BIOCHEMISTRY

BIOCHEMISTRY

SECOND EDITION

THE MOLECULAR BASIS

OF CELL STRUCTURE AND FUNCTION

ALBERT L. LEHNINGER

THE JOHNS HOPKINS UNIVERSITY

SCHOOL OF MEDICINE

WORTH PUBLISHERS, INC.

For Jan

BIOCHEMISTRY, Second Edition

by Albert L. Lehninger

Printed in the United States of America

Library of Congress Catalog Card No. 75-11082

ISBN: 0-87901-047-9

Designed by Malcolm Grear Designers, Inc.

Second printing, May 1976

Worth Publishers, Inc.

444 Park Avenue South

New York, N.Y. 10016

PREFACE TO THE SECOND EDITION

Someone long ago said that to be happy at one's work, one must be fit for it, have a sense of success in it, and not do too much of it. In the rush to finish the second edition, there was often too much to do, but even in the lowest moments my spirits were buoyed by the hundreds of heartening letters I received from students who used the first edition.

I am tempted to call this second edition a brand new text, since nearly every paragraph was rewritten and there are so many additions. But I have kept the organization and style that students and teachers found useful the first time around. I have also tried to hold the focus on central biochemical concepts and on the impact of new discoveries in this exciting field, rather than dwell on encyclopedic detail.

I continue to believe that biochemistry is best approached by starting with a set of organizing principles rather than scattered facts and hypotheses. Biochemistry has a structure, a set of unifying themes, all the wonderful consequences of genetic control over the shape and function of proteins. The compelling sense of this pattern of organizing principles and their implications is set forth in an Introduction (page 3), which I call "The Molecular Logic of Living Organisms." As with the first edition, the rest of the book builds upon this introductory essay, beginning with the structures and properties of biomolecules, moving to a study of energy-yielding processes and energy-requiring processes, and ending with the culmination of our journey — the replication, transcription, and translation of genetic information.

The book is now cross-referenced with page numbers, not just chapter numbers. All enzymatic reactions are balanced, the metabolic pathway illustrations are essentially complete, and all enzyme names are included. (I am using the recommended enzyme names of the 1972 report of the Enzyme Commission.) The problems have been completely reworked and there are more of them than before.

There is so much other new material in the second edition that I can only take room here to mention the most salient. There are many more applications of biochemical knowledge to problems of human health and disease. There are two new chapters particularly relevant to human biochemistry—"Biochemical Aspects of Hormone Action" and "Organ Interrelationships in the Metabolism of Mammals." Among other things, the hormone chapter deals with the organization of the endocrine system, the relationship of the hypothalamus and pituitary, and the role of hormone receptors and intracellular messengers, such as cyclic AMP. The chapter on integration of metabolism includes an account of the metabolic characteristics of each organ as well as the metabolic interplay between organs, metabolic rate, transport of nutrients and gases by the blood, the function of the kidneys in adjustment of urine composition, and the special role of the liver and adipose tissue in processing and distributing fuel supplies. I have also added a discussion of starvation and diabetes mellitus to show how the human organism adjusts its metabolism in response to stress.

In the area of proteins and enzymes, I have included detailed discussion of the conformation of proteins, the kinetics and thermodynamics of protein folding and unfolding, and the sequential and all-or-none models of allosteric transitions. There is also a new chapter on the art and science of protein purification and characterization. The enzyme chapters are expanded with new material on enzyme reaction mechanisms and kinetics, including bisubstrate reactions. And I have written a new chapter on vitamins and coenzymes.

The most completely rewritten portion of the book is Part 4. The chapter on DNA includes new material on the structure of bacterial and viral genomes as well as much new information on eukaryotic chromosomes, including repetititve sequences and inverted repetitions. The chapter on DNA replication and transcription has also been revised and updated to include the important new advances of the last few years, as have the chapters on protein synthesis and the amino acid code. The chapter on regulation of gene expression has received particular attention because of the great strides made in this important area of cell biology. The material on morphogenesis and self-assembly, as well as that on the origin of life, has also been substantially revised.

All told, I found the writing of this second edition to be a renewed biochemical education for myself. I can only hope I may have made the way a little easier for others.

Many of the chapters have been reviewed by experts in the fields concerned: Bruce M. Alberts, Daniel I. Arnon, William Beranek, Jr., Clanton C. Black, Jr., Ralph A. Bradshaw, Alexander C. Brownie, Richard M. Caprioli, Waldo E. Cohn, Stanley Dagley, Julian E. Davies, Richard E. Dickerson, John T. Edsall, Irving Geis, Govindjee, Franklin M. Harold, William F. Harrington, Jens G. Hauge, Bernard L. Horecker, William P. Jencks, Thomas H. Jukes, Walter Kauzmann, Daniel E. Koshland, Jr., Robert Lehman, John M. Lowenstein, Alan H. Mehler, Alton Meister, Daniel J. O'Kane, Peter L. Pedersen, Norman S. Radin, William J.

Rutter, Carl Sagan, Norman G. Sansing, Irwin H. Segel, Thomas P. Singer, David B. Sprinson, Donald F. Steiner, Jack L. Strominger, Charles Sweeley, Thomas E. Thompson, Gordon M. Tomkins, John L. Westley, and William B. Wood. I owe them much for their critical advice and encouragement. Moreover, many teachers and students took the time to send me useful suggestions based on their experience with the first edition. As before, I will greatly appreciate comments from those who use this new book.

For help in preparing this second edition I must also express my deepest thanks to two fine teachers of biochemistry: Herbert Friedmann of the University of Chicago, who read every chapter at least twice, examined each statement with microscopic attention, and ruthlessly skewered many bad phrases, and Richard Mintel, who enthusiastically developed many new problems and their solutions, made useful suggestions for teaching certain topics, and, not least, contributed many hours to patient proofreading, always with cheerful ebullience. I also wish to thank Daniel DiMaio, a Johns Hopkins medical student, for introducing me to his "Cell Game" and for formulating many of the problems in Part 4. In addition, I thank Linda Hansford, who laid out the pages and backstopped our proofreading. And I must thank my secretary Peggy Ford for again keeping everything flying while typing manuscript against impossible deadlines. My thanks to Chris Mello, too.

Again, it was a pleasure to work with all the good people of Worth Publishers, Inc.; they produced this book with their usual care and high standards. But above all, I thank my wife, who has, on top of everything else, endured my writing six books—she is an everlasting source of encouragement.

ALBERT L. LEHNINGER

Sparks, Maryland
May, 1975

PREFACE TO THE FIRST EDITION

This book is written for students who are taking their first and perhaps their only course in biochemistry, whether as undergraduates or as graduate or medical students. I undertook this task because I want to convey to students my picture of what this science has recently become. Biochemistry is no longer a mere catalog of the biological occurrence and enzymatic reactions of a large number of organic compounds. In the last few years it has acquired, along with many new facts, a set of organizing principles which have made it a much simpler field to comprehend, and, at the same time, a more powerful way of analyzing many important problems in biology.

How has this come about? Each field of scientific study at some time in its evolution undergoes a profound transition in which a collection of widely scattered facts and hypotheses crystallizes into a logical pattern, unified by a few basic concepts. Biochemistry has been undergoing such a transition, stimulated by new experimental findings and new insights. Among these are the recognition of the principles of energy transfer in cells, the mechanisms by which the major metabolic pathways are regulated, the importance of membranes, ribosomes, and other ultrastructural elements of cells in their molecular activities, and the far-reaching conclusion that the amino acid sequence determines the three-dimensional conformation of protein molecules and thus their biological functions. The new knowledge of the molecular basis of genetics, which has transformed all of biology, has had the most profound influence. Because of these developments biochemistry now has a central story, a *leitmotiv*, which I have tried to express in simple terms in the *Introduction*.

This book is concerned primarily with biochemistry at the cell level, where its organizing principles are most clearly evident. Central concepts are emphasized rather than an encyclopedic treatment of biochemical details. There are four major parts in the book:

1. Biomolecules
2. Energy-yielding processes
3. Energy-requiring processes
4. Transfer of genetic information

These are subdivided into what I believe is a logical progression of chapters, each of which is a manageable "package" for both students and teachers, equivalent to the content of one lecture or discussion period. I agree with many teachers that the structure and properties of some biomolecules may best be taught together with their metabolism. This approach is quite feasible using this book, although for student convenience I have chosen to collect most of the material on the structure, chemistry, and occurrence of the various types of biomolecules into one section. I believe this makes for easy reference, while still allowing for flexibility of approach.

Biochemistry has many new frontiers today. I have tried to sketch out some of the most promising in chapters on the regulation of protein synthesis and its role in cell differentiation, the molecular basis of self-assembly and morphogenesis, and the origin of life. These chapters may well be out-of-date soon, but I hope they will serve to acquaint students with some of the biochemistry of the future.

Acknowledgments

Many may think it foolhardy for a single author to attempt to write a comprehensive textbook of biochemistry. However, my publishers have made it possible to enlist the criticism and advice of a number of chemists, biochemists, and biologists expert in research and/or teaching in the areas covered by the book. Each chapter has been read and criticized by at least one and often several authorities. To them I owe a great deal, not only for kind encouragement and sometimes deservedly blunt criticism, but also for the insight and perspective that only the real expert can convey. It is perhaps inevitable that some errors of fact, interpretation, or emphasis will be found, but I trust no one will attribute these to anyone but me. I will greatly appreciate receiving from students and teachers alike their comments, criticisms, notice of errors, and advice about improvements that can be made in later printings or editions.

To the following reviewers I give my most sincere thanks: Jay Martin Anderson, Christian B. Anfinsen, Robert E. Beyer, R. G. S. Bidwell, Rodney L. Biltonen, Konrad E. Bloch, Benjamin Bouck, Daniel Branton, Robert H. Burris, Melvin Calvin, Roderick K. Clayton, Helena Curtis, Robert E. Davies, Bernard D. Davis, John T. Edsall, Paul T. Englund, Allan H. Fenselau, J. Lawrence Fox, Richard Goldsby, Ursula Johnson Goodenough, Guido Guidotti, Gordon G. Hammes, William F. Harrington, Edward C. Heath, Harold G. Hempling, Donald P. Hollis, Lloyd L. Ingraham, Andre T. Jagendorf, William P. Jencks, Daniel E. Koshland, Jr., Sir Hans A. Krebs, Myron Ledbetter, William J. Lennarz, Richard C. Lewontin, Julius Marmur, Daniel Nathans, Leslie Orgel, Peter L. Pedersen, Keith R. Porter, David Prescott, John Sinclair,

Gunther Stent, Jack L. Strominger, Maurice Sussman, Serge N. Timasheff, and William B. Wood. My thanks go to many others, acknowledged at the end of the book, who generously gave me permission to use drawings, electron micrographs, and other illustrative material.

I am also grateful to the officers and staff of Worth Publishers for their genuine interest in the needs of both students and teachers, their appreciation of the struggles of a university author, and above all, their desire to produce an educationally useful book.

My colleagues in the Department of Physiological Chemistry of The Johns Hopkins School of Medicine furnished much advice and also took on many responsibilities which the gestation of this book had forced me to neglect. I also owe a great deal to Johns Hopkins medical students, who taught me whatever I have learned about teaching. Two of them, Bill Scott and Penny Pate, gave me much help in the early stages of preparation of the manuscript; one happy outcome is that they are now Mr. and Mrs. William Wallace Scott, Jr. To Linda Hansford I am particularly indebted for invaluable help with proof-reading, indexing, checking of problems, and collection of data and references. Thanks also to Ronald Garrett, who photographed the molecular models and to my secretary Peggy Ford, who not only effectively marshalled my time and attention among teaching, research, departmental administration, and book-writing, but also typed many chapters of the manuscript.

Finally, I want to express my deep appreciation to my family, who patiently endured the many weekends and evenings that were devoted to writing and who gave encouragement when it was most needed.

ALBERT L. LEHNINGER

Sparks, Maryland
March, 1970

CONTENTS IN BRIEF

CONTENTS

Contents

PART 3

BIOSYNTHESIS AND THE UTILIZATION OF PHOSPHATE-BOND ENERGY   617

CHAPTER 23

The Biosynthesis of Carbohydrates   623

CHAPTER 24

The Biosynthesis of Lipids   659

Contents

INTRODUCTION **THE MOLECULAR LOGIC OF LIVING ORGANISMS**

Living things are composed of lifeless molecules. These molecules, when isolated and examined individually, conform to all the physical and chemical laws that describe the behavior of inanimate matter. Yet living organisms possess extraordinary attributes not shown by collections of inanimate matter. If we examine some of these special properties, we can approach the study of biochemistry with a better understanding of the fundamental questions it seeks to answer.

The Identifying Characteristics of Living Matter

Perhaps the most conspicuous attribute of living organisms is that they are complicated and highly organized. The cells of which they are composed possess intricate internal structures containing many kinds of complex molecules. Furthermore, living organisms occur in an enormous number of different species. In contrast, the inanimate matter around us, as represented by soil, water, and rocks, usually consists of random mixtures of simple chemical compounds, with comparatively little structural organization.

Second, each component part of a living organism appears to have a specific purpose or function. This is true not only of such macroscopic, visible structures as wings, eyes, flowers, or leaves but also of intracellular structures, such as the nucleus and the cell membrane. Moreover, individual chemical compounds in the cell, e.g., lipids, proteins, and nucleic acids, also have specific functions. In living organisms it is quite legitimate to ask what the function of a given molecule is. However, to ask such questions about molecules in collections of inanimate matter is irrelevant and meaningless.

Third, living organisms have the capacity to extract and transform energy from their environment, which they use to build and maintain their own intricate structures from simple raw materials. They can also carry out other forms of purposeful work, such as the mechanical work of locomotion. Inanimate matter cannot utilize external energy to maintain its own structural organization. In fact, inanimate matter usually decays to a more random state when it absorbs external energy in the form of heat or light.

The most extraordinary attribute of living organisms is their capacity for precise self-replication, a property that can be regarded as the very quintessence of the living state. In contrast, collections of inanimate matter show no apparent ability to reproduce themselves in forms identical in mass, shape, and internal structure, through "generation" after "generation."

Biochemistry and the Living State

We may now ask: If living organisms are composed of molecules that are intrinsically inanimate, why does living matter differ so radically from nonliving matter, which also consists of intrinsically inanimate molecules? Why does the living organism appear to be more than the sum of its inanimate parts? Early philosophers would have answered that living organisms are endowed with a mysterious and divine life force. But this doctrine, called vitalism, is nothing more than superstition, and it has been rejected by modern science. Today the central goal of the science of biochemistry is to determine how the collections of inanimate molecules found in living organisms interact with each other to constitute, maintain, and perpetuate the living state.

Biochemistry is a very young science. Until the last few decades only a very few universities recognized it as a science in its own right. There are two parent lines in the genealogy of present-day biochemistry. One line arose from medicine and physiology, a by-product of early inquiries into the chemical composition of blood, urine, and the tissues and their variation in health and disease. The other lineage traces from organic chemistry, from early studies of the structure of naturally occurring organic compounds. For a long time biochemistry was simply regarded as a branch of physiology or as a branch of chemistry. It did not really emerge as a full-fledged science in its own right, with powerful experimental methods and predictive insight into biological phenomena, until the last quarter century. Two major developments brought this about. One was the recognition of multienzyme systems as catalytic units in the major metabolic pathways and the development of a unifying hypothesis for the transfer of energy in living cells. The other, which has had a most pervasive and profound influence, was the recognition that heredity, one of the most fundamental aspects of biology, has a rational molecular basis. Today biochemistry is making exciting probes into a number of fundamental areas of biology—the differentiation of cells and organisms, the origin of life and evolution, behavior and memory, and human disease—probes which have demonstrated that these basic problems can be fruitfully approached by biochemical methods.

Indeed, the success of biochemistry in explaining many cellular phenomena has been so great that many scientists have come to the conclusion that biology is chemistry. Some biologists do not share this view; they maintain that the essence or gestalt of complex living organisms cannot be reduced, now or ever, to the level of molecules and molecular interactions. But this is a minority view. Today it is perhaps more logical to assume, as a working philosophy, that all biological phenomena are ultimately molecular in basis and to abandon this view only when it is no longer useful for designing critical experiments or for explaining experimental data. We must not, however, look upon biology as merely another branch of classical chemistry, such as organic chemistry, physical chemistry, or inorganic chemistry. If biology is chemistry, it must be a kind of "superchemistry" which includes but at the same time transcends classical chemistry. This is because the molecules found in living organisms not only conform to all the familiar physical and chemical principles governing the behavior of all molecules but, in addition, interact with each other in accordance with another set of principles that we shall refer to as the *molecular logic of the living state*. These principles do not necessarily involve any new or as yet undiscovered physical laws or forces. Rather, they should be regarded as a set of ground rules that govern the nature, function, and interactions of the specific types of molecules found in living organisms, that endow them with the capacity for self-organization and self-replication. Not all the principles comprised by the molecular logic of the living state have yet been identified; indeed, some are only dimly perceived. In fact, it is perhaps more appropriate to speak of these principles as axioms, since some of them are intuitive and not yet provable.

Now let us see if we can identify some of the important axioms in the molecular logic of the living state. We shall begin with a brief survey of the structure and function of the molecules found in living matter, which we shall henceforth call *biomolecules*.

Biomolecules

The chemical composition of living organisms is qualitatively quite different from that of the physical environment in which they live. Most of the chemical components of living organisms are organic compounds of carbon, in which the carbon is relatively reduced, or hydrogenated; many organic biomolecules also contain nitrogen. In contrast, the elements carbon and nitrogen are rather scarce in nonliving matter; moreover, they occur in the atmosphere and the earth's crust only in such simple inorganic forms as carbon dioxide, molecular nitrogen, carbonates, and nitrates.

The organic compounds present in living matter occur in extraordinary variety, and most of them are extremely complex. Even the simplest and smallest cells, the bacteria, contain a very large number of different organic molecules. The bacterium *Escherichia coli* is estimated to contain about 5,000 different kinds of compounds, including some 3,000 different kinds of proteins and 1,000 different kinds of nucleic acids. Moreover, most of the organic matter in living cells consists of *macromolecules,* with very large molecular weights, including not only the proteins and nucleic acids but also such polymeric substances as starch and cellulose.

If we turn to larger and more complex organisms, the higher animals and plants, we find that they contain proteins and nucleic acids in much greater variety. In the human organism, for example, there may be as many as 100,000 different kinds of proteins, compared with about 3,000 in E. coli. Although some of the proteins in cells of E. coli function in ways quite similar to some of the proteins in human cells, none of the protein molecules of E. coli is identical with any of the proteins found in man. Indeed, each species of organism has its own chemically distinct sets of protein molecules and of nucleic acid molecules. Since there are over 1.5 million species of living organisms, it may be calculated that all living species together must contain somewhere between 10^{10} and 10^{12} different kinds of protein molecules and about 10^{10} different kinds of nucleic acids. If we compare these figures with the total number of all organic compounds whose structure is known to the organic chemist today, which is only about 1 million, or 10^6, it is clear that we know the precise structure of only a trivially small fraction of all the organic molecules that are believed to exist in living matter. Therefore, for biochemists to attempt to isolate, identify, and synthesize all the different organic molecules present in living matter might appear to be a hopeless undertaking.

Paradoxically, however, the immense diversity of organic molecules in living organisms is ultimately reducible to a surprising simplicity. We now know that cell macromolecules are composed of simple, small building-block molecules strung together in long chains. Starch and cellulose consist of long strings of covalently linked glucose molecules. The different types of proteins consist of long covalently linked chains of amino acids, small organic compounds of known structure. Only 20 different kinds of amino acids are found in proteins, but they are arranged in many different sequences to form proteins of many different kinds. Thus, all 3,000 or more proteins in the E. coli cell are built from only 20 different kinds of small molecules. Similarly, the 1,000 or more nucleic acids of the E. coli cell, of which there are two kinds, deoxyribonucleic acid (DNA) and ribonucleic acid (RNA), are constructed from a total of eight different building blocks, the _nucleotides_, four of which are building blocks of DNA and four the building blocks of RNA. Moreover, the 20 different amino acid building blocks of proteins and the eight different nucleotide components of nucleic acids are identical in all living species. Even though we have precise knowledge of the covalent structure of fewer than 100 proteins today, the techniques of protein chemistry are sufficiently well developed for it now to be within the capability of biochemistry to elucidate the structure of any protein from any species of organism.

The few simple building-block molecules from which all macromolecules are constructed have another striking characteristic. Each serves more than one function in living cells; indeed, some are extremely versatile and play a number of roles. The amino acids serve not only as building blocks of protein molecules but also as precursors of hormones, alkaloids, porphyrins, pigments, and many other biomolecules. Various nucleotides serve not only as building blocks of nucleic acids but also as coenzymes and as energy-carrying molecules. It therefore appears probable that the building-block biomolecules were selected during the course of biological evolution for their capacity to serve several functions. So far as we know, living organisms normally contain no functionless compounds, although there are some biomolecules whose functions are not yet understood.

Now we can see emerging some of the axioms in the molecular logic of the living state. Since the thousands of different macromolecules present in cells are constructed from only a few simple building-block molecules, we can formulate the first axiom: *There is an underlying simplicity in the molecular organization of the cell.* Since the building-block biomolecules are identical in all known species, we can infer another: *All living organisms have a common ancestor.* Because each organism has its own distinctive sets of nucleic acids and proteins another axiom emerges: *The identity of each species of organism is preserved by its possession of characteristic sets of nucleic acids and proteins.* Furthermore, from the functional versatility of the building-block biomolecules, we can make another inference: *There is an underlying principle of molecular economy in living organisms.* Perhaps living cells contain the simplest possible molecules in the least number of different types, just sufficient to endow them with the attribute of life and with species identity, under the environmental conditions in which they exist.

Energy Transformations in Living Cells

The molecular complexity and the orderliness of structure of living organisms, in contrast to the randomness of inanimate matter, have profound implications to the physical scientist. The second law of thermodynamics, the branch of physics dealing with energy and its transformations, states that all physical and chemical processes always proceed with an increase in the disorder or randomness in the world, i.e., its entropy. How is it, then, that living organisms can create and maintain their intricate orderliness in an environment that is relatively disordered and becoming more so with time?

Living organisms do not constitute exceptions to the laws of thermodynamics. Their high degree of molecular orderliness must be paid for in some way, since it cannot arise spontaneously from disorder. The first law of thermodynamics states that energy can be neither created nor destroyed. Living organisms thus cannot consume or use up energy; they can only transform one form of energy into another. They absorb from their environment a form of energy that is useful to them under the special conditions of temperature and pressure in which they live and then return to the environment an equivalent amount of energy in some other, less useful form. The useful form of energy that cells take in is called *free energy*, which may be simply defined as that type of energy which can do work at constant temperature and pressure. The less useful type of energy that cells return to their environment consists of heat and other forms that quickly become randomized in the environment and thus increase its disorder, or entropy. We may now state another important axiom in the molecular logic of the living state: *Living organisms create and maintain their essential orderliness at the expense of their environment, which they cause to become more disordered and random.*

The environment of living organisms is absolutely essential to them, not only as a source of free energy but also as a source of raw materials. In the language of thermodynamics, living organisms are *open* systems because they exchange both energy and matter with their environment and, in so doing, transform it. It is characteristic of open systems that they are not in equilibrium with their environment. Although living organisms may appear to be in equilibrium, because they may not change visibly as we observe them over a period of time, actually they usually exist in a *steady state*, that condition of an open system in which the rate of transfer of matter and energy from the environment into the system is exactly balanced by the rate of transfer of matter and energy out of the system. It is therefore part of the molecular logic of the living state that the cell is a nonequilibrium open system, a machine for extracting free energy from the environment, which it causes to increase in randomness. Moreover, and this is another reflection of the principle of maximum economy, living cells are highly efficient in handling energy and matter. They greatly exceed most man-made machines in the efficiency with which they convert input energy into work performed.

The energy-transforming machinery of living cells is built entirely of relatively fragile and unstable organic molecules that are unable to withstand high temperatures, strong electric currents, or extremely acid or basic conditions. The living cell is also essentially isothermal; at any given time, all parts of the cell have essentially the same temperature. Furthermore, there are no significant differences in pressure between one part of the cell and another. For these reasons, cells are unable to use heat as a source of energy, since heat can do work at constant pressure only if it passes from a zone of higher temperature to a zone of lower temperature. Living cells therefore do not resemble heat engines or electric engines, the types of engines with which we are most familiar. Instead they follow another important axiom in the molecular logic of the living state: *Living cells function as isothermal chemical engines.* The energy that cells absorb from their environment is transformed into chemical energy, which is then used to carry out the chemical work involved in the biosynthesis of cell components, the osmotic work required to transport materials into the cell, and the mechanical work of contraction and locomotion; all of these transformations take place at essentially constant temperature.

Among familiar man-made machines, we see very few capable of using chemical energy to do work at constant temperature. In fact, engineering technology has yet to produce a useful engine that can convert chemical energy isothermally into mechanical energy, yet this type of energy conversion is familiar to all of us in the contraction of muscles.

Chemical Reactions in Living Cells

Cells can function as chemical engines because they possess *enzymes*, catalysts capable of greatly enhancing the rate of specific chemical reactions. Enzymes are highly specialized protein molecules, made by cells from simple amino acids. Each type of enzyme can catalyze only one specific type of chemical reaction; nearly 2,000 different kinds of enzymes are known. Enzymes far exceed man-made catalysts in their reaction specificity, their catalytic efficiency, and their capacity to operate under mild conditions of temperature and hydrogen-ion concentration. They can catalyze in seconds complex sequences of reactions that would require days, weeks, or months of work in the chemical laboratory.

One especially remarkable property of chemical reactions in living cells ultimately makes their efficient function as chemical engines possible; enzyme-catalyzed reactions proceed with a 100 percent yield; there are no by-products. In contrast, the reactions of organic chemistry carried out in the laboratory with man-made catalysts are nearly always accompanied by the formation of one or more by-products, so that yields are usually much less than 100 percent and intensive purification of the product is required at each step. Because enzymes can enhance a single reaction pathway of a given molecule without enhancing its other possible reactions, living organisms can carry out, simultaneously, many different individual reactions without bogging down in a morass of useless by-products. The great specificity of enzymes results from the operation of another fundamental axiom in the molecular logic of the living state: *The specificity of molecular interactions in cells results from the structural complementarity of the interacting molecules.* Enzyme molecules combine with their substrates during the catalytic cycle in such a way that the active site of the enzyme molecule fits the substrate with a near-perfect lock-and-key complementarity. We shall see that the principle of structural complementarity underlies the specificity of many different types of molecular interactions in cells.

The hundreds of enzyme-catalyzed chemical reactions in the cell do not take place independently of each other but are linked into sequences of consecutive reactions having common intermediates, so that the product of the first reaction becomes the substrate or reactant of the second, and so on. Such linked, or coupled, sequences, which may have anywhere from 2 to 20 or more reaction steps, are in turn connected into networks of converging or diverging pathways. This arrangement has several important biological implications. One is that such systems of sequential reactions provide for the channeling of chemical reactions along

specific routes to specific end products. Another is that
sequential reactions make the transfer of chemical energy
possible. Energy transfer cannot take place between two
chemical reactions under conditions of constant temperature and pressure unless the two reactions have a common
intermediate. If two independent reactions such as

$$A \longrightarrow B$$

$$C \longrightarrow D$$

occur in the same container under constant temperature and
pressure, each will proceed with a decrease in free energy
regardless of the presence of the other. However, in two consecutive reactions, such as

$$A \longrightarrow B$$

$$B \longrightarrow C$$

some of the chemical energy of A may be transferred to C by
the common intermediate B.

Living cells can be divided into two great classes according to the type of energy they obtain from their environment. *Photosynthetic cells* utilize sunlight as their main
source of energy; the radiant energy is absorbed by the pigment chlorophyll and transformed into chemical energy.
Heterotrophic cells obtain energy from the degradation of
highly reduced, energy-rich organic molecules, such as glucose, which they require as nutrients from the environment.
Most cells of the animal world are heterotrophic. In heterotrophic cells glucose is oxidized to carbon dioxide and
water; in this process some of the free energy of the glucose
molecule is conserved and employed to carry out various
types of cellular work.

Although these two classes of organisms obtain energy
from their environment in different forms, both transform it
into chemical energy, largely in the form of the compound
adenosine triphosphate (ATP). ATP functions as the major
carrier of chemical energy in the cells of all living species.
As it transfers its energy to other molecules, it loses its terminal phosphate group and becomes *adenosine diphosphate*
(ADP), which is the discharged, or energy-poor, counterpart of ATP. In turn, ADP can accept chemical energy again
by regaining a phosphate group to become ATP, at the expense of solar energy (in photosynthetic cells) or chemical
energy (in heterotrophic cells). ATP serves as a common
intermediate or connecting link between two large networks
of enzyme-catalyzed reactions in the cell. One of these networks conserves chemical energy derived from the environment by causing the phosphorylation of the energy-poor ADP
to the energy-rich ATP. The other network utilizes the energy of ATP to carry out the biosynthesis of cell components
from simple precursors, with simultaneous breakdown of the
ATP to ADP. We may now state another axiom in the molecular logic of cells: *Consecutively linked sequences of
enzyme-catalyzed reactions provide the means for transfer-*

ring chemical energy from energy-yielding to energy-requiring processes.

Self-Regulation of Cell Reactions

There is another important result of the fact that all chemical reactions in the cell are enzyme-catalyzed and are linked by common intermediates. A simple bacterial cell like *E. coli* simultaneously synthesizes all its thousands of different complex molecular components from just three simple precursors—glucose, ammonia, and water. Here the living cell employs a kind of chemical logic which is still beyond the current state of the art of synthetic chemistry in the laboratory. If a chemist were confronted with the problem of synthesizing two products, say an amino acid and a lipid, he would never dream of synthesizing them from the same precursors simultaneously in the same reaction vessel. He would start each synthesis with different precursors and use different sequences of reactions. He would carry out the two syntheses independently, in separate vessels, and probably at different times. Yet, in living cells, the synthesis of hundreds and thousands of widely different molecules is carried out simultaneously, literally in the same vessel, starting from only a few common precursors. The linking of enzyme-catalyzed reactions into sequences of consecutive reactions makes possible the orderly channeling of the thousands of chemical reactions taking place in cells, so that all the specific biomolecules required in cell structure and function are produced in the right proportions and rates.

A bacterial cell synthesizes simultaneously perhaps 3,000 or more different kinds of protein molecules in specific molar ratios to each other. Each of these protein molecules contains a minimum of 100 amino acid units in a chain; most contain many more. Yet at 37°C the bacterial cell requires only a few seconds to complete the synthesis of any single protein molecule. In contrast, the synthesis of a protein by man in the laboratory, a feat which was accomplished for the first time only in 1969, required the work of highly skilled chemists, many expensive reagents, hundreds of separate operations, complex automated equipment, and months of time in preparation and execution. Not only can the bacterial cell make individual protein molecules very rapidly, but it can make 3,000 or more different kinds of proteins simultaneously, in the precise molar ratios required to constitute a living, functioning cell.

The linkage of enzyme-catalyzed reactions into consecutive sequences makes the regulation of metabolism possible and endows it with self-adjusting properties. In the simplest case, the overaccumulation of an end product of metabolism, such as an amino acid, can inhibit the rate-determining step in the sequence of reactions by which it was formed, a type of control known as *feedback inhibition*. Moreover, living cells possess the power to regulate the synthesis of their own catalysts. Thus the cell can "turn off" the synthesis of the enzymes required to make a given product from its precursors whenever that product is available, ready-made, from the environment. We have then another important principle: *Cells are capable of regulating their metabolic reactions and the biosynthesis of their enzymes to achieve maximum efficiency and economy.*

The Self-Replication of Living Organisms

The most remarkable of all the properties of living cells is their capacity to reproduce themselves with nearly perfect fidelity, not just once or twice, which would be remarkable enough, but for hundreds and thousands of generations. Three features immediately stand out. First, some living organisms are so immensely complex that the amount of genetic information transmitted seems out of all proportion to the minute size of the cells that must carry it, namely, the single sperm cell and the single egg. But we know today that virtually all the genetic information is present in the chromosomes, coded in the form of the specific sequence of nucleotide building blocks in a very small amount of DNA, which in a human sperm or egg cell weighs no more than about 6 picograms (1 pg $= 1 \times 10^{-12}$ g). Modern research in the biochemical aspects of genetics thus has led to another axiom in the molecular logic of the living state: *The symbols in which the genetic information is coded in DNA are submolecular in dimension.*

A second remarkable characteristic of the self-replicating property of living organisms is the extraordinary stability of the genetic information stored in DNA. Very few early historical records prepared by man have survived for long, even though they have been etched in copper or stone and preserved against the elements. The Dead Sea scrolls and the Rosetta stone, for example, are only a few thousand years old. But there is good reason to believe that modern bacteria have nearly the same size, shape, and internal structure and contain the same kinds of building-block molecules and the same kinds of enzymes as those which lived hundreds of millions of years ago, despite the fact that bacteria, like all organisms, have been undergoing constant evolutionary change. Genetic information is preserved, not on a copper scroll or engraved in stone, but in the form of DNA, an organic molecule so fragile that when isolated in solution, it will break into many pieces if the solution is merely stirred or pipetted.

The capacity of living cells to preserve their genetic information is the result of the operation of the principle of *structural complementarity*. One DNA strand serves as the template for the enzymatic replication of a structurally complementary DNA strand. In fact, the DNA-synthesizing enzymes of the cell cannot make DNA without a template. It is altogether remarkable that, even in the intact cell, a DNA strand may break frequently, but it is quickly and automatically repaired by specific enzymes. Errors or mutations occur only infrequently, but even these are not always deleterious, since they may possess advantages in allowing a given species of organism to change its identity gradually, in order to adapt itself better to changes in its environment, during the course of evolution.

There is a third remarkable characteristic of genetic-information transfer in living organisms. Genetic information is encoded in the form of a specific sequence of four different nucleotide building blocks in the linear DNA molecule. But living cells are three-dimensional in structure, and they have three-dimensional parts or components. Here we come to a very crucial axiom in the molecular logic of the living state, one that provides the connecting link between the simple linear chemistry of DNA and all the three-dimensional attributes of the great variety of multicellular organisms: *The one-dimensional information of DNA is translated into three-dimensional macromolecular and supramolecular components of living organisms by translation of DNA structure into protein structure.* The specific linear base sequence of DNA is translated into a corresponding linear amino acid sequence of a polypeptide chain in the process of protein synthesis. However, unlike a DNA molecule, a polypeptide is not stable in extended linear form. It spontaneously curls up and folds into a specific, stable, three-dimensional structure, the precise geometry of which is determined by the particular sequence of amino acids in its polypeptide chain. Each type of polypeptide chain will assume its own specific three-dimensional conformation, which in turn endows it with a specific type of biological activity. Moreover, the many different kinds of protein molecules that serve as the components of such biostructures as membranes, ribosomes, and organelles can recognize each other and group themselves spontaneously into precisely reproducible three-dimensional assemblies because they fit each other in only one specific way—again, according to the principle of structural complementarity.

We have now described a number of the characteristic interactions and interrelationships of biomolecules that together constitute the molecular logic of the living state. We may summarize these principles by the following statements: *A living cell is a self-assembling, self-regulating, self-replicating isothermal open system of organic molecules operating on the principle of maximum economy of parts and processes; it promotes many consecutive, linked organic reactions for the transfer of energy and for the synthesis of its own components by means of organic catalysts that it produces itself.* At no point in our examination of the molecular logic of living cells have we encountered any violation of known physical laws, nor has it been necessary to define new ones. The machinery of living cells functions within the same set of laws that governs the operation of man-made machines, but the chemical reactions and processes of cells have been refined far beyond the present capabilities of chemical engineering.

In this orienting survey we have sketched the central and most fundamental goal of biochemistry today, namely, to determine in detail the molecular logic of the living cell. This is not, however, the only goal of biochemistry or even the ultimate one. The biochemistry of the cell, the theme of this book, is but the starting point for molecular study of many other problems of biology. Most fundamental, perhaps, is to deduce how, in the obscure early history of the earth, certain inanimate organic compounds first "found" each other, "learned" to interact with each other, and ultimately organized themselves into the first "living" structures. Another is to learn how the first cells underwent evolutionary development into the remarkable panoply of plant and animal species we see around us today. A further goal is a molecular description of the interactions of cells within tissues and of specialized functions such as muscular contraction. Still another goal is the biochemical analysis of neurofunction from the level of simple intercellular communication upward to integration, to memory, to behavior, and ultimately to thought—indeed, all the profound questions posed by man's quest for understanding of his nature. And along with these fundamental questions, biochemistry also is developing increasing insight into human disease and its alleviation, into plant life and agriculture, and into the ecological balance of the biosphere.

As we begin the study of biochemistry, the organizing principles that constitute the molecular logic of cells should serve as a frame of reference. This book begins with a description of the various classes of biomolecules (Part 1). It then proceeds to analyze the isothermal, self-adjusting, consecutively linked, enzyme-catalyzed reactions that constitute the open system through which both energy and matter flow, i.e., the process of metabolism. Metabolism, as we have seen, consists of two networks of reactions. The network that yields chemical energy as ATP will be the subject of Part 2. The other network, that which utilizes ATP for cell synthesis and performance of cell work, is developed in Part 3. Finally, in Part 4, we shall consider the molecular basis of the self-replication of cells and the assembly of cell components. The book will end by returning to the origin of life and its molecular logic.

PART **1** THE MOLECULAR COMPONENTS OF CELLS

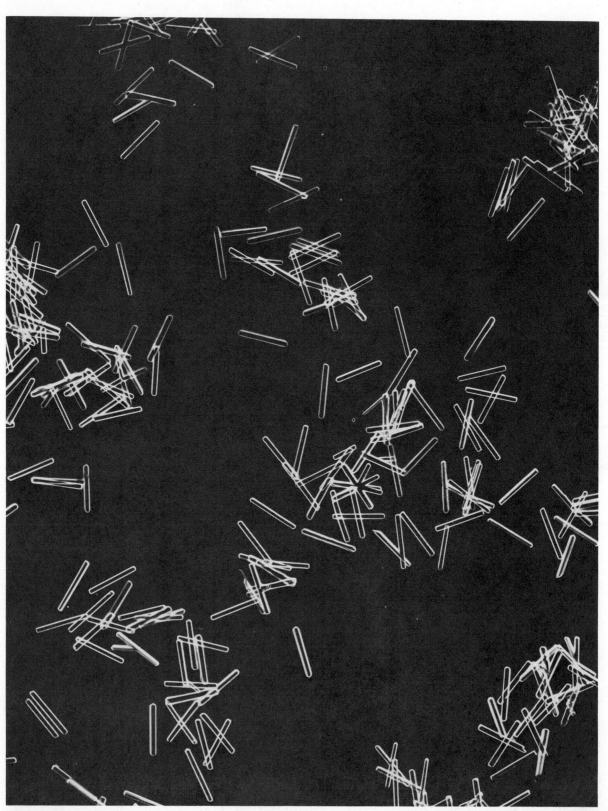

Crystals of bovine trypsin.

CHAPTER **1** BIOMOLECULES AND CELLS

Table 1-1 The bioelements

The following elements are essential in the nutrition of one or more species, but not all are essential for every species

The elements of organic matter	Trace elements
O	Mn
C	Fe
N	Co
H	Cu
P	Zn
S	B
	Al
	V
	Mo
Monoatomic ions	I
Na$^+$	Si
K$^+$	Sn
Mg^{2+}	Ni
Ca^{2+}	Cr
Cl$^-$	F
	Se

Table 1-2 Relative abundance of the major chemical elements in the earth's crust and in the human body as percent of total number of atoms

Earth's crust		Human body	
O	47	H	63
Si	28	O	25.5
Al	7.9	C	9.5
Fe	4.5	N	1.4
Ca	3.5	Ca	0.31
Na	2.5	P	0.22
K	2.5	Cl	0.08
Mg	2.2	K	0.06
Ti	0.46	S	0.05
H	0.22	Na	0.03
C	0.19	Mg	0.01

In this chapter, the first in a series devoted to the structures and properties of the major classes of biomolecules, we shall develop the idea that biomolecules should be studied from two points of view. Of course we must examine their structure and properties as we would those of nonbiological molecules, by the principles and approaches used in classical chemistry. But we must also examine them in the light of the hypothesis that biomolecules are the products of evolutionary selection, that they may be the fittest possible molecules for their biological function, and that they interact with each other in the specific relationships that we have called the molecular logic of the living state. Moreover, we must also consider the size and shape of biomolecules, which, as we shall see, determine not only the specificity of biological interactions but also the dimensions and ultrastructure of living cells and their component organelles.

The Biological Fitness of Organic Compounds

Some chemical elements appear to be more "fit" than others to make up the molecules of living organisms, since only 27 of the 90 natural chemical elements in the earth's crust have been found to be essential components in various living organisms (Table 1-1). Moreover, the chemical elements in living organisms are not distributed in proportion to their occurrence in the earth's crust (Table 1-2). The four most abundant elements in the earth's crust are oxygen, silicon, aluminum, and iron. In contrast, the four most abundant elements in living organisms are hydrogen, oxygen, carbon, and nitrogen, which make up about 99 percent of the mass of most cells. We may therefore presume that compounds of these four elements possess unique molecular fitness for the processes that collectively constitute the living state.

Carbon, hydrogen, nitrogen, and oxygen possess a common property: they readily form covalent bonds by electron-pair sharing. Hydrogen needs one electron, oxygen two, nitrogen three, and carbon four to complete their outer electron shells and thus form stable covalent bonds. These four elements thus react with each other to form a large number of different covalent compounds. Furthermore, three of these elements (C, N, and O) can share either one or two electron pairs to yield either single or double bonds, a capacity that endows them with considerable versatility of chemical bonding. Carbon, nitrogen, hydrogen, and oxygen are uniquely fit in yet another way: they are the lightest elements capable of forming covalent bonds. Moreover, since the strength of a covalent bond is inversely related to the atomic weights of the bonded atoms, these four elements are capable of forming very strong covalent bonds.

Carbon atoms possess another and most significant property, the capacity to bond with each other. Since a carbon atom may either accept or donate four electrons to complete an outer octet, it can form covalent bonds with four other carbon atoms. In this way covalently linked carbon atoms can form linear or branched or cyclic backbones for an immense variety of different organic molecules. Moreover, since carbon atoms also form covalent bonds with oxygen, hydrogen, nitrogen, and sulfur, many different kinds of functional groups can be introduced into the structure of organic molecules.

Organic compounds of carbon have yet another distinctive feature. Because of the tetrahedral configuration of electron pairs around singly bonded carbon, many different three-dimensional structures can be achieved by carbon-carbon bonding. No other chemical element can form molecules of such widely different sizes and shapes or with such a variety of functional groups.

The Hierarchy of the Molecular Organization of Cells

The biomolecules of living organisms are ordered into a hierarchy of increasing molecular complexity, as shown in Figure 1-1. All organic biomolecules are ultimately derived from very simple, low-molecular-weight _precursors_ obtained from the environment, particularly carbon dioxide, water, and atmospheric nitrogen. These precursors are converted by living matter, via sequences of metabolic _intermediates_, into the _building-block biomolecules_, organic compounds of somewhat larger molecular weight. These building-block molecules are then linked to each other covalently to form the _macromolecules_ of the cell, which have relatively high molecular weights. Thus, the amino acids are the building blocks of proteins, nucleotides are the building blocks of nucleic acids, monosaccharides the building blocks of polysaccharides, and fatty acids the building blocks of most lipids. [Although the molecular weights of individual lipids are low (mol wt 750 to 1500) compared with those of proteins, nucleic acids, and polysaccharides, some of the lipids spontaneously associate into structures of high particle weight and usually function in macromolecular systems. We may arbitrarily include them for present purposes among the macromolecules.]

At the next higher level of organization of cells, macromolecules of different classes associate with each other to form

Figure 1-1
The hierarchy in the molecular organization
of cells.

	The Cell
Organelles	Nucleus
	Mitochondria
	Chloroplasts
	Golgi bodies
Supramolecular assemblies (particle weight 10^6–10^9)	Ribosomes
	Enzyme complexes
	Contractile systems
	Microtubules
Macromolecules (mol wt 10^3–10^9)	Nucleic acids
	Proteins
	Polysaccharides
	Lipids
Building blocks (mol wt 100–350)	Nucleotides
	Amino acids
	Monosaccharides
	Fatty acids
	Glycerol
Metabolic intermediates (mol wt 50–250)	Pyruvate
	Citrate
	Malate
	Glyceraldehyde 3-phosphate
Precursors from the environment (mol wt 18–44)	Carbon dioxide
	Water
	Ammonia
	Nitrogen

supramolecular systems, such as lipoproteins, which are complexes of lipids and proteins, and ribosomes, which are complexes of nucleic acids and proteins. For example, each ribosome of a bacterial cell contains 3 different ribonucleic acid molecules and about 50 different protein molecules. However, there is now a distinctive difference in the way the components are assembled. In supramolecular complexes the component macromolecules are not covalently bonded to each other. The nucleic acid and protein components of ribosomes are held together by relatively weak noncovalent forces, such as hydrogen bonds, hydrophobic interactions, and van der Waals interactions. However, because they are held together by a large number of such weak associations, supramolecular complexes like ribosomes are surprisingly stable under biological conditions. Moreover, the non-covalent association of macromolecules into supramolecular complexes is very specific, the result of a precise geometrical fit, or complementarity, between the component parts. For example, as we shall see later (Chapter 33), the three-dimensional structure of ribosomes is highly ordered and specific, in accord with their complex function in the translation of genetic information into protein structure.

Finally, at the highest level of organization in the hierarchy of cell structure, various supramolecular complexes and systems are further assembled into cell organelles —nuclei, mitochondria, and chloroplasts—and into other intracellular structures and inclusions—lysosomes, microbodies, and vacuoles. Here again, so far as is known, the various macromolecular and supramolecular components are held together by noncovalent interactions.

The relative amounts of the major classes of biomolecules in the common bacterium Escherichia coli are shown in Table 1-3. The proteins (Greek proteios, "first") are the most prominent macromolecules in the cell, making up over 50

Table 1-3 Major molecular components of an E. coli cell

Component	Percent total weight	Approximate number of molecular species
Water	70	1
Proteins	15	3,000
Nucleic acids		
DNA	1	1
RNA	6	1,000
Carbohydrates	3	50
Lipids	2	40
Building-block molecules and intermediates	2	500
Inorganic ions	1	12

percent of the dry weight of cells. The *E. coli* cell may contain over 3,000 different kinds of protein molecules. The second most abundant set of macromolecules in *E. coli* are the nucleic acids, followed by the carbohydrates and lipids. Actually, all living cells contain about the same proportions of the major classes of biomolecules as shown for the *E. coli* cell if we exclude from consideration such relatively inert parts of living organisms as the exoskeleton, the mineral portion of bone, extracellular materials like hair or feathers, and such intracellular storage materials as starch and fat, whose content may vary widely from one cell or organism to another. Thus the "living" portion of all types of cells has about the same gross molecular composition.

The major classes of biomolecules have identical functions in all species of cells. The nucleic acids serve universally to store and transmit genetic information. In all cells the proteins are the direct products and effectors of gene action. Some have specific catalytic activity and function as enzymes; others serve as structural elements. Proteins are the most versatile class of macromolecules; this is why they are the subject of 7 of the 13 chapters in this book devoted to the structure and properties of biomolecules. The polysaccharides have two major functions in all cells. Some, e.g., starch, serve as storage forms of energy-yielding fuel for cell activity, and others, e.g., cellulose, serve as extracellular structural elements. The lipids in turn also play the same roles in all cells, either as major structural components of membranes or as a storage form of energy-rich fuel.

One other point requires comment. There is an important and fundamental difference between the nucleic acids and proteins on the one hand and the polysaccharides and lipids on the other (Figure 1-2). Nucleic acids and proteins are *informational macromolecules*. Each nucleic acid molecule contains four or more types of nucleotides arranged in a specific information-rich sequence. Similarly, each protein molecule contains a specific information-rich sequence of some 20 different amino acids. On the other hand, the polysaccharides and lipids do not have an information-carrying function. For example, the recurring building blocks of polysaccharides either are all identical, as in starch, a polymer of glucose, or they consist of regularly alternating building-block components.

The Primordial Biomolecules

Table 1-3 shows that there are perhaps 3,000 different types of proteins and 1,000 types of nucleic acids in an *E. coli* cell. All the different proteins are made from only 20 different amino acids, and the nucleic acids are largely made from only 8 different nucleotides. Similarly, the polysaccharides of *E. coli* are made from only a few simple sugar molecules. Since the macromolecules in all living organisms are made from only a relatively few, simple building-block molecules, it has been suggested that the first cells to have arisen on earth may have been built from only two or three dozen different organic molecules (Figure 1-3). This set of primordial

Figure 1-2

Informational macromolecules of the cell. The letters A, T, G, and C symbolize the four nucleoside building blocks of DNA. The amino acid building blocks of proteins are symbolized by their three-letter symbols Arg, Ala, Trp, etc.

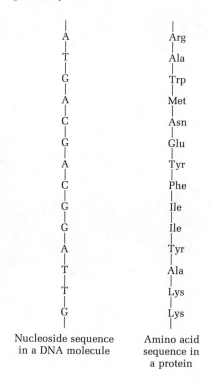

| Nucleoside sequence in a DNA molecule | Amino acid sequence in a protein |

Figure 1-3

The primordial biomolecules (right). Although they are somewhat arbitrarily chosen, the compounds shown may be regarded as the simplest ancestors from which all other organic biomolecules have been derived during the course of biochemical evolution. They are grouped according to their chemical structure and are shown in un-ionized form. The amino acids are building blocks of the proteins; the pyrimidines, purines, and D-ribose are precursors of nucleic acids; D-glucose is the precursor of many polysaccharides; and glycerol, choline, and palmitic acid are building blocks of lipids.

The Primordial Biomolecules

The amino acids (in un-ionized form)

$HCHCOOH$ | NH_2 — Glycine

$CH_3CHCOOH$ | NH_2 — Alanine

$CH_3CHCHCOOH$ with CH_3 and NH_2 — Valine

$CH_3CHCH_2CHCOOH$ with CH_3 and NH_2 — Leucine

$CH_3CH_2CHCHCOOH$ with CH_3 and NH_2 — Isoleucine

$HOCH_2CHCOOH$ | NH_2 — Serine

$CH_3—S—CH_2CH_2CHCOOH$ | NH_2 — Methionine

$CH_3CHCHCOOH$ with OH and NH_2 — Threonine

Phenyl—$CH_2CHCOOH$ | NH_2 — Phenylalanine

HO—phenyl—$CH_2CHCOOH$ | NH_2 — Tyrosine

Indole—$C—CH_2CHCOOH$ | NH_2 — Tryptophan

$HS—CH_2CHCOOH$ | NH_2 — Cysteine

Proline

$HOOCCH_2CHCOOH$ | NH_2 — Aspartic acid

$H_2N—CCH_2CHCOOH$ with O and NH_2 — Asparagine

$HOOCCH_2CH_2CHCOOH$ | NH_2 — Glutamic acid

$H_2N—CCH_2CH_2CHCOOH$ with O and NH_2 — Glutamine

Histidine: $HC=C—CH_2CHCOOH$, N, NH, NH_2, CH

Arginine: $H_2N—C—NH—CH_2CH_2CH_2CHCOOH$, NH, NH_2

Lysine: $H_2N—CH_2CH_2CH_2CH_2CHCOOH$ | NH_2

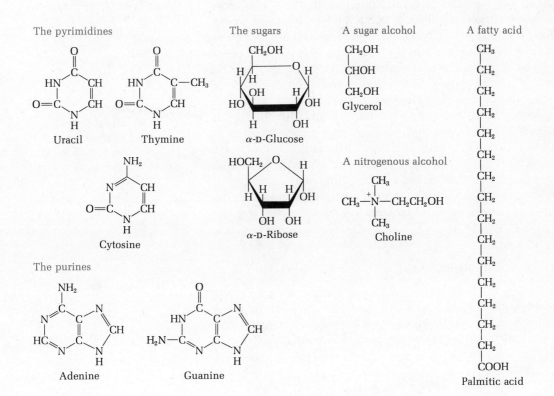

The pyrimidines: Uracil, Thymine, Cytosine

The purines: Adenine, Guanine

The sugars: α-D-Glucose, α-D-Ribose

A sugar alcohol: CH_2OH — $CHOH$ — CH_2OH — Glycerol

A nitrogenous alcohol: Choline

A fatty acid: Palmitic acid

biomolecules may have included 20 amino acids, 5 nitrogenous bases, 1 or more fatty acids, 2 sugars, the alcohol glycerol, and the amine choline. Whatever their precise number or identity, this set of primordial biomolecules may be regarded as the ancestors of all other biomolecules; they are the first alphabet of living matter. We should look upon this group of simple organic substances with some awe and wonder, since an extraordinary and unique relationship exists between them.

The Specialization and Differentiation of Biomolecules

As living organisms evolved into more highly differentiated and complex forms, new biomolecules of greater complexity and variety are believed to have evolved from the primordial biomolecules. For example, over 150 different biologically occurring amino acids are known today, but nearly all of them are derived from the basic 20 amino acids used for the construction of proteins. Similarly, dozens of different nucleotides and nucleotide derivatives are known, all containing descendants of the five primordial nitrogenous bases. Over 70 simple sugars are biologically derived from glucose, and from them a large variety of polysaccharides are formed in different organisms. There are many different fatty acids, all descended from one or a few primordial fatty acids. Some examples of the specialized descendants of the primordial biomolecules are given in Table 1-4.

Many specialized biomolecules are extremely complex and at first glance appear to bear little resemblance to the 30 primordial biomolecules. Among these are various pigments, odor-bearing essential oils, hormones, antibiotics, alkaloids, and various structural molecules, such as the lignin of wood. Nevertheless, recent research on the biogenesis of such complex substances shows that they are also ultimately derived from one of the primordial biomolecules. For example, the *terpenes,* a large class of biomolecules that includes some of the vitamins, many essential oils, plant pigments, and such complex natural products as rubber, are all ultimately built from acetic acid, the major breakdown product of glucose and of fatty acids. Hundreds of different *alkaloids* are known, most of which are derived from amino acids. Thus the primordial biomolecules are the biological ancestors of the vast number of different organic compounds found in the various species of living organisms today.

The Origin of Biomolecules

Apart from their existence in living matter, organic compounds occur only in traces in the earth's crust. How, then, did the first living organisms acquire their characteristic organic building blocks, the primordial biomolecules?

Table 1-4 Specialized derivatives of some of the primordial biomolecules

Arginine
  Ornithine
  Citrulline

Proline
  3-Hydroxyproline
  4-Hydroxyproline
  4-Hydroxymethylproline
  4-Methyleneproline
  4-Ketoproline

Leucine
  β-Hydroxyleucine
  δ-Hydroxyleucine
  γ,δ-Dihydroxyleucine
  γ-Hydroxyleucine
  N-Methylleucine

Guanine
  1-Methylguanine
  2-Methylguanine
  2-Dimethylguanine
  2-O-Methylguanine
  7-Methylguanine

D-Glucose
  D-Mannose
  D-Fructose
  D-Galactose
  N-Acetylglucosamine
  D-Glucuronic acid
  D-Glucose 6-phosphate
  Ascorbic acid
  Inositol
  Sucrose
  Maltose
  Lactose

Palmitic acid
  Oleic acid
  Stearic acid
  Lauric acid
  Palmitoleic acid
  Palmitaldehyde
  Stearaldehyde

Figure 1-4
Spark-discharge apparatus for demonstrating abiotic formation of organic compounds under primitive-atmosphere conditions.

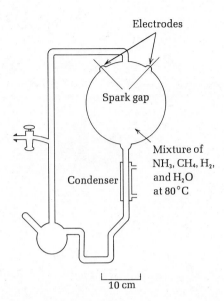

Electrodes

Spark gap

Condenser

Mixture of
NH_3, CH_4, H_2,
and H_2O
at 80°C

10 cm

Table 1-5   Some organic compounds generated by spark discharges under simulated primitive atmosphere conditions†

Glycine
Alanine
Sarcosine
β-Alanine
α-Aminobutyric acid
N-Methylalanine
Aspartic acid
Glutamic acid
Iminodiacetic acid
Iminoacetopropionic acid
Formic acid
Acetic acid
Glycolic acid
Lactic acid
α-Hydroxybutyric acid
Succinic acid
Urea
Methylurea

† Many others have been found; see Chapter 37.

Recent research suggests that early in the history of the earth many different organic compounds occurred in relatively high concentrations in the surface waters of the ocean. From this warm "soup" of organic compounds the first living cells somehow arose (see Chapter 37). It is now generally believed that the earth is approximately 4,800 million (4.8×10^9) years old. Living organisms may have arisen as early as 4,000 million years ago. The fossil remains of bacteria similar to those known today have been dated with some certainty to be at least 3,300 million years old.

In the 1920s, A. I. Oparin, a biochemist in the Soviet Union, suggested that natural chemical and physical processes could have led to the spontaneous formation of simple organic compounds, such as amino acids and sugars, from methane, ammonia, and water vapor, which he postulated to be components of the primitive atmosphere. According to his theory, these gases were activated by the radiant energy of sunlight or by lightning discharges to react with each other. The simple organic products so formed condensed and dissolved in the primitive ocean, which gradually became enriched in a large variety of organic compounds. Oparin postulated that the first living cell arose spontaneously from this warm, concentrated solution of organic compounds, a view independently put forward by J. B. S. Haldane in England. These views were attended by much controversy and remained an untested speculation for over 20 years.

That the gaseous components now known to be present in the primitive atmosphere can be the precursors of organic compounds is today well supported by laboratory studies. Among the early experiments on the abiotic origin of organic molecules were those carried out in 1953 by Stanley Miller. He subjected mixtures of the gases methane, ammonia, water, and hydrogen, then believed to be predominant in the primitive atmosphere, in a closed flask at 80°C to electric sparking across a pair of electrodes, to simulate lightning, for periods of a week or more (Figure 1-4). Then he collected and analyzed the contents of the system. The gas phase contained carbon monoxide, carbon dioxide, and nitrogen, which were evidently formed from the gases initially introduced. In the chilled, dark-colored condensate he found significant amounts of water-soluble organic substances, which he separated by chromatographic methods. Among the compounds Miller identified were a number of α-amino acids, including some known to be present in proteins, i.e., glycine, alanine, aspartic acid, and glutamic acid. He also found several of the simple organic acids known to occur in living organisms, such as formic, acetic, propionic, lactic, and succinic acids (Table 1-5).

Miller postulated that the various organic compounds formed in these experiments arose by sequences of reactions like those shown in Table 1-6. He proposed that hydrogen cyanide was formed from methane and ammonia and that the electric discharge converted methane into ethylene and other hydrocarbons. The hydrogen cyanide could react with

Table 1-6 Chemical reactions in spark discharges

$CH_4 + NH_3 \longrightarrow HCN + 3H_2$	(1)
$C_2H_4 + HCN \longrightarrow CH_3CH_2CN$	(2)
A nitrile	
$CH_3CH_2CN + 2H_2O \longrightarrow CH_3CH_2COOH + NH_3$	(3)
Propionic acid	
$CH_3CHOHCN + NH_3 \longrightarrow CH_3CHNH_2CN + H_2O$	(4)
An aminonitrile	
$CH_3CHNH_2CN + 2H_2O \longrightarrow CH_3CHNH_2COOH + NH_3$	(5)
Alanine	

ethylene to form a nitrile [reaction (2)], which could then undergo hydrolysis to propionic acid [reaction (3)]. Similarly, α-hydroxy nitriles could react with ammonia to form α-amino nitriles [reaction (4)], which on hydrolysis could form α-amino acids like alanine [reaction (5)].

Miller's experiments were carried out in a system rich in the reduced compounds methane and ammonia, but later experiments with mixtures containing nitrogen, hydrogen, carbon monoxide, and carbon dioxide (but no methane or ammonia) exposed to radiant energy again formed amino acids and other organic molecules, which showed that highly reduced precursors such as ammonia and methane are not essential for the abiotic formation of organic molecules.

Many different forms of energy or radiation lead to organic compounds from such simple gas mixtures, including visible light, ultraviolet light, x-rays, gamma radiation, sparking and silent electric discharges, ultrasonic waves, shock waves, and α and β particles. Several hundred different organic compounds have been formed in such experiments, including representatives of all the important types of molecules found in cells as well as many not found in cells. All the common amino acids present in proteins, the nitrogenous bases adenine, guanine, cytosine, uracil, and thymine, which serve as the building blocks of nucleic acids, and many biologically occurring organic acids and sugars have been detected among the products of such primitive-earth-simulation experiments. It appears quite likely that the primitive ocean was indeed rich in dissolved organic compounds, which may have included many or all of the basic building-block molecules we recognize in living cells today.

The Fitness of Biomolecules

Why should living organisms have selected the specific types of organic molecules they now possess? Why should 20 α-amino acids be the building blocks of all proteins in all organisms? Why not only 10? Why not 40? Why are they all α-amino acids? Couldn't we equally well construct large "protein" molecules from amino acids having their amino groups in the β positions? Why are the purines adenine and guanine and the pyrimidines cytosine and thymine, out of the dozens of purine and pyrimidine derivatives known, the essential building blocks of DNA in all species? Much evidence supports the concept that the biomolecules we know today were selected from a much larger number of available organic compounds. Actually, several hundred different organic compounds have been isolated from simulated primitive-earth experiments on the abiotic origin of organic molecules like those described above. Since only a small number of different organic compounds may have been required to form the earliest biostructures capable of survival, it appears very likely that a process of selection took place.

Another argument for the fitness concept can be made from the fact that over 150 different amino acids occur biologically, yet all proteins in all species are built from the same set of 20 primordial amino acids. If any of the other amino acids had been more fit as components of protein molecules than the primordial amino acids, ample evolutionary time was available for living organisms to have acquired the ability to use them.

The Dimensions and Shapes of Biomolecules

In Part 1 we examine the structures and properties of the major classes of biomolecules. As we do this, we must also take special notice of the size and shape of each type of biomolecule, since these attributes are of great significance in biochemistry and molecular biology. We have already seen (in the Introduction) that the complementary fit between the substrate and the active site of an enzyme is so precise that it makes possible the great efficiency and selectivity of enzymatic catalysis and the absence of by-products. If only a very small change is made in some critical dimension of a substrate molecule, it may no longer fit the active site and thus fail to be acted upon by the enzyme. Moreover, the great accuracy with which the genetic information of DNA is replicated also depends on the precision of fit of specific biomolecules with each other. It is therefore essential to become familiar with the dimensions and the shape of biomolecules and how they relate to the dimensions of various intracellular structures. Table 1-7 shows the units of mass and length commonly used in connection with molecular and cell dimensions.

Molecular dimensions were formerly given in units of angstroms, cell dimensions in microns, and wavelengths in millimicrons, where 1 micron = 10^{-3} mm. However, by international agreement the metric units nanometers, micrometers, and millimeters are now recommended. Accordingly, in this book we shall give wavelengths of light and molecular dimensions in nanometers (nm) and cell dimensions in nanometers (nm) or micrometers (μm).

Table 1-8 shows the standard prefixes used to indicate powers of 10 in the metric system. These prefixes are repeated for convenience in the appendix, where other ab-

Table 1-7 Some units of mass and length

Mass
  1 dalton   = mass of one hydrogen atom
              = 1.67×10^{-24} g
  1 picogram = 1×10^{-12} g

Length
  1 nanometer (nm) = 10^{-9} m
                   = 10 angstroms (Å)
  1 micrometer (μm) = 10^{-6} m
                   = 1,000 nm
                   = 10,000 angstroms (Å)

Table 1-8 Prefixes for powers of 10 for use with SI units[†]

Value	Prefix	Abbreviation
10^6	mega	M
10^3	kilo	k
10^{-1}	deci	d
10^{-2}	centi	c
10^{-3}	milli	m
10^{-6}	micro	μ
10^{-9}	nano	n
10^{-12}	pico	p
10^{-15}	femto	f
10^{-18}	atto	a

† Combinations of prefixes are no longer allowed, so that n- is used instead of mμ and p- instead of $\mu\mu$-.

breviations of the SI (Système International d'Unités, International System of Units) are listed.

The planar two-dimensional projections in which the structures of organic molecules are necessarily shown on the printed page are quite insufficient to describe the true three-dimensional configuration of biomolecules. For this reason biochemists often construct three-dimensional models of biomolecules when confronted with problems of molecular specificity. There are two classes of molecular models (Figure 1-5). _Crystallographic_ models show the covalent skeleton with the correct bond angles and lengths, but such models do not indicate the actual space occupied by the molecule. _Space-filling_ models (Figures 1-5 and 1-6), on the other hand, show few details of bond angles and distances in the backbone, but they do show the van der Waals contour, or surface, of the molecule. While both types of model are useful in studying the structure of biomolecules, it is the space-filling model that represents the molecule as it is "seen" by the cell or by one of its specific components, such as an enzyme. Actually, an enzyme sees much more than the three-dimensional shape of its substrate. It sees the location and sign of the electric charges and the precise distance between charged groups. It sees the positions of uncharged polar groups, such as hydroxyl, carbonyl, and amide groups, which can potentially enter into hydrogen-bond formation. It sees the size and shape of the nonpolar or hydrocarbon areas on the surface of the biomolecule, which may provide important contact areas with other molecules.

Three-dimensional shape and surface topography are especially important for macromolecules. Protein molecules usually have only one characteristic three-dimensional conformation under normal intracellular conditions, called the _native conformation_, which is indispensable for their biological activity. For example, only the native conformation of an enzyme molecule has catalytic activity. However, we cannot easily deduce the three-dimensional conformation of a macromolecule from a two-dimensional structure on the printed page; nor can we reconstruct it unambiguously from ordinary space-filling atomic models. Complex physical methods, particularly _x-ray diffraction analysis_, are required to establish the precise conformation of biological macromolecules. Indeed, charting the three-dimensional structure of macromolecules by x-ray analysis and correlating their structure with their biological activity are major objectives of biochemistry and molecular biology.

Biomolecules, Supramolecular Structures, and Cell Organelles

The size and shape of biomolecules are of crucial importance in another way. We have seen that in living cells there is a hierarchy of molecular organization (Figure 1-1); the simple biomolecules are the building blocks of macromolecules, macromolecules are components of supramolecular complexes, supramolecular complexes are assembled into cell

Figure 1-5
Different representations and models of the structure of alanine, shown in un-ionized form.

Figure 1-6
Space-filling models of some biologically important atoms (to scale).

Empirical formula

$$C_3H_7O_2N$$

Structural formula

$$CH_3-\underset{\underset{NH_2}{|}}{\overset{\overset{H}{|}}{C}}-COOH$$

Crystallographic models

0.1 nm

Carbon

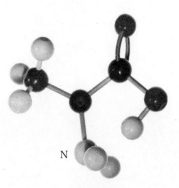

Ball-and-stick model

Hydrogen

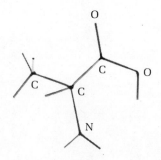

Dreiding model

Oxygen

Space-filling model

Nitrogen

Sulfur

Phosphorus

organelles, and organelles and other structures are organized into cells. It is clear that the dimensions, shape, and physical properties of the simple building-block biomolecules must determine the dimensions and properties of macromolecules, whose shape and surface topography must in turn determine how they fit together to form supramolecular structures, which in turn determine the structure of cell organelles and ultimately of the cell itself.

A dramatic example of how the size, shape, and properties of a relatively small building-block biomolecule can influence the size, shape, and biological behavior of an entire cell is given by the human genetic disease _sickle-cell anemia_. The red blood cells of patients with this disease have a normal biochemical composition except for the oxygen-carrying protein hemoglobin. This protein is composed of nearly 600 amino acid units linked into four polypeptide chains. The hemoglobin molecules from patients with this disease differ very slightly in composition from normal hemoglobin as the result of a genetic mutation: two molecules of glutamic acid in normal hemoglobin are replaced by two molecules of the amino acid valine in the sickle-cell hemoglobin. This slight change, affecting only 2 of the nearly 600 amino acid residues, alters the structure of the sickle hemoglobin molecule so that it "stacks" improperly with neighboring molecules. This defect in turn causes a profound change in the shape of the entire red blood cell, which assumes the shape of a sickle, or crescent, whereas normal red blood cells are flat disks. As a consequence, sickled red blood cells tend to aggregate in small blood vessels, blocking the circulation and causing other serious disturbances. Thus, only a very small difference in the structure of a simple, small amino acid molecule (weight about 100 daltons; length about 0.7 nm) can result in a profound change not only in the hemoglobin molecule of which it is a component but also in the very much larger structure of the entire red blood cell (weight about 1×10^{14} daltons; diameter 7,000 nm).

To provide some orientation regarding the relative size of various biostructures Table 1-9 gives some data on the weight and the dimensions of representative building-block molecules, cell macromolecules, supramolecular assemblies, organelles, viruses, and cells.

The Structural Organization of Cells

Throughout this book we shall relate the structure and dynamic function of each type of biomolecule to the structure and biological role of various cell components, e.g., cell walls, membranes, ribosomes, chloroplasts, contractile systems, endoplasmic reticulum, and the cell nucleus. Therefore, before we begin the detailed study of biomolecules, it is important to review the major structural features of different types of cells, the dimensions and molecular composition of their internal organelles, and the compart-

Table 1-9   Approximate dimensions and weights of some biomolecules and cell components

		Weight	
	Long dimension, nm	Daltons	Picograms
Alanine	0.5	89	
Glucose	0.7	180	
Phospholipid	3.5	750	
Myoglobin (a small protein)	3.6	16,900	
Hemoglobin (a medium-sized globular protein)	6.8	65,000	
Myosin (a large rod-shaped protein)	160	470,000	
Ribosome of *E. coli*	18	2,800,000	
Bacteriophage ϕX174 of *E. coli*	25	6,200,000	
Tobacco mosaic virus (a rod)	300	40,000,000	6.68×10^{-4}
Mitochondrion (liver cell)	1,500		1.5
E. coli cell	2,000		2
Chloroplast (spinach leaf)	8,000		60
Liver cell	20,000		8,000

mentation and division of vital cell functions among these internal structures. This review is presented schematically in Figures 1-7 to 1-9, which represent three types of cells. The first, the bacterium *Escherichia coli,* is the best-known member, biochemically and genetically speaking, of the great class of prokaryotic cells (Figure 1-7). The second is the hepatocyte, or liver cell, of the rat, a well-studied example of the other great class of cells, eukaryotic cells (Figure 1-8). Both the *E. coli* cell and the hepatocyte obtain their energy from the oxidation of organic nutrient molecules acquired from the environment. Figure 1-9 shows an example of a photosynthetic cell, which obtains energy from sunlight. The cell chosen is from a green leaf of a higher plant and is also a eukaryotic cell.

Biochemistry today is increasingly concerned with the structure of cells and their organelles. Some of the most illuminating recent progress has come from combined biochemical and morphological studies of cellular processes. As we shall see, it is a fundamental goal of modern biochemistry not only to identify the nature and mechanism of the enzymatic reactions of intermediary metabolism and replication and transfer of genetic information but also to determine where these events take place in the cell and how biochemical events taking place in different parts of a cell are coordinated, both spatially and temporally. The dividing lines between biochemistry and cell biology are becoming more and more difficult to identify, since these fields of cell science truly form a logical continuum. Thus the application of exact chemical and physical methods to the analysis of the structure of cell components is yielding significant new insights into the functions of biomolecules and their dynamic interactions with each other in living cells.

Figure 1-7
The structural organization of prokaryotic cells.
Prokaryotes are very small, relatively simple cells having only a single membrane, the cell membrane, which is usually surrounded by a rigid cell wall. Since they have no other membranes, they contain no nucleus and no membranous organelles such as mitochondria or endoplasmic reticulum. The prokaryotes include the eubacteria, the blue-green algae, the spirochetes, the rickettsiae, and the mycoplasma or pleuropneumonialike organisms. They contain only one chromosome, which consists of a single molecule of double helical DNA, densely coiled to form the nuclear zone; prokaryotes reproduce largely by asexual division. Prokaryotes were the first cells to arise in biological evolution.

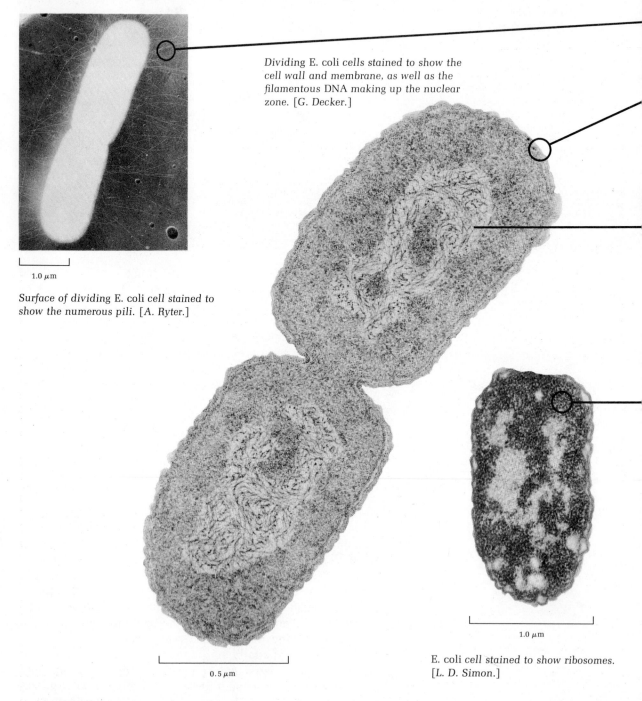

Dividing E. coli cells stained to show the cell wall and membrane, as well as the filamentous DNA making up the nuclear zone. [G. Decker.]

1.0 μm

Surface of dividing E. coli cell stained to show the numerous pili. [A. Ryter.]

1.0 μm

0.5 μm

E. coli cell stained to show ribosomes. [L. D. Simon.]

Electron micrographs of the bacterium E. coli. This aerobic organism is a member of the coliform group of bacteria, typically found in the human intestinal tract. The mature cell is a cylindrical rod about 2 μm long and 1 μm in diameter; it weighs about 2 pg. E. coli cells multiply rapidly on a simple medium containing glucose as carbon source and ammonium ions as nitrogen source; the division time may be as short as 20 min at 37°C. Most of our knowledge of the molecular basis of genetics has arisen from the study of various strains and mutants of E. coli and E. coli bacteriophages. Although more is known about the biochemistry and genetics of E. coli than for any other cell, we are still very far from a complete molecular description.

Schematic drawing	Molecular composition	Properties and functions
Cell wall and membrane Cell membrane · Cell wall Protein molecule · Lipid bilayer · 9 nm · 20 nm · Pili	The cell wall contains a rigid framework of polysaccharide chains cross-linked with short peptide chains. Its outer surface is coated with lipopolysaccharide. The pili, not found in all bacteria, are extensions of the cell wall. The cell membrane contains about 45% lipid and 55% protein; the lipids form a continuous non-polar phase. Infoldings of the cell membrane are called mesosomes.	The cell wall protects bacteria against swelling in hypotonic media. It is porous and allows most small molecules to pass. Some of the pili are hollow and serve to transfer DNA during sexual conjugation. The membrane is a selectively permeable boundary which allows water, certain nutrients, and metal ions to pass freely. Enzymes responsible for conversion of nutrient energy into ATP are located in the membrane.
Nuclear zone	The genetic material is a single chromosome of double-helical DNA 2 nm in diameter and about 1.2 mm long, which is tightly coiled.	DNA is the carrier of genetic information. During division, each strand is replicated to yield two daughter double-helical molecules. From one strand of DNA the genetic message is transcribed to form messenger RNA.
Ribosomes 18 nm · 50S · 30S	Each *E. coli* cell contains about 15,000 ribosomes. Each ribosome has a large and a small subunit. Each subunit contains about 65% RNA and 35% protein.	Ribosomes are the sites of protein synthesis. Messenger RNA binds in the groove between the subunits and specifies the sequence of amino acids in the growing polypeptide chains.
Storage granules	*E. coli* and many other bacteria contain storage granules that are polymers of sugars. Some bacteria contain granules of poly-β-hydroxybutyric acid.	When needed as fuel, these polymers are enzymatically degraded to yield free glucose or free β-hydroxybutyric acid.
Cytosol	The soluble portion of the cytoplasm is highly viscous; the protein concentration is very high, exceeding 20%.	Most of the proteins of the cytosol are enzymes required in metabolism. The cytosol also contains metabolic intermediates and inorganic salts.

Figure 1-8
The structural organization of eukaryotic cells.
Eukaryotic cells are much larger and much more complex than prokaryotic cells. The cell volume of most eukaryotes is from 1,000 to 10,000 times larger than that of typical prokaryotes. The cells of all higher organisms in both plant and animal worlds are eukaryotic, as are those of fungi, protozoa, and most algae. Eukaryotes contain a membrane-surrounded nucleus. The genetic material is divided into several or many chromosomes, which undergo mitosis during cell division. Eukaryotes also contain internal membranes surrounding organelles such as the mitochondria and Golgi bodies, as well as an endoplasmic reticulum. Many of their metabolic reactions are segregated within structural compartments. Eukaryotes are more recent in evolutionary origin than prokaryotes; presumably they were derived from the latter.

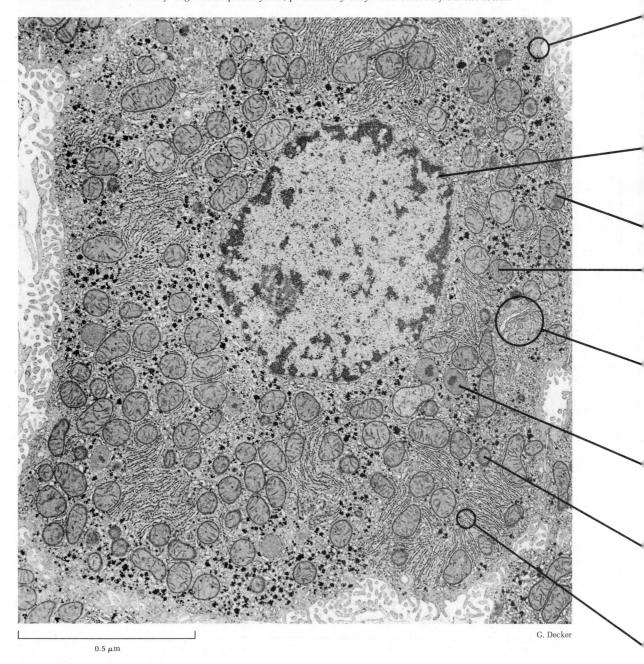

0.5 μm

G. Decker

Electron micrograph of a thin section of a rat-liver cell (hepatocyte) fixed in osmium tetroxide. Rat-liver hepatocytes are polyhedral and about 20 μm in diameter. They are metabolically versatile cells whose most important function is the biochemical processing and distribution of foodstuff molecules brought to the liver from the intestinal tract. They store glucose as glycogen, prepare nitrogenous wastes for excretion, and synthesize blood-plasma proteins and lipids. Hepatocytes can carry out all the major metabolic activities of cells. They are perhaps the most thoroughly studied animal cells because they are readily available in quantity and can be easily fractionated after homogenization to yield nuclei, mitochondria, endoplasmic reticulum (the "microsome fraction"), and other subcellular fractions by means of differential centrifugation.

Schematic drawing	Molecular composition	Properties and functions
Cell membrane Cell membrane Protein molecule Lipid bilayer Cell coat (glycocalyx)	The cell coat of hepatocytes is flexible and sticky. It is composed of acid mucopolysaccharides, glycolipids, and glycoproteins. The plasma membrane is about 9 nm thick and contains about equal amounts of lipids and proteins; the lipids are arranged in a bilayer. It contains a greater variety of lipids than bacterial membranes.	The adhesive properties of cell coats are specific and play an important role in cell-cell recognition and thus tissue organization. The plasma membrane is selectively permeable. It contains active-transport systems for Na^+ and K^+, glucose, amino acids, and other nutrients, as well as a number of important enzymes.
Nucleus Perinuclear envelope Nucleolus	The nucleus, about 4–6 μm in diameter, is surrounded by a perinuclear envelope. The DNA within is combined with histones and organized into chromosomes. The nucleolus is rich in RNA.	During mitosis, chromosomes undergo replication of their DNA and separation into daughter chromosomes.
Mitochondrion Cristae Granules Matrix	There are about 800 mitochondria in each hepatocyte. They are globular and a little over 1 μm in diameter, occupying about 20% of the cytoplasmic volume. Their outer and inner membranes differ in lipid composition and in enzymatic activity. The matrix is rich in enzymes.	The mitochondria are the power plants of the cell, where carbohydrates, lipids, and amino acids are oxidized to CO_2 and H_2O by molecular oxygen, and the energy set free is converted into the energy of ATP. The enzymes of electron transport and energy conversion are located in the inner membrane.
Golgi complex Vacuole	The Golgi complex consists of flattened, single-membrane vesicles, which are often stacked. Small vesicles arise peripherally by a pinching-off process. Some become vacuoles in which secretory products are concentrated.	The Golgi apparatus functions in the secretion of cell products, such as proteins, to the exterior. It also helps to form the plasma membrane and the membranes of lysosymes.
Microbody (peroxisome) Crystalline array	Microbodies are single-membrane vesicles about 0.5 μm in diameter. They contain catalase, D-amino acid oxidase, urate oxidase, and other oxidative enzymes often present in crystalline arrays.	Microbodies participate in the oxidation of certain nutrients. Hydrogen peroxide, the reduction product of oxygen in these organelles, is decomposed to form water and oxygen.
Lysosome 	Lysosomes are single-membrane vesicles, 0.25–0.5 μm in diameter, containing hydrolytic enzymes, such as ribonuclease and phosphatase.	Lysosomes function in the digestion of materials brought into the cell by phagocytosis or pinocytosis. They also serve to digest cell components after cell death.
Endoplasmic reticulum and ribosomes Cisternae Ribosomes	The endoplasmic reticulum consists of flattened, single-membrane vesicles whose inner compartments, the cisternae, interconnect to form channels throughout the cytoplasm. The rough-surfaced portion is studded with ribosomes, which are larger than those of prokaryotes.	Proteins synthesized by the adhering ribosomes cross the membrane of the endoplasmic reticulum and appear in the intracisternal space, which forms a highly ramified channel for intracellular transport to the periphery of the cell. Protein synthesis by unattached ribosomes also occurs, as in prokaryotes.

Figure 1-9
Structural organization of a photosynthetic leaf cell of a higher plant.
The parenchymal cells in the mesophyll of leaves of higher plants are active in photosynthesis. They contain most of the distinctive organelles and structures observed in eukaryotic cells of animals, such as nucleus, mitochondria, Golgi apparatus, endoplasmic reticulum, and ribosomes. In addition, leaf cells contain three other major structures usually absent in animal cells: plastids (including chloroplasts), large vacuoles, and thick, rigid cell walls. Green leaf cells are rich in the pigment chlorophyll, which is localized in the chloroplasts. The major biochemical activities of the parenchymal cell are photosynthetic formation of glucose from CO_2 and H_2O and the storage of glucose as starch. In the dark, photosynthetic cells oxidize glucose and other fuels at the expense of atmospheric oxygen.

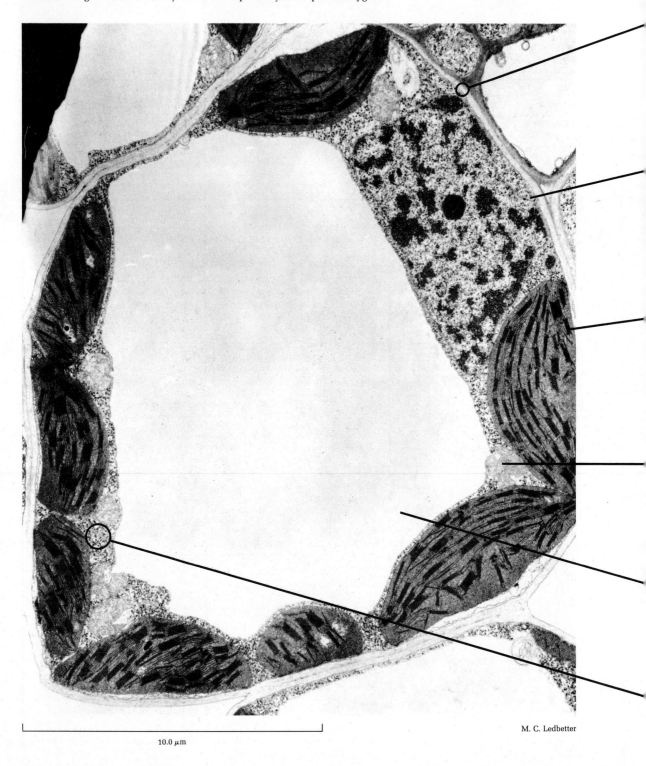

M. C. Ledbetter

10.0 μm

Electron micrograph of a section through a parenchymal leaf cell of Phleum pratense (timothy).

Schematic drawing	Molecular composition	Properties and functions
Cell wall and membrane 	The plant cell wall is thick, rigid, and boxlike. It consists of cellulose fibrils encased in a cement of polysaccharides and proteins. The cell membrane of plants is generally similar in thickness, structure, and composition to animal cell membranes, although lipid components differ somewhat.	The rather porous cell wall protects the cell membrane from mechanical or osmotic rupture, firmly fixes the position of the cell, and confers physical shape and strength upon plant tissue. The cell membrane of plant cells is selective in permeability, containing active-transport systems for specific nutrients and inorganic ions and also certain enzymes.
Nucleus 	The nucleus, nucleolus, and perinuclear membrane of plant cells are grossly similar in structure and composition to those of animal cells.	Chromosomes in plant cells undergo replication of their DNA, as in animal cells.
Chloroplast 	The cells of higher plants characteristically contain plastids, membrane-surrounded organelles some of which possess a distinctive DNA. Those containing chlorophyll are called chloroplasts. Chloroplasts are relatively large compared to mitochondria. There may be one, several, or many chloroplasts per cell, depending on the species; they may assume different forms.	Chloroplasts are receptors of light energy, which they convert into the chemical energy of ATP for the biosynthesis of glucose and other organic biomolecules from carbon dioxide, water, and other precursors. Oxygen is generated during plant photosynthesis. Chloroplasts are the main source of energy of photosynthetic cells in the light.
Mitochondrion 	Mitochondria are found in all plant cells, including photosynthetic cells. Their structural organization is similar to that of animal-cell mitochondria, as is their molecular and enzymatic composition. They also contain a specific type of DNA.	Mitochondria in plant cells promote oxidation of nutrients and conversion of energy into ATP, as in animal cells. In nonphotosynthetic plant cells the mitochondria are the main source of energy via respiration. In photosynthetic cells mitochondrial respiration is the main source of energy in the dark.
Vacuole 	Vacuoles are characteristic of plant cells. They are small in young cells and increase greatly in size with age, often causing the cytoplasm to become compressed against the cell wall. They contain dissolved sugars, salts of organic acids, proteins, mineral salts, pigments, oxygen, and carbon dioxide.	Vacuoles segregate waste products of plant cells and remove salts and other solutes, which gradually increase in concentration during the lifetime of the cell. Sometimes certain solutes crystallize within vacuoles.
Endoplasmic reticulum 	The endoplasmic reticulum of plant cells is similar in structure to that in animal cells, but the ribosomes of plant cells are slightly different in size and chemical composition from those in animal cells.	Ribosomes are the site of synthesis of proteins in plant cells. The endoplasmic reticulum serves to channel protein products through the cytoplasm.

Summary

Living matter requires only 27 of the 90 common chemical elements found in the crust of the earth; the four elements carbon, hydrogen, nitrogen, and oxygen make up 99 percent of the total mass of most living organisms. Nearly all the nonaqueous portion of living cells consists of organic compounds of carbon, which are otherwise very sparse on the earth's surface. Carbon appears to be uniquely fit for the backbone structure of biomolecules because of its capacity to form stable covalent bonds with hydrogen, oxygen, and nitrogen, and, above all, other carbon atoms.

There is a hierarchy in the molecular organization of cells. Simple precursors obtained from the environment, such as carbon dioxide, water, and ammonia, are used to form the building-block molecules, such as amino acids, nucleotides, sugars, and fatty acids. These in turn are joined covalently to form various macromolecules, the proteins, nucleic acids, polysaccharides, and lipids. Macromolecules are noncovalently bound together into supramolecular complexes, which ultimately are assembled into cell organelles.

The first primitive cells may have been formed from a relatively small number of primordial biomolecules, which probably included twenty different α-amino acids, five purine and pyrimidine bases, two sugars, a fatty acid, glycerol, and choline. From these primordial biomolecules are descended hundreds of other biomolecules, serving more specialized and differentiated functions in various organisms. The primordial biomolecules probably had an abiotic origin, arising by interaction of the components of the primitive atmosphere under the influence of radiant energy or lightning discharges. The primitive sea, it is believed, contained a large number of simple organic compounds. From this primordial soup were selected those molecules most suited for the formation and survival of the first living organisms. Presumably the biomolecules are the simplest, most versatile, and most fit molecules for their multiple functions in cells. The size, shape, and surface characteristics of biomolecules are exceedingly important for the specificity of their biological interactions and their function as building blocks of macromolecules.

References

Books

FAWCETT, D. W.: *The Cell: An Atlas of Fine Structure,* Saunders, Philadelphia, 1966. Excellent electron micrographs of cells and their organelles and inclusions.

HENDERSON, L. J.: *The Fitness of the Environment,* Macmillan, New York, 1927; reprinted 1958. A classic statement.

HENDRICKSON, J. B.: *The Molecules of Nature,* Benjamin, New York, 1965. A paperback review of the structural interrelationships among alkaloids, terpenes, acetogenins, and other natural products.

LEDBETTER, M. C., and K. R. PORTER: *Introduction to the Fine Structure of Plant Cells,* Springer-Verlag, New York, 1970.

LOEWY, A., and SIEKEVITZ, P.: *Cell Structure and Function,* 2d ed., Holt, New York, 1970.

OPARIN, A. I.: *Life: Its Nature, Origin, and Development,* Academic, New York, 1962.

ORGEL, L.: *The Origins of Life: Molecules and Natural Selection,* Chapman & Hall, London, 1973. Up-to-date summary of modern research on the origin of biomolecules.

PORTER, K. R., and BONNEVILLE, M. A.: *Fine Structure of Cells and Tissues,* Lea & Febiger, Philadelphia, 1972.

SPEAKMAN, J. C.: *Molecules,* McGraw-Hill, New York, 1966. Paperback on the structure and properties of molecules in relation to biology.

WOLFE, S. L.: *Biology of the Cell,* Wadsworth, Belmont, Calif., 1972. Another textbook of cell biology.

Articles

FRIEDEN, E.: "The Chemical Elements of Life," *Sci. Am.,* 227: 52–64 (1972). An interesting account of the role of various chemical elements in biology.

PALADE, G. E.: "The Organization of Living Matter," pp. 179–203, in *The Scientific Endeavor,* Rockefeller Institute Press, New York, 1964.

WALD, G.: "The Origins of Life," pp. 113–134, in *The Scientific Endeavor,* Rockefeller Institute Press, New York, 1964.

Water not only makes up 70 to 90 percent of the weight of most forms of life, it also represents the continuous phase of living organisms. Because it is familiar and ubiquitous, water is often regarded as a bland, inert liquid, a mere space filler in living organisms. Actually, however, it is a highly reactive substance with unusual properties that distinguish it strikingly from most other common liquids. We now recognize that water and its ionization products, hydronium and hydroxide ions, are important determinants of the characteristic structure and biological properties of proteins and nucleic acids, as well as membranes, ribosomes, and many other cell components.

Physical Properties and Hydrogen Bonding of Water

Water has a higher melting point, boiling point, heat of vaporization, heat of fusion, and surface tension than such comparable hydrides as H_2S or NH_3 or, for that matter, than most common liquids. All these properties indicate that the forces of attraction between the molecules in liquid water, and thus its internal cohesion, are relatively high. For example, Table 2-1 shows that the heat of vaporization of water is considerably higher than that of any of the other common liquids listed. The heat of vaporization is a direct measure of the amount of energy required to overcome the attractive forces between adjacent molecules in a liquid so that individual molecules can escape from each other and enter the gaseous state.

The strong intermolecular forces in liquid water are caused by the specific distribution of electrons in the water molecule. Each of the two hydrogen atoms shares a pair of electrons with the oxygen atom, through overlap of the $1s$ orbitals of the hydrogen atoms with two hybridized sp^3 orbitals of the oxygen atom. From spectroscopic and x-ray analyses the precise H—O—H bond angle is 104.5°, and the average hydrogen-oxygen interatomic distance is 0.0965 nm (Figure 2-1). This arrangement of electrons in the water mole-

Table 2-1 Heat of vaporization of some common liquids at their boiling point (1.0 atm)

Liquid	ΔH_{vap}, cal g^{-1}
Water	540
Methanol	263
Ethanol	204
n-Propanol	164
Acetone	125
Benzene	94
Chloroform	59

cule gives it electrical asymmetry. The highly electronegative oxygen atom tends to withdraw the single electrons from the hydrogen atoms, leaving the hydrogen nuclei bare. As a result, each of the two hydrogen atoms has a local partial positive charge (designated δ^+). The oxygen atom, in turn, has a local partial negative charge (designated δ^-) located in the zone of the unshared orbitals. Thus, although the water molecule has no net charge, it is an electric dipole.

When two water molecules approach each other closely, electrostatic attraction occurs between the partial negative charge on the oxygen atom of one water molecule and the partial positive charge on a hydrogen atom of an adjacent water molecule. This is accompanied by a redistribution of the electronic charges in both molecules which greatly enhances their interaction. A complex electrostatic union of this kind is called a _hydrogen bond_. Because of the nearly tetrahedral arrangement of the electrons about the oxygen atom, each water molecule is potentially capable of hydrogen-bonding with four neighboring water molecules (Figure 2-2). It is this property that is responsible for the great internal cohesion of liquid water.

Hydrogen bonds are relatively weak compared with covalent bonds. The hydrogen bonds in liquid water are estimated to have a bond energy of only about 4.5 kcal mol^{-1}, compared with 110 kcal mol^{-1} for the covalent H—O bonds in the water molecule (note that bond energy is the energy required to break a bond). Another important property of hydrogen bonds is that they are strongest when the two interacting groups are oriented to yield maximum electrostatic

Figure 2-1
The water molecule. The outline of the space-filling model represents the border at which van der Waals attractions are counterbalanced by repulsive forces. (Below) Within the outline of the space-filling model is shown a ball-and-stick, or crystallographic, model of the water molecule, giving the bond angle and length.

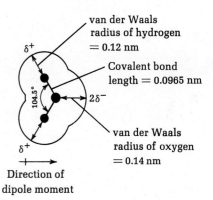

Figure 2-2
(Below) _Tetrahedral hydrogen bonding around a water molecule in ice. Molecules 1 and 2 and the central molecule are in the plane of the page; molecule 3 is above it, and molecule 4 is behind it. (Right) Schematic diagram of the lattice of water molecules in ice._ [_J. D. Watson, The Molecular Biology of the Gene, p. 120, W. A. Benjamin, Inc., New York, 1965._]

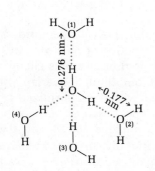

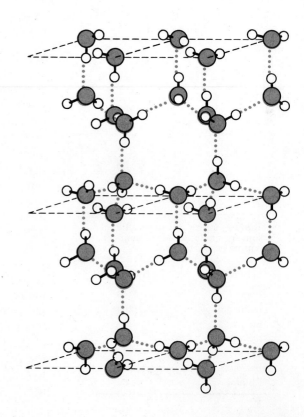

Figure 2-3
Directionality of the hydrogen bond. (Above) When the O—H bond is linear with the acceptor atom, the hydrogen bond formed will have maximum stability. (Below) When the acceptor oxygen is not on the electrical vector of the O—H bond, a weaker hydrogen bond is formed.

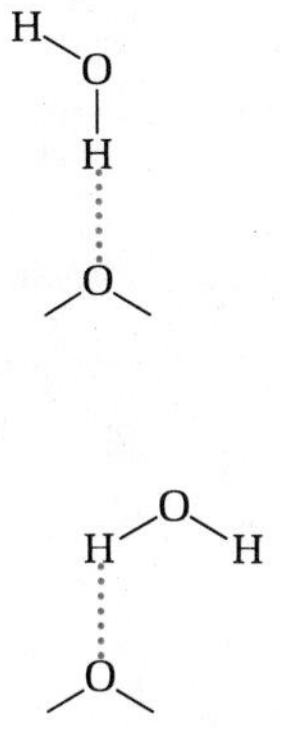

attraction (Figure 2-3). Hydrogen bonds also have a characteristic bond length, which differs from one type of hydrogen bond to another, according to the structural geometry and the electron distribution in the bonded molecules. In ice, for example, the length of each hydrogen bond is 0.177 nm (Figure 2-2).

The Structure of Liquid Water

Hydrogen bonding between water molecules occurs not only in liquid water but also in ice and in water vapor. In the most common crystalline form of ice, called ice I, each water molecule is hydrogen-bonded with exactly four nearest neighbors in a regular lattice having an average oxygen-oxygen distance of 0.276 nm. In liquid water at 0°C, each water molecule is hydrogen-bonded at any given time with an average of about 3.6 other water molecules; the average oxygen-oxygen distance is only slightly greater than in ice, about 0.29 nm at 15°C and 0.305 nm at 83°C. It has been estimated from the heat of fusion of ice that only a small fraction, perhaps about 10 percent, of the hydrogen bonds in ice are broken when it is melted to water at 0°C. Thus liquid water has considerable short-range order but no long-range structure. Liquid water is still highly hydrogen-bonded at 100°C, as indicated by its high heat of vaporization and dielectric constant.

The small difference in degree of hydrogen bonding between ice and liquid water may appear surprising in view of the rigidity of ice and the fluidity of liquid water. Part of the explanation lies in the very high rate at which hydrogen bonds in liquid water are made and broken. Although at any given time most of the molecules in liquid water are hydrogen-bonded, the half-life of each hydrogen bond is only about 10^{-11} s. The short-range structure of liquid water is therefore statistical, since it is averaged over both space and time. Consequently, liquid water is at the same time fluid and highly hydrogen-bonded.

Many models have been proposed for the structure of liquid water, but none has been completely verified experimentally. The simplest models suggest that liquid water consists of icelike clusters of water molecules in labile equilibrium with free water molecules. Other models propose that liquid water contains three or more types of hydrogen-bonded components. Yet another, the *continuum model,* suggests that although the great majority of the hydrogen bonds between water molecules in ice at 0°C remain unbroken when ice is melted, they become distorted, i.e., bent at different angles from the most stable linear configuration shown in Figure 2-3. The higher the temperature of liquid water, the greater the amount of distortion and the greater the instability. In this model the individual short-lived domains of water molecules, although highly hydrogen-bonded, deviate significantly from the regular latticelike structure of ice. Clearly, continued research into the structure of liquid water will be of great significance to biochemistry and molecular biology.

Other Properties of Hydrogen Bonds

Hydrogen bonds are not unique to water. They tend to form between a small, highly electronegative atom, e.g., oxygen, nitrogen, or fluorine, and a hydrogen atom covalently bonded to another electronegative atom. Hydrogen bonds may form between two molecules or between two parts of the same molecule. Some biologically important hydrogen bonds are shown in Figure 2-4.

When two molecules are joined by a single hydrogen bond, the bond will be very weak in an aqueous system because the surrounding water molecules compete to form hydrogen bonds with the solute molecules. However, when it is possible for two or more hydrogen bonds to form between two solute molecules, for geometrical reasons establishment of the first hydrogen bond greatly increases the probability that the second bond will form. Once the second bond forms, the probability is increased that a third will form, and so on, leading to a very strong association between the two solute molecules, which can overcome the competing effects of water molecules. Such enhancement of the strength of the attraction between two molecules by the cooperation of many weak bonds is called *cooperativity*.

Cooperative hydrogen bonding is a characteristic of both protein (page 128) and nucleic acid (page 864) molecules, which may contain dozens, hundreds, or even thousands of cooperative hydrogen bonds. Cooperativity of hydrogen bonding in biological macromolecules depends on the relative positions of the functional groups capable of forming hydrogen bonds. It is something like putting together a ladder. Once one or two rungs are in place, so that the side rails are lined up, all the rest of the rungs can be nailed on quickly.

Hydrogen bonds form and break in aqueous systems much faster than most covalent bonds. This fact, together with their geometrical specificity and directionality, endows hydrogen bonds with a great biological advantage over covalent bonds in biomolecular phenomena that must occur at very high rates, such as the folding of proteins into their native conformations (pages 143 and 144).

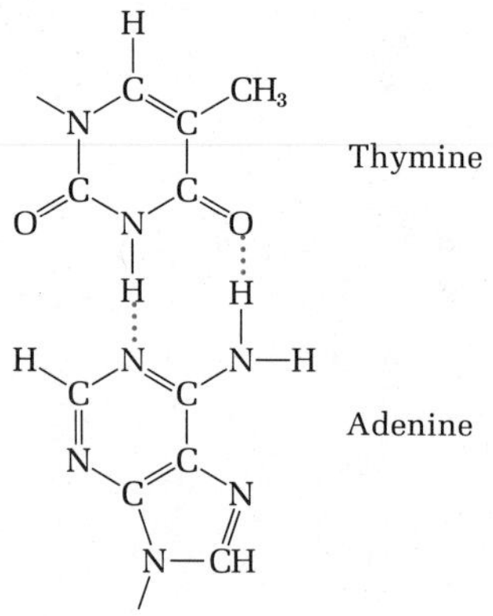

Solvent Properties of Water

Water is a much better solvent than most common liquids. Many crystalline salts and other ionic compounds readily dissolve in water but are nearly insoluble in nonpolar liquids like chloroform or benzene. Since the crystal lattice of salts, e.g., sodium chloride, is held together by very strong electrostatic attractions between alternating positive and negative ions, considerable energy is required to pull these ions away from each other. Water, however, readily dissolves crystalline sodium chloride because the strong electrostatic attraction between water dipoles and the Na^+ and Cl^- ions, which leads to the very stable hydrated Na^+ and Cl^- ions, greatly exceeds the tendency of Na^+ and Cl^- to attract each other.

Figure 2-4
Some hydrogen bonds of biological importance.

Between a hydroxyl group and H_2O

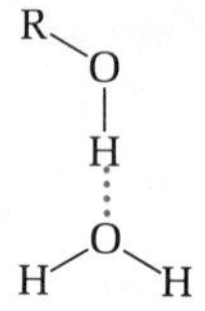

Between a carbonyl group and H_2O

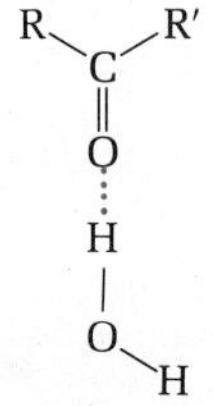

Between two peptide chains

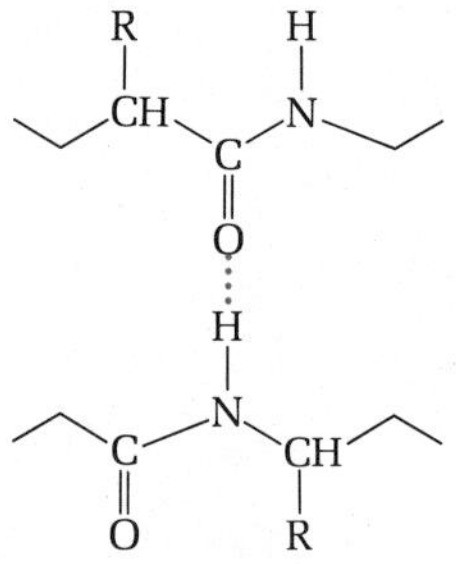

Between complementary base pairs in DNA

Thymine

Adenine

Table 2-2 Dielectric constant D of some liquids at 20°C

Liquid	D
Water	80
Methanol	33
Ethanol	24
Acetone	21.4
Benzene	2.3
Hexane	1.9

Figure 2-5
Formation of a soap micelle in water. The nonpolar tails of the sodium oleate are hidden from the water, whereas the negatively charged carboxyl groups are exposed.

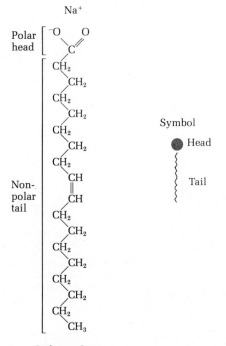

Sodium oleate

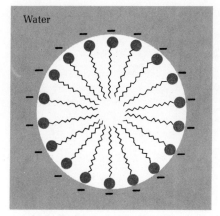

Sodium oleate micelle

Water also tends to oppose the electrostatic attraction between positive and negative ions. This tendency is given by the *dielectric constant D*, defined by the relationship

$$F = \frac{e_1 e_2}{D r^2}$$

where F is the attractive force between two ions of opposite charge, e_1 and e_2 are the charges on the ions, and r is the distance between them. As Table 2-2 shows, water has a relatively high dielectric constant and benzene a very low one. The attractive force between Na^+ and Cl^- ions at a given distance in water is only one-fortieth that in benzene, greatly favoring dissolution of the crystal lattice by water.

A second large class of substances readily dissolved by water includes the nonionic but polar compounds—sugars, simple alcohols, aldehydes, and ketones. Their solubility is due to the propensity of polar functional groups, such as the hydroxyl groups of sugars and alcohols and the carbonyl oxygen atom of aldehydes and ketones, to form hydrogen bonds with water molecules (Figure 2-4).

Hydrophobic Interactions

Water also disperses or solubilizes in the form of *micelles* many compounds which contain both strongly nonpolar and strongly polar groups. Such molecules are called *amphipathic*.

An example of an amphipathic biomolecule that tends to form micelles is the sodium salt of the long-chain fatty acid *oleic acid*. This molecule has a single carboxyl group, which is polar and thus tends to hydrate readily, and a long hydrocarbon tail, which is nonpolar and intrinsically insoluble in water. Because of this long, hydrophobic tail there is very little tendency for sodium oleate (a soap) to dissolve in water to yield a true ionic solution. However, it readily disperses in water to form micelles, in which the negatively charged carboxylate groups are exposed and form hydrogen bonds with water molecules and the nonpolar, insoluble hydrocarbon chains, which do not hydrogen-bond with water, are hidden within (Figure 2-5). Soap micelles have a net negative charge and remain suspended because of mutual electrostatic repulsion. Such micelles, which may contain hundreds or even thousands of soap molecules, form spontaneously because water "likes" water (and also carboxyl groups) more than it "likes" nonpolar structures. Within micelles additional attractive forces between adjacent hydrocarbon structures are provided by van der Waals interactions.

It must be emphasized that there is no true stoichiometric bonding between the hydrocarbon tails in micelles. For this reason we use the term *hydrophobic interaction* rather than "hydrophobic bond" to refer to the clustering or aggregation of the hydrophobic portions of amphipathic molecules out of contact with water. Hydrophobic interactions have relatively little directionality, compared with hydrogen bonds, but they tend to produce systems of high stability.

As we shall see later, many cell components are amphipathic and tend to form structures in which the nonpolar, hydrophobic parts are hidden from water—in particular, the proteins (page 142), the nucleic acids (page 865), and the polar lipids (page 287).

Effect of Solutes on Water Properties

The presence of dissolved solutes causes the structure and properties of liquid water to change. For example, when a salt such as NaCl is dissolved in water, the Na^+ and Cl^- ions become surrounded by shells of water dipoles. In these ion hydrates the geometry of water molecules differs from that of the clusters of hydrogen-bonded water molecules in pure water: the molecules are more highly ordered and more regular in structure. Table 2-3 gives the average interionic distance in aqueous NaCl solutions as a function of the concentration of NaCl. We see that at 0.15 M NaCl, the approximate concentration of NaCl in blood plasma (and of K^+ salts in the cytoplasm of cells), Na^+ and Cl^- ions are separated by only about 1.9 nm on the average. Since each hydrated Na^+ and Cl^- ion is 0.5 to 0.7 nm in diameter and a tetrahedral cluster of five water molecules is about 0.5 nm in diameter, the three-dimensional structure of liquid water is significantly altered when NaCl is dissolved in a concentration approximating that occurring in biological fluids. Dissolved salts thus tend to "break" the normal structure of liquid water.

The effect of a solute on the solvent is manifest in another set of properties, namely, the *colligative properties* of solutions, which depend on the number of solute particles per unit volume of solvent. Solutes produce such characteristic effects in the solvent as depression of the freezing point, elevation of the boiling point, and depression of the vapor pressure. They also endow a solution with the property of osmotic pressure. One gram molecular weight of an ideal, nondissociating, nonassociating, nonvolatile solute dissolved in 1,000 g of water at a pressure of 760 mm Hg depresses the freezing point by 1.86°C and elevates the boiling point by 0.543°C. Such a solution also yields an osmotic pressure of 22.4 atm when measured in an appropriate apparatus (Chapter 7). Since aqueous solutions usually deviate from ideal behavior, the proportionality between molal concentration and freezing-point depression, for example, holds exactly only in very dilute solution, i.e., on extrapolation to zero concentration of solute.

Ionization of Water

Because the mass of the hydrogen atom is so small and because the atom's single electron is tightly held by the oxygen atom, there is a finite tendency for a hydrogen ion to dissociate from the oxygen atom to which it is covalently bound in one water molecule and "jump" to the oxygen atom of the adjacent water molecule to which it is hydrogen-

Table 2-3 Average interionic distance in solutions of NaCl

Concentration, M	Distance, nm
0.001	9.4
0.010	4.4
0.10	2.0
0.150	1.9
1.00	0.94

Figure 2-6
*Hydrated form of hydronium ion ($H_9O_4^+$).
The hydration shell is stable to 100°C.*

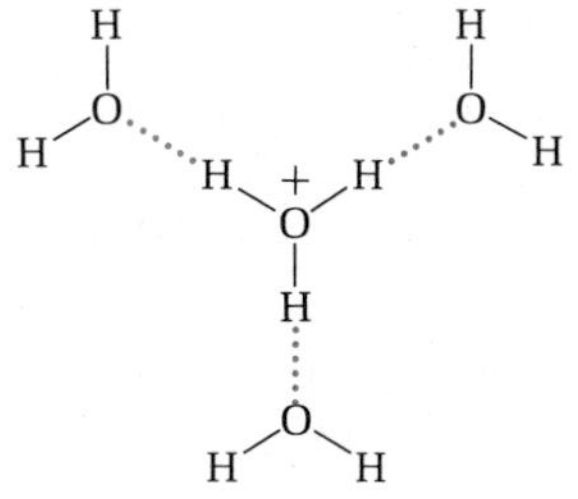

Table 2-4   Electrical mobility of some cations at infinite dilution (25°C)

Ion	Mobility, $cm^2\ V^{-1}\ s^{-1}$
H^+	36.3×10^{-4}
Na^+	5.2×10^{-4}
K^+	7.6×10^{-4}
NH_4^+	7.6×10^{-4}
Mg^{2+}	5.4×10^{-4}
Li^+	4.0×10^{-4}

Figure 2-7
*Proton jumps. The curved arrows show the
path taken by protons in successive jumps
from a hydronium ion to a water molecule.
In liquid water the jumps are random in
space; in ice they occur along the hydrogen-
bonded lattice. The last H_2O molecule
(color) becomes a hydronium ion at the end
of the series of proton jumps.*

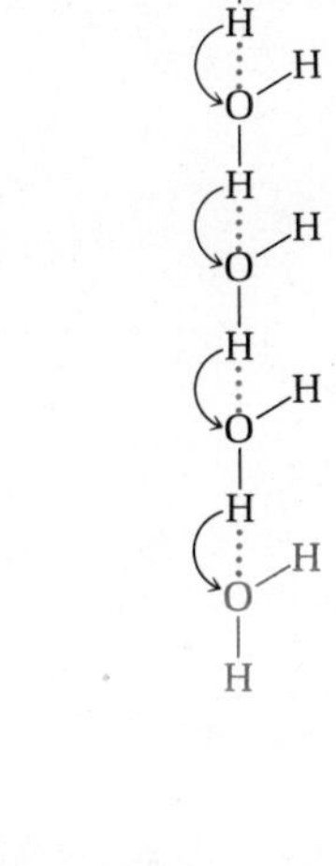

bonded, provided that the internal energy of each molecule is favorable:

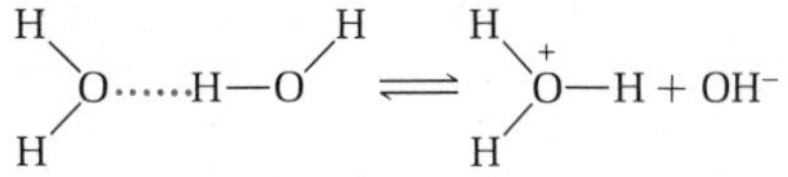

In this reaction two ions are produced, the hydronium ion (H_3O^+) and the hydroxide ion (OH^-). In a liter of pure water at 25°C at any given time there is only 1.0×10^{-7} mol of H_3O^+ ions and an equal amount of OH^- ions, as shown by electrical-conductivity measurements.

Although it has become the convention, for brevity, to use the symbol H^+ to designate the hydronium ion, it must be strongly emphasized that protons or hydrogen ions do not exist "bare" in water to any significant extent; they occur only in hydrated form. Moreover, the H_3O^+ or hydronium ion is itself further hydrated through additional hydrogen bonding with water to form the $H_9O_4^+$ ion, as well as more highly hydrated forms (Figure 2-6). The hydroxide ion is also hydrated in liquid water.

Table 2-4 shows that the apparent rate of migration of H_3O^+ ions in an electric field is many times greater than that of the univalent cations Na^+ and K^+. This anomaly results because a proton can jump very rapidly from a hydronium ion to a neighboring water molecule to which it is hydrogen-bonded. Thus a positive electric charge can move a given distance from one molecule of water to another with little movement of the water molecules themselves. A series of such proton jumps has the effect of translocating protons at a rate that is much higher than the rate of diffusive or bulk movement of H_3O^+ ions per se (Figure 2-7). Proton jumps along immobilized water molecules in the crystal lattice of ice are responsible for the fact that ice, despite its rigid structure, has about the same electrical conductivity as liquid water. Conduction of protons through hydrogen-bonded water molecules, called <u>tunneling</u>, may be an important phenomenon in biological systems.

The Ion Product of Water: The pH Scale

The dissociation of water is an equilibrium process:

$$H_2O \rightleftharpoons H^+ + OH^-$$

for which we can write the equilibrium constant

$$K_{eq} = \frac{[H^+][OH^-]}{[H_2O]}$$

where the brackets indicate concentration in moles per liter. The magnitude of the equilibrium constant at any given temperature can be calculated from conductivity measurements on pure distilled water. Since the concentration of water in pure water is very high (it is equal to the number of grams of H_2O in a liter divided by the gram molecular weight of water, or $1,000/18 = 55.5\ M$) and since the concentrations of H^+ and OH^- ions are very low in comparison ($1 \times 10^{-7}\ M$ at 25°C), the molar concentration of water is not significantly changed by its very slight ionization. The equilibrium-constant expression may thus be simplified to

$$55.5K_{eq} = [H^+][OH^-]$$

and the term $55.5K_{eq}$ can then be replaced by a lumped constant K_w, called the *ion product* of water,

$$K_w = [H^+][OH^-]$$

The value of K_w at 25°C is 1.0×10^{-14}. In an acid solution, the H^+ concentration is relatively high and the OH^- concentration correspondingly low; in a basic solution, the situation is reversed.

K_w, the ion product of water, is the basis for the *pH scale* (Table 2-5), a means of designating the actual concentration of H^+ (and thus of OH^-) ions in any aqueous solution in the acidity range between $1.0\ M\ H^+$ and $1.0\ M\ OH^-$. The pH scale was devised by the Danish biochemist S. P. L. Sørensen as a means of avoiding cumbersome numbers like 0.0000001 or 1.0×10^{-7} to express the low hydrogen-ion concentrations in biological fluids. He defined the term pH as

$$pH = \log_{10} \frac{1}{[H^+]} = -\log_{10} [H^+]$$

In a precisely neutral solution at 25°C

$$[H^+] = [OH^-] = 1.0 \times 10^{-7}\ M$$

The pH of such a solution is

$$pH = \log \frac{1}{1 \times 10^{-7}} = 7.0$$

The value of 7.0 for the pH of a precisely neutral solution is thus not an arbitrarily chosen figure; it is derived from the absolute value of the ion product of water at 25°C. It is important to note that the higher the pH number, the lower the hydrogen-ion concentration, and vice versa. Note that the pH scale is logarithmic, not arithmetic. To say that two solutions differ in pH by 1 pH unit means only that one solution has 10 times the hydrogen-ion concentration of the other. Table 2-6 lists the pH of some fluids.

Measurement of pH

Measurement of pH is one of the most common and useful analytical procedures in biochemistry since the pH deter-

Table 2-5 The pH scale

$[H^+]$, M	pH	$[OH^-]$, M
1.0	0	10^{-14}
0.1	1	10^{-13}
0.01	2	10^{-12}
0.001	3	10^{-11}
0.0001	4	10^{-10}
0.00001	5	10^{-9}
10^{-6}	6	10^{-8}
10^{-7}	7	10^{-7}
10^{-8}	8	10^{-6}
10^{-9}	9	10^{-5}
10^{-10}	10	10^{-4}
10^{-11}	11	0.001
10^{-12}	12	0.01
10^{-13}	13	0.1
10^{-14}	14	1.0

Table 2-6 pH of some fluids

Fluid	pH
Seawater (varies)	7.5
Blood plasma	7.4
Interstitial fluid	7.4
Intracellular fluids	
Muscle	6.1
Liver	6.9
Gastric juice	1.2–3.0
Pancreatic juice	7.8–8.0
Saliva	6.35–6.85
Cow's milk	6.6
Urine	5–8
Tomato juice	4.3
Grapefruit juice	3.2
Soft drink (cola)	2.8
Lemon juice	2.3

mines many important aspects of the structure and activity of biological macromolecules and thus of the behavior of cells and organisms. The primary standard for measurement of hydrogen-ion concentration (and thus of pH) is the *hydrogen electrode,* a specially treated platinum electrode immersed in the solution whose pH is to be measured. The solution is in equilibrium with gaseous hydrogen at a known pressure and temperature. The electromotive force at the electrode responds to the equilibrium

$$H_2 \rightleftharpoons 2H^+ + 2e^-$$

The potential difference between the hydrogen electrode and a reference electrode of known emf, e.g., a calomel electrode, is measured and used to calculate the hydrogen-ion concentration.

The hydrogen electrode proved too cumbersome for general use and has been replaced by the *glass electrode,* which responds directly to hydrogen-ion concentration in the absence of hydrogen gas. The response of the glass electrode must be calibrated against buffers of precisely known pH. Another way of measuring pH is to use acid-base indicators (see below).

Acids and Bases

The most general and comprehensive definitions of acids and bases, applicable to both nonaqueous and aqueous systems, are those of G. N. Lewis. A Lewis acid is a potential *electron-pair acceptor,* and a Lewis base a potential *electron-pair donor.* However, the formalism introduced by J. N. Brönsted and T. M. Lowry is more widely used in describing acid-base reactions in dilute aqueous systems. According to the Brönsted-Lowry concepts, an acid is a *proton donor* and a base is a *proton acceptor* (Figure 2-8). An acid-base reaction always involves a *conjugate acid-base pair,* made up of a proton donor and the corresponding proton acceptor. For example, acetic acid (CH_3COOH) is a proton donor, and the acetate anion (CH_3COO^-) is the corresponding proton acceptor; together they constitute a conjugate acid-base pair.

The equation for the dissociation or ionization of an acid (HA) in dilute aqueous solution involves the transfer of a proton from the acid to water, which itself can act as a proton acceptor to yield the acid H_3O^+:

$$HA + H_2O \rightleftharpoons H_3O^+ + A^-$$

Each conjugate base has a characteristic affinity for a proton relative to the proton affinity of OH^-. Acids that have only a slight tendency to give up protons to water are *weak acids;* acids that readily give up their protons are *strong acids.* The tendency of any given acid to dissociate is given by its *dissociation constant* at a given temperature

$$K = \frac{[H^+][A^-]}{[HA][H_2O]}$$

Figure 2-8
Some conjugate acid-base pairs.

Proton donor	Proton acceptor
$CH_3COOH \rightleftharpoons$	$H^+ + CH_3COO^-$
$H_2PO_4^- \rightleftharpoons$	$H^+ + HPO_4^{2-}$
$NH_4^+ \rightleftharpoons$	$H^+ + NH_3$
$HOH \rightleftharpoons$	$H^+ + OH^-$

where the brackets indicate concentrations in moles per liter. It is conventional to simplify this expression by eliminating the water required for hydration of the proton:

$$K = \frac{[H^+][A^-]}{[HA]}$$

It is also the convention in biochemistry to employ dissociation constants based on the analytically measured concentrations of reactants and products under a given set of experimental conditions, i.e., at a given total concentration and ionic strength and with other solutes specified. Such a constant, called an *apparent* or *concentration dissociation constant*, is designated K' to distinguish it from the *true* or *thermodynamic dissociation constant* K employed by the physical chemist, which is corrected for deviation of the system from ideal behavior caused by such factors as concentration and ionic strength.

The apparent dissociation constants of some acids and bases are given in Table 2-7. Note that in the Brönsted-Lowry formalism, acids and bases are treated alike, i.e., solely in terms of the tendency of protons to dissociate from the proton-donor species. (So-called basic dissociation constants, such as K_b for the dissociation reaction $NH_4OH \rightleftharpoons NH_4^+ + OH^-$, are not employed. In fact, in the Brönsted-Lowry formalism, NH_4OH is neither an acid nor a base.) Table 2-7 also gives values for the expression pK', which is a logarithmic transformation of K', just as the term pH is a logarithmic transformation of $[H^+]$:

$$pK' = \log \frac{1}{K'} = -\log K'$$

The pK' values are less cumbersome to handle than K' values, just as the pH numbers are less cumbersome than actual hydrogen-ion molarities. Strong acids have low pK' values

Table 2-7 Apparent dissociation constant and pK' of some acids (25°C)

Acid (proton donor)	K', M	pK'
HCOOH	1.78×10^{-4}	3.75
CH_3COOH	1.74×10^{-5}	4.76
CH_3CH_2COOH (propionic acid)	1.35×10^{-5}	4.87
$CH_3CHOHCOOH$ (lactic acid)	1.38×10^{-4}	3.86
$COOHCH_2CH_2COOH$ (succinic acid)	6.16×10^{-5}	4.21
$COOHCH_2CH_2COO^-$	2.34×10^{-6}	5.63
H_3PO_4	7.25×10^{-3}	2.14
$H_2PO_4^-$	6.31×10^{-8}	7.20
HPO_4^{2-}	3.98×10^{-13}	12.4
H_2CO_3	1.70×10^{-4}	3.77
HCO_3^-	6.31×10^{-11}	10.2
NH_4^+	5.62×10^{-10}	9.25
$CH_3NH_3^+$	2.46×10^{-11}	10.6

and strong bases have high pK' values. Note that water itself may be considered to be a very weak acid of pK' about 14; its conjugate base, the hydroxide ion, is obviously a very strong base with a high affinity for a proton.

Figure 2-9 shows the titration curves of some weak acids titrated with sodium hydroxide. The pH resulting after each increment of NaOH is plotted against the equivalents of OH⁻ added. The shapes of such titration curves are very similar from one acid to another; the important difference is that the curves are displaced vertically along the pH scale. The pH intercept at the midpoint of the titration is numerically equal

Figure 2-9
(Right) *Acid-base titration curves of some acids, showing the major ionic species at the beginning, midpoint, and end of the titration.* (Below) *The relative buffering power of these acids plotted against pH. Maximum buffering power is given at pH = pK', at which there is minimum change in pH following addition of a given increment of acid or base.*

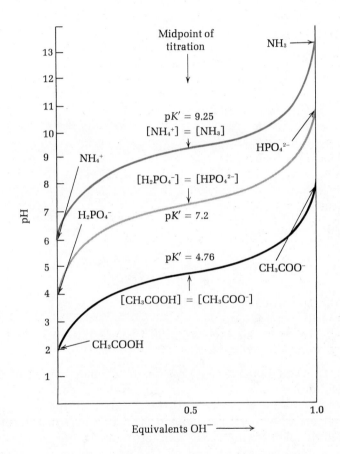

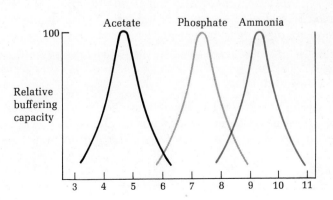

to the pK' of the acid titrated. At the midpoint, equimolar concentrations of proton-donor (HA) and proton-acceptor species (A⁻) of the acid are present. In fact, the pK' of an acid can be calculated from the pH at any point on the titration curve of an acid if the concentrations of the proton-donor and proton-acceptor species at this point are known. The shape of the titration curve can be expressed by the _Henderson-Hasselbalch equation,_ which is a logarithmic transformation of the expression for the dissociation constant. It is derived as follows:

$$K' = \frac{[H^+][A^-]}{[HA]}$$

Solve for H⁺:

$$H^+ = K' \frac{[HA]}{[A^-]}$$

Take the negative logarithm of both sides:

$$-\log[H^+] = -\log K' - \log \frac{[HA]}{[A^-]}$$

Substitute pH for $-\log [H^+]$ and pK' for $-\log K'$:

$$pH = pK' - \log \frac{[HA]}{[A^-]}$$

If we now change signs, we obtain the Henderson-Hasselbalch equation:

$$pH = pK' + \log \frac{[A^-]}{[HA]}$$

which in more general form is

$$pH = pK' + \log \frac{[\text{proton acceptor}]}{[\text{proton donor}]}$$

This equation makes it possible to calculate the pK' of any acid from the molar ratio of proton-donor and proton-acceptor species at a given pH, to calculate the pH of a conjugate acid-base pair of a given pK' and a given molar ratio, and to calculate the molar ratio of proton donor and proton acceptor given the pH and pK'. Note that when the concentrations of proton donor and proton acceptor are equal, the observed pH is numerically equal to the pK'. The Henderson-Hasselbalch equation is fundamental to quantitative treatment of all acid-base equilibria in biological systems.

Acid-Base Indicators

The pH of a solution can be determined by using indicator dyes, most of which are weak acids (designated HInd). Such an indicator dissociates according to the equilibrium

$$HInd \rightleftharpoons H^+ + Ind^-$$

Let us suppose that the species HInd is colorless and the species Ind⁻ is colored. The equilibrium position of this dissociation, and thus the amount of light absorbed by the species Ind⁻ at its characteristic wavelength, is determined by the ambient hydrogen-ion concentration. In a strongly acid solution, the equilibrium is shifted to the left, and little light will be absorbed at the wavelength of maximum absorption of Ind⁻. In a basic solution, the species Ind⁻ is favored since excess OH⁻ ions react with H⁺ ions to form H_2O, thus pulling the equilibrium to the Ind⁻ side. In this case, much light will be absorbed.

Buffers

Figure 2-9 shows that the titration curve of each acid has a relatively flat zone extending about 1.0 pH unit on either side of its midpoint. In this zone, the pH of the system changes relatively little when small increments of H⁺ or OH⁻ are added. This is the zone in which the conjugate acid-base pair acts as a *buffer,* a system which tends to resist change in pH when a given increment of H⁺ or OH⁻ is added. At pH values outside this zone there is less capacity to resist changes in pH. The buffering power is maximum at the pH of the exact midpoint of the titration curve, at which the concentration of the proton acceptor equals that of the proton donor and pH = pK′ (Figure 2-9). Buffering power decreases as the pH is raised or lowered from this point, a direct consequence of the change in ratio of the proton-acceptor and proton-donor species. Each conjugate acid-base pair has a characteristic pH at which its buffering capacity is greatest, namely, the point at which pH = pK′.

Intracellular and extracellular fluids of living organisms contain conjugate acid-base pairs which act as buffers at the normal pH of these fluids. The major intracellular buffer is the conjugate acid-base pair $H_2PO_4^-$–HPO_4^{2-} (pK′ = 7.2). Organic phosphates such as glucose 6-phosphate and ATP also contribute buffering power in the cell. The major extracellular buffer in the blood and interstitial fluid of vertebrates is the bicarbonate buffer system. The extraordinary buffering power of blood plasma can be shown by the following comparison. If 1 ml of 10 N HCl is added to 1.0 l of neutral physiological saline, i.e., about 0.15 M NaCl, the pH of the saline will fall to pH 2.0, since NaCl solutions have no buffering power. However, if 1 ml of 10 N HCl is added to 1 l of blood plasma, the pH will decline only slightly, from pH 7.4 to about pH 7.2. Buffer action is best appreciated by solving problems with the Henderson-Hasselbalch equation.

The bicarbonate buffer system (H_2CO_3–HCO_3^-) has some distinctive features. While it functions as a buffer in the same way as other acid-base pairs, the pK′ of H_2CO_3, a relatively strong acid, is about 3.8 (Table 2-7), which is far lower than the normal range of blood pH. The question therefore arises why an acid having such a low pK′ is capable of serving as a physiological buffer at pH near 7.0. In a bicarbonate buffer system, the proton-donor species, carbonic acid, is in reversible equilibrium with dissolved CO_2:

$$H_2CO_3 \rightleftharpoons CO_2(aq) + H_2O$$

If such an aqueous system is in contact with a gas phase, the dissolved CO_2 will in turn equilibrate between the gaseous and aqueous phases:

$$CO_2(aq) \rightleftharpoons CO_2(g)$$

Since by Henry's law the solubility of a gas in water is proportional to its partial pressure, the pH of the bicarbonate buffer system is a function of the partial pressure of CO_2 in the gas phase over the buffer solution. If the CO_2 pressure is increased, all other variables remaining constant, the pH of the bicarbonate buffer declines, and vice versa. The bicarbonate system can buffer blood plasma effectively near pH 7.0, at which the proton-acceptor/proton-donor ratio is very high, because a small amount of proton donor H_2CO_3 is in labile equilibrium with a relatively large reserve capacity of gaseous CO_2 in the lungs. Under any conditions in which the blood must absorb excess OH^-, the H_2CO_3 which is used up and converted to HCO_3^- is quickly replaced from the large pool of gaseous CO_2 in the lungs.

There is another distinctive feature of the bicarbonate buffer system. CO_2 is a major end product of the aerobic combustion of fuel molecules and in mammals is ultimately eliminated via the lungs. The steady-state ratio of $[HCO_3^-]$ / $[H_2CO_3]$ in the blood is a reflection of the rate of CO_2 production during tissue oxidation and the rate of loss of CO_2 by expiration.

The pH of blood plasma in mammals is held at remarkably constant values. The blood plasma of man normally has a pH of 7.40. Should the pH-regulating mechanisms fail, as may happen in disease, and the pH of the blood fall below 7.0 or rise above 7.8, irreparable damage may occur. We may ask: What molecular mechanisms in cells are so extraordinarily sensitive that a change in hydrogen-ion concentration of as little as 3×10^{-8} M (approximately the difference between blood plasma at pH 7.4 and at pH 7.0) can be lethal? Although many aspects of cell structure and function are influenced by pH, the catalytic activity of enzymes is especially sensitive. The typical curves in Figure 2-10 (see also page 196) show that enzymes have maximal activity at a characteristic pH, called the *optimum pH*, and that their activity declines sharply on either side of the optimum. Thus biological control of the pH of cells and body fluids is of central importance in all aspects of intermediary metabolism and cellular function.

The Fitness of the Aqueous Environment for Living Organisms

Living organisms have effectively adapted to their aqueous environment and have even evolved means of exploiting the

Figure 2-10
The effect of pH on the activity of some enzymes. Each enzyme has a characteristic pH-activity profile. The pH may influence the degree of ionization not only of enzymes but also of coenzyme and substrate molecules.

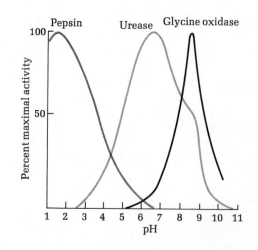

unusual properties of water. The high specific heat of water is useful to large terrestrial animals, because body water acts as a heat buffer, allowing the temperature of the organism to remain relatively constant as the air temperature fluctuates. Furthermore, the high heat of evaporation of water is exploited by vertebrates as an effective means of losing heat by evaporation of sweat. The high degree of internal cohesion of liquid water, due to hydrogen bonding, is exploited by higher plants to transport dissolved nutrients from the roots up to the leaves by transpiration. Even the fact that ice has a lower density than liquid water and therefore floats has important biological consequences in the ecology of aquatic organisms. But most fundamental to all living organisms is that many important biological properties of cell macromolecules, particularly the proteins and nucleic acids, derive from their interactions with water molecules of the surrounding medium, as we shall see.

Summary

Water is the most abundant compound in living organisms. Its relatively high freezing point, boiling point, heat of vaporization, and surface tension are the result of strong intermolecular attractions in the form of hydrogen bonding between neighboring water molecules. Liquid water has considerable short-range order although individual hydrogen bonds have a very short half-life.

The polarity and hydrogen-bonding properties of the water molecule make it a potent solvent for many ionic compounds and neutral molecules. Water also disperses amphipathic molecules, such as soaps, to form micelles, clusters of molecules in which the hydrophobic groups are hidden from exposure to water and the hydrophilic, or polar, groups are located on the external surface, exposed to water. The formation of micelles results from the tendency of surrounding water molecules to engage in maximum hydrogen bonding with each other and with externally located polar groups.

Water ionizes very slightly to form hydronium (H_3O^+) and hydroxide (OH^-) ions. Protons may readily jump from H_3O^+ to a neighboring hydrogen-bonded H_2O molecule. Such proton jumps account for the high electrical mobility of protons in water and ice. In dilute aqueous solutions, the concentrations of H^+ and OH^- ions are inversely related by the expression $K_w = [H^+][OH^-] = 1 \times 10^{-14}$ (25°C). The hydrogen-ion concentration of biological systems is expressed in terms of pH, defined as $pH = -\log [H^+]$. The pH of aqueous solutions is measured with the glass electrode or indicators.

Acids are defined as proton donors and bases as proton acceptors. A conjugate acid-base pair consists of a proton donor (HA) and the corresponding proton acceptor (A^-). The tendency of an acid to donate protons in aqueous solutions is expressed by its dissociation constant K' or by the function pK', defined as $-\log K'$. There is a quantitative relationship between the pH of a solution of a weak acid, its pK', and the ratio of the concentrations of its proton-acceptor and proton-donor species. This relationship is the Henderson-Hasselbalch equation. A conjugate acid-base pair can act as a buffer and resist changes in pH; its capacity to do so is greatest at the pH numerically equal to its pK'. The most important biological buffer pairs are H_2CO_3–HCO_3^- and $H_2PO_4^-$–HPO_4^{2-}. The catalytic activity of enzymes is strongly influenced by pH.

References

General reference works in biochemistry and physical chemistry are given in Appendix B.

Books

DAVENPORT, H. W.: *ABC of Acid-Base Chemistry*, 5th ed., University of Chicago Press, Chicago, 1969.

DAWES, E. A.: *Quantitative Problems in Biochemistry*, 5th ed., Williams & Wilkins, Baltimore, 1968. Succinct treatment of quantitative aspects of biochemistry; problem solving.

DICK, D. A. T.: *Cell Water*, Butterworth, Washington, D.C., 1966. Structure, properties, movement, and control of cellular water.

EDSALL, J. T., and J. WYMAN: *Biophysical Chemistry*, vol. 1, Academic, New York, 1958. Detailed treatment of water and solutions of electrolytes, including amino acids.

EISENBERG, D., and W. KAUZMANN: *The Structure and Properties of Water*, Oxford University Press, Fair Lawn, N.J., 1969. Most recent and authoritative monograph.

HENDERSON, L. J.: *The Fitness of the Environment*, Macmillan, New York, 1927; reprinted 1958. A classic statement which is still absorbing reading.

MONTGOMERY, R., and C. A. SWENSON: *Quantitative Problems in Biochemical Sciences*, Freeman, San Francisco, 1969. Problems in acid-base chemistry and buffer action, as well as in many other aspects of biochemistry.

MORRIS, J. G.: *A Biologist's Physical Chemistry*, Addison-Wesley, Reading, Mass., 1968. An excellent primer; chapters 5 and 6 are especially useful.

SEGEL, I. H.: *Biochemical Calculations*, Wiley, New York, 1968. Another useful book on quantitative aspects of biochemistry.

Articles

NARTEN, A. H., and H. A. LEVY: "Observed Diffraction Pattern and Proposed Models of Liquid Water," *Science*, 165: 447–454 (1969). A discussion of different hypotheses for the structure of liquid water.

WICKE, E.: "Structure, Formation, and Molecular Mobility in Water and Aqueous Solutions," *Angew. Chem. Int. Ed. (Engl.)*, 5: 106–112 (1966).

Problems

A large number of acid-base and buffer problems, which are fundamental to many operations in biochemical research, are given for practice.

1. Calculate:
   (a) The pH of 10^{-5} N HCl
   (b) The pOH of 5×10^{-3} N NaOH
   (c) The pH of 3×10^{-6} N KOH
   (d) The pH of 10^{-9} N HCl
   (e) The pOH of 3×10^{-4} N HCl
   (f) The pH of 7 mM H_2SO_4

2. Calculate the following (see Table 2-6):
   (a) The $[H^+]$ of blood plasma
   (b) The $[H^+]$ of the intracellular fluid of muscle
   (c) The $[OH^-]$ of tomato juice
   (d) The $[H^+]$ of gastric juice (pH 1.4)
   (e) The $[OH^-]$ of saliva (pH 6.5)
   (f) The $[H^+]$ of seawater

3. The pH of a 0.01 M solution of a given acid, HA, is 3.80. Calculate (a) K', (b) pK' of the acid.

4. The degree of dissociation, α, of an acid, HA, is defined by the relationship

$$\alpha = \frac{[A^-]}{[A^-] + [HA]}$$

Calculate the degree of dissociation for acetic acid ($pK' = 4.76$) at (a) 0.1 M and (b) 0.01 M, assuming that [HA] is not appreciably diminished by the dissociation.

5. Show that the degree of dissociation is also given by

$$\alpha = \frac{K'}{K' + [H^+]}$$

6. Calculate the degree of dissociation of acetic acid at (a) pH 3.0, (b) pH 4.0, and (c) pH 4.76.

7. Conductivity measurements show that a 0.1 M solution of propionic acid (CH_3CH_2COOH) is ionized 1.16 percent at 25°C. Calculate (a) the dissociation constant and (b) the pK' of propionic acid.

8. (a) Calculate the $[HCO_3^-]$ / $[H_2CO_3]$ ratio in blood plasma at pH 7.4 (see Table 2-7).
   (b) Calculate the $[HPO_4^{2-}]$ / $[H_2PO_4^-]$ ratio in blood plasma.
   (c) Which of these two conjugate acid-base pairs is the more effective buffer in a sample of blood plasma in a closed flask with no gas space?

9. The internal pH of a muscle cell is 6.8. Calculate the $[H_2PO_4^-]$ / $[HPO_4^{2-}]$ ratio in the cell. The second dissociation constant of phosphoric acid is 6.31×10^{-8} M.

10. (a) What volume of 0.10 N HCl must be added to 20.0 ml of 0.04 M phosphate buffer pH 6.50 containing urease in order to diminish its enzymatic activity by exactly 50 percent when assayed under conditions in which pH is the rate-limiting factor?
   (b) What will the hydrogen-ion concentration in such a solution be? (Use data in Figure 2-10; neglect buffering capacity of urease.)

11. Calculate (a) the pH and (b) the concentration of CO_3^{2-} ions in a solution 0.01 M in H_2CO_3.

12. On the basis of your answer to Problem 11, make a generalization relating the concentration of A^{2-} ions to pK_2' for acids of the type H_2A.

13. You have available 0.1 N NaOH and 0.1 N solutions of H_2SO_4, acetic acid ($pK' = 4.76$), lactic acid ($pK' = 3.86$), phosphoric acid ($pK' = 7.2$), and ammonium chloride ($pK' = 9.25$).
   (a) Assuming that the above solutions are used without dilution by addition of water, how would you prepare 50 ml of a buffered medium to keep the pH essentially constant at 5.40 in an enzyme experiment in which acid will be produced?

(b) If 0.1 milliequivalent of acid is produced in the 50 ml of buffered medium during the experiment, what will be the final pH of the buffer mixture prepared as in part (a)?

14. Calculate to three significant figures the pH of a 0.001 M solution of formic acid ($pK' = 3.75$)

    (a) Assuming that the concentration of hydrogen ion is very small compared to the concentration of formic acid.

    (b) Assuming that the concentration of hydrogen ion is not very small compared to the concentration of formic acid.

15. Repeat the calculations of Problem 14 for a 10^{-5} M solution of formic acid.

16. Calculate the pH of a 0.01 M solution of sodium acetate.

17. Calculate (a) the pH and (b) the concentration of acetate ion resulting from mixture of 400 ml of 0.1 M formic acid and 100 ml of 0.1 M sodium acetate. This mixture constitutes a two-component buffer. Such buffers are frequently used in the laboratory; moreover, cell and body fluids contain two or more buffer systems.

18. The following equilibria are involved in the dissociation of citric acid, a tribasic acid of the type H_3A, and one of several tricarboxylic acids important in metabolism:

	pK'
$H_3A \rightleftharpoons H^+ + H_2A^-$	3.13
$H_2A^- \rightleftharpoons H^+ + HA^{2-}$	4.76
$HA^{2-} \rightleftharpoons H^+ + A^{3-}$	6.40

Calculate the concentrations of each of the following forms in a solution at pH 5.0 in which the total concentration of the acid and its anion forms is 0.01 M: (a) H_3A, (b) H_2A^-, (c) HA^{2-}, and (d) A^{3-}.

CHAPTER 3 **PROTEINS AND THEIR BIOLOGICAL FUNCTIONS: A BIRD'S-EYE VIEW**

In this chapter we shall take a bird's-eye view of the chemical nature and biology of proteins. Our purpose is to provide orientation in this complex field and to define some essential terms and concepts, in preparation for the much more detailed treatment of protein structure and function that follows in Chapters 4 to 9.

Proteins are the most abundant organic molecules in cells, constituting 50 percent or more of their dry weight. They are found in every part of every cell, since they are fundamental in all aspects of cell structure and function. There are many different kinds of proteins, each specialized for a different biological function. Moreover, most of the genetic information is expressed by proteins. For this reason we must also survey the general nature of the genetic relationship between deoxyribonucleic acid and the structure of proteins, as well as the effect of mutations on protein structure. The structure of protein molecules and its relationship to their biological function and activity are central problems in biochemistry today.

Composition of Proteins

Hundreds of different proteins have been isolated in pure crystalline form (Figure 3-1). All contain carbon, hydrogen, nitrogen, and oxygen; nearly all contain sulfur. Some proteins contain additional elements, particularly phosphorus, iron, zinc, and copper. The molecular weights of proteins are very high, but on acid hydrolysis they all yield a group of simple organic compounds of low molecular weight, the _α-amino acids_ (Figures 1-3 and 3-2). These building-block molecules contain at least one carboxyl group and one α-amino group but differ from each other in the structure of their R groups, or side chains. Twenty different α-amino acids are commonly found as the building blocks of proteins (Chapter 4).

In protein molecules the amino acid residues are covalently linked to form very long, unbranched chains. They are united in a head-to-tail arrangement through substituted

Figure 3-1
Crystals of horse cytochrome c, a protein functioning in electron transport.

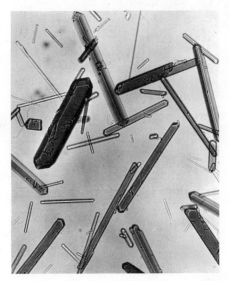

E. Margoliash

amide linkages called peptide bonds (Figure 3-2) that arise by elimination of the elements of water from the carboxyl group of one amino acid and the α-amino group of the next. These macromolecules, called polypeptides, may contain hundreds of amino acid units. Some proteins contain only one polypeptide chain; others contain two or more. The polypeptide chains of proteins are not random polymers of indefinite length; each polypeptide chain has a definite molecular weight, chemical composition, sequential order of its amino acid building blocks, and three-dimensional shape (Chapter 5).

Proteins are divided into two major classes on the basis of their composition: simple and conjugated. Simple proteins are those which on hydrolysis yield only amino acids and no other major organic or inorganic hydrolysis products. They usually contain about 50 percent carbon, 7 percent hydrogen, 23 percent oxygen, 16 percent nitrogen, and from 0 to 3 percent sulfur. Conjugated proteins are those yielding not only amino acids but also other organic or inorganic components. The non-amino acid portion of a conjugated protein is called its prosthetic group. Conjugated proteins may be classified on the basis of the chemical nature of their prosthetic groups (Table 3-1); thus we have nucleoproteins and lipoproteins, which contain nucleic acids and lipids, respectively, as well as phosphoproteins, metalloproteins, and glycoproteins.

The Size of Protein Molecules

By physical methods to be described in Chapter 7 the molecular weights of proteins can be determined. Some characteristic values are given in Table 3-2; they range from about 5,000, which is arbitrarily the lower limit, to 1 million or more. However, even among proteins having the same type of function we cannot make generalizations about size. Different enzymes, for example, vary in molecular weight from about 12,000 to over 1 million. The upper limit of the molecular weight of proteins can be set only arbitrarily, since it depends on how we define the terms protein and molecule, as we shall see.

Table 3-2 also shows that many proteins having molecular weights above 36,000 contain two or more polypeptide chains. The individual polypeptide chains of most proteins of known structure contain from 100 to 300 amino acid residues (mol wt 12,000 to 36,000). The single polypeptide chains of ribonuclease, cytochrome c, and myoglobin, which are among the best-known small proteins, contain between 100 and 155 amino acid residues. However, some proteins have much longer chains, such as serum albumin (approximately 550 residues) and myosin (approximately 1,800 residues).

The Conformation of Proteins

In its native state each type of protein molecule has a characteristic three-dimensional shape, referred to as its conformation (defined more rigorously in Chapter 6). Depending on

Figure 3-2
Amino acids, their R groups, and the structure of peptides.

General structural formula for the α-amino acids found in proteins. The shaded portion is common to all amino acids; the R groups are distinctive.

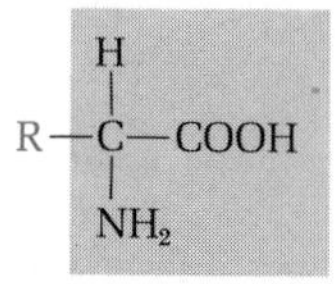

R groups of representative amino acids. The structures of all the amino acids of proteins are shown on pages 73 to 75.

R group	Name of amino acid
H—	Glycine
CH$_3$—	Alanine
HOCH$_2$—	Serine
CH$_2$—	Phenylalanine

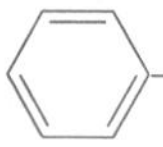

Structure of a tetrapeptide.

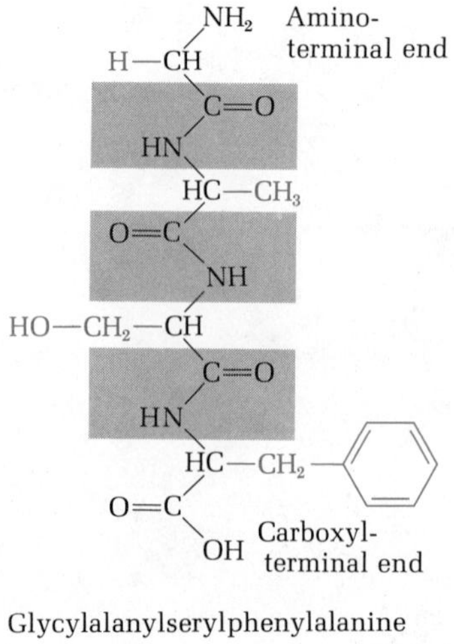

Glycylalanylserylphenylalanine

Table 3-1   Some conjugated proteins

Class	Prosthetic group components	Approximate percentage of weight
Nucleoprotein systems		
Ribosomes	RNA	50–60
Tobacco mosaic virus	RNA	5
Lipoproteins		
Plasma β_1-lipoproteins	Phospholipid, cholesterol, neutral lipid	79
Glycoproteins		
γ-Globulin	Hexosamine, galactose, mannose, sialic acid	2
Plasma orosomucoid	Galactose, mannose, N-acetylgalactosamine, N-acetylneuraminic acid	40
Phosphoproteins		
Casein (milk)	Phosphate esterified to serine residues	4
Hemoproteins		
Hemoglobin	Iron protoporphyrin	4
Cytochrome c	Iron protoporphyrin	4
Catalase	Iron protoporphyrin	3.1
Flavoproteins		
Succinate dehydrogenase	Flavin adenine dinucleotide	2
D-Amino acid oxidase	Flavin adenine dinucleotide	2
Metalloproteins		
Ferritin	$Fe(OH)_3$	23
Cytochrome oxidase	Fe and Cu	0.3
Alcohol dehydrogenase	Zn	0.3
Xanthine oxidase	Mo and Fe	0.4

Table 3-2   Molecular weights of some proteins

Protein	Molecular weight	No. of chains
Insulin (bovine)	5,700	2
Ribonuclease I (bovine pancreas)	12,600	1
Lysozyme (egg white)	13,900	1
Myoglobin (horse heart)	16,900	1
Chymotrypsinogen (bovine pancreas)	23,200	1
β-Lactoglobulin (bovine)	35,000	2
Hemoglobin (human)	64,500	4
Hexokinase (yeast)	102,000	2
Tryptophan synthetase (E. coli)	159,000	4
Aspartate transcarbamoylase (E. coli)	310,000	12
Glycogen phosphorylase (bovine liver)	370,000	4
Glutamine synthetase (E. coli)	592,000	12
Pyruvate dehydrogenase complex (bovine kidney)	7,000,000	160
Tobacco mosaic virus	40,000,000	2,130

their conformation, proteins can be placed in two major classes, fibrous and globular (Figure 3-3). The *fibrous proteins* consist of polypeptide chains arranged in parallel along a single axis, to yield long fibers or sheets. Fibrous proteins are physically tough and are insoluble in water or dilute salt solutions. They are the basic structural elements in the connective tissue of higher animals. Examples are *collagen* (page 135) of tendons and bone matrix, *α-keratin* (page 126) of hair, horn, skin, nails, and feathers, and *elastin* of elastic connective tissue.

In *globular proteins,* on the other hand, the polypeptide chains are tightly folded into compact spherical or globular shapes (Figure 3-3). Most globular proteins are soluble in aqueous systems. They usually have a mobile or dynamic function in the cell. Of the nearly 2,000 different enzymes known to date, nearly all are globular proteins, as are the antibodies, a number of hormones, and many proteins having a transport function, e.g., serum albumin and hemoglobin. Some proteins fall between the fibrous and globular types, resembling fibrous proteins in their long rodlike structures and the globular proteins in their solubility in aqueous salt solutions. Examples are *myosin*, an important structural element of muscle, and *fibrinogen,* the precursor of fibrin, the structural element of blood clots.

Specific terms commonly used to refer to different levels of protein structure (Figure 3-3) will be defined more fully and precisely in Chapter 5. *Primary structure* refers to the covalent backbone of the polypeptide chain and the sequence of its amino acid residues. *Secondary structure* refers to a regular, recurring arrangement in space of the polypeptide chain along one dimension. Secondary structure is particularly evident in the fibrous proteins, where the polypeptide chains have an extended or longitudinally coiled conformation; it also occurs in segments of the polypeptide chains in globular proteins. *Tertiary structure* refers to how the polypeptide chain is bent or folded in three dimensions, to form the compact, tightly folded structure of globular proteins (Figure 3-3).

Quaternary structure refers to how individual polypeptide chains of a protein having two or more chains are arranged in relation to each other. Most larger proteins contain two or more polypeptide chains, between which there are usually no covalent linkages (Figure 3-3). The more general term *conformation* is often used to refer to the combined secondary, tertiary, and quaternary structure of a protein (Chapter 6).

Proteins with two or more polypeptide chains are known as *oligomeric* proteins; their component chains are called *subunits* or *protomers.* A well-known example of an oligomeric protein is *hemoglobin,* the respiratory pigment of the red blood cell, which consists of four polypeptide chains fitting together tightly to form a compact, globular assembly of considerable stability, despite the lack of covalent linkages between them. Oligomeric proteins usually contain an even number of polypeptide chains, which may be identical or different in length or amino acid sequence. There may be

Figure 3-3
Fibrous and globular proteins.

Fibrous proteins

The backbone of the polypeptide chain in a typical fibrous protein, α-keratin. The term secondary structure refers to regularly coiled or zigzag arrangements of polypeptide chains along one dimension.

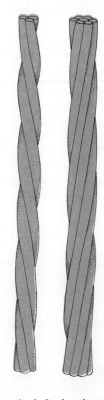

α-Helical coil

Supercoiling of α-helical coils to form ropes

Globular proteins

The polypeptide chain is folded into a compact globular shape, called the tertiary structure. Short lengths of the polypeptide chain of globular proteins may also have regular coiled or zigzag secondary structure. In oligomeric proteins the three-dimensional packing arrangement of the polypeptide chains is referred to as the quaternary structure.

The tertiary structure
of a single-chain
globular protein

The quaternary structure of a multichain
or oligomeric globular protein

anywhere from two to twelve subunit chains in the smaller oligomeric proteins.

Since oligomeric proteins contain two or more polypeptide chains, usually not covalently attached to each other, it may appear improper or at least ambiguous to refer to such proteins as "molecules" and to speak of their "molecular weight." However, in most oligomeric proteins the separate chains are so tightly associated that the complete particle behaves in solution like a single molecule. Moreover, all the component subunits of oligomeric proteins are necessary for their biological function.

Supramolecular Assemblies of Proteins

Sometimes a set of protein molecules functioning together occurs in cells as a cluster or complex that can be isolated in homogeneous or even crystalline form. An example of a cluster of functionally related macromolecules, called a _supramolecular assembly_ or _complex,_ is the _fatty acid synthetase complex,_ which contains one molecule of each of the seven different enzymes required for the biosynthesis of fatty acids (page 660). This complex can be isolated from yeast cells in homogeneous form (Table 3-2). The largest supramolecular protein complexes are the _viruses,_ complexes of proteins and nucleic acids; some viruses also contain lipids and metal ions. _Tobacco mosaic virus_ (Figure 3-4), one of the smaller viruses, has a particle weight of nearly 40 million, of which about 5 percent, or 2 million, consists of ribonucleic acid. The remaining 38 million is contributed by the protein portion, consisting of some 2,200 identical polypeptide chains. However, virus particles behave like single homogeneous structures having a definite molecular weight because their subunit components stick together very tightly.

Denaturation

Most protein molecules retain their biological activity only within a very limited range of temperature and pH. Exposing soluble or globular proteins to extremes of pH or to high temperatures for only short periods causes most of them to undergo a physical change known as _denaturation,_ in which the most visible effect is a decrease in solubility. Since no covalent bonds in the backbone of the polypeptide chain are broken during this relatively mild treatment, the primary structure remains intact. Most globular proteins undergo denaturation when heated above 60 to 70°C. Formation of an insoluble white coagulum when egg white is boiled is a common example of protein denaturation. But the most significant consequence of denaturation is that the protein usually loses its characteristic biological activity; e.g., heating usually destroys the catalytic ability of enzymes.

Denaturation is the unfolding of the characteristic native folded structure of the polypeptide chain of globular protein molecules (Figure 3-5). When thermal agitation causes the native folded structure to uncoil or unwind into a randomly looped chain, the protein loses its biological ac-

Figure 3-4
Portion of a tobacco mosaic virus particle, a supramolecular assembly containing 2,200 polypeptide chains and a molecule of RNA.

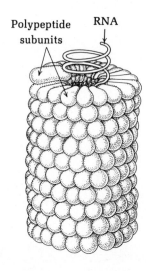

Figure 3-5
Denaturation and renaturation of a globular protein. After the polypeptide chain has been unfolded (by heating, by exposure to low pH, or by treatment with urea), it will often spontaneously refold to the native form.

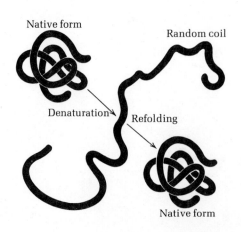

tivity. Although each type of protein has an amino acid composition and sequence fixed during biosynthesis, the amino acid sequence as such does not *directly* endow a protein with its biological function or activity. However, we shall see that the amino acid sequence *ultimately* determines the biological activity of a protein because it determines the native conformation, or folded state, of the protein molecule, through interactions of the amino acid side chains with each other, with the solvent, and with other solutes. This conclusion follows from the discovery that denaturation, or unfolding, of native proteins into randomly coiled, biologically inactive forms is not irreversible, as was once thought. Many cases have now been observed in which an unfolded protein molecule spontaneously returns to its native biologically active form in the test tube, a process called renaturation (Figure 3-5). If the denatured protein was an enzyme, its catalytic activity returns on renaturation, without change in the specificity of the reaction catalyzed. However, renaturation of a denatured protein cannot evoke any biological activity that was not present in the original protein. These facts therefore indicate that the sequence of amino acids in the polypeptide chain contains the information required to specify its native folded conformation and that this native conformation determines its biological activity (Chapter 6).

The Functional Diversity of Proteins

Proteins have many different biological functions. Table 3-3 gives some representative types of proteins, classified according to function. The enzymes represent the largest class. Nearly 2,000 different kinds of enzymes are known, each catalyzing a different kind of chemical reaction. Enzymes have extraordinary catalytic power, far beyond that of man-made catalysts. They are highly specific in their function. The enzyme hexokinase catalyzes transfer of a phosphate group from adenosine triphosphate (ATP) to glucose, the first step in glucose metabolism. Other enzymes dehydrogenate fuel molecules. Still others, e.g., cytochrome *c*, transfer electrons toward molecular oxygen during respiration or, like DNA polymerase and amino acid–activating enzymes, participate in the biosynthesis of cell components. Each type of enzyme molecule contains an active site, to which its specific substrate is bound during the catalytic cycle. Many enzymes contain a single polypeptide chain; others contain two or more. Some enzymes, called regulatory or allosteric enzymes, are further specialized to serve a regulatory function in addition to their catalytic activity. Virtually all enzymes are globular proteins, as defined above. How enzymes catalyze chemical reactions is a major concern of modern biochemistry.

Another major class of proteins has the function of storing amino acids as nutrients and as building blocks for the growing embryo, e.g., ovalbumin of egg white, casein of milk, and gliadin of wheat seeds.

Some proteins have a transport function; they are capable of binding and transporting specific types of molecules via the blood. Serum albumin binds free fatty acids tightly and

Table 3-3   Classification of proteins by biological function

Type and examples	Occurrence or function
Enzymes	
Hexokinase	Phosphorylates glucose
Lactate dehydrogenase	Dehydrogenates lactate
Cytochrome *c*	Transfers electrons
DNA polymerase	Replicates and repairs DNA
Storage proteins	
Ovalbumin	Egg-white protein
Casein	A milk protein
Ferritin	Iron storage in spleen
Gliadin	Seed protein of wheat
Zein	Seed protein of corn
Transport proteins	
Hemoglobin	Transports O_2 in blood of vertebrates
Hemocyanin	Transports O_2 in blood of some invertebrates
Myoglobin	Transports O_2 in muscle cells
Serum albumin	Transports fatty acids in blood
β_1-Lipoprotein	Transports lipids in blood
Iron-binding globulin	Transports iron in blood
Ceruloplasmin	Transports copper in blood
Contractile proteins	
Myosin	Thick filaments in myofibril
Actin	Thin filaments in myofibril
Dynein	Cilia and flagella
Protective proteins in vertebrate blood	
Antibodies	Form complexes with foreign proteins
Complement	Complexes with some antigen-antibody systems
Fibrinogen	Precursor of fibrin in blood clotting
Thrombin	Component of clotting mechanism
Toxins	
Clostridium botulinum toxin	Causes bacterial food poisoning
Diphtheria toxin	Bacterial toxin
Snake venoms	Enzymes that hydrolyze phosphoglycerides
Ricin	Toxic protein of castor bean
Gossypin	Toxic protein of cottonseed
Hormones	
Insulin	Regulates glucose metabolism
Adrenocorticotrophic hormone	Regulates corticosteroid synthesis
Growth hormone	Stimulates growth of bones
Structural proteins	
Viral-coat proteins	Sheath around nucleic acid
Glycoproteins	Cell coats and walls
α-Keratin	Skin, feathers, nails, hoofs
Sclerotin	Exoskeletons of insects
Fibroin	Silk of cocoons, spider webs
Collagen	Fibrous connective tissue (tendons, bone, cartilage)
Elastin	Elastic connective tissue (ligaments)
Mucoproteins	Mucous secretions, synovial fluid

thus serves to transport these molecules between adipose tissue and other tissues or organs in vertebrates. The *lipoproteins* of blood plasma transport lipids between the intestine, liver, and adipose (fatty) tissues. *Hemoglobin* of vertebrate erythrocytes transports oxygen from the lungs to the tissues. Invertebrates have other types of oxygen-carrying protein molecules, such as the *hemocyanins*.

Other types of proteins function as essential elements in contractile and motile systems. *Actin* and *myosin* are the two major protein elements of the contractile system of skeletal muscle. Actin is a long, filamentous protein composed of many globular polypeptide chains arranged like a string of beads; myosin is a long rodlike molecule containing two helically intertwined polypeptide chains (Chapter 27). In muscles these proteins are arranged in parallel arrays and slide along each other during contraction.

Some proteins have a protective or defensive function. The blood proteins *thrombin* and *fibrinogen* participate in blood clotting and thus prevent the loss of blood from the vascular system of vertebrates, but the most important protective proteins are the *antibodies,* or *immune globulins,* which combine with and thus neutralize foreign proteins and other substances that happen to gain entrance into the blood or tissues of a given vertebrate. Indeed, the study of antibodies has led to the conclusion that each species of organism has its own specific set of protein molecules (page 112).

Toxins, i.e., substances that are extremely toxic to higher animals in very small amounts, represent another group of proteins and include *ricin* of the castor bean, *gossypin* of cottonseed, *diphtheria toxin,* and the toxin of the anaerobic bacterium *Clostridium botulinum,* which is responsible for some types of food poisoning.

Among the most interesting proteins are those functioning as hormones, such as growth hormone, or *somatotrophin,* a hormone of the anterior pituitary gland. *Insulin,* secreted by certain specialized cells of the pancreas, is a hormone regulating glucose metabolism; its deficiency in man causes the disease diabetes mellitus.

Yet another class of proteins comprises those serving as structural elements. In vertebrates the fibrous protein *collagen* is the major extracellular structural protein in connective tissue and bone. Collagen fibrils, by forming a structural continuum, also help bind a group of cells together to form a tissue. Two other fibrous proteins in vertebrates are *elastin,* of yellow elastic tissue, and *α-keratin,* mentioned above. Cartilage contains not only collagen but also *glycoproteins,* which endow mucous secretions and synovial fluid in the joints of vertebrates with a slippery, lubricating quality.

Besides these major classes of proteins others have unusual functions. Spiders and silkworms secrete a thick solution of the protein *fibroin,* which quickly solidifies into an insoluble thread of exceptional tensile strength used to form webs or cocoons. The blood of some fishes living in subzero Antarctic waters contains a protein that keeps the blood from freezing, aptly called "antifreeze" protein. Monellin is a sweet-tasting protein found in some fruits; when it is denatured, it no longer tastes sweet.

It is extraordinary that all proteins, including those having intense biological or toxic effects, are built from the same 20 amino acids, which by themselves have little or no biological activity or toxicity. Its three-dimensional conformation gives each type of protein its specific biological activity; its conformation is in turn determined by the specific sequence of the amino acids in its polypeptide chain(s) (Chapter 6).

Antibodies and the Immune Response; The Species Specificity of Proteins

Among the many different proteins in living organisms the antibodies, or immune globulins, have been of the utmost importance in demonstrating that proteins are specific for each species of organism. Antibody molecules appear in the blood serum and certain cells of a vertebrate in response to the introduction of a protein or some other macromolecule foreign to that species; such a species-foreign macromolecule is called the antigen. The specific antibody molecules generated in this manner can combine with the antigen which elicited their formation to form an antigen-antibody complex (Figure 3-6). This reaction, called the immune response, is the basis for the whole field of immunology. Immunity to a specific infectious disease can often be conferred by injecting very small amounts of certain macromolecular components (i.e., the antigenic components), of the causative microorganism or virus. A specific antibody or immune globulin is formed in response to the foreign antigen and may persist in the blood for a long time. If the causative microorganism should later gain access to the blood or lymph, these specific antibodies can inactivate or kill it by combining with its antigenic components. The immune response is given only by vertebrates and thus is a fairly recent product of biological evolution.

Antibody molecules have binding sites that are specific for and complementary to the structural features of the antigen that induced their formation. Usually the antibody molecule has two binding sites, making possible the formation of a three-dimensional lattice of alternating antigen and antibody molecules; since it ultimately precipitates from the serum, it is called the precipitin (Figure 3-6). The structure and origin of immune globulins are described in more detail in Chapter 35.

Antibodies are highly specific for the foreign proteins that evoke their formation. An antibody formed by a rabbit to injected hen's-egg albumin, for example, will combine with the latter but not with unrelated proteins such as human hemoglobin. Moreover, it is specific for the three-dimensional structure of native hen's-egg albumin, so that if the albumin is heated or denatured to unfold its polypeptide chains or is

Figure 3-6
The antigen-antibody reaction. The colored sites on the antigen are the determinants of antigenic specificity. The colored sites on the antibody molecule are structurally complementary to the determinant sites of the antigen. The antibody is divalent, but the antigen may be multivalent.

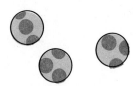

Antigen molecules

Antibody molecule

Antigen-antibody complex

Insoluble antigen-antibody lattice
(precipitin)

chemically modified, the antibody will no longer combine with it.

Applying the highly specific antigen-antibody reaction to the study of various proteins from different species of organisms has yielded several important conclusions. The first is that functionally different proteins from any single species lead to the formation of different antibodies. Thus, when a rabbit is immunized against horse hemoglobin, the antibody formed will precipitate horse hemoglobin but not (usually) any other horse proteins.

The second major conclusion has far-reaching biological implications: homologous proteins of different species are not immunologically identical. Homologous proteins are those with the same or similar functions, e.g., the hemoglobins of different vertebrate species. Although hemoglobins of different mammals have not only the same function but also about the same molecular weight, with four iron atoms, four porphyrin rings, and four polypeptide chains, nevertheless they are immunologically different molecules. The antibodies produced by the rabbit after immunization with horse hemoglobin, for example, react maximally with horse hemoglobin but are much less reactive with hemoglobins from other vertebrates.

A third conclusion is that antibody specificity reflects phylogenetic relationships. Homologous proteins of closely related species are more similar to each other than those of widely separated species (Table 3-4). Thus, while antibodies generated by the rabbit to horse hemoglobin react best with horse hemoglobin, they also react strongly with the hemoglobins of species closely related to the horse, e.g., zebra, cow, pig, and other ungulates, but are far less reactive with the hemoglobins from rodents, birds, and amphibians.

The structural differences between homologous proteins of different species, which were first revealed by the antigen-antibody reaction, have since been found to be the result of differences in their amino acid sequences (Chapter 5). The more closely related two species of organism are, the more nearly identical the amino acid sequence of their homologous proteins. For this reason the amino acid sequence of homologous proteins can give us valuable information about the evolution of different organisms and their phylogenetic relationships (page 113).

Table 3-4 Reactivity of serum albumin from different species with rabbit antibody to bovine serum albumin†

Species	Reactivity in precipitin test
Cow	(100)
Sheep	76
Pig	32
Cat	25
Horse	16
Man	15
Hamster	15
Rat	14
Dog	14
Mouse	9

† Data from W. O. Weigle, *J. Immunol.,* 88:9 (1962).

Sequence Isomerism in Polypeptide Chains

Earlier (in the Introduction) we estimated that the total number of different kinds of proteins in all species of living organisms is of the order of 10^{10} to 10^{12}. Can this enormous number of different proteins, each having its own specific amino acid sequence, be made from only 20 different amino acids? It is possible to answer from mathematical considerations alone. In a dipeptide containing two different amino acids A and B, two sequence isomers are possible, that is, A–B and B–A. In a tripeptide having three different amino acids, six sequential arrangements of the three amino acids are possible: A–B–C, A–C–B, B–A–C, B–C–A, C–A–B, and C–B–A. In permutation theory, the general expression for the number of possible sequential arrangements of different objects is given by the expression n! (read "n factorial"), where n is the number of different objects. For a tetrapeptide having four different amino acids $4! = 4 \times 3 \times 2 \times 1 = 24$ different sequences are possible. For a polypeptide of 20 different amino acids, in which each of the amino acids occurs only once, the number of possible sequential arrangements is given by 20!, or $20 \times 19 \times 18 \times 17 \times 16 \times \cdots$, which comes out to the startling figure of about 2×10^{18}. But this is only a small polypeptide chain (mol wt $\approx 2{,}600$) in which each amino acid occurs only once. For a protein of molecular weight 34,000, containing 12 different amino acids in equal numbers, some 10^{300} sequence isomers are possible. Just as the English alphabet of 26 letters can be used to make an enormously large number of written words, so the 20 different amino acids can be used to make an almost limitless number of different proteins. Unraveling the amino acid sequences of different proteins and relating each sequence to the properties and functions of the protein and the phylogenetics of the organism are major objectives of contemporary biochemistry.

The Genetic Coding of Amino Acid Sequences in Proteins

Because the structure and function of proteins are ultimately reflections of their amino acid sequences, we can hardly discuss proteins and their biological activities without at least a rudimentary knowledge of the nature of the molecular relationships between genes and proteins. These relationships, discussed in much more detail in Part 4, have yielded penetrating insight into the comparative biochemistry of proteins in different species and into the evolution of protein molecules. Genetic information is stored in deoxyribonucleic acid (DNA), the informational macromolecule of the chromosomes. This information instructs each cell to produce a characteristic set of proteins, in accordance with the central statement of molecular genetics; i.e., genetic information flows in the direction DNA → RNA → protein.

It is the *sequence* of amino acids in the polypeptide chain of each type of protein that is ultimately specified or coded by the *sequence* of nucleotide residues in deoxyribonucleic acid (DNA). The segment of a DNA molecule specifying one complete polypeptide chain is called a *cistron* or *gene*. Each amino acid is coded for by three successive nucleotide residues in DNA, called a *coding triplet*. However, genes normally remain in the chromosomes and do not directly serve as the coding templates during the biosynthesis of proteins, which takes place on the ribosomes. Instead the genetic message in the gene is first enzymatically transcribed to form a specific type of ribonucleic acid called *messenger RNA* (mRNA), whose nucleotide sequence is complementary to that of the DNA of the gene. In the mRNA coding triplets, or *codons*, complementary to those in DNA, serve as the im-

Figure 3-7
Colinearity of the nucleotide sequences of DNA, mRNA, and the amino acid sequence of polypeptide chains. The triplets of nucleotide units in DNA determine the sequence of amino acids in proteins through the intermediary formation of RNA, which has nucleotide triplets (codons) complementary to those of DNA. The letters symbolize the distinctive base components of the nucleotide residues in DNA and RNA. A, adenine; T, thymine; G, guanine; C, cytosine; U, uracil.

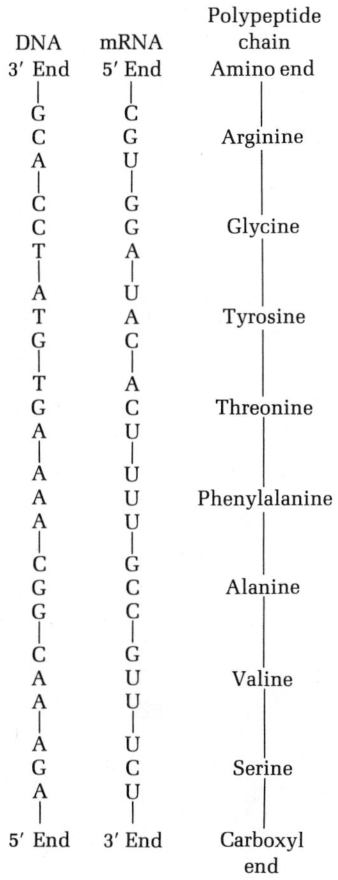

DNA	mRNA	Polypeptide chain
3' End	5' End	Amino end
G	C	
C	G	Arginine
A	U	
C	G	
C	G	Glycine
T	A	
A	U	
T	A	Tyrosine
G	C	
T	A	
G	C	Threonine
A	U	
A	U	
A	U	Phenylalanine
A	U	
C	G	
G	C	Alanine
G	C	
C	G	
A	U	Valine
A	U	
A	U	
G	C	Serine
A	U	
5' End	3' End	Carboxyl end

mediate template and provide the genetic information specifying the sequence of amino acids during protein biosynthesis on the ribosomes.

The coding relationship between the sequence of nucleotides in DNA, in the complementary RNA, and in the sequence of amino acids in the resulting polypeptide chain is illustrated schematically in Figure 3-7. The sequence of coding triplets in DNA corresponds in a linear or sequential manner; i.e., it is colinear with the amino acid sequence of the polypeptide chain for which it codes, through the intermediate template function of the RNA.

Mutations

Sometimes a gene specifying a given protein undergoes a chemical change, as the result of some physical agency, e.g., x-radiation or radioactivity, or the action of certain chemical agents, so that one of the three bases in a coding triplet for an amino acid residue is chemically changed or deleted; sometimes an extra nucleotide is inserted. As a result, the normal continuous, or *commaless*, sequence of coding triplets in DNA is altered and produces a corresponding alteration in the nucleotide sequence of the mRNA, which in turn codes for an altered polypeptide chain. In an abnormal polypeptide chain one or perhaps many amino acid residues in its specific sequence may be replaced by others; as a consequence it may be defective in its biological function. The experimental study of mutationally altered proteins is of great importance since, for one thing, it can reveal which amino acid residues in a polypeptide chain are essential for the structure and function of a protein (Chapters 6 and 8).

With this outline of the structure and biology of proteins to serve as orientation, we may now proceed to examine protein structure in more detail.

Summary

Proteins are made up of one or more polypeptide chains, each consisting of many α-amino acid residues covalently linked by peptide bonds. The molecular weights of proteins vary from about 5,000 to 1 million or more. All proteins, regardless of function or species of origin, are constructed from a basic set of 20 amino acids, arranged in various specific sequences. Simple proteins yield only α-amino acids on hydrolysis, whereas conjugated proteins contain a metal or an organic prosthetic group. Proteins are classified according to their three-dimensional conformation. Fibrous proteins occur as rods or sheets and have parallel, relatively extended polypeptide chains. They are insoluble and serve as structural elements. Globular proteins have tightly folded polypeptide chains and are spherical or globular; they have dynamic functions.

The primary structure of a protein is its specific amino acid sequence. The secondary structure is the extended or helical arrangement of the polypeptide chain along a single long axis, as in fibrous proteins. The tertiary structure is the three-dimensional folding of the polypeptide chain to form globular proteins. In oligomeric proteins, which contain two or more polypeptide chains, the term "quaternary structure" refers to how the individual polypeptide chains are clustered together. Ultimately, it is the amino acid sequence that determines the three-dimensional conformation of protein molecules. Proteins are denatured or unfolded, without cleavage of the peptide-chain backbone, by extremes of pH or temperature and by other agencies. Denaturation causes proteins to lose their biological activity; denaturation is sometimes reversible.

Proteins serve many diverse functions: as catalysts, as structural elements, in contractile systems, for nutrient storage, as vehicles of transport, as hormones, as toxins, and as protective agents. In this last category are the immune globulins or antibodies formed by vertebrates in response to antigens, i.e., substances foreign to the species. Antigen-antibody specificity studies have led to the conclusion that homologous proteins of different species are species-specific. The amino acid sequence of a protein is specified during its biosynthesis by a colinear sequence of consecutive triplets of nucleotides in RNA, which is in turn complementary to the nucleotide sequence in DNA. The segment of DNA that codes one polypeptide chain is called a gene.

References

References to more specific aspects of protein biochemistry are given at the ends of Chapters 4 to 9.

Books

DAVIS, B. D., R. DULBECCO, H. N. EISEN, H. S. GINSBERG, and W. B. WOOD, JR.: *Principles of Microbiology and Immunology*, 2d ed., Harper & Row, New York, 1973. A textbook that includes an excellent account of the immune response and the biology of the antigen-antibody reaction.

NEURATH, H.: *The Proteins*, 3d ed., vols. 1–4, Academic, New York, 1963–1966. Authoritative and comprehensive monograph.

WATSON, J. D.: *Molecular Biology of the Gene*, 3d ed., Benjamin, Menlo Park, Calif., 1975. Elementary account of the genetic background of protein synthesis.

We now consider the physical and chemical properties of the amino acids in some detail, since they are the alphabet of protein structure and determine many of the important properties of proteins. The first amino acid isolated from a protein hydrolyzate was glycine, obtained in 1820 from gelatin by H. Braconnot. The most recently discovered of the 20 amino acids commonly found in proteins is threonine, first isolated from hydrolyzates of fibrin by W. C. Rose in 1935. Besides the 20 amino acids found as building blocks of proteins, many additional biologically occurring amino acids serve other functions in cells.

Although much important information on the structure, synthesis, optical properties, and chemical reactions of amino acids arose from early investigations many years ago, full appreciation of the role of amino acids in determining protein structure has come only in the last two decades.

The Common Amino Acids of Proteins

Figure 4-1
The general structural formula for the α-amino acids found in proteins. The shaded portion is common to all α-amino acids.

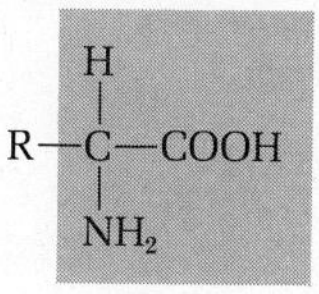

Figure 4-1 shows the general structural formula of the 20 α-amino acids commonly found in proteins, also called _standard_ amino acids. All except proline have as common denominators a free carboxyl group and a free unsubstituted amino group on the α carbon atom. They differ from each other in the structure of their distinctive side chains, called the R groups.

Various ways of classifying the amino acids on the basis of their R groups have been proposed. The most meaningful is based on their polarity. There are four main classes of amino acids: those with (1) nonpolar or hydrophobic R groups, (2) neutral (uncharged) polar R groups, (3) positively charged R groups, and (4) negatively charged R groups (at pH 6.0 to 7.0, the zone of intracellular pH). Within any single class of amino acids there are considerable variations in the size, shape, and properties of the R groups. Later we shall see that this way of classifying amino acids may bear a relationship to the genetic code words for the different amino acids found in proteins (page 962).

Amino acids are ordinarily designated by three-letter symbols, but a set of one-letter symbols has also been adopted to facilitate comparative display of amino acid sequences of homologous proteins (Table 4-1).

Amino Acids with Nonpolar (Hydrophobic) R Groups

Figure 4-2 shows the structural formulas and space-filling models of the eight amino acids having nonpolar R groups, together with their symbols. This family includes five amino acids with aliphatic hydrocarbon R groups (*alanine, leucine, isoleucine, valine,* and *proline*), two with aromatic rings (*phenylalanine* and *tryptophan*), and one containing sulfur (*methionine*). As a group, these amino acids are less soluble in water than the amino acids with polar R groups. The least hydrophobic member of this class is alanine, which is thus near the border line between nonpolar amino acids and those with uncharged polar R groups (below).

Proline differs from all the other standard amino acids in actually being an α-imino acid; it may be regarded as an α-amino acid in which the R group is a substituent in the amino group.

Amino Acids with Uncharged Polar R Groups

These amino acids (Figure 4-3) are relatively more soluble in water than those with nonpolar R groups. Their R groups contain neutral (uncharged) polar functional groups which can hydrogen-bond with water. The polarity of *serine, threonine,* and *tyrosine* is contributed by their hydroxyl groups; that of *asparagine* and *glutamine* by their amide groups; and that of *cysteine* by its sulfhydryl (—SH) group. *Glycine*, the borderline member of this group, is sometimes classified as a nonpolar amino acid (see above), but its R group, a single hydrogen atom, is too small to influence the high degree of polarity of the α-amino and α-carboxyl groups.

Asparagine and glutamine are the amides of aspartic acid and glutamic acid (below). Since these amino acids are easily hydrolyzed by acid or base to aspartic acid and glutamic acid, respectively, the symbols Asx and Glx are used to designate these chemically related pairs of amino acids when the amounts of the amide forms are not known.

Cysteine and tyrosine have the most polar substituents of this class of amino acids, namely, the thiol and phenolic hydroxyl groups, respectively. These groups tend to lose protons by ionization far more readily than the R groups of other amino acids of this class, although they are only slightly ionized at pH 7.0. Cysteine often occurs in proteins in its oxidized form *cystine*, in which the thiol groups of two molecules of cysteine have been oxidized to a disulfide group to provide a covalent cross-linkage between them. The structure of cystine is given on page 83.

Amino Acids with Positively Charged (Basic) R Groups

The basic amino acids (Figure 4-4), in which the R groups have a net positive charge at pH 7.0, all have six carbon

Table 4-1 Amino acid symbols

Amino acid	Three-letter symbol	One-letter symbol
Alanine	Ala	A
Arginine	Arg	R
Asparagine	Asn	N
Aspartic acid	Asp	D
Asn and/or Asp	Asx	B
Cysteine	Cys	C
Glutamine	Gln	Q
Glutamic acid	Glu	E
Gln and/or Glu	Glx	Z
Glycine	Gly	G
Histidine	His	H
Isoleucine	Ile	I
Leucine	Leu	L
Lysine	Lys	K
Methionine	Met	M
Phenylalanine	Phe	F
Proline	Pro	P
Serine	Ser	S
Threonine	Thr	T
Tryptophan	Trp	W
Tyrosine	Tyr	Y
Valine	Val	V

Figure 4-2
Amino acids with nonpolar R groups. The three-letter and one-letter symbols and the molecular weights are also given. The amino acids are shown in the ionized forms predominating at pH 6.0 to 7.0.

	R groups	**R groups**

Alanine
Ala
A

Mol wt 89

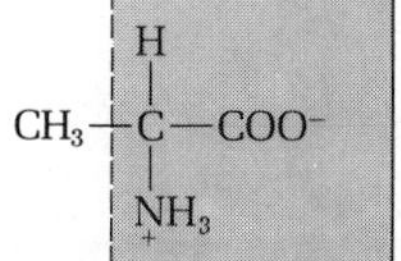

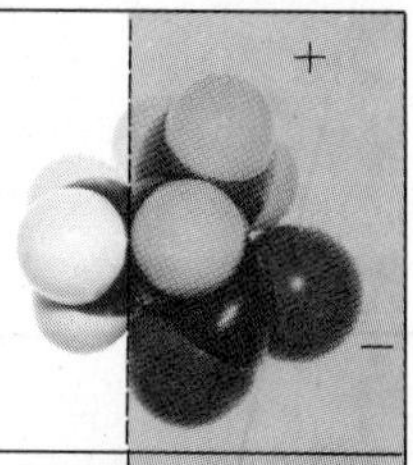

$$CH_3 - \underset{\underset{+}{\overset{|}{NH_3}}}{\overset{\overset{H}{|}}{C}} - COO^-$$

Valine
Val
V

Mol wt 117

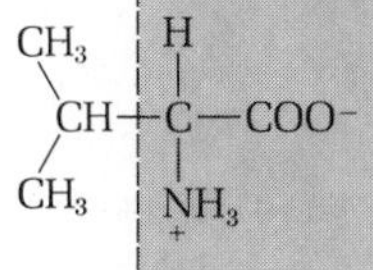

$$\underset{CH_3}{\overset{CH_3}{\diagdown}} CH - \underset{\underset{+}{NH_3}}{\overset{\overset{H}{|}}{C}} - COO^-$$

Leucine
Leu
L

Mol wt 131

$$\underset{CH_3}{\overset{CH_3}{\diagdown}} CH - CH_2 - \underset{\underset{+}{NH_3}}{\overset{\overset{H}{|}}{C}} - COO^-$$

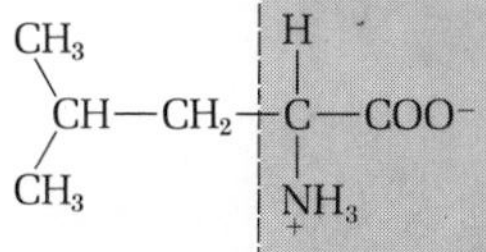

Isoleucine
Ile
I

Mol wt 131

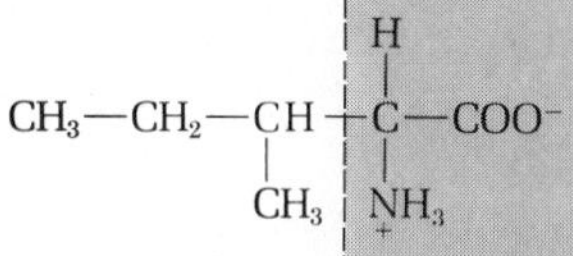

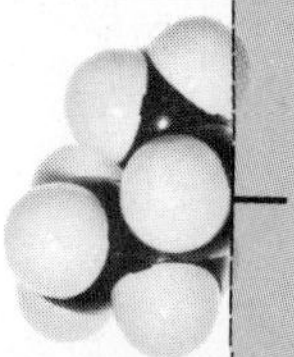

$$CH_3 - CH_2 - \underset{CH_3}{\overset{}{CH}} - \underset{\underset{+}{NH_3}}{\overset{\overset{H}{|}}{C}} - COO^-$$

Proline
Pro
P

Mol wt 115

$$\begin{array}{c} H_2 \\ C \\ H_2C \diagup \diagdown \\ | \hspace{1.2em} C - COO^- \\ H_2C \diagdown \diagup \hspace{0.3em} \vdots \\ N \hspace{1.5em} H \\ | \\ H \end{array}$$

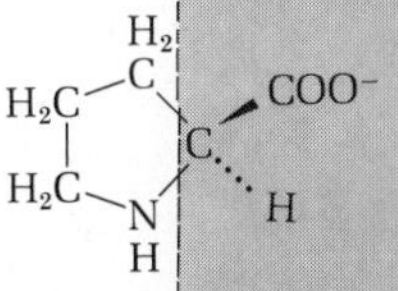

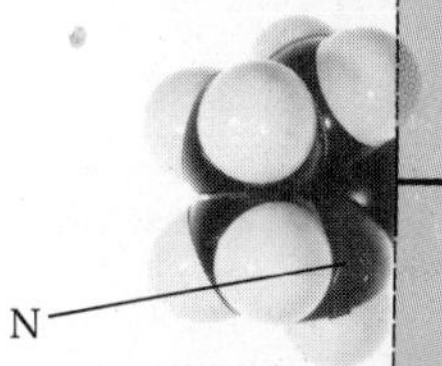

Phenylalanine
Phe
F

Mol wt 165

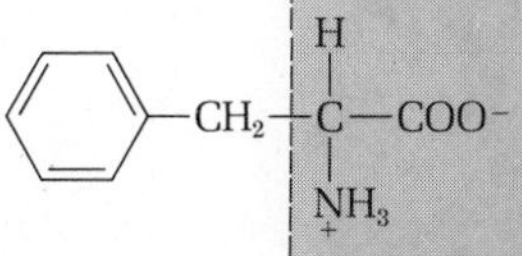

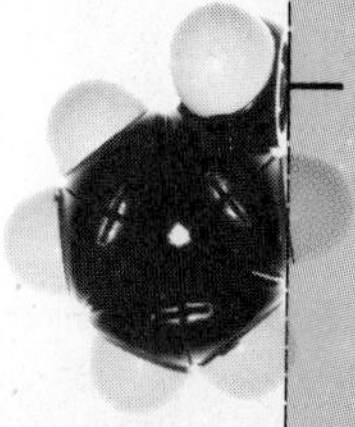

$$\text{⬡} - CH_2 - \underset{\underset{+}{NH_3}}{\overset{\overset{H}{|}}{C}} - COO^-$$

Tryptophan
Trp
W

Mol wt 204

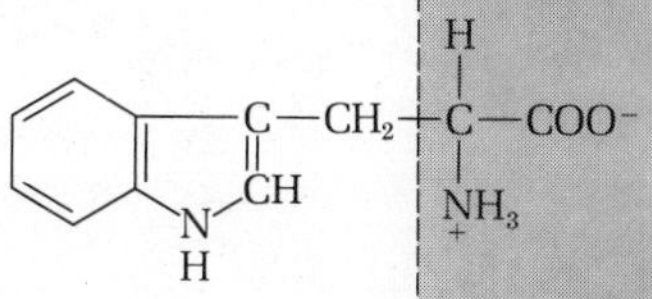

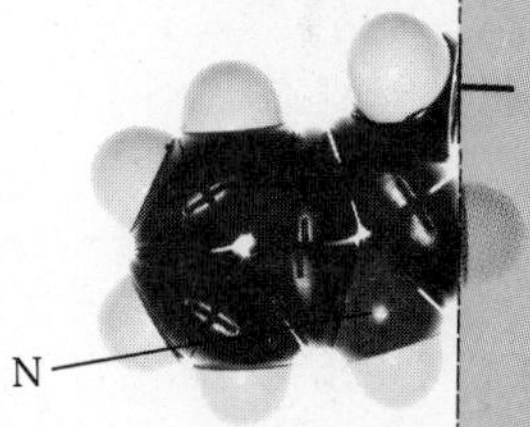

$$\begin{array}{c} \text{⬡⬠} \\ C - CH_2 - \underset{\underset{+}{NH_3}}{\overset{\overset{H}{|}}{C}} - COO^- \\ CH \\ N \\ | \\ H \end{array}$$

Methionine
Met
M

Mol wt 149

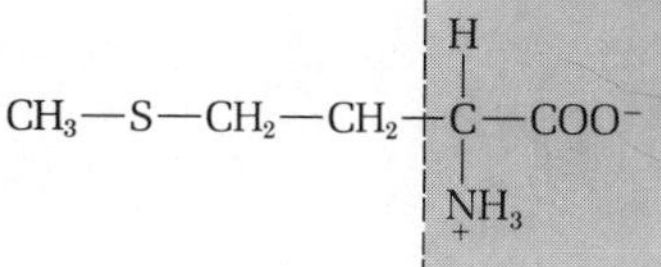

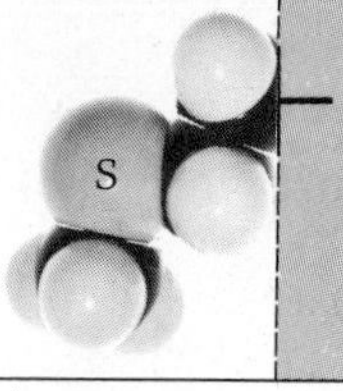

$$CH_3 - S - CH_2 - CH_2 - \underset{\underset{+}{NH_3}}{\overset{\overset{H}{|}}{C}} - COO^-$$

0.5 nm

Figure 4-3
Amino acids with uncharged polar R groups.

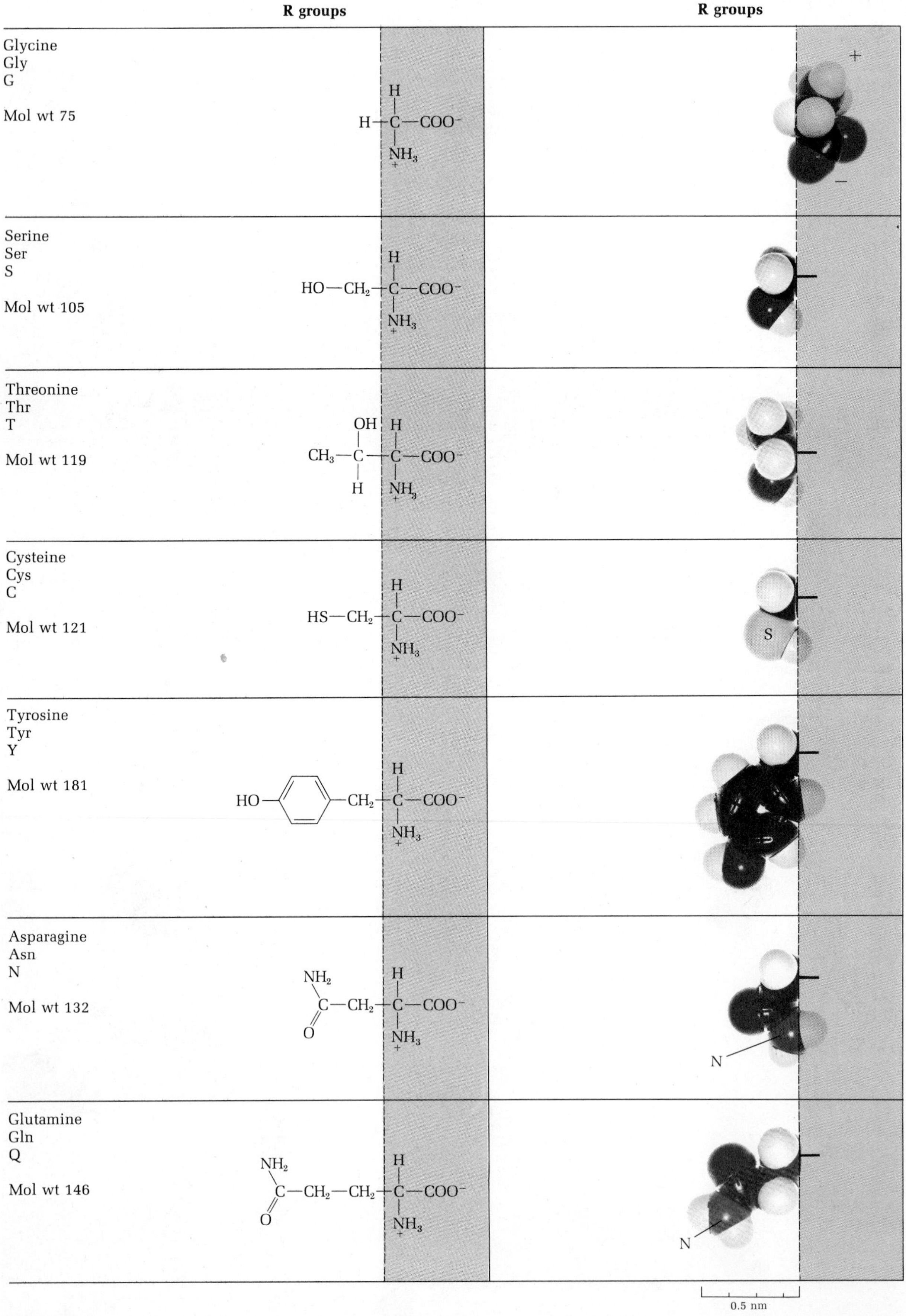

R groups		**R groups**	
Glycine Gly G Mol wt 75	$\begin{array}{c} H \\ \| \\ H-C-COO^- \\ \| \\ \overset{+}{N}H_3 \end{array}$		
Serine Ser S Mol wt 105	$\begin{array}{c} H \\ \| \\ HO-CH_2-C-COO^- \\ \| \\ \overset{+}{N}H_3 \end{array}$		
Threonine Thr T Mol wt 119	$\begin{array}{c} OH \quad H \\ \| \quad\; \| \\ CH_3-C-C-COO^- \\ \| \quad\; \| \\ H \quad\; \overset{+}{N}H_3 \end{array}$		
Cysteine Cys C Mol wt 121	$\begin{array}{c} H \\ \| \\ HS-CH_2-C-COO^- \\ \| \\ \overset{+}{N}H_3 \end{array}$		
Tyrosine Tyr Y Mol wt 181	$\begin{array}{c} H \\ \| \\ HO-\bigcirc-CH_2-C-COO^- \\ \| \\ \overset{+}{N}H_3 \end{array}$		
Asparagine Asn N Mol wt 132	$\begin{array}{c} NH_2 \qquad\; H \\ \| \qquad\quad \| \\ C-CH_2-C-COO^- \\ \| \qquad\quad \| \\ O \qquad\quad \overset{+}{N}H_3 \end{array}$		
Glutamine Gln Q Mol wt 146	$\begin{array}{c} NH_2 \qquad\qquad\quad H \\ \| \qquad\qquad\qquad \| \\ C-CH_2-CH_2-C-COO^- \\ \| \qquad\qquad\qquad \| \\ O \qquad\qquad\qquad \overset{+}{N}H_3 \end{array}$		

0.5 nm

Figure 4-4
Amino acids with charged polar groups at pH 6.0 to 7.0.

Acidic amino acids (negatively charged at pH 6.0)

R groups		**R groups**
Aspartic acid Asp D Mol wt 133	$\overset{-O}{\underset{O}{C}}\!-\!CH_2\!-\!\overset{\overset{H}{\mid}}{\underset{\underset{+}{NH_3}}{C}}\!-\!COO^-$	
Glutamic acid Glu E Mol wt 147	$\overset{-O}{\underset{O}{C}}\!-\!CH_2\!-\!CH_2\!-\!\overset{\overset{H}{\mid}}{\underset{\underset{+}{NH_3}}{C}}\!-\!COO^-$	

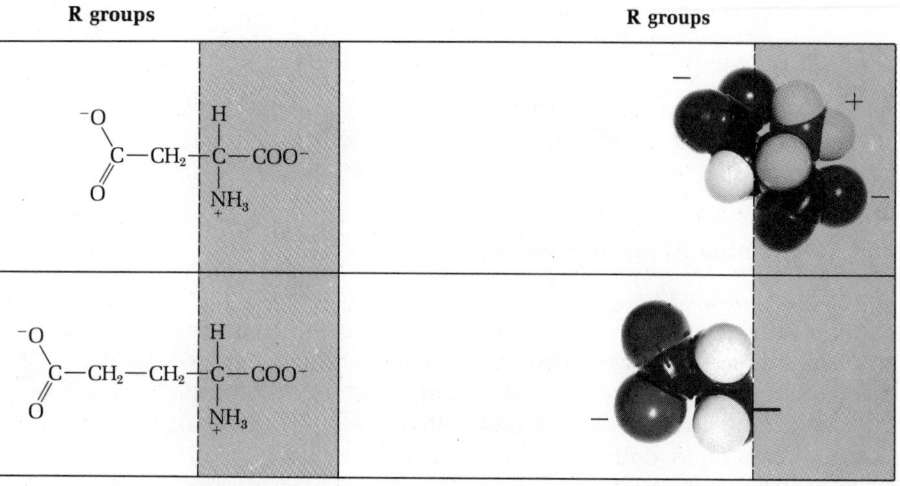

Basic amino acids (positively charged at pH 6.0)

R groups		**R groups**
Lysine Lys K Mol wt 146	$H_3\overset{+}{N}\!-\!CH_2\!-\!CH_2\!-\!CH_2\!-\!CH_2\!-\!\overset{\overset{H}{\mid}}{\underset{\underset{+}{NH_3}}{C}}\!-\!COO^-$	
Arginine Arg R Mol wt 174	$H_2N\!-\!\overset{\overset{\displaystyle \shortparallel}{C}}{\underset{\underset{+}{NH_2}}{}}\!-\!NH\!-\!CH_2\!-\!CH_2\!-\!CH_2\!-\!\overset{\overset{H}{\mid}}{\underset{\underset{+}{NH_3}}{C}}\!-\!COO^-$	
Histidine (at pH 6.0) His H Mol wt 155	$HC\!=\!\overset{}{C}\!-\!CH_2\!-\!\overset{\overset{H}{\mid}}{\underset{\underset{+}{NH_3}}{C}}\!-\!COO^-$	

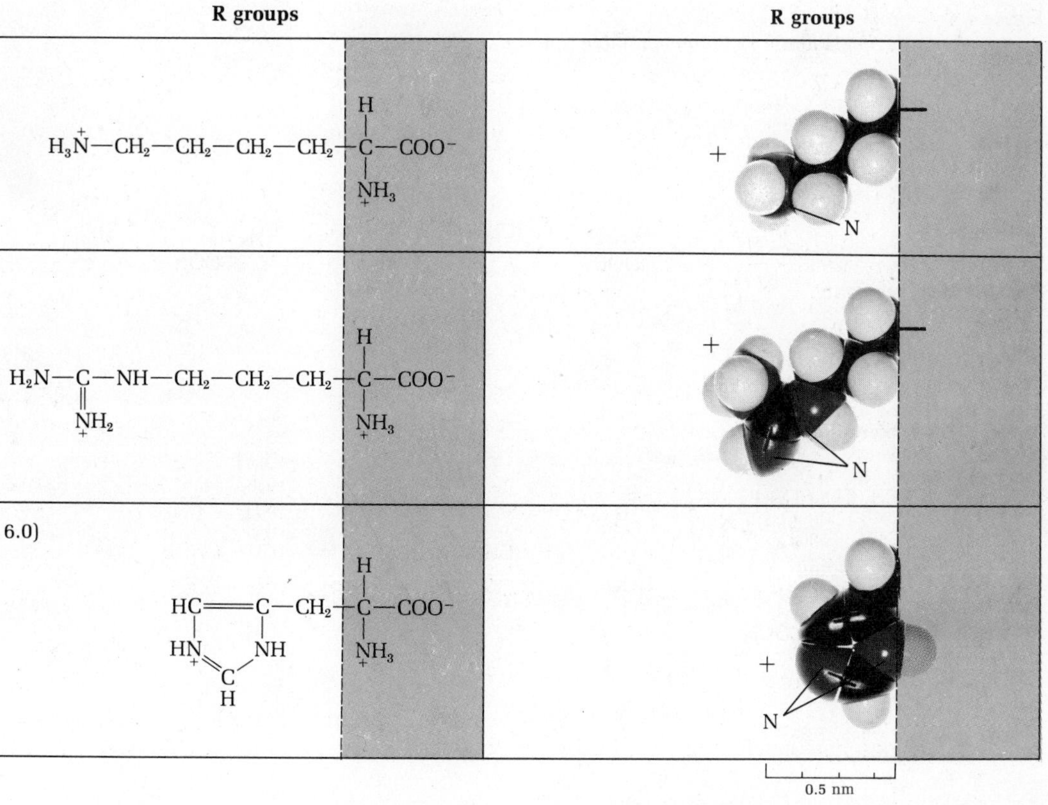

0.5 nm

atoms. They consist of *lysine*, which bears a positively charged amino group at the ε position on its aliphatic chain; *arginine*, which bears the positively charged guanidinium group; and *histidine*, which contains the weakly basic imidazolium function. Histidine is borderline in its properties. At pH 6.0 somewhat over 50 percent of histidine molecules possess a protonated, positively charged R group, but at pH 7.0, less than 10 percent have a positive charge. It is the only amino acid whose R group has a pK' near 7.0.

Amino Acids with Negatively Charged (Acidic) R Groups

The two members of this class are *aspartic acid* and *glutamic acid*, each with a second carboxyl group which is fully ionized and thus negatively charged at pH 6 to 7 (Figure 4-4).

The Rare Amino Acids of Proteins

In addition to the 20 standard amino acids, several others of relatively rare occurrence have been isolated from hydrolyzates of some specialized types of proteins. All are derivatives of some standard amino acid. Among them is *4-hydroxyproline*, a derivative of proline found in some abundance in the fibrous protein collagen (page 135) and in some plant proteins (Figure 4-5). *Hydroxylysine*, the 5-hydroxy derivative of lysine, is present in collagen. *Desmosine* and *isodesmosine* occur in the fibrous protein elastin. Their rather extraordinary structures can be visualized as formed from four lysine molecules with their R groups joined to form a substituted pyridine ring. Possibly this structure permits des-

Figure 4-5
Some rare amino acids found in fibrous proteins (un-ionized forms).

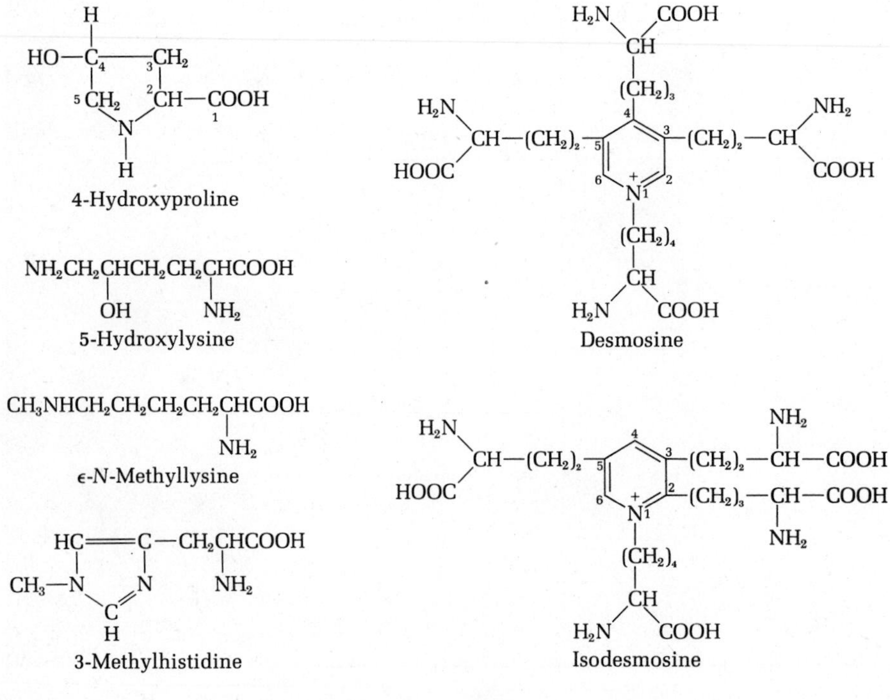

4-Hydroxyproline

5-Hydroxylysine

ε-N-Methyllysine

3-Methylhistidine

Desmosine

Isodesmosine

mosine and isodesmosine to connect four polypeptide chains in a radial array; elastin differs from other fibrous proteins in that it is capable of undergoing two-way stretch. The unusual amino acids ϵ-N-methyllysine, ϵ-N-trimethyllysine, and methylhistidine, methyl derivatives of standard amino acids (page 75), have been found in certain muscle proteins.

Although it is likely that additional rare amino acids will be discovered in proteins, on genetic grounds we can assume that they will be few in number, will be derivatives of the presently known standard amino acids, and will be limited in occurrence to specific types of proteins. The rare amino acids in proteins are genetically distinctive since there are no triplet code words for them (page 962). In all known cases, they arise by enzymatic modification after their parent amino acids have already been inserted into the polypeptide chain.

Nonprotein Amino Acids

In addition to the 20 common and several rare amino acids of proteins, over 150 other amino acids are known to occur biologically in free or combined form but never in proteins. Most are derivatives of the L-α-amino acids found in proteins, but β-, γ-, and δ-amino acids are also known (Figure 4-6).

Some nonprotein amino acids are important precursors or intermediates in metabolism. Thus, *β-alanine* is a building block of the vitamin pantothenic acid; *homocysteine* and *homoserine* are intermediates in amino acid metabolism; *citrulline* and *ornithine* are intermediates in the synthesis of arginine. Other nonprotein amino acids function as chemical agents for the transmission of nerve impulses, such as *γ-aminobutyric acid*. Some nonprotein amino acids have the D configuration, for example, D-*glutamic acid*, found in substantial amounts in the cell walls of many bacteria (page 269), D-*alanine*, found in the larvae or pupae of some insects, and D-*serine*, found in the earthworm.

Fungi and higher plants contain an extraordinary variety of nonprotein amino acids, some with very curious structures. The metabolic functions of most of these specialized plant amino acids are not yet understood; some, e.g., *canavanine*, *djenkolic acid*, and *β-cyanoalanine*, are toxic to certain other forms of life.

The Acid-Base Properties of Amino Acids

A knowledge of the acid-base properties of amino acids is extremely important in understanding and analyzing the properties of proteins. Furthermore, much of the art of separating, identifying, and quantitating the different amino acids and determining their sequence in proteins is based on their acid-base behavior.

First, let us consider the common ionic species of amino acids. Crystalline amino acids have relatively high melting or decomposition points, usually above 200°C. They are much more soluble in water than in nonpolar solvents. These properties are precisely those to be expected if the lattice of amino acid molecules in the crystalline state is stabilized by

Figure 4-6
Some naturally occurring amino acids not found in proteins.

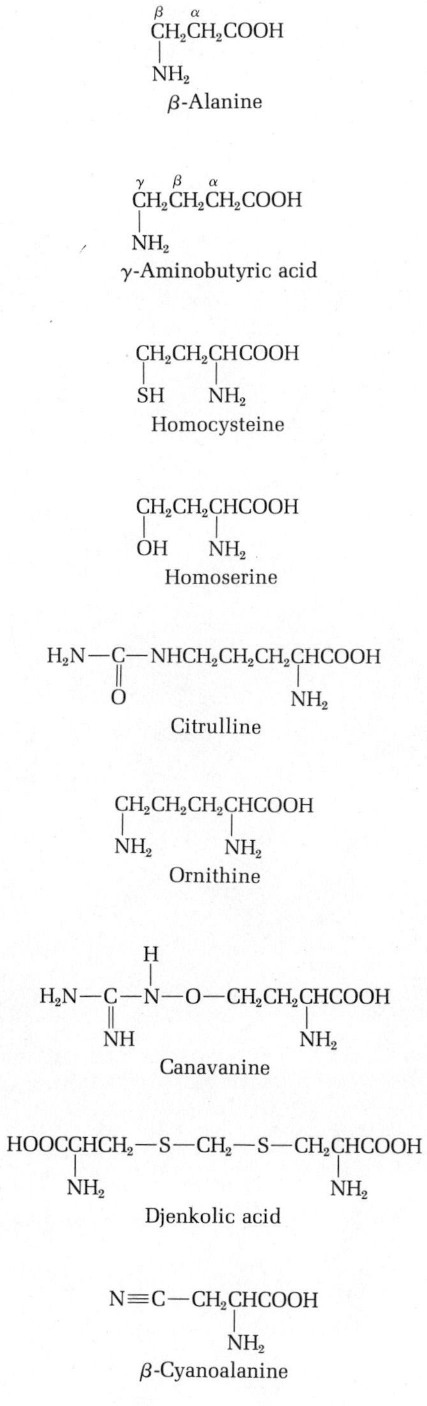

$$\overset{\beta}{C}H_2\overset{\alpha}{C}H_2COOH$$
$$|$$
$$NH_2$$
β-Alanine

$$\overset{\gamma}{C}H_2\overset{\beta}{C}H_2\overset{\alpha}{C}H_2COOH$$
$$|$$
$$NH_2$$
γ-Aminobutyric acid

$$CH_2CH_2CHCOOH$$
$$|\qquad\quad|$$
$$SH\quad\;\; NH_2$$
Homocysteine

$$CH_2CH_2CHCOOH$$
$$|\qquad\quad|$$
$$OH\quad\;\; NH_2$$
Homoserine

$$H_2N-\overset{}{\underset{\underset{O}{\|}}{C}}-NHCH_2CH_2CH_2\overset{}{\underset{\underset{NH_2}{|}}{C}}HCOOH$$
Citrulline

$$CH_2CH_2CH_2CHCOOH$$
$$|\qquad\qquad\quad|$$
$$NH_2\qquad\;\; NH_2$$
Ornithine

$$H_2N-\overset{\overset{H}{|}}{\underset{\underset{NH}{\|}}{C}}-N-O-CH_2CH_2\overset{}{\underset{\underset{NH_2}{|}}{C}}HCOOH$$
Canavanine

$$HOOCCHCH_2-S-CH_2-S-CH_2CHCOOH$$
$$|\qquad\qquad\qquad\qquad\qquad\;\; |$$
$$NH_2\qquad\qquad\qquad\qquad\quad NH_2$$
Djenkolic acid

$$N{\equiv}C-CH_2CHCOOH$$
$$|$$
$$NH_2$$
β-Cyanoalanine

electrostatic forces of attraction between oppositely charged groups, like the high-melting crystal lattice of salts such as sodium chloride. If amino acids crystallized in a nonionic form, they would be stabilized by the much weaker van der Waals forces and would have low melting points. This and many other points of evidence have led to the conclusion that amino acids occur in, and crystallize from, neutral aqueous solutions as *dipolar* ions, or *zwitterions* (Figure 4-7), rather than as undissociated molecules. That amino acids exist as dipolar ions in neutral aqueous solutions is also indicated by their high dielectric constants and their large dipole moments, which are reflections of the occurrence of both positive and negative charges in the same molecule.

When a crystalline zwitterionic amino acid, say alanine, is dissolved in water, it can act either as an acid (proton donor) or as a base (proton acceptor):

As an acid: $H_3\overset{+}{N}CH(CH_3)COO^- \rightleftharpoons H^+ + H_2NCH(CH)_3COO^-$

As a base: $H^+ + H_3\overset{+}{N}CH(CH_3)COO^- \rightleftharpoons H_3\overset{+}{N}CH(CH_3)COOH$

Substances having this property are *amphoteric* (Greek *amphi*, "both") and are called *ampholytes,* abbreviated from "amphoteric electrolytes."

The acid-base behavior of ampholytes is most simply formalized in terms of the Brönsted-Lowry theory of acids and bases (page 47). A simple monoamino monocarboxylic α-amino acid like alanine is considered to be a dibasic acid in its fully protonated form, which can donate two protons during its complete titration with a base. The course of such a two-stage titration with NaOH can be represented in the following equations, which indicate each ionic species involved:

$$H_3\overset{+}{N}CHRCOOH + OH^- \longrightarrow H_3\overset{+}{N}CHRCOO^- + H_2O$$

$$H_3\overset{+}{N}CHRCOO^- + OH^- \longrightarrow H_2NCHRCOO^- + H_2O$$

Figure 4-8 shows the biphasic titration curve of alanine, which may be compared with the titration curves of simple weak acids in Figure 2-9 (page 49). The pK' values of the two stages of ionization of alanine are wide enough apart to yield two clearly separate legs. Each leg has a midpoint where there is minimal change in pH as increments of OH$^-$ are added. The apparent pK' values for the two dissociation steps can be determined from the midpoints of each stage; they are $pK_1' = 2.34$ and $pK_2' = 9.69$. At pH 2.34, the midpoint of the first step, equimolar concentrations of proton-donor ($H_3\overset{+}{N}CHRCOOH$) and proton-acceptor ($H_3\overset{+}{N}CHRCOO^-$) species are present. At pH 9.69, equimolar concentrations of $H_3\overset{+}{N}CHRCOO^-$ and $H_2NCHRCOO^-$ are present. Each of the two legs of the biphasic curve can be expressed mathematically to a very close approximation by the Henderson-Hasselbalch equation (page 50); this means that we can calculate the ratios of ionic species of an amino acid at any pH, given the values for pK_1' and pK_2'.

Figure 4-7
Un-ionized and zwitterion forms of amino acids.

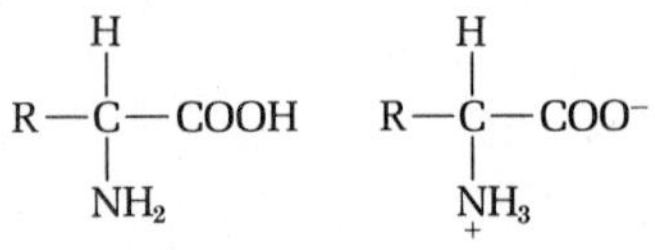

Undissociated form   Zwitterion or dipolar form

Figure 4-8
Titration curve of alanine. The predominant ionic species at each cardinal point in the titration is given in color; R is the methyl group of alanine.

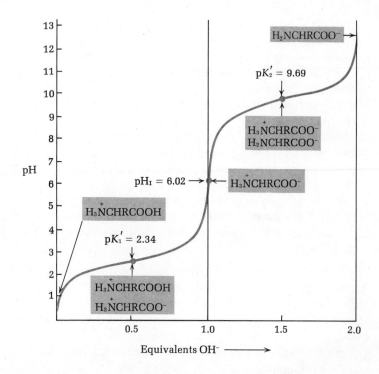

At pH 6.02 there is a point of inflection between the two separate legs of the titration curve of alanine. There is no net electric charge on the molecule at this pH, and it will not move in an electric field. This is the *isoelectric pH* (symbolized pH_I), which is the arithmetic mean of pK_1' and pK_2', that is, $pH_I = \frac{1}{2}(pK_1' + pK_2')$. These relationships are sufficiently exact for most purposes. However, a small but finite amount of the uncharged form ($H_2NCHRCOOH$) is also present in equilibrium with the charged form. All mono-amino monocarboxylic acids show essentially the same behavior.

Table 4-2 gives the pK' values for the ionizing groups of some amino acids. A number of important generalizations follow from these data:

1 The α-carboxyl group of monoamino monocarboxylic acids is a stronger acid than the carboxyl group of comparable aliphatic acids such as acetic acid ($pK' = 4.76$). The increased acid strength of this group is caused by the presence of the electron-withdrawing α-ammonium group and its positive charge, which produces a strong field effect, thus increasing the tendency of the carboxyl hydrogen to dissociate as a proton.

2 The α-amino group of monoamino monocarboxylic acids is a stronger acid (or weaker base) than the amino group of comparable aliphatic amines.

3 All the monoamino monocarboxylic amino acids with uncharged R groups have nearly identical pK_1' values and nearly identical pK_2' values.

4 None of the monoamino monocarboxylic amino acids has significant buffering capacity at the physiological pH zone, pH 6.0 to 8.0. They do show buffering capacity in the zones near their pK' values, i.e., pH 1.3 to 3.3 and 8.6 to

Table 4-2 The pK' values for the ionizing groups of some amino acids at 25°C

Amino acid	pK_1' α-COOH	pK_2' α-NH$_3^+$	pK_R' R group
Glycine	2.34	9.6	
Alanine	2.34	9.69	
Leucine	2.36	9.60	
Serine	2.21	9.15	
Threonine	2.63	10.43	
Glutamine	2.17	9.13	
Aspartic acid	2.09	9.82	3.86
Glutamic acid	2.19	9.67	4.25
Histidine	1.82	9.17	6.0
Cysteine	1.71	10.78	8.33
Tyrosine	2.20	9.11	10.07
Lysine	2.18	8.95	10.53
Arginine	2.17	9.04	12.48

10.6. The only amino acid with significant buffering capacity at pH 6 to 8 is histidine.

5 The β-carboxyl group of aspartic acid and the γ-carboxyl group of glutamic acid, although fully ionized at pH 7.0, have pK′ values considerably higher than the pK′ values of the α-carboxyl groups and more nearly equal to that of simple carboxylic acids such as acetic acid.

6 The thiol or sulfhydryl group (—SH) of cysteine and the p-hydroxyl group of tyrosine are only very weakly acidic. At pH 7.0, the former is about 8 percent ionized and the latter about 0.01 percent ionized.

7 The ε-amino group of lysine and the guanidinium group of arginine are strongly basic; they lose their protons only at a very high pH. At pH 7.0, these amino acids have a net positive charge.

The titration curves of amino acids with R groups that ionize, e.g., histidine, lysine, and glutamic acid, are more complex, since they are a composite of the curve corresponding to the R-group dissociation and the curves for the α-amino and α-carboxyl groups (Figure 4-9).

Formaldehyde in excess readily combines with the free, i.e., unprotonated, amino groups of amino acids to give methylol derivatives. This reaction causes an isoelectric amino acid to lose a proton from the $^+NH_3$— group of the zwitterion form:

$$\overset{+}{H_3}NCHRCOO^- \rightleftharpoons H_2NCHRCOO^- + H^+$$

$$H_2NCHRCOO^- + 2HCHO \longrightarrow (HOCH_2)_2NCHRCOO^-$$

Dimethylol amino acid

The proton so liberated can be titrated directly with NaOH to pH 8, the end point of phenolphthalein (Figure 4-10). Titration of amino acids or amino acid mixtures in the presence of excess formaldehyde (the *formol titration*) is a useful analytical method for following the formation of free amino acids during hydrolysis of proteins by proteolytic enzymes.

The Stereochemistry of Amino Acids

With the single exception of glycine, all amino acids obtained from hydrolysis of proteins under sufficiently mild conditions show optical activity; they can rotate the plane of polarization of plane-polarized light when examined in a polarimeter. Optical activity is shown by all compounds capable of existing in two forms that are nonsuperimposable mirror images of each other; such compounds, which can exist in right-handed and left-handed forms, are called *chiral* compounds (Greek for "hand"). The phenomenon of stereoisomerism, also called *chirality* ("handedness"), occurs in all compounds having an asymmetric carbon atom, i.e., one with four different substituents. Because of the tetrahedral nature of the sp^3 orbitals of the carbon atom, the four different substituent groups can occupy two different arrangements in space around the carbon atom to yield two different *stereoisomers*, or *enantiomers*. Glycine has no asymmetric

Figure 4-9
Titration curves of glutamic acid, lysine, and histidine. The pK′ of the R group is designated pK′ᴿ.

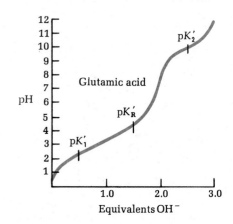

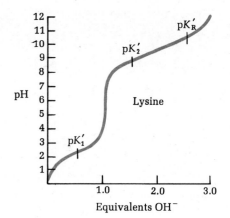

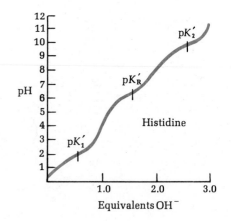

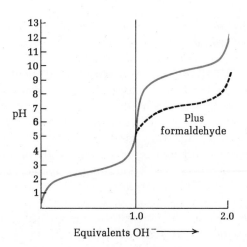

Figure 4-10
Alanine is titrated in the absence and presence of formaldehyde to show the release of titrable H⁺ when formaldehyde combines with the free amino group.

Table 4-3 Specific rotation of some amino acids isolated from proteins (L-stereoisomers) in aqueous solution

Amino acid	Specific rotation $[\alpha]_D^{25}$
L-Alanine	+1.8
L-Arginine	+12.5
L-Leucine	−11.0
L-Isoleucine	+12.4
L-Phenylalanine	−34.5
L-Glutamic acid	+12.0
L-Histidine	−38.5
L-Aspartic acid	+5.0
L-Methionine	−10.0
L-Lysine	+13.5
L-Serine	−7.5
L-Proline	−86.2
L-Threonine	−28.5
L-Tryptophan	−33.7
L-Valine	+5.6

Figure 4-11
Effect of pH on the optical rotation of amino acids.

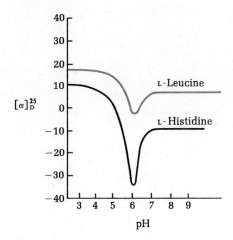

carbon atom. All the other amino acids commonly found in proteins have one asymmetric carbon, except threonine and isoleucine, which have two. The number of possible stereo-isomers of any given compound is 2^n, where n is the number of asymmetric carbon atoms.

Optical activity is expressed quantitatively as the *specific rotation* $[\alpha]_D^{25}$:

$$[\alpha]_D^{25} = \frac{\text{observed rotation, deg} \times 100}{\text{optical path length, dm} \times \text{concentration, g/100 ml}}$$

The temperature (usually 25°C) and the wavelength of the light employed (usually the D line of sodium 589.3 nm) must be specified. The data in Table 4-3 show that some α-amino acids isolated from proteins are *dextrorotatory* (Ala, Ile, Glu, etc.) whereas others are *levorotatory* (Trp, Leu, Phe) when measured at pH 7.0. Dextrorotatory compounds are designated with the symbol (+) and levorotatory compounds with (−). Figure 4-11 shows that the specific rotation of an amino acid varies with the pH at which it is measured; in general, a monoamino monocarboxylic amino acid is at its most levorotatory when it is in its isoelectric form. From the data in Table 4-3, we can also conclude that the specific rotation of an amino acid depends on the nature of its R group.

The stereochemistry of the amino acids normally found in proteins is best discussed not in terms of specific-rotation measurements, like those described above, but in terms of the absolute configuration of the four different substituents in the tetrahedron around the asymmetric carbon atom. All optically active centers can be related stereochemically (by means of appropriate reaction sequences carried out in such a way that optical activity is not lost) to the optically active center of a single parent compound arbitrarily chosen to serve as a standard of reference for stereoisomers. This is the three-carbon sugar *glyceraldehyde*, the smallest sugar to have an asymmetric carbon atom (page 252).

By convention, the two possible stereoisomers of glyceraldehyde are designated L and D (note the use of small capital letters). Figure 4-12 shows the conventions by which the structural formulas of stereoisomers are designated, and Figure 4-13 shows how L- and D-alanine are related to L- and D-glyceraldehyde, respectively. We see that the amino group on the asymmetric carbon atom of alanine can be sterically related to the substituent hydroxyl group on the asymmetric carbon atom of glyceraldehyde, that the carboxyl group of the amino acid can be related to the aldehyde group of glyceraldehyde, and that the R group of the amino acid can be related to the —CH_2OH group of glyceraldehyde. The stereoisomers of all the naturally occurring amino acids can be structurally related to the two stereoisomers of glyceraldehyde in this manner. Isomers stereochemically related to L-glyceraldehyde are designated L, and those related to D-glyceraldehyde are designated D, *regardless of the direction of rotation of plane-polarized light* given by the isomers. The symbols D and L thus refer to *absolute configuration*, not direction of rotation. It has been recommended that the prefixes *d-* and *l-*, which indicate direction of rotation, be replaced by the signs (+) and (−) to eliminate confusion and ambiguity. Whenever the absolute configuration of a compound having an asymmetric carbon atom is known, it is the convention to designate it by D or L; specifying the direction of rotation then becomes unnecessary. If the absolute configuration of an optically active compound has not been established, by convention such compounds may be designated (+) or (−) to indicate direction of rotation but the conditions of measurement must be specified. Throughout this text, the correct configuration about a single asymmetric carbon atom is indicated by a projection formula (Figure 4-12), that is, when the carbon atom is shown with four single bonds joining it and the four dissimilar substituents. Whenever the four single bonds are not explicitly shown, the stereochemistry is not specified.

The D and L stereoisomers of any given compound, e.g. alanine, have identical physical properties and identical chemical reactivities, with two exceptions: (1) they rotate the plane of plane-polarized light equally but in opposite directions; (2) they react at different rates with reagents that are themselves asymmetric. Most enzymes acting upon amino acids have asymmetric binding sites and are thus capable of discriminating completely between the D and L forms of amino acids.

All naturally occurring amino acids found in proteins belong to the L stereochemical series. However, as the examples in Table 4-3 show, some are levorotatory and some are dextrorotatory when dissolved in water. The optical activity of amino acids can be preserved without racemization if the hydrolysis of the protein is carried out under appropriate conditions, but an amino acid synthesized by simple organic chemical reactions in the laboratory is usually obtained in an optically inactive form. Designated a *racemate*, it consists of an equimolar mixture of the D and L stereoisomers, symbolized by the prefix DL. Optically active amino acids are racemized, i.e., converted into DL mixtures, during any chemical reaction in which the asymmetric carbon atom passes through a symmetrical intermediate state, e.g., by

Figure 4-12
Structural formulas of stereoisomers. Shown below are three different representations of a pair of stereoisomers containing an asymmetric carbon atom. In perspective formulas it is the convention that the wedge-shaped bonds project above the plane of the paper, the dotted bonds behind. In ordinary projection formulas the horizontal bonds are assumed to project above the plane of the paper, the vertical bonds behind. This is a convention that is used in this textbook to designate stereochemical relationships. However, projection formulas are sometimes loosely used without reference to stereochemical configuration. In Fischer projections the carbon symbol is usually but not always omitted, the horizontal bonds projecting above the plane of the paper, the vertical bonds behind, and the principal carbon chain appearing vertically, with the lowest-numbered carbon at the top.

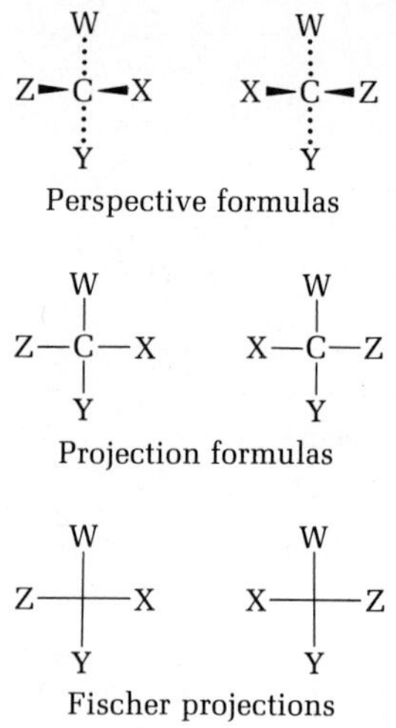

Perspective formulas

Projection formulas

Fischer projections

Figure 4-13
Configurations of the stereoisomers of glyceraldehyde and alanine.

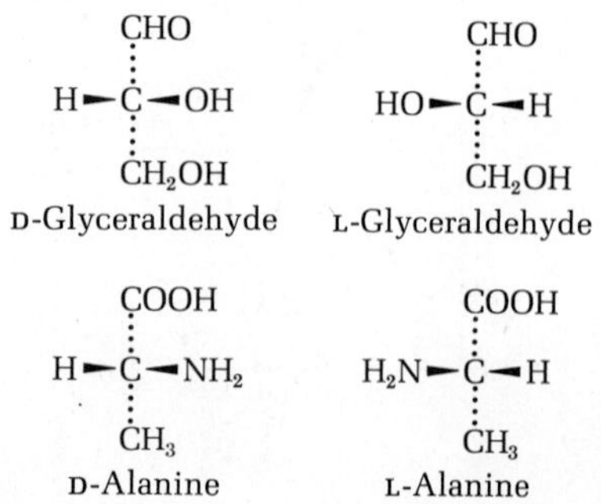

D-Glyceraldehyde   L-Glyceraldehyde

D-Alanine   L-Alanine

Figure 4-14
Stereoisomers of threonine.

```
        COOH                    COOH
         |                       |
  H₂N—C—H              H₂N—C—H
         |                       |
    H—C—OH                HO—C—H
         |                       |
        CH₃                     CH₃

   L-Threonine           L-allo-Threonine


        COOH                    CCOH
         |                       |
    H—C—NH₂               H—C—NH₂
         |                       |
   HO—C—H                 H—C—OH
         |                       |
        CH₃                     CH₃

   D-Threonine           D-allo-Threonine
```

Figure 4-15
Stereoisomers of cystine.

```
        COOH                    COOH
         |                       |
  H₂N—C—H              H₂N—C—H
         |                       |
        CH₂————S—S————CH₂

              L-Cystine


        COOH                    COOH
         |                       |
  H₂N—C—H              H—C—NH₂
         |                       |
        CH₂————S—S————CH₂

            meso-Cystine


        COOH                    COOH
         |                       |
    H—C—NH₂               H—C—NH₂
         |                       |
        CH₂————S—S————CH₂

              D-Cystine
```

Figure 4-16
The ultraviolet absorption spectra of
tryptophan, tyrosine, and phenylalanine.

boiling in strong base. Little or no racemization of amino acids occurs when they are heated with strong acids.

The amino acids with two asymmetric carbon atoms, threonine and isoleucine, have four stereoisomers. The form of threonine isolated from protein hydrolyzates is by convention designated L; its mirror image is the D form. The other two stereoisomers are *diastereoisomers*, or *allo*, forms, also mirror images of each other (Figure 4-14). It is the configuration about the α carbon atom that is the basis for configurational assignment. Cystine, the oxidized form of cysteine (page 86), which contains two asymmetric carbon atoms, one in each half of the molecule, may occur not only in L and D forms but also as an isomer in which the asymmetric carbon atoms are mirror images of each other. This *internally compensated* isomer, which does not occur biologically, is called a *meso* form (Figure 4-15). Obviously, naming stereoisomers of compounds with more than one asymmetric carbon atom can lead to ambiguities. These and other difficulties are avoided by a new system of designating stereoisomers, the Cahn-Ingold-Prelog convention, usually called the RS system. Although it is not yet widely used in biochemistry, it provides the only unambiguous way of designating absolute configurations of molecules having two or more asymmetric centers (see References).

Although only L-amino acids are present in protein molecules, many different D-amino acids are found in living cells in other chemical forms, e.g., in the cell walls of certain microorganisms or as part of the structure of peptide antibiotics like *gramicidin* (page 802) and *actinomycin D* (page 920).

Absorption Spectra

Although none of the 20 amino acids found in proteins absorbs light in the visible range, three amino acids—tyrosine, tryptophan, and phenylalanine—absorb light significantly in the ultraviolet (Figure 4-16). Since most proteins contain tyrosine residues, measurement of light absorption at 280 nm in a spectrophotometer is an extremely rapid and convenient means of estimating the protein content of a solution. Cystine absorbs weakly at 240 nm due to its disulfide group. All amino acids absorb in the far ultraviolet (< 220 nm).

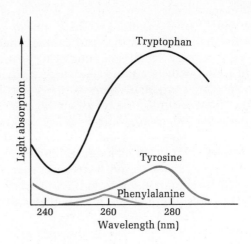

The Chemical Reactions of Amino Acids

The characteristic organic reactions of amino acids are those of their functional groups, i.e., the carboxyl groups, the α-amino groups, and the functional groups present in the different side chains. Knowledge of these reactions is useful in several important aspects of protein chemistry: (1) identification and analysis of amino acids in protein hydrolyzates, (2) identification of the amino acid sequence in protein molecules, (3) identification of the specific amino acid residues of native proteins that are required for their biological function, (4) chemical modification of amino acid residues in protein molecules to produce changes in their biological activity or other properties, and (5) chemical synthesis of polypeptides.

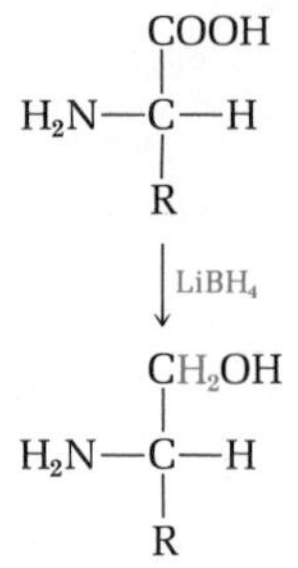

Reactions of Carboxyl Groups

The carboxyl groups of all amino acids undergo well-known organic reactions leading to the formation of amides, esters, and acid halides. These reactions need not be detailed here. Esterification of amino acids with ethanol or benzyl alcohol is often used as a means of protecting the carboxyl group of amino acids in the chemical synthesis of peptides (page 117). Another reaction of the carboxyl group of special usefulness in the study of amino acid sequence (page 104) is its reduction in an anhydrous medium with the potent reducing agent lithium borohydride to yield the corresponding primary alcohol (Figure 4-17).

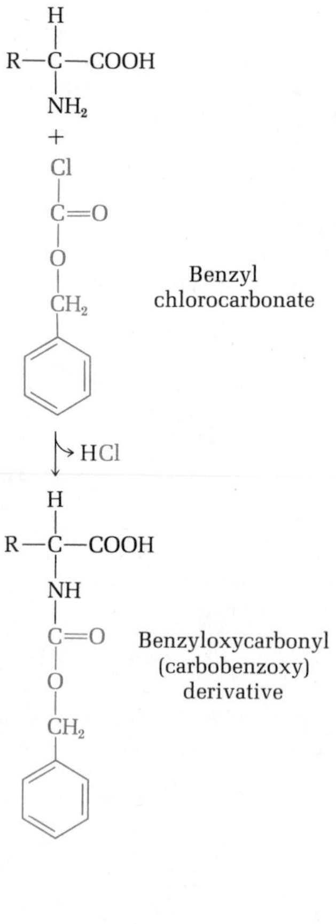

Reactions of Amino Groups

The α-amino group of amino acids can be acylated by treatment with acid halides or anhydrides; this procedure is commonly used to block or protect the α-amino group in the chemical synthesis of peptides (page 117). A widely used reagent for this purpose is _benzylchlorocarbonate_, which yields the corresponding _benzyloxycarbonyl_ derivative of the amino acid (Figure 4-18), often called a _carbobenzoxy_ derivative. When acylation is carried out under mild conditions, the stereochemical integrity of the α-carbon atom is retained, but under drastic conditions, i.e., if heat is applied, racemization may occur.

A very widely used reaction of the α-amino group, the _ninhydrin reaction_, can be utilized to estimate amino acids quantitatively in very small amounts. On heating, an α-amino acid reacts with two molecules of ninhydrin to yield an intensely colored product (Figure 4-19). A purple color is given in the ninhydrin reaction by all amino acids and peptides having a free α-amino group, whereas proline and hydroxyproline, in which the α-amino group is substituted, yield derivatives with a characteristic yellow color.

The α-amino groups of amino acids react reversibly with aldehydes to form compounds called _Schiff's bases_ (Figure 4-20). Schiff's bases appear to be intermediates in a number

Figure 4-17
Reduction of the carboxyl group of amino acids.

Figure 4-18
Formation of the benzyloxycarbonyl (carbobenzoxy) derivative of amino acids.

Figure 4-19
The ninhydrin reaction. This strong oxidizing agent brings about the oxidative decarboxylation of the amino acid. The ammonia and hydrindantin so formed react
with a second molecule of ninhydrin to
yield a purple pigment, of which only the
nitrogen atom arises from the amino acid.

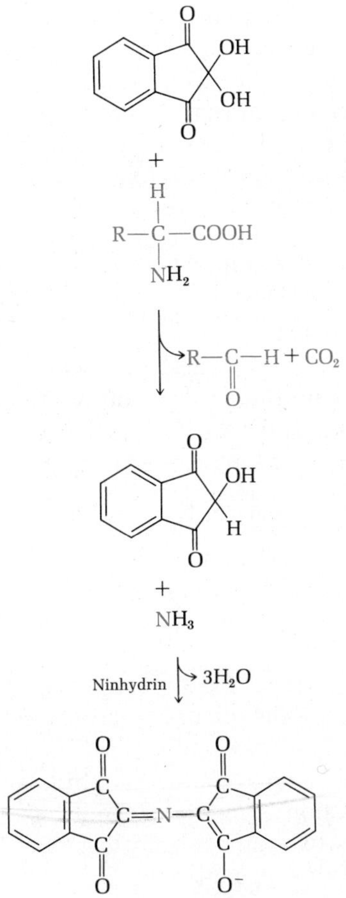

Ninhydrin

Amino acid

$R-C-H + CO_2$

Hydrindantin

$+$

NH_3

Ninhydrin $3H_2O$

Purple pigment

Figure 4-20
Formation of a Schiff's base.

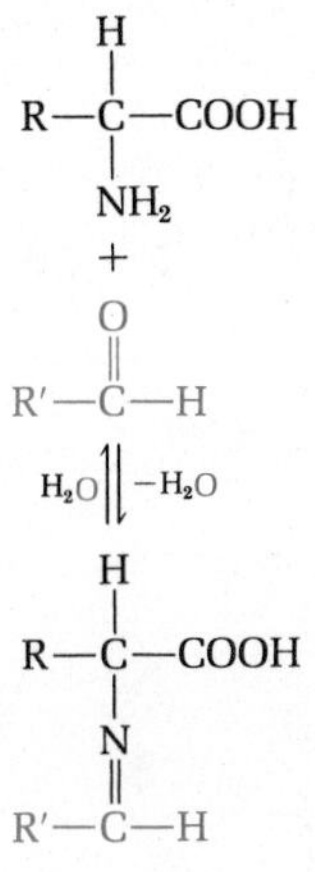

Amino acid

Aldehyde

A Schiff's base

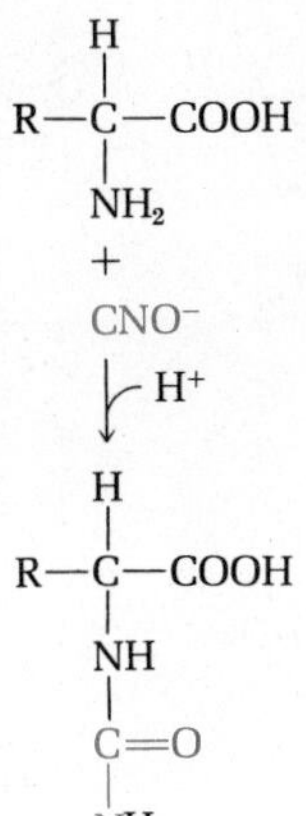

of enzymatic reactions involving interaction of the enzyme
with amino or carbonyl groups of the substrate (pages 221,
425, and 469). Another important reaction of amino groups is
with cyanate, to yield carbamoyl derivatives (Figure 4-21).
This reaction has been used to modify the properties of
sickle-cell hemoglobin to make it more nearly like normal
adult hemoglobin (page 149).

In addition to these reactions of the amino group, several
others are particularly useful in labeling the amino acid residue at that end of a polypeptide chain at which the amino
group is free or uncombined, i.e., the amino-terminal end.
These important reactions will be described in detail in
Chapter 5, pages 102 to 105, in connection with methods for
establishing the amino acid sequence of peptides.

Figure 4-21
Reaction of amino acids with cyanate.

Cyanate ion

Carbamoylamino
acid

Reactions of the R Groups

Amino acids also show qualitative color reactions typical of
certain functions present in their R groups, e.g., the thiol
group of cysteine, the phenolic hydroxyl group of tyrosine,
and the guanidinium group of arginine. Although these color
reactions are sometimes useful for spot tests or qualitative
identification, amino acids are more accurately identified

and measured in very low concentrations by means of chromatographic methods (see below). However, certain other functional groups in the side chains of amino acids play important roles in the biological activity of proteins, among them the thiol or sulfhydryl group of cysteine. This very reactive group is highly susceptible to oxidation to the disulfide by atmospheric oxygen in the presence of iron salts or by other mild oxidizing agents. The oxidation product is cystine (Figure 4-22), in which the disulfide bond constitutes a covalent bridge between two residues of cysteine. Another important and characteristic reaction of the thiol group of cysteine is with heavy metals such as Hg^{2+} and Ag^+, which form mercaptides (Figure 4-23). The thiol groups of cysteine and cysteine residues in peptides and proteins can be measured by a quantitative reaction with Ellman's reagent (Figure 4-24), yielding a product that can be measured colorimetrically. We shall see later (page 224) that the thiol groups of specific cysteine residues in certain enzyme molecules play an important role in their catalytic activity; modification of such thiol groups by reactions like those described above often inactivates the enzyme.

Cystine, the oxidized form of cysteine, plays a special role in protein structure since its disulfide group serves as a covalent cross-link between two polypeptide chains or between two points in a single chain. Disulfide cross-links can be cleaved by the action of reducing agents, e.g., mercaptoethanol, which yields two molecules of cysteine from cystine (Figure 4-25). Cystine can also be oxidized by such agents as performic acid, to yield two molecules of cysteic acid. This important reaction will be discussed later (page 100). Both the sulfhydryl and disulfide groups are destroyed in strongly alkaline solution, with the formation of a number of products.

Analysis of Amino Acid Mixtures

The quantitative separation and estimation of each amino acid in a complex mixture like the hydrolyzate of a protein is a formidable problem when attacked by such classical separation methods as fractional precipitation, crystallization, or distillation. Twenty-five years ago the quantitative analysis of but one amino acid in a mixture might have taken months of work. It was not until chromatographic methods were systematically applied to the analysis of amino acid mixtures that any significant progress was achieved in the detailed study of protein structure. Since then analytical methods based on these principles have been vastly refined and are capable of great speed, precision, and sensitivity; they have also been automated.

Chromatographic methods are applicable not only to separation, identification, and quantitative analysis of amino acid mixtures but also of peptides, proteins, nucleotides, nucleic acids, lipids, and carbohydrates. Because of the universal importance in biochemistry of the many different forms of

Figure 4-22
Oxidation of cysteine to cystine.

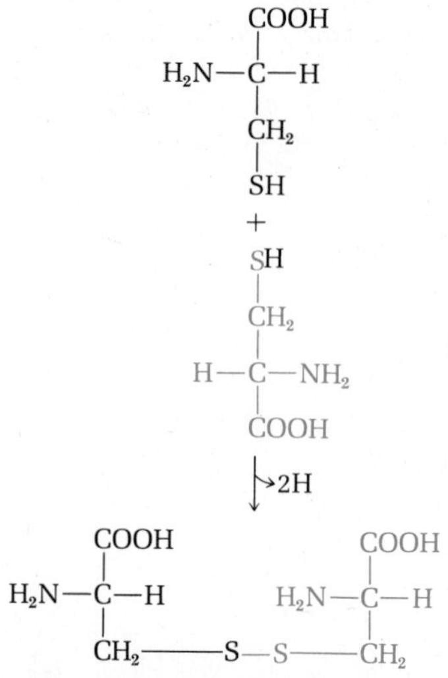

Figure 4-23
Reaction of cysteine with Ag^+.

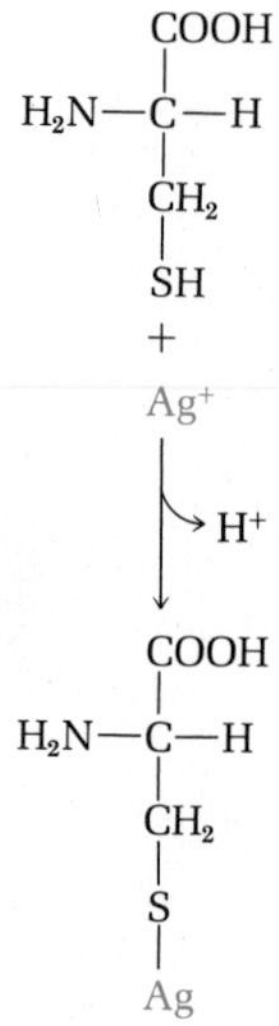

Figure 4-24
Reaction of cystine with Ellman's reagent. This reaction causes release of a molecule of thionitrobenzoic acid, which at pH 8.0 has an intense absorption at 412 nm.

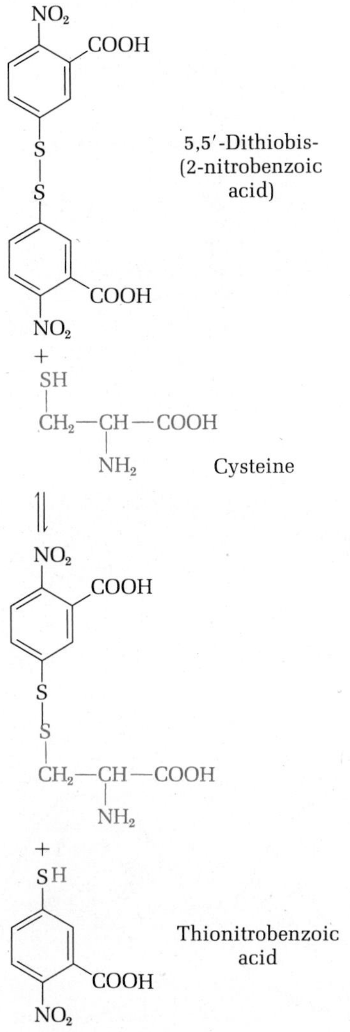

5,5'-Dithiobis-(2-nitrobenzoic acid)

Cysteine

Thionitrobenzoic acid

Figure 4-25
Reduction of cystine to cysteine.

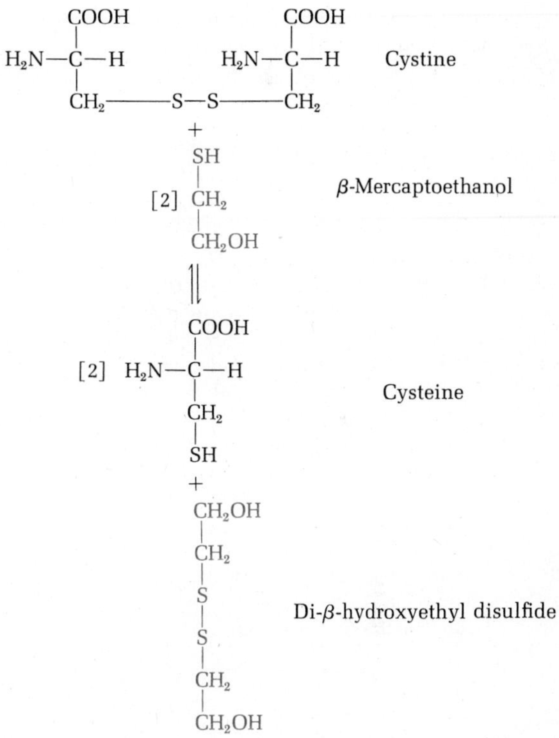

Cystine

β-Mercaptoethanol

Cysteine

Di-β-hydroxyethyl disulfide

chromatography employed today, we shall briefly review their basic physical principles. It will become evident that the art of amino acid analysis by chromatographic and electrophoretic methods is entirely based on knowing the relative solubility and acid-base behavior of the different amino acids.

The Partition Principle: Partition Chromatography

When a solute is allowed to distribute itself between equal volumes of two immiscible liquids, the ratio of the concentrations of the solute in the two phases at equilibrium at a given temperature is called the *partition coefficient.* Amino acids can be partitioned in this manner between two liquid phases, e.g., the pairs phenol-water or n-butanol–water; each amino acid has a distinctive partition coefficient for any given pair of immiscible solvents. A mixture of substances with different partition coefficients can be quantitatively separated by a technique known as *countercurrent distribution,* first developed by L. C. Craig, in which many repetitive partition steps take place. The principle of countercurrent distribution is shown in a simple example in Figure 4-26.

Partition chromatography is the chromatographic separation of mixtures essentially by the countercurrent-partition

Figure 4-26
Countercurrent distribution. Countercurrent distribution involves many repetitive partitions of a mixture of substances between two immiscible solvents. The principle is illustrated with a single substance Y having a partition coefficient $(K = [Y]_A/[Y]_B)$ equal to 1.0 between the solvents A and B. Solute Y (64 units) dissolved in one volume of solvent B is present in tube 1; tubes 2 to 7 contain equal volumes of pure solvent B. One volume of solvent A is added to tube 1 and shaken to equilibrium; the result is the partition of solute Y between solvents A and B in equal amounts, that is, 32 units in each. Solvent A (the top layer) from tube 1 is now transferred to tube 2 and fresh solvent A added to tube 1. Tubes 1 and 2 are shaken to equilibrium with the result that each solvent layer in both tubes contains 16 units of Y, according to the partition coefficient $K = 1.0$. After each partition step the top layer of each tube is removed and added to the next tube to the right; each time a fresh layer of solvent A is added to tube 1. The diagram shows the results of seven partition steps; in actual laboratory practice 100 or more partition steps are employed.

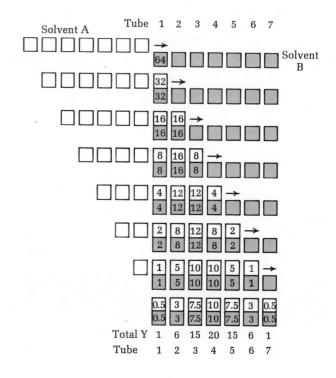

The total amount of solute Y in each of the seven tubes is shown in a plot. For comparison the distribution of two other solutes X and Z (64 arbitrary units) carried through the same set of countercurrent distributions is shown. Solute X has a lower partition coefficient $(K_X = [X]_A/[X]_B = 0.33)$ and solute Z a higher partition coefficient $(K_Z = [Z]_A/[Z]_B = 3.0)$ than solute Y. The higher the partition coefficient the greater the fraction of solute moved per transfer. Since each solute moves independently of the other in accordance with its partition coefficient, a mixture of solutes X, Y, and Z will separate completely in three peaks after a much larger number of transfers.

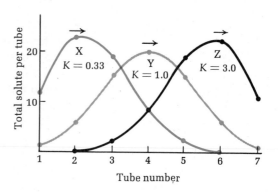

principle described above. It was first developed as a powerful method for separating amino acids by A. J. P. Martin and R. L. M. Synge in England and has since been applied to an enormous number of different substances. The separation is achieved in a huge number of separate partition steps, which take place on microscopic granules of a hydrated insoluble inert substance, such as starch or silica gel, packed in a column about 10 to 100 cm long. Starch or silica gel granules are hydrophilic and are surrounded by a layer of tightly bound water, which serves as a stationary aqueous phase, past which flows a moving phase of an immiscible solvent containing the mixture to be separated. The mixture of solutes undergoes a microscopic partition process between the fixed water layer and the flowing solvent. This process, which occurs on the surface of each granule, is similar to the partition steps in countercurrent distribution. Individual solute partition processes occurring in a starch or silica gel column are not necessarily complete to the point of equilibrium, but the total number of partition steps in the column is so great that the different amino acids in a mixture move down the column at different rates as the moving liquid phase flows through it. The liquid appearing at the bottom of the column, called the _eluate_, is caught in small fractions with an automatic fraction collector and analyzed by means of the quantitative ninhydrin reaction (page 85). A plot of the amount of amino acid in each tube versus the tube or fraction number will show a series of peaks, each corresponding to a different amino acid.

Precisely the same principle is involved in _filter-paper chromatography_ of amino acids. The cellulose of the filter-paper fibers is hydrated. As a solvent containing an amino acid mixture ascends in the vertically held paper by capillary action (or descends, in descending chromatography), many microscopic distributions of the amino acids occur between the flowing phase and the stationary water phase bound to the paper fibers. At the end of the process, the different amino acids have moved different distances from the origin. The paper is dried, sprayed with ninhydrin solution, and heated in order to locate the amino acids. In the important refinement of _two-dimensional paper chromatography_, the mixture of amino acids is chromatographed in one direction; then the paper is dried and subjected to chromatography with a different solvent system in a direction at right angles to the first. A two-dimensional map of the different amino acids results. Two-dimensional paper chromatography, or paper electrophoresis (see below) in one direction followed by chromatography at right angles, is widely used for separating amino acids and peptides (page 107).

Ion-Exchange Chromatography

The partition principle has been further refined in _ion-exchange chromatography_, first developed by W. Cohn. In this method solute molecules are sorted out by the differences in their acid-base behavior. For this process a column is filled with granules of a synthetic resin containing fixed charged groups. There are two major classes of ion-exchange resins: cation exchangers and anion exchangers. Amino acids are usually separated on cation-exchange columns filled with solid granules of a sulfonated polystyrene resin previously equilibrated with an NaOH solution in order to charge its sulfonic acid groups with Na^+. This form of the resin is called the sodium form; the resin may also be prepared in the protonated, or hydrogen, form by washing it with acid. To the washed Na⁺ form of the resin an acid solution (pH = 3.0) of the amino acid mixture is added; at pH 3.0, amino acids are largely cations with net positive charge. The cationic amino acids tend to displace some of the bound Na⁺ ions from the resin particles; the amount of displacement will vary slightly among different amino acids because of small differences in degree of ionization. At pH 3.0, the most basic amino acids (lysine, arginine, and histidine) are bound tightest to the resin by electrostatic forces, and the most acid (glutamic and aspartic acid) are bound the least. As the pH and the NaCl concentration of the eluting aqueous medium are gradually increased, the amino acids move down the column at different rates and the eluate is collected in many small fractions. These are analyzed quantitatively by means of the ninhydrin reaction. The most anionic, e.g., glutamic acid, appear first, and the most cationic, e.g., lysine, appear last. An elution curve is constructed from the data (Figure 4-27). The entire analytical procedure has been automated, so that elution, collection of fractions, analysis of each fraction, and recording data are performed automatically by servomechanisms in an apparatus called an *amino acid analyzer*. This technique of amino acid analysis was pioneered by W. Stein and S. Moore at the Rockefeller Institute, New York City.

To reduce the time required for precise chromatographic measurement of amino acid composition, particularly in hydrolyzates of proteins, very short ion-exchange columns are used and the eluting buffers are forced through the column under pressure. These and other refinements make it possible to determine the type and amounts of all the amino acids in a mixture in little more than 2 h; only a fraction of a milligram of sample is required.

Figure 4-27
Automatically recorded chromatographic analysis of amino acids on an ion-exchange resin. The elution is carried out with different buffers of successively higher pH. The effluent is caught in small volumes, and the amino acid content of each tube is automatically analyzed. The area under each peak is proportional to the amount of each amino acid in the mixture. [Redrawn from D. H. Spackman, W. H. Stein, and S. Moore, Anal. Chem., 30: 1190 (1958).]

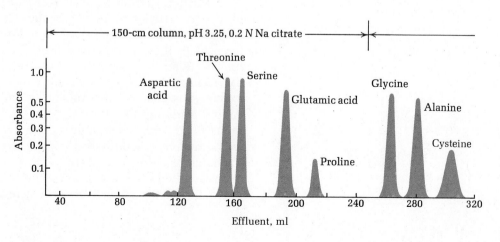

In another method of separating amino acids, *paper electrophoresis*, a drop of a solution of the amino acid mixture is placed on a filter-paper sheet moistened with a buffer of a given pH. The ends of the sheet dip into electrode vessels, and an electric field is applied. Because of their different pK' values, the amino acids migrate in different directions and at different rates, depending on the pH of the system and the emf applied. For example, at pH 1, histidine, arginine, and lysine have a charge of +2 whereas all other amino acids have a charge of +1. At pH 11, aspartate, glutamate, cysteine, and tyrosine have a charge of −2 and all others −1. Knowing the acid-base properties of amino acids makes it possible to select conditions to achieve separation of any given mixture of amino acids. Amino acids can also be separated by *thin-layer chromatography*, a refinement of partition chromatography which will be described in Chapter 11 (page 285).

Summary

The 20 amino acids commonly found as hydrolysis products of proteins are alike in containing an α-carboxyl group and an α-amino group but differ in the chemical nature of the R groups substituted on the α-carbon atom. They are classified on the basis of the polarity of their R groups. The nonpolar (hydrophobic) class includes alanine, leucine, isoleucine, valine, proline, phenylalanine, tryptophan, and methionine. The polar neutral class includes glycine, serine, threonine, cysteine, tyrosine, asparagine, and glutamine. The positively charged (basic) class contains arginine, lysine, and histidine, and the negatively charged (acidic) class contains aspartic acid and glutamic acid. In a few specialized proteins, other amino acids may occur, such as hydroxyproline, hydroxylysine, and desmosine.

Monoamino monocarboxylic amino acids are dibasic acids ($H_3N^+CHRCOOH$) at low pH. As the pH is raised to about 6, the proton is lost from the carboxyl group to form the dipolar, or zwitterion, species $H_3\overset{+}{N}CHRCOO^-$, which is electrically neutral. Further increase in pH causes loss of the second proton, to yield the ionic species $NH_2CHRCOO^-$. The pK' of the first ionization, i.e., of the α-carboxyl group, is about 2.0 to 2.5; the pK' of the second step is about 9 to 10. Amino acids possessing ionizable R groups may exist in additional ionic species, depending on the pK' values of their R groups. The α-carbon atom of the amino acids (except glycine) is asymmetric and thus can exist in at least two stereoiso-

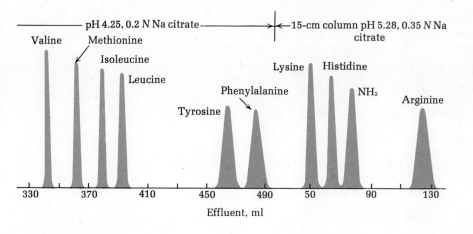

meric forms; only the L stereoisomers, which are related to L-glyceraldehyde, are found in proteins.

All α-amino acids form analytically useful colored derivatives with ninhydrin. The α-amino group is also readily acylated; benzylchlorocarbonate yields benzyloxycarbonyl derivatives of amino acids. The α-amino group forms Schiff's bases with aldehydes and carbamoyl derivatives with cyanate. The carboxyl groups and R groups of amino acids also show characteristic reactions. The thiol group of cysteine is readily oxidized to yield cystine, which can in return be reduced to cysteine. Such reactions are extremely useful for identification and analysis of particular amino acids.

Complex mixtures of amino acids can be separated, identified, and estimated by chromatography on paper or ion-exchange columns or by paper electrophoresis. These methods exploit differences in the solubility and/or acid-base behavior of the different amino acids.

References

Books   .

See also references to Chapters 3, 5, 6, and 9.

ALWORTH, W. L.: *Stereochemistry and Its Application to Biochemistry*, Wiley-Interscience, New York, 1972. A valuable book on a complex subject. The Cahn-Ingold-Prelog, or RS, system of designating the absolute configuration of stereoisomers is given in detail; the principles of the RS system are also described in textbooks of organic chemistry.

BLACKBURN, S.: *Amino Acid Determination*, Dekker, New York, 1968. Comprehensive treatise on analytical methods.

EDSALL, J. T., and J. WYMAN: *Biophysical Chemistry*, vol. 1, Academic, New York, 1958. Excellent treatment of amino acids as electrolytes.

GREENSTEIN, J. P., and M. WINITZ: *Chemistry of the Amino Acids*, 3 vols., Wiley, New York, 1961.

MEISTER, A.: *Biochemistry of the Amino Acids*, 2d ed., 2 vols., Academic, New York, 1965. Authoritative and comprehensive treatment of structure, occurrence, and metabolism of protein and nonprotein amino acids.

Articles

FOWDEN, L.: "The Non-Protein Amino Acids of Plants," in L. Reinhold and Y. Liwschitz (eds.), *Progress in Phytochemistry*, vol. 2, Interscience, London, 1970.

FOWDEN, L., D. LEWIS, and H. TRISTRAM: "Toxic Amino Acids: Their Actions as Antimetabolites," *Adv. Enzymol.*, 29: 89–163 (1968).

HIRS, C. H. W. (ed.): "Amino Acid Analysis and Related Procedures," Sec. 1 in *Methods in Enzymology*, vol. II, Academic, New York, 1967.

SPACKMAN, D. H., W. H. STEIN, and S. MOORE: "Automatic Recording Apparatus for Use in the Chromatography of Amino Acids," *Anal. Chem.*, 30: 1190–1206 (1958). A description of the automatic amino acid analyzer.

Problems

1. Name an amino acid found in proteins that can be converted by treatment with strong base into another amino acid found in proteins.

2. Name a naturally occurring amino acid with an isoelectric point greater than 8.0 that would be expected to form a derivative with molecular weight 118 upon treatment with lithium borohydride.

3. Calculate the pH_I values of glycine, alanine, serine, threonine, and glutamic acid from their pK' values (Table 4-2).

4. Referring to Table 4-2 for the necessary pK' values, indicate the probable net charge (−, 0, or +) of glycine, aspartic acid, lysine, and histidine at (a) pH 1.0, (b) pH 2.10, (c) pH 4.0, (d) pH 10.

5. Paper electrophoresis at pH 6.0 was carried out on a mixture of glycine, alanine, glutamic acid, lysine, arginine, and serine.
   (a) Which compound(s) moved fastest toward the anode?
   (b) Which moved fastest toward the cathode?
   (c) Which remained at or near the origin?

6. A mixture of lysine, arginine, aspartic acid, glutamic acid, tyrosine, and alanine is added to an anion exchange resin at high pH. Predict the order of elution upon treatment of the resin with solutions successively decreasing in pH.

7. One-directional filter-paper chromatography of a mixture of serine, alanine, glutamic acid, and isoleucine is carried out in a solvent system containing n-butanol, water, and acetic acid. Predict the relative mobilities of the amino acids in the system described, giving first the one which would be expected to move the greatest distance from the origin.

8. How many grams of NaOH would you have to add to 500 ml of 0.01 M histidine in its fully protonated form to yield a buffer of pH 7.0?

9. To 1.0 l of a 1.0 M solution of glycine at the isoelectric pH is added 0.3 mol of HCl.
   (a) What will be the pH of the resultant solution?
   (b) What would be the pH if 0.3 mol of NaOH were added instead?

10. A solution of L-alanine (400 ml) is brought to pH 8.0 and then treated with an excess of formaldehyde. The resultant solution requires a total of 250 ml of 0.2 M NaOH solution to titrate it back to pH 8.0. How many grams of L-alanine did the original solution contain?

11. Using data in Table 4-3, calculate the optical rotation of a 1.10 M solution of alanine in water at 25°C in a 25-cm polarimeter tube with the sodium D line as light source.

12. A solution containing only L-isoleucine and L-phenylalanine gives an observed rotation of −1.97° in a 25-cm polarimeter tube at 25°C. When a 100-ml portion of the solution is brought to pH 8.0 and treated with excess formaldehyde, 66.8 ml of 0.5 N NaOH is required to titrate it back to pH 8.0. What is the molar concentration of (a) isoleucine and (b) phenylalanine in the original sample?

13. Glycine is 4 times as soluble in solvent A as in solvent B, whereas phenylalanine is only one-half as soluble in solvent A as in solvent B. Solvents A and B are immiscible; solvent B is

heavier than solvent A. A mixture of 1.0 mg of glycine and 1.0 mg of phenylalanine is shaken with equal volumes of solvent A and B in a separatory funnel until equilibrium is reached. The top layer is then removed and transferred to another separatory funnel, and an equal volume of fresh solvent B is added. The contents are then shaken to equilibrium. Calculate the amount of glycine and phenylalanine in solvent A after the second equilibration.

14. Amino acid X is 4 times as soluble in solvent A as in solvent B. Amino acid Y is only $1\frac{1}{2}$ times as soluble in A as in B. Counter-current distribution between solvents A and B was used to separate X and Y. Starting with 100 mg of X and 100 mg of Y: (a) Which tube of a four-tube countercurrent system (see Figure 4-26) contains the most X and which the most Y? (b) How many milligrams of each amino acid are in each of the tubes?

15. Suppose that the countercurrent distribution of Problem 14 is extended to a total of 20 tubes. How many milligrams of (a) X and (b) Y are present in tube 20?

16. To 400 ml of 0.1 M glycine at its isoelectric point is added 100 ml of 0.5 M acetic acid. Calculate (a) the pH of the mixture and (b) an approximate concentration for the predominant isoelectric form of glycine in the mixture.

17. A solution of 0.1 M glutamic acid is at its isoelectric point.
(a) Calculate an approximate concentration for the major isoelectric form.
(b) Calculate an approximate concentration for the completely protonated form.
(c) A total of how many forms is present in solution, even at vanishingly small concentrations?

In this chapter we examine various aspects of the *primary structure* of proteins, which we have defined (page 60) as the covalent backbone structure of polypeptide chains, including the sequence of amino acid residues. We begin by considering the properties of simple peptides. Then we examine three major problems: (1) the determination of amino acid sequence in polypeptide chains, (2) the significance of variations in the amino acid sequences of different proteins in different species, and (3) the laboratory synthesis of polypeptide chains. We shall use various terms and concepts already defined in Chapter 3, which may be referred to for orientation.

The Structure of Peptides

Simple peptides containing two, three, four, or more amino acid residues, i.e., dipeptides, tripeptides, tetrapeptides, etc., joined covalently through peptide bonds, are formed on partial hydrolysis of much longer polypeptide chains of proteins. Many hundreds of different peptides have been isolated from such hydrolyzates or synthesized by chemical procedures (page 117). Peptides are also formed in the gastrointestinal tract during the digestion of proteins by *proteases*, enzymes that hydrolyze peptide bonds. Peptides are named from their component amino acid residues in the sequence beginning with the amino-terminal (abbreviated N-terminal) residue (Figure 5-1).

Much evidence supports the conclusion that the peptide bond is the sole covalent linkage between amino acids in the linear backbone structure of proteins. This evidence comes not only from chemical- and enzymatic-degradation studies, but also from various physical measurements. For example, proteins have absorption bands in the far ultraviolet (180 to 220 nm) and infrared regions that are similar to those given by authentic peptides. Furthermore, x-ray diffraction analysis (Chapter 6) directly shows the presence of peptide bonds in native proteins. There is only one other major covalent linkage between amino acids: the disulfide bond of cystine (page 86) serves in some proteins as a cross-link

Figure 5-1
Structure of a pentapeptide. Peptides are named beginning with the N-terminal residue. The peptide bonds are shaded in color.

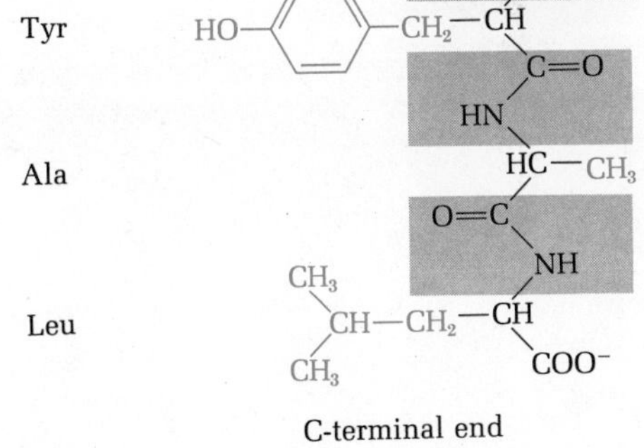

N-terminal end

Ser

Gly

Tyr

Ala

Leu

C-terminal end

Serylglycyltyrosinylalanylleucine
(Ser-Gly-Tyr-Ala-Leu)

between two separate polypeptide chains (interchain disulfide bond) or between loops of a single chain (intrachain disulfide bond).

Peptides may be regarded as substituted amides. Like the amide group, the peptide bond shows a high degree of resonance stabilization. The C—N single bond in the peptide linkage has about 40 percent double-bond character and the C=O double bond about 40 percent single-bond character. This fact has two important consequences: (1) The imino (—NH—) group of the peptide linkage has no significant tendency to ionize or protonate in the pH range 0 to 14. (2) The C—N bond of the peptide linkage (Figure 5-1) is relatively rigid and cannot rotate freely, a property of supreme importance with respect to the three-dimensional conformation of polypeptide chains, as we shall see in Chapter 6.

Peptides of Nonprotein Origin

In addition to the large number of different short peptides identified as partial hydrolysis products of proteins, many peptides not derived from proteins have been found in living matter (Figure 5-2). Such nonprotein peptides usually differ structurally from those derived from proteins. For example, the tripeptide *glutathione*, found in all cells of higher animals, contains a glutamic acid residue joined in an unusual peptide linkage involving its γ-carboxyl rather than the α-carboxyl group. The muscle dipeptide *carnosine* contains a β-amino acid. Some nonprotein peptides contain D-amino acids, such as the antibiotic *tyrocidin A* (Figure 5-2). β-Amino acids, γ-peptide bonds, and D-amino acids do not occur in proteins; presumably these variations in structure protect these specialized peptides from the action of proteases, which are generally specific for peptides having L-α-amino acids in normal peptide linkage. Among the nonprotein peptides are some having hormonal activity, e.g., the hypothalamic regulatory factor, for release of thyrotropic hormone from the anterior pituitary gland; the posterior pituitary hormones *oxytocin* and *vasopressin;* and the nonapeptide *bradykinin* of blood plasma, which participates in regulation of blood pressure (see page 810). Many antibiotics are peptides or derivatives of peptides; including *gramicidin* and the cyclic decapeptide *valinomycin* (page 802).

Acid-Base Properties of Peptides

Peptides usually have high melting points, indicating that they crystallize from neutral solutions in an ionic lattice as dipolar ions, like the amino acids (page 77). Since none of the α-carboxyl groups and none of the α-amino groups that are combined in peptide linkages can ionize in the pH zone 0 to 14, the acid-base behavior of peptides is contributed by the free α-amino group of the N-terminal residue, the free α-carboxyl group of the carboxy-terminal (abbreviated C-terminal) residue, and those R groups of the residues in intermediate positions which can ionize (page 79). In long polypeptide chains the ionizing R groups necessarily greatly

Figure 5-2
Some biological peptides of nonprotein origin. Such peptides often contain amino acids other than those found in proteins. The arrows, where used, show the polarity or direction of the peptide bonds; they point from the N-terminal toward the C-terminal end of the chain.

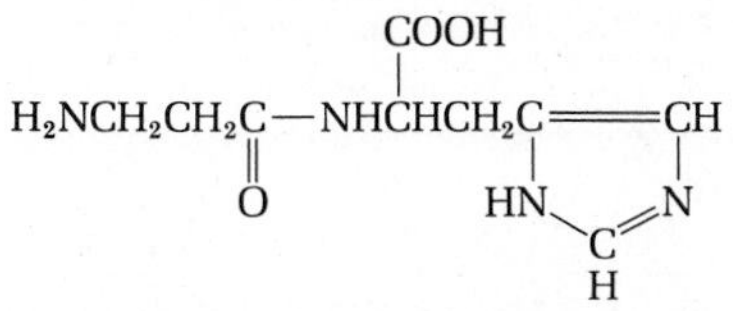

Carnosine (β-alanylhistidine)

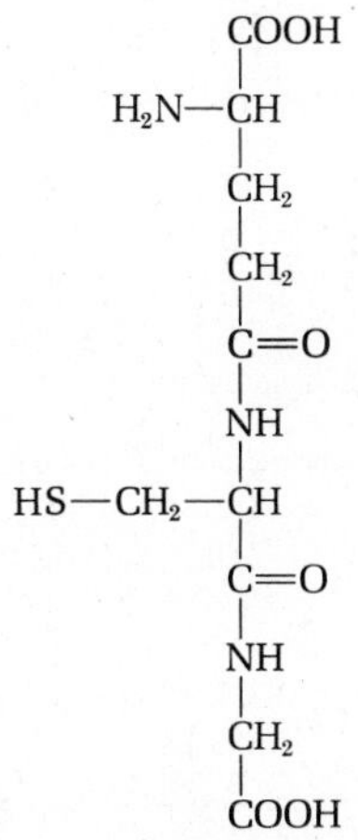

Glutathione (γ-glutamylcysteinylglycine)

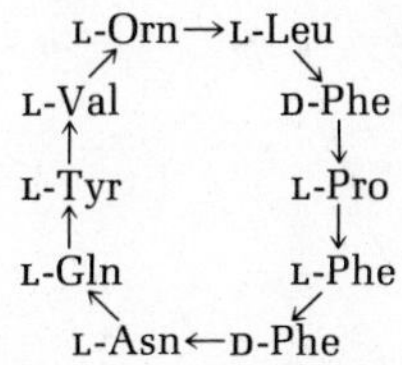

Tyrocidin A [Orn is the symbol for
ornithine (page 77)]

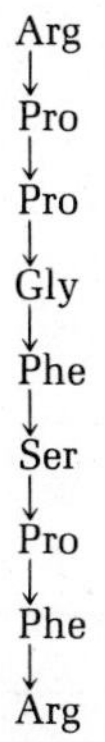

$$\text{Arg} \downarrow \text{Pro} \downarrow \text{Pro} \downarrow \text{Gly} \downarrow \text{Phe} \downarrow \text{Ser} \downarrow \text{Pro} \downarrow \text{Phe} \downarrow \text{Arg}$$

Bradykinin

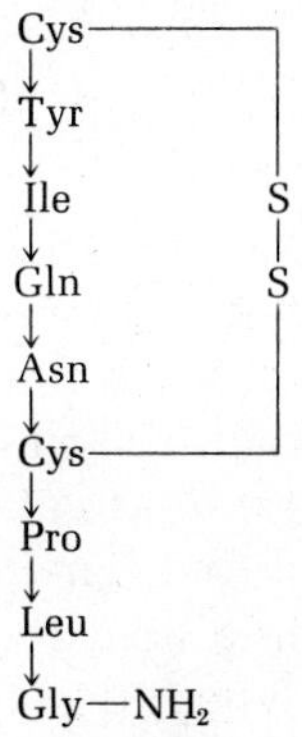

Bovine
oxytocin

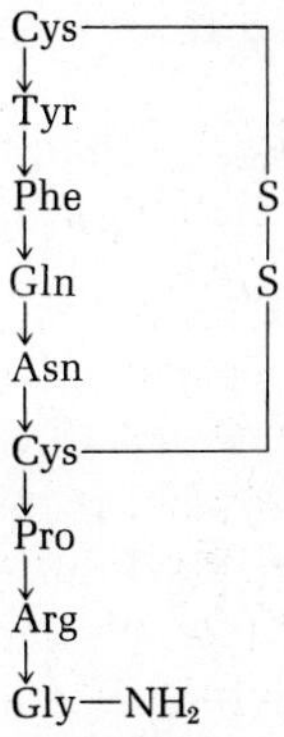

Bovine
vasopressin

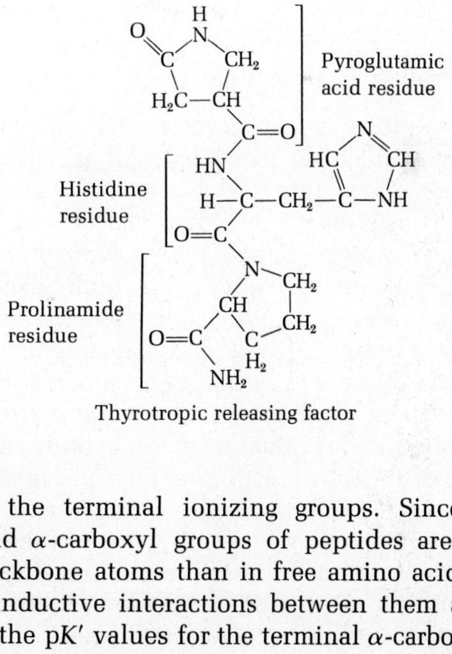

Thyrotropic releasing factor

outnumber the terminal ionizing groups. Since the free α-amino and α-carboxyl groups of peptides are separated by more backbone atoms than in free amino acids, electrostatic and inductive interactions between them are diminished; thus the pK' values for the terminal α-carboxyl groups are somewhat higher and those for the α-amino groups somewhat lower than in free α-amino acids (Table 5-1). The pK' values for the R groups in short peptides are close to those of the corresponding free amino acids.

The acid-base titration curves of short peptides are very similar to those of free α-amino acids. The predominant ionic species at different stages in the titration curves of peptides are comparable to those for free amino acids (Figure

Table 5-1 pK' values for some amino acids and peptides at 25°C

	pK$_1'$ α-COOH	pK$_2'$ α-NH$_3^+$	pK$_R'$ R group	pH$_I$
Gly	2.34	9.6	—	5.97
Gly-Gly	3.06	8.13	—	5.59
Gly-Gly-Gly	3.26	7.91	—	5.58
Ala	2.34	9.69	—	6.02
Ala-Ala-Ala-Ala	3.42	7.94	—	5.68
Ala-Ala-Lys-Ala	3.58	8.01	10.58	~9.3
Gly-Asp	2.81	8.60	4.45	~3.6

5-3). Peptides also have a characteristic isoelectric pH, which can be calculated from their pK' values (Table 5-1).

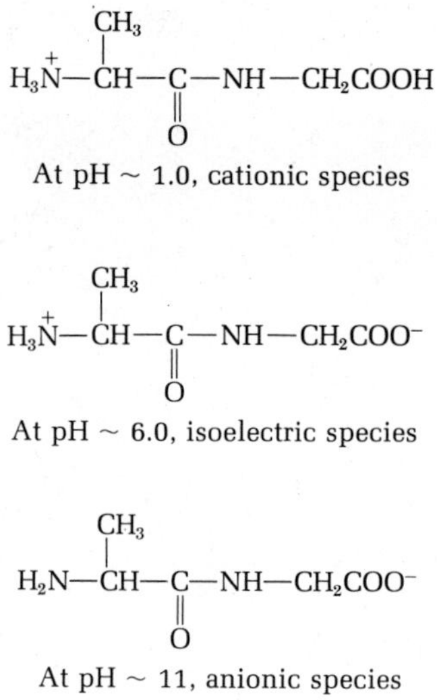

Optical Properties of Peptides

If partial hydrolysis of a protein is carried out under sufficiently mild conditions, so that no racemization of the asymmetric α-carbon atom occurs, the peptides formed are optically active, since they contain only L-amino acid residues. In relatively short peptides, the total observed optical activity is approximately an additive function of the optical activities of the component amino acid residues. However, the optical activity of long polypeptide chains of proteins in their native conformation (page 131) is much less than additive, a fact of great significance with regard to the secondary and tertiary structure of proteins, as we shall see in Chapter 6.

Chemical Properties of Peptides

The free N-terminal amino groups of peptides undergo the same kinds of chemical reactions as those given by the α-amino groups of free amino acids, such as acylation and carbamoylation (page 84). The N-terminal amino acid residue of peptides also reacts quantitatively with _ninhydrin_ (page 85) to form colored derivatives; the ninhydrin reaction is widely used for detection and quantitative estimation of peptides in electrophoretic and chromatographic procedures. Similarly, the C-terminal carboxyl group of a peptide may be esterified or reduced. Moreover, the various R groups of the different amino acid residues found in peptides usually yield the same characteristic reactions as free amino acids.

One widely employed color reaction of peptides and proteins that is not given by free amino acids is the _biuret reaction._ Treatment of a peptide or protein with Cu^{2+} and alkali yields a purple Cu^{2+}–peptide complex, which can be measured quantitatively in a spectrophotometer.

Steps in the Determination of Amino Acid Sequence

With this information on the properties of simple peptides as background, we can examine the general strategy used to determine the amino acid sequence of peptides and proteins devised by Frederick Sanger in 1953 in his epoch-making determination of the amino acid sequence of the polypeptide

Figure 5-3
The major ionic species of the dipeptide alanylglycine.

$$H_3\overset{+}{N}-\underset{|}{\overset{\overset{\displaystyle CH_3}{|}}{CH}}-\underset{\overset{\displaystyle ||}{O}}{C}-NH-CH_2COOH$$

At pH ~ 1.0, cationic species

$$H_3\overset{+}{N}-\underset{|}{\overset{\overset{\displaystyle CH_3}{|}}{CH}}-\underset{\overset{\displaystyle ||}{O}}{C}-NH-CH_2COO^-$$

At pH ~ 6.0, isoelectric species

$$H_2N-\underset{|}{\overset{\overset{\displaystyle CH_3}{|}}{CH}}-\underset{\overset{\displaystyle ||}{O}}{C}-NH-CH_2COO^-$$

At pH ~ 11, anionic species

chains of insulin, the first protein for which the complete covalent structure became known. Although each protein offers special problems, the following sequence of steps is generally used:

1 If the protein contains more than one polypeptide chain, the individual chains are first separated and purified.
2 All the disulfide groups are reduced and the resulting sulfhydryl groups alkylated.
3 A sample of each polypeptide chain is subjected to total hydrolysis, and its amino acid composition is determined.
4 On another sample of the polypeptide chain the N-terminal and C-terminal residues are identified.
5 The intact polypeptide chain is cleaved into a series of smaller peptides by enzymatic or chemical hydrolysis.
6 The peptide fragments resulting from step 5 are separated, and their amino acid composition and sequence are determined.
7 Another sample of the original polypeptide chain is partially hydrolyzed by a second procedure to fragment the chain at points other than those cleaved by the first partial hydrolysis. The peptide fragments are separated and their amino acid composition and sequence determined (as in steps 5 and 6).
8 By comparing the amino acid sequences of the two sets of peptide fragments, particularly where the fragments from the first partial hydrolysis overlap the cleavage points in the second, the peptide fragments can be placed in the proper order to yield the complete amino acid sequence.
9 The positions of the disulfide bonds and the amide groups in the original polypeptide chain are determined.

We shall now describe the methods used in carrying out the individual steps in the overall strategy.

Cleavage of Disulfide Bonds and Separation of Polypeptide Chains

Before analyzing the amino acid sequence of a protein, the investigator must determine whether the protein contains more than one polypeptide chain. The number of chains is usually deduced from the number of N-terminal amino acid residues per molecule of protein, by methods to be described below (see also page 102). Clearly, the number of polypeptide chains will be equal to the number of N-terminal amino acid residues per molecule of protein. If the polypeptide chains have no covalent cross-linkages, they can be separated by treating the protein with acid, base, or high concentrations of salt or a denaturing agent (page 150)

If the polypeptide chains are covalently cross-linked by one or more disulfide bonds between half residues of cystine (page 86), these cross-linkages must be cleaved by appropriate chemical reactions. The commonest procedure is to reduce the disulfide bond to sulfhydryl groups with an excess of mercaptoethanol (page 87). An alkylating agent like *iodoacetate* is then used to alkylate the sulfhydryl group of the cysteine residues to yield their *S-carboxymethyl* derivatives (Figure 5-4). When the polypeptide chain is subsequently hydrolyzed, these residues appear as _S-carboxymethylcysteine_, which is easily identified by the chromatographic procedures used for amino acid analysis. Alkylation of cysteine residues is desirable because the sulfhydryl group of cysteine is relatively unstable and tends to undergo oxidation. Other reagents such as _iodoacetamide_ and _ethyleneimine_ are also employed for alkylation of sulfhydryl groups. An older but less common method for cleaving disulfide cross-linkages, first developed by Sanger, is to oxidize the disulfide group to yield _cysteic acid_ residues from the half-cysteines (Figure 5-5). Once the interchain disulfide bonds have been cleaved, the individual polypeptide chains are separated, usually by electrophoresis (page 165).

Even if the protein to be examined contains but a single polypeptide chain, its intrachain disulfide bonds, if any, must be cleaved and all cysteine residues alkylated to the more stable S-carboxymethyl derivatives.

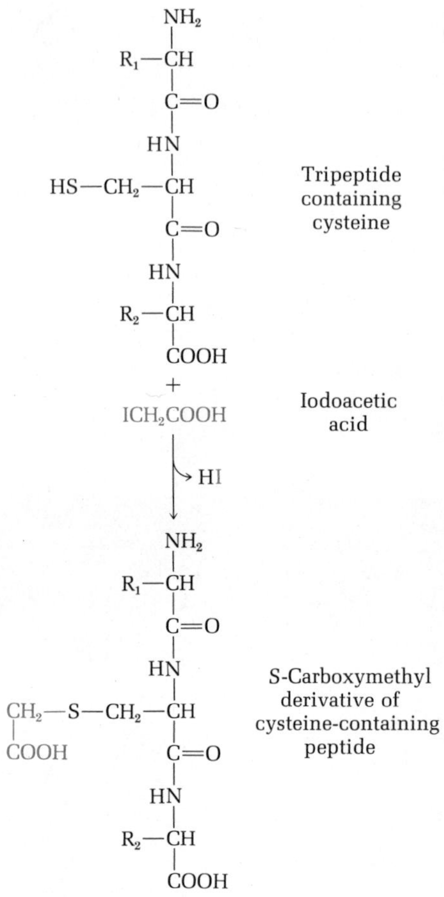

Complete Hydrolysis of Polypeptide Chains and Determination of Amino Acid Composition

Once the polypeptide chain to be examined has been obtained in homogeneous form, with no remaining disulfide cross-links or free sulfhydryl groups, it is completely hydrolyzed and its amino acid composition determined. Peptide bonds are readily hydrolyzed by heating with either acid or base. Heating polypeptides with excess 6 N hydrochloric acid at 100 to 120°C for 10 to 24 h, usually in an evacuated, sealed tube, is the usual procedure for complete hydrolysis. Little or no racemization of the amino acids takes place under these conditions. However, not all the amino acids are recovered quantitatively following acid hydrolysis; tryptophan is usually destroyed by this treatment, which also causes some loss of serine and threonine. Moreover, the amide groups of asparagine and glutamine (page 74) undergo complete hydrolysis in acid, to yield free aspartic and glutamic acids, respectively, plus free ammonium ions.

Figure 5-4
S-alkylation of a peptide containing a cysteine residue.

Tripeptide containing cysteine

Iodoacetic acid

S-Carboxymethyl derivative of cysteine-containing peptide

Figure 5-5
Oxidative cleavage of disulfide cross-linkages by oxidation with performic acid.

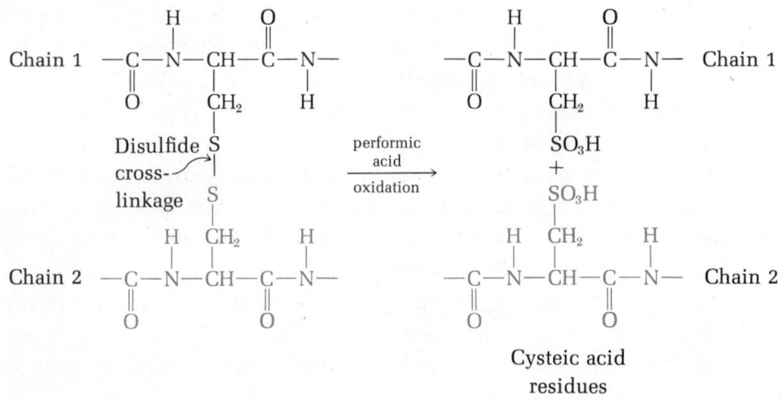

Cysteic acid residues

Polypeptides can also be hydrolyzed by boiling with strong sodium hydroxide solutions, but alkaline hydrolysis causes destruction of cysteine, cystine, serine, and threonine and racemization of all the amino acids. Alkaline hydrolysis is normally used only for the separate estimation of tryptophan, which is unstable to acid but stable to base.

The amino acid composition of hydrolyzates of polypeptides and proteins is determined by automated ion-exchange chromatography in an amino acid analyzer (page 90). The first pure protein for which the complete amino acid composition was deduced was β-lactoglobulin of milk. This analysis, which required several years of work by older methods, was completed in 1947. Today, the amino acid analyzer determines the complete amino acid composition of a protein hydrolyzate within 2 to 4 h. Only very small samples are required.

At this point it is instructive to consider the amino acid composition of representative pure proteins (Table 5-2).

Table 5-2 Amino acid composition of some representative proteins as number of residues per molecule. The code letters of the proteins are arbitrary.

| | Protein code letters (see below for key) | | | | | | | | | | |
	A	B	C	D	E	F	G	H	I	J	K
Nonpolar											
Ala	12	15	9	9	22	14	6	45	19	28	12
Val	6	9	11	7	23	14	3	9	17	39	9
Leu	8	21	17	8	19	12	6	2	20	25	2
Ile	6	10	11	4	10	9	6	1	10	24	3
Pro	2	8	17	4	9	8	4	7	17	20	4
Met	2	4	5	0	2	0	2	0	2	9	4
Phe	3	4	8	2	6	8	4	4	11	18	3
Trp	6	2	2	1	8	3	1	1	6	2	0
Polar (uncharged)											
Gly	12	3	9	6	23	6	12	74	16	38	3
Ser	10	7	16	7	28	16	0	17	30	26	15
Thr	7	8	5	8	23	16	10	2	14	24	10
Cys	8	5	0	5	10	1	2	2	1	14	8
Tyr	3	4	10	4	4	4	4	23	8	4	6
Asn	13	5	8	2	14	10	5	1	17	8	10
Gln	3	9	14	4	10	9	3	0	9	8	7
Negatively charged											
Asp	8	11	7	11	9	8	3	4	14	17	5
Glu	2	16	25	9	5	7	9	4	13	21	5
Positively charged											
Lys	6	14	14	4	14	2	19	5	18	30	10
Arg	11	3	6	1	4	11	2	3	7	12	4
His	1	2	5	1	2	0	3	2	11	7	4
Percent nonpolar	35	46	40	36	40	43	31	33	39	44	30
Total residues	129	160	199	97	245	158	104	206	260	374	124

A = chicken lysozyme
C = bovine α-casein
E = bovine chymotrypsinogen
G = equine cytochrome c
I = human carbonic anhydrase
K = bovine ribonuclease

B = bovine β-lactoglobulin
D = spinach ferredoxin
F = coat protein of tobacco mosaic virus
H = silk fibroin of *Bombyx mori*
J = equine alcohol dehydrogenase

Some generalizations may be made from these and other available data:

1 Not all proteins contain all the 20 amino acids normally found in proteins; e.g., ribonuclease lacks tryptophan. Fibrous proteins, e.g., silk fibroin and collagen, lack several amino acids.
2 Some amino acids occur much less frequently in proteins than others. For example, in most proteins there are relatively few histidine, tryptophan, and methionine residues, as is evident from Tables 5-2 and 5-3.
3 In most proteins 30 to 40 percent of the residues are amino acids with nonpolar R groups. Membrane proteins tend to have a somewhat higher content. Over 90 percent of the amino acid residues of the insoluble fibrous protein elastin are nonpolar.
4 In some proteins, such as lysozyme, cytochrome c, and the histones, the positively charged R groups predominate (at pH 7.0); such proteins are basic. In others, the negatively charged R groups of glutamic or aspartic acid predominate, as in pepsin, which is highly acidic.

Table 5-3   Relative frequency of occurrence of amino acids in *E. coli* proteins based on Ala = 100

Amino acid	Relative frequency
Ala	100
Glx	83
Asx	76
Leu	60
Gly	60
Lys	54
Ser	46
Val	46
Arg	41
Thr	35
Pro	35
Ile	34
Met	29
Phe	25
Tyr	17
Cys	14
Trp	8
His	5

Identification of the N-Terminal Residue of a Peptide

Very important in the procedure for establishing amino acid sequence are methods for identifying the terminal amino acid residues. The first useful method for the N-terminal residue of polypeptides was described by Sanger, who found that the free unprotonated α-amino group of peptides reacts with 2,4-dinitrofluorobenzene (DNFB) to form yellow 2,4-dinitrophenyl derivates (Figure 5-6). When such a derivative of a peptide, regardless of its length, is subjected to hydrolysis with 6 N HCl, all the peptide bonds are hydrolyzed, but the bond between the 2,4-dinitrophenyl group and the α-amino group of the N-terminal amino acid is relatively stable to acid hydrolysis. Consequently, the hydrolyzate of such a dinitrophenyl peptide contains all the amino acid residues of the peptide chain as free amino acids except the N-terminal one, which appears as the yellow 2,4-dinitrophenyl derivative. This labeled residue can easily be separated from the unsubstituted amino acids and identified by chromatographic comparison with known dinitrophenyl derivatives of the different amino acids.

Sanger's method has been largely supplanted by more sensitive and efficient procedures. One employs the labeling reagent 1-dimethylaminonaphthalene-5-sulfonyl chloride (abbreviated *dansyl chloride*), as shown in Figure 5-7. Since the dansyl group is highly fluorescent, dansyl derivatives of the N-terminal amino acid can be detected and measured in minute amounts by fluorimetric methods. The dansyl procedure is 100 times more sensitive than the Sanger method.

The most important and most widely used labeling reaction for the N-terminal residue is that designed by P. Edman

Figure 5-6
Identification of the N-terminal amino acid residue of a tetrapeptide by means of the Sanger reaction.

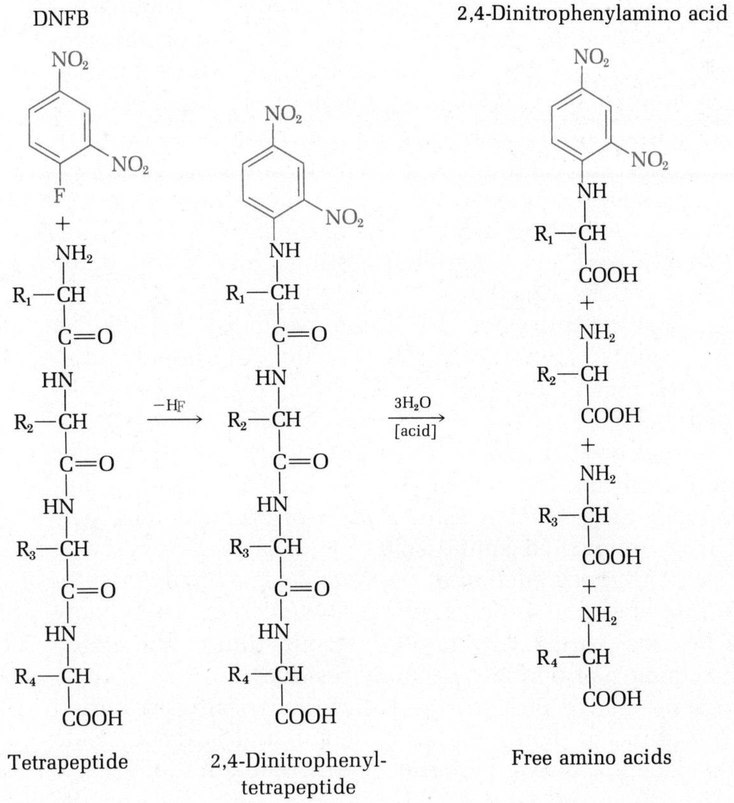

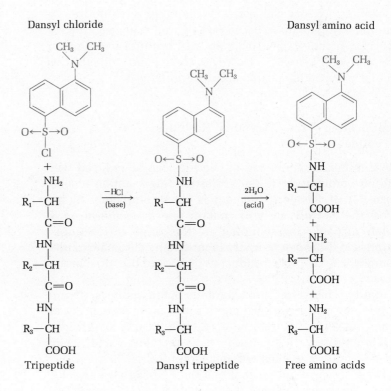

Figure 5-7
Identification of the N-terminal residue of a tripeptide as the dansyl derivative.

(Figure 5-8). In the Edman procedure phenylisothiocyanate reacts quantitatively with the free amino group of a peptide to yield the corresponding phenylthiocarbamoyl peptide. On treatment with anhydrous acid the N-terminal residue is split off as a phenylthiocarbamoyl amino acid, leaving the rest of the peptide chain intact. The phenylthiocarbamoyl amino acid is then cyclized to the corresponding *phenylthiohydantoin* derivative, which can be separated and identified, usually by gas-liquid chromatography (page 283). Alternatively, the N-terminal residue removed as the phenylthiocarbamoyl derivative can be identified simply by determining the amino acid composition of the peptide before and after removal of the N-terminal residue; this is called the *subtractive* Edman method.

The great advantage of the Edman method is that the rest of the peptide chain after removal of the N-terminal amino acid is left intact for further cycles of this procedure; thus the Edman method can be used in a sequential fashion to identify several or even many consecutive amino acid residues starting from the N-terminal end. This great advantage has been further exploited by Edman and G. Begg, who have perfected an automated amino acid "sequenator" for carrying out sequential degradation of peptides by the phenylisothiocyanate procedure. Automated amino acid sequencers, now widely used, permit very rapid determination of the amino acid sequence of peptides up to 20 residues.

In some native proteins the N-terminal residue is buried deep within the tightly folded molecule and is inaccessible to the labeling reagent; in such cases, denaturation of the protein can render it accessible. In other proteins, e.g., the tobacco mosaic virus coat protein, the α-amino group of the N-terminal amino acid is acetylated and hence not reactive to labeling reagents. Some natural peptides have no free N-terminal α-amino group because they are cyclic; e.g., the antibiotic tyrocidin A has 10 amino acid residues in a circular arrangement (Figure 5-2). However, there is no evidence that circular polypeptide chains occur in proteins.

Identification of the C-Terminal Residues of Peptides

The C-terminal amino acid of peptides can be reduced with lithium borohydride to the corresponding α-amino alcohol (page 84). If the peptide chain is then completely hydrolyzed, the hydrolyzate will contain one molecule of an α-amino alcohol corresponding to the original C-terminal amino acid. This can be easily identified by chromatographic methods; all the other residues will be found as free amino acids.

Another important procedure is *hydrazinolysis* (Figure 5-9), which cleaves all the peptide bonds by converting all except the C-terminal amino acid residues into hydrazides. The C-terminal residue appears as a free amino acid, which can be readily identified chromatographically.

The C-terminal amino acid of a peptide can also be selectively removed by action of the enzyme *carboxypeptidase*,

Figure 5-8
Identification of the N-terminal residue by the Edman degradation. Note that the peptide chain remains intact after removal of the N-terminal amino acid.

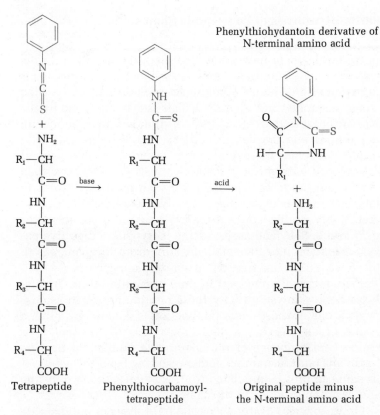

Phenylisothiocyanate

Phenylthiohydantoin derivative of N-terminal amino acid

base →

acid →

Tetrapeptide

Phenylthiocarbamoyl-tetrapeptide

Original peptide minus the N-terminal amino acid

Figure 5-9
Identification of the C-terminal residue by reaction of a peptide with hydrazine (NH₂NH₂) (Akabori procedure).

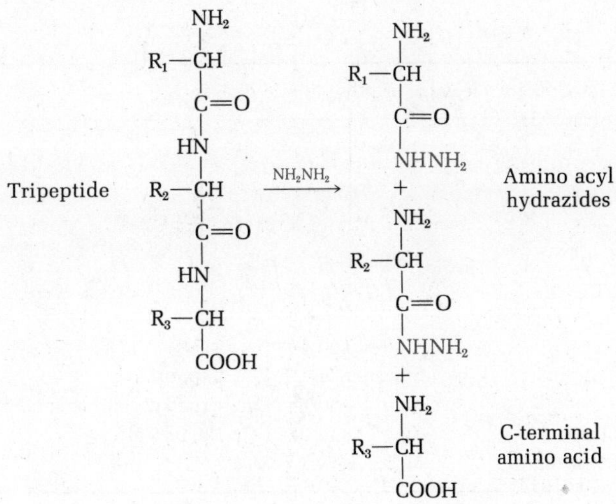

Tripeptide

NH₂NH₂ →

Amino acyl hydrazides

C-terminal amino acid

which specifically attacks C-terminal peptide bonds. A drawback is that the enzyme, after removal of the terminal residue, proceeds to attack the new C-terminal peptide bond. It is therefore necessary to measure the rate of liberation of different amino acids from the peptide by carboxypeptidase in order to identify the C-terminal residue unequivocally.

Partial Hydrolysis of Polypeptide Chains

Once the N-terminal and C-terminal amino acid residues of a polypeptide chain have been identified, the next step in the grand strategy for determining the sequence of amino acids is to fragment the chain to yield a set of short peptides which can be separated and identified. This is accomplished by the partial or selective hydrolysis of the polypeptide chain. Sometimes this can be achieved by partial hydrolysis with dilute acid, since the peptide bonds between certain pairs of amino acids are more susceptible to acid hydrolysis than others, but the method of choice for partial hydrolysis is to use _proteases,_ enzymes that hydrolyze peptide bonds. Several highly purified proteases have been used for this purpose (Table 5-4). The most specific is _trypsin,_ a digestive enzyme secreted into the small intestine from the pancreas in the form of its inactive precursor _trypsinogen_ (page 560). Trypsin is readily obtained in pure crystalline form. It catalyzes the hydrolysis of only those peptide bonds in a polypeptide chain whose carbonyl function is donated by either a lysine or an arginine residue, regardless of the length or amino acid sequence of the chain. The number of peptide fragments (and free amino acids) resulting from the action of trypsin can thus be predicted from the total number of lysine and arginine residues in the chain.

Other enzymes useful for partial hydrolysis of polypeptide chains are _chymotrypsin_ (pages 218 and 230), _pepsin_ (pages 243 and 560), and _thermolysin._ Their peptide-bond specificities are shown in Table 5-4. Although they are less specific than trypsin, they may be very useful, depending on the amino acid composition of the polypeptide chain under study. Thermolysin, a heat-stable bacterial protease, can hydrolyze peptide bonds in which the amino function is contributed by the nonpolar amino acids leucine, isoleucine,

---

Table 5-4 Specific cleavage of polypeptide chains

$$-N-\underset{\underset{R_1}{|}}{\overset{\overset{H}{|}}{C}}-\underset{O}{\overset{H}{\overset{|}{C}}}\Big\downarrow N-\underset{\underset{H}{|}}{\overset{\overset{H}{|}}{C}}-\underset{O}{\overset{R_2}{\overset{|}{C}}}-$$

Amino acid 1   Amino acid 2

Method	Peptide bonds cleaved
Trypsin	Amino acid 1 = Lys or Arg
Chymotrypsin	Amino acid 1 = Phe, Trp, or Tyr
Pepsin	Amino acid 1 = Phe, Trp, Tyr, and several others
Thermolysin	Amino acid 2 = Leu, Ile, or Val
Cyanogen bromide	Amino acid 1 = Met

Figure 5-10
Cleavage of a polypeptide chain at a methionine residue (color) by cyanogen bromide. The methionine residue becomes a C-terminal homoserine lactone residue.

Cyanogen bromide

Methyl thiocyanate

Peptidyl homoserine lactone

Amino acyl peptide

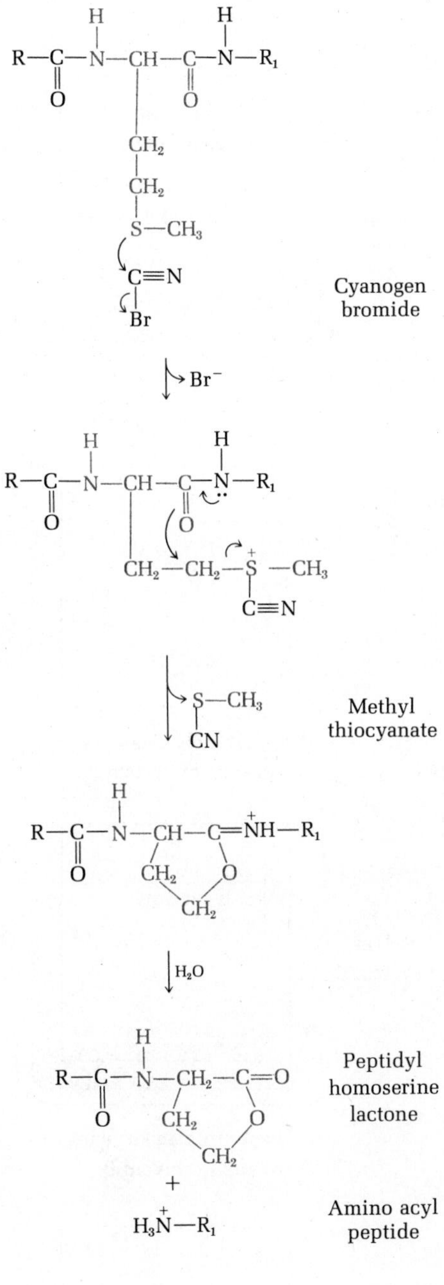

and valine. It is especially useful for partial hydrolysis of polypeptides containing no arginine or lysine and therefore refractory to cleavage by trypsin. Thermolysin has also been useful in establishing the sequence of amino acids in the *protamines*, basic proteins of the cell nucleus that contain so many arginine and lysine residues that cleavage by trypsin yields no useful information about sequence.

Specific chemical methods have also been developed for cleaving polypeptide chains at specific amino acid residues. The most successful involves reaction of the polypeptide with *cyanogen bromide*, which cleaves peptide bonds whose carbonyl function is contributed by a methionine residue. The methionine residue is converted into a C-terminal *homoserine lactone* residue (Figure 5-10). The number of fragments produced from a polypeptide by cyanogen bromide can be predicted from the number of methionine residues in the chain.

Part of the strategy of amino acid sequence analysis is to fragment the polypeptide chain in at least two different ways, so that the small peptide fragments resulting from one procedure overlap with those resulting from the other (see below). For example, if trypsin is used for the first cleavage, the second cleavage may be carried out with some other protease, such as chymotrypsin, pepsin, or thermolysin, or by the cyanogen bromide reaction. Sometimes a third or fourth partial hydrolysis is required to give the necessary overlapping peptides.

Separation and Analysis of Peptides

Complex mixtures of peptides resulting from partial hydrolysis of proteins are much more difficult to separate than mixtures of amino acids (page 90) because the number of possible peptides is very much greater than the number of amino acids present in proteins. Two general approaches are possible, paper techniques, e.g., paper chromatography and paper electrophoresis, and (more effective and more widely used) column chromatography. Both approaches exploit differences in the acid-base behavior of peptides.

When paper techniques are used, two-dimensional methods (page 89) are most effective. The peptide mixture is first subjected to electrophoresis in one direction of the paper, followed by chromatography (or electrophoresis at some other pH) at right angles to the first (Figure 5-11). This two-step process results in a two-dimensional *peptide map*. The peptides can be located on the dried paper upon spraying with ninhydrin, followed by application of heat. From an unstained paper obtained under identical conditions, the peptide-containing spots can be cut out and eluted to recover the individual peptides. If two or more peptides form an overlapping spot, they can be eluted and rechromatographed in a different solvent system. Peptide maps, also called peptide fingerprints, are particularly useful in identifying differences between homologous proteins of different species (page 112) and in locating the sites of amino acid replacements in mutant proteins (page 115).

107

Column chromatography of peptide mixtures is usually carried out on ion-exchange columns. The elution is carried out with gradients of pH or salt concentration, as in amino acid analysis. This method has a great advantage in that it can be automated and carried out in an amino acid analyzer (page 90). Another very effective way of separating peptide mixtures is _exclusion chromatography_ or _gel filtration,_ performed on columns of the polymerized carbohydrate derivative known as Sephadex. This type of chromatography separates peptides on the basis of their molecular weight rather than their acid-base properties. The physical principles underlying gel-exclusion chromatography are given in detail in Chapter 7 (page 159). A combination of automated ion-exchange and gel-exclusion chromatography is extremely effective in separating complex peptide mixtures.

Sequence Analysis of Peptide Fragments

Once all the peptide fragments resulting from the partial hydrolysis of a polypeptide chain have been separated, a sample of each is completely hydrolyzed and its amino acid content determined. Then, on other samples, the amino acid sequence of each peptide is determined by sequential Edman degradation. If the peptide is too long to be sequenced by this method, its N-terminal and C-terminal residues are identified by methods outlined above and the peptide is then fragmented further by partial hydrolysis with some method other than that used in the primary fragmentation. The resulting smaller peptides can then be analyzed as just described.

Ordering the Peptide Fragments

Once two sets of peptide fragments have been obtained by two different procedures for cleavage of the original polypeptide chain and the amino acid sequence of all the fragments has been established, it is possible to deduce the complete amino acid sequence from overlaps of the peptide sequences. The principle is shown in Figure 5-12. Often, however, unequivocal overlaps are not established for all portions of a given chain, and a third or even a fourth type of partial

Figure 5-11
Two-dimensional paper chromatogram. The mixture of peptides is spotted and run in one dimension with solvent system A. The paper is then dried, turned 90°, and allowed to develop with solvent system B, spreading the peptide spots over the entire sheet. See Figure 5-19 for an example.

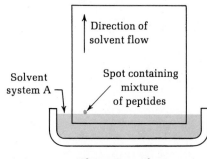

Chromatography in
first solvent system

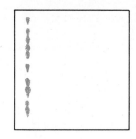

Paper after chromatography
in solvent system A

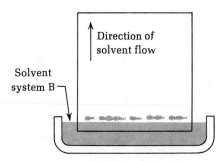

Paper rotated 90° and
run in solvent B

Complete two-dimensional
chromatogram

Figure 5-12
_Determination of amino acid sequence
from overlapping peptides._

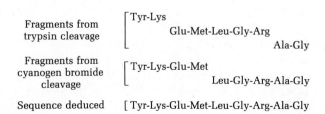

Fragments from trypsin cleavage	[Tyr-Lys
	Glu-Met-Leu-Gly-Arg
	Ala-Gly
Fragments from cyanogen bromide cleavage	[Tyr-Lys-Glu-Met
	Leu-Gly-Arg-Ala-Gly
Sequence deduced	[Tyr-Lys-Glu-Met-Leu-Gly-Arg-Ala-Gly

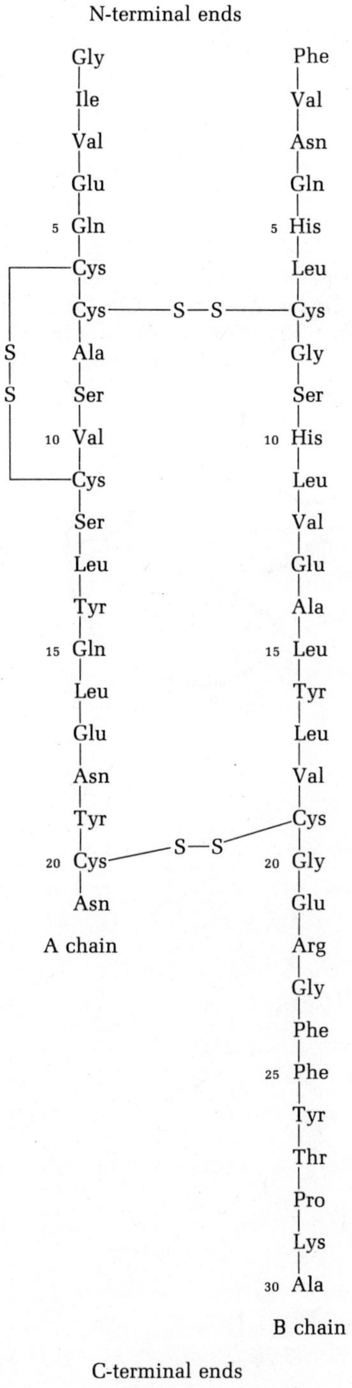

Figure 5-13
The amino acid sequence of bovine insulin and the positions of the —S—S— cross-linkages.

N-terminal ends

Gly — Phe
Ile — Val
Val — Asn
Glu — Gln
5 Gln — 5 His
Cys — Leu
Cys —S—S— Cys
Ala — Gly
Ser — Ser
10 Val — 10 His
Cys — Leu
Ser — Val
Leu — Glu
Tyr — Ala
15 Gln — 15 Leu
Leu — Tyr
Glu — Leu
Asn — Val
Tyr — Cys
20 Cys —S—S— 20 Gly
Asn — Glu
A chain — Arg
— Gly
— Phe
— 25 Phe
— Tyr
— Thr
— Pro
— Lys
— 30 Ala
B chain

C-terminal ends

hydrolysis of the original polypeptide is necessary to obtain the required overlaps.

Assignment of the Position of Disulfide Cross-Linkages

Once the complete amino acid sequences of the polypeptide chains of a protein have been deduced, the next step is to establish which of the cysteine residues are paired to form intrachain and interchain disulfide cross-linkages, if there are any. This is usually done by fragmenting a sample of the protein with the cross-linkages still intact, usually by partial hydrolysis with trypsin. The peptides containing the disulfide bridge are isolated, their disulfide bonds reduced, and the resulting sulfhydryl groups alkylated, as described before (page 99). The two peptides derived from each of the disulfide-containing fragments are then identified by comparison with the tryptic peptides from the original reduced polypeptide chain.

Assignment of Amide Positions

The amide groups of asparagine and glutamine residues are lost as ammonia during acid or basic hydrolysis, to yield the corresponding aspartic acid and glutamic acid residues. Although the number of ammonia molecules formed per mole of protein during acid hydrolysis can give the sum of the asparagine and glutamine residues, the positions of amide groups, particularly in proteins containing both aspartic and glutamic acids and their amides, cannot be deduced from peptide fragments obtained following acid hydrolysis; however, they are established from peptide fragments obtained by the action of specific proteases that do not hydrolyze the amide groups.

The Amino Acid Sequence of Some Peptides and Proteins

Figure 5-13 shows the complete amino acid sequence of bovine insulin, the first protein whose amino acid sequence was established. It consists of two polypeptide chains: the A chain, with 21 amino acid residues, and the B chain, with 30. The two polypeptide chains of insulin are cross-linked by two disulfide bridges; in addition, there is one intrachain disulfide cross-link. Sanger cleaved both the interchain and the intrachain disulfide bonds and then separated the A and B chains. After establishing the amino acid sequence of each, by the strategy already outlined, he located the position of interchain and intrachain cross-links.

Sanger's work, completed in 1953, opened the door for sequence analysis of longer polypeptide chains. Soon two different research groups reported the sequence of amino acids in adrenocorticotrophin, a hormone of the anterior

Figure 5-14
The amino acid sequence of human adreno-corticotrophin.

Ser-Tyr-Ser-Met-Glu-His-Phe-Arg-Trp-Gly-Lys-Pro-Val-Gly-Lys-Lys-Arg-Arg-Pro-Val$_{20}$-

Lys-Val-Tyr-Pro-Asp-Ala-Gly-Glu-Asp-Gln-Ser-Ala-Glu-Ala-Phe-Pro-Leu-Glu-Phe $_{39}$

Figure 5-15
The amino acid sequence of bovine ribonuclease. Shown below are the positions of the intrachain disulfide bridges.

Lys-Glu-Thr-Ala-Ala-Ala-Lys-Phe-Glu-Arg-Gln-His-Met-Asp-Ser-Ser-Thr-Ser-Ala-Ala-Ser-Ser-Ser-Asn-Tyr-Cys-Asn-Gln-Met$_{29}$-
Met-Lys-Ser-Arg-Asn-Leu-Thr-Lys-Asp-Arg-Cys-Lys-Pro-Val-Asn-Thr-Phe-Val-His-Glu-Ser-Leu-Ala-Asp-Val-Gln-Ala-Val-Cys$_{58}$-
Ser-Gln-Lys-Asn-Val-Ala-Cys-Lys-Asn-Gly-Gln-Thr-Asn-Cys-Tyr-Gln-Ser-Tyr-Ser-Thr-Met-Ser-Ile-Thr-Asp-Cys-Arg-Glu-Thr$_{87}$-
Gly-Ser-Ser-Lys-Tyr-Pro-Asn-Cys-Ala-Tyr-Lys-Thr-Thr-Gln-Ala-Asn-Lys-His-Ile-Ile-Val-Ala-Cys-Glu-Gly-Asn-Pro-Tyr-Val$_{116}$-
Pro-Val-His-Phe-Asp-Ala-Ser-Val$_{124}$

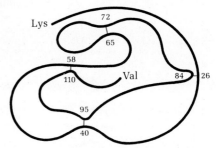

pituitary gland that stimulates the adrenal cortex (Figure 5-14). This hormone consists of a single chain of 39 residues with a molecular weight of about 4,600. Several years later, the first sequence analysis of an enzyme protein, ribonuclease, was achieved by S. Moore and W. Stein in New York, with important independent studies by C. B. Anfinsen in Bethesda. Ribonuclease has 124 amino acid residues in a single chain; it contains four intrachain disulfide cross-linkages (Figure 5-15).

The next important landmark, the identification of the amino acid sequences of the two types of polypeptide chains in hemoglobin (Figure 5-16), was the first sequence analysis of a large oligomeric (multichain) protein, carried out by two groups in the United States and another in Germany. Hemoglobin contains four polypeptide chains, two identical α-chains (141 residues) and two identical β-chains (146 residues). The α and β chains have identical amino acid residues in many positions (Figure 5-16); i.e., the two chains possess what is known as *sequence homology*. The single polypeptide chain of the muscle protein myoglobin, which also contains an iron-porphyrin group and resembles hemoglobin in its ability to bind oxygen, also has sequence homology with the hemoglobin chains. The significance of sequence homologies in functionally similar proteins or in homologous proteins of different species will be discussed in Chapter 6 (page 145).

The longest polypeptide chains for which complete amino acid sequences have been deduced to date are those of horse alcohol dehydrogenase (374 residues) and bovine glutamate dehydrogenase (500 residues).

Figure 5-16
The amino acid sequence of the α and β chains of normal hemoglobin. Residues identical in both chains are in color. Residues identical in both chains and in the single chain of human myoglobin are shaded.

N-terminal ends

#	α	β		#	α	β		#	α	β		#	α	β
	Val	Val			Thr	Arg			Met	Leu			Phe	Phe
		His			Tyr	Phe			Pro	Lys			Thr	Thr
	Leu	Leu			Phe	Phe			Asn	Gly			Pro	Pro
	Ser	Thr			Pro	Glu			Ala	Thr		120	Ala	Pro
	Pro	Pro			His	Ser		80	Leu	Phe			Val	Val
	Ala	Glu			Phe	Phe			Ser	Ala			His	Gln
	Asp	Glu				Gly			Ala	Thr			Ala	Ala
	Lys	Lys			Asp	Asp			Leu	Leu			Ser	Ala
	Thr	Ser			Leu	Leu			Ser	Ser			Leu	Tyr
	Asn	Ala			Ser	Ser			Asp	Glu			Asp	Gln
10	Val	Val		50	His	Thr			Leu	Leu			Lys	Lys
	Lys	Thr			Gly	Pro			His	His			Phe	Val
	Ala	Ala			Ser	Asp			Ala	Cys			Leu	Val
	Ala	Leu			Ala	Ala			His	Asp		130	Ala	Ala
	Trp	Trp				Val		90	Lys	Lys			Ser	Gly
	Gly	Gly				Met			Leu	Leu			Val	Val
	Lys	Lys				Gly			Arg	His			Ser	Ala
	Val	Val				Asn			Val	Val			Thr	Asp
	Gly	Asn				Pro			Asp	Asp			Val	Ala
	Ala				Gln	Lys			Pro	Pro			Leu	Leu
20	His				Val	Val			Val	Glu			Thr	Ala
	Ala	Val			Lys	Lys			Asn	Asn			Ser	His
	Gly	Asp			Gly	Ala			Phe	Phe			Lys	Lys
	Glu	Glu			His	His			Lys	Arg		140	Tyr	Tyr
	Tyr	Val			Gly	Gly		100	Leu	Leu			Arg	His
	Gly	Gly		60	Lys	Lys			Leu	Leu	
	Ala	Gly			Lys	Lys			Ser	Gly	
	Glu	Glu			Val	Val			His	Asn	
	Ala	Ala			Ala	Leu			Cys	Val	
	Leu	Leu			Asp	Gly			Leu	Leu	
30	Glu	Gly			Ala	Ala			Leu	Val	
	Arg	Arg			Leu	Phe			Val	Cys	
	Met	Leu			Thr	Ser			Thr	Val	
	Phe	Leu			Asn	Asp			Leu	Leu	
	Leu	Val			Ala	Gly		110	Ala	Ala	
	Ser	Val		70	Val	Leu			Ala	His	
	Phe	Tyr			Ala	Ala			His	His	
	Pro	Pro			His	His			Leu	Phe	
	Thr	Trp			Val	Leu			Pro	Gly	
	Thr	Thr			Asp	Asp			Ala	Lys	
40	Lys	Gln		75	Asp	Asn		116	Glu	Glu	

C-terminal ends

From analysis of the amino acid sequences of many globular proteins, a few cautious generalizations can be made. To date, no periodic, frequently recurring sequences such as ABABABABAB · · · or ABCDABCDABCD · · · have been found in globular proteins. Only rarely does a single amino acid occur more than three times in a row, except in the protamines, which contain sequences of four, five, and six consecutive arginine residues. There is, therefore, little obvious regularity in the amino acid sequence. Examination of the known sequences of globular proteins shows that all or nearly all the possible short sequences of two and three amino acids do occur but with no apparent periodic pattern. Nor is there any apparent regularity in the presence or the position of intrachain—S—S— cross-linkages; some proteins have none, others may have several.

On the other hand, in some fibrous proteins certain amino acids do appear in periodic sequences. Collagen has a preponderance of glycine, proline, and hydroxyproline residues, which occur in the periodic sequence -Gly-X-Y-, where Y is frequently proline or hydroxyproline. Nearly 80 percent of the amino acid residues of silk fibroin are contributed by alanine, glycine, and serine (Chapter 6). The antifreeze proteins of certain Antarctic fish (page 274) have the repeating sequence -Ala-Ala-Thr-, with a disaccharide esterified to the threonine residue.

Species Variations in Sequence of Homologous Proteins

Comprehensive information on the amino acid sequences of homologous proteins from different species has showed that some amino acid residues in specific positions of homologous proteins are relatively *invariant*; i.e., in all species they are identical or replaced only infrequently. Homologous proteins also contain *variable* residues, usually the majority of the residues in the chain, which vary much more widely from one species to another and in which several different amino acids may replace each other.

The complete amino acid sequences of insulins isolated from many different vertebrate species are now known. These insulins have virtually the same specific hormonal activity and molecular weight. The A chain of the insulins of man, pig, dog, rabbit, and sperm whale are identical, and the B chains of the cow, pig, dog, sperm whale, sheep, goat, and horse insulins are identical. The B chains of human and elephant insulins are also identical. In the A chain, the amino acid replacements from one species to another usually occur at positions 8, 9, and 10, that is, the positions between the two half-cystine residues that form the intrachain cross-link. However, replacements have also been observed at positions 4, 13, 14, 15, and 18. Further study of insulin sequences in different species shows that at position A8, alanine, histidine, and threonine may replace each other; at A9, serine, arginine, lysine, asparagine, and glycine replace each other; and at A10, valine, isoleucine, proline, and threonine replace each other. We shall see that such specific

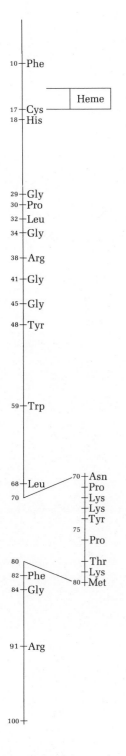

Figure 5-17
The 27 invariant amino acid residues in eukaryotic cytochrome c's. Fifty different cytochrome c's were examined, from 14 mammals, 4 birds, 2 reptiles, 1 amphibian, 5 fish, 4 insects, 5 fungi, 11 higher plants, and 2 baker's yeasts. As more species are investigated, the number of invariant residues may become smaller. [Modified from E. Margoliash in B. Chance and R. Estabrook (eds.), Hemes and Hemoproteins, p. 373, Academic Press Inc., New York, 1966.]

amino acid replacements are related to the constitution of the genetic code (pages 961 to 965).

Insulin is found only in vertebrates and is apparently absent in lower animals, plants, and bacteria. For this reason and also because it is a relatively small molecule, insulin does not lend itself well to examination of possible relationships between amino acid sequence and taxonomy. Much better in this respect is the electron-transferring protein *cytochrome c*, which occurs in all animals, plants, and aerobic microorganisms. Cytochrome c has a single chain of 104 residues in terrestrial vertebrates, 103 or 104 in fishes, 107 in insects, 111 or 112 in green plants, and 107 to 109 in yeasts and molds. E. Margoliash and his colleagues have carried out a comprehensive study of amino sequence variations in the cytochrome c isolated from over 50 different species. In all these species, 27 amino acid residues are absolutely invariant (Figure 5-17). The invariant residues are irregularly spaced along the polypeptide chain, although 7 of them occur in positions 70 to 80. In all species but one, cysteine residues 14 and 17, to which the porphyrin ring is attached, are invariant; the single species in which this does not hold has an alanine residue at 14. Clearly these cysteine residues are important for the structure and function of cytochrome c. Besides the absolutely invariant residues in this group of 50 cytochrome c's, several other residues are replaced only infrequently and then only by one other amino acid. The number of residue differences between species is roughly in proportion to their phylogenetic differences; 48 residues differ in the cytochrome c molecules from the widely different species horse and yeast, whereas only 2 residues differ in the cytochrome c's of the closely related duck and chicken. The cytochrome c molecule is identical in the chicken and turkey; it is also identical in the pig, cow, and sheep.

The number of residue differences in the cytochrome c's from many species has been used to construct a phylogenetic tree that not only shows the course of biological evolution of cytochrome c but also makes it possible to estimate the probable times when the major genera and species of living organisms diverged (Figure 5-18).

Evolution of Related Proteins

Amino acid sequence studies reveal that certain sets of proteins may have arisen from a common evolutionary ancestor. We have already seen that the single polypeptide chain of the muscle protein myoglobin has considerable sequence homology with the four polypeptide chains of adult hemoglobin. Both proteins have the function of binding oxygen reversibly to their heme groups, the former in muscle and the latter in erythrocytes. Later we shall see (page 137) that the three-dimensional conformation of the polypeptide chain of myoglobin is very similar to that of the α and β chains of hemoglobin. It is probable that both myoglobin and hemoglobin were derived from a common ancestor, which in all probability possessed a single chain and a single heme group.

Figure 5-18
Phylogenetic tree showing the evolution of cytochrome c. It was constructed
by computer from the amino acid sequences of the cytochrome c's of many
species. Each circle represents the sequence of a cytochrome c deduced to
be ancestral to all species higher in the branches leading from that circle.
The small figures alongside each branch indicate the number of amino acid
residue differences, per 100 residues, from the ancestor. Thus, mung bean
cytochrome c is different from its ancestor by 5 residues per 100, whereas
sesame cytochrome c is 2 residues per 100 different. (Adapted from M. O.
Dayhoff, C. M. Park, and P. J. McLaughlin in M. O. Dayhoff (ed.), Atlas of
Protein Sequence and Structure, vol. 5, p. 8, National Biomedical Research
Foundation, Washington, D.C., 1972.)

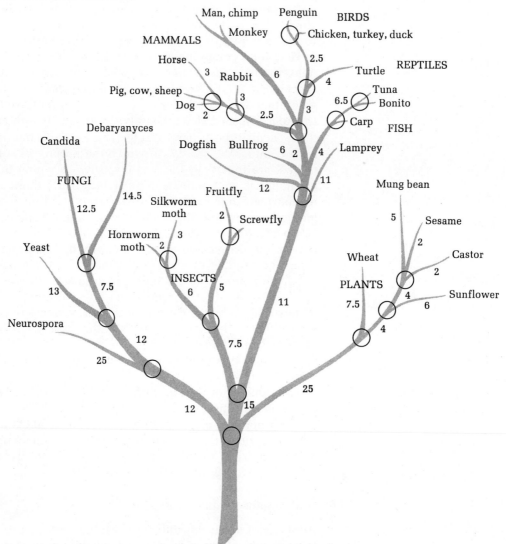

Another well-known pair of proteins that are cousins are
trypsin and chymotrypsin; H. Neurath and his colleagues
have pointed out the extensive similarity in their amino acid
sequences. Trypsin and chymotrypsin are both digestive enzymes synthesized by the pancreas and secreted in the form
of their inactive zymogens trypsinogen and chymotrypsinogen; however, as we have seen, they have very different
substrate specificities. Of all the amino acid residues of chymotrypsin 41 percent are present in homologous positions in
trypsin; moreover, four of the five disulfide bridges of chymotrypsin occur at the same positions as four of the six
bridges in trypsin.

Sometimes two proteins of apparently different function and quite different location show unmistakable sequence homology. R. L. Hill and his colleagues found that 54 of the 124 residues of α-lactalbumin, a milk protein, are identical to those in the enzyme lysozyme of hen's egg white. These seemingly unrelated proteins do share a common denominator: α-lactalbumin is a regulatory subunit of the enzyme lactose synthetase (page 644), which forms lactose (milk sugar) from its component monosaccharides, and lysozyme is an enzyme capable of hydrolyzing polysaccharide chains in the bacterial cell wall (page 231). Thus both enzymes are capable of activating carbohydrate groups.

Yet another interesting case is the sequence homology between the hormone insulin (Figure 5-13, page 109) and nerve growth factor, a protein in the submaxillary gland of mammals that greatly enhances the growth of certain nerve ganglia. This observation suggests a possible evolutionary and functional relationship between these endocrine and secretory glands and the mode of action of their secretions.

Gene Duplication

Studies of amino acid sequence have revealed another important principle in the evolution of proteins, the duplication of genes. The simplest case is the iron-containing electron-carrying protein ferredoxin of bacteria (page 606), which contains 55 amino acid residues in its polypeptide chain. The two halves of this polypeptide chain show distinct amino acid homologies with each other, suggesting that the gene for the entire molecule arose during evolution by doubling of a single gene coding for a half molecule of about 28 amino acid residues. Moreover, the polypeptide chains of higher-plant ferredoxins are twice as long as those in the bacteria, suggesting that further elongation and fusion of genes took place during evolution of these higher forms.

Gene duplication has also occurred in the myoglobin and hemoglobin family of proteins, which may have had a common single-chain ancestor. In accordance with this idea is the discovery that the hemoglobin in the blood of the modern lamprey has but a single polypeptide chain and a single heme group; it is the only known species with a single-chain hemoglobin. Presumably the gene for the single-chain ancestor of hemoglobin became duplicated, leading to separate genes for the α and β chains of modern hemoglobins, which show considerable sequence homology (Figure 5-16). Gene duplication also occurred in the evolution of the polypeptide chains of the immune globulins, to be discussed in Chapter 35.

Mutational Changes in Amino Acid Sequence within a Species

In the human disease sickle-cell anemia the erythrocytes tend to sickle at low oxygen tensions, i.e., assume a crescent shape instead of the flat, disk conformation of normal erythrocytes (Figure 5-19). The electrophoretic mobility of the hemoglobin from sickle cells differs slightly from that of normal hemoglobin.

Chemical studies have shown that sickle-cell hemoglobin (hemoglobin S) differs from normal hemoglobin (hemoglobin A) in but a single amino acid residue. The α chains of the two forms are identical, but the glutamic acid residue at position 6 in the β chain of normal hemoglobin is replaced by a valine residue in hemoglobin S. The two valines at positions 1 and 6 form a hydrophobic association, leading the hemoglobin S molecule to assume a conformation that stacks in such a way that the shape of the erythrocyte itself is distorted. Sickle-cell anemia is thus a molecular disease of genetic origin; the amino acid replacement is the result of a mutation in the DNA molecule that codes for the synthesis of the hemoglobin β chain.

Hemoglobin S is inherited as a simple mendelian character. The majority of people possessing hemoglobin S in their erythrocytes are _heterozygotes,_ who carry one gene for normal hemoglobin and one gene for hemoglobin S and whose erythrocytes contain both normal hemoglobin and hemoglobin S; such persons are spoken of as possessing _sickle-cell trait._ The minority, called _homozygotes,_ carry two genes for hemoglobin S; such persons suffer from _sickle-cell anemia;_ their erythrocytes contain only hemoglobin S. Although 1 out of every 10 Americans of African ancestry is a heterozygote, only 1 in 400 is homozygous for hemoglobin S. Most homozygotes die of sickle-cell anemia before the age of thirty. Sickled cells last for only half as long as normal cells, resulting in anemia, and they clump together, especially when deoxygenated, because of the abnormal tertiary structure of hemoglobin S. The clumping blocks capillaries carrying the blood supply to vital regions and produces a clinical crisis. On the other hand, the heterozygotes live fairly normal lives. It is of some interest that the red cells of persons with sickle-cell trait are resistant to invasion by malarial parasites, thus giving them an advantage in Africa, where malaria is still common. Many African infants with normal hemoglobin die of cerebral malaria, but those with sickle-cell trait are resistant.

Altogether, about 150 different kinds of mutant hemoglobins have been found in human beings. The specific amino acid replaced in a mutant protein can be determined very simply by application of the peptide-map technique (Figure 5-11). Nearly all the genetic changes observed in mutant hemoglobins are due to a single amino acid replacement, which may be in either the α or the β chain. Table 5-5 lists some of the many mutations detected; the names of these abnormal forms are often derived from the location of their discovery. In a study in Taiwan, 165 individuals in over 100,000 whose hemoglobin was tested by electrophoresis were found to have abnormal hemoglobins. Since only an average of one in three types of abnormal hemoglobins can be detected by this test, it has been concluded that about 5 out of every 1,000 human beings carries a mutant hemoglobin.

Figure 5-19
Photomicrograph of normal (disklike) and sickle (crescent-shaped) blood cells. [Walter Dawn, from National Audubon Society.]

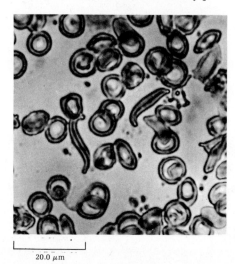

20.0 μm

Peptide maps of trypsin peptides of hemoglobin A and sickle-cell hemoglobin. Only one peptide spot (color) differs in location; it contains the genetically replaced amino acid. [Redrawn from C. Baglioni, Biochem. Biophys. Acta, 48: 392 (1961).]

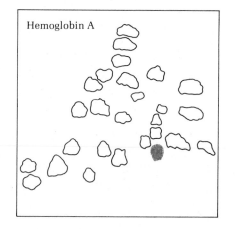

Hemoglobin A

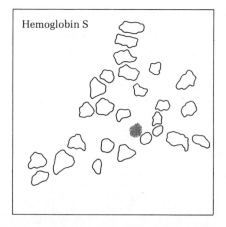

Hemoglobin S

Table 5-5   Amino acid replacements in human hemoglobins

Abnormal hemoglobin	Position and normal residue	Replacement
α chain		
I	16 Lys	Glu
G$_{Honolulu}$	30 Glu	Gln
Norfolk	57 Gly	Asp
M$_{Boston}$	58 His	Tyr
G$_{Philadelphia}$	68 Asn	Lys
O$_{Indonesia}$	116 Glu	Lys
β chain		
C	6 Glu	Lys
S	6 Glu	Val
G$_{San Jose}$	7 Glu	Gly
E	26 Glu	Lys
M$_{Saskatoon}$	63 His	Tyr
Zürich	63 His	Arg
M$_{Milwaukee}$	67 Val	Glu
D$_{Punjab}$	121 Glu	Gln

Some hemoglobin mutations are lethal; i.e., the patient may never live to maturity because the amino acid replacement results in a functionally defective hemoglobin molecule. On the other hand, in some mutations the physiological function of the hemoglobin is less seriously impaired, and in still others the mutations are apparently harmless. Such mutations are not limited to hemoglobin; it is very likely that all types of proteins in all organisms undergo changes by mutation, some of which are advantageous and some deleterious. Many harmless changes apparently pass into proteins by evolutionary drift.

Laboratory Synthesis of Polypeptide Chains

The chemical synthesis of long polypeptide chains is of special interest because of the formidable technical difficulties involved and the ingenious methods that are being applied to the problem. The basic difficulty is that the reagents required to form peptide bonds can also react with other functional groups of amino acids that are not involved in the peptide linkage, e.g., the free amino group of the N-terminal residue, the free carboxyl group of the C-terminal residue, or certain R·groups, such as the thiol group of cysteine. These sensitive groups must therefore be blocked or protected by appropriate reactions first, to shield them from the peptide-forming reagent (Figure 5-20). After the peptide bond has been formed, the protective blocking groups must be removed. For this reason, the addition of each amino acid residue during the synthesis of a polypeptide chain requires several individual reaction steps for attaching and removing the blocking groups. Obviously, the yield of each chemical reaction step must be rather high if a polypeptide chain of any length is to be synthesized.

Polypeptide chains having as many as 16 residues were synthesized early in this century by Emil Fischer, but no naturally occurring polypeptide was synthesized until 1953, when V. du Vigneaud and his colleagues succeeded in the synthesis of the posterior pituitary hormones *oxytocin* and *vasopressin*, which contain nine amino acid residues (Figure 5-2). Since then, a number of other active polypeptides, such as the blood-pressure-regulating hormone *bradykinin* (9 resi-

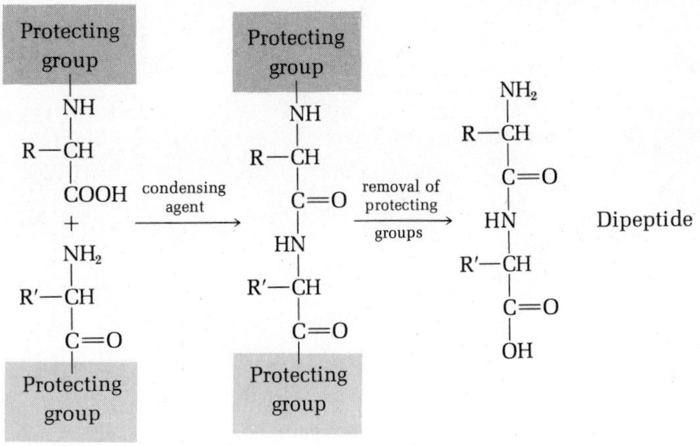

Figure 5-20
Condensation of an amino-protected and a carboxyl-protected amino acid to yield a protected dipeptide. The protecting groups are then removed to yield the free dipeptide.

dues), *α-melanocyte-stimulating hormone* (24 residues), and *adrenocorticotrophin* (39 residues) (Figure 5-14), have been chemically synthesized without use of enzymes. Moreover, the two chains of insulin have been synthesized by groups in Germany, the United States, and China.

R. B. Merrifield and his colleagues have developed an ingenious method, the *solid-phase technique*, for chemical synthesis of polypeptides. In this procedure the polypeptide chain is built residue by residue starting from the C-terminal amino acid, which is covalently anchored to an insoluble solid resin particle large enough to be separated from a

Figure 5-21

Steps in the solid-phase synthesis of a tripeptide. The chain is built by adding successive amino acyl residues to the C-terminal residue, which is attached to a resin particle. The amino groups are protected by the tert-butyloxycarbonyl group (color), which is easily removed in the form of volatile products on acidification. By repetition of these steps very long polypeptide chains can be built. The finished chain is then cleaved from the resin particle.

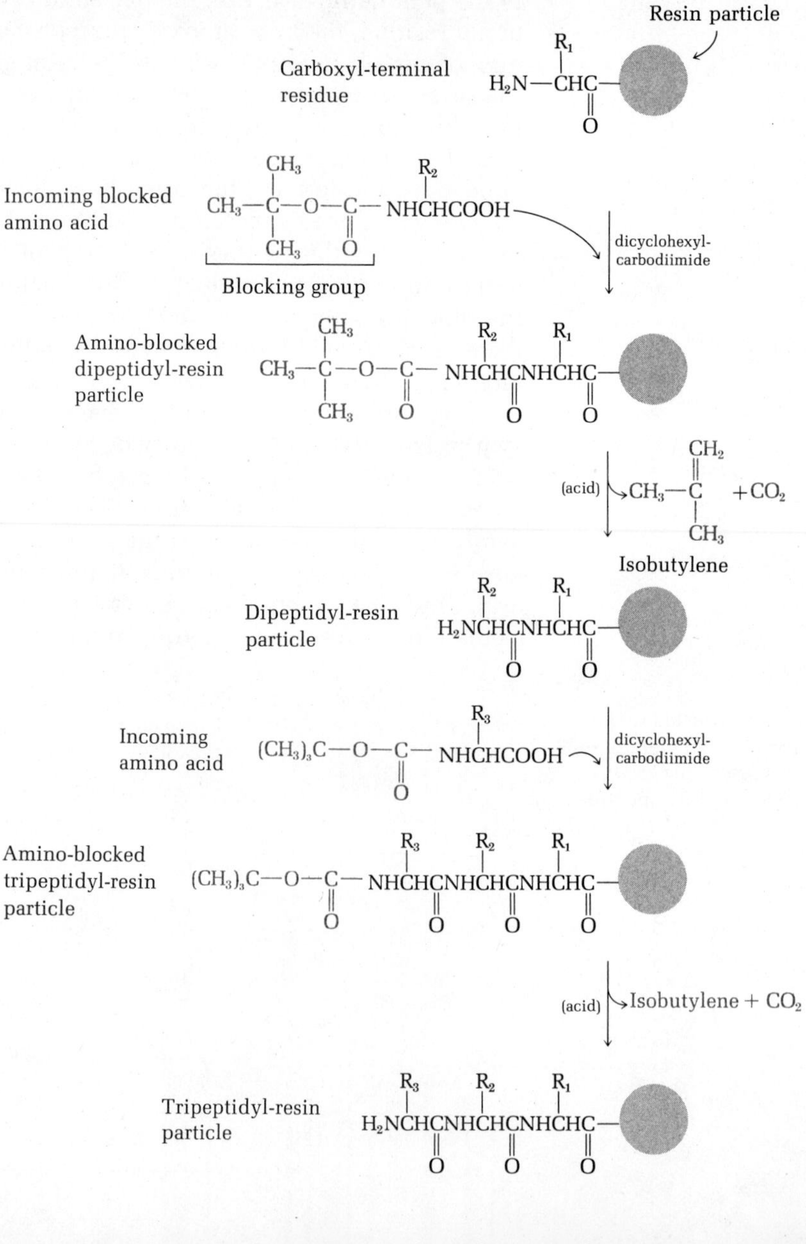

Figure 5-22
tert-Butyloxycarbonyl chloride, a blocking reagent for amino groups. It is readily removed to yield gaseous products (see Figure 5-21).

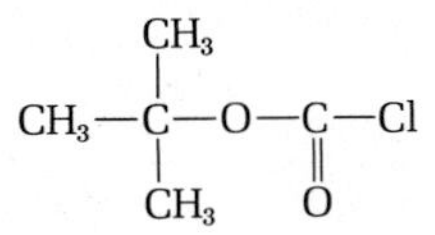

$$CH_3-\underset{\underset{CH_3}{|}}{\overset{\overset{CH_3}{|}}{C}}-O-\underset{O}{\overset{||}{C}}-Cl$$

Figure 5-23
Dicyclohexylcarbodiimide, a powerful condensing agent. It becomes hydrated to dicyclohexylurea when it removes the elements of water from two amino acid residues.

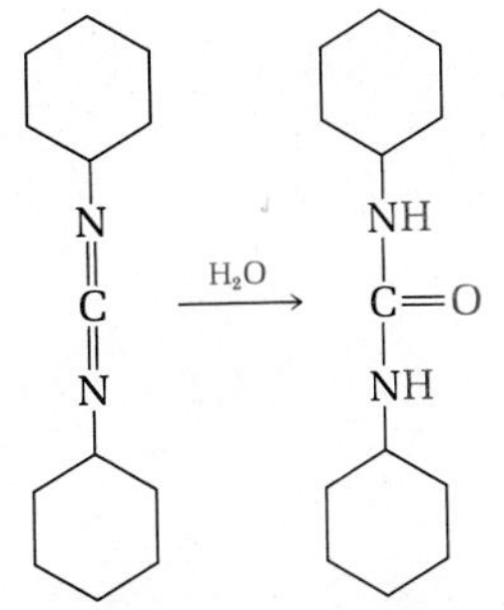

Dicyclohexyl-   Dicyclohexylcarbodiimide   urea

liquid phase by filtration. Excess reagents at the many repetitive steps are removed by filtration and thorough washing of the resin particles with appropriate solvents—simple operations that are carried out by an automatically programmed machine. All the reactions take place in a single reaction chamber, with reagents added automatically from reservoirs by means of measuring pumps. When the chain is complete it is cleaved from its resin anchor by a reaction that does not attack the newly formed peptide linkages (Figure 5-21). The simple filtration and washing operations between reaction steps and eliminating the need to isolate the intermediate products in pure form mean that this procedure has very high yields at each step and requires relatively little time.

The synthesis of the polypeptide chain (Figure 5-21) is begun by attaching the carboxyl group of the C-terminal amino acid residue of the chain to be built to the insoluble resin particle. The next amino acid to be introduced, following blocking of its amino group with the reagent *tert*-butyloxycarbonyl chloride (Figure 5-22), is allowed to react with the free amino group of the C-terminal residue in the presence of the condensing agent *dicyclohexylcarbodiimide* (Figure 5-23). This reaction forms an amino-blocked dipeptide covalently attached to the insoluble resin particle via the C-terminal carboxyl group. The amino-blocking group is then removed by acidification; it subsequently decomposes into gaseous carbon dioxide and isobutylene. These steps are repeated cyclically many times.

Merrifield and his colleagues synthesized the nonapeptide bradykinin by this automated procedure, with an overall yield of 85 percent, in 27 h; the average rate of synthesis was thus about 3 h per peptide bond. The two chains of insulin were also made in this way; the A chain (21 residues) required only 8 days and the B chain (30 residues) 11 days. They also synthesized bovine pancreatic ribonuclease (124 residues), the first protein to be made chemically from its amino acid components. The overall yield was 18 percent. At the same time a second group of investigators succeeded in synthesizing ribonuclease by a somewhat different approach; they built segments of the ribonuclease chain first and then spliced them together.

Amino Acid Homopolymers

Extremely long polypeptide chains can be synthesized quite easily if they contain only one type of amino acid residue. Such chains are called *homopolypeptides*; examples are polyglycine, polyalanine, and polyglutamic acid. The procedure involves using relatively simple polymerization reactions that become self-sustaining. The length of the homopolymer can be controlled by the nature of the reaction initiator, the temperature, and the solvent. Although homopolymers of amino acids are not found in nature, they are exceedingly valuable model compounds for studying the various parameters influencing the structure and behavior of polypeptide chains; e.g., they have led to important insight into the relationship between the optical rotation of peptides and their secondary structure (Chapter 6).

Summary

The peptide bond is the only covalent linkage between successive amino acids in the backbone of polypeptide chains. The acid-base behavior of a polypeptide is a function of its N-terminal amino group, its C-terminal carboxyl group, and R groups that ionize. Partial hydrolysis of polypeptide chains by boiling with acid or by the action of enzymes yields mixtures of smaller peptides, which can be separated by chromatography or electrophoresis. Complete hydrolysis of a protein yields all its amino acids in free form. Not all proteins contain all the amino acids; nor do the amino acids occur with equal frequency.

To determine the sequence of amino acids in a protein, its polypeptide chains are first separated, following reduction of the disulfide bridges to sulfhydryl groups. Next the cysteine residues are alkylated. The N-terminal and C-terminal residues of the chain are then identified and the total amino acid composition of the chain determined on a complete hydrolyzate. One sample of the chain is partially cleaved by using trypsin, which hydrolyzes those peptide bonds in which lysine or arginine contributes the carbonyl carbon atom. The tryptic peptides are separated, their amino acid content determined, and their sequence established, usually by the Edman degradation. Another intact sample of the polypeptide chain is then cleaved by a second method, using chymotrypsin, pepsin, thermolysin, or cyanogen bromide, to yield a different set of fragments, which are separated and analyzed as in the first cleavage. From the overlaps between the two sets of fragments, the sequence of the fragments in the original protein can be deduced.

Each protein has a characteristic amino acid sequence, and all molecules of a given type are identical in sequence. Functionally homologous proteins of different species, e.g., hemoglobins and cytochromes, possess the same amino acid residues at certain invariant positions in the chain; other residues may vary from one species to another. The more distant two species are, the greater the number of amino acid differences in the variant positions, a fact which permits construction of phylogenetic trees. Some proteins of different function are related to each other through sequence homologies, e.g., trypsin and chymotrypsin. During evolution of proteins their genes often undergo doubling in length or duplication. Efficient methods have been worked out for the chemical synthesis of polypeptides. Several polypeptides and an enzyme have been synthesized chemically. Their biological activity confirms the correctness of the structure of these proteins deduced by analytical methods.

References

Books

BAILEY, J. L.: *Techniques in Protein Chemistry*, 2d ed., American Elsevier, New York, 1967. A valuable compendium of laboratory procedures.

DAYHOFF, M. O.: *Atlas of Protein Sequence and Structure 1972*, National Biomedical Research Foundation, Washington, D.C., 1972. A complete atlas of amino acid sequences of proteins, with much interesting information on species differences, evolutionary relationships and the genetic code for amino acids, as well as transfer RNA sequences.

DICKERSON, R. E., and I. GEIS: *The Structure and Action of Proteins*, Benjamin, Menlo Park, Calif., 1969. An illuminating and brilliantly illustrated short account.

HASCHEMEYER, R. H., and A. E. V. HASCHEMEYER: *Proteins: A Guide to Study by Physical and Chemical Methods*, Wiley, New York, 1973. Convenient, short handbook of methodology.

MEANS, G. E., and R. E. FEENEY: *Chemical Modification of Proteins*, Holden-Day, San Francisco, 1971.

NEURATH, H. (ed.): *The Proteins: Composition, Structure and Function*, 2d ed., vols. 1–5, Academic, New York, 1963–1970. Collection of detailed reviews.

STEWART, J. M., and J. D. YOUNG: *Solid-Phase Peptide Synthesis*, Freeman, San Francisco, 1969. A short handbook of laboratory methods.

Articles

EDELMAN, G. M., B. A. CUNNINGHAM, W. E. GALL, P. D. GOTTLIEB, V. RUTISHAUSER, and M. J. WARDEL: "The Covalent Structure of an Entire γG Immunoglobulin Molecule," *Proc. Natl. Acad. Sci. U.S.*, 63: 78–85 (1969).

EDMAN, P., and G. BEGG: "A Protein Sequenator," *Eur. J. Biochem.*, 1: 80–91 (1967). Automated equipment for sequence determination.

FRAZIER, W. A., R. H. ANGELETTI, and R. A. BRADSHAW: "Nerve Growth Factor and Insulin," *Science*, 176: 482–488 (1972).

GUTTE, B., and R. B. MERRIFIELD: "The Total Synthesis of an Enzyme with Ribonuclease A Activity," *J. Am. Chem. Soc.*, 91: 501–502 (1969).

HALL, D. O., R. CAMMACK, and K. K. RAO: "The Plant Ferredoxins and Their Relationship to the Evolution of Ferredoxins," *Pure Appl. Chem.* 34: 553–577 (1973).

HARTLEY, B. S.: "Strategy and Tactics in Protein Chemistry," *Biochem. J.*, 119: 805–822 (1970). A personal account of the problems encountered in analysis of amino sequence of proteins.

MARGOLIASH, E.: "The Molecular Variations of Cytochrome *c* as a Function of the Evolution of Species," *Harvey Lect.*, ser. 66, p. 177, Academic, New York, 1972.

MOON, K., D. PISZKIEWICZ, and E. L. SMITH: "Glutamate Dehydrogenase: Amino Acid Sequence of the Bovine Enzyme and Comparison with That from Chicken Liver," *Proc. Natl. Acad. Sci. U.S.*, 69: 1380–1383 (1972).

MOORE, S., and W. H. STEIN: "Chemical Structures of Pancreatic Ribonuclease and Deoxyribonuclease," *Science*, 180: 458–464 (1973). Combined text of the Nobel prize lectures by the authors.

NEURATH, H., K. A. WALSH, and W. P. WINTER: "Evolution of Structure and Function of Proteases," *Science*, 158: 1638–1644 (1967). Evolutionary relationships between proteolytic enzymes secreted into the digestive tract.

SANGER, F. and E. O. P. THOMPSON: "The Amino Acid Sequence in the Glycyl Chain of Insulin," *Biochem. J.*, 53: 353–374 (1963). This paper and the one below describe the classical work in protein sequencing.

SANGER, F., and H. TUPPY: "The Amino Acid Sequence in the Phenylalanyl Chain of Insulin," *Biochem. J.*, 49: 463–490 (1961).

Problems

1. (a) Predict the action of trypsin on the following peptides. Each
   of the resulting fragments is then treated with 2,4-dini-
   trofluorobenzene, followed by hydrolysis of the peptide linkages. (b) List the resulting 2,4-dinitrophenyl (2,4-DNP) amino
   acids.
   (1) Lys-Asp-Gly-Ala-Ala-Glu-Ser-Gly
   (2) Ala-Ala-His-Arg-Glu-Lys-Phe-Ile
   (3) Tyr-Cys-Lys-Ala-Arg-Arg-Gly
   (4) Phe-Ala-Glu-Ser-Ala-Gly

2. (a) List the peptides formed when the following polypeptide is
   treated with chymotrypsin.
   Val-Ala-Lys-Glu-Glu-Phe-Val-Met-Tyr-Cys-Glu-Trp-Met-Gly-
   Gly-Phe
   (b) Suppose the peptides resulting from chymotrypsin treatment
   are then reacted with cyanogen bromide; list the products.

3. List the peptides formed (a) when the following polypeptide is
   treated with trypsin and (b) when the resulting fragments are
   reacted with thermolysin.
   Leu-Met-His-Tyr-Lys-Arg-Ser-Val-Cys-Ala-Lys-Asp-Gly-Ile-Phe-
   Ile

4. (a) What treatments could you apply to the following polypeptide to obtain two sets of peptides with appropriate overlaps in
   order to establish the complete amino acid sequence? (b) Write
   the amino acid sequence of the peptides obtained by the two
   treatments.
   Gly-Leu-Ser-Pro-Phe-His-Thr-Asp-Val-Ser-Ala-Ala-Trp-Gly-Glu-
   Val-Gly-Ala-His-Leu-Gly-Glu-Tyr-Gly-Ala-Glu-Ala-Thr-Glu

5. Reaction of a tetrapeptide with 2,4-dinitrofluorobenzene, followed by hydrolysis in 6 N HCl, yielded the 2,4-dinitrophenyl
   derivative of valine and three other amino acids. Hydrolysis of
   another sample of the tetrapeptide with trypsin gave two fragments. One was reduced with LiBH$_4$ and then hydrolyzed. In
   the hydrolyzate, the amino alcohol corresponding to glycine
   was detected, together with an amino acid forming a yellow
   reaction product with ninhydrin. Give a possible amino
   acid sequence for the tetrapeptide.

6. Predict the direction of migration, i.e., stationary (0), toward
   cathode (C), or toward anode (A), of the following peptides
   during paper electrophoresis at pH 1.9, 3.0, 6.5, and 10.0:
   (a) Lys-Gly-Ala-Gly
   (b) Lys-Gly-Ala-Glu
   (c) His-Gly-Ala-Glu
   (d) Glu-Gly-Ala-Glu
   (e) Gln-Gly-Ala-Lys

7. If the five peptides of Problem 6 were present in a mixture,
   predict the relative order of elution of each peptide from an
   ion-exchange resin column containing Dowex-1 (an anion exchanger) when the column is eluted with a buffer system whose
   pH is continuously changed from an initial value of 10 to a
   final value of 1.0.

8. Indicate how the following abnormal hemoglobins will differ in electrophoretic mobility toward the anode from normal human hemoglobins at pH 7.0:
   (a) HbS
   (b) HbI
   (c) HbE
   (d) HbM$_{Milwaukee}$
   (e) Hb$_{Zurich}$

9. The following data were collected on the structure of a polypeptide chain. Total hydrolysis gave Gly, Ala, Val$_2$, Leu$_2$, Ile, Cys$_4$, Asp$_2$, Glu$_4$, Ser$_2$, Tyr$_2$. Treatment of the chain with 2,4-dinitrofluorobenzene followed by acid hydrolysis gave 2,4-dinitrophenylglycine; C-terminal analysis gave aspartate. Partial acid hydrolysis gave the following oligopeptides, among others: Cys-Cys-Ala, Glu-Asp-Tyr, Glu-Glu-Cys, Glu-Leu-Glu, Cys-Asp, Tyr-Cys, Ser-Val-Cys, Glu-Cys-Cys, Ser-Leu-Tyr, Leu-Tyr-Glu, Gly-Ile-Val-Glu-Glu. Cleavage with pepsin yielded a peptide that on hydrolysis gave Ser-Val-Cys and Ser-Leu. Give a structure consistent with the data.

10. A peptide A of the composition Lys, His, Asp, Glu$_2$, Ala, Ile, Val, and Tyr gave 2,4-dinitrophenylaspartate on N-terminal analysis with 2,4-dinitrofluorobenzene and free valine as the first product with carboxypeptidase. Digestion of A with trypsin yielded two peptides. One contained Lys, Asp, Glu, Ala, and Tyr. The other (His, Glu, Ile, Val) gave 2,4-dinitrophenylhistidine on N-terminal analysis with 2,4-dinitrofluorobenzene. Cleavage of the latter with thermolysin yielded, among other products, free histidine. Two chymotryptic peptides were also formed from A. One contained Asp, Ala, and Tyr, and the other contained Lys, His, Glu$_2$, Ile, and Val. Deduce a structure for peptide A.

11. Acid hydrolysis of 1 mmol of a hypothetical pentapeptide yields 2 mmol of glutamic acid, 1 mmol of lysine, and quantitative recovery of no other amino acid. Trypsin splits the original pentapeptide into two fragments. Upon electrophoresis at pH 7.0, one of the tryptic fragments moves toward the anode and the other toward the cathode. Treatment of one of the tryptic fragments with DNFB followed by acid hydrolysis yields DNP–glutamic acid. Treatment of the original pentapeptide with chymotrypsin yields two dipeptides and free glutamic acid. Give an amino acid sequence for the pentapeptide that is consistent with the foregoing data.

12. Acid hydrolysis of a hypothetical hexapeptide gives quantitative recovery of glycine and one other amino acid. Treatment of the hexapeptide with a lysine-blocking reagent and then trypsin yields a product that migrates as a single spot in a variety of chromatographic and electrophoretic systems but whose migration differs from that of the untreated hexapeptide. Treatment of the original hexapeptide with chymotrypsin yields, among other products, a free amino acid that absorbs ultraviolet light. Give an amino acid sequence for the hexapeptide that is consistent with the foregoing data.

We now come to a problem central in the study of proteins and fundamental to the molecular logic of living cells. We have seen that genetic information is stored in a linear, one-dimensional code, in the form of a specific base sequence in the long, chainlike deoxyribonucleic acid molecule. Proteins also consist of long chains, in which amino acids occur in specific linear sequences. Yet we know that in each type of protein the polypeptide chain is folded into a specific three-dimensional conformation, which is required for its specific biological function or activity. How is the linear, or one-dimensional, information inherent in the amino acid sequence of polypeptide chains translated into the characteristic three-dimensional conformation of native protein molecules?

The answer to this question comes from some of the most significant advances in modern biological research. These discoveries, made possible by the application of physical-chemical measurements to pure proteins, have illuminated the function and comparative biology of proteins.

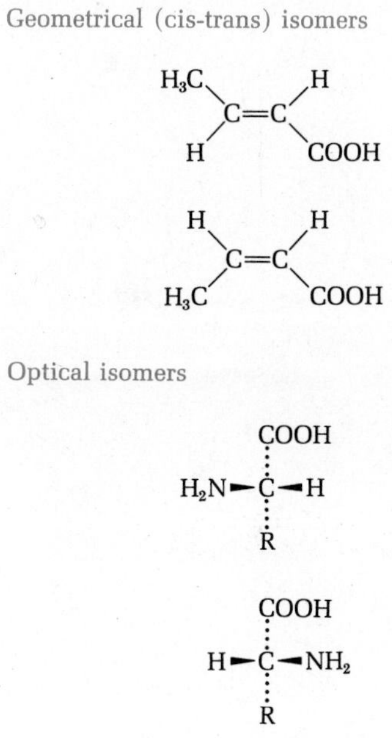

Configuration and Conformation

First, we must clarify two terms often confused. *Configuration* denotes the arrangement in space of substituent groups in stereoisomers (Figure 6-1); such structures cannot be interconverted without breaking one or more covalent bonds. *Conformation* refers to the spatial arrangement of substituent groups that are free to assume many different positions, without breaking bonds, because of rotation about the single bonds in the molecule. In the hydrocarbon ethane, for example, one might expect complete freedom of rotation around the C—C single bond to yield an infinite number of conformations of the molecule. However, the *staggered* conformation (Figure 6-2) is more stable than all others and thus predominates, whereas the *eclipsed* form is least stable.

The covalent backbone of a polypeptide chain is formally single-bonded. We would therefore expect the backbone of a

Figure 6-1
Configuration of stereoisomers. Such isomers cannot be interconverted without breaking covalent bonds.

Geometrical (cis-trans) isomers

Optical isomers

polypeptide chain to have an infinite number of possible conformations and the conformation of any given polypeptide to undergo constant change because of thermal motion. However, it is now known that the polypeptide chain of a protein has only one conformation (or a very few) under normal biological conditions of temperature and pH. This *native* conformation (page 62), which confers biological activity, is sufficiently stable so that the protein can be isolated and retained in its native state. This fact therefore implies that the single bonds in the backbone of native proteins cannot rotate freely.

Fibrous Proteins

We shall consider the conformation of fibrous proteins first. Not only are they very abundant, particularly in higher animals, but they also have simpler conformations than the globular proteins, since their polypeptide chains are usually arranged or coiled along a single dimension, often in parallel bundles. As a result the conformation of the polypeptide chains in some fibrous proteins has been easier to examine experimentally; actually, the fibrous proteins gave the first important clues to the constraints on the freedom of rotation of the single bonds in the polypeptide-chain backbone of proteins.

Two major classes of fibrous proteins, the *keratins* and *collagens*, will be considered here. Study of the keratins has been especially important in revealing the most prevalent conformations of the polypeptide chains in native proteins, namely, the *α helix* and the *β conformation*.

The Keratins

The keratins are fibrous, insoluble proteins of animals derived from ectodermal (skin) cells. They include the structural protein elements of skin (leather is almost pure keratin) as well as the biological derivatives of ectoderm, such as hair, wool, scales, feathers, quills, nails, hoofs, horns, and silk. There are two classes of keratins. The *α-keratins* are relatively rich in cystine residues and thus contain many disulfide cross bridges (page 99); in addition, they contain most of the common amino acids. The α-keratins include the hard, brittle proteins of horns and nails, which have a very high content of cystine (up to 22 percent), as well as the softer, more flexible keratins of skin, hair, and wool, which contain about 10 to 14 percent cystine. The *β-keratins*, on the other hand, contain no cysteine or cystine but are rich in amino acids with small side chains, particularly glycine, alanine, and serine. The β-keratins are found in the fibers spun by spiders and silkworms and in the scales, claws, and beaks of reptiles and birds. Another important difference is that the α-keratins stretch when heated; hair, for example, stretches to almost double its length when exposed to moist heat but contracts to its normal length on cooling. The β-keratins do not stretch under these conditions.

Electron microscopy has revealed that hair and wool fibers contain bundles of macrofibrils, each made up of

Figure 6-2
Conformations of ethane. Different conformational forms are rapidly interconvertible by rotation around the single bond; they cannot be separated from each other.

Overhead views

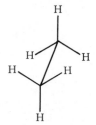

Staggered

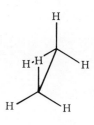

Eclipsed

End views

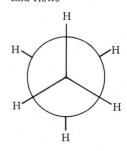

Staggered

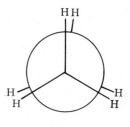

Eclipsed

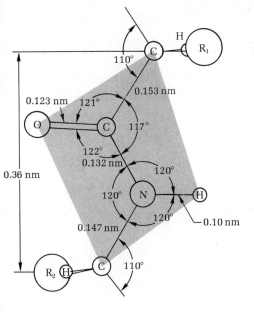

Figure 6-3

Dimensions of the peptide group from x-ray data. The six atoms in the shaded zone lie in a plane. Because the central C—N bond has some double-bond character, this plane tends to be rigid. Bond lengths are given in nanometers. [Redrawn from T. P. Bennett, Graphic Biochemistry, vol. 1, The Macmillan Company, New York, 1968.]

thinner fibrils consisting in turn of parallel bundles of protein filaments arranged along a single axis. This structural feature allows them to be examined readily by x-ray diffraction analysis.

X-Ray Analysis of Keratins

The spacing of regularly repeating atomic or molecular units in crystals can be determined by studying the angles and intensities at which x-rays of a given wavelength are scattered, or diffracted, by the electrons that surround each atom. Atoms with the highest electron density, such as heavy-metal atoms, diffract x-rays most; atoms with the lowest electron density (hydrogen atoms) diffract x-rays least. X-ray analysis of crystals of salts like NaCl is relatively simple because only two different kinds of atoms are involved and they are regularly spaced. In principle, crystals of complex organic molecules, even very large ones such as proteins, can also be analyzed by x-ray diffraction methods, but the mathematical analysis of the diffraction patterns is very complex because the large number of atoms in the molecule may yield thousands of diffraction spots.

In the early 1930s W. Astbury in England carried out the first pioneering x-ray studies of proteins. Hair and wool and certain other fibrous proteins of the α-keratin class gave similar x-ray diffraction patterns, which indicated that these proteins possess a major periodicity, or _repeat unit_, of about 0.5 to 0.55 nm along their long axes. His observations suggested further that the polypeptide chains in this family of fibrous proteins are not fully extended but twisted or coiled in some regular way, since a fully extended polypeptide chain could not give the observed spacings. On the other hand, _fibroin_, the β-keratin of silk fibers, has a distinctly different x-ray diffraction pattern compatible with a repeat unit of 0.7 nm. Significantly, when hair or wool was stretched after steaming, it assumed an x-ray pattern resembling that of β-keratins, with a periodicity of between 0.65 and 0.70 nm. Astbury concluded that the polypeptide chains in α- and β-keratins are twisted or coiled in different ways.

The next stage in the development of our knowledge of the structure of keratins came from the work of L. Pauling and R. B. Corey in the United States, who recorded the x-ray diffraction patterns of crystals of amino acids and of simple dipeptides and tripeptides and from them deduced the precise structure of the peptide bond. They found that the C—N bond of the peptide linkage is shorter than most other C—N bonds; they concluded that it has some double-bond character and thus cannot rotate freely. They further deduced that the four atoms of the peptide group and the two α carbon atoms lie in a single plane, in such a way that the oxygen atom of the carbonyl group and the hydrogen atom of the —NH— group are *trans* to each other (Figure 6-3). This planar arrangement, which is rigid, is the result of resonance stabilization of the peptide bond.

From these findings the backbone of a polypeptide chain may be pictured as a series of relatively rigid planes sepa-

Figure 6-4
Restricted rotation around the single bonds of a polypeptide chain. Only the single bonds to the α carbon atoms are free to rotate; the C—N single bonds of the planar peptide groups are rigid.

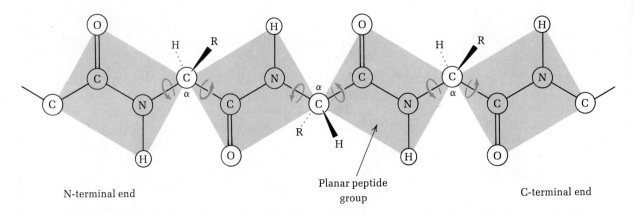

N-terminal end

Planar peptide group

C-terminal end

rated by substituted methylene groups (—CHR—) (Figure 6-4). In the backbone of a polypeptide chain, one-third of all the formal single bonds are C—N bonds and cannot rotate because they have double-bond character. Thus the peptide bonds impose significant constraints upon the number of conformations that can be assumed by a polypeptide chain. We shall see later that there are further constraints upon the number of conformations of a polypeptide chain, since even the carbon-carbon single bonds in the backbone are not always free to rotate.

The α Helix and the Structure of α-Keratins

Pauling and Corey used precisely constructed models to study all the possible ways of twisting or coiling the backbone of the polypeptide chain along one axis, in view of the constraint imposed by the planar peptide bonds, to account for the observed repeat units of 0.50 to 0.55 nm in α-keratins. The simplest arrangement they found is the helical structure shown in Figure 6-5. In this structure, the *α helix*, the backbone is arranged in a helical coil having about 3.6 amino acid residues per turn. The R groups of the amino acids extend outward from the rather tight helix formed by the backbone. In such a structure the *repeat unit*, consisting of a single complete turn of the helix, extends about 0.54 nm (5.4 Å) along the long axis, corresponding closely to the major periodicity of 0.50 to 0.55 nm deduced from the x-ray pattern of natural α-keratins. The rise per residue is about 0.15 nm, corresponding to the minor periodicity of 0.15 nm also observed in the diffraction patterns. Such an α helix permits the formation of intrachain hydrogen bonds between successive coils of the helix, parallel to the long axis of the helix and extending between the hydrogen atom attached to the electronegative nitrogen of one peptide bond and the carbonyl oxygen of the third amino acid beyond it (Figure 6-5).

Figure 6-5
The α helix. [Top left and right and bottom right models from G. H. Haggis, D. Michie, A. R. Muir, K. B. Roberts, and P. M. B. Walker, Introduction to Molecular Biology, John Wiley & Sons, Inc., New York, 1964.]

Formation of a right-handed α helix. The planes of the rigid peptide bonds are parallel to the long axis of the helix.

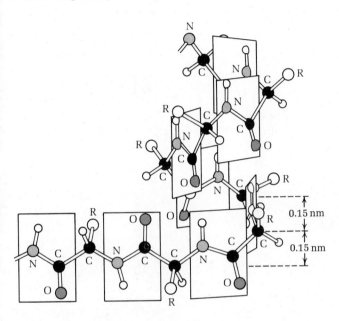

The average dimensions of the α helix. The pitch and the rise per residue correspond to the major and minor periodicities of 0.54 and 0.15 nm, respectively. [From L. Pauling and R. B. Corey, Proc. Int. Wool Text. Res. Conf., B: 249 (1955).]

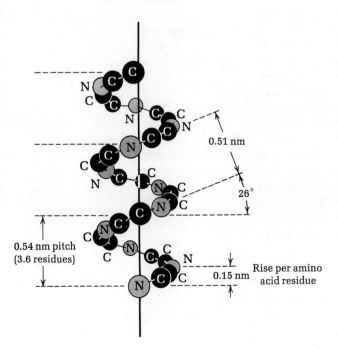

Ball-and-stick model of α helix, showing intrachain hydrogen bonds (colored dots).

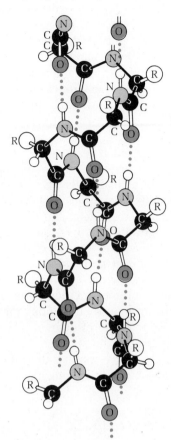

Space-filling model of α helix.

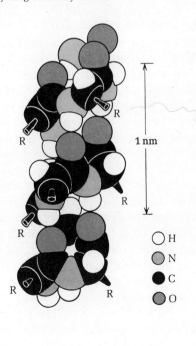

○ H
○ N
● C
○ O

The electrical vectors (page 40) of these hydrogen bonds are so oriented that they give nearly maximal bond strength. But especially significant is that the α-helical arrangement allows every peptide bond of the chain to participate in intrachain hydrogen bonding. Although other kinds of helical coils of polypeptide chains can be formed, such as a π helix (4.4 residues per turn), they cannot account for the characteristic spacing of the repeat units in the α-keratin family of proteins, nor would they be as stable as the α helix.

An α helix may form with either L- or D-amino acids, but a helix cannot form from a polypeptide chain containing a mixture of L and D residues. Furthermore, starting from the naturally occurring L-amino acids, either right-handed or left-handed helical coils can be built; however, the right-handed helix is significantly more stable. In all native proteins examined to date, the α helix is right-handed. From these structural considerations Pauling and Corey proposed that the α-keratins consist of polypeptide chains in right-handed α-helical coils. In the α-keratins of hair and wool, three or seven such α helixes may be coiled around each other to form three-stranded or seven-stranded ropes (Figure 6-6), held together by disulfide cross-linkages.

To return to a term defined in Chapter 3 (page 60), the α helix represents the *secondary structure* of α-keratins, i.e., the regular, coiled conformation of their polypeptide chains around and along their long axis.

Helix-Forming and Helix-Destabilizing Amino Acids

The extent to which any given polypeptide chain can exist in the form of a stable α helix is a reflection of its amino acid composition and sequence; not all polypeptide chains can form a stable α helix (Table 6-1). Especially important information on this point has come from study of the *polyamino acids* (page 119), polypeptides in which all the amino acid residues are identical. Although polyalanine, whose R groups are small and uncharged, spontaneously forms α-helical coils in aqueous solution at pH 7.0, polylysine does not form an α helix at pH 7.0 but exists in an irregular random form, in which the flexible backbone undergoes continuous change as a result of thermal motion. This is because at pH 7.0 the R groups of polylysine all have a positive charge and repel each other so strongly that they overcome the tendency for intrachain hydrogen bonds to form. However, at pH 12, the lysine R groups bear no charge and thus do not repel each other; at this pH, polylysine spontaneously forms an α helix (Figure 6-7). Similarly, polyglutamic acid is a random structure at pH 7.0 because its R groups at that pH are all negatively charged, but at pH 2.0, where its R groups are protonated, it forms an α helix (Figure 6-7).

Not only the presence of charge interactions but also the size or bulk of the side chains determines whether any given polypeptide chain can form an α helix. Polyisoleucine fails to form an α helix because of the steric hindrance of its bulky

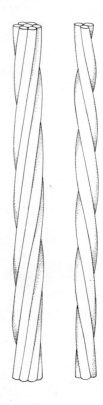

Figure 6-6
Models for supercoiling of α-helical coils in hair and wool keratins.

Table 6-1   Helix-forming and helix-destabilizing amino acids

Allow stable α helix
Alanine
Leucine
Phenylalanine
Tyrosine
Tryptophan
Cysteine
Methionine
Histidine
Asparagine
Glutamine
Valine

Destabilize α helix
Serine
Isoleucine
Threonine
Glutamic acid
Aspartic acid
Lysine
Arginine
Glycine

Break α helix
Proline
Hydroxyproline

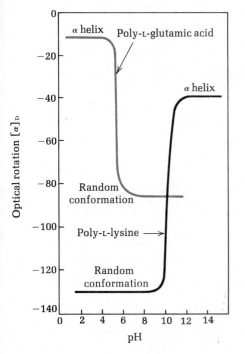

Figure 6-7
The effect of pH *on the transition between the random conformation and α-helical forms of poly-L-lysine and poly-L-glutamic acid. The unfolding of the α helix is accompanied by a large change in optical rotation to a more negative value. [Redrawn from P. Doty in J. L. Oncley (ed.), Biophysical Science, p. 108, John Wiley & Sons, Inc., New York, 1959.]*

R groups next to the α carbon atoms. In polyproline, the α nitrogen atoms are part of the rigid R-group rings (page 72), and no rotation of either the ring N—C bonds or the peptide C—N bonds is possible. Moreover, no intrachain hydrogen bonds can form in polyproline. Whenever proline (or hydroxyproline) occurs in a polypeptide chain, it interrupts the α helix and creates a rigid kink or bend. Polyglycine can form an α helix, but it prefers another type of backbone conformation, the *β conformation*, in which the chains are relatively extended (page 133). As a result of model studies with different synthetic polyamino acids the various amino acids have been classified with respect to their potential for forming α-helical coils (Table 6-1).

Optical Properties of the α Helix

Native proteins are significantly more dextrorotatory than the sum of the individual rotations contributed by each asymmetric α-carbon atom of the L-amino acid residues. Such extra dextrorotatory power is maximal in the α-helix form, whereas randomly coiled polypeptide chains show only simple additivity of rotatory power. For example, the specific rotation of a given length of poly-L-glutamic acid at pH 2 (at which it is an α helix) is about −15°, whereas its specific rotation at pH 7 (at which it is a random form) is about −85° (Figure 6-7). Nearly all globular proteins become more levorotatory when they are denatured.

The ability to rotate the plane of plane-polarized light is shown by compounds containing one or more asymmetric carbon atoms, but other forms of molecular asymmetry may show optical activity, even without an asymmetric carbon. The basic criterion is whether the molecule can exist in two different nonsuperimposable mirror-image forms. In an α helix of L-amino acids, the total asymmetry of the molecule is the sum of the asymmetry contributed by the asymmetric carbon atoms and that contributed by the α-helical coil, which is asymmetric since it can exist in right-handed or left-handed forms. The optical-rotatory power before and after denaturation, i.e., conversion to a random form, has been used as a measure of the approximate amount of α-helical coiling in any given polypeptide.

Allowed and Disallowed Conformations of Polypeptide Chains: The Ramachandran Plot

We have seen (Figure 6-4) that every third bond in the backbone of a polypeptide chain has some double-bond character and is unable to rotate freely. Moreover, we have also seen that the side chains of some amino acids such as isoleucine or lysine (at pH 7.0) are not compatible with the formation of stable α-helical coils. These observations have led to attempts to predict more quantitatively the conformation of the backbone of a given polypeptide chain starting from knowledge of its amino acid sequence. A very useful set of relationships has been developed by G. N. Ramachandran and

Figure 6-8

Ball-and-stick model showing the φ and ψ angles between two adjacent peptide groups. The planar peptide groups are separated by the α carbon atom, which bears the R group of the right-hand amino acid residue. In this representation the planes of the peptide groups are in the plane of the page; this is the extended, or open, form of the peptide bond. The angle φ is the angle subtended when the left-hand plane is rotated in the direction given by the arrow. The ψ angle is that subtended when the right-hand plane is rotated in the direction shown. Both may be rotated independently, and both φ and ψ may have values ranging from −180° to +180°. By convention, the φ and ψ angles in the extended form depicted below are assigned the maximum permissible value of +180°, i.e., φ = ψ = +180°. When the φ angle approaches 0° and the ψ angle is +180°, the carbonyl oxygen atoms (dashed outline) overlap and prevent further rotation. Similarly, when φ is +180° and ψ approaches 0°, the hydrogen atoms (dotted outlines) attached to the peptide nitrogens overlap. When φ is approximately +120° and ψ is +180° the N—H bond is trans to the C_α—R bond, and when φ is +180° and ψ is +120° the C_α—H bond is trans to the C—O bond.

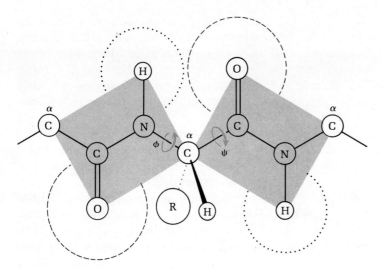

his colleagues. Figure 6-8 shows the backbone of a tripeptide with its two planar peptide bonds. The C—N backbone bonds in each peptide group have some double-bond character and do not rotate. The other backbone bonds, designated C_α—N and C_α—C, are theoretically free to rotate, since they are true single bonds, but if the R group attached to the central α carbon atom is large enough, it will prevent complete rotation around the C_α—N and C_α—C bonds. Moreover, if these bonds are rotated in relation to each other, angles will be found where the two H atoms (or O atoms) of the peptide bonds would overlap each other and obstruct free rotation. Thus each pair of successive peptide bonds has two kinds of constraints on the freedom of rotation of the C_α—C and the C_α—N single bonds.

The angle of rotation of the C_α—N bond is called the φ (phi) angle and that of the C_α—C bond the ψ (psi) angle (Figure 6-8). From estimates based on theoretical considerations or accurate models of peptides, Ramachandran has constructed a plot of the allowed and disallowed φ and ψ angles of adjacent C_α—C and C_α—N single bonds (Figure 6-9). The number of possible stable conformations of the C_α—C and C_α—N single bonds is rather limited; only conformations

Figure 6-9

Ramachandran plot of φ angles vs. ψ angles. The colorless (white) areas indicate the φ and ψ angles of the stable allowed conformations. The brown areas indicate conformations of lesser stability. All other areas of the plot correspond to disallowed φ and ψ angles. White or brown areas within the orange zones are compatible with right-handed α helixes and within the gray zones with left-handed α helixes. α_R = right-handed α helix; α_L = left-handed α helix; C = collagen helix; β_A = antiparallel β-pleated sheet; β_P = parallel β-pleated sheet. The plot is for an L-alanyl residue.

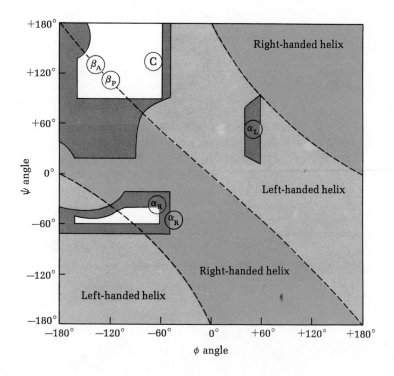

falling within the relatively small white areas in Figure 6-9 are stable and allowed.

For example, the α helix is an allowed conformation in the Ramachandran plot since all the pairs of successive peptide bonds in an α-helical polypeptide chain fall within the permitted area. The most stable angles for a right-handed α helix are $\phi = -57°$ and $\psi = -47°$. We may note that if the φ angle is increased significantly without change in the ψ angle, we leave the allowed zone; or if the ψ angle is increased or decreased significantly without change in the φ angle, we also leave the zone of allowed structures. Thus, for any given pair of φ and ψ angles we can predict whether the conformation is allowed. All naturally occurring conformations of the backbone of native proteins fall within the allowed zones of the Ramachandran plot (Figure 6-9).

β-Keratins: The β Conformation and the Pleated Sheet

We have seen that x-ray study of α-keratins led to our present knowledge of the α helix; in a similar way, x-ray studies of β-keratins have revealed important clues to the β conformation of the polypeptide chain. We recall (page 126) that when fibers of α-keratin are subjected to moist heat, they can be stretched to almost double their original length. In this stretched condition they yield x-ray diffraction patterns resembling that of silk fibroin, an example of a β-keratin. Pauling and Corey concluded that the transition from α-keratin to β-keratin structure when hair or wool is steamed is caused by the thermal breakage of the intrachain hydrogen bonds that normally stabilize the α helix and the consequent

Figure 6-10
*The β conformation of the polypeptide
chain.*

*Schematic representation of three parallel
chains in β structure, showing the pleated-
sheet arrangement. All the R groups project
above or below the plane of the page. [Re-
drawn from T. P. Bennett, Graphic Biochemistry, vol. 1, The Macmillan Company,
New York, 1968.]*

*Ball-and-stick models. Note the maximal
hydrogen bonding between the chains to
form a sheet in antiparallel arrangement.
(Redrawn from H. D. Springall, The Structural Chemistry of Proteins, p. 64, Academic
Press Inc., New York, 1954.)*

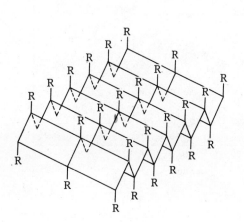

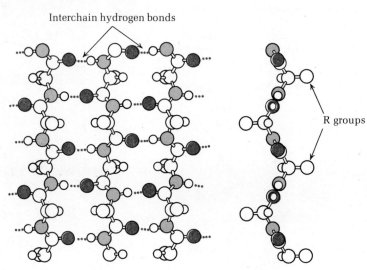

Interchain hydrogen bonds

R groups

Top view   Edge view

stretching of the relatively tight α helix into a more extended, zigzag conformation of the polypeptide chain,
characteristic of β-keratins generally, which they designated
the *β conformation*. Side-by-side polypeptide chains in the β
conformation are arranged in *pleated sheets*, which are
cross-linked by interchain hydrogen bonds (Figure 6-10). All
the peptide linkages participate in this cross-linking and
thus lend the structure great stability; the R groups lie above
or below the zigzagging planes of the pleated sheet. This is
the type of secondary structure found in fibroin secreted by
the silkworm *Bombyx mori*. In most types of fibroin every
other amino acid is glycine, so that all the R groups on one
side of the pleated sheet are hydrogen atoms. Since alanine
makes up most of the rest of the amino acids of fibroin, most
of the R groups on the other side of the sheet are methyl
groups. Fibroin and other β-keratins are rich in amino acids
having relatively small R groups, particularly glycine and
alanine. If the R groups are bulky or have like charges, the
pleated sheet cannot exist because of R-group interactions.
This is why the stretched form of α-keratins is unstable and
reverts spontaneously to the α-helical form; the R groups of
α-keratins are bulkier and more highly charged than those of
silk fibroin (see Table 5-2, page 101).

There are two other differences between α-keratins and
native β-keratins. In the α forms all the polypeptide chains
are *parallel*, i.e., run in the same N-terminal to C-terminal
direction, whereas in fibroin, the adjacent polypeptide
chains are *antiparallel*, i.e., run in opposite directions. Also,

Figure 6-11
Electron micrograph of collagen fibrils of connective tissue. Note periodicity of cross striations, which have a repeat distance of 70 nm.

K. R. Porter

1.0 μm

Figure 6-12
Conformation of polypeptide chains in triple-stranded tropocollagen molecule. Each chain is a coil with many repeating sequences of Gly-X-Y.

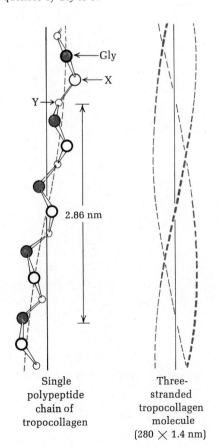

Gly

X

Y

2.86 nm

Single polypeptide chain of tropocollagen

Three-stranded tropocollagen molecule
(280 × 1.4 nm)

α-keratin contains many cystine residues so arranged as to provide interchain —S—S— cross-linkages between adjacent polypeptide chains. In contrast, the β-keratins, such as fibroin, have no —S—S— cross-linkages.

If we return to the Ramachandran plot (Figure 6-9), we note that the ϕ and ψ angles of the peptide bonds of both the parallel and antiparallel pleated sheets fall within the area of allowed structures.

Collagen

Another major type of fibrous protein in higher animals, the collagen of connective tissues, is the most abundant of all proteins in higher vertebrates, making up one-third or more of the total body protein. The larger and heavier the animal, the greater the fraction of its total proteins contributed by collagen. It has been aptly said, for example, that a cow is largely held together by the collagen fibrils in its hide, tendons, bones, and other connective tissues. Collagen fibrils are arranged in different ways, depending on the biological function of the particular type of connective tissue. In tendons collagen fibers are arranged in parallel bundles to yield structures of great strength but little or no capacity to stretch. In the hide of the cow the collagen fibrils form an interlacing network laid down in sheets. The organic material of the cornea of the eye is almost pure collagen. Whatever the arrangement of collagen fibrils in connective tissue, the fibrils always show a characteristic cross-striated appearance under the electron microscope (Figure 6-11), in which the repeat distance is between 60 and 70 nm, depending on the type of collagen and species of organism. Boiling in water converts collagen into gelatin, a mixture of polypeptides.

Although collagens of different species differ somewhat in amino acid sequence, most contain about 35 percent glycine and 11 percent alanine, resembling the β-keratins in this respect. Collagens are distinctive in containing about 12 percent proline (page 72) and 9 percent hydroxyproline (page 76), an amino acid rarely found in proteins other than collagen.

Collagens also have a distinctive x-ray diffraction pattern, different from those of α- and β-keratins. From comparisons of the x-ray patterns of collagen and of polyproline (page 119) it has been deduced that the secondary structure of collagen is that of a triple helix of polypeptide chains (Figure 6-12). Each of the chains is a left-handed three-residue helix; the chains are held together by hydrogen bonds. The frequent proline residues determine the distinctive type of helical arrangement of the chain, whereas the smaller R groups of the glycine residues, which occur in every third position, allow the chains to intertwine. The complete amino acid sequence of the collagen chains is not yet known, but -Gly-X-Pro-, -Gly-Pro-X-, and -Gly-X-Hyp- are frequently occurring sequences, in which X may be any amino acid. No proteins other than the collagens appear to contain similar triple-helical chains.

Collagen is built of recurring subunit structures, triple-stranded _tropocollagen_ molecules, having distinctive "heads." These subunits are arranged head to tail in many parallel bundles, but the heads are staggered (Figure 6-13), thus accounting for the characteristic 60- to 70-nm spacing of the repeat units in collagen fibrils from different species. The polypeptide chains of tropocollagen are covalently cross-linked by _dehydrolysinonorleucine_ residues (Figure 6-13), formed by an enzymatic reaction between two lysine residues of adjacent tropocollagen subunits.

The secondary structure of the polypeptide chains in other fibrous proteins is not yet known. Studies are under way on _elastin_ of the elastic connective tissue of ligaments and on _sclerotin_, the structural protein of the light, rigid exoskeleton of insects. Elastin is especially interesting since its polypeptide chains are covalently connected to form a stretchy, two-dimensional sheet resembling a trampoline net. The polypeptide chains are joined through covalent attachment to residues of _desmosine_ and _isodesmosine_ (page 76). Another structural protein of great interest, _resilin_, found in the wing hinges of some insects, is remarkable for its perfectly reversible elastic properties.

Tertiary Structure of Globular Proteins

We now turn from the fibrous proteins, which have relatively simple structures, to the far more complex globular proteins, which have polypeptide chains tightly folded into compact three-dimensional structures with many different kinds of specialized biological activities.

Until x-ray analysis of crystalline globular proteins became feasible, next to nothing could be learned about how their polypeptide chains are folded in three dimensions. In fact, only the barest outlines of the shape of globular proteins can be deduced from other physical methods, e.g., measurements of viscosity, sedimentation, and diffusion, which allow calculation of the axial ratio of protein molecules but can give no information on their internal structure.

Interpretation of the x-ray diffraction patterns is far more difficult for globular than for fibrous proteins because the polypeptide chains of globular proteins are not arranged along one axis but are irregularly and compactly folded into nearly spherical shapes. However, the introduction of intensely diffracting, electron-dense heavy-metal atoms into the molecules of globular proteins to provide reference points for the mathematical interpretation of the diffraction patterns has made it possible to determine the three-dimensional structures of a number of globular proteins to a resolution of 0.6 nm, and in some cases 0.2 nm. Among the globular proteins whose tertiary structures are now well known are myoglobin, hemoglobin, lysozyme, ribonuclease, chymotrypsin, carboxypeptidase A, cytochrome _c_, lactate dehydrogenase, and subtilisin, a proteolytic enzyme from a bacterium. Although the conformations of only a few proteins are known in detail, the results have already yielded some

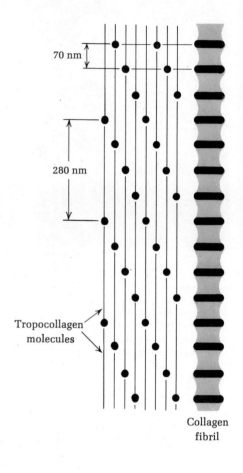

Figure 6-13
The staggered alignment of tropocollagen molecules, which possess heads, is responsible for the 70-nm repeat units in hydrated collagen fibers. Below is shown the structure of dehydrolysinonorleucine, which forms a covalent cross-link between parallel collagen chains.

70 nm

280 nm

Tropocollagen molecules

Collagen fibril

Dehydrolysinonorleucine

important generalizations that are probably applicable to many globular proteins.

Myoglobin

The first important breakthrough came from the x-ray studies by J. C. Kendrew and his colleagues in England of sperm whale myoglobin. Myoglobin is a relatively small globular protein containing a single polypeptide chain of 153 amino acid residues, whose sequence is now known. It contains an iron-porphyrin, or heme, group (page 490) identical with that of hemoglobin; myoglobin is also capable of reversible oxygenation and deoxygenation. It is, in fact, a functional and structural relative of hemoglobin, which contains four polypeptide chains and four hemes and has four times the molecular weight of myoglobin. Myoglobin is found in skeletal-muscle cells and is particularly abundant in diving mammals like the whale, seal, and walrus, whose muscles are so rich in myoglobin that they are deep brown. Myoglobin not only stores oxygen but also enhances its rate of diffusion through the cell.

The x-ray diffraction pattern of crystalline myoglobin, which contains about 2,500 atoms, consists of nearly 25,000 reflections. Analysis of the structure of myoglobin took place in two stages. In the first, completed in 1957, the results were calculated to 0.6 nm resolution, a feat that required precise analysis of 400 of the diffraction spots (Figure 6-14). This degree of resolution was insufficient to reveal the exact positions of individual atoms, but it did indicate how the polypeptide chain backbone is folded in the myoglobin molecule. In the second stage the x-ray analysis of myoglobin was carried out to 0.2 nm resolution, requiring the analysis of some 10,000 reflections. High-speed electronic computing methods were used. This level of resolution was sufficiently high to identify the sequence of most of the R groups, which agreed with the amino acid sequence determined by chemical methods. The backbone of the myoglobin molecule consists of eight relatively straight segments set off by bends (Figure 6-14). Each straight segment is a length of α helix, the longest consisting of 23 amino acids and the shortest only 7; all are right-handed. Some 70 percent of the amino acids in the molecule are in these straight α-helical regions; this figure confirmed the results of optical-rotation measurements on myoglobin solutions (page 131). Although the three-dimensional structure of myoglobin appears irregular and asymmetric, it is not at all random. All myoglobin molecules have the same conformation; otherwise myoglobin could not crystallize and yield reproducible x-ray diffraction patterns.

Some other important features of the myoglobin structure were found:

1 The molecule is very compact; in its interior there is room for only four molecules of water.
2 All the polar R groups of the amino acid residues are located on the outer surface of the molecule and are hydrated.

Figure 6-14
The structure of myoglobin.

X-ray diffraction pattern of myoglobin (sperm whale). [From J. C. Kendrew, Scientific American, December 1961.]

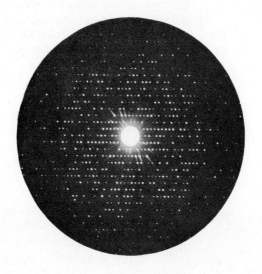

Backbone structure of myoglobin deduced from high-resolution (0.2 nm) x-ray data. [Redrawn from R. E. Dickerson in H. Neurath (ed.), The Proteins, vol. 2, p. 634, Academic Press Inc., New York, 1964.]

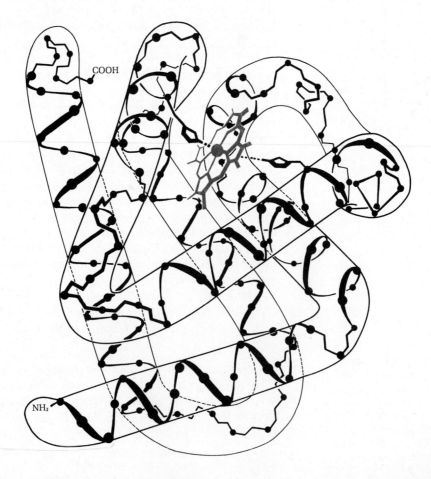

3 Nearly all the nonpolar or hydrophobic R groups are in the interior of the molecule, hidden from exposure to water.

4 Proline residues occur only at the bends, which also contain some amino acids known not to form α-helical coils readily, such as isoleucine and serine.

5 The gross conformation of the polypeptide chain is apparently identical in myoglobins of all species examined although they differ somewhat in amino acid composition. Thus the conserved or invariant (page 112) residues in the sequence may be involved in determining the position of the bends and the directions of the straight segments.

The Structures of Other Globular Proteins

X-ray analysis of the structure of a number of other globular proteins shows that each type of protein has a distinctive folding pattern or tertiary structure. All globular proteins examined to date share three common denominators with myoglobin: (1) compact folding, with little or no internal space for water molecules, (2) external location of nearly all the hydrophilic R groups, and (3) internal location of half or more of the hydrophobic R groups. Although over 70 percent of the polypeptide chain of myoglobin is in the form of α-helical segments, this is an unusually high fraction. Most globular proteins examined so far have much less α-helical content. Some contain segments in β conformation, which is absent in myoglobin.

The distinctive structural features of a few other globular proteins can be outlined. Lysozyme is an enzyme in egg white that catalyzes the hydrolysis of certain glycosidic linkages in the complex polysaccharides found in the cell walls of some bacteria (page 268). It has 129 residues in a single chain and four intrachain cystine cross-links. Only about 25 percent of its residues are in the form of α-helical regions. On the other hand, lysozyme resembles myoglobin in being compactly folded, with nearly all its polar R groups on the outside of the molecule and nearly all the hydrophobic R groups on the inside. Lysozyme has a large crevice in its surface, which has been identified as its active site, into which the substrate molecule fits (page 232). This crevice is lined with α-helical coils. The role of the three-dimensional structure of lysozyme in its catalytic activity will be considered in Chapter 9 (page 231).

Cytochrome c has also been studied at high resolution (Figure 6-15). Although it is a heme protein, it shows little or no structural similarity to myoglobin. The cytochrome c molecule has a few segments of α-helix and a number of regions in which the polypeptide chain has an extended conformation. The latter segments are wrapped and packed around the heme group, which is nearly buried in a crevice. This structure gives the heme group a hydrophobic environment, except where one edge of the heme is exposed to the external medium.

Ribonuclease also has a long, shallow crevice, which appears to be the active site into which the long substrate molecule fits. Ribonuclease also has relatively little α helix but a considerable amount of β structure. X-ray analysis

Figure 6-15
Conformation of the polypeptide chain backbone of cytochrome c with the heme shown in color. [From R. E. Dickerson, "The Structure and History of an Ancient Protein," Scientific American, April 1972. Drawing by Irving Geis. Copyright 1972 by Dickerson and Geis.]

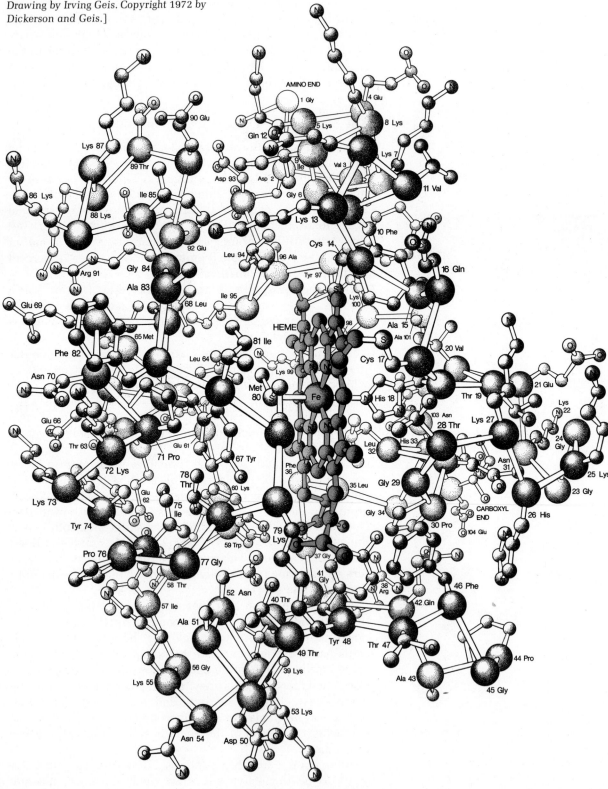

shows chymotrypsin to be a roughly spherical molecule with many parallel loops of the backbone in β conformation; only about 18 of its 241 residues are in α-helical segments.

Most of the globular proteins whose structure has been determined by the x-ray method are enzymes. Until some representative proteins with other biological functions have been analyzed it is premature to generalize further regarding the tertiary structure of globular proteins, but we may ask: Is the three-dimensional structure of a protein molecule revealed by x-ray diffraction analysis of protein crystals an accurate picture of the structure of that protein in aqueous solution, in which it must perform its biological functions?

In one approach taken to answer this question, F. M. Richards and his colleagues found that when the individual molecules in crystals of ribonuclease are prevented from going into solution by covalently cross-linking them into the crystal lattice, they still retain catalytic activity, indicating that the crystallization has not significantly distorted the molecule into an inactive conformation. Similarly, chymotrypsin and carboxypeptidase have been found to be catalytically active in the crystalline state. From this and other approaches it appears likely that globular proteins have the same general conformation in solution as in the crystalline state, but other experimental evidence (see page 150) clearly shows that some proteins, particularly enzymes, exhibit a certain amount of flexibility in their normal function in aqueous solution. Thus protein molecules in solution are not entirely rigid, but may "breathe" a little.

Specification of the Tertiary Structure of Globular Proteins by Their Amino Acid Sequence

Native globular proteins undergo denaturation to yield unfolded, random conformations of their polypeptide chains on heating, on treatment with acids or bases, or on exposure to strong solutions of urea or guanidine hydrochloride (page 150). Although this change is accompanied by loss of their biological activity, some denatured proteins spontaneously recover their biological activity and thus their original native conformation, sometimes very rapidly. Classical experiments carried out by F. White and C. B. Anfinsen and their colleagues on ribonuclease first showed the importance of amino acid sequence in the determination of native conformation. Treatment of native ribonuclease with 8 M urea in the presence of the reducing agent β-mercaptoethanol (page 87) caused complete unfolding of the ribonuclease molecule, to yield a random form (Figure 6-16). In this process the four intrachain disulfide bridges contributed by the cystine residues of ribonuclease were cleaved by the β-mercaptoethanol, converting them into eight cysteine residues. The combined unfolding and cleavage of the cross-links caused complete loss of enzymatic activity, but when the urea and β-mercaptoethanol were slowly removed from the ribonuclease solution by dialysis (page 158), the enzymatic activity of the ribonuclease gradually returned, indicating

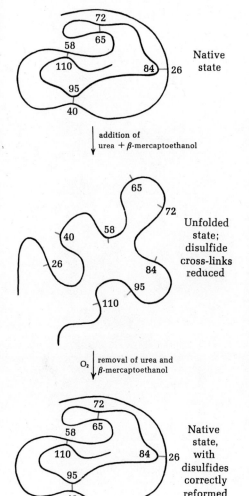

Figure 6-16
Renaturation of unfolded, reduced ribonuclease, with reestablishment of correct disulfide cross-links.

Native state

addition of urea + β-mercaptoethanol

Unfolded state; disulfide cross-links reduced

O₂ | removal of urea and β-mercaptoethanol

Native state, with disulfides correctly reformed

that even after complete unfolding, the polypeptide chain of ribonuclease still contains the information necessary to fold spontaneously back into its catalytically active tertiary structure. In this process the eight cysteine residues became reoxidized by atmospheric oxygen to reestablish four disulfide cross bridges. What is particularly significant is that these four cystine cross bridges were the "correct" ones, involving the same pairs of cysteine residues as the native molecule. Probability calculations show that on a random basis eight cysteine residues in a single polypeptide chain can form 105 sets of four different disulfide pairs. Yet only the unique set present in the native ribonuclease molecule was formed, indicating that the amino acid sequence of ribonuclease accurately and precisely determines the arrangement in space of the polypeptide chain in the native molecule as well as the exact fit of the proper —SH groups to yield the correct disulfide cross-links.

Some proteins lacking disulfide cross-links can also refold into the native, active configuration spontaneously and quickly after denaturation. Anfinsen and his colleagues have shown that a nuclease from *Staphylococcus* cells loses all its biological activity and its native conformation when acidified to pH 3.0; when the pH is restored to 7, the activity is quickly regained. From these and many similar experiments on renaturation of other globular proteins, it is now quite firmly established that the amino acid sequence specifies the distinctive tertiary structure of globular proteins, which is a reflection of different kinds of constraints upon the freedom of rotation around the single bonds of the polypeptide chain. To summarize, these constraints include the rigid, planar nature of the peptide bonds, the allowed ϕ and ψ angles of the C_α—C and C_α—N bonds of successive residues, the number and location of hydrophobic and hydrophilic residues in the sequence, and the number and location of positive and negatively charged R groups. The conformation of the polypeptide chain thus results from the adjustment of each single bond in the backbone to various local and long-range constraints, yielding a single characteristic tertiary conformation having a specific biological activity.

The Stabilization of Tertiary Structure of Globular Proteins

Once the native tertiary structure of a globular protein has formed, four major types of weak interactions or bonds cooperate in stabilizing it: (1) hydrogen bonds between peptide groups, as in α-helical or β-pleated sheets; (2) hydrogen bonds between R groups; (3) hydrophobic interactions (page 43) between nonpolar R groups; and (4) ionic bonds between positively charged and negatively charged groups, such as the —COO$^-$ of aspartate or glutamate R groups and the —NH$_3^+$ of lysine R groups. From studies on the relative contribution of each of these four types of weak bond to the total conformational stability of native protein molecules it is now clear that hydrophobic interactions between the nonpolar R groups are by far the most important. Most proteins contain from 30 to 50 percent of amino acids with nonpolar R groups (page 72); x-ray analysis shows that nearly all these R groups are in the interior of native globular proteins (page 139), shielded from exposure to water.

To fully understand the important role of hydrophobic interactions in stabilizing protein structure, we must first ask a fundamental question: Why does a denatured, randomly coiled polypeptide chain tend to fold spontaneously into a highly ordered, biologically active conformation, a process that apparently *decreases* the entropy of the polypeptide chain? Is protein folding a violation of the second law of thermodynamics (page 390), which states that all processes proceed in that direction which maximizes entropy, or randomness? The answer to this dilemma is found in a balance of forces. One force is the tendency of the polypeptide chain to seek its own conformation of maximum randomness or entropy. The opposing force is the tendency of the surrounding water molecules to seek *their* position of maximum randomness or entropy. The critical factor in this balance of forces is represented by the nonpolar R groups. When nonpolar groups are inserted into water, a new interface is created, which requires the adjacent water molecules to assume a more ordered arrangement than they would have in pure liquid water; thus *input* of energy is required to force a nonpolar R group into water. A random polypeptide chain, with its nonpolar R groups exposed, will thus tend to assume a conformation in which the nonpolar R groups are shielded from exposure to water. It is the tendency of the surrounding water molecules to relax into their maximum-entropy state that brings about the transition of the polypeptide chain from a random unfolded state to a highly ordered tertiary conformation. At equilibrium, when the random chain is fully folded, the *increase* in the entropy of the surrounding water molecules is greater than the *decrease* in the entropy of the now correctly coiled polypeptide chain. The second law has not been violated because the combination of the system (the polypeptide) and the surroundings (the water) has undergone a net *increase* in entropy.

However, much evidence suggests that the folded, native conformation is more stable than the unfolded, or denatured, conformation by only a relatively small margin. The stability of a native globular protein is thus the result of a delicate balance between two relatively massive and opposing forces: (1) the tendency of the polypeptide chain to unfold into a more random arrangement and (2) the tendency of the surrounding water molecules to seek their most random state.

We have assumed in this discussion that the native conformation of a globular protein is more stable, i.e., has less free energy, than the random-coil form under biological conditions, but this assumption may not be true for all proteins; it is the subject of much debate and study. Proteins that spontaneously refold into their native form may indeed be more stable than their denatured forms under specific conditions of pH, ionic strength, and temperature (Figure 6-17). On the other hand, the unfolded form of some proteins may have less free energy than the native form. In such cases the

Figure 6-17
Relative free energy of native and unfolded (denatured) polypeptide chains. (Top) A protein whose native conformation has a lower energy content than the unfolded form; the unfolded polypeptide refolds very quickly and spontaneously if the activation-energy barrier is low. (Bottom) A protein whose denatured form has a lower energy content than the native form. In this case the native form is stable under biological conditions because of the large activation-energy barrier for unfolding, which is overcome only by heat or other agencies.

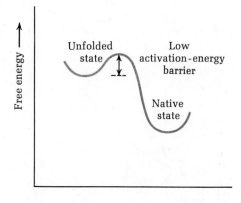

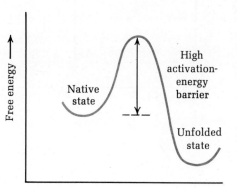

transition from the native to the random state may have a very high activation-energy barrier, thus locking the polypeptide chain into its native conformation. In this case, once the native form is unfolded, the polypeptide chain will not spontaneously refold into the native conformation.

These considerations have great biological significance. When proteins are synthesized from their constituent amino acids by living cells, they are made residue by residue starting from the N-terminal end (page 932). However, we know very little about the mechanism and characteristics of the process by which newly forming polypeptide chains undergo folding into their specific biologically active conformations.

The Kinetics of Protein Folding

A particularly critical aspect of the folding of polypeptide chains into their native conformation is the very high rate at which it must occur biologically. If we assume that a protein molecule has n amino acid residues, that each residue has 2 bonds capable of rotation, and that there are 3 possible conformations (ϕ or ψ angles) for each rotatable bond in the backbone, the maximum number of possible conformations is 3^{2n}, which is approximately equal to 10^n. Since each single bond can rotate completely in about 10^{-13} s, the total time required for *every* formal single bond in the backbone to rotate once is about 2×10^{-13} s. Therefore the time required for a polypeptide chain to try out every possible conformation it can assume is $t = 10^n(2n \times 10^{-13})$. For a polypeptide chain of six residues t is in the range of microseconds, for a chain of 11 residues, about 0.2 s, but for a chain of 100 amino acid residues it would be about 2×10^{89} s, or longer than the age of the earth. Yet staphylococcal nuclease, which has 149 residues, requires at most 0.1 to 0.2 s to regain its native, enzymatically active conformation after complete denaturation with acid. Actually, it is a biological necessity that protein molecules fold into their native conformations very rapidly and through some kind of favored route, since the biosynthesis of proteins in cells is very fast, occurring at a rate exceeding one amino acid residue per second.

How is it possible, then, for a long polypeptide chain to fold so quickly into its native conformation, without trying out *all* its possible conformations? Although the full answer is by no means known, it appears likely that this process is aided by the principle of *cooperativity* (page 42). That is, once a few weak bonds (hydrogen bonds or hydrophobic interactions) have correctly formed in part of the polypeptide chain, they greatly increase the probability of the formation of further correct bonds without requiring the chain to try out all possible conformations (page 42). It is as though the randomly coiled polypeptide underwent a nucleation process resembling that in a supersaturated solution of a salt, which crystallizes instantly when it is nucleated, or seeded, with a tiny crystal. What is not known is whether the folding process starts at one end of the chain or the other or in the middle. Thermal unfolding of proteins shows rather sharp "melting" curves (Figure 6-18), which suggest that a polypep-

Figure 6-18
Thermal "melting" curve of a protein. The steep transition to the unfolded form over a very short temperature interval indicates that unfolding or denaturation is a cooperative process, in which rupture of the first few weak bonds greatly increases the probability that the remainder rupture.

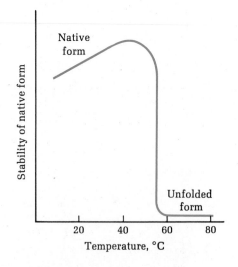

tide chain exists in only two states, completely folded or completely unfolded. However, recent studies of protein unfolding over extremely short time intervals indicate that intermediate conformational states may form in some proteins. Perhaps these will reveal how unfolding and folding of polypeptide chains is initiated and propagated under biological conditions.

The Quaternary Structure of Oligomeric Proteins

We have seen that some globular proteins are *oligomeric*, i.e., contain two or more separate polypeptide chains or subunits (Table 3-2, page 59). The *quaternary structure* (page 60) designates the characteristic manner in which the individual, folded polypeptide chains fit each other in the native conformation of an oligomeric protein. Among the simplest oligomeric proteins is hemoglobin, which has four polypeptide chains (page 110).

Because oligomeric proteins have relatively high molecular weights, and because they contain multiple chains, each of which may have a characteristic conformation, their three-dimensional conformation is far more difficult to analyze by x-ray methods than that of single-chain proteins.

Hemoglobin

Hemoglobin was the first oligomeric protein for which the complete tertiary and quaternary structure became known from x-ray analysis. This achievement, accomplished by M. F. Perutz and his colleagues in England, culminated some 25 years of detailed study of the structure of this important protein. Because of the similarity of function and the homology of amino acid sequence of the polypeptide chains of myoglobin and hemoglobin (page 111), a number of extremely important relationships have developed from concurrent investigations of the structure of these two proteins, which were carried out in the same laboratory.

Hemoglobin contains two α chains (141 residues) and two β chains (146 residues), to each of which is bound a heme residue in noncovalent linkage. The molecule was examined in its oxygenated form, which has a compact spheroidal structure of dimensions 6.4 by 5.5 by 5.0 nm. Figure 6-19 shows the low-resolution outlines of the hemoglobin chains, and Figure 6-20 shows how the chains fit together in an approximately tetrahedral arrangement. Each chain has an irregularly folded conformation, in which lengths of pure α-helical regions are separated by bends. Both the α and β chains have about 70 percent α-helical character, as is true for myoglobin. The α and β chains are very similar to each other in their tertiary structure, which consists of similar lengths of α helix with bends of about the same angles and directions. But most remarkable is that the tertiary structure of the α and β chains is very similar to that of the single chain of myoglobin, consonant with the similar biological

Figure 6-19
Tertiary conformation of the α and β chains of hemoglobin. [From A. F. Cullis, H. Muirhead, A. C. T. North, M. F. Perutz, and M. G. Rossmann, Proc. R. Soc. Lond., A265: 161 (1962).]

α chains

β chains

function of these two proteins, namely, their capacity to bind oxygen reversibly, myoglobin in muscle and hemoglobin in blood.

In hemoglobin there is very little contact between the two α chains and between the two β chains, but there are numerous R-group contacts between the pairs of unlike chains. Of special interest is the location of the four heme groups, one in each subunit, that bind the four molecules of oxygen. These heme groups, flat molecules in which the iron atoms form square-planar coordination complexes, are quite far apart from each other and are situated at different angles from each other (Figure 6-20). Each is partially buried in a pocket lined with nonpolar R groups. The fifth coordination bond of each iron atom is to an imidazole nitrogen of a histidine residue; the sixth position is available for coordination with an oxygen molecule. There is a central cavity within the hemoglobin molecule, lined with polar R groups.

The amino acid sequences of hemoglobin chains of many species have been compared. Although only nine of the residues in each chain are absolutely invariant, the amino acid replacements in many other positions suggest that the polypeptide chain subunits of the hemoglobins from nearly all species have the same tertiary structure. Moreover, in nearly all hemoglobins a histidine R group coordinates with the iron atom of the heme group (page 490).

The quarternary conformation of other oligomeric proteins has now been established, in particular the enzyme lactate dehydrogenase, which also has four polypeptide chains. Its structure and that of aspartate transcarbamoylase, which has 12 chains, are discussed elsewhere (pages 483 and 237).

Oxygenation and Deoxygenation of Hemoglobin

X-ray analysis of deoxyhemoglobin shows that the tertiary structure of the four subunit chains is identical to that of oxyhemoglobin, but there is a significant change in the quaternary structure, i.e., in how the chains are oriented in relation to each other. On deoxygenation the α chains undergo a rotation of some 9° and the β chains of about 7°, but about different axes, causing a change in the contact points between the four subunits and the formation of new ionic bonds between them. The two α hemes come closer together by 0.1 nm and the two β hemes separate by about 0.65 nm. Thus the binding of four oxygen molecules, each having only a relatively small diameter, can cause a profound change in the quaternary structure of hemoglobin, the significance of which will now be considered.

The Binding Equilibria of Hemoglobin

Considerable importance attaches to the characteristic manner in which the hemoglobin molecule binds oxygen in its function as a carrier of oxygen in the erythrocyte, from the oxygen-rich gas phase of the lungs to the oxygen-poor peripheral tissues.

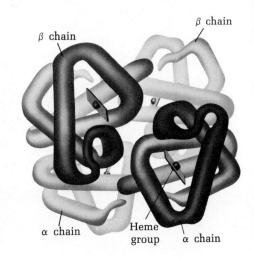

Figure 6-20
The quaternary structure of hemoglobin. [*Adapted with permission from R. E. Dickerson and I. Geis,* The Structure and Action of Proteins, *published by W. A. Benjamin, Inc., Menlo Park, California. Copyright 1969 by Dickerson and Geis. Coordinates courtesy of M. F. Perutz, Cambridge, England.*]

Hemoglobin constitutes about 90 percent of the total protein in erythrocytes, where it is concentrated in the cytoplasm. Because of the ability of the hemoglobin to bind oxygen, whole blood can absorb some 21 ml of gaseous oxygen per 100 ml, whereas blood plasma alone can absorb only about 0.3 ml of oxygen by physical solution. A plot of percent saturation of hemoglobin against oxygen partial pressure is sigmoidal (Figure 6-21), whereas a similar plot for myoglobin of muscle, which has but a single heme group and a single polypeptide chain, shows a hyperbolic curve. The sigmoidal oxygen binding curve of hemoglobin means that hemoglobin has relatively low affinity for binding the first one or two oxygen molecules but once they are bound, the binding of subsequent oxygen molecules is greatly enhanced. Conversely, the loss of one oxygen molecule from fully oxygenated hemoglobin causes the rest to dissociate more readily when the oxygen pressure is decreased.

The position of the hemoglobin-oxygen equilibrium is also affected by the pH (Figure 6-21). The higher the pH of the hemoglobin solution at a given partial pressure of oxygen, the greater the percent saturation with oxygen. This reversible effect results because when hemoglobin is oxygenated, it ionizes, to set free one proton for each oxygen bound, according to the equation

$$HHb^+ + O_2 \rightleftharpoons HbO_2 + H^+$$

in which HHb^+ is a protonated subunit of a deoxyhemoglobin molecule. Since this reaction is freely reversible, increasing the hydrogen-ion concentration will cause the equilibrium to shift to the left, toward decreased saturation, whereas decreasing the hydrogen-ion concentration will cause the equilibrium to shift to the right, toward increased saturation. This effect of pH on the oxygen-hemoglobin equilibrium is called the _Bohr effect_.

The partial pressure of oxygen and the pH are the two most important factors regulating the function of hemoglobin in the transport of oxygen. In the lungs, where the partial pressure of oxygen is high (about 100 mm Hg) and the pH also relatively high, hemoglobin tends to become almost maximally saturated with oxygen, about 96 percent (see Figure 6-21). In the interior of the peripheral tissues, where the oxygen tension is low (about 45 mm Hg) and the pH also low (due to the high concentration of CO_2 formed as the end product of respiration), the hemoglobin binds oxygen less strongly and will thus unload some of its oxygen to the respiring cell mass, until the hemoglobin is only about 65 percent saturated. Hemoglobin thus cycles between 65 and 96 percent saturation with oxygen in its function as oxygen carrier.

The sigmoid binding curve for oxygen has been the subject of much research and speculation, since it evidently reflects a biological adaptation of the hemoglobin molecule that allows it to function with maximum molecular efficiency. The equilibria involved in the stepwise oxygenation

Figure 6-21

The oxygen-saturation curves of hemoglobin and myoglobin. The sigmoid saturation curve of hemoglobin and its response to change in pH allow hemoglobin to become nearly fully saturated in the lungs (partial pressure 100 mm Hg and pH 7.4) and to unload over 30 percent of its oxygen in the tissues (partial pressure 45 mm Hg and pH 7.2). Myoglobin, whose oxygen-saturation curve is a rectangular hyperbola, would allow only about 2 to 3 percent unloading of oxygen under the same conditions. In human erythrocytes the metabolite 2,3-diphosphoglycerate is tightly bound by hemoglobin but only slightly by oxyhemoglobin. This ligand has the effect of further facilitating the unloading of oxygen, by lowering the affinity of hemoglobin for oxygen. 2,3-Diphosphoglycerate is thus a modulator of the activity of hemoglobin in transporting oxygen.

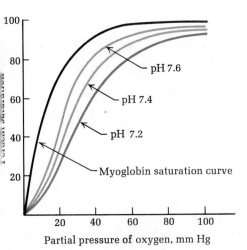

Partial pressure of oxygen, mm Hg

of hemoglobin are shown in the following reactions (for simplicity, the proton equilibria are not included).

$$Hb + O_2 \rightleftharpoons HbO_2$$

$$HbO_2 + O_2 \rightleftharpoons Hb(O_2)_2$$

$$Hb(O_2)_2 + O_2 \rightleftharpoons Hb(O_2)_3$$

$$Hb(O_2)_3 + O_2 \rightleftharpoons Hb(O_2)_4$$

The problem is to explain how the subunits in the hemoglobin molecule are coupled or linked so that oxygenation of one of them is signaled to the others and can increase their oxygen affinity. Two general models accounting for the subunit interactions during the oxygenation of hemoglobin have been proposed (Figure 6-22). Both assume that each subunit of the hemoglobin molecule can exist in two different conformations, one having a high affinity and the other a lower affinity for oxygen. The first model, suggested by G. Adair many years ago but recently refined and restated by D. E. Koshland, Jr., and his colleagues, is called the _sequential model_ (Figure 6-22). As an oxygen molecule binds to a given subunit, the subunit changes its conformation to the high-affinity form, which, because of its molecular contacts with its neighbors, increases the probability that the next subunit will switch to the high-affinity form and bind the next oxygen atom, and so on, until all four are bound and all four subunits are in the high-affinity conformation. The hallmark of the sequential model is that the molecule passes through a series of _intermediate_ conformational states.

The other general model, the _symmetry model_ (Figure 6-22) proposed by J. Monod, J. Wyman, and J.-P. Changeux, postulates that the hemoglobin molecule exists in only two forms, one with all subunits in the low-affinity form and one with all in the high-affinity form. In this model the hemoglobin molecule is always symmetrical, i.e., all the subunits are either in one state or the other.

Although both models can account for the sigmoidal oxygenation curve of hemoglobin, the evidence at present favors the second, or symmetry, model, particularly the x-ray evidence of the difference in structure between the oxygenated and deoxygenated forms. Perutz has postulated that the first two oxygens bind to the α subunits, changing their ionic interactions with the β subunits and causing release of two of the Bohr-effect protons. All four subunits then switch to the high-affinity form. The remaining two subunits bind oxygen with high affinity and then release the other two Bohr protons.

Hemoglobin is a model or prototype of many oligomeric proteins that bind two or more ligand molecules to each of two or more polypeptide chain subunits. Among such oligomeric proteins are the group of regulatory enzymes known as _allosteric enzymes_, discussed in Chapter 9 (pages 234 to 241). The response of some allosteric enzymes to changes in their substrate concentration resembles the response of hemoglobin to changes in oxygen partial pressure; e.g., when

Sequential model

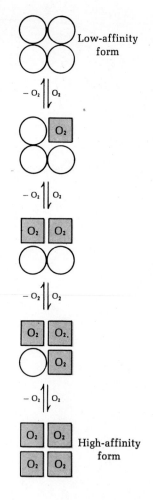

Symmetry model

In the absence of oxygen, there are two forms in equilibrium, with the equilibrium far in the direction of the low-affinity form.

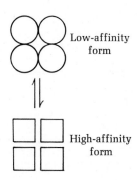

Low-affinity form

High-affinity form

When oxygen is present, it binds preferentially to the high-affinity form, pulling the equilibrium toward the latter.

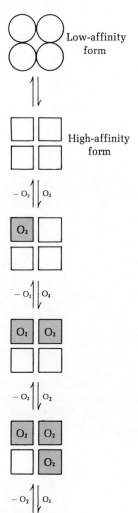

Low-affinity form

High-affinity form

substrate concentration is increased, a sigmoidal rise in enzyme activity may occur. Both the sequential and symmetry models (Figure 6-22) have been applied to the analysis of allosteric enzyme function; in many cases the sequential model appears more appropriate (page 240).

The Conformation of Sickle Hemoglobin

In sickle-cell hemoglobin (hemoglobin S) the glutamic acid residues in the 6 position of the β chains are replaced by valine, a change that causes the red blood cells in patients with sickle-cell anemia to become crescent- or sickle-shaped (Figure 5-19, page 28) at low oxygen tensions, whereas normal red blood cells retain their disklike shape on deoxygenation. These facts suggest that the conformation of the hemoglobin S molecule differs from that of normal hemoglobin A, in such a way that the packing of the hemoglobin S molecules within the sickle erythrocyte causes it to change shape on deoxygenation.

A. Cerami and J. M. Manning have found that treating erythrocytes from sickle-cell anemia patients with sodium cyanate in vitro largely prevents sickling of the cells when the oxygen tension is diminished. Such cells also survive longer and apparently function better than untreated sickle cells when returned to the patient. The treatment with cyanate results in the carbamoylation (page 85) of the free amino group in the N-terminal valine residues of hemoglobin S (Figure 6-23). This reaction has the effect of removing the positive electric charge normally present on this amino group. Thus it appears that alteration of but a single charged group in the hemoglobin molecule can "correct" its conformation so that the molecule is again oxygenated and deoxygenated normally without significant sickling. Moreover, cyanate appears to be very selective in this effect, causing no alteration of other amino groups in hemoglobins. This effect of cyanate is now being examined to see whether it may be useful in the clinical treatment of the crises that occur in sickle-cell anemia (page 115).

Figure 6-23
Formation of carbamoyl derivative of N-terminal valine residue of β chain of sickle-cell hemoglobin.

$$CH_3-CH-CH-\overset{\displaystyle O}{\overset{\|}{C}}- \quad + \quad H-N{=}C{=}O \quad \longrightarrow \quad CH_3-CH-CH-\overset{\displaystyle O}{\overset{\|}{C}}-$$

N-terminal valine residue   Isocyanic acid   Carbamoyl derivative of N-terminal valine

Dissociation and Denaturation of
Oligomeric Proteins

When proteins with two or more polypeptide chain subunits are subjected to agencies causing denaturation, e.g., high acidity, heat, or high concentrations of urea or guanidine, they undergo conformational changes to more random forms in two stages: (1) dissociation of the chains from each other and (2) unfolding of the separated chains into random coils. If the treatment is carried out very carefully, the polypeptide chains of an oligomeric protein can sometimes be separated from each other without damage to their normal tertiary structure. Hemoglobin, for example, can be dissociated by salts into two half molecules, i.e., two $\alpha\beta$ subunits. The separated subunits reassociate to reconstitute the fully active oligomeric protein when the excess salt is dialyzed away.

If, however, more drastic conditions are employed, the polypeptide-chain subunits will not only separate but also unfold to form random coils. Although it would be expected that fully denatured oligomeric proteins are less likely to refold spontaneously back into the native form than simple proteins having a single polypeptide chain, a number of oligomeric proteins are capable of completely reversible denaturation. For example, acidification of the enzyme aldolase (page 425) causes its four subunits to separate and to unfold; when the pH is raised to about 7.0, the subunits spontaneously refold into their native tertiary structures and then reassociate, to form the catalytically active oligomeric enzyme. These facts indicate that the amino acid sequences of the polypeptide chains in oligomeric proteins specify not only their secondary and tertiary structure but also the geometry of the contact sites that make possible the exact fit of the subunit polypeptide chains and their very tight quaternary association with each other. These features thus endow some oligomeric proteins, such as aldolase and hemoglobin, with the capacity for self-assembly.

Probes of Protein Conformation

Protein molecules often undergo changes in conformation during the course of their biological function, particularly the enzymes. Such changes can be detected by certain physical measurements. We have seen (page 131) that optical-rotation measurements on polypeptides can tell us whether they are present as random coils or as α-helical forms, since the α helix has intrinsic asymmetry. In another type of optical measurement, called *optical-rotatory dispersion*, the optical rotation of a protein molecule is scanned over a range of wavelengths of incident light. A typical optical-rotatory-dispersion plot of a native protein is shown in Figure 6-24; there is a deep valley at about 234 nm. Such a sharp change in the magnitude of the optical rotation as a function of the wavelength is known as a *Cotton effect*; most proteins show a Cotton effect at about 225 to 235 nm. The optical rotation of a protein at or near such inflection points is an extremely sensitive indicator of changes in its conformation. Another

Figure 6-24

Optical-rotatory-dispersion curve of a typical protein, showing Cotton effect at 234 nm. Cotton effects are also produced when proteins bind ligand molecules; usually they occur at longer wavelengths than the intrinsic Cotton effect shown here.

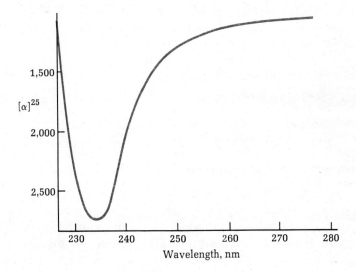

sensitive probe of conformational changes is measurement of the <u>circular dichroism</u> of proteins. When circularly or elliptically polarized light passes through a protein solution, the transmitted light may be plane-polarized by unequal absorption of the right and left-handed components of the incident light.

Another sensitive method of detecting changes in the three-dimensional conformation of a protein in solution is by measuring changes in fluorescence. When the three aromatic amino acids (phenylalanine, tyrosine, and tryptophan) are excited by visible or ultraviolet light, they fluoresce, i.e., emit light at a wavelength other than that of the incident light. The ratio of the fluorescent light emitted to the incident light absorbed, called the <u>quantum yield</u>, is exceedingly sensitive to short-range changes in the environment surrounding the R groups of the aromatic amino acids. In a related method, called <u>fluorescence polarization</u>, the protein is excited with plane-polarized light, and the degree of polarization of the emitted fluorescent light is measured, yielding valuable information on the size, shape, and rotational movement of the protein.

<u>Nuclear magnetic resonance (NMR) spectroscopy</u> has also become a very powerful tool for detecting which amino acid residues in a native protein molecule participate in conformational changes. The protein is placed in an oscillating magnetic field of varying strength. Some of the electromagnetic radiation is absorbed by atomic nuclei having nonzero angular-spin moments, particularly those of hydrogen and deuterium atoms. Since the protons of different types of functional groups ($-OH$, $=CH-$, $-CH_2-$, $-CH_3$) have different and characteristic magnetic resonances, changes in the local environment of such groups can be detected by NMR measurements. Of particular importance is a refinement of NMR spectroscopy which senses the angular momentum of the ^{13}C isotope, present in small amounts in the natural carbon of proteins. With this powerful method it is

possible to measure changes in the environment of the carbon atoms in peptide bonds, α carbon atoms, and unsaturated and aromatic R groups. With these new methods the specific portions of the polypeptide chain undergoing conformational changes can be detected.

Displays or drawings of the three-dimensional conformation of the polypeptide-chain backbone of protein molecules may be generated by electronic computers (Figure 6-25).

Prediction of Tertiary Structure

If the secondary and tertiary conformations of protein molecules are the reflection of residue-by-residue adjustment of the polypeptide chain to produce local minimum-free-energy states, it might seem theoretically possible to predict the complete tertiary structure of a given protein from its amino acid sequence, starting from such considerations as the constraints imposed by the ϕ and ψ angles at each peptide bond. Attempts are being made to carry out such calculations with the aid of electronic computers. For each amino acid residue the minimum-free-energy conformation must be calculated from data on electrostatic, van der Waals, and steric interactions of the adjacent R groups. Although it is not yet possible to predict complete protein conformations from amino acid sequence, because of the great complexity of the short-range and long-range factors that must be taken into account, significant advances are being made toward this end.

Summary

Each protein has at least one three-dimensional conformation in which it is stable and active under biological conditions of temperature and pH, the native conformation. X-ray analysis of the α-keratins, fibrous proteins of hair and wool, show that their polypeptide chains are right-handed α-helical coils, having 3.6 amino acid residues per turn, held together by maximal intrachain hydrogen bonding. Each peptide carbonyl oxygen atom in the α helix is hydrogen-bonded to the —NH— of the peptide bond three residues removed. The disulfide cross-linkages of cystine hold together the parallel α-helical coils in α-keratins. Silk fibroin, a β-keratin, has a pleated-sheet structure, in which the polypeptide backbones are in the extended or β conformation of the polypeptide chain and are cross-linked by interchain hydrogen bonds. The three chains of the fibrous protein collagen consist of three kinked, left-handed helixes, which intertwine to form the basic tropocollagen repeat unit of a collagen fibril. Collagen is rich in glycine, proline, and hydroxyproline.

X-ray analysis of globular proteins, such as myoglobin, shows that their chains are compactly folded, leaving little space in the interior for water molecules. All or nearly all the polar R groups of globular proteins are on the surface and are hydrated; the hydrophobic residues remain shielded inside. Depending on their amino acid sequence, globular proteins contain widely varying amounts of α helix or β conformation. Proline residues make bends in α-helical coils.

The conformation of the polypeptide-chain backbone is automatically determined by (1) the rigid planar peptide groups, (2) limitations upon the angles of rotation (the ϕ and ψ angles) about the

Figure 6-25
Computer representations of the backbone
structure of a polypeptide chain as the
model is rotated around its vertical axis.
[From C. Levinthal, Scientific American,
November 1966.]

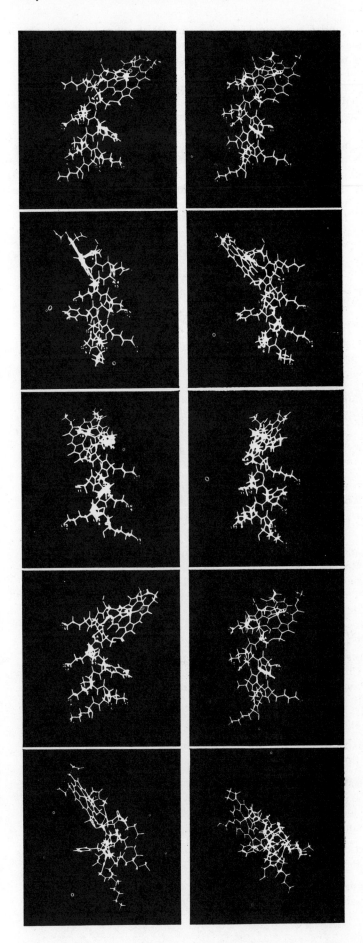

C_α—C and C_α—N single bonds, and (3) the size, polarity, and charge of the R groups. Some denatured or unfolded proteins spontaneously refold into their native conformation very quickly, in the face of the fact that the time required to try out all possible conformations is almost infinite. Cooperative interactions take place in spontaneous refolding.

The quaternary structure of oligomeric proteins, such as hemoglobin, is also determined by the primary amino acid sequence of the component polypeptide chains. Hemoglobin can bind four molecules of oxygen in a relationship indicating that the subunits interact cooperatively. Oxygen binding is accompanied by conformational changes in the subunits. Changes in conformational state with ligand binding occurs in many oligomeric proteins. It can be explained either by sequential transitions of successive subunits or by symmetrical all-or-none transitions of the whole oligomeric molecule.

References

Books

DICKERSON, R. E., and I. GEIS: *The Structure and Action of Proteins*, Benjamin, Menlo Park, Calif., 1969. With a stereo viewing supplement.

HASCHEMEYER, R. H., and A. E. V. HASCHEMEYER: *Proteins: A Guide to Study by Physical and Chemical Methods*, Wiley, New York, 1973.

NEURATH, H.: *The Proteins*, 2d ed., 4 vol., Academic, New York, 1964–1966.

TANFORD, C.: *The Hydrophobic Effect*, Wiley-Interscience, New York, 1973. A short, valuable book on the physical chemistry of hydrophobic interactions, stability of proteins, and formation of lipid micelles and membranes.

WILSON, H. R.: *Diffraction of X-rays by Proteins, Nucleic Acids and Viruses*, St. Martin's, New York, 1966. An elementary account of the x-ray technique.

Articles

ANFINSEN, C. B.: "Principles That Govern the Folding of Polypeptide Chains," *Science*, 181: 223–230 (1973). A Nobel prize lecture recapitulating the early history and recent progress.

BLOW, D. M., and T. A. STEITZ: "X-Ray Diffraction Studies of Enzymes," *Ann. Rev. Biochem.*, 39: 63–100 (1970).

CERAMI, A., J. M. MANNING, and others: "Effect of Cyanate on Red Blood Cell Sickling," *Fed. Proc.*, 32: 1668–1672 (1973).

DARNALL, D. W., and I. M. KLOTZ: "Subunit Constitution of Proteins: A Table," *Arch. Biochem. Biophys.*, 166: 651–682 (1975). Valuable compilation of the number and molecular weight of subunits of proteins.

DICKERSON, R. E.: "X-Ray Studies of Protein Mechanisms," *Ann. Rev. Biochem.*, 41: 815–842 (1972).

GALLUP, P. M., O. O. BLUMENFELD, and S. SEIFTER: "Structure and Metabolism of Connective Tissue Proteins," *Ann. Rev. Biochem.*, 41: 617–672 (1972).

HARTE, R. A., and J. A. RUPLEY: "Three-Dimensional Pictures of Molecular Models," *J. Biol. Chem.,* 243: 1664–1669 (1968). Three-dimensional "xographs" of lysozyme.

HEWITT, J. A., J. V. KILMARTIN, L. F. TEN EYCK, and M. F. PERUTZ: "Noncooperativity of the $\alpha\beta$ Dimer in the Reaction of Hemoglobin with Oxygen," *Proc. Natl. Acad. Sci. (U.S.),* 69: 203–207 (1972).

KENDREW, J. C.: "The Three-Dimensional Structure of a Protein Molecule," *Sci. Am.,* 205: 96–110 (December 1961).

KOSHLAND, D. E., JR., G. NEMETHY, and D. FILMER: "Comparison of Experimental Binding Data and Theoretical Models in Proteins Containing Subunits," *Biochemistry,* 5: 365–385 (1971).

LEVINTHAL, C.: "Molecular Model-Building by Computer," *Sci. Am.,* 214: 42–52 (June 1966). Computer representation of protein conformations.

RAMACHANDRAN, G. N., and V. SASISEKHARAN: "Conformation of Polypeptides and Proteins," *Adv. Protein Chem.,* 23: 283–437 (1968).

TANFORD, C.: "Protein Denaturation," *Adv. Protein Chem.,* 23: 121–282 (1968); 24: 1–95 (1970).

TANIUCHI, H., and A. N. SCHECHTER: "Chemical and Physical Factors Involved in Protein Folding as Exemplified by Staphylococcal Nuclease," *Panam. Assoc. Biochem. Soc. Rev.,* 1: 419–494 (1973).

WETLAUFER, D. B., and S. RISTOW: "Acquisition of Three-Dimensional Structure of Proteins," *Ann. Rev. Biochem.,* 42: 135–158 (1973).

Problems

1. Calculate the length (in nanometers) of a polypeptide chain containing 105 amino acid residues if (*a*) it exists entirely in α-helical form, or (*b*) if the backbone bonds are fully extended and linear.

2. In the following polypeptide, (*a*) which sections would you expect to have α-helical conformation at pH 7.0, (*b*) where might bend points occur, and (*c*) where might cross-linkages be formed?

   ```
    1    2    3    4    5    6    7    8    9   10   11   12   13   14
   Ile-Ala-His-Thr-Tyr-Gly-Pro-Phe-Glu-Ala-Ala-Met-Cys-Lys-
   15   16   17   18   19   20   21   22   23   24   25   26   27   28
   Trp-Glu-Glu-Glu-Pro-Asp-Gly-Met-Glu-Cys-Ala-Phe-His-Arg
   ```

3. The protein sheath of the tobacco mosaic virus contains 2,130 identical subunit polypeptide chains of 130 amino acids each. If each error in insertion of an amino acid results in a defective subunit which is rejected, estimate the maximum frequency of errors possible (number per 1,000,000 residues) if the final yield of intact active viral particles is to be 50 percent of all the viral polypeptide made. Repeat the calculation assuming that the entire coat is constructed of but one polypeptide chain.

4. The polypeptide chain of a given protein is α-helical in some segments and has the β-conformation in others. The protein has a mol wt of 240,000 and a contour length of 5.06×10^{-5} cm. Calculate the fraction of the molecule that exists in the α-helical configuration.

CHAPTER 7 PROTEINS: PURIFICATION AND CHARACTERIZATION

We have seen (Chapters 3 to 6) that each type of cell may contain thousands of different proteins, that each species of organism contains a distinctive set of proteins, chemically different from those of other organisms, and that proteins are relatively fragile molecules, retaining their biological activity only within relatively narrow ranges of pH and temperature. The isolation in pure form of a given protein from a given cell or tissue thus may appear to be a difficult task, particularly since any given protein may exist in only a very low concentration in the cell, along with thousands of others. Yet despite these difficulties a great many different proteins have been isolated in pure form. Moreover, current methods for separating proteins have exceptionally high resolving power.

This chapter describes the physical principles underlying techniques for separating proteins, the strategy employed in their purification, and some of the methods for determining their molecular weight. Although there is little discussion of the biology of proteins in this chapter, much of what we now know about their biology depended on the availability of highly purified preparations of proteins, whose isolation has required an enormous amount of painstaking and largely unsung effort.

The Behavior of Proteins in Solution

Although the proteins have been known for over a century, our present understanding of their behavior as solutes in aqueous systems has been gained only in relatively recent years and on the basis of the most intensive physicochemical studies. For many years before their structure became known, proteins were regarded as substances of mysterious properties, quite different in their behavior as solutes from molecules of other kinds. Proteins in solution were long thought to consist of colloidal micelles of indefinite and variable molecular weight. As late as 1916, Emil Fischer, who did more than any other chemist to transform the study of biomolecules into an exact, rigorous science, vigorously denied that proteins had molecular weights exceeding 5,000;

he believed they aggregated into micellelike complexes. Moreover, proteins in solution were found to undergo extraordinary changes in solubility in the presence of neutral salts, acids, or bases, changes which seemed quite different from the effects produced by these agents on simple, small organic molecules.

Two generations of physicochemical research were required to establish our present view that proteins are macromolecules of well-defined molecular weight, that they form true molecular solutions, and that they are electrolytes whose behavior conforms to the same physical principles as small electrolytes. Through the efforts of many investigators, the study of the behavior of proteins in solution and the separation of proteins has become nearly an exact science.

In the sections that follow we consider how various characteristic properties of globular proteins in solution can be exploited to separate mixtures of proteins, based on their (1) molecular size, (2) solubility, (3) electric charge, (4) differences in adsorption characteristics, and (5) biological affinity for other molecules.

Separation Procedures Based on Molecular Size

The most striking characteristic of proteins is their large size, which makes possible simple methods for separation of proteins from small molecules, as well as methods for resolving mixtures of proteins.

Dialysis and Ultrafiltration

Globular proteins in solution can easily be separated from low-molecular-weight solutes by *dialysis* (Figure 7-1), which utilizes a semipermeable membrane to retain protein molecules and allow small solute molecules and water to pass through. Another way of separating proteins from small molecules is by *ultrafiltration* (Figure 7-2), in which pressure or centrifugal force is used to filter the aqueous medium and small solute molecules through a semipermeable membrane, which retains the protein molecules. Cellophane and other synthetic materials are commonly used as the membrane in such procedures.

Density-Gradient (Zonal) Centrifugation

Because proteins in solution tend to sediment at high centrifugal fields, thus overcoming the opposing tendency of diffusion (page 174), it is possible to separate mixtures of proteins by centrifugal methods. *Density-gradient* or *zonal centrifugation* is a widely used and versatile procedure for separating not only proteins and other types of macromolecules but also organelles and viruses. In the most common procedure (Figure 7-3) a continuous density gradient of sucrose is first prepared in a plastic centrifuge tube by a device that mixes concentrated sucrose solution and water in decreasing ratio as the tube is filled, so that the density of the medium is greatest at the bottom of the tube. The mix-

Figure 7-1.
Dialysis. Since the membrane enclosing the protein solution is semipermeable, water and small solutes, such as glucose or ammonium sulfate, pass through the membrane freely but proteins do not. By replacing the outer aqueous phase with distilled H_2O several times, the concentration of small solute molecules in the protein solution can be decreased to a vanishingly small amount.

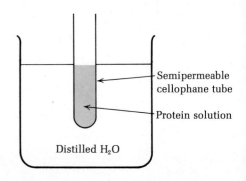

Semipermeable
cellophane tube

Protein solution

Distilled H_2O

Figure 7-2
Ultrafiltration of a protein solution. By applying positive pressure above (or a vacuum below) the membrane the protein can be concentrated by filtration of water and dissolved salts.

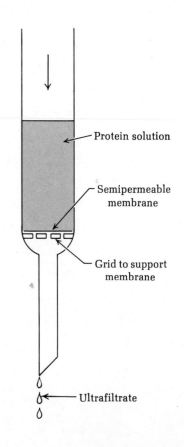

Protein solution

Semipermeable
membrane

Grid to support
membrane

Ultrafiltrate

Figure 7-3
Separation of proteins by centrifugation in a sucrose density gradient. The individual proteins band according to their size, shape, and density.

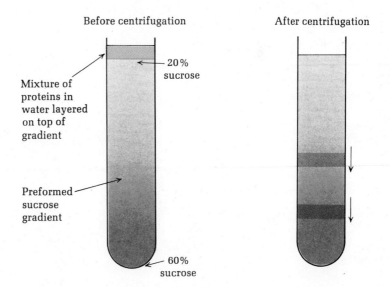

Before centrifugation

After centrifugation

Mixture of proteins in water layered on top of gradient

20% sucrose

Preformed sucrose gradient

60% sucrose

ture of macromolecules to be resolved is layered on top of the gradient. Centrifugation of the tube in a horizontal position in a rotor at a high speed causes each type of macromolecule to sediment down the density gradient at its own rate, determined primarily by its particle weight but also by its density and shape, in the form of separate bands or zones. Usually centrifugation is stopped before equilibrium is reached. The positions of the protein bands can be located optically or by draining off the contents of the tube carefully through a pinhole in the bottom and analyzing successive small samples. Alternatively, the plastic tube can be frozen solid and then cut into thin slices for analysis.

Molecular-Exclusion Chromatography

One of the most useful and powerful tools for separating proteins from each other on the basis of size is molecular-exclusion chromatography, also known as gel-filtration or molecular-sieve chromatography. It differs from ion-exchange chromatography, which separates solutes on the basis of their electric charge and acid-base properties. In molecular-exclusion chromatography the mixture of proteins, dissolved in a suitable buffer, is allowed to flow by gravity down a column packed with beads of an inert, highly hydrated polymeric material that has previously been washed and equilibrated with the buffer alone. Common column materials are Sephadex, the commercial name of a polysaccharide derivative; Bio-Gel, a commercial polyacrylamide derivative; and agarose, another polysaccharide—all of which can be prepared with different degrees of internal porosity. In the column proteins of different molecular size penetrate into the internal pores of the beads to different degrees and thus travel down the column at different rates (Figure 7-4). Very large protein molecules cannot enter the pores of the beads; they are said to be excluded and thus remain in the excluded volume of the column, defined as the

Figure 7-4
Separation of two proteins of different size on a Sephadex column.

Magnification showing the exclusion process.

Small solute molecules penetrate into the pores of the Sephadex beads and are retarded.

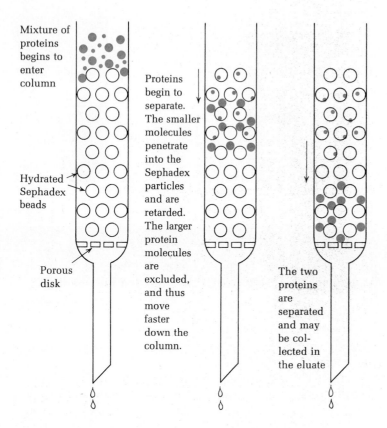

Mixture of proteins begins to enter column

Hydrated Sephadex beads

Porous disk

Proteins begin to separate. The smaller molecules penetrate into the Sephadex particles and are retarded. The larger protein molecules are excluded, and thus move faster down the column.

The two proteins are separated and may be collected in the eluate

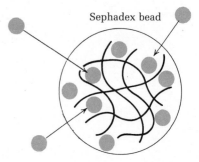

Sephadex bead

Large solute molecules cannot penetrate and are excluded.

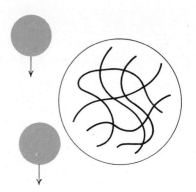

volume of the aqueous phase outside the beads. On the other hand, very small proteins can enter the pores of the beads freely (Figure 7-4). Small proteins are retarded by the column while large proteins pass through rapidly, since they cannot enter the hydrated polymer particles. Proteins of intermediate size will be excluded from the beads to a degree that depends on their size; hence the term *exclusion* chromatography. From measurements of the protein concentration in small fractions of the eluate an elution curve can be constructed (Figure 7-5).

Molecular-exclusion chromatography can also be used to separate mixtures of other kinds of macromolecules, as well as very large biostructures, e.g., viruses, ribosomes, cell nuclei, or even bacteria, simply by using beads or gels with different degrees of internal porosity. The resolving power of molecular-exclusion chromatography is so great that this simple method is now widely used as a way of determining the molecular weight of proteins (Figure 7-5).

Separation Procedures Based on Solubility Differences

Proteins in solution show profound changes in solubility as a function of (1) pH, (2) ionic strength, (3) the dielectric properties of the solvent, and (4) temperature. These

Figure 7-5
Elution of proteins from a molecular-exclusion column and determination of molecular weight. To calibrate the column, proteins (A, B, and C) of known molecular weight are allowed to pass through the column, and their peak elution volumes are plotted against the logarithm of the molecular weight. From such a graph the molecular weight of an unknown protein can be extrapolated, given its elution volume. This relationship holds true only for spherical proteins. For nonspherical particles, the elution volume is directly related to the Stokes radius, i.e., the radius of a spherical particle of equivalent hydrodynamic properties.

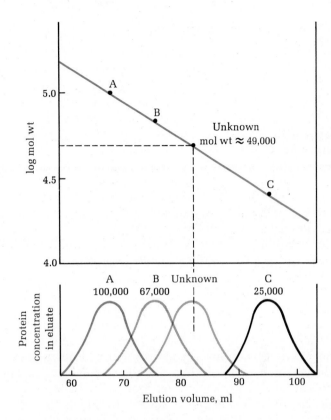

variables — reflections of the fact that proteins are electrolytes of very large molecular weight — can be used to separate mixtures of proteins, since each protein has a characteristic amino acid composition, which determines its behavior as an electrolyte.

Isoelectric Precipitation

The solubility of most globular proteins is profoundly influenced by the pH of the system. Figure 7-6 shows that the solubility of β-lactoglobulin, a milk protein, is at a minimum at pH 5.2 to 5.3, regardless of the concentration of sodium chloride present. On either side of this critical pH, the solubility rises very sharply. Nearly all globular proteins show a solubility minimum, although the pH at which it occurs varies from one protein to another.

The pH at which a protein is least soluble is its *isoelectric pH*, defined as that pH at which the molecule has no net electric charge and fails to move in an electric field (Table 7-1). Under these conditions there is no electrostatic repulsion between neighboring protein molecules, and they tend to coalesce and precipitate. However, at pH values above or below the isoelectric point, all the protein molecules have a net charge of the same sign. They therefore repel each other, preventing coalescence of single molecules into insoluble aggregates. Some proteins are virtually insoluble at their isoelectric pH.

Since different proteins have different isoelectric pH values, because their content of amino acids with ionizable R

Figure 7-6
Effect of pH and salt concentration on the solubility of β-lactoglobulin at 25°C. Figures give the concentration of NaCl.

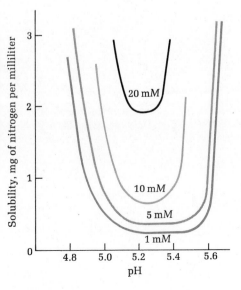

groups differs, they can often be separated from each other by *isoelectric precipitation*. When the pH of a protein mixture is adjusted to the isoelectric pH of one of its components, much or all of that component will precipitate, leaving behind in solution proteins with isoelectric pH values above or below that pH. The precipitated isoelectric protein remains in its native conformation and can be redissolved in a medium having an appropriate pH and salt concentration.

For any given protein the isoelectric pH will vary somewhat, depending on the ionic composition of the medium, since proteins can bind certain anions and/or cations. When a protein solution is thoroughly dialyzed against distilled H_2O to remove all small ions other than H^+ and OH^-, the pH of the resulting solution is known as the *isoionic pH*. The isoionic pH is a constant for any given protein.

Table 7-1   Isoelectric point of some proteins

	Isoelectric pH
Pepsin	~ 1.0
Egg albumin	4.6
Serum albumin	4.9
Urease	5.0
β-Lactoglobulin	5.2
γ_1-Globulin	6.6
Hemoglobin	6.8
Myoglobin	7.0
Ribonuclease	9.6
Chymotrypsinogen	9.5
Cytochrome c	10.6
Lysozyme	11.0

Salting-in and Salting-out of Proteins

Neutral salts have pronounced effects on the solubility of globular proteins, as shown in Figures 7-6 and 7-7. In low concentration, salts increase the solubility of many proteins, a phenomenon called *salting-in*. Salts of divalent ions, such as $MgCl_2$ and $(NH_4)_2SO_4$, are far more effective at salting-in than salts of monovalent ions, such as NaCl, NH_4Cl, and KCl. The ability of neutral salts to influence the solubility of proteins is a function of their *ionic strength* (Figure 7-7), a measure of both the concentration and the number of electric charges on the cations and anions contributed by the salt. Salting-in effects are caused by changes in the tendency of dissociable R groups on the protein to ionize.

On the other hand, as the ionic strength is increased further, the solubility of a protein begins to decrease (Figure 7-7). At sufficiently high ionic strength a protein may be almost completely precipitated from solution, an effect called *salting-out*. The physicochemical basis of salting-out is rather complex; one factor is that the high concentration of salt may remove water of hydration from the protein molecules, thus reducing their solubility, but other factors are also involved. Whatever their physical basis, salting-in and salting-out are important procedures in the separation of protein mixtures, since different proteins vary in their response to the concentration of neutral salts. Proteins precipitated by salting-out retain their native conformation and can be dissolved again, usually without denaturation. Ammonium sulfate is preferred for salting out proteins because it is so soluble in water that very high ionic strengths can be attained.

Figure 7-7
Effect of a neutral salt (K_2SO_4) on the solubility of carbon monoxide hemoglobin at its isoelectric pH. The ionic strength of a solution μ is given by $\frac{1}{2}\Sigma c_i z_i^2$, in which c is the concentration and z is the charge. At low ionic strength, the protein is salted in, i.e., increases in solubility. At high salt concentration, it is salted out.

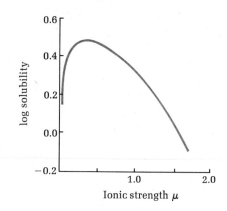

Solvent Fractionation

The addition of water-miscible neutral organic solvents, particularly ethanol or acetone, decreases the solubility of most globular proteins in water to such an extent that they precipitate out of solution. Quantitative study of this effect shows that protein solubility at a fixed pH and ionic strength is a function of the dielectric constant of the medium. Since ethanol has a lower dielectric constant than water (Table 2-2, page 43), its addition to an aqueous protein solution increases the attractive force between opposite charges (page 43), thus decreasing the degree of ionization of the R groups of the protein. As a result, the protein molecules tend to aggregate and precipitate. Mixtures of proteins can be separated on the basis of quantitative differences in their solubility in cold ethanol-water or acetone-water mixtures. A disadvantage of this method is that since such solvents can denature proteins at higher temperatures, the temperature must be kept rather low.

Effect of Temperature on Solubility of Proteins

Within a limited range, from about 0 to about 40°C, most globular proteins increase in solubility with increasing temperature, although there are some exceptions, as there are for simple electrolytes. Above 40 to 50°C, most proteins become increasingly unstable and begin to denature (page 141), ordinarily with a loss of solubility at the neutral pH zone. Protein-fractionation procedures are usually carried out at 0°C or refrigerator temperatures, since most proteins are stable at low temperatures; however, there are exceptions. Some proteins are most stable and maximally soluble at room temperature or the temperature of their normal cellular surroundings.

Utilizing these four basic parameters of protein solubility, namely, pH, ionic strength, dielectric constant, and temperature, E. J. Cohn and J. T. Edsall and their colleagues at Harvard Medical School designed successful procedures during World War II for the large-scale isolation of various proteins from human blood plasma, e.g., serum albumin, used to restore blood volume in patients with blood loss or shock, serum γ-globulins, or antibodies (page 1002), useful for immunization against measles, mumps, and other diseases, and fibrinogen, which promotes blood clotting. These solubility parameters are still widely used, particularly in the early stages of protein purification, but they cannot give the high resolution of more recently developed methods.

Separation Procedures Based on Electric Charge

Separation of proteins on the basis of their electric charge depends ultimately on their acid-base properties, which are largely determined by the number and types of ionizable R groups in their polypeptide chains. Since proteins differ in amino acid composition and sequence, each protein has distinctive acid-base properties. The principles involved in the electrophoretic separation of proteins are best understood if we first consider the acid-base titration curve of a globular protein of known amino acid content. Figure 7-8 shows the titration curve of native ribonuclease with 124 amino acids, of which 34 possess ionizing R groups, in addition to the N-terminal and C-terminal groups. The titration curve of ribonuclease is thus a composite reflecting the ionization of many groups. The approximate contribution of each type of

Figure 7-8
Titration of ribonuclease. It is assumed that the titration is begun from the isoelectric point of ribonuclease (pH 9.6) and proceeds (to the left) by addition of acid or (to the right) by addition of base. The addition of H^+ brings the protein into the pH zone where it has a net positive charge; the addition of base brings it into the zone of net negative charge. The shape of the titration curve reflects the number and type of ionizing groups (see Table 7-2).

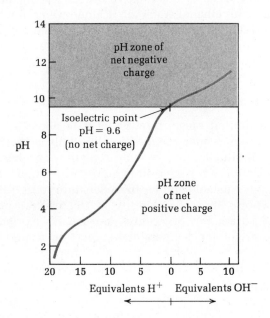

ionizing R group can be determined (Table 7-2). These data show that the average pK' values for some types of R groups in ribonuclease may vary somewhat from the pK's of these groups in free amino acids, due to the effects of neighboring charges. All the R groups of ribonuclease are accessible to acid-base titration, in agreement with the generalization from x-ray studies (Chapter 6, page 137) that nearly all the ionizing R groups of native globular proteins are on the outer surface of the molecule. Some native proteins, however, possess one or more ionizing groups that are not accessible to titration, presumably because they are hidden or participate in hydrogen bonding. On denaturation of the native protein (page 62) such hidden R groups become accessible. For example, in myoglobin, the R groups of 5 of the 11 histidine residues are inaccessible to titration until the protein is denatured.

The titration curve of ribonuclease (Figure 7-8) also shows its *isoelectric pH*, at which the molecule carries no *net* electric charge and fails to migrate in an electric field. The isoelectric pH is determined by the number and pK' of the ionizing R groups. It will be relatively high, above pH 7.0, if the protein has a relatively high content of basic amino acids (lysine, arginine), as is the case with ribonuclease, which has an isoelectric pH of 9.6 (Figure 7-8). The isoelectric pH will be relatively low if the protein has a preponderance of acidic residues (aspartic and glutamic acids), as is the case with pepsin. Most globular proteins have isoelectric points between pH 4.5 and 6.5 (Table 7-1).

Table 7-2 Titratable groups of ribonuclease†

Group	From amino acid composition	From titration curve	Approximate pK′	pK′ of group in free amino acid
α-COOH	1		4.7	2.3
R—COOH (Glu, Asp)	10		4.7	4.0
Imidazole (His)	4	} 5	6.5	6.0
α-Amino	1		7.8	9.7
Phenolic OH (Tyr)	6	} 16	9.95	10.0
ε-Amino (Lys)	10		10.2	10.5
Guanidinyl (Arg)	4	4	12	12.5

Number per molecule spans the "From amino acid composition" and "From titration curve" columns.

† For titration curve see Figure 7-8.

The titration curve of a protein also indicates the sign and magnitude of its net electric charge at any given pH. At any pH above the isoelectric point, a protein has a net negative charge and will move toward the anode. Its negative charge increases in magnitude as the pH is increased, in accordance with the shape of the titration curve. Similarly, at any pH below the isoelectric point, the protein has a net positive charge and will move toward the cathode (Figure 7-8). Knowledge of the acid-base properties of a given protein thus makes it possible to predict its behavior in an electric field.

Both the shape of the titration curve and the isoelectric pH of a protein may change significantly in the presence of neutral salts, which influence the degree of ionization of the different types of R groups. Proteins also may bind cations such as Ca^{2+} and Mg^{2+} or anions such as Cl^- or HPO_4^{2-} (page 162). For these reasons, the observed isoelectric pH values for proteins depend somewhat on the nature of the medium in which the protein is dissolved; the *isoionic point* (page 162) is characteristically constant for each protein.

The characteristic acid-base properties of proteins are directly exploited in two widely used general methods for separating and analyzing protein mixtures, electrophoresis and ion-exchange chromatography.

Electrophoretic Methods

There are a number of different forms of *electrophoresis*, also called *ionophoresis*, useful for analyzing and separating mixtures of proteins. The prototype of all modern methods is *free*, or *moving-boundary, electrophoresis*, first developed by A. Tiselius in Sweden in the 1930s. The mobility μ in square centimeters per volt-second of a molecule in an electric field is given by the ratio of the velocity of migration v, in centimeters per second, to electric field strength E, in volts per centimeter:

$$\mu = \frac{v}{E}$$

For small ions, such as chloride, μ is between 4 and 9×10^{-4}

$cm^2 V^{-1} s^{-1}$ (25°C); for proteins, it is about 0.1 to 1.0×10^{-4} $cm^2 V^{-1} s^{-1}$. Proteins thus migrate much more slowly in an electric field than small ions such as Na^+ or Cl^-, simply because they have a much smaller ratio of charge to mass. In free electrophoresis a buffered solution of the protein mixture is placed in a U-shaped observation cell, with pure buffer layered over the protein solution (Figure 7-9). The cell is immersed in a constant-temperature bath insulated from vibrations and an electric field is generated between the electrodes; negatively charged proteins move toward the anode and positively charged proteins toward the cathode. In order to get a complete picture of all the proteins in a mixture the pH is usually chosen so that most or all of the proteins will have the same charge but different mobilities. As the negatively charged protein molecules move toward the anode, they migrate from the protein solution into the zone of protein-free buffer and form a _front_, or _boundary_. The refractive index of the solution changes sharply at this boundary because the index of refraction of the protein molecules is different from that of the pure buffer. Optical measurements of the refractive-index changes along the electrophoresis cell yield electrophoretic patterns (called schlieren patterns) that show the direction and relative rate of migration of the major proteins in the mixture. Figure 7-9 shows such a pattern; each peak in the pattern corresponds to the position of the moving boundary of a specific protein (it does not represent the peak of protein concentration). If the electrophoretic mobility of a given protein is determined at several different pH values, the isoelectric pH of the protein can be extrapolated. Actually, the titration curve of a protein (page 164) is an approximate measure of its electrophoretic mobility as a function of pH. For many years moving-boundary electrophoresis was the most valuable method for quantitative analysis of complex mixtures of proteins, e.g., those in blood plasma.

Free electrophoresis has been largely supplanted by various forms of _zone electrophoresis_, which are much simpler, have much greater resolving power, and require smaller samples. In zone electrophoresis the aqueous protein solution is immobilized in a solid matrix or support, a hydrated porous material that has mechanical rigidity, eliminating convection and vibration disturbance. The most widely used supports are filter paper or cellulose acetate strips, relatively inert materials that do not interact with migrating proteins or retard them. The electrophoretic process is allowed to continue until the major protein components separate into discrete zones; hence the name zone electrophoresis. The position and amount of the proteins in the separated zones are determined by applying a protein stain; the density of staining, which is proportional to the amount of protein, can be estimated with a scanning densitometer (Figure 7-10). Zone electrophoresis is capable of significantly higher resolution than free electrophoresis. This method is often used in hospital laboratories to measure the amounts of the major proteins in blood plasma.

Figure 7-9
Free electrophoresis.

Schematic view of Tiselius moving-boundary electrophoresis apparatus.

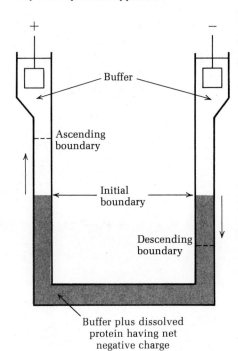

Electrophoretic pattern of human blood plasma proteins (pH 8.6). A = serum albumin; ϕ = fibrinogen; α_1, α_2, β, and γ are various globulins. [Redrawn from R. Alberty, J. Chem. Educ., 25: 619 (1948).]

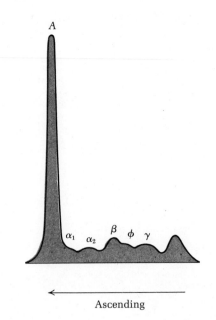

Figure 7-10
Zone electrophoresis on a cellulose acetate strip. After staining, the strip is scanned to yield the tracing of the protein peaks shown below.

Stained strip after electrophoresis

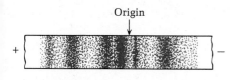

Densitometer tracing

Figure 7-11
Separated zones of proteins revealed by staining of the gel after disc electrophoresis. [Redrawn from K. Linderstrom-Lang and S. O. Nielsen in M. Bier (ed.), Electrophoresis, p. 139, Academic Press Inc., New York, 1967.]

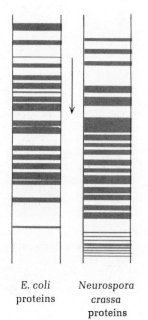

E. coli
proteins

Neurospora
crassa
proteins

Much higher electrophoretic resolution is possible when the support or matrix material can retard or exclude protein molecules on the basis of molecular size, like the materials used in molecular-exclusion chromatography (page 159). This form of zone electrophoresis can separate a protein mixture on the basis of *both* electric charge and molecular size. For this purpose, gels of potato starch or polyacrylamide are commonly used. By this technique the protein components of blood plasma can be resolved into 15 or more bands, compared with about 5 or 6 observed with free or zone electrophoresis of the simple kind outlined above. Polyacrylamide gel electrophoresis can be scaled up to carry out isolation of larger quantities of purifed proteins.

In yet another variation of zone electrophoresis, called <u>disc electrophoresis,</u> the protein mixture to be analyzed is subjected to an electric field in a retarding gel support that is separated into two sections differing in porosity and buffered at different pHs. The protein mixture migrates from the more porous into the less porous gel, a process accompanied by a change in pH. As a result, each protein species becomes concentrated into a very thin, sharp band, producing much higher resolution than can be achieved in a continuous buffer. This form of zone electrophoresis is called *disc* (not *disk*) electrophoresis because of the discontinuous buffer employed and the discoid appearance of the protein zones (Figure 7-11).

Perhaps the most ingenious and effective electrophoretic method for separating proteins is <u>isoelectric focusing</u> or <u>electrofocusing,</u> invented by H. Svensson in Sweden, in which the mixture of proteins is subjected to an electric field in a gel support in which a pH gradient has first been generated. Each protein then migrates toward, and is "focused" at, that portion of the pH gradient where the pH is equal to its isoelectric pH and forms a sharp stationary band there (Figure 7-12). The power of isoelectric focusing is extraordinary: it can resolve the proteins of human blood plasma into 40 or more bands. Isoelectric focusing is normally used as an analytical tool but can also be used on a large scale to prepare purified proteins.

Ion-Exchange Chromatography

A second general way of utilizing the acid-base behavior of proteins as a basis for separation is *ion-exchange chromatography,* which also has a number of variations. The same basic principles that make the separation and analysis of amino acid or peptide mixtures on columns of ion-exchange resins feasible (pages 89 and 90) were first successfully applied to the separation of protein mixtures by H. Sober and E. Peterson in the United States in the 1950s. The most commonly used materials for chromatography of proteins are synthetically prepared derivatives of cellulose. <u>Diethylaminoethylcellulose</u> (abbreviated DEAE-cellulose) contains positively charged groups at pH 7.0 and is therefore an anion exchanger (Figure 7-13). *Carboxymethylcellulose* (abbreviated CM-cellulose) contains negatively charged groups at neutral pHs and is a cation exchanger. Protein mixtures are resolved and the individual components successively eluted from DEAE-cellulose columns by passing a series of buffers of decreasing pH or a series of salt solutions of increasing ionic strength, which have the effect of decreasing the binding of anionic proteins. The composition of the eluting solution may also be varied gradually and continuously during chromatography; this process is called *gradient elution*. Gradients used for this purpose may be linear with the volume of the eluting solution or varied in some other relationship by using devices capable of mixing liquids in varying ratios. The protein concentration in the eluate, which is collected in small fractions, is estimated optically by its capacity to absorb light in the ultraviolet region (page 83).

In an important variation of this method the ion-exchange and molecular-exclusion principles have been combined. The diethylaminoethyl derivative of the molecular-exclusion material Sephadex (DEAE–Sephadex) is widely used for resolving protein mixtures on the basis of both molecular size and electric charge.

Separation of Proteins by Selective Adsorption

Proteins can be adsorbed to, and selectively eluted from, columns of finely divided, relatively inert materials with a very large surface area in relation to particle size. They include nonpolar substances, e.g., charcoal, and polar substances, e.g., silica gel or alumina. The precise nature of the forces binding the protein to such adsorbents is not known, but presumably van der Waals and hydrophobic interactions prevail with nonpolar adsorbents, whereas ionic attractions and/or hydrogen bonding are the main forces with polar adsorbents.

Perhaps the most widely used and effective adsorbent for protein purification is a form of crystalline calcium phosphate, *hydroxyapatite*, the same mineral found in bone. Presumably negatively charged groups in protein molecules bind to the Ca^{2+} ions in the hydroxyapatite crystal lattice. Proteins can be eluted from hydroxyapatite columns with phosphate buffers.

Separations Based on Ligand Specificity: Affinity Chromatography

Some proteins can be isolated from a very complex mixture and brought to a high degree of purification, often in a single step, by *affinity chromatography*. This method is based on a biological property of some proteins, namely, their capacity for specific, noncovalent binding of another molecule, called the *ligand*. For example, some enzymes bind their specific coenzymes (page 185) very tightly through noncovalent forces. In order to separate such an enzyme from other proteins by affinity chromatography, its specific coenzyme is covalently attached, by means of an appropriate chemical

Figure 7-12
Resolution of the isozymes (different molecular forms) of crystalline L-amino acid oxidase from rattlesnake venom by electrofocusing. Disc gel electrophoresis (left) separates the isozymes into only three bands, whereas electrofocusing (right) resolves 18 different molecular forms of the enzyme. [From M. B. Hayes and D. Wellner, J. Biol. Chem., 244: 6636 (1969)].

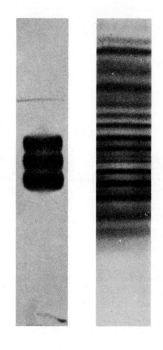

Figure 7-13
Ion-exchange materials for chromatography of proteins. Each cellulose particle contains a large number of the ion-exchange groups, which are covalently attached to hydroxyl groups of the cellulose.

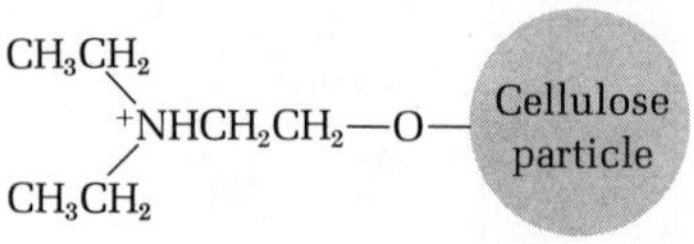

$$CH_3CH_2$$
$$^+NHCH_2CH_2\!-\!O\!-\!\text{Cellulose particle}$$
$$CH_3CH_2$$

Diethylaminoethylcellulose particle
(anion exchanger)

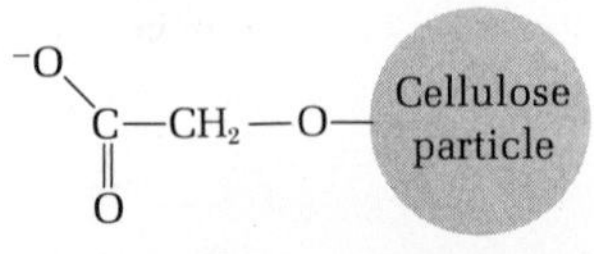

$$^-O$$
$$\underset{\underset{O}{\|}}{C}\!-\!CH_2\!-\!O\!-\!\text{Cellulose particle}$$

Carboxymethylcellulose particle
(cation exchanger)

reaction, to a functional group on the surface of large hydrated particles of a porous column material, e.g., the polysaccharide agarose, which otherwise allows protein molecules to pass freely (Figure 7-14). When a mixture of proteins containing the enzyme to be isolated is added to such a column, the enzyme molecule, which is capable of binding tightly and specifically to the immobilized ligand molecule, adheres to the ligand-derivatized agarose particles, whereas all the other proteins, which lack a specific binding site for that particular ligand molecule, will pass through. Similarly, one might use a substrate or a competitive inhibitor (page 197) of an enzyme as the specific ligand derivatized to the column material. This method thus depends on the biological affinity of the protein for its characteristic ligand. The protein specifically bound to the column particles in this manner can then be eluted, often with a solution of the free ligand molecule.

Affinity chromatography is used to isolate not only enzymes (page 172) but also the receptor molecules in cell membranes that bind specific hormones. For example, the insulin receptor protein of the plasma membrane of certain animal cells (page 821) has been separated and greatly purified by affinity chromatography on a column material to which insulin molecules had been covalently attached.

The Extraction and Purification of Proteins

The great number of different proteins, their great variety of biological activities, and the chemical difference between homologous proteins of different organisms make the extrac-

Figure 7-14
Principle of affinity chromatography.

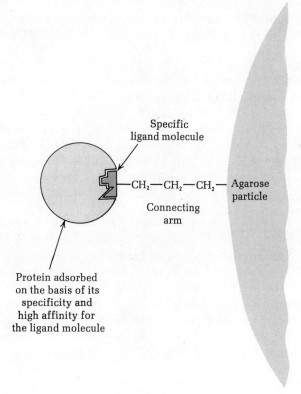

Specific
ligand molecule

—CH$_2$—CH$_2$—CH$_2$— Agarose
particle

Connecting
arm

Protein adsorbed
on the basis of its
specificity and
high affinity for
the ligand molecule

tion, purification, and characterization of proteins central to all research in biochemistry. Over a thousand different enzymes have been at least partially purified, and 200 or more have been obtained in pure crystalline form. In addition, hundreds of proteins other than enzymes have been isolated with a high degree of purity.

Early methods for isolation of proteins were empirical, slow, and very laborious. However, with the new methods now available, some of which were described above, protein isolation has become a fine art. There is no single procedure or set of procedures by which any and every protein can be isolated, but for any protein it is usually possible to choose a sequence of separation steps that will result in a high degree of purification and a high yield. The general objective is to increase the purity or biological activity of the desired protein per unit weight, by ridding it of inactive or unwanted proteins while at the same time maximizing the yield.

The first requirement is a specific and sensitive method to distinguish and measure quantitatively the particular protein to be isolated. If it is an enzyme, a quantitative assay system capable of estimating its catalytic activity is required (page 207). If the protein is a hormone, a suitable bioassay must be available. If the protein has a distinctive chemical component, e.g., a trace metal such as copper, a sensitive analytical method for that component can be utilized.

Also necessary is a procedure for liberating the protein in soluble form from the intact cell or tissue structure without causing loss of activity. Usually mechanical blending or homogenization of animal tissues (for example, see page 380) is used to break the cell membranes and release the cell contents, which may then be assayed for their content of the desired protein. In bacteria, yeasts, and many plant cells much more vigorous procedures are necessary to break the tough cell walls, e.g., sonic radiation, grinding with sand, or fragmentation in a high-pressure press. Sometimes the cell wall can be loosened or lysed by treatment with certain enzymes (page 270).

Next, it is customary to determine whether the protein in question is localized in one of the major subcellular organelles, such as the nucleus, the mitochondria, or the soluble portion of the cytoplasm (pages 30 to 35). This can be accomplished by carrying out separation of the cell organelles by *differential centrifugation* (see page 381). If the protein is found in one of the major cell fractions, a substantial degree of purification can be achieved by using that cell fraction as the starting point for the next stage of purification. If the desired protein happens to be associated with a membrane or membranous organelle, it must be extracted therefrom in soluble form, which often can be done either by simple extraction with water, by mechanical or sonic disruption of the membranes, or by use of detergents to disaggregate membrane structure (see page 303).

Once the desired protein has been obtained in soluble form, the fractionation methods described earlier in this chapter can be applied to separate it from contaminating proteins. By direct assay it can be determined in which fraction the desired protein appears and whether it has been selectively enriched with an increase in its specific activity. Since the starting cell or tissue extract may contain hundreds of different proteins, the purification of a given protein may require many steps, in order to remove a large number of other proteins from that to be purified. A variety of procedures is used in an empirically chosen sequence.

Among the separation procedures used as early steps are isoelectric precipitation (page 161), fractionation by salting-out (page 162), or solvent precipitation (page 162). At each fractionation step the enrichment factor and the yield of the desired protein are determined. These early steps, which do not have much resolving power, are usually followed by chromatographic procedures. Often molecular-exclusion chromatography and ion-exchange chromatography are used, in either sequence, to obtain fractions enriched in the desired protein on the basis of molecular size and electric charge, respectively. When affinity chromatography is possible, it is a very powerful method and may be used after or instead of other forms of chromatography. Sometimes, as a last purification step, usually carried out on a small scale, the protein is subjected to one or another form of zone electrophoresis, disc electrophoresis, or isoelectric focusing, to bring about high-resolution separation of the protein from the remaining impurities.

If the overall yield of the purified protein following a sequence of such purification steps has been sufficiently high, it is often possible to crystallize the protein. One procedure is very slowly to approach the salt concentration or pH required for salting-out or isoelectric precipitation, respectively. Another method is to reverse this procedure; the protein is salted-out and diluted just enough to redissolve it, and the solution is allowed to stand to lose water by evaporation. However, crystallization is not necessarily a sign of complete purity, since protein crystals often contain trapped contaminants. Throughout procedures of the kind described, the pH must be carefully controlled with appropriate buffers and the temperature held at its optimal level, which for most proteins is near 0°C.

Table 7-3 shows the data obtained in a recently published procedure for the isolation of the enzyme *acetylcholinesterase* from the electric tissue of the electric eel, *Electrophorus electricus*. This enzyme, which catalyzes the reaction

$$\text{Acetylcholine} + \text{H}_2\text{O} \longrightarrow \text{acetic acid} + \text{choline}$$

was quantitatively assayed by measuring the increase in acidity (decrease in pH) as the substrate acetylcholine is hydrolyzed with formation of acetic acid. The specific activity of each enzyme fraction is shown in micromoles of acetylcholine hydrolyzed per minute per milligram of protein (see page 208 for a discussion of standard enzyme units). At each purification step the specific activity increases, from a

Table 7-3 Purification of acetylcholinesterase by standard procedures[†]

Step	Specific activity, μmol min^{-1} mg^{-1}	Yield, %
Fresh tissue homogenate	16.7	
Extraction and ammonium sulfate precipitation	520	100
DEAE-cellulose	2,330	52
Concentration and dialysis	2,420	50
Sephadex G-200	4,170	43
Cellex-P (a cation exchanger)	6,830	25
DEAE-cellulose	7,910	16
DEAE–Sephadex	8,330	12

† Data from T. L. Rosenberry, H. W. Chang, and Y. Y. Chen, "Purification of Acetylcholinesterase by Affinity Chromatography and Determination of Active Site Stoichiometry," *J. Biol. Chem.*, 247: 1555–1565 (1972). The authors' original data have been recalculated in terms of standard enzyme units, described on page 208.

Table 7-4 Purification of acetylcholinesterase by affinity chromatography[†]

Step	Specific activity, μmol min^{-1} mg^{-1}	Yield, %
Ammonium sulfate precipitate (see Table 7-3)	470	(100)
Affinity chromatography (see Figure 7-15)	9,750	70

† See footnote in Table 7-3.

specific activity of 16.7 μmol mg^{-1} in the first tissue homogenate to 8,330 μmol mg^{-1} in the final product. Simultaneously, there were losses in the activity recovered after each step; at the end of the procedure only 12 percent of the starting activity was recovered. This is not an unusually low yield, considering the large number of steps involved.

To show the great power of affinity chromatography, Table 7-4 gives the specific activity and yield of acetylcholinesterase when affinity chromatography was used to purify the enzyme from the starting material. In only a single step this procedure yielded a product with a significantly higher specific activity than the sequence of procedures in Table 7-3, and in a much higher yield. The structure of the specific ligand attached to the column material in order to bind acetylcholinesterase molecules selectively is shown in Figure 7-15. It is a very specific competitive inhibitor (page 197) of acetylcholinesterase, which resembles the natural substrate in structure but is not cleaved by the enzyme.

The Characterization of Protein Molecules

After a given protein has been isolated in highly purified form, its homogeneity must be established. For this purpose free electrophoresis and sedimentation analysis in the ultracentrifuge were once common, but these expensive and relatively insensitive methods have been largely supplanted by simpler methods, e.g., molecular-exclusion chromatography, electrophoresis on polyacrylamide gels, and isoelectric focusing, which have much higher resolving power and can easily detect the presence of minor protein impurities.

Once the homogeneity of the protein is established, it can be characterized in a succession of approaches to ascertain (1) its molecular weight, (2) whether it contains a single or multiple polypeptide chains, (3) the molecular weight of the polypeptide chains, (4) their amino acid composition, and (5) their amino acid sequence. We have already seen how amino acid composition (page 100) and sequence (page 98) can be ascertained. Here we shall consider briefly the principles of different methods commonly used to establish the molecular weight and subunit composition of globular proteins.

Determination of Minimum Molecular Weight from Chemical Composition

Since each molecule of a given protein must contain at least one molecule of its prosthetic group or at least one residue of any of its component amino acids, the mass of the protein in daltons containing one such residue is equal to the minimum molecular weight. For example, myoglobin contains 0.335 percent iron. The minimum molecular weight can be calculated as

$$\text{Minimum mol wt} = \frac{\text{at w iron}}{\text{\% iron}} \times 100$$

$$= \frac{55.8}{0.335} \times 100 = 16{,}700$$

The true molecular weight is n times the minimum molecular weight, where n is the number of iron atoms per molecule. Since n = 1 in myoglobin, its true molecular weight is 16,700. Hemoglobin also contains iron, but it has four iron atoms per molecule. Thus, n = 4, and the true molecular weight is 4 times the minimum molecular weight calculated from the iron content. Such calculations are most accurate if the residue or element used as the basis for calculation has a small value for n.

Determination of the Molecular Weight from Osmotic-Pressure Measurements

When a semipermeable membrane separates a solution of a protein from pure water, the water moves across the membrane into the compartment containing the solute, a process called osmosis. Osmosis is a reflection of the tendency of the water to move in whatever direction will make its thermodynamic activity uniform throughout all parts of the system available to it. The osmotic pressure is the force that must be applied to counterbalance the force of such osmotic flow (Figure 7-16). Osmotic pressure is one of the colligative properties of solutions; it is a function of the number of solute particles per unit volume but is independent of the molecular nature of the solute or its shape. The molecular weight of a protein can be determined from measurements of the osmotic pressure of a solution of a known concentration of protein by the relationship

$$M = \frac{c}{\pi} RT$$

where M is molecular weight, c the concentration in grams per liter, R the gas constant (0.082 liter-atm mol^{-1} K^{-1}), T the absolute temperature in kelvins, and π the osmotic pressure in atmospheres. Expressed another way, this relationship states that a 1.0 M solution of an ideal nondissociating solute in an ideal solvent gives an osmotic pressure of 22.4 atm at 0°C. However, in practice, this relationship holds only for very dilute solutions. Usually osmotic-pressure measurements are made at several concentrations of solute and then extrapolated to zero protein concentration. The osmotic-pressure method has important theoretical advantages; e.g., it requires no knowledge of the shape of the protein. However, since the osmotic pressure depends on the number of molecules in solution, an impermeant molecule of small mass has the same effect as a molecule of large mass. The method is therefore highly susceptible to errors caused by low-molecular-weight impurities and is seldom used today.

Determination of Molecular Weight by Sedimentation Analysis

The ultracentrifuge, invented by Svedberg in 1925, can yield centrifugal fields exceeding 250,000 times the force of gravity. Such a high centrifugal field causes protein mole-

Figure 7-16
Osmosis and osmotic pressure. In B, water has moved into the protein solution. At equilibrium, the hydrostatic pressure h of the column of protein solution just counterbalances the osmotic flow of water. The osmotic pressure (C) is equal to the hydrostatic pressure of head h.

A. Initial state

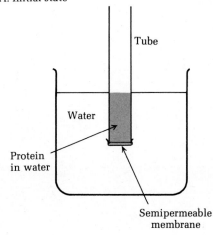

B. Final state

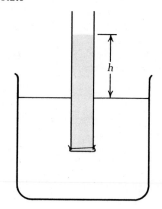

C. Osmotic pressure

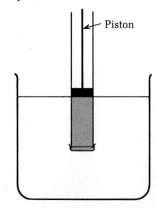

Figure 7-17.
Principle of the ultracentrifuge, showing
how optical measurements are made while
sample is undergoing sedimentation.

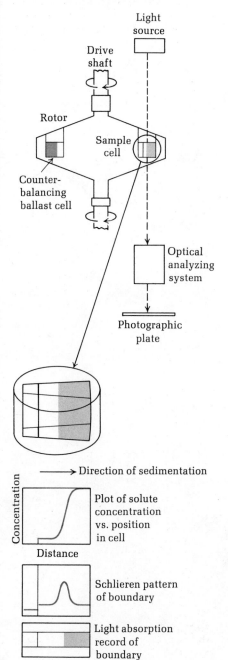

A schlieren pattern for a mixture of three
sedimenting proteins, each peak indicating
the position of a boundary.

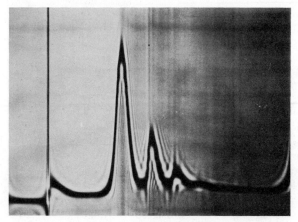

Spinco Division, Beckman Instruments, Inc.

cules to sediment from solution, opposing the force of
diffusion (see below), which normally keeps them evenly
dispersed in solution. Three types of sedimentation measurements are used to determine the molecular weight of
proteins: *sedimentation velocity*, *sedimentation equilibrium*,
and *approach to equilibrium*.

First, we shall consider the *sedimentation-velocity
method*. If the centrifugal force exerted on protein molecules
in a solution greatly exceeds the opposing diffusion force,
the molecules will sediment down from the surface of the
solvent (the *meniscus*), leaving behind pure solvent. A sharp
boundary is thus formed. The rate of movement of this boundary down the centrifuge cell is observed by optical measurements of the index of refraction at different positions
along the cell (Figure 7-17). The measurements are made
photographically at timed intervals during the centrifugation
while the rotor is spinning. The drive system of the centrifuge is engineered to produce constant speeds without
vibration.

When the sedimenting boundary of protein moves at a
constant rate, the centrifugal force just counterbalances the
frictional resistance of the solvent. The *sedimentation coefficient* s of the protein is given by the equation

$$s = \frac{dx/dt}{\omega^2 x}$$

where x is the distance of the boundary from the center of rotation in centimeters, t is the time in seconds, and ω is the
angular velocity in radians per second. Proteins have sedimentation coefficients (denoted $s_{20,w}$, where the temperature
is 20°C and the medium water) in the range between
1×10^{-13} and 200×10^{-13} seconds (Table 7-5). A sedimentation coefficient of 1×10^{-13} seconds is called a *Svedberg unit*
or simply a Svedberg (S). Thus, a sedimentation coefficient
of 8×10^{-13} seconds is denoted 8S.

175

Table 7-5 Physical constants of some proteins

Protein	Mol wt	Diffusion coefficient $D_{20,w} \times 10^7$, cm^2 s^{-1}	Sedimentation coefficient $s_{20,w} \times 10^{13}$, s
Cytochrome c (bovine heart)	13,370	11.4	1.71
Myoglobin (horse heart)	16,900	11.3	2.04
Chymotrypsinogen (bovine pancreas)	23,240	9.5	2.54
β-Lactoglobulin (goat milk)	37,100	7.48	2.85
Serum albumin (human)	68,500	6.1	4.6
Hemoglobin (human)	64,500	6.9	4.46
Aldolase	149,100	4.63	7.35
Catalase (horse liver)	221,600	4.3	11.2
Urease (jack bean)	482,700	3.46	18.6
Fibrinogen (human)	339,700	1.98	7.63
Myosin (cod)	524,800	1.10	6.43
Tobacco mosaic virus	40,590,000	0.46	198

Although the sedimentation coefficient increases with molecular weight, it is not proportional to the molecular weight since it is also influenced by the frictional resistance of the solvent and by the shape of the protein. Nevertheless, with some additional data, the molecular weight M of a protein can be calculated from the sedimentation coefficient by means of the *Svedberg equation*, which is derived by equating the centrifugal force with the opposing frictional force, the condition existing when the rate of sedimentation is constant. This equation is

$$M = \frac{RTs}{D(1 - \overline{v}\rho)}$$

where R is the gas constant (8.31×10^7 ergs mol^{-1} K^{-1}), T the absolute temperature in kelvins, s the sedimentation coefficient, $\overline{v}$ the partial specific volume of the protein, ρ the density of the solvent, and D the diffusion coefficient (page 177). The partial specific volume is the increase in volume when 1.0 g of dry solute is added to an infinitely large volume of solvent; for most proteins in water it is about 0.74 cm^3 g^{-1}. With experimentally determined values for the diffusion coefficient of the protein, obtained as described below (page 178), the molecular weight of the protein can be calculated from this equation. For the most accurate results, values of the sedimentation coefficient s and the diffusion coefficient D must be obtained from measurements made at several different protein concentrations and extrapolated to infinite dilution. Sedimentation-velocity measurements can also give valuable information about the state of purity of a protein and the composition of a protein mixture, since different proteins sediment at different rates (Figure 7-17).

The *sedimentation-equilibrium method* for determination of molecular weight has two important advantages over the sedimentation-velocity method: it does not require knowledge of the diffusion coefficient or the shape of the protein molecule. In this method, the ultracentrifuge is operated at a relatively low speed, just high enough for the system to come to an equilibrium state in which the rate of sedimentation of the protein is exactly balanced by the opposing diffusion force. At equilibrium no pure solvent region is present at the surface meniscus; instead a gradient of protein molecules is formed down the centrifuge tube, in which the bottom layer may have a protein concentration about twice that of the top layer. By measuring the concentration of the protein as a function of the distance from the center of rotation, data can be obtained to calculate the molecular weight from the equation

$$M = \frac{2RT \ln (c_2/c_1)}{\omega^2(1 - \bar{v}\rho) (x_2{}^2 - x_1{}^2)}$$

in which R and T have their usual meanings, c_1 and c_2 are the concentrations of the protein at two points in the tube at distances x_1 and x_2 from the center of rotation, ω is the angular velocity, ρ is the density of solvent, and $\bar{v}$ is the partial specific volume of the protein. Although the sedimentation-equilibrium method is the most accurate of the sedimentation methods, it may require several days of centrifugation to attain equilibrium, a difficulty solved in part by using cells with only a very short column of protein solution (1 to 2 mm). Further, the protein must be quite pure and homogeneous.

The _approach-to-equilibrium method_ represents a compromise, in which some of the accuracy of the equilibrium method is sacrificed to make a more rapid measurement of molecular weight possible. In the approach-to-equilibrium method, the rotor speed is brought to approximately the equilibrium speed over a 1- to 2-h period by a series of adjustments. Each time, measurements of protein concentration are made near the bottom of the tube. From these, the molecular weight can be extrapolated.

Diffusion and the Diffusion Coefficient

The sedimentation-velocity method for determination of the molecular weight of a protein requires knowledge of its diffusion coefficient. Since diffusion has broad biological significance, a brief discussion of diffusion and the diffusion coefficient is warranted. In a solution of a protein at equilibrium the distribution of the solute is statistically uniform throughout the solution, although the protein molecules are in constant thermal motion. If a concentration gradient of the protein is formed, e.g., by carefully layering pure water over a solution of the protein in water, the protein molecules will tend to move from the region of high concentration in the lower layer to the region of low concentration in the upper layer. At equilibrium the protein molecules will be uniformly randomized throughout the system. Such a net movement of solute molecules in response to a concentration gradient is called _diffusion_.

The rate of diffusion is given by Fick's first law of diffusion: the amount of solute ds diffusing across the area A in
a period of time dt is proportional to the concentration gradient dc/dx at that point:

$$\frac{ds}{dt} = -DA\,\frac{dc}{dx}$$

The proportionality constant D is the *diffusion coefficient*; it
is defined as the quantity of solute diffusing per second
across a surface area of 1.0 cm² when there is a concentration
gradient of unity. Since diffusion is in the direction of the
lower concentration, the sign of the expression is negative.
The diffusion coefficient is a function of the size and shape
of the molecule and the frictional resistance offered by the
viscosity of the solvent. For spherical macromolecules the
diffusion coefficient is inversely proportional to the cube
root of the molecular weight.

The diffusion coefficient of a protein can be determined
by measuring its upward-migration rate after pure solvent is
layered over a protein solution of known concentration. The
change with time in the concentration of the protein at a specific point in the cell (Figure 7-18) is followed optically. The
diffusion coefficient of proteins decreases with increasing
molecular weight (Table 7-5). However, it will be noted that
serum albumin, which has the same shape but twice the
molecular weight of β-lactoglobulin, has a diffusion coefficient that is only 23 percent less, in agreement with the fact
that the diffusion coefficient is inversely proportional to the

Figure 7-18
Diffusion of proteins.

*The protein distribution in the diffusion cell
at zero time and at 100 h.*

*Graphical representation showing the con-
centration of protein in the cell at different
time intervals.*

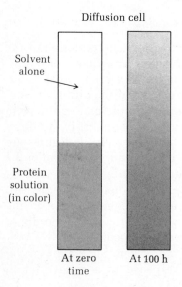

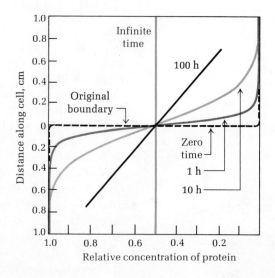

cube root of the molecular weight. Diffusion is opposed by the frictional resistance of the solvent, a very sensitive function of the radius of the particle. For this reason, the diffusion coefficient alone is not a useful measure of the molecular weight of a protein, but combined with sedimentation-rate measurements, it can yield quite accurate values for spherical proteins.

Diffusion is a fundamental process in all cellular transport activities. The rate of diffusion and the diffusion-path length of various metabolites and enzymes are believed to set physical limits on the size and volume of the metabolizing mass of living cells and their organelles.

Determining Molecular Weight by Light Scattering

When a beam of light is passed through a protein solution in a darkened room, the path of the beam can be seen because the light is scattered by the protein molecules. This is called the *Tyndall effect*. From the wavelength of the incident radiation, the intensity of the scattered light, the refractive index of the solvent and solute, and the concentration of the solute, the molecular weight of the protein can be calculated. Since the measurements can be made instantaneously and recorded with time, the method can be used to study rapid changes in molecular weight, like those occurring during dissociation or polymerization of a protein. No other method for measuring molecular weight permits such continuous measurements; however, extraneous dust particles produce very large errors and must be removed by careful filtration.

Determining Molecular Weight by Molecular-Exclusion Chromatography

We have seen that protein mixtures can be sorted out on the basis of molecular weight by molecular-exclusion chromatography (page 159). This simple method, which requires no complex equipment, can yield surprisingly accurate determinations of the molecular weight of a protein (Figure 7-5). Molecular-exclusion columns measure not the true molecular weight of an unknown protein but its *Stokes radius*, which is most simply defined as the radius of a perfect unhydrated sphere having the same rate of passage through the column as the unknown protein in question. If the unknown and marker proteins are spherical, the method yields the molecular weight directly.

Molecular-exclusion chromatography has another unique advantage: it can yield the Stokes radius or approximate molecular weight of a given protein even in very complex mixtures, providing the protein has a characteristic biological activity or property that can be measured. For example, although a crude cell extract may contain hundreds of different enzymes, it is often possible to determine the approximate molecular weight of a single type of enzyme in this extract without isolating it, simply by passing the extract through a Sephadex column and determining the position of the peak of the enzyme's catalytic activity in the eluates. The presence of other proteins is irrelevant since each protein passes through the column independent of the others, at a rate determined by its Stokes radius. Molecular-exclusion columns are also extremely useful for measuring association and dissociation of protein molecules.

Determining the Number and Molecular Weight of Subunits: SDS-Gel Electrophoresis

Many proteins are oligomeric and contain more than one polypeptide chain. By a variation of zone electrophoresis called *SDS-gel electrophoresis,* an oligomeric protein can be dissociated into its subunits and the molecular weight of the subunits determined. The method also can be used to determine the molecular weight of single-chain proteins. The purified protein is treated with the detergent *sodium dodecyl sulfate* (SDS), which dissociates the protein into subunits and completely unfolds each polypeptide chain to form a long, rodlike SDS-polypeptide complex. In this complex the polypeptide chain is coated with a layer of SDS molecules in such a way that their hydrocarbon chains are in tight hydrophobic association with the polypeptide chain and the charged sulfate groups of the detergent are exposed to the aqueous medium. Such complexes contain a constant ratio of SDS to protein (about 1.4:1 by weight) and differ only in mass. When an SDS-treated single-chain protein is subjected to electrophoresis in a molecular-sieve gel containing SDS, its rate of migration is determined primarily by the mass of the SDS-polypeptide particle, through the molecular-exclusion principle. The electric field simply supplies the driving force for the molecular sieving. To calibrate a given gel system, marker proteins of known molecular weight are run for comparison.

Many other specialized procedures are available for characterization of protein molecules according to molecular weight, shape, and conformation, to which the References will provide an introduction.

Summary

Proteins can be separated from each other on the basis of differences in (1) size, (2) solubility properties, (3) electric charge, (4) adsorption behavior, and (5) the biological affinity of a protein for a specific ligand. Separation of proteins on the basis of molecular size can be carried out by different forms of density-gradient centrifugation and by molecular-exclusion chromatography. Proteins can also be separated on the basis of differential solubility, depending on the four variables of pH, ionic strength, dielectric constant of the medium, and temperature. Salting-out and isoelectric precipitation are especially useful. Separation of proteins on the basis of electric charge depends on their acid-base properties, which are a reflection of the ionizable R groups of the polypeptide chain(s). Each protein possesses a characteristic isoelectric pH, at which it will not move in an electric field. Above the isoelectric pH, it has a net negative charge; below it, a net positive charge.

Mixtures of proteins can be separated on the basis of their relative rates of movement in an electric field, either by free electrophoresis in aqueous solution or by zone electrophoresis in a gel or semisolid support. Disc electrophoresis and isoelectric focusing provide especially great resolving power. Proteins can also be separated by ion-exchange chromatography.

Protein purification involves (1) availability of a specific assay method, (2) a method to release the protein from the cell, either in solution or associated with a subcellular organelle, (3) extraction of the protein from the organelle, if needed, and (4) use of a sequence of different fractionation procedures until maximum and constant specific activity of the protein is attained and homogeneity is established by physicochemical criteria, such as gel electrophoresis or isoelectric focusing.

The molecular weight of a protein can be determined from knowledge of its chemical composition, from osmotic-pressure measurements, from measurement of sedimentation velocity or equilibrium, or by means of gel-exclusion chromatography. The subunit polypeptide chains of proteins can be separated and their molecular weight determined by gel electrophoresis in the presence of detergents such as sodium dodecyl sulfate.

References

Books

BAILEY, J. L.: *Techniques in Protein Chemistry*, 2d ed., Elsevier, New York, 1967.

COHN, E. J., and J. T. EDSALL: *Proteins, Amino Acids, and Peptides*, Reinhold, New York, 1941; reprinted 1958. The development of the modern theory of proteins as electrolytes.

EDSALL, J. T., and J. WYMAN: *Biophysical Chemistry*, vol. 1, Academic, New York, 1958. Detailed treatment of acid-base and electrolyte properties of amino acids, peptides, and proteins.

HASCHEMEYER, R. H., and A. E. V. HASCHEMEYER: *Proteins: A Guide to Study by Physical and Chemical Methods*, Wiley, New York, 1973. A valuable short book. Provides up-to-date references in all aspects of protein separation and characterization.

JAKOBI, W. (ed.): *Enzyme Purification and Related Techniques*, vol. 22 of *Methods in Enzymology*, Academic, New York, 1971.

MORRIS, C. J. O. R., and P. MORRIS: *Separation Methods in Biochemistry*, Pitman, London, 1963.

SCHACHMAN, H. K.: *Ultracentrifugation in Biochemistry*, Academic, New York, 1959. Classic treatise.

TANFORD, C.: *Physical Chemistry of Macromolecules*, Wiley, New York, 1961. The authoritative textbook in this field.

VAN HOLDE, K. E.: *Physical Biochemistry*, Prentice-Hall, Englewood Cliffs, N.J., 1971. An excellent, clearly written short treatment of fundamentals.

WORK, T. S., and E. WORK: *Laboratory Techniques in Biochemistry and Molecular Biology*, vol. 1, North-Holland, London, 1969. Gel electrophoresis and immunoelectrophoresis are described.

Articles

ACKERS, G. K.: "Analytical Gel Chromatography of Proteins," *Adv. Protein Chem.*, 24: 343 (1970).

ANDERSEN, N. G.: "Preparative Particle Separation in Density Gradients," *Quart. Rev. Biophys.*, 1: 217 (1968).

CUATRECASAS, P., and C. B. ANFINSEN: "Affinity Chromatography," *Ann. Rev. Biochem.*, 40: 259–278 (1971).

WEBER, K., and M. OSBORN: "The Reliability of Molecular Weight Determinations by Dodecyl Sulfate-Polyacrylamide Gel Electrophoresis," *J. Biol. Chem.*, 244: 4406 (1969).

WELLNER, D., "Electrofocusing in Gels," *Anal. Chem.*, 43: 597 (1971).

Problems

1. A pure heme protein was found to contain 0.426 percent iron. Calculate its minimum molecular weight.

2. A pure enzyme was found to contain 1.65 percent leucine and 2.48 percent isoleucine by weight. Calculate its minimum molecular weight.

3. Calculate the molecular weight of a pure isoelectric protein if a 1 percent solution gives an osmotic pressure of 46 mm H_2O at 0°C. Assume that it yields an ideal solution.

4. A protein has a partial specific volume of 0.707 cm^3 g^{-1} and a diffusion coefficient of 13.1×10^{-7} cm^2 s^{-1} corrected to water at 20°C. It has a sedimentation coefficient of 2.05S. The density of water at 20°C is 0.998 g cm^{-3}. Calculate its molecular weight using the Svedberg equation.

5. A solution contains 1 mg ml^{-1} of myosin and 10^{14} latex particles per milliliter. When a given volume of this solution is dried on a grid and viewed under the electron microscope, a typical field contains 122 protein molecules and 10 latex particles. Calculate the molecular weight of myosin.

6. In what direction, i.e., toward anode (A), toward cathode (C), or stationary (0), will the following proteins migrate in an electric field at the pH indicated? (Use data in Table 7-1.)
   (a) Egg albumin at pH 5.0
   (b) β-Lactoglobulin at pH 5.0; at pH 7.0
   (c) Chymotrypsinogen at pH 5.0, 9.5, and 11

7. Electrophoresis at what pH would be most effective in separating the following protein mixtures? (Use data in Table 7-1.)
   (a) Serum albumin and hemoglobin
   (b) Myoglobin and chymotrypsinogen
   (c) Egg albumin, serum albumin, and urease

8. Predict the sequence in which the following proteins would be eluted from a molecular-exclusion column with a protein-fractionation range of 5,000 to 400,000: myoglobin, catalase, cytochrome c, myosin, chymotrypsinogen, and serum albumin (see Table 7-5 for molecular weights).

9. What is the molecular weight of an unknown protein if the elution volumes of cytochrome c, β-lactoglobulin, the unknown protein, and hemoglobin from a molecular-exclusion column as described in Problem 8 are 118, 58, 37, and 24 ml, respectively? Assume that all the proteins are spherical and fall within the protein-fractionation range of the column.

CHAPTER 8 ENZYMES: KINETICS AND INHIBITION

Enzymes are proteins specialized to catalyze biological reactions. They are among the most remarkable biomolecules known because of their extraordinary specificity and catalytic power, which are far greater than those of man-made catalysts.

Much of the history of biochemistry is the history of enzyme research. The name enzyme ("in yeast") was not used until 1877, but much earlier it was suspected that biological catalysts are involved in the fermentation of sugar to form alcohol (hence the earlier name "ferments"). The first general theory of chemical catalysis, published in 1835 by J. J. Berzelius, included an example of what is now known as an enzyme, diastase of malt, and pointed out that hydrolysis of starch is more efficiently catalyzed by diastase than by sulfuric acid.

Although Louis Pasteur recognized that fermentation is catalyzed by enzymes, he postulated in 1860 that they are inextricably linked with the structure and life of the yeast cell. It was therefore a major landmark in the history of enzyme research when, in 1897, Eduard Buchner succeeded in extracting from yeast cells the enzymes catalyzing alcoholic fermentation. This achievement clearly demonstrated that these important enzymes, which catalyze a major energy-yielding metabolic pathway, can function independently of cell structure. However, it was not until many years later that an enzyme was first isolated in pure crystalline form. This was accomplished by J. B. Sumner in 1926 for the enzyme urease, isolated from extracts of the jack bean. Sumner presented evidence that the crystals consist of protein, and he concluded, contrary to prevailing opinion, that enzymes are proteins. His views were not immediately accepted, however, and it was not until the period 1930 to 1936, during which J. Northrop crystallized the enzymes pepsin, trypsin, and chymotrypsin (Figure 8-1), that the protein nature of enzymes was firmly established. Today nearly 2,000 different enzymes are known. Many have been isolated in pure homogeneous form, and at least 200 have been crystallized.

Figure 8-1
Crystals of bovine chymotrypsin.

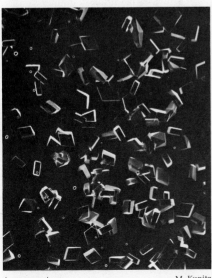

M. Kunitz

0.1 mm

Although most of the enzymes concerned with the basic metabolic housekeeping of the cell have been identified, many important problems remain to be solved, including the genetic control of enzyme synthesis, the molecular mechanisms by which enzyme activity is regulated, and the role of multiple forms of certain enzymes in development and differentiation. Above all, we still do not know in molecular terms how enzymes catalyze chemical reactions with such efficiency, precision, and specificity.

In this and the following chapter no attempt is made to catalog and describe the large number of different enzymes known today. Instead the properties and characteristics common to most enzymes will be examined. Specific enzymes participating in various metabolic cycles will be discussed in more detail in succeeding chapters.

Naming and Classification of Enzymes

Many enzymes have been named by adding the suffix -*ase* to the name of the *substrate*, i.e., the molecule on which the enzyme exerts catalytic action. For example, *urease* catalyzes hydrolysis of urea to ammonia and CO_2, *arginase* catalyzes the hydrolysis of arginine to ornithine and urea, and *phosphatase* the hydrolysis of phosphate esters. However, this nomenclature has not always been practical, with the result that many enzymes have been given chemically uninformative names, e.g., pepsin, trypsin, and catalase. For this reason and because the number of newly discovered enzymes is increasing rapidly, a systematic classification of enzymes has been adopted on the recommendation of an international enzyme commission. The new system divides enzymes into six major classes and sets of subclasses, according to the type of reaction catalyzed (Table 8-1). Each enzyme is assigned a *recommended name*, usually short and appropriate for everyday use, a *systematic name*, which identifies the reaction it catalyzes, and a *classification number*, which is used where accurate and unambiguous identification of an enzyme is required, as in international research journals, abstracts, and indexes. An example is given by the enzyme catalyzing the reaction

$$\text{ATP} + \text{creatine} \rightleftharpoons \text{ADP} + \text{phosphocreatine}$$

The recommended name of this enzyme, that normally used, is *creatine kinase*, and the systematic name, based on the reaction catalyzed, is *ATP:creatine phosphotransferase*. Its classification number is EC 2.7.3.2, where EC stands for Enzyme Commission, the first digit (2) for the class name (transferases), the second digit (7) for the subclass (phosphotransferases), the third digit (3) for the sub-subclass (phosphotransferases with a nitrogenous group as acceptor), and the fourth digit (2) designates creatine kinase (see Table 8-1). In this book we shall use the recommended names of enzymes, as listed in the 1973 edition of *Enzyme Nomenclature* (see References), with a few exceptions.

Table 8-1 International classification of enzymes (class names, code numbers, and types of reactions catalyzed)

1. Oxido-reductases (oxidation-reduction reactions)

   1.1 Acting on $\diagdown$CH—OH

   1.2 Acting on $\diagdown$C=O

   1.3 Acting on $\diagdown$C=CH—

   1.4 Acting on $\diagdown$CH—NH$_2$

   1.5 Acting on $\diagdown$CH—NH—

   1.6 Acting on NADH; NADPH

2. Transferases (transfer of functional groups)

   2.1 One-carbon groups

   2.2 Aldehydic or ketonic groups

   2.3 Acyl groups

   2.4 Glycosyl groups

   2.7 Phosphate groups

   2.8 S-containing groups

3. Hydrolases (hydrolysis reactions)

   3.1 Esters

   3.2 Glycosidic bonds

   3.4 Peptide bonds

   3.5 Other C—N bonds

   3.6 Acid anhydrides

4. Lyases (addition to double bonds)

   4.1 $\diagdown$C=C$\diagup$

   4.2 $\diagdown$C=O

   4.3 $\diagdown$C=N—

5. Isomerases (isomerization reactions)

   5.1 Racemases

6. Ligases (formation of bonds with ATP cleavage)

   6.1 C—O

   6.2 C—S

   6.3 C—N

   6.4 C—C

Enzyme Cofactors

Some enzymes depend for activity only on their structure as proteins, while others also require one or more nonprotein components, called *cofactors*. The cofactor may be a <u>metal ion</u> or an organic molecule called a <u>coenzyme</u>; some enzymes require both. Cofactors are generally stable to heat, whereas most enzyme proteins lose activity on heating. The catalytically active enzyme-cofactor complex is called the *holoenzyme*. When the cofactor is removed, the remaining protein, which is catalytically inactive by itself, is called an *apoenzyme*.

Table 8-2 lists some enzymes requiring metal ions as cofactors. In such enzymes the metal ion may serve as (1) the primary catalytic center; (2) a bridging group, to bind substrate and enzyme together through formation of a coordination complex; or (3) an agent stabilizing the conformation of the enzyme protein in its catalytically active form. Enzymes requiring metal ions are sometimes called <u>metalloenzymes</u>. In some metalloenzymes the metal component alone already possesses primitive catalytic activity, which is greatly enhanced by the enzyme protein; e.g., the iron-porphyrin enzyme <u>catalase</u>, which catalyzes very rapid decomposition of hydrogen peroxide to water and oxygen. Simple iron salts also catalyze this reaction but at a much lower rate.

Table 8-3 summarizes the principal coenzymes and the types of enzymatic reactions in which they participate. Each of the coenzymes listed contains as part of its structure a molecule of one or another of the *vitamins*, trace organic substances that are vital to the function of all cells and required in the diet of certain species. The vitamins and their coenzyme forms are treated in Chapter 13 (page 335). Coenzymes usually function as intermediate carriers of functional groups, of specific atoms, or of electrons that are transferred in the overall enzymatic reaction. When the coenzyme is very tightly bound to the enzyme molecule, it is usually called a *prosthetic group*, e.g., the biocytin group of acetyl-CoA carboxylase, which is covalently incorporated in the

Table 8-2 Some enzymes containing or requiring metal ions as cofactors

Zn^{2+}
  Alcohol dehydrogenase
  Carbonic anhydrase
  Carboxypeptidase

Mg^{2+}
  Phosphohydrolases
  Phosphotransferases

Mn^{2+}
  Arginase
  Phosphotransferases

Fe^{2+} or Fe^{3+}
  Cytochromes
  Peroxidase
  Catalase
  Ferredoxin

Cu^{2+} (Cu^+)
  Tyrosinase
  Cytochrome oxidase

K^+
  Pyruvate kinase (also requires Mg^{2+})

Na^+
  Plasma membrane ATPase (also requires K^+ and Mg^{2+})

Table 8-3 Coenzymes in group-transferring reactions

Coenzyme	Entity transferred
Nicotinamide adenine dinucleotide	Hydrogen atoms (electrons)
Nicotinamide adenine dinucleotide phosphate	Hydrogen atoms (electrons)
Flavin mononucleotide	Hydrogen atoms (electrons)
Flavin adenine dinucleotide	Hydrogen atoms (electrons)
Coenzyme Q	Hydrogen atoms (electrons)
Thiamin pyrophosphate	Aldehydes
Coenzyme A	Acyl groups
Lipoamide	Acyl groups
Cobamide coenzymes	Alkyl groups
Biocytin	Carbon dioxide
Pyridoxal phosphate	Amino groups
Tetrahydrofolate coenzymes	Methyl, methylene, formyl or formimino groups

polypeptide chain (page 345). In some cases, however, the coenzyme is only loosely bound and essentially functions as one of the specific substrates of that enzyme.

Chemical Kinetics

Before we examine the catalysis of reactions by enzymes, some relationships and terms used in measuring and expressing the rates of chemical reactions must be outlined. Chemical reactions may be classified on the basis of the number of molecules that must ultimately react to form the reaction products. Thus, we have _monomolecular_, _bimolecular_, and _termolecular_ reactions, in which one, two, or three molecules, respectively, undergo reaction.

Chemical reactions are also classified on a kinetic basis, by _reaction order_, and we have zero-order, first-order, second-order, and third-order reactions, depending on how the reaction rate is influenced by the concentration of the reactants under a given set of conditions.

First-order reactions are those which proceed at a rate exactly proportional to the concentration of _one_ reactant (Figure 8-2). The simplest example is when the rate of the reaction

$$A \longrightarrow P$$

is exactly proportional to the concentration of A. Then the reaction rate at any time t is given by the _first-order rate equation_

$$\frac{-d[A]}{dt} = k[A]$$

where [A] is the molar concentration of A and $-d[A]/dt$ is the rate at which the concentration of A decreases. The proportionality constant k is called the _rate constant_ or _specific reaction rate_. First-order rate constants have the dimensions of reciprocal time, usually s^{-1}.

The integrated form of this equation, which is more useful for carrying out kinetic calculations, is

$$\log \frac{[A_0]}{[A]} = \frac{kt}{2.303}$$

in which $[A_0]$ is the concentration of A at zero time and [A] is the concentration at time t.

In first-order reactions, the half-time $t_{1/2}$ of the reaction is given by

$$t_{1/2} = \frac{0.693}{k}$$

a relationship that is simply derived. In first-order reactions the half-time is independent of the initial concentration of substrate.

Second-order reactions are those in which the rate is proportional to the product of the concentrations of _two_ reac-

Figure 8-2
Plot of the course of a first-order reaction. The half-time $t_{1/2}$ is the time required for one-half of the initial reactant to be consumed.

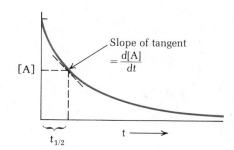

tants or to the second power of a single reactant. The simplest example is the reaction

$$A + B \longrightarrow P$$

The rate of this reaction, which may be designated as $-d[A]/dt$, $-d[B]/dt$, or $+d[P]/dt$, is proportional to the product of the concentrations of A and B, as given by the *second-order rate equation*

$$\frac{-d[A]}{dt} = k[A][B]$$

where k is the *second-order rate constant*. If the reaction has the form

$$2A \longrightarrow P$$

and its rate is proportional to the product of the concentration of the two reacting molecules, the second-order rate equation is

$$\frac{-d[A]}{dt} = k[A][A] = k[A]^2$$

The rate constants of second-order reactions have the dimensions $1/(\text{concentration} \times \text{time})$, or M^{-1} s^{-1}. The integrated form of the second-order rate equation is

$$t = \frac{2.303}{k([A_0] - [B_0])} \log \frac{[B_0][A]}{[A_0][B]}$$

where $[A_0]$ and $[B_0]$ are initial concentrations and $[A]$ and $[B]$ the concentrations at time t.

For second-order reactions in which the initial concentrations of the reactants are equal, the half-time is equal to $1/C_0 k$, where C_0 is the initial concentration of reactants and k the second-order rate constant.

It is important to note that a second-order reaction such as

$$A + B \longrightarrow P$$

may under some conditions *appear* to be a first-order reaction. For example, if the concentration of B is very high and that of A very low, this reaction might appear to be first-order because its rate will be nearly proportional to the concentration of only one reactant, namely, A. Under these special conditions the reaction is an *apparent-* or *pseudo-first-order reaction*.

Third-order reactions, which are relatively rare, are those whose velocity is proportional to the product of three concentration terms. Some chemical reactions are independent of the concentration of any reactant; these are called *zero-order* reactions. Many catalyzed reactions are zero order with respect to the reactants. When this is true, the rate of reaction depends on the concentration of the catalyst or on some factor other than the concentration of the molecular species undergoing reaction. Reaction rates need not necessarily be pure first order or pure second order; often reactions are of mixed order.

The Free Energy of Activation and the Effects of Catalysts

A chemical reaction such as A $\longrightarrow$ P takes place because a certain fraction of the population of A molecules at any given instant possesses enough energy to attain an activated condition, called the *transition state*, in which the probability is very high that a chemical bond will be made or broken to form the product P. This transition state is at the top of the energy barrier separating the reactants and products (Figure 8-3). The rate of a given chemical reaction is proportional to the concentration of this transition-state species. The *free energy of activation* $\Delta G^{\ddagger}$ (the symbol $\ddagger$ designates the activation process) is the amount of energy required to bring all the molecules in 1 mol of a substance at a given temperature to the transition state at the top of the activation barrier.

There are two general ways in which the rate of a chemical reaction may be accelerated. A rise in temperature, because it increases thermal motion and energy, increases the number of molecules capable of entering the transition state and thus accelerates the rate of chemical reactions. In many reactions, the reaction rate is approximately doubled by a 10°C rise in temperature. The rate of a chemical reaction

Figure 8-3
Energy diagram for a chemical reaction, uncatalyzed and catalyzed.

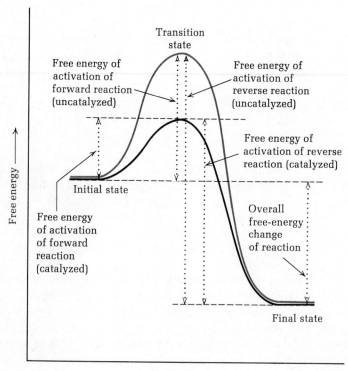

Table 8-4   Estimated free energy of activation ΔG^+ for the decomposition of hydrogen peroxide at 20°C; catalase accelerates the rate of the reaction by more than 10^8-fold

Conditions	ΔG^+ kcal mol^{-1}
Uncatalyzed	18
Catalyzed by colloidal platinum	13
Catalyzed by catalase	7

can also be accelerated by addition of a catalyst. Catalysts combine transiently with the reactants to produce a transition state having a lower energy of activation than the transition state of the uncatalyzed reaction. Thus they accelerate chemical reactions by lowering the energy of activation (Figure 8-3). When the reaction products are formed, the free catalyst is regenerated. Table 8-4 shows the energy of activation for the decomposition of hydrogen peroxide under noncatalyzed and catalyzed conditions.

Kinetics of Enzyme-Catalyzed Reactions: The Michaelis-Menten Equation

The general principles of chemical-reaction kinetics apply to enzyme-catalyzed reactions, but they also show a distinctive feature not usually observed in nonenzymatic reactions, _saturation_ with substrate. In Figure 8-4 we see the effect of the substrate concentration on the rate of the enzyme-catalyzed reaction A ⟶ P. At a low substrate concentration, the initial reaction velocity v_0 is nearly proportional to the substrate concentration, and the reaction is thus approximately first order with respect to the substrate. However, as the substrate concentration is increased, the initial rate increases less, so that it is no longer nearly proportional to the substrate concentration; in this zone, the reaction is mixed order. With a further increase in the substrate concentration, the reaction rate becomes essentially independent of substrate concentration and asymptotically approaches a constant rate. In this range of substrate concentrations the reaction is essentially zero order with respect to the substrate and the enzyme is spoken of as being _saturated_ with its substrate. All enzymes show the saturation effect, but they vary widely with respect to the substrate concentration required to produce it. This saturation effect led some early investigators, particularly A. J. Brown and also V. Henri, to the hypothesis that the enzyme and substrate react reversibly to form a complex, as an essential step in the catalyzed reaction.

In 1913 a general theory of enzyme action and kinetics was developed by L. Michaelis and M. L. Menten, which was later extended by G. E. Briggs and J. B. S. Haldane. This theory, which is basic to the quantitative analysis of all as-

Figure 8-4
Effect of substrate concentration on the rate of an enzyme-catalyzed reaction.

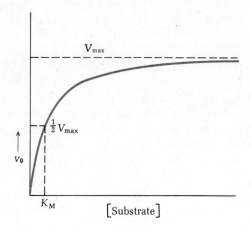

pects of enzyme kinetics and inhibition, is best developed for the simple case of a reaction in which there is only one substrate. The Michaelis-Menten theory assumes that the enzyme E first combines with the substrate S to form the enzyme-substrate complex ES; the latter then breaks down in a second step to form free enzyme and the product P:

$$E + S \underset{k_{-1}}{\overset{k_{+1}}{\rightleftharpoons}} ES \tag{1}$$

$$ES \underset{k_{-2}}{\overset{k_{+2}}{\rightleftharpoons}} E + P \tag{2}$$

These reactions are assumed to be reversible; the rate constants for the forward and reverse directions respectively have a positive and a negative subscript.

We now derive the Michaelis-Menten equation, which expresses the mathematical relationship between the initial rate of an enzyme-catalyzed reaction, the concentration of the substrate, and certain characteristics of the enzyme. The Michaelis-Menten equation is the _rate equation_ for reactions catalyzed by enzymes having a single substrate. In this derivation, that of Briggs and Haldane, [E] represents the concentration of the free or uncombined enzyme, [ES] the concentration of the enzyme-substrate complex, and [E_T] the total enzyme concentration (the sum of the free and combined forms). [S] represents the substrate concentration, which is assumed to be far greater than [E], so that the amount of S bound by E at any given time is negligible compared with the total concentration of S.

It is the purpose of this derivation to define a general expression for v_0, the initial velocity of an enzyme-catalyzed reaction, assuming that enzyme-catalyzed reactions take place in two steps, as shown in reactions (1) and (2). The initial velocity is of course equal to the rate of breakdown of the enzyme-substrate complex ES, according to equation (2), for which we can write the first-order rate equation

$$v_0 = k_{+2}[ES] \tag{3}$$

However, since neither k_{+2} nor [ES] can be determined directly, we must find an alternative expression for v_0 in terms of other variables that can be measured more readily. To do this we first write the second-order rate equation for the _formation_ of ES from E and S [see reaction (1)]:

$$\frac{d[ES]}{dt} = k_{+1}([E_T] - [ES])\,[S] \tag{4}$$

in which k_{+1} is the second-order rate constant. Although ES can also be formed from E and P by reversal of reaction (2), the rate of this back reaction may be neglected, since we are considering the beginning of the reaction in the forward direction, when [S] is very high and [P] is zero or close to zero.

Next we may write the rate equation for the _breakdown_ of ES by the sum of two reactions; first, the reaction yielding the product (forward direction) and, second, the reaction yielding E + S [the reverse direction of equation (1)]. We then have

$$\frac{-d[ES]}{dt} = k_{-1}[ES] + k_{+2}[ES] \tag{5}$$

When the rate of *formation* of ES is equal to its rate of *breakdown*, i.e., when the reaction system has entered the <u>steady state</u>, defined as the condition in which the concentration of ES remains constant, then

$$k_{+1}([E_T] - [ES])[S] = k_{-1}[ES] + k_{+2}[ES] \tag{6}$$

Figure 8-5 illustrates the course with time of the various participants.

Figure 8-5
Time-course of the formation of an enzyme-substrate complex and initiation of the steady state, as derived from computer solutions of data obtained in an actual experiment on a typical enzyme. The portion shaded in the top graph is shown in magnified form on the lower graph.

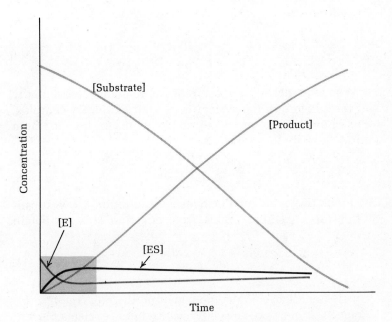

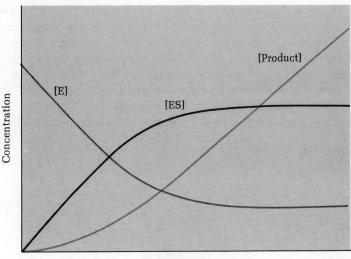

Rearranging equation (6), we obtain

$$\frac{[S]([E_T] - [ES])}{[ES]} = \frac{k_{-1} + k_{+2}}{k_{+1}} = K_M \tag{7}$$

The lumped constant K_M, which replaces the term $(k_{-1} + k_{+2})/k_{+1}$, is called the *Michaelis-Menten constant*.

From this equation the steady-state concentration of the ES complex can be obtained by solving for [ES]:

$$[ES] = \frac{[E_T][S]}{K_M + [S]} \tag{8}$$

We have seen (above) that the initial rate v_0 of an enzymatic reaction is

$$v_0 = k_{+2}\,[ES] \tag{3}$$

We can now substitute for the term [ES] in equation (3) its value from equation (8):

$$v_0 = k_{+2}\,\frac{[E_T][S]}{K_M + [S]} \tag{9}$$

When the substrate concentration is so high that essentially all the enzyme in the system is present as the ES complex, i.e., when the enzyme is *saturated*, we reach the maximum initial velocity V_{max}, given by

$$V_{max} = k_{+2}[E_T] \tag{10}$$

in which $[E_T]$ is the total enzyme concentration. Now, substituting for $k_{+2}\,[E_T]$ its value from equation (10) we obtain

$$v_0 = \frac{V_{max}\,[S]}{K_M + [S]} \tag{11}$$

This is the Michaelis-Menten equation, the <u>rate equation</u> for a one-substrate enzyme-catalyzed reaction. It relates the initial velocity, the maximum velocity, and the initial substrate concentration through the Michaelis-Menten constant. It is important to note that although the Michaelis-Menten equation *appears* to have no term for enzyme concentration, it is actually contained in the term V_{max}, which we have seen is equal to $k_{+2}\,[E_T]$.

An important numerical relationship emerges from the Michaelis-Menten equation in the special case when the initial reaction rate is exactly one-half the maximum velocity, i.e., when $v_0 = \frac{1}{2}V_{max}$ (Figure 8-4).

$$\frac{V_{max}}{2} = \frac{V_{max}\,[S]}{K_M + [S]}$$

If we divide by V_{max}, we obtain

$$\frac{1}{2} = \frac{[S]}{K_M + [S]}$$

On rearranging, this becomes

$$K_M + [S] = 2[S]$$

$$K_M = [S]$$

Thus we see that K_M, *the Michaelis-Menten constant, is equal to the substrate concentration at which the initial reaction velocity is half maximal.* K_M for a one-substrate reaction usually has the dimensions moles per liter and is independent of the enzyme concentration.

The value of K_M for any given enzyme can easily be approximated from a series of simple experiments in which the initial reaction velocity is measured at different initial concentrations of the substrate with a fixed concentration of enzyme. The approximate value of K_M is obtained graphically from a plot of initial velocity vs. initial substrate concentration (Figure 8-4), which has the form of a rectangular hyperbola. At very low substrate concentrations, the initial velocity v_0 is nearly proportional to [S]; that is, the reaction shows essentially first-order behavior. At very high substrate concentrations the reaction rate approaches V_{max} asymptotically and is essentially zero order, i.e., nearly independent of substrate concentration. A few enzymes, such as catalase, appear not to show saturation with substrate; this is because the rate of decomposition of the ES complex to form products is so fast that it cannot easily be made rate-limiting.

Table 8-5 gives the K_M values for a number of enzymes. Note that K_M is not a fixed value but may vary with the structure of the substrate, with pH, and with temperature. For enzymes having more than one substrate, each substrate has a characteristic K_M. Under intracellular conditions, enzymes are not necessarily saturated with their substrates. The maximum velocity V_{max}, which, we recall, is equal to $k_{+2} [E_T]$, also varies widely from one enzyme to another for a given enzyme concentration. V_{max} also varies with the structure of the substrate (Table 8-6), with pH, and with temperature.

The Michaelis constant of an enzyme is an important and useful characteristic, fundamental not only to the mathematical description of enzyme kinetics but also to the quantitative assay of enzyme activity in tissues and enzyme purification. Moreover, the substrate concentration yielding half-maximal velocity provides a useful index for the analysis of some enzyme regulatory mechanisms (page 236). A striking example from recent medical research shows the usefulness of K_M in another way. Some types of animal and human leukemia (a form of cancer in which white blood cells proliferate abnormally) can be suppressed by intravenous administration of the enzyme *asparaginase*, which catalyzes the reaction

$$\text{Asparagine} + H_2O \rightleftharpoons \text{aspartate} + NH_4^+$$

This finding led to the conclusion that asparagine present in the blood is an essential nutrient for the growth of the malignant white cells; intravenous asparaginase causes hydrolysis of asparagine to aspartate, which cannot satisfy the require-

Table 8-5 K_M for some enzymes

Enzyme and substrate	K_M, mM
Catalase	
H_2O_2	25
Hexokinase	
Glucose	0.15
Fructose	1.5
Chymotrypsin	
N-Benzoyltyrosinamide	2.5
N-Formyltyrosinamide	12.0
N-Acetyltyrosinamide	32
Glycyltyrosinamide	122
Carbonic anhydrase	
HCO_3^-	9.0
Glutamate dehydrogenase	
Glutamate	0.12
α-Ketoglutarate	2.0
NH_4^+	57
NAD_{ox}	0.025
NAD_{red}	0.018
Aspartate aminotransferase	
Aspartate	0.9
α-Ketoglutarate	0.1
Oxaloacetate	0.04
Glutamate	4.0

Table 8-6 Effect of substrate structure on V_{max} for D-amino acid oxidase (relative to D-alanine = 100)

Substrate	Relative V_{max}
D-Tyrosine	297
D-Proline	231
D-Methionine	125
D-Alanine	100
D-Valine	55
D-Histidine	9.7
Glycine	0.0

ment for asparagine. During a search for sources of asparaginase suitable for treatment of leukemia, the puzzling
discovery was made that not all asparaginases are effective in
suppressing experimental leukemia. The reason was finally
found: asparaginases from different animal, plant, and bacterial sources differ widely in their K_M for asparagine. Since
the concentration of asparagine in the blood is very low, the
administration of an asparaginase from another species can
be therapeutically effective only if its K_M value is low enough
to hydrolyze asparagine rapidly at the low concentration at
which it is present in the blood.

The kinetic behavior of most enzymes is more complex
than that of the simple, idealized one-substrate reaction we
have just discussed. For one thing, our formulation has assumed that in a one-substrate reaction there is only one
enzyme-substrate complex, but it now appears likely that
many one-substrate reactions may involve two or three complexes, as indicated in the sequence

$$E + S \rightleftharpoons ES \rightleftharpoons EZ \rightleftharpoons EP \rightleftharpoons E + P$$

where EZ is an intermediate complex and EP an enzyme-
product complex. Moreover, only a minority of enzymatic
reactions have a single substrate; most enzymes have two or
more substrates and may have two or more products. Kinetic
analysis of enzymatic reactions involving multiple reactants
and multiple products is complex; nevertheless, the Michaelis-Menten relationship remains the starting point for
analysis of the kinetics of all enzymatic reactions.

The Michaelis Constant K_M and the Substrate Constant K_S

The Michaelis constant, as we noted above, is an experimentally determined, operationally defined quantity: the substrate concentration at which the reaction velocity is half
maximal. In the idealized case used in the derivation above
it is represented by

$$K_M = \frac{k_{-1} + k_{+2}}{k_{+1}} \tag{7}$$

but in some enzymatic reactions k_{-1} is very large compared
with k_{+2}, in which case the rate constant k_{+2} becomes negligibly small and equation (7) simplifies to the expression

$$K_M \approx \frac{k_{-1}}{k_{+1}}$$

where K_M is approximately equal to the dissociation constant of the enzyme-substrate complex K_S, also called the
substrate constant:

$$K_S = \frac{[E][S]}{[ES]}$$

Unfortunately, K_M and K_S are frequently but wrongly
regarded as synonymous. K_M should not be regarded as the

dissociation constant of the ES complex unless specific information is available that k_{+2} is very small compared with k_{-1}.

Transformations of the Michaelis-Menten Equation

The Michaelis-Menten relationship [equation (11)] can be algebraically transformed into other forms that are more useful in plotting experimental data. One common transformation is derived simply by taking the reciprocal of both sides of the Michaelis-Menten equation (11):

$$\frac{1}{v_0} = \frac{K_M + [S]}{V_{max} [S]}$$

Rearranging, we have

$$\frac{1}{v_0} = \frac{K_M}{V_{max}[S]} + \frac{[S]}{V_{max}[S]}$$

which reduces to

$$\frac{1}{v_0} = \frac{K_M}{V_{max}} \frac{1}{[S]} + \frac{1}{V_{max}} \tag{12}$$

Equation (12) is the *Lineweaver-Burk equation*. When $1/v_0$ is plotted against $1/[S]$, a straight line is obtained. This line will have a slope of K_M/V_{max}, an intercept of $1/V_{max}$ on the $1/v_0$ axis, and an intercept of $-1/K_M$ on the $1/[S]$ axis (Figure 8-6). Such a *double-reciprocal plot* has the advantage of allowing a much more accurate determination of V_{max}, which can only be approximated as a limiting value at infinite substrate concentration from a simple plot of v_0 vs. $[S]$, as seen in Figure 8-4. The double-reciprocal plot can also give valuable information on enzyme inhibition, as we shall see later (page 199).

Another useful transformation of the Michaelis-Menten equation is obtained by multiplying both sides of equation (12) by V_{max} and rearranging to yield

$$v_0 = -K_M \frac{v_0}{[S]} + V_{max}$$

A plot of v_0 against $v_0/[S]$, called the *Eadie-Hofstee plot* (Figure 8-7), not only yields V_{max} and K_M in a very simple way but also magnifies departures from linearity which might not be apparent in a double-reciprocal plot.

Effect of pH on Enzymatic Activity

Most enzymes have a characteristic pH at which their activity is maximal; above or below this pH the activity declines. Although the pH-activity profiles of many enzymes are bell-shaped, they may vary considerably in form (Figure 8-8). The pH-activity relationship of any given enzyme depends on the acid-base behavior of enzyme and substrate, as well as many other factors that are usually difficult to ana-

Figure 8-6
A double-reciprocal (Lineweaver-Burk) plot.

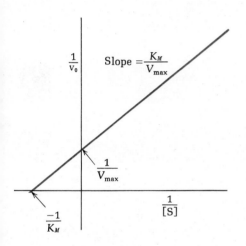

Figure 8-7
An Eadie-Hofstee plot.

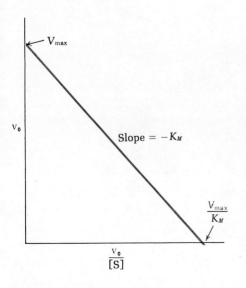

Figure 8-8
The pH-*activity profiles of some enzymes.*

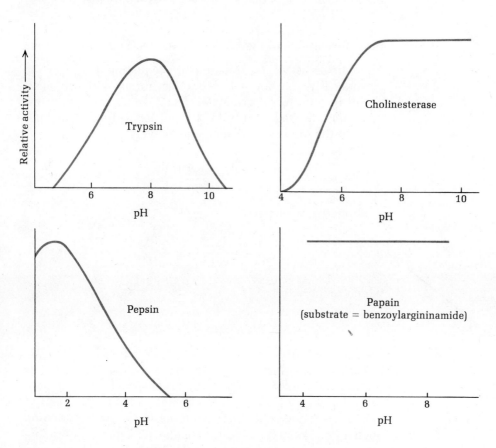

lyze quantitatively. The shape of the pH-activity profile usually varies with substrate concentration, since the K_M of most enzymes changes with pH. Such curves are most meaningful if the enzyme is kept saturated with substrate at all the pH values tested. In most studies of enzyme kinetics the pH is held constant at or near the optimum pH.

The optimum pH of an enzyme is not necessarily identical with the pH of its normal intracellular surroundings, which may be on the ascending or descending slope of its pH-activity profile. This suggests that the pH-activity relationship of an enzyme may be a factor in intracellular control of its activity.

Effect of Temperature on Enzymatic Reactions

As is true for most chemical reactions, the rate of enzyme-catalyzed reactions generally increases with temperature, within the temperature range in which the enzyme is stable and retains full activity. The rate of most enzymatic reactions approximately doubles for each 10°C rise in temperature ($Q_{10} \approx 2.0$). However, the temperature coefficient Q_{10} varies somewhat from one enzyme to another, depending on the energy of activation of the catalyzed reaction, i.e., the height of the energy barrier to the transition state (Figure 8-3).

Although enzyme-catalyzed reactions often *appear* to

Figure 8-9
Effect of temperature on the activity of an enzyme. Enzymes differ with respect to their thermal stability. The descending portion of the curve is due to thermal denaturation.

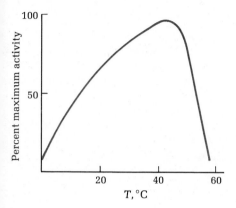

have an optimum temperature (Figure 8-9), the peak in such a plot of catalytic activity vs. temperature results because enzymes, being proteins, are denatured by heat and become inactive as the temperature is raised beyond a certain point. The apparent temperature "optimum" is thus the resultant of two processes: (1) the usual increase in reaction rate with temperature and (2) the increasing rate of thermal denaturation of the enzyme above a critical temperature. Although most enzymes are inactivated at temperatures above about 55 to 60°C, some are quite stable and retain activity at much higher temperatures, e.g., enzymes of various species of thermophilic bacteria inhabiting hot springs, which are still active at temperatures exceeding 85°C.

Some enzymes, such as ribonuclease, lose activity on heating but quickly regain it on cooling, indicating that their unfolded polypeptide chain quickly snaps back into its native conformation (page 62).

Enzyme Inhibition

From the study of enzyme inhibitors valuable information has been obtained on the mechanism and pathway of enzyme catalysis, the substrate specificity of enzymes, the nature of the functional groups at the active site, and the participation of certain functional groups in maintaining the active conformation of the enzyme molecule. Moreover, inhibition of certain enzymes by specific metabolites is an important element in the regulation of intermediary metabolism.

The three major types of reversible enzyme inhibition, *competitive*, *uncompetitive*, and *noncompetitive*, can be experimentally distinguished by the effects of the inhibitor on the reaction kinetics of the enzyme, which may be analyzed in terms of the basic Michaelis-Menten rate equation. For valid kinetic analysis the inhibitor must combine rapidly *and reversibly* with the enzyme or enzyme-substrate complex. Inhibitors that react with the enzyme slowly and/or irreversibly are considered elsewhere (pages 201 and 220).

Competitive Inhibition

The hallmark of competitive inhibition is that the inhibitor can combine with the free enzyme in such a way that it competes with the normal substrate for binding at the active site. A competitive inhibitor reacts reversibly with the enzyme to form an *enzyme-inhibitor complex* (EI), analogous to the enzyme-substrate complex:

$$E + I \rightleftharpoons EI$$

The inhibitor molecule is not chemically changed by the enzyme. Following the Michaelis-Menten formalism, we can define the *inhibitor constant* K_I as the dissociation constant of the enzyme-inhibitor complex:

$$K_I = \frac{[E][I]}{[EI]}$$

The inhibitor constant K_I is thus comparable to K_S, the dissociation constant of the enzyme-substrate complex.

Competitive inhibition is easily recognized experimentally because the percent inhibition at a fixed inhibitor concentration is decreased by increasing the substrate concentration. For quantitative kinetic analysis, the effect of varying the substrate concentration [S] on the initial velocity v_0 is determined at a fixed concentration of inhibitor. This experiment is then repeated with a different concentration of inhibitor; often several series of such experiments are carried out, each at a different concentration of inhibitor. Plots of $1/v_0$ vs. $1/[S]$ are then prepared, one for each concentration of inhibitor. These plots characteristically give a family of straight lines intersecting at a common intercept on the $1/v_0$ axis (Figure 8-10). The presence of a competitive inhibitor thus increases the apparent K_M of the enzyme for substrate, i.e., causes it to require a higher substrate concentration to achieve its maximum velocity. The apparent K_M for the substrate will be greater than the true K_M by the increase in the intercept on the $1/[S]$ axis. Since the slope of the plot of the uninhibited reaction is K_M/V_{max} (Figure 8-10; Table 8-7) and the slope for the inhibited reaction is $K_M/V_{max}(1 + [I]/K_I)$, the slope is increased by a factor of $1 + [I]/K_I$. From this relationship K_I can be calculated. On the other hand, a competitive inhibitor characteristically does not affect V_{max}, indicating that it does not interfere with the rate of breakdown of the enzyme-substrate complex.

The classic example of competitive competition is the inhibition of succinate dehydrogenase (page 459) by malonate and other dicarboxylate anions (Figure 8-11). Succinate dehydrogenase is a member of the group of enzymes responsible for the reactions of the tricarboxylic acid cycle (page 444). It catalyzes the removal of two hydrogen atoms from the two methylene carbon atoms of succinate (the nature of the hydrogen acceptor is not relevant to this discussion). The competitive inhibitor malonate resembles succinate in having two ionized carboxyl groups at pH 7.0 but differs in not being itself dehydrogenated by succinate dehydrogenase. If sufficient malonate is added to inhibit the dehydrogenation of a given concentration of succinate by, say, 50 percent, increasing the succinate concentration will reduce the percent inhibition by malonate.

Table 8-7   Summary of the effects of inhibitors on Lineweaver-Burk plots $1/V_0$ vs. $1/[S]$

	Slope	Intercept on ordinate
No inhibitor	$\dfrac{K_M}{V_{max}}$	$\dfrac{1}{V_{max}}$
Competitive	$\dfrac{K_M}{V_{max}}\left(1 + \dfrac{[I]}{K_I}\right)$	$\dfrac{1}{V_{max}}$
Uncompetitive	$\dfrac{K_M}{V_{max}}$	$\dfrac{1}{V_{max}}\left(1 + \dfrac{[I]}{K_I}\right)$
Noncompetitive (both K_I's identical)	$\dfrac{K_M}{V_{max}}\left(1 + \dfrac{[I]}{K_I}\right)$	$\dfrac{1}{V_{max}}\left(1 + \dfrac{[I]}{K_I}\right)$

Figure 8-10
Double-reciprocal plots showing the effect of competitive, uncompetitive, and non-competitive inhibition of enzymes.

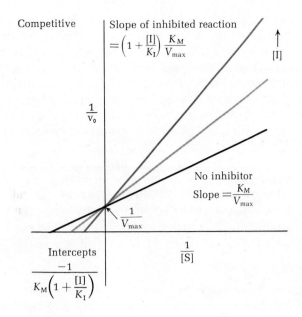

Competitive

Slope of inhibited reaction
$$= \left(1 + \frac{[I]}{K_I}\right)\frac{K_M}{V_{max}}$$

$\frac{1}{v_0}$

[I]

No inhibitor
Slope $= \dfrac{K_M}{V_{max}}$

$\dfrac{1}{V_{max}}$

Intercepts
$$\frac{-1}{K_M\left(1 + \dfrac{[I]}{K_I}\right)}$$

$\dfrac{1}{[S]}$

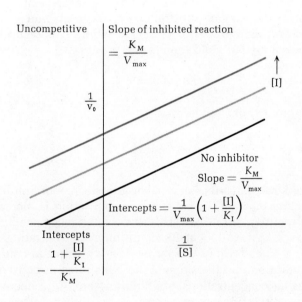

Uncompetitive

Slope of inhibited reaction
$$= \frac{K_M}{V_{max}}$$

$\frac{1}{v_0}$

[I]

No inhibitor
Slope $= \dfrac{K_M}{V_{max}}$

Intercepts $= \dfrac{1}{V_{max}}\left(1 + \dfrac{[I]}{K_I}\right)$

Intercepts
$$-\frac{1 + \dfrac{[I]}{K_I}}{K_M}$$

$\dfrac{1}{[S]}$

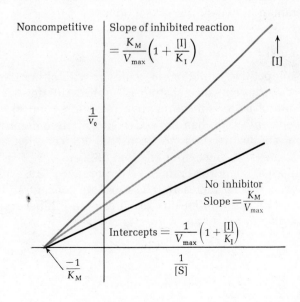

Noncompetitive

Slope of inhibited reaction
$$= \frac{K_M}{V_{max}}\left(1 + \frac{[I]}{K_I}\right)$$

$\frac{1}{v_0}$

[I]

No inhibitor
Slope $= \dfrac{K_M}{V_{max}}$

Intercepts $= \dfrac{1}{V_{max}}\left(1 + \dfrac{[I]}{K_I}\right)$

$\dfrac{-1}{K_M}$

$\dfrac{1}{[S]}$

Besides malonate, anions of certain other dibasic acids capable of assuming the proper distance between the two anionic groups may act as competitive inhibitors of succinate dehydrogenase, e.g., the pyrophosphate anion, leading to the conclusion that the catalytic site of succinate dehydrogenase has two appropriately spaced positively charged groups capable of attracting the two negatively charged carboxylate groups of the substrate. The catalytic site thus shows complementarity to the structure of the substrate.

From the relationship between the molecular structure of a competitive inhibitor and its affinity to the enzyme, as expressed by K_I, valuable information about the structure and geometry of the active site can be obtained. This is an important approach for mapping enzyme active sites.

Uncompetitive Inhibition

In uncompetitive inhibition, which is not very aptly named, the inhibitor does not combine with the free enzyme or affect its reaction with its normal substrate; however, it does combine with the enzyme-substrate complex to give an inactive enzyme-substrate-inhibitor complex, which cannot undergo further reaction to yield the normal product:

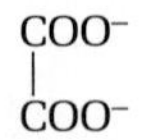

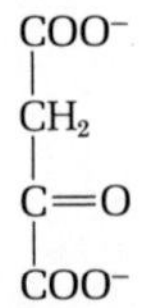

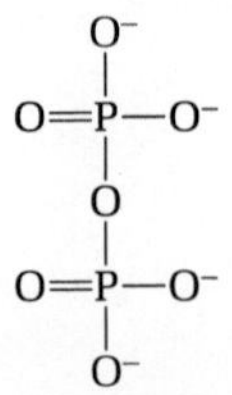

$$ES + I \rightleftharpoons ESI$$

The inhibitor constant is thus

$$K_I = \frac{[ES][I]}{[ESI]}$$

These relationships show that the degree of inhibition may increase when the substrate concentration is increased. Uncompetitive inhibition is most easily recognized from plots of $1/v_0$ vs. $1/[S]$ at fixed inhibitor concentrations. As Figure 8-10 and Table 8-7 show, it is typical of uncompetitive inhibition that the slope of the plots remains constant at increasing concentrations of inhibitor, but V_{max} decreases. Uncompetitive inhibition is rare in one-substrate reactions but common in two-substrate reactions.

Noncompetitive Inhibition

A noncompetitive inhibitor can combine with either the free enzyme or the enzyme-substrate complex, interfering with the action of both. Noncompetitive inhibitors bind to a site on the enzyme other than the active site, often to deform the enzyme, so that it does not form the ES complex at its normal rate and, once formed, the ES complex does not decompose at the normal rate to yield products. These effects are not reversed by increasing the substrate concentration. In noncompetitive inhibition the reaction with inhibitor yields two inactive forms, EI and ESI:

$$E + I \rightleftharpoons EI$$

$$ES + I \rightleftharpoons ESI$$

Figure 8-11
Competitive inhibition of succinate dehydrogenase.

The succinate dehydrogenase reaction.

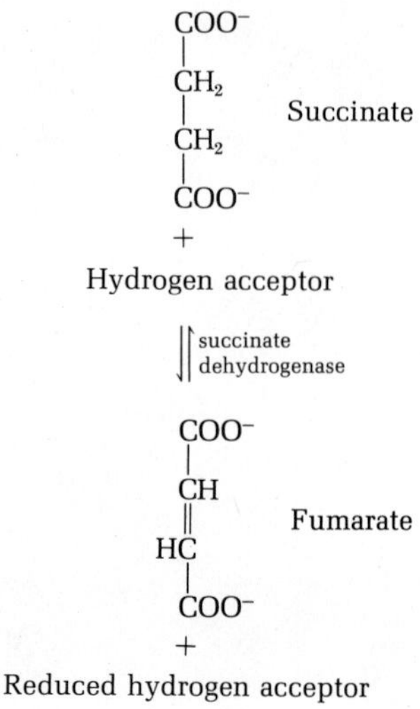

COO⁻
|
CH₂
| Succinate
CH₂
|
COO⁻
+
Hydrogen acceptor

succinate dehydrogenase

COO⁻
|
CH
‖   Fumarate
HC
|
COO⁻
+
Reduced hydrogen acceptor

Some competitive inhibitors of succinate dehydrogenase. Note that all contain two anionic groups whose spacing resembles that in succinate.

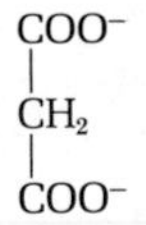

COO⁻
|
CH₂
|
COO⁻
Malonate

COO⁻
|
COO⁻
Oxalate

COO⁻
|
CH₂
|
C=O
|
COO⁻
Oxaloacetate

O⁻
|
O=P—O⁻
|
O
|
O=P—O⁻
|
O⁻
Pyrophosphate

200

for which there are two inhibitor constants:

$$K_I^{EI} = \frac{[E][I]}{[EI]}$$

$$K_I^{ESI} = \frac{[ES][I]}{[ESI]}$$

which may or may not be equal. Noncompetitive inhibition is also most easily recognized from plots of $1/v_0$ vs. $1/[S]$ in the presence of different fixed concentrations of inhibitor (Figure 8-10; Table 8-7). The plots differ in slope but do not share a common intercept on the $1/v_0$ axis. The intercept on the $1/v_0$ axis is greater for the inhibited than the uninhibited enzyme, indicating that V_{max} is decreased by the inhibitor and cannot be restored regardless of how high the substrate concentration may be.

The most common type of noncompetitive inhibition is given by reagents that can combine reversibly with some functional group of the enzyme (outside the active site) that is essential for maintaining the catalytically active three-dimensional conformation of the enzyme molecule. Some (but not all) enzymes possessing an essential —SH group are noncompetitively inhibited by heavy-metal ions (page 86), suggesting that such —SH groups must be intact for the enzyme to retain its normal active conformation.

Some enzymes that require metal ions for activity are inhibited noncompetitively by agents capable of binding the essential metal. For example, the chelating agent *ethylenediamine tetraacetate* (EDTA) reversibly binds Mg^{2+} and other divalent cations and thus noncompetitively inhibits some enzymes requiring such ions for activity (Figure 8-12).

Figure 8-12
A chelate of ethylenediamine tetraacetate with a divalent metal cation (Me²⁺). The shaded portion represents the plane of the coordination bonds.

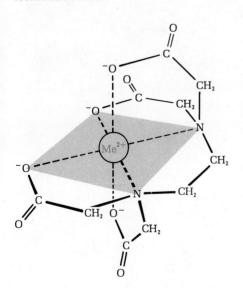

Irreversible Inhibition: Enzyme Modification

In *reversible* inhibition of enzymes, discussed above, the inhibitor participates in a rapidly established, easily reversible equilibrium with the enzyme or enzyme-substrate complex, which can be analyzed in terms of the Michaelis-Menten formalism. However, some enzymes undergo irreversible inactivation when they are treated with agents capable of covalently and permanently modifying a functional group required for catalysis, making the enzyme molecule inactive. This type of inhibition cannot be treated by Michaelis-Menten principles, which assume *reversible* formation of EI or ESI complexes. Often such an irreversible inhibition sets in slowly compared with the normal reaction kinetics of the enzyme, so that the inhibition is incomplete at first but continuously increases with time because chemical modification of an increasing fraction of the enzyme molecules takes place.

Later, in Chapter 9 (page 220), the irreversible inactivation of enzymes by covalent modification is considered in greater detail, since it has yielded important information on the identity of the catalytic functional groups at the active site.

Kinetics of Enzymatic Reactions Having
Two or More Substrates

Most enzymes catalyze reactions with two interacting substrates; such reactions show much more complex kinetics than the simple one-substrate reactions considered above. Enzymes catalyzing bisubstrate reactions include the large class of transferases, which catalyze the transfer of a specific functional group from one of the substrates to the other (Table 8-1).

The kinetic analysis of two-substrate reactions, symbolized by the equation

$$A + B \overset{E}{\rightleftharpoons} P + Q$$

is more complicated than for one-substrate reactions because they may have several enzyme-substrate complexes, such as the binary complexes EA, EB, EP, and EQ, and the ternary complexes EAB, EPQ, EAQ, and EPB. Determination of K_M and V_{max} for bisubstrate systems is similar to the approach used for one-substrate reactions. The concentration of one substrate, say B, is fixed, usually at a saturating level, and the concentration of substrate A is varied to determine its effect on the initial reaction rate and thus to give the Michaelis-Menten constant for substrate A, namely, K_M^A. Three or more fixed concentrations of B are usually employed to arrive at the value for K_M^A. Then the experimental arrangement is reversed: the concentration of substrate A is held constant at the saturation level, and the effect of varying the concentration of substrate B on the initial reaction velocity is determined, to yield the K_M for substrate B, namely, K_M^B. The values of K_M^A and K_M^B are most conveniently obtained from double-reciprocal plots. Bisubstrate reactions therefore reduce to the single-substrate case when one of the substrates is present at a saturating concentration. Table 8-5 shows the K_M values for the two substrates and two products of aspartate aminotransferase (see page 562).

Most two-substrate reactions can be put into one of two classes, *single-displacement reactions* and *double-displacement reactions*, which can generally be distinguished by kinetic analysis, as summarized in Figure 8-13. However, kinetic analysis is not infallible, and other experimental approaches must often be applied for diagnosis of the reaction pathway. In the following we shall examine the characteristics of idealized cases.

Single-Displacement Reactions

In single-displacement reactions both substrates A and B must be present on the enzyme active site simultaneously to yield a ternary complex EAB in order for the reaction to proceed. Single-displacement reactions occur in two forms, *random* and *ordered*, which differ in the sequence in which the two substrates bind to the enzyme. In random bisubstrate reactions, either substrate may bind to the enzyme first, indicating that the ternary complex (also called the *central complex*) EAB can be formed equally well in two different ways

$$E + A \rightleftharpoons EA$$
$$EA + B \rightleftharpoons EAB \tag{13}$$

or

$$E + B \rightleftharpoons EB$$
$$EB + A \rightleftharpoons EAB \tag{14}$$

Random single-displacement reactions are catalyzed by many of the *phosphotransferases*. An example is the reaction catalyzed by *creatine kinase* (page 767):

$$ATP + creatine \rightleftharpoons ADP + phosphocreatine$$

In this reaction both ATP and creatine are bound to the active site, in either sequence, to form a ternary complex. Following transfer of the phosphate group from the bound ATP to the bound creatine both products leave, in either sequence.

In *ordered* single displacements there is a *compulsory* sequence of reaction, so that one specific substrate, the *leading* substrate, must be bound first, before the second, or *following*, substrate can be bound, as shown in the reactions

$$E + A \rightleftharpoons EA \tag{15}$$
$$EA + B \rightleftharpoons EAB \tag{16}$$

where A is the leading substrate. Many dehydrogenases utilizing nicotinamide adenine dinucleotide (NAD^+) as coenzyme to accept electrons from their substrates (page 481) catalyze ordered bisubstrate reactions. For example, *malate dehydrogenase* (page 460), which catalyzes the reaction

$$Malate + NAD^+ \rightleftharpoons oxaloacetate + NADH + H^+ \tag{17}$$

must first bind NAD^+ to yield the E–NAD^+ complex, with which malate then combines to form the ternary complex E–NAD^+–malate.

Random bisubstrate reactions can be distinguished from ordered reactions experimentally. If the *last* reaction product inhibits the overall reaction by competing with *only* the *first* or *leading* substrate, the reaction is ordered. For example, the malate dehydrogenase reaction (17), which is an ordered reaction, is inhibited by excess NADH, which competes with the normal leading substrate NAD^+ for binding to the enzyme; NADH does not however compete with malate.

Under certain conditions the ternary complex in some ordered bisubstrate reactions is present in vanishingly low concentrations, so that the products of the reaction *appear* to result directly from the reaction of the binary complex of the enzyme with the leading substrate, namely, EA, and the second substrate B, without apparently forming an EAB com-

Figure 8-13
Summary of the characteristics of bisubstrate reactions of the type $A + B \rightleftharpoons P + Q$. *The different types of bisubstrate reactions can often be distinguished by the nature of double-reciprocal plots of $1/v_0$ vs. $1/[S]$ with the other substrate held constant at a fixed concentration. Random single displacements usually give a pattern of intersecting straight lines but are sometimes more complex and plots are not shown here. Ordered single displacements usually give converging plots as seen below, while double displacements usually yield parallel double-reciprocal plots.*

Single-displacement reactions

Random: Either substrate A or substrate B may combine with the free enzyme E to yield a binary complex. Addition of the other substrate follows to yield the ternary complex EAB. As shown here, the discharge of the products P and Q is in random sequence, but that need not necessarily be the case.

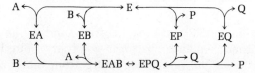

Ordered: The free enzyme E must first combine with the leading substrate A to form a binary complex. Addition of the following substrate B then yields the ternary complex EAB. Ordered release of products is indicated, but again that need not necessarily be the case.

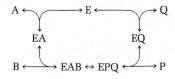

General rate equation for ordered reactions, where A is the leading substrate

$$v_0 = \frac{V_{max}}{\dfrac{K_S^A\, K_M^B}{[A][B]} + \dfrac{K_M^A}{[A]} + \dfrac{K_M^B}{[B]} + 1}$$

In slope-intercept (double-reciprocal) form with respect to A

$$\frac{1}{v_0} = \frac{1}{V_{max}}\left(K_M^A + \frac{K_S^A\, K_M^B}{[B]}\right)\left(\frac{1}{[A]}\right) + \frac{1}{V_{max}}\left(1 + \frac{K_M^B}{[B]}\right)$$

Plots of $1/v_0$ vs. $1/[A]$ at various concentrations of the fixed substrate B (assuming no interactions between the binding sites of A and B)

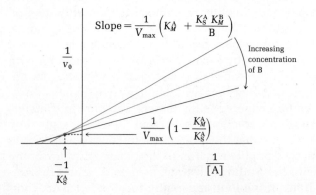

Double-displacement (ping-pong) reactions

Substrate A combines with free enzyme E to yield the complex EA from which product P departs, leaving the covalently substituted enzyme E*. Substrate B then combines with E* to give the complex E*B, which decomposes to yield product Q and free enzyme E.

General rate equation

$$v_0 = \frac{V_{max}}{1 + \dfrac{K_M^A}{[A]} + \dfrac{K_M^B}{[B]}}$$

In slope-intercept form

$$\frac{1}{v_0} = \frac{K_M^A}{V_{max}}\left(\frac{1}{[A]}\right) + \left(1 + \frac{K_M^B}{[B]}\right)\left(\frac{1}{V_{max}}\right)$$

Plots of $1/v_0$ vs. $1/[A]$ at various concentrations of fixed substrate B

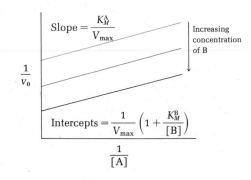

plex. This type of behavior, first discovered and analyzed by H. Theorell and B. Chance, occurs under certain experimental conditions with the NAD-linked lactate and alcohol dehydrogenases (page 483).

Double-Displacement (Ping-Pong) Reactions

In bisubstrate reactions of the double-displacement type (Figure 8-13), one substrate must be bound and one product released *before* the entry of the second substrate and the departure of the second product. In such reactions the first substrate reacts with the enzyme to yield a chemically modified form of the latter, usually by transfer of a functional group. In the second step this functional group is transferred from the enzyme to the second substrate. An example of such a double-displacement reaction is that catalyzed by aspartate aminotransferase, also called aspartate transaminase, which catalyzes the reaction

$$\text{Aspartate} + \alpha\text{-ketoglutarate} \rightleftharpoons \text{oxaloacetate} + \text{glutamate}$$

Aspartate, the leading substrate, combines with the enzyme first. The amino group of aspartate is then transferred to the tightly bound prosthetic group of the enzyme, pyridoxal phosphate (page 344). The first product, oxaloacetate, now leaves the active site and is replaced by the second substrate, α-ketoglutarate, to which the amino group is then transferred from the enzyme to form glutamate, which then leaves the active site. These two-step bisubstrate reactions have been given the descriptive name *ping-pong reactions* (Figure 8-13).

Equilibrium Exchanges in Bisubstrate Reactions

Although single- and double-displacement bisubstrate reactions can often be distinguished by their characteristic kinetic behavior, as outlined above and in Figure 8-13, kinetic analysis alone does not always lead to infallible conclusions. There is another important criterion by which they can sometimes be distinguished. We shall illustrate it by comparing two formally similar enzymatic reactions, the *sucrose phosphorylase* reaction, a double displacement, and the *maltose phosphorylase* reaction, a single displacement.

Sucrose phosphorylase catalyzes the reaction

$$\text{Sucrose} + P_i \rightleftharpoons \alpha\text{-D-glucose 1-phosphate} + \beta\text{-D-fructose}$$

in which the disaccharide sucrose, containing α-D-glucose and β-D-fructose (page 261), undergoes reversible phosphorolytic cleavage to yield α-D-glucose 1-phosphate and free D-fructose. The same enzyme, functioning in the reverse direction, participates in the biosynthesis of sucrose. Detailed study of this enzyme shows that it can catalyze an exchange reaction in which free inorganic phosphate exchanges with the phosphate group of glucose 1-phosphate when neither sucrose nor fructose is present, without any net transformation of phosphate or glucose 1-phosphate. The course of this equilibrium exchange reaction is followed by starting with inorganic phosphate labeled with the radioactive isotope ^{32}P (indicated by color) and unlabeled glucose 1-phosphate:

Phosphate + α-D-glucose 1-phosphate $\rightleftharpoons$
phosphate + α-D-glucose 1-phosphate

The glucose 1-phosphate becomes labeled with the isotope as the exchange proceeds. Similarly, sucrose phosphorylase also catalyzes an equilibrium exchange reaction between isotopically labeled fructose (in color) and the fructose portion of unlabeled sucrose, in the absence of phosphate or glucose 1-phosphate:

Fructose + glucose-fructose $\rightleftharpoons$ fructose + glucose-fructose
         (Sucrose)   (Sucrose)

From these and other facts it has been deduced that the reaction catalyzed by sucrose phosphorylase is a double-displacement reaction occurring in two distinct steps and involving a covalent enzyme-glucose intermediate. These steps are

E + sucrose $\rightleftharpoons$ E–glucose + D-fructose

E–glucose + P_i $\rightleftharpoons$ E + α-D-glucose 1-phosphate

The equilibrium exchange of free labeled fructose with the fructose portion of sucrose is brought about by the first reaction, which may proceed independently of the second, and the exchange of isotopic phosphate with the 1-phosphate group of α-D-glucose 1-phosphate takes place by the second reaction, which can proceed independently of the first. This formulation of the sucrose phosphorylase reaction is supported by the direct demonstration of the formation of a covalent enzyme-glucose intermediate by the enzyme.

In striking contrast is the *maltose phosphorylase* reaction, an example of a single displacement:

Maltose + P_i $\rightleftharpoons$ D-glucose + β-D-glucose 1-phosphate

Maltose, a disaccharide of two molecules of D-glucose, undergoes a phosphorolytic cleavage formally resembling the sucrose phosphorylase reaction. However, maltose phosphorylase does not catalyze an isotopic exchange between phosphate and β-D-glucose 1-phosphate in the absence of maltose and D-glucose, nor does it catalyze an exchange between isotopic free D-glucose and a glucose residue of maltose in the absence of phosphate and glucose 1-phosphate. Moreover, no covalent enzyme-sugar intermediate has been detected in the action of maltose phosphorylase. It has been concluded from this and other evidence that the maltose phosphorylase reaction is a single-displacement reaction, in which both maltose and phosphate must be present at the active site simultaneously. When both kinetic tests and equilibrium-exchange tests are in agreement, a reliable decision can be made as to whether a given enzymatic reaction is a single- or double-displacement reaction.

Quantitative Assay of Enzymatic Activity

The amount of an enzyme in a given solution or tissue extract can be assayed quantitatively in terms of the catalytic effect it produces. For this purpose, it is necessary to know:

1 The overall stoichiometry of the reaction catalyzed
2 Whether the enzyme requires the addition of cofactors such as metal ions or coenzymes
3 Its dependence on substrate and cofactor concentrations, i.e., the K_M for both substrate and cofactor
4 Its optimum pH
5 A temperature zone in which it is stable and has high activity
6 A simple analytical procedure for determining the disappearance of the substrate or the appearance of the reaction products

Where possible, enzymes are assayed in test systems in which the pH is optimum and the substrate concentration is above the saturating level, so that the initial reaction rate is zero order for substrate. Under these conditions, the initial reaction rate is proportional to enzyme concentration alone. When enzymes require cofactors, such as metal ions or coenzymes, they must also be added in concentrations that exceed saturation so that the true rate-limiting factor in the system is the enzyme concentration. Usually measurement of the rate of formation of the reaction product is more accurate than measurement of the disappearance of the substrate, since the substrate must often be present at relatively high concentration to preserve zero-order kinetics.

The quantitative assay of enzyme activity is carried out most quickly and conveniently when the substrate or the product is colored or absorbs light in the ultraviolet region because the rate of appearance or disappearance of a light-absorbing product or substrate can be followed with a spectrophotometer, giving a continuous record of the progress of the reaction on a strip chart. Other light-absorbing or light-scattering components must be absent or subtracted by appropriate blank measurements. An example is the measurement of the activity of the enzyme *lactate dehydrogenase*, which transfers electrons from lactate to the oxidized form of the coenzyme nicotinamide adenine dinucleotide, NAD^+ (Table 8-3; see also page 483), to yield pyruvate, the reduced coenzyme NADH, and a proton:

$$\text{Lactate} + NAD^+ \rightleftharpoons \text{pyruvate} + \text{NADH} + H^+$$

The reduced form of the coenzyme, NADH, absorbs light in the ultraviolet at 340 nm (page 482), whereas the oxidized form NAD^+, the lactate, and the pyruvate do not. Thus the progress of the reaction in the forward direction can be followed by measuring the increase in light absorption of the system at 340 nm in a spectrophotometer—much simpler than chemical determination of any of the substrates or products.

Optical assays are so simple and convenient that they are often used to follow the time course of an enzymatic reaction in which neither the substrate(s) nor product(s) have any characteristic light-absorption maxima, by coupling the reaction to some other enzymatic reaction which has an easily measured optical change. An example is the reaction between phosphoenolpyruvate and ADP to yield pyruvate and ATP by transfer of a phosphate group, catalyzed by *pyruvate kinase*:

$$\text{Phosphoenolpyruvate} + \text{ADP} \rightleftharpoons \text{pyruvate} + \text{ATP}$$

Although the substrates and products of this reaction do not absorb light in the zone 300 to 400 nm, this reaction is easy to measure if a large excess of the enzyme lactate dehydrogenase and NADH is added to the system to yield the following coupled reactions, which have the common intermediate pyruvate:

$$\text{Phosphoenolpyruvate} + \text{ADP} \rightleftharpoons \text{pyruvate} + \text{ATP}$$
$$\text{Pyruvate} + \text{NADH} + \text{H}^+ \rightleftharpoons \text{lactate} + \text{NAD}^+$$

In the presence of excess lactate dehydrogenase and NADH, the formation of pyruvate in the first reaction is followed by the very rapid reduction of pyruvate to lactate in the second. For each molecule of pyruvate so formed and reduced, a molecule of NADH is oxidized to NAD^+, causing a decrease in light absorption at 340 nm, the absorption maximum for NADH. In such a *coupled assay* the enzyme whose activity is to be measured is made the rate-limiting component by appropriate adjustment of the reaction system.

Enzyme-Activity Units

The most widely used unit of enzyme activity is defined as that amount which causes transformation of 1.0 μmol (10^{-6} mol) of substrate per minute at 25°C under optimal conditions of measurement. The *specific activity* is the number of enzyme units per milligram of protein. It is a measure of enzyme purity, increasing during purification of an enzyme and becoming maximal and constant when the enzyme is in the pure state. The *molar* or *molecular activity*, formerly called the *turnover number*, is the number of substrate molecules transformed per minute by a single enzyme molecule (or by a single active site) when the enzyme is the rate-limiting factor (Table 8-8). The molar activity of an enzyme with a single active site can be calculated from its V_{max} value and its molecular weight. The enzyme carbonic anhydrase has the highest molar activity of any known enzyme, 36,000,000 min^{-1} per molecule.

A new international unit for enzyme activity recommended by the Enzyme Commission (see References) is the *katal* (abbreviated kat), defined as the amount of enzyme activity that transforms 1 mol s^{-1} of substrate. The new unit is in accord with the dimensions of rate constants in chemical kinetics, which are based on the second rather than the

Table 8-8 Molecular activity[†] of some enzymes in the range 20 to 38°C

Enzyme	Molecular activity[†]
Carbonic anhydrase C	36,000,000
Δ^5-3-Ketosteroid isomerase	17,100,000
Catalase	5,600,000
β-Amylase	1,100,000
β-Galactosidase	12,500
Phosphoglucomutase	1,240
Succinate dehydrogenase	1,150

[†] The molecular activity is the number of substrate molecules transformed by a single molecule of enzyme per minute under optimal conditions.

minute. Activities can also be given in microkatals (μkat), nanokatals (nkat), or picokatals (pkat). One old enzyme unit (see above) is thus equal to $\frac{1}{60}$ μkat or 16.67 nkat. In the new convention *specific activity* is given as katals per kilogram of protein, equivalent to microkatals per milligram of protein, and the *molar activity* as katals per mole of enzyme.

Enzyme Purification

By applying quantitative assay methods, like those described above, to determine the amount of catalytic activity present in different protein fractions obtained from cells or tissues by the methods described in detail in Chapter 7 (page 157), it is possible to purify enzymes and ultimately obtain them in homogeneous form. The approaches and procedures used have been illustrated (page 172) by some actual data obtained during the purification of the enzyme acetylcholinesterase. It should be kept in mind that a crude extract of bacterial cells or of a plant or animal tissue may contain several hundred different proteins, most of them enzymes, whose solubility and acid-base properties do not differ widely. Therefore purification of a single enzyme from such an extract requires an empirical choice of a sequence of separation procedures to exploit differences in properties between the enzyme to be isolated and all the other proteins present.

Enzyme-Substrate Complexes and Covalent Enzyme-Substrate Compounds

As we have seen, kinetic evidence strongly suggests that all enzymes combine transiently with their substrates to form enzyme-substrate complexes during their catalytic cycles. But because of the instability of enzyme-substrate complexes, which by definition decompose rapidly, it might be expected that proof of their formation is difficult to obtain. Nevertheless, evidence for the formation of such complexes has come from physical and chemical measurements. One approach is spectroscopic observation of transient spectral changes occurring when the substrate is mixed with the enzyme. The heme enzymes *catalase* and *peroxidase*, which have distinctive absorption spectra due to their heme prosthetic groups, show transient changes in their spectra on mixing with hydrogen peroxide, their substrate, which reflect the formation and decomposition of their enzyme-substrate complexes. Similarly, some enzymes on mixing with their substrates undergo changes in optical rotation at certain wavelengths (page 150), which are interpreted to be manifestations of conformation changes of the enzyme accompanying the formation of the enzyme-substrate complex.

Especially convincing is the demonstration of rather stable enzyme-substrate complexes when some enzymes react with a sluggish substrate, i.e., one that readily forms a noncovalent enzyme-substrate complex but decomposes very slowly compared with the normal biological substrate. Such sluggish enzyme-substrate complexes have been observed for lysozyme (page 233), D-amino acid oxidase, and several other enzymes. Indeed, a sluggish enzyme-substrate complex of lysozyme has been crystallized and subjected to x-ray analysis, yielding some revealing insights into the structure of the active site (page 232). It is also possible to demonstrate the formation of stable complexes of some enzymes with competitive inhibitors, which resemble the substrate in structure and bind to the substrate site. Formation of enzyme-substrate complexes can also be directly observed in certain two-substrate reactions when one of the substrates is added in the absence of the other. For example, alcohol dehydrogenase forms a complex with oxidized nicotinamide adenine dinucleotide in the absence of ethanol, its other substrate.

Besides the reversible, noncovalent Michaelis-Menten type of enzyme-substrate complex, which presumably is formed by all enzymes and which, as it decomposes, becomes the rate-limiting step when the substrate concentration is saturating, some enzymes form covalent enzyme-substrate compounds as intermediates, particularly those catalyzing double-displacement reactions. A classical case is the enzyme _fructose-diphosphate aldolase_ (page 425), which catalyzes the reversible reaction.

Fructose 1,6-diphosphate $\rightleftharpoons$
dihydroxyacetone phosphate + glyceraldehyde 3-phosphate

Addition of the strong reducing agent sodium borohydride to a mixture of the aldolase and isotopically labeled dihydroxyacetone phosphate results in the formation of an isotopically labeled, covalent aldolase-dihydroxyacetone phosphate compound, which is catalytically inactive. After complete hydrolysis of this compound with acid, a stable ϵ-N-glyceryl derivative of a single lysine residue was found in the hydrolyzate (Figure 8-14). This derivative did not form in the absence of the reducing agent. From the structure of the labeled product it was concluded that aldolase combines covalently but reversibly with dihydroxyacetone phosphate to form a labile Schiff's base (page 426) between the ϵ-amino group of a lysine residue in the active site of the enzyme and the carbonyl group of the substrate. Treatment with borohydride reduces this very labile intermediate to a stable but inactive covalent derivative. In this way very labile covalent enzyme-substrate compounds of a number of enzymes acting on substrates with either an amino or carbonyl function have been successfully trapped in their stable reduced forms.

In some double-displacement bisubstrate reactions an enzyme-substrate compound may be isolated if the other substrate is absent from the system. For example, an enzyme-glucose compound is formed when sucrose phosphorylase acts on sucrose in the absence of phosphate (page 205).

Enzymes and Substrates in Living Cells

Enzymes in living cells may function under conditions very different from those in test systems, like the in vitro kinetic tests described in this chapter. If a liver cell contains 1,000

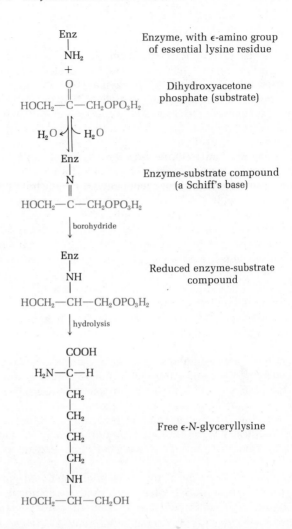

different enzymes, and if their average molecular weight is about 140,000, then on the average each enzyme is present in the cytoplasm in a concentration of a little less than 10^{-6} M (1.0 μM). Some enzymes, particularly those catalyzing reactions in the central pathways, may have much higher concentrations; e.g., hexokinase, which catalyzes the phosphorylation of glucose by ATP, is present in muscle in excess of 10^{-4} M (100 μM). On the other hand, some enzymes, e.g., those catalyzing biosynthesis of the coenzymes, are probably present in concentrations as low as 10^{-8} M (0.01 μM).

Measurements have also been made of the cytoplasmic concentration of various metabolites acting as substrates. Most are present in a total concentration of anywhere from 5 to 500 μM. In many, if not most, cases the concentration of the substrate in intact cells is insufficient to saturate the enzyme; indeed, in some cases the concentration of the substrate is not much greater than the concentration of the enzyme. Clearly, enzymes in the intact cell do not necessarily exhibit classical Michaelis-Menten kinetic behavior, which assumes that the enzyme concentration is negligibly small compared with the substrate concentration. Quantitative analysis of enzyme kinetics in the intact cell is an infant field of enzymology, which is of the greatest importance in understanding the biological regulation of enzyme activity (Chapter 9, page 234).

Summary

Enzymes are classified on the basis of the reaction catalyzed. Some enzymes are simple proteins; others are conjugated and contain prosthetic groups of metal ions, coenzymes, or both. Coenzymes and prosthetic groups serve as intermediate carriers of specific functional groups, atoms, or electrons. Many coenzymes contain a molecule of a given vitamin, an organic nutrient required in trace amounts for normal cell function.

In enzyme-catalyzed reactions an increase in the concentration of the substrate increases the reaction rate until a point is reached where the reaction rate becomes independent of substrate concentration. At this point the enzyme is saturated, and the reaction becomes zero order with respect to substrate. Each enzyme has a characteristic substrate concentration (K_M, the Michaelis-Menten constant) at which the reaction velocity is one-half maximal. The quantitative relationship between the initial reaction rate, the substrate concentration K_M, and the maximum velocity of an enzyme is given by the Michaelis-Menten equation. Its derivation is based on the assumption that an enzyme-substrate complex is formed reversibly as an essential step in catalysis. Enzymes have an optimum pH and a temperature range in which they are stable and active.

Competitive enzyme inhibitors are those reacting reversibly with the free enzyme in competition with the substrate to form an enzyme-inhibitor complex; their action can be reversed by increasing the substrate concentration. Uncompetitive inhibitors do not react with free enzyme but combine reversibly with the enzyme-substrate complex and prevent formation of the products. Noncompetitive inhibitors react reversibly with both free enzyme and the enzyme-substrate complex. Irreversible inhibitors cause a permanent, chemical modification of some essential functional group in the enzyme molecule. Kinetic tests are used to distinguish the various types of reversible enzyme inhibition; double-reciprocal plots are especially useful in analyzing these kinetic data.

In random single-displacement bisubstrate reactions the enzyme forms a ternary complex with both substrates, and the substrates may add in either order. In ordered single-displacement reactions there is a compulsory sequence of addition of the substrates to form the ternary complex. In double-displacement, or ping-pong, reactions one substrate reacts with the enzyme, and its corresponding product leaves before the other substrate combines.

Enzymes are usually assayed by measuring the initial reaction rate under conditions in which the enzyme is saturated with substrate and the pH is optimal. The occurrence of enzyme-substrate complexes and covalent compounds has been deduced from kinetic studies, from trapping experiments, and from spectroscopic measurements.

References

Books

BARMAN, T.: *Enzyme Handbook*, vol. 1, Springer, New York, 1969. Valuable summary of the major properties of enzymes, classified according to the international rules.

BENDER, M. L., and L. J. BRUBACHER: *Catalysis and Enzyme Action*, McGraw-Hill, New York, 1973. An introductory paperback.

BOYER, P. D.: *The Enzymes*, 3d ed., vols. 1–10, Academic, New York, 1970–1974. Extremely valuable collection of comprehensive reviews of many aspects of enzyme biochemistry.

COLOWICK, S. P., and N. O. KAPLAN (eds.): *Methods in Enzymology*, Academic, New York. A continuing and very valuable series of volumes on experimental techniques of enzyme assay, isolation, and substrate preparation.

DIXON, M., and E. C. WEBB: *Enzymes*, 2d ed., Longmans, London, 1964. Still a classic text on general enzyme properties.

Enzyme Nomenclature, American Elsevier, New York, 1973. The 1972 recommendations of the Commission on Enzyme Nomenclature, including units and kinetic symbols.

GUTFREUND, H.: *Enzymes: Physical Principles*, Wiley-Interscience, New York, 1972. Physical chemistry and kinetics of enzymes. Excellent and readable short treatment.

HALDANE, J. B. S.: *The Enzymes*, MIT Press, Cambridge, Mass., 1965. A reprinted classic.

LAIDLER, K. J.: *The Chemical Kinetics of Enzyme Action*, 2d ed., Oxford, New York, 1973.

MEISTER, A. (ed.): *Advances in Enzymology*, Academic, New York. Annual volumes containing reviews of special topics in enzyme biochemistry.

PLOWMAN, K.: *Enzyme Kinetics*, McGraw-Hill, New York, 1972.

WEBB, J. L.: *Enzyme and Metabolic Inhibitors*, 3 vols., Academic, New York, 1963–1966.

WESTLEY, J.: *Enzymic Catalysis*, Harper & Row, New York, 1969. Complex kinetic phenomena are given precise verbal description as well as detailed mathematical treatment.

Articles

CLELAND, W. W.: "Steady State Kinetics," in P. D. Boyer (ed.), *The Enzymes*, 3d ed., vol. 2, pp. 1–65, Academic, New York, 1970.

Problems

Because the kinetics of enzyme reactions is central to much biochemical research, a large assortment of commonly encountered problems follows.

1. In the first-order reaction A $\longrightarrow$ B, the concentration of A at time 0 is 0.50 mM. After 2 s, it is 0.25 mM. What will it be after 5 s?

2. In the second-order reaction A + B $\longrightarrow$ C, the concentrations at time 0 are, for reactant A, 5.0 mM and, for reactant B, 4.0 mM. After 1 s, the concentration of A is 4.0 mM and that of B is 3.0 mM. After 3 s, what will be the ratio of the concentration of A to that of B?

3. (a) If the half-time of a first-order reaction is 0.3 s, what is its rate constant k?
   (b) How long will it take for 95 percent of the reactant to disappear?

4. The following data were reported for the reaction

$$RSO_2S^- + CN^- \longrightarrow SCN^- + RSO_2^-$$

t, s	30	60	120
$[SCN^-], \times 10^4 \ M^{-1}$	2.92	5.89	10.79

Calculate a pseudo-first-order rate constant for the reaction, given that the initial CN^- concentration was 0.125 M and the initial RSO_2S concentration 0.0035 M.

5. What relationship exists between K_M and [S] when an enzyme-catalyzed reaction proceeds at 80 percent V_{max}?

6. Show that for competitive inhibition of an enzymic reaction, the intercepts on the horizontal axis of a plot of $1/v_0$ vs. $1/[S]$ at different inhibitor concentrations are equal to

$$\frac{-1}{K_M(1 + [I]/K_1)}$$

7. To a 10.0-ml solution of a pure enzyme containing 1.0 mg protein per milliliter is added just enough $AgNO_3$ to completely inactivate the enzyme. A total of 0.342 μmol of $AgNO_3$ was required. Calculate a minimum molecular weight for the enzyme.

8. Transaminase catalyzes the reaction

$$\text{Glutamate} + \text{oxaloacetate} \longrightarrow \alpha\text{-ketoglutarate} + \text{aspartate}$$

Pyridoxal phosphate (PP) acts as a coenzyme in this catalytic process. Calculate K_M for the apoenzyme-coenzyme complex from the following data, obtained when the concentration of PP was varied while concentrations of glutamate and oxaloacetate and other conditions were held constant:

Glutamate disappearing per minute, mg	0.17	0.27	0.43	0.65	0.73	0.78	0.79	0.81
PP added, μM	0.30	0.50	1.0	2.0	3.0	4.0	5.0	10.0

9. Salicylate inhibits the catalytic action of glutamate dehydrogenase. (a) Determine the type of inhibition by graphical analysis of the following data. Assume that the salicylate concentration is held constant at 40 mM. Also calculate (b) K_M for the substrate and (c) K_1, the dissociation constant for the enzyme-inhibitor complex.

Substrate concentration, mM	Product per minute, mg	
	Without salicylate	With salicylate
1.5	0.21	0.08
2.0	0.25	0.10
3.0	0.28	0.12
4.0	0.33	0.13
8.0	0.44	0.16
16.0	0.40	0.18

10. From the following data on an enzymatic reaction, determine (a) the type of inhibition, (b) K_M for the substrate, and (c) K_1 for the inhibitor-enzyme complex.

Substrate concentration, mM	Product per hour, μg	
	No inhibitor	6 mM inhibitor
2.0	139	88
3.0	179	121
4.0	213	149
10.0	313	257
15.0	370	313

11. Glycerokinase catalyzes the reaction

$$\text{Glycerol} + \text{ATP} \longrightarrow \text{glycerol phosphate} + \text{ADP}$$

The following results were obtained upon the addition to the reaction mixture of the chromium salt of ATP:

Chromium ATP concentration, μM	v_0 (arbitrary units)				
	0.100	0.050	0.033	0.025	mM *Glycerol*
0	50.0	40.0	33.3	33.3	
10	6.25	6.06	6.06	5.88	
20	3.45	3.13	3.08	3.23	
30	2.38	2.27	2.35	2.33	

(a) Classify the inhibition according to type.
(b) Calculate a value for the inhibitor constant, K_I.

12. For an ordered single-displacement reaction in which A is the leading and B the following substrate, give algebraic expressions for (a) the slope, (b) the intercept, (c) the horizontal coordinate of the point of intersection, and (d) the vertical coordinate of the point of intersection when $1/v_0$ is plotted as a function of $1/[B]$ at different fixed concentrations of A.

13. Nucleoside diphosphate kinase catalyzes the transfer of a terminal phosphate from ATP to an acceptor nucleoside diphosphate. Kinetic studies involving the reaction

$$\text{ATP} + \text{UDP} \longrightarrow \text{ADP} + \text{UTP}$$

have been reported with the following results. Plots of $1/v_0$ vs. $1/[\text{ATP}]$ at different fixed concentrations of UDP yield a pattern of parallel lines, as do plots of $1/v_0$ vs. $1/[\text{UTP}]$ at different fixed concentrations of ADP. ADP and UDP were found to be mutually competitive inhibitors, as were ATP and UTP. The enzyme was found to catalyze isotopic exchange reactions between UTP and UDP and between ATP and ADP. Propose a mechanism consistent with these observations.

14. Glycerol dehydrogenase catalyzes the reaction

$$\text{Glycerol} + \text{NAD}^+ \rightleftharpoons \text{dihydroxyacetone} + \text{NADH} + \text{H}^+$$

The following kinetic properties of the enzyme from *Aerobacter aerogenes* have been reported: Plots of $1/v_0$ vs. $1/[\text{glycerol}]$ at different fixed concentrations of NAD^+ yield lines which intersect to the left of the $1/v_0$ axis. Similarly, plots of $1/v_0$ vs. $1/[\text{NADH}]$ at different fixed concentrations of dihydroxyacetone show intersection to the left of the $1/v_0$ axis. NADH was found to be a competitive inhibitor with respect to NAD^+, and dihydroxyacetone was uncompetitive with respect to NAD^+, as was glycerol with respect to NADH at high concentrations of the fixed substrates. Give a mechanism consistent with these observations.

15. Citrate synthase catalyzes the reaction

$$\text{Acetyl-CoA} + \text{oxaloacetate} \rightleftharpoons \text{CoA} + \text{citrate}$$

When kinetic properties of the enzyme from rat brain were studied, the following results were obtained. Plots of $1/v_0$ vs.

1/[acetyl-CoA] at different fixed concentrations of oxaloacetate yielded a pattern of straight lines intersecting to the left of the $1/v_0$ axis. Similar results were obtained when oxaloacetate, coenzyme A, and citrate each served as the varied substrate. Inhibition studies showed that acetyl-CoA and coenzyme A were mutually competitive while citrate was a competitive inhibitor of both oxaloacetate and acetyl-CoA. Give a kinetic mechanism that is consistent with these observations.

CHAPTER 9 ENZYMES: MECHANISM, STRUCTURE, AND REGULATION

One of the most remarkable attributes of enzymes is their specificity of action, so that only certain substrates are acted upon and only a single type of reaction takes place, without side reactions or by-products. The organic chemist considers himself fortunate if he can realize a 90 percent yield in carrying out a reaction in the laboratory. However, if the yield of each product in a 10-step metabolic sequence were as poor as 90 percent, only about a third of the starting material would be recovered as the final end product. Clearly, if it were not for the specificity of enzymes, cells would soon be swamped with side reactions and by-products.

Another remarkable attribute of enzymes is their enormous catalytic power. Although enzymes are proteins, and thus relatively fragile molecules, they bring about their extraordinary catalytic effects in dilute aqueous solution at biological pH and moderate temperature, in sharp contrast to the rather extreme conditions often required to accelerate chemical reactions in the organic laboratory.

We shall consider in this chapter not only the specificity and mechanism of enzyme action but also the properties and structure of different classes of regulatory enzymes, which serve as pacemakers of multienzyme systems in cells and as elements in the integration of metabolic pathways.

Substrate Specificity of Enzymes

One of the first important studies on the specificity of enzymes was carried out by Emil Fischer, who found that enzymes capable of hydrolyzing glycosides can distinguish between their stereoisomeric forms. In 1894 this observation led him to enunciate the principle that the substrate molecule fits the active site of the enzyme in a lock-and-key, or complementary, relationship.

Although enzymes are in general very specific compared to man-made catalysts, they vary considerably in their degree of specificity. Some have nearly absolute specificity for a given substrate and will not attack even very closely related

molecules, whereas others will attack a whole class of molecules sharing a common denominator of structure but at
widely different rates. Aspartate ammonia-lyase, commonly
known as aspartase, is an example of an enzyme with nearly
absolute substrate specificity (Figure 9-1). Moreover, aspartase also has strict stereospecificity; thus, it will not de-
aminate D-aspartate, nor will it add ammonia to maleate, the
cis geometrical isomer of fumarate. In fact, it is the absolute
stereospecificity of many enzymes that is especially remarkable, since man-made catalysts are generally unable to select
one of a pair of stereoisomers. Other stereospecific enzymes
are lactate dehydrogenase, specific for the L stereoisomer of
lactate, and D-amino acid oxidase, specific for the D stereoisomer of various amino acids. Some enzymes, like aconitase
(page 454), are even capable of distinguishing between two
identical substituents on a nonasymmetric carbon atom.

At the other end of the spectrum are enzymes with relatively broad specificity, capable of acting on a number of different structurally related substrates, but at widely different
rates. In this group are _alkaline phosphatase_, which hydrolyzes many different esters of phosphoric acid, _carboxyes-
terase_, which hydrolyzes esters of various carboxylic acids,
and _carboxypeptidase_, which catalyzes hydrolysis of the C-
terminal peptide bond of peptides, regardless of the length of
the peptide chain or the identity of the C-terminal amino
acid residue.

Studies of the substrate specificity of enzymes show that
substrate molecules generally reflect, by the principle of
complementarity, the structure of the active site of the enzyme in two distinctive structural features: (1) the substrate
must have a susceptible chemical bond that can be attacked
by the enzyme; (2) it usually has some other structural feature required for its binding to the enzyme active site, presumably to position the substrate molecule in the proper
geometrical relationship so that the susceptible bond can be
attacked. For example, substrate specificity studies on _ace-
tylcholinesterase_ (page 171) have shown that although the
susceptible bond is the ester linkage between choline and the
acetyl group, the part of the molecule required for its positioning on the active site is the positively charged quaternary
ammonium group adjoining a nonpolar group (Figure 9-2).
This positioning group is also required in the structure of
competitive inhibitors of acetylcholinesterase (page 197). Indeed, analysis of the structural requirements for competitive
inhibition has given much valuable information on the active
sites of enzymes, since competitive inhibitors can bind to the
active site but are not acted upon.

The substrate specificity of chymotrypsin, mapped in detail by M. Bergmann and J. S. Fruton, presents an especially
instructive case (Figure 9-3). Because chymotrypsin is secreted into the small intestine, it was first thought to be specific for the hydrolysis of relatively long polypeptides
formed by the action of pepsin on ingested proteins in the
stomach, but later work revealed that chymotrypsin can also
attack short peptides. Tests of various synthetic peptides as

Figure 9-1
_The aspartase reaction. Aspartase has rela-
tively strict substrate specificity. It is unable
to catalyze addition of ammonia to methyl-
fumarate, to esters or amides of fumaric
acid, to α,β-unsaturated monocarboxylic
acids, or to maleate, nor does it catalyze
deamination of aminomalonic, glutamic, or
various α-amino monocarboxylic acids._

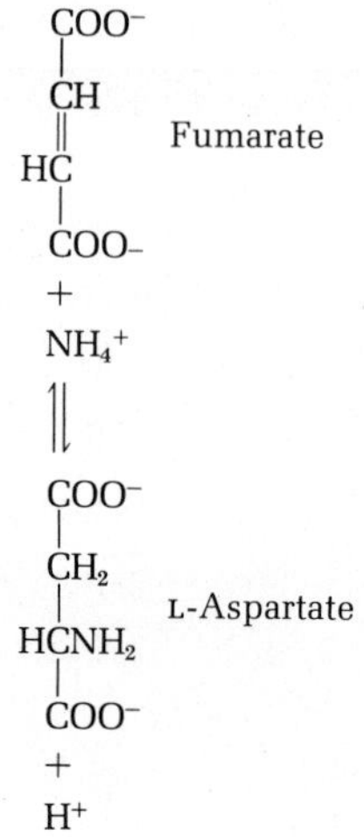

$$
\begin{array}{c}
COO^- \\
| \\
CH \\
\| \\
HC \quad\quad \text{Fumarate} \\
| \\
COO^- \\
+ \\
NH_4^+
\end{array}
$$

$$
\begin{array}{c}
COO^- \\
| \\
CH_2 \\
| \quad\quad \text{L-Aspartate} \\
HCNH_2 \\
| \\
COO^- \\
+ \\
H^+
\end{array}
$$

Figure 9-2
_The positioning group and the susceptible
bond in acetylcholine, the substrate of
acetylcholinesterase._

$$
CH_3 - \overset{+}{\underset{|}{\underset{CH_3}{\overset{CH_3}{N}}}} - CH_2CH_2O - \overset{O}{\underset{\|}{C}} - CH_3
$$

Binding or   Susceptible
positioning group   bond

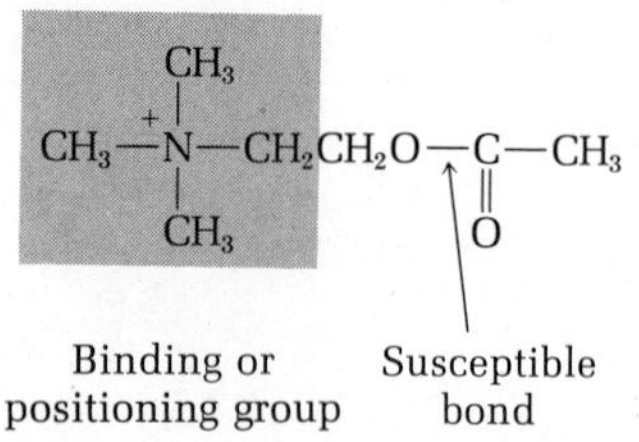

Figure 9-3
The substrate specificity of chymotrypsin. From specificity studies on synthetic substrates (right) it has been concluded that chymotrypsin is a transferase for hydrophobic acyl groups rather than strictly a peptidase. Its minimum structural requirements are shown below.

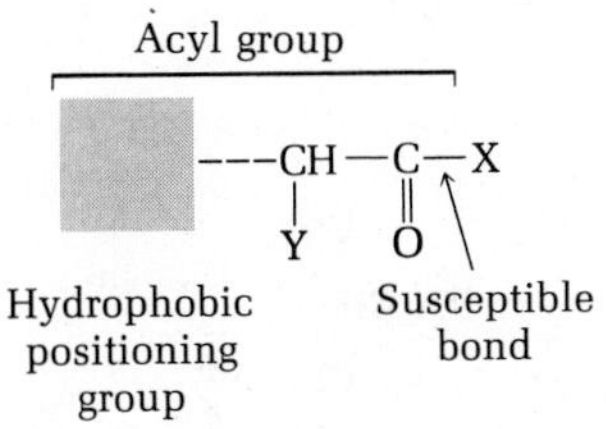

Some compounds hydrolyzed by chymotrypsin. The points of cleavage are shown by colored arrows.

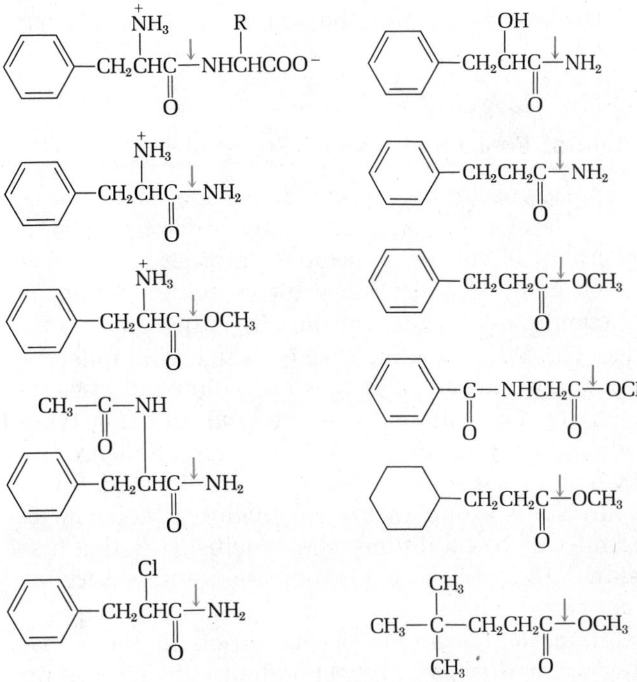

substrates showed that the enzyme is an *endopeptidase*; i.e., it can split certain types of peptide linkages wherever they occur in a peptide chain, in contrast to the *exopeptidases*, which can split only terminal peptide bonds. Moreover, chymotrypsin is specific for those peptide linkages in which the carbonyl function is contributed by aromatic amino acid residues, e.g., tyrosine, tryptophan, and phenylalanine; it also hydrolyzes the amides of these amino acids.

It was a considerable surprise when later work on synthetic substrates revealed that chymotrypsin can hydrolyze not only peptides and amides of aromatic amino acids but also their esters. In fact, esters of tyrosine are the most active known substrates for chymotrypsin. Another unanticipated finding was that chymotrypsin catalyzes the transfer of the aromatic acyl group to acyl acceptors other than water, e.g., ammonia, other amino acids, or alcohols. Moreover, the aromatic rings traditionally thought to be required by chymotrypsin are also dispensable, since replacement of the benzene ring of a phenylalanine residue with a cyclohexyl ring or some other bulky hydrocarbon group causes no great decrease in activity. Furthermore, not even the α-amino group of the aromatic amino acid is required, since it may be replaced by a hydrogen atom, a hydroxyl group, or a chlorine atom (Figure 9-3). Strictly speaking, then, chymotrypsin is not a peptidase at all; it could more accurately be designated as a "hydrophobic acyl-group transferase." Its active site has two distinct features, a hydrophobic zone for binding and positioning the substrate on the active site and a catalytic portion for removing and transferring the acyl group (Figure 9-3). From such studies of substrate specificity and of the structural requirements of competitive inhibitors (page 197) it has become possible to map the active sites of many enzymes.

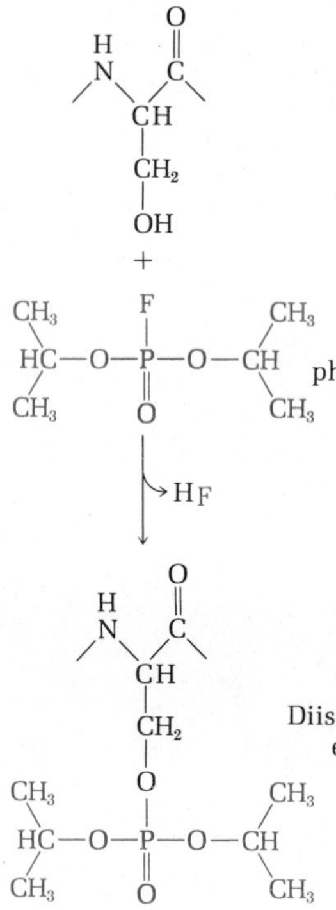

Identification of Functional Groups Essential for Catalysis

Another approach useful in mapping active sites is the use of reagents capable of covalently modifying different types of functional groups in enzyme molecules, in order to establish whether such groups are necessary for catalytic activity. A classical example is the action of the alkylating agent iodoacetate (page 99). When ribonuclease is treated with iodoacetate at pH 5.5, the enzyme undergoes alkylation and loses its catalytic activity. Two different inactive forms of the enzyme are formed: in one the imidazole ring of histidine residue 119 is alkylated; in the other, histidine 12 is alkylated. Since no other functional group in the ribonuclease molecule is alkylated under these conditions, the conclusion is that histidine residues 12 and 119 of ribonuclease are necessary for catalytic activity.

Another important example is the action of the phosphorylating agent *diisopropylphosphofluoridate* on certain enzymes to yield inactive derivatives in which the hydroxyl group of a specific serine residue is phosphorylated (Figure 9-4). This reagent is one of a group of toxic organophosphorus compounds sometimes called *nerve poisons* since they combine with and completely inactivate the enzyme *acetylcholinesterase*, which functions in the activity of the nervous system. Some of these compounds are used as insecticides. Diisopropylphosphofluoridate inhibits not only acetylcholinesterase but also other enzymes having an essential serine residue at their active sites, e.g., chymotrypsin, trypsin, phosphoglucomutase, and several esterases. In chymotrypsin, diisopropylphosphofluoridate selectively phosphorylates the serine residue at position 195, thus identifying this residue as essential for catalysis. When an enzyme phosphorylated by this reagent is subjected to partial hydrolysis, the phosphoserine residue remains intact and can be found in one of the peptide fragments. Chemical analysis of such phosphorylated peptides has given important information on the amino acid sequences near the active sites in this group of enzymes, sometimes called the *serine enzymes* (Table 9-1). Serine enzymes inactivated in this manner can sometimes be slowly reactivated by reagents with a higher reactivity for the phosphorylating agent than the enzyme itself.

It is highly significant that those functional groups of enzymes required for catalytic activity are usually much more accessible or reactive than similar groups elsewhere in the molecule that are not directly involved in catalysis. For example, ribonuclease contains many functional groups ca-

Figure 9-4
Formation of an inactive phosphorylated derivative of an enzyme of the serine class.

Table 9-1 Amino acid sequences around reactive serine residues of some enzymes

Chymotrypsin
  -Gly-Asp- Ser-Gly-Gly-

Trypsin
  -Gly-Asp-Ser-Gly-Pro-

Thrombin
   -Asp-Ser-Gly-

Elastase
  -Gly-Asp-Ser-Gly-

Phosphoglucomutase
  -Thr-Ala- Ser-His-Asp-

Phosphorylase
  -Glu-Ile- Ser-Val-Arg-

pable of reacting with iodoacetate, but the imidazole groups of histidine residues 12 and 119 are far more reactive than all the others. Similarly, chymotrypsin contains many serine residues, but only that at position 195 is phosphorylated by diisopropyl phosphofluoridate under mild conditions.

In *affinity labeling*, another way of identifying essential functional groups in enzyme active sites, the enzyme is allowed to react with a molecule that is synthetically prepared to resemble the true substrate, so that it is specifically bound to the active site like the true substrate, but contains in addition a functional group capable of rapid covalent reaction with some specific group of the enzyme on or near the active site. Figure 9-5 gives the structure of an affinity label for chymotrypsin, showing the similarity of much of its molecule to the structure of the normal substrate for chymotrypsin, as well as its reactive chemical group, which combines with histidine residue 57 of chymotrypsin. Thus histidine 57 and serine 195 (see above) of chymotrypsin have been identified as participants in its catalytic activity.

Other chemical tricks have been used to identify specific amino acid residues of various enzymes involved in catalytic activity. For example, in the preceding chapter (page 210) we saw that reduction of an enzyme-substrate complex of aldolase with the powerful reducing agent sodium borohydride traps the complex in a stable reduced form. Partial hydrolysis of the reduced enzyme-substrate compound yields a peptide containing a glyceryl derivative of the specific lysine residue with which the substrate forms a Schiff's base intermediate.

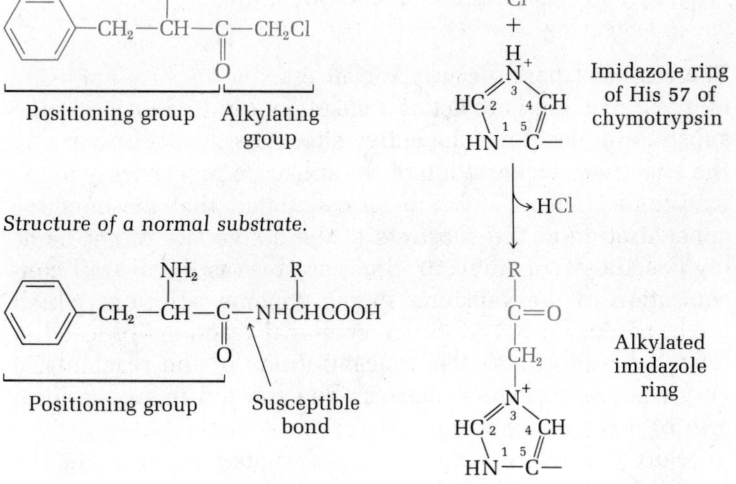

Figure 9-5
Reaction of an affinity labeling agent with chymotrypsin. The affinity label, N-tosyl-L-phenylalanylchloromethyl ketone (TPCK), was designed to resemble a normal substrate for chymotrypsin. However, instead of the usual susceptible structure of the substrate (—CO—NH—), TPCK contains the —CO—CH₂Cl group, a potent alkylating agent. Incubation of TPCK with chymotrypsin causes it to bind to the active site in the same way as the normal substrate. However, instead of hydrolysis of the substrate, alkylation of the essential His 57 residue occurs, indicating that the latter is close to the susceptible bond normally undergoing hydrolysis. The alkylated His residue can be isolated following complete hydrolysis of all the peptide bonds of the enzyme.

Structure of TPCK.

Positioning group   Alkylating group

Structure of a normal substrate.

Positioning group   Susceptible bond

Reaction of TPCK with chymotrypsin (R designates the positioning group of TPCK).

TPCK

Imidazole ring of His 57 of chymotrypsin

Alkylated imidazole ring

Factors Contributing to the Catalytic Efficiency of Enzymes

How efficient are enzymes as catalysts? By how much do they actually enhance the reaction rate? Quantitative data have been obtained for the hydrolysis of urea,

$$NH_2CONH_2 + 2H_2O + H^+ \longrightarrow 2NH_4^+ + HCO_3^-$$

catalyzed by urease isolated from the jack bean. The apparent first-order rate constant of this reaction in water in the absence of enzymes is 3×10^{-10} s^{-1} at pH 8.0 and 20°C. In contrast, the first-order rate constant of the breakdown of the urea-urease complex to yield the products is 3×10^4 s^{-1} under the same conditions. Thus the enzyme accelerates the uncatalyzed reaction some 10^{14}-fold. Similar calculations indicate that enzyme-catalyzed reactions proceed at rates that are anywhere from 10^8 to 10^{20} times faster than the corresponding uncatalyzed reactions. Only a few man-made catalysts begin to approach the activity of enzymes, and most are far less efficient.

Four major factors appear to contribute to the large rate accelerations produced by enzymes:

1 The enzyme may bind the substrate molecule in such a way that the susceptible bond is (a) in close proximity to the catalytic group on the active site and (b) so oriented in relation to the catalytic group that the transition state is readily formed.
2 Some enzymes may combine with the substrate to form an unstable covalent intermediate that more readily undergoes reaction to form the products.
3 By providing functional groups capable of acting as proton donors or proton acceptors the enzyme may bring about general-acid or general-base catalysis.
4 The enzyme may induce strain or distortion in the susceptible bond of the substrate molecule, making the bond easier to break.

These factors will be discussed in turn.

Proximity and Orientation of the Substrate; Orbital Steering

The simplest possible way for an enzyme to accelerate the rate of a reaction is to act as a means of binding or fixing the substrate molecule at the active site, thus greatly increasing the effective concentration of the substrate in a sharply localized zone. Indeed, it has been calculated that the effective concentration of the substrate at the active site might be as high as 100 M, or some 10^5 times as great as the overall concentration of the substrate in the enzyme solution, which might be only 0.001 M. Since chemical reactions proceed at rates proportional to the concentrations of the reactants, a very large rate enhancement can be expected in such a local area of high concentration.

Many interesting experiments designed to evaluate the factor of proximity have been carried out with organic model

Figure 9-6
Effect of proximity of the catalytic group on reaction rate. In these intramolecular models the general base —COO$^-$ is the catalyst, and the ester linkage is the substrate. R is a p-bromophenyl group. The cleavage of the ester is accompanied by cyclization to yield the acid anhydride. As the freedom of the COO$^-$ group to assume many different positions in relation to the ester group is restricted in this series of compounds, its catalytic activity is enhanced, showing the importance of the proximity factor. (Adapted from T. C. Bruice and S. J. Benkovic, Bio-organic Mechanisms, Vol. 1, p. 178, W. A. Benjamin, Inc., New York, 1966.)

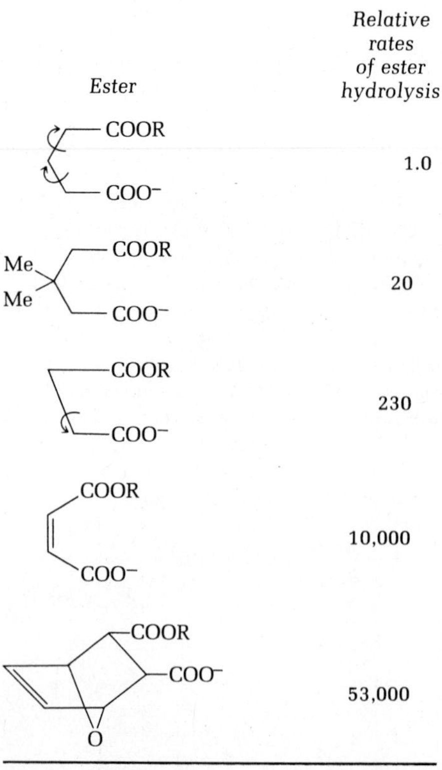

Ester	Relative rates of ester hydrolysis
—COOR / —COO$^-$	1.0
Me, Me —COOR / —COO$^-$	20
—COOR / —COO$^-$	230
COOR / COO$^-$	10,000
—COOR / —COO$^-$ (O)	53,000

reactions. In one such model two reacting groups are covalently built into the same molecule, so that they are necessarily brought close together. The reaction is then *intramolecular* rather than intermolecular and is said to be promoted by *anchimeric assistance*. An especially well-known case, studied by T. C. Bruice and his colleagues, is the intramolecular catalysis of the hydrolysis of monophenyl esters of dicarboxylic acids, in which the free carboxylate group functions as the catalyst (Figure 9-6). In this model the greater the proximity of the catalytic carboxylate group to the susceptible ester bond, the greater the reaction rate. In the most dramatic case enhancement of the reaction rate was 53,000-fold.

It may well be asked whether such rate enhancements are due to proximity alone or are the result of specific orientation of the reacting molecules in relation to each other. D. R. Storm and D. E. Koshland, Jr., have postulated that a major function of the active site of an enzyme is to produce *orbital steering*, i.e., the precise orientation of the substrate and the catalytic group of the enzyme with respect to each other, so that their bonding orbitals are aligned into the specific relationship required to bring the enzyme-substrate complex directly into the transition state (Figure 9-7). This hypothesis has provoked much heated debate; we shall return to this matter later (page 230).

Covalent Catalysis

Another way in which some enzymes may enhance the rate of a chemical reaction is to form a highly reactive covalent intermediate between the enzyme and the substrate which has a high probability of entering the transition state and thus permits the substrate to find a lower pass over the activation-energy barrier (page 188).

Many enzymes form covalent enzyme-substrate intermediates (Table 9-2). One case, the formation of a covalent Schiff's base intermediate in the action of fructose-diphosphate aldolase (page 210), which has already been described. Chymotrypsin also forms a covalent intermediate. When this enzyme is mixed with an equimolar amount of the ester p-nitrophenyl acetate, a relatively "slow" substrate, free p-nitrophenol is liberated at a higher rate than acetate, the other product. Quantitative study of this effect led to the conclusion that the hydrolysis of p-nitrophenyl acetate takes place in two steps:

Chymotrypsin + p-nitrophenyl acetate $\longrightarrow$
$$\text{acetyl-chymotrypsin} + p\text{-nitrophenol} \qquad (1)$$

$$\text{Acetyl-chymotrypsin} + H_2O \longrightarrow \text{acetate} + \text{chymotrypsin} \qquad (2)$$

The first reaction is relatively fast, but the second, which results in the release of free acetate, is rather slow. The acetyl-chymotrypsin intermediate is fairly stable at low pHs and can be isolated. When an ester of trimethylacetic acid is used as the substrate for chymotrypsin, trimethylacetyl-chymotrypsin is formed, which is so stable that it can be

Figure 9-7
Schematic representation of the orbital-steering hypothesis, which proposes that the substrate and the catalytic group on the enzyme must not only be brought into proximity but also into the proper alignment so that the relevant orbitals overlap. In this way the transition state can be attained with a high probability. [Adapted from A. Dafforn and D. E. Koshland, Jr., Biochem. Biophys. Res. Commun., 52: 780 (1973).]

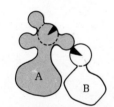

Unfavorable orientation,
unfavorable proximity

Favorable proximity,
unfavorable orientation

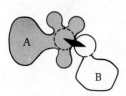

Favorable proximity,
favorable orientation

Table 9-2 Some enzymes forming covalent enzyme-substrate compounds

Enzyme	Type of covalent intermediate
Serine class	
Phosphoglucomutase	Phosphoenzyme
Acetylcholinesterase	Acyl-enzyme
Trypsin	Acyl-enzyme
Chymotrypsin	Acyl-enzyme
Elastase	Acyl-enzyme
Cysteine class	
Glyceraldehyde-phosphate dehydrogenase	Acyl-enzyme
Papain	Acyl-enzyme
Acetyl-CoA acetyltransferase	Acyl-enzyme
Histidine class	
Glucose 6-phosphatase	Phosphoenzyme
Succinyl-CoA synthetase	Phosphoenzyme
Lysine class	
Fructose-diphosphate aldolase	Schiff's base
Transaldolase	Schiff's base
D-Amino-acid oxidase	Schiff's base

crystallized. The functional group of chymotrypsin that becomes acylated during its catalytic cycle is the hydroxyl group of serine residue 195, which, as we have seen, becomes phosphorylated on treatment with diisopropyl phosphofluoridate.

Enzymes functioning via covalent enzyme-substrate intermediates are classified according to the type of amino acid residue with which the substrate reacts (Table 9-2). The *serine* class includes acetylcholinesterase, chymotrypsin, trypsin, and phosphoglucomutase. In these the hydroxyl group of a specific serine residue participates in the formation of an intermediate ester, either with an acyl group, as in chymotrypsin, to form an *acyl-enzyme*, or with a phosphoric group, as in phosphoglucomutase (page 435), to form a *phosphoenzyme*. In the *cysteine* class of enzymes, which includes glyceraldehyde-phosphate dehydrogenase and the proteolytic enzyme papain, a covalent thioester bond is formed between an acyl group of the substrate and the sulfhydryl group of a specific cysteine residue in the active site of the enzyme. The *histidine* class includes certain phosphate-transferring enzymes, e.g., succinyl-CoA synthetase, in which the imidazole group of a specific histidine residue of the enzyme becomes phosphorylated. In the *lysine* class, represented by fructose-diphosphate aldolase, an intermediate Schiff's base is formed between the ε-amino group of a lysine residue of the enzyme and a carbonyl group of the substrate. Many enzymes forming covalent intermediates show the characteristic kinetic behavior of double-displacement, or ping-pong, reactions (pages 204 and 205).

How does the intermediate formation of a covalent

Figure 9-8
Important nucleophilic groups of proteins.

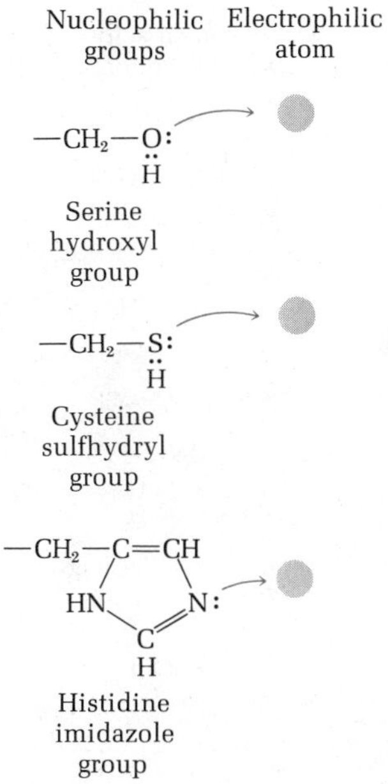

Nucleophilic   Electrophilic
groups   atom

$-CH_2-O:$
$\quad\quad\;\; H$

Serine
hydroxyl
group

$-CH_2-S:$
$\quad\quad\;\; H$

Cysteine
sulfhydryl
group

$-CH_2-C=CH$
$HN\quad\quad N:$
$\quad\; C$
$\quad\; H$

Histidine
imidazole
group

enzyme-substrate compound make a large rate enhancement of the overall reaction possible? To answer this we must first review a few principles of organic reaction mechanisms. The most common reaction pattern in covalent catalysis involves the attack of a _nucleophilic_ group of the catalyst on an electrophilic carbon atom of the substrate. Nucleophilic (nucleus-seeking) groups contain electron-rich atoms capable of donating electrons. Nucleophiles are very effective and versatile catalysts; for example, they promote the hydrolysis of simple organic esters and the transfer of acyl groups from one compound to another. Enzyme molecules contain at least three kinds of nucleophilic groups potentially capable of functioning catalytically, the imidazole group of histidine, the hydroxyl group of serine, and the sulfhydryl group of cysteine (Figure 9-8). Moreover, many coenzymes have nucleophilic centers.

A nucleophilic group can catalyze an acyl transfer reaction, of which ester hydrolysis is a special case because it attacks the molecule containing the acyl group, called the acyl-group donor, to yield an acyl derivative of the nucleophilic group. The acyl derivative of the catalyst functions as a reaction intermediate; in a second step the acyl group is transferred from the nucleophilic catalyst to the ultimate acyl-acceptor molecule, which may be some other alcohol or simply water. As shown in the schematic examples in Figure 9-9, in order for nucleophilic catalysis to take place the substrate must react more rapidly with the nucleophilic catalyst than it would with the final acyl-group acceptor in the absence of catalyst; moreover, the acylated catalyst so formed must also react more rapidly with the final acyl acceptor than the original substrate would (in the absence of catalyst). In this way the formation of a covalent intermediate can lower the activation-energy barrier of reactions involving acyl-group transfer.

Figure 9-9
Reaction-rate enhancement by nucleophilic catalysis. Note that the nucleophilic catalyst Y is regenerated at the end of the catalytic cycle.

Uncatalyzed reaction:

$$RX + H_2O \xrightarrow{slow} ROH + X^- + H^+$$

Catalysis by the nucleophilic agent Y:

$$RX + Y \xrightarrow{fast} RY + X^-$$

$$RY + H_2O \xrightarrow{fast} ROH + Y + H^+$$

Sum:

$$RX + H_2O \xrightarrow{fast} ROH + X^- + H^+$$

Acid-Base Catalysis

Acids and bases, defined in a broad sense, are the most versatile and universal catalysts of organic reactions. There are two major types of acid-base catalysis: _specific_ acid-base catalysis and _general_ acid-base catalysis. Specific-acid and specific-base catalysis are defined as rate enhancements proportional only to the concentration of H^+ or OH^- ions, respectively. General-acid and general-base catalysis are rate enhancements that are proportional to the concentration of general acids and general bases, i.e., proton donors and proton acceptors, respectively, as defined by J. N. Brönsted and T. M. Lowry (page 47).

Specific acid-base catalysis by H^+ or OH^- is of relatively limited importance in enzymatic reactions, but general acid-base catalysis is much more likely to be involved in the action of enzymes. Many types of organic reactions occurring in cells — the addition of water to carbonyl groups, the hydrolysis of carboxylic and phosphoric esters, the elimination of water from double bonds, various molecular rearrangements, and many substitution reactions—are subject to this type of catalysis. Moreover, enzyme molecules are known to contain several kinds of functional groups capable of acting as general acids or general bases, e.g., amino, carboxyl, sulfhydryl, phenolic hydroxyl, and imidazole groups (Figure 9-10).

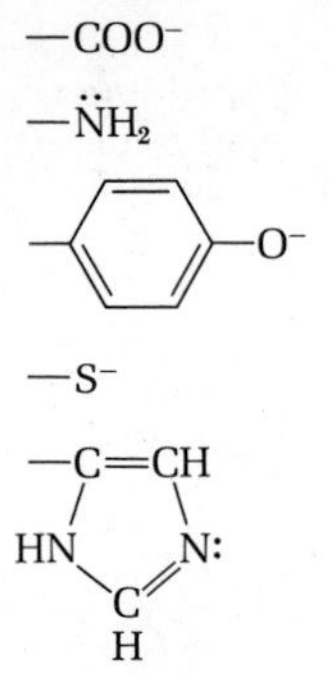

In reactions catalyzed by general acids or bases the catalyst at some point in the catalytic cycle functions as an acceptor or donor of protons. Usually the critical proton-transfer step is to or from a carbon atom of the transition-state species. For example, the rate of some enzymatic reactions is greatly decreased when they are carried out in deuterium oxide (D_2O) rather than in water (H_2O), indicating that a protonation step is rate-limiting, since the deuterium ion (D^+) reacts much more slowly than the proton (H^+) with a general base. Sometimes both a general acid and a general base are involved as catalysts in *concerted* proton transfers. In such reactions the general acid donates a proton to the transition-state species and the general base accepts one.

Two important factors affect the rate of reactions catalyzed by a general acid or general base. The first is the strength of the general acid or base, i.e., its proton dissociation constant. Among the most active general acid-base catalysts is the imidazole group of histidine, whose pK' is about 6.0 (page 75), enabling it to act both as a proton donor and as a proton acceptor at a pH near that of biological fluids. The second important factor is the rate at which the acid or base can donate or accept protons. Here again the imidazole group is particularly effective, since its rates of protonation and deprotonation are not only nearly equal near neutral pH but also very fast, with a half-time of less than 0.1 ns. In fact, the imidazole group of specific histidine residues is involved in the catalytic cycle of many different enzymes. The imidazole group, as we have seen, is also a potent nucleophile. Since there are relatively few histidine residues in most proteins, this amino acid may have been selected during the course of biological evolution to play a special role in enzyme catalysis.

General acid-base catalysis provides a means for catalyzing at neutral pH, where the concentrations of H^+ and OH^- are very low, chemical reactions that would otherwise require very high concentrations of H^+ or OH^-. For example, in the absence of enzymes hydrolysis of peptide bonds requires very high concentrations of H^+ or OH^-, long reaction periods, and high temperatures (page 100), whereas hydrolysis of peptide bonds by chymotrypsin proceeds rapidly at neutral pH, promoted by general acid-base catalysis at the active site.

The Factor of Strain in Enzyme Catalysis: The Relationship of Enzyme Conformation to Catalytic Activity

Two questions have long confronted enzyme biochemists: Why is the native three-dimensional conformation of the enzyme molecule generally required for catalytic activity? Why are enzyme molecules so large in relation to the susceptible

Figure 9-10
Functional groups of proteins capable of acting as general acids and general bases.

Some general acid groups in
proteins (proton donors)

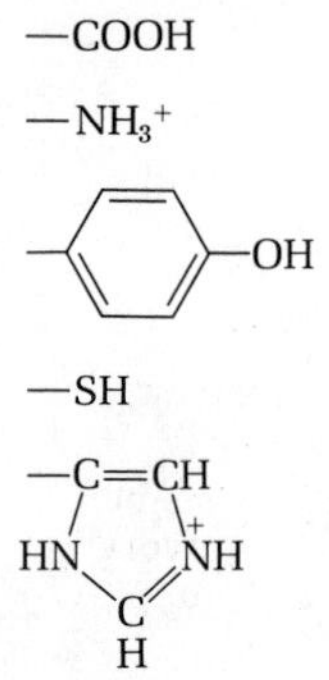

—COOH

—NH_3^+

Some general base groups
(proton acceptors)

—COO^-

—$\ddot{N}H_2$

Figure 9-11
Requirement of S-peptide and S-protein for ribonuclease activity. The fragments need not be covalently joined for restoration of activity-merely mixed.

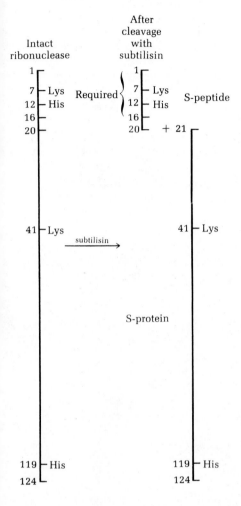

structures of the substrate molecule? Research on ribonuclease and other enzymes has given some clues. We have seen that histidine residues 12 and 119 are required for catalytic activity of ribonuclease, suggesting that the polypeptide chain is folded in such a way that these two histidines, which are separated by 107 residues in the backbone, are brought close together at the active site. This view is confirmed by some interesting experiments by F. M. Richards and his colleagues on the effect of cleaving the polypeptide-chain backbone of ribonuclease. The bacterial protease subtilisin cleaves the 124-residue chain of ribonuclease between residues 20 and 21 (Figure 9-11). The long fragment, called the *S-protein*, contains histidine residue 119, and the short fragment, the *S-peptide*, contains histidine 12. The S-protein and the S-peptide are catalytically inactive when tested singly, but when they are simply mixed again at pH 7.0, enzymatic activity is restored, even though no covalent linkage is formed between the pieces. The S-peptide evidently binds to the S-protein through weak forces, such as hydrogen bonding and hydrophobic interactions, in such a way as to bring the two essential histidine residues together at the active site even though the polypeptide chain extending between them has been broken.

A similar example is provided by chymotrypsin, which is secreted as an inactive precursor, or zymogen, *chymotrypsinogen*. Chymotrypsinogen has a single polypeptide chain of 245 residues held together by five intrachain disulfide bridges (Figure 9-12). It is converted into active α chymotrypsin by the enzymatic hydrolysis of four peptide linkages, by sequential action of trypsin and chymotrypsin, with the release of two dipeptides. The active α chymotrypsin produced this way consists of three polypeptide chains held together by two —S—S— bonds. Thus the two specific residues essential for catalytic activity, histidine 57 and serine 195, are present in two different chains. However, it has been directly established by x-ray analysis of the tertiary structure of chymotrypsin that these residues are actually very close to each other in the conformation of the native enzyme (Figures 9-12 and 9-13).

It has long been suspected that a conformational change in the tightly folded loops of the polypeptide chain may somehow contribute to the catalytic process. Indeed, many enzymes undergo a change in conformation during their catalytic cycle, as determined by optical-rotation measurements (page 150). Moreover, x-ray diffraction analysis shows that the conformation of free carboxypeptidase is significantly different from that of carboxypeptidase saturated with a sluggish substrate like glycyltyrosine.

There are several different views about the significance of these conformational differences between the free and combined enzyme. One view holds that once the substrate is bound to the active site in a lock-and-key relationship, a conformational change occurs which imposes a strain on the substrate molecule, something like a medieval torture rack. The drawback to this view is that if the active site has a rigid

Figure 9-12
Linear representation of the conversion of chymotrypsinogen to chymotrypsin. After excision of the two dipeptides Ser-Arg and Thr-Asn, the A, B, and C chains of chymotrypsin are connected only by —S—S— bridges. The catalytically active residues (color) come from two chains.

Folded representation of chymotrypsin, showing how the essential residues may be brought together in the active site, although they are in separate chains.

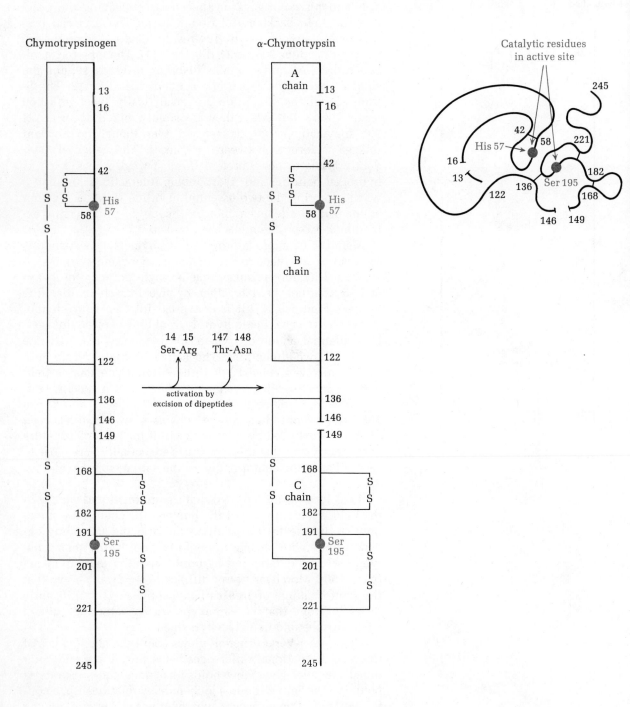

Figure 9-13
Three-dimensional model of the polypeptide-chain backbone of chymotrypsin as determined by x-ray analysis. Residues His 57 and Ser 195 function in the catalytic cycle. A, B, and C refer to the three polypeptide chains. See Figure 9-12. [Redrawn from B. W. Matthews, P. B. Sigler, R. Henderson, and D. M. Blow, Nature, 214: 652 (1967).]

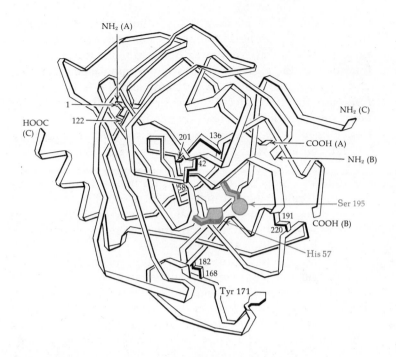

structure, as implied by the lock-and-key analogy, it cannot fit *both* the substrate and the products of a reversible reaction in an optimum fashion. This dilemma has led to other hypotheses. One is that the active site of the free enzyme exactly fits neither the substrate nor the products but only the transition-state species. Evidence for this view is that certain enzymes are very profoundly inhibited by compounds specifically resembling the transition state rather than the free substrate molecule.

Another idea, the *induced-fit hypothesis* of Koshland, was originally postulated to explain a number of anomalies in enzyme action. For example, many enzymes are unable to act upon smaller homologs of their normal substrate. This point is best illustrated by <u>hexokinase</u>, catalyzing the reaction

$$ATP + \text{D-glucose} \rightleftharpoons ADP + \text{glucose 6-phosphate}$$

Because hexokinase can phosphorylate the 6-hydroxyl group of several different hexoses, one might expect to be able to utilize substrate molecules smaller than glucose. Yet hexokinase cannot phosphorylate lower homologs of glucose such as glyceraldehyde or simpler alcohols such as glycerol, ethanol, or even water, which, because of its very small size and very high concentration in an aqueous system, some 55 *M* (page 46), could be expected to penetrate readily to the active site and function as a very effective phosphate-group acceptor from ATP. These and other considerations led Koshland to postulate that the essential functional groups on the active site of the free enzyme are not in their optimal positions for promoting catalysis when the active site is unoccupied but that when the substrate molecule is bound by the enzyme, the binding affinity forces the enzyme molecule into a conformation in which the catalytic groups assume a favorable geometrical position to form the transition state; i.e., there is an _induced fit_ (Figure 9-14). The enzyme molecule is unstable in this active conformation and tends to revert to its free form in the absence of substrate. A poor substrate, i.e., one bound to the active site with low affinity, is less able to force the enzyme molecule into the maximally active form because it is either too small or has incorrect steric properties. This explains why water at 55 _M_ concentration usually cannot compete with the normal substrate glucose in the hexokinase reaction.

Induced conformational changes occurring when the substrate molecule is bound may also be a significant factor in the enhancement of reaction rate by enzymes. Presumably they distort both the enzyme and the substrate, causing them to enter the transition state more readily.

Now that we have surveyed the four major factors that may contribute to the catalytic efficiency of enzymes, can we assess the contribution of each more quantitatively? Probably no single factor can account for the entire catalytic activity of all enzymes, and it is more likely that for each enzyme a specific combination of factors is responsible for the overall acceleration of the reaction rate. Calculations of W. P. Jencks and his colleagues indicate that a combination of the proximity and orientation factors might provide a rate enhancement as high as 10^8, which comes very close to accounting for the catalytic efficiency of some enzymes. By bringing the substrate and the catalytic group close together and by orienting them at precise angles to each other their freedom to undergo translational and rotational movement, i.e., their entropy, can be reduced sufficiently to allow them to enter the transition state readily. Covalent catalysis and acid-base catalysis are very important in enzyme action but probably contribute relatively small enhancements, perhaps no greater than 10^3.

In addition to the major factors discussed above, it has also been observed that in some enzymes the active site cavity is relatively nonpolar, so that the catalytic group is surrounded by a low-dielectric environment. This has the effect of polarizing not only the catalytic group but also the susceptible bond of the substrate, rendering both more reactive. This feature may also contribute to the total rate enhancement given by some enzymes.

Some Reaction Mechanisms at Enzyme Active Sites

Although the mechanism of catalysis is not yet known in detail for any single enzyme, great progress in this direction is being made through kinetic and specificity studies, chemical analysis of enzyme active sites, and x-ray comparison of the structure of certain enzymes and their complexes with sluggish substrates or competitive inhibitors.

Chymotrypsin is a well-studied example. Its activity depends on the His 57 and Ser 195 residues, which are located near each other in the active site. Moreover, a covalent acyl-enzyme intermediate is formed, involving the hydroxyl

Figure 9-14
The induced-fit hypothesis. The active site is postulated to be flexible and to adjust its conformation to that of the substrate molecule.

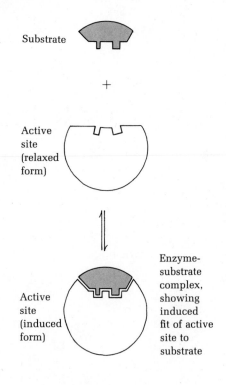

Substrate

+

Active site (relaxed form)

Active site (induced form)

Enzyme-substrate complex, showing induced fit of active site to substrate

group of Ser 195. This and other information has been used to formulate a reaction mechanism for the hydrolysis of a peptide bond by chymotrypsin. According to one hypothesis (Figure 9-15), the imidazole group of the His 57, acting as a general base, promotes the nucleophilic attack of the hydroxyl group of Ser 195 on the carboxyl carbon atom of the aminoacyl group of the substrate, thus displacing the amino group of the other amino acid, the leaving group, and causing the formation of the aminoacyl derivative of the hydroxyl group of Ser 195. In the second step of this double displacement, which is the reverse of the first, the imidazole group of His 57 promotes the transfer of the aminoacyl group from the serine hydroxyl group of the enzyme to the second substrate, the acyl-group acceptor, which is water in this case but may also be an alcohol, an amino acid, or an amine (page 219). Although this reaction mechanism (Figure 9-15) is supported by much evidence, it is not sufficient as it stands to account for the high catalytic rates yielded by chymotrypsin; presumably the factors of proximity, orientation, and strain must be superimposed on the general-base-catalyzed nucleophilic displacement mechanism.

Lysozyme is another enzyme whose catalytic mechanism is being studied. X-ray analysis of the crystalline enzyme shows that it contains a long cleft as its active site (Figure 9-16), which can accommodate its normal substrate, the

Figure 9-15
A proposed mechanism for the action of chymotrypsin on a dipeptide. This is only one of many hypotheses for the covalent catalysis steps, which are undoubtedly accompanied by other effects leading to rate enhancement.

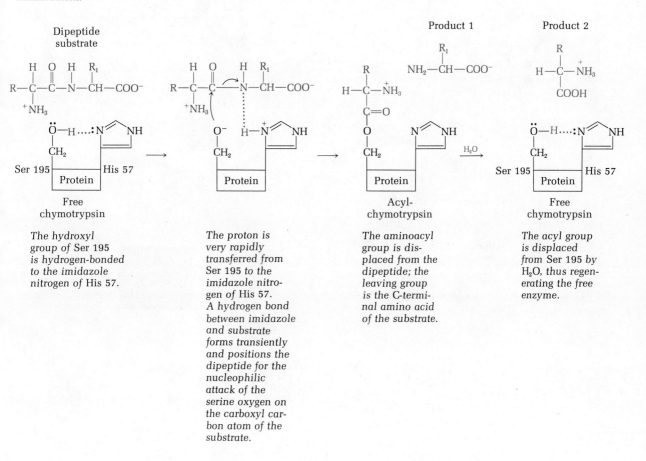

The hydroxyl group of Ser 195 is hydrogen-bonded to the imidazole nitrogen of His 57.

The proton is very rapidly transferred from Ser 195 to the imidazole nitrogen of His 57. A hydrogen bond between imidazole and substrate forms transiently and positions the dipeptide for the nucleophilic attack of the serine oxygen on the carboxyl carbon atom of the substrate.

The aminoacyl group is displaced from the dipeptide; the leaving group is the C-terminal amino acid of the substrate.

The acyl group is displaced from Ser 195 by H_2O, thus regenerating the free enzyme.

Figure 9-16
Structure and postulated action of the lysozyme-substrate complex. The drawing below shows the backbones of the lysozyme (open and color screened lines) and substrate (solid color) molecules in the crystalline complex. [The drawings on this page and the facing page are reproduced by permission from R. E. Dickerson and I. Geis, The Structure and Action of Proteins, published by W. A. Benjamin, Inc., Menlo Park, California. Copyright 1969 by Dickerson and Geis. Coordinates by courtesy of D. C. Phillips, Oxford.]

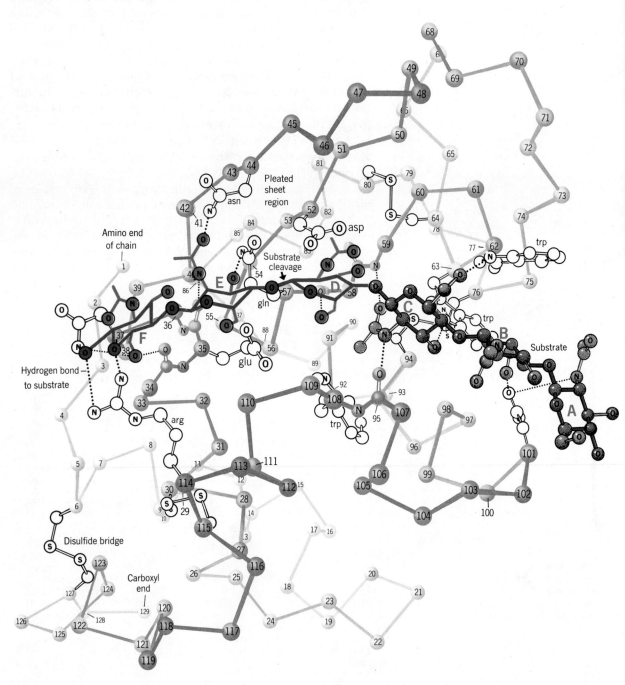

The magnification below shows the postulated mechanism of action of lysozyme. Only part of the substrate molecule is shown (in color). Ring D is a residue of N-acetylmuramic acid (page 269) and ring E is a residue of N-acetylglucosamine. In the top figure a hydrogen atom of Glu 35 attacks the glycosidic oxygen between the two sugar residues. In the middle figure, carbon atom 1 of ring D becomes a carbonium ion, which is stabilized by the neighboring carboxyl group of Asp 52; ring D twists into a half-chair conformation. A hydroxide ion (from a water molecule) attacks the carbonium ion and completes the cleavage (bottom figure).

long-chain peptidoglycan of bacterial cell walls. Lysozyme forms enzyme-substrate complexes with certain sluggish substrates or competitive inhibitors (page 197) structurally similar to the normal substrate; these are specifically bound to the cleft but not attacked by the enzyme. One enzyme-substrate complex of lysozyme has been crystallized and its structure analyzed by x-ray methods. The structure of the active site, in relation to the susceptible bond of the substrate molecule, is clearly visualized (Figure 9-16). Most significant is the identification of the carboxyl group of glutamic residue 35 and the carboxyl group of aspartic residue 52, which lie in the cleft near the bond to be broken, as the proton-donor and proton-acceptor groups that appear to function in a concerted general-acid–general-base catalytic mechanism. It is also of interest that glutamic residue 35 is surrounded by nonpolar R groups, which could be expected to enhance the proton transfers.

Carboxypeptidase, an exopeptidase that catalyzes the hydrolysis of the C-terminal peptide bond of peptides, has an active site in the form of a hole or cavity, into which the C-terminal residue of the polypeptide substrate fits, end on. This cavity contains a zinc atom, as well as the characteristic R groups of residues Arg 145, Tyr 248, and Glu 270. These groups are so located in the active site, as deduced by x-ray analysis, that the hydrolysis of the C-terminal peptide bond of the substrate is facilitated (Figure 9-17).

Figure 9-17
Schematic representation of the active site of carboxypeptidase. It consists of a cleft capable of accommodating the six C-terminal residues of the polypeptide substrate and a pocket that accepts the R group of the C-terminal residue. Tyr 248, Arg 145, Glu 270, and the bound Zn^{2+} of the enzyme position the substrate molecule properly.

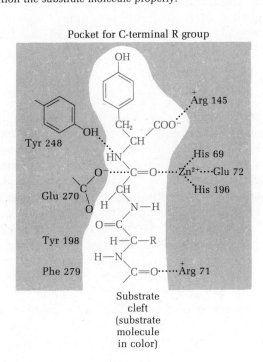

Although nearly all the enzymes whose structures have been determined by x-ray methods are hydrolases, the structure of lactate dehydrogenase, an intracellular enzyme catalyzing an ordered bisubstrate oxidation-reduction reaction, has now been established. The general properties of this enzyme, the structure of its active site, and a postulated mechanism are given elsewhere (pages 244 and 484).

Regulatory Enzymes

All enzymes exhibit various features that could conceivably be elements in the regulation of their activity in living cells. All have a characteristic optimum pH, which makes possible alteration of their catalytic rates with changes in intracellular pH. The rates of all enzymatic reactions also depend on the substrate concentration, which may vary significantly under intercellular conditions. Moreover, many enzymes require either metal ions, such as Mg^{2+} or K^+, or coenzymes (page 337) for activity, suggesting that fluctuations in the concentration of these metals or coenzymes in the cell can regulate enzyme activity. However, over and above these properties of all enzymes, some enzymes possess other properties that specifically endow them with regulatory roles in metabolism. Such more highly specialized forms are called regulatory enzymes. There are two major types of regulatory enzymes: (1) allosteric enzymes, whose catalytic activity is modulated through the noncovalent binding of a specific metabolite at a site on the protein other than the catalytic site, and (2) covalently modulated enzymes, which are interconverted between active and inactive forms by the action of other enzymes. Some of the enzymes in the second class also respond to noncovalent allosteric modulators. These two types of regulatory enzymes are responsive to alterations in the metabolic state of a cell or tissue on a relatively short time scale—allosteric enzymes within seconds and covalently regulated enzymes within minutes. We shall now examine the properties of these classes of regulatory enzymes.

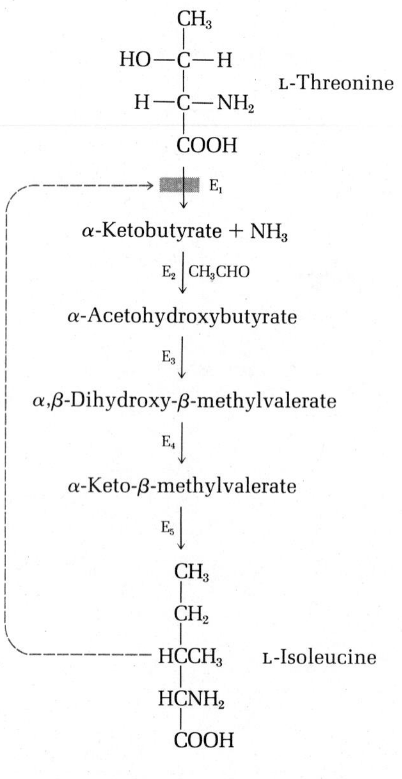

Allosteric Enzymes

In many multienzyme systems the end product of the reaction sequence may act as a specific inhibitor of an enzyme at or near the beginning of the sequence, with the result that the rate of the entire sequence of reactions is determined by the steady-state concentration of the end product. The classical example is the multienzyme sequence catalyzing the conversion of L-threonine to L-isoleucine, which occurs in five enzyme-catalyzed steps (Figure 9-18). The first enzyme of the sequence, L-threonine dehydratase, is strongly inhibited by L-isoleucine, the end product, but not by any other intermediate in the sequence. The kinetic characteristics of the inhibition by isoleucine are atypical; the inhibition is neither competitive with the substrate L-threonine, nor is it noncompetitive or uncompetitive. Isoleucine is quite specific as an inhibitor; other amino acids or related compounds do

Figure 9-18
Feedback inhibition of the formation of isoleucine from threonine. Isoleucine, the end product of the sequence, inhibits the first enzyme E_1 (threonine dehydratase). E_2, E_3, E_4, and E_5 symbolize enzymes catalyzing the intermediate steps. The intermediate metabolites in the sequence, i.e., α-ketobutyrate, α-acetohydroxybutyrate, etc., do not inhibit E_1. Feedback inhibition is designated by a dashed line (color) leading from the end product to the colored bar across the reaction arrow for the reaction being modulated. This notation is used throughout this book.

Figure 9-19

Patterns of allosteric modulation. In linear pathways the end product usually inhibits the first enzyme in the sequence. Sometimes the precursor S may act as a positive modulator and stimulate the first reaction, as shown by the colored arrow. In branched pathways the metabolite at the branch point is often the feedback inhibitor of the first enzyme, whereas the two end products of the branches (P₁ and P₂) often act as feedback inhibitors of the first enzymes after the branch point.

Linear pathways

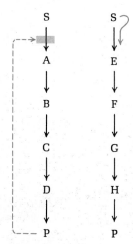

Feedback   Feedforward
inhibition   stimulation

A branched pathway

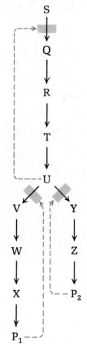

not inhibit. This type of inhibition is variously called *end-product inhibition*, *feedback inhibition*, or *retroinhibition*. The first enzyme in this sequence, that which is inhibited by the end product, is called an *allosteric enzyme*, a name proposed by J. Monod, J.-P. Changeux, and F. Jacob, of the Pasteur Institute in Paris, who first developed a comprehensive theory for the function of this type of regulatory enzyme. The term allosteric denotes "another space" or "another structure"; allosteric enzymes possess, in addition to the catalytic site, the "other space," to which the specific *effector* or *modulator* is reversibly and noncovalently bound. In general, the allosteric site is as specific for binding the modulator as the catalytic site is for binding the substrate. Some modulators, e.g., L-isoleucine for threonine dehydratase (Figure 9-18), are inhibitory and therefore called *inhibitory* or *negative* modulators. Other allosteric enzymes may have stimulatory, or *positive*, modulators. When an allosteric enzyme has only one specific modulator, it is said to be *monovalent*. Some allosteric enzymes respond to two or more specific modulators, each bound to a specific site on the enzyme; they are *polyvalent*. Moreover, a given allosteric enzyme may have both positive and negative modulators. Two or more multienzyme systems may be connected by one or more polyvalent enzymes in a control network (Figure 9-19), examples of which are described elsewhere (Chapters 23 to 26).

The first step in a multienzyme reaction sequence, i.e., the step catalyzed by the allosteric enzyme, is usually irreversible under intracellular conditions. It is often called the *committing reaction*; once it occurs, all the ensuing reactions of the sequence takes place. Clearly, it is good strategy for the cell to regulate a metabolic pathway at its first step, to achieve maximum economy of metabolites.

Allosteric enzymes are usually much larger in molecular weight, more complex, and often more difficult to purify than ordinary enzymes because nearly all known allosteric enzymes are oligomeric and thus have two or more polypeptide chain subunits, usually in an even number; some contain many chains. Allosteric enzymes show a number of anomalous properties. Some are unstable at 0°C but stable at room or body temperature, unlike single-chain enzymes, which do not show cold lability. Most allosteric enzymes exhibit atypical dependence of initial reaction velocity on substrate concentration and fail to conform to the classical Michaelis-Menten relationship (pages 189ff). Moreover, the inhibition produced by the modulating metabolite does not always conform to the simple, well-known prototypes of competitive, uncompetitive, and noncompetitive inhibition.

Allosteric enzymes show two different types of control, *heterotropic* and *homotropic*, depending on the nature of the modulating molecule. *Heterotropic* enzymes are stimulated or inhibited by an effector or modulator molecule other than their substrates. For the heterotropic enzyme threonine dehydratase (Figure 9-18) the substrate is threonine and the modulator is L-isoleucine. In *homotropic* enzymes, on the other hand, the substrate also functions as the modulator. Homotropic enzymes contain two or more binding sites for the substrate; modulation of these enzymes depends on how many of the substrate sites are occupied. However, a great many (if not most) allosteric enzymes are of mixed homotropic-heterotropic type, in which *both* the substrate and some other metabolite(s) may function as modulators.

Kinetics of Allosteric Enzymes

Allosteric enzymes do not usually show classical Michaelis-Menten kinetic relationships between substrate concentration, V_{max}, and K_M because their kinetic behavior is greatly altered by variations in the concentration of the allosteric modulator. Many allosteric enzymes, particularly homotropic ones, show a sigmoid curve relating initial velocity to substrate concentration, rather than the rectangular hyperbola yielded by the Michaelis-Menten relationship (Figure 9-20). The sigmoid curve implies that binding the first substrate molecule to the enzyme enhances the binding of subsequent substrate molecules to the other substrate sites, just as binding one oxygen molecule to hemoglobin enhances the binding of subsequent oxygen molecules (page 146). Sigmoid curves are sometimes very steep, so that a rather small increase in substrate concentration can cause a very large acceleration of the rate of catalysis, much more than is afforded by a simple, nonregulatory enzyme obeying the hyperbolic Michaelis-Menten relationship. Such a sigmoid relationship is an example of *positive cooperativity*, since the binding of one substrate molecule at one site enhances the binding of subsequent molecules at the other sites. It should be made clear, however, that not all allosteric enzymes exhibit sigmoid curves of v_0 vs. [S]; moreover, not all enzymes showing such sigmoid curves are necessarily allosteric enzymes.

Some allosteric enzymes show the opposite behavior: binding one substrate molecule appears to *decrease* the binding of subsequent substrate molecules. This is *negative cooperativity*, which results in a rather flattened plot of initial velocity vs. substrate concentration (Figure 9-20). In such a case the enzyme is far less sensitive to small changes in substrate concentration and approaches saturation more slowly than nonregulatory or positively cooperative enzymes.

Some allosteric enzymes respond to the binding of a modulator with a change in the *apparent K_M* for the substrate, without change in V_{max} (Figure 9-20). A negative modulator will then produce an increase in the apparent K_M and a positive modulator a decrease in the apparent K_M. [Here the term "apparent K_M," also designated $S_{0.5}$, refers to the substrate concentration giving half-maximal velocity; it cannot be used to calculate the initial velocity of an allosteric enzyme since the v_0-vs.-[S] relationship is not a rectangular hyperbola, as required by the Michaelis-Menten assumptions and equations (page 189)]. Thus a negative modulator will decrease the reaction rate at a fixed, nonsaturating concentration of substrate, whereas a positive modulator will increase the rate. Allosteric enzymes that respond to positive or negative modulators with a change in the apparent K_M but without a change in V_{max} are sometimes

Figure 9-20
Comparison of idealized plots of percent saturation of the substrate sites vs. substrate concentration for (A) a nonregulatory enzyme, (B) a regulatory enzyme showing positive cooperativity, and (C) a regulatory enzyme showing negative cooperativity.

A. Nonregulatory enzyme obeying Michaelis-Menten rate equation (rectangular hyperbola). An 81-fold increase in [S] is required to bring activity from 10 to 90 percent maximal.

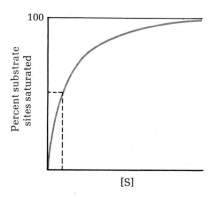

B. Regulatory enzyme showing sigmoidal curve (positive cooperativity). Only a ninefold increase in [S] is required to raise the activity from 10 to 90 percent maximal.

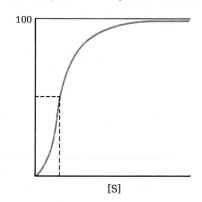

C. Regulatory enzyme showing negative cooperativity. An increase in substrate concentration of over 6,000-fold is required to raise the activity from 10 to 90 percent maximal.

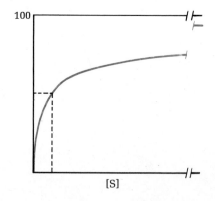

called *K enzymes* (Figure 9-21). Conversely, allosteric enzymes showing a change in V_{max} in response to the modulator, without a change in apparent K_M, are called *M enzymes*; there are fewer examples of this class. However, not all allosteric enzymes can be placed in one of these two classes. Generally speaking, interpretation of the regulatory properties of allosteric enzymes from plots of initial velocity vs. substrate concentration is biologically significant only at the lower range of substrate concentration, since the concentration of metabolites in intact cells is usually far below saturation levels for the enzymes acting on them (page 211).

The capacity of an allosteric enzyme for activation or inhibition by its specific modulators can sometimes be abolished without damaging its catalytic activity, often by gentle heating, exposure to urea, or treating the enzyme with an agent that produces a chemical modification of the modulator site. The enzyme is then said to be *desensitized*. A desensitized allosteric enzyme not only loses its sensitivity to modulators but may even show normal Michaelis-Menten kinetics. Genetic mutations sometimes result in the biosynthesis of a defective allosteric enzyme that is insensitive to its normal modulator, presumably because of a nonfunctional replacement of a critical amino acid residue.

Aspartate Transcarbamoylase: Kinetics and Inhibition

Although many allosteric enzymes are known, the best understood is the *aspartate transcarbamoylase* of *E. coli*, often abbreviated as ATCase (its recommended EC name is *aspartate carbamoyltransferase*). It is not to be regarded as typical of all allosteric enzymes but as representative of one class.

ATCase catalyzes the reaction

Carbamoyl phosphate + L-aspartate $\longrightarrow$

$$\text{N-carbamoyl-L-aspartate} + P_i$$

an early step in the enzymatic biosynthesis of the pyrimidine nucleotide cytidine triphosphate (CTP), to be discussed further elsewhere (page 736). CTP, the end product of this biosynthetic sequence, is the specific negative modulator of ATCase; cytidine diphosphate (CDP) and cytidine monophosphate (CMP) have no activity. ATCase also has a positive modulator, namely, ATP, which reverses the inhibitory effect of CTP. Figure 9-22 shows the effect of aspartate concentration on the activity of ATCase, alone and in the presence of its modulators, ATP and CTP. CTP increases the apparent K_M of the enzyme, whereas ATP decreases it.

The ATCase of *E. coli* has been isolated and studied in great detail by J. C. Gerhart, A. B. Pardee, and H. K. Schachman in the United States and K. Weber in Germany, among others. It has a molecular weight of about 310,000 but can be dissociated by treatment with mercurial reagents to yield two identical *catalytic subunits* and three identical *regulatory subunits*. Each of the two catalytic subunits has a molecular weight of about 100,000 and contains three poly-

Figure 9-21
*Effect of positive (accelerating) and
negative (inhibitory) modulators on the
v_0-vs.-[S] curves of the K and M classes of
allosteric enzymes.*

A. K enzymes. *These enzymes respond to
increasing concentrations of accelerating
modulators by a decrease in the apparent
K_M and to increasing concentrations of
inhibitory modulators by an increase in
the apparent K_M, so that at a fixed, nonsaturating concentration of substrate the
reaction rate increases in the presence of an
accelerating modulator and decreases in the
presence of an inhibitory modulator. V_{max}
of K enzymes remains constant.*

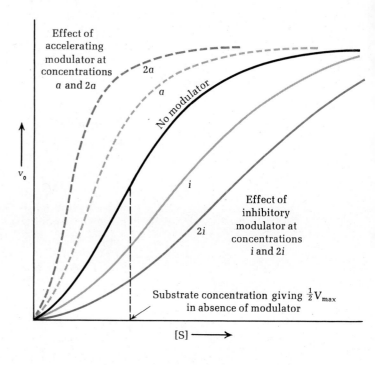

B. M enzymes. *These enzymes undergo
changes in V_{max} but not in apparent K_M in
the presence of accelerating or inhibitory
modulators, as shown. The terms K_M and
V_{max} cannot be strictly applied to allosteric
enzymes not obeying the hyperbolic Michaelis-Menten relationship; they are used here
only to designate the concentration of
substrate giving half-maximal velocity and
the maximal velocity, respectively.*

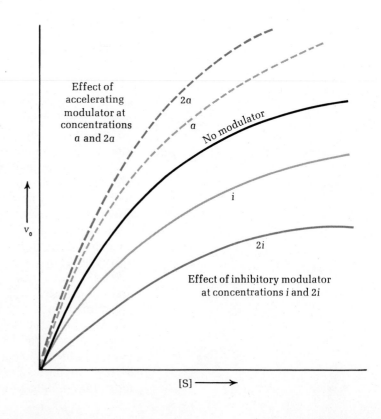

Figure 9-22
Properties of aspartate transcarbamoylase.

Plot of the initial velocity v_0 vs. aspartate concentration. If the concentration of aspartate is kept at a constant level, well below the saturating concentration, the addition of the negative modulator CTP decreases the activity of the enzyme, since the apparent K_M is increased. Conversely, when the positive modulator ATP is added, the enzyme activity increases, since the apparent K_M is lowered.

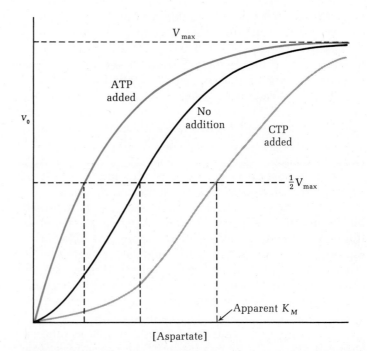

Schematized representation of the subunit structure of the enzyme which dissociates on treatment with Hg^{2+} into two trimeric catalytic subunits and three dimeric regulatory subunits (color). These can be dissociated further to give a total of six monomeric C chains and six monomeric R chains. Each of the subunits may exist in two conformations, indicated by circles and squares. The active conformation of ATCase is given by circles, and the inactive conformation, induced by the binding of CTP to the regulatory subunits, is indicated by squares.

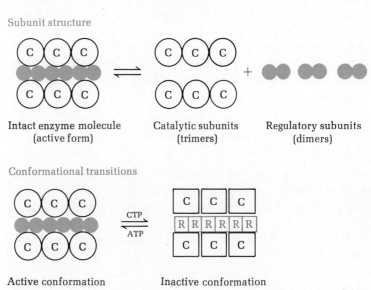

peptide chains of molecular weight 34,000, called C' chains. Each catalytic subunit contains three binding sites for the substrate aspartate, one on each of the three C chains. The catalytic subunits have enzymatic activity but are not sensitive to the modulator CTP. Each of the three regulatory subunits, which have no catalytic activity, contains two polypeptide chains (R chains) of molecular weight 17,000; one atom of Zn^{2+} is bound to each of the R chains. Each regulatory subunit can bind two molecules of the modulator CTP, one on each of the two R chains.

When the native ATCase molecule is dissociated into a mixture of its catalytic and regulatory subunits, the mixture retains catalytic activity but the reaction is not inhibited by CTP so long as the catalytic and regulatory subunits are detached from each other. For inhibition by CTP to occur, the catalytic and regulatory subunits must be physically associated in the intact oligomeric ATCase molecule. Thus binding CTP to the regulatory subunits diminishes the enzymatic activity of the associated catalytic subunits. This signal is believed to be transmitted by a conformational change; when CTP binds to its specific sites, the regulatory subunits are thought to undergo a change in conformation that is sterically transmitted to the catalytic subunits. They in turn undergo a cooperative transition from a catalytically active conformation to another conformation having less activity at the same substrate concentration. Figure 9-22 summarizes the kinetic behavior of ATCase and its modulation by the negative effector CTP and the positive effector ATP through cooperative conformational changes. ATCase may be representative of those allosteric enzymes in which the substrate and modulator binding sites are on different subunits. There are other allosteric enzymes in which both sites are present on the same subunit. Still other allosteric enzymes appear to have but a single polypeptide chain.

It is an important feature of ATCase, as well as other allosteric enzymes, that the manner of its allosteric regulation is species-specific. Although the ATCase of E. coli and certain other bacteria is strongly inhibited by CTP and stimulated by ATP, the corresponding ATCase of animal tissues is not affected by these nucleotides (page 737). Homologous allosteric enzymes from different species of organisms may have quite different allosteric modulators.

Mechanism of the Regulatory Activity
of Allosteric Enzymes

Much attention has been focused on the molecular mechanisms by which the binding of the modulator to the regulatory site of an allosteric enzyme can change the activity of the catalytic site. One of the most fruitful avenues of research and speculation has been evoked by the striking functional similarities between allosteric enzymes and hemoglobin (page 145), in particular, the similarity in the sigmoid relationship between the concentration of substrate and the initial velocity of some allosteric enzymes on the one hand and the concentration (partial pressure) of oxygen and its binding to hemoglobin on the other.

We have seen (page 148) that two types of models have been proposed for the cooperative interactions of the subunits of hemoglobin, the *sequential* model proposed by D. E. Koshland and the *symmetry* model of J. Monod and his colleagues. These models are also applicable to allosteric enzymes having multiple subunits, which are assumed to occur in two different conformations. Figure 9-23 shows the application of these models to a homotropic allosteric enzyme having four subunits, each capable of binding a substrate molecule, with a sigmoid relationship between substrate concentration and activity (see Figure 9-20). According to the symmetry model, binding the first substrate molecule increases the tendency of the remaining subunits to undergo transition to the high-affinity form through an all-or-none effect, all subunits being either in their low-affinity or their high-affinity forms (Figure 9-23).

The sequential model also postulates that the catalytic subunits have two conformational states, but it differs from the symmetry model in postulating that the subunits may un-undergo *individual* sequential changes in conformation; between the all-on and all-off states there may be many intermediate conformational states of the enzyme molecule, each having its own intrinsic catalytic activity (Figure 9-23). The sequential model thus offers the possibility of a finer tuning or modulation of allosteric-enzyme activity than the symmetry model.

Other types of models have been proposed to account for the characteristic modulation and kinetics of various allosteric enzymes. Some investigators believe that no single model will be found to explain the behavior of all allosteric enzymes; others support the view that a single model may suffice. To account for the very complex kinetic behavior of some allosteric enzymes it has been postulated that their subunits undergo rather slow conformational interactions.

Figure 9-23
Two general models for the interconversion of inactive and active forms of allosteric enzymes. In the symmetry, or all-or-none, model (see also Figure 6-23) all the subunits are postulated to be in the same conformation, either all ○ (low affinity) or all □ (high affinity). Depending on the equilibrium, K_1, between ○ and □ forms, the binding of one or more substrate (S) molecules will pull the equilibrium toward the □ form. A possible pathway is given by the gray shading. In the sequential model each individual subunit can be in either the ○ or □ form of the oligomeric protein. A very large number of conformations is thus possible, but the shaded pathway, shown by the diagonal arrows, is the most probable route.

Symmetry model

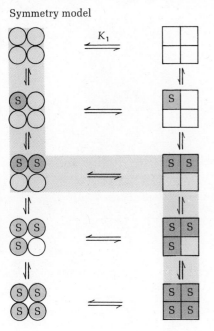

Sequential model

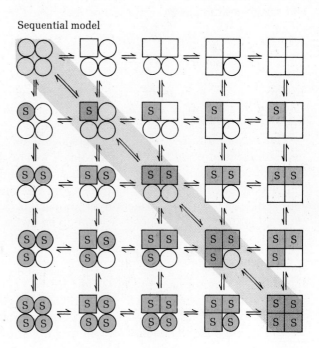

Covalently Modulated Regulatory Enzymes

In the second class of regulatory enzymes the active and inactive forms are interconverted by covalent modifications of their structures that are catalyzed by other enzymes. The classical example of this type of regulatory enzyme is *glycogen phosphorylase* of animal tissues, which catalyzes the breakdown of the storage polysaccharide glycogen, a polymer of glucose (page 266), to yield glucose 1-phosphate:

$$(\text{Glucose})_n + P_i \Longleftrightarrow (\text{glucose})_{n-1} + \text{glucose 1-phosphate}$$

Glycogen   Shortened
                   glycogen
                   molecule

This enzyme, whose biological function is developed elsewhere (pages 433 and 647), occurs in two forms, phosphorylase *a*, the more active form, and phosphorylase *b*, the less active form. Phosphorylase *a* is an oligomeric protein with four major subunits (Figure 9-24). Each subunit contains a serine residue that is phosphorylated at the hydroxyl group; these phosphate groups are required for maximum catalytic activity. The phosphate groups in phosphorylase *a* can be hydrolytically removed by the enzyme *phosphorylase phosphatase*,

$$\text{Phosphorylase } a + 4H_2O \xrightarrow[\text{phosphatase}]{\text{phosphorylase}} 2 \text{ phosphorylase } b + 4P_i$$

Removal of the phosphate groups causes phosphorylase *a* to dissociate into two half molecules, phosphorylase *b*, which are much less active in cleaving glycogen than phosphorylase *a*.

The relatively inactive phosphorylase *b* can be reactivated to form phosphorylase *a* by the enzyme *phosphorylase kinase*, which catalyzes the enzymatic phosphorylation of the serine residues at the expense of ATP:

$$4\text{ATP} + 2 \text{ phosphorylase } b \xrightarrow[\text{kinase}]{\text{phosphorylase}} 4\text{ADP} + \text{phosphorylase } a$$

In this way the activity of glycogen phosphorylase is regulated by the action of two enzymes that shift the balance between its active and inactive forms (Figure 9-24).

The second striking attribute of glycogen phosphorylase and similar regulatory enzymes modulated by covalent modification is that they can greatly amplify a chemical signal. All enzymes can bring about amplification; i.e., one enzyme molecule can catalyze formation of thousands of product molecules from a given substrate in a given period of time. But here an enzyme acts upon another enzyme as its substrate. One molecule of phosphorylase kinase can convert thousands of molecules of the inactive phosphorylase *b* into the active phosphorylase *a*, which in turn can catalyze the production of thousands of molecules of glucose 1-phosphate molecules from glycogen. Phosphorylase kinase and phosphorylase thus constitute an *amplification cascade* with two steps. Elsewhere we shall see (page 813) that these

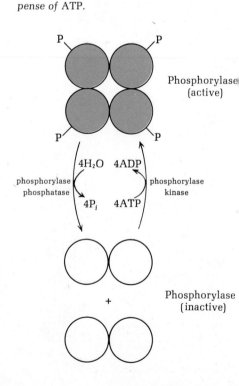

Figure 9-24
Modulation of glycogen phosphorylase activity by covalent modification. The active form of the enzyme, phosphorylase a, a tetramer, is enzymatically dephosphorylated by phosphorylase phosphatase to yield the inactive phosphorylase b, a dimer. By the action of phosphorylase kinase, phosphorylase a is regenerated by the phosphorylation of four specific serine residues at the expense of ATP.

two enzymes actually are elements in a longer cascade, with two further amplification stages, whose biological function is to amplify a very small signal produced by a few molecules of the hormone epinephrine into a relatively massive breakdown of glycogen in the tissue.

So far two types of chemical modification are known to be involved in covalently modulated regulatory enzymes. Phosphorylase and certain other enzymes are modulated by transfer of a phosphate group from ATP to the enzyme to yield a phosphoenzyme. The other type of covalent modification is through transfer of an adenylyl group from ATP to the enzyme, as in the regulation of *glutamine synthetase* of *E. coli*, which catalyzes the reaction

$$\text{ATP} + \text{glutamate} + \text{NH}_3 \rightleftharpoons \text{ADP} + \text{glutamine} + \text{P}_i$$

Glutamine synthetase is converted from its relatively active form to its less active form by the transfer of adenylyl groups from ATP to specific tyrosine residues in each of the 12 subunits of the enzyme, to yield covalent adenylyl derivatives of the phenolic hydroxyl groups of the tyrosines. The enzyme may also be enzymatically de-adenylylated to its active form. The complex regulatory properties of this enzyme are described in more detail elsewhere (page 695).

Covalent Activation of Zymogens

A somewhat different type of regulation of enzyme activity is the enzyme-catalyzed activation of inactive precursors of enzymes (zymogens) to yield the catalytically active forms. The classical examples are the digestive enzymes pepsin, trypsin, and chymotrypsin (page 106), which are synthesized as the inactive zymogens *pepsinogen*, *trypsinogen*, and *chymotrypsinogen*, respectively. When these enzymes are secreted into the gastrointestinal tract, they are converted into their active forms by the selective hydrolytic cleavage of one or more specific peptide bonds in the zymogen molecule. Pepsinogen is converted into active pepsin in the stomach by the action of free pepsin at low pH, which causes removal of 42 amino acid residues as a mixture of peptides from the N-terminal end of pepsinogen:

$$\text{Pepsinogen} \xrightarrow[\text{H}^+]{\text{pepsin}} \text{pepsin} + \text{peptides}$$

Trypsinogen is converted into active trypsin by removal of a hexapeptide from the N-terminal end, through the action of the enzyme *enterokinase*:

$$\text{Trypsinogen} \xrightarrow{\text{enterokinase}} \text{trypsin} + \text{hexapeptide}$$

Chymotrypsinogen is converted into active chymotrypsin by the action of trypsin, which causes removal of two dipeptide fragments, as described earlier (page 228):

$$\text{Chymotrypsinogen} \xrightarrow{\text{trypsin}} \text{chymotrypsin} + 2 \text{ dipeptides}$$

These zymogens are kept from exerting proteolytic activity on intracellular proteins so long as they remain within the cells in which they are made. They are turned on to generate the active form only after they are secreted into the gastrointestinal tract. However, this type of covalent regulation is one-way: there are no known enzymatic reactions which can transform these three enzymes back into their respective zymogens.

Isozymes

Another type of regulation of metabolic activity is through the participation of *isozymes*, multiple forms of a given enzyme that occur within a single species of organism or even in a single cell. Such multiple forms can be detected and separated by gel electrophoresis of cell extracts; since they are coded by different genes, they differ in amino acid composition and thus in their isoelectric pH values.

Lactate dehydrogenase, one of the first enzymes in this class to be studied intensively, occurs as five different isozymes in the tissues of the rat and other vertebrates (Figure 9-25). All the rat isozymes have been isolated. They all catalyze the same overall reaction,

$$\text{Lactate} + \text{NAD}^+ \rightleftharpoons \text{pyruvate} + \text{NADH} + \text{H}^+$$

in which NAD^+ and NADH symbolize the oxidized and reduced forms, respectively, of *nicotinamide adenine dinucleotide* (page 340). All five isozymes have the same molecular weight, about 134,000, and all contain four polypeptide chains, each of molecular weight 33,500. The five isozymes consist of five different combinations of two different kinds of polypeptide chains, designated M and H. The isozyme predominating in skeletal muscle has four identical M chains and is designated M_4; another, which predominates in heart, has four identical H chains and is designated H_4. The other three isozymes have the composition M_3H, M_2H_2, and MH_3. Single M and H chains have been isolated and found to differ significantly in amino acid content and sequence. When single M and H chains, which are inactive, are mixed in appropriate proportions, all the different isozymes of lactate dehydrogenase can be made to form spontaneously in the test tube and all have full catalytic activity.

Genetic research indicates that the amino acid sequences of the two different polypeptide chains M and H of lactate dehydrogenase are coded by two different genes. The biosynthesis of the two types of chains and thus the relative amounts of the lactate dehydrogenase isozymes present in a given cell are under genetic regulation. Figure 9-25 shows that the isozymes of lactate dehydrogenase exist in different proportions in different tissues. Moreover, the relative proportions of the lactate dehydrogenase isozymes in a tissue may change during embryological development. They are also important in diagnosis of heart and liver disease (Figure 9-25).

Careful kinetic study of the lactate dehydrogenase isozymes has revealed that although they all catalyze the same

Figure 9-25
Isozymes of lactate dehydrogenase. Separation of the isozymes of rat skeletal muscle and of rat heart by zone electrophoresis. The extracts are applied at the origin line. Four of the isozymes migrate to the positive electrode, the other to the negative. The size and density of the spots reflect the amount of each isozyme. [Redrawn from I. H. Fine, N. O. Kaplan, and P. Kuftinec, Biochemistry, 2: 116 (1963).]

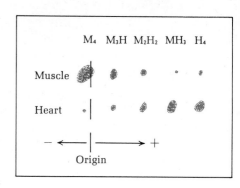

Relative content of lactate dehydrogenase isozymes in human tissues. The blood-serum isozyme levels are used in clinical diagnosis; the H_4 and MH_3 isozymes are elevated in liver disease.

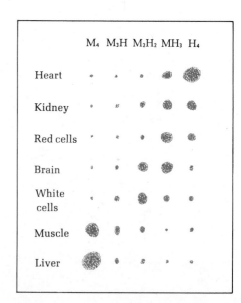

reaction, they differ significantly in their K_M values for their substrates, particularly for pyruvate, as well as their V_{max} values when pyruvate is the substrate. The isozyme M_4, characteristic of skeletal muscle and embryonic tissues, has a relatively low K_M for pyruvate and a relatively high rate at which it can reduce pyruvate to lactate. The H_4 isozyme, characteristic of the heart and other red muscles, has a relatively high K_M for pyruvate and reduces it at a relatively low rate. Moreover, the dehydrogenation of lactate catalyzed by the H_4 isozyme is strongly inhibited by pyruvate. The other lactate dehydrogenase isozymes have kinetic properties intermediate between those of the M_4 and H_4 isozymes, in proportion to their relative content of M and H chains.

These kinetic characteristics, when compared with the metabolic features of the tissues in which the M_4 and H_4 isozymes predominate, have provided some insight into the function of the lactate dehydrogenase isozymes. Skeletal muscle and embryonic tissue tend to utilize glucose anaerobically and break it down to form lactate in the process of glycolysis (page 417). Lactate dehydrogenase catalyzes the last step in glycolysis, the reduction of pyruvate to lactate. Thus the skeletal muscle or M_4 isozyme of lactate dehydrogenase, which has a low K_M and a high V_{max} for pyruvate, is well adapted to convert pyruvate rapidly into lactate. On the other hand, heart muscle does not normally form lactate from glucose; it oxidizes pyruvate to carbon dioxide aerobically without going through an intermediate formation of lactate. The H_4 isozyme is well adapted for this different route of glucose metabolism since it has a high K_M for pyruvate and a low V_{max} for reducing pyruvate to lactate and is inhibited by excess pyruvate, factors which greatly diminish its activity in the direction of lactate formation. Although heart muscle does not ordinarily use lactate dehydrogenase in the oxidation of glucose, in emergencies when the oxygen supply is low, the presence of lactate dehydrogenase allows the heart to obtain its energy from the conversion of glycogen to lactate (pages 838–839).

Isozymes are now known for a great many different enzymes. They usually consist of tightly associated mixtures of different kinds of polypeptide chains, in which are blended the specific kinetic and binding properties contributed by each type of chain. Many allosteric enzymes occur as two or more isozymes that vary in sensitivity to their allosteric modulators (pages 699 and 711). In other cases different isozymes are present in different intracellular compartments.

The study of isozymes has grown into one of fundamental significance in the investigation of the molecular basis of cellular differentiation and morphogenesis. Many proteins in cells, not only those with catalytic activity, may occur in multiple forms.

Summary

The active site of enzymes can be mapped by studying their substrate specificity, by using specific chemical reagents capable of bringing about covalent modification of functional groups essential for catalysis, by affinity labeling, and by analyzing crystalline enzyme-substrate or enzyme-inhibitor complexes with x-rays. Serine, histidine, lysine, and cysteine R groups of enzyme proteins are frequently involved in catalysis at the active site.

Enzyme-catalyzed reactions are from 10^8 to 10^{20} times faster than the corresponding uncatalyzed reactions. A major part of the rate enhancement probably comes from the exact positioning of the substrate in the proper proximity and orientation to the catalytic group, so that it enters the transition state readily. A smaller degree of rate enhancement is provided in some enzymes by the occurrence of covalent catalysis, in which a covalent enzyme-substrate compound is quickly formed and quickly broken down. Rate enhancement is also made possible through general-acid or general-base catalysis promoted by proton-donating and proton-accepting groups in the active site. An unknown degree of rate enhancement may be provided by conformational changes occurring when enzyme and substrate combine. The mechanism of action of a few enzymes, such as chymotrypsin and lysozyme, is beginning to be understood.

Some enzymes are biologically adapted to serve a regulatory as well as a catalytic function. The allosteric enzymes are modulated by noncovalent binding of some specific metabolite. They usually catalyze the first reaction of a multienzyme sequence and are often inhibited by the end product of the sequence, which binds to a specific regulatory or allosteric site on the enzyme molecule. Some allosteric enzymes are stimulated by their modulator, which may be the substrate itself. Allosteric enzymes show atypical kinetics which do not appear to follow the classical Michaelis-Menten rate equation. Some allosteric enzymes show sigmoidal plots of initial velocity vs. substrate concentration, whereas others show non-rectangular hyperbolic curves. The modulator molecule, when bound to its specific site, changes the apparent K_M or V_{max} of the enzyme. Nearly all known allosteric enzymes have multiple subunits, which in some cases are of two types, catalytic and regulatory. Several models have been proposed to explain the mechanism of allosteric regulation. The symmetry model proposes that the allosteric enzyme molecule occurs in only one of two possible conformations, active or inactive, whereas the sequential model postulates that the subunits change their conformation in sequence, not simultaneously, so that intermediate states of differing catalytic activity occur.

Another class of regulatory enzymes undergoes interconversion between active and inactive forms by covalent modification of some specific group in the enzyme molecule by other enzymes. An example is glycogen phosphorylase, which is converted into its inactive b form by enzymatic hydrolysis of its phosphorylated serine residues and dissociation of its tetrameric structure into a dimeric form; the latter can be converted back into active phosphorylase a by enzymatic phosphorylation.

Some enzymes occur in multiple forms, called isozymes, within a given species or cell type. They contain different proportions of two or more types of polypeptide chains, which cause the isozyme forms to differ in K_M or V_{max}.

References

Other references for enzymes are given at the end of Chapter 8 (pages 212 and 213).

Books

BRUICE, T. C., and S. J. BENKOVIC: *Bioorganic Mechanisms*, 2 vols., Benjamin, Menlo Park, Calif., 1966. Comprehensive review of organic model reactions for various types of enzymes.

JENCKS, W. P.: *Catalysis in Chemistry and Enzymology*, McGraw-Hill, New York, 1969. A definitive treatment of catalytic mechanisms.

Articles

BELL, R. M., and D. E. KOSHLAND, JR.: "Covalent Enzyme-Substrate Intermediates," *Science*, 172: 1253–1256 (1971). A brief analysis of covalent catalysis and a list of 60 enzymes in which covalent enzyme-substrate intermediates are probably formed.

BLOW, D. M., and T. A. STEITZ: "X-Ray Diffraction Studies of Enzymes," *Ann. Rev. Biochem.*, 39: 63–100 (1970).

BRUICE, T. C.: "Proximity Effects and Enzyme Catalysis," pp. 217–280 in P. D. Boyer (ed.), *The Enzymes*, 3d ed., vol. 2, Academic, New York, 1970.

HOLZER, H., and W. DUNTZE: "Metabolic Regulation by Chemical Modification of Enzymes," *Ann. Rev. Biochem.*, 40: 345–374 (1971).

JACOBSON, G. R., and G. R. STARK: "Aspartate Transcarbamylases," in P. D. Boyer (ed.), *The Enzymes*, 3d ed., vol. 9, pt B, Academic, New York, 1973.

JENCKS, W. P.: "Approximation, Chelation, and Enzymic Catalysis," *PAABS Rev.*, 2: 235–319 (1973). An important review and collection of original papers on orientation, orbital steering, and the catalytic power of enzymes.

JENCKS, W. P., and M. I. PAGE: "On the Importance of Togetherness in Enzymatic Catalysis," in Enzymes: Structure and Function, *Proc. 8th FEBS Meet., 1972*, American-Elsevier, New York.

KIEFER, H. C., W. I. CONGDON, I. S. SCARPA, and I. M. KLOTZ: "Catalytic Accelerations of 10^{12}-fold by an Enzyme-like Synthetic Polymer," *Proc. Natl. Acad. Sci. U.S.*, 69: 2155–2159 (1972).

KIRSCH, J.: "Mechanism of Enzyme Action," *Ann. Rev. Biochem.*, 42: 205–234 (1973).

KOSHLAND, D. E., JR.: "The Molecular Basis for Enzyme Regulation," pp. 341–396 in P. D. Boyer (ed.), *The Enzymes*, 3d ed., vol. 1, Academic, New York, 1970. Theory and models of allosteric regulation.

KOSHLAND, D. E., JR.: "Protein Shape and Biological Control," *Sci. Am.*, 229: 52–64 (1973). A well-written, easily understood analysis of the significance of conformational changes in enzyme catalysis and its regulation.

KOSHLAND, D. E., JR., G. NEMETHY, and D. FILMER: "Comparison of Experimental Binding Data and Theoretical Models in Proteins Containing Subunits," *Biochemistry*, 5: 365–387 (1966). The sequential and symmetry models compared.

LIENHARD, G. E.: "Enzymatic Catalysis and Transition-State Theory," *Science*, 180: 149–154 (1973). The importance of the enzyme in inducing and binding the transition-state species.

MONOD, J., J. WYMAN, and J.-P. CHANGEUX: "On the Nature of Allosteric Transitions: A Plausible Model," *J. Mol. Biol.*, 12: 88–118 (1965).

PANAGOU, D., M. D. ORR, J. R. DUNSTONE, and R. L. BLAKLEY,: "A Monomeric Allosteric Enzyme with a Single Polypeptide Chain:

Ribonucleotide Reductase of *L. leichmannii*," *Biochemistry*, 11: 2378–2388 (1972).

QUIOCHO, F. A., and W. N. LIPSCOMB: "Carboxypeptidase A: A Protein and an Enzyme," *Adv. Protein Chem.*, 25: 1–59 (1971). A comprehensive account of the amino acid sequence, x-ray analysis, and mechanism of action.

SEGAL, H. L.: "Enzymatic Interconversion of Active and Inactive Forms of Enzymes," *Science*, 180: 25–32 (1973).

SHAW, E.: "Chemical Modification by Active-Site-Directed Reagents," pp. 91–147 in P. D. Boyer (ed.), *The Enzymes*, 3d ed., vol. 1, Academic, New York, 1970.

STADTMAN, E. R.: "Mechanisms of Enzyme Regulation in Metabolism," pp. 397–459 in P. D. Boyer (ed.), *The Enzymes*, 3d ed., vol. 1, Academic, New York, 1970.

CHAPTER 10 SUGARS, STORAGE POLYSACCHARIDES, AND CELL WALLS

The carbohydrates, or saccharides, are most simply defined as polyhydroxy aldehydes or ketones and their derivatives. Many have the empirical formula $(CH_2O)_n$, which originally suggested they were "hydrates" of carbon. _Monosaccharides_, also called simple sugars, consist of a single polyhydroxy aldehyde or ketone unit. The most abundant monosaccharide is the six-carbon D-glucose; it is the parent monosaccharide from which most others are derived. D-Glucose is the major fuel for most organisms and the basic building block of the most abundant polysaccharides, such as starch and cellulose.

Oligosaccharides (Greek _oligo_, "few") contain from two to ten monosaccharide units joined in glycosidic linkage. _Polysaccharides_ contain many monosaccharide units joined in long linear or branched chains. Most polysaccharides contain recurring monosaccharide units of only a single kind or two alternating kinds.

Polysaccharides have two major biological functions, as a storage form of fuel and as structural elements. In the biosphere there is probably more carbohydrate than all other organic matter combined, thanks largely to the abundance in the plant world of two polymers of D-glucose, _starch_ and _cellulose_. Starch is the chief form of fuel storage in most plants, whereas cellulose is the main extracellular structural component of the rigid cell walls and the fibrous and woody tissues of plants. Glycogen, which resembles starch in structure, is the chief storage carbohydrate in animals. Other polysaccharides serve as major components of the cell walls of bacteria and of the soft cell coats in animal tissues.

Families of Monosaccharides

Monosaccharides have the empirical formula $(CH_2O)_n$, where $n = 3$ or some larger number. The carbon skeleton of the common monosaccharides is unbranched and each carbon atom except one contains a hydroxyl group; at the remaining carbon there is a carbonyl oxygen, which, as we shall see, is often combined in an acetal or ketal linkage. If the carbonyl group is at the end of the chain, the monosaccharide is an aldehyde derivative and called an _aldose_; if it is at any other position, the monosaccharide is a ketone derivative and called a _ketose_. The simplest monosaccharides are the three-carbon _trioses_ glyceraldehyde and dihydroxyacetone (Figure 10-1). Glyceraldehyde is an _aldotriose_; dihydroxyacetone is a _ketotriose_. Also among the monosaccharides are the _tetroses_ (four carbons), _pentoses_ (five carbons), _hexoses_ (six carbons), _heptoses_ (seven carbons), and _octoses_ (eight carbons). Each exists in two series, i.e., aldotetroses and ketotetroses, aldopentoses and ketopentoses, aldohexoses and ketohexoses, etc. The structures of D aldoses and D ketoses are shown in Figures 10-2 and 10-3. In both classes of monosaccharides the hexoses are by far the most abundant. However, aldopen-

Figure 10-1
The trioses.

Figure 10-2
The family of D aldoses having from three to six carbon atoms shown in open-chain structural formulas. To conserve space, horizontal bonds are not shown in this and subsequent figures, but these representations of sugars should be regarded as projection formulas (page 82).

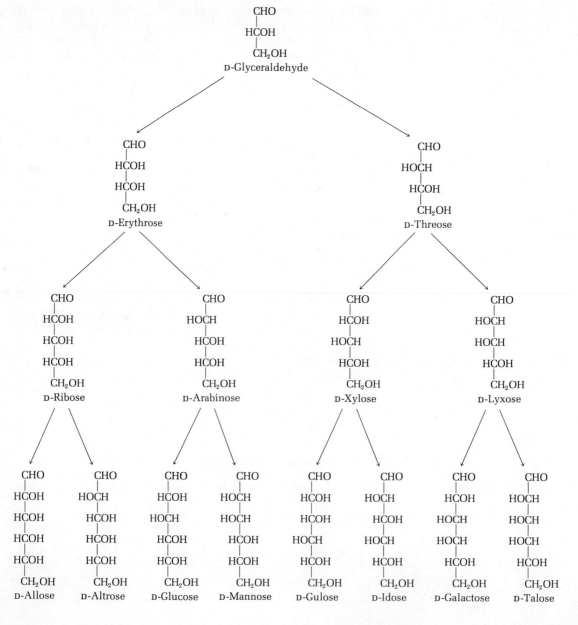

toses are important components of nucleic acids and various polysaccharides; derivatives of trioses and heptoses are important intermediates in carbohydrate metabolism. All the simple monosaccharides are white crystalline solids that are freely soluble in water but insoluble in nonpolar solvents. Most have a sweet taste.

Stereoisomerism of Monosaccharides

All the monosaccharides except dihydroxyacetone contain one or more asymmetric carbon atoms and thus are chiral molecules. Glyceraldehyde contains only one asymmetric

Figure 10-3
D ketoses (open-chain form).

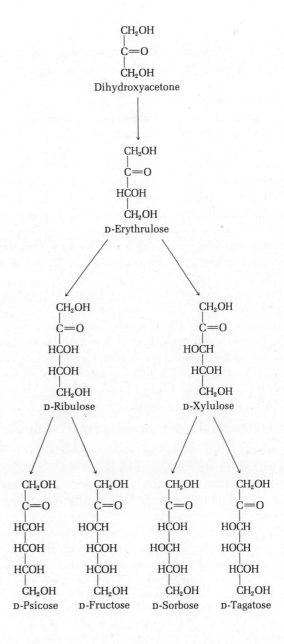

251

carbon atom and therefore can exist as two different stereo-isomers (Figure 10-4). It will be recalled that D- and L-glyceraldehyde are the reference, or parent, compounds for designating the absolute configuration of all stereoisomeric compounds (page 82). Aldotetroses have two asymmetric carbon atoms and aldopentoses three. The aldohexoses have four asymmetric carbon atoms and thus exist in the form of $2^n = 2^4 = 16$ different stereoisomers, 8 of which are shown in Figure 10-2. As expected, the monosaccharides with asymmetric carbon atoms are optically active. For example, the usual form of glucose found in nature is dextrorotatory ($[\alpha]_D^{20} = +52.7°$), and the usual form of fructose is levorotatory ($[\alpha]_D^{20} = -92.4°$), but both are members of the D series since their absolute configurations are related to D-glyceraldehyde (page 82). For sugars having two or more asymmetric carbon atoms, the convention has been adopted that the prefixes D and L refer to the asymmetric carbon atom *farthest removed from the carbonyl carbon atom.*

Figure 10-2 shows the projection formulas (page 82) of the D aldoses. All have the same configuration at the asymmetric carbon atom farthest from the carbonyl carbon, but because most have two or more asymmetric carbon atoms, a number of isomeric D aldoses exist, most important biologically being D-glyceraldehyde, D-ribose, D-glucose, D-mannose, and D-galactose.

Figure 10-3 shows the projection formulas of the D ketoses; all share the same configuration at the asymmetric carbon atom farthest from the carbonyl group. Ketoses are sometimes designated by inserting *ul* into the name of the corresponding aldose; e.g., D-ribulose is the ketopentose corresponding to the aldopentose D-ribose. The most important ketoses biologically are dihydroxyacetone, D-ribulose, and D-fructose.

Aldoses and ketoses of the L series are mirror images of their D counterparts, as shown in Figure 10-5. L sugars are found in nature, but they are not so abundant as D sugars. Among the most important are L-*fucose*, L-*rhamnose* (page 260), and L-*sorbose.*

Two sugars differing only in the configuration around one specific carbon atom are called *epimers* of each other. Thus, D-glucose and D-mannose are epimers with respect to carbon atom 2, and D-glucose and D-galactose are epimers with respect to carbon atom 4 (see Figure 10-2).

Mutarotation and the Anomeric Forms of D-Glucose

In aqueous solution many monosaccharides act as if they had one more asymmetric center than is given by the open-chain structural formulas in Figures 10-2 and 10-3. D-Glucose may exist in two different isomeric forms differing in specific rotation, α-D-glucose, for which $[\alpha]_D^{20} = +112.2°$, and β-D-glucose, for which $[\alpha]_D^{20} = +18.7°$. Both have been isolated in pure form. Although they do not differ in elementary composition, their physical and chemical properties differ (Table 10-1). When the α and β isomers of D-glucose are dissolved in water, the optical rotation of each gradually changes with

Figure 10-4
The stereoisomers of glyceraldehyde, showing projection formulas (top) and perspective formulas (bottom). See Figure 4-12, page 82.

$$\begin{array}{cc}
\text{CHO} & \text{CHO} \\
| & | \\
\text{H—C—OH} & \text{HO—C—H} \\
| & | \\
\text{CH}_2\text{OH} & \text{CH}_2\text{OH}
\end{array}$$

$$\begin{array}{cc}
\text{CHO} & \text{CHO} \\
\vdots & \vdots \\
\text{H—C—OH} & \text{HO—C—H} \\
\vdots & \vdots \\
\text{CH}_2\text{OH} & \text{CH}_2\text{OH} \\
\text{D-Glyceraldehyde} & \text{L-Glyceraldehyde}
\end{array}$$

Figure 10-5
D- and L-glucose.

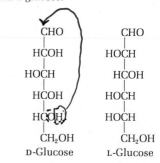

$$\begin{array}{cc}
\text{CHO} & \text{CHO} \\
\text{HCOH} & \text{HOCH} \\
\text{HOCH} & \text{HCOH} \\
\text{HCOH} & \text{HOCH} \\
\text{HCOH} & \text{HOCH} \\
\text{CH}_2\text{OH} & \text{CH}_2\text{OH} \\
\text{D-Glucose} & \text{L-Glucose}
\end{array}$$

Figure 10-6
The α and β forms of D-glucopyranose. The structure of pyran is shown in a comparable form. Below is a space-filling model of α-D-glucopyranose. Note that the α form of D-glucopyranose has the anomeric hydroxyl group to the right in the projection formula, on the same side as the hydroxyl group of carbon atom 5, denoting that the sugar is of the D series; in the β form the anomeric hydroxyl group is on the left. Also note that the anomeric carbon atom is distinguished from the others in being linked to two oxygen atoms.

HCOH
|
HCOH
| O
HOCH
|
HCOH
|
HC——
|
CH₂OH

α-D-Glucopyranose

HOCH
|
HCOH
| O
HOCH
|
HCOH
|
HC——
|
CH₂OH

β-D-Glucopyranose

CH
‖
CH
| O
CH₂
|
CH
‖
HC——

Pyran

α-D-Glucopyranose

Table 10-1 Properties of α- and β-D-glucose

Property	α-D-Glucose	β-D-Glucose
Specific rotation $[\alpha]_D^{20}$	+112.2°	+18.7°
Melting point, °C	146	150
Solubility in H_2O, g per 100 ml	82.5	178
Relative rate of oxidation by glucose oxidase	100	< 1.0

time and approaches a final equilibrium value of $[\alpha]_D^{20} = +52.7°$. This change, called *mutarotation*, is due to the formation of an equilibrium mixture consisting of about one-third α-D-glucose and two-thirds β-D-glucose at 20°C. From various chemical considerations it has been deduced that the α and β isomers of D-glucose are not open-chain structures, as in Figure 10-2, but six-membered ring structures formed by the reaction of the alcoholic hydroxyl group at carbon atom 5 with the aldehydic carbon atom 1 (Figure 10-6). The six-membered ring forms of sugars are called *pyranoses* because they are derivatives of the heterocyclic compound *pyran*. The systematic name for the ring form of α-D-glucose is *α-D-glucopyranose*.

The formation of pyranoses is a special case of a more general type of reaction between an aldehyde and an alcohol to form a *hemiacetal* (Figure 10-7), which contains an asymmetric carbon atom and therefore can exist in two stereoisomeric forms. A *hemiketal* is an analogous product formed by reaction of a ketone with an alcohol. D-Glucopyranose is an intramolecular hemiacetal in which the hydroxyl group at carbon atom 5 has reacted with the aldehydic carbon atom 1, rendering it asymmetric. D-Glucopyranose therefore can exist as two different stereoisomers, designated α and β (Figure 10-6). Isomeric forms of monosaccharides that differ from each other only in configuration about the carbonyl carbon atom are *anomers*, and the carbonyl carbon atom is called the *anomeric carbon*. All aldoses with five or more carbon atoms form stable pyranose rings and can exist in anomeric forms.

Figure 10-7
Hemiacetals and hemiketals. Note the asymmetric carbon atoms.

Formation of a hemiacetal

O
‖
R—C—H   Aldehyde
+
OH
|
R'   Alcohol

‖

OH
|
R—C—H   Hemiacetal
|
OR'

Formation of a hemiketal

O
‖
R—C—R'   Ketone
+
OH
|
R''   Alcohol

‖

OH
|
R—C—R'   Hemiketal
|
OR''

Figure 10-8
The furanose ring.

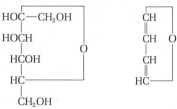

β-D-Fructofuranose   Furan   β-D-Fructofuranose

Ketoses of five or more carbon atoms also occur in α and β anomeric forms. In these compounds, the alcoholic hydroxyl group on carbon atom 5 has been joined with the carbonyl group at carbon atom 2 to yield a hemiketal in the form of a five-membered _furanose ring_, a derivative of _furan_ (Figure 10-8). The common ring form of D-fructose is _β-D-fructofuranose_. Aldohexoses may also exist in furanose forms, but since the six-membered aldopyranose ring is much more stable than the furanose ring, it predominates in aldohexose solutions. The open-chain form of hexoses occurs in only minute amounts in aqueous solution but is an intermediate in the interconversion of α and β forms during mutarotation.

Although we earlier defined carbohydrates as polyhydroxy aldehydes or ketones (page 249), most naturally occurring monosaccharides do not have free carbonyl groups and are more properly defined as polyhydroxy acetals or ketals.

Haworth projection formulas may be used to indicate the ring forms of monosaccharides; the edge of the ring nearest the reader is represented by bold lines (Figure 10-9). The Haworth projections are somewhat misleading, however, since they suggest that the five- and six-membered furanose and pyranose rings are planar, which is not the case. The pyranose ring exists in two conformations, the _chair_ form and the _boat_ form. (Recall that conformation denotes the arrangement in space of the atoms in a molecule which can be achieved by rotations about single bonds.) The chair form of the pyranose ring, which is relatively rigid and much more stable than the boat form, predominates in aqueous solutions of hexoses. The substituent groups in the chair form are not geometrically and chemically equivalent; they fall into two classes, _axial_ and _equatorial_ (Figure 10-10). The equatorial hydroxyl groups of pyranoses are more readily esterified than axial groups.

Action of Acids and Bases on Monosaccharides

Monosaccharides are stable to hot dilute mineral acids. Concentrated acids, however, cause dehydration of sugars to yield _furfurals_, aldehyde derivatives of furan. For example,

Figure 10-9
Haworth projection formulas.

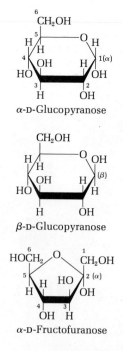

α-D-Glucopyranose

β-D-Glucopyranose

α-D-Fructofuranose

Figure 10-10
Boat and chair forms of six-membered rings.
The chair form is more stable.

The axial and equatorial bonds
of cyclohexane. Axial bonds are
parallel to the axis of symmetry.

The stable conformation (chair
form) of α-ᴅ-glucopyranose.

Axis of symmetry

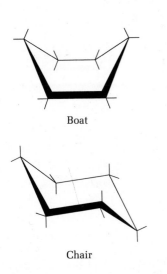

Boat

Chair

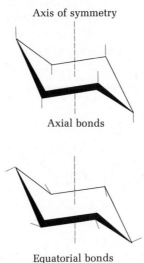

Axial bonds

Equatorial bonds

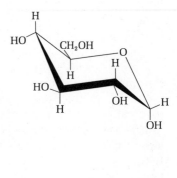

ᴅ-glucose heated with strong HCl yields 5-_hydroxymethylfur-
fural_ (Figure 10-11). Furfurals condense with phenols, such
as orcinol, to give characteristic colored products often used
for colorimetric analysis of sugars.

Dilute aqueous bases at room temperature cause rearrangements about the anomeric carbon atom and its adjacent
carbon atom without affecting substituents at other carbon
atoms. For example, treatment of ᴅ-glucose with dilute alkali
yields an equilibrium mixture of ᴅ-glucose, ᴅ-fructose, and
ᴅ-mannose (Figure 10-12). This reaction involves intermediate enol forms, called _enediols_, of the hydroxyaldehyde
and hydroxyketone structures of carbon atoms 1 and 2
(Figure 10-12).

At high temperatures or high concentrations, alkalis cause
free monosaccharides to undergo further rearrangement, fragmentation, or polymerization, but, as we shall see later,
glycosides and some polysaccharides are stable to alkali.

Figure 10-11
Furfural formation.

HC=O
HCOH
HOCH   ᴅ-Glucose
HCOH
HCOH
CH₂OH

↓ H⁺

HC=O
C
HC   O   5-Hydroxymethyl-
HC   furfural
C
CH₂OH
+
3H₂O

Figure 10-12
The isomerization of ᴅ-glucose by dilute
base.

HC=O	HOCH	HOCH₂	HOCH	O=CH
HCOH	COH	C=O	HOC	HOCH
HOCH	HOCH	HOCH	HOCH	HOCH
HCOH	HCOH	HCOH	HCOH	HCOH
HCOH	HCOH	HCOH	HCOH	HCOH
CH₂OH	CH₂OH	CH₂OH	CH₂OH	CH₂OH
ᴅ-Glucose	trans-Enediol	ᴅ-Fructose	cis-Enediol	ᴅ-Mannose

Important Derivatives of Monosaccharides

Glycosides

Aldopyranoses readily react with alcohols in the presence of a mineral acid to form anomeric α- and β-glycosides. The glycosides are asymmetric mixed acetals formed by the reaction of the anomeric carbon atom of the intramolecular hemiacetal or pyranose form of the aldohexose with a hydroxyl group furnished by an alcohol. This is called a _glycosidic bond_. The anomeric carbon in such glycosides is asymmetric. D-Glucose yields, with methanol, _methyl α-D-glucopyranoside_ ($[\alpha]_D^{20} = +158.9°$) and _methyl β-D-glucopyranoside_ ($[\alpha]_D^{20} = -34.2°$), shown in Figure 10-13.

The glycosidic linkage is also formed by the reaction of the anomeric carbon of a monosaccharide with a hydroxyl group of another monosaccharide to yield a _disaccharide_ (page 261). Oligosaccharides and polysaccharides are chains of monosaccharides joined by glycosidic linkages. The glycosidic linkage is stable to bases but is hydrolyzed by boiling with acid to yield the free monosaccharide and free alcohol. Glycosides are also hydrolyzed by enzymes called _glycosidases_, which differ in their specificity according to the type of glycosidic bond (α or β), the structure of the monosaccharide unit(s), and the structure of the alcohol.

Whether a given glycoside exists in furanose or pyranose form can be ascertained by oxidative degradation with periodic acid, which cleaves 1,2-dihydroxy compounds. Treatment of methyl α-D-glucopyranoside with periodate cleaves the pyranose ring to yield a dialdehyde and formic acid (Figure 10-14). Periodate cleavage of methyl α-D-arabinofuranoside yields the same dialdehyde but no formic acid.

Figure 10-13
The anomeric methyl D-glucosides in Haworth projections (above) and conformational formulas (below).

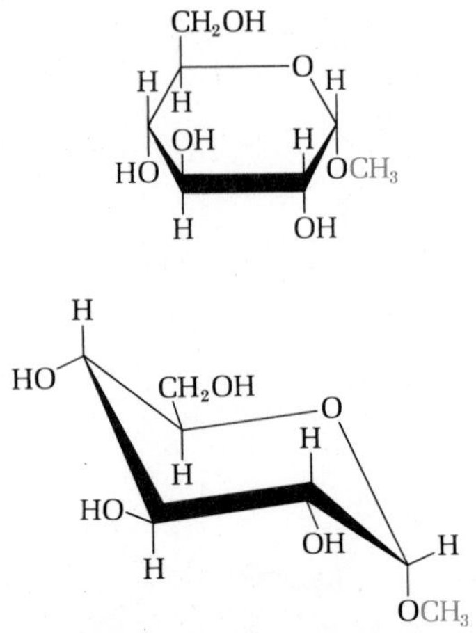

Methyl α-D-glucopyranoside

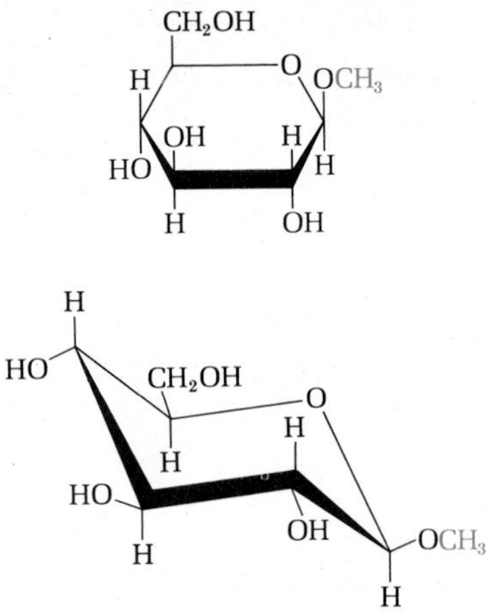

Methyl β-D-glucopyranoside

Figure 10-14
Action of periodate. Furanosides can be distinguished from pyranosides, since the latter yield formic acid.

Cleavage of
a furanoside.

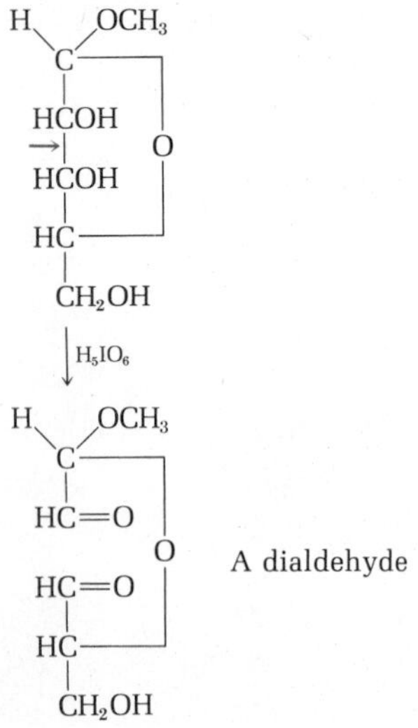

Cleavage of
a pyranoside.

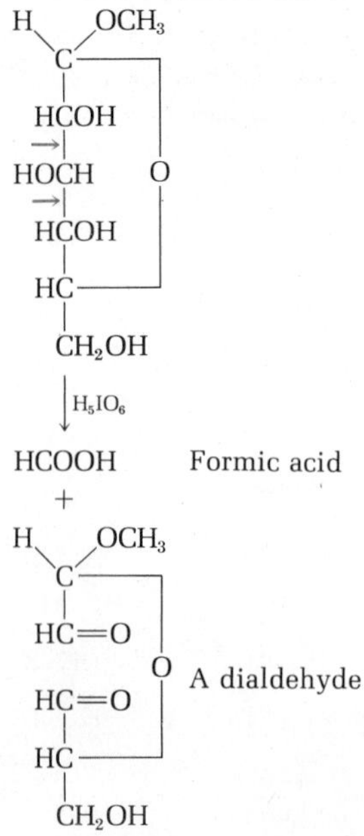

Figure 10-15
Two N-glycosylamines.

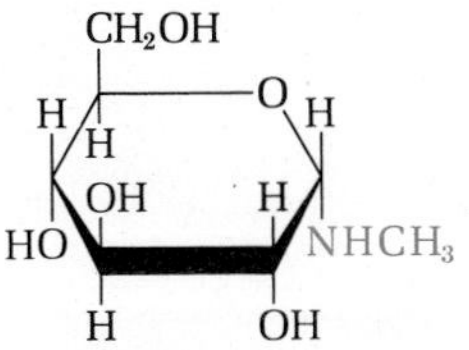

N-(α-D-glucopyranosyl)methylamine

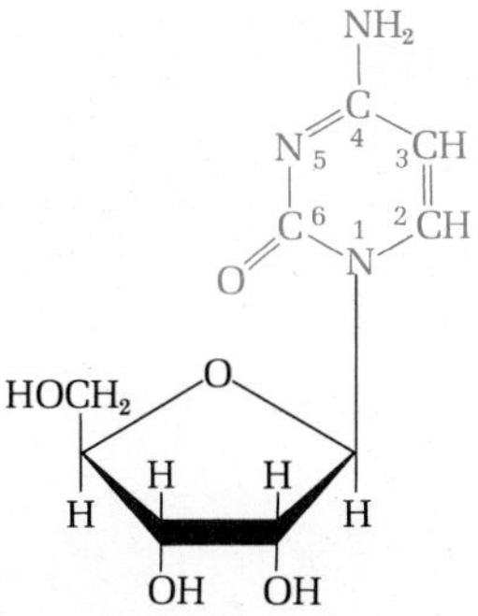

N¹-(β-D-ribofuranosyl)cytosine (cytidine)

N-Glycosylamines (N-Glycosides)

Aldoses and ketoses react with amines in an appropriate solvent to form N-glycosylamines, also called N-glycosides. (Figure 10-15). Such compounds are very important biologically. In the nucleotides and nucleic acids (Chapter 12) ring nitrogen atoms of purine or pyrimidine bases form N-glycosylamine linkages with carbon atom 1 of D-ribose or 2-deoxy-D-ribose.

O-Acyl Derivatives

The free hydroxyl groups of monosaccharides and polysaccharides can be acylated to yield O-acyl derivatives, which are useful in determining structure. For example, treatment of α-D-glucose with excess acetic anhydride yields penta-O-acetyl-α-D-glucose (Figure 10-16). All the hydroxyl groups of a monosaccharide can be acylated, although they differ somewhat in reactivity. The resulting esters can be hydrolyzed again.

O-Methyl Derivatives

The hydroxyl groups of monosaccharides can also be methylated. The hydroxyl group on the anomeric carbon atom reacts readily with methanol in the presence of acid to yield methyl glycosides, which are acetals, as we have seen. The remaining hydroxyl groups of monosaccharides require much more drastic conditions for methylation, e.g., treatment with dimethyl sulfate or methyl iodide plus silver oxide. In this case, methyl ethers are formed, not methyl acetals. Methyl acetals are easily hydrolyzed by boiling with acid; methyl ethers are not. Methylation of all the free hydroxyl groups of a carbohydrate is called *exhaustive methylation*. Exhaustive methylation is commonly used to establish the positions of substituents, e.g., amino groups, phosphate groups, or glycosidic linkages (see below), in which the hydroxyl group is no longer free to form an ether and also to determine whether a given monosaccharide is a furanose or a pyranose. Methylation of methyl α-D-glucopyranoside, for example, yields *methyl-2,3,4,6-tetra-O-methyl-D-glucopyranoside* (Figure 10-17).

Figure 10-16
Penta-O-acetyl-α-D-glucose.

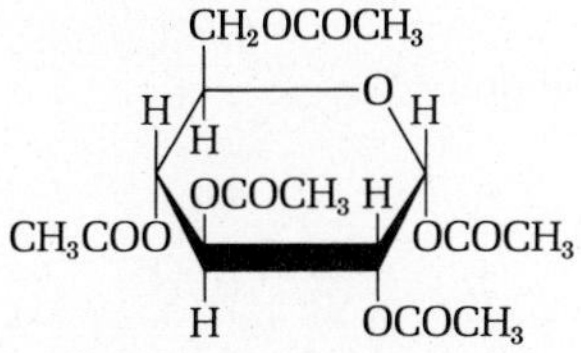

Figure 10-17
Exhaustive methylation.

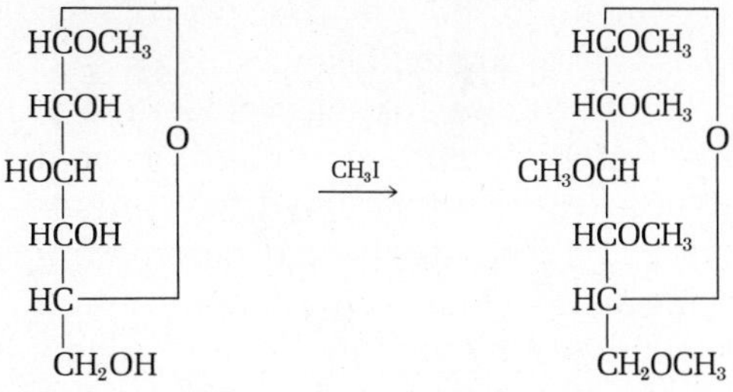

Methyl-α-D-glucopyranoside   Methyl-2,3,4,6-tetra-O-methyl-
                                                          D-glucopyranoside

Osazones

Monosaccharides in slightly acid solution at 100°C react with excess phenylhydrazine to form *phenylosazones*, which are insoluble in water and easily crystallized. The structure of the phenylosazone of D-glucose is given in Figure 10-18. Glucose, fructose, and mannose yield the same osazones since the differences in structure and configuration about carbon atoms 1 and 2 are abolished. Osazones are used to identify sugars.

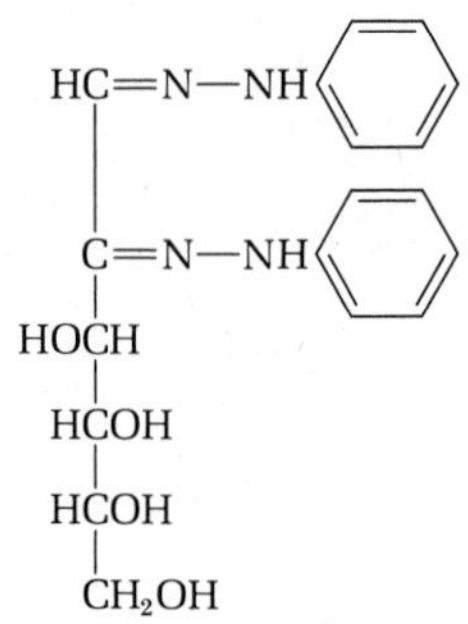

Sugar Alcohols

The carbonyl group of monosaccharides can be reduced by H_2 gas in the presence of metal catalysts or by sodium amalgam in water to form the corresponding sugar alcohols. D-Glucose, for example, yields the sugar alcohol D-*glucitol*, also formed by reduction of L-sorbose and often called L-*sorbitol*. D-Mannose yields D-*mannitol*. Such reductions can also be carried out by enzymes.

Two other sugar alcohols occur in nature in some abundance. One is *glycerol*, an important component of some lipids. The other is the fully hydroxylated cyclohexane derivative *inositol*, which can exist in several stereoisomeric forms. One of the stereoisomers of inositol, *myo-inositol*, is found not only in the lipid *phosphatidylinositol* (page 289) but also in *phytic acid*, the hexaphosphoric ester of inositol. The calcium-magnesium salt of phytic acid is called *phytin*; it is abundant in the extracellular supporting material in higher-plant tissues. The structures of some sugar alcohols are shown in Figure 10-19.

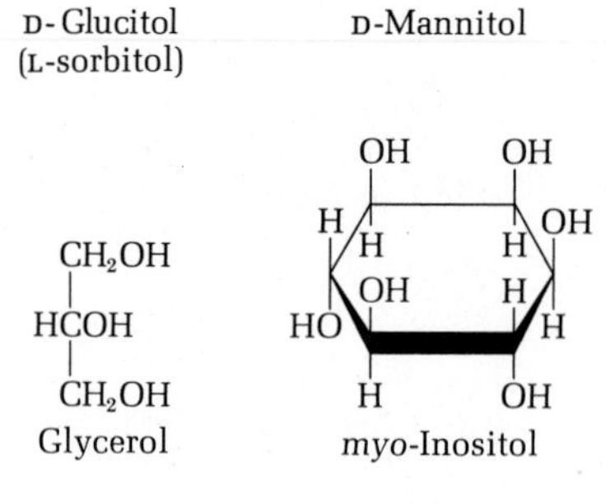

Sugar Acids

There are three important types of sugar acids: *aldonic, aldaric,* and *uronic* acids (Figure 10-20). The aldoses are oxidized at the aldehydic carbon atom by weak oxidizing agents, e.g., sodium hypoiodite, or by specific enzymes to form the corresponding carboxylic acids, which are called generically *aldonic acids*. D-Glucose, for example, yields D-gluconic acid, which in phosphorylated form is an important intermediate in carbohydrate metabolism (page 468).

If a stronger oxidizing agent is employed, e.g., nitric acid, both the aldehydic carbon atom and the carbon atom bearing the primary hydroxyl group are oxidized to carboxyl groups, yielding *aldaric acids* (also called *saccharic acids*). With D-glucose the product is called D-*glucaric acid*. Aldaric acids are sometimes useful for the identification of sugars, but they are of no great biological significance.

However, the third class of sugar acids, the *uronic acids*, are biologically very important. In uronic acids, only the carbon atom bearing the primary hydroxyl group is oxidized, to a carboxyl group. The uronic acid derived from D-glucose is D-*glucuronic acid*. Other important uronic acids are D-*galacturonic acid* and D-*mannuronic acid*. The uronic acids are components of many polysaccharides (page 272). Aldonic and uronic acids usually exist in lactone forms if a

Figure 10-18
D-*Glucose phenylosazone.*

Figure 10-19
Some sugar alcohols. Myo-inositol is the most abundant stereoisomer of inositol.

D-Glucitol
(L-sorbitol)

D-Mannitol

Glycerol

myo-Inositol

Figure 10-20
Sugar acids and lactones. In the lactones the carbon atoms are given Greek locants starting from the carbon adjacent to the carboxyl group.

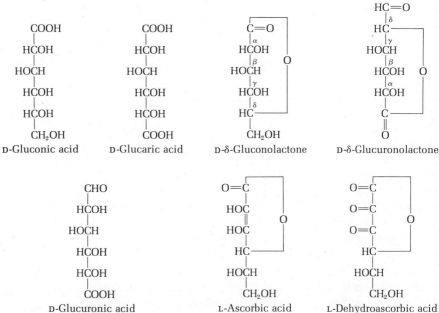

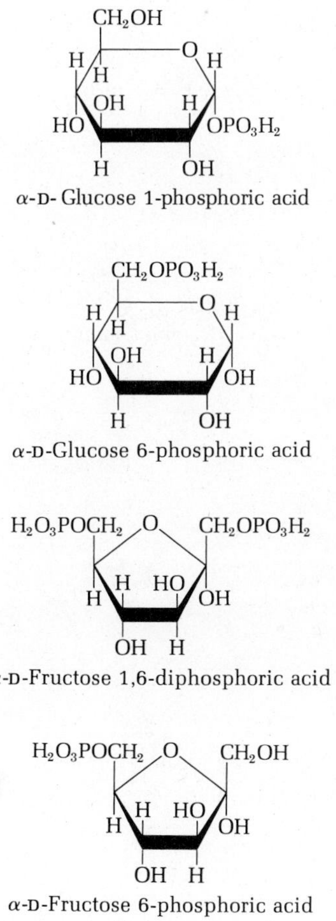

five- or six-membered ring can form. Examples are D-δ-gluconolactone and D-δ-glucuronolactone.

One of the most important sugar acids is <u>ascorbic acid</u>, or vitamin C, the γ-lactone of a hexonic acid having an enediol structure at carbon atoms 2 and 3 (Figure 10-20). This is a very unstable compound and readily undergoes oxidation to <u>dehydroascorbic acid</u>. Prolonged lack of ascorbic acid in the diet of human beings results in scurvy (page 350); less severe deficiency of ascorbic acid produces alterations in connective-tissue structure and may also cause decreased resistance to some infections. Ascorbic acid is present in large amounts in citrus fruit and tomatoes.

Monosaccharides can be detected or estimated quantitatively on the basis of their oxidation in alkaline solution by Cu^{2+}, Ag^+, or ferricyanide; a mixture of sugar acids results. Sugars capable of reducing such oxidizing agents are called <u>reducing sugars</u>.

Sugar Phosphates

Phosphate derivatives of monosaccharides are found in all living cells, in which they serve as important intermediates in carbohydrate metabolism (page 421). Representative sugar phosphates are shown in Figure 10-21.

Deoxy Sugars

Several deoxy sugars are found in nature (Figure 10-22). The most abundant is 2-*deoxy*-D-*ribose*, the sugar component of deoxyribonucleic acid. L-*Rhamnose* (6-deoxy-L-mannose) and L-*fucose* (6-deoxy-L-galactose) are important components of some bacterial cell walls (page 271).

Figure 10-21
Some hexose phosphates.

α-D- Glucose 1-phosphoric acid

α-D-Glucose 6-phosphoric acid

α-D-Fructose 1,6-diphosphoric acid

α-D-Fructose 6-phosphoric acid

Figure 10-22
Some deoxy sugars (open-chain forms).

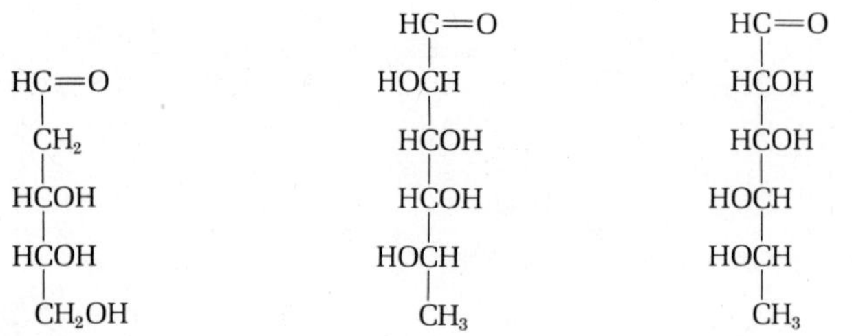

2-Deoxy-D-ribose   L-Fucose   L-Rhamnose

Amino Sugars

Two amino sugars of wide distribution are D-*glucosamine* (2-amino-2-deoxy-D-glucose) and D-*galactosamine* (2-amino-2-deoxy-D-galactose), in which the hydroxyl group at carbon atom 2 is replaced by an amino group (Figure 10-23). D-Glucosamine occurs in many polysaccharides of vertebrate tissues and is also a major component of *chitin*, a structural polysaccharide found in the exoskeletons of insects and crustaceans. D-Galactosamine is a component of glycolipids (page 287) and of the major polysaccharide of cartilage, chondroitin sulfate (page 272).

Muramic Acid and Neuraminic Acid

These sugar derivatives are important building blocks of the structural polysaccharides found in the cell walls of bacteria (page 268) and the cell coats of higher-animal cells (page 271), respectively. Both are nine-carbon amino sugar derivatives; they may be visualized as consisting of a six-carbon amino sugar linked to a three-carbon sugar acid; the amino group is usually acetylated. N-*Acetylmuramic acid* (Figure 10-24) is a major building block of the polysaccharide backbone of bacterial cell walls (page 269). It consists of N-acetyl-D-glucosamine in ether linkage with the three-carbon D-lactic acid. N-*Acetylneuraminic acid* is derived from N-acetyl-D-mannosamine and pyruvic acid. It is an important building

Figure 10-23
Common amino sugars and an amino sugar derivative (open-chain forms).

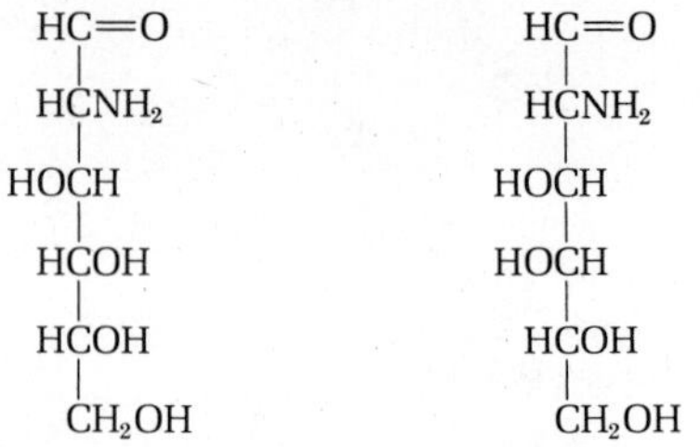

D-Glucosamine   D-Galactosamine

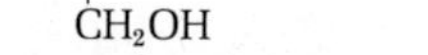

N-Acetyl-D-glucosamine

Figure 10-24
N-acetylmuramic and N-acetylneuraminic acids. The portions in color are derived from three-carbon acids.

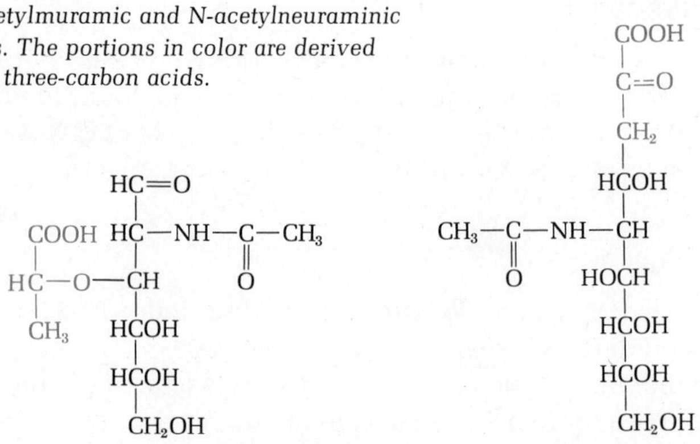

N-Acetylmuramic acid   N-Acetylneuraminic acid

block of the oligosaccharide chains found in the glycoproteins and glycolipids of the cell coats and membranes of animal tissues (page 649). N-Acyl derivatives of neuraminic acid are generically called *sialic acids*. The sialic acids found in human tissues contain an N-acetyl group; in some other species they contain an N-glycolyl group.

Disaccharides

Disaccharides consist of two monosaccharides joined by a glycosidic linkage. The most common disaccharides are maltose, lactose, and sucrose (Figure 10-25). *Maltose*, which is formed as an intermediate product of the action of amylases on starch, contains two D-glucose residues. It is a mixed acetal of the anomeric carbon atom 1 of D-glucose; one hydroxyl group is furnished intramolecularly by carbon atom 5 and the other by carbon atom 4 of a second D-glucose molecule. Both glucose moieties are in pyranose form, and the configuration at the anomeric carbon atom in glycosidic linkage is α. Maltose may therefore be called *O-α-D-glucopyranosyl-(1 → 4)-α-D-glucopyranose*. The second glucose residue of maltose has a free anomeric carbon atom capable of existing in α and β forms; both the α and β forms are products of enzyme action. The first glucose residue cannot undergo oxidation, but the second residue can; it is called the reducing end. The position of the glycosidic linkage between the two glucose residues is symbolized 1 → 4. Exhaustive methylation of all the free hydroxyl groups, followed by hydrolysis of the glycosidic linkage, has proved that the glycosidic linkage in maltose involves carbon atom 1 of the first residue and carbon atom 4 of the second glucose unit. The resulting methylated fragments were 2,3,4,6-tetra-O-methyl-D-glucose and 2,3,6-tri-O-methyl-D-glucose.

Two other common disaccharides that contain two D-glucose units are cellobiose and gentiobiose. *Cellobiose*, the repeating disaccharide unit of cellulose, has a $\beta(1 → 4)$ glycosidic linkage; its full name is thus *O-β-D-glucopyranosyl-(1 → 4)-β-D-glucopyranose*. In *gentiobiose*, the glycosidic linkage is $\beta(1 → 6)$. Since both these disaccharides have a free anomeric carbon, they are reducing sugars.

The disaccharide *lactose* [*O-β-D-galactopyranosyl-(1 → 4)-β-D-glucopyranose*] is found in milk but otherwise does not occur in nature. It yields D-galactose and D-glucose on hydrolysis. Since it has a free anomeric carbon on the glucose residue, lactose is a reducing disaccharide (Figure 10-25).

Sucrose, or cane sugar (Figure 10-25), is a disaccharide of glucose and fructose [*O-β-D-fructofuranosyl-(2 → 1)-α-D-glucopyranoside*]. It is extremely abundant in the plant world and is familiar as table sugar. Unlike most disaccharides and oligosaccharides, sucrose contains no free anomeric carbon atom; the anomeric carbon atoms of the two hexoses are linked to each other. For this reason sucrose does not undergo mutarotation, does not react with phenylhydrazine to form osazones, and does not act as a reducing sugar. It is much more readily hydrolyzed than other disaccharides. The

Figure 10-25
Important disaccharides.

Maltose (β form)
(O-α-D-glucopyranosyl-(1 ⟶ 4)-β-D-glucopyranose)

Haworth projection

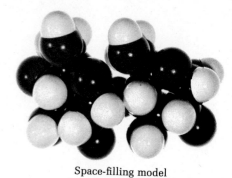

Space-filling model

Conformational formula

Lactose (β form)
(O-β-D-galactopyranosyl-(1 ⟶ 4)-β-D-glucopyranose)

Sucrose
(O-β-D-fructofuranosyl-(2⟶1)-α-D-glucopyranoside)

Space-filling model

hydrolysis of sucrose ($[\alpha]_D^{20} = +66.5°$) to D-glucose ($[\alpha]_D^{20} = +52.5°$) and D-fructose ($[\alpha]_D^{20} = -92°$) is often called _inversion_ since it is accompanied by a net change in optical rotation from dextro to levo as the equimolar mixture of glucose and fructose is formed; this mixture is often called _invert sugar_. The hydrolysis of sucrose, which is also catalyzed by the enzyme invertase, can therefore be followed with a polarimeter.

Trehalose [O-α-D-glucopyranosyl-(1 → 1)-α-D-glucopyranoside], which contains two D-glucose residues, is another example of a nonreducing disaccharide in which the two anomeric carbon atoms are joined; it is the major sugar found in the hemolymph of many insects.

Trisaccharides

A number of trisaccharides occur free in nature. Raffinose [O-α-D-galactopyranosyl-(1 → 6)-O-α-D-glucopyranosyl-(1 → 2)-β-D-fructofuranoside] is found in abundance in sugar beets and many other higher plants. Melezitose [O-α-D-glucopyranosyl-(1 → 3)-O-β-D-fructofuranosyl-(2 → 1)-α-D-glucopyranoside] is found in the sap of some coniferous trees.

Identification and Analysis of Monosaccharides and Oligosaccharides

Chromatographic procedures have revolutionized the art of isolating, separating, and identifying sugars and sugar derivatives. Paper chromatography and thin-layer chromatography are widely employed. For most effective separation and analysis, sugar mixtures are chromatographed on columns of ion-exchange materials in the presence of excess boric acid, which converts neutral sugars into their weakly acid borate complexes and thus allows them to be separated on the basis of differences in acid-base properties. The structure of these borate complexes is not known with certainty; they dissociate readily to form the free sugar and boric acid. Automated systems similar to amino acid analyzers (page 90) have been devised for analysis of sugar mixtures resulting from hydrolysis of oligosaccharides and polysaccharides.

Polysaccharides (Glycans)

Most of the carbohydrates found in nature occur as polysaccharides of high molecular weight. On complete hydrolysis with acid or specific enzymes, these polysaccharides yield monosaccharides and/or simple monosaccharide derivatives. D-Glucose is the most prevalent monosaccharide unit in polysaccharides, but polysaccharides of D-mannose, D-fructose, D- and L-galactose, D-xylose, and D-arabinose are also common. Monosaccharide derivatives commonly found as structural units of natural polysaccharides are D-glucosamine, D-galactosamine, D-glucuronic acid, N-acetylmuramic acid, and N-acetylneuraminic acid.

Polysaccharides, which are also called *glycans*, differ in the nature of their recurring monosaccharide units, in the length of their chains, and in the degree of branching. They are divided into *homopolysaccharides*, which contain only one type of monomeric unit, and *heteropolysaccharides*, which contain two or more different monomeric units. Starch, which contains only D-glucose units, is a homopolysaccharide. Hyaluronic acid consists of alternating residues of D-glucuronic acid and N-acetyl-D-glucosamine and is thus a heteropolysaccharide. Homopolysaccharides are given class names indicating the nature of their building blocks. For example, those containing D-glucose units, e.g., starch and glycogen, are called *glucans* and those containing mannose units are *mannans*. The important polysaccharides are best described in terms of their biological function.

Storage Polysaccharides

These polysaccharides, of which starch is the most abundant in plants and glycogen in animals, are usually deposited in the form of large granules in the cytoplasm of cells. Glycogen or starch granules can be isolated from cell extracts by differential centrifugation. In times of glucose surplus glucose units are stored by undergoing enzymatic linkage to the ends of starch or glycogen chains; in times of metabolic need they are released enzymatically for use as fuel.

Starch

Starch occurs in two forms, *α-amylose* and *amylopectin*. α-Amylose consists of long unbranched chains in which all the D-glucose units are bound in $\alpha(1 \to 4)$ linkages. The chains are polydisperse and vary in molecular weight from a few thousand to 500,000. Amylose is not truly soluble in water but forms hydrated micelles, which give a blue color with iodine. In such micelles, the polysaccharide chain is twisted into a helical coil (Figure 10-26). *Amylopectin* is highly branched; the average length of the branches is from 24 to 30 glucose residues, depending on the species. The backbone glycosidic linkage is $\alpha(1 \to 4)$, but the branch points are $\alpha(1 \to 6)$ linkages (Figure 10-27). Amylopectin yields colloidal or micellar solutions, which give a red-violet color with iodine. Its molecular weight may be as high as 100 million.

The major components of starch can be enzymatically hydrolyzed in two different ways. Amylose can be hydrolyzed by *α-amylase* [$\alpha(1 \to 4)$-glucan 4-glucanohydrolase], which is present in saliva and pancreatic juice and participates in the digestion of starch in the gastrointestinal tract. It hydrolyzes $\alpha(1 \to 4)$ linkages at random to yield a mixture of glucose and free maltose; the latter is not attacked. Amylose can also be hydrolyzed by *β-amylase* [$\alpha(1 \to 4)$-glucan maltohydrolase]. This enzyme, which occurs in malt, cleaves away successive maltose units beginning from the nonreducing end to yield maltose quantitatively. The α- and β-amylases also attack amylopectin (Figure 10-28). The polysaccharides of intermediate chain length that are formed

Figure 10-26
The helical coil of amylose.

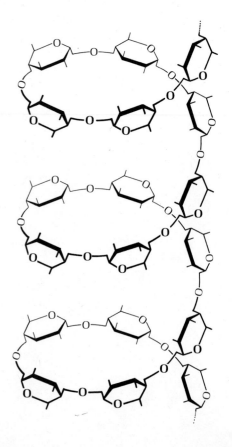

Figure 10-27
An α(1 → 6) branch point in amylopectin.

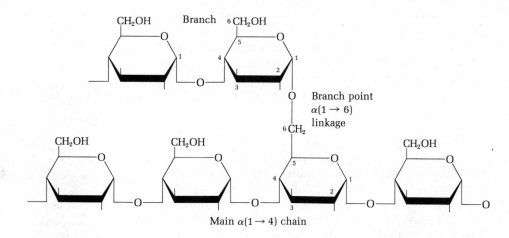

Main α(1 → 4) chain

from starch components by the action of amylases are called _dextrins_. Neither α- nor β-amylases can hydrolyze the α(1 → 6) linkages at the branch points of amylopectin. The end product of exhaustive β-amylase action on amylopectin is a large, highly branched core, or _limit dextrin_, so called because it represents the limit of the attack of β-amylase (Figure 10-28). A debranching enzyme [α(1 → 6)-glucan 6-glucanohydrolase, also called α(1 → 6)-glucosidase] can hydrolyze the α(1 → 6) linkages at the branch points. The combined action of a β-amylase and an α(1 → 6)-glucosidase can therefore completely degrade amylopectin to maltose and glucose.

Figure 10-28
Action of β-amylase on amylopectin. Successive maltose residues are hydrolyzed until the α(1 → 6) branch points are reached. The remaining core (color), which represents about 40 percent of the molecule, is the limit dextrin.

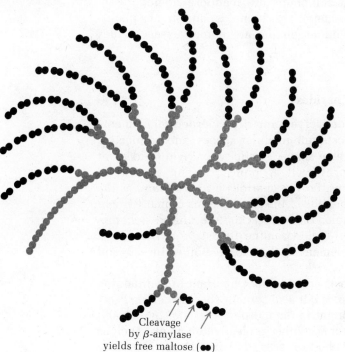

Cleavage
by β-amylase
yields free maltose (●●)

Glycogen

Glycogen is the main storage polysaccharide of animal cells, the counterpart of starch in plant cells. Glycogen is especially abundant in the liver, where it may attain up to 10 percent of the wet weight. It is also present to about 1 to 2 percent in skeletal muscle. In liver cells the glycogen is found in large granules, which are themselves clusters of smaller granules composed of single, highly branched molecules with a molecular weight of several million (Figure 10-29).

Like amylopectin, glycogen is a polysaccharide of D-glucose in $\alpha(1 \rightarrow 4)$ linkage. However, it is a more highly branched and more compact molecule than amylopectin; the branches occur about every 8 to 12 glucose residues. The branch linkages are $\alpha(1 \rightarrow 6)$. Glycogen can be isolated from animal tissues by digesting them with hot KOH solutions, in which the nonreducing $\alpha(1 \rightarrow 4)$ and $\alpha(1 \rightarrow 6)$ linkages are stable. Glycogen is readily hydrolyzed by α- and β-amylases to yield glucose and maltose, respectively; the action of β-amylase also yields a limit dextrin. Glycogen gives a red-violet color with iodine.

Other Storage Polysaccharides

Dextrans, too, are branched polysaccharides of D-glucose, but they differ from glycogen and starch in having backbone linkages other than $\alpha(1 \rightarrow 4)$. Found as storage polysaccharides in yeasts and bacteria, they vary in their branch points, which may be $1 \rightarrow 2$, $1 \rightarrow 3$, $1 \rightarrow 4$, or $1 \rightarrow 6$ in different species. Dextrans form highly viscous, slimy solutions. *Fructans* (also called levans) are homopolysaccharides composed of D-fructose units; they are found in many plants. *Inulin*, found in the artichoke, consists of D-fructose residues in $\beta(2 \rightarrow 1)$ linkage. *Mannans* are mannose homopolysaccharides found in bacteria, yeasts, molds, and higher plants. Similarly, *xylans* and *arabinans* are homopolysaccharides found in plant tissues.

Structural Polysaccharides

Many polysaccharides serve primarily as structural elements in cell walls and coats, intercellular spaces, and connective tissue, where they give shape, elasticity, or rigidity to plant and animal tissues as well as protection and support to unicellular organisms. Polysaccharides also are found as the major organic compounds of the exoskeletons of many invertebrates. For example, the polysaccharide *chitin*, a homopolymer of N-acetyl-D-glucosamine in $\beta(1 \rightarrow 4)$ linkage, is the major organic element in the exoskeleton of insects and crustacea.

Cell walls and coats are not only important in maintaining the structure of tissues but also contain specific cell-cell recognition sites important in the morphogenesis of tissues and organs. They also contain other protective elements, such as the cell-surface antibodies of vertebrate tissues. For this

Figure 10-29
Electron micrograph of glycogen granules in a liver cell of the hamster. They consist of clusters of many smaller granules, about 15 to 30 nm in diameter. They are usually found near the smooth-surfaced endoplasmic reticulum, as shown here.

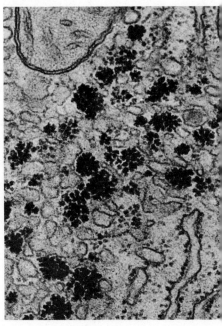

D. Fawcett

0.1 μm

Figure 10-30

Electron micrograph of the cell wall of an alga (Chaetomorpha). The wall consists of successive layers of cellulose fibrils in parallel arrangement. [D. M. Jones, Adv. Carbohydr. Chem., 19: 219 (1964).]

1.0 μm

reason we shall examine the structural polysaccharides in the context of the molecular organization of cell walls and coats.

Plant Cell Walls

Since plant cells must be able to withstand the large osmotic-pressure difference between the extracellular and intracellular fluid compartments, they require rigid cell walls to keep from swelling. In larger plants and trees the cell walls not only must contribute physical strength or rigidity to stems, leaves, and root tissues but must also be able to sustain large weights.

The most abundant cell-wall and structural polysaccharide in the plant world is *cellulose*, a linear polymer of D-glucose in $\beta(1 \rightarrow 4)$ linkage. Cellulose is the major component of wood and thus of paper; cotton is nearly pure cellulose. Cellulose is also found in some lower invertebrates. It is almost entirely of extracellular occurrence.

On complete hydrolysis with strong acids, cellulose yields only D-glucose, but partial hydrolysis yields the reducing disaccharide *cellobiose* (page 261), in which the linkage between the D-glucose units is $\beta(1 \rightarrow 4)$. When cellulose is exhaustively methylated and then hydrolyzed, it yields only 2,3,6-tri-O-methylglucose, showing not only that all its glycosidic linkages are $1 \rightarrow 4$ but also that there are no branch points. The only chemical difference between starch and cellulose, both homopolysaccharides of D-glucose, is that starch has $\alpha(1 \rightarrow 4)$ linkages and cellulose $\beta(1 \rightarrow 4)$. Cellulose is not attacked by either α- or β-amylase. In fact, enzymes capable of hydrolyzing the $\beta(1 \rightarrow 4)$ linkages of cellulose are not secreted in the digestive tract of most mammals, and they cannot use cellulose for food. However, the ruminants, e.g., the cow, are an exception: they can utilize cellulose as food since bacteria in the rumen form the enzyme *cellulase*, which hydrolyzes cellulose to D-glucose.

The minimum molecular weight of cellulose from different sources has been estimated to vary from about 50,000 to 2,500,000 in different species, equivalent to 300 to 15,000 glucose residues. X-ray diffraction analysis indicates that cellulose molecules are organized in bundles of parallel chains to form fibrils (Figure 10-30). Although cellulose has a high affinity for water, it is completely insoluble in it.

In the cell walls of plants densely packed cellulose fibrils surround the cell in regular parallel arrays, often in crisscross layers (Figure 10-30). These fibrils are cemented together by a matrix of three other polymeric materials: *hemicellulose*, *pectin*, and *extensin*. *Hemicelluloses* are not related structurally to cellulose but are polymers of pentoses, particularly D-xylans, polymers of D-xylose in $\beta(1 \rightarrow 4)$ linkage with side chains of arabinose and other sugars. *Pectin* is a polymer of methyl D-galacturonate. *Extensin*, a complex glycoprotein (page 273), is attached covalently to the cellulose fibrils. Extensin resembles its animal-tissue counterpart *collagen* (page 135) in being rich in hydroxyproline residues; it also contains many side chains with arabinose and galactose residues. The cell walls of higher plants can be compared to cases of reinforced concrete, in which the cellulose fibrils correspond to the steel rods and the matrix material to the concrete. These walls are capable of withstanding enormous weights and physical stress. Wood contains another polymeric substance, _lignin_, which makes up nearly 25 percent of its dry weight. Lignin is a polymer of aromatic alcohols (page 709).

Other polysaccharides serving as cell-wall or structural components in plants include _agar_ of seaweeds, which contains D- and L-galactose residues, some of which are esterified with sulfuric acid; _alginic acid_ of algae and kelp, which contains D-mannuronic acid units; and _gum arabic_, a vegetable gum, which contains D-galactose and D-glucuronic acid residues, as well as arabinose and rhamnose.

Bacterial Cell Walls

The cell walls of bacteria are rigid, porous, boxlike structures that provide physical protection to the cell. Since bacteria have a high internal osmotic pressure, and since they are often exposed to a quite variable and sometimes hypotonic external environment, they must have a rigid cell wall to prevent swelling and rupture of the cell membrane. The structure and biosynthesis of bacterial cell walls have been intensively studied, because they contain specific antigens useful in diagnosing infectious diseases and because the biosynthesis of bacterial cell walls is inhibited by penicillin and other medically useful antibiotics.

Bacteria are classically divided into _Gram-positive_ and _Gram-negative_ organisms, according to their reaction to the Gram stain, an empirical procedure in which the cells are treated successively with the dye crystal violet and with iodine, then decolorized and treated with safranine. In Gram-positive cells the walls contain very little lipid, e.g., _Streptococcus albus_ and _Micrococcus lysodeikticus_, whereas in Gram-negative cells the walls are rich in lipid, e.g., _E. coli_. Cell walls of Gram-negative bacteria can be isolated by shaking thick suspensions of the cells with very fine glass beads to rupture the walls; the bacterial contents can then be washed out. Cell walls appear relatively thin but are highly dense; they make up about 20 to 30 percent of the dry weight of Gram-negative bacteria.

Gram-positive and Gram-negative cell walls have one molecular feature in common: in both the rigid structural framework consists of parallel polysaccharide chains covalently cross-linked by peptide chains. This framework makes up 50 percent or more of the weight of the cell wall. To it are attached characteristic accessory components, which are different in Gram-negative and Gram-positive cells, as we shall see below. The covalently linked framework of the cell wall actually may be regarded as a single large sacklike molecule. It is called a _peptidoglycan_ or _murein_ (Latin _murus_, wall). The structure (page 618) and enzymatic biosynthesis of the peptidoglycan were largely elucidated during investigations on the mechanism of action of the antibiotic penicillin by J.

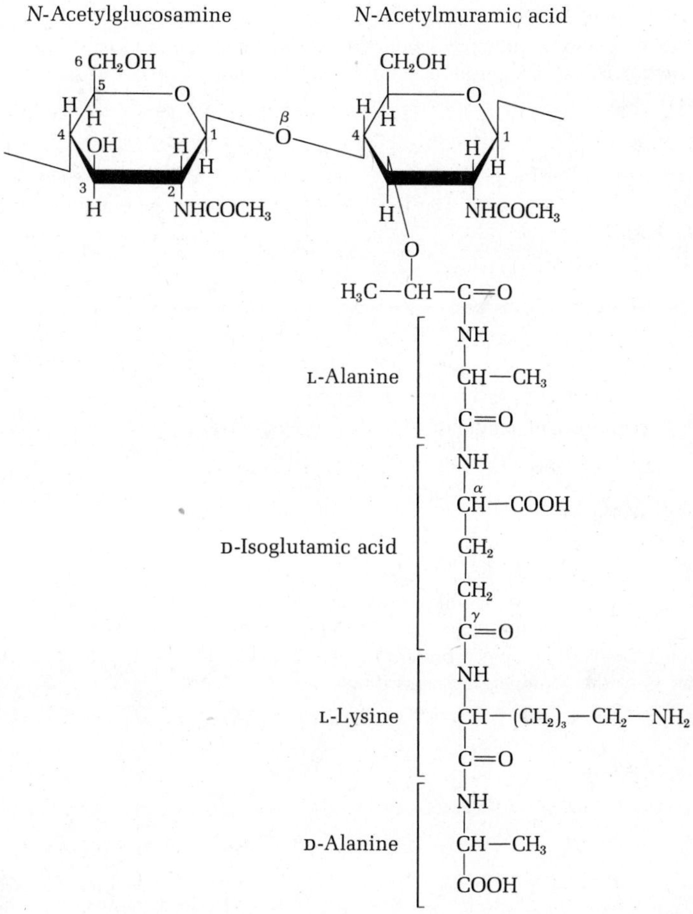

Figure 10-31

Recurring muropeptide of bacterial cell walls. It consists of a disaccharide to which a tetrapeptide side chain is attached. The linkage between successive units is $\beta(1 \to 4)$. The D-glutamic acid residue is linked to L-lysine by its γ-carboxyl group and is therefore designated D-isoglutamic acid. In some species, e.g., Staphylococcus aureus (Figure 10-32), the α-carboxyl group of the D-glutamic acid residue is substituted with an amino group; this residue is then called D-isoglutamine.

N-Acetylglucosamine   N-Acetylmuramic acid

Strominger and his colleagues (page 652). The basic recurring unit in the peptidoglycan structure is the muropeptide (Figure 10-31). It is a disaccharide of N-acetyl-D-glucosamine and N-acetylmuramic acid (page 260) in $\beta(1 \to 4)$ linkage. The backbone may be regarded as a substituted chitin (page 266) with D-lactic acid substituted on alternating residues. To the carboxyl group of the N-acetylmuramic acid residues of the backbone are attached tetrapeptide side chains, each containing L-alanine, D-alanine, D-glutamic acid or D-glutamine, and either *meso*-diaminopimelic acid, L-lysine, L-hydroxylysine, or ornithine (page 77) depending on the bacterial species.

The parallel polysaccharide chains of the cell wall are cross-linked through their peptide side chains. The terminal D-alanine residue of the side chain of one polysaccharide chain is joined covalently with the peptide side chain of an adjacent polysaccharide chain, either directly, as in *E. coli,* or through a short connecting peptide, e.g., the pentaglycine in *Staphylococcus aureus* (Figure 10-32). The peptidoglycan forms a completely continuous, covalent structure around the cell; Gram-positive bacteria are encased by up to 20 layers of cross-linked peptidoglycan.

The peptidoglycan structure of the bacterial cell wall is resistant to the action of peptide-hydrolyzing enzymes, which do not attack peptides containing D-amino acids. How-

Figure 10-32
Manner of cross-linking of peptidoglycan chains in S. aureus cell walls.
[Redrawn from J. L. Strominger and J. M. Ghuysen, Science, 156: 213–221 (1967).]

The backbones are in color. X is an N-acetylglucosamine residue, and Y an N-acetylmuramic acid residue. The tetrapeptide side chains of each disaccharide unit are cross-linked to those of adjacent chains by a glycine pentapeptide.

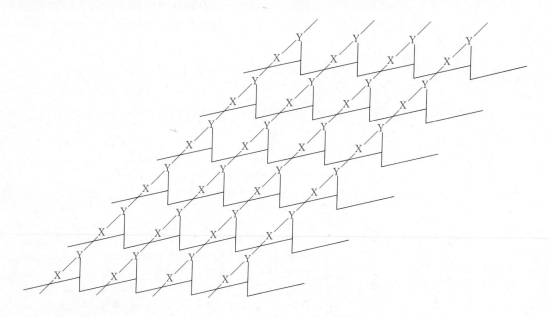

The arrangement of several parallel peptidoglycan chains into a tight, cross-linked structure. Each murein chain is about 12 disaccharide units long. The vertical lines represent the tetrapeptide side chains; the horizontal lines the pentapeptide cross-links.

ever, the enzyme <u>lysozyme</u> (page 231), found in tears and in egg white, causes lysis of Gram-positive bacteria, such as *Micrococcus lysodeikticus*, by hydrolyzing the $\beta(1 \rightarrow 4)$ glycosidic bonds of the polysaccharide backbone of the peptidoglycan (the name *lysodeikticus* derives from Greek words meaning "lysis meter"; this organism is often used to assay preparations of lysozyme for biological activity). The products of lysozyme action are disaccharides of N-acetyl-D-glucosamine and N-acetylmuramic acid to which the peptide side chains are still attached. Once the backbones have been broken in this manner, the cell swells, with rupture of the

Figure 10-33
Segment of a teichoic acid containing alternating residues (color) of D-alanine and N-acetylglucosamine.

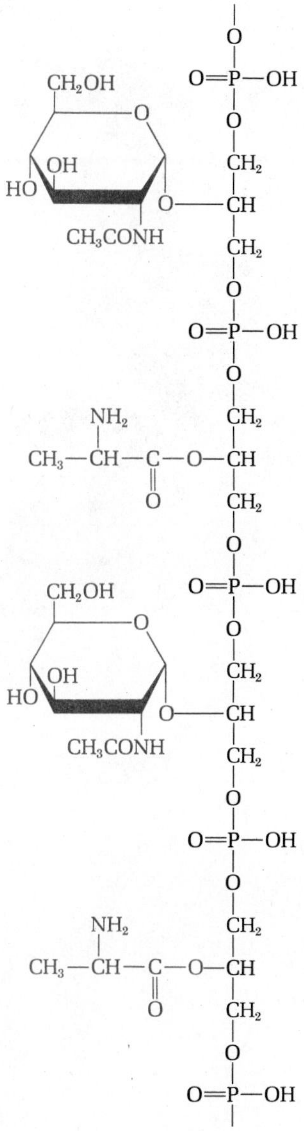

membrane and loss of cell contents. However, when Gram-positive bacteria are treated with lysozyme in the presence of high concentrations of the impermeant solute sucrose (0.8 M), the cell is protected against osmotic swelling after the wall is removed. The naked bacterial cell, surrounded only by its membrane, is called a *protoplast*. Protoplasts are very fragile and remain intact only so long as the medium is isotonic, to prevent swelling and subsequent rupture of the membrane.

In addition to the peptidoglycan framework, bacterial cell walls contain a number of accessory polymers, which make up almost 50 percent of the weight of the wall. These accessory components differ from one species to another. There are three types of accessory polymers: (1) teichoic acids, (2) polysaccharides, and (3) polypeptides or proteins. The *teichoic acids* (Greek *teichos*, "city wall"), which make up from 20 to 40 percent of the dry weight of the cell walls of Gram-positive bacteria, are polymeric chains of glycerol or ribitol molecules linked to each other by phosphodiester bridges. In one type of teichoic acid (Figure 10-33) alternating free hydroxyl groups of the glycerol phosphate backbone are occupied by D-alanine and either D-glucose or N-acetyl-D-glucosamine residues. The accessory polysaccharides contain rhamnose, glucose, galactose, or mannose (or their amines). Depending on the species, they may be slimy or form hard, tough capsules. Both teichoic acids and the polysaccharides of bacterial cell walls are antigenic (page 66); this property is useful in the taxonomic classification of bacteria.

The walls of Gram-negative cells, such as E. coli, are much more complex than those of Gram-positive cells. Their accessory components consist of polypeptides, lipoproteins, and, particularly, a very complex *lipopolysaccharide* whose structure is just beginning to be understood. It has a trisaccharide backbone repeating unit, consisting of two heptose (seven-carbon) sugars and *octulosonic acid* (an eight-carbon sugar acid). To this backbone are attached oligosaccharide side chains and the fatty acid *β-hydroxymyristic acid*, which gives this complex structure its lipid character. The lipopolysaccharide forms an outer lipid membrane and contributes to the complex antigenic specificity of Gram-negative cells. The cell wall of Gram-positive organisms can be thought of as a rigid, brittle box, like the shell of a crustacean, whereas the cell wall of Gram-negative organisms has an outer lipid-rich skin, with the rigid peptidoglycan skeleton buried underneath.

Cell Coats and Ground Substance in Animal Tissues

The cells in the tissues of vertebrates are not surrounded by rigid walls, presumably because they do not require protection against large changes in external osmotic pressure. However, they do have an outer coat that is in some respects comparable to a cell wall. Electron microscopy often reveals the presence of a filamentous or fuzzy coat outside the cell membrane in some animal cells (Figure 10-34). These cell coats are soft and flexible and have adhesive properties.

Although the chemical nature of some of the major components of animal cell coats is known, there is still little information about how these components are arranged on the outer surface of the cell. This is an important field of study since the cells in animal tissues engage in specific contacts with their neighbors. For example, when kidney cells are separated from each other and then grown in tissue culture, they will seek each other out from a mixture of liver and kidney cells and reassociate, presumably because their surfaces contain specific cell-cell recognition sites. Furthermore, the growth of normal mammalian cells in tissue culture is subject to the phenomenon of _contact_ or _density inhibition_. Normal cells grow in an orderly manner, so that the division of any given cell is inhibited when most of its surface is in near contact with neighboring cells. On the other hand, when cancer cells are cultured, they are not subject to contact inhibition, a condition leading to a disorderly piling up of the cells. Moreover, the cell coats of cancer cells contain specific antigens not found in normal cells. Cell coats also are the site of cell and tissue antibodies, like those involved in the rejection of grafted or transplanted tissues by a given organism. Cell surfaces are involved in many other types of intercellular reactions, e.g., the synaptic contacts between neurons of the nervous system.

The major components of the cell coats of higher organisms consist of (1) _glycosphingolipids_, to be discussed in Chapter 11 (page 294), (2) _acid mucopolysaccharides_, and (3) _glycoproteins_.

Acid Mucopolysaccharides

The acid mucopolysaccharides are a group of related heteropolysaccharides usually containing two types of alternating monosaccharide units of which at least one has an acidic group, either a carboxyl or sulfuric group (Table 10-2). When they occur as complexes with specific proteins, they are called _mucins_ or _mucoproteins_; in this class of glycoproteins (page 273) the polysaccharide makes up most of the weight. Mucoproteins are jellylike, sticky, or slippery substances; some provide lubrication, and some function as a flexible intercellular cement.

Figure 10-34
Electron micrograph of the absorptive surface of an epithelial cell in the small intestine of the cat, showing the polysaccharide cell coat, or glycocalyx, on the tips of the villi.

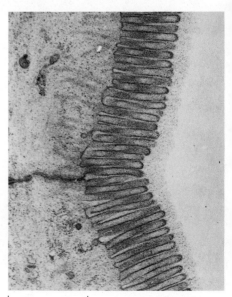

1.0 μm

D. Fawcett

Table 10-2 Acid mucopolysaccharides

Polysaccharide	Constituents	Occurrence
Hyaluronic acid	Glucuronic acid, N-acetyl-D-glucosamine	Synovial fluid
Chondroitin	Glucuronic acid, N-acetyl-D-galactosamine	Cornea
Chondroitin 4-sulfate	Glucuronic acid, N-acetyl-D-galactosamine 4-sulfate	Cartilage
Dermatan sulfate	Iduronic acid, N-acetyl-D-galactosamine 4-sulfate	Skin
Keratan sulfate	Galactose, galactose 6-sulfate, N-acetyl-D-glucosamine 6-sulfate	Cornea
Heparin	Glucosamine 6-sulfate, glucuronic acid 2-sulfate, iduronic acid	Lung

The most abundant acid mucopolysaccharide is *hyaluronic acid*, present in cell coats and in the extracellular ground substance of the connective tissues of vertebrates; it also occurs in the synovial fluid in joints and in the vitreous humor of the eye. The repeating unit of hyaluronic acid is a disaccharide composed of D-glucuronic acid and N-acetyl-D-glucosamine in $\beta(1 \rightarrow 3)$ linkage (Figure 10-35). Since each disaccharide unit is attached to the next by $\beta(1 \rightarrow 4)$ linkages, hyaluronic acid contains alternating $\beta(1 \rightarrow 3)$ and $\beta(1 \rightarrow 4)$ linkages. Hyaluronic acid is a linear polymer. Because its carboxyl groups are completely ionized and thus negatively charged at pH 7.0, hyaluronic acid is soluble in water, in which it forms highly viscous solutions. The enzyme *hyaluronidase* catalyzes hydrolysis of the $\beta(1 \rightarrow 4)$ linkages of hyaluronic acid; this hydrolysis is accompanied by a decrease in viscosity.

Another acid mucopolysaccharide is *chondroitin*, which is nearly identical in structure to hyaluronic acid; the only difference is that it contains N-acetyl-D-galactosamine instead of N-acetyl-D-glucosamine residues. Chondroitin itself is only a minor component of extracellular material, but its sulfuric acid derivatives, *chondroitin 4-sulfate* (chondroitin A) and *chondroitin 6-sulfate* (chondroitin C), are major structural components of cell coats, cartilage, bone, cornea, and other connective-tissue structures in vertebrates. *Dermatan sulfate* and *keratan sulfate* (Table 10-2) are acid mucopolysaccharides found in skin, cornea, and bony tissues. A related acid mucopolysaccharide is *heparin*, which prevents coagulation of blood. It is found in the lungs and in the walls of arteries.

Recently it has been discovered that some acid mucopolysaccharides contain the element silicon, which is essential in the nutrition of rats and chicks. The silicon is bound to the polysaccharides in covalent form.

Glycoproteins

Among the several different classes of conjugated proteins (page 58), the *glycoproteins*, which contain carbohydrate groups attached covalently to the polypeptide chain, represent a large group of wide distribution and considerable biological significance. In fact, on closer study many proteins once thought to be simple proteins, i.e., containing only

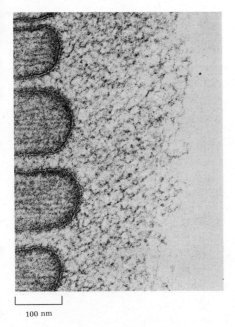

The higher magnification below shows its filamentous meshlike structure, often called a "fuzzy coat." [S. Ito, Fed. Proc., 28:12 (1969).]

100 nm

Figure 10-35
Repeating unit of hyaluronic acid.

D-Glucuronic acid — N-acetyl-D-glucosamine

amino acid residues, have been found to contain carbohydrate groups. Table 10-3 shows the biological distribution of some major types of glycoproteins. The percent by weight of carbohydrate groups in different glycoproteins may vary from less than 1 percent in ovalbumin to as high as 80 percent in the mucoproteins. Glycoproteins having a very high content of carbohydrate are called *proteoglycans*.

Glycoproteins are found in all forms of life. In vertebrates most but not all of the glycoproteins are extracellular in occurrence and function or are secreted from cells; it has accordingly been suggested that one purpose of the attached sugar residues is to label the protein for export from the cell. Among the glycoproteins having extracellular location or function are the cell-coat glycoproteins, the blood glycoproteins, the circulating forms of some protein hormones, the antibodies, various digestive enzymes secreted into the intestine, the mucoproteins of mucous secretions, and the glycoproteins of extracellular basement membranes.

Many different monosaccharides and monosaccharide derivatives have been found in glycoproteins. The linear or branched side chains of glycoproteins may contain from two to dozens of monosaccharide residues, usually of two or more kinds. Often the terminal monosaccharide unit is a negatively charged residue of *N-acetylneuraminic acid*, a sialic acid (page 650; Figure 10-36).

The oligosaccharide groups of most glycoproteins are covalently attached to the R groups of specific amino acid residues in the polypeptide chain. Three different types of linkages have been found (Figure 10-36). In some glycoproteins, e.g., ovalbumin and the immunoglobulins, the oligosaccharide is attached via a glycosylamine linkage between N-acetyl-D-glucosamine of the oligosaccharide to the amide nitrogen of an asparagine residue in the polypeptide chain (Figure 10-36). In a second class of glycoproteins, including the submaxillary mucoprotein, there is a glycosidic bond between N-acetyl-D-galactosamine of the side-chain oligosaccharide and the hydroxyl group of a serine or threonine residue. The submaxillary mucins have recurring units of about 28 amino acid residues; each recurring unit contains three oligosaccharide side chains. In the third class of glycoproteins, represented by collagen (page 135), the oligosaccharide side chains are attached to the hydroxyl groups of hydroxylysine residues. The precise sequences of residues in the oligosaccharide side chains are known in only a few cases.

The antifreeze proteins found in the blood plasma of Antarctic fishes (pages 66 and 112) are particularly interesting. Their backbones consist of the recurring amino acid sequence Ala-Ala-Thr; the disaccharide galactosyl-N-acetylgalactosamine is attached to every threonine residue. The molecular weights of these proteins vary from 10,000 to 23,000. The antifreeze proteins have a flexible, expanded structure in water which presumably interferes with the formation of the crystal lattice of ice.

The human blood-group proteins contain oligosaccharide side chains with residues of L-fucose, D-galactose, N-acetyl-D-galactosamine, and N-acetyl-D-glucosamine; these side chains determine the blood-group specificity.

Table 10-3 Some glycoproteins, grouped according to biological occurrence. Note that most glycoproteins are extracellular.

Blood plasma
   Fetuin
   α_1-Acid glycoprotein
   Fibrinogen
   Immune globulins
   Thyroxine-binding protein
   Blood-group proteins

Urine
   Urinary glycoprotein

Hormones
   Chorionic gonadotrophin
   Follicle-stimulating hormone
   Thyroid-stimulating hormone

Enzymes
   Ribonuclease B
   β-Glucuronidase
   Pepsin
   Serum cholinesterase

Egg white
   Ovalbumin
   Avidin
   Ovomucoid

Mucus secretions
   Submaxillary glycoproteins
   Gastric glycoproteins

Connective tissue
   Collagen

Cell membranes
   Glycophorin of erythrocyte membrane

Extracellular membranes
   Basement-membrane glycoprotein
   Lens-capsule glycoprotein

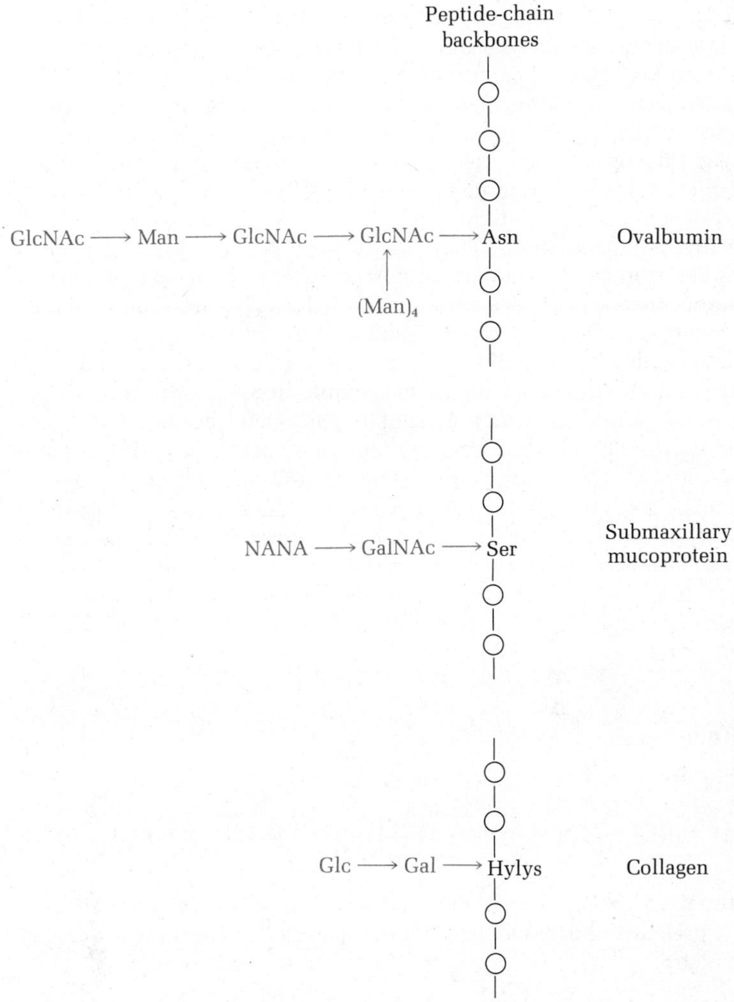

Figure 10-36
Oligosaccharide side chains (color) of three representative glycoproteins. The symbols used are GlcNAc, N-acetylglucosamine; GalNAc, N-acetylgalactosamine; Glc, glucose; Gal, galactose; Man, mannose; NANA, N-acetylneuraminic acid; and Hylys, hydroxylysine.

Peptide-chain backbones

GlcNAc ⟶ Man ⟶ GlcNAc ⟶ GlcNAc ⟶ Asn   Ovalbumin

(Man)$_4$

NANA ⟶ GalNAc ⟶ Ser   Submaxillary mucoprotein

Glc ⟶ Gal ⟶ Hylys   Collagen

Summary

Carbohydrates are polyhydroxylic acetals or ketals with the empirical formula $(CH_2O)_n$. Pentoses $(C_5H_{10}O_5)$ and hexoses $(C_6H_{12}O_6)$ are the most abundant simple carbohydrates, or sugars. Sugars have one or more asymmetric carbon atoms and thus exist in the form of stereoisomers; most naturally occurring sugars, e.g., ribose, glucose, fructose, and mannose, are in the D series. Sugars with five or more carbon atoms can exist in two anomeric forms, which are stereoisomeric cyclic hemiacetals formed between a hydroxyl group of the sugar and the carbonyl group. The five- and six-membered rings thus formed are called furanoses and pyranoses, respectively. Pyranoses may occur in boat and chair conformations.

Anomeric glycosides result from the reaction of pentoses and hexoses with alcohols. The free hydroxyl groups of sugars can be completely acetylated or methylated; their rings can also be cleaved by periodic acid. Sugars can be reduced to sugar alcohols; they can be oxidized at the aldehydic carbon to aldonic acids, at the primary hydroxyl group to uronic acids, and at both terminal carbon atoms to aldaric acids.

Disaccharides consist of two monosaccharides joined in glycosidic linkage. Maltose contains two glucose residues in $\alpha(1 \rightarrow 4)$ linkage, cellobiose contains two glucose residues, lactose contains galactose and glucose, and sucrose contains glucose and fructose. In sucrose, the anomeric carbon atoms of both monosaccharides are linked together and cannot undergo oxidation.

Polysaccharides (glycans) are classified chemically as homopolysaccharides, which contain a single recurring monosaccharide unit (e.g., glycogen, a polymer of glucose) and heteropolysaccharides, which contain two or more recurring monosaccharide units (e.g., hyaluronic acid, an alternating polymer of D-glucuronic acid and N-acetyl-D-glucosamine). They are also classified functionally as either storage or structural polysaccharides. The most important storage polysaccharides are starch and glycogen; these are branched glucose polymers with $\alpha(1 \rightarrow 4)$ linkages in the chains and $\alpha(1 \rightarrow 6)$ linkages at branch points. The most important structural polysaccharide is cellulose, with D-glucose units in $\beta(1 \rightarrow 4)$ linkages.

The walls of bacterial cells contain peptidoglycans (mureins), heteropolysaccharides of N-acetylmuramic acid and N-acetylglucosamine, with short cross-linking peptides containing D-amino acids. The cell walls in higher plants contain cellulose, other polysaccharides, and protein. Animal cells possess flexible cell coats containing acid mucopolysaccharides coupled to proteins. There are three classes of glycoproteins, distinguished by the amino acid residues to which the oligosaccharide side chains are attached.

References

Books

Advances in Carbohydrate Chemistry, 1945 to present. Annual review series.

BAILEY, R. W.: *Oligosaccharides*, Macmillan, New York, 1965. Comprehensive reference work on structure, preparation, and chemistry.

DAVIDSON, E. A.: *Carbohydrate Chemistry*, Holt, New York, 1967. An excellent survey.

FLORKIN, M., and E. H. STOTZ (eds.): *Carbohydrates*, sec. II, vol. 5 of *Comprehensive Biochemistry*, American Elsevier, New York, 1963. Reference treatise.

GOTTSCHALK, A. (ed.): *Glycoproteins: Composition, Structure, and Function*, 2d ed., 2 vols., Elsevier, Amsterdam, 1972. Comprehensive reference work.

JEANLOZ, R. W., and E. A. BALASZ (eds.): *Amino Sugars*, vols. 1 and 2, Academic, New York, 1965.

PIGMAN, W. W., and D. HORTON (eds.): *The Carbohydrates*, vols. 1A and 1B, Academic, New York, 1972.

QUINTARELLI, G. (ed.): *The Chemical Physiology of Mucopolysaccharides*, Little, Brown, Boston, 1967.

Articles

BADDILEY, J.: "Teichoic Acids in Cell Walls and Membranes of Bacteria," *Essays Biochem.*, 8: 35–78 (1972).

DEVRIES, A. L., J. VANDENHEEDE, and R. E. FEENEY: "Primary Structure of Freezing-Point Depressing Glycoproteins," *J. Biol. Chem.*, 246: 305–308 (1971).

HEATH, E. C.: "Complex Polysaccharides," *Ann. Rev. Biochem.*, 40: 29–56 (1971).

MARSHALL, R. D.: "Glycoproteins," *Ann. Rev. Biochem.*, 41: 673–702 (1972).

SPIRO, R. G.: "Glycoproteins: Their Biochemistry, Biology, and Role in Human Disease," *NE J. Med.*, 281: 991–1001, 1043–1056 (1969). Excellent short account.

Problems

1. If 80 ml of a freshly prepared 10 percent solution of α-D-glucose is mixed with 20 ml of a freshly prepared solution of β-D-glucose, estimate (a) the initial specific rotation $[\alpha]_D^{20}$ of the resulting solution and (b) the rotation after several hours have elapsed. (c) Do the same for a mixture of 50 ml of methyl-α-D-glucoside and 50 ml of methyl-β-D-glucoside.

2. Name the products of (a) treatment of α-D-galactose with excess acetic anhydride and (b) treatment of α-D-glucose with dimethyl sulfate, followed by gentle hydrolysis.

3. Name the products of (a) treatment of D-galactose with sodium amalgam in water, (b) oxidation of L-mannose with nitric acid, and (c) exhaustive methylation of D-galactosamine.

4. (a) Lactose is exhaustively methylated and then hydrolyzed. Name the products. (b) What are the products of the exhaustive methylation of sucrose followed by hydrolysis?

5. On acid hydrolysis, a trisaccharide yields D-glucose and D-galactose in a 2:1 ratio. Exhaustive methylation, followed by hydrolysis yields 2,3,6-tri-O-methylgalactose, 2,3,4,6-tetra-O-methylglucose, and 2,3,4-tri-O-methylglucose. Name the trisaccharide.

6. After exhaustive methylation and hydrolysis, a polysaccharide yields equimolar amounts of 2,3,4-tri-O-methylglucose and 2,3,6-tri-O-methylglucose. The polysaccharide has one reducing terminus. Indicate its structure.

7. A 10.0-g sample of glycogen yields 6 mmol of 2,3-di-O-methylglucose on methylation and hydrolysis. (a) What percentage of the glucose residues occur at $1 \rightarrow 6$ branch points? (b) What is the average number of glucose residues per branch? (c) How many millimoles of 2,3,6-tri-O-methylglucose were formed? (d) If the molecular weight of the polysaccharide is 2×10^6, how many glucose residues does it contain?

8. On treatment with periodate, a 100-mg sample of cellulose gave 0.0015 mmol of formic acid. What is the approximate chain length?

CHAPTER 11 LIPIDS, LIPOPROTEINS, AND MEMBRANES

Lipids are water-insoluble organic biomolecules that can be extracted from cells and tissues by nonpolar solvents, e.g., chloroform, ether, or benzene. There are several different families or classes of lipids but all derive their distinctive properties from the hydrocarbon nature of a major portion of their structure. Lipids have several important biological functions, serving (1) as structural components of membranes, (2) as storage and transport forms of metabolic fuel, (3) as a protective coating on the surface of many organisms, and (4) as cell-surface components concerned in cell recognition, species specificity, and tissue immunity. Some substances classified among the lipids have intense biological activity; they include some of the vitamins and hormones.

Although lipids are a distinct class of biomolecules, we shall see that they often occur combined, either covalently or through weak bonds, with members of other classes of biomolecules to yield hybrid molecules such as *glycolipids*, which contain both carbohydrate and lipid groups, and *lipoproteins*, which contain both lipids and proteins. In such biomolecules the distinctive chemical and physical properties of their components are blended to fill specialized biological functions.

Classification of Lipids

Lipids have been classified in several different ways. The most satisfactory classification is based on their backbone structures (Table 11-1). The *complex lipids*, which characteristically contain fatty acids as components, include the *acylglycerols*, the *phosphoglycerides*, the *sphingolipids*, and the *waxes*, which differ in the backbone structures to which the fatty acids are covalently joined. They are also called *saponifiable lipids* since they yield soaps (salts of fatty acids) on alkaline hydrolysis. The other great group of lipids consists of the *simple lipids*, which do not contain fatty acids and hence are nonsaponifiable.

Let us first consider the structure and properties of fatty acids, characteristic components of all the complex lipids.

Table 11-1 Classification of lipids

Lipid type	Backbone
Complex (saponifiable)	
Acylglycerols	Glycerol
Phosphoglycerides	Glycerol 3-phosphate
Sphingolipids	Sphingosine
Waxes	Nonpolar alcohols of high molecular weight
Simple (nonsaponifiable)	
Terpenes	
Steroids	
Prostaglandins	

Fatty Acids

Although fatty acids occur in very large amounts as building-block components of the saponifiable lipids, only traces occur in free (unesterified) form in cells and tissues. Well over 100 different kinds of fatty acids have been isolated from various lipids of animals, plants, and microorganisms. All possess a long hydrocarbon chain and a terminal carboxyl group (Figure 11-1). The hydrocarbon chain may be saturated, as in *palmitic acid*, or it may have one or more double bonds, as in *oleic acid*; a few fatty acids contain triple bonds. Fatty acids differ from each other primarily in chain length and in the number and position of their unsaturated bonds. They are often symbolized by a shorthand notation that designates the length of the carbon chain and the number, position, and configuration of the double bonds. Thus palmitic acid (16 carbons, saturated) is symbolized 16:0 and oleic acid [18 carbons and one double bond (cis) at carbons 9 and 10] is symbolized $18:1^{\Delta 9}$. It is understood that the double bonds are cis (see below) unless indicated otherwise. Table 11-2 gives the structures and symbols of some important saturated and unsaturated fatty acids and a few with unusual structures.

Some generalizations can be made on the different fatty acids of higher plants and animals. The most abundant have an even number of carbon atoms with chains between 14 and 22 carbon atoms long, but those with 16 or 18 carbons predominate. The most common among the saturated fatty acids are palmitic acid (C_{16}) and stearic acid (C_{18}) and among the unsaturated fatty acids oleic acid (C_{18}). Unsaturated fatty acids predominate over the saturated ones, particularly in higher plants and in animals living at low temperatures. Unsaturated fatty acids have lower melting points than saturated fatty acids of the same chain length (Table 11-2). In most monounsaturated (*monoenoic*) fatty acids of higher organisms there is a double bond between carbon atoms 9 and 10. In most polyunsaturated (*polyenoic*) fatty acids one double bond is between carbon atoms 9 and 10; the additional double bonds usually occur between the 9,10 double bond and the methyl-terminal end of the chain. In most types of polyunsaturated fatty acids the double bonds

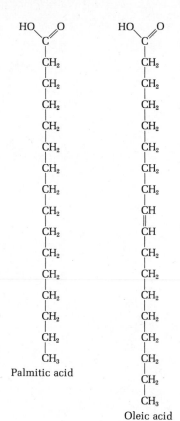

Figure 11-1
Two common fatty acids.

Palmitic acid

Oleic acid

Table 11-2   Some naturally occurring fatty acids

Symbol	Structure	Systematic name	Common name	m.p., °C
	Saturated fatty acids			
12:0	$CH_3(CH_2)_{10}COOH$	n-Dodecanoic	Lauric	44.2
14:0	$CH_3(CH_2)_{12}COOH$	n-Tetradecanoic	Myristic	53.9
16:0	$CH_3(CH_2)_{14}COOH$	n-Hexadecanoic	Palmitic	63.1
18:0	$CH_3(CH_2)_{16}COOH$	n-Octadecanoic	Stearic	69.6
20:0	$CH_3(CH_2)_{18}COOH$	n-Eicosanoic	Arachidic	76.5
24:0	$CH_3(CH_2)_{22}COOH$	n-Tetracosanoic	Lignoceric	86.0
	Unsaturated fatty acids			
$16:1^{\Delta 9}$	$CH_3(CH_2)_5CH{=}CH(CH_2)_7COOH$		Palmitoleic	-0.5
$18:1^{\Delta 9}$	$CH_3(CH_2)_7CH{=}CH(CH_2)_7COOH$		Oleic	13.4
$18:2^{\Delta 9,12}$	$CH_3(CH_2)_4CH{=}CHCH_2CH{=}CH(CH_2)_7COOH$		Linoleic	-5
$18:3^{\Delta 9,12,15}$	$CH_3CH_2CH{=}CHCH_2CH{=}CHCH_2CH{=}CH(CH_2)_7COOH$		Linolenic	-11
$20:4^{\Delta 5,8,11,14}$	$CH_3(CH_2)_4(CH{=}CHCH_2)_3CH{=}CH(CH_2)_3COOH$		Arachidonic	-49.5
	Some unusual fatty acids			
$16:1^{\Delta 9,trans}$	$CH_3(CH_2)_5CH{=}CH(CH_2)_7COOH$ (trans)		trans-Hexadecenoic	
$18:1^{\Delta 9,trans}$	$CH_3(CH_2)_7CH{=}CH(CH_2)_7COOH$ (trans)		Elaidic	
	$CH_3(CH_2)_5HC{-}{-}CH(CH_2)_9COOH$ $\diagdown\diagup$ CH_2		Lactobacillic	
	$CH_3(CH_2)_7CH(CH_2)_8COOH$ $\mid$ CH_3		Tuberculostearic	
	OH $\mid$ $CH_3(CH_2)_{21}CHCOOH$		Cerebronic	

are separated by one methylene group, for example, —CH=CH—CH₂—CH=CH—; only in a few types of plant fatty acids are the double bonds in conjugation, that is, —CH=CH—CH=CH—. The double bonds of nearly all kinds of naturally occurring unsaturated fatty acids are in the cis geometrical configuration; only a very few are trans.

Bacteria contain fewer and simpler types of fatty acids than higher organisms, namely, C_{12} to C_{18} saturated acids, some of which have a branched methyl group (see Table 11-2), and also C_{16} and C_{18} monounsaturated acids. Fatty acids with more than one double bond have not been found in bacteria.

Fatty acids with an odd number of carbon atoms occur only in trace amounts in terrestrial animals but occur in significant amounts in many marine organisms.

Essential Fatty Acids

When weanling or immature rats are placed on a fat-free diet, they grow poorly, develop a scaly skin, lose hair, and ultimately die with many pathological signs. When linoleic acid is present in the diet, these conditions do not develop. Linolenic acid and arachidonic acid (Table 11-2) also prevent these symptoms. Saturated and monounsaturated fatty acids are inactive. It has been concluded that mammals can synthesize saturated and monounsaturated fatty acids from other precursors but are unable to make linoleic and γ-linolenic acids. Fatty acids required in the diet of mammals are called *essential fatty acids*. The most abundant essential fatty acid in mammals is linoleic acid, which makes up from 10 to 20 percent of the total fatty acids of their triacylglycerols and phosphoglycerides. Linoleic and γ-linolenic acids cannot be synthesized by mammals but must be obtained from plant sources, in which they are very abundant. Linoleic acid is a necessary precursor in mammals for the biosynthesis of arachidonic acid, which is not found in plants.

Although the specific functions of essential fatty acids in mammals were a mystery for many years, one function has been discovered. Essential fatty acids are necessary precursors in the biosynthesis of a group of fatty acid derivatives called *prostaglandins* (page 300), hormonelike compounds which in trace amounts have profound effects on a number of important physiological activities.

Physical and Chemical Properties of Fatty Acids

Saturated and unsaturated fatty acids have quite different conformations. In saturated fatty acids, the hydrocarbon tails are flexible and can exist in a very large number of conformations because each single bond in the backbone has complete freedom of rotation. The fully extended form shown in Figure 11-2, the minimum-energy form, is the most probable

Figure 11-2
Space-filling models of a saturated, a mono-unsaturated, and diunsaturated fatty acid (anionic forms).

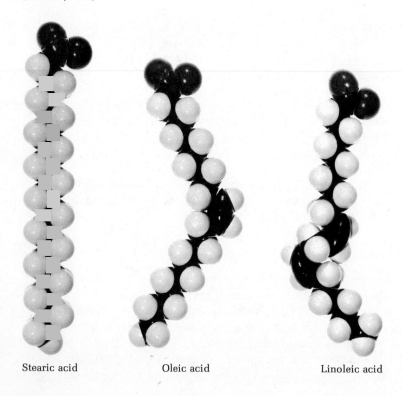

Stearic acid   Oleic acid   Linoleic acid

Figure 11-3
Geometry of double bonds in fatty acids.

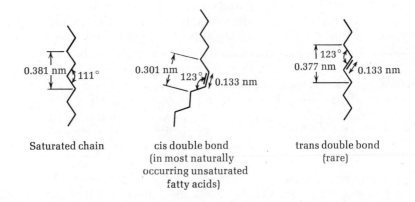

Saturated chain

cis double bond
(in most naturally
occurring unsaturated
fatty acids)

trans double bond
(rare)

conformation of saturated fatty acids. Unsaturated fatty acids, on the other hand, show one or more rigid kinks contributed by the nonrotating double bond(s). The cis configuration of the double bonds produces a bend of about 30° in the aliphatic chain (Figure 11-2), whereas the trans configuration more nearly resembles the extended form of saturated chains (Figure 11-3).

The cis forms of unsaturated fatty acids can be converted into trans forms by heating with certain catalysts. In this way oleic acid can be readily converted to its trans isomer *elaidic acid*, which has a much higher melting point (Table 11-2). Although elaidic acid is not a naturally occurring fatty acid, it is formed in appreciable amounts in the catalytic hydrogenation of liquid vegetable oils, a step in the manufacture of semisolid cooking fats and margarine. Elaidic acid has been found in the lipids of human tissues, presumably as a result of the consumption of such commercially hydrogenated products.

Unsaturated fatty acids undergo addition reactions at their double bonds. Quantitative titration with halogens, e.g., iodine or bromine, can yield information on the relative number of double bonds in a given sample of fatty acids or lipid.

Gas-Liquid Chromatography of Fatty Acids

Analysis of complex fatty acid mixtures obtained on hydrolysis of natural lipids was once an extremely difficult problem, but now precise and very sensitive analysis of fatty acid mixtures like that shown in Table 11-3 can be carried out by *gas-liquid chromatography*. In this procedure the fatty acids are first converted into a more volatile form, usually their methyl esters. An inert carrier gas such as nitrogen is used as the moving phase for partition chromatography of the vaporized mixture of methyl esters between the moving gas phase and a stationary liquid phase of a high-melting polyester or silicone polymer coated on particles of diatomaceous earth or on the inner surface of a long, heated capillary tube. The methyl esters of the various fatty acids partition themselves between the moving gas phase and the stationary liquid phase according to their individual gas-liquid partition coefficients. The separated methyl esters in

Table 11-3   Fatty acid composition (percent) of lipids of mouse liver

	Phospho-glycerides	Triacyl-glycerols
Saturated		
Myristic	0	0
Palmitic	28	24
Stearic	20	4
Unsaturated		
Palmitoleic	4	6
Oleic	17	43
Linoleic	12	20
Linolenic	1	1
Arachidonic	18	2

the gas phase leaving the column can be measured by a variety of extremely sensitive detectors. In one, the flame-ionization detector, the carrier gas stream containing the fatty acid esters is mixed with a stream of hydrogen and air and burned in a high-voltage electric field. The current generated by the flow of ionized fragments of the fatty acid in the flame is automatically recorded on a chart, which shows a series of separate peaks. Each peak corresponds to a separate fatty acid, and the area under the peak is proportional to the amount. Very complex mixtures of fatty acids can be sorted out in this fashion and quantitated; the amount of sample required for analysis is only a fraction of a milligram. Gas-liquid chromatography can also be used to analyze mixtures of sterols and hydrocarbons, as well as other compounds that are volatile at reasonable temperatures (up to 350°C) or can be converted chemically into volatile derivatives.

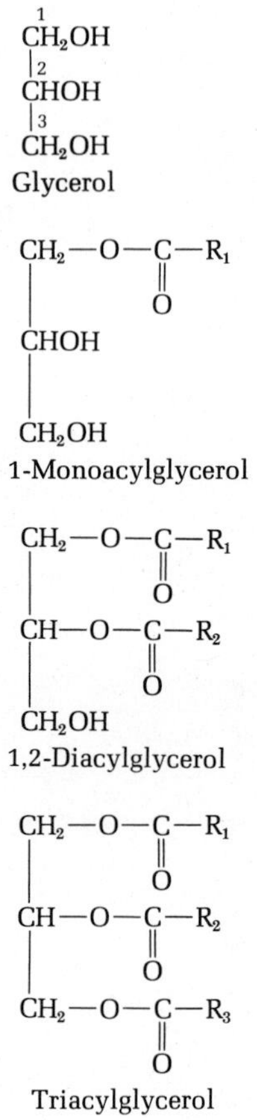

Triacylglycerols (Triglycerides)

Fatty acid esters of the alcohol glycerol (Figure 11-4) are called *acylglycerols* or *glycerides;* they are sometimes referred to as "neutral fats," a term that has become archaic. When all three hydroxyl groups of glycerol are esterified with fatty acids, the structure is called a *triacylglycerol* (Figures 11-4 and 11-5). (Although the name "triglyceride" has been traditionally used to designate these compounds, an international nomenclature commission has recommended that this chemically inaccurate term no longer be used.) Triacylglycerols are the most abundant family of lipids and the major components of depot or storage lipids in plant and animal cells. Triacylglycerols that are solid at room temperature are often referred to as "fats" and those which are liquid as "oils." *Diacylglycerols* (also called diglycerides) and *monoacylglycerols* (or monoglycerides) are also found in nature, but in much smaller amounts.

Triacylglycerols occur in many different types, according to the identity and position of the three fatty acid components esterified to glycerol. Those with a single kind of fatty acid in all three positions, called *simple* triacylglycerols, are named after the fatty acids they contain. Examples are *tristearoylglycerol, tripalmitoylglycerol,* and *trioleoylglycerol;* the trivial and more commonly used names are *tristearin, tripalmitin,* and *triolein,* respectively. *Mixed* triacylglycerols contain two or more different fatty acids. Triacylglycerols containing two different fatty acids A and B can exist in six different isomeric forms, BBA, AAB, ABA, ABB, BAA, and BAB, of which four (AAB, BAA, ABB, BBA) are stereoisomers (see below). The naming of mixed triacylglycerols can be illustrated by the example of 1-palmitoyldistearoylglycerol (trivial name, 1-palmitodistearin). Most natural fats are extremely complex mixtures of simple and mixed triacylglycerols.

Although there have been many attempts to discover the biological ground rules that determine the mode of distribution of different fatty acids in natural triacylglycerols, no simple, all-encompassing generalizations can yet be made.

Figure 11-4
Glycerol and mono-, di-, and triacylglycerols.

$$\overset{1}{C}H_2OH$$
$$\overset{2}{C}HOH$$
$$\overset{3}{C}H_2OH$$
Glycerol

$$CH_2-O-\underset{\underset{O}{\|}}{C}-R_1$$
$$CHOH$$
$$CH_2OH$$
1-Monoacylglycerol

$$CH_2-O-\underset{\underset{O}{\|}}{C}-R_1$$
$$CH-O-\underset{\underset{O}{\|}}{C}-R_2$$
$$CH_2OH$$
1,2-Diacylglycerol

$$CH_2-O-\underset{\underset{O}{\|}}{C}-R_1$$
$$CH-O-\underset{\underset{O}{\|}}{C}-R_2$$
$$CH_2-O-\underset{\underset{O}{\|}}{C}-R_3$$
Triacylglycerol

Figure 11-5
Space-filling model of 1-myristoyldipalmitoylglycerol, a mixed triacylglycerol.

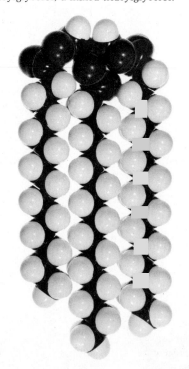

Table 11-4 Occurrence of simple
and mixed triacylglycerols in
depot fat of the rat (S = saturated;
U = unsaturated)

Type	Symbol	Mol %
Simple	SSS	0.3
	UUU	61.8
Mixed	SSU	4.1
	SUS	1.6
	SUU	19.5
	USU	12.8

Table 11-4 shows the distribution of various types of triacylglycerols in the depot fat of the rat. The fatty acid composition of depot fat in part reflects the composition of the
ingested lipids.

Properties of Triacylglycerols

The melting point of triacylglycerols is determined by their
fatty acid components. In general, the melting point increases with the number and length of the saturated fatty
acid components. For example, tripalmitin and tristearin are
solids at body temperature, whereas triolein and trilinolein
are liquids. All triacylglycerols are insoluble in water and do
not tend by themselves to form highly dispersed micelles.
However, diacylglycerols and monoacylglycerols have appreciable polarity because of their free hydroxyl groups and
thus can form micelles (pages 43 and 300). Diacyl- and
monoacylglycerols find wide use in the food industry in the
production of more homogeneous and more easily processed
foods; they are completely digestible and utilized biologically. Acylglycerols are soluble in ether, chloroform, benzene,
and hot ethanol. Their specific gravity is lower than that of
water.

Although glycerol itself is optically inactive, carbon
atom 2 becomes asymmetric whenever the fatty acid substituents on carbon atoms 1 and 3 are different. Naturally occurring triacylglycerols with an asymmetric carbon atom are
by convention named as if they were derived from L-
glyceraldehyde.

Acylglycerols undergo hydrolysis when boiled with acids
or bases or by the action of lipases, e.g., those present in
pancreatic juice (Figure 11-6). Hydrolysis with alkali (page
279), called *saponification*, yields a mixture of soaps and
glycerol.

Thin-Layer Chromatography

Triacylglycerols are separated and identified by the technique of thin-layer chromatography (Figure 11-7). A glass
plate about 10 by 10 cm is covered with an aqueous slurry of
an inert adsorbent material, such as silica gel or cellulose;
the slurry also contains a binder such as plaster of paris.
Sometimes silver nitrate is added, because Ag^+ forms weak
bonds with unsaturated molecules, causing unsaturated
acylglycerols to move more slowly than saturated ones. The

Figure 11-6
Hydrolysis of triacylglycerols.

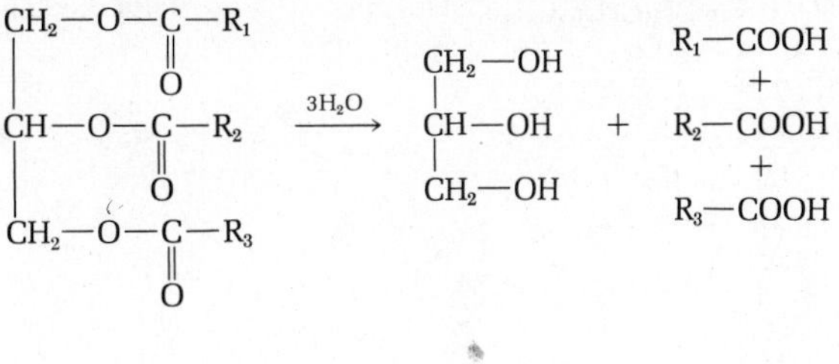

plate is air-dried and then baked to remove the remaining water, leaving a thin, uniform layer of firmly bound absorbent. The mixture to be analyzed is spotted at the bottom of the plate, and the lower edge of the plate is dipped into a pool of a suitable solvent in a closed chamber. The solvent rises by capillary action, as in paper chromatography, and the mixture of triacylglycerols is resolved into discrete spots. When the solvent front approaches the top, which takes only 20 to 30 min, the plate is dried and the positions of the separated components are located by spraying with a suitable indicator. The separated lipids can also be recovered by elution from patches of adsorbent scraped off the plate. This method can separate minute quantities of acylglycerols. Thin-layer chromatography is also useful for separating and identifying other types of lipids, as well as mixtures of amino acids, nucleotides, carbohydrates, and other cell components.

Alkyl Ether Acylglycerols

In addition to triacylglycerols, in which the three hydroxyl groups of glycerol are esterified with fatty acids, there is another family of closely related glycerol lipids, the *alkyl ether acylglycerols*, which are much less abundant than the triacylglycerols but occur widely. They contain fatty acids esterified to two of the hydroxyl groups of glycerol; the remaining hydroxyl group is joined in ether linkage with a long alkyl or alkenyl chain (Figure 11-8). These lipids are difficult to separate from the triacylglycerols; indeed, they escaped detection until the advent of refined chromatographic methods. Mild alkaline or enzymatic hydrolysis of the alkyl ether acylglycerols removes the fatty acyl groups to yield the *glyceryl ethers*, such as *chimyl* and *batyl* alcohols (Figure 11-8), hexadecyl and octadecyl ethers of glycerol, respectively.

Figure 11-7
Thin-layer chromatography of acylglycerols on silica gel impregnated with AgNO$_3$. A = synthetic mixture, B = lard, C = cocoa butter, D = cottonseed oil, E = peanut oil. The spots are (1) tristearin, (2) 2-oleodistearin, (3) 1-oleodistearin, (4) 1-stearodiolein, (5) 1-linoleodistearin, (6) triolein, (7) trilinolein, and (8) monostearin.

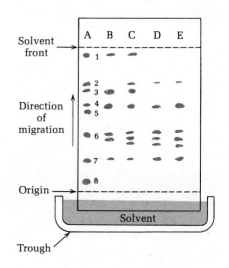

Figure 11-8
Alkyl ether acylglycerols and glyceryl ethers.

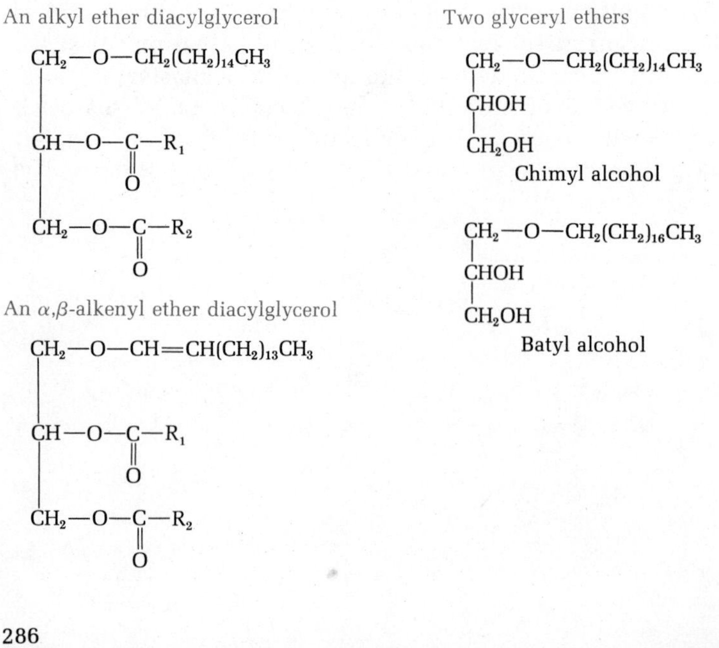

An alkyl ether diacylglycerol

CH$_2$—O—CH$_2$(CH$_2$)$_{14}$CH$_3$

CH—O—C—R$_1$
‖
O

CH$_2$—O—C—R$_2$
‖
O

An α,β-alkenyl ether diacylglycerol

CH$_2$—O—CH=CH(CH$_2$)$_{13}$CH$_3$

CH—O—C—R$_1$
‖
O

CH$_2$—O—C—R$_2$
‖
O

Two glyceryl ethers

CH$_2$—O—CH$_2$(CH$_2$)$_{14}$CH$_3$

CHOH

CH$_2$OH
Chimyl alcohol

CH$_2$—O—CH$_2$(CH$_2$)$_{16}$CH$_3$

CHOH

CH$_2$OH
Batyl alcohol

Figure 11-9
A glycosyldiacylglycerol.

Polar head

Nonpolar tails

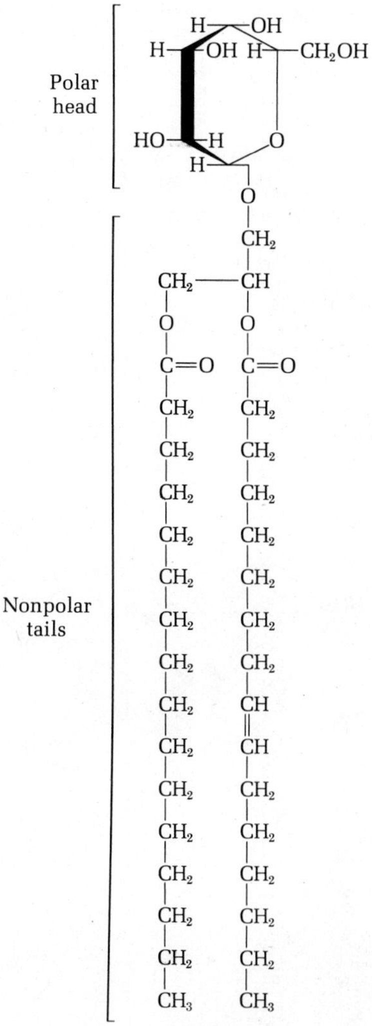

Monogalactosyl
diacylglycerol

Glycosylacylglycerols

Another family of acylglycerols includes the *glycosyldiacylglycerols*, which contain a sugar in glycosidic linkage with the unesterified 3-hydroxyl group of diacylglycerols. A common example is *galactosyldiacylglycerol* (Figure 11-9), found in higher plants and also in neural tissue of vertebrates. Similar glycolipids containing di- and trisaccharides are also known; a dimannosyldiacylglycerol has been isolated from bacteria.

Phosphoglycerides

The second large class of complex lipids consists of the *phosphoglycerides*, also called *glycerol phosphatides*. They are characteristic major components of cell membranes; only very small amounts of phosphoglycerides occur elsewhere in cells. Phosphoglycerides are also loosely referred to as *phospholipids* or *phosphatides*, but it should be noted that not all phosphorus-containing lipids are phosphoglycerides; e.g., sphingomyelin (page 292) is a phospholipid because it contains phosphorus, but it is better classified as a sphingolipid because of the nature of the backbone structure to which the fatty acid is attached.

In phosphoglycerides one of the primary hydroxyl groups of glycerol is esterified to phosphoric acid; the other hydroxyl groups are esterified to fatty acids. The parent compound of the series is thus the phosphoric ester of glycerol. This compound has an asymmetric carbon atom and can be designated as either D-glycerol 1-phosphate or L-glycerol 3-phosphate. Because of this ambiguity, the stereochemistry of glycerol derivatives is based on the *stereospecific numbering* (sn) of the carbon atoms, as shown in Figure 11-10. The isomer of glycerol phosphate found in natural phosphoglycerides is called sn-glycerol 3-phosphate; it belongs to the L-stereochemical series. In addition to the two fatty acid residues esterified to the hydroxyl groups at carbon atoms 1 and 2, phosphoglycerides contain a polar head group, namely, an alcohol designated X—OH, whose hydroxyl group is esterified to the phosphoric acid.

Because phosphoglycerides possess a polar head in addition to their nonpolar hydrocarbon tails (Figure 11-11), they are called *amphipathic* (page 43) or *polar* lipids. The different types of phosphoglycerides differ in the size, shape,

Figure 11-10
Stereochemical configuration of phosphoglycerides.

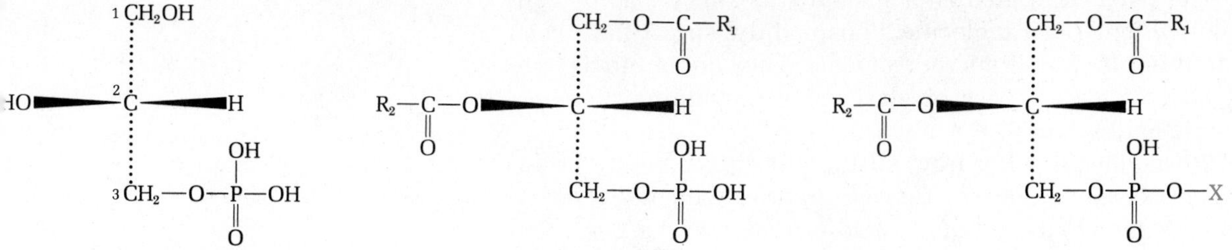

L-Glycerol 3-phosphoric acid
(sn-glycerol 3-phosphoric acid)

L-Phosphatidic acid
(3-sn-phosphatidic acid)

General structure of phosphoglycerides.
The moiety X is contributed by an alcohol.

and electric charge of their polar head groups (Table 11-5 and Figures 11-11 and 11-12). Each type of phosphoglyceride can exist in many different chemical species differing in their fatty acid substituents. Usually there is one saturated and one unsaturated fatty acid, the latter in the 2 position of glycerol.

The parent compound of the phosphoglycerides is *phosphatidic acid* (Figure 11-10), which contains no polar alcohol head group. It occurs in only very small amounts in cells, but it is an important intermediate in the biosynthesis of the phosphoglycerides. The most abundant phosphoglycerides in higher plants and animals are *phosphatidylethanolamine* and *phosphatidylcholine* (Table 11-5 and Figure 11-12), which contain as head groups the amino alcohols *ethanolamine* and *choline*, respectively. (The new names recommended for these phosphoglycerides are *ethanolamine phosphoglyceride* and *choline phosphoglyceride*, but they have not yet gained wide use. The old trivial names are cephalin and lecithin, respectively.) These two phosphoglycerides are major components of most animal cell membranes.

In *phosphatidylserine*, the hydroxyl group of the amino acid L-serine is esterified to the phosphoric acid. In *phosphatidylinositol*, the head group is the six-carbon cyclic sugar alcohol inositol. In *phosphatidylglycerol*, the head group is a molecule of glycerol. Phosphatidylglycerol is often found in bacterial membranes as an amino acid derivative, particularly of L-lysine, which is esterified at the 3′ position of the glycerol head group. This type of amino acid–containing lipid is called a *lipoamino acid* or, more accurately, an *O-aminoacylphosphatidylglycerol* (see Table 11-5).

Closely related to phosphatidylglycerol is the more complex lipid *cardiolipin*, also called *diphosphatidylglycerol*, which consists of a molecule of phosphatidylglycerol in which the 3′-hydroxyl group of the second glycerol moiety is esterified to the phosphate group of a molecule of phosphatidic acid (Figure 11-12 and Table 11-5). The backbone of cardiolipin thus consists of three molecules of glycerol joined by two phosphodiester bridges; the two hydroxyl groups of both external glycerol molecules are esterified with fatty acids. Phosphatidylglycerol, O-aminoacylphosphatidylglycerol, and cardiolipin are therefore structurally related. They are characteristically abundant in the cell membranes of bacteria. Cardiolipin is also present in large amounts in the inner membrane of mitochondria; it was first isolated from heart muscle, in which mitochondria are abundant.

The polar head groups of phosphatides may also be contributed by a sugar molecule. Phosphatidyl sugars have been found in plants and microorganisms. They are not to be confused with other types of glycolipids containing no phosphoric acid.

Plasmalogens differ from all the other phosphoglycerides described above. One of the two hydrocarbon tails is contributed by a long-chain fatty acid esterified to the 2 position of the glycerol, but the other is a long aliphatic chain in cis

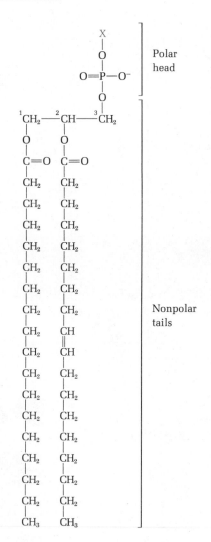

Figure 11-11
General structure of phosphoglycerides in a form emphasizing their amphipathic nature. Usually the fatty acid in the 2 position is unsaturated.

Table 11-5   Polar head groups of the phosphoglycerides

The head alcohols are shown in color. The open bonds on the phosphoric residues are to position 3 of 1,2-diacylglycerol.

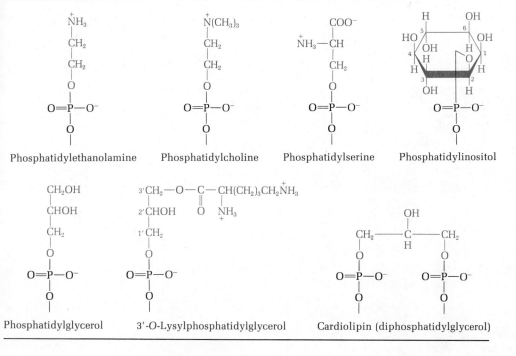

Phosphatidylethanolamine   Phosphatidylcholine   Phosphatidylserine   Phosphatidylinositol

Phosphatidylglycerol   3'-O-Lysylphosphatidylglycerol   Cardiolipin (diphosphatidylglycerol)

Figure 11-12
Space-filling models of three phospho-glycerides, showing location of the charged groups.

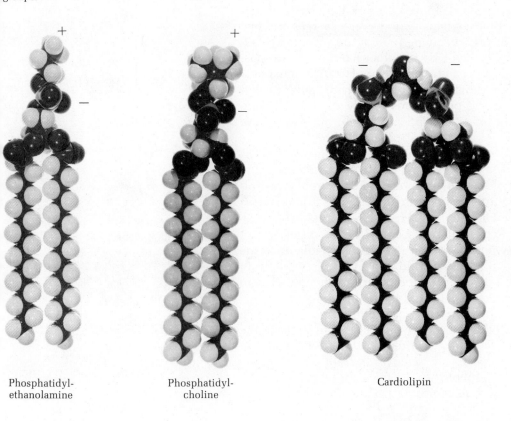

Phosphatidyl-
ethanolamine

Phosphatidyl-
choline

Cardiolipin

α,β-unsaturated ether linkage at the 1 position (Figure 11-13). Thus plasmalogens are phosphoglyceride analogs of the alkyl ether acylglycerols described above (page 286); ethanolamine is the most common polar head group. They are especially abundant in the membranes of muscle and nerve cells.

Properties of Phosphoglycerides

Pure phosphoglycerides are white waxy solids. On exposure to air they darken and undergo complex chemical changes because of the tendency of their polyunsaturated fatty acid components to be peroxidized by atmospheric oxygen, which results in polymerization. Phosphoglycerides are soluble in most nonpolar solvents containing some water and are best extracted from cells and tissues with chloroform-methanol mixtures. They are not readily soluble in anhydrous acetone. When phosphoglycerides are placed in water, they appear to dissolve, but only very minute amounts go into true solution; most of the "dissolved" lipid is in the form of micelles (see below).

All phosphoglycerides have a negative charge at the phosphate group at pH 7; the pK' of this group is in the range of 1 to 2. The head groups of phosphatidylinositol, phosphatidylglycerol, and the phosphatidyl sugars have no electric charge, but they are quite polar because of their high content of hydroxyl groups. The head groups of phosphatidylethanolamine and phosphatidylcholine have a positive charge at pH 7; thus at this pH these two phosphoglycerides are dipolar zwitterions with no net electric charge. The head group of phosphatidylserine contains an α-amino group (pK' = 10) and a carboxyl group (pK' = 3); the phosphatidylserine molecule thus contains two negative charges and one positive charge at pH 7.0, giving it a net negative charge. O-Lysylphosphatidylglycerol, on the other hand, with two positive charges and one negative charge at pH 7.0 has a net positive charge. These variations in the size, shape, polarity, and electric charge of the polar heads (Table 11-5 and Figure 11-12) presumably play a significant role in the structure of various types of cell membranes (page 302).

Mild alkaline hydrolysis of phosphoglycerides yields the fatty acids as soaps but leaves the glycerol–phosphoric acid–alcohol portion of the molecule intact. For example, hydrolysis of phosphatidylcholine under these conditions yields glycerol 3-phosphorylcholine. Hydrolysis of phosphoglycerides with strong alkali causes hydrolytic cleavage not only of the fatty acids but also the head alcohol; since the linkage between phosphoric acid and glycerol is relatively stable to alkaline hydrolysis, the other product is glycerol phosphate, which can be cleaved by acid hydrolysis.

Phosphoglycerides can also be hydrolyzed by specific phospholipases, which have become important tools in the determination of phosphoglyceride structure (Figure 11-14). Phospholipase A_1 specifically removes the fatty acid from the 1 position and phospholipase A_2 from the 2 position. Removal of one fatty acid molecule from a phosphoglyceride

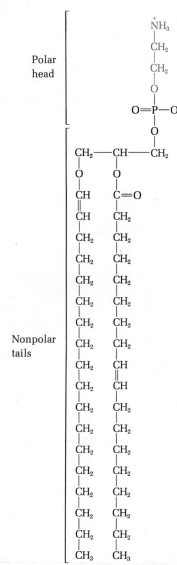

Figure 11-13
A plasmalogen

Figure 11-14
Sites of action of phospholipases on phosphatidylcholine.

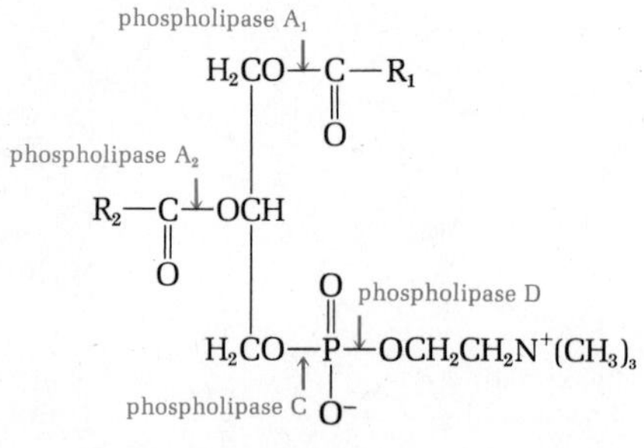

Figure 11-15
Sphingosine bases and ceramide. In the structure of the latter the sphingosine moiety is in color.

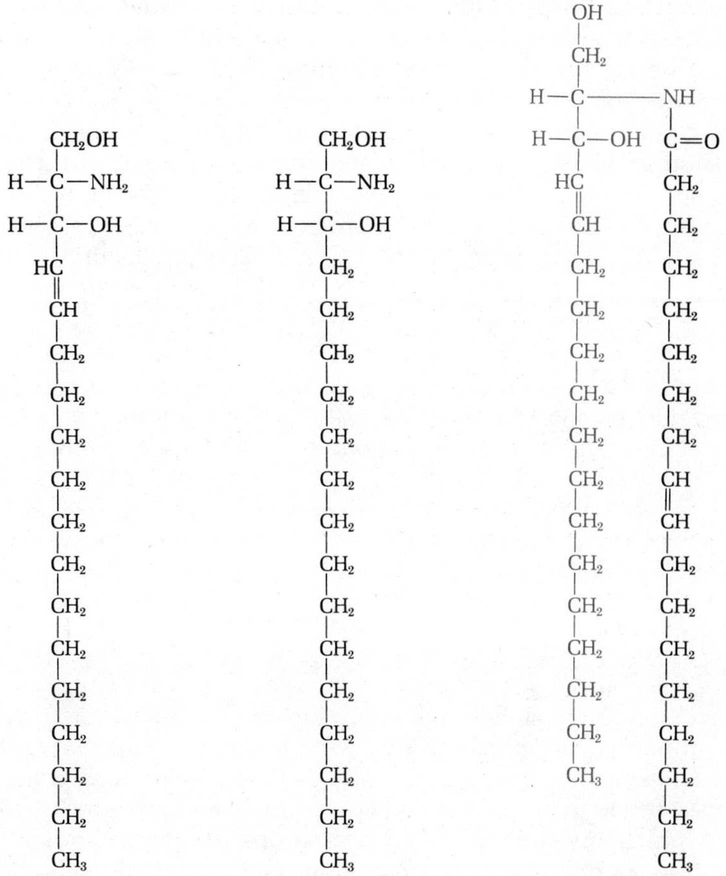

Sphingosine (4-sphingenine)   Dihydrosphingosine (sphinganine)   Ceramide

yields a *lysophosphoglyceride*, e.g., *lysophosphatidylethanolamine*. Lysophosphoglycerides are intermediates in phosphoglyceride metabolism but are found in cells or tissues in only very small amounts; in high concentrations they are toxic and injurious to membranes. Phospholipase B, a mixture of phospholipases A_1 and A_2, can bring about successive removal of the two fatty acids of phosphoglycerides. Phospholipase C hydrolyzes the bond between phosphoric acid and glycerol, while phospholipase D removes the polar head group to leave a phosphatidic acid.

Phosphoglycerides, like other complex lipids described below, are readily separated and identified by thin-layer chromatography or by chromatography on silicic acid columns.

Sphingolipids

Sphingolipids, complex lipids containing as their backbone sphingosine or a related base (Figure 11-15), are important membrane components in both plant and animal cells. They are present in especially large amounts in brain and nerve tissue. Only trace amounts of sphingolipids are found in depot fats. All sphingolipids contain three characteristic building-block components: one molecule of a fatty acid, one molecule of sphingosine or one of its derivatives, and a polar head group, which in some sphingolipids is very large and complex.

Sphingosine (Figure 11-15) is one of 30 or more different long-chain amino alcohols found in sphingolipids of various species. In mammals *sphingosine* (4-sphingenine) and *dihydrosphingosine* (sphinganine) are the major bases of sphingolipids, in higher plants and yeast *phytosphingosine* (4-hydroxysphinganine) is the major base, and in marine invertebrates doubly unsaturated bases such as *4,8-sphingadiene* are common. The sphingosine base is connected at its amino group by an amide linkage to a long saturated or monounsaturated fatty acid of 18 to 26 carbon atoms. The resulting compound, which has two nonpolar tails and is called a *ceramide* (Figure 11-15), is the characteristic parent structure of all sphingolipids. Different polar head groups are attached to the hydroxyl group at the 1 position of the sphingosine base.

Sphingomyelins

The most abundant sphingolipids in the tissues of higher animals are *sphingomyelins*, which contain phosphorylethanolamine or phosphorylcholine as their polar head groups, esterified to the 1-hydroxyl group of ceramide (Figure 11-16). Sphingomyelins have physical properties very similar to those of phosphatidylethanolamine and phosphatidylcholine; they are zwitterions at pH 7.0.

Neutral Glycosphingolipids

A second class of sphingolipids contains one or more neutral sugar residues as their polar head groups and thus has no electric charge; they are called neutral glycosphingolipids. The simplest of these are the *cerebrosides*, which contain as their polar head group a monosaccharide bound in β glycosidic linkage to the hydroxyl group of ceramide (Figure 11-17). The cerebrosides of the brain and nervous system contain D-galactose and are therefore called *galactocerebrosides*. Cerebrosides are also present in much smaller amounts in nonneural tissues of animals, where, because they usually contain D-glucose instead of D-galactose, they are called *glucocerebrosides*. Sulfate esters of galactocerebrosides (at the 3 position of the D-galactose) are also present in brain tissue; they are called *sulfatides*. Cerebrosides and sulfatides usually contain fatty acids with 22 to 26 carbon atoms. A common fatty acid component of cerebrosides is *cerebronic acid* (Table 11-2), which has a D-hydroxyl group at carbon atom 2. When the fatty acid is cleaved from a cerebroside by alkaline hydrolysis, the remaining glycosylsphingosine compound is called a *psychosine*.

Neutral glycosphingolipids with disaccharides as their polar head groups are called *dihexosides*. Also known are *trihexosides* and *tetrahexosides* (Table 11-6), containing tri-

Figure 11-16
Structure of a representative sphingomyelin.

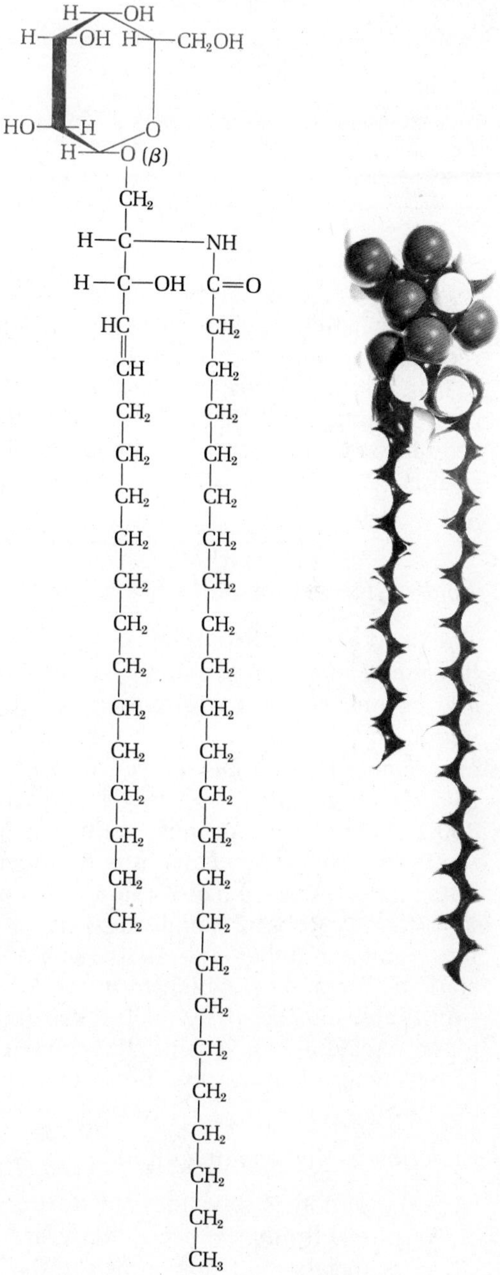

Figure 11-17
A galactocerebroside containing lignoceric acid (C_{24}) as its fatty acid component.

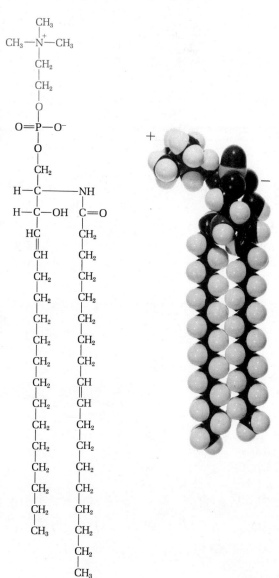

saccharide and tetrasaccharide head groups, respectively. The monosaccharide units found in these glycosphingolipids include D-glucose, D-galactose, N-acetyl-D-glucosamine, and N-acetyl-D-galactosamine. The neutral glycosphingolipids are important cell-surface components in animal tissues. Their nonpolar tails presumably penetrate into the lipid bilayer structure of cell membranes (see page 305), whereas the polar heads protrude outward from the surface. Some of the neutral glycosphingolipids are found on the surface of red blood cells and give them blood-group specificity; they

Table 11-6 Major classes of neutral glycosphingolipids. The symbols are Glc, D-glucose; Gal, D-galactose; GalNAc, N-acetyl-D-galactosamine.

Glucosylceramides

Monohexoside (glucocerebroside)	Glc 1$\xrightarrow{\beta}$ceramide
Dihexoside	Gal 1$\xrightarrow{\beta}$4 Glc 1$\xrightarrow{\beta}$ceramide
Trihexoside	Gal 1$\xrightarrow{\alpha}$4 Gal 1$\xrightarrow{\beta}$4 Glc 1$\xrightarrow{\beta}$ceramide
Tetrahexoside	GalNAc 1$\xrightarrow{\beta}$3 Gal 1$\xrightarrow{\alpha}$4 Gal 1$\xrightarrow{\beta}$4 Glc 1$\xrightarrow{\beta}$ceramide

Galactosylceramides

Galactocerebroside	Gal 1$\xrightarrow{\beta}$ceramide
Dihexoside	Gal 1$\xrightarrow{\beta}$4 Gal 1$\xrightarrow{\beta}$ceramide

are responsible, in part, for the need to match donor and recipient blood for compatibility.

The neutral glycosphingolipids are classified on the basis of the identity of the sugar attached to the ceramide unit, the sequence of the sugars, and the length of the oligosaccharide chains. Examples of some glycosphingolipids are shown in Table 11-6.

Acidic Glycosphingolipids (Gangliosides)

The third and most complex group of glycosphingolipids are the gangliosides; they contain in their oligosaccharide head groups one or more residues of a sialic acid (page 650), which gives the polar head of the gangliosides a net negative charge at pH 7.0 (Table 11-7). The sialic acid usually found in human gangliosides is N-acetylneuraminic acid (page 260). Gangliosides are most abundant in the gray matter of the brain, where they constitute 6 percent of the total lipids, but small amounts are also found in nonneural tissues. Over 20 different types of gangliosides have been identified, differing in the number and relative positions of the hexose and sialic acid residues, which form the basis of their classification (Table 11-7). Nearly all the known gangliosides have a glucose residue in glycosidic linkage with ceramide; residues of D-galactose and N-acetyl-D-galactosamine are also present.

Function of Glycosphingolipids

Much attention is now focused on the biochemistry of the glycosphingolipids. Although they are only minor constituents of membranes, they appear to be extremely important in a number of specialized functions. Because gangliosides are especially abundant in nerve endings, it has been suggested that they function in the transmission of nerve impulses across synapses. They are also believed to be present at receptor sites for acetylcholine and other neurotransmitter substances. Some of the cell-surface glycosphingolipids are concerned not only in blood-group specificity but also in organ and tissue specificity. These complex lipids are also involved in tissue immunity and in cell-cell recognition sites fundamental to the development and structure of tissues. Cancer cells, for example, have characteristic glycosphingolipids different from those in normal cells. Ganglioside G_{M2}

Table 11-7 Structures of some gangliosides†

	Symbol
NANA 2→3 Gal 1$\xrightarrow{\beta}$4 Glc 1$\xrightarrow{\beta}$ceramide	G_{M3}

GalNAc 1$\xrightarrow{\beta}$4 Gal 1$\xrightarrow{\beta}$4 Glc 1$\xrightarrow{\beta}$ceramide G_{M2}
3
↑
2 NANA

3 GalNAc 1$\xrightarrow{\beta}$4 Gal 1$\xrightarrow{\beta}$4 Glc 1$\xrightarrow{\beta}$ceramide G_{M1}
↑β   3
1 Gal   ↑
    2 NANA

3 GalNAc 1→4 Gal 1$\xrightarrow{\beta}$4 Glc 1→ceramide G_{D1}
↑β   3
1 Gl   ↑
3   2 NANA
↑
NANA 2

3 GalNAc 1$\xrightarrow{\beta}$4 Gal 1$\xrightarrow{\beta}$4 Glc 1$\xrightarrow{\beta}$ceramide G_{T1}
↑β   3
1 Gal   ↑
3   2 NANA   8
↑   ↑
NANA 2   2 NANA

† Glc = D-glucose, Gal = D-galactose, GalNAc = N-acetyl-D-galactosamine, NANA = N-acetylneuramic acid (sialic acid). In this nomenclature of gangliosides, devised by L. Svennerholm, the subscript letters indicate the number of sialic acid groups (M = monosialo, D = disialo, and T = trisialo). The numeral in the subscript is $5 - n$, where n is the number of neutral sugar residues. The sialic acid residues are in color.

accumulates in the brain in Tay-Sachs disease, due to genetic lack of the enzyme required for its degradation (page 678). Several other genetic deficiency diseases result in abnormal accumulation of different glycosphingolipids (page 678).

Waxes

Waxes are water-insoluble, solid esters of higher fatty acids with long-chain monohydroxylic fatty alcohols or with sterols (see below). They are soft and pliable when warm but hard when cold. Waxes are found as protective coatings on skin, fur, and feathers, on leaves and fruits of higher plants, and on the exoskeleton of many insects. The major components of beeswax are palmitic acid esters of long-chain fatty alcohols with 26 to 34 carbon atoms. Lanolin, or wool fat, is a mixture of fatty acid esters of the sterols lanosterol and agnosterol (see below).

Simple (Nonsaponifiable) Lipids

The lipids discussed up to this point contain fatty acids as building blocks, which can be released on alkaline hydrolysis. The simple lipids contain no fatty acids. They occur in smaller amounts in cells and tissues than the complex lipids, but they include many substances having profound biological activity — vitamins, hormones, and other highly specialized fat-soluble biomolecules.

There are two major classes of nonsaponifiable lipids, the *terpenes* and the *steroids*. Although it is convenient to consider them as two distinct classes, they are closely related structurally, since both ultimately derive from five-carbon building blocks.

Terpenes

Terpenes are constructed of multiples of the five-carbon hydrocarbon *isoprene* (2-methyl-1,3-butadiene) (Figure 11-18). Terpenes containing two isoprene units are called *monoterpenes*, those containing three isoprene units are called *sesquiterpenes*, and those containing four, six, and eight units are called *diterpenes*, *triterpenes*, and *tetraterpenes*, respectively. Terpenes may be either linear or cyclic molecules; some terpenes contain structures of both types. The successive isoprene units of terpenes are usually linked in a head-to-tail arrangement, particularly in the linear segments, but sometimes the isoprene units are in tail-to-tail arrangement. The double bonds in the linear segments of most terpenes are in the stable trans configuration, but in some, particularly vitamin A and its precursor β-carotene (below), one or more of the double bonds are cis.

Of the very large number of terpenes identified in plants, many have characteristic odors or flavors and are major components of essential oils derived from such plants. Thus the monoterpenes *geraniol*, *limonene*, *menthol*, *pinene*, *camphor*, and *carvone* are major components of oil of geranium, lemon oil, mint oil, turpentine, camphor oil, and caraway oil, respectively. *Farnesol* is an example of a sesquiterpene. The diterpenes include *phytol*, a linear terpenoid alcohol, which is a component of the photosynthetic pigment chlorophyll (page 595). The triterpenes include *squalene*, an important precursor in the biosynthesis of cholesterol. Other higher terpenes include the *carotenoids*, a class of tetraterpene hydrocarbons and their oxygen-containing derivatives in which the head-to-tail arrangement of the isoprene units is characteristically reversed at the center of the molecule (Figure 11-19). An important carotenoid is *β-carotene*, the hydrocarbon precursor of vitamin A. Natural rubber and gutta-percha are *polyterpenes*; they consist of long hydrocarbon chains containing hundreds of isoprene units in regular linear order.

Among the most important terpenes are three members of the group of *fat-soluble vitamins*, namely, vitamins A, E, and K. Although these substances, which are required in trace amounts in the diet of mammals, may be classified among the lipids, their biological functions are so distinctive that their structure and function will be considered separately in Chapter 13, page 351.

Another important class of terpenes is represented by the *polyprenols*, long-chain linear polyisoprenoid compounds with a terminal primary alcohol group. The most important of these is *undecaprenyl alcohol*, also called *bactoprenol*, which contains 11 isoprene units and thus has 55 carbon atoms (Figure 11-19). *Dolichol* is the corresponding analog in

Figure 11-18
Isoprene units in the structure of some simple terpenes.

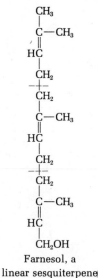

Head-to-tail or regular arrangement of isoprene units

Tail-to-tail or irregular arrangement of isoprene units

Geraniol, a linear monoterpene

Farnesol, a linear sesquiterpene

Limonene, a cyclic monoterpene

Figure 11-19
Some higher terpenes. Terpene structures
are often shown in shorthand notation.

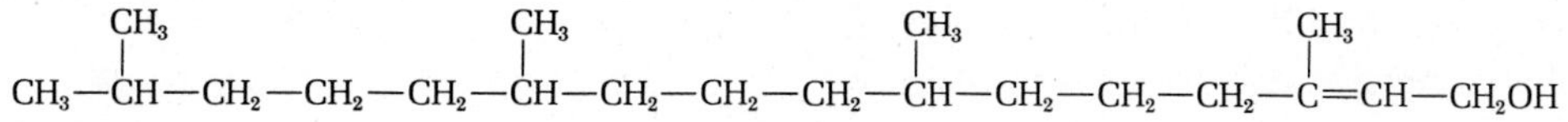

$$CH_3-\overset{\overset{\displaystyle CH_3}{|}}{CH}-CH_2-CH_2-CH_2-\overset{\overset{\displaystyle CH_3}{|}}{CH}-CH_2-CH_2-CH_2-\overset{\overset{\displaystyle CH_3}{|}}{CH}-CH_2-CH_2-CH_2-\overset{\overset{\displaystyle CH_3}{|}}{C}=CH-CH_2OH$$

Phytol

$$_3-\overset{\overset{\displaystyle CH_3}{|}}{C}=CH-CH_2-CH_2-\overset{\overset{\displaystyle CH_3}{|}}{C}=CH-CH_2-CH_2-\overset{\overset{\displaystyle CH_3}{|}}{C}=CH-CH_2-CH_2-CH=\overset{\underset{\underset{\displaystyle CH_3}{|}}{}}{C}-CH_2-CH_2-CH=\overset{\underset{\underset{\displaystyle CH_3}{|}}{}}{C}-CH_2-CH_2-CH=\overset{\underset{\underset{\displaystyle CH_3}{|}}{}}{C}-CH_3$$

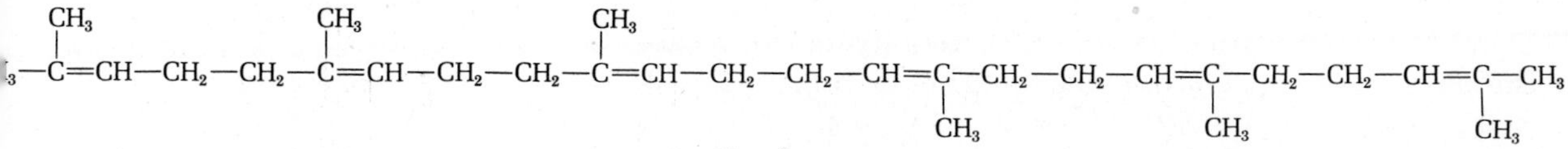

Squalene

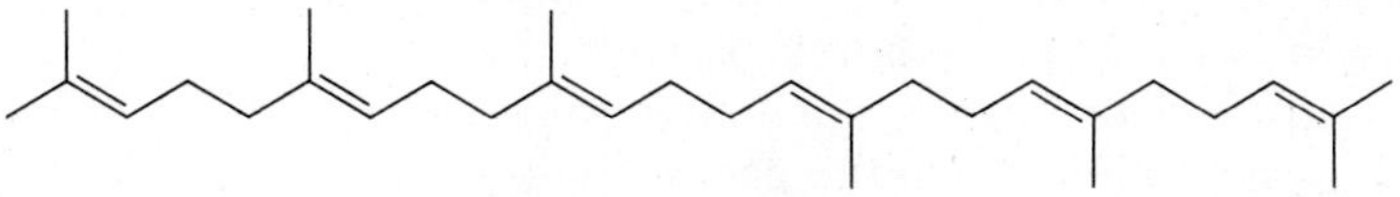

Squalene (shorthand form)

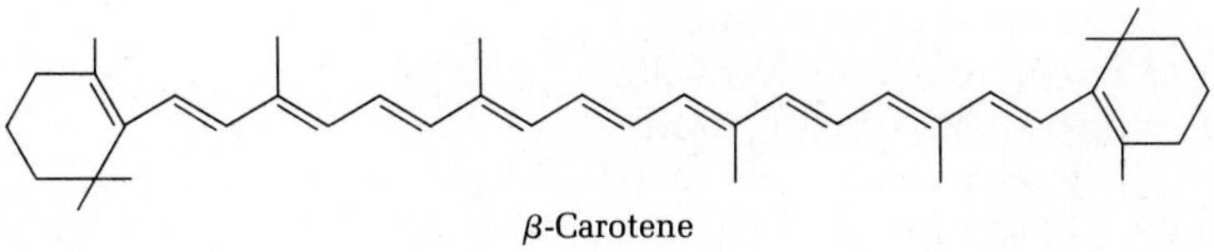

β-Carotene

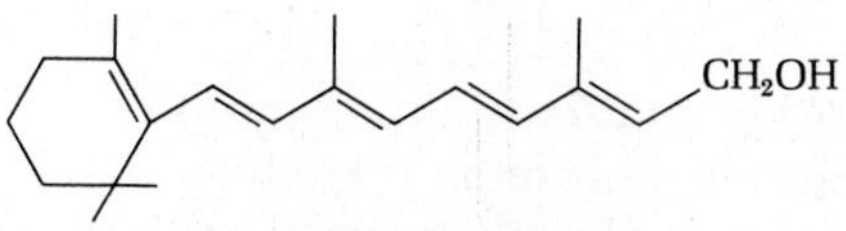

Vitamin A$_1$ (retinol)

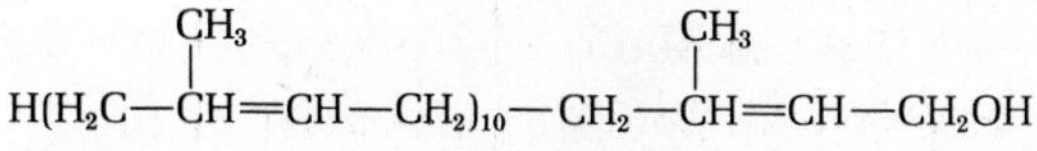

$$H(H_2C-\overset{\overset{\displaystyle CH_3}{|}}{CH}=CH-CH_2)_{10}-CH_2-\overset{\overset{\displaystyle CH_3}{|}}{CH}=CH-CH_2OH$$

Bactoprenol

animal tissues; it contains 19 isoprene units (95 carbon atoms). These polyprenols, in the form of their phosphate esters, *undecaprenyl phosphate* and *dolichyl phosphate,* respectively, have a coenzymelike function in the enzymatic transfer of sugar groups from the cytoplasm to the outer surface of the cell during the synthesis of cell-surface and cell-wall lipopolysaccharides, peptidoglycans, teichoic acids, and glycoproteins (pages 268, 273, and 649). In this process the long, nonpolar hydrocarbon chain of the polyprenols is believed to be anchored within the nonpolar core of the membrane (page 302), whereas the polar end of the molecule serves as an arm for the transfer of the covalently bound sugar groups across the membrane.

Still another class of terpenoid compounds functioning as coenzymes is the *ubiquinone* or *coenzyme Q* family of compounds, which function as hydrogen carriers for biological oxidations in the mitochondria (page 493). They contain a substituted quinone ring, which can be reduced and then reoxidized, and a long isoprenoid side chain, whose length differs with the organism. Analogous compounds, called *plastoquinones*, are found in chloroplasts, where they function in photosynthesis (page 607).

Steroids

Steroids are derivatives of the saturated tetracylic hydrocarbon *perhydrocyclopentanophenanthrene* (Figure 11-20). A great many different steroids, each with a distinctive function or activity, have been isolated from natural sources. Steroids differ in the number and position of double bonds, in the type, location, and number of substituent functional groups, in the configuration (α or β) of the bonds between the substituent groups and the nucleus, and in the configuration of the rings in relation to each other, since the parent hydrocarbon has six centers of asymmetry. The main points of substitution are carbon 3 of ring A, carbon 11 of ring C, and carbon 17 of ring D. All steroids originate from the linear triterpene *squalene* (Figure 11-19), which cyclizes readily (page 683). The first important steroid product of this cyclization is *lanosterol*, which in animal tissues is the precursor of *cholesterol*, the most abundant steroid in animal tissues. Cholesterol and lanosterol are members of a large subgroup of steroids called the *sterols*. They are steroid alcohols containing a hydroxyl group at carbon 3 of ring A and a branched aliphatic chain of eight or more carbon atoms at carbon 17. They occur either as free alcohols or as long-chain fatty acid esters of the hydroxyl group at carbon 3; all are solids at room temperature. *Cholesterol* melts at 150°C and is insoluble in water but readily extracted from tissues with chloroform, ether, benzene, or hot alcohol. Cholesterol occurs in the plasma membranes of many animal cells and in the lipoproteins of blood plasma. *Lanosterol* (Figure 11-20) was first found in the waxy coating of wool in esterified form before it was established as an important intermediate in the biosynthesis of cholesterol in animal tissues (page 683). Cholesterol occurs only rarely in higher plants, which contain other types of sterols known collectively as *phytosterols*. Among these are *stigmasterol* and *sitosterol*. Fungi and yeasts contain still other types of sterols, the *mycosterols*. Among these is *ergosterol*, which is converted to vitamin D on irradiation by sunlight. Sterols are not present in bacteria.

Cholesterol is the precursor of many other steroids in animal tissues, including the *bile acids*, detergentlike compounds that aid in emulsification and absorption of lipids in the intestine; the *androgens*, or male sex hormones; the *estrogens*, or female sex hormones; the progestational hormone *progesterone*; and the *adrenocortical hormones* (Figure 11-21). The biological activity of some of the steroid hormones will be discussed in Chapter 29, pages 823 and 824. Among the most important steroids are a group of compounds having vitamin D activity; their structure and function will be discussed in Chapter 13 (page 355).

Figure 11-20
The structures of the steroid nucleus, cholesterol, and lanosterol. The ring designations and numbering of the carbon atoms of steroids are also shown.

Perhydrocyclopentanophenanthrene nucleus

Cholesterol

Space-filling model of cholesterol

Lanosterol

Figure 11-21
Some important steroids. The two bile acids
usually occur as amides of glycine and
taurine.

Two bile acids

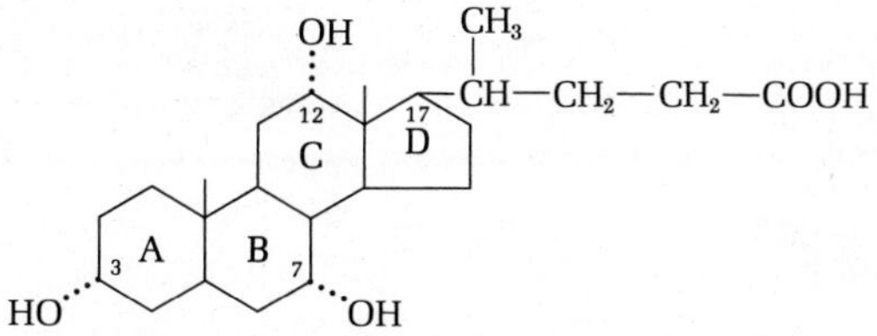

Cholic acid

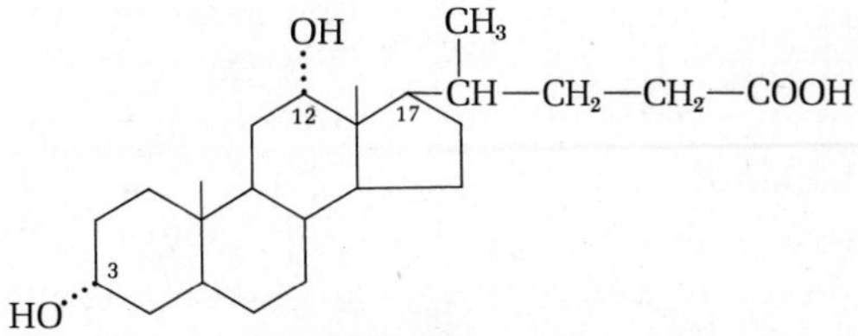

Deoxycholic acid

Two adrenocortical steroids

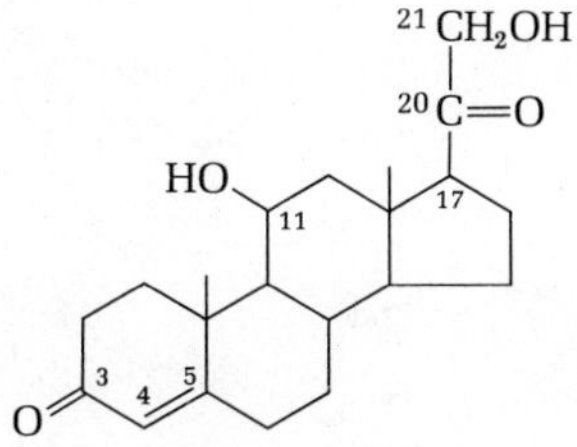

Corticosterone

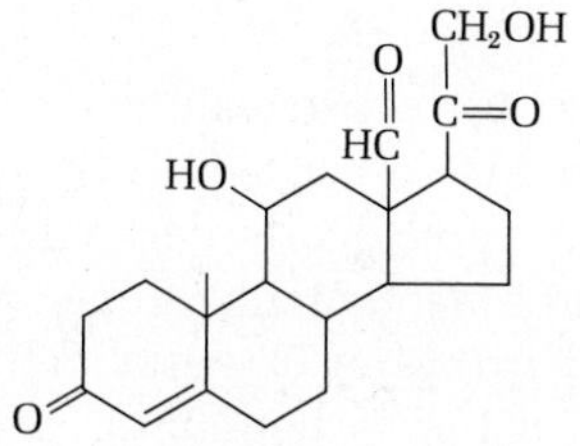

Aldosterone

Two estrogens (female sex hormones)

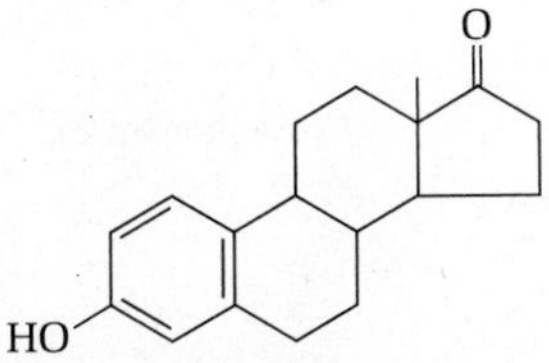

Estrone

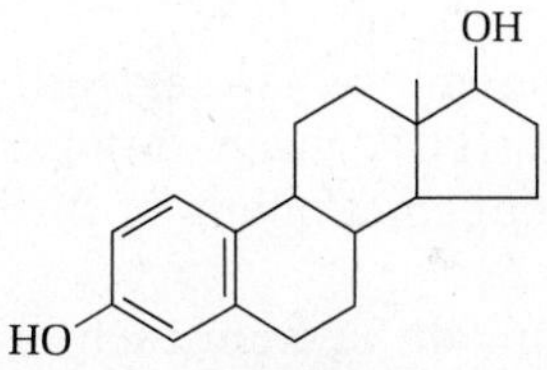

β-Estradiol

An androgen
(male sex hormone)

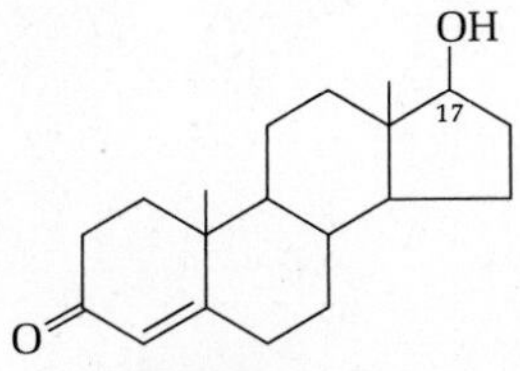

Testosterone

A progestational hormone

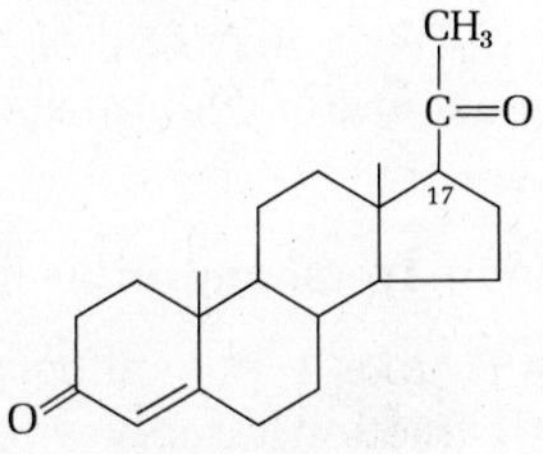

Progesterone

An insect molting hormone

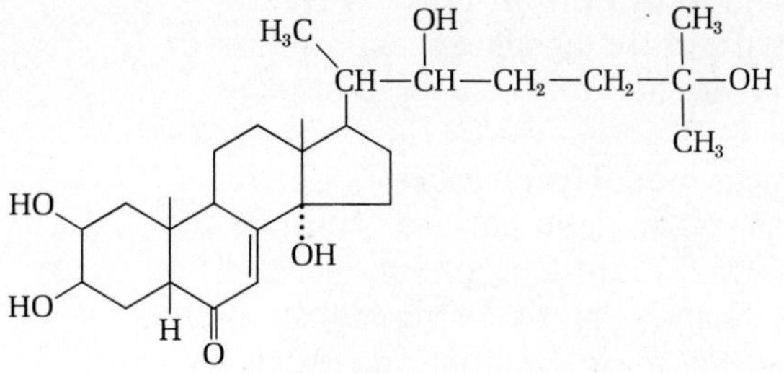

Ecdysone

Prostaglandins

Prostaglandins are a family of fatty acid derivatives which have a variety of potent biological activities of a hormonal or regulatory nature. The name prostaglandin was first given in the 1930s by the Swedish physiologist U. S. von Euler to a lipid-soluble acidic substance found in the seminal plasma, the prostate gland, and the seminal vesicles. In very small amounts this material was found to lower blood pressure and to stimulate certain smooth muscles to contract. At first prostaglandin was thought to be a single substance, characteristically secreted by the male genital tract, but more recent research has shown that there are many different prostaglandins which function as regulators of metabolism in a number of tissues and in a number of ways. At least 14 prostaglandins occur in human seminal plasma, and many others have been found in other tissues or prepared synthetically in the laboratory.

The structure of prostaglandins was established by S. Bergström and his colleagues in Sweden. All the natural prostaglandins are biologically derived by cyclization of 20-carbon unsaturated fatty acids, such as arachidonic acid, which is formed from the essential fatty acid linoleic acid (page 281). Five of the carbon atoms of the fatty acid backbone (carbons 8 through 12) are looped to form a five-membered ring (Figure 11-22). The prostaglandins are named according to their ring substituents and the number of additional side-chain double bonds, which have the cis configuration. The best known are prostaglandins E_1, $F_{1\alpha}$, and $F_{2\alpha}$, abbreviated as PGE_1, $PGF_{1\alpha}$, and $PGF_{2\alpha}$, respectively. These in turn are the parent compounds of further biologically active prostaglandins.

The prostaglandins differ from each other with respect to their biological activity, although all show at least some activity in lowering blood pressure and inducing smooth muscle to contract. Some, like PGE_1, antagonize the action of certain hormones. PGE_2 and $PGE_{2\alpha}$ may find clinical use in inducing labor and bringing about therapeutic abortion.

Lipid Micelles, Monolayers, and Bilayers

When a polar lipid, like a phosphoglyceride, is added to water, only a small fraction dissolves to form a true molecular solution. Above the *critical micelle concentration* the polar lipids associate into various types of aggregates resembling the micelles formed from soaps (page 43). In such structures (Figure 11-23) the hydrocarbon tails are hidden from the aqueous environment and form an internal hydrophobic phase whereas the hydrophilic heads are exposed on the surface. Triacylglycerols do not form such aggregates since they have no polar heads.

Phosphoglycerides also form monolayers on air-water interfaces as well as bilayers separating two aqueous compartments. Liposomes (Figure 11-23) are completely closed, vesicular bilayer structures formed by exposing phosphoglyceride-water suspensions to sonic oscillation. Bilayer systems of this sort have been extensively studied as models

Figure 11-22
Some prostaglandins. The parent compound is prostanoic acid. The dotted bonds (color) project behind the plane of the page.

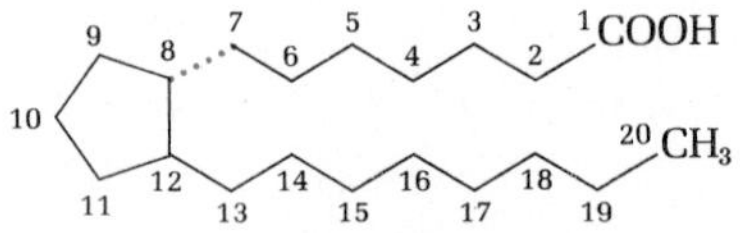

Prostanoic acid

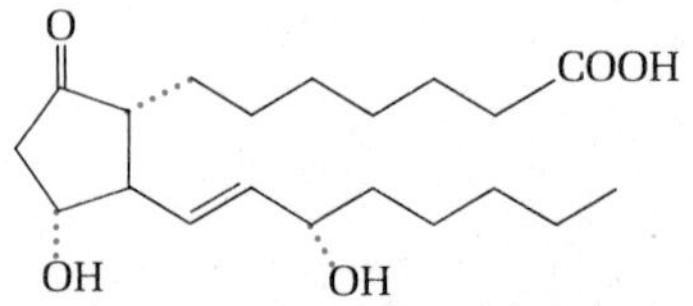

Prostaglandin E_1 (PGE_1)

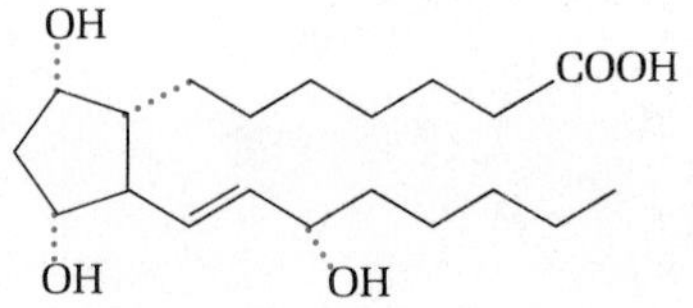

Prostaglandin $F_{1\alpha}$ ($PGF_{1\alpha}$)

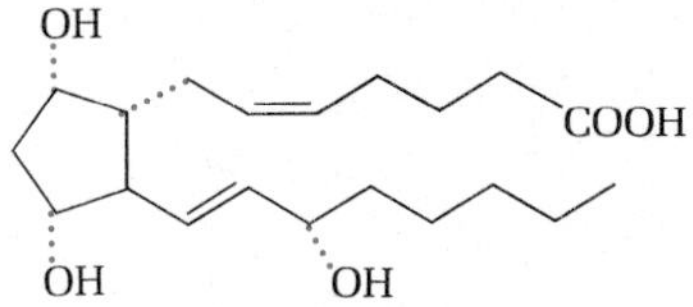

Prostaglandin $F_{2\alpha}$ ($PGF_{2\alpha}$)

Figure 11-23
Stable phosphoglyceride-water systems.

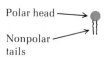

Polar head

Nonpolar
tails

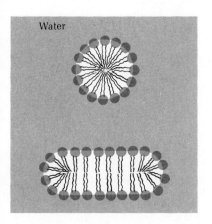

Micelles in water

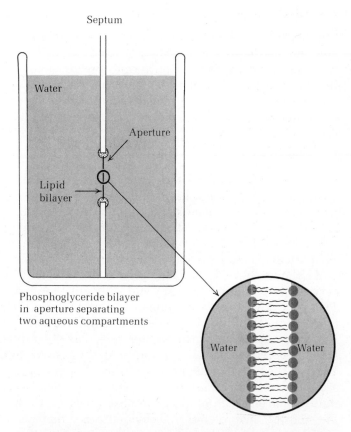

Septum

Water

Aperture

Lipid
bilayer

Phosphoglyceride bilayer
in aperture separating
two aqueous compartments

Water   Water

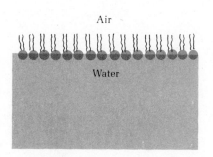

Air

Water

Monolayer at air-water interface

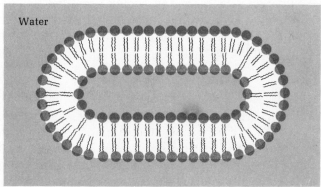

Water

Cross section through a liposome. The phospholipid bilayer forms
a completely closed vesicle. Bilayers, like natural membranes,
have self-sealing properties.

of natural membranes, which appear to contain polar phospholipid bilayers as their continuous phase (page 305).

Lipoprotein Systems

Certain lipids associate with specific proteins to form
lipoprotein systems in which the specific physical properties
of these two classes of biomolecules are blended. There are
two major types, *transport lipoproteins* and *membrane
systems*. In these systems the lipids and proteins are not
covalently joined but are held together largely by hydrophobic interactions (page 43) between the nonpolar portions of the lipid and the protein components.

Transport Lipoproteins of Blood Plasma

The plasma lipoproteins are complexes in which the lipids and proteins occur in a relatively fixed ratio. They carry water-insoluble lipids between various organs via the blood, in a form with a relatively small and constant particle diameter and weight. Human plasma lipoproteins occur in four major classes that differ in density as well as particle size (Table 11-8). They are physically distinguished by their relative rates of flotation in high gravitational fields in the ultracentrifuge. All four lipoprotein classes have densities less than 1.21 g ml^{-1}, whereas the other plasma proteins, such as albumin and γ-globulin, have densities in the range 1.33 to 1.35 g ml^{-1}. The characteristic flotation rates in Svedberg flotation units (S_f) of the lipoproteins are determined in an NaCl medium of density 1.063 g ml^{-1} at 26°C, in which lipoproteins float upward and simple proteins sediment.

As shown in Table 11-8, the plasma lipoproteins contain varying proportions of protein and different types of lipid. The very low-density lipoproteins contain four different types of polypeptide chains having distinctive amino acid sequences. The high-density lipoproteins have two different types of polypeptide chains, of molecular weight 17,500 and 28,000. The polypeptide chains of the plasma lipoproteins are believed to be arranged on the surface of the molecules, thus conferring hydrophilic properties. However, in the very low-density lipoproteins and chylomicrons, there is insufficient protein to cover the surface; presumably the polar heads of the phospholipid components also contribute hydrophilic groups on the surface, with the nonpolar triacylglycerols in the interior.

Membranes

Membranes are a conspicuous feature of cell structure (pages 31 and 35); in some eukaryotic cells (page 33) the different membrane systems may make up as much as 80 percent of the total dry cell mass. Membranes serve not only as barriers separating aqueous compartments with different sol-

Table 11-8   Major classes of human plasma lipoproteins

	Chylomicrons	Very low density lipoproteins (VLDL)	Low-density lipoproteins (LDL)	High-density lipoproteins (HDL)
Density, g ml^{-1}	< 0.94	0.94–1.006	1.006–1.063	1.063–1.21
Flotation rate, S_f	> 400	20–400	0–20	(Sediment)
Particle size, nm	75–1,000	30–50	20–22	7.5–10
Protein, % of dry weight	1–2	10	25	45–55
Triacylglycerols, % of dry weight	80–95	55–65	10	3
Phospholipids, % of dry weight	3–6	15–20	22	30
Cholesterol, free, % of dry weight	1–3	10	8	3
Cholesterol, esterified, % of dry weight	2–4	5	37	15

Figure 11-24
Lipid composition of erythrocyte membranes of different mammals. Note that the proportions of cholesterol and phosphatidylethanolamine are approximately constant but that the ratio of phosphatidylcholine to sphingomyelin varies greatly with the species.

Key

C = cholesterol
PE = phosphatidylethanolamine
PC = phosphatidylcholine
SP = sphingomyelin

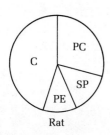

Rat

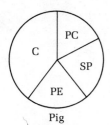

Pig

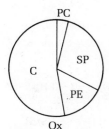

Ox

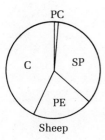

Sheep

ute composition but also as the structural base to which certain enzymes and transport systems are firmly bound. They are very thin (about 8 nm) and flexible.

Membrane Components

Most membranes contain about 40 percent lipid and 60 percent protein, but there is considerable variation. At one extreme the inner mitochondrial membrane contains only about 20 to 25 percent lipid, and at the other the myelin membrane surrounding certain nerves may contain up to 75 percent lipid. The lipids of membranes are largely polar; phosphoglycerides predominate, with much smaller amounts of sphingolipids. In fact, nearly all the polar lipids of cells are localized in their membranes. Endoplasmic reticulum and organelle membranes contain relatively little cholesterol or triacylglycerol, whereas the plasma membrane of some cells of higher animals contains much cholesterol, both free and esterified. The ratio of different kinds of polar lipids in membranes is characteristic of the type of membrane system, the organ, and the species. As an example, Figure 11-24 shows the distribution of the major lipids in the plasma membranes of the erythrocytes of different animal species. The molar ratio of the different types of lipids in a membrane appears to be genetically determined; it cannot be altered, for example, by feeding vertebrates different mixtures of lipids. However, the fatty acid components of the individual lipids are not fixed and vary with nutritional state and environmental temperature. Within any given species of cell, the lipid composition of the different types of membranes are not necessarily identical; e.g., the lipid composition of the plasma membrane, the mitochondrial membrane, and the endoplasmic reticulum membrane of rat liver cells differs significantly.

Each type of membrane contains several or many kinds of proteins or polypeptides. Membrane proteins can be classified in two categories. The *extrinsic*, or *peripheral*, proteins are only loosely attached to the membrane surface and can easily be removed in soluble form by mild extraction procedures. The *intrinsic*, or *integral*, proteins, which make up 70 percent or more of the total membrane protein, are very tightly bound to the lipid portion and can be removed only by drastic treatment. The intrinsic proteins are highly insoluble in neutral aqueous systems but can be extracted by detergents such as sodium dodecyl sulfate or by unfolding agents such as 6 *M* guanidine hydrochloride. When the latter reagent was used to extract erythrocyte membranes, 17 different polypeptide chains having molecular weights from 27,000 to 220,000 were obtained. Among them is *glycophorin,* a glycoprotein that extends completely across the membrane. The inner mitochondrial membrane is one of the most complex membranes; it probably contains over 100 different kinds of polypeptide chains.

Various physical methods have been used to ascertain the arrangement of the lipid and protein molecules in the structure of membranes. Electron microscopy has revealed that membranes have a trilaminar structure (Figure 11-25) with a total thickness between 7.0 and 9.0 nm, depending on the type of membrane. Optical-rotatory-dispersion and circular-dichroism measurements (page 150) indicate that the protein molecules in membranes have a rather high content of right-handed α helix. The lipid molecules of membranes appear to be arranged in a bilayer structure. This follows not only from the high electrical resistance of natural membranes, which indicates the presence of a continuous hydrocarbon phase, but also from experiments with the spin-label technique. In this ingenious method, first developed by H. McConnell and his colleagues, a fatty acid or lipid bearing a *spin label* is incorporated into the membrane. A widely used spin label is the nitroxyl group (Figure 11-26); its $N \rightarrow O$ bond contains an unpaired electron whose spin has a specific direction in relation to the long axis of the fatty acid chain. The directional orientation and motion of the spin-labeled fatty acid or lipid in a bilayer or membrane can be determined by *electron spin-resonance spectroscopy* (ESR), also called *electron paramagnetic-resonance spectroscopy* (EPR), which detects the paramagnetism generated by the unpaired electron spin. With this method it has been shown that the phospholipids in natural membranes move freely in the plane of the membrane but do not cross readily from one side to the other. Moreover, by placing the spin label at different distances along the hydrocarbon chain of a fatty acid molecule it has been found that the nonpolar end of the chain wiggles readily, whereas the end near the carboxyl group is relatively rigid, nearly perpendicular to the plane of the membrane.

Membrane Structure

The first important hypothesis of the structure of biological membranes was proposed by H. Davson and J. Danielli in 1935. An important feature of their hypothesis is the proposal that membranes contain a continuous hydrocarbon phase contributed by the lipid components of the membrane. Some years later this hypothesis was modified and refined, particularly by J. D. Robertson, into the unit-membrane hypothesis. The unit membrane was proposed to consist of a bilayer of mixed polar lipids, with their hydrocarbon chains oriented inward to form a continuous hydrocarbon phase and their hydrophilic heads oriented outward. Each surface was thought to be coated with a monomolecular layer of protein molecules, with the polypeptide chains in extended form. The total thickness of the unit membrane was suggested to be about 8.0 to 9.0 nm, the thickness of the lipid bilayer about 6.0 to 7.0 nm. Later other investigators proposed globular or subunit models, in which membranes were viewed as consisting of sheets of recurring lipoprotein subunits of diameter 4.0 to 9.0 nm, resembling the subunit structure of some oligomeric proteins or the coats of some viruses (page 62). However, globular models have failed to account satisfactorily for many properties of membranes.

The most satisfactory model of membrane structure to date appears to be the *fluid-mosaic* model, postulated by S. J.

Figure 11-25
High-magnification electron micrograph of erythroctye plasma membrane, showing the trilaminar image yielded by fixation in osmium tetroxide. Usually the two electron-dense lines separate a clear space of about 2.5 nm.

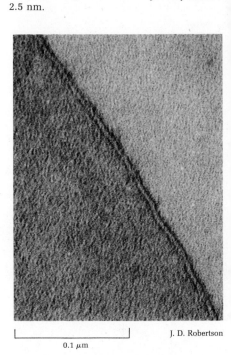

J. D. Robertson

0.1 μm

Figure 11-26
A fatty acid bearing the nitroxyl group as a spin label.

$$CH_3(CH_2)_m - C - (CH_2)_n - COOH$$

Figure 11-27

The fluid-mosaic model of membrane structure. The membrane consists of a fluid phospholipid bilayer with globular protein molecules penetrating into either side or extending entirely through the membrane. There is no long-range regularity in the spacing of the protein molecules, but some may be organized into complexes. Presumably the membrane is asymmetric. The outer surface of the plasma membrane of eukaryotic cells has oligosaccharide chains protruding from glycolipids and glycoproteins (see Figure 10-36, page 275). [Modified from S. J. Singer and G. L. Nicolson, Science, 175: 720–731 (1972).]

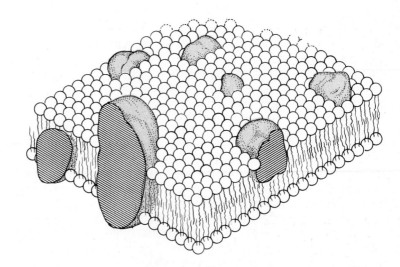

Singer and G. L. Nicolson in 1972. This model (Figure 11-27) postulates that the phospholipids of membranes are arranged in a bilayer to form a fluid, liquid-crystalline matrix or core. In this bilayer individual lipid molecules can move laterally, endowing the bilayer with fluidity, flexibility, and a characteristically high electrical resistance and relative impermeability to highly polar molecules. The fluid-mosaic model postulates that the membrane proteins are globular, to account for their high content of α helix. Some of the proteins are partially embedded in the membrane, penetrating into the lipid phase from either side, and others completely span the membrane. To what extent a given globular protein penetrates into the lipid phase would be determined by the amino acid sequence of the protein and the location on its surface of the nonpolar amino acid R groups. Thus the various membrane proteins would form a mosaiclike structure in the otherwise fluid phospholipid bilayer. This mosaic is not fixed or static, since the proteins are free to diffuse laterally in two dimensions, at least in some membranes. The relative viscosity of the lipid bilayer is thought to be from 100 to 1,000 times that of water.

The fluid-mosaic model accounts satisfactorily for many features and properties of biological membranes. It provides for membranes with widely different protein content, depending on the number of different protein molecules per unit area of membrane; it provides for the varying thickness of different types of membranes; it can account for the asymmetry of natural membranes, since it permits proteins of different types to be arranged on the two surfaces of the lipid bilayer; it accounts for the electrical properties and permeability of membranes; and it also accounts for the observation that some protein components of cell membranes move in the plane of the membrane at a rather high rate.

Most membranes appear to be asymmetric, with different types and/or numbers of specific proteins on each surface. In particular, the plasma membrane of eukaryotic cells often has an outer coat or *glycocalyx* (page 271) made up of the hydrophilic oligosaccharide side chains of membrane glycoproteins and the oligosaccharide head groups of membrane glycolipids.

Summary

Lipids are water-insoluble components of cells that can be extracted by nonpolar solvents. The complex, or saponifiable, lipids contain fatty acids, usually with an even number of carbon atoms, 12 to 22 carbon atoms long. The double bonds of unsaturated fatty acids usually have the cis configuration. In most unsaturated fatty acids, one double bond is at the 9,10 position. Fatty acids can be separated and analyzed by gas-liquid partition chromatography.

Triacylglycerols (triglycerides) contain three fatty acid molecules esterified to the three hydroxyl groups of glycerol. Triacylglycerols serve primarily to store fuel in the form of fat droplets in cells. The phosphoglycerides contain two fatty acid molecules esterified to the two free hydroxyl groups of glycerol 3-phosphate and an alcohol esterified to the phosphoric acid. Their polar head groups differ in polarity and charge. They occur mainly in membranes. Sphingolipids contain no glycerol but have two long hydrocarbon chains, one contributed by a fatty acid and the other by sphingosine, a long-chain aliphatic amino alcohol. Sphingomyelin is the only sphingolipid containing phosphoric acid. The neutral glycosphingolipids contain a carbohydrate head group; cerebrosides, the simplest, contain either D-glucose or D-galactose. Gangliosides are acidic glycosphingolipids containing one or more residues of N-acetyl-neuraminic acid; they are important elements in cell surfaces. Waxes are fatty acid esters of high-molecular-weight alcohols.

The simple, or nonsaponifiable, lipids include the terpenes and the steroids. Terpenes are linear or cyclic compounds built of two or more isoprene units. Steroids are derived from the terpene squalene. The sterols are steroid alcohols; cholesterol is the most abundant sterol in animal tissues. Other steroids include sex hormones, adrenocortical hormones, and bile acids. The prostaglandins, cyclic derivatives of 20-carbon unsaturated fatty acids, function in biological regulation.

Polar lipids spontaneously form micelles, monolayers, and bilayers. Lipids are transported in the blood by the plasma lipoproteins, of which there are four different classes differing in density.

Most membranes contain about 50 to 60 percent protein and 40 to 50 percent lipid. The lipids are present in fixed molar ratios, which are probably genetically determined. Several models of membrane structure have been proposed. Much evidence supports the fluid-mosaic model, which consists of a liquid-crystalline phosphoglyceride bilayer, into which globular proteins penetrate partially or completely. Some plasma membrane proteins contain oligosaccharide side chains, which protrude from the cell surface.

References

Books

ANSELL, G. B., J. N. HAWTHORNE, and R. M. C. DAWSON: *Form and Function of Phospholipids*, 2d ed., Elsevier, London, 1973. Up-to-date, comprehensive treatise covering all aspects of phospholipid biochemistry.

CHAPMAN, D.: *The Structure of Lipids by Spectroscopy and X-Ray Techniques*, Methuen, London, 1965. The physicochemical aspects of lipids and membranes.

CRC Handbook of Chromatography, Chemical Rubber Co., Cleveland, 1972. Detailed methods and data on lipid chromatography.

GURR, M. I., and A. T. JAMES: *Lipid Biochemistry: An Introduction,* Cornell University Press, Ithaca, N.Y., 1971. A short, excellent introduction.

HEFTMANN, E.: *Steroid Biochemistry,* Academic, New York, 1969.

HORTON, E. W.: *Prostaglandins,* vol. 7, Monographs on Endocrinology, Springer-Verlag, New York, 1971. A recent treatise on all aspects of prostaglandin research.

ROTHFIELD, L. I. (ed.): *Structure and Function of Biological Membranes,* Academic, New York, 1971. A collection of papers dealing with the biochemical aspects of membranes.

Articles

BERGSTRÖM, S., and B. SAMUELSSON: "The Prostaglandins," *Endeavour,* 27: 109–113 (1968). An excellent brief summary.

BRETSCHER, M. S.: "Membrane Structure: Some General Principles," *Science,* 181: 622–629 (1973).

COLEMAN, R.: "Membrane-Bound Enzymes and Membrane Ultrastructure," *Biochim. Biophys. Acta,* 300: 1–30 (1973).

IUPAC-IUB COMMISSION OF BIOCHEMICAL NOMENCLATURE: "The Nomenclature of Lipids," *Biochem. J.,* 105: 897–902 (1967).

SINGER, S. J., and G. L. NICOLSON: "The Fluid Mosaic Model of the Structure of Membranes," *Science,* 175: 720–731 (1972).

STOECKENIUS, W., and D. M. ENGELMAN: "Current Models for the Structure of Biological Membranes," *J. Cell Biol.,* 42: 613–646 (1967). An excellent review of the unit membrane and globular (subunit) models.

VAN DEENEN, L. L. M.: "Phospholipids and Biomembranes," *Prog. Chem. Fats Lipids,* 8: 1–127 (1966). Comprehensive review of the chemical relationship of phospholipids to membrane structure.

Problems

1. Indicate the structures of all possible isomers of a triacylglycerol containing palmitic, stearic, and oleic acids. Include both positional isomers and stereoisomers.

2. A mixture of triacylglycerols when hydrolyzed yielded oleic (O), palmitic (P), and stearic (S) acids.
    (a) Indicate the structure of all possible molecular species in the original mixture.
    (b) If only L-stereoisomers are present in the mixture, how many species would be present?

3. A sample (5 g) of the triglycerides extracted from an avocado required 36.0 ml of 0.5 *M* KOH for complete hydrolysis and conversion of its fatty acids into soaps. Calculate the average chain length of the fatty acids in the sample.

4. A mixture of 1-palmitoyl-2-stearoyl-3-lauroylglycerol and phosphatidic acid in benzene is shaken with an equal volume of water. After the two phases are allowed to separate, which lipid will be in higher concentration in the aqueous phase? Why?

5. Electrophoresis at pH 7.0 was carried out on a mixture of lipids containing (a) cardiolipin, (b) phosphatidylglycerol, (c) phosphatidylethanolamine, (d) phosphatidylserine, and (e) O-lysylphosphatidylglycerol. Indicate how you would expect these compounds to move: toward the anode (A), toward the cathode (C), or remain at origin (O).

6. Name the products of hydrolysis with dilute sodium hydroxide of (a) 1-stearoyl-2,3-dipalmitoylglycerol, (b) 1-stearoyl-2-elaidoylphosphatidylinositol, (c) 1-palmitoyl-2-oleyl phosphatidylcholine.

7. Name the products of the following: (a) hydrolysis of 1-stearoyl-2-oleyl phosphatidylserine by strong base, followed by acid hydrolysis, (b) treatment of 1-palmitoyl-2-linoleyl phosphatidylcholine with phospholipase D.

8. Most membranes of animal cells contain about 60 percent by weight of protein and 40 percent by weight phosphoglycerides. (a) Calculate the average density of a membrane, assuming that protein has a density of 1.33 g cm^{-3} and phosphoglyceride a density of 0.92 g cm^{-3}.
    (b) If a sample of membrane material were centrifuged in NaCl solution of density 1.05 g cm^{-3}, would it sediment or float?

9. If a membrane contains 60 percent by weight of proteins and 40 percent of phosphoglycerides, calculate the molar ratio of phosphoglyceride to protein. Assume that the lipid molecules have an average molecular weight of 800 and the proteins an average molecular weight of 50,000.

CHAPTER 12 NUCLEOTIDES AND THE COVALENT STRUCTURE OF NUCLEIC ACIDS

Deoxyribonucleic acid (DNA) and ribonucleic acid (RNA) are chainlike macromolecules that function in the storage and transfer of genetic information. They are major components of all cells, together making up from 5 to 15 percent of their dry weight. Nucleic acids are also present in viruses, infectious nucleic acid–protein complexes capable of directing their own replication in specific host cells. Although nucleic acids are so named because DNA was first isolated from cell nuclei, both DNA and RNA also occur in other parts of cells.

Just as the amino acids are the building blocks, or monomeric units, of polypeptides, the nucleotides are the monomeric units of nucleic acids. The analogy between proteins and nucleic acids may be taken further. Just as one type of protein molecule is distinguished from another by the sequence of the characteristic side chains or R groups of the amino acid monomers, each type of nucleic acid is distinguished by the sequence of the characteristic heterocyclic bases of its nucleotide monomers.

In this chapter we examine first the structure and properties of nucleotides, which serve not only as building blocks of nucleic acids but also have important functions in intermediary metabolism. Then we shall examine the covalent backbone structure of DNA and RNA and the problem of deducing the sequence of nucleotide units in nucleic acids. Finally, we shall examine some supramolecular, particulate structures containing nucleic acids, especially ribosomes and viruses. The three-dimensional structure and the biological function of nucleic acids in storing and transferring genetic information is developed in Part 4, beginning on page 853.

General Structure of the Nucleotides

The monomeric units of DNA are called _deoxyribonucleotides_; those of RNA are _ribonucleotides_. Each nucleotide contains three characteristic components: (1) a nitrogenous heterocyclic base, which is a derivative of either pyrimidine or purine; (2) a pentose; and (3) a molecule of phosphoric acid. The major nucleotides are shown in Figure 12-1.

Four different deoxyribonucleotides serve as the major components of DNAs (Figure 12-1); they differ from each other only in their nitrogenous base components, after which they are named. The four bases characteristic of the deoxyribonucleotide units of DNA are the purine derivatives _adenine_ and _guanine_ and the pyrimidine derivatives _cytosine_ and _thymine_ (Figure 12-2). Similarly, four different ribonucleotides are the major components of RNAs (Figure 12-1); they contain the purine bases _adenine_ and _guanine_ and the pyrimidine bases _cytosine_ and _uracil_ (Figure 12-2). Thus thymine, which is 5-methyluracil, is characteristically present in DNA but not usually in RNA, whereas uracil is normally present in RNA but only rarely in DNA.

The other difference in the composition between these two kinds of nucleic acids is that deoxyribonucleotides contain as their pentose component _2-deoxy-D-ribose_, whereas ribonucleotides contain _D-ribose_. Both sugars occur in their furanose forms in nucleotides. The chemical and physical properties of these and other sugars are discussed in Chapter 10 (page 249). The pentose is joined to the base by a β-N-glycosyl bond between carbon atom 1 of the pentose and nitrogen atom 9 of purine bases or nitrogen atom 1 of pyrimidine bases. The phosphate group of nucleotides is in ester linkage with carbon atom 5 of the pentose.

When the phosphate group of a nucleotide is removed by hydrolysis, the structure remaining is called a _nucleoside._

Figure 12-1
The major ribonucleotides and deoxyribonucleotides. (Right) Space-filling model of adenylic acid.

Ribonucleoside
5′-monophosphates

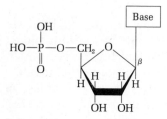

General structure

Names

Adenosine 5′-phosphoric acid
(adenylic acid; AMP)

Guanosine 5′-phosphoric acid
(guanylic acid; GMP)

Cytidine 5′-phosphoric acid
(cytidylic acid; CMP)

Uridine 5′-phosphoric acid
(uridylic acid; UMP)

2′-Deoxyribonucleoside
5′-monophosphates

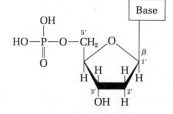

General structure

Names

Deoxyadenosine 5′-phosphoric acid
(deoxyadenylic acid; dAMP)

Deoxyguanosine 5′-phosphoric acid
(deoxyguanylic acid; dGMP)

Deoxycytidine 5′-phosphoric acid
(deoxycytidylic acid; dCMP)

Deoxythymidine 5′-phosphoric acid
(deoxythymidylic acid; dTMP)

Space-filling model of
adenylic acid (anionic form)

Thus nucleotides are the 5′-phosphates of the corresponding nucleosides (Figure 12-1). Later we shall see that cells also contain the 5′-diphosphates and 5′-triphosphates of the common nucleosides; hence the inclusion of M for mono in the abbreviations for the nucleotides (Figure 12-1).

We shall now examine the properties of the components of nucleotides more closely, beginning with the nitrogenous bases, since they give the monomeric nucleotide units of DNA and RNA their chemical individuality.

The Pyrimidines and Purines

The parent compounds of the two classes of nitrogenous bases found in nucleotides are the heterocyclic compounds <u>pyrimidine</u> and <u>purine</u> (Figure 12-2), which have pronounced aromatic character. Purine may itself be regarded as

Figure 12-2
Structures and space-filling models of the common purine and pyrimidine bases.

Purine, the parent compound

Pyrimidine, the parent compound

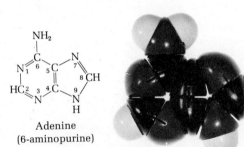

Adenine
(6-aminopurine)

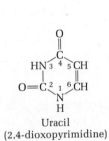

Uracil
(2,4-dioxopyrimidine)

Guanine
(2-amino-6-oxopurine)

Thymine
(5-methyl-2,4-dioxopyrimidine)

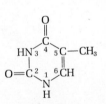

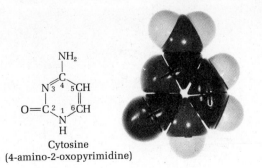

Cytosine
(4-amino-2-oxopyrimidine)

a derivative of pyrimidine; it consists of a pyrimidine ring and an imidazole ring fused together. Three pyrimidine derivatives, *uracil*, *thymine*, and *cytosine*, and two purine derivatives, *adenine* and *guanine*, constitute the major nitrogenous bases found in nucleotides (Figure 12-2). In free or uncombined form these bases occur only in traces in most cells, usually as products of the enzymatic hydrolysis of nucleic acids and nucleotides.

The precise three-dimensional structure of various pyrimidines and purines has been deduced by x-ray diffraction analysis. Pyrimidines are planar molecules; purines are very nearly planar, with a slight pucker. The exact dimensions of adenine are shown in Figure 12-3 and space-filling models of the purines and pyrimidines in Figure 12-2. Not only the dimensions of the bases but also their capacity for hydrogen bonding are crucial to the biological function of nucleic acids (Chapter 31, page 864). The important functional groups involved in formation of hydrogen bonds are the amino groups of adenine, guanine, and cytosine, the ring —NH— groups at position 1 of adenine and guanine and position 3 of the pyrimidine bases, and the strongly electronegative oxygen atoms at position 2 of the pyrimidines and position 6 of guanine (Figure 12-2).

Free pyrimidine and purine bases are relatively insoluble in water. They are weakly basic compounds that may exist in two or more tautomeric forms depending upon the pH. Uracil, for example, occurs in *lactam* and *lactim* forms (Figure 12-4); at pH 7.0, the lactam form of uracil predominates. The structures of the other purines and pyrimidines shown in Figure 12-2 are the tautomers predominating at pH 7.0. These are also the forms responsible for the observed hydrogen bonding between bases in native DNA molecules, as we shall see later (Chapter 31, page 864).

In addition to the common bases listed above, a large number of other purine and pyrimidine derivatives, called

Figure 12-3
Dimensions of the adenine molecule.

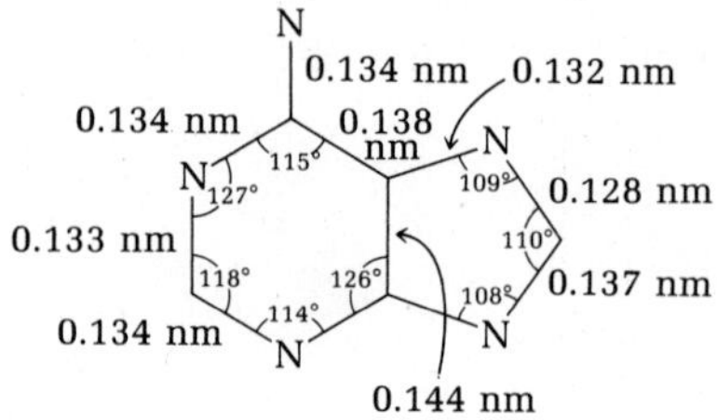

Figure 12-4
Tautomeric forms of uracil.

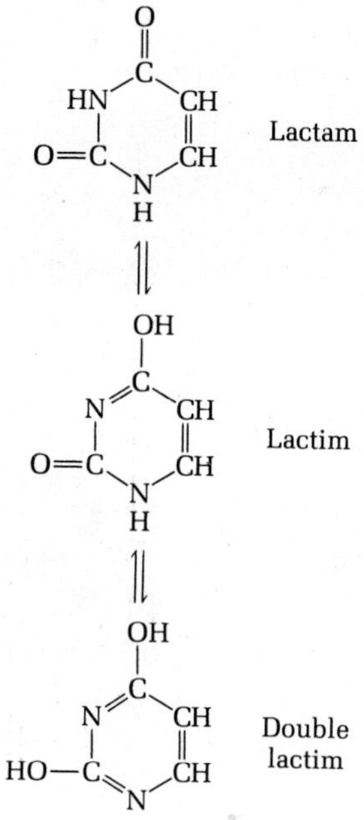

Lactam

Lactim

Double lactim

Figure 12-5
Some rare bases.

Two minor purines

6-Methyladenine

2-Methylguanine

Two minor pyrimidines

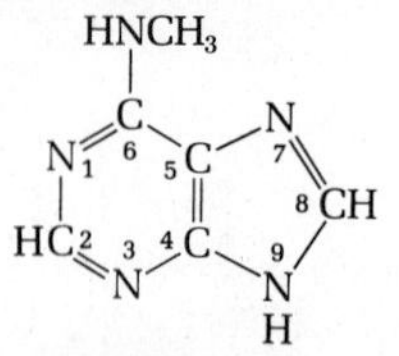

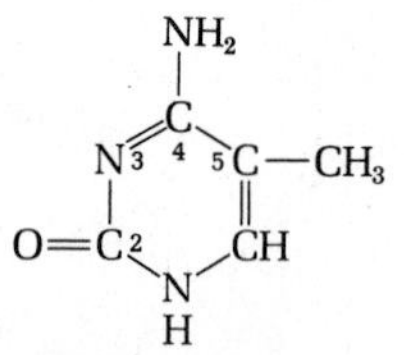

5-Methylcytosine

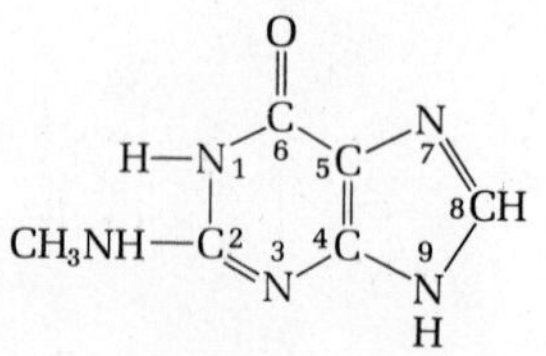

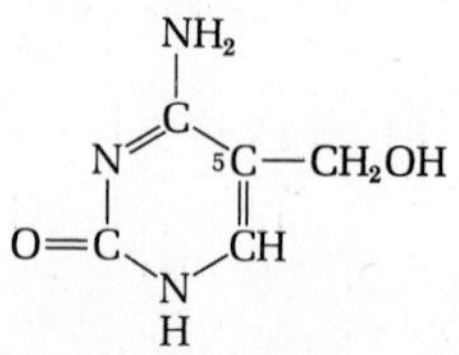

5-Hydroxymethylcytosine

Table 12-1   Other rare bases
found in nucleic acids

5,6-Dihydrouracil
1-Methyluracil
3-Methyluracil
5-Hydroxymethyluracil
2-Thiouracil
N^4-Acetylcytosine
3-Methylcytosine
5-Methylcytosine
5-Hydroxymethylcytosine
1-Methyladenine
2-Methyladenine
7-Methyladenine
N^6-Methyladenine
N^6,N^6-Dimethyladenine
N^6-(Δ^2-Isopentenyl)adenine
1-Methylguanine
7-Methylguanine
N^2-Methylguanine
N^2,N^2-Dimethylguanine

the rare or minor bases, occur in small amounts in some nucleic acids. Among the rare pyrimidines are 5-methylcytosine and 5-hydroxymethylcytosine; the minor purines include 6-methyladenine and 2-methylguanine (Figure 12-5). These and other minor bases are listed in Table 12-1; most of them are methyl derivatives of the major bases, but some contain acetyl, isopentenyl, or hydroxymethyl groups. Rare bases are especially prominent in transfer RNAs (page 321), which characteristically contain up to 10 percent of these unusual components. Over 30 different kinds of rare bases have been found in tRNAs.

All the purine and pyrimidine bases of nucleic acids strongly absorb ultraviolet light in the region 250 to 280 nm. This property is very useful in the detection and quantitative analysis not only of the free bases but also of nucleosides and nucleotides. Free purine and pyrimidine bases are easily separated by chromatographic or electrophoretic methods.

Nucleosides

There are two series of nucleosides: the ribonucleosides, which contain D-ribose as the sugar component, and the deoxyribonucleosides, which contain 2-deoxy-D-ribose (Figure 12-6). Like the free purines and pyrimidines, free nucleosides occur only in trace amounts in most cells, the products of chemical or enzymatic hydrolysis of nucleotides. Nucleosides are much more soluble in water than their corresponding free bases. They are readily separated and identified by chromatographic methods. Like the glycosides (page 256), the nucleosides are relatively stable in alkali. The purine nucleosides are rather easily hydrolyzed by acid to yield the free base and the pentose. However, the pyrimidine nucleosides are resistant to acid hydrolysis. Both types of nucleosides are hydrolyzed by specific nucleosidases.

Nucleotides

Ribonucleotides and deoxyribonucleotides occur in the free form in cells in significant amounts. Their phosphoric

Figure 12-6
Nucleosides of adenine.

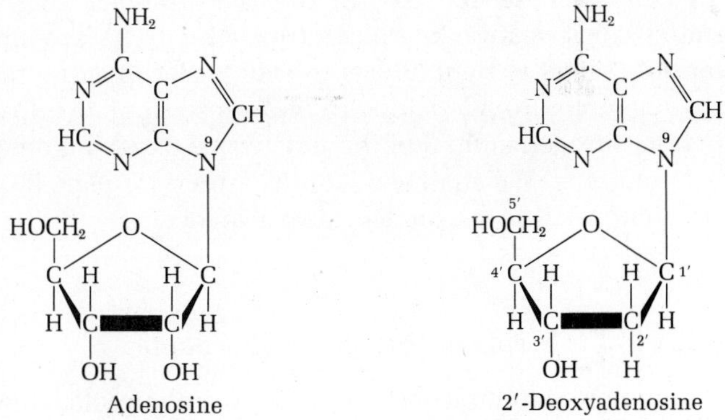

Adenosine
(9-β-D-ribofuranosyladenine)

2'-Deoxyadenosine
(9-β-2'-deoxy-D-ribofuranosyladenine)

Figure 12-7
The Lambert-Beer law, the absorption spectra of the common nucleotides,
and their molar absorption coefficients at 260 nm.

Measurement of light absorption is an important tool for analysis of nucleotides and nucleic acids (pages 340, 874, and 875). The fraction of the incident light absorbed by a solution at a given wavelength is related to the
thickness of the absorbing layer and the concentration of the absorbing
species. These two relationships are combined into the Lambert-Beer law,
given in integrated form as

$$\log\frac{I_0}{I} = \epsilon c l$$

where I_0 is the intensity of the incident light, I is the intensity of the transmitted light, ϵ is the molar absorption coefficient (in units of liters per mole-
centimeter), c the concentration of the absorbing species in moles per liter,
and l the thickness of the light-absorbing sample in centimeters. The
Lambert-Beer law assumes that the incident light is parallel and monochromatic and that the solvent and solute molecules are randomly oriented.
The expression $\log (I_0/I)$ is called the absorbance, designated A.

It is important to note that each millimeter thickness of absorbing solution in a 1.0 cm cell absorbs not a constant amount but a constant fraction
of the incident light. However, with an absorbing layer of fixed thickness,
the absorbance A is directly proportional to the concentration of the
absorbing solute.

The molar absorption coefficient varies with the nature of the absorbing
compound, the solvent, the wavelength, and also with pH if the light-absorbing species is one that is in equilibrium with another species having a
different spectrum through gain or loss of protons. At the right are shown
the spectra (pH 7.0) of the purine ribonucleotides (top) and the pyrimidine
nucleotides (bottom). The spectra of the corresponding ribo- and deoxyribonucleotides, as well as the nucleosides, are essentially identical.
When mixtures of nucleotides are being measured the wavelength 260 nm
(dashed vertical lines) is ordinarily used.

The table gives values of ϵ at 260 nm and pH 7.0.

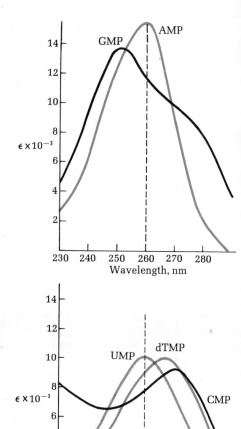

groups are relatively strong acids; at pH 7.0 the free nucleotides thus exist primarily in the form $N-O-PO_3^{2-}$, where N
is the nucleoside group. Due to the presence of a pyrimidine
or purine base, all the nucleotides show strong ultraviolet absorption in the region 250 to 280 nm, which is very useful
for quantitative analysis. Figure 12-7 gives the characteristic
ultraviolet spectra of some nucleotides. Nucleotides are
easily separated and quantitated by ion-exchange chromatography.

Nucleosides and nucleotides contain two nearly planar
rings, that of the base and that of the ribofuranose. In the
most stable conformation of nucleotides the rings are not
coplanar but almost at right angles to each other, placing the
2' hydrogen or hydroxyl of the ribofuranose ring in close
proximity to nitrogen atom 3 of the purines or oxygen atom 2
of the pyrimidines. We shall see later (Chapter 31, page 863)
that this conformation of nucleotides also occurs in intact
nucleic acid molecules.

Nucleoside 5'-Diphosphates and 5'-Triphosphates

All the common ribonucleosides and deoxyribonucleosides
occur in cells not only as the 5'-monophosphates, as

Molar absorption coefficient of nucleotides	
	ϵ_{260}
AMP	15,400
GMP	11,700
CMP	7,500
UMP	9,900
dTMP	9,200

Figure 12-8
General structure of nucleoside 5'-mono-, 5'-di-, and 5'-triphosphates (NMPs, NDPs, and NTPs). In the corresponding deoxyribonucleoside phosphates (dNMPs, dNDPs, and dNTPs) the pentose is 2-deoxy-D-ribose. At pH 7.0 the ionized form shown predominates.

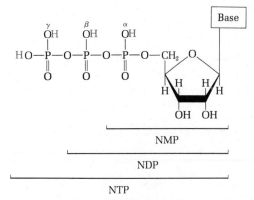

NMP

NDP

NTP

Abbreviations of ribonucleoside 5'-phosphates			
Base	Mono-	Di-	Tri-
Adenine	AMP	ADP	ATP
Guanine	GMP	GDP	GTP
Cytosine	CMP	CDP	CTP
Uracil	UMP	UDP	UTP

Abbreviations of deoxyribonucleoside 5'-phosphates			
Base	Mono-	Di-	Tri-
Adenine	dAMP	dADP	dATP
Guanine	dGMP	dGDP	dGTP
Cytosine	dCMP	dCDP	dCTP
Thymine	dTMP	dTDP	dTTP

described above, but also as the 5'-diphosphates and the 5'-triphosphates, i.e., as the 5'-pyrophosphoric and the 5'-triphosphoric esters of the nucleosides (Figure 12-8). Thus we have three series of 5' phosphorylated nucleosides; e.g., for adenosine they are _adenosine 5'-monophosphate_ (AMP), _adenosine 5'-diphosphate_ (ADP), and _adenosine 5'-triphosphate_ (ATP). The phosphoric residues of these compounds are designated by the symbols α, β, and γ (Figure 12-8).

The nucleoside 5'-diphosphoric acids and 5'-triphosphoric acids, generically designated as NDPs and NTPs, are relatively strong acids and dissociate three and four protons, respectively, from their condensed phosphoric groups. The condensed phosphoric groups of the NDPs and NTPs form complexes with divalent cations such as Mg^{2+} and Ca^{2+}. Because of the relatively high concentration of Mg^{2+} in the cytoplasm, the nucleoside 5'-di- and 5'-triphosphates exist primarily as the Mg^{2+} complexes in the intact cell. The significance of these complexes will be discussed in Chapter 15 (page 402). The terminal phosphate group of the NDPs and NTPs can be selectively removed by specific enzymes without cleavage of other bonds. Moreover, the β- and γ-phosphate groups of NTPs and the β-phosphate group of NDPs are hydrolyzed to yield inorganic phosphate by heating at 100°C in 1 N HCl for 7 min; in contrast, the α- or 5′-phosphate group of NMPs is stable to this treatment. This difference permits simple analysis of the sum of NTPs and NDPs in mixtures with NMPs.

The NTPs have a number of important functions. ATP is a carrier of phosphate and pyrophosphate in several important enzymatic reactions involved in the transfer of chemical energy (Chapter 15). After the dephosphorylation of ATP in such reactions, the ADP formed is rephosphorylated to ATP during respiration. Although the ATP-ADP system is the primary or mainline system for transferring phosphate groups in the cell, the other NTPs, namely, GTP, UTP, and CTP, also channel chemical energy into specific biosynthetic pathways (Chapter 15).

A second major function of the NTPs and NDPs is to serve as coenzymelike, energized carriers of specific types of building-block molecules. For example, uridine diphosphate (UDP) is a specific carrier of sugar residues in the biosynthesis of polysaccharides; thus, _uridine diphosphate glucose_ (Figure 12-9) is the specific donor of glucose residues in the enzymatic biosynthesis of glycogen. Similarly, _cytidine diphosphate choline_ is a donor of phosphocholine in the enzymatic biosynthesis of choline-containing phosphoglycerides (Figure 12-9).

The third major function of NTPs and dNTPs is to serve as energy-rich precursors of mononucleotide units in the enzymatic biosynthesis of DNA and RNA. During these reactions, the dNTPs and NTPs lose their terminal pyrophosphate groups to become the nucleoside monophosphate residues of the nucleic acids. In all three functions of the NTPs and dNTPs the chemical energy inherent in the β- and γ-phosphate groups is utilized to help form new covalent bonds (Chapters 23 to 26 and 31).

Figure 12-9
Uridine diphosphate glucose and cytidine diphosphate choline. The glucose and choline moieties are in color.

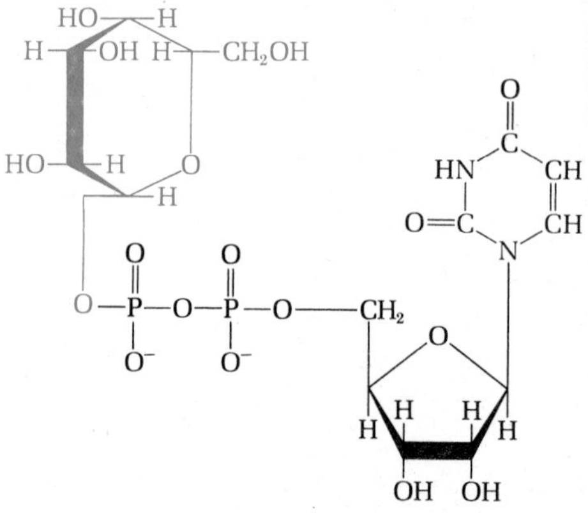

Uridine diphosphate glucose
(UDPG)

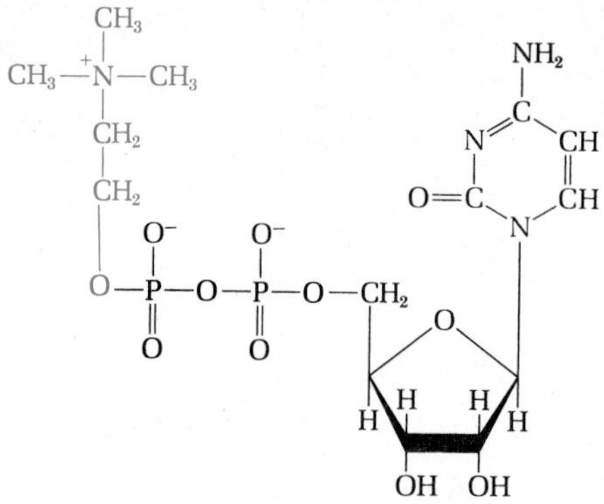

Cytidine diphosphate choline

Other Nucleotides

In addition to the nucleoside 5′-phosphates just described, nucleotides having their phosphate groups in other positions occur biologically (Figures 12-10 and 12-11). _Ribonucleoside 2′,3′-cyclic phosphates_ are intermediates and _ribonucleoside 3′-phosphates_ are end products of the hydrolysis of certain ribonucleotide linkages of RNA by the action of some ribonucleases. These compounds are also formed during the hydrolysis of RNA by alkali, as are ribonucleoside 2′-phosphates (Figure 12-10).

Two very important nucleotides play a key role in the biochemical action of a number of hormones: _adenosine 3′,5′-cyclic phosphate_ (abbreviated cyclic AMP or cAMP) and _guanosine 3′,5′-cyclic phosphate_ (abbreviated cyclic GMP or cGMP) (Figure 12-11). Cyclic AMP arises in eukaryotic cells from ATP by the action of an enzyme located in the cell membrane, _adenylate cyclase_, that is stimulated by certain hormones arriving from the bloodstream (Chapter 29, page 812). Cyclic AMP is called a second messenger

Figures 12-10

Some adenosine monophosphates. Adenosine 2'-, 3'-, and 2',3'-cyclic phosphates are intermediates in the hydrolysis of ribonucleic acid by alkali. See also Figure 12-11.

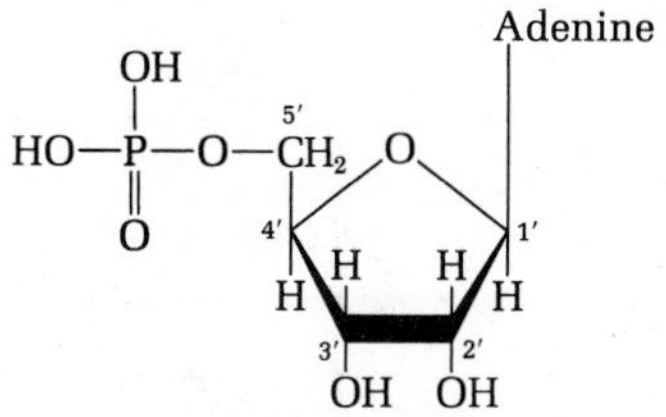

Figure 12-11

Three nucleotides active in regulatory mechanisms.

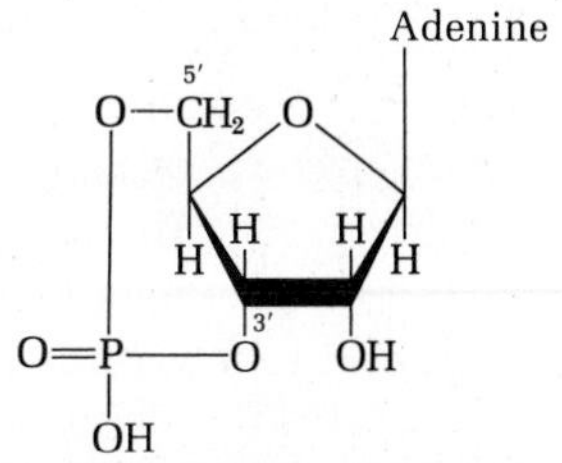

Adenosine 5'-phosphoric acid

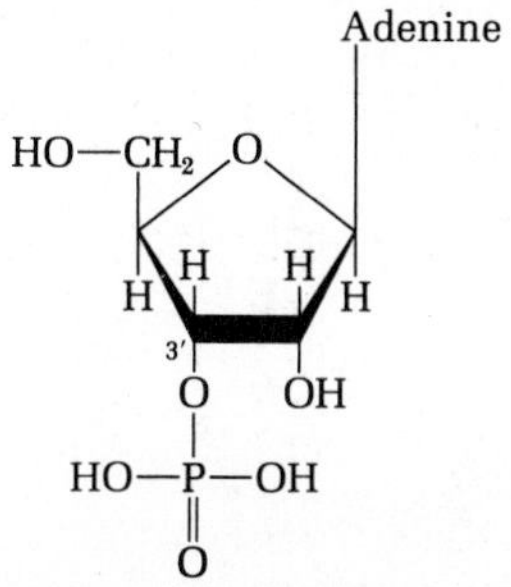

Adenosine 3'-phosphoric acid

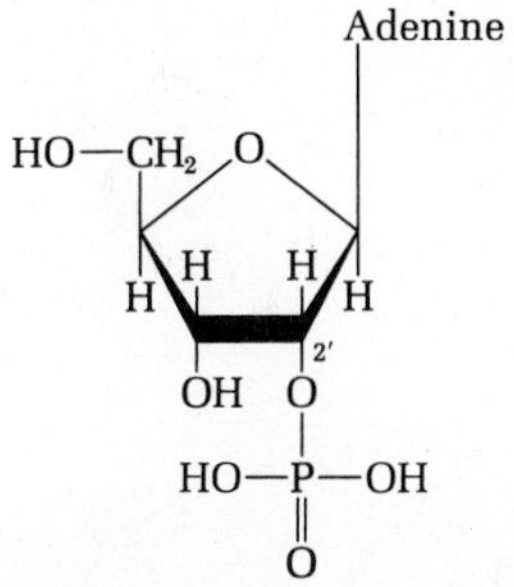

Adenosine 2'-phosphoric acid

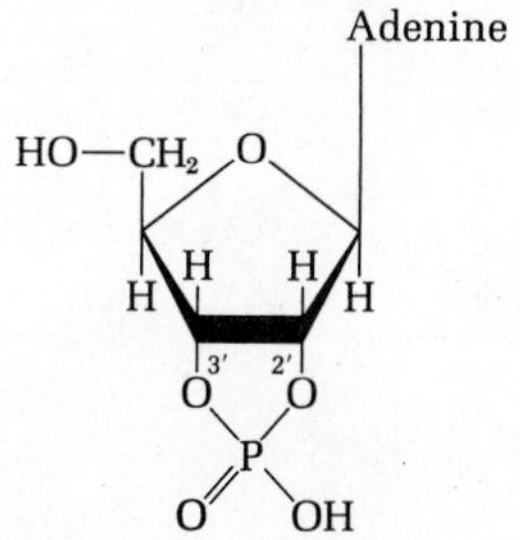

Adenosine 2',3'-cyclic phosphoric acid

Adenosine 3',5'-cyclic phosphoric acid
(cyclic AMP; cAMP)

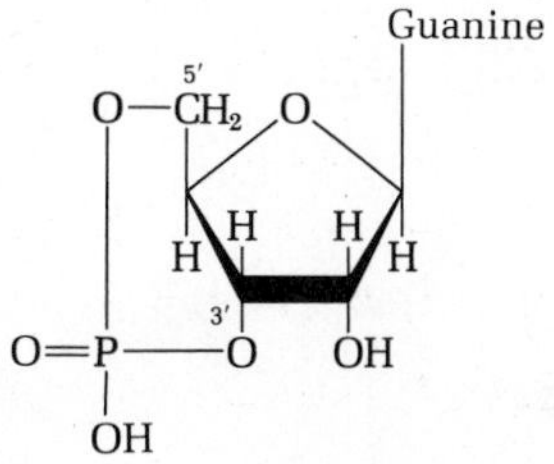

Guanosine 3',5'-cyclic phosphoric acid
(cyclic GMP; cGMP)

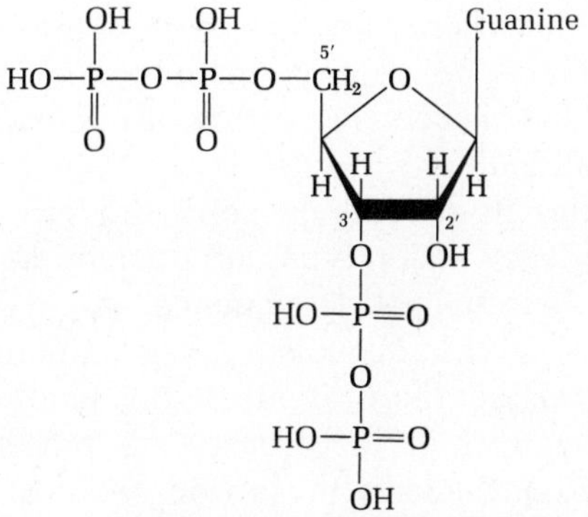

Guanosine 5'-diphosphate 3'-diphosphate
(also called guanosine tetraphosphate)

because it transmits and amplifies within the cell the chemical signals delivered via the blood by hormones, the first messengers. Two other important nucleotides, now known to participate in the regulation of gene transcription in bacteria, are *guanosine 5'-diphosphate 3'-diphosphate* (abbreviated ppGpp) (Figure 12-11) and *guanosine 5'-triphosphate 3'-diphosphate* (pppGpp) (Figure 12-11).

Many coenzymes are nucleotides or derivatives of nucleotides; they are discussed in the next chapter (page 335).

317

Nucleic Acids

Deoxyribonucleic acid (DNA) consists of covalently linked chains of deoxyribonucleotides, and ribonucleic acid (RNA) consists of chains of ribonucleotides. DNA and RNA share a number of chemical and physical properties because in both of them the successive nucleotide units are covalently linked in identical fashion by phosphodiester bridges formed between the 5'-hydroxyl group of one nucleotide and the 3'-hydroxyl group of the next (Figure 12-12). Thus the backbone of both DNA and RNA consists of alternating phosphate and pentose groups, in which phosphodiester bridges provide the covalent continuity. The purine and pyrimidine bases of the nucleotide units are not present in the backbone structure but constitute distinctive side chains, just as the R groups of amino acid residues are the distinctive side chains of polypeptides.

Before examining further the chemical properties of the backbone structure of the nucleic acids, we must briefly sketch the classification, chemical composition, and function of the major types of nucleic acids; these matters are further developed in Part 4 (page 853).

DNA

DNA was first isolated (from pus cells and salmon sperm) and intensively studied by Friedrich Miescher, a Swiss, in a series of remarkable investigations beginning in 1869. He named it "nuclein" from its occurrence in cell nuclei. Over 70 years of research were required before the major building-block units and the backbone structure of nucleic acids were completely identified.

DNA molecules from different cells and viruses vary in the ratio of the four major types of nucleotide monomers (Table 12-2), in their nucleotide sequence, and in their molecular weight. Besides the four major bases (adenine, guanine thymine, and cytosine) found in all DNAs, small amounts of methylated derivatives of these bases are present in some DNA molecules, particularly those from viruses (page 328). The DNAs isolated from different organisms and viruses normally have two strands in complementary double-helical arrangement (Chapter 31, page 864). In most cells the DNA molecules are so large that they are not easily isolated in intact form. In prokaryotic cells (page 30), which contain only a single chromosome, essentially all the DNA is present as a single double-helical, i.e., two-stranded, macromolecule exceeding 2×10^9 in molecular weight. In eukaryotic cells, which contain several or many chromosomes, there are, correspondingly, several or many DNA molecules. In bacteria, the DNA molecule, which makes up about 1 percent of the cell weight, is found in the nuclear zone (page 31); it is usually attached, apparently at a single point, to an infolding of the cell membrane (page 411) called a mesosome. In bacteria no protein is associated with the DNA. Sometimes small molecules of extrachromosomal DNA occur in the cytoplasm of bacteria; such DNA molecules, which carry only a

Figure 12-12
The covalent backbone structure of nucleic acid chains.

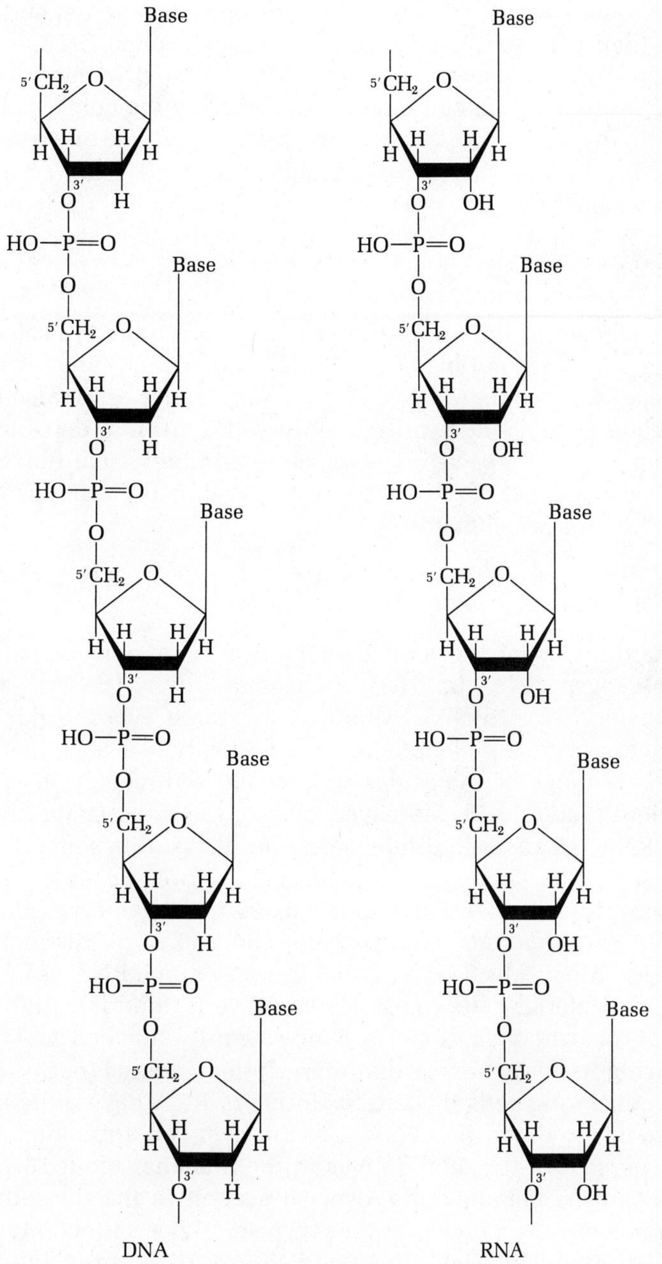

DNA   RNA

Table 12-2   Mole percent of bases in some nucleic acid preparations

	Adenine	Guanine	Cytosine	Thymine	Uracil
		DNAs			
Human	30.9	19.9	19.8	29.4	—
E. coli	24.7	26.0	25.7	23.6	—
Bacteriophage λ	21.3	28.6	27.2	22.9	—
		RNAs			
Ox liver (total)†	17.1	27.3	33.9	—	21.7
E. coli mRNA†	25.1	27.1	24.1	—	23.7
Tobacco mosaic virus	29.8	25.4	18.5	—	26.7

† Mixture.

few genes, are called either *plasmids* or *episomes*, depending on their genetic relationship to the chromosomal DNA.

In diploid eukaryotic cells nearly all the DNA molecules are present in the cell nucleus, where they are combined in ionic linkage with basic proteins called *histones*. In addition to the nuclear DNA, diploid eukaryotic cells also contain very small amounts of DNA in the mitochondria (page 870); it differs in its base composition and molecular weight from nuclear DNA. Mitochondrial DNA (designated mtDNA) has a molecular weight of about 10 million; it accounts for 0.1 to 0.2 percent of the total cellular DNA. Chloroplasts also contain a distinctive type of small DNA molecule. Many viruses (page 328) contain DNA, which may range in molecular weight from about 2 million to over 100 million, depending upon the viral species. The structure and biological function of various types of DNA are further developed in Chapters 31 and 32 (pages 859 and 891).

RNA

The three major types of ribonucleic acid in cells are called *messenger RNA* (mRNA), *ribosomal RNA* (rRNA), and *transfer RNA* (tRNA). Although all three types occur as single polyribonucleotide strands, each type has a characteristic range of molecular weight and sedimentation coefficient (Table 12-3). Moreover, each of the three major kinds of RNA occurs in multiple molecular forms. Ribosomal RNA of any given biological species exists in three or more major forms, transfer RNA in as many as 60 forms, and messenger RNA in hundreds and perhaps thousands of distinctive forms. Most cells contain 2 to 8 times as much RNA as DNA.

In bacterial cells, most of the RNA is found in the cytoplasm, although some is noncovalently attached to DNA during its formation in the transcription process (page 917). In eukaryotic cells the various forms of RNA have a distinctive intracellular distribution. In the liver cell approximately 11 percent of the total RNA is in the nucleus, about 15 percent in the mitochondria, over 50 percent in the ribosomes, and about 24 percent in the cytosol. Like mitochondrial DNA, mitochondrial rRNAs and tRNAs differ from the extramitochondrial forms (page 951). RNA is present in some viruses discussed later in this chapter.

Table 12-3 Properties of *E. coli* RNAs

Type	S	Mol wt	No. of nucleotide residues	Percent of total cell RNA
mRNA	6–25	25,000–1,000,000	75–3,000	~2
tRNA	~4	23,000–30,000	75–90	16
rRNA	5	~35,000	~100	
	16	~550,000	~1,500	82
	23	~1,100,000	~3,100	

Messenger RNA

Messenger RNA contains only the four major bases. It is synthesized in the nucleus during the process of transcription, in which the sequence of bases in one strand of the chromosomal DNA is enzymatically transcribed in the form of a single strand of mRNA; some mRNA is also made in the mitochondria. The sequence of bases of the mRNA strand so formed is complementary (page 918) to that of the DNA strand being transcribed. After transcription, the mRNA passes into the cytoplasm and then to the ribosomes, where it serves as the template for the sequential ordering of amino acids during the biosynthesis of proteins (pages 931 and 958). Although mRNA makes up only a very small part of the total RNA of the cell (Table 12-3), it occurs in many distinctive forms which vary greatly in molecular weight and base sequence. Each of the thousands of different proteins synthesized by the cell is coded by a specific mRNA or segment of an mRNA molecule.

Messenger RNAs of eukaryotic cells are distinctive in containing a long sequence of about 200 successive adenylate residues at the 3′ end, which apparently plays a role in the processing or transport of mRNA from the nucleus to the ribosomes (page 919).

Transfer RNAs

Transfer RNAs are relatively small molecules that act as carriers of specific individual amino acids during protein synthesis on the ribosomes. They have molecular weights in the range of 23,000 to 28,000 and sedimentation coefficients (page 175) of about 4S. They contain from 75 to 90 nucleotide units. Each of the 20 amino acids found in proteins has at least one corresponding tRNA, and some have multiple tRNAs. For example, there are five distinctly different tRNAs specific for the transfer of leucine in E. coli cells. Moreover, there are different types of tRNAs for a given amino acid in the mitochondria and in the cytoplasm of eukaryotic cells. Many different tRNAs have been isolated from different kinds of cells.

Besides the major purine and pyrimidine bases transfer RNAs characteristically contain a rather large number of the rare bases, up to 10 percent of the total. In addition, tRNAs also contain some unusual nucleotides, e.g., pseudouridylic acid and ribothymidylic acid (Figure 12-13). Transfer RNA molecules share other identifying features. At one end of the polynucleotide chain all tRNAs contain a terminal guanylic acid residue; at the other end all tRNAs contain the terminal sequence cytidylic-cytidylic-adenylic (C-C-A). The 5′-hydroxyl group of the terminal adenylic residue is linked to the 3′-hydroxyl of the preceding cytidylic residue by a phosphodiester bridge. A free hydroxyl group of the terminal adenylic residue is enzymatically acylated with its specific α-amino acid to yield the charged form of tRNA, namely, aminoacyl-tRNA. This amino acid residue is enzymatically transferred to the end of the growing polypeptide chain on

Figure 12-13
Two unusual nucleotides found in tRNAs. In pseudouridylic acid the N-glycosyl linkage is at position 5 of uracil, rather than the usual position 1. Ribothymidylic acid is unusual in that thymine is normally present in DNA but not in RNA.

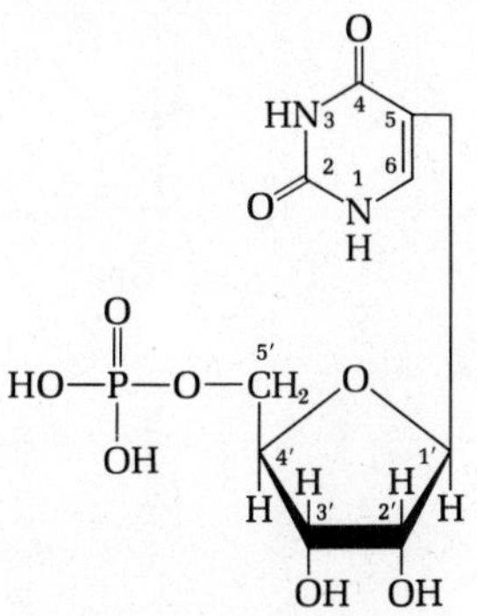

Pseudouridylic acid

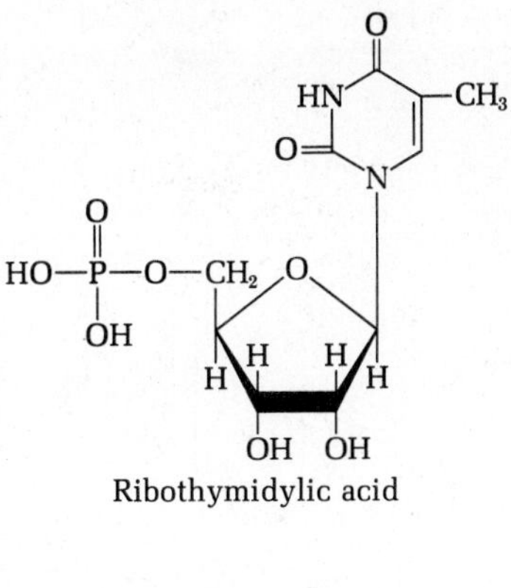

Ribothymidylic acid

the surface of the ribosome during protein biosynthesis (Chapter 33).

Further details of the structure and function of tRNAs are provided below and in Chapter 33 (page 929).

Ribosomal RNA

Ribosomal RNA (rRNA) constitutes up to 65 percent of the mass of ribosomes. It can be obtained from E. coli ribosomes in the form of linear, single-stranded molecules that appear in three characteristic forms, sedimenting at 23S, 16S, and 5S, respectively; these three forms differ in base ratios and sequences. In eukaryotic cells, which have larger ribosomes than prokaryotes, there are four types of rRNA; 5S, 7S, 18S, and 28S. Although rRNAs make up a large fraction of total cellular RNA, their function in ribosomes is not yet clear. A few of the bases in rRNAs are methylated.

Shorthand Representation of Nucleic Acid Backbones

The covalent structure and base sequence of polynucleotide chains is often schematized as shown in Figure 12-14. These diagrams are also useful in indicating the specific bonds cleaved during chemical or enzymatic hydrolysis of nucleic acids, as we shall see below. In addition, an international convention is frequently used to indicate the nucleoside sequences of polynucleotides. The nucleosides of RNA are symbolized by A, U, G, and C; those of DNA by dA, dT, dG, and dC. The letter p designates a terminal phosphate group, a hyphen an internal phosphate group. When p appears to the left of a nucleoside symbol, the phosphate is esterified to the 5' position; when it appears at the right of the nucleoside symbol, the phosphate is esterified to the 3' position. Thus pA is adenosine 5'-phosphate and Ap is adenosine 3'-phosphate. Oligonucleotides (Greek oligo, "few") are conveniently symbolized as shown in the examples in Figure 12-14. To symbolize a DNA sequence the base symbols are prefixed by d, as in dA-T-G-Cp. The 3' terminus of an oligonucleotide is that end at which the terminal nucleoside is attached by its 5' carbon to the phosphoric group of the preceding nucleotide in the main chain.

Hydrolysis of Nucleic Acids by Acids and Bases

Gentle acid hydrolysis of DNA at pH 3.0 causes selective hydrolytic removal of all its purine bases without affecting the pyrimidine-deoxyribose bonds or the phosphodiester bonds of the backbone. The resulting DNA derivative, which is devoid of purine bases, is called an apurinic acid. Selective removal of the pyrimidine bases, accomplished by somewhat different chemical conditions, produces apyrimidinic acid.

DNA is not hydrolyzed by dilute alkali, whereas RNA is because of the 2'-hydroxyl groups it contains. Dilute sodium hydroxide produces from RNA a mixture of nucleoside

Figure 12-14
Schematic representation and shorthand notation of polynucleotide structure. The vertical lines represent the pentose backbone, the numbers 3' and 5' the carbon atoms of the pentose, and P the phosphoric group. The diagrams always show the 3',5' phosphodiester linkage going from left to right. The shorthand notation for oligodeoxyribonucleotides may include the prefix d, if needed.

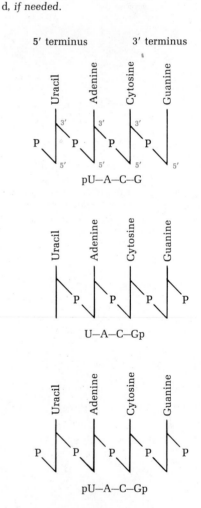

The a (3') and b (5') linkages (color) of the phosphodiester internucleotide bonds.

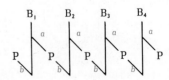

2'- and 3'-phosphates (Figure 12-10). Cyclic 2',3'-monophosphates are the first products of the action of alkali on RNA and are obligatory intermediates; they are then further hydrolyzed by the alkali, which attacks either one of the two P—O—C linkages to yield a mixture of 2'- and 3'-nucleoside monophosphates (Figure 12-10). The discovery of this mechanism explains why DNA is not hydrolyzed by base: DNA has no 2'-hydroxyl groups and therefore cannot form the necessary 2',3'-cyclic monophosphate intermediates. Nucleoside 2',3'-cyclic monophosphates are also intermediates in the action of some ribonucleases (see below).

Selective hydrolytic cleavage of polynucleotides by enzymatic or chemical methods is used in determining the base sequence of nucleic acids, which in principle is approached with the same logic as that employed in the sequence analysis of proteins (page 98).

Enzymatic Hydrolysis of Nucleic Acids

Nucleic acids ingested by animals undergo enzymatic hydrolysis in the intestine by nucleases secreted by the pancreas. These and other enzymes capable of hydrolyzing nucleic acids are important tools in analyzing nucleotide sequence. The phosphodiester bridges of DNA and RNA are attacked by two classes of enzymes, designated a and b (or 3' and 5') in Figure 12-14 and Table 12-4, depending on which side of the phosphodiester bridge is attacked. The a, or 3', enzymes specifically hydrolyze the ester linkage between the 3' carbon and the phosphoric group, and the b, or 5', enzymes hydrolyze the ester linkage between the phosphoric group and the 5' carbon of the phosphodiester bridge (Figure 12-14). The best known of the class a enzymes is a phosphodiesterase from the venom of the rattlesnake or Russell's viper, which hydrolyzes all the 3' bonds in either RNA or

Table 12-4   Specificity of some enzymes acting on nucleic acids

Enzyme	Nucleic acid	Specificity
Class a (3') nucleases		
Exonuclease		
Snake-venom phosphodiesterase	DNA and RNA	Starts from 3' end
Endonuclease		
Deoxyribonuclease I	DNA	Some 3' linkages
Class b (5') nucleases		
Exonuclease		
Spleen phosphodiesterase	DNA and RNA	Starts from 5' end
Endonucleases		
Deoxyribonuclease II	DNA	Some 5' linkages
Ribonuclease I (pancreas)	RNA	5' linkages in which the 3' linkage is to pyrimidine nucleotide
Ribonuclease T_1 (mold)	RNA	5' linkages in which the 3' linkage is to a guanine nucleotide

DNA, liberating nearly all the nucleotide units as nucleoside 5'-phosphates. The enzyme requires a free 3'-hydroxyl group on the terminal nucleotide residue and proceeds stepwise from that end of the polynucleotide chain. This enzyme and all other nucleases attacking only at the ends of polynucleotide chains are called exonucleases. The class b enzymes are represented by a phosphodiesterase from bovine spleen, also an exonuclease, which hydrolyzes all the b, or 5', linkages of both DNA and RNA and thus liberates only nucleoside 3'-phosphates. It begins its attack at the end of the chain having a free 5'-hydroxyl group.

Endonucleases do not require a free 3'- or 5'-hydroxyl group at the end of the chain; they attack certain 3' or 5' linkages wherever they occur in the polynucleotide chain. Deoxyribonuclease I of bovine pancreas, an endonuclease, catalyzes hydrolysis of some of the a, or 3', linkages of DNA to yield oligonucleotides containing on the average about four nucleotide residues. Another endonuclease, deoxyribonuclease II, isolated from spleen, thymus, or various bacteria, hydrolyzes some of the b, or 5', linkages (Table 12-4).

RNA can similarly be degraded by RNA-specific nucleases. Crystalline ribonuclease I from bovine pancreas (page 227) is an endonuclease; it hydrolyzes those 5' linkages of RNA in which the 3' linkage is attached to a pyrimidine nucleotide (Figure 12-14 and Table 12-4). Thus the end products of ribonuclease action are pyrimidine nucleoside 3'-phosphates and oligonucleotides terminating in pyrimidine 3'-phosphate residues. A cyclic pyrimidine 2',3'-phosphate residue is an obligatory intermediate, which is then selectively hydrolyzed to yield the terminal pyrimidine 3'-phosphate residues. Ribonuclease T_1 from the mold *Aspergillus oryzae* hydrolyzes b, or 5', linkages between 3'-guanylate residues and the 5' carbon of the adjacent residues. A number of other nucleases with different types of nucleotide specificity are important tools for fragmenting nucleic acids and determining their nucleotide sequences.

Analysis of Nucleotide Sequence in Nucleic Acids

Since the nucleotide sequence of various types of nucleic acids is the primary means of storing and transmitting genetic information, the experimental identification of such sequences is a central problem in genetic biochemistry.

Just as selective hydrolysis of certain peptide bonds by specific enzymes and chemical procedures has made possible the directed fragmentation of polypeptides for analysis of amino acid sequence (Chapter 5, page 106), selective hydrolysis of internucleotide linkages allows directed fragmentation of nucleic acids. However, sequence analysis is generally more difficult for nucleic acids. Since there are only four major nucleotides in nucleic acids whereas proteins contain 20 amino acids, recognition of distinctive sequences of nucleotides is more difficult because the chances of ambiguity are greater.

Figure 12-15
(Opposite) *Analysis of the nucleotide sequence of yeast alanine tRNA. Yeast alanine tRNA was first isolated from all other tRNAs by countercurrent distribution. The intact tRNA chain (77 residues) was then fragmented into two series of oligonucleotides by action of two ribonucleases, one from pancreas (a, b, c, d) the other from a mold (e, f, g). These fragments were separated by column chromatography. Each fragment was next hydrolyzed completely to determine its base content. To analyze the base sequence of the single fragment shown, snake-venom phosphodiesterase was used for removal of nucleotide units from one end of the chain. Chromatography of the resulting complex mixture (lower right) permitted identification of the base sequence in this fragment. Many repetitive cleavage and chromatographic steps were required.*
[R. Holley, Scientific American, 214: 30 (1966).]

The problem was first tackled with the tRNAs, since they have relatively short chains and can be isolated in pure form. In 1965, R. W. Holley and his colleagues deduced the first complete RNA sequence, a remarkable experimental achievement. The approach taken was, in principle, that introduced by Sanger for determination of the amino acid sequence of polypeptide chains (Figure 12-15). The 77-member polynucleotide chain of yeast alanine tRNA was cleaved into a series of small oligonucleotide fragments by the action of specific nucleases. These were separated, their sequence determined by the action of exonucleases (Figure 12-15), and the sequence of the entire tRNA chain then pieced together by using a second fragmentation method to provide overlaps. Alanine tRNA contains nine nucleotide residues with minor bases, which served as distinctive markers of different parts of the polynucleotide chain. The complete base sequence of this particular yeast alanine tRNA is given in Figure 12-16.

In the last few years the complete nucleotide sequences of over 40 different tRNA molecules have been determined, for a number of different amino acids and from various orga-

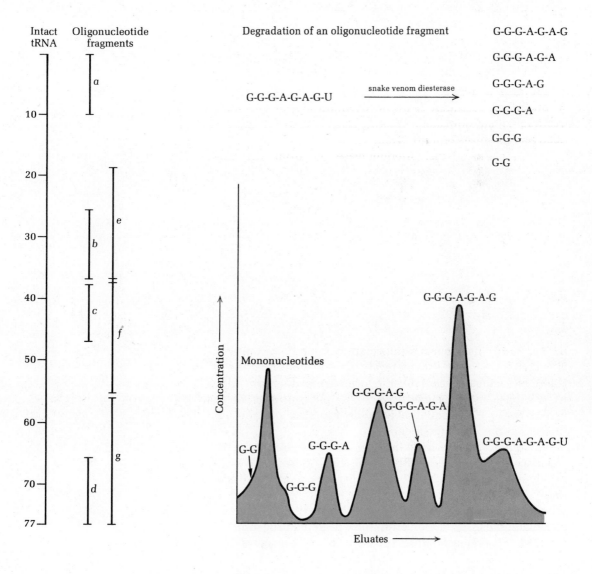

Figure 12-16
The nucleoside sequence in yeast alanine tRNA₁. In addition to A, G, U, and C, the following symbols are used for the nucleosides: ψ = pseudouridine, I = inosine, T = ribothymidine, hU = 5,6-dihydrouridine, m'I = 1-methylinosine, m'G = 1-methylguanosine, m₂G = N²-dimethylguanosine. The symbols for the rare nucleosides are in color. The dots between the parallel sections show the points where hydrogen bonding occurs between complementary bases (page 865). All tRNAs have a similar cloverleaf structure. The specific anticodon shown is the nucleotide triplet capable of recognizing a codon for alanine in the mRNA molecule. Other features of tRNA structure are given in Chapter 32 (page 935). [Redrawn from R. W. Holley, J. Apgar, G. A. Everett, J. T. Madison, M. Marguisee, S. H. Merrill, J. R. Penswick, and A. Zamir, Science, 147: 1462 (1965).]

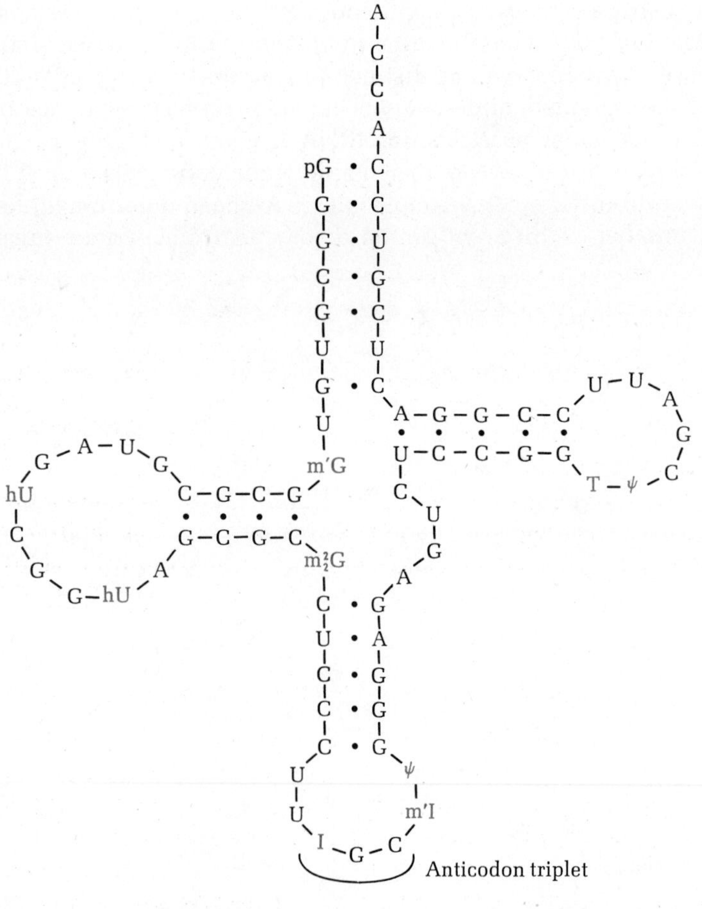

Anticodon triplet

nisms. From this sequence information some important conclusions have been drawn regarding the conformation of tRNA molecules (page 935), the structure of the anticodons for various amino acids (page 969), the evolution of the tRNAs (page 971), and the possible function of their distinctive minor bases (page 969).

Not long after the breakthrough on tRNA structure, the sequence of 5S ribosomal RNA of *E. coli* ribosomes was determined by F. Sanger and his colleagues, a more difficult problem, since it is longer than the tRNAs and contains no minor bases (Figure 12-17). They utilized two-dimensional chromatography as a powerful tool in separating oligonucleotide fragments. S. Spiegelman and his colleagues have

Figure 12-17
The nucleoside sequence of 5S
RNA from E. coli. It contains 120
nucleotide residues.

pU-G-C-C-U-G-G-C-G-G-C-C-G-U-A-G-C-G-C-G-G-U-G-G-U-C-C-C-A-C-C-U-G-A-C-C-C-C-A-U-G-C-C-G-A-A-C-U-C-A-G-A-A-G-U-

G-A-A-A-C-G-C-C-G-U-A-G-C-G-C-C-G-A-U-G-G-U-A-G-U-G-U-G-G-G-G-U-C-U-C-C-C-C-A-U-G-C-G-A-G-A-G-U-A-G-G-G-A-A-

C-U-G-C-C-A-G-G-C-A-U$_{\overline{120}}$OH

reported the nucleotide sequence of an RNA containing 218 nucleotide residues.

Determination of the nucleotide sequence of DNAs is much more difficult, since even the smallest DNA molecules contain at least 5,000 nucleotide units and thus are very much longer than tRNA and 5S RNA. Moreover, the deoxyribonucleases are much less specific for certain internucleotide linkages than the ribonucleases. However, some important beginnings are being made. A newly discovered class of deoxyribonucleases, called restriction endonucleases, has been found very useful for specific fragmentation of DNA. These species-specific bacterial enzymes can cleave, at certain specific points, DNA molecules other than those naturally present in the cells from which they derived (page 881). Restriction endonucleases function biologically to protect a given organism from the deleterious effects of a foreign DNA introduced into the cell. Another approach to the sequencing of DNA specimens is made possible by chemical modification of DNA, e.g., to yield apurinic or apyrimidinic acids (page 322).

Nucleic Acid–Protein Supramolecular Complexes

Certain nucleic acids occur in cells in noncovalent association with specific proteins as supramolecular complexes (page 19). Of these nucleic acid–protein systems, which have very complex structure and biological functions, the ribosomes and the viruses are by far the best understood. Perhaps the most complex nucleic acid–protein systems are the chromosomes of eukaryotic cells (page 869).

Ribosomes

Ribosomes are ribonucleoprotein particles found in all types of cells (page 19); they are essential in the biosynthesis of proteins (page 929). The ribosomes of prokaryotic cells have a diameter of about 18 nm, a mass of about 2.8 megadaltons, and a sedimentation coefficient of 70S; they consist of 60 to 65 percent rRNA and about 35 to 40 percent protein (Figure 12-18). An E. coli cell contains about 15,000 ribosomes, which accounts for some 25 percent of the dry weight of the cell. The cytoplasmic ribosomes of eukaryotic cells are larger and vary somewhat in size in different organisms; they have a diameter of 20 to 22 nm and a sedimentation coefficient of

73S to 80S. Most of the ribosomes are found in the cytoplasm, either in free form or bound to the surface of the endoplasmic reticulum (pages 33 and 949). In eukaryotic cells ribosomes are also found in the cell nucleus and in organelles such as mitochondria and chloroplasts; mitochondrial ribosomes are smaller than cytoplasmic ribosomes (page 951). Many of the ribosomes in both prokaryotes and eukaryotes are associated in beadlike strings called polyribosomes or simply polysomes, which are formed by the attachment of a number of ribosomes to a single molecule of mRNA during protein synthesis (page 947).

The ribosomes of all cells are constructed according to the same architectural plan. They have two subunits of unequal size (Figure 12-18). In E. coli ribosomes the subunits have sedimentation coefficients of 30S and 50S (particle weights of 1.0 and 1.8 megadaltons, respectively). The 50S subunit contains a molecule of 23S rRNA and a molecule of 5S rRNA; the 30S subunit contains one molecule of 16S rRNA. Both subunits contain a large number of different polypeptide chains. Eukaryotic ribosomes contain larger rRNA molecules and more polypeptide chains than prokaryotic ribosomes (Figure 12-18). Ribosomes are stable in solutions containing relatively high Mg^{2+} concentration but may dissociate into their subunits when the Mg^{2+} concentration is decreased. Ribosomal structure and function are considered in more detail in Chapters 33 and 36.

Much is now known about the structural organization and the manner of assembly of prokaryotic ribosomes. Exposure of ribosomal subunits to strong salt solutions sequentially extracts a series of specific protein components. Removal of these proteins yields a core containing the rRNA and a number of more tightly bound proteins. The complete ribosome can then be reconstituted from these cores by restoring the proteins removed by extraction. The complete assembly of bacterial ribosomes is described in Chapter 36 (page 1023).

Viruses

Viruses, which have been aptly described as structures "at the threshold of life," are particles composed principally of nucleic acid and a number of specific protein subunits; they have the ability to direct their own replication when they gain entry to a specific host cell. Viruses infect animal, plant, and bacterial cells. The bacterial viruses are also called bacteriophages. In man, many diseases, e.g., poliomyelitis, influenza, and the common cold, are transmitted by viruses. Much evidence also suggests that some types of cancer are caused or transmitted by viruses, at least in some animal species.

Although viruses have extremely large particle weights, they can be isolated in homogeneous form and many have been crystallized. Virus particles, or virions, are capable of attaching themselves to the surface of specific host cells and delivering their nucleic acid, the infectious portion of the virion, into the cytoplasm. The viral nucleic acid then mo-

Figure 12-18

Structure of ribosomes from prokaryotic and eukaryotic cells. The latter contain two distinctive types of ribosomes: cytoplasmic, shown below, and mitochondrial. The latter more nearly resemble prokaryotic ribosomes.

Prokaryotic ribosome [E. coli]

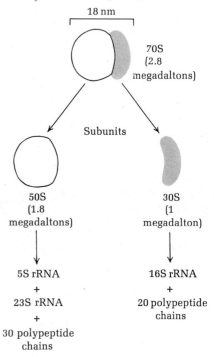

18 nm

70S
(2.8 megadaltons)

Subunits

50S
(1.8 megadaltons)

30S
(1 megadalton)

5S rRNA
+
23S rRNA
+
30 polypeptide chains

16S rRNA
+
20 polypeptide chains

Eukaryotic ribosomes (cytoplasmic)

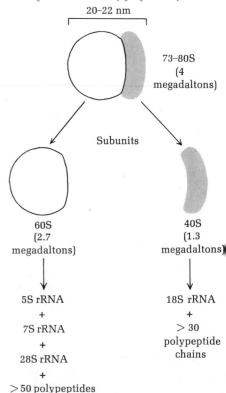

20–22 nm

73–80S
(4 megadaltons)

Subunits

60S
(2.7 megadaltons)

40S
(1.3 megadaltons)

5S rRNA
+
7S rRNA
+
28S rRNA
+
>50 polypeptides

18S rRNA
+
>30 polypeptide chains

Table 12-5   Composition of some viruses

	Virion weight, megadaltons	Nucleic acid	No. of chains	Percent nucleic acid	Shape	Long dimension, nm
E. coli bacteriophages						
T2, T4, T6	~220	DNA	2	61	Tadpole	18
T7	38	DNA	2	41	Tadpole	6
ϕX174	6	DNA	1 or 2	26	Polyhedral	15
λ	50	DNA	2	64	Tadpole	20
MS2	3.6	RNA	1	32	Polyhedral	17.5
Plant viruses						
Tobacco mosaic	40	RNA	1	5	Rod	300
Tomato bushy stunt	10.6	RNA	1	15	Polyhedral	28
Tobacco necrosis	1.97	RNA	1	20	Polyhedral	21
Animal viruses						
Poliomyelitis	6.7	RNA	1	28	Polyhedral	30
Polyoma	21	DNA	2	13.4	Polyhedral	45
Adenovirus	200	DNA	2	5.0	Polyhedral	70
Vaccinia	2,000	DNA	?	7.5	Brick	230

nopolizes the biosynthetic machinery of the host cell, forcing it to synthesize the molecular components of virus molecules rather than the normal host-cell nucleic acids and proteins. The DNA of DNA-containing viruses, after gaining entry into the host, serves as the template for transcription of complementary mRNA molecules, which in turn can usurp the host ribosomal apparatus and cause it to synthesize viral proteins and the enzymes required to synthesize viral DNA. The RNA of RNA viruses is bound to the host-cell ribosomes in preference to the host-cell mRNA molecules. It then acts as a template for the synthesis of viral coat proteins and for the additional enzymes required to replicate the viral RNA itself. In some cases, viral RNA causes the synthesis of a DNA that codes for the synthesis of certain viral proteins.

Viruses vary considerably in size, shape, and chemical composition (Table 12-5). One of the smaller viruses is the *E. coli* bacteriophage ϕX174, whose DNA has about 5,000 nucleotide residues. In the larger viruses, e.g., the *E. coli* bacteriophage T2, the DNA may contain over 200,000 residues. In the simpler viruses, e.g., tobacco mosaic, only a single type of protein subunit is present, but in bacteriophage T2 there are at least 50 different kinds of protein subunits.

All plant viruses contain RNA and are either rodlike helixes, as in *tobacco mosaic virus*, or eicosahedral (20-sided), as in *tomato bushy stunt virus*. Animal viruses contain either DNA or RNA. Bacterial viruses are most convenient for genetic and biochemical study because they replicate in large numbers in bacterial suspensions and are easily isolated. The most widely studied bacterial viruses are those of *E. coli* cells, such as the DNA bacteriophages T2, T4, T6, ϕX174, and *lambda* (λ), and the RNA bacteriophages Qβ, MS2, and R17.

The structure of viruses has been elucidated in great detail by electron microscopy and x-ray diffraction. Electron mi-

Figure 12-19
Electron micrographs of viruses.

Tobacco mosaic virus

Bacteriophage T4

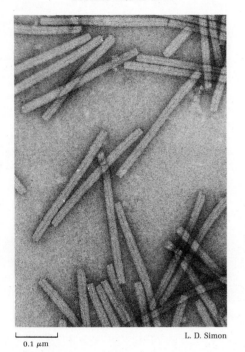

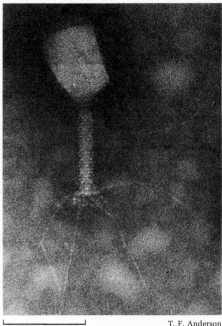

0.1 μm

L. D. Simon

0.1 μm

T. F. Anderson

crographs of three viruses are shown in Figure 12-19. Other aspects of viruses are discussed in Part 4, including their genetic constitution (page 866), replication (pages 904 and 922), and assembly (page 1017).

Summary

Nucleic acids contain recurring monomeric units called nucleotides. The monomeric units of ribonucleic acid, the ribonucleotides, are composed of one molecule of a purine or pyrimidine base, one of D-ribose, and one of phosphoric acid. The monomeric units of deoxyribonucleic acid, the deoxyribonucleotides, contain 2-deoxy-D-ribose instead of D-ribose. The purine or pyrimidine bases are covalently bound in β-N-glycosyl linkage to carbon atom 1 of D-ribose or 2-deoxy-D-ribose. The resulting compounds are known as nucleosides. The most abundant nucleotides contain phosphate in ester linkage at the 5'-hydroxyl group of the pentose. The common ribonucleotides contain either one of the pyrimidines cytosine and uracil or one of the purines adenine and guanine. The common deoxyribonucleotides contain either one of the pyrimidines cytosine and thymine or one of the purines adenine and guanine. Also important are the 5'-diphosphoric and 5'-triphosphoric derivatives of nucleosides, called nucleoside (or deoxyribonucleoside) 5'-diphosphates and 5'-triphosphates. Nucleotides serve as building blocks of nucleic acids but also have other functions. ATP serves to carry phosphate groups, UDP and GDP serve to carry sugar groups, and GDP carries amine groups in enzyme-catalyzed reactions.

Shope papilloma virus

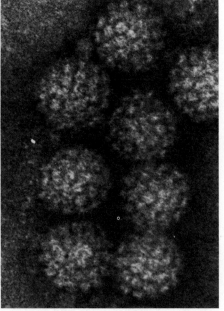

0.1 μm

C. Breedis, L. Berwick,
and T. F. Anderson

In nucleic acids the nucleotide monomers are joined by phosphodiester bridges between the 3'-hydroxyl group of one nucleotide unit and the 5'-hydroxyl group of the next. Deoxyribonucleic acids serve to store and transmit genetic information. Most have an extremely high molecular weight; they vary in base composition and sequence. There are at least three major types of ribonucleic acids: messenger RNA, transfer RNA, and ribosomal RNA. RNAs function in the expression of the genetic information of DNA.

The internucleotide linkages of RNA are split by dilute alkali to yield a mixture of nucleoside 2'- and 3'-phosphates; DNA is not split by bases. Both DNA and RNA may be hydrolyzed by snake-venom phosphodiesterase in such a way as to yield only 5'-phosphorylated nucleosides. Others, such as spleen phosphodiesterase, cleave the other side of the phosphodiester linkage to yield only 3'-phosphorylated nucleosides. Both these enzymes are exonucleases. Nucleases specific for the linkages between certain pairs of mononucleotides also are known; they are important tools in degrading nucleic acids. Pancreatic ribonuclease hydrolyzes RNA in such a way as to yield pyrimidine-containing nucleoside 3'-phosphates and oligonucleotides terminating in them. Dilute acids cleave purines from DNA to yield apurinic acids. The nucleoside sequences of nucleic acids can be deduced by fragmentation of the chain, determination of the sequence in the fragments, and the use of overlaps to put the fragments in the proper order. The base sequences of many RNAs have been established.

Nucleic acids are associated with specific proteins in supramolecular complexes such as ribosomes and viruses.

References

Many other references to nucleic acid structure and function are given at the ends of Chapters 31 to 35 (pages 888, 925, 953, 973, and 1007).

Books

DAVIDSON, J. N.: The Biochemistry of Nucleic Acids, Academic, New York, 1972. Elementary paperback on all aspects of nucleic acid biochemistry.

FRAENKEL-CONRAT, H.: Design and Function at the Threshold of Life: The Viruses, Academic, New York, 1972. Classic elementary account of virus structure and function.

KORNBERG, A.: DNA Synthesis, Freeman, San Francisco, 1974. Valuable, well-illustrated account of DNA biochemistry.

MICHELSON, A. M.: The Chemistry of Nucleosides and Nucleotides, Academic, New York, 1963. Structure, laboratory synthesis, and chemical and physical properties are considered in detail. Progress in Nucleic Acid Research and Molecular Biology. A series of annual review volumes.

STEWART, P. R., and D. S. LETHAM (eds.): The Ribonucleic Acids, Springer, New York, 1973. Comprehensive, up-to-date résumé of the biochemistry of RNAs.

Articles

BROWNLEE, G. G., F. SANGER, and B. G. BARRELL: "The Sequence of 5S Ribosomal Ribonucleic Acid," J. Mol. Biol., 34: 379–412 (1968).

HOLLEY, R. W., J. APGAR, G. A. EVERETT, J. T. MADISON, M. MARGUISEE, S. H. MERRILL, J. R. PENSWICK, and A. ZAMIR: "The Base Sequence of Yeast Alanine Transfer RNA," *Science*, 147: 1462–1465 (1965). The first determination of the base sequence of an RNA.

HOLMQUIST, R., T. H. JUKES, and S. PANGBURN: "Evolution of Transfer RNA," *J. Mol. Biol.*, 78: 91–116 (1973). Comparison of the nucleotide sequences in tRNAs for different amino acids and from different species.

MILLS, D. R., F. R. KRAMER, and S. SPIEGELMAN: "Complete Nucleotide Sequence of a Replicating RNA Molecule," Science, 180: 916–927 (1973). Interesting account of the sequence analysis and evolutionary significance of an RNA molecule capable of replicating itself by action of a virus-induced RNA replicase.

MIRSKY, A. E.: "The Discovery of DNA," *Sci. Am.*, June 1968, p. 78. An absorbing account.

TS'O, P. O. P.: "Monomeric Units of Nucleic Acids: Bases, Nucleosides, and Nucleotides," p. 49 in *Fine Structure of Proteins and Nucleic Acids*, G. Fasman and S. Timasheff (eds.), Dekker, New York, 1970. A valuable comprehensive review of the structure and physical properties of the basic units of nucleic acids.

Problems

1. Calculate $A_{260\,nm}$, the absorbance, of the following aqueous solutions in 1.0-cm cells at pH 7.0. Use the molar absorption coefficients in Figure 12-7.
   (a) 3.2×10^{-5} M AMP
   (b) 47.5 μM CMP
   (c) 6.0 μM UMP
   (d) A mixture of 4.8×10^{-5} M AMP and 3.2×10^{-5} M UMP

2. Calculate the molar concentration of the following aqueous solutions at pH 7.0, using the data in Figure 12-7.
   (a) A GMP solution of $A_{260\,nm} = 0.325$ (optical path 1.0 cm)
   (b) A dTMP solution of $A_{260\,nm} = 0.090$ (optical path 1.0 cm)

3. A solution containing a mixture of AMP and GMP gave an absorbancy index of 0.652 at 260 nm and 0.284 at 280 nm. Given the molar absorbancy indexes below, calculate the concentrations of AMP and GMP in the solution.

   AMP: $A_{260\,nm} = 15.4 \times 10^3$ M^{-1} cm^{-1}
   $\quad\quad\ A_{280\,nm} = 2.5 \times 10^3$ M^{-1} cm^{-1}
   GMP: $A_{260\,nm} = 11.7 \times 10^3$ M^{-1} cm^{-1}
   $\quad\quad\ A_{280\,nm} = 7.7 \times 10^3$ M^{-1} cm^{-1}

4. Indicate the products of the action of (a) snake-venom phosphodiesterase on A-U-A-A-C-U, (b) bovine pancreatic ribonuclease on A-U-A-A-C-U, (c) snake-venom phosphodiesterase on dT-A-G-G-Cp, (d) mild acid hydrolysis on dT-A-C-G-G-C-A, (e) mild basic hydrolysis on dT-A-C-G-G-C-A.

5. An oligoribonucleotide of base composition $A_2C_4G_2U$ was incubated with various enzymes. Treatment with pancreatic ribonuclease yielded 2 mol of Cp, 1 mol of a dinucleotide con-

taining adenine and uracil, 1 mol of a dinucleotide containing guanine and cytosine, and 1 mol of a trinucleotide containing adenine, cytosine, and guanine. Treatment with takadiastase yielded one molecule each of free cytosine, Ap and pGp, a trinucleotide containing cytosine, guanine, and uracil, and one other product. (Takadiastase is an enzyme preparation from a mold which hydrolyzes those b linkages of oligonucleotides in which the a linkage is attached to purine-containing nucleotides.) Snake-venom phosphodiesterase treatment of the original oligonucleotide for a limited time yielded some pC. Deduce a sequence for the oligonucleotide which is consistent with these data.

6. An oligoribonucleotide of base composition $A_2C_4G_4U_2$ was treated with certain enzymes and the base composition of the products determined following their hydrolysis. Pancreatic ribonuclease gave 3 mol of Cp, 1 mol of a dinucleotide containing adenine and uracil, 1 mol of a dinucleotide containing only guanine, a dinucleotide containing guanine and uracil, and a trinucleotide containing adenine, cytosine, and guanine. Takadiastase gave Ap, two Gp, and three trinucleotides, one containing adenine, cytosine, and uracil, the second containing cytosine and guanine, and the third containing cytosine, guanine, and uracil. Deduce a sequence for the oligonucleotide which is consistent with these data.

7. A sample is known to contain 5′-AMP, 5′-CMP, cytidine, and cytosine. A portion of the sample is placed in a cell with a 1.0-cm light path and is read in a spectrophotometer with the following results: $A_{260\,nm} = A_{280\,nm} = 0.500$. Digestion of a 10-ml aliquot of the original sample with alkaline phosphatase gave 3.07×10^{-7} mol inorganic phosphate. A total ribose determination on a 1.0-ml portion of the original sample gave 3.87×10^{-8} mol ribose. Calculate the molar concentrations of (a) 5′-AMP, (b) 5′-CMP, (c) cytidine, and (d) cytosine in the original sample. Use the molar-absorbancy values provided below, and assume that the same values apply to bases, nucleosides, and nucleotides.

AMP: $A_{260\,nm} = 15.4 \times 10^3\ M^{-1}\ cm^{-1}$
$A_{280\,nm} = 2.5 \times 10^3\ M^{-1}\ cm^{-1}$
CMP: $A_{260\,nm} = 6.47 \times 10^3\ M^{-1}\ cm^{-1}$
$A_{280\,nm} = 13.0 \times 10^3\ M^{-1}\ cm^{-1}$

CHAPTER 13 VITAMINS AND COENZYMES

In addition to their bulk components—proteins, nucleic acids, carbohydrates, and lipids—living cells also contain certain organic substances that function in trace amounts, the *vitamins*. Although these substances are vital for many forms of life, their biological importance was first recognized because some organisms cannot synthesize them and must therefore acquire them from exogenous sources.

The story of the vitamins, one of the most important episodes in the history of biochemistry, has touched profoundly on man's health and well-being and on our understanding of the catalytic processes taking place in the metabolism of living organisms. That diet is related to disease was recognized in antiquity. Thus, liver was early known to cure night-blindness. In the eighteenth century cod liver oil was first used to treat rickets and later the juice of limes was discovered to prevent the symptoms of scurvy among seamen of the British navy (henceforth, British sailors came to be known as "limeys"). But it was not until 1912 that F. G. Hopkins, in England, proved experimentally that animals require more than protein, fat, and carbohydrate in the diet for normal growth. He postulated that one or more "accessory factors" present in certain natural foods are also necessary in the nutrition of animals. In the same year Casimir Funk obtained a concentrate of an amine from rice husks and polishings that alleviated the symptoms of the disease beriberi, prevalent among Japanese sailors limited to a diet of milled or polished rice. He coined the name *vitamine*, denoting an amine essential for life (today spelled vitamin, since many of the substances in this class are not amines). Shortly later it became clear that there must be several vitamins, when E. V. McCollum, in the United States, discovered that young rats require both fat-soluble and water-soluble growth factors.

These early discoveries ultimately led to the development of a systematic approach to the isolation of trace factors required in the nutrition of the rat and other laboratory animals. Synthetic diets, prepared from specific proportions of chemically known components, such as proteins or amino acids, fats, carbohydrates, and minerals, were fed to young experimental animals. Although these components alone do not suffice to give normal growth rates, when the ration is supplemented with natural foods rich in growth factors, increases in growth rates are observed. By systematically fractionating the natural source of the growth factor(s) and assaying the potency of the concentrates on the rate of growth of the animals, the growth factors were ultimately obtained in pure form for chemical identification. In addition to growth rate, other criteria and/or deficiency symptoms, such as scaly skin, matted fur, altered gait, etc., have also been used to assay the potency of vitamin concentrates. Determining the precise amounts of exogenous vitamin required by experimental animals or man is often complicated by the fact that some or all of the requirement may be met through the presence of intestinal microorganisms which can synthesize some of the vitamins.

A dramatic breakthrough came in the mid-1930s, when some of the vitamins were first isolated and their molecular structures were established, a laboratory feat of heroic proportions at the time, since only milligram quantities of pure substance could be obtained, often from hundreds of pounds—indeed tons—of starting material. Within a few years _thiamin_ (_vitamin B₁_) the antiberiberi substance, _riboflavin_ (_vitamin B₂_), and _nicotinic acid_, the antipellagra factor, were identified. Soon afterward it was discovered that these vitamins function as essential building-block components of certain coenzymes. For example, the enzyme now called _pyruvate decarboxylase_, which catalyzes the decarboxylation of pyruvic acid to acetaldehyde and CO_2, a step in the alcoholic fermentation of sugar by yeast, requires a heat-stable organic cofactor, called _cocarboxylase_, which K. Lohmann and P. Schuster succeeded in isolating in 1936. Within months they found that cocarboxylase contains a molecule of thiamin, or vitamin B_1. Very soon riboflavin and nicotinic acid were identified as important components of other coenzymes required in the enzymatic oxidation of carbohydrate. Not only did these discoveries demonstrate the biological role of many vitamins and show why these substances function in only trace amounts but they have since pointed the way to a molecular understanding of the mechanism by which coenzymes and enzymes promote the rate of chemical reactions.

In this chapter we describe the major vitamins known today, the nature of their biological effects, the coenzymes in which the vitamins are essential components, and examples of coenzyme function in specific enzyme-catalyzed metabolic reactions.

Classification of Vitamins

Vitamins are classified into two large groups, _water-soluble_ and _fat-soluble_ (Table 13-1). The coenzyme function of all of the water-soluble vitamins is reasonably well known, with the exception of vitamin C. In addition several other water-soluble substances are necessary as growth factors for some

Table 13-1 Vitamins and their coenzyme forms

Type	Coenzyme or active form	Function promoted
Water-soluble		
Thiamin	Thiamin pyrophosphate (TPP)	Aldehyde-group transfer
Riboflavin	Flavin mononucleotide (FMN)	Hydrogen-atom (electron) transfer
	Flavin adenine dinucleotide (FAD)	Hydrogen-atom (electron) transfer
Nicotinic acid	Nicotinamide adenine dinucleotide (NAD)	Hydrogen-atom (electron) transfer
	Nicotinamide adenine dinucleotide phosphate (NADP)	Hydrogen-atom (electron) transfer
Pantothenic acid	Coenzyme A (CoA)	Acyl-group transfer
Pyridoxine	Pyridoxal phosphate	Amino-group transfer
Biotin	Biocytin	Carboxyl transfer
Folic acid	Tetrahydrofolic acid	One-carbon-group transfer
Vitamin B_{12}	Coenzyme B_{12}	1,2 shift of hydrogen atoms
Lipoic acid	Lipoyllysine	Hydrogen-atom and acyl-group transfer
Ascorbic acid	—	Cofactor in hydroxylation
Fat-soluble		
Vitamin A	11-cis-Retinal	Visual cycle
Vitamin D	1,25-Dihydroxycholecalciferol	Calcium and phosphate metabolism
Vitamin E	—	Antioxidant
Vitamin K	—	Prothrombin biosynthesis

organisms, but they occur in more than trace amounts; this group includes _inositol_, _choline_, and _carnitine_.

The fat-soluble vitamins include vitamins A, D, E, and K. Only higher animals appear to require them from exogenous sources; an essential role of the fat-soluble vitamins in plants or microorganisms has not been clearly established. They do not appear to serve as components of coenzymes but function in other ways requiring only trace amounts.

Thiamin (Vitamin B_1) and Thiamin Pyrophosphate

Thiamin is necessary in the diet of most vertebrates and some microorganisms. Its deficiency causes beriberi in man and polyneuritis in birds. After Funk's early success in concentrating the antiberiberi factor, it was crystallized in 1925 by B. C. P. Jansen; its molecular structure was finally established by the American biochemist R. R. Williams and his colleagues in 1935. Thiamin consists of a substituted pyrimidine joined by a methylene bridge to a substituted thiazole (Figure 13-1). Thiamin occurs in cells largely as its active coenzyme form _thiamin pyrophosphate_, formerly called _cocarboxylase_ (page 338).

Thiamin pyrophosphate serves as the coenzyme for two classes of enzyme-catalyzed reactions in the mainstream of carbohydrate metabolism in which aldehyde groups are removed and/or transferred: (1) the decarboxylation of α-keto acids and (2) the formation or degradation of α-ketols. In these reactions the thiazole ring of the thiamin pyrophosphate serves as a transient carrier of a covalently bound "active" aldehyde group. Mg^{2+} is also required as a cofactor.

Our present view of the mechanism by which thiamin pyrophosphate functions as a coenzyme arose from the discovery that thiamin alone promotes nonenzymatic decar-

Figure 13-1
Thiamin and thiamin pyrophosphate. The
substituted thiazole (color) is the active
portion.

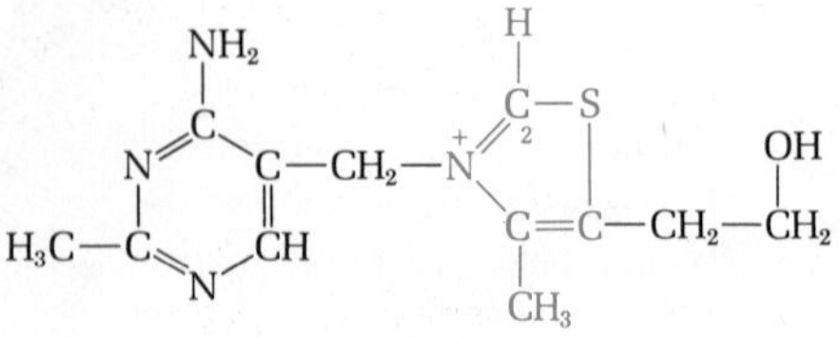

Thiamin (vitamin B₁)

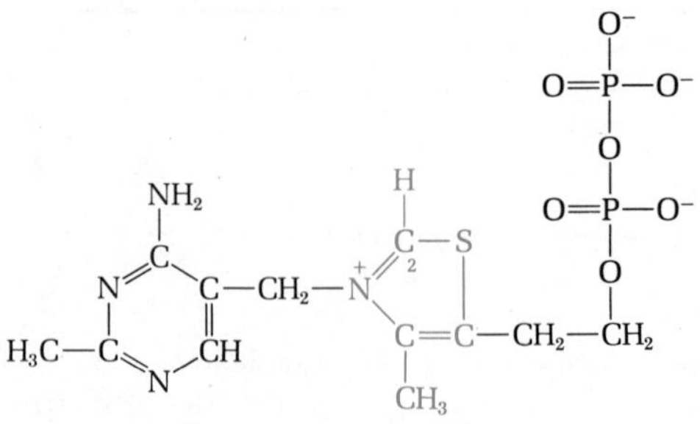

Thiamin pyrophosphate

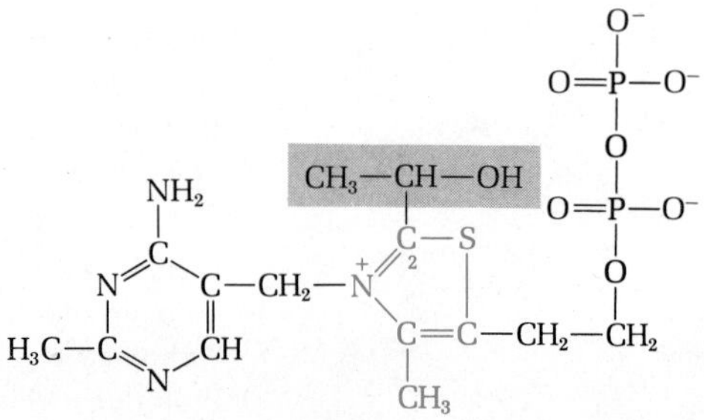

2(α-Hydroxyethyl)thiamin pyrophosphate

Figure 13-2
Steps in the action of thiamin pyrophosphate in decarboxylation of pyruvate by
pyruvate decarboxylase.

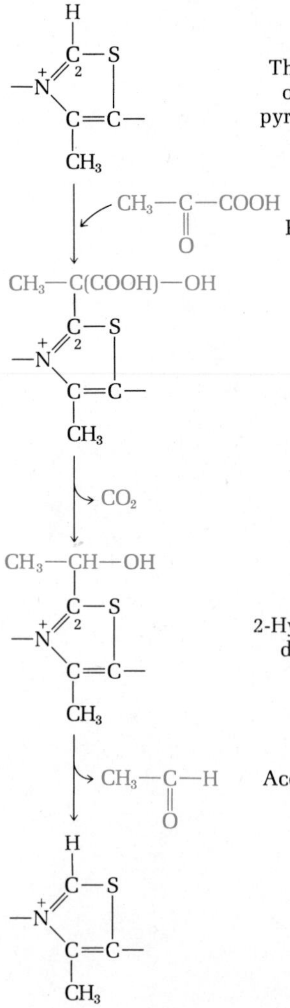

Thiazole ring
of thiamin
pyrophosphate

$CH_3—C—COOH$
Pyruvate

CO_2

2-Hydroxyethyl
derivative

$CH_3—C—H$ Acetaldehyde

boxylation of pyruvate to yield acetaldehyde and CO_2.
Studies of this model reaction disclosed that the hydrogen
atom at position 2 of the thiazole ring (Figure 13-1) ionizes
readily to yield a carbanion, which reacts with the carbonyl
carbon atom of pyruvate at elevated temperatures to yield
CO_2 and the hydroxyethyl derivative of the thiazole ring. The
hydroxyethyl group may then undergo hydrolysis, to yield
acetaldehyde, or become oxidized, to yield an acyl group
(Figure 13-2). Following the prediction from this model reaction, it has been found that hydroxyethylthiamin pyrophosphate (Figure 13-1) is formed when yeast pyruvate decarboxylase acts on pyruvate in the presence of thiamin
pyrophosphate.

Elsewhere further details are given on the role of thiamin
pyrophosphate in the decarboxylation of pyruvate (page
450), the oxidation of α-keto acids (page 457), and the formation and utilization of α-ketols by transketolases (page 468).

When thiamin is withheld from the diet of experimental
animals, the ability of the different tissues to utilize pyruvate
does not decline uniformly, indicating that in some tissues
the bound thiamin pyrophosphate is tenaciously retained
whereas in others it is more readily lost, through metabolic
degradation or excretion. For example, in pigeons deprived
of thiamin, the capacity for pyruvate utilization is most conspicuously impaired in the brain, due to loss of the coenzyme,
whereas the activity of muscle tissue is only mildly affected.
This tissue difference explains why the earliest symptoms of
thiamin deficiency in pigeons are defects in nervous-system
activity concerned with posture and coordination; hence the
term polyneuritis applied to thiamin deficiency in birds.
Similar considerations apply to the other water-soluble vitamins, the most deprivation-sensitive tissue differing from

Figure 13-3
Riboflavin and the flavin enzymes.

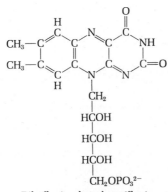

Riboflavin

Riboflavin phosphate (flavin mononucleotide; FMN)

Space-filling model of riboflavin phosphate

Riboflavin

Flavin adenine dinucleotide

Reduced form of the isoalloxazine ring of flavin nucleotides. R indicates the remainder of the coenzyme molecule.

one vitamin to another and in many cases from one animal species to another.

Symptoms of thiamin deficiency can be induced by feeding certain analogs of thiamin, such as *pyrithiamin*, in which the thiazole group is replaced by pyridine, and *oxythiamin*, in which the amine group of the pyrimidine moiety is replaced by a hydroxyl group. Presumably these analogs, which have no vitamin activity themselves, compete with or displace thiamin pyrophosphate from the enzymes to which it is normally bound. The enzyme *thiaminase*, present in intestinal microorganisms, inactivates thiamin by cleaving it between the rings. This enzyme is responsible for *Chastek paralysis*, a disease of ranch foxes and minks fed entrails of stock animals. In general, the human intake of thiamin tends to be marginal, and some foods, e.g., white bread, are often supplemented with the vitamin.

Riboflavin (Vitamin B₂) and the Flavin Nucleotides

Yellow pigments now known to be identical or related to riboflavin were first isolated from animal tissues, eggs, and milk. The one isolated from milk, first called *lactoflavin* and later *riboflavin*, was found to be a necessary growth factor for the dog and other mammals. In 1935 the structure of riboflavin was established as a derivative of *isoalloxazine* by the European chemists R. Kuhn and P. Karrer (Figure 13-3). Riboflavin is made by all plants and many microorganisms but not by higher animals.

That yellow flavin pigments function as coenzymes had earlier been deduced by H. Theorell, of Sweden, and O. Warburg, of Germany, who found that an enzyme participating in the oxidation of reduced pyridine nucleotides (page 486) contains a yellow prosthetic group, later identified by Theorell as riboflavin 5'-phosphate, also called *flavin mononucleotide* (FMN). Later, in 1938, Warburg found a second coenzyme form of riboflavin, *flavin adenine dinucleotide* (FAD). FMN is not a true nucleotide since it contains no pentose sugar; instead, like FAD, it contains the sugar alcohol ribitol.

The flavin nucleotides function as prosthetic groups of oxidation-reduction enzymes known as *flavoenzymes* or *flavoproteins*. These enzymes function in the oxidative degradation of pyruvate (page 450), fatty acids (page 548), and amino acids (page 566), and also in the process of electron transport (page 486). In most flavoenzymes the flavin nucleotide is tightly but noncovalently bound to the protein; an exception is succinate dehydrogenase (page 459), in which the flavin nucleotide FAD is covalently bound to a histidine residue of the polypeptide chain. The *metalloflavoproteins* contain one or more metals as additional cofactors. Flavin nucleotides undergo reversible reduction of the isoalloxazine ring in the catalytic cycle of flavoproteins, to yield the reduced nucleotides, symbolized $FMNH_2$ and $FADH_2$.

The oxidized forms of different flavoenzymes are intensely colored; they are characteristically yellow, red, or green due to strong absorption bands in the visible range. On reduction, flavoenzymes undergo bleaching, with a characteristic change in the absorption spectrum (Figure 13-4) that is useful in measuring their activity.

Nicotinic Acid (Niacin) and the Pyridine Nucleotides

Deficiency of nicotinic acid (Figure 13-5) leads to the disease *pellagra* in man and *blacktongue* in dogs. Pellagra was endemic among children in the southern United States in the early 1900s. It was finally recognized during 1915–1920 as a dietary deficiency disease through classical nutritional studies carried out on children in orphan asylums by Joseph Goldberger, a physician of the U.S. Public Health Service. Goldberger showed that the disease was noninfectious, was associated with low-grade starchy diets, and could be prevented by supplementing the diet with meat, eggs, and milk. However, the responsible dietary factor was not recognized as nicotinic acid until 1937, following clues derived from an entirely independent line of biochemical investigation that began in 1904, when the British biochemists A. Harden and W. J. Young discovered that a heat-stable, dialyzable cofactor is required for the alcoholic fermentation of sugar by yeast extracts (pages 420 and 427). This cofactor, first called cozymase or coenzyme I, was finally isolated by H. von Euler in Sweden in 1933. A closely related coenzyme, then called coenzyme II, was isolated by Warburg and W. Christian in 1934. Shortly later they identified *nicotinamide* (Figure 13-5), the amide of nicotinic acid, as a component of both coenzymes. This discovery quickly led to the identification of nicotinic acid as the nutritional factor preventing blacktongue in dogs and pellagra in man by the American biochemists C. A. Elvehjem and D. W. Woolley in 1937. Nicotinic acid is so named because it is a component of the toxic alkaloid nicotine of tobacco; it has since been given the trivial nonassociative name *niacin* for lay use.

Actually, plants and most animals can make nicotinic acid from other precursors, particularly the amino acid tryp-

Figure 13-4
Absorption spectrum of flavin nucleotides (oxidized form). Reduction causes the 450-nm peak to disappear.

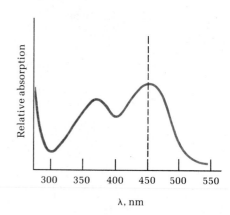

Figure 13-5
Nicotinic acid and the nicotinamide
nucleotides.

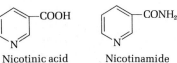

Nicotinic acid
(niacin)

Nicotinamide

Nicotinamide adenine dinucleotide
(NAD). The nicotinamide moiety,
which is the portion undergoing reversible reduction, is shown in color.
In nicotinamide adenine
dinucleotide phosphate (NADP)
the 2'-hydroxyl indicated by
the colored arrow is esterified
with phosphoric acid.

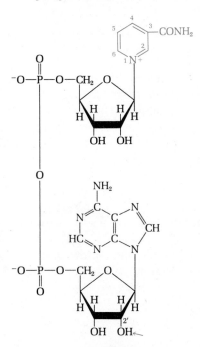

Space-filling model of extended
form of NAD

Table 13-2 Pyridine-linked
dehydrogenases

NAD-linked
    Isocitrate
    D-β-Hydroxybutyrate
    Glyceraldehyde 3-phosphate
    Dihydrolipoamide
    L-β-Hydroxyacyl-CoA
    Ethanol
    Lactate
    Glycerol 3-phosphate
    L-Malate
NADP-linked
    Isocitrate
    Glucose 6-phosphate
NAD or NADP
    L-Glutamate

tophan (page 717). If animals are amply supplied with proteins rich in tryptophan, they show no deficiency when nicotinic acid is withheld from the diet. However, if the protein
content of the diet is low, insufficient tryptophan is available
for the manufacture of niacin and a deficiency results. Although pellagra is due largely to lack of nicotinic acid, it is
usually compounded by deficiencies of other vitamins and
of the essential fatty acids.

Figure 13-5 shows the structures of the two coenzymes
containing nicotinamide as an essential component, _nico-
tinamide adenine dinucleotide_ (NAD), also called _diphos-
phopyridine nucleotide_ (DPN), and _nicotinamide adenine
dinucleotide phosphate_ (NADP), also called _triphospho-
pyridine nucleotide_ (TPN). [The older names coenzyme I
(or cozymase) and coenzyme II are no longer used.] These
two coenzymes are also referred to as _pyridine coenzymes_
or _pyridine nucleotides_, since nicotinamide is a derivative
of pyridine.

The pyridine nucleotides function as the coenzymes of a
large number of oxidoreductases (page 481), collectively
called _pyridine-linked dehydrogenases_ (Table 13-2). These
coenzymes are bound to the dehydrogenase protein relatively loosely during the catalytic cycle and therefore serve
more as substrates than as prosthetic groups (page 481). They
act as electron acceptors during the enzymatic removal of
hydrogen atoms from specific substrate molecules. One hy-

drogen atom from the substrate is transferred as a hydride ion to the nicotinamide portion of the oxidized forms of these coenzymes (symbolized NAD$^+$ and NADP$^+$), to yield the reduced coenzymes (symbolized NADH and NADPH, respectively); the other hydrogen atom from the substrate becomes a hydrogen ion (Figure 13-6). An example of a reaction catalyzed by a pyridine-linked dehydrogenase is the oxidation of malate to oxaloacetate, catalyzed by *malate dehydrogenase* (Figure 13-6), an important step in the oxidation of carbohydrates via the tricarboxylic acid cycle (page 460). Malate dehydrogenase from the liver is specific for NAD$^+$ as coenzyme and will not accept NADP$^+$. In general, pyridine-linked dehydrogenases are specific for either NAD$^+$ or NADP$^+$, but a few, such as glutamate dehydrogenase, will function with both. In general, the reactions catalyzed by pyridine nucleotide dehydrogenases are reversible (page 481). The reduced forms of NAD$^+$ and NADP$^+$ characteristically have an absorption maximum at 340 nm, which is used for following the course of pyridine nucleotide–linked reactions (pages 207 and 482). Figure 13-6 shows that the 1 and 4 positions of the pyridine ring of NAD$^+$ and NADP$^+$ become reduced by transfer of a hydride ion from the substrate.

Other important properties of pyridine nucleotides and pyridine-linked dehydrogenases are detailed in Chapters 17 (page 456) and 18 (page 481).

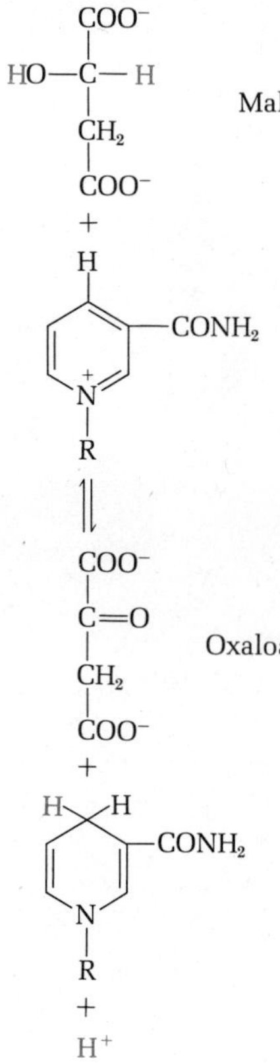

Pantothenic Acid and Coenzyme A

Pantothenic acid (Figure 13-7) was first recognized by R. J. Williams as a growth factor for yeast. It is formed by plants and many bacteria but is required in the diet of vertebrates. Soon after its discovery pantothenic acid was found to occur in tissues in a low-molecular-weight bound or combined form, but it was not until 1948 that the latter was identified as *coenzyme A* (A for acetyl) by N. O. Kaplan and F. Lipmann, who had earlier found that certain enzymatic acetylation reactions require a heat-stable coenzyme.

The function of coenzyme A is to serve as a carrier of acyl groups in enzymatic reactions involved in fatty acid oxidation (page 544), fatty acid synthesis (page 661), pyruvate oxidation (page 450), and biological acetylations (page 343). The precise chemical mechanism by which coenzyme A carries acyl groups was established by the German biochemist F. Lynen in 1951. He isolated an "active" form of acetate from yeast and showed it to consist of a thioester of acetic acid with the thiol or sulfhydryl group of coenzyme A (Figure 13-7). Coenzyme A is abbreviated CoA (or CoA-SH where its thiol function is being emphasized).

Acetyl–coenzyme A (acetyl-CoA) is formed during the enzymatic oxidation of pyruvate or fatty acids; it may also be generated from free acetate in the presence of the enzyme *acetyl-CoA synthetase*:

ATP + coenzyme A + acetate $\rightleftharpoons$

AMP + acetyl-CoA + pyrophosphate

Figure 13-6
The malate dehydrogenase reaction, showing the path of the hydrogen atoms removed by the enzyme. One is transferred as a hydride ion to the 4 position of the pyridine ring; the other appears as a hydrogen ion in the medium.

Figure 13-7
*Panthothenic acid and coenzyme A. The
thiol group (color) of coenzyme A becomes
esterified with an acyl group to yield a thioester during the activity of enzymes utilizing
coenzyme A as acyl group acceptor. Shown
below is the thioester bond in acetyl-CoA.*

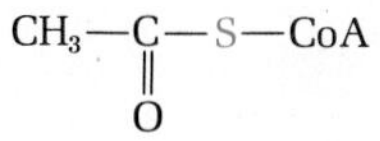

$$CH_3—CH_2—\underset{\underset{CH_3}{|}}{\overset{\overset{CH_3}{|}}{C}}—\underset{OH}{\overset{|}{CH}}—\overset{\overset{O}{\parallel}}{C}—\underset{H}{\overset{|}{N}}—CH_2—CH_2—COOH$$

Pantothenic acid

$$CH_3—\underset{\parallel}{\overset{}{C}}—S—CoA$$
$$O$$

Acetyl–coenzyme A

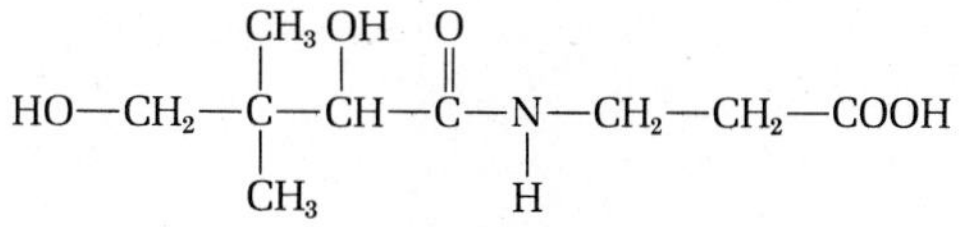

β-Mercaptoethylamine

Pantothenic
acid

Adenine

Ribose 3′-phosphate

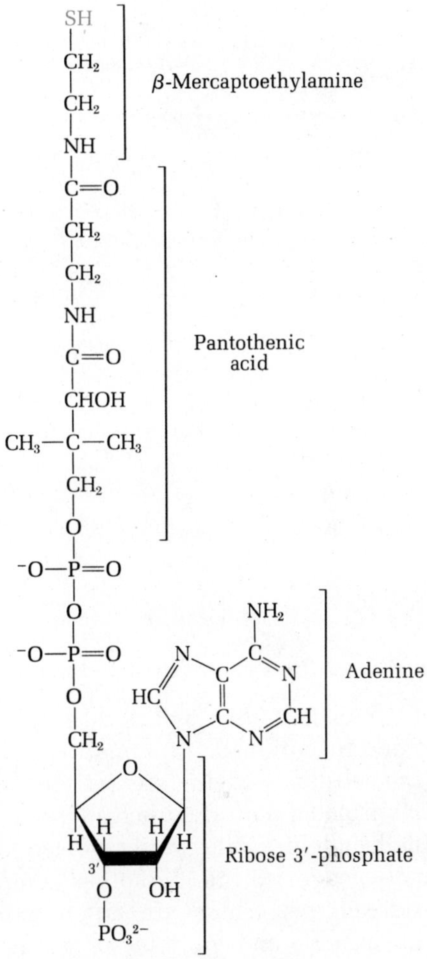

Coenzyme A

The acetyl-CoA so formed may then react enzymatically with an acyl-group acceptor such as choline, to yield acetyl-choline (page 218), or with oxaloacetic acid, to yield citric acid, as in the reactions

$$\text{Acetyl-CoA} + \text{choline} \xrightarrow[\text{acetylase}]{\text{choline}} \text{acetylcholine} + \text{CoA-SH}$$

$$\text{Acetyl-CoA} + \text{oxaloacetic acid} + H_2O \xrightarrow[\text{synthase}]{\text{citrate}} \text{citrate} + \text{CoA-SH}$$

In nonenzymatic model reactions thioesters like acetyl-CoA are more effective acylating agents than the corresponding oxygen esters, since the thioester linkage shows less reso-nance stabilization than the oxyester linkage. Enzymatic reactions involving coenzyme A are described in detail else-where (pages 450, 546, 561, and 661).

343

Figure 13-8
Pyridoxine (vitamin B_6) and its coenzyme forms.

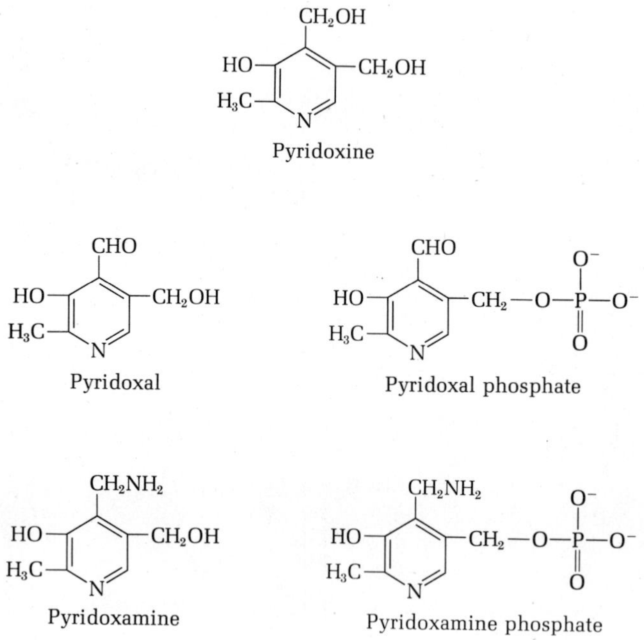

Figure 13-9
The function of pyridoxal phosphate in aminotransferases. The amino group of the α-amino acid reacts with the carbonyl group of the pyridoxal phosphate, tightly bound to the enzyme, to yield a Schiff's base intermediate. This is transformed to its tautomeric form, which then undergoes hydrolysis to yield the corresponding α-keto acid, leaving the enzyme with the amino group covalently bound as pyridoxamine phosphate. By reversal of this sequence of reactions the aminated form of the aminotransferase can then transfer its amino group to another α-keto acid, to yield the corresponding amino acid.

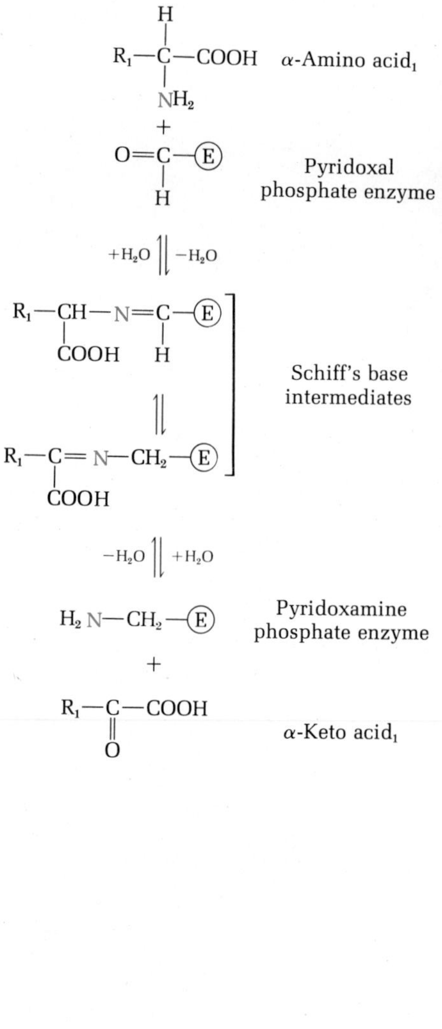

Vitamin B_6 and the Pyridoxine Coenzymes

Vitamin B_6 was first identified as essential in the nutrition of the rat for prevention of a dermatitis called _acrodynia_. A crystalline substance having this activity, _pyridoxine_ (Figure 13-8), was isolated and identified in 1938. Subsequent work of E. Snell and his colleagues revealed that pyridoxine is biologically converted into two other compounds, _pyridoxal_ and _pyridoxamine_, which are much more potent growth factors for bacteria and more direct precursors of the active forms of this vitamin. The active coenzyme forms of vitamin B_6 are _pyridoxal phosphate_ and _pyridoxamine phosphate_ (Figure 13-8).

The pyridoxine coenzymes are extremely versatile, functioning in a large number of different enzymatic reactions in which amino acids or amino groups are transformed or transferred. The most common type of enzymatic reaction requiring pyridoxal phosphate as a coenzyme is _transamination_, the transfer of the α-amino group of an amino acid to the α-carbon atom of an α-keto acid. Enzymes catalyzing such reactions are called _transaminases_ or _aminotransferases_. Our present understanding of the role of these coenzymes arose from model experiments of Snell and his associates, in which pyridoxal was found to react nonenzymatically at 100°C with glutamic acid to yield pyridoxamine and α-ketoglutaric acid. This and other observations led Snell, as well as A. Braunstein in the Soviet Union, to propose that pyridoxal phosphate functions as a coenzyme by virtue of the ability of its aldehyde group to react with the α-amino group of the amino acid substrate to yield a Schiff's

Figure 13-10

Biotin and the carboxylated form of the biotinyl derivative of a lysine residue of the enzyme, which can transfer its carboxyl group (color) to acetyl-CoA or other acceptors.

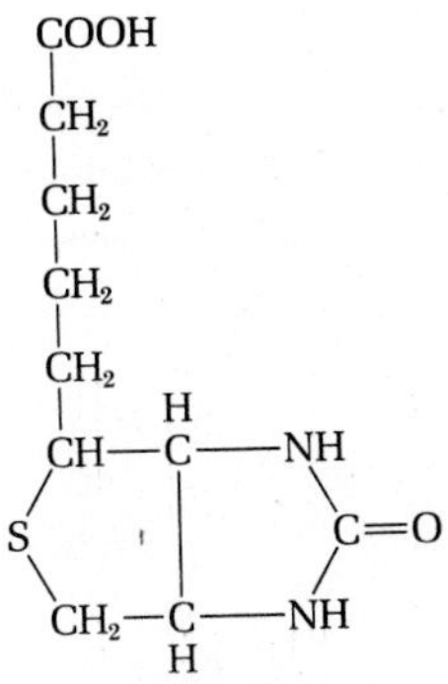

Biotin

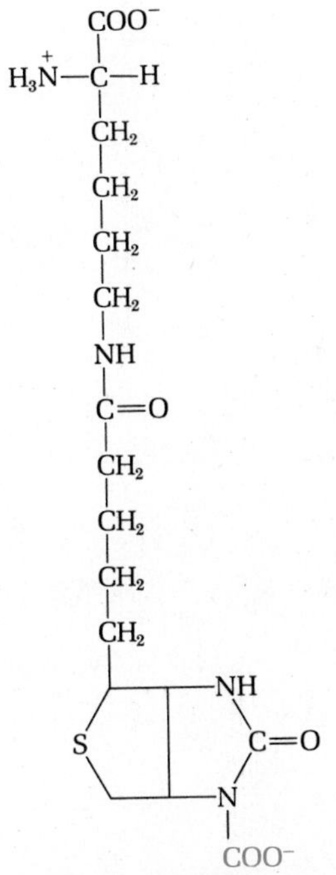

N-Carboxybiotinyllysine

base between the enzyme-bound pyridoxal phosphate and the amino acid (Figure 13-9). The amino group then detaches from the amino acid, converting it into an α-keto acid; the resulting bound pyridoxamine phosphate on the enzyme then reacts with another α-keto acid, called the amino-group acceptor, in a reaction that is the reverse of those described above, to yield a new amino acid and the pyridoxal phosphate–enzyme. It will be recalled that the transaminase reaction is an example of a double-displacement reaction (page 205).

Several specific examples and further details of the coenzyme action of pyridoxal phosphate are described elsewhere (pages 563, 574, 699, and 701).

Biotin and Biocytin

Biotin (Figure 13-10) was first isolated by the Dutch biochemist F. Kögl in 1935 from a liver concentrate known to contain growth factors for yeast. Its structure was solved by V. du Vigneaud and his colleagues. Biotin contains fused imidazole and thiophene rings. Biotin protects animals against a peculiar toxicity produced by feeding them raw egg white. Because biotin is made by intestinal bacteria, a biotin deficiency cannot easily be produced merely by withholding the vitamin from the diet. However, raw egg white induces a biotin deficiency because it contains a protein, avidin, which specifically binds biotin very tightly and prevents its absorption from the intestine. Biotin deficiency has been found in people who consume large amounts of raw eggs.

The mode of action of biotin was obscure for many years, but it was ultimately found to function in the enzymatic transfer or incorporation of carbon dioxide. M. D. Lane and Lynen found that biotin is present in the covalently bound prosthetic group of propionyl-CoA carboxylase. The biotin molecule is combined as a substituted amide with the ε-amino group of a specific lysine residue of the enzyme protein, as the compound *biotinyllysine* or *biocytin* (Figure 13-10). Bound biotin serves as an intermediate carrier of carbon dioxide during the action of certain carboxylating enzymes, e.g., propionyl-CoA carboxylase and acetyl-CoA carboxylase, in the form of a very labile *carboxybiotin* derivative (Figure 13-10). The overall course of such carboxylation reactions is indicated by the equations

$$\text{ATP} + \text{HCO}_3^- + \text{biotinyl-enzyme} \rightleftharpoons$$
$$\text{ADP} + \text{P}_i + \text{carboxybiotinyl-enzyme}$$

$$\text{Carboxybiotinyl-enzyme} + \text{substrate} \rightleftharpoons$$
$$\text{biotinyl-enzyme} + \text{carboxylated substrate}$$

Folic Acid and Its Coenzyme Forms

Folic acid (Latin *folium*, "leaf"), first found in spinach leaves, is broadly distributed in plants; its deficiency in mammals results in failure to grow and in various forms of anemia. It contains three characteristic building blocks: (1) a

Figure 13-11
Structure of folic acid and tetrahydrofolic acid. The four hydrogen atoms added to form tetrahydrofolic acid are shown in color. The N^5 and N^{10} nitrogen atoms participate in the transfer of one-carbon groups.

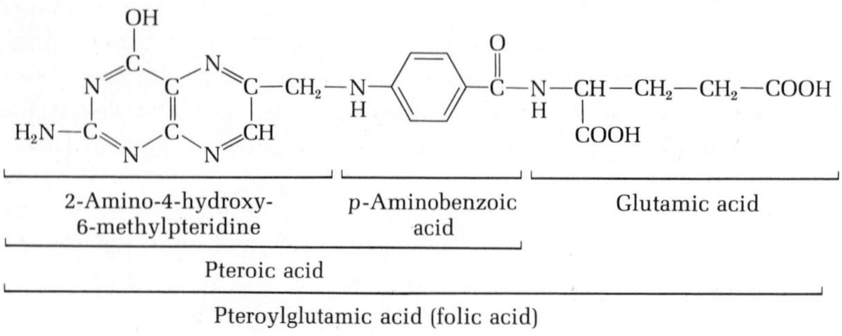

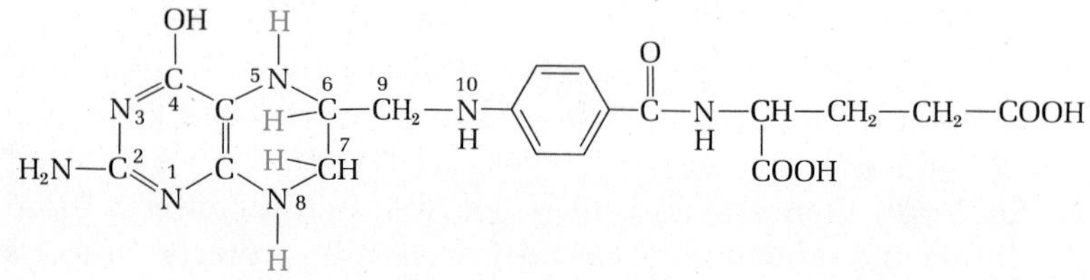

2-Amino-4-hydroxy-
6-methylpteridine

p-Aminobenzoic
acid

Glutamic acid

Pteroic acid

Pteroylglutamic acid (folic acid)

Tetrahydrofolic acid

substituted *pteridine*, (2) *p-aminobenzoic acid*, and (3) *glutamic acid*. Folic acid is also known as *pteroylglutamic acid* or *folacin* (Figure 13-11). Pteridine is the bicyclic nitrogenous parent compound of the *pterins*, which are derivatives of 2-amino-4-hydroxypteridine. Some pterins, such as *xanthopterin* (Figure 13-12), serve as eye and wing pigments in insects. The pterin present in folic acid is 6-*methylpterin*. In some species folic acid is attached by the γ-carboxyl group of the glutamic acid to one or more additional glutamic acid residues joined in peptide linkages involving the γ-carboxyl groups of the glutamic residues. Some organisms require only the p-aminobenzoic acid portion of folic acid; they can synthesize folic acid if p-aminobenzoic acid is available.

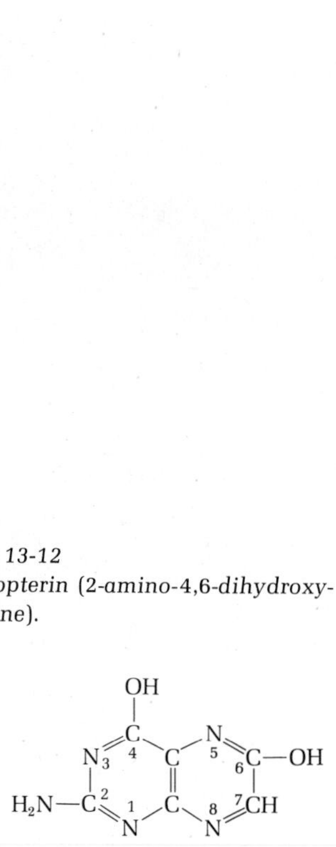

The most conspicuous biochemical symptom of folic acid deficiency is impaired biosynthesis of purines and the pyrimidine thymine. Enzyme and metabolic studies show that the coenzyme form of folic acid functions in the transfer of certain one-carbon groups utilized in these and other biosynthetic pathways. Folic acid is converted by reduction in two steps into its coenzyme form *tetrahydrofolic acid* (Figure 13-11). Tetrahydrofolate, often abbreviated FH_4, serves as an intermediate carrier of hydroxymethyl ($-CH_2OH$), formyl ($-CHO$), or methyl ($-CH_3$) groups in a large number of enzymatic reactions in which such groups are transferred from one metabolite to another or are interconverted. The structures of N^5,N^{10}-methylenetetrahydrofolate and other types of enzymatically active one-carbon derivatives of tetrahydrofolate are given in Figure 13-13. Some of these complex one-carbon transfer reactions are described in connection with the intermediary metabolism of amino acids (pages 567 and 714), purines (page 731), and pyrimidines (page 739).

Figure 13-12
Xanthopterin (2-amino-4,6-dihydroxy-pteridine).

Figure 13-13
One-carbon derivatives of tetrahydrofolate
(for its complete structure see Figure 13-11).

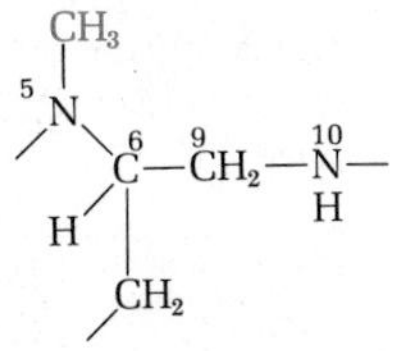

N⁵-Methyltetrahydrofolate

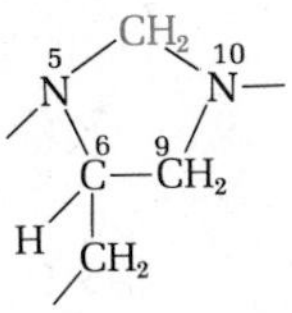

N⁵,N¹⁰-Methylenetetrahydrofolate

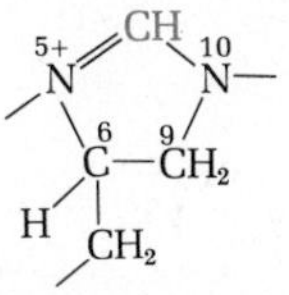

N⁵,N¹⁰-Methenyltetrahydrofolate

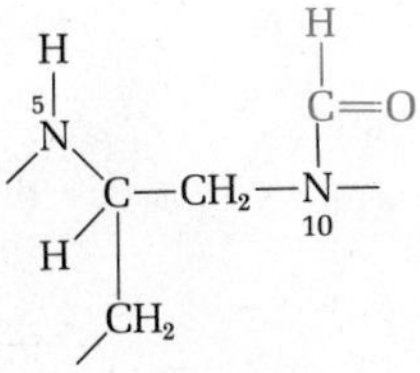

N¹⁰-Formyltetrahydrofolate

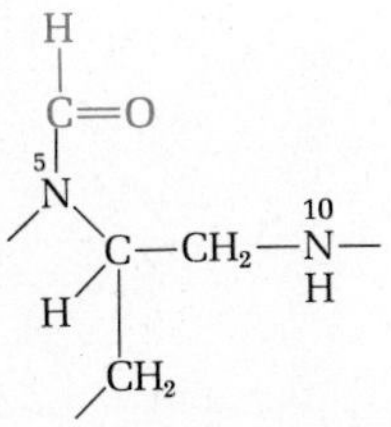

N⁵-Formyltetrahydrofolate

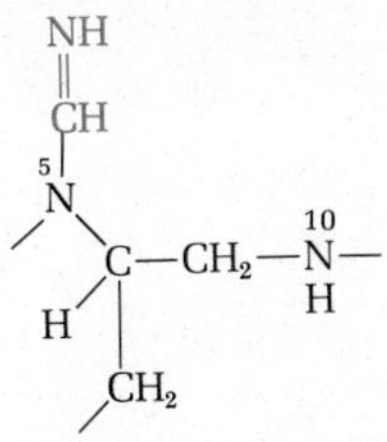

N⁵-Formiminotetrahydrofolate

Following is an example of a sequence of enzymatic reactions in which the hydroxymethyl group of the amino acid serine is enzymatically removed to form the N^5,N^{10}-methylene derivative of tetrahydrofolate, which is reduced to the N^5-methyl derivative. The latter then donates its methyl group to homocysteine to yield methionine. Tetrahydrofolate thus serves as a shuttle to which the one-carbon group is covalently but transiently attached:

$$\text{Serine} + FH_4 \rightleftharpoons \text{glycine} + N^5,N^{10}\text{-methylene-}FH_4 + H_2O$$

$$N^5,N^{10}\text{-methylene-}FH_4 + NADH + H^+ \rightleftharpoons N^5\text{-methyl-}FH_4 + NAD^+$$

$$N^5\text{-methyl-}FH_4 + \text{homocysteine} \rightleftharpoons FH_4 + \text{methionine}$$

Lipoic Acid

Lipoic acid, sometimes called *thioctic acid*, was first isolated in crystalline form in 1953 from extracts of liver by the American biochemists L. J. Reed and I. C. Gunsalus and their colleagues; only a few milligrams were obtained from tons of liver. There are two forms of this rather simple eight-carbon vitamin (Figure 13-14), lipoic acid, a cyclic disulfide, and its reduced open-chain form *dihydrolipoic acid*, which has two sulfhydryl groups, in the 6 and 8 positions. These two forms are readily interconverted by oxidation-reduction reactions.

Figure 13-14
Lipoic acid and its derivatives. The thiol and disulfide groups are in color.

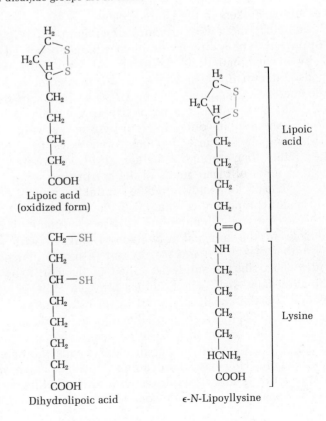

Lipoic acid
(oxidized form)

Dihydrolipoic acid

Lipoic acid

Lysine

ε-N-Lipoyllysine

347

Lipoic acid functions as one of the coenzymes in the oxidative decarboxylation of pyruvate and other α-keto acids, complex reactions in which several coenzymes participate (page 450). The pyruvate first undergoes loss of the carboxyl group (Figure 13-15), to yield the hydroxyethyl derivative of enzyme-bound thiamin pyrophosphate (page 338). The latter then reacts with lipoic acid bound to the enzyme *dihydrolipoyl transacetylase,* with transfer of electrons and the acyl group to yield 6-acetyldihydrolipoic acid. In subsequent reactions, described elsewhere (page 450), lipoic acid is regenerated following transfer of the acetyl group to coenzyme A and reoxidation of the thiol groups by transfer of electrons to NAD$^+$, to yield the oxidized or cyclic disulfide form of lipoic acid. Lipoic acid is covalently bound via an amide linkage to the ε-amino group of a specific lysine residue in dihydrolipoyl transacetylase; the lipoyllysine residue is also known as *lipoamide* (Figure 13-14).

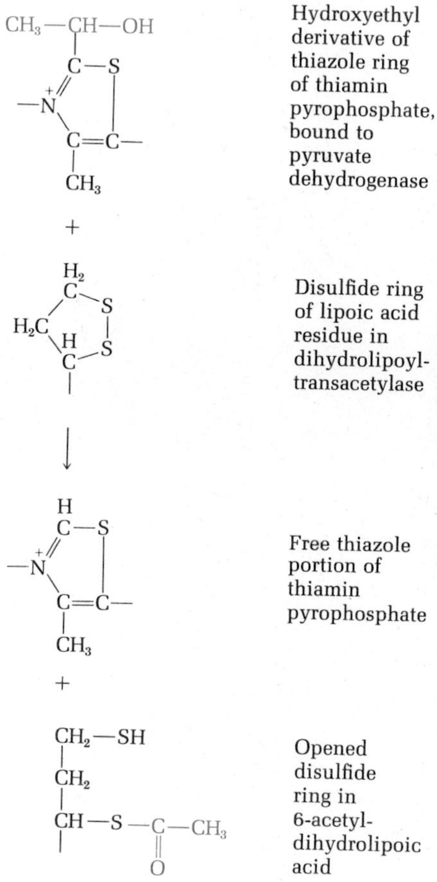

Vitamin B$_{12}$ and the B$_{12}$ Coenzymes

The history of vitamin B$_{12}$ is of extraordinary interest since the isolation, structure, and mode of action of this vitamin have been exceedingly challenging problems. In 1926 the American physicians G. R. Minot and W. P. Murphy discovered that liver in the diet cures patients suffering from pernicious anemia. Despite many attempts over the years to isolate the liver factor, little progress was made, since the disease could not be produced in experimental animals and patients with pernicious anemia had to be used to assay the potency of liver-factor concentrates. However, in 1948, the liver factor, called vitamin B$_{12}$, was finally isolated in crystalline form by the groups of E. L. Smith in England and E. Rickes and K. Folkers in the United States.

The elucidation of the structure of vitamin B$_{12}$ (Figure 13-16) proved to be a difficult problem, ultimately solved in 1957 by a combination of chemical and x-ray-diffraction methods. Vitamin B$_{12}$, also known as *cyanocobalamin,* has two characteristic components. The larger is the *corrin* ring system, which resembles the porphyrin ring system of hemoglobin (page 490) in containing four pyrrole-type rings but in which a pair of these five-membered rings is joined directly rather than through a methene bridge. Coordinated to the four inner nitrogen atoms of the corrin ring system is an atom of cobalt, long known to be essential for growth. The second major component of vitamin B$_{12}$ is a ribonucleotide, which is exceptional in containing as base 5,6-*dimethylbenzimidazole* in the unusual α-N-glycosyl linkage with D-ribose, rather than the β linkage present in most other nucleotides. This ribonucleotide is joined to the corrin by a coordination bond between the other nitrogen atom of the nucleotide and the cobalt atom and by an ester linkage between the 3′-phosphate group of the ribonucleotide and a side chain of the corrin ring. Cyanide occupies one of the coordination positions of the cobalt atom; hence the name cyanocobalamin. However, the cyanide is present as an artifact of isolation; similar complexes with nitrite, sulfite, and hydroxide ions are known. In *coenzyme* B$_{12}$ the cyanide

Figure 13-15
Transfer of hydrogen atoms and an acetyl group from thiamin pyrophosphate to lipoic acid. This reaction step occurs in the oxidative decarboxylation of pyruvate (page 450), a complex reaction requiring the sequential action of three enzymes. See also Figures 13-1 and 13-2 (page 338).

Hydroxyethyl derivative of thiazole ring of thiamin pyrophosphate, bound to pyruvate dehydrogenase

Disulfide ring of lipoic acid residue in dihydrolipoyl-transacetylase

Free thiazole portion of thiamin pyrophosphate

Opened disulfide ring in 6-acetyl-dihydrolipoic acid

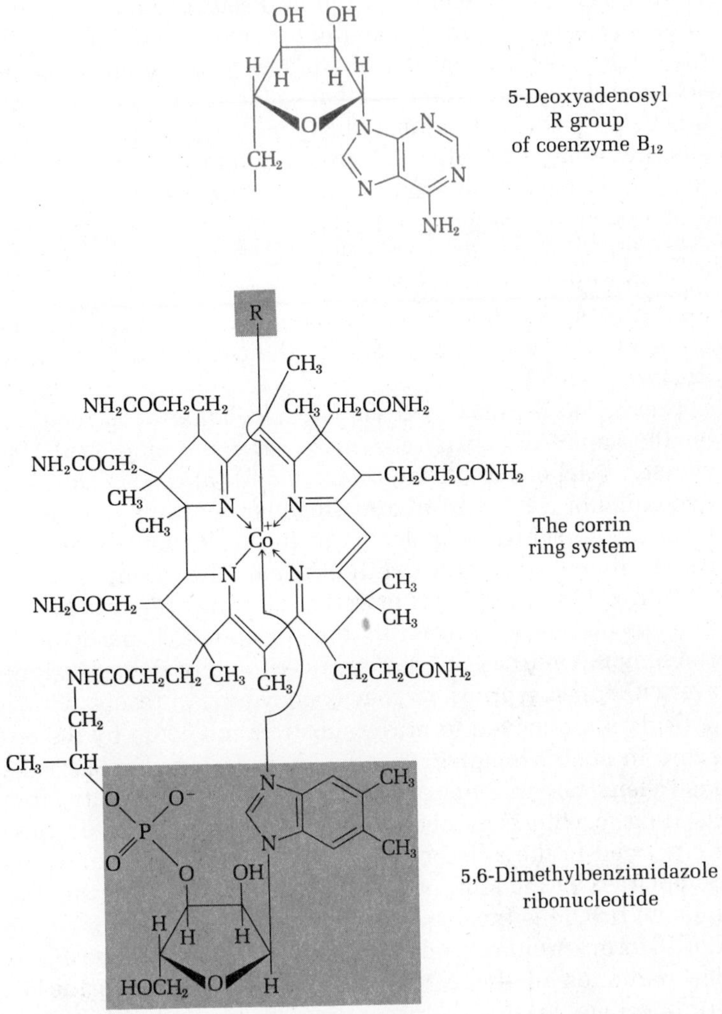

Figure 13-16

Vitamin B$_{12}$ (cobalamin) and its derivatives. In cyanocobalamin R = cyanide. Cobalamin forms similarly named complexes with sulfite, hydroxide, and nitrite ions. In deoxyadenosylcobalamine (coenzyme B$_{12}$) R is the 5'-deoxyadenosyl group (color).

5-Deoxyadenosyl
R group
of coenzyme B$_{12}$

The corrin
ring system

5,6-Dimethylbenzimidazole
ribonucleotide

ligand is replaced by the 5-deoxyadenosyl group (Figure 13-16). There is another form of vitamin B$_{12}$, *pseudovitamin B$_{12}$*, which also occurs as a coenzyme, to be described below.

Neither animals nor plants can synthesize vitamin B$_{12}$, which is manufactured only by certain microorganisms. Vitamin B$_{12}$ is required only in traces by animals; normal human blood contains only 0.0002 μg of vitamin B$_{12}$ per milliliter. Pernicious anemia is not simply the result of a deficiency of vitamin B$_{12}$ in the diet but is caused by the failure of the patient to absorb vitamin B$_{12}$ from ingested food, due to the lack of a specific glycoprotein in gastric juice called the *intrinsic factor*. This protein binds one molecule of vitamin B$_{12}$ and carries it into the intestinal cells, from which it is transported, bound to other proteins called *transcobalamins*, to the peripheral tissues.

Vitamin B$_{12}$ is essential for the normal maturation and development of erythrocytes. The molecular basis of its action was obscure until 1958, when H. A. Barker discovered that the 5'-deoxyadenosyl derivative of pseudovitamin B$_{12}$ is an essential cofactor in the enzymatic conversion of glutamic acid into β-methylaspartic acid in the anaerobic mud bacterium *Clostridium tetanomorphum.* Pseudovitamin B_{12} contains adenine instead of 5,6-dimethylbenzimidazole in its ribonucleotide; this variant of vitamin B_{12} is found in *C. tetanomorphum* and certain other microorganisms that are unable to make 5,6-dimethylbenzimidazole but can make adenine. The derivative of pseudovitamin B_{12} found by Barker to have coenzyme activity contains the 5-deoxyadenosyl group as a ligand to the cobalt atom. The coenzyme forms of pseudovitamin B_{12} and vitamin B_{12} each have *two* nucleotide moieties; the former contains two adenine nucleotides and the latter a benzimidazole and an adenine nucleotide, in each case bound to the cobalt atom via coordination bonds.

Coenzyme B_{12} or *5'-deoxyadenosylcobalamin* is required for the action of several enzymes (pages 555 and 739). Enzymatic reactions requiring coenzyme B_{12} have as a common denominator a 1,2 shift of a hydrogen atom from one carbon atom of the substrate molecule to the next, usually accompanied by a reverse or 2,1 shift of some other group, such as a hydroxyl, amino, alkyl, or carboxyl group (Figure 13-17). The precise mechanism of the reaction is particularly challenging and has not yet been completely defined. However, it appears probable that coenzyme B_{12} reacts with a hydride ion removed from the substrate molecule by the enzyme in such a manner that the hydride ion displaces the methylene carbon atom of the 5'-deoxyadenosyl group from its linkage with the cobalt atom. The hydride ion is then transferred to the adjacent carbon atom of the substrate, accompanied by the shift of the other group being transferred. In such reactions the cobalt atom presumably remains in the Co(II) form. Another important reaction of coenzyme B_{12} is the reduction of the 2' carbon atom of ribonucleoside 5'-triphosphate to the corresponding 2'-deoxyribonucleoside triphosphate (page 739).

In a second class of enzymatic reactions vitamin B_{12} functions as a coenzyme with the sixth coordination position of the cobalt atom filled by a methyl group rather than the 5'-deoxyadenosyl group; the resulting compound is *methylcobalamin*. In these reactions the methylcobalamin functions as a carrier of a methyl group from N^5-methyltetrahydrofolate (page 347) to certain acceptor molecules, particularly homocysteine, which is methylated to become methionine (page 701). Enzymatic reactions involving the vitamin B_{12} coenzymes are described in more detail elsewhere (pages 555, 701, and 739).

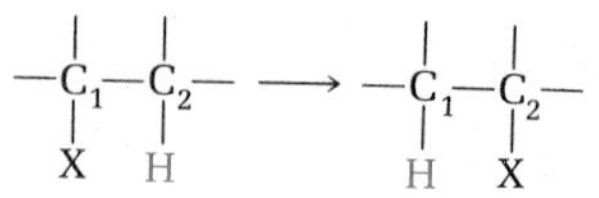

Vitamin C

Vitamin C, or ascorbic acid (page 259), was first isolated in pure crystalline form from lemon juice by the American biochemists C. G. King and W. A. Waugh in 1932. It is one of the simplest vitamins in structure, being a lactone of a sugar acid (Figure 13-18). Ascorbic acid is required in the diet of only a few vertebrates—man, monkeys, the guinea pig, the Indian fruit bat, and certain fishes. Some insects and other invertebrates also require ascorbic acid, but most other

Figure 13-17
The 1,2 shift of a hydrogen atom in exchange with the group X, promoted by enzymes dependent on coenzyme B_{12}.

Figure 13-18
Ascorbic acid and its derivatives.

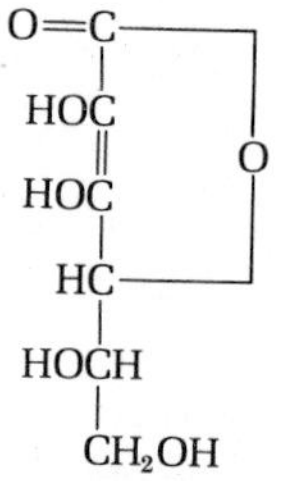

L-Ascorbic acid

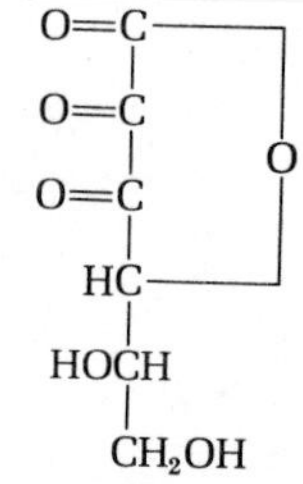

L-Dehydroascorbic acid (active)

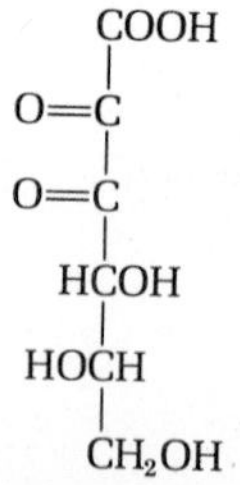

L-Diketogulonic acid (inactive)

higher animals and plants can synthesize ascorbic acid from glucose or other simple precursors. Ascorbic acid is not present in microorganisms, nor does it seem to be required.

Ascorbic acid is a strong reducing agent, readily losing hydrogen atoms to become *dehydroascorbic acid*, which also has vitamin C activity. However, vitamin activity is lost when the lactone ring of dehydroascorbic acid is hydrolyzed to yield diketogulonic acid (Figure 13-18). Ascorbic acid in food is largely destroyed by cooking.

In animal and plant tissues rather large concentrations of ascorbic acid are present, in comparison with other water-soluble vitamins; e.g., human blood plasma contains about 1 mg of ascorbic acid per 100 ml. Ascorbic acid is especially abundant in citrus fruits and tomatoes. Although the symptoms of scurvy in man can be prevented by as little as 20 mg of ascorbic acid per day, there is some evidence that far larger amounts may be required for completely normal physiological function and well-being.

Despite the relatively high concentration of ascorbic acid in the tissues and its rather simple structure, its physiological function is not yet known. It acts as a cofactor in the enzymatic hydroxylation of proline to hydroxyproline (page 696) and in other hydroxylation reactions (page 570), but it is not specific in these reactions and can be replaced by other reducing agents without antiscorbutic activity.

Other Water-Soluble Growth Factors

In addition to the water-soluble vitamins described above, several other factors are essential as nutrients but are required in the diet by only a few species or only under special circumstances. The cyclic sugar alcohol myo-*inositol* (Figure 13-19) is required in the diet for normal hair growth in mice and to prevent a "spectacled-eye" condition in rats. It is curious that inositol is readily made from glucose by the rat but apparently in insufficient quantities. Inositol is not a building block of any known coenzyme, but it is found in fairly large amounts in animal tissues as a component of the *inositol phosphoglycerides* (page 288).

Choline (Figure 13-19) is also a component of a class of phosphoglycerides, in this case the *choline phosphoglycerides* (page 288). Although it has no known coenzyme function, choline is required in the diet of rats fed inadequate amounts of amino acids, particularly methionine, normally a precursor of choline. Deficiency of choline causes fatty infiltration of the liver and hemorrhagic kidneys.

Carnitine is required as an essential growth factor by the mealworm *Tenebrio molitor* but not by mammals, which can make it. It is especially abundant in muscle tissue. Carnitine is a necessary component for shuttling fatty acids across the mitochondrial membrane, preparatory to their enzymatic oxidation (page 545).

Fat-Soluble Vitamins

The four fat-soluble vitamins, A, D, E, and K, are all isoprenoid compounds (page 296). Although they have been

Figure 13-19
Some other growth factors.

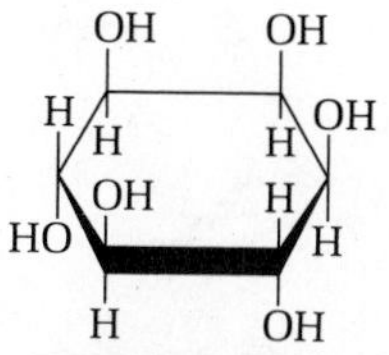

myo-Inositol

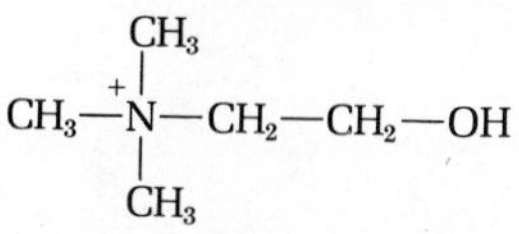

Choline

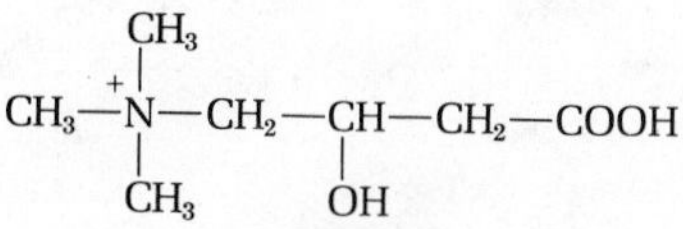

Carnitine

351

studied intensively and widely used in human nutrition, we know less about their specific biological function than about the water-soluble vitamins. No specific coenzyme function has yet been found for any of the fat-soluble vitamins, and only for vitamins A and D are we at all close to an understanding of the molecular basis of their action.

Vitamin A

Vitamin A occurs in two common forms, *vitamin A₁*, or *retinol* (Figure 13-20), the form most common in mammalian

Figure 13-20
Vitamin A₁ (retinol) and β-carotene, its precursor. Oxidative cleavage (colored arrow) yields two molecules of vitamin A. The isoprene units are set apart by colored dashes. Vitamin A₂ has a second double bond, between carbon atoms 3 and 4 in the ring; otherwise it is identical with vitamin A₁.

Vitamin A₁

Model of vitamin A₁

β-carotene

352

tissues and marine fishes, and *vitamin A₂*, or *retinol₂*, common in freshwater fishes. Both are isoprenoid compounds containing a six-membered carbocyclic ring and an eleven-carbon side chain. Vitamin A activity in mammals is given not only by the retinols but also by certain carotenoids widely distributed in plants, particularly α-, β-, and γ-carotene. The carotenes have no intrinsic vitamin A activity per se but are converted into vitamin A by enzymatic reactions in the intestinal mucosa and the liver. β-Carotene, a symmetrical molecule, is cleaved in its center to yield two molecules of retinol (Figure 13-20). Retinol occurs in the tissues of mammals and is transported in the blood in the form of esters of long-chain fatty acids.

Vitamin A deficiency was first recognized in the rat, but all mammals, including man, appear to be susceptible, with little variation in the symptoms. In vitamin A deficiency young animals fail to grow, the bones and nervous system fail to develop properly, the skin becomes dry and thickened, the kidneys and various glands degenerate, and both males and females become sterile. Although all tissues appear to be disturbed by vitamin A deficiency, the eyes are most conspicuously affected. In infants and young children the condition known as *xerophthalmia* ("dry eyes") is an early symptom of deficiency and is a common cause of blindness in some tropical areas where nutrition is generally poor. In adults an early sign of vitamin A deficiency is nightblindness, a deficiency in dark adaptation, which is often used as a diagnostic test. Young animals are most susceptible to vitamin A deficiency, which is not readily produced in adults because the liver can store sufficient vitamin A to last for months or even years.

The vitamin A requirement of man, less than a milligram per day, is met in large part by green and yellow vegetables, such as lettuce, spinach, sweet potatoes, and carrots, which are rich in carotenes. Fish-liver oils are particularly rich in vitamin A. However, excessive intake of vitamin A is toxic and leads to easily fractured, fragile bones in children, as well as abnormal development of the fetus.

Although the more general biological function of vitamin A is not known, detailed information is available on its role in the *visual cycle* in vertebrates, the sequence of molecular events involved in (1) the absorption of light energy by a pigment in the photoreceptor cells in the retina, to yield a specific photochemical product, (2) the initiation of the nerve impulse by the photoproduct, and (3) the regeneration of the light-sensitive form of the visual pigment (Figure 13-21).

The human retina, like that of other mammals, contains two types of light-sensitive photoreceptor cells. *Rod cells* are adapted to sensing low light intensities, but not colors; they are the cells involved in night vision, whose function is impaired by vitamin A deficiency. *Cone cells*, which sense colors, are adapted for high light intensities. The visual cycle in rod cells has been intensively investigated, particularly by G. Wald of Harvard University, whose research has developed the outlines of our present knowledge of the role of vitamin A in this function.

Figure 13-21
The visual cycle in rod cells.

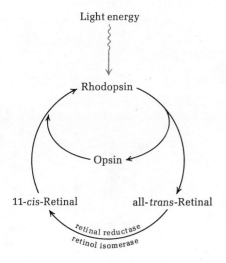

353

Retinal rod cells contain many stacked, disklike membrane vesicles, parallel to the light-receiving surface of the retina; these vesicles serve as light receptors. About one-half of the protein in the membrane of these vesicles consists of the light-absorbing conjugated protein _rhodopsin_ (earlier called visual purple), which has a molecular weight of about 28,000. Rhodopsin is insoluble in water but can be extracted from the vesicles with detergents. The absorption spectrum of purified rhodopsin in the visible range is very similar to the light-sensitivity curve of the eye in dim light, indicating that rhodopsin is the major light receptor of the rod cells. Rhodopsin consists of a protein, _opsin_, and tightly bound 11-_cis_-retinal, the aldehyde of vitamin A, which contains four carbon-carbon double bonds in its side chain; three are in trans form and the fourth, at position 11,12, is cis (Figure 13-22). It will be recalled that the double bonds in most isoprenoid compounds are trans (page 296). When rhodopsin is exposed to light, the bound 11-_cis_-retinal undergoes transformation into all-_trans_-retinal, which causes a substantial change in the configuration of the retinal molecule (Figure 13-22). This reaction is nonenzymatic; it is a purely photochemical reaction which can take place at liquid-nitrogen temperatures. The isomerization of retinal is followed by a series of other molecular changes, ending in the dissociation of the bleached rhodopsin to yield free opsin and all-_trans_-retinal, which functions as a trigger setting off the nerve impulse (below).

In order for rhodopsin to be regenerated from opsin and all-_trans_-retinal, the latter must undergo isomerization back to 11-_cis_-retinal. This appears to occur in a sequence of enzymatic reactions catalyzed by two enzymes:

$$\text{all-}trans\text{-Retinal} + \text{NADH} + \text{H}^+ \xrightarrow{\overset{\text{retinal}}{\text{reductase}}} \text{all-}trans\text{-retinol} + \text{NAD}^+$$

$$\text{all-}trans\text{-Retinol} \xrightarrow{\overset{\text{retinol}}{\text{isomerase}}} 11\text{-}cis\text{-retinol}$$

$$11\text{-}cis\text{-Retinol} + \text{NAD}^+ \xrightarrow{\overset{\text{retinal}}{\text{reductase}}} 11\text{-}cis\text{-retinal} + \text{NADH} + \text{H}^+$$

The 11-_cis_-retinal so formed now recombines with opsin to yield rhodopsin, thus completing the visual cycle.

The crucial part of the visual cycle is the mechanism by which the bleaching of rhodopsin initiates the nerve impulse

Figure 13-22
Light-induced isomerization of 11-cis-retinal.

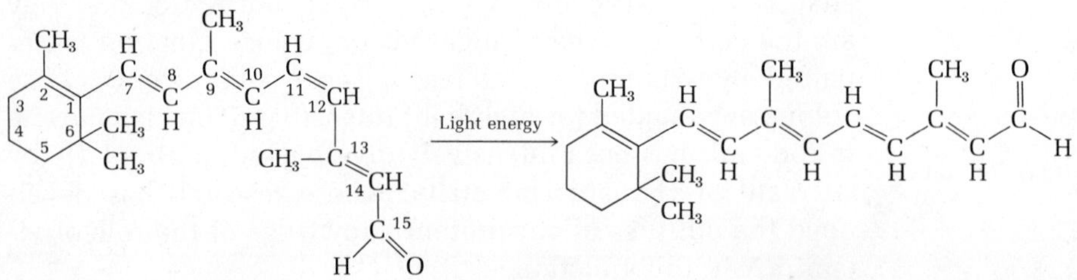

11-_cis_-Retinal   all-_trans_-Retinal

in the retina, so that the light is actually perceived by the brain. Recent research indicates that the absorption of light by the disklike flattened membrane vesicles in the rod cells causes the rhodopsin molecules to undergo a change of conformation, accompanying the conversion of 11-*cis*-retinal to all-*trans*-retinal. This change alters the permeability of the vesicle membrane, across which there is normally a potential difference, so that Ca^{2+} ions are allowed to flow out of the vesicles, thus triggering the nerve impulse. Ca^{2+} serves as a messenger or mediator for coupling the exciting stimulus to the function of the receptor system, just as Ca^{2+} couples the exciting nerve impulse to the contractile system of muscle (page 762).

Since vitamin A deficiency affects all tissues of mammals, not the retina alone, the role of retinal in the visual cycle does not represent the entire action of vitamin A. It appears possible that vitamin A may play a general role in the transport of Ca^{2+} across certain membranes, by analogy with its role in the rod cells; such a more general role might explain the effects of vitamin A deficiency and excess on bony and connective tisues.

Vitamin D

It has long been known that rickets, a disease of growing bone causing bowlegs and pigeon breast, is common in areas with long winters where children are exposed to little sunlight and that ingestion of fish-liver oils can prevent the symptoms of rickets. These clues led to the isolation and identification of several compounds having vitamin D or potential vitamin D activity. Most important are vitamin D_2, or *ergocalciferol,* and vitamin D_3, or *cholecalciferol* (Figure 13-23); the form normally found in mammals. These com-

Figure 13-23
Formation of vitamin D from its precursors.

Formation of vitamin D_3 in animals.

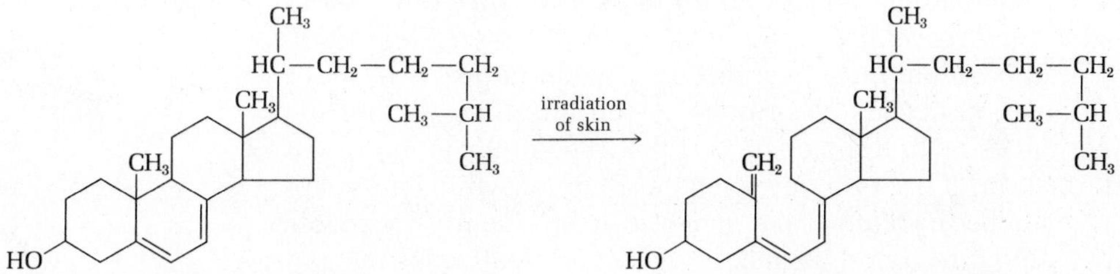

7-Dehydrocholesterol   Vitamin D_3 (cholecalciferol)

Formation of vitamin D_2 from ergosterol.

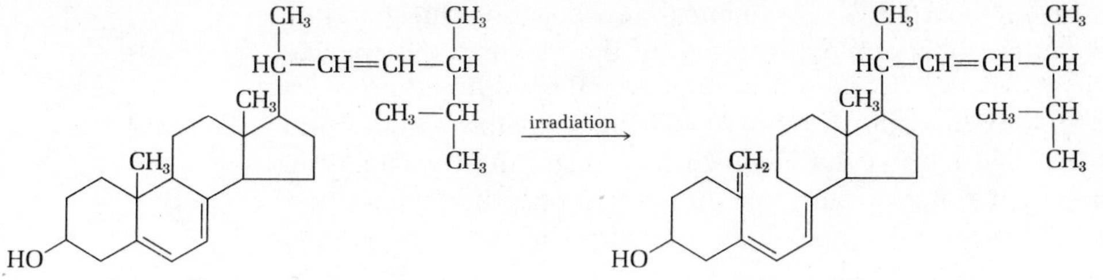

Ergosterol   Vitamin D_2 (ergocalciferol)

pounds may be regarded as steroids (page 298) in which the B ring has been ruptured. Since irradiation of certain foodstuffs with ultraviolet light also produces substances with vitamin D activity, a search for precursors of vitamin D revealed that 7-dehydrocholesterol, common in animal tissues, is converted by irradiation into cholecalciferol. Similarly, the yeast sterol ergosterol is converted by irradiation into ergocalciferol (Figure 13-23). From these findings it is now known that 7-dehydrocholesterol in the skin is the natural precursor of cholecalciferol in man; the conversion requires irradiation of the skin by sunlight. On a normal unsupplemented diet this is the major route by which people usually acquire vitamin D. Most natural foods contain little if any vitamin D; preformed vitamin D in the diet comes largely from fish-liver oils or from irradiated natural sources. Vitamin D preparations available commercially are products of the ultraviolet irradiation of ergosterol from yeast. About 20 μg of vitamin D is required by an adult daily. The vitamin can be stored in sufficient amounts in the liver for a single dose to suffice for some weeks. As with vitamin A, excessive intake of vitamin D causes the bones to become fragile and to undergo multiple fractures, suggesting that both vitamins play a role in biological transport and deposition of calcium.

Although it has long been known that vitamin D is concerned with the proper calcification of bones, its specific biochemical role is only now emerging as a result of recent research, particularly by H. F. DeLuca and his colleagues. The first important clues came from a study of the metabolic fate of synthetically prepared radioactive cholecalciferol of very high specific activity, injected into animals in the microgram quantities in which it is physiologically active. A labeled derivative of cholecalciferol, 25-hydroxycholecalciferol, appeared in the blood and tissues (Figure 13-24). This product is more active biologically than cholecalciferol and it has since been found to be the main circulating form of vitamin D in animals, formed in the liver. DeLuca and his colleagues then injected radioactive 25-hydroxycholecalciferol into animals and found it to be metabolized further to 1,25-dihydroxycholecalciferol (Figure 13-24). This compound is still more active; its administration produces rapid stimulation of Ca^{2+} absorption by the intestine. Further experiments by E. Kodicek proved that the kidney is the site of formation of 1,25-dihydroxycholecalciferol, which now appears to be the biologically active form of vitamin D, capable of acting directly on its major targets, the small intestine and the bones.

DeLuca has postulated that 1,25-dihydroxycholecalciferol is actually a hormone, in accordance with the classical definition of a hormone as a chemical messenger produced in one organ or gland and transported by the blood to the specific target tissue(s) where it exerts its regulatory function. Thus 1,25-dihydroxycholecalciferol is formed in the kidneys and is transported via the blood to its targets, the intestine and bones. It differs from other hormones only in that its precursors must be supplied from dietary sources if the animal

Figure 13-24
Conversion of cholecalciferol (vitamin D₃) into 25-hydroxycholecalciferol and then into 1,25-dihydroxycholecalciferol, the active form.

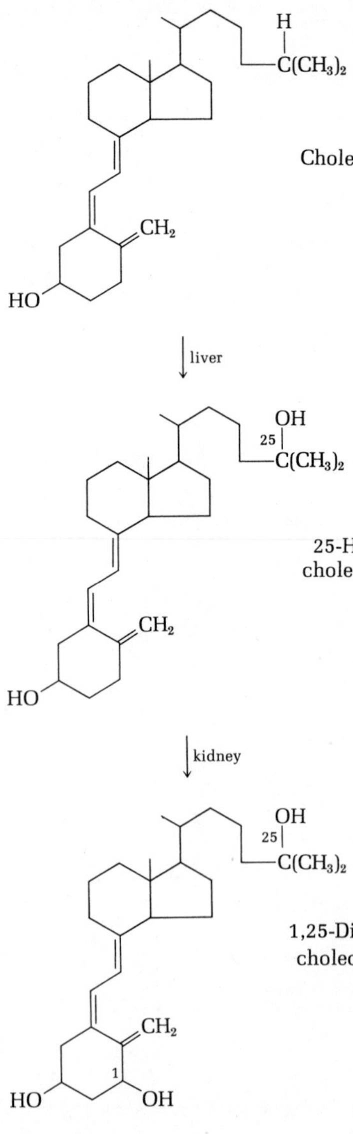

Cholecalciferol

25-Hydroxy-cholecalciferol

1,25-Dihydroxy-cholecalciferol

Figure 13-25
Summary of the precursors, metabolism, and function of vitamin D_3.

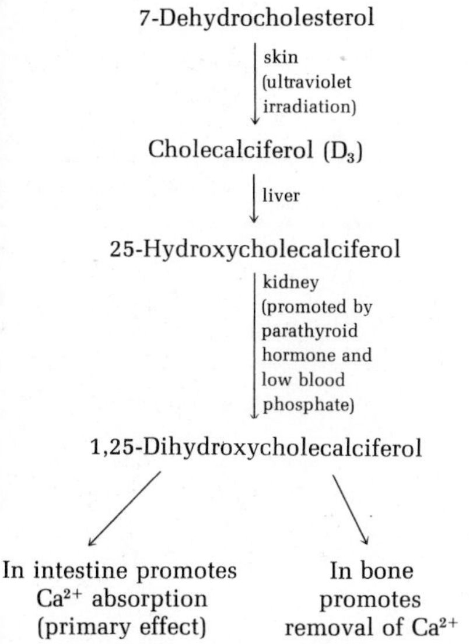

7-Dehydrocholesterol

| skin
(ultraviolet
irradiation)

Cholecalciferol (D_3)

| liver

25-Hydroxycholecalciferol

| kidney
(promoted by
parathyroid
hormone and
low blood
phosphate)

1,25-Dihydroxycholecalciferol

In intestine promotes
Ca²⁺ absorption
(primary effect)

In bone
promotes
removal of Ca²⁺

Figure 13-26
Vitamin E and vitamin K. Both are isoprenoid compounds; the isoprene units are set off by colored dashes.

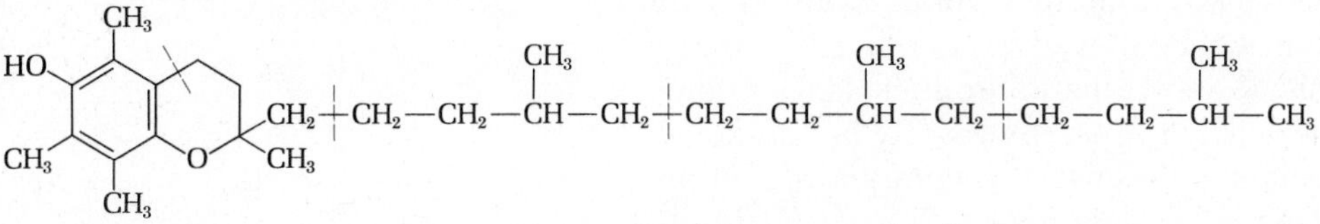

is unable to convert endogenous 7-hydroxycholesterol into cholecalciferol.

Other research has shown that 1,25-dihydroxycholecalciferol promotes absorption of Ca^{2+} from the intestine into the blood, through its ability to stimulate the biosynthesis of specific protein(s) that participate in transport or binding of Ca^{2+} in the intestinal mucosa. This role of 1,25-dihydroxycholecalciferol is integrated with the action of _parathyroid hormone_. Whenever the Ca^{2+} concentration of the blood becomes lower than normal, the parathyroid glands secrete larger amounts of parathyroid hormone, a polypeptide of 84 amino residues. This hormone acts on the kidney, stimulating it to excrete more phosphate in the urine and to produce more 1,25-dihydroxycholecalciferol from its precursor 25-hydroxycholecalciferol. Parathyroid hormone may thus be a _tropic hormone_ for 1,25-dihydroxycholecalciferol; a tropic hormone is one that stimulates the synthesis or secretion of another hormone (page 808). Figure 13-25 summarizes these relationships.

Vitamin E

Vitamin E was first recognized as a factor in vegetable oils that restores fertility in rats grown on cow's milk alone and otherwise incapable of bearing young. It was isolated from wheat germ and was given the name _tocopherol_ (Greek _tokos,_ "childbirth"). Several different tocopherols having vitamin E activity have been found in plants; the most active and abundant is _α-tocopherol_ (Figure 13-26).

The biological function of tocopherol is still obscure. One reason is that its deficiency produces many other symptoms besides infertility in male and female rats, e.g., degeneration

Vitamin E (α-tocopherol)

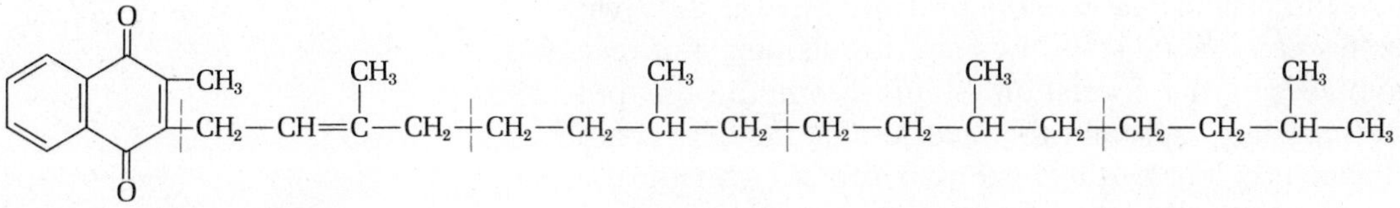

Vitamin K₁

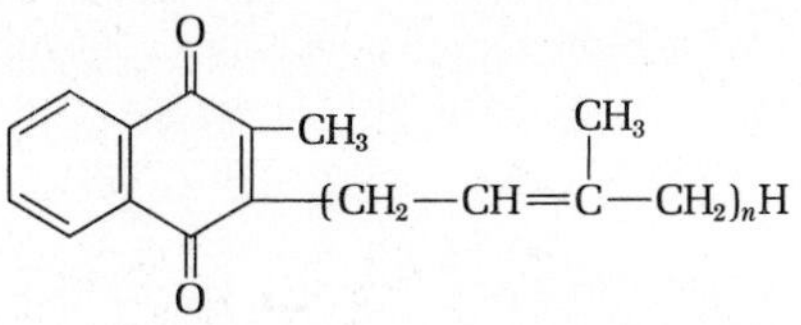

Vitamin K₂ (n may be 6, 7, 8, 9, or 10,
depending on the species)

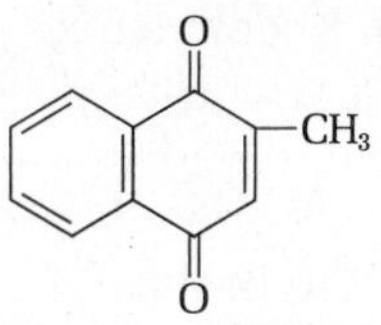

Vitamin K₃ (menadione)

of the kidneys, the deposition of brown pigments in lipid depots, necrosis of the liver, and dystrophy, or wasting, of skeletal muscles, especially in herbivorous animals such as the guinea pig. It is not known whether tocopherol deficiency causes infertility in man.

Tocopherols have been found to have *antioxidant* activity; i.e., they prevent the autoxidation of highly unsaturated fatty acids (page 281) when they are exposed to molecular oxygen. Such autoxidation results in the polymerization of unsaturated fatty acids, a process similar to that occurring in the "drying" of the linseed oil in paint, to produce a hard, tough, insoluble polymeric product. Some of the symptoms of tocopherol deficiency in animals can in fact be prevented by other compounds with antioxidant activity, some of which, like N,N-dimethyl-p-phenylenediamine, have no obvious structural relationship to tocopherol and do not occur biologically. One of the functions of tocopherol may be to protect highly unsaturated fatty acids in the lipids of biological membranes against the deleterious effects of molecular oxygen. Normally, autoxidation products of unsaturated fats do not occur in the tissues, but in tocopherol deficiency they are detectable in the fat depots, liver, and other organs. However, it appears unlikely that this is the only biological action of tocopherol. Coenzyme Q and some of its derivatives (page 493) also have tocopherol-like activity.

Vitamin K

Vitamin K (K for Danish, *koagulation*) was first discovered by H. Dam in Denmark as a nutritional factor required for normal blood-clotting time in chicks fed a diet producing a tendency to hemorrhage. It was isolated and its structure determined by E. A. Doisy and his colleagues in the United States in 1939. At least two forms of vitamin K are known (Figure 13-26); vitamin K_2 is believed to be the active form. Menadione, or vitamin K_3, a synthetic product, lacks a long side chain. Vitamin K deficiency cannot readily be produced in rats and other mammals because the vitamin is synthesized by intestinal bacteria.

The only known result of vitamin K deficiency is a failure in the biosynthesis of the enzyme *proconvertin* in the liver. This enzyme catalyzes a step in a complex sequence of reactions involved in the formation of *prothrombin*, the precursor of *thrombin*, a protein that accelerates the conversion of *fibrinogen* into *fibrin*, the insoluble protein constituting the fibrous portion of blood clots. The compound *dicumarol*, an analog of vitamin K (Figure 13-27), produces symptoms in animals resembling vitamin K deficiency; it is believed to block the action of vitamin K. Dicumarol is used in clinical medicine to prevent clotting in blood vessels.

Since vitamin K is produced by many microorganisms and most plants and is found in the tissues of all organisms, the question has arisen whether it does not have some other more general biological activity than as a factor in blood clot-

Figure 13-27
Dicumarol [3,3'-methylenebis(4-hydroxy-1,2-benzopyrone)], an antagonist of vitamin

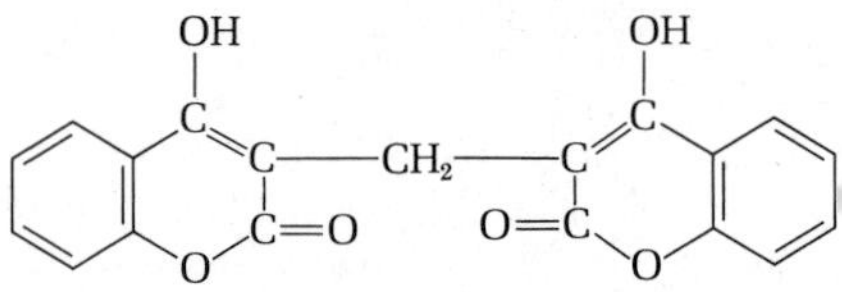

ting. Some evidence indicates that it may function as a coenzyme in a specialized route of electron transport in animal tissues; since vitamin K is a quinone which can be reduced reversibly to a quinol, it may serve as an electron carrier.

Summary

The vitamins are trace organic substances essential in the function of most forms of life but which some organisms are unable to synthesize and must obtain from exogenous sources. They are classified as water-soluble or fat-soluble.

Most of the water-soluble vitamins function as necessary building-block components of a number of different coenzymes important in central metabolic pathways. Thiamin (vitamin B_1), a deficiency of which causes beriberi in man, is the active component of thiamin pyrophosphate, a coenzyme necessary for the enzymatic decarboxylation of α-keto acids. Riboflavin (vitamin B_2) is a component of the coenzymes flavin mononucleotide (FMN) and flavin adenine dinucleotide (FAD), which function as hydrogen-carrying prosthetic groups of certain oxidative enzymes. Nicotinic acid, a deficiency of which causes pellagra, is a component of the nicotinamide adenine dinucleotides (NAD and NADP), which serve as carriers of electrons for the pyridine-linked dehydrogenases. Pantothenic acid is an essential component of coenzyme A, which functions as an acyl-group carrier during the enzymatic oxidation and synthesis of fatty acids. Vitamin B_6 (pyridoxine) is an essential precursor of the coenzyme pyridoxal phosphate, which serves as the prosthetic group of transaminases and other amino acid–transforming enzymes. Biotin functions as the prosthetic group of certain carboxylases; it serves as a carrier of carbon dioxide. Folic acid is the precursor of tetrahydrofolic acid, a coenzyme functioning in the enzymatic transfer of certain one-carbon compounds. Lipoic acid serves as a carrier of hydrogen atoms and acyl groups in the oxidative decarboxylation of α-keto acids. Vitamin B_{12} or cobalamin, essential in the nutrition of most higher organisms, has a complex corrin ring system. Its 5'-deoxyadenosyl derivative, called coenzyme B_{12}, functions in the enzymatic catalysis of 1,2 shifts of hydrogen atoms between adjacent carbon atoms. Methylcobalamin functions in a number of methyl-group transfer reactions.

The fat-soluble vitamins appear not to function as components of coenzymes but to serve other important roles. Vitamin A, whose precursor is β-carotene, functions as a light-sensitive receptor pigment in the visual cycle of rod cells in vertebrates. Vitamin D, a steroid derivative formed from 7-dehydrocholesterol by irradiation, is the major biological precursor of 1,25-dihydroxycholecalciferol, which has a hormonelike action in the activation of Ca^{2+} binding and transport in the intestine and bones. Deficiency of vitamin E, or tocopherol, in the diet yields a variety of symptoms in vertebrates. Vitamin K, a quinone with an isoprenoid side chain, is concerned in biosynthesis of components of the blood-clotting mechanism and probably has other functions as well.

References

Books

BLAKLEY, R. L.: *The Biochemistry of Folic Acid and Related Pteridines*, North-Holland, Amsterdam, 1969.

DELUCA, H. F., and J. W. SUTTIE (eds.): *The Fat-Soluble Vitamins,* University of Wisconsin Press, Madison, 1970.

DYKE, S. F.: *The Chemistry of the Vitamins,* Interscience, New York, 1965.

FLORKIN, M., and E. H. STOTZ (eds.): *Metabolism of Vitamins and Trace Elements,* vol. 2 of *Comprehensive Biochemistry,* Elsevier, Amsterdam, 1970. Detailed reviews, with much valuable background.

HUTCHINSON, D. W.: *Nucleotides and Coenzymes,* Wiley, New York, 1964.

SEBRELL, W. H., JR., and R. S. HARRIS, (eds.): *The Vitamins,* 2d ed., vols. 1–5, Academic, New York, 1967–1972. Important reference work.

SMITH, E. L.: *Vitamin B_{12},* Wiley, New York, 1965.

WAGNER, A. F., and K. FOLKERS: *Vitamins and Coenzymes,* Interscience, New York, 1964.

Articles

BARKER, H. A.: "Corrinoid-Dependent Enzymic Reactions," *Ann. Rev. Biochem.,* 41: 55–90 (1972).

BURNS, J. J.: "Ascorbic Acid," pp. 394–411, in D. M. Greenberg (ed.), *Metabolic Pathways,* 3d ed., Academic, New York, 1967.

CHATTERJEE, I. B.: "Evolution and the Biosynthesis of Ascorbic Acid," *Science,* 182: 1271–1272 (1973).

KNAPPE, J.: "Mechanism of Biotin Action," *Ann. Rev. Biochem.,* 39: 757 (1970).

WASSERMAN, R. H., and A. N. TAYLOR: "Metabolic Roles of Fat-Soluble Vitamins D, E, and K," *Ann. Rev. Biochem.,* 41: 179 (1972).

PART 2 CATABOLISM AND THE GENERATION OF PHOSPHATE-BOND ENERGY

0.5 μm

M. C. Ledbetter

Thin section of a chloroplast of Phleum pretense (timothy).

CHAPTER 14 METABOLIC AND ENERGY-TRANSFER PATHWAYS: A SURVEY OF INTERMEDIARY METABOLISM

Intermediary metabolism is often briefly defined as the sum total of all the enzymatic reactions occurring in the cell. Although this is not an inaccurate statement, it is incomplete as a definition since it does not indicate that metabolism is a highly coordinated, purposeful activity in which many sets of interrelated multienzyme systems participate, exchanging both matter and energy between the cell and its environment. Metabolism has four specific functions: (1) to obtain chemical energy from fuel molecules or from absorbed sunlight, (2) to convert exogenous nutrients into the building blocks, or precursors, of macromolecular cell components, (3) to assemble such building blocks into proteins, nucleic acids, lipids, and other cell components, and (4) to form and degrade biomolecules required in specialized functions of cells.

Although intermediary metabolism involves hundreds of different enzyme-catalyzed reactions and is often depicted in highly detailed charts or metabolic maps, which give an impression of hopeless complexity, the form and function of the central metabolic pathways are not difficult to understand. Moreover, the central pathways of metabolism are remarkably similar in most forms of life.

In this chapter we shall survey the sources of nutrients and energy for cellular life, the major routes by which cell components are synthesized and degraded, the pathway of cellular energy transfers, and some experimental approaches used in the study of cell metabolism.

Sources of Carbon and Energy for Cellular Life

Cells can be divided into two large groups on the basis of the chemical form of carbon they require from the environment. *Autotrophic* ("self-feeding") cells can utilize carbon dioxide as the sole source of carbon and construct from it the carbon skeletons of all their organic biomolecules. *Heterotrophic* ("feeding on others") cells, on the other hand, cannot utilize carbon dioxide and must obtain carbon from their environment in a relatively complex reduced form, such as glucose. Autotrophs are relatively self-sufficient cells, whereas heterotrophs, with their requirements for carbon in a fancier form, must subsist on the products formed by other cells. Photosynthetic cells and some bacteria are autotrophic, whereas the cells of higher animals and most microorganisms are heterotrophic.

The second criterion by which cells may be classified is the nature of their energy source. Cells using light as energy source are *phototrophs*; those using oxidation-reduction reactions are *chemotrophs*. Chemotrophs can be further subdivided on the basis of the nature of the electron donors they oxidize to obtain energy. Recall that oxidation-reduction reactions are those in which electrons are transferred from an electron donor (reducing agent) to an electron acceptor (oxidizing agent). Chemotrophs requiring complex organic molecules as their electron donors, such as glucose, are called *chemoorganotrophs*. Organisms that can employ simple inorganic electron donors, such as hydrogen, hydrogen sulfide, ammonia, or sulfur, are called *chemolithotrophs* (Greek *lithos*, "stone"). Table 14-1 shows the classification of all organisms into four major groups: chemoorganotrophs, chemolithotrophs, photoorganotrophs, and photolithotrophs, according to their sources of energy and carbon.

The great majority of organisms are either photolithotrophs or chemoorganotrophs. Although the other two groups of organisms include relatively few species, they should not be regarded as rare curiosities. Some of them play extremely important roles in the biosphere, particularly the soil microorganisms that fix molecular nitrogen or oxidize ammonia to nitrate. Recall also that over half the living matter on the earth is microbial and that most microbial life occurs in the soil and seas.

Heterotrophs can be divided into two major classes, *aerobes*, which use molecular oxygen as the ultimate acceptor of electrons from their organic electron donors, and *anaerobes*, which instead of oxygen use some other molecule as electron acceptor. Many cells can live either aerobically or

Table 14-1 The metabolic classification of organisms

Type of organism	Carbon source	Energy source	Electron donors	Examples
Photolithotroph	CO_2	Light	Inorganic compounds: H_2O, H_2S, S	Green cells of higher plants (in the light), blue-green algae, photosynthetic bacteria
Photoorganotroph	Organic compounds	Light	Organic compounds	Nonsulfur purple bacteria
Chemolithotroph	CO_2	Oxidation-reduction reactions	Inorganic compounds: H_2, S, H_2S, Fe(II), NH_3	Hydrogen, sulfur, iron, and denitrifying bacteria
Chemoorganotroph	Organic compounds	Oxidation-reduction reactions	Organic compounds: glucose	All higher animals, most microorganisms, nonphotosynthetic plant cells, also photosynthetic cells in the dark

anaerobically; these are called *facultative* organisms. They use oxygen when it is available, and when it is not, they can use certain organic compounds as electron acceptors. Organisms that cannot utilize oxygen at all are called strict anaerobes; indeed, many strict anaerobes are poisoned by oxygen. Most heterotrophic cells, particularly those of higher organisms, are facultative, and if oxygen is available to them, they prefer to use it since it allows more economical use of fuel molecules.

It is important to note that not all cells of a given organism are of the same class and that some types of cells have great metabolic flexibility. For example, in higher plants the green chlorophyll-containing cells of leaves are photosynthetic autotrophs, whereas the root cells are heterotrophs. Moreover, most green leaf cells function as photosynthetic autotrophs in the sunlight but as heterotrophs in the dark.

The Carbon and Oxygen Cycles

Living organisms in nature are nutritionally interdependent in many ways. Most fundamental are the cycles of carbon and oxygen in the biosphere, in which photosynthetic cells and aerobic heterotrophic cells literally feed each other, a relationship called *syntrophy*. Photosynthetic cells produce organic compounds, such as glucose, from atmospheric CO_2 and water at the expense of solar energy. Heterotrophic cells utilize the organic compounds produced by photosynthetic cells as fuels and as building blocks; the carbon dioxide formed as the end product of their metabolism is then returned to the atmosphere to be used again by photosynthetic cells (Figure 14-1). Accompanying the cycling of carbon is the exchange of oxygen between photosynthetic and heterotrophic organisms (Figure 14-1). Most photosynthetic organisms produce oxygen, which is in turn utilized by heterotrophs to oxidize fuels. The magnitude of the carbon cycle in the biosphere is enormous: it has been estimated to turn over 3.5×10^{11} tons of carbon dioxide annually.

Nutritional interdependence between living organisms in nature, exemplified here by the carbon cycle, occurs at many levels in the biosphere, from the global level just described to the microscopic level; it is characteristic of all ecological systems.

The Nitrogen Cycle

Nitrogen, a component element of proteins, nucleic acids, and other important biomolecules, also cycles through the living organisms of the biosphere. As in the carbon cycle, we find a nutritional and metabolic interdependence of different types of organisms. Although molecular nitrogen (N_2) occurs in vast amounts in the atmosphere, it is relatively inert chemically and cannot be used by most forms of life. The great majority of living organisms must obtain their nitrogen in some combined form, e.g., nitrate, ammonia, or more complex compounds like amino acids. However, such combined forms of nitrogen are very scarce in surface water

Figure 14-1
The carbon and oxygen cycles in the biosphere.

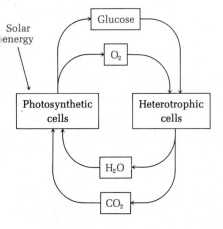

365

Figure 14-2
The nitrogen cycle.

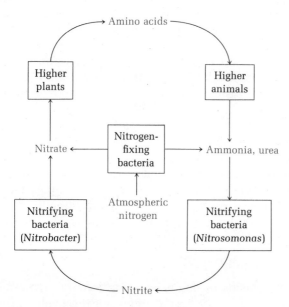

and soil, and they undergo continuous turnover (Figure 14-2). Most plants obtain their nitrogen from the soil as nitrate, which they reduce to form ammonia, amino acids, and other reduced products. These products are built up to form nitrogenous cell components, such as proteins. Heterotrophic organisms then utilize plant proteins as nutrients and return nitrogen to the soil as excretory end products or as products of decay after their death, usually as ammonia. Soil microorganisms in turn oxidize ammonia to form nitrite and nitrate, which can be utilized again by plants. Only a few forms of life, such as the nitrogen-fixing bacteria, can reduce atmospheric nitrogen and thus supplement the biologically available supply of combined nitrogen in the biosphere.

The most self-sufficient cells known are the nitrogen-fixing photosynthetic blue-green algae, prokaryotes found in soil, fresh water, and the oceans. These organisms obtain their energy from sunlight, their carbon from carbon dioxide, their nitrogen from atmospheric nitrogen, and their electrons for reduction of carbon dioxide from water. Blue-green algae are believed to be the first microorganisms that colonized land during evolution, a hypothesis given support by an interesting observation made many years ago. After the eruption of Mount Krakatoa in 1883 had completely destroyed all life in a large surrounding oceanic area, the first microorganisms to reestablish themselves were the nitrogen-fixing blue-green algae.

The Flow of Energy in the Biosphere

There is a massive flow of energy in the biosphere which is very largely coupled to the carbon cycle. Photosynthetic organisms trap solar energy and convert it into the chemical

Figure 14-3
The flow of energy in the biosphere. Solar energy is the origin of all cellular energy. Glucose and other photosynthetic products are utilized in both plants and animals to provide energy for vital cell activities. Ultimately solar energy is dissipated in useless forms, such as heat.

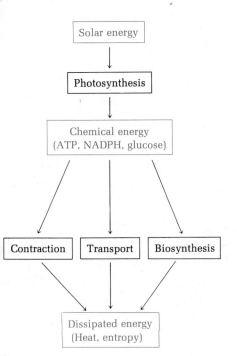

energy of glucose and other organic products. Heterotrophic organisms then utilize these products as precursors of their structural biomolecules and as energy-rich fuels to perform their energy-requiring activities (Figure 14-3). Solar energy is thus the ultimate source of energy for nearly all organisms, whether they are autotrophic or heterotrophic. However, it is important to note that energy is not cycled in the biosphere; rather it flows in a one-way direction. This flow begins with solar energy captured by photosynthetic cells and converted into the chemical energy of the photosynthetic products. These products then are used by heterotrophs to carry out various forms of cell work. In the course of these processes chemical energy is degraded into biologically useless forms of energy, such as heat, which becomes dissipated in the environment.

The flow of energy in the biosphere is of enormous proportions. Annually some 10^{19} kcal of solar energy is used to convert carbon dioxide into biomass by photosynthetic organisms in the biosphere. This biological energy flow is some twentyfold greater than the flow of energy through all the man-made machines on the face of the earth.

Nutritional Interdependence of Organisms and Cells

Besides such bulk nutrients as carbohydrates and fats many organisms require specific organic substances formed by other organisms. Although the bacterium *E. coli* can utilize ammonia as sole nitrogen source for the manufacture of all its amino acids, nucleotides, and other nitrogenous components, the lactic acid–forming bacterium *Leuconostoc mesenteroides* requires a total of 32 different nitrogenous nutrients, including most of the common amino acids as well as a number of vitamins (Table 14-2).

Many vertebrates, including man, also lack the ability to manufacture a number of amino acids and therefore require these amino acids preformed in the diet; they are called the *essential amino acids* (page 693). Moreover, many microorganisms and animals lack the capacity to synthesize one or more of the vitamins (page 335), which must also be obtained from exogenous sources. The nutritional requirements of the albino rat, which are probably identical to those of man, are given in Table 14-3. Syntrophy as an organizing principle in ecosystems thus extends beyond the bulk exchange of carbon and nitrogen; the metabolic partnership between organisms of different species may involve highly specialized trace organic molecules.

Even within a given multicellular organism, one type of cell is often dependent on the products of metabolism of another type of cell. Indeed, such nutritional interdependence is an organizing principle in the evolution of multicellular organisms. In general, the cells of higher animals require many exogenous components for growth; in fact, the complex nutritional requirements for culturing some mammalian cells in vitro are not yet known.

Table 14-2 Nutritional requirements of the lactic acid bacterium *Leuconostoc mesenteroides*

Glucose

NH$_4^+$

Amino acids

Ala	Glu	Lys	Thr
Arg	Gly	Met	Trp
Asn	His	Phe	Tyr
Asp	Ile	Pro	Val
Cys	Leu	Ser	

Bases
  Adenine
  Guanine
  Uracil
  Xanthine

Vitamins
  Thiamin
  Pyridoxine
  Pantothenic acid
  Riboflavin
  Nicotinic acid
  p-Aminobenzoic acid
  Biotin
  Folic acid
Na$^+$
K$^+$
Mg^{2+}
HPO$_4^{2-}$
SO$_4^{2-}$
Fe^{2+}
Mn^{2+}

Table 14-3 Organic nutrients required by the albino rat

Amino acids
  Arginine
  Histidine
  Isoleucine
  Leucine
  Lysine
  Methionine
  Phenylalanine
  Threonine
  Tryptophan
  Valine

Vitamins
  Water-soluble
    Thiamin
    Riboflavin
    Nicotinic acid
    Pantothenic acid
    Pyridoxine
    Biotin
    Vitamin B$_{12}$
    Folic acid
  Fat-soluble
    Vitamin A
    Vitamin D
    Vitamin E
    Vitamin K

Other organic compounds
  Polyunsaturated fatty acids
  Choline
  Inositol

Flexibility and Economy in Intermediary Metabolism

Living organisms have considerable metabolic flexibility and can adjust to the type and amount of the various nutrients available from the environment. For example, all *E. coli* cells are chemoorganotrophs, but within this category they are metabolically versatile. They can use as sole carbon source not only glucose but also various other sugars, glycerol, amino acids, ethanol, or acetate. This flexibility is possible because all these carbon sources are converted by enzymes in *E. coli* cells into compounds that can be accepted as fuels by the central metabolic pathways.

E. coli cells can also utilize sources of nitrogen other than ammonia, namely, amino acids, pyrimidines, choline, and other nitrogenous compounds. Actually, *E. coli* cells grow much faster if instead of ammonia the medium contains a complete mixture of the different amino acids, purines, and pyrimidines required in synthesis of proteins and nucleic acids. Under these circumstances the cells are spared the job of making all these building blocks from ammonia. Indeed, when *E. coli* cells subsisting on ammonia as their nitrogen source are provided with an ample supply of exogenous amino acids, they stop using the ammonia and use the preformed amino acids instead. The availability of the latter acts as a signal to turn off the biosynthesis of the enzymes required to catalyze the biosynthesis of the amino acids. Thus the cell is spared the metabolic work of making these now superfluous enzymes. However, when the concentration of preformed amino acids in the environment falls below a critical level, the necessary enzymes are made again and the cell begins to manufacture its own amino acids using ammonia as nitrogen source. We shall see repeated examples of the principle of maximum economy operating in the use of energy and nutrients by cells, made possible by various types of regulatory systems.

Catabolism and Anabolism

Metabolism is divided into two major phases, catabolism and anabolism. *Catabolism* is the degradative phase of metabolism, in which relatively large and complex nutrient molecules (carbohydrates, lipids, and proteins), coming either from the environment of the cell or from its own nutrient storage depots, are degraded to yield smaller, simpler molecules, such as lactic acid, acetic acid, CO_2, ammonia, or urea. Catabolism is accompanied by release of the chemical energy inherent in the structure of organic nutrient molecules and its conservation in the form of the energy-transferring molecule adenosine triphosphate (ATP).

Anabolism is the building-up or biosynthetic phase of metabolism, the enzymatic biosynthesis of such molecular components of cells as nucleic acids, proteins, polysaccharides, and lipids from their simple building-block precursors. Biosynthesis of organic molecules from simple precursors requires input of chemical energy, which is furnished by the ATP generated during catabolism. Catabolism and anabolism take place concurrently and simultaneously in cells, but they are independently regulated, as we shall see. Because metabolism proceeds in a stepwise manner through many intermediates, the term *intermediary metabolism* is often used to denote the chemical pathways of metabolism. The intermediates of metabolism are also called *metabolites*.

A characteristic energy change accompanies each enzyme-catalyzed reaction of metabolism. At specific steps in the catabolic pathways the chemical energy of metabolites is conserved in the form of ATP, whereas at certain steps in biosynthetic pathways the energy of ATP is utilized. Therefore, as we examine any given metabolic pathway, we must consider not only the successive enzymatic reaction steps by which the covalent structure of the precursor molecule is altered to form the end product but also the chemical mechanism by which energy is either conserved or utilized during the metabolic conversion.

We shall see that each catabolic and anabolic pathway consists of a sequence of consecutive enzyme-catalyzed reactions; sometimes there are as many as 20 steps. It may well be asked whether the same end product could not be formed in fewer steps or perhaps only one step. One answer is that multiple sequential steps are more versatile and flexible than a single, virtually irreversible reaction step for providing interconnections in the metabolic network. Another and more fundamental reason for multiple steps is that a specific, fixed amount of free energy is required to form a molecule of ATP from ADP and phosphate during catabolism; conversely, there is a maximum limit to the amount of free energy that can be delivered by a molecule of ATP in a biosynthetic pathway. Metabolic pathways have many steps so that those which provide and those which require chemical energy conform to the dimensions of the energy currency of the cell, namely, the packet, or "quantum," of free energy inherent in the terminal phosphate group of an ATP molecule (page 398).

Multienzyme Systems

Enzymes are the catalytic units of intermediary metabolism. They usually function sequentially by catalyzing consecutive reactions linked by common intermediates, so that the product of the first enzyme becomes the substrate of the next, and so on. Multienzyme systems may involve anywhere from 2 to 20 or more enzymes acting in sequence. Most of the consecutive reactions of intermediary metabolism involve the enzymatic transfer of hydrogen atoms, water molecules, or such specific functional units as amino, acetyl, phosphate, methyl, formyl, carboxyl, or adenylyl groups.

Three levels of complexity can be discerned in the organization of multienzyme systems. In the simplest multienzyme systems, the individual enzymes are in solution in the cytoplasm as independent molecules not directly associated with each other at any time during their action. The intermediates in such an enzyme system, which are generally much smaller molecules than the enzymes and thus have much higher rates of diffusion, diffuse very rapidly from one enzyme molecule to the next in the sequence (Figure 14-4).

Other multienzyme systems are more highly organized, so that the individual enzymes are physically associated and function together as _multienzyme complexes_ (Figure 14-4). For example, the fatty acid synthetase system of yeast catalyzes the biosynthesis of fatty acids from small precursor molecules, a process that requires the sequential cooperation of seven different enzymes. One molecule of each of these enzymes is present in a tightly bound cluster or complex, which does not readily dissociate; in fact, the separated enzyme molecules are inactive. Such an arrangement of sequentially active enzyme molecules into a nondissociating complex is biologically advantageous in that it limits the distance through which the substrate molecules must diffuse during the course of the reaction sequence. Actually, in the fatty acid synthetase system of yeast the intermediates in the sequence are covalently bound and never leave the complex (page 659).

Figure 14-4
Types of multienzyme systems.

A soluble or dissociated multienzyme system with diffusing intermediates B, C, D, and E.

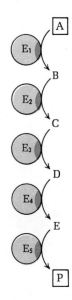

A multienzyme complex. Often the intermediates in the reaction sequence catalyzed by an enzyme complex are covalently bound and do not diffuse away from the complex.

A membrane-bound enzyme system.

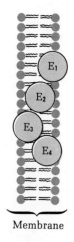

Membrane

The most complicated and highly organized multienzyme systems are those associated with large supramolecular structures such as membranes or ribosomes (Figure 14-4). An important example is the chain of electron-carrying enzymes responsible for transferring electrons from substrates to oxygen in heterotrophic cells. These enzymes are attached to the inner membrane of the mitochondria and actually form part of its structure (pages 493 and 511). The more complicated the enzyme system, the more likely it is to be associated with some organelle or other intracellular structure.

The kinetic behavior of a multienzyme system is obviously much more difficult to analyze quantitatively than the kinetics of a single enzyme. Each member of a multienzyme sequence not only has its own characteristic K_M for each substrate and cofactor but also a characteristic V_{max} and pH dependence (page 189). The rate of each individual step is determined by the steady-state concentration of each intermediate in the metabolic sequence, as well as by the concentration of each enzyme. Many times, it is the first reaction in a multienzyme sequence that sets the rate for the whole system; it is this reaction that is usually catalyzed by a regulatory enzyme (page 234). Often the kinetics of a multienzyme system can be simulated with electronic computers, starting from known data on the K_M and V_{max} values for each enzyme, the steady-state concentrations of intermediates, and the characteristics of the regulatory enzyme participating in the sequence.

Catabolic, Anabolic, and Amphibolic Pathways

The enzymatic degradation of each of the bulk nutrients of cells, namely, polysaccharides, lipids, and proteins, proceeds through a number of consecutive enzymatic reactions organized into three major stages (Figure 14-5). (Although the nucleic acids represent a substantial fraction of cell dry weight, they are not utilized as a source of energy by most organisms and are therefore not included in this discussion.) In stage I of catabolism large nutrient molecules are degraded to their major building blocks. Thus, polysaccharides are degraded to yield hexoses or pentoses; lipids to yield fatty acids, glycerol, and other components; and proteins to yield their component amino acids. In stage II of catabolism, the many different products of stage I are collected and converted into a smaller number of still simpler intermediates. Thus the hexoses, pentoses, and glycerol are degraded via the three-carbon intermediate pyruvic acid to yield a single two-carbon species, the acetyl group of acetyl-CoA. Similarly, the various fatty acids and amino acids are broken down to form acetyl-CoA and a few other end products. Finally, the acetyl groups of acetyl-CoA, as well as other products of stage II, are channeled into stage III, the final common catabolic pathway, in which they are ultimately oxidized to carbon dioxide and water.

Biosynthesis also takes place in three stages (Figure 14-5). Small precursor molecules are generated in stage III, then

Figure 14-5
The three stages of metabolism. The catabolic pathways (gray arrows, downward) converge to common end products and lead to ATP synthesis in stage III. The anabolic (biosynthetic) pathways (colored arrows, upward) start from a few precursors in stage III and utilize ATP energy to yield many different cell components.

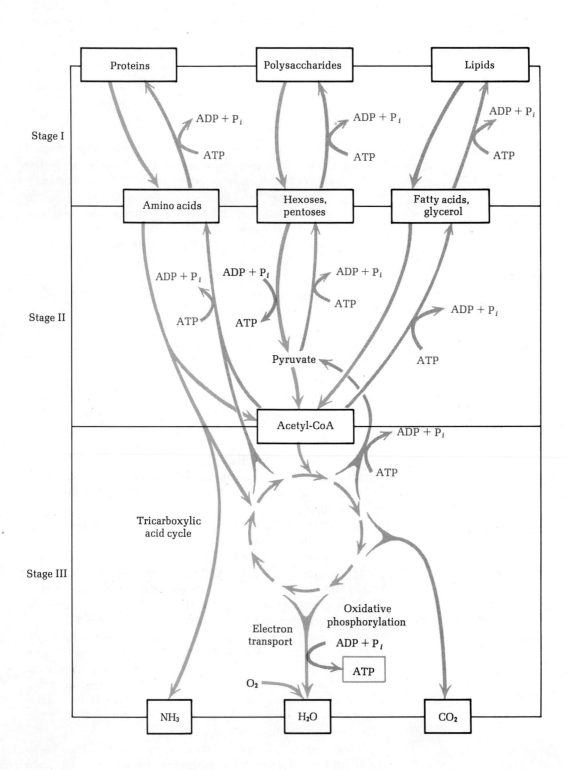

Figure 14-6
Convergence and divergence of metabolic
pathways.

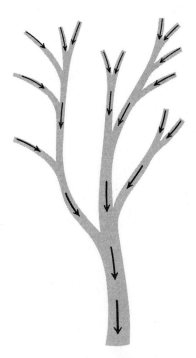

Converging pathways
of catabolism

Diverging pathways
of biosynthesis

converted in stage II into building-block molecules, which are finally assembled into macromolecules in stage I. For example, biosynthesis of proteins begins in stage III with the formation of certain α-keto acids, which are precursors of the α-amino acids. In stage II the α-keto acids are aminated by amino-group donors to form α-amino acids. Finally, in stage I, the amino acids are assembled into polypeptide chains.

Note that the catabolic pathways of metabolism, which have diffuse beginnings, starting from many different polysaccharides, lipids, and proteins, converge into a final common route in stage III (Figure 14-6). In contrast, biosynthetic pathways diverge; they start from a few precursors in stage III. As these paths proceed through stage II and stage I, they branch and diverge, leading to formation of many different kinds of biomolecules (Figure 14-6).

We come now to another important point. The catabolic and anabolic pathways between a given precursor and a given product are not the reverse of each other. For example, a sequence of 12 well-characterized enzymes catalyzes the sequential steps responsible for the degradation of glycogen into lactic acid. It might seem logical and economical for the biosynthesis of glycogen from lactic acid to occur by simple reversal of these 12 enzymatic steps. However, the biosynthesis of glycogen from lactic acid involves reversal of only 9 of the 12 enzymatic steps involved in degradation. Energetic considerations preclude simple reversal of the remaining 3 steps, and therefore bypass routes involving different enzymes and intermediates are used in the direction of biosynthesis. The catabolic and anabolic pathways between proteins and amino acids are also dissimilar, as are those between fatty acids and acetyl-CoA. Although it may seem wasteful to have two sets of metabolic pathways, one for catabolism and one for anabolism, this arrangement has important advantages. Indeed, such parallel routes for energy-yielding vs. energy-requiring pathways are a necessity, since the pathway taken in catabolism is energetically impossible for anabolism. Degradation of a complex organic molecule is energetically a "downhill" process, whereas its synthesis is an "uphill" process. As is shown in Figure 14-5, the catabolic pathways cause the formation of ATP from ADP and phosphate, at the expense of the free energy yielded during the degradation of various fuel molecules, especially in the process of oxidative phosphorylation in stage III. Conversely, the anabolic or biosynthetic pathways, which are uphill, require the input of ATP and are accompanied by its breakdown to ADP and phosphate.

Another advantage of independent catabolic and anabolic routes is that they are independently regulated. For example, the rate of breakdown of glycogen to lactic acid is controlled by different regulatory enzymes than the reverse process, the conversion of lactic acid to glycogen. For this reason the metabolic traffic between glycogen and lactic acid can be very effectively regulated in both directions; if one route is turned on, the other is turned off. Dual regulation is a general principle characteristic of all corresponding or parallel anabolic and catabolic pathways.

Catabolic and anabolic pathways may differ in another important respect: they often take place in different locations in eukaryotic cells. For example, the oxidation of fatty acids to the stage of acetyl-CoA takes place by the action of a set of enzymes localized in the mitochondria, whereas the biosynthesis of fatty acids from acetyl-CoA takes place by a different set of enzymes located in the extramitochondrial cytoplasm. Separation of opposite catabolic and anabolic pathways in different compartments of the cell allows these processes to take place independently yet simultaneously.

Although the corresponding pathways of catabolism and anabolism are not identical, stage III (Figure 14-5) constitutes a central meeting ground or pathway that is accessible to both. This common central route, sometimes called an *amphibolic pathway*, has a dual function (Greek *amphi*, "both"). The amphibolic route can be used catabolically to bring about completion of the degradation of small molecules derived from stage II of catabolism, or it can be used anabolically to furnish small molecules as precursors in biosynthetic reactions.

The Energy Cycle in Cells

Complex organic molecules, like glucose, contain much potential energy because of their high degree of structural order; they have relatively little randomness, or entropy (defined more exactly in Chapter 15). When the glucose molecule is oxidized by molecular oxygen to form six molecules of CO_2 and six of water, its carbon atoms undergo an increase in randomness; they become separated from each in the form of CO_2 and thus can assume many different positions in relation to each other. As a result of this transformation, which means an increase in the degrees of freedom of its constituent atoms, the glucose molecule undergoes a loss of free energy, i.e., that form of energy capable of doing work under conditions of constant temperature and pressure (page 390).

Biological oxidations are in essence flameless or low-temperature combustions. As we have seen (Introduction), heat cannot be used as energy source by living organisms since they are essentially isothermal; heat can do work at constant pressure only when it can flow from one body to another at a lower temperature. Instead, the free energy of cellular fuels is conserved as the chemical energy inherent in the covalent bonding structure of the terminal phosphate groups in the adenosine triphosphate molecule, as we have seen in Figure 14-5. ATP (page 398) is enzymatically generated from adenosine diphosphate (ADP) and inorganic phosphate in enzymatic phosphate-group transfer reactions that are chemically coupled to specific oxidative steps during catabolism. The ATP so formed can now diffuse to those sites in the cell where its energy is required; it is a transport form of free energy. Some of the chemical energy transported by ATP is transferred, along with the terminal phosphate group of ATP, to certain specific acceptor molecules, which become "energized" and can then function as precursors of larger biomol-

Figure 14-7
The ATP-ADP cycle. The high-energy phosphate bonds of ATP are used in coupled reactions for carrying out energy-requiring functions; ultimately, inorganic phosphate is released. ADP is rephosphorylated to ATP during energy-yielding reactions of catabolism.

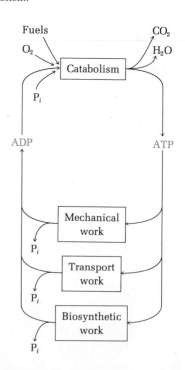

Figure 14-8
Transfer of reducing power via the nicotinamide adenine dinucleotide phosphate (NADP) cycle. Other electron-carrying coenzymes, such as flavin nucleotides, also participate in reductive biosynthesis.

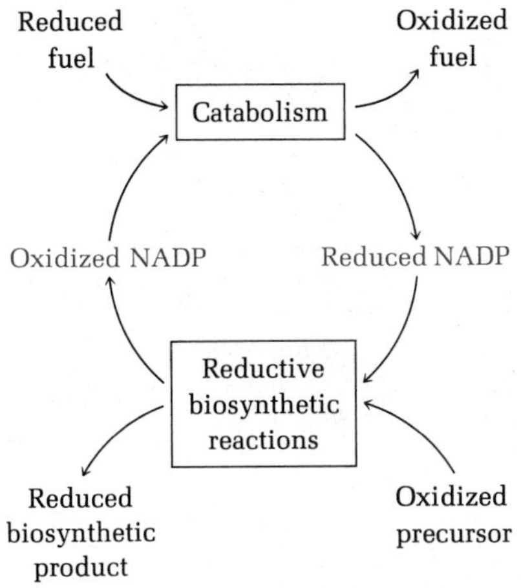

ecules (Figure 14-5). The ATP cycle is shown in Figure 14-7 and is discussed in more detail in Chapter 15 (page 407).

Electrons provide another vehicle for transfer of chemical energy from the energy-yielding reactions of catabolism to the energy-requiring reactions of biosynthesis. In the biosynthesis of some hydrogen-rich biomolecules, e.g., fatty acids and cholesterol, electrons or hydrogen atoms are required for the reduction of double bonds to single bonds. Electrons are transported enzymatically from the electron-yielding oxidations of catabolism to such electron-requiring groups as carbon-carbon or carbon-oxygen double bonds by means of electron-carrying coenzymes, the most important of which is _nicotinamide adenine dinucleotide phosphate_ (NADP). NADP (page 341) thus serves as a carrier of energy-rich electrons from catabolic reactions to electron-requiring anabolic reactions (Figure 14-8), just as ATP is a carrier of phosphate groups and energy from reactions of catabolism to those of anabolism.

Metabolic Turnover: The Dynamic State of Cell Constituents

For many years it was widely believed that cell proteins remained intact and stable for the lifetime of the cell. This view is not unreasonable, since considerable chemical work is required to synthesize a protein molecule from its component amino acids. However, this concept was overthrown in the 1930s, particularly by the research of R. Schoenheimer and his colleagues. When amino acids enriched with the heavy isotope ^{15}N (the normal, most abundant isotope of nitrogen is ^{14}N) were fed to animals, the labeled amino acids became incorporated into the polypeptide chains of liver proteins at a high rate even though the total amount of protein in the liver did not change. When the labeled amino acids were removed from the diet and replaced by normal, or ^{14}N amino acids, the previously labeled liver proteins rapidly lost their isotopic nitrogen, again with no change in the total amount of liver protein. Schoenheimer concluded that the proteins of the liver cell exist in a _dynamic steady state_, in which a relatively high rate of synthesis is exactly counterbalanced by a relatively high rate of degradation. Thus liver proteins undergo _metabolic turnover_. The proteins of rat liver were found to have a half-life of about 5 to 6 days, whereas the liver cell itself has a lifetime of several months (Table 14-4). On the other hand, the proteins of muscle tissue turn over very slowly, as do the lipids of the brain. As Table 14-4 shows, in each organ the major molecular components have a characteristic rate of turnover. Metabolic turnover is particularly rapid in cells or tissues (like the liver and the intestinal mucosa) that must adapt themselves rapidly to changes in the chemical composition of their exogenous nutrients. These tissues rapidly synthesize specific enzymes from amino acids to catalyze the utilization of the new organic components in the changing nutrient intake; such enzymes are then degraded after they are no longer needed,

Table 14-4 Turnover of some components of rat tissues

Tissue	Half-life, days
Liver	
Total protein	5–6
Glycogen	0.5–1.0
Phosphoglycerides	1–2
Triacylglycerols	1–2
Cholesterol	5–7
Mitochondrial proteins	5–6
Muscle	
Total protein	30
Glycogen	0.5–1.0
Brain	
Triacylglycerols	10–15
Phospholipid	200
Cholesterol	>100

to be replaced by other enzymes required as the nutrient intake changes again.

Experimental Approaches to Intermediary Metabolism

There are two major objectives in the experimental analysis of a given metabolic sequence. The first is the identification of the chemical pathway, the stoichiometry, and the mechanism of each reaction step in the sequence. This phase involves the identification of each intermediate in the pathway and the extraction of each enzyme of the sequence, its purification, and investigation of its specificity, kinetics, mechanism, and inhibition. It culminates in the reconstruction of the enzyme system in the test tube, starting from highly purified known components. The second major goal is the identification of the mechanisms by which the rate of the pathway is regulated, which involves the identification of the regulatory enzymes in the pathway and their response to specific modulating metabolites.

The experimental elucidation of a metabolic pathway is approached experimentally by a succession of methods beginning with the intact organism.

Metabolic Studies on Intact Organisms

The beginning and end of some major metabolic pathways have been identified from balance-sheet studies of the metabolic input and output of intact organisms. In this way carbon dioxide, which can be quantitatively recovered in the expired air and the urine of animals, was recognized to be the final end product of oxidation of carbohydrate. Similarly, ethanol and carbon dioxide were found to be the major products of glucose fermentation in yeast, and urea was identified in the urine of some mammals as a major nitrogenous end product of protein degradation. However, the intermediate steps in the catabolism of major nutrients cannot be recognized by such balance-sheet approaches.

Although the intermediates of metabolism normally are present in low steady-state concentrations in cells, specific metabolites may sometimes accumulate under the stress or imbalance created by dysfunction or disease and yield important clues to the chemical nature of a given metabolic pathway or some step in it. For example, it was early observed that during repeated muscular contractions lactic acid appears in large amounts in the blood as the glycogen of the muscle disappears, indicating that the six-carbon glucose molecule is cleaved to yield three-carbon fragments. Other examples are furnished by studies of metabolism in fasting or diabetic animals, in which glucose utilization is impaired and fatty acids become a major energy source. In such animals the _ketone bodies_ (β-hydroxybutyric acid and acetoacetic acid) appear in the blood and urine in large amounts and were thus recognized as intermediates in fatty acid oxidation. Moreover, in diabetic animals glucose excretion is increased when certain specific amino acids, such as alanine or glutamic acid, are fed, showing that the carbon skeletons

Figure 14-9
Warburg-Barcroft apparatus for measurement of oxygen consumption of sliced or minced tissue suspensions.

The incubation is carried out in a special reaction vessel attached to a manometer for recording the decrease in oxygen pressure as respiration takes place.

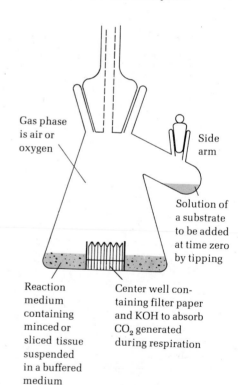

Gas phase is air or oxygen

Side arm

Solution of a substrate to be added at time zero by tipping

Reaction medium containing minced or sliced tissue suspended in a buffered medium

Center well containing filter paper and KOH to absorb CO_2 generated during respiration

of these amino acids can serve as metabolic precursors of glucose. On the other hand, tyrosine and phenylalanine do not increase the excretion of glucose but increase the formation of acetoacetic acid, showing that the carbon skeletons of these two amino acids are metabolically converted into acetoacetate.

Treatment of intact organisms with certain drugs or inhibitors may also cause a specific metabolite to accumulate in the blood and tissues. For example, administration of *fluorocitrate* causes citrate to accumulate in the liver of some animals because it competitively inhibits the enzymatic oxidation of citrate.

Perfusion of the vascular system of isolated organs such as the liver or kidney with blood or buffered saline containing a metabolic precursor, followed by chemical analysis of the perfusate, is another approach that can give valuable information on metabolic pathways. With this method it was established that the liver is the major site of formation of the ketone bodies, of urea, and of the conversion of certain amino acids into glucose.

Surviving-Slice and Manometric Methods

Another technique widely used in early studies of intermediary metabolism in animal or plant tissues is the surviving-slice method, introduced by O. Warburg in the 1920s. Solid animal or plant tissues are sectioned into thin slices in which most of the cells remain intact. Such slices can be made sufficiently thin (< 0.4 mm) to ensure that the rate of diffusion of oxygen and metabolites into and out of the cells from the aqueous suspending medium is not rate-limiting for the metabolic exchanges occurring within the cells. The tissue slices are incubated in a buffered medium with a given metabolite to study its conversion into metabolic products, which accumulate in the medium. The rate of many metabolic events, e.g., the conversion of glucose to lactic acid by animal tissues, can be determined by chemical measurements of the changes occurring in the medium in which tissue slices are suspended. It is also possible to determine the rate of oxygen consumption by surviving slices of tissues. The decrease in partial pressure of oxygen over a suspension of tissue slices is measured in a manometric device called the Warburg-Barcroft apparatus (Figure 14-9). This manometric technique was very important, for example, in elucidation of the reactions of the tricarboxylic acid cycle (page 444).

Genetic Defects in Metabolism; Auxotrophic Mutants

Another important approach to the study of intermediary metabolism in intact cells and organisms is afforded by genetic mutants of an organism in which there is a defect in the biosynthesis of a given enzyme. Such genetic defects, if they are not lethal, often result in the accumulation and excretion of the substrate of the defective enzyme. In the normal or unmutated organism, of course, the intermediate

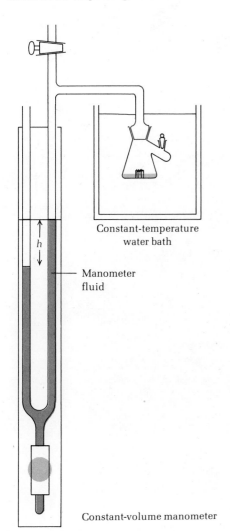

The reaction vessel is shaken in a constant-temperature bath. The carbon dioxide generated during respiration is continuously absorbed from the gas phase by KOH in the center well. In this way, the partial pressure of CO_2 in the gas phase is kept at or near zero, and the pressure changes observed are due only to changes in oxygen pressure. Consumption or production of metabolites can be determined by chemical measurements on the suspending medium.

Constant-temperature water bath

h

Manometer fluid

Constant-volume manometer

would not accumulate since it would undergo further metabolic conversion. Identification of metabolites excreted by such mutant organisms can lead to valuable information about a metabolic pathway. The abnormal excretion of _homogentisic acid_ in the urine (page 569) is due to one such genetic defect in human patients, called _alkaptonuria_ (page 570). Since excretion of this acid is increased by feeding phenylalanine or tyrosine but not by other amino acids, it was concluded that homogentisic acid is formed as an intermediate during the catabolism of phenylalanine and tyrosine. Many other genetic defects in human amino acid metabolism are known and have provided important evidence for the nature of the intermediate reaction steps in amino acid catabolism (page 570).

Far more useful for study of metabolism are specific genetic defects induced in microorganisms by various mutagenic agents, such as x-radiation. Mutant microorganisms deficient in a specific pathway or enzyme can literally be produced at will and have become powerful tools for studying intermediary metabolism. In 1941 the "one gene–one enzyme" relationship was postulated by G. W. Beadle and E. L. Tatum on the basis of their studies of mutants of the mold _Neurospora crassa_. The approaches they developed were extremely important not only for study of the gene-enzyme relationship but also for the analysis of the pathways of intermediary metabolism. Wild-type, i.e., unmutated, _Neurospora_ can grow on a simple medium containing glucose as sole carbon source and only ammonia as nitrogen source. However, exposure of _Neurospora_ spores to x-rays yields some mutant cells no longer capable of growing on this simple medium. Such mutant cells will grow normally if the medium is supplemented with the specific metabolite whose biosynthesis was impaired by the mutation. For example, some mutants of _Neurospora_ are unable to grow unless the medium contains arginine, suggesting that an enzyme required in the synthesis of arginine from ammonia is genetically defective in these mutants. For lack of arginine such mutant cells cannot manufacture their proteins. The mutant cells can utilize arginine for protein biosynthesis and show normal growth only when this amino acid is supplied in the medium. Further studies have revealed that not all mutants of _Neurospora_ defective in the capacity to make arginine are identical; they differ with respect to the specific step in the pathway of arginine biosynthesis that is genetically defective. Thus it was found that the growth of _arginineless_ mutant I (Figure 14-10) can be sustained only by adding arginine to the medium and not by adding any known precursor of arginine such as metabolites A, B, C, or D, indicating that this mutant is defective in enzyme E_4. However, _arginineless_ mutant II will grow when the medium is supplied with precursor D or with arginine but not when precursors A, B, or C are added, indicating that arginine synthesis and thus growth is blocked because enzyme E_3 is defective. Such mutants, blocked at different points in a metabolic pathway, can be used to test suspected precursors

Figure 14-10

Auxotrophic mutants of Neurospora crassa *with defective enzymes (color) at different points in the biosynthesis of arginine from precursor A.*

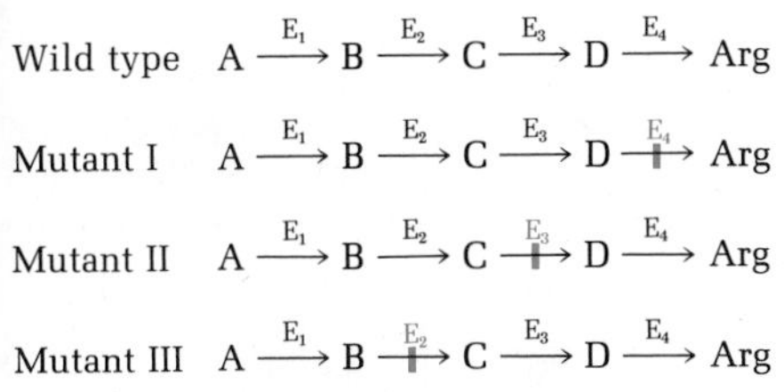

Wild type $A \xrightarrow{E_1} B \xrightarrow{E_2} C \xrightarrow{E_3} D \xrightarrow{E_4} Arg$

Mutant I $A \xrightarrow{E_1} B \xrightarrow{E_2} C \xrightarrow{E_3} D \xrightarrow{E_4} Arg$

Mutant II $A \xrightarrow{E_1} B \xrightarrow{E_2} C \xrightarrow{E_3} D \xrightarrow{E_4} Arg$

Mutant III $A \xrightarrow{E_1} B \xrightarrow{E_2} C \xrightarrow{E_3} D \xrightarrow{E_4} Arg$

Table 14-5 Some isotopes useful as tracers†

Isotope	Relative natural abundance, %	Type of radiation	Half-life
$^{2}_{1}H$	0.0154		Stable
$^{3}_{1}H$		β^-	12.1 years
$^{13}_{6}C$	1.1		Stable
$^{14}_{6}C$		β^-	5,700 years
$^{15}_{7}N$	0.365		Stable
$^{18}_{8}O$	0.204		Stable
$^{24}_{11}Na$		β^-,γ	15 h
$^{32}_{15}P$		β^-	14.3 days
$^{35}_{16}S$		β^-	87.1 days
$^{36}_{17}Cl$		β^-	3.1×10^5 years
$^{42}_{19}K$		β^-	12.5 h
$^{45}_{20}Ca$		β^-	152 days
$^{59}_{26}Fe$		β^-,γ	45 days
$^{131}_{53}I$		β^-,γ	8 days

† The superscript before the symbol of the element designates the mass number, the subscript the atomic number. β^- radiation is due to negative electrons. Radioactivity is expressed in terms of the *curie* (Ci), the quantity undergoing the same number of disintegrations per second as 1.0 g of radium (3.7 × 10^{10} s⁻¹); the millicurie and microcurie are more convenient units. Stable isotopes are measured in atoms percent excess over the natural abundance.

of a given product and to determine the sequence in which such precursors are converted into the product.

Such sets of mutants can be used in another way to identify intermediates in a biochemical sequence. For example, when mutant I (Figure 14-10) is grown in the presence of small, limiting amounts of arginine, the blocked precursor D will accumulate in the growth medium since it cannot be converted into arginine. After removal of the mutant I cells by filtration, this medium will support the growth of mutant II in the absence of arginine since it supplies the precursor D. The filtrate of mutant II, however, will not support the growth of mutant I. This type of experiment is known as *cross-feeding*. The precursor generated by mutant I that can feed mutant II can then be isolated and identified; the growth rate of mutant II can be used as a bioassay for the precursor. Mutants which are defective in a biosynthetic pathway but which revert to normal growth when they have been provided with the normal product of the pathway are called *auxotrophic mutants*.

Mutants can also be used to analyze catabolic pathways. For example, wild-type *E. coli* cells are able to grow on lactose, glucose, or galactose as carbon source. In some *E. coli* mutants the capacity to grow on lactose is lost, but the cells are still able to grow on the other sugars. The genetic defect in this case involves the enzyme β-galactosidase, which can hydrolyze lactose (page 978) to its components glucose and galactose.

Through such genetic approaches employing mutant microorganisms, many intermediates in metabolic pathways have been identified.

Isotopic-Labeling Methods

Another powerful method applicable to study of metabolism in intact organisms is the use of a metabolite labeled so that its metabolic fate can be traced. This approach was first used as long ago as 1904, when the German biochemist F. Knoop found that a phenyl group substituted in the terminal methyl carbon atom of a fatty acid remains attached to that atom throughout the oxidative degradation of the fatty acid. Animals fed phenyl-substituted long-chain fatty acids excreted in the urine phenyl-substituted degradation products from whose structure Knoop deduced that fatty acids are oxidized by successive removal of two-carbon fragments (page 544). This primitive chemical-labeling method has been superseded by the use of either stable or radioactive isotopes to label certain atoms of a given metabolite so that for all practical purposes it is chemically and biologically indistinguishable from its normal analog. Because of the presence of the isotope, the metabolite and its transformation products can be detected and measured (Table 14-5). An extraordinary range of important observations has been made with the isotope tracer technique applied to intact animals, plants, and microorganisms, as well as isolated organs or tissue extracts. Among these is the discovery that the carbon atoms of cholesterol (page 679) are derived from acetate and that glycine is a precursor in the synthesis of purines (page 729) and porphyrins (page 718). Study of many aspects of metabolism (including metabolic turnover rates, biosynthesis of proteins and nucleic acids, and photosynthesis) would have been quite impossible without isotopically labeled metabolites.

The isotope tracer method can also be used to determine what the rate of metabolic processes is in intact organisms and whether, for a given metabolite, a given metabolic pathway is the predominant route. It has also been used to establish whether a metabolic pathway postulated by study of isolated enzymes in the test tube actually occurs in the intact cell (see Chapter 17).

Cell-Free Systems

The most direct experimental approach to identification of the individual reaction steps in a given metabolic sequence is the study of cell or tissue dispersions in which the cell membrane is broken and the cell contents released. Such dispersions or soluble extracts prepared by centrifugation can often carry out an entire metabolic sequence. Thus our present knowledge of the details of the conversion of glucose to ethanol and carbon dioxide began with the discovery by E. Buchner in 1897 that cell-free extracts of yeast catalyze the alcoholic fermentation of glucose to yield ethanol and CO_2. Similarly, it was shown later that conversion of glucose to lactic acid occurs in cell-free extracts of muscle. Once such preparations were obtained, metabolic intermediates could be made to accumulate by use of specific enzyme inhibitors (page 419), by inactivation of specific enzymes, or by removal from the extract of essential coenzymes. Chemical identification of an intermediate that accumulates after such a treatment then makes it possible to identify the enzymatic reaction by which the intermediate is formed and the enzymatic reaction by which it is utilized. Ultimately the enzyme may be isolated from the cell or tissue extract. The final goal is the reconstruction of part or all of the enzyme system in vitro, starting from highly purified enzymes, coenzymes, and other components.

If the cell membrane is disrupted by mild homogenization in isotonic sucrose solution, such subcellular organelles as nuclei, mitochondria, and lysosomes (page 18) and such supramolecular structures as ribosomes remain intact and can be isolated by differential centrifugation from the homogenate (Figure 14-11). These fractions can then be tested in vitro for their capacity to catalyze a given metabolic sequence. In this way it was shown that isolated mitochondria catalyze the entire sequence of reactions constituting the tricarboxylic acid cycle. Extracts prepared from isolated mitochondria formed the starting point for tracing the enzymatic steps in fatty acid oxidation and for isolating the enzymes participating in this process. Similarly, isolated ribosomes are used to study the pathway and mechanism of protein biosynthesis.

Figure 14-11
Isolation of intracellular structures from rat liver by differential centrifugation. The cell membranes are ruptured by the shearing forces developed by the rotating homogenizer pestle. Following removal of connective tissue and fragments of blood vessels and bile ducts by a stainless steel sieve, the cell extract is centrifuged at a series of increasing rotor speeds.

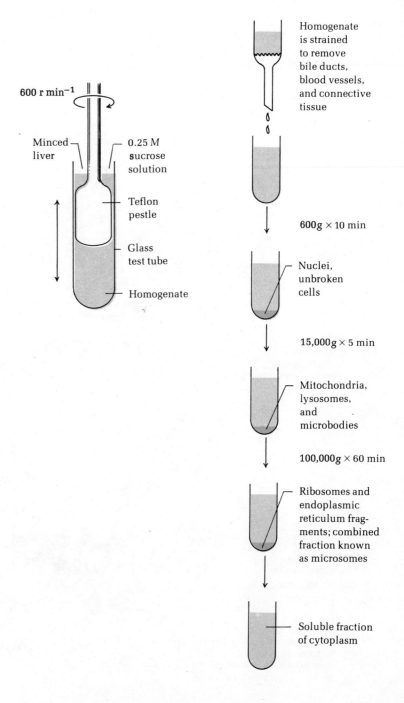

Intracellular Compartmentation of Enzymes and Enzyme Systems

Different enzymes and enzyme systems are characteristically located in one or another organelle or intracellular structure of eukaryotic cells (Figure 14-12). The entire glycolytic enzyme system is located in the soluble portion of the cytoplasm, the _cytosol_, whereas the enzymes concerned in oxidation of pyruvate, fatty acids, and some amino acids via the tricarboxylic acid cycle are located in the mitochrondria, as are the enzymes of electron transport and oxidative phosphorylation of ADP.

Figure 14-12
Compartmentation of some important enzymes and metabolic sequences in the liver cell of the rat. The electron micrograph from which the cell drawing was traced is shown in Figure 1-8.

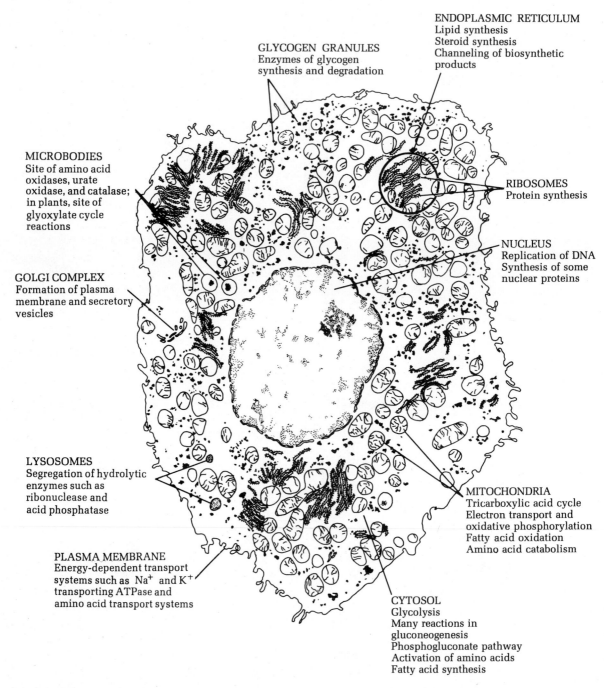

ENDOPLASMIC RETICULUM
Lipid synthesis
Steroid synthesis
Channeling of biosynthetic
products

GLYCOGEN GRANULES
Enzymes of glycogen
synthesis and degradation

MICROBODIES
Site of amino acid
oxidases, urate
oxidase, and catalase;
in plants, site of
glyoxylate cycle
reactions

RIBOSOMES
Protein synthesis

GOLGI COMPLEX
Formation of plasma
membrane and secretory
vesicles

NUCLEUS
Replication of DNA
Synthesis of some
nuclear proteins

LYSOSOMES
Segregation of hydrolytic
enzymes such as
ribonuclease and
acid phosphatase

MITOCHONDRIA
Tricarboxylic acid cycle
Electron transport and
oxidative phosphorylation
Fatty acid oxidation
Amino acid catabolism

PLASMA MEMBRANE
Energy-dependent transport
systems such as Na^+ and K^+
transporting ATPase and
amino acid transport systems

CYTOSOL
Glycolysis
Many reactions in
gluconeogenesis
Phosphogluconate pathway
Activation of amino acids
Fatty acid synthesis

Compartmentation of enzyme systems also permits control and integration of some intracellular activities. For example, the biosynthesis of glucose from pyruvate involves a complex interplay of a series of enzymes, some located in the mitochondria and some in the soluble phase of the cytoplasm. The rate of this overall reaction depends not only on the activity of regulatory enzymes in both compartments but also on the rates of exchange of essential intermediates across the mitochondrial membranes.

Cellular Regulation of Metabolic Pathways

In general, the rate of catabolism is controlled not by the concentration of nutrients available in the environment of the cell but by the cell's second-to-second needs for energy in the form of ATP. In short, cells consume their fuels only as fast as necessary to furnish the energy required for their activities at any given instant. Similarly, the rate of biosynthesis of cell components is also adjusted to immediate needs. For example, cells synthesize amino acids just fast enough to keep pace with the rate of their utilization. Thus the principle of maximum economy (Introduction) pervades all aspects of cellular metabolism.

The regulation of a metabolic pathway may occur at several levels. The reaction rate of each enzymatic reaction is a function of the pH and the intracellular concentrations of its substrate(s), product(s), and cofactor, which are primary elements in the regulation of enzymatic activity.

The second level of control of metabolic sequences is through the action of regulatory enzymes (page 234). Many regulatory enzymes are inhibited by the end product of the sequence in which they function. For example, in enzyme systems concerned with the generation of ATP coupled to catabolic reactions, ATP, the end product, may be a specific inhibitor of an allosteric enzyme early in the catabolic sequence. In biosynthetic enzyme systems the end product often functions as an allosteric inhibitor. In addition, some regulatory enzymes are activated or stimulated by specific metabolites, sometimes by their own substrates (page 236). Some allosteric enzymes are located at branch points in a metabolic pathway. Such enzymes are often multivalent; i.e., they can respond to two or more activators or inhibitors which may be the products of the two branching pathways. In this way multivalent allosteric enzymes can integrate the rates of two or more enzyme systems.

The third level at which metabolic regulation is exerted is through genetic control of the rate of enzyme synthesis. The rate of a given metabolic sequence must depend on the concentration of the active form of each enzyme in a sequence, which in turn is the result of a balance between the rate of its synthesis and the rate of its degradation. Enzymes that are always present in nearly constant amounts in a given cell are called *constitutive enzymes*. Others that are synthesized only in response to the presence of certain substrates are called *induced enzymes*. The genes specifying the synthesis of inducible enzymes are usually under repression and come into play, i.e., undergo derepression, only in response to the presence of the inducing agent. An entire sequence of enzymes may be repressed or induced as a group; their synthesis is coded by a set of consecutive genes in DNA, called an *operon*, which can be repressed or derepressed together. A more complete description of the regulation of enzyme synthesis is given in Chapter 35 (page 977).

Metabolic regulation is exerted at still another level in higher multicellular organisms with endocrine systems. Hormones elaborated by an endocrine gland are chemical messengers that pass via the blood to certain "target" tissues, where they stimulate or inhibit specific metabolic activities. Deficiency of insulin secretion by the pancreas results in impaired transport of glucose into cells, which leads to a number of secondary metabolic effects, such as a decrease in the biosynthesis of fatty acids from glucose and an excessive formation of ketone bodies by the liver. Administration of trace amounts of insulin repairs these defects. Opposed to the action of insulin is glucagon, another hormone secreted by the pancreas. The action of hormones is considered in detail elsewhere (pages 807 to 828).

Summary

Organisms can be classified on the basis of their exogenous carbon requirements. Autotrophs require only carbon dioxide, whereas heterotrophs require carbon in a more complex reduced form, such as glucose. Organisms can also be classified by their energy source: phototrophs obtain their energy from light, whereas chemotrophs obtain energy from oxidation-reduction reactions. Cells using inorganic reductants as electron donors are called lithotrophs; those using organic molecules are organotrophs. Photosynthetic autotrophs and chemoorganotrophs "feed" each other: photosynthetic cells use atmospheric carbon dioxide and solar energy to synthesize glucose and evolve oxygen, whereas the chemoorganotrophic cells of animals oxidize glucose and other products of photosynthesis at the expense of molecular oxygen to yield energy and release carbon dioxide. Solar energy is the ultimate source of energy for all forms of life.

In the biological nitrogen cycle plants obtain their nitrogen as nitrate and reduce it to ammonia and amino acids; the latter are then utilized by animals and returned to the soil as urea or ammonia, which are reoxidized to nitrate by soil bacteria. Only nitrogen-fixing bacteria can utilize the molecular nitrogen of the atmosphere. Some organisms require exogenous or preformed amino acids, purines, or pyrimidines as nitrogen sources.

Intermediary metabolism can be divided into catabolic pathways, responsible for the degradation of energy-rich nutrient molecules; anabolic pathways, for the biosynthesis of cellular components; and the central amphibolic pathway, which serves in both capacities. Each pathway is promoted by a sequence of specific enzymes catalyzing consecutive reactions. The anabolic and catabolic pathways to and from a given nutrient, such as glucose, are not the exact reverse of each other; they are chemically and enzymatically different. Moreover, they are independently regulated and often located in different parts of the cell. Catabolism of nutrient molecules is accompanied by conservation of some of the energy of the nutrient in the form of the phosphate-bond energy of adenosine triphosphate (ATP). Conversely, the chemical energy required for biosynthetic pathways is provided by the dephosphorylation of ATP. Chemical energy is also carried from catabolic to anabolic pathways in the form of reduced coenzymes, especially NADPH.

Metabolic pathways in intact cells can be investigated by means of input-output studies on normal, pathological, stressed, or poisoned organisms, by perfusion of intact organs, and by studying surviving slices of tissue. Genetically defective microorganisms, called auxotrophs, have also been very useful for analysis of metabolic pathways. The isotope tracer technique is an especially powerful tool for studying the pathways, rates, and various disturbances of metabolism. Once a given metabolic conversion can be demonstrated in a cell-free system, the individual enzymes catalyzing the separate reactions can be isolated and identified. Metabolism is regulated through the intrinsic kinetic properties of enzymes, through the action of regulatory or allosteric enzymes, through changes in the rate of biosynthesis of enzymes, and by hormonal action.

References

Books

BIRNIE, G. D. (ed.): *Subcellular Components: Preparation and Fractionation,* 2d ed., University Park Press, Baltimore, 1972.

CHASE, G. D., and J. L. RABINOWITZ: *Principles of Radioisotope Methodology,* 2d ed., Burgess, Minneapolis, 1962.

COLOWICK, S. P., and N. O. KAPLAN: *Methods in Enzymology,* Academic, New York, 1955 to present. Multivolume treatise on experimental methods in study of intermediary metabolism.

DAGLEY, S., and D. E. NICHOLSON: *Metabolic Pathways,* Wiley, New York, 1970. Comprehensive compilation of "metabolic maps," showing pathways and names of enzymes. Very useful for reference.

GREEN, D. E., and R. F. GOLDBERGER: *Molecular Insights into the Living Process,* Academic, New York, 1967.

GREENBERG, D. M. (ed.): *Metabolic Pathways,* Academic, New York, 3d ed., 1967 to present. Valuable series of volumes with articles on different aspects of intermediary metabolism.

GUNSALUS, I. C., and R. Y. STANIER (eds.): *The Bacteria,* vols. II and III, Academic, New York, 1961. The metabolic classification of microorganisms is described in detail, as are many metabolic pathways.

KREBS, H. A., and H. L. KORNBERG: *Energy Transformation in Living Matter,* Springer-Verlag, Berlin, 1957. A classical and readable analysis of metabolism in terms of energy exchanges; it has had a wide influence.

MOROWITZ, H. J.: *Energy Flow in Biology,* Academic, New York, 1968. A theoretical analysis of some profound aspects of bioenergetics.

NOVIKOFF, A. B., and E. HOLTZMAN: *Cells and Organelles,* Holt, New York, 1970.

ROODYN, D. B. (ed.): *Enzyme Cytology,* Academic, New York, 1967. The intracellular location of various enzymes and enzyme systems.

STANBURY, J. B., J. B. WYNGAARDEN, and D. S. FREDRICKSON: *The Metabolic Basis of Inherited Disease,* 3d ed., McGraw-Hill, New York, 1972. Comprehensive treatment of genetic aberrations in human metabolism, with excellent reviews of the relevant metabolic pathways.

Article

SIEKEVITZ, P.: "The Turnover of Proteins and the Usage of Information," *J. Theoret. Biol.,* 37: 321–334 (1972).

Problems

1. The decay or disintegration of radioisotopes is a first-order process (page 186) given by the relationship

$$\log \frac{N_0}{N_t} = \frac{\lambda t}{2.303}$$

where N_0 is the number of atoms of the radioisotope present at $t = 0$, N_t is the number present at time t, and λ is the disintegration constant, which is expressed in units of t^{-1}. The half-life, $t_{1/2}$, of a radioisotope is the time taken for the number of atoms of the radioisotope to decrease by exactly 50 percent. Give an expression for $t_{1/2}$ in terms of λ.

2. Direct determination of the number of atoms of a radioisotope in a given sample is not feasible. Instead, radioisotopes are assayed by measurement of a proportional quantity, the number of disintegrations per unit time. Such measurements are made with a Geiger counter or scintillation counter and are frequently reported as counts per unit time. If a sample containing ^{32}P gives 1.5×10^6 counts per minute in a given counting system, how many counts per minute will it give exactly 1 week later? (Refer to the data of Table 14-5 in this and subsequent problems.)

3. A sample containing a hypothetical radioisotope gave 10,000 counts per minute and then 6,000 counts per minute 24 h later. What is the half-life of the radioisotope?

4. How long will it take for a sample of ^{35}S to lose 95 percent of its initial radioactivity?

5. A sample containing ^{32}P and ^{35}S gave 3.52×10^5 counts per minute initially and 2.21×10^5 counts per minute exactly 30 days later. What fraction of the radioactivity in the original sample was due to ^{32}P, and what fraction was due to ^{35}S?

6. The specific activity of a substance may be defined as the number of counts per minute per unit of that substance. To a solution of D-galactose was added 0.1 mg of ^{14}C-labeled galactose with a specific activity of 1.00×10^6 counts per minute per milligram. The specific activity of the resulting mixture was determined to be 1.37×10^3 counts per minute per milligram. What was the concentration of galactose in the original sample?

7. Volumes of distribution of substances ("pool size") can be estimated by dilution of radioisotopes. Suppose that 10 μl of a solution containing 3.00×10^6 counts per minute of ^{24}Na is injected into the bloodstream of an experimental animal. Thirty minutes later, assay of 50 μl of the blood yields 126 counts per minute. If it is assumed that the Na^+ is not excreted and that it does not penetrate into cells, what is the volume of distribution of the Na^+?

CHAPTER 15 **BIOENERGETIC PRINCIPLES AND THE ATP CYCLE**

In the molecular logic of the living state (Introduction, page 7) cells may be regarded as chemical engines capable of operating under conditions of essentially constant temperature, pressure, and volume. Like man-made engines, all living organisms must obtain their energy supply from the surrounding environment. Photosynthetic organisms utilize the radiant energy of sunlight, whereas heterotrophic organisms utilize the energy inherent in the structure of organic nutrient molecules obtained from their surroundings. These forms of energy are transformed by cells into the chemical energy of adenosine triphosphate (ATP), which functions as a carrier of energy to those processes in cells that require energy input.

In this chapter we shall develop the chemical and thermodynamic principles that underlie the function of the ATP system in the energetics of living cells. These principles are fundamental to all aspects of intermediary metabolism and will be repeatedly emphasized in subsequent chapters.

The ATP Cycle

ATP was first discovered in muscle extracts by C. Fiske and Y. Subbarow in the United States and independently by K. Lohmann in Germany in 1929; for some time thereafter ATP was thought to be primarily concerned in muscular contraction. It was not until the early 1940s that the full scope of the role of ATP in all living cells became apparent, when a number of earlier observations crystallized into a meaningful picture. Some of these may be enumerated. In the 1930s it became clear, particularly from the research of the German biochemists Otto Warburg and Otto Meyerhof, that ATP is generated from ADP in coupled enzymatic reactions during the anaerobic breakdown of glucose to lactic acid in muscle (page 419). Later H. Kalckar in Denmark and V. Belitser in the Soviet Union proved that ATP is also generated from ADP during aerobic oxidations in animal tissues, in the process of oxidative phosphorylation (page 514), in accordance with an earlier hypothesis of V. A. Engelhardt. The recognition that ATP is in turn utilized for energy-requiring functions came when Engelhardt and M. N. Lyubimova discovered that ATP is hydrolyzed to ADP and phosphate by myosin, the major contractile protein of muscle, and when C. F. Cori and G. T. Cori in the United States, as well as other investigators, recognized that ATP is required to phosphorylate glucose and thus "energize" it for the biosynthesis of glycogen.

These and many other observations were assembled in 1941 by Fritz Lipmann into a general hypothesis for energy transfer in living cells, to which Kalckar also made important contributions. Lipmann postulated that ATP functions in a cyclic manner as a carrier of chemical energy from the degradative or catabolic reactions of metabolism, which yield

Figure 15-1
The ATP cycle. At the top is shown the cycle as depicted by Lipmann in 1941. In his words, "The metabolic dynamo generates ~P-current. This is brushed off by adenylic acid, which likewise functions as the wiring system distributing the current. Creatine ~P, when present, serves as a ~P-accumulator." [F. Lipmann, Adv. Enzymol., 1: 122 (1941).] Below is a version showing the major uses of ATP energy.

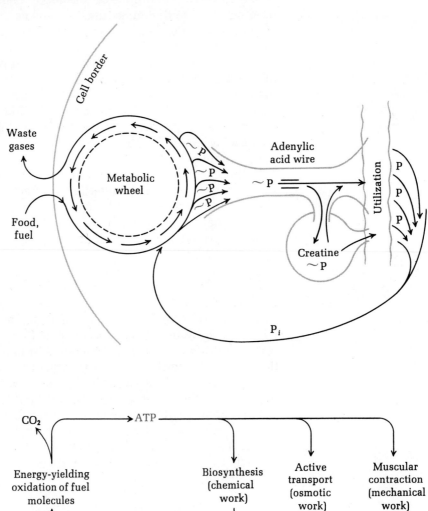

chemical energy, to the various cellular processes that require an energy input (Figure 15-1). ATP is generated from ADP by coupled or linked phosphorylation reactions at the expense of energy yielded by degradation of fuel molecules. The ATP so generated was postulated to donate its terminal phosphate group to specific acceptor molecules, to energize them for carrying out various energy-requiring functions in the cell, e.g., the biosynthesis of cell macromolecules (chemical work), the active transport of inorganic ions and cell nutrients across membranes against gradients of concentration (osmotic work), and the contraction of muscles (mechanical work). As the energy of ATP is delivered to these energy-requiring processes, the ATP undergoes cleavage to ADP and inorganic phosphate. The ADP is then rephosphorylated at the expense of energy-yielding oxidation of fuels to yield ATP, thus completing the cellular energy cycle. The terminal phosphate group of the ATP was thus visualized as undergoing constant turnover, being continuously transferred to acceptor molecules and continuously replaced by phosphate groups that become energized during the catabolic degradation of cell fuels.

To understand how ATP functions as a molecular linking agent in the flow of chemical energy from the oxidation of fuels to various energy-requiring processes we must first examine some chemical properties of ATP and then consider the physical principles that govern the transfer of energy by chemical reactions.

Occurrence and Properties of ATP, ADP, and AMP

The structure of ATP (page 314 and Figure 15-2), first deduced by Lohmann in 1930, was ultimately confirmed through total chemical synthesis by Alexander Todd and his colleagues in 1948. ATP, ADP, and AMP have been found in all forms of life examined, in both the animal and plant worlds. The sum of their concentrations in the aqueous phase of living cells remains relatively constant and amounts to between 2 to 10 mM, depending on the species. In actively metabolizing cells the concentration of ATP usually greatly

Figure 15-2
Space-filling model and structure of ATP.
See also Figure 12-8.

exceeds the sum of the concentrations of ADP and AMP. These nucleotides are present not only in the soluble cytoplasm but also such organelles as mitochondria and nuclei. We shall see later that the intracellular compartmentation of ATP is an important feature in cellular regulation of metabolism (page 536).

At pH 7.0 both ATP and ADP are highly charged anions. ATP has four ionizable protons in its condensed phosphate groups; ADP has three. Three of the four protons in ATP are fully ionized at pH 7.0; the fourth has a pK' of 6.95 and is thus about 50 percent dissociated at pH 7.0. We shall see that the high concentration of closely spaced negative charges around the triphosphate group of ATP is an important factor in its energy-transferring function. In intact cells ATP and ADP are largely present as the 1:1 MgATP^{2-} and MgADP^{-} complexes, because of the high affinity of the pyrophosphate groups for divalent cations and the relatively high concentrations of Mg^{2+} in intracellular fluid. The affinity of ATP for Mg^{2+} is about 10 times that of ADP. In most enzymatic reactions in which ATP participates as phosphate donor, its active form is the MgATP^{2-} complex (Figure 15-3).

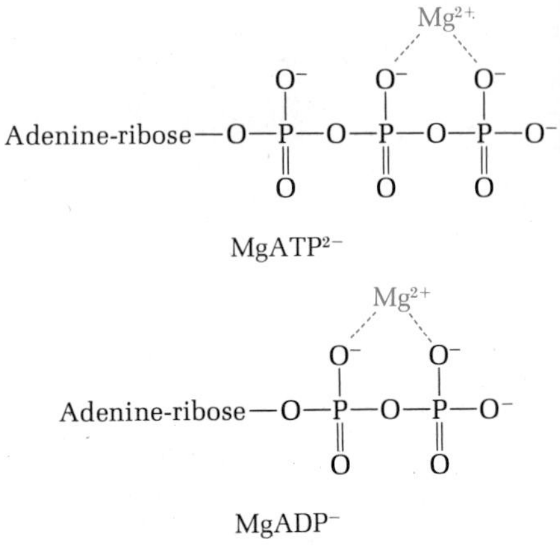

Free Energy

A description of the physicochemical basis of the function of ATP in the energy cycle of the cell requires consideration of some principles of equilibrium thermodynamics as applied to chemical reactions.

The first law of thermodynamics is the principle of energy conservation: in any process the total energy of the system plus surroundings remains constant. Although energy is neither created nor destroyed during a given process, it may undergo transformation from one form to another; for example, chemical energy may be transformed into thermal, radiant, electric, or mechanical energy.

The second law of thermodynamics states that in all processes the _entropy_ of the system plus the surroundings, i.e., the entropy of the "universe," always increases until equilibrium is attained, at which point the entropy is the maximum possible under the prevailing conditions of temperature and pressure. Entropy is a term that may be regarded as _disorder_ or _randomness_. Put in another way, the second law says that the ultimate driving force of all chemical and physical processes is the tendency for the entropy of the universe to be maximized.

However, entropy changes are of limited usefulness in predicting the direction and the equilibrium position of chemical reactions. For one thing, entropy cannot always be directly measured or calculated in chemical processes. Moreover, we must have information about the entropy change not only in the system under study, but also in its surroundings. A more useful criterion than the entropy change has been derived for predicting the direction and the equilibrium position of chemical reactions, namely, the change in _free energy_, that form of energy capable of doing work under

Figure 15-3
Metal complexes of ATP and ADP. In the Mg^{2+} complexes the two terminal phosphate groups are the ligands.

MgATP^{2-}

MgADP^{-}

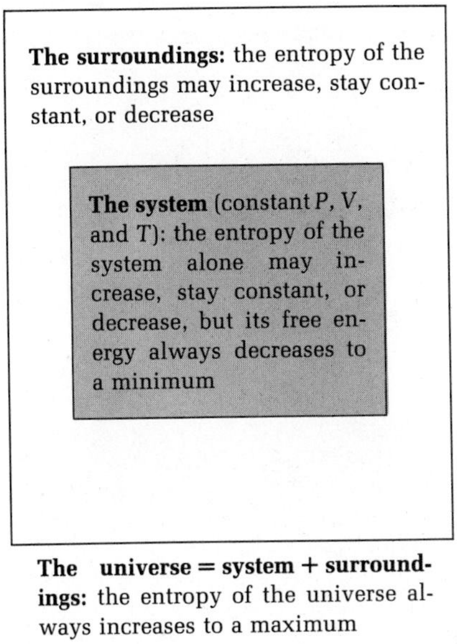

Figure 15-4
Summary of free-energy and entropy rela-
tionships between a system and its sur-
roundings when the temperature, pressure,
and volume of the system are constant. The
entropy of the system + surroundings or the
free energy of the system alone is the crite-
rion for predicting the direction of chemical
reactions under these conditions.

The surroundings: the entropy of the surroundings may increase, stay constant, or decrease

The system (constant P, V, and T): the entropy of the system alone may increase, stay constant, or decrease, but its free energy always decreases to a minimum

The   universe = system + surroundings: the entropy of the universe always increases to a maximum

conditions of constant temperature and pressure. The relationship between the free-energy change of a reacting system and the entropy under conditions of constant temperature and pressure, like those existing in living cells, is summarized in the equation

$$\Delta G = \Delta H - T\Delta S$$

in which ΔG is the free-energy change of the system, ΔH is the change in enthalpy, T is the absolute temperature, and ΔS is the change in entropy. The energy terms ΔG and ΔH are in calories, ΔS is in cal deg^{-1}, and the temperature is in kelvins. The free-energy change ΔG can be defined as that portion of the total energy change which is available to do work as the system proceeds toward equilibrium at constant temperature, pressure, and volume. These relationships are summarized in Figure 15-4.

The Standard-Free-Energy Change of Chemical Reactions

We shall now consider the fundamental relationship between the free-energy change of a chemical reaction and its equilibrium constant, a relationship basic to all aspects of biochemical energetics. For the generalized reaction

$$a\text{A} + b\text{B} \rightleftharpoons c\text{C} + d\text{D} \tag{1}$$

where a, b, c, and d are the number of molecules of A, B, C, and D participating in the reaction, the free-energy change ΔG at constant temperature and pressure is given by

$$\Delta G = \Delta G^\circ + RT \ln\frac{[\text{C}]^c[\text{D}]^d}{[\text{A}]^a[\text{B}]^b} \tag{2}$$

in which the brackets denote molal concentrations, R is the gas constant, T is the absolute temperature in kelvins, and ΔG° is the *standard-free-energy change* of the reaction.

We note from equation (2) that the free-energy change of a given chemical reaction is the summation of two terms. The first term, ΔG° or the standard-free-energy change, is a fixed constant whose value is characteristic for any given chemical reaction. It is a measure of the decrease in free energy of the given reaction when it is allowed to take place under arbitrarily defined standard conditions; it will be more precisely defined below. The second term is a variable; it reflects the concentrations of the reactants and the products of the given reaction. Thus we can calculate the free-energy change for any chemical reaction at a given temperature if we know its characteristic standard-free-energy change and the concentrations of the reactants and products.

If we now allow reaction (1) to proceed toward equilibrium, the free energy of the system will decrease during the drive toward equilibrium, and it is in this process that the reaction will theoretically be able to do work at constant temperature and pressure. Finally the system will reach its equilibrium point, at which no further net chemical change is occurring. At this point the free energy of the system is at a minimum for the given set of conditions. ΔG is now zero, and the system is no longer capable of doing work.

At the equilibrium point equation (2) becomes

$$0 = \Delta G^\circ + RT \ln \frac{[C]^c[D]^d}{[A]^a[B]^b} \tag{3}$$

which on rearrangement is

$$\Delta G^\circ = -RT \ln \frac{[C]^c[D]^d}{[A]^a[B]^b} \tag{4}$$

Since the equilibrium constant K'_{eq} (see page 47) for equation (2) is

$$K'_{eq} = \frac{[C]^c[D]^d}{[A]^a[B]^b} \tag{5}$$

we can substitute K'_{eq} in equation (4) and obtain the general equation

$$\Delta G^\circ = -RT \ln K'_{eq}$$

or

$$\Delta G^\circ = -2.303RT \log K'_{eq} \tag{6}$$

When the gas constant R is given in calories as energy units ($R = 1.98$ cal mol^{-1} K^{-1}), ΔG° is obtained in calories per mole. When the gas constant is given in joules as energy units ($R = 8.31$ J mol^{-1} K^{-1}), ΔG° is obtained in joules per mole. Equation (6) can thus be used to calculate ΔG°, the standard-free-energy change, from the equilibrium constant K'_{eq} for any given reaction at any given temperature.

Now let us define the standard-free-energy change more precisely. We shall do this in two different ways, which may help to clarify its true meaning.

1 The standard-free-energy change of a given chemical reaction is the *difference* between the sum of the free energies of the products and the sum of the free energies of the reactants, each reactant and product being present in its standard state. The standard state for the components of reactions in aqueous solution is by convention defined as a concentration of 1.0 m, a temperature of 25°C or 298° K, and a pressure of 1.0 atm. [In biochemical energetics the distinction between molal (m) and molar (M) solutions is not strictly observed; hereafter, we shall assume *molar* concentrations in all thermodynamic derivations and calculations.] Each chemical compound has a characteristic, intrinsic free energy by virtue of its molecular structure. The standard-free-energy change ΔG° of a chemical reaction can therefore be expressed by the general equation

$$\Delta G^\circ = \Sigma\ G^\circ_{prod} - \Sigma\ G^\circ_{react}$$

For the special case of reaction (1) $\Delta G°$ is given by

$$\Delta G° = (cG_C° + dG_D°) - (aG_A° + bG_B°)$$

2 The standard-free-energy change may be defined in a second way. For reaction (1) the standard-free-energy change $\Delta G°$ is that amount of free energy absorbed or lost per mole when A and B are transformed into C and D under such conditions that the concentrations of A, B, C, and D remain at 1 M while the conversion is being carried out at standard temperature and pressure.

It is extremely important to understand clearly the difference between $\Delta G°$, the *standard*-free-energy change, and ΔG, the *actual* or observed free-energy change. This difference may be explained by analogy. $\Delta G°$ is a constant for any given chemical reaction at a given temperature, just as the pK' of a weak acid is a fixed constant at a given temperature (page 47). On the other hand, ΔG varies with the concentrations of the reactants and products, just as the pH of a solution of a weak acid varies with the concentrations of its proton-donor and proton-acceptor forms (page 45). ΔG equals $\Delta G°$ only when all reactants and products are present at 1.0 M concentration, in the same way that the pH of a solution of a weak acid is equal to its pK' when the proton-donor and proton-acceptor species are 1.0 M.

It is ΔG that determines whether a chemical reaction will occur in the direction written, starting from given concentrations of reactants and products. Remember that a chemical reaction will occur only if ΔG is negative in sign, i.e., only if the free energy of the system decreases. On the other hand, a chemical reaction whose standard-free-energy change $\Delta G°$ is positive can still go forward as written, provided the concentrations of the reactants and products are such that ΔG will be negative.

When a chemical reaction at constant temperature and pressure proceeds with a decline in free energy, it can theoretically do an amount of work that is energetically equivalent to the decrease in free energy. Actually, however, a chemical reaction performs work only if it can be harnessed in some way to utilize the energy. The $\Delta G°'$ of a chemical reaction represents the <u>maximum</u> work it can theoretically perform; the amount actually performed may be much less or zero, depending on the efficiency of the machine or device used to harness it.

Exergonic and Endergonic Reactions

Equation (6) enables us to calculate the standard-free-energy change $\Delta G°$ of any chemical reaction from its equilibrium constant, which in turn can be estimated from analytical measurements. If the equilibrium constant for a reaction is 1.0, then $\Delta G° = 0.0$ and no change in free energy occurs when 1 mol of reactant(s) is completely converted to product(s), all at a concentration of 1.0 M. If the equilibrium constant is greater than 1.0, the standard-free-energy change $\Delta G°$

is negative. If the equilibrium constant is less than 1.0, $\Delta G°$ is positive. Chemical reactions with a negative standard-free-energy change are termed _exergonic_; such reactions proceed spontaneously in the direction written starting from 1.0 M concentrations of all components. Reactions with a positive standard-free-energy change are called _endergonic_; they do not proceed spontaneously in the direction written starting from 1.0 M concentrations of all reactants and products; instead they tend to go in the reverse direction. Table 15-1 shows the quantitative relationship between the standard-free-energy change $\Delta G°$ and the magnitude of the equilibrium constant K'_{eq}.

Table 15-1 The relationship between the equilibrium constant and the standard-free-energy change at 25°C

K'_{eq}	$\Delta G°$, cal mol^{-1}
0.001	+4,092
0.01	+2,728
0.1	+1,364
1.0	0
10.0	−1,364
100.0	−2,728
1,000.0	−4,092

Conventions in Biochemical Energetics

Some conventions relevant to thermodynamic analysis of biochemical systems must now be specified.

1 Whenever water is a reactant or product in a dilute aqueous system, its thermodynamic activity or concentration is arbitrarily set at 1.0, even though the molar concentration of water in dilute aqueous systems is actually about 55.5 M.

2 In biochemical energetics, pH 7.0 is usually designated as the reference state, rather than pH 0.0 (a hydrogen-ion concentration of 1.0 M) as normally used in physical chemistry. The standard-free-energy change at pH 7.0 is designated by $\Delta G°'$; that at pH 0.0 by $\Delta G°$.

3 The $\Delta G°'$ values used in biochemical energetics assume that the standard state of each reactant and product capable of ionization is that mixture of its un-ionized and ionized forms which exists at pH 7.0. Therefore $\Delta G°'$ values based on pH = 7.0 may not necessarily be used at pH values other than 7.0, because the extent of ionization of one or more components may change with pH. Moreover, a change in pH may lead to a difference in the number of H^+ and OH^- ions taken up or released during a reaction. The variation of $\Delta G°$ with pH for some biochemical reactions is quite large and sometimes difficult to calculate.

4 Standard-free-energy changes of biochemical systems have in the past been expressed in units of calories (or kilocalories), but an international commission has recently recommended that they hereafter be expressed in joules or kilojoules per mole (1 cal is equivalent to 4.184 J; 1 kcal is thus equivalent to 4.184 kJ). We shall give standard-free-energy changes in kilocalories per mole in the text and in both kilocalories and kilojoules in tabulations of free-energy data.

A Sample Calculation of $\Delta G°'$

A sample calculation of the standard-free-energy change can now be made from equilibrium data on the enzyme _phosphoglucomutase_, which catalyzes the reversible reaction

$$\text{Glucose 1-phosphate} \rightleftharpoons \text{glucose 6-phosphate} \qquad (7)$$

Chemical analysis shows that if we start with 0.020 M glucose 1-phosphate, add the enzyme, and allow the reaction to go in the forward direction, or if we start with 0.020 M glucose 6-phosphate and go in the reverse direction, the final equilibrium mixture in either case will contain 0.001 M glucose 1-phosphate and 0.019 M glucose 6-phosphate at 25°C and pH 7.0. We can then calculate the equilibrium constant:

$$K'_{eq} = \frac{[\text{glucose 6-phosphate}]}{[\text{glucose 1-phosphate}]} = \frac{0.019}{0.001} = 19$$

From this value of K'_{eq}, the standard-free-energy change $\Delta G^{\circ\prime}$ is calculated from equation (6):

$$\begin{aligned} \Delta G^{\circ\prime} &= -RT \ln K'_{eq} \\ &= -1.987 \times 298 \ln 19 \\ &= -1.987 \times 298 \times 2.303 \log 19 \\ &= -1{,}745 \text{ cal mol}^{-1} \end{aligned}$$

If instead of the value 1.987 cal mol^{-1} K^{-1} for the gas constant R, we use the equivalent value 8.314 J mol^{-1} K^{-1}, equation (6) will give us the standard-free-energy change directly in joules, namely, $-7{,}301$ J mol^{-1}.

The magnitude of the standard-free-energy changes of biochemical reactions is such that they are more conveniently given in kilocalories or kilojoules than calories or joules; $\Delta G^{\circ\prime}$ for the phosphoglucomutase reaction above is thus -1.745 kcal mol^{-1} or -7.301 kJ mol^{-1}. Since the sign of the standard-free-energy change we have calculated for equation (8) is negative, the conversion of glucose 1-phosphate to glucose 6-phosphate is an exergonic process.

Additive Nature of the Standard-Free-Energy Change

The standard-free-energy changes of chemical reactions are additive in any sequence of consecutive reactions. As an example, we may consider the following consecutive reactions, which are coupled, or linked, by common intermediates; that is, B, the product of the first reaction, is the reactant in the second reaction; C, the product of the second reaction, is the reactant in the third reaction:

Reaction	Standard-free-energy change
A $\longrightarrow$ B	$\Delta G^{\circ\prime}_1$
B $\longrightarrow$ C	$\Delta G^{\circ\prime}_2$
C $\longrightarrow$ D	$\Delta G^{\circ\prime}_3$

The sum of these reactions is A $\rightarrow$ D, whose standard-free-energy change $\Delta G^{\circ\prime}_s$ is the algebraic sum of the $\Delta G^{\circ\prime}$ values of the individual steps, each being given its proper sign:

$$\Delta G^{\circ\prime}_s = \Delta G^{\circ\prime}_1 + \Delta G^{\circ\prime}_2 + \Delta G^{\circ\prime}_3$$

This property is extremely useful for calculating the standard-free-energy change of a reaction when its equilibrium constant cannot be determined directly. In such a case the reaction may be coupled to one or more other reactions of known equilibrium constants, to yield a sequence whose overall equilibrium can be measured more readily. In the above example, if we know the values of $\Delta G_1^{\circ\prime}$, $\Delta G_2^{\circ\prime}$, and $\Delta G_s^{\circ\prime}$, we can calculate $\Delta G_3^{\circ\prime}$.

Calculation of $\Delta G^{\circ\prime}$ from Standard Free Energies of Formation

Another way of arriving at the $\Delta G^{\circ\prime}$ for a given chemical reaction is to calculate it from the *standard free energy of formation* of its substrates and products. The results of many thermodynamic measurements have shown that each type of organic compound has a characteristic standard free energy of formation ΔG_f°, defined as the decrease in free energy as 1 mol of the compound is formed from its elements, each in its standard state and in the proper stoichiometric ratio. The standard free energies of formation for many biological compounds have been calculated (Table 15-2).

The ΔG° of a chemical reaction is equal to the sum of the standard free energies of formation of the products minus the sum of the standard free energies of formation of the reactants, taking into account the actual stoichiometry of the reaction:

$$\Delta G^{\circ} = \Sigma \, \Delta G_{f,\text{prod}}^{\circ} - \Sigma \, \Delta G_{f,\text{react}}^{\circ}$$

Let us now calculate with this relationship the standard-free-energy change $\Delta G^{\circ\prime}$ of the following reaction, catalyzed by the enzyme *fumarase*, using the data in Table 15-2:

$$\text{Fumarate} + H_2O \rightleftharpoons \text{malate}$$

$$\begin{aligned}\Delta G^{\circ\prime} &= \Delta G_{f,\text{malate}}^{\circ} - (\Delta G_{f,\text{fumarate}}^{\circ} + \Delta G_{f,H_2O}^{\circ}) \\ &= -201.98 - (-144.41 - 56.69) \\ &= -0.88 \text{ kcal mol}^{-1}\end{aligned}$$

Although the error in such calculations can be rather large because the value for $\Delta G^{\circ\prime}$ usually represents a small difference between two large terms, this approach often represents the only convenient way of obtaining the $\Delta G^{\circ\prime}$ of a reaction. Note that the $\Delta G^{\circ\prime}$ for the fumarase reaction calculated by this procedure differs from this value of -0.75 kcal mol^{-1} obtained from another type of measurement, as listed in Table 15-3.

Table 15-3 gives the standard-free-energy changes for a number of biologically important reactions calculated from either equilibrium measurements or standard free energies of formation. Note that the oxidation of organic molecules by molecular oxygen proceeds with especially large decreases of free energy. Such oxidations serve as the main source of energy for aerobic cells.

Table 15-2 Standard free energies of formation for 1 M aqueous solutions at pH 7.0 and 25°C

Substance	$\Delta G_f^{\circ\prime}$	
	kcal mol^{-1}	kJ mol^{-1}
Acetate$^-$	-88.99	-372.3
cis-Aconitate^{3-}	-220.51	-922.61
L-Alanine	-88.75	-371.3
Ammonium ion	-19.00	-79.50
L-Aspartate$^-$	-166.99	-698.69
Bicarbonate ion	-140.33	-587.14
Carbon dioxide (gas)	-94.45	-395.2
Ethanol	-43.39	-181.6
Fumarate^{2-}	-144.41	-604.21
α-D-Glucose	-219.22	-917.21
Glycerol	-116.76	-488.64
Hydrogen ion	-9.55	-39.96
Hydroxide ion	-37.60	-157.3
α-Ketoglutarate^{2-}	-190.62	-797.56
Lactate$^-$	-123.76	-517.81
L-Malate^{2-}	-201.98	-845.08
Oxaloacetate^{2-}	-190.53	-797.18
Pyruvate$^-$	-113.44	-474.63
Succinate^{2-}	-164.97	-690.23
Water (liquid)	-56.69	-237.2

Table 15-3 Standard-free-energy changes of some chemical reactions in dilute aqueous systems at pH 7.0 and 25°C

Reaction	$\Delta G°'$	
	kcal mol^{-1}	kJ mol^{-1}
Hydrolysis:		
Acid anhydrides:		
Acetic anhydride + $H_2O \longrightarrow$ 2 acetate	−21.8	−91.2
Pyrophosphate + $H_2O \longrightarrow$ 2 phosphate	−8.0	−33.4
Esters:		
Ethyl acetate + $H_2O \longrightarrow$ ethanol + acetate	−4.7	−19.7
Glucose 6-phosphate + $H_2O \longrightarrow$ glucose + phosphate	−3.3	−13.8
Amides:		
Glutamine + $H_2O \longrightarrow$ glutamate + NH_4^+	−3.4	−14.2
Glycylglycine + $H_2O \longrightarrow$ 2 glycine	−2.2	−9.2
Glycosides:		
Sucrose + $H_2O \longrightarrow$ glucose + fructose	−7.0	−29.3
Maltose + $H_2O \longrightarrow$ 2 glucose	−4.0	−16.7
Esterification:		
Glucose + phosphate $\longrightarrow$ glucose 6-phosphate + H_2O	+3.3	+13.8
Rearrangement:		
Glucose 1-phosphate $\longrightarrow$ glucose 6-phosphate	−1.7	−7.11
Fructose 6-phosphate $\longrightarrow$ glucose 6-phosphate	−0.4	−1.67
Elimination:		
Malate $\longrightarrow$ fumarate + H_2O	+0.75	+3.14
Oxidation:		
Glucose + $6O_2 \longrightarrow 6CO_2 + 6H_2O$	−686	−2,870
Palmitic acid + $23O_2 \longrightarrow 16CO_2 + 16H_2O$	−2338	−9,782

The Standard Free Energy of Hydrolysis of Phosphate Compounds

The remainder of this chapter deals primarily with the energy relationships between various phosphorylated compounds that participate in the transfer of chemical energy in the cell. For this purpose a standard yardstick is used to express the thermodynamic tendency or potential of a phosphate group on one molecule to be transferred to another molecule in an enzyme-catalyzed reaction, so that we can predict in which direction such a phosphate-group transfer will take place and at what point the reaction will come to equilibrium. The thermodynamic yardstick is furnished by a comparison of the tendency of the phosphate group in different phosphorylated compounds to be transferred to a standard, arbitrarily chosen acceptor molecule, namely, water. The transfer of a phosphate group from a donor molecule $R—O—PO_3^{2-}$ to water is given by the reaction

$$R—O—PO_3^{2-} + HOH \rightleftharpoons R—OH + HO—PO_3^{2-}$$

We recognize that this reaction actually represents the hydrolysis of the phosphate ester $R—O—PO_3^{2-}$. Thus, in order to obtain some idea of the relative transfer tendency of phosphate groups in various biological phosphate compounds we must determine their standard free energies of hydrolysis. It must be made clear at the outset that biological phosphate compounds do not normally undergo simple hydrolysis reactions in the intact cell; the choice of water as the standard phosphate acceptor for thermodynamic comparison is wholly arbitrary.

Table 15-4 shows the standard free energy of hydrolysis of a number of important phosphate compounds found in cells, including ATP. We note that ATP has a $\Delta G^{\circ\prime}$ value of -7.3 kcal mol^{-1} (-30.5 kJ mol^{-1}) and stands about midway on the energy scale of phosphate compounds. Let us see how this value has been determined.

The Standard Free Energy of Hydrolysis of ATP

In principle, the simplest way to arrive at the value of $\Delta G^{\circ\prime}$ for the reaction

$$ATP + HOH \rightleftharpoons ADP + phosphate \qquad (8)$$

is to determine its equilibrium constant at pH 7.0 and calculate $\Delta G^{\circ\prime}$ from the relationship [equation (6)]

$$\Delta G^{\circ\prime} = -2.303RT \log K'_{eq}$$

However, direct measurement of the equilibrium constant for the hydrolysis of ATP is not practical, since the reaction at equilibrium has gone so far in the direction of hydrolysis that the available analytical methods are not sensitive enough to determine exactly when equilibrium has been reached and what the exact equilibrium concentrations of ATP, ADP, and phosphate are. In fact, this is a serious practical problem in determining $\Delta G^{\circ\prime}$ for any reaction having a large negative (or a large positive) $\Delta G^{\circ\prime}$ value. Actually, the $\Delta G^{\circ\prime}$ for ATP hydrolysis, which we shall designate $\Delta G^{\circ\prime}_{ATP}$, can be arrived at by exploiting the additive nature of the $\Delta G^{\circ\prime}$ values of consecutive reactions. In principle, the relatively

Table 15-4 Standard free energy of hydrolysis of some phosphorylated compounds

| Compound | $\Delta G^{\circ\prime}$ | | Phosphate-group transfer potential† |
	kcal mol^{-1}	kJ mol^{-1}	
Phosphoenolpyruvate	-14.80	-61.9	14.8
3-Phosphoglyceroyl phosphate	-11.80	-49.3	11.8
Phosphocreatine	-10.30	-43.1	10.3
Acetyl phosphate	-10.10	-42.3	10.1
Phosphoarginine	-7.70	-32.2	7.7
ATP ($\longrightarrow$ ADP + P$_i$)	-7.30	-30.5	7.3
Glucose 1-phosphate	-5.00	-20.9	5.0
Fructose 6-phosphate	-3.80	-15.9	3.8
Glucose 6-phosphate	-3.30	-13.8	3.3
Glycerol 1-phosphate	-2.20	-9.2	2.2

† Defined as $-\Delta G^{\circ\prime}$ (kcal mol^{-1}).

large standard-free-energy change for the hydrolysis of ATP is broken up into two or more energetically smaller steps that can be measured more easily. As an example, ATP may first be allowed to react with glucose in the presence of the enzyme hexokinase to yield ADP and glucose 6-phosphate. The equilibrium constant of this reaction is measured, and from it the standard-free-energy change is calculated:

$$\text{ATP} + \text{glucose} \xrightarrow{\text{hexokinase}} \text{ADP} + \text{glucose 6-phosphate} \qquad (9)$$

$$K'_{eq} = 661 \qquad \Delta G_1^{\circ\prime} = -4.0 \text{ kcal mol}^{-1}$$

This is followed by a measurement of the equilibrium constant and the calculation of the standard-free-energy change $\Delta G^{\circ\prime}$ for the hydrolysis of glucose 6-phosphate to yield glucose and phosphate, catalyzed by glucose-6-phosphatase:

$$\text{Glucose 6-phosphate} + \text{H}_2\text{O} \xrightarrow[\text{phosphatase}]{\text{glucose-6-}} \text{glucose} + \text{phosphate} \qquad (10)$$

$$K'_{eq} = 171 \qquad \Delta G_2^{\circ\prime} = -3.3 \text{ kcal mol}^{-1}$$

The sum of reactions (9) and (10) is the equation for the hydrolysis of ATP,

$$\text{ATP} + \text{H}_2\text{O} \rightleftharpoons \text{ADP} + \text{phosphate}$$

Since the $\Delta G^{\circ\prime}$ values of the two reactions are additive, the standard free energy of hydrolysis of ATP can be calculated:

$$\Delta G_{\text{ATP}}^{\circ\prime} = \Delta G_1^{\circ\prime} + \Delta G_2^{\circ\prime} = -4.0 + (-3.3) = -7.3 \text{ kcal mol}^{-1}$$

Another approach to the determination of $\Delta G^{\circ\prime}$ for the hydrolysis of ATP to ADP and phosphate is afforded by measurements on the equilibrium of the reaction

$$\text{ATP} + \text{glutamate} + \text{NH}_3 \rightleftharpoons \text{ADP} + \text{phosphate} + \text{glutamine} \qquad (11)$$

which is catalyzed by the enzyme glutamine synthetase. This reaction can be visualized as having two components. The exergonic component, which yields free energy, is the reaction

$$\text{ATP} + \text{H}_2\text{O} \rightleftharpoons \text{ADP} + \text{phosphate} \qquad (12)$$

The endergonic component, which requires input of free energy, is

$$\text{Glutamate} + \text{NH}_3 \rightleftharpoons \text{glutamine} + \text{H}_2\text{O} \qquad (13)$$

The sum of these two reactions is of course reaction (11). As it happens, the standard-free-energy change $\Delta G_2^{\circ\prime}$ for reaction (13) is known; it is $+3.4$ kcal mol^{-1}. If the equilibrium constant and the standard-free-energy change $\Delta G_s^{\circ\prime}$ of the overall reaction (11) can be measured, we can calculate the standard-free-energy change $\Delta G_1^{\circ\prime}$ for reaction (12), using the principle of the additivity of standard-free-energy changes. It has been found that $\Delta G_s^{\circ\prime}$ for the overall reaction (11) is -3.9 kcal mol^{-1}. The standard-free-energy change $\Delta G_1^{\circ\prime}$ for the hydrolysis of ATP is then calculated as follows:

$$\Delta G_s^{\circ\prime} = \Delta G_1^{\circ\prime} + \Delta G_2^{\circ\prime}$$
$$-3.9 = \Delta G_1^{\circ\prime} + 3.4$$
$$\Delta G_1^{\circ\prime} = -7.3 \text{ kcal mol}^{-1}$$

The published values for $\Delta G^{\circ\prime}$ for the hydrolysis of ATP to ADP and phosphate in various textbooks and research articles vary somewhat, due in part to analytical difficulties in obtaining precise values for the equilibrium constants of the reactions employed, as mentioned above, and in part to the fact that measurements in different laboratories have not always been made under precisely the same conditions of temperature, pH, and Mg^{2+} concentration, each of which can influence the free energy of hydrolysis of ATP very markedly, as will be seen below. The most accurate values for $\Delta G^{\circ\prime}$ for the hydrolysis of ATP to ADP and phosphate at pH 7.0, 25°C, and in the presence of 20 mM Mg^{2+} appear to lie between -7 and -8 kcal mol^{-1}. We shall use the value $\Delta G_{\text{ATP}}^{\circ\prime} = -7.3$ kcal mol^{-1} (-30.5 kJ mol^{-1}) throughout this book (see References). In any case, what is really important in the function of ATP as an intermediate carrier of chemical energy is not so much the absolute value for $\Delta G_{\text{ATP}}^{\circ\prime}$ but its value relative to the values for $\Delta G^{\circ\prime}$ of hydrolysis of the biologically important donors of phosphate groups to ADP and the values for $\Delta G^{\circ\prime}$ of the phosphorylated compounds formed by transfer of the terminal phosphate of ATP to various acceptors during the ATP cycle (Table 15-4), as we shall see below.

The terminal phosphate group of ADP also has a relatively large standard free energy of hydrolysis at pH 7.0:

$$ADP + H_2O \rightleftharpoons AMP + P_i$$
$$\Delta G^{\circ\prime} = -7.3 \text{ kcal mol}^{-1}$$

However, the single phosphate group of AMP has a much lower value:

$$AMP + H_2O \rightleftharpoons \text{adenosine} + P_i$$
$$\Delta G^{\circ\prime} = -3.40 \text{ kcal mol}^{-1}$$

Recall (page 315) that the bonds between the adjacent phosphate groups of ATP and ADP are anhydride linkages, whereas the bond between phosphoric acid and ribose in AMP is an ester linkage. In general, anhydride bonds have a much larger (negative) standard free energy of hydrolysis than ester linkages.

The Structural Basis of the Free-Energy Change during Hydrolysis of ATP

What structural features of the ATP molecule give its terminal phosphate group a more negative standard free energy of hydrolysis at pH 7.0 ($\Delta G^{\circ\prime} = -7.3$ kcal mol^{-1}) than other phosphate compounds below it on the thermodynamic scale (Table 15-4)? Why is it more negative, say, than glucose 6-phosphate, whose standard free energy of hydrolysis to yield glucose and phosphate is only -3.3 kcal mol^{-1}? This is tantamount to asking why the equilibrium of hydrolysis of the phosphate group lies farther in the direction of completion for ATP than it does for glucose 6-phosphate.

Because the standard free energy of hydrolysis is a measure of the *difference* between the free energy of the products and the free energy of the reactants, the answer to this question lies in the properties of the products and reactants. The phosphate ion at pH near 7.0 is a resonance hybrid, in which the formal double bond has significant single-bond character and the opposite formal single bond has considerable double-bond character. Such resonance hybrids are more stable (i.e., contain less free energy) than their formal structure indicates, by an amount called the resonance stabilization energy. The phosphate groups in ATP, ADP, AMP, and various other phosphorylated compounds are all resonance hybrids, stabilized by different amounts of resonance energy, depending upon the electronic configuration of the adjacent functional groups. When such compounds undergo hydrolysis there is a significant change in the resonance energy, which is reflected in different ways. As it happens, the difference in resonance energy between the reactants and products is greater for hydrolysis of ATP than for hydrolysis of glucose 6-phosphate. Moreover, at pH 7.0 ATP molecules have on the average about 3.5 closely spaced negative charges, which repel each other strongly. When the terminal phosphate bond is hydrolyzed, some of this electrical stress within the complex triphosphate group is removed by separation of the resulting products, the anions HPO_4^{2-} and ADP^{3-}. Because these anions are both negatively charged, they will repel each other and thus will not readily recombine to form ATP. In contrast, when glucose 6-phosphate undergoes hydrolysis, glucose, one of the products, has no net charge; since the glucose and phosphate molecules do not repel each other, they have a greater tendency to recombine. Both terminal phosphorus atoms of ATP have a strong electron-withdrawing tendency, thus making the phosphoric anhydride linkage more prone to hydrolysis than the simple phosphate ester glucose 6-phosphate. Another consequence of the difference in the electronic configuration of reactants and products is that ADP and phosphate are more strongly hydrated than ATP itself, again a factor tending to pull ATP hydrolysis more nearly to completion.

In general, acid anhydrides, e.g., the pyrophosphate group of ATP and the acyl phosphate group of 3-phosphoglyceroyl phosphate, are high-energy compounds with a large negative $\Delta G^{\circ\prime}$ of hydrolysis. Other classes of compounds having a large negative $\Delta G^{\circ\prime}$ include enol esters (phosphoenolpyruvate), thioesters (acetyl-CoA), and phosphoguanidine compounds (phosphocreatine). In all cases it is the special electron configuration in or near the bond undergoing hydrolysis that ultimately is responsible for the strongly negative values for the $\Delta G^{\circ\prime}$ of hydrolysis.

The term *phosphate-bond energy* sometimes used by biochemists is not to be confused with the term _bond energy_ used by the physical chemist, which denotes the energy required to _break_ a bond between two atoms. Actually, a relatively large amount of energy is required to break a covalent chemical bond, which would not exist if it were not quite stable. Although we do break a P—O bond during the hydrolysis of phosphate esters, a new P—O bond is formed. Phosphate-bond energy specifically denotes the _difference_ in the free energy of the products and the free energy of the reactants when a phosphorylated compound undergoes hydrolysis.

High-energy phosphate bonds, i.e., those whose hydrolysis proceeds with a negative $\Delta G^{\circ\prime}$, are sometimes symbolized by the squiggle ~ , high-energy phosphate groups being written as ~P. For example, ATP can be symbolized as A—R—P~P~P, ADP as A—R—P~P, and phosphocreatine as P~Cr.

Conditions Affecting the ΔG° of Hydrolysis of ATP

Several factors influence the magnitude of the free energy of hydrolysis of ATP. Data in Figure 15-5 show that ΔG_{ATP} rises sharply as pH increases. This is a reflection of the fact that the pK's of the last ionization steps of ATP, ADP, and phosphate are dissimilar:

$$\text{HATP}^{3-} \rightleftharpoons \text{ATP}^{4-} + \text{H}^+ \qquad \text{pK}_1' = 6.95$$
$$\text{HADP}^{2-} \rightleftharpoons \text{ADP}^{3-} + \text{H}^+ \qquad \text{pK}_2' = 6.88$$
$$\text{H}_2\text{PO}_4^- \rightleftharpoons \text{HPO}_4^{2-} + \text{H}^+ \qquad \text{pK}_3' = 6.78$$

The presence of Mg^{2+} will also alter the value of ΔG_{ATP}. The ionic species ATP^{4-}, ADP^{3-}, and HPO_4^{2-} form complexes with Mg^{2+} in the reversible reactions

$$\text{Mg}^{2+} + \text{ATP}^{4-} \rightleftharpoons \text{MgATP}^{2-}$$
$$\text{Mg}^{2+} + \text{ADP}^{3-} \rightleftharpoons \text{MgADP}^-$$
$$\text{Mg}^{2+} + \text{HPO}_4^{2-} \rightleftharpoons \text{MgHPO}_4$$

Moreover, the ionic species HATP^{3-}, HADP^{2-}, and H_2PO_4^- also form complexes with Mg^{2+}. Since the affinity of Mg^{2+} for each of the six species of phosphate compounds differs, and since this affinity in turn increases with pH, it is clear that the effects of pH and Mg^{2+} concentration on ΔG_{ATP} can be quite complex. Figure 15-6 shows the effect of increasing Mg^{2+} concentration on ΔG_{ATP} at pH 7.0. The value of ΔG_{ATP} can be obtained for any given pH, Mg^{2+} concentration, and ionic strength from graphs prepared by computer methods (see References).

Finally, the concentrations of ATP, ADP, and phosphate in intact cells are far from the standard concentrations of 1.0 M on which $\Delta G^{\circ\prime}$ values for all reactions are based. After appropriate corrections are made for intracellular pH, Mg^{2+} concentration, and the actual steady-state concentrations of ATP, ADP, and phosphate in the intracellular water phase, the free energy of hydrolysis of ATP in intact cells is in the neighborhood of -12.5 kcal mol^{-1}. For consistency in com-

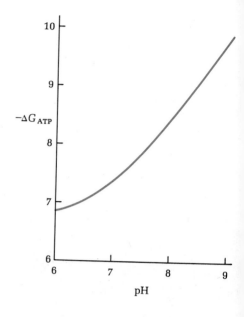

Figure 15-5
Effect of pH on the free energy of hydrolysis of ATP (ΔG_{ATP}).

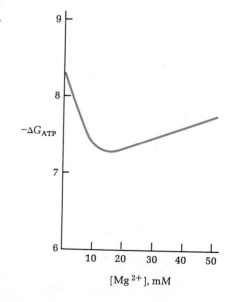

Figure 15-6
Effect of Mg^{2+} concentration on the free energy of hydrolysis of ATP at pH 7.0.

paring various reactions the *standard*-free-energy change $\Delta G^{\circ\prime} = -7.3$ kcal mol^{-1} must be used, but it is quite clear that rather large corrections must be applied to such calculations if they are to reflect actual intracellular conditions and processes. Furthermore, it is also evident that the free energy of hydrolysis of ATP in the cell is not necessarily constant; it can vary from time to time or even from place to place in the cell, depending on the pH, Mg^{2+} concentration, and the concentration of ATP, ADP, and phosphate at any given time and place.

The Transfer Potential of the Phosphate Group

The phosphorylated compounds found in cells are often classified as high-energy or low-energy compounds, according to the magnitude of the $\Delta G^{\circ\prime}$ values for their hydrolysis. ATP, for example, is usually classed as a high-energy phosphate. However, the $\Delta G^{\circ\prime}$ values for various phosphate compounds in Table 15-4, which are arranged in order of decreasingly negative values, show that there is no sharp division of phosphate compounds into two classes. Moreover, several phosphorylated compounds have a much more negative $\Delta G^{\circ\prime}$ value than ATP itself, which has an intermediate value in this thermodynamic scale. Compounds higher on the scale, such as 3-phosphoglyceroyl phosphate, tend to be more completely hydrolyzed at equilibrium than those lower in the scale and thus have a higher equilibrium constant for hydrolysis.

Any compound above ATP on this scale would tend to lose its phosphate group to a phosphate-acceptor molecule lower on the scale, such as ADP, provided a catalyst is available to promote its transfer. Similarly, ATP would tend to donate its phosphate group to an acceptor lower on the scale, such as D-glucose. Thus this scale tells us about the thermodynamic tendency or potential of the phosphate group of different phosphate compounds to undergo transfer. The term *phosphate-group transfer potential* is often used to denote the phosphate-group "pressure"; it is sometimes given as a dimensionless number with a positive sign, as shown in Table 15-4.

The fact that ATP has an intermediate value on this scale is of special significance, since ATP serves as a _common intermediate_ in most enzyme-catalyzed phosphate-group transfer reactions in the cell: ADP can accept a phosphate group from a phosphorylated compound higher on the scale, and the ATP so formed can then donate it to an acceptor molecule to yield a phosphate compound lower on the scale.

The Common-Intermediate Principle: Coupled Reactions

In two consecutive reactions in which a product of the first is a substrate of the second, like the consecutive reactions

$$A + B \longrightarrow C + D$$
$$D + E \longrightarrow F + G$$

the reactions are said to be *coupled* through a *common intermediate,* in this case the component D. The only way chemical energy can be transferred from one reaction to another under isothermal conditions (apart from fluorescence phenomena) is for the two reactions to have such a common reaction intermediate. Nearly all metabolic reactions in the cell proceed in consecutive sequences of this sort. In the consecutive reactions responsible for energy transfers via ATP, chemical energy is transferred from a high-energy phosphate-group donor to ADP and is conserved in the form of ATP as a reaction product. In the succeeding reaction, in which ATP is a substrate, its terminal phosphate group is transferred to an acceptor molecule, causing the latter to be changed to a compound with a higher energy content. ATP is thus a common intermediate that serves to link or couple enzymatic reactions involving the transfer of phosphate groups and is thus a vehicle for the transfer of chemical energy. Actually, the transfer of chemical energy through common intermediates is a general attribute of consecutive chemical reactions and does not necessarily require either phosphate groups or ATP. In fact, we shall see that many chemical groups other than phosphate, e.g., hydrogen atoms, amino groups, and acetyl groups, are enzymatically transferred by means of consecutive reactions having common intermediates. However, cells utilize primarily phosphate groups for the purpose of transferring energy in the central metabolic pathways.

Now let us examine more closely the changes in free energy occurring in (1) the enzymatic transfer of a phosphate group from a donor molecule to ADP and (2) the enzymatic transfer of the terminal phosphate group from ATP to a phosphate acceptor, in order to see in more quantitative terms how the ATP–ADP system functions as a common intermediate in the mainstream of the transfer of chemical energy in the cell.

Enzymatic Transfer of Phosphate Groups to ADP

From the viewpoint of metabolism there are two general classes of high-energy phosphate compounds for which $\Delta G^{\circ\prime}$ of hydrolysis is more negative than that of ATP and which therefore can serve as phosphate donors to ADP (Table 15-4). Members of the first group, which includes *3-phosphoglyceroyl phosphate* and *phosphoenolpyruvate,* are generated during the energy-yielding enzymatic breakdown, or catabolism, of fuel molecules. Members of the second group, which includes *phosphocreatine* and *phosphoarginine,* serve as storage reservoirs of chemical energy in muscles. Let us examine the enzymatic reactions by which such compounds donate phosphate groups to ADP.

Both 3-phosphoglyceroyl phosphate and phosphoenolpyruvate are generated during *glycolysis* (page 417), the breakdown of glucose to lactic acid, a process that furnishes muscle and other cells with considerable energy. These compounds are not actually hydrolyzed in the cell; rather, their phosphate groups are transferred to ADP. For 3-

Figure 15-7
Transfer of a phosphate group from 3-phosphoglyceroyl phosphate to ADP.

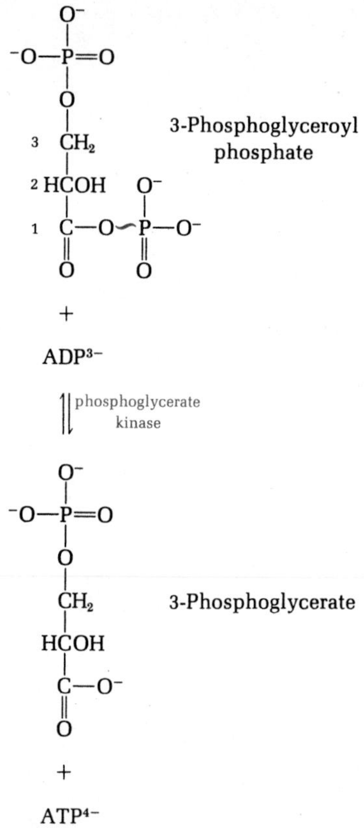

phosphoglyceroyl phosphate this reaction, catalyzed by 3-phosphoglycerate kinase, is

3-Phosphoglyceroyl phosphate + ADP $\rightleftharpoons$

3-phosphoglycerate + ATP   (14)

The standard-free-energy change of this reaction has been calculated from its equilibrium constant ($K' = 2{,}070$) to be -4.50 kcal mol^{-1}. From the very high value of the equilibrium constant and the rather large negative value for $\Delta G^{\circ\prime}$, it is clear that this reaction, starting from 1 M reactants and products, will proceed strongly to the right; at its equilibrium well over 99.9 percent of the ADP will have been converted to ATP by transfer of phosphate groups from 3-phosphoglyceroyl phosphate. In this process there is a transfer of chemical energy from 3-phosphoglyceroyl phosphate to ADP; this energy is now conserved in the ATP formed. Note from the data in Table 15-4 that the value for $\Delta G^{\circ\prime}$ of reaction (14), namely, -4.50 kcal mol^{-1}, is equal to the difference between the $\Delta G^{\circ\prime}$ of hydrolysis of 3-phosphoglyceroyl phosphate and $\Delta G^{\circ\prime}$ for the hydrolysis of ATP, or $-11.8 - (-7.3) = -4.5$ kcal mol^{-1}. Thus $4.5/11.8 \times 100 = 38$ percent of the overall free energy of hydrolysis of 3-phosphoglyceroyl phosphate has been used to "push" its phosphate group to ADP, and the remaining 62 percent is now stored in the ATP formed. It is clear from this example that if we know the $\Delta G^{\circ\prime}$ for the hydrolysis of a phospate compound, we can predict the $\Delta G^{\circ\prime}$ for the reaction by which its phosphate group is transferred to ADP, or vice versa. The compound 3-phosphoglyceroyl phosphate owes its high phosphate-group transfer potential to the fact that it is an acid anhydride (Figure 15-7).

In a very similar way, phosphoenolpyruvate, which is also formed during breakdown of glucose to lactic acid (page 431), donates its phosphate group to ADP in a reaction catalyzed by pyruvate kinase:

Phosphoenolpyruvate + ADP $\rightleftharpoons$ pyruvate + ATP   (15)

Actually, this reaction takes place in two steps (Figure 15-8). In the first the enolate form of pyruvate is formed, after transfer of the phosphate group to ADP. At pH 7.0 this becomes protonated and rapidly reverts to the keto form, presumably nonenzymatically; this step is responsible for much of the thermodynamic push. The overall reaction has a very large positive equilibrium constant and thus a large negative $\Delta G^{\circ\prime}$ value, namely, -7.5 kcal mol^{-1}, indicating that at equilibrium nearly all the phosphenolpyruvate molecules have donated their phosphate groups to ADP. Again, we may note that the $\Delta G^{\circ\prime}$ for this reaction is equal to the difference between the $\Delta G^{\circ\prime}$ for the hydrolysis of phosphoenolpyruvate and that for hydrolysis of ATP, namely, $-14.8 - (-7.3) = -7.5$ kcal mol^{-1}. Thus, through these two phosphate-group transfer reactions much of the energy of the breakdown of glucose to lactate is transferred to ADP and conserved as ATP, ready to be used for energy-requiring reactions.

Figure 15-8
Transfer of a phosphate group from phosphoenolpyruvate to ADP. A major cause of the large negative $\Delta G^{\circ\prime}$ for this reaction is the nonenzymatic conversion of enolpyruvate, the immediate product, into the keto form that predominates at pH 7, shown below.

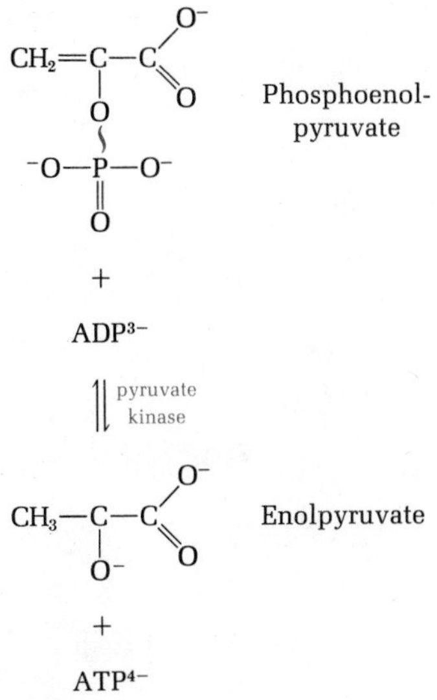

Nonenzymatic conversion of enol to keto form of pyruvate.

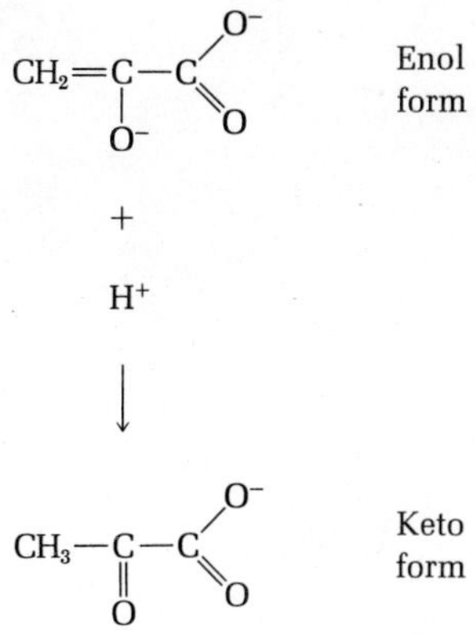

The high-energy phosphate compounds serving as reservoirs of phosphate-bond energy are often called *phosphagens*; the most important are *phosphocreatine* (Figure 15-9), present in vertebrate muscle and nerve tissue, and *phosphoarginine* (Figure 15-10), found in invertebrates. In both the phosphorus atom is bonded directly to a nitrogen atom of a guanido group. The phosphate group is enzymatically transferred from these compounds to ADP by the enzymes *creatine kinase* and *arginine kinase*, respectively. For phosphocreatine the reaction is

$$\text{Phosphocreatine} + \text{ADP} \rightleftharpoons \text{creatine} + \text{ATP}$$
$$\Delta G^{\circ\prime} = -3.0 \text{ kcal mol}^{-1}$$

The significantly large negative value for $\Delta G^{\circ\prime}$ indicates that this reaction also tends to go to the right, starting from 1 M concentrations of reactants and products. Again, note that the $\Delta G^{\circ\prime}$ of this reaction could have been predicted from the $\Delta G^{\circ\prime}$ values for the hydrolysis of phosphocreatine and ATP.

Many experiments on the mechanism and pathway of enzymatic phosphate-transferring reactions between ATP and various phosphate acceptors, often using isotopic oxygen as tracer, have proved that it is not the phosphoric group

$(-O-P\substack{O^- \\ \diagup \\ \diagdown \\ O^-}=O)$ that is transferred but the phospho group

$(-P\substack{O^- \\ \diagup \\ \diagdown \\ O^-}=O)$, as shown in Figure 15-11.

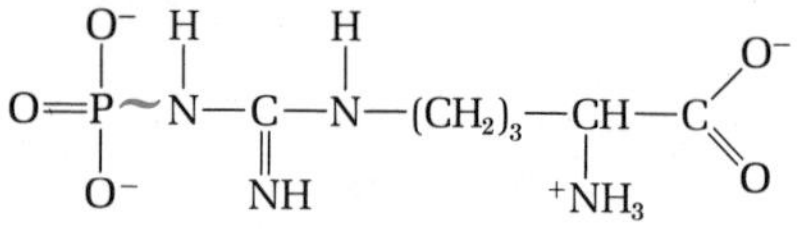

Transfer of Phosphate Groups from ATP to Various Acceptors

In the second stage of its function as energy carrier ATP can donate its terminal phosphate group to a large variety of phosphate-acceptor molecules in reactions catalyzed by specific enzymes. Among these acceptors are D-glucose and

Figure 15-9
Structure and space-filling model of phosphocreatine.

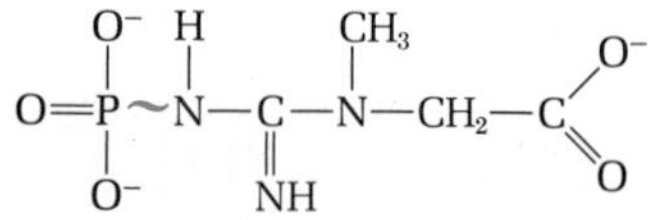

Figure 15-10
Structure of phosphoarginine.

Figure 15-11
Enzymatic transfer of the terminal phosphate group of ATP to an acceptor ROH. Such reactions usually occur by a nucleophilic displacement on the phosphorus atom. The group transferred is —PO₃²⁻.

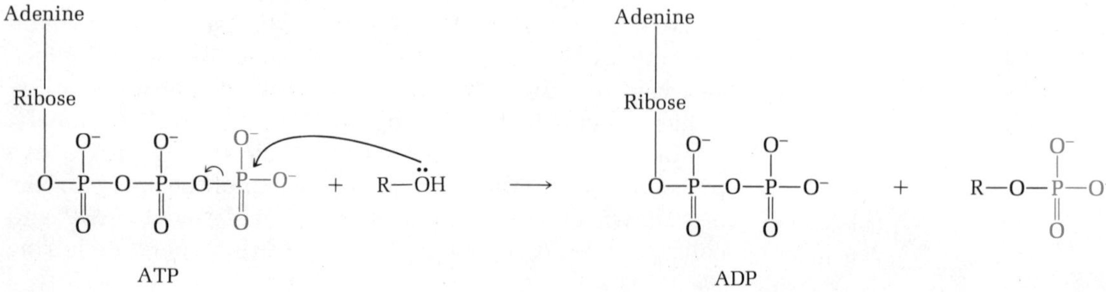

ATP + R—ÖH ⟶ ADP + R—O—P—O⁻

glycerol. The enzyme *hexokinase* catalyzes the reaction

$$ATP + \text{D-glucose} \rightleftharpoons ADP + \text{D-glucose 6-phosphate} \quad (16)$$
$$\Delta G^{\circ\prime} = -4.00 \text{ kcal mol}^{-1}$$

Note that the $\Delta G^{\circ\prime}$ for this reaction is strongly negative, indicating that phosphate groups will tend to be transferred to glucose if we start from 1 M concentrations of all reactants and products. Thus, in this reaction the direction of phosphate group transfer is *away* from ATP, whereas in reactions (14) and (15) it was *toward* ADP, to make ATP. The difference is that the phosphorylated product resulting from the transfer of phosphate from ATP to glucose, namely, glucose 6-phosphate, has a *lower* (less negative) standard free energy of hydrolysis than ATP (see Table 15-4), whereas 3-phosphoglyceroyl phosphate and phosphoenolpyruvate have higher (more negative) $\Delta G^{\circ\prime}$ values than ATP. Again, note that the $\Delta G^{\circ\prime}$ value for reaction (16) can be predicted from the $\Delta G^{\circ\prime}$ values for hydrolysis of ATP and of glucose 6-phosphate (Table 15-4). In this way the terminal phosphate group of ATP can be transferred to a number of different building-block molecules, a process that "energizes" them and prepares them for subsequent biosynthetic reactions.

The Enzymatic Pathways of Phosphate Transfers

Figure 15-12 is a flow sheet of enzymatic phosphate-transfer reactions in the cell. An important feature is that the ATP–ADP system is the primary connecting link between high- and low-energy phosphate compounds. Phosphate groups are first transferred by action of specific phosphotransferases from high-energy compounds to ADP, as in

$$\text{Phosphoenolpyruvate} + ADP \xrightarrow[\text{kinase}]{\text{pyruvate}} \text{pyruvate} + ATP$$

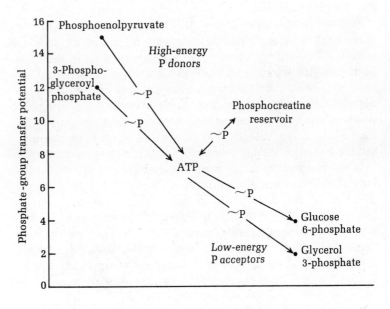

Figure 15-12
Flow of phosphate groups from high-energy phosphate donors to low-energy acceptors via ATP–ADP system. The direction of flow is toward compounds having a low phosphate-group potential, assuming standard conditions, with all reactants and products at 1 M.

The ATP so formed then becomes the specific phosphate donor in a second enzymatic reaction to form a low-energy phosphate compound, as in

$$\text{ATP} + \text{D-glucose} \xrightarrow{\text{hexokinase}} \text{ADP} + \text{D-glucose 6-phosphate}$$

The net result is the transfer of a phosphate group from a high-energy donor to a low-energy acceptor through the ATP–ADP system as mediator. The overall reaction is then

Phosphoenolpyruvate + D-glucose $\rightleftharpoons$

pyruvate + D-glucose 6-phosphate

As a result of these two coupled reactions glucose has been converted by its phosphorylation to a compound with a higher energy content; glucose 6-phosphate may be regarded as an energized form of glucose, which can now serve as a precursor in the energy-requiring biosynthesis of glycogen and starch (page 645).

In nearly all phosphate-transfer reactions between metabolites ADP serves as intermediate phosphate acceptor and ATP as intermediate phosphate donor in reactions similar to those above.

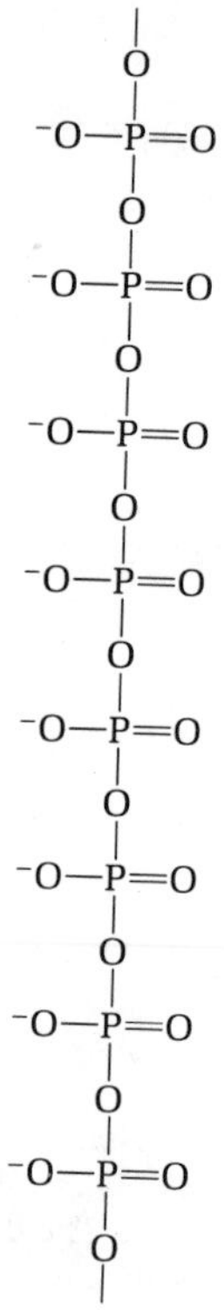

The Standard-Free-Energy Change and the Biological Reversibility of Enzymatic Reactions

In the preceding discussion we have seen that whenever the $\Delta G^{\circ\prime}$ for a reaction has a relatively large negative value, it tends to go to the right as written if we start with 1 M concentrations of all reactants and products, the exact position of equilibrium being predictable from the value of $\Delta G^{\circ\prime}$. Thus we have shown that phosphate groups will tend to be transferred from, say, phosphoenolpyruvate to ADP and then from ATP to glucose, since each of these steps proceeds with a negative value of $\Delta G^{\circ\prime}$, the first with $\Delta G^{\circ\prime} = -7.5$ kcal mol^{-1} and the second with $\Delta G^{\circ\prime} = -4.0$ kcal mol^{-1}. This is, in fact, the direction of phosphate-group transfer in the intact cell.

However, we cannot necessarily conclude that because a given enzymatic reaction has a strongly negative value for its standard-free-energy change that it is irreversible in the cell. Some enzymatic reactions that have a $\Delta G^{\circ\prime}$ value of -4.0 kcal mol^{-1}, e.g., that catalyzed by hexokinase, are known from independent evidence to be essentially irreversible in the cell, whereas other reactions having about the same value for $\Delta G^{\circ\prime}$ may go in either direction in the cell. As has been pointed out (page 391), the $\Delta G^{\circ\prime}$ values for chemical reactions assume 1 M concentrations of all reactants and products, which are very different from the concentrations actually existing in the cell. Thus it is quite possible for a given enzyme-catalyzed reaction A $\rightleftharpoons$ B having a $\Delta G^{\circ\prime}$ value of -4.0 kcal mol^{-1} to proceed toward the left in the cell, providing A is present at a very low concentration and B at a very high concentration. For example, if A is constantly being removed at a high rate by another reaction, thus keeping its concentra-

Figure 15-13
Structure of polymetaphosphate.

tion vanishingly small, then B can be quite rapidly converted into A. Thus the thermodynamic analysis of the individual reactions of metabolism requires knowledge of the actual concentrations of the relevant metabolites in the cell, as well as their rates of formation and utilization.

Reservoirs of High-Energy Phosphate Groups

It must be strongly emphasized that ATP does not function primarily as a reservoir of chemical energy; instead, it acts as a transmitter or carrier of energy. The amount of ATP in the cell at any given time is sufficient for only a short period. However, some cells do have phosphate compounds that function as reservoirs of energy, e.g., phosphocreatine. Figure 15-12 shows the reservoir role of phosphocreatine, which is formed by direct enzymatic transfer of a phosphate group from ATP to creatine whenever ATP is at a high concentration; there is no other pathway for the formation of phosphocreatine. Furthermore, the only known pathway for the dephosphorylation of phosphocreatine is the reversal of the reaction by which it is formed. The phosphocreatine reservoir is therefore filled with phosphate groups whenever ATP is present at high concentrations. Whenever the ATP concentration falls, thus raising the concentration of ADP, phosphate groups are transferred back to ADP from phosphocreatine. The phosphocreatine system is especially important in skeletal muscle, where it can provide the chemical energy required for several minutes of contractions (Chapter 27). It is also found in smooth muscle and nerve cells, but only in very small amounts in liver, kidney, and other mammalian tissues and not at all in bacteria. Phosphoarginine functions in a similar way in muscles of some invertebrates, e.g., the crab and lobster.

Some microorganisms store high-energy phosphate groups in the form of insoluble granules containing *polymetaphosphate,* a linear polymer of indefinite size (Figure 15-13). These granules stain in a characteristic way with basic dyes; they are often called *volutin* granules. Phosphate groups can be released from polymetaphosphate by specific enzymes.

Channeling of Phosphate Groups via Other Nucleoside 5'-Triphosphates

Although the ATP–ADP system is the obligatory phosphate-group carrier in the mainstream of energy transfers in the cell, the 5'-diphosphates and 5'-triphosphates of other ribonucleosides and 2-deoxyribonucleosides (Chapter 12) also participate in cellular energy transfers. The 5'-di- and triphosphates of the various ribonucleosides not only serve as energized precursors in RNA synthesis but also channel high-energy phosphate groups into other specific biosynthetic reactions (Figure 15-14). These channels all connect with ATP through the action of the enzyme *nucleoside diphosphate kinase,* which is found in mitochondria and in the soluble cytoplasm of cells; it catalyzes reversible reac-

Figure 15-14
Channeling of high-energy phosphate groups into different biosynthetic pathways via the ribonucleoside and deoxyribonucleoside 5'-triphosphates.

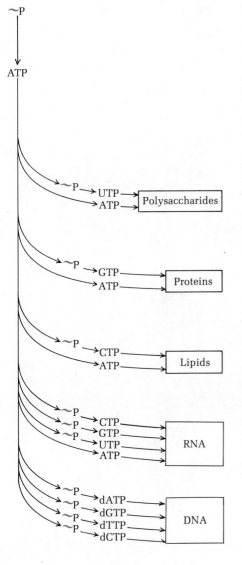

tions of the type shown in Figure 15-15. This enzyme is relatively nonspecific with respect to its substrates. It not only transfers phosphate between ATP and any NDP but also between any NTP and any NDP, although the rates of such reactions depend on the nature of the substrates. The equilibrium constant for all such phosphate-group transfers is approximately 1.0 at pH 7.0, since the standard free energy of hydrolysis of the terminal phosphate group of all the ribonucleoside and deoxyribonucleoside 5'-triphosphates is approximately the same.

Figure 15-14 also shows that each type of nucleoside 5'-triphosphate has a specialized function. Uridine triphosphate (UTP) is the immediate phosphate donor (and thus the energy donor) for many reactions in animal tissues leading to polysaccharide synthesis (pages 643 and 645). Similarly, cytidine triphosphate (CTP) is the energy donor for several reactions of lipid biosynthesis (page 671). Figure 15-14 also shows that deoxyribonucleoside 5'-triphosphates serve to channel high-energy phosphate groups into the biosynthesis of DNA.

Figure 15-15
Reactions catalyzed by nucleoside diphosphate kinase.

$$ATP + UDP \rightleftharpoons ADP + UTP$$

$$ATP + GDP \rightleftharpoons ADP + GTP$$

$$ATP + CDP \rightleftharpoons ADP + CTP$$

$$GTP + UDP \rightleftharpoons GDP + UTP$$

$$ATP + dCDP \rightleftharpoons ADP + dCTP$$

$$GTP + dADP \rightleftharpoons GDP + dATP$$

The Role of AMP and Pyrophosphate

We have seen that ADP is the product of many ATP-utilizing reactions in the cell and that it is also the direct phosphate acceptor in the energy-yielding reactions of glycolysis and of oxidative phosphorylation. However, in many ATP-utilizing reactions in the cell the two terminal phosphate groups of ATP are enzymatically removed in one piece as pyrophosphate (PP_i), leaving AMP as the other product. An example is the enzymatic activation of a fatty acid to form its ester with coenzyme A, an important "energized" intermediate in the biosynthesis of triacylglycerols and phosphoglycerides (page 546):

$$ATP + RCOOH + CoA—SH \rightleftharpoons AMP + PP_i + RCO—S—CoA$$

Fatty acid   Fatty acyl–CoA ester

This reaction proceeds by what is termed a *pyrophosphate cleavage* of ATP, in contrast to the usual *orthophosphate cleavage*, in which ATP loses a single phosphate group. The two types of cleavage may be symbolized as follows:

Orthophosphate cleavage:

$$A—R—P{\sim}P{\sim}P \rightleftharpoons A—R—P{\sim}P + Pi$$

Pyrophosphate cleavage:

$$A—R—P{\sim}P{\sim}P \rightleftharpoons A—R—P + P{\sim}P$$

The decrease in free energy when ATP undergoes pyrophosphate cleavage is substantially greater than in orthophosphate cleavage. The standard-free-energy change $\Delta G^{\circ\prime}$ for the reaction

$$ATP + H_2O \rightleftharpoons AMP + PP_i$$

is -10.0 kcal mol^{-1}. Pyrophosphate cleavage of ATP thus yields an extra thermodynamic "push" in certain enzymatic reactions, as we shall see.

The question now arises: By what mechanism can ATP be regenerated from AMP and pyrophosphate, the products of pyrophosphate cleavage? Two important auxiliary enzymes allow AMP and pyrophosphate to return to the mainstream of phosphate-group transfer via the ATP-ADP cycle, _inorganic pyrophosphatase_ and _adenylate kinase_. The former catalyzes the hydrolysis of inorganic pyrophosphate (PP_i) to form two molecules of inorganic orthophosphate (P_i):

$$PP_i + H_2O \rightleftharpoons 2P_i \qquad \Delta G^{\circ\prime} = -4.6 \text{ kcal mol}^{-1}$$

This reaction, which proceeds with a large decrease in free energy, may appear to be wasteful of energy, but we shall see (pages 546 and 640) that the hydrolysis of pyrophosphate may sometimes help assure the completeness of certain biosynthetic reactions. The orthophosphate so formed can then be utilized in the regeneration of ATP from ADP. _Adenylate kinase_, which is sometimes called _myokinase_, catalyzes the rephosphorylation of AMP to ADP in the reaction

$$ATP + AMP \rightleftharpoons ADP + ADP$$

Nucleoside monophosphate kinases make possible the formation of other NDPs from the corresponding NMPs and ATP:

$$ATP + NMP \rightleftharpoons ADP + NDP$$

Note, however, that AMP and the other NMPs cannot be rephosphorylated unless some ATP is available from other sources.

Dynamics of Phosphate-Group Turnover in the Cell

In intact living cells in the steady state the concentrations of ATP, ADP, and AMP over short periods are relatively constant. Normally ATP is present in much higher concentration than ADP and AMP, indicating that the adenylate system is normally nearly "full" of phosphate groups. If a sudden work load is placed on a cell, so that it must for a time utilize ATP at a much higher rate than usual, the concentration of ATP in the cell will at first decrease and that of ADP will rise. This change is a signal that causes acceleration of the ATP-generating reactions of glycolysis and respiration, which then proceed at higher rates to keep pace with the rate of ATP dephosphorylation (page 536). When the work load is suddenly removed from the cell, the ATP concentration abruptly increases and the ADP concentration falls, thus signaling the ATP-yielding reactions to slow down. In general, then, the rate of ATP production in cells is adjusted to the rate of ATP utilization in a dynamic steady state. We shall see elsewhere (page 424) that the regulation of ATP synthesis is made possible by the action of allosteric enzymes whose modulators are ATP, ADP, or AMP.

These considerations imply that the terminal phosphate group of ATP must undergo very rapid turnover in the cell; i.e., it must be rapidly removed and equally rapidly replaced. This prediction has been verified by use of the radioactive isotope of phosphorus as a tracer. Introduction of inorganic phosphate labeled with ^{32}P into the cell is followed by extremely rapid incorporation of the labeled phosphate into the terminal phosphate group of the intracellular ATP; in a very short time the specific radioactivity of the terminal phosphate group and the inorganic phosphate pool become identical. In fact, the turnover rate of the terminal phosphate group of the intracellular ATP is so high that it cannot be measured easily. The half-time of turnover of ATP in a rapidly respiring bacterial cell, such as *E. coli*, has been calculated to be a matter of only seconds and that of the larger, more slowly respiring eukaryotic cells, such as a liver cell, about 1 to 2 min.

Energetics of Open Systems

The principles of classical or equilibrium thermodynamics we have used in analyzing energy changes in isolated chemical reactions in this chapter are applicable only to _closed systems_, systems that do not exchange matter with their surroundings. Analysis of closed systems is relatively simple since we need consider only the initial and final states of a given system or collection of matter, after it has come to equilibrium. From such approaches much important information has become available on the energetics of individual enzymatic reactions. But when we now attempt to apply this information to analysis of energy exchanges in intact living cells, we face great difficulties since living cells are _open systems_; they _do_ exchange matter with their surroundings. Furthermore, cells are never in equilibrium. A living cell at any given moment exists in a steady state in which the rate of input of matter equals the rate of output of matter. In this steady state the concentration of all components of the cell, including the ATP, remains constant; under these conditions the rate of formation of each component is exactly equal to its rate of utilization.

The analysis of the magnitude and the efficiency of energy exchanges in steady-state or open systems is much more complex than in closed systems, but an extension of thermodynamic theory, called _nonequilibrium_ or _irreversible thermodynamics_, has been developed for the analysis of open systems. The application of these principles is beyond the scope of this book, but at least two general attributes of open systems in steady states have considerable significance in biology. An open system in the steady state is capable of doing work, precisely because it is away from the condition of equilibrium; systems already at equilibrium can do no work. Moreover, only a process away from equilibrium can be regulated. But the most important implication is this: in the formalism of nonequilibrium thermodynamics, the steady state, which is a characteristic of all smoothly running machinery, may be considered to be the *orderly* state of an open system, the state in which the rate of entropy production is at a minimum and in which the system is operating with maximum efficiency under the prevailing conditions. The significance of this relationship has been aptly commented on by A. Katchalsky, a pioneer in the application of nonequilibrium thermodynamics to biology.[1]

> This remarkable conclusion sheds new light on the wisdom of living organisms. Life is a constant struggle against the tendency to produce entropy. The synthesis of large and information-rich macromolecules, the formation of intricately structured cells, the development of organization, all these are powerful antientropic forces. But since there is no possibility of escaping the entropic doom imposed on all natural phenomena under the Second Law of thermodynamics, living organisms choose the least evil—they produce entropy at a minimum rate by maintaining a steady-state.

Summary

Energy changes of chemical reactions can be analyzed quantitatively in terms of the first and second laws of thermodynamics. Chemical reactions proceed in such a direction that at equilibrium the entropy S of the system plus surroundings is at a maximum and the free energy G of the system alone is at a minimum. Every chemical reaction has a characteristic standard-free-energy change $\Delta G^{\circ\prime}$ at standard temperature and pressure with all reactants and products at $1\ M$ concentration and pH = 7.0. $\Delta G^{\circ\prime}$ can be calculated from the equilibrium constant K'_{eq} by the equation $\Delta G^{\circ\prime} = -2.303RT \log K'_{eq}$. Standard-free-energy changes can also be calculated from equilibrium data of a consecutive series of reactions or from the difference in the standard free energy of formation of reactants and products. The $\Delta G^{\circ\prime}$ of hydrolysis of ATP to ADP and phosphate is -7.30 kcal mol^{-1} at pH 7.0 and 25°C in the presence of 20 mM Mg^{2+}. This relatively negative value is the result of electrostatic repulsion between the products of hydrolysis, ADP^{3-} and HPO_4^{2-}, their resonance stabilization and tendency to undergo extensive hydration, and the electron-withdrawing properties of the phosphoric groups in ATP. ΔG_{ATP} varies with pH and Mg^{2+} concentration. Under intracellular conditions it is approximately -12.5 kcal mol^{-1}. Some phosphorylated compounds, e.g., phosphoenolpyruvate and 3-phosphoglyceroyl phosphate, which are generated in glycolysis, have much more negative $\Delta G^{\circ\prime}$ values for their hydrolysis than ATP, whereas others, e.g., glucose 6-phosphate, have more positive values. The intermediate position of ATP in the thermodynamic scale of phosphate-bond energy and the specificity of phosphate-transferring enzymes for ADP or ATP as phosphate acceptor or donor, respectively, mean that the ADP–ATP system is the obligatory common intermediate or carrier of phosphate groups from high-energy phosphate compounds generated during catabolism to certain phosphate acceptors, which thus become energized. Phosphocreatine, phosphoarginine, and polymetaphosphate are reservoirs of high-energy phosphate groups.

ATP may undergo loss of either an orthophosphate or a pyrophosphate group during its utilization in biosynthetic reactions, to

[1] A. Katchalsky, "Non-Equilibrium Thermodynamics," p. 194 in R. Colbern (ed.), *Modern Science and Technology*, D. Van Nostrand Company, Inc., New York, 1965.

form ADP or AMP, respectively. AMP is rephosphorylated to ADP by the adenylate kinase reaction, $ATP + AMP \rightleftharpoons 2ADP$. Other nucleoside 5'-triphosphates such as GTP, UTP, CTP, dATP, dTTP, etc., also participate as carriers of high-energy phosphate groups, which they channel into specific biosynthetic routes. The terminal phosphate group of ATP undergoes extremely rapid replacement by inorganic phosphate in intact respiring cells. Living cells are open systems which exchange both matter and energy with their surroundings; they exist in steady states, far from equilibrium.

References

Books

BLUM, H. F.: *Time's Arrow and Evolution*, Harper, New York, 1962. An interesting discussion of entropy in biology.

BRAY, H. G., and K. WHITE: *Kinetics and Thermodynamics in Biochemistry*, 2d ed., Academic, New York, 1966.

FLORKIN, M., and E. H. STOTZ (eds.): *Bioenergetics*, vol. 22 of *Comprehensive Biochemistry*, American Elsevier, New York, 1967.

KALCKAR, H. M.: *Biological Phosphorylations: Development of Concepts*, Prentice-Hall, Englewood Cliffs, N.J., 1969. A collection of reprinted papers describing classical investigations in bioenergetics, with an accompanying narrative.

KAPLAN, N. O., and E. P. KENNEDY (eds.): *Current Aspects of Biochemical Energetics*, Academic, New York, 1966. A volume of essays and papers dedicated to Fritz Lipmann.

KATCHALSKY, A., and P. F. CURRAN: *Non-Equilibrium Thermodynamics in Biophysics*, Harvard University Press, Cambridge, Mass., 1965.

KREBS, H. A., and H. L. KORNBERG: *Energy Transformations in Living Matter*, Springer-Verlag, Berlin, 1957. A classical analysis of the energetics of metabolism.

LEHNINGER, A. L.: *Bioenergetics*, 2d ed., Benjamin, Menlo Park, Calif., 1972. Elementary treatment stressing biochemical aspects.

LIPMANN, F.: *Wanderings of a Biochemist*, Wiley-Interscience, New York, 1971. Scientific autobiography and some biochemical essays.

VAN HOLDE, K. E.: *Physical Biochemistry*, Prentice-Hall, Englewood Cliffs, N.J., 1971.

WALL, F. T.: *Chemical Thermodynamics*, Freeman, San Francisco, 1965. A standard textbook of equilibrium thermodynamics.

WOOD, W. B., J. H. WILSON, R. M. BENBOW, and L. E. HOOD: *Biochemistry: A Problems Approach*, Benjamin, Menlo Park, Calif., 1974. Chapter 8 contains additional problems and review in bioenergetics.

Articles

ALBERTY, R. A.: "Effect of pH and Metal Ion Concentration on the Equilibrium Hydrolysis of Adenosine Triphosphate to Adenosine Diphosphate," *J. Biol. Chem.*, 243: 1337–1343 (1968). Important paper on the effect of pH and Mg^{2+} on $\Delta G^{o\prime}$ of ATP hydrolysis.

BENZINGER, T. H.: "Thermodynamics, Chemical Reactions, and Molecular Biology," *Nature*, 229: 100–102 (1971). Important theoretical article.

INGRAHAM, L. L., and A. B. PARDEE: "Free Energy and Entropy in Metabolism," in D. M. Greenberg (ed.), *Metabolic Pathways*, 3d ed., vol. 1, pp. 2–45, Academic, New York, 1967. Excellent review article.

JENCKS, W. P.: "Free Energies of Hydrolysis and Decarboxylation," p. J181 in *Handbook of Biochemistry*, 2d ed., Chemical Rubber Co., Cleveland (1970). Authoritative compilation of thermodynamic data, the source of many values used in this book.

LIPMANN, F.: "Metabolic Generation and Utilization of Phosphate Bond Energy," *Adv. Enzymol.*, 18: 99–162 (1941). Classical statement of the ATP–ADP cycle.

ROSING, J., and E. C. SLATER: "The Value of $\Delta G^{\circ\prime}$ for the Hydrolysis of ATP," *Biochim. Biophys. Acta*, 267: 275–290 (1972).

SHIKAMA, K., and K-I. NAKAMURA: "Standard Free Energy Maps for the Hydrolysis of ATP as a Function of pH and Metal Ion Concentration," *Arch. Biochem. Biophys.*, 157: 457–463 (1973).

Problems

1. Calculate the percent dissociation of (a) the third ionizable proton of ADP and (b) the fourth ionizable proton of ATP at the pH of the intracellular fluid of muscle (pH 6.0).

2. Calculate the $\Delta G^{\circ\prime}$ value for the alcoholic fermentation of glucose (D-glucose $\rightleftharpoons$ 2 ethanol $+$ $2CO_2$) from data on the standard free energies of formation of reactants and products (Table 15-2).

3. Calculate the equilibrium constants for the following reactions at pH $= 7.0$ and $T = 25°C$, using the $\Delta G^{\circ\prime}$ values of Table 15-3:

   (a) Glucose 6-phosphate $+$ H_2O $\rightleftharpoons$ glucose $+$ phosphate
   (b) Glutamine $+$ H_2O $\rightleftharpoons$ glutamate $+$ NH_4^+

4. Calculate the standard-free-energy changes of the following reactions at 25°C from the equilibrium constants given (pH 7.0):

   (a) Glutamate $+$ oxaloacetate $\rightleftharpoons$ aspartate $+$ α-ketoglutarate
   $$K_{eq}' = 6.8$$
   (b) H_2O $\rightleftharpoons$ H^+ $+$ OH^- $K_{eq}' = 6.31 \times 10^{-15}\ M$
   (c) Isopropanol $+$ NAD^+ $\rightleftharpoons$ acetone $+$ NADH $+$ H^+
   $$K_{eq}' = 7.20 \times 10^{-9}\ M$$
   (d) Malate $+$ NAD^+ $\rightleftharpoons$ oxaloacetate $+$ NADH $+$ H^+
   $$K_{eq}' = 7.50 \times 10^{-13}\ M$$

5. Calculate $\Delta G'$ (pH 7.0; $T = 25°$) for the hydrolysis of ATP to ADP and phosphate, assuming that ATP and ADP are present in equimolar concentrations and that the phosphate concentration is (a) 1.0 M, (b) 0.1 M, (c) 0.01 M, and (d) 1.0 mM.

6. Calculate the free energy of hydrolysis of ATP under conditions existing in a resting muscle cell, namely, [ATP] $= 5.0$ mM, [ADP] $= 0.5$ mM, [P_i] $= 1.0$ mM, pH $= 6.0$, and $T = 25°C$. Start from data in Figure 15-5.

7. Glucose 1-phosphate is converted to fructose 6-phosphate in two successive reactions

    Glucose 1-phosphate $\rightleftharpoons$ glucose 6-phosphate
    Glucose 6-phosphate $\rightleftharpoons$ fructose 6-phosphate

    Using the $\Delta G^{o\prime}$ values of Table 15-3, determine the $\Delta G^{o\prime}$ value for the overall reaction.

8. At what minimum concentration must malate be present to make the fumarase reaction (malate $\rightleftharpoons$ fumarate $+ H_2O$) proceed to the right at $pH = 7.0$ and $T = 25°C$ if the fumarate is present at a concentration of 1 mM?

9. The standard-free-energy change for the reaction phosphoenolpyruvate $+ ADP \rightleftharpoons$ pyruvate $+ ATP$ is -7.50 kcal. If phosphoenolpyruvate and ADP are originally present at 10 mM concentrations but no ATP or pyruvate are present, what will the equilibrium concentrations of the products and reactants be? Repeat the calculation, assuming the reaction is initiated with 6.0 mM ADP, 6.0 mM phosphoenolpyrurate, and 6.0 mM ATP.

10. A mixture 30 mM in 3-phosphoglyceroyl phosphate and 10 mM in pyruvate is incubated at pH 7.0 and 25°C with phosphoglycerate kinase, pyruvate kinase, and a catalytic amount of ATP until equilibrium occurs. Calculate the concentration of phosphoenolpyruvate at equilibrium.

11. Adenylate kinase catalyzes the reaction

    $$2ADP \rightleftharpoons ATP + AMP$$
    $$\Delta G^{o\prime} = +0.486 \text{ kcal mol}^{-1}$$

    (a) Suppose the concentration of ATP is 4.90 mM. Calculate the concentrations of ADP and AMP assuming the adenylate kinase reaction to be at equilibrium and the total adenine nucleotide concentration to be 5.00 mM.

    (b) If the ATP concentration is decreased by 10 percent from that in part (a), what will be the percentage change in AMP concentration from that in part (a)?

CHAPTER **16** GLYCOLYSIS

We now begin consideration of the multienzyme systems that catalyze the degradation of fuel molecules and the recovery of part of their chemical energy as ATP. We shall first examine *glycolysis*, the anaerobic degradation of glucose to yield lactic acid. Glycolysis is one of several catabolic pathways, known generically as *anaerobic fermentations*, by which many organisms extract chemical energy from various organic fuels in the absence of molecular oxygen. Since living organisms first arose in an atmosphere lacking oxygen (page 23), anaerobic fermentation is the most ancient type of biological mechanism for obtaining energy from nutrient molecules. Most higher organisms have retained the capacity for anaerobic degradation of glucose to lactate, which has become a preparatory pathway in aerobic glucose catabolism. Moreover, in most animals glycolysis serves as an important emergency mechanism capable of yielding energy for short periods when oxygen is not available.

The glycolytic pathway was the first major enzyme system to be elucidated; it engaged the attention of some of the greatest biochemists of the first half of the twentieth century. The lessons learned in the study of glycolysis offered new insights and opened new approaches to the study of enzymology and intermediary metabolism. We shall examine glycolysis in some detail since it will serve as prototype for the study of other multienzyme systems and metabolic pathways.

Fermentation and Respiration

All heterotrophic organisms ultimately obtain their energy from oxidation-reduction reactions, i.e., reactions in which electrons are transferred from one compound, the *electron donor*, or reducing agent, to an *electron acceptor*, or oxidizing agent. Aerobic organisms obtain most of their energy from *respiration*, defined as the oxidation of organic fuels by molecular oxygen; oxygen thus serves as the final electron acceptor in respiration. Anaerobic heterotrophs also obtain most of their energy from oxidation-reduction reactions, but in this case electrons pass from one organic intermediate of sugar breakdown, the electron donor, to some other organic intermediate in the fermentation process, which serves as the electron acceptor. In anaerobic fermentation processes, however, there is no net oxidation of the fuel.

Organisms that can live anaerobically and thus employ fermentation as a source of energy are divided into two classes. The *obligate*, or *strict*, *anaerobes* cannot use oxygen at all and, indeed, often cannot tolerate it. They include bacteria living in marine mud, the denitrifying bacteria of the soil responsible for reducing nitrate to nitrogen, the methane-producing bacteria, which form marsh gas, and some organisms that are pathogenic for man, such as the soil bacterium *Clostridium perfringens*, which causes gas gangrene in wound infections, and *Clostridium botulinum*, responsible for a deadly form of food poisoning. The *facultative anaerobes*, on the other hand, can live either in the absence or presence of oxygen. When they live anaerobically, they obtain energy from a fermentation process. When they live aerobically, they often continue to degrade their fuel by the anaerobic pathway and then oxidize the products of the anaerobic pathway at the expense of molecular oxygen (Figure 16-1).

The most common fuels for anaerobic fermentation are the sugars, particularly D-glucose, but some bacteria can obtain their metabolic energy by carrying out anaerobic fermentation of such fuel molecules as fatty acids, amino acids, purines, or pyrimidines, depending on the species. In fact, the taxonomic classification of microorganisms is in part based on their characteristic organic fuels and their fermentation products.

Of the many kinds of glucose fermentation, two closely related types predominate (Figure 16-2). In glycolysis, sometimes called *homolactic fermentation*, the six-carbon glucose molecule is degraded to two molecules of the three-carbon lactic acid as sole end product. This type of glucose breakdown occurs in many microorganisms and in the cells of most higher animals and plants. In *alcoholic fermentation*, characteristic of many yeasts, the glucose molecule is broken down into two molecules of the two-carbon compound ethanol (C_2H_5OH) and two molecules of CO_2. Alcoholic fermentation occurs by the same enzymatic pathway as glycolysis but requires two different enzymatic steps at the end (Figure 16-2). Most of the other types of glucose fermentation are variations on the basic pathway of glycolysis (page 438).

The Balance Sheets for Glycolysis and Alcoholic Fermentation

The pathways for glycolysis and alcoholic fermentation schematized in Figure 16-2 describe only the fate of the carbon atoms of the glucose and tell us nothing about the energetics of this process. Actually, during both glycolysis and alcoholic fermentation ATP is generated from ADP and phosphate. The complete balanced equations for glycolysis and

Figure 16-1
The pattern of glucose utilization in facultative organisms, including most higher plants and animals. The fermentation pathway is common to both the anaerobic (A) and the aerobic (B) pathways of glucose utilization.

A. *Under anaerobic conditions.*

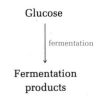

B. *Under aerobic conditions.*

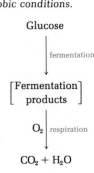

Figure 16-2
Path of carbon atoms in (A) glycolysis and (B) alcoholic fermentation. The numbers in parentheses indicate the carbon atoms of D-glucose from which the carbon atoms of the products were derived.

A. Glycolysis.

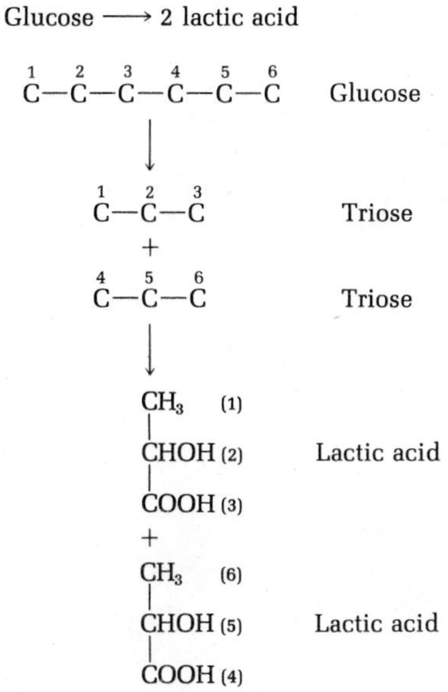

Glucose $\longrightarrow$ 2 lactic acid

B. Alcoholic fermentation.

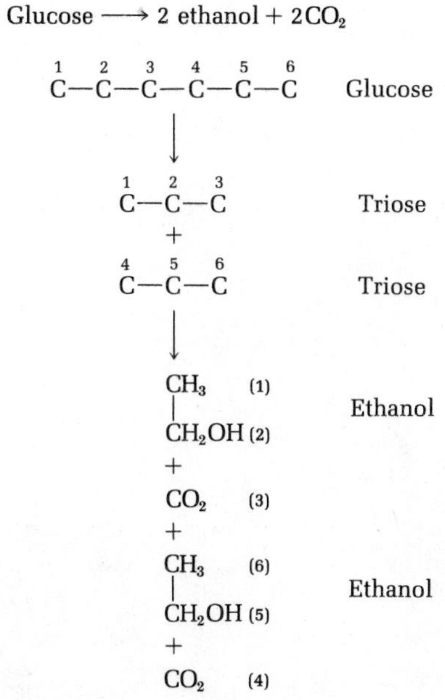

Glucose $\longrightarrow$ 2 ethanol + 2CO$_2$

alcoholic fermentation, including the energy-conserving steps, are

Glycolysis:

$$C_6H_{12}O_6 + 2P_i + 2ADP \longrightarrow 2CH_3CHOHCOOH + 2ATP + 2H_2O$$
$$\text{Lactic acid}$$

Alcoholic fermentation:

$$C_6H_{12}O_6 + 2P_i + 2ADP \longrightarrow 2CH_3CH_2OH + 2CO_2 + 2ATP + 2H_2O$$
$$\text{Ethanol}$$

To analyze the energetics of glycolysis under standard conditions we can resolve the overall equation into two processes, the conversion of glucose to lactate, which is exergonic, and the formation of ATP from ADP and phosphate, which is endergonic:

Exergonic process:

$$\text{Glucose} \longrightarrow 2 \text{ lactate}$$
$$\Delta G_1^{\circ\prime} = -47.0 \text{ kcal mol}^{-1}$$

Endergonic process:

$$2P_i + 2ADP \longrightarrow 2ATP + 2H_2O$$
$$\Delta G_2^{\circ\prime} = 2 \times 7.30 = +14.6 \text{ kcal mol}^{-1}$$

Sum:

$$\text{Glucose} + 2P_i + 2ADP \longrightarrow 2 \text{ lactate} + 2ATP + 2H_2O$$
$$\Delta G_s^{\circ\prime} = \Delta G_1^{\circ\prime} + \Delta G_2^{\circ\prime}$$
$$= -47.0 + 14.6$$
$$= -32.4 \text{ kcal mol}^{-1}$$

From the standard-free-energy changes shown it is clear that the breakdown of glucose to lactate ($\Delta G^{\circ\prime} = -47.0$ kcal mol^{-1}) provides more than sufficient energy to cause the phosphorylation of two molecules of ADP to ATP ($\Delta G^\circ = +14.6$ kcal mol^{-1}). Therefore, 14.6/47.0 × 100, or about 31 percent, of the free-energy decrease during breakdown of glucose to lactate is conserved in the form of ATP. Actually if we adjust such calculations, which are based on 1.0 M standard concentrations, to take the actual intracellular concentrations of reactants and products into account, the true efficiency of glycolysis is much higher than 31 percent (page 432).

The overall process of glycolysis, even after making allowance for the coupled formation of ATP, still proceeds with a very large net decrease in free energy, -32.4 kcal mol^{-1}. Glycolysis is thus an essentially "irreversible" reaction, with its equilibrium overwhelmingly in the direction of lactate formation. We shall see later, however, that most of its reaction steps have a relatively small standard-free-energy change and are also employed in the reverse direction for the biosynthesis of glucose from lactate and other precursors (page 624).

Experimental History

The enzymatic pathway of glycolysis and alcoholic fermentation was elucidated over the course of many years of research. Some important landmarks will illustrate the experimental and conceptual approaches taken in the investigation of this fundamental metabolic pathway.

We recall (page 380) E. Buchner's discovery in 1897 that an extract of yeast, freed of intact cells by filtration, retains the ability to ferment glucose to ethanol. This observation demonstrated that the enzymes of fermentation can function independently of cell structure, contrary to Pasteur's earlier dictum (page 183). A second major landmark was the discovery by A. Harden and W. J. Young in England (1905) that alcoholic fermentation in yeast extracts requires phosphate and that a hexose diphosphate accumulates in fermenting yeast extracts under some conditions but is utilized in others, suggesting that it is an intermediate in the overall fermentation process. This intermediate was later identified as fructose 1,6-diphosphate (page 424). Harden and Young also found that two fractions of yeast extracts are required for alcoholic fermentation to take place, a heat-labile fraction, called zymase, presumably containing the enzymes required for the process, and a heat-stable fraction (cozymase) required for activity of zymase. The heat-stable fraction was later shown to contain two essential components, the oxidation-reduction coenzyme nicotinamide adenine dinucleotide, or NAD (page 340), discussed further below, and a mixture of the adenine nucleotides ADP and ATP (pages 314 and 389).

Another important set of observations revealed that in the presence of the inhibitor fluoride, fermenting yeast extracts showed an accumulation of two phosphate esters, 3-phosphoglycerate and 2-phosphoglycerate. On the other hand, the inhibitor iodoacetate caused an accumulation of fructose 1,6-diphosphate. Once these intermediates were identified, it became possible to study the enzymatic reactions by which they were formed and utilized.

These basic observations on yeast extracts, as well as the later discovery that muscle extracts can catalyze glycolysis of glucose to lactate, served as the starting point for more intensive investigations by German biochemists in the 1930s. Among the most important contributors to this phase were Gustav Embden, who postulated the manner of cleavage of fructose 1,6-diphosphate and the overall pattern of the subsequent steps, and Otto Meyerhof, who verified the major features of Embden's hypothesis and also studied the energetics of glycolysis. The sequence of reactions from glucose to pyruvate is often called the Embden-Meyerhof pathway. Very important contributions were also made by Otto Warburg in Berlin, by C. F. Cori and G. T. Cori in the United States, and by J. Parnas in Poland.

Although the individual steps of glycolysis have been known since about 1940, research on this pathway has by no means ceased. In fact, intensive research on the enzymes of glycolysis is now revealing some of the mechanisms by which certain enzymes in the sequence participate in the regulation of glycolysis in the intact cell. Although most of the classical research on the mechanism of fermentation and glycolysis was carried out on yeast and muscle extracts, these pathways occur with only minor variations in most forms of life, an indication of the evolutionary survival value of this ancient energy-yielding pathway.

There are sometimes differences in the properties of the homologous enzymes of the glycolytic sequence from one species or cell type to another. Such variations presumably relate to differences in the regulation of this pathway as a reflection of species or tissue differentiation.

The Stages of Glycolysis

Glycolysis is catalyzed by the consecutive action of a group of 11 enzymes, most of which have been crystallized and thoroughly studied. Since they are easily extracted in soluble form from cells, they are believed to be localized in the soluble portion of the cytoplasm. It is also thought that the individual enzymes catalyzing the steps in glycolysis have no physical dependence on each other; i.e., they appear not to be associated into a stable multienzyme complex. There is some evidence, however, that in different types of cells certain individual enzymes of glycolysis may be loosely associated with the plasma membrane, with myofibrils, or with the mitochondria.

All the intermediates of glycolysis between glucose and pyruvate are phosphorylated compounds. Their phosphate groups appear to have three functions. They provide each intermediate with a polar, negatively charged group, rendering it impermeant through the cell membrane, which generally does not allow highly polar molecules to pass (pages 314 and 777). In fact, most intermediates of metabolism are ionic compounds at pH 7, a fact which prevents them from leaking out of cells by simple diffusion. The phosphate groups of the glycolytic intermediates also serve as binding or recognition groups in the formation of enzyme-substrate complexes. Most important, however, is the function of the phosphate groups of the glycolytic intermediates in conservation of energy, since they ultimately become the terminal phosphate groups of ATP in the course of glycolysis.

There are two major stages of anaerobic glycolysis (Figure 16-3). In the first stage glucose is primed or prepared for its catabolism by its phosphorylation and then cleaved to form the three-carbon sugar *glyceraldehyde 3-phosphate*; in the second stage glyceraldehyde 3-phosphate is converted into lactate. The first stage of glycolysis serves as a preparatory or collection phase, in which a number of different hexoses, after phosphorylation by ATP, enter the glycolytic sequence and are converted into a common product, glyceraldehyde 3-phosphate. In this stage two ATP molecules are expended to phosphorylate the 1 and 6 positions of the hexose, analogous to priming a pump. The second stage of glycolysis is the common pathway for all sugars; in it occur the oxidoreduction steps and the energy-conserving mechanisms by which ADP is phosphorylated to ATP. In the second stage four molecules of ATP are formed, so that the net yield, after subtracting the priming ATPs, is two molecules of ATP per molecule of glucose degraded to lactate.

Figure 16-3
*The two stages of glycolysis. The inputs into the system are shown in color
and the outputs in colored boxes. The pathway from glucose 6-phosphate is
identical in glycolysis and alcoholic fermentation (page 437). Although the
term glycolysis originally referred to the breakdown of glucose to lactate
and is so used in this chapter, it is often employed more loosely to refer to
the pathway to the stage of pyruvate.*

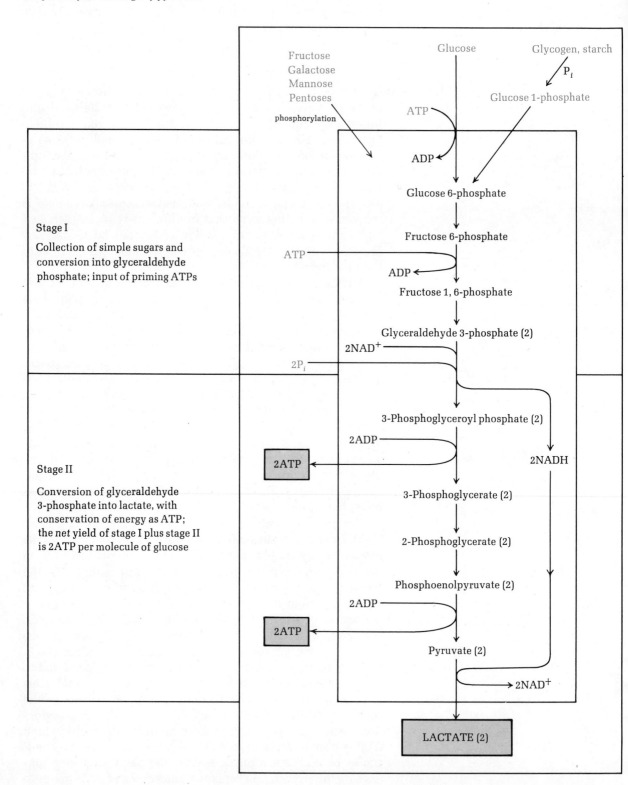

Figure 16-4
The structure and space-filling model of α-D-glucose 6-phosphate.

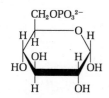

Three different types of chemical transformation take place during glycolysis; their pathways are interconnected:

1 The sequence of reactions by which the carbon skeleton of glucose is degraded to form lactate, i.e., the pathway of the carbon atoms
2 The sequence of reactions by which inorganic phosphate becomes the terminal phosphate group of ATP, i.e., the pathway of phosphate
3 The sequence of oxidoreductions, i.e., the pathway of electrons

The Enzymatic Steps in the First Stage of Glycolysis

Phosphorylation of D-Glucose by ATP

This is the first of the two priming steps of glycolysis, in which ATP is utilized. In it the neutral D-glucose molecule is prepared for the subsequent enzymatic steps by its phosphorylation to a negatively charged molecule at the expense of ATP. There is relatively little free D-glucose in cells, most intracellular glucose existing in phosphorylated form. The phosphorylation of D-glucose at the 6 position by ATP to yield D-glucose 6-phosphate (Figure 16-4) is catalyzed by two types of enzyme, *hexokinase* and *glucokinase*, which differ in their sugar specificity and affinity for D-glucose. The reaction for both enzymes is

$$\text{ATP} + \alpha\text{-D-glucose} \xrightarrow{\text{Mg}^{2+}} \text{ADP} + \alpha\text{-D-glucose 6-phosphate}$$
$$\Delta G^{\circ\prime} = -4.0 \text{ kcal mol}^{-1}$$

Hexokinase is the more widely distributed and is the enzyme normally employed by most cells. It catalyzes the phosphorylation not only of D-glucose but also of many other hexoses and hexose derivatives, including D-fructose, D-mannose, and D-glucosamine; it has a higher affinity for aldohexoses than for ketohexoses. Hexokinases are found in yeast and bacteria and in many animal and plant tissues. Yeast hexokinase has been crystallized (mol wt 96,000). The hexokinase of animal tissues occurs in the form of three isozymes (page 244) that differ in their affinity for glucose. The hexokinase of animal tissues is a regulatory enzyme; it is inhibited by its own product, glucose 6-phosphate (page 234). Whenever the cell has a high concentration of glucose 6-phosphate and requires no more for its energy demands, hexokinase is inhibited, thus preventing the formation of further glucose 6-phosphate.

The second type of glucose-phosphorylating enzyme, glucokinase, phosphorylates only D-glucose and does not act on other hexoses. Glucokinase has a much higher K_M for D-glucose ($K_M = 10$ mM) and thus requires a much higher glucose concentration to become fully active than hexokinase ($K_M = 100$ μM). It differs from hexokinase in another respect: it is not inhibited by glucose 6-phosphate. Glucokinase is present in liver, where it predominates over hexokinase, but it is absent in muscles. It normally comes into play when the blood glucose concentration is temporarily high, as it is following a meal rich in sugar (Figure 16-5). However, this en-

Figure 16-5
Comparison of hexokinase and glucokinase. At normal concentrations of blood glucose (about 5 mM) hexokinase is fully saturated. When blood glucose concentration becomes very high, the glucokinase of the liver becomes significantly more active.

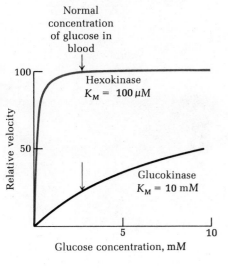

zyme is deficient in patients suffering from *diabetes mellitus,* in which there is a high blood-sugar concentration as a consequence of failure to secrete the pancreatic hormone insulin (page 849). Both hexokinase and glucokinase require a divalent cation (Mg^{2+} or Mn^{2+}), which first combines with ATP to form the true substrate, $MgATP^{2-}$ or $MnATP^{2-}$. Hexokinase is inhibited by certain sulfhydryl reagents. The phosphorylation of glucose by either hexokinase or glucokinase is not reversible under intracellular conditions.

The enzymatic dephosphorylation of D-glucose 6-phosphate to regenerate free D-glucose (a reaction of great importance in the liver since it delivers free glucose to the blood) occurs through the action of an entirely different enzyme, *glucose 6-phosphatase* (page 628):

$$\text{D-Glucose 6-phosphate} + H_2O \longrightarrow \text{D-glucose} + \text{phosphate}$$
$$\Delta G^{\circ\prime} = -3.30 \text{ kcal mol}^{-1}$$

Conversion of Glucose 6-Phosphate to Fructose 6-Phosphate

Glucose phosphate isomerase, which has been isolated from muscle tissue in highly purified form, catalyzes isomerization of glucose 6-phosphate to fructose 6-phosphate (Figure 16-6):

$$\alpha\text{-D-Glucose 6-phosphate} \rightleftharpoons \alpha\text{-D-fructose 6-phosphate}$$
$$\Delta G^{\circ\prime} = +0.4 \text{ kcal mol}^{-1}$$

The reaction proceeds readily in either direction and is reversible in the cell. Glucose phosphate isomerase is specific for glucose 6-phoshate and fructose 6-phosphate.

Phosphorylation of D-Fructose 6-Phosphate to Fructose 1,6-Diphosphate

In this, the second of the two priming reactions of glycolysis, a second molecule of ATP is required to phosphorylate fructose 6-phosphate in the 1 position to yield fructose 1,6-diphosphate (Figure 16-7) by the action of *6-phosphofructokinase:*

$$\text{ATP} + \text{D-fructose 6-phosphate} \longrightarrow$$
$$\text{ADP} + \text{D-fructose 1,6-diphosphate}$$
$$\Delta G^{\circ\prime} = -3.40 \text{ kcal mol}^{-1}$$

Mg^{2+} is required, presumably because the true substrate is $MgATP^{2-}$. Although fructose 6-phosphate is the specific phosphate acceptor in the reaction, UTP and ITP may replace ATP as phosphate donors.

Phosphofructokinase is an allosteric enzyme; the phosphorylation of fructose 6-phosphate is the most important control point in the glycolytic sequence (pages 536 to 539). Like many allosteric enzymes, it has a rather high molecular weight (380,000), contains a number of subunits, and shows

Figure 16-6
Structure of α-D-fructose 6-phosphate.

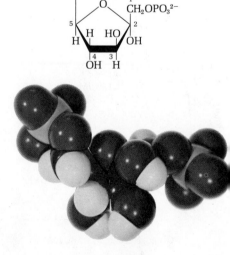

Figure 16-7
Structure and space-filling model of α-D-fructose 1,6-diphosphate. An international nomenclature commission has recommended that the prefix di be replaced by bis; this compound would thus be α-D-fructose 1,6-bisphosphate.

a complex dependence of its reaction velocity on the concentration of its substrates. Phosphofructokinase has multiple allosteric modulators. It is inhibited by high concentrations of ATP, citrate, and long-chain fatty acids but is stimulated by ADP or AMP. Therefore, whenever the cell has a high concentration of ATP, or whenever other fuels such as fatty acids or citrate are available, 6-phosphofructokinase is inhibited and turns off glycolysis. Conversely, whenever the ATP concentration is low and AMP and ADP thus predominate, or whenever the concentration of other fuels such as citrate or fatty acids is low, 6-phosphofructokinase activity is stimulated. Thus the kinetic behavior of 6-phosphofructokinase, though very complex, is extraordinarily well adapted for the regulation of this important step of glycolysis. The positive and negative allosteric modulators of this enzyme vary from one type of cell to another. The role of 6-phosphofructokinase in regulating the rate of glycolysis will be described in more detail later (pages 433 and 536). The 6-phosphofructokinase reaction is essentially irreversible in the cell; it will be recalled that most regulatory enzymes catalyze irreversible reactions (page 234).

By a separate enzymatic pathway D-fructose 1,6-diphosphate may be converted back to fructose 6-phosphate, through a hydrolytic reaction catalyzed by <u>hexose diphosphatase,</u> also an allosteric enzyme whose role in the regulation of glucose biosynthesis will be considerd later (page 627):

$$\text{Fructose 1,6-diphosphate} + H_2O \longrightarrow \text{fructose 6-phosphate} + P_i$$
$$\Delta G^{\circ\prime} = -4.0 \text{ kcal mol}^{-1}$$

Cleavage of Fructose 1,6-Diphosphate to Dihydroxyacetone Phosphate and Glyceraldehyde 3-Phosphate

This reaction is catalyzed by the well-studied enzyme <u>fructose diphosphate aldolase,</u> which is easily isolated in crystalline form from rabbit-muscle extracts. The reaction catalyzed is a reversible aldol condensation, yielding two different triose phosphates:

D-Fructose 1,6-diphosphate $\rightleftharpoons$
   dihydroxyacetone phosphate + D-glyceraldehyde 3-phosphate
$$\Delta G^{\circ\prime} = +5.73 \text{ kcal mol}^{-1}$$

The structures of the products are shown in Figure 16-8.

Although the $\Delta G^{\circ\prime}$ of this reaction is strongly positive, under intracellular conditions it readily proceeds in the forward direction (see Problems 9 and 10 for this chapter).

Typical of the fructose diphosphate aldolases found in higher animals and plants, often called <u>class I aldolases</u>, is the one isolated from skeletal muscle, which has a molecular weight of 160,000 and contains four subunits. It also contains a number of free —SH groups, some of which are essential for catalytic activity. Incubation of crystalline muscle aldolase with dihydroxyacetone phosphate in the presence of the reducing agent sodium borohydride causes a stable covalent bond to form between the enzyme molecule and

Figure 16-8
The triose phosphates. The numbers at the left of the stuctures refer to the carbon atoms of the triose phosphates. The numbers in parentheses at the right refer to the carbon atoms of the fructose 1,6-diphosphate molecule from which the triose carbons directly derived.

3 CH$_2$OPO$_3^{2-}$ (1)
|
2 C=O   (2)
|
1 CH$_2$OH   (3)

Dihydroxyacetone phosphate

1 HC=O   (4)
|
2 HCOH   (5)
|
3 CH$_2$OPO$_3^{2-}$ (6)

D-Glyceraldehyde
3-phosphate

dihydroxyacetone phosphate. Hydrolysis of the trapped enzyme-substrate compound, which is catalytically inactive, yields a glyceryl derivative of the ε-amino group of a lysine residue of the enzyme (page 210). These facts have led to the conclusion that muscle aldolase forms a covalent enzyme-substrate compound, a Schiff's base or ketimine (page 84) between the ε-amino group of a lysine residue of the enzyme and the carbonyl group of the dihydroxyacetone phosphate (Figure 16-9). The Schiff's base loses a proton to form a carbanion, which attacks the aldehydic carbon atom of the glyceraldehyde 3-phosphate to yield fructose 1,6-diphosphate and free enzyme, following uptake of a proton. These events are reversible.

Class 1 fructose diphosphate aldolases occur in different isoenzyme forms. In the tissues of the rabbit there are three major forms, aldolase A, predominant in muscle, aldolase B in the liver, and aldolase C in the brain. All contain four polypeptide subunits which differ in amino acid composition.

The fructose diphosphate aldolases found in bacteria, yeasts, and fungi (class II aldolases) differ from class I forms in containing a specific divalent metal ion, usually Zn^{2+}, Ca^{2+}, or Fe^{2+}; they also require K^+. Their molecular weight is about 65,000, or less than one-half that of the animal enzymes. They function by a different mechanism than class I aldolases; they do not form Schiff's base intermediates.

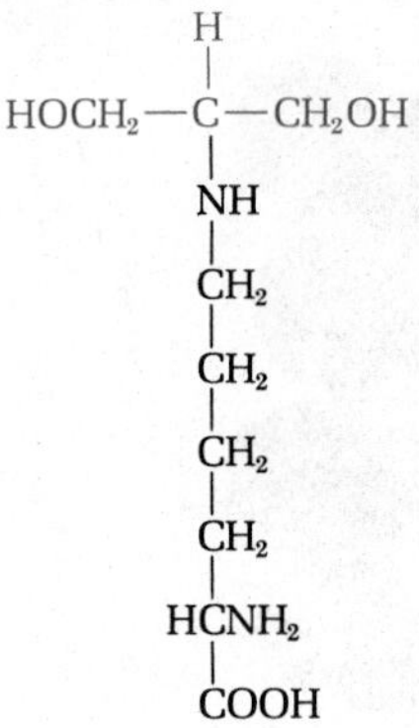

The Interconversion of the Triose Phosphates

Only one of the two triose phosphates, namely, glyceraldehyde 3-phosphate, can be directly degraded in the further reactions of glycolysis. However, the other, dihydroxyacetone phosphate, is reversibly converted into glyceraldehyde 3-phosphate by the enzyme triosephosphate isomerase:

$$\text{Dihydroxyacetone phosphate} \rightleftharpoons \text{D-glyceraldehyde 3-phosphate}$$
$$\Delta G^{\circ\prime} = +1.83 \text{ kcal mol}^{-1}$$

Note that by this reaction carbon atoms 1 and 6 of the starting glucose now become carbon 3 of D-glyceraldehyde 3-phosphate. Similarly, carbons 2 and 5 of glucose become carbon 2, and carbons 3 and 4 become carbon 1 of glyceraldehyde 3-phosphate. Dihydroxyacetone phosphate constitutes over 90 percent of the equilibrium mixture of the two triose phosphates.

This reaction completes the first stage of glycolysis, in which the glucose molecule has been prepared for the second stage by two priming phosphorylation steps and cleavage. The "collecting" reactions, by which glycogen, starch, and sugars other than glucose are fed into the first stage, are described later.

The Second Stage of Glycolysis

This stage (Figure 16-3) includes the oxidoreduction steps, as well as the phosphorylation steps in which ATP is generated

Figure 16-9
Enzyme-substrate compound of aldolase.

The structure of the Schiff's base between dihydroxyacetone phosphate (color) and a specific lysyl residue of the enzyme protein.

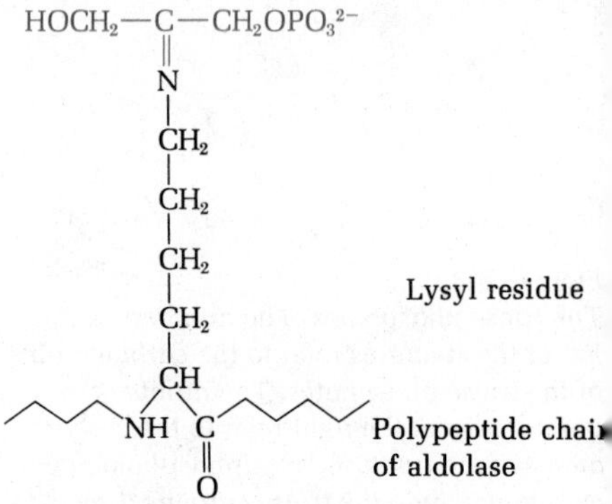

Lysine derivative of dihydroxyacetone (color) isolated from the hydrolyzate of the reduced aldolase-substrate compound.

from ADP. Since one molecule of glucose forms two of glyceraldehyde 3-phospate, both halves of the glucose molecule follow the same pathway.

Oxidation of Glyceraldehyde 3-Phosphate to 3-Phosphoglyceroyl Phosphate

This is one of the most important steps of the glycolytic sequence, since it conserves the energy of oxidation of the aldehyde group of glyceraldehyde 3-phosphate in its oxidation product 3-phosphoglycerol phosphate (page 404). The elucidation of the pathway of this and the following reaction by Warburg and his colleagues in 1938–1939 is considered one of the most important achievements in modern biology since it was the first demonstration of an enzymatic and chemical mechanism by which energy yielded by the oxidation of an organic molecule could be conserved in the form of ATP.

At this point we may interject some clarification of what may appear to be inconsistencies of nomenclature between the various phosphorylated three-carbon intermediates of glycolysis. Although we have used the official name *glyceraldehyde 3-phosphate* up to this point, it may be helpful in the following discussion to use the alternative and equally correct name *3-phosphoglyceraldehyde* because it emphasizes that the ester linkage between phosphate and carbon atom 3 of 3-phosphoglyceraldehyde remains unchanged in the next two intermediates of glycolysis, namely 3-phosphoglyceroyl phosphate (in which the other phosphate is on carbon atom 1) and 3-phosphoglycerate.

The enzyme catalyzing the oxidation of 3-phosphoglyceraldehyde, usually called *glyceraldehydephosphate dehydrogenase* (3-phosphoglyceraldehyde dehydrogenase is also correct), is easily isolated in crystalline form (mol wt 140,000) from rabbit muscle or yeast. It contains four identical subunits, each consisting of a single polypeptide chain of some 330 residues, the amino acid sequence of which has been deduced. The overall reaction catalyzed by the enzyme is

$$\text{D-3-Phosphoglyceraldehyde} + \text{NAD}^+ + P_i \rightleftharpoons$$
$$\text{D-3-phosphoglyceroyl phosphate} + \text{NADH} + \text{H}^+$$
$$\Delta G^{\circ\prime} = +1.5 \text{ kcal mol}^{-1}$$

In this reaction the aldehyde group of D-3-phosphoglyceraldehyde is oxidized to the oxidation level of a carboxyl group. However, instead of a free carboxylic acid, the reaction yields a mixed anhydride of the carboxyl group of 3-phosphoglyceric acid and phosphoric acid, namely, *3-phosphoglyceroyl phosphate* (Figure 16-10), which, as we have seen (page 398), is a high-energy phosphate compound having a more negative standard free energy of hydrolysis than ATP.

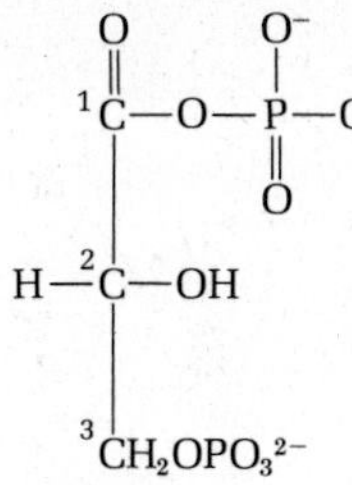

The other important component of this reaction is NAD^+, the oxidized form of nicotinamide adenine dinucleotide, which accepts electrons from the aldehyde group of D-3-

Figure 16-10
Structure of 3-phosphoglyceroyl phosphate.

427

phosphoglyceraldehyde. The structure, properties, and reactions of NAD are more fully described elsewhere (pages 340 and 481). NAD is now known to be one of the components of cozymase, the heat-stable fraction required for alcoholic fermentation found in the early experiments of Harden and Young (page 420). NAD serves as a carrier of electrons from the electron donor D-3-phosphoglyceraldehyde to pyruvate, which is formed later in the glycolytic sequence.

The overall reaction has a small positive value of $\Delta G^{\circ\prime}$ and thus proceeds readily in either direction depending on the concentration of the reactants and products. The forward reaction can now be broken down into two separate processes for analysis of the energy changes, using RCHO to designate 3-phosphoglyceraldehyde; RCOOH, 3-phosphoglyceric acid; and $RCOOPO_3H_2$, 3-phosphoglyceroyl phosphate:

Exergonic partial reaction:

$$RCHO + H_2O + NAD^+ \rightleftharpoons RCOOH + NADH + H^+$$
$$\Delta G_1^{\circ\prime} = -10.3 \text{ kcal mol}^{-1}$$

Endergonic partial reaction:

$$RCOOH + H_3PO_4 \rightleftharpoons RCOOPO_3H_2 + H_2O$$
$$\Delta G_2^{\circ\prime} = +11.8 \text{ kcal mol}^{-1}$$

Sum:

$$RCHO + H_3PO_4 + NAD^+ \rightleftharpoons RCOOPO_3H_2 + NADH + H^+$$
$$\Delta G_s^{\circ\prime} = \Delta G_1^{\circ\prime} + \Delta G_2^{\circ\prime}$$
$$= -10.3 + 11.8$$
$$= +1.5 \text{ kcal mol}^{-1}$$

The oxidation of RCHO by NAD^+ is a highly exergonic process that would normally proceed far in the direction of completion as written, starting from $1\ M$ concentrations of reactants, whereas the formation of 3-phosphoglyceroyl phosphate from 3-phosphoglycerate and phosphate is highly endergonic and would not proceed as written. However, in the overall enzymatic reaction, the endergonic process is obligatorily coupled to the exergonic process, so that the energy released on oxidation of the aldehyde is conserved in the form of the high-energy acyl phosphate group of 3-phosphoglyceroyl phosphate.

The mechanism of this important oxidoreduction has been studied in detail (Figure 16-11). Each of the enzyme's four identical subunits contains an active catalytic site to which is bound a molecule of NAD^+. 3-Phosphoglyceraldehyde dehydrogenase is inhibited by heavy metals as well as by alkylating agents, such as iodoacetate. For this and other reasons, it has been concluded that a sulfhydryl group in the active site is essential for catalytic activity. The enzyme binds the oxidized form of the coenzyme NAD^+ first, in a reaction in which the essential sulfhydryl group becomes sterically masked (Figure 16-11). In the next step the aldehyde group of the substrate forms a _thiohemiacetal_ linkage with the

Figure 16-11
Postulated mechanism of action of D-glyceraldehyde 3-phosphate dehydrogenase. This enzyme catalyzes a 3-substrate reaction, in which the substrates must bind to the active site in a compulsory ordered sequence as shown. The bound NADH formed is displaced from the active site by free NAD^+ from the medium.

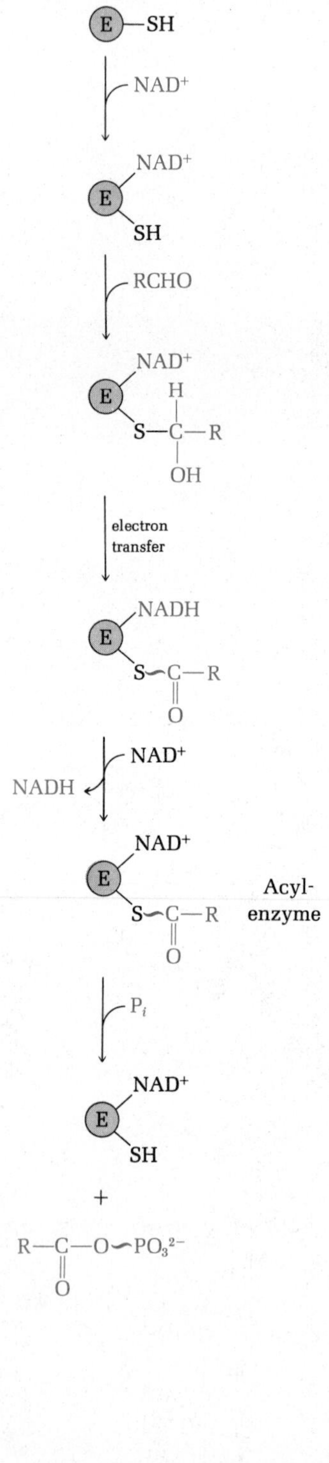

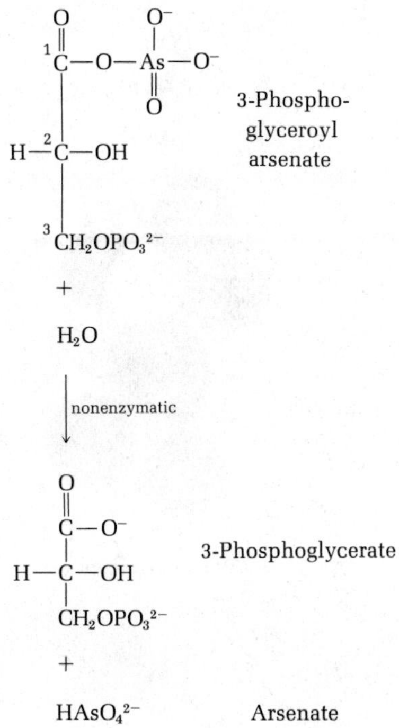

Figure 16-12
Uncoupling action of arsenate.

sulfhydryl group. The enzyme then catalyzes hydrogen transfer from the covalently bound 3-phosphoglyceraldehyde to the bound NAD^+, forming a thioester between the enzyme sulfhydryl group and the carboxyl group of the substrate; this form of the enzyme is called the *acyl-enzyme*. The NADH then leaves the enzyme active site in exchange for a molecule of free NAD^+ from the medium. The acyl group is then transferred from the sulfhydryl group of the enzyme to inorganic phosphate to form 3-phosphoglyceroyl phosphate, the oxidation product. The free oxidized form of the enzyme is now ready for another catalytic cycle.

The enzyme requires NAD^+ specifically as oxidant. Although it is most active with D-3-phosphoglyceraldehyde, it also oxidizes D- and L-glyceraldehyde and even acetaldehyde, but at very low rates. The enzyme can also utilize arsenate instead of phosphate, presumably forming *3-phosphoglyceroyl arsenate* (Figure 16-12), a highly unstable compound that immediately and spontaneously decomposes into 3-phosphoglycerate and arsenate in aqueous systems. Note that in the presence of arsenate, no high-energy phosphate compound is generated by the dehydrogenase, although the overall oxidoreduction takes place. In this way arsenate can uncouple oxidation from phosphorylation.

Glyceraldehyde 3-phosphate dehydrogenase is an allosteric enzyme; its major effector is NAD^+, which is also one of its substrates. The enzyme has four binding sites for NAD^+. Binding of the first molecule of NAD^+ diminishes the affinity of the other subunits for NAD^+ but enhances their intrinsic activity. This is an example of an allosteric enzyme showing negative cooperativity (page 236.)

Transfer of Phosphate from 3-Phosphoglyceroyl Phosphate to ADP

Warburg and his colleagues showed that 3-phosphoglyceroyl phosphate formed in the preceding reaction now reacts enzymatically with ADP, with transfer of the acyl phosphate group to ADP and formation of 3-phosphoglycerate (Figure 16-13), catalyzed by *phosphoglycerate kinase*:

3-Phosphoglyceroyl phosphate + ADP $\rightleftharpoons$
$$3\text{-phosphoglycerate} + \text{ATP}$$
$$\Delta G^{\circ\prime} = -4.50 \text{ kcal mol}^{-1}$$

This reaction is highly exergonic and serves to "pull" the preceding reaction toward completion. The phosphate-transferring enzyme has an extremely high affinity for 3-phosphoglyceroyl phosphate. The overall equation for the two reactions, the first involving oxidation of glyceraldehyde 3-phosphate to 3-phosphoglyceroyl phosphate by the action of 3-phosphoglyceraldehyde dehydrogenase and the second involving transfer of the acyl phosphate group to ADP catalyzed by phosphoglycerate kinase, is

Glyceraldehyde 3-phosphate + P_i + ADP + NAD^+ $\rightleftharpoons$
$$3\text{-phosphoglycerate} + \text{ATP} + \text{NADH} + \text{H}^+$$
$$\Delta G^{\circ\prime} = -3.0 \text{ kcal mol}^{-1}$$

Through these two consecutive reactions the energy of oxidation of an aldehyde group to a carboxylate group has been conserved in the form of ATP.

Conversion of 3-Phosphoglycerate to 2-Phosphoglycerate

This reaction is catalyzed by the enzyme phosphoglyceromutase:

$$\text{3-Phosphoglycerate} \rightleftharpoons \text{2-phosphoglycerate}$$
$$\Delta G^{\circ\prime} = +1.06 \text{ kcal mol}^{-1}$$

Mg^{2+} is essential for this reaction, which involves transfer of the phosphate group from the 3 to the 2 position of glyceric acid (Figure 16-13). The reaction has only a small standard-free-energy change and is freely reversible in the cell. There are two forms of this enzyme; the form in animal tissues appears to require 2,3-diphosphoglycerate as an intermediate (see page 435), according to the equation

$$\text{2,3-Diphosphoglycerate} + \text{3-phosphoglycerate} \rightleftharpoons$$
$$\text{2-phosphoglycerate} + \text{2,3-diphosphoglycerate}$$

Dehydration of 2-Phosphoglycerate to Phosphoenolpyruvate

The conversion of 2-phosphoglycerate to phosphoenolpyruvate (page 404) is the second reaction of the glycolytic sequence in which a high-energy phosphate compound is generated (Figure 16-14). It is catalyzed by enolase:

$$\text{2-Phosphoglycerate} \rightleftharpoons \text{phosphoenolpyruvate} + H_2O$$
$$\Delta G^{\circ\prime} = +0.44 \text{ kcal mol}^{-1}$$

Enolase has been obtained in pure crystalline form from several sources (mol wt 85,000). It has an absolute requirement for a divalent cation (Mg^{2+} or Mn^{2+}), which makes a complex with the enzyme before the substrate is bound. The enzyme is strongly inhibited by fluoride, particularly if phosphate is present, the inhibitory species being the phosphofluoridate ion, which forms a complex with Mg^{2+}. Although the reaction catalyzed by enolase is formally an elimination of a molecule of water from carbon atoms 2 and 3 of 2-phosphoglycerate, it may also be regarded as an intramolecular oxidoreduction, since the removal of water causes carbon atom 2 to become more oxidized and carbon atom 3 more reduced. Despite the relatively small standard-free-energy change in this reaction, there is a very large change in the standard free energy of hydrolysis of the phosphate group of the reactant and product, that of 2-phosphoglycerate being about -4.2 kcal mol^{-1} and that of phosphoenolpyruvate about -14.8 kcal mol^{-1}. Evidently there is a large change in the distribution of energy within the 2-phosphoglycerate molecule when it is dehydrated to phosphoenolpyruvate (page 405).

Figure 16-13
Structures and space-filling models of phosphorylated forms of D-glycerate.

$$\begin{array}{c} COO^- \\ | \\ H-C-OH \\ | \\ CH_2OPO_3^{2-} \end{array}$$

3-Phosphoglycerate

$$\begin{array}{c} COO^- \\ | \\ H-C-OPO_3^{2-} \\ | \\ CH_2OH \end{array}$$

2-Phosphoglycerate

$$\begin{array}{c} COO^- \\ | \\ H-C-OPO_3^{2-} \\ | \\ CH_2OPO_3^{2-} \end{array}$$

2,3-Diphosphoglycerate

Figure 16-14
Structure and space-filling model of phosphoenolpyruvate.

$$
\begin{array}{c}
\text{COO}^- \\
|\\
\text{C}=\text{O} \\
|\\
\text{CH}_3
\end{array}
$$

Pyruvate

$$
\begin{array}{c}
\text{COO}^- \\
|\\
\text{HO}-\text{C}-\text{H} \\
|\\
\text{CH}_3
\end{array}
$$

L-Lactate

Transfer of Phosphate from Phosphoenolpyruvate to ADP

The transfer of the phosphate group from phosphoenolpyruvate to ADP, yielding free pyruvate (Figure 16-15), is catalyzed by the enzyme *pyruvate kinase*,

$$\text{Phosphoenolpyruvate} + \text{ADP} \longrightarrow \text{pyruvate} + \text{ATP}$$
$$\Delta G^{\circ\prime} = -7.5 \text{ kcal mol}^{-1}$$

which has been obtained in pure crystalline form (mol wt 250,000). The reaction is highly exergonic and it has been found to be irreversible under intracellular conditions. The enzyme requires Mg^{2+} or Mn^{2+}, with which it must form a complex before binding the substrate. Ca^{2+} competes with Mn^{2+} or Mg^{2+} and forms an inactive complex. The enzyme also requires an alkali-metal cation, which may be K^+, Rb^+, or Cs^+; K^+ is the physiological activator. It is believed that the binding of K^+ causes a conformational change of the enzyme to produce a more active form. Pyruvate kinase in mammals is a regulatory enzyme and occurs in different forms in various tissues. The L, or liver, form is activated by fructose 1,6-diphosphate and by high concentrations of phosphoenolpyruvate but is inhibited by ATP, AMP, citrate, and alanine. It is also inhibited by long-chain fatty acids and acetyl-CoA. The M, or muscle, form is not activated by fructose 1,6-diphosphate but is inhibited by phenylalanine. Like hexokinase (page 423) and 6-phosphofructokinase (page 424), pyruvate kinase "turns off" whenever the ATP concentration in the cell is relatively high or when other fuels, such as fatty acids, citrate, acetyl-CoA, or alanine, are available. It "turns on" whenever there is a buildup of the preceding glycolytic intermediates, particularly fructose 1,6-diphosphate and phosphoenolpyruvate.

Reduction of Pyruvate to Lactate

In the last step of glycolysis, pyruvate is reduced to lactate (Figure 16-15) at the expense of electrons originally donated by 3-phosphoglyceraldehyde. These electrons are carried by NADH. The reaction is catalyzed by *lactate dehydrogenase:*

$$\text{Pyruvate} + \text{NADH} + \text{H}^+ \rightleftharpoons \text{lactate} + \text{NAD}^+$$
$$\Delta G^{\circ\prime} = -6.0 \text{ kcal mol}^{-1}$$

The overall equilibrium of this reaction is far to the right, as suggested by the large negative value of $\Delta G^{\circ\prime}$. Lactate dehydrogenase exists in at least five different molecular forms, or isozymes, in higher animals (page 244), which differ in the rate at which they bring about reduction of pyruvate at low pyruvate concentrations. Their regulatory role in glycolysis is described in detail elsewhere (pages 244 and 630). This reaction completes the internal oxidoreduction cycle of glycolysis.

Lactate, the end product of the glycolytic sequence under anaerobic conditions, diffuses through the cell membrane to the surroundings as waste. When muscles of higher animals must function anaerobically during short bursts of exceptionally vigorous activity, the lactate that escapes from the muscle cells into the blood is salvaged by the liver and rebuilt to form blood glucose (pages 624 and 767).

The Overall Balance Sheet

A balance sheet for glycolysis can now be constructed to account for the fate of the carbon skeleton of glucose, the oxidoreduction reactions, and the input and output of phosphate, ADP, and ATP. The left-hand part of the following equation shows all the inputs of the glycolytic sequence and the right, all the outputs, adjusted for the fact that each molecule of glucose yields two molecules of glyceraldehyde 3-phosphate:

Glucose + 2ATP + 2NAD$^+$ + 2P$_i$ + 4ADP + 2NADH + 2H$^+$ $\longrightarrow$
2 lactate + 2ADP + 2NADH + 2H$^+$ + 4ATP + 2NAD$^+$ + 2H$_2$O

By canceling out common terms on both sides of the equation we get

Glucose + 2P$_i$ + 2ADP $\longrightarrow$ 2 lactate + 2ATP + 2H$_2$O

In the overall process D-glucose is converted to two molecules of lactate, two molecules of ADP and phosphate are converted to ATP, and four electrons have been transferred from glyceraldehyde 3-phosphate to pyruvate via 2NADH + 2H$^+$. Two molecules of ATP must be fed into the scheme to prime it (page 421), and four molecules of ATP are produced, resulting in a net yield of two ATPs per molecule of glucose transformed.

Energetics of Glycolysis in the Intact Cell

Much interest attaches to the energetics of glycolysis as it occurs in intact cells. Such an analysis has been accomplished for human red blood cells, which differ from most cells in the body in obtaining all their energy from glycolysis. The actual steady-state concentrations of all the intermediates of glycolysis in red blood cells have been measured (Table 16-1). From these values and the known equilibrium constants for the glycolytic reactions, the free-energy change (ΔG, not $\Delta G^{\circ\prime}$) of each step in the intact erythrocyte has been calculated from the general equation (page 391)

$$\Delta G = \Delta G^{\circ\prime} + RT \ln\frac{[C]^c [D]^d}{[A]^a [B]^b}$$

adjusted to take appropriate account of the stoichiometry of the reaction. From these results a free-energy profile of glycolysis in the intact erythrocyte in the steady state has been constructed (Figure 16-16). It shows that eight of the reactions of glycolysis are at or very close to equilibrium, since their ΔG values are close to zero whereas three reactions occur with large decreases in ΔG and thus are far from equi-

Table 16-1 Steady-state concentrations of the intermediates of glycolysis in the human erythrocyte

Intermediate	Concentration, μM
Glucose	5,000
Glucose 6-phosphate (G6P)	83
Fructose 6-phosphate (F6P)	14
Fructose 1,6-diphosphate (FDP)	31
Dihydroxyacetone phosphate (DHP)	138
Glyceraldehyde 3-phosphate (GAP)	18.5
3-Phosphoglycerate (3PG)	118
2-Phosphoglycerate (2PG)	29.5
Phosphoenolpyruvate (PEP)	23
Pyruvate (Pyr)	51
Lactate (Lact)	2,900
ATP	1,850
ADP	138
Phosphate	1,000

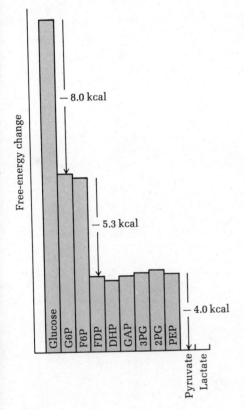

Figure 16-16
Free-energy profile of glycolysis in the
human erythrocyte. All the reactions are at
or near equilibrium except those catalyzed
by hexokinase, phosphofructokinase, and
pyruvate kinase, at which large decreases in
ΔG occur. In the cell, all of the steps of
glycolysis must proceed with either no
change or a decline in free energy. The slight
increases in free energy depicted here for
several of the steps of glycolysis must be
regarded as deriving from errors in experi-
mental measurement.

librium, namely, the hexokinase, phosphofructokinase, and pyruvate kinase reactions. From these data it has also been calculated that the free energy of hydrolysis of ATP in the erythrocyte is about -13.3 kcal mol^{-1}; the actual efficiency of energy recovery in erythrocyte glycolysis is about 53 percent, or much greater than the theoretical efficiency of 31 percent calculated from standard-free-energy data (page 419). In muscle the actual efficiency appears to be even higher.

There is now good evidence that the phosphofructokinase reaction is the major rate-limiting reaction in glycolysis. But there are also at least two secondary control points that may become rate-limiting under special circumstances, the reactions catalyzed by hexokinase (page 423) and pyruvate kinase (page 431), which are allosteric enzymes. Glyceraldehyde 3-phosphate also appears to have the characteristics of a regulatory enzyme but its role in the regulation of glycolysis in vivo is not clear. The regulation of the rate of glucose utilization is discussed further in Chapters 19, 29, and 30.

Entry of Other Carbohydrates into the Glycolytic Sequence

The storage polysaccharides glycogen and starch and simple sugars other than D-glucose are channeled into the first stage of glycolysis by feeder pathways catalyzed by auxiliary enzymes whose action will now be described.

Glycogen and Starch

The D-glucose units of glycogen and starch gain entrance into the glycolytic sequence through the sequential action of two enzymes, *glycogen phosphorylase* (or *starch phosphorylase* in plants) and *phosphoglucomutase*. Glycogen phosphorylase and starch phosphorylase are members of a general class of enzymes designated as $\alpha(1 \rightarrow 4)$-glucan phosphorylases. Widely distributed in animal, plant, and microbial cells, they catalyze the general reaction shown, in which (glucose)$_n$ designates the glucan chain and (glucose)$_{n-1}$ the shortened glucan chain:

$$\text{(Glucose)}_n + \text{HPO}_4^{2-} \rightleftharpoons \text{(glucose)}_{n-1} + \text{glucose 1-phosphate}$$
$$\Delta G^{\circ\prime} = +0.73 \text{ kcal mol}^{-1}$$

In this reaction the terminal $\alpha(1 \rightarrow 4)$ glycosidic linkage at the nonreducing end of a glycogen side chain undergoes *phosphorolysis*. Just as hydrolysis brings about cleavage of a molecule by introducing the elements of water, phosphorolysis brings about cleavage of a molecule by introducing the components of phosphoric acid. The cleavage of the terminal glycosidic bond results in the removal of the terminal glucose as glucose 1-phosphate, leaving behind a glycogen chain with one less glucose unit (Figure 16-17). The enzyme acts repetitively on the nonreducing ends of glycogen chains until it meets the $\alpha(1 \rightarrow 6)$ branch points, which it cannot attack. Exhaustive action of glycogen phosphorylase can thus produce a *limit dextrin* (page 265), which may be further degraded by glycogen phosphorylase after the action of a

Figure 16-17
Phosphorolytic removal of a glucose residue (color) from the nonreducing end of a glycogen chain by phosphorylase.

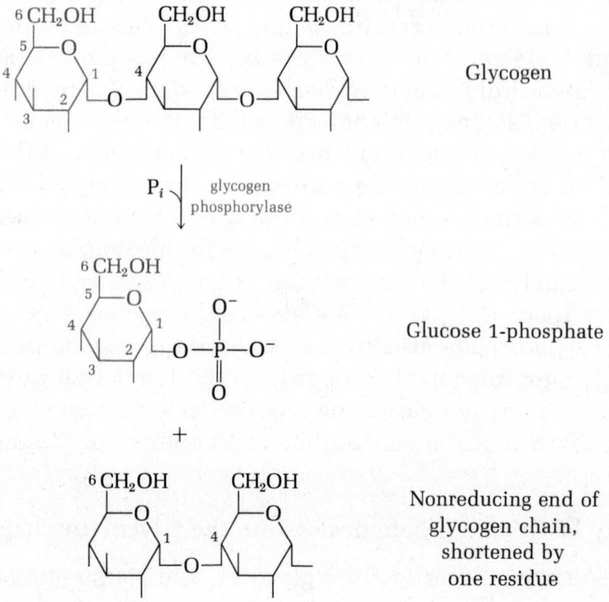

debranching enzyme, a hydrolytic *amylo-1,6-glucosidase* (page 265), which hydrolyzes the $1 \rightarrow 6$ linkage at the branch point, thus making another length of the polysaccharide chain available to the action of glycogen phosphorylase.

Although glycogen phosphorylase, discovered and studied in detail by Cori and Cori and their colleagues, seemed at first to be responsible for both formation and breakdown of glycogen in the cell, it has since been found that its primary function is to catalyze the breakdown of glycogen. Under intracellular conditions, in which the inorganic phosphate concentration is relatively high and the glucose 1-phosphate concentration relatively low, the phosphorylase equilibrium greatly favors the formation of glucose 1-phosphate. A different enzyme, *glycogen synthase* (page 645), is responsible for formation of glycogen from glucose phosphate units.

Phosphorylase is situated at a strategically important point between the fuel reservoir (glycogen or starch) and the glycolytic sequence that utilizes the fuel. Its activity in muscle and liver is under the regulation of an elaborate set of controls. The glycogen phosphorylase of skeletal muscle occurs in two forms, the active form (phosphorylase *a*) and a much less active form (phosphorylase *b*) (Chapter 9, page 242), both of which have been crystallized. Phosphorylase *a* has a molecular weight of 380,000 and consists of four identical subunits. Each subunit contains a phosphoserine residue that is essential for catalytic activity and a molecule of pyridoxal phosphate (page 344), which is covalently bound to a lysine residue. The function of the pyridoxal phosphate is still unknown. The active form of phosphorylase *a* can be attacked by the hydrolytic enzyme *phosphorylase phospha-*

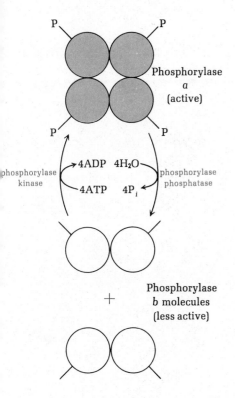

Figure 16-18
Conversion of phosphorylase a to phosphorylase b by phosphorylase phosphatase and reactivation of phosphorylase b by phosphorylase kinase.

Phosphorylase *a* (active)

phosphorylase kinase   4ADP   4H$_2$O   phosphorylase phosphatase

4ATP   4P$_i$

+

Phosphorylase *b* molecules (less active)

Figure 16-19
Diagrammatic representation of the role of glucose 1,6-diphosphate in the phosphoglucomutase reaction.

E—P + [1 P, 6] ⇌ E + [1 P, 6 P]

E + [1 P, 6 P] ⇌ E—P + [1, 6 P]

tase, which removes the phosphate groups from the serine phosphate residues (Figure 16-18). This reaction causes phosphorylase *a* to dissociate into two molecules of phosphorylase *b*, the less active form. Phosphorylase *b* is converted back to active phosphorylase *a*, not by simple reversal of the above reaction, which is irreversible, but by an alternate pathway in which four molecules of ATP act on two molecules of phosphorylase *b* in the presence of the enzyme *phosphorylase kinase* (page 242). By the action of phosphorylase phosphatase and phosphorylase kinase the ratio of the active phosphorylase *a* to the less active phosphorylase *b* in the cell can be controlled, thus varying the rate of conversion of glycogen into glucose 1-phosphate. Actually, the activity of phosphorylase kinase is itself under regulation, since it in turn occurs in active and inactive forms whose ratio is controlled by certain hormones (pages 647 and 811).

Liver also contains active and inactive forms of phosphorylase, the ratio of which can also be controlled by the rate of the interconversion between the active and inactive forms. However, in the liver these enzymes have a distinctly different architecture from those in muscle, and they are controlled in a different fashion, in harmony with the differences in the dynamics of glucose metabolism in liver and muscle (page 813).

Glucose 1-phosphate, the end product of the glycogen and starch phosphorylase reactions, is converted into glucose 6-phosphate by the interesting enzyme *phosphoglucomutase*, which has been obtained in pure form from many sources. It catalyzes the readily reversible reaction

$$\text{Glucose 1-phosphate} \rightleftharpoons \text{glucose 6-phosphate}$$
$$\Delta G^{\circ\prime} = -1.74 \text{ kcal mol}^{-1}$$

Although it can also catalyze the conversion of D-mannose 1-phosphate to D-mannose 6-phosphate, the rate of this reaction is only one-hundredth that of the conversion of D-glucose 1-phosphate.

Phosphoglucomutase contains a serine residue that is essential for catalytic activity; during the catalytic cycle the hydroxyl group of serine becomes esterified with phosphoric acid, which transits between the 1 and 6 positions of glucose (see below). The hydroxyl group of the serine residue can also be esterified by the inhibitor *diisopropyl phosphofluoridate* (page 220) to yield an inactive form of the enzyme. Phosphoglucomutase is a member of the serine class of enzymes (page 220). It requires not only Mg^{2+} but also an organic cofactor; *glucose 1,6-diphosphate*. Much evidence suggests that the latter is an intermediate in the action of the enzyme, as indicated by the following sequence of steps and by Figure 16-19:

Phosphoenzyme + glucose 1-phosphate ⇌
  dephosphoenzyme + glucose 1,6-diphosphate

Dephosphoenzyme + glucose 1,6-diphosphate ⇌
  phosphoenzyme + glucose 6-phosphate

Sum:

$$\text{Glucose 1-phosphate} \rightleftharpoons \text{glucose 6-phosphate}$$

The role of glucose 1,6-diphosphate in the phosphoglucomutase reaction is thus similar to the role of 2,3-diphosphoglycerate in the phosphoglyceromutase reaction (page 430).

Glucose 1,6-diphosphate can also be generated by another reaction,

$$\text{Glucose 1-phosphate} + \text{ATP} \longrightarrow \text{glucose 1,6-diphosphate} + \text{ADP}$$

catalyzed by the enzyme *phosphoglucokinase*. Glucose 6-phosphate formed as the product of the phosphoglucomutase reaction now may enter the glycolytic cycle.

Entry of Disaccharides

Disaccharides ingested by higher animals are usually hydrolyzed to their monosaccharide components before absorption in the intestine. The most important reactions are

$$\text{Sucrose} + H_2O \xrightarrow{\beta\text{-fructofuranosidase}} \text{D-glucose} + \text{D-fructose}$$

$$\text{Maltose} + H_2O \xrightarrow{\alpha\text{-glucosidase}} 2\text{D-glucose}$$

$$\text{Lactose} + H_2O \xrightarrow{\beta\text{-galactosidase}} \text{D-glucose} + \text{D-galactose}$$

The enzyme lactase, a β-galactosidase secreted into the small intestine, is low in activity or lacking in many Arabian, Jewish, Bantu, Japanese, American Indian, and Filipino adults. Such individuals fail to absorb lactose normally, a condition known as *lactose intolerance*. Monosaccharides other than glucose that are formed in these reactions gain entry into glycolysis by reactions described below.

Entry of Monosaccharides Other Than Glucose

In the liver of vertebrates, fructose (page 251) gains entry into glycolysis by the action of *fructokinase*, which catalyzes the phosphorylation of fructose at carbon atom 1:

$$\text{D-Fructose} + \text{ATP} \longrightarrow \text{D-fructose 1-phosphate} + \text{ADP}$$

The resulting fructose 1-phosphate is then cleaved into D-glyceraldehyde and dihydroxyacetone phosphate,

Fructose 1-phosphate $\longrightarrow$
$$\text{D-glyceraldehyde} + \text{dihydroxyacetone phosphate}$$

by aldolase. The free D-glyceraldehyde so formed is phosphorylated to glyceraldehyde 3-phosphate in the reaction

D-Glyceraldehyde + ATP $\longrightarrow$
$$\text{D-glyceraldehyde 3-phosphate} + \text{ADP}$$

Figure 16-20

Epimerization of D-*galactose* 1-*phosphate.*

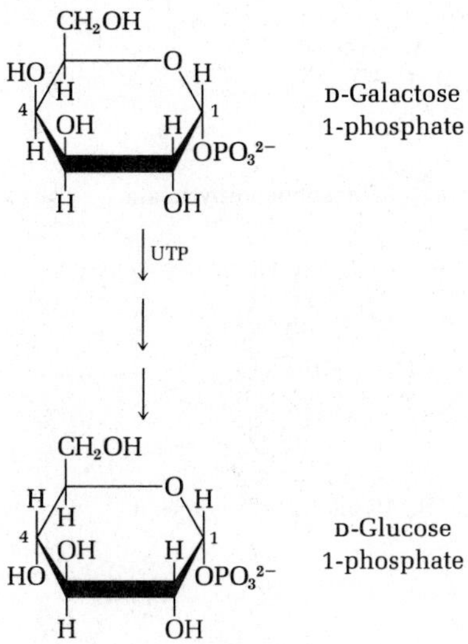

The dihydroxyacetone phosphate and glyceraldehyde 3-phosphate so formed are of course intermediates in glycolysis and are ultimately degraded to pyruvate.

The hexose D-galactose (page 250), a component of milk sugar (D-lactose), enters the glycolytic cycle following phosphorylation at the expense of ATP by *galactokinase*:

$$\text{ATP} + \text{D-galactose} \longrightarrow \text{ADP} + \text{D-galactose 1-phosphate}$$

D-Galactose 1-phosphate is now converted into its epimer at carbon atom 4, namely, D-glucose 1-phosphate (Figure 16-20), through a sequence of reactions requiring as a coenzyme uridine triphosphate (UTP). The details of this and other UTP-requiring sugar transformations are described elsewhere (pages 640 to 651).

D-Mannose (page 250) is phosphorylated at the 6 position by hexokinase, which is relatively nonspecific:

$$\text{D-Mannose} + \text{ATP} \longrightarrow \text{D-mannose 6-phosphate} + \text{ADP}$$

D-Mannose 6-phosphate is reversibly isomerized into D-fructose 6-phosphate by the action of *mannose phosphate isomerase*:

$$\text{D-Mannose 6-phosphate} \rightleftharpoons \text{D-fructose 6-phosphate}$$

Pentoses may also enter the glycolytic cycle, following phosphorylation and conversion to hexose and triose phosphates by mechanisms described on pages 467 to 472. Glycerol and L-glycerol 3-phosphate, derived from triacylglycerols and phosphoglycerides, respectively (pages 284 and 287), may also be utilized. Free glycerol is phosphorylated at the expense of ATP by *glycerol kinase*:

$$\text{ATP} + \text{glycerol} \longrightarrow \text{ADP} + \text{L-glycerol 3-phosphate}$$

L-Glycerol 3-phosphate then undergoes oxidation to dihydroxyacetone phosphate, either by *cytoplasmic glycerol 3-phosphate dehydrogenase*, an enzyme that requires NAD^+ as electron acceptor,

$$\text{L-Glycerol 3-phosphate} + NAD^+ \rightleftharpoons$$
$$\text{dihydroxyacetone phosphate} + \text{NADH} + H^+$$

or by *mitochondrial glycerol 3-phosphate dehydrogenase*, a flavoprotein (page 535). The dihydroxyacetone phosphate formed in these reactions may then be enzymatically converted into glyceraldehyde 3-phosphate to enter the second stage of glycolysis.

Alcoholic Fermentation

In organisms like brewers' yeast, which ferment glucose to ethanol and CO_2 rather than to lactic acid, the fermentation pathway is identical to that described for glycolysis except for the terminal step catalyzed by lactate dehydrogenase, which is replaced by two other enzymatic steps (Figure 16-21).

In the first step, pyruvate is decarboxylated to acetaldehyde and CO_2 by the enzyme *pyruvate decarboxylase*, which is not present in animal tissues:

$$Pyruvate \longrightarrow acetaldehyde + CO_2$$

The decarboxylation of pyruvate to form acetaldehyde and CO_2 is essentially irreversible. Pyruvate decarboxylase requires Mg^{2+} and has a tightly bound coenzyme, *thiamin pyrophosphate* (page 337). The decarboxylation of pyruvate proceeds through a series of intermediates covalently bound to the thiamin pyrophosphate (Figure 16-22).

In the final step of alcoholic fermentation, acetaldehyde is reduced to ethanol, with $NADH + H^+$ furnishing the reducing power, through the enzyme *alcohol dehydrogenase*:

$$Acetaldehyde + NADH + H^+ \rightleftharpoons ethanol + NAD^+$$

Ethanol and CO_2 are thus the end products of alcoholic fermentation. The overall equation of alcoholic fermentation can therefore be written

$$Glucose + 2P_i + 2ADP \longrightarrow 2\ ethanol + 2CO_2 + 2ATP + 2H_2O$$

The energy-conserving steps leading to ATP formation are identical in both glycolysis and alcoholic fermentation.

Other Types of Anaerobic Fermentation

Homolactic and alcoholic fermentations are the simplest common fermentation mechanisms. Other pathways are known, some of which are variations of the Embden-Meyerhof scheme. In *heterolactic*, or mixed, lactic fermentations, which are predominantly found in microorganisms, one molecule each of lactic acid, ethanol, and carbon dioxide constitute the end products. In other types of bacterial fermentations propionic acid, butyric acid, succinic acid, and acetone are end products. Microorganisms are often used in chemical industry to form such products as acetone, ethanol, butanol, and many other important chemicals by fermentation of sugars or other natural carbohydrate sources.

Summary

Anaerobic fermentation is the most ancient pathway for obtaining energy from fuels such as glucose. In anaerobic cells it is the sole energy-producing process. In most facultative cells it is an obligatory first stage in glucose catabolism, which is followed by aerobic oxidation of the fermentation products. The two most common types of fermentation are glycolysis and alcoholic fermentation. Both utilize identical energy-conserving mechanisms and differ only in their terminal steps. The overall equation for glycolysis is glucose $+ 2ADP + 2P_i \rightarrow 2$ lactic acid $+ 2ATP + 2H_2O$ and for alcoholic fermentation it is glucose $+ 2ADP + 2P_i \rightarrow 2$ ethanol $+ 2CO_2 + 2ATP + 2H_2O$. Both fermentations are essentially irreversible.

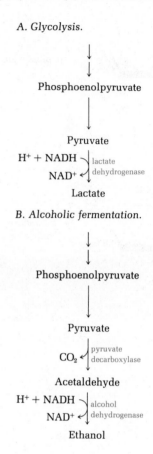

Figure 16-21
Comparison of the concluding steps in (A) glycolysis and (B) alcoholic fermentation.

A. Glycolysis.

Phosphoenolpyruvate

Pyruvate

$H^+ + NADH$ — lactate dehydrogenase
NAD^+

Lactate

B. Alcoholic fermentation.

Phosphoenolpyruvate

Pyruvate

CO_2 — pyruvate decarboxylase

Acetaldehyde

$H^+ + NADH$ — alcohol dehydrogenase
NAD^+

Ethanol

Figure 16-22
Steps in the action of enzyme-bound
thiamin pyrophosphate in decarboxylation
of pyruvate (see page 338).

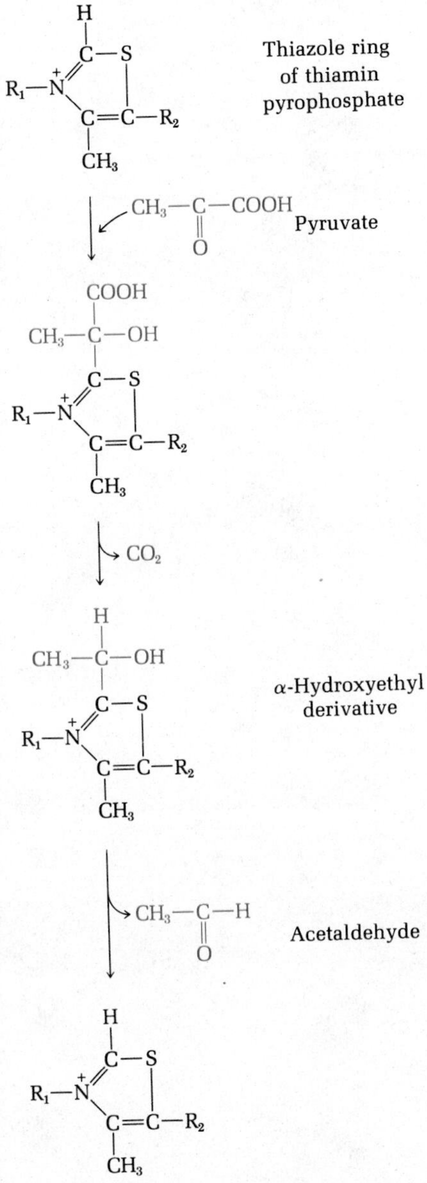

Thiazole ring
of thiamin
pyrophosphate

CH₃—C—COOH Pyruvate

Pyruvate

α-Hydroxyethyl
derivative

CO₂

Acetaldehyde

Glycolysis takes place in two stages. In the first, D-glucose is enzymatically phosphorylated by ATP and ultimately cleaved to yield two molecules of D-glyceraldehyde 3-phosphate. Other hexoses, pentoses, and glycerol are also collected and converted into glyceraldehyde 3-phosphate, following their phosphorylation.

In the second stage of glycolysis, the glyceraldehyde 3-phosphate is oxidized by NAD^+, with uptake of inorganic phosphate, by the action of glyceraldehyde phosphate dehydrogenase, to form 3-phosphoglyceroyl phosphate. The latter donates its acyl phosphate group to ADP to yield ATP and 3-phosphoglycerate, which is then isomerized to 2-phosphoglycerate. After dehydration of the latter by enolase, the phosphoenolpyruvate formed donates its phosphate group to ADP. The other product, free pyruvate, is reduced to lactate by NADH formed in the dehydrogenation of glyceraldehyde 3-phosphate. Two molecules of ATP enter the first stage of glycolysis, and four are formed from ADP in the second stage, giving a net yield of two ATPs from one molecule of glucose. The efficiency of energy recovery by glycolysis in the intact erythrocyte is over 50 percent. There are three essentially irreversible steps in glycolysis, catalyzed by hexokinase, 6-phosphofructokinase, and pyruvate kinase. The reaction catalyzed by phosphofructokinase, a regulatory enzyme, is the major rate-limiting step of glycolysis. The entry of glucose residues of glycogen and starch into glycolysis is made possible by glycogen (starch) phosphorylase and phosphoglucomutase. Glycogen phosphorylase, which catalyzes conversion of glycogen to glucose 1-phosphate, is a regulatory enzyme existing in active (phosphorylase a) and less active (phosphorylase b) forms. Hexoses other than glucose, such as fructose and galactose, are enzymatically converted into intermediates of the glycolytic pathway, as are certain pentoses. In alcoholic fermentation, the reaction sequence is identical up to the stage of pyruvate, but instead of being reduced to lactate, pyruvate is decarboxylated to acetaldehyde, which is then reduced to ethanol.

References

See also references to Chapter 15.

Books

COLOWICK, S. P., and N. O. KAPLAN (eds.): *Carbohydrate Metabolism,* vol. 9 of *Methods in Enzymology,* Academic, New York, 1966. Compendium of experimental methods and approaches.

DICKENS, F., P. J. RANDLE, and W. J. WHELAN: *Carbohydrate Metabolism and Its Disorders,* 2 vols., Academic, New York, 1968. A series of papers on metabolism and control mechanisms.

FLORKIN, M., and E. H. STOTZ (eds.): *Carbohydrate Metabolism,* vol. 17 of *Comprehensive Biochemistry,* American Elsevier, New York, 1967.

NEWSHOLME, E. A., and C. START: "Regulation in Metabolism," Wiley, New York, 1974. An excellent and indispensible account of the regulatory steps of a number of metabolic processes.

Articles

AXELROD, B.: "Glycolysis," pp. 112–145 in D. M. Greenberg (ed.), *Metabolic Pathways,* 3d ed., vol. 1, Academic, New York, 1967.

COLOWICK, S. P.: "The Hexokinases," pp. 1–48 in P. D. Boyer (ed.), *The Enzymes*, 3d ed., vol. 9, Academic, New York, 1973.

EVERSE, J., and N. O. KAPLAN: "Lactate Dehydrogenases: Structure and Function," *Adv. Enzymol.*, 37: 61–134 (1973).

FISCHER, E. H., A. POCKER, and J. C. SAARI: "The Structure, Function, and Control of Glycogen Phosphorylase," *Essays Biochem.*, 6: 23–68 (1970). Excellent and comprehensive article on this complex enzyme.

GRISOLIA, S.: "Phosphoglyceromutases," in N. V. Thoai and J. Roche (eds.), *Homologous Enzymes and Biological Evolution*, Gordon and Breach, New York, 1971.

LAI, C. Y., and B. L. HORECKER: "Aldolase: A Model for Enzyme Structure-Function Relationships," *Essays Biochem.*, 8: 149–178 (1972).

MANSOUR, T.: "Studies on Heart Phosphofructokinase: Purification, Inhibition, and Activation," *J. Biol. Chem.*, 238: 2285–2292 (1963).

MINIKAMI, S., and H. YOSHIKAWA: "Thermodynamic Considerations of Erythrocyte Glycolysis," *Biochem. Biophys. Res. Commun.*, 18: 345–349 (1965). Determination of the steady-state concentrations of intermediates and ΔG of ATP hydrolysis in intact cell.

PASSONNEAU, J. V., and O. H. LOWRY: "Phosphofructokinase and the Pasteur Effect," *Biochem. Biophys. Res. Commun.*, 7: 10–15 (1962). This and the following paper were important in demonstrating the important regulatory role of this enzyme.

PASSONNEAU, J. V., and O. H. LOWRY: "Phosphofructokinase and the Control of the Citric Acid Cycle," *Biochem. Biophys. Res. Commun.*, 13: 372–379 (1963).

PURICH, D. L., H. J. FROMM, and F. B. RUDOLPH: "The Hexokinases," *Adv. Enzymol.*, 39: 249–326 (1973).

ROLLESTON, F. S., and E. A. NEWSHOLME: "Control of Glycolysis in Cerebral Cortex Slices," *Biochem. J.*, 104: 524–533 (1967). Provides valuable data on the concentrations of glycolytic intermediates under a variety of conditions and includes a very useful discussion of possible regulatory steps.

STADTMAN, E. R.: "Allosteric Regulation of Enzyme Activity," *Adv. Enzymol.*, 28: 42–154 (1966). Pages 71 to 117 deal with enzymes of carbohydrate metabolism.

VILLAR-PALASI, C., and J. LARNER: "Glycogen Metabolism and Glycolytic Enzymes," *Ann. Rev. Biochem.*, 39: 639–672 (1970). Comprehensive review and guide to recent research literature.

Problems

1. Predict whether the addition of significant quantities of the compounds specified below to a yeast extract fermenting glucose might result in an increase (+), a decrease (−), or no change (0) in the rate of conversion of glucose to ethanol: (a) iodoacetate, (b) ATP, (c) ADP, (d) AMP, (e) phosphate, (f) fluoride, (g) bisulfite (reacts with aldehydes), (h) citrate, (i) arsenate.

2. Write a balanced equation for the conversion of D-fructose to lactic acid in the liver (a) assuming that the fructose is phosphorylated by hexokinase and (b) assuming that the fructose is

phosphorylated by fructokinase. Include all associated phosphorylation steps in the overall equation.

3. Write balanced equations for the following processes, including all associated phosphorylation steps:
   (a) Conversion of glycerol to lactic acid
   (b) Conversion of L-glycerol 3-phosphate to ethanol and CO_2
   (c) Conversion of D-mannose to phosphoenolpyruvate
   (d) Fermentation of sucrose to ethanol and CO_2
   (e) Fermentation of glyceraldehyde to ethanol and CO_2

4. Write the overall equation for the conversion of fructose 1,6-diphosphate to phosphoenolpyruvate in the presence of NAD^+, phosphate, ADP, and arsenate.

5. A yeast extract containing all the enzymes required in the alcoholic fermentation of glucose is incubated with 200 mM D-glucose, 20 mM ATP, 2.0 mM NAD^+, and 20 mM phosphate. The incubation is continued until the system comes to equilibrium.
   (a) Predict the concentration of glucose and the concentration of ethanol at equilibrium.
   (b) How can the fermentation be made to go to completion so that essentially all the glucose is converted into ethanol?

6. Name the compounds present at equilibrium when pure muscle aldolase acts on a mixture of fructose 1,6-diphosphate, D-glyceraldehyde, and acetaldehyde.

7. (a) Calculate the overall free-energy change when glycolysis occurs under the following set of conditions, which are similar to those existing in the intact cell: glucose = 5 mM, phosphate = 1.0 mM, ADP = 0.5 mM, ATP = 3.0 mM, and lactate = 3.0 mM.
   (b) What will the free-energy change be when the lactate concentration is raised to 100 mM?

8. The concentration of 3-phosphoglycerate in the intracellular fluid of cerebral cortex tissue is 0.283 mM, whereas that of 2-phosphoglycerate is 0.026 mM. Calculate the free-energy change ΔG at 25°C for the phosphoglyceromutase reaction under these conditions.

9. Calculate what percentage of the starting concentration of fructose 1,6-diphosphate is cleaved by pure aldolase at equilibrium, when the initial concentration of the fructose 1,6-diphosphate is (a) 1.0 M, (b) 0.1 M, (c) 0.01 M, (d) 0.001 M, (e) 0.0001 M. Assume that $\Delta G°'$ for the aldolase reaction is +5.73 kcal mol^{-1}.

10. The concentration of fructose 1,6-diphosphate in the intracellular fluid of cerebral cortex tissue is 0.146 mM while the total concentration of the two triose phosphates is 0.0942 mM. Calculate the free-energy change at 25°C for the aldolase reaction under these conditions, assuming the reaction catalyzed by triose phosphate isomerase to be at equilibrium.

11. Glucokinase from rat liver has a K_M of 10 mM and an approximate maximum catalytic activity of 1.5 μmol min^{-1} g^{-1} of fresh tissue whereas hexokinase from rat liver has a K_M of approximately 0.1 mM and a maximum catalytic activity of 0.1 μmol min^{-1} g^{-1}. Give the ratios of glucokinase to hexokinase activities at the following glucose concentrations: (a) 0.1 mM, (b) 1.0 mM, (c) 5.0 mM (normal), (d) 10 mM, (e) 30 mM (diabetic level).

We now begin consideration of _respiration_, the process by
which aerobic cells obtain energy from the oxidation of fuel
molecules by molecular oxygen. In this chapter we shall
examine the tricarboxylic acid cycle, the common central
pathway for the degradation of the two-carbon acetyl residues derived not only from carbohydrates but also from fatty
acids and amino acids. The chapter to follow will describe
the final stage of respiration, namely, electron transport and
oxidative phosphorylation.

The tricarboxylic acid cycle is a cyclic sequence of reactions of almost universal occurrence in aerobic organisms. It
is catalyzed by a multienzyme system that accepts the acetyl
group of acetyl–coenzyme A as fuel and dismembers it to
yield carbon dioxide and hydrogen atoms. The latter are then
led via a sequence of electron-carrying proteins to molecular
oxygen, which is reduced to form water.

In this chapter we shall also outline a second pathway for
oxidation of glucose, the _phosphogluconate_ pathway, which
serves as one of the sources of reducing power for biosynthetic reactions and is responsible for the synthesis and degradation of pentoses and other sugars.

The Energetics of Fermentation and Respiration

Glycolysis releases only a very small fraction of the chemical
energy potentially available in the structure of the glucose
molecule. Much more energy is released when the glucose
molecule is oxidized completely to CO_2 and H_2O, as is
shown by comparison of the standard-free-energy changes
for anaerobic glycolysis (page 419) and for complete oxidation of glucose:

Glycolysis:

$$Glucose \longrightarrow 2 \text{ lactate}$$
$$\Delta G^{\circ\prime} = -47.0 \text{ kcal mol}^{-1}$$

Complete oxidation:

$$Glucose + 6O_2 \longrightarrow 6CO_2 + 6H_2O$$
$$\Delta G^{\circ\prime} = -686.0 \text{ kcal mol}^{-1}$$

When cells degrade glucose anaerobically via the glycolytic sequence, the lactate formed, which cannot be utilized further, still contains most of the energy of the original glucose molecule. On the other hand, under aerobic conditions glucose degradation does not stop at the stage of lactate but continues further, so that the products of glycolysis are oxidized completely to CO_2 and H_2O, thus releasing the remainder of the available energy of the glucose molecule.

The Flow Sheet of Respiration

The flow sheet of respiration is shown in Figure 17-1. Acetyl-CoA derived from the oxidation of carbohydrates, fatty acids, and amino acids in stage II of catabolism (Chapter 14, page 371) now enters the tricarboxylic acid cycle in stage III, the final common pathway of oxidation of all fuel molecules in aerobic cells. It is in this cycle that the acetyl group of acetyl-CoA is enzymatically degraded to form two molecules of CO_2 and four pairs of hydrogen atoms (in bound form). The latter (or the corresponding electrons) are then fed into the _respiratory chain_, a series of electron carriers. The ensuing process of _electron transport_ to molecular oxygen proceeds with a very large decline in free energy, much of which is conserved by the phosphorylation of ADP to yield ATP, in the process of _oxidative phosphorylation_.

The cyclic nature of the tricarboxylic acid cycle contrasts with the linear reaction sequence of glycolysis. In each turn around the tricarboxylic acid cycle (Figure 17-2) one molecule of acetic acid (two carbon atoms) enters as acetyl-CoA and condenses with a molecule of the four-carbon compound oxaloacetic acid to form the six-carbon tricarboxylic compound citric acid. The citric acid is then degraded by a reaction sequence that yields two molecules of CO_2 and regenerates the four-carbon oxaloacetic acid. Another turn of the cycle may now start by reaction of the oxaloacetic acid with another molecule of acetyl-CoA. Thus, in each turn of the cycle one molecule of acetic acid enters, two molecules of CO_2 are formed, and a molecule of oxaloacetate is utilized to form citrate but is regenerated at the end of the cycle. There is therefore no _net_ disappearance of oxaloacetate in the operation of the cycle. One molecule of oxaloacetate, if not removed by side reactions, suffices to bring about oxidation of an unlimited number of acetate molecules. The tricarboxylic acid cycle is thus catalytic in two senses. Each separate step of the cycle is of course catalyzed by a specific enzyme, as is true in all enzyme systems, but superimposed on this level of catalysis is the catalytic effect of the cycle intermediates themselves: one molecule of oxaloacetate or any of its precursors in the cycle can promote the oxidation of many acetate molecules.

The Discovery of the Tricarboxylic Acid Cycle

The tricarboxylic acid cycle was first postulated by H. A. Krebs in 1937 under its original name "citric acid cycle." It was the outcome of a brilliant piece of reasoning and experi-

Figure 17-1
The flow sheet of respiration. Acetyl-CoA is mobilized and collected in stage II of catabolism (page 371). It then enters stage III, which consists of the tricarboxylic acid cycle and electron transport with its coupled oxidative phosphorylations. The end products of respiration are indicated in color.

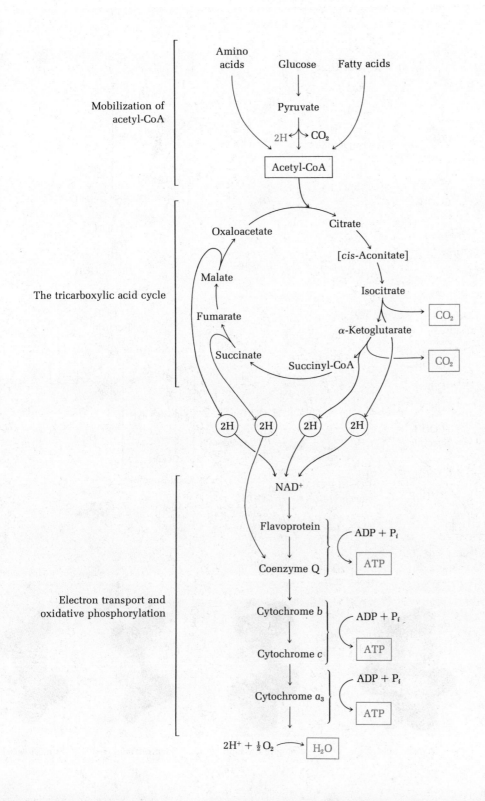

Figure 17-2
The tricarboxylic acid cycle. The intermediates are shown as their free acids. The end products (two CO_2, four pairs of H atoms, and GTP) are in boxes. The carbon atoms entering as acetyl-CoA are in color; their position in the cycle intermediates is given to the stage of succinyl-CoA. The succinate formed, because of its symmetry, will yield fumarate, malate, and oxaloacetate containing carbon from acetyl-CoA in equal amounts in all positions, assuming no side reactions. Space-filling models of the intermediates are shown below.

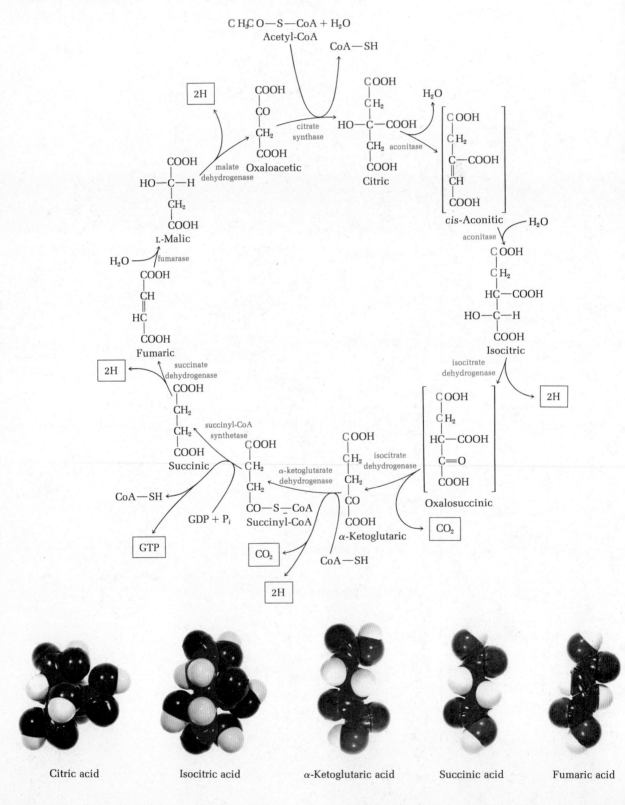

Citric acid   Isocitric acid   α-Ketoglutaric acid   Succinic acid   Fumaric acid

Figure 17-3
The oxidized and reduced forms of methyl-
ene blue. Since the reduced form is rapidly
reoxidized by molecular oxygen, the reduc-
tion of methylene blue by tissue suspensions
must be followed under anaerobic condi-
tions in order to detect the color change that
accompanies reduction.

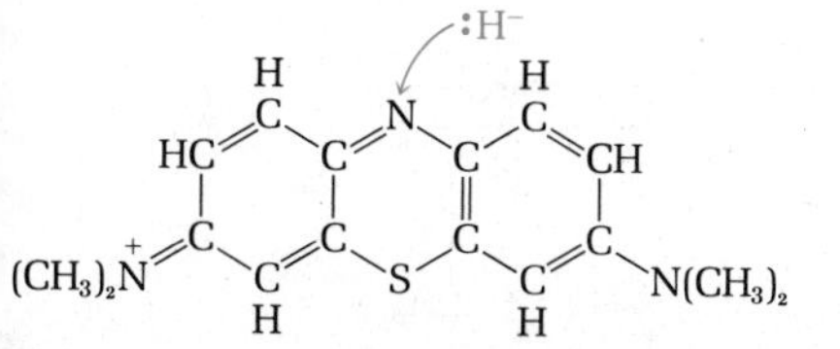

Oxidized form (blue)

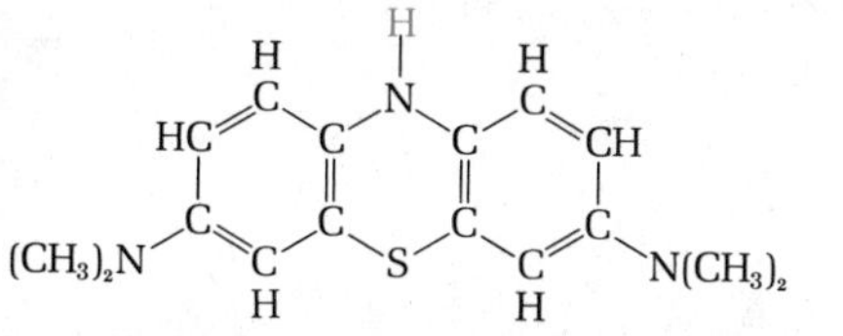

Reduced form (colorless)

Table 17-1   Catalytic effect of malate on the
respiration of minced pigeon-breast muscle

Malate added, μmol	Oxygen uptake calculated for oxidation of malate added, μatoms	Total oxygen uptake observed, μatoms
1.0	6.0	47.2
2.0	12.0	69.4

L-Malic acid

Oxaloacetic acid

mentation which ranks among the classic investigations of modern biology. The outlines of the research that led to the cycle will be reconstructed briefly since it was an influential and instructive model for the analysis of other metabolic pathways.

First some background must be sketched. It was known from the work of T. Thunberg and of F. Battelli and L. S. Stern in the period 1910–1920 that anaerobic suspensions of minced animal tissues catalyze the transfer of hydrogen atoms from certain organic acids known to occur in cells—especially succinic, malic, and citric acids—to the reducible dye methylene blue, giving its colorless reduced form (Figure 17-3). Enzymes catalyzing such reactions were called *dehydrogenases*. By the early 1930s several investigators using manometric measurement of the oxygen-utilization rate of minced-tissue suspensions (page 376) found that succinate, fumarate, malate, and citrate are rapidly oxidized to carbon dioxide by molecular oxygen.

In 1935 A. Szent-Györgyi, working in Hungary, assembled these and other observations into a sequence of enzymatic reactions for the oxidation of succinate:

$$\text{Succinate} \longrightarrow \text{fumarate} \longrightarrow \text{malate} \longrightarrow \text{oxaloacetate}$$

Especially significant was Szent-Györgyi's observation that adding small amounts of oxalacetate or malate to minced muscle suspensions evokes the utilization of an amount of oxygen far beyond that required to oxidize the added dicarboxylic acid to CO_2 and water (Table 17-1). From this and other experiments Szent-Györgyi concluded that these acids stimulate the oxidation of some endogenous substrate in the tissue, presumably glycogen, one molecule of malate or oxaloacetate promoting the oxidation of many molecules of the endogenous substrate.

Somewhat later C. Martius and F. Knoop, in Germany, found that citrate is enzymatically oxidized to succinate by animal tissues in the sequence

$$\text{Citrate} \longrightarrow \alpha\text{-ketoglutarate} \longrightarrow \text{succinate}$$

In 1936, with these observations as background, Krebs began to study the interrelationships in the oxidative metabolism of various dicarboxylic and tricarboxylic acids in suspensions of minced flight muscles of pigeons; these muscles have a very high rate of respiration. In particular, he sought the biological significance of these acids in the oxidation of glucose. The observations and reasoning which led Krebs to the postulation of the tricarboxylic acid cycle are summarized in the following paragraphs.

1 First, Krebs pointed out that muscle suspensions oxidize at a very high rate only certain dicarboxylic acids, namely, succinic, fumaric, malic, oxaloacetic, and α-ketoglutaric acids, and only certain tricarboxylic acids, namely, citric, isocitric, and *cis*-aconitic acids. (At neutral pH these acids are of course present as their anions.) Other organic acids,

447

e.g., tartaric acid and maleic acid, are not readily oxidized by muscle.

2 The oxidation of endogenous carbohydrate or of added pyruvate by muscle suspensions was catalytically stimulated not only by small amounts of succinate, fumarate, malate, and oxaloacetate, as Szent-Györgyi had shown, but also by citrate, cis-aconitate, isocitrate, and α-ketoglutarate.

3 Malonate (Figure 17-4; page 198) completely prevented the stimulation of the oxidation of pyruvate by any of the catalytically active tricarboxylic and dicarboxylic acids enumerated above. Since malonate is a specific competitive inhibitor of succinate dehydrogenase (page 198) and does not inhibit the dehydrogenases attacking the other dicarboxylic and tricarboxylic acids, Krebs concluded that the oxidation of succinate to fumarate by succinate dehydrogenase must be an essential link in a chain of reactions involving all the tricarboxylic and dicarboxylic acids capable of stimulating oxidation of pyruvate. (At this point it is useful to refer to Figure 17-5.)

4 When oxaloacetate and pyruvate were incubated with a muscle suspension under anaerobic conditions, citrate was formed. Krebs therefore postulated that the condensation of pyruvate and oxaloacetate to form citrate was the missing link by which the known enzymatic reactions of the dicarboxylic and tricarboxylic acids could be arranged into a cyclic sequence (Figure 17-5). The following experiments put this hypothesis to critical test.

5 When malonate was added to muscle suspensions to block succinate dehydrogenase, Krebs found that succinate accumulated quantitatively following the oxidation of added citrate, isocitrate, cis-aconitate, or α-ketoglutarate, as expected from the cyclic formulation in Figure 17-5.

6 In malonate-poisoned muscle the oxidation of fumarate, malate, or oxaloacetate also led to quantitative accumulation of succinate. This important observation clearly established that a pathway must exist for the oxidative conversion of fumarate into succinate, one that does not involve succinate dehydrogenase. These observations thus strongly supported Krebs' hypothesis of an oxidative pathway from fumarate to succinate via oxaloacetate and citrate and established that the overall sequence of reactions was cyclic.

7 Krebs then found that when pyruvate utilization was blocked by malonate, the inhibition could be relieved or overcome by raising the concentration of oxaloacetate. Under these circumstances, one molecule of oxaloacetate disappeared for each molecule of pyruvate consumed. He explained this finding as follows. In the uninhibited cycle, one molecule of oxaloacetate can stimulate the oxidative removal of many molecules of pyruvate, as would be expected if oxaloacetate is regenerated at each turn of the cycle. However, when the cycle is poisoned by malonate, oxaloacetate can no longer be regenerated. Under these circumstances one molecule of oxaloacetate is required to

Figure 17-4
Malonate, a competitive inhibitor of succinate dehydrogenase (see also Figure 8-11, page 200). Note the similarity in structure of malonate and succinate.

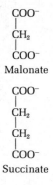

Malonate

Succinate

Figure 17-5
Position of the malonate block. When succinate dehydrogenase is blocked, the oxidation of citrate, cis-aconitate, isocitrate, or α-ketoglutarate leads to accumulation of succinate. Inhibition by malonate also blocks the reverse action of succinate dehydrogenase. Since fumarate is converted into succinate in the presence of malonate, Krebs postulated that oxaloacetate, the oxidative end product of fumarate, reacts with pyruvate to form citrate, a precursor of succinate.

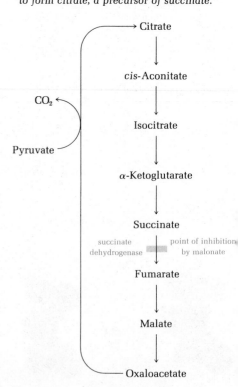

remove each molecule of pyruvate to yield citrate, which is oxidized to succinate. The latter now accumulates as a consequence of the malonate block. In malonate-poisoned muscle the following overall reaction therefore takes place:

$$\text{Fumarate} + \text{pyruvate} + 2O_2 \longrightarrow \text{succinate} + 3CO_2 + H_2O$$

8 Krebs demonstrated that all the individual enzymatic reactions of the postulated cycle take place at a rate high enough to account for the total rate of pyruvate and oxygen utilization by the tissue. He therefore concluded that this series of reactions is the major if not sole pathway for the oxidation of pyruvate in muscle.

From these simple yet elegant experiments and arguments Krebs postulated the citric acid cycle as the main pathway for oxidation of carbohydrate in muscle. Because there was uncertainty for some years whether citric acid was the first tricarboxylic acid formed in the reaction between pyruvate and oxaloacetate, the name of the cycle was changed to tricarboxylic acid cycle. As it turned out, citric acid is in fact the first tricarboxylic acid formed, but over 10 years were to pass before this was proved conclusively. Today the names "tricarboxylic acid cycle" and "citric acid cycle" are used synonymously.

Intracellular Location of the Enzymes of the Tricarboxylic Acid Cycle

In 1948 E. P. Kennedy and A. L. Lehninger found that isolated mitochondria (page 514), obtained from rat-liver homogenates by differential centrifugation (page 381) and suspended in a buffered medium containing phosphate, adenine nucleotides, and Mg^{2+}, catalyze the oxidation of pyruvate and all the intermediates of the tricarboxylic acid cycle at the expense of molecular oxygen. The overall rates of oxygen consumption and pyruvate utilization by the isolated liver mitochondria accounted for the rate of respiration of the intact liver cell. On the other hand, the nuclei, the microsomes, and the soluble fraction of the cytoplasm (page 381) were inactive. The liver mitochondria thus contain all the enzymes required for the tricarboxylic acid cycle, as well as the components required for electron transport. In all animal and plant cells examined to date the mitochondria are the sites of the tricarboxylic acid cycle reactions.

The enzymatic reactions of the tricarboxylic acid cycle take place within the inner compartment of the mitochondrion (pages 511 and 529). Some of the cycle enzymes occur in the soluble matrix of the inner compartment, whereas others are attached to the inner mitochondrial membrane (pages 489 and 511) and are not readily extracted in soluble form. Some of the enzymes of the tricarboxylic acid cycle, particularly aconitate hydratase, NADP-specific isocitrate dehydrogenase, fumarase, and malate dehydrogenase (page 481), also occur in the cytosol of some tissues.

The Oxidation of Pyruvate to Acetyl-CoA

After the tricarboxylic cycle was postulated in 1937, the precise pathway of citrate formation from pyruvate and oxaloacetate became the subject of much research. It was not until 1948–1950 that the problem was solved. Pyruvate is first oxidized, with loss of CO_2, to acetyl-CoA, which then reacts enzymatically with oxaloacetate to form citrate.

The oxidation of pyruvate to acetyl-CoA, catalyzed by the *pyruvate dehydrogenase complex*, is a very complicated process, whose details were unraveled by L. J. Reed and his colleagues. The overall equation is

$$\text{Pyruvate} + NAD^+ + CoA \longrightarrow \text{acetyl-CoA} + NADH + H^+ + CO_2$$
$$\Delta G^{\circ\prime} = -8.0 \text{ kcal mol}^{-1}$$

This reaction, which is irreversible in animal tissues, is not itself part of the tricarboxylic acid cycle but is obligatory for the entry of all carbohydrates (via pyruvate) into the tricarboxylic acid cycle.

The oxidative decarboxylation of pyruvate to acetyl-CoA and CO_2 requires three different enzymes and five different coenzymes organized into a multienzyme complex. The reaction steps promoted by this complex are schematized in Figure 17-6. Step I is catalyzed by *pyruvate dehydrogenase*, whose prosthetic group is the coenzyme thiamin pyrophosphate (page 337). Pyruvate undergoes decarboxylation to yield CO_2 and the α-hydroxyethyl derivative of the thiazole ring of thiamin pyrophosphate (page 338), in a reaction that is identical or similar to that involved in the nonoxidative decarboxylation of pyruvate during alcoholic fermentation (page 438). In step II the hydroxyethyl group is dehydrogenated, and the resulting acetyl group is transferred to the sulfur atom at carbon 6 of lipoic acid (page 347), which constitutes the covalently bound prosthetic group of the second enzyme of the complex, *dihydrolipoyl transacetylase* (official name, *lipoate acetyltransferase*). The transfer of a pair of hydrogen atoms (or equivalent electrons) from the hydroxyethyl group of thiamin pyrophosphate to the disulfide bond of lipoic acid converts the latter into its reduced, or dithiol, form, dihydrolipoic acid (page 347). In step III, the acetyl group is enzymatically transferred from the lipoyl group of dihydrolipoic acid to the thiol group of coenzyme A; the acetyl-CoA so formed then leaves the enzyme complex in free form. In step IV the dithiol form of the lipoyl group of dihydrolipoyl transacetylase is reoxidized to its disulfide form by transfer of hydrogen atoms to the third enzyme of the complex, usually known as *dihydrolipoyl dehydrogenase* (official name, *lipoamide dehydrogenase*), whose reducible prosthetic group is tightly bound flavin adenine dinucleotide (FAD) (page 339). The resulting $FADH_2$, which remains bound to the enzyme, is reoxidized in step V by NAD^+, with formation of NADH (Figure 17-6).

The pyruvate dehydrogenase complex of heart and kidney mitochondria has a particle weight of over 7 million and consists of an icosahedron, a 20-sided polyhedron. It contains a core of dihydrolipoyl transacetylase, to which mole-

Figure 17-6
Steps in the oxidation of pyruvate to acetyl-CoA by the pyruvate dehydrogenase complex; the fate of pyruvate is traced in color. The structure of thiamin pyrophosphate and its α-hydroxyethyl derivative is given in Figure 13-1 (page 338).

Key:

E_1 = pyruvate dehydrogenase
TPP = thiamin pyrophosphate
TPP—CHOH—CH_3 = α-hydroxyethylthiamin pyrophosphate
E_2 = dihydrolipoyl transacetylase
E_3 = dihydrolipoyl dehydrogenase

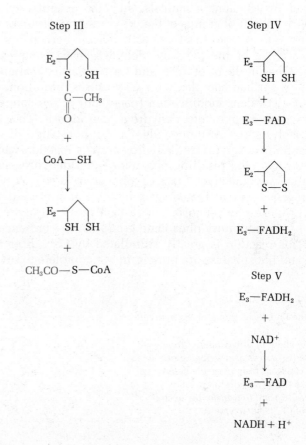

Step I

E_1—TPP

+

$CH_3COCOOH$

$\searrow CO_2$

E_1—TPP—CHOH—CH_3

Step II

E_1—TPP—CHOH—CH_3

+

E_2
S—S

↓

E_1—TPP

+

E_2
S   SH
C—CH_3
O

Step III

E_2
S   SH
C—CH_3
O

+

CoA—SH

↓

E_2
SH SH

+

CH_3CO—S—CoA

Step IV

E_2
SH SH

+

E_3—FAD

↓

E_2
S—S

+

E_3—FADH$_2$

Step V

E_3—FADH$_2$

+

NAD$^+$

↓

E_3—FAD

+

NADH + H$^+$

cules of pyruvate dehydrogenase and dihydrolipoyl dehydrogenase are attached. The core (mol wt 3.1 million) consists of 60 identical polypeptide chains, each containing a molecule of covalently bound lipoic acid. Twenty molecules of pyruvate dehydrogenase (mol wt 154,000) are attached to the core, as well as five or six molecules of dihydrolipoyl dehydrogenase (mol wt 110,000). Also firmly attached to the dihydrolipoyl transacetylase core are five molecules of *pyruvate dehydrogenase kinase* (mol wt ≈ 62,000); less firmly attached are several molecules of *pyruvate dehydrogenase phosphatase* (mol wt 100,000). The latter two enzymes play a regulatory role; their function is described below. The pyruvate dehydrogenase complex is localized in the mitochondrial matrix.

In *E. coli* cells the pyruvate dehydrogenase complex may have a somewhat different composition; an electron micrograph of the *E. coli* enzyme complex is shown in Figure 17-7.

In the pyruvate dehydrogenase complex the long lipoyllysyl side chain of dihydrolipoyl transacetylase serves as a "swinging arm" to transfer the hydroxyethyl group from the bound thiamin pyrophosphate of pyruvate dehydrogenase to the active site of the dihydrolipoyl transacetylase, where the hydroxyethyl group is oxidized. The resulting acetyl group is then transferred, again by the swinging arm, to the active site of dihydrolipoyl dehydrogenase (Figure 17-8).

The pyruvate dehydrogenase complex is characteristically inhibited by trivalent arsenicals, such as arsenite, which react with both thiol groups of the dihydrolipoyl group of the transacetylase to yield inactive cyclic arsenic derivatives.

The activity of the pyruvate dehydrogenase complex is regulated by the level of ATP and Ca^{2+} ions. Incubation of the highly purified complex with ATP causes inhibition, the result of a covalent modification (page 220) of the complex by the action of *pyruvate dehydrogenase kinase*. This enzyme catalyzes the ATP-dependent phosphorylation of one of the subunits of pyruvate dehydrogenase, a reaction which requires Mg^{2+}. The resulting phosphorylated dehydrogenase is catalytically inactive. Inactive phosphorylated pyruvate dehydrogenase is reactivated again by the action of *pyruvate dehydrogenase phosphatase*, which catalyzes hydrolytic removal of the inhibitory phosphate groups in the presence of Mg^{2+}. This reaction is greatly stimulated by Ca^{2+}. Both the kinase and phosphatase are present in the complete pyruvate

Figure 17-7
Electron micrograph of the pyruvate dehydrogenase complex of E. coli, showing its subunit structure. Below are shown models of the complex.

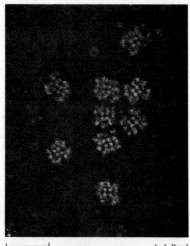

100 nm   L. J. Reed

Figure 17-8
Role of the lipoyllysyl group in the pyruvate dehydrogenase complex. The long lipoyl-lysyl side chain of dihydrolipoyl trans-acetylase (E₂) serves as a swinging arm to transfer electrons from pyruvate dehydrogenase (E₁) to dihydrolipoyl dehydrogenase (E₃) and to transfer the acetyl group from E₁ to coenzyme A.

Top view

Side view

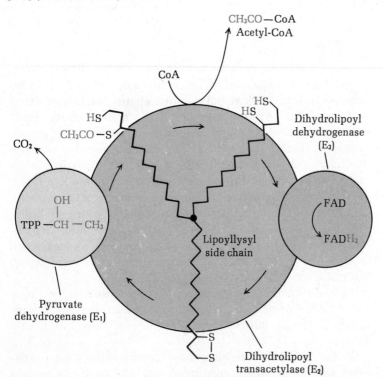

dehydrogenase complex and thus function as regulatory subunits (page 242).

Regulation of the activity of the pyruvate dehydrogenase complex may be summarized as follows. Whenever ATP, the energy-rich end product of the tricarboxylic acid cycle and oxidative phosphorylation, accumulates to high levels, the pyruvate dehydrogenase complex is turned off through its phosphorylation by the kinase, slowing down the rate of formation of acetyl-CoA, the fuel of the cycle, and thus the subsequent formation of ATP. However, whenever the ADP concentration is high and ample pyruvate is available, the pyruvate dehydrogenase complex is turned on by dephosphorylation of the inactive dehydrogenase, a reaction that is enhanced by Ca^{2+}.

We now proceed to describe the fate of the acetyl-CoA formed as the end product of the oxidative decarboxylation of pyruvate.

The Reactions of the Tricarboxylic Acid Cycle

The tricarboxylic acid cycle as originally postulated in 1937 was a skeleton. It has since been filled in with many details from the study of highly purified preparations of the enzymes catalyzing the individual steps. Moreover, much attention has been given the stereochemistry of the cycle reactions. Figure 17-2 shows the cycle as it is known today.

Citrate Synthase

Citric acid, the first tricarboxylic intermediate of the cycle, is formed by the condensation of acetyl-CoA with oxaloacetate:

$$\text{Acetyl-CoA} + \text{oxaloacetate} + H_2O \longrightarrow \text{citrate} + \text{CoA}$$
$$\Delta G^{\circ\prime} = -7.7 \text{ kcal mol}^{-1}$$

This reaction is catalyzed by *citrate synthase*, discovered and first called *condensing enzyme* by S. Ochoa. This enzyme (mol wt 100,000) catalyzes an aldol condensation between the methyl group of acetyl-CoA and the carbonyl group of oxaloacetate, with hydrolysis of the thioester bond and formation of free CoA—SH (Figure 17-9). Citroyl-CoA is believed to be formed as a nondissociating intermediate on the active site; synthetic citroyl-CoA is hydrolyzed by the enzyme to yield citrate and CoA.

The citrate synthase reaction proceeds far in the direction of citrate formation because of the exergonic hydrolysis of the high-energy thioester linkage of citroyl-CoA ($\Delta G^{\circ\prime} = -7.7$ kcal). Citrate synthase also catalyzes formation of *monofluorocitrate* from monofluoracetyl-CoA. This is an example of lethal synthesis, since fluoracetate is not itself toxic whereas fluorocitrate is a potent inhibitor of *aconitase*, the next enzyme in the tricarboxylic acid cycle.

The citrate synthase reaction is the primary pacemaker step of the tricarboxylic acid cycle (page 465); its rate is largely determined by the availability of acetyl-CoA and oxaloacetate and by the concentration of succinyl-CoA (see below), which competes with acetyl-CoA and inhibits citrate

Figure 17-9
The citrate synthase reaction. The carbon atoms originating from acetyl-CoA are in color.

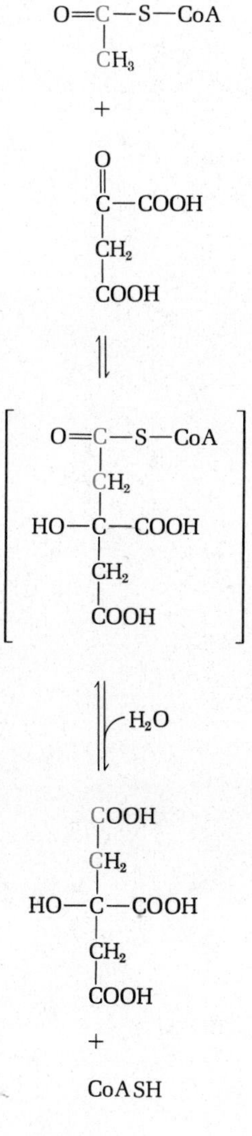

Acetyl-CoA

Oxaloacetic acid

Citroyl-CoA
(enzyme-bound
intermediate)

Citric acid

CoASH

synthase. Although the enzyme is also inhibited by ATP, NADH, and by long-chain fatty acyl-CoA esters, it is not certain whether these effects have physiological significance.

Conversion of Citrate to Isocitrate

The enzyme *aconitate hydratase,* more commonly known as *aconitase,* catalyzes the reversible interconversion of citrate and isocitrate via the enzyme-bound intermediate *cis*-aconitate (Figure 17-10):

$$\text{Citrate} \rightleftharpoons [\textit{cis}\text{-aconitate}] \rightleftharpoons \text{isocitrate}$$

The equilibrium mixture at pH 7.4 and 25°C contains about 93 percent citrate and only 7 percent isocitrate. However, isocitrate is very quickly oxidized in the next step of the cycle, thus pulling the aconitase reaction toward isocitrate formation. Aconitase contains Fe(II) and requires a thiol such as cysteine or reduced glutathione (page 714). The enzyme catalyzes the reversible addition of H_2O to the double bond of *cis*-aconitic acid in two directions, one leading to citric and the other to isocitric acid. *cis*-Aconitate normally functions as an enzyme-bound intermediate; it leaves the active site only very slowly. From cleverly designed experiments with deuterium-labeled substrates, it has been deduced that H— and —OH are always added to the double bond of *cis*-aconitate trans to each other (Figure 17-11), in the formation of either citrate or isocitrate. The stereochemistry of aconitase action is developed further below.

Figure 17-10
The aconitase reaction. The carbon atoms derived from acetyl-CoA are in color.

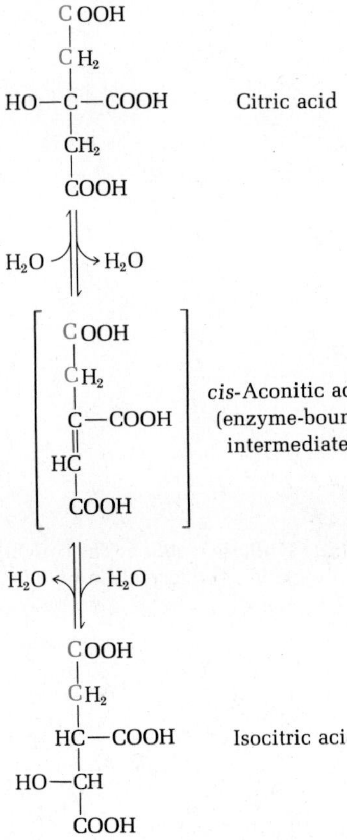

Figure 17-11
The stereospecific trans addition of the elements of water (color) to cis-aconitic acid. The six central atoms in the cis-aconitic acid molecule form a plane (color). The elements of water add from in front of the plane and behind the plane in two different ways, as shown, to form citric acid or isocitric acid. [Adapted from J. P. Glusker, in P. D. Boyer (ed.), The Enzymes, vol. 5, p. 421, Academic Press Inc., New York, 1971.]

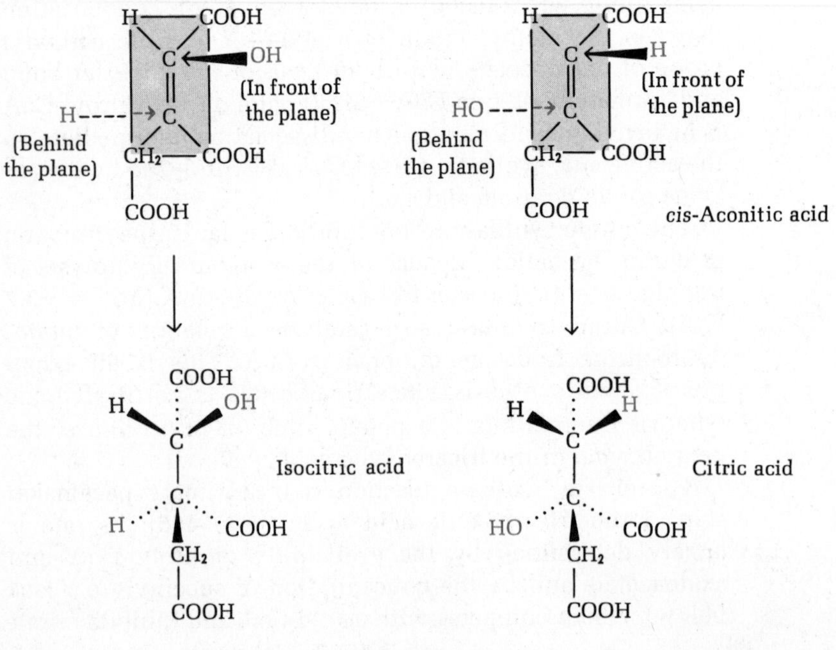

Figure 17-12
The "ferrous wheel" mechanism postulated for aconitase. The cis-aconitate is assumed to be bound to the enzyme designated E at three points and to the essential Fe(II) atom (color) at the active site. The plane of the double bond of cis-aconitate (see Figure 17-11) is shown by dashed lines enclosing color. The addition of OH from above by a partial turn of the ferrous wheel and H from below leads to citrate, whereas addition of OH from below (dotted arrow) and H from above, after a partial turn of the ferrous wheel in the other direction, leads to formation of isocitrate. This hypothesis was derived from results of x-ray analysis of the Fe(II) complexes of the tricarboxylic acids. [Adapted from J. P. Glusker, in P. D. Boyer (ed.), The Enzymes, vol. 5, p. 434, Academic Press Inc., New York, 1971.]

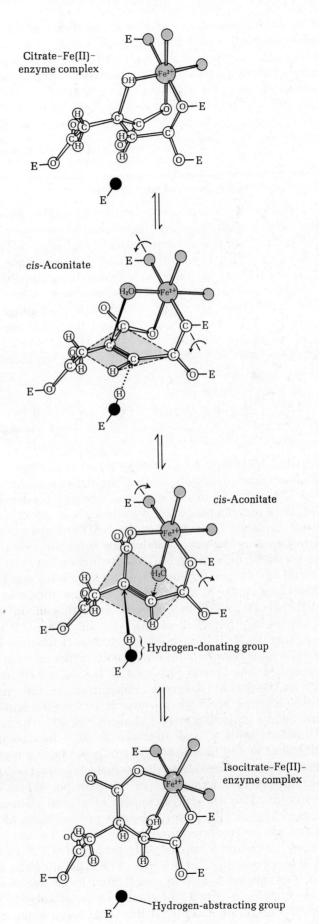

The mechanism of the aconitase reaction has attracted much attention because the Fe(II) ion, which is known to form a stable chelate complex with citric acid, is required in the action of the enzyme. The "ferrous wheel" hypothesis for the action of aconitase is shown in Figure 17-12.

Aconitase is present in animal tissues in two isozyme forms, one in the mitochondria and the other in the cytosol fraction.

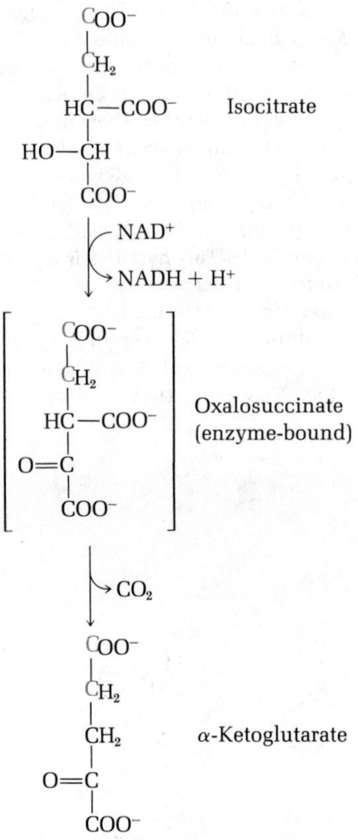

Oxidation of Isocitrate to α-Ketoglutarate

The oxidation of isocitrate to α-ketoglutarate (Figure 17-13) is of special interest because of its unusual experimental history. Most microorganisms and tissues of higher animals and plants contain two types of isocitrate dehydrogenase. For years a controversy has raged over which is responsible for the oxidation of isocitrate to α-ketoglutarate in the tricarboxylic acid cycle. One type of isocitrate dehydrogenase requires NAD^+ as electron acceptor, and the other requires $NADP^+$ (pages 340 and 481). The overall reactions catalyzed by the two types of isocitrate dehydrogenase are identical:

$$\text{Isocitrate} + NAD^+ (NADP^+) \rightleftharpoons$$
$$\alpha\text{-ketoglutarate} + CO_2 + NADH (NADPH) + H^+$$
$$\Delta G^{\circ\prime} = -5.0 \text{ kcal mol}^{-1}$$

Both the NAD-linked and NADP-linked isocitrate dehydrogenases occur in mitochondria of animal tissues, but the former is found only in mitochondria whereas the latter is present both in mitochondria and in the cytosol. Most of the available evidence indicates that the NAD-linked isocitrate dehydrogenase is the major catalyst for isocitrate oxidation in the tricarboxylic acid cycle. This distinction between the NAD- and NADP-specific isocitrate dehydrogenases was long obscured because the NAD-linked enzyme is an allosteric enzyme that requires ADP as a specific activating modulator. Until the stimulating effect of ADP was discovered, it was thought that the NAD-linked dehydrogenase, as measured in extracts of mitochondria, was only feebly active and unable to account for the known high rate of isocitrate oxidation.

The NAD-specific isocitrate dehydrogenase of mitochondria requires Mg^{2+} for activity. It has eight identical subunits and a molecular weight of about 380,000. The reaction proceeds with a large decrease in $\Delta G^{\circ\prime}$ because the simultaneous loss of the β-carboxyl group as CO_2 is a highly exergonic process. Possibly oxalosuccinate (Figure 17-13) is an enzyme-bound intermediate in the reaction. When the NAD-specific isocitrate dehydrogenase of animal tissues is stimulated by ADP, the latter undergoes no enzymatic alteration and can be recovered again at the end of the reaction. No other nucleoside 5'-diphosphate (except dADP) can activate the enzyme, nor can AMP or ATP. The allosteric nature of the enzyme is also apparent from its characteristic reaction kinetics. Figure 17-14 shows a plot of velocity vs. isocitrate concentration of the rat-liver enzyme as a function

Figure 17-13
Oxidation of isocitrate to α-ketoglutarate. Oxalosuccinate (in brackets) is believed to be an enzyme-bound intermediate.

Figure 17-14

Allosteric stimulation of isocitrate dehydrogenase by ADP. Increasing concentrations of ADP cause the apparent K_M to decrease, without change in V_{max}, thus activating the enzyme when the substrate concentration is low.

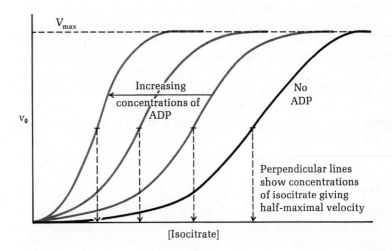

of ADP concentration. In similar experiments, it has been found that increasing the concentration of ADP also increases the apparent affinity of the enzyme for NAD^+. On the other hand, isocitrate dehydrogenase is strongly inhibited by the negative modulators NADH and ATP. In any metabolic condition in which the ADP concentration rises in the cell, presumably because of an excessively high rate of breakdown of ATP in energy-requiring reactions, an increase in the rate of isocitrate oxidation follows. As the concentration of ATP in the cell increases again, that of ADP must decrease, a change which tends to turn off isocitrate dehydrogenase. Furthermore, any accumulation of NADH in the mitochondria also can turn off oxidation of isocitrate. Isocitrate dehydrogenase does not appear to be the primary rate-controlling step in the cycle under most conditions; however, it may be involved in a secondary regulatory role.

The NAD-linked isocitrate dehydrogenase reaction is stereospecific for one of the four possible stereoisomers of isocitrate, namely, threo-D_s-isocitrate, and also for the NAD^+, since the enzyme transfers a hydrogen atom from isocitrate to only one side of the nicotinamide ring of NAD^+, namely, the A side (page 483).

The NADP-linked isocitrate dehydrogenase may under some circumstances participate in oxidation of isocitrate by mitochondria; it does not appear to be an allosteric enzyme.

Oxidation of α-Ketoglutarate to Succinyl-CoA

The oxidation of α-ketoglutarate to succinyl-CoA, which is biologically irreversible in animal cells, is carried out by the α-ketoglutarate dehydrogenase complex:

$$\alpha\text{-Ketoglutarate} + NAD^+ + CoA \rightleftharpoons$$
$$\text{succinyl-CoA} + CO_2 + NADH + H^+$$
$$\Delta G^{\circ\prime} = -8.0 \text{ kcal mol}^{-1}$$

This reaction is analogous to the oxidation of pyruvate to acetyl-CoA and CO_2 (page 450) and occurs by the same

457

mechanism, with thiamin pyrophosphate, lipoic acid, CoA, FAD, and NAD$^+$ participating as coenzymes. The α-ketoglutarate dehydrogenase complex is very similar in structure and properties to the pyruvate dehydrogenase complex (page 450). It has been isolated from animal tissues and from E. coli, in which it has a particle weight of 2.1×10^6 daltons.

The product of the α-ketoglutarate oxidation step is succinyl-CoA, a high-energy thioester of one carboxyl group of succinic acid.

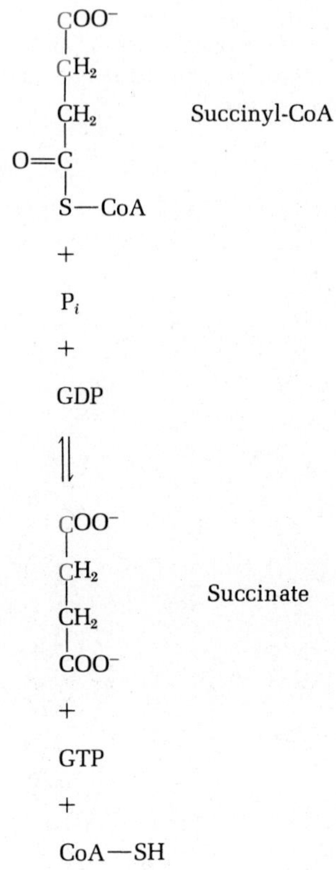

Deacylation of Succinyl-CoA

Succinyl-CoA undergoes loss of its CoA group, not by a simple hydrolysis, but by an energy-conserving reaction with guanosine diphosphate (GDP) and phosphate:

$$\text{Succinyl-CoA} + P_i + \text{GDP} \rightleftharpoons \text{succinate} + \text{GTP} + \text{CoA—SH}$$
$$\Delta G^{\circ\prime} = -0.7 \text{ kcal mol}^{-1}$$

The enzyme catalyzing this reaction, *succinyl-CoA synthetase*, causes the formation of the high-energy phosphate bond of GTP from GDP and P_i at the expense of the high-energy thioester bond of succinyl-CoA (Figure 17-15). The enzyme from animal tissues is specific for GDP as phosphate acceptor; that from E. coli utilizes ADP. Studies by P. D. Boyer and his colleagues have revealed that a covalent phosphoenzyme is formed as an intermediate step in this reaction. A histidine residue of the enzyme protein becomes phosphorylated when the enzyme is incubated with $^{32}P_i$, succinyl-CoA, and Mg^{2+} in the absence of GDP or with ^{32}P-labeled GTP in the absence of succinate. The 3-phosphohistidine residue (Figure 17-16) is believed to donate its phosphate group to GDP in the last step of the overall reaction. This and other evidence suggests the following sequence, in which E designates the enzyme protein:

$$\text{Succinyl-CoA} + P_i + E \rightleftharpoons \text{E–succinyl phosphate} + \text{CoA}$$

$$\text{E–succinyl phosphate} \rightleftharpoons E \sim P + \text{succinate}$$

$$E \sim P + \text{GDP} \rightleftharpoons E + \text{GTP}$$

In this formulation succinyl phosphate, a mixed anhydride of succinic and phosphoric acids, is postulated to be formed on the active site and to transfer its phosphate group to position 3 of the imidazole ring of a histidine residue of the enzyme. The enzyme-bound succinyl phosphate does not exchange readily with added free succinyl phosphate.

The GTP formed in this reaction then donates its terminal phosphate group to ADP to form ATP in the nucleoside diphosphate kinase reaction (page 409):

$$\text{GTP} + \text{ADP} \rightleftharpoons \text{GDP} + \text{ATP}$$

GTP and ATP have approximately the same standard free energy of hydrolysis.

Figure 17-15
Deacylation of succinyl-CoA.

Figure 17-16
Structure of 3-phosphohistidine.

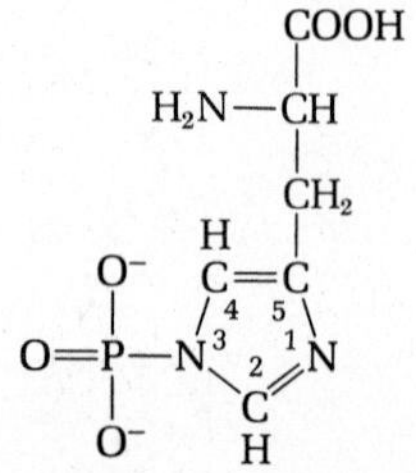

Figure 17-17
Dehydrogenation of succinate. E designates
the succinate dehydrogenase protein.

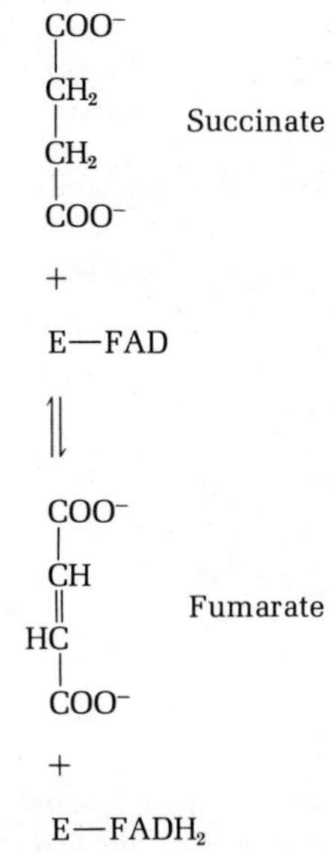

COO⁻
|
CH₂
| Succinate
CH₂
|
COO⁻

+

E—FAD

⇃↾

COO⁻
|
CH
‖   Fumarate
HC
|
COO⁻

+

E—FADH₂

The formation of GTP (or ATP) coupled to deacylation of succinyl-CoA is called a *substrate-level phosphorylation*, to distinguish it from phosphorylations linked to the respiratory chain (page 514). The formation of ATP coupled to oxidation of 3-phosphoglyceraldehyde during glycolysis is another example of a substrate-level phosphorylation. The formation of GTP during the oxidation of α-ketoglutarate to succinate is not inhibited by 2,4-dinitrophenol, the characteristic uncoupling agent for oxidative phosphorylation (page 519).

Succinyl-CoA can be deacylated to succinate by certain other reactions important in auxiliary functions of the tricarboxylic acid cycle; they are described elsewhere (pages 554 and 719). Note from Figure 17-2 that the succinate resulting from the α-ketoglutarate dehydrogenase reaction contains carbon atoms from the acetyl-CoA fed into the cycle.

Succinate Dehydrogenase

Succinate is oxidized to fumarate (Figure 17-17) by the flavoprotein *succinate dehydrogenase*, which contains covalently bound flavin adenine dinucleotide (page 339). This enzyme, tightly bound to the inner mitochondrial membrane, is rather difficult to extract from the membrane in soluble form. Many years of intensive research have been required to analyze its composition, properties, and mechanism. Its reducible coenzyme FAD functions as hydrogen acceptor in the reaction

$$\text{Succinate} + \text{E—FAD} \rightleftharpoons \text{fumarate} + \text{E—FADH}_2$$

The reduced enzyme can donate electrons to various artificial electron acceptors, e.g., reducible dyes; the normal electron acceptor is not known with certainty. Other properties of this and other flavin dehydrogenases are given elsewhere (pages 486 to 488).

As extracted from beef-heart mitochondria with solutions of sodium perchlorate, which disrupts the hydrogen-bonded structure of water, succinate dehydrogenase has a molecular weight of about 100,000 and contains one molecule of FAD, eight atoms of iron, and eight acid-labile sulfur atoms. The highly purified enzyme appears to have two subunits, of 30,000 and 70,000 molecular weight. The larger subunit of succinate dehydrogenase contains the FAD (pages 339 and 486), four atoms of iron, and four of acid-labile sulfur. The smaller subunit is an iron-sulfur protein (page 488) containing four iron atoms and four acid-labile sulfur atoms. The FAD is covalently bound and can be released on tryptic digestion of the large subunit. The FAD is attached via the 8-methyl group of riboflavin to imidazole ring nitrogen 3 of a histidine residue of the protein. The amino acid sequence of a 23-residue tryptic peptide containing the FAD has been established. Probably the iron atoms of both subunits of succinate dehydrogenase undergo Fe(II)–Fe(III) valence changes during electron transfer from succinate to the respiratory chain. A number of iron-sulfur proteins are known to function in electron-transfer reactions (page 489).

The stereochemistry of the removal of hydrogen atoms from succinate by succinate dehydrogenase has been studied in detail. An important clue has come from analysis of the action of the enzyme on L-chlorosuccinate, which it dehydrogenates, and D-chlorosuccinate, which it does not. From this fact, as well as isotopic experiments with deuterium (^{2}H) as tracer, it has been concluded that the dehydrogenase removes trans hydrogen atoms from the methylene carbon atoms of succinate (Figure 17-18).

Succinate dehydrogenase has some of the attributes of an allosteric enzyme: it is activated by succinate, phosphate, ATP, and by reduced coenzyme Q (page 493); it is inhibited by very low concentrations of oxaloacetate (page 201). However, it is not certain that these effects play a role in setting the overall rate of the tricarboxylate cycle, since succinate dehydrogenase activity in mitochondria is usually far greater than the activity of the other enzymes in the cycle and greater than the activity of the electron-transport chain. This matter is discussed further elsewhere (page 465).

Hydration of Fumarate

The reversible hydration of fumarate to L-malate is catalyzed by the enzyme *fumarate hydratase,* more commonly known as *fumarase:*

$$\text{Fumarate} + \text{H}_2\text{O} \rightleftharpoons \text{L-malate}$$

$$\Delta G^{\circ\prime} \approx 0$$

This enzyme has been obtained in crystalline form from pig heart. The reaction is freely reversible in vivo. Fumarase has a molecular weight of about 200,000 and contains four polypeptide-chain subunits, which are inactive in separated form. It requires no coenzyme. ATP decreases the apparent affinity of the enzyme for fumarate. Some very important experimental and mathematical approaches to analysis of enzyme kinetics arose from studies of the factors affecting the rate of fumarase activity.

Fumarase also acts stereospecifically, since it forms (and dehydrates) only the L-stereoisomer of malate. Stereochemical studies on the mechanism, using water labeled with deuterium, have revealed that fumarase catalyzes trans addition of H— and —OH to the double bond of fumarate. Although fumarate is a symmetrical molecule, the —OH group can be added to only one side of the double bond, to yield the L stereoisomer of malate (Figure 17-19).

Oxidation of Malate to Oxaloacetate

In the last reaction of the cycle the NAD-linked L-malate dehydrogenase catalyzes the oxidation of L-malate to oxaloacetate:

$$\text{L-Malate} + \text{NAD}^+ \rightleftharpoons \text{oxaloacetate} + \text{NADH} + \text{H}^+$$

$$\Delta G^{\circ\prime} = +7.1 \text{ kcal mol}^{-1}$$

Figure 17-18
Stereochemistry of the succinate dehydrogenase reaction.

Projection formula of succinic acid. The equivalent carbons are marked by asterisks and circles. Succinate dehydrogenase removes either the two $\boxed{H}$ *atoms or the two* ⓗ *atoms, at equal rates. (Modified from W. L. Alworth, Stereochemistry and Its Application in Biochemistry, p. 13, Interscience Publishers, a division of John Wiley & Sons, Inc., New York, 1973.)*

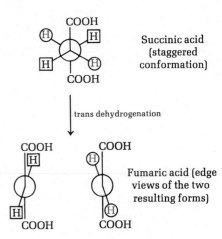

End view of succinic acid. The dehydrogenase removes either the two $\boxed{H}$ *or the two* ⓗ *to yield the two forms of fumaric acid shown. The four single bonds extending from the central carbon-carbon double bond lie in a plane.*

Succinic acid (staggered conformation)

trans dehydrogenation

Fumaric acid (edge views of the two resulting forms)

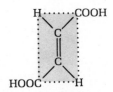

Front view, showing planarity of fumaric acid

Figure 17-19
Stereochemistry of the fumarase reaction. The D— and —OD groups of deuterium oxide (D₂O) approach the double bond of fumaric acid, whose four substituent groups are in the plane of the page, from behind and from in front of the plane, respectively. Since only one stereoisomer is formed, the —OD group can be added to only one side of the double bond, to yield the stereoisomer shown.

Front view. The plane of the fumaric acid molecule is perpendicular to the line of view.

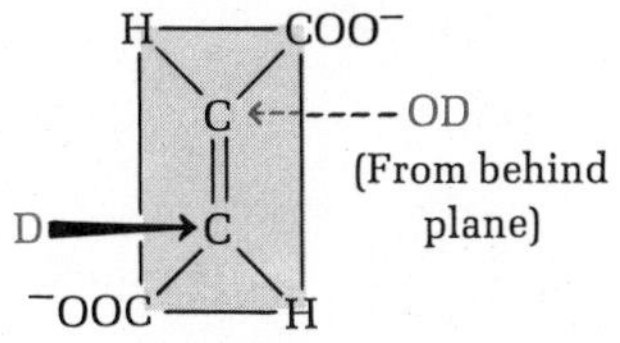

Edge view, with the plane of the fumaric acid molecule nearly parallel to the line of view.

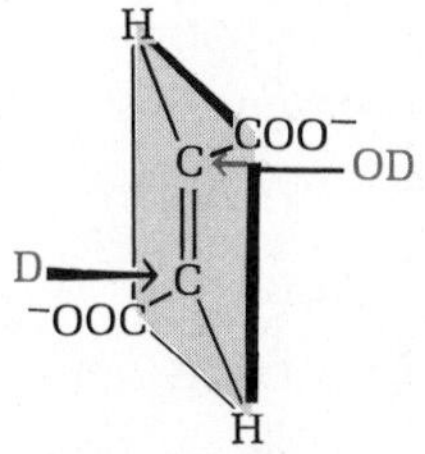

The product, erythro-3-monodeutero-L-malate. The deuterium entering as —OD undergoes exchange with hydrogen ions of the medium; the deuterium bonded to carbon is stable.

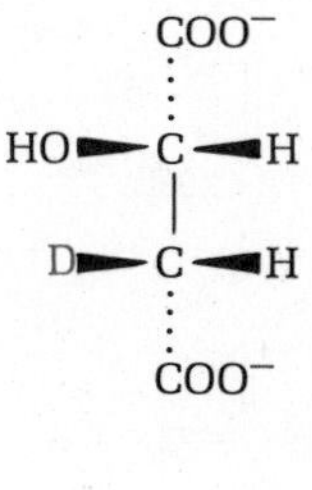

Although the reaction is endergonic as written (page 393), it goes in the forward direction very readily in the cell because of the rapid removal of the reaction products oxaloacetate and NADH in subsequent steps. NADP⁺ is only feebly reduced by the enzyme. The malate dehydrogenase reaction is strictly stereospecific for the L-stereoisomer of malate and for the A side of the pyridine ring of NAD⁺ (page 483). The cells of higher animals contain two forms of L-malate dehydrogenase, one in the mitochondria and the other in the extramitochondrial cytoplasm.

We can now sum up the output of one turn around the tricarboxylic acid cycle. For each acetyl group entering, two carbon atoms appear as carbon dioxide; however, these carbon atoms are not the same ones that entered as acetyl groups. Four pairs of hydrogen atoms are yielded by enzymatic dehydrogenation; three pairs have been used to reduce NAD⁺ and one pair to reduce the bound FAD of succinate dehydrogenase. These four pairs of hydrogen atoms become H⁺ ions; the corresponding electrons combine with oxygen, following their transport down the respiratory chain.

Isotopic Tests of the Tricarboxylic Acid Cycle

That the tricarboxylic acid cycle actually takes place in intact cells and can account quantitatively for the oxidation of carbohydrate, fatty acids, and amino acids has been verified by stringent isotopic tests employing precursors and intermediates of the cycle in which specific carbon atoms were isotopically labeled with either the ¹³C or ¹⁴C isotopes. However, some of the earliest isotopic experiments produced an unexpected result, which aroused considerable controversy about the pathway and mechanism of the cycle reactions. For example, acetate labeled with isotopic carbon (indicated by a star) in the carboxyl group (CH₃ČOOH) was incubated with a tissue suspension carrying out the oxidations of the cycle. Since acetate can be enzymatically converted into acetyl-CoA in animal tissues (page 546), the pathway of the carboxyl carbon atom of acetyl-CoA in the cycle reactions could be followed. Various intermediates of the tricarboxylic acid cycle were isolated from the tissue suspension following a period in which the acetate was being oxidized. The intermediates were then degraded to establish the positions of the isotopic carbon in each intermediate. Condensation of unlabeled oxaloacetate with carboxyl-labeled acetate would be expected to produce citrate labeled in one carboxyl group (Figure 17-20). Because citric acid *appears* to be a symmetrical molecule, without an asymmetric carbon atom, it was expected that the two terminal carboxyl groups of citrate would be chemically indistinguishable. Therefore, half of the labeled citrate molecules formed from labeled acetate were expected to yield α-ketoglutarate with the isotope in the α-carboxyl group and the other half α-ketoglutarate with the isotope in the γ-carboxyl group (Figure 17-20). However, contrary to this expectation, isotopic analysis of the labeled α-ketoglutarate isolated from the tissue suspension showed that the isotope originally introduced as the carboxyl carbon of acetate was found *only* in the γ-carboxyl group of the α-ketoglutarate, *not* in both α- and γ-carboxyl carbons (Figure 17-20). It was therefore concluded that citric acid itself or any other apparently symmetrical molecule could not possibly be an intermediate in the pathway from acetate to α-ketoglutarate. Hence it was postulated that an asymmetrical tricarboxylic acid, presumably *cis*-aconitic acid or isocitric acid, had to be the first condensation product formed from acetate and oxaloacetate.

In 1948, however, A. Ogston pointed out that although citric acid has no asymmetric carbon atom, the active sites of citrate synthase and aconitase may be asymmetric. Specifically, he suggested that the active site of aconitase may have three binding loci for citrate, asymmetrically arranged so that the citrate must make a "three-point landing" on the active site. Three different functional groups of citrate could then interact specifically with three complementary groups on the active site (Figure 17-21). Today, however, we recognize that the asymmetric utilization of citrate by aconitase follows from a stereochemical property of the citric acid molecule itself.

We have seen that molecules having an asymmetric carbon atom are <u>chiral</u> molecules (page 80) since they exist in left-handed and right-handed forms. However, modern stereochemistry recognizes another class of molecules in which there are no asymmetric carbon atoms but which are *potentially* capable of reacting asymmetrically; these are called <u>prochiral</u> molecules. Citric acid is a prochiral molecule because its two halves (see Figure 17-22) possess a non-superimposable mirror-image relationship to each other and thus are stereochemically different. Just as some enzymes can differentiate between *separate* enantiomeric molecules, such as D- and L-lactate (page 431) or D- and L-amino acids (page 566), some enzymes can differentiate between the enantiomerically different parts of a single prochiral molecule.

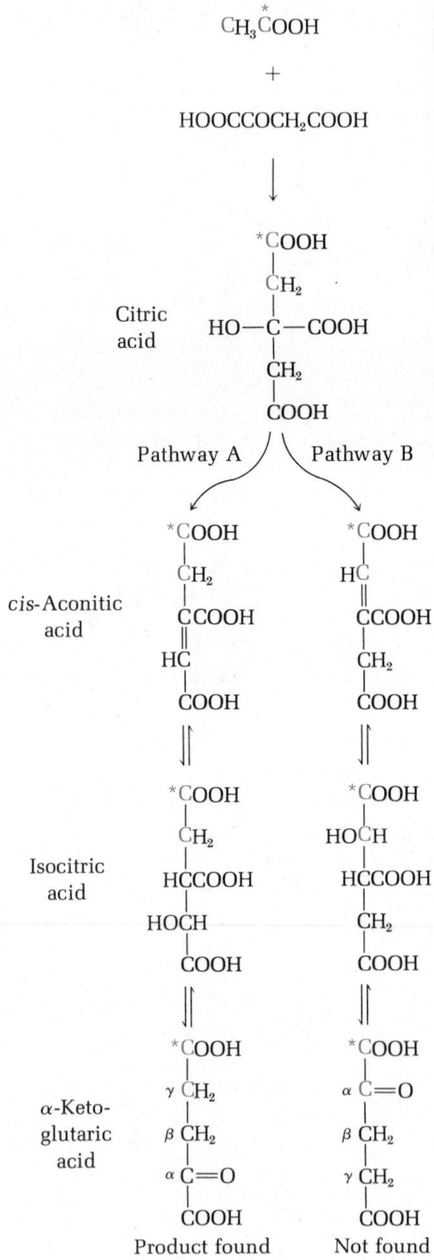

Figure 17-21
Ogston's explanation for the asymmetric action of aconitase on citrate. However, a three-point landing of citrate is not necessary to explain the action of aconitase on citrate. See Figure 17-22.

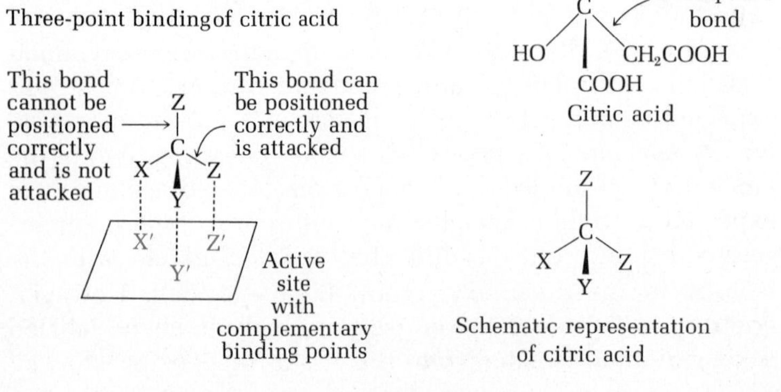

Three-point binding of citric acid

This bond cannot be positioned correctly and is not attacked

This bond can be positioned correctly and is attacked

Active site with complementary binding points

Schematic representation of citric acid

Figure 17-20
Incorporation of the carbon atoms (color) of the acetyl group of acetyl-CoA into α-ketoglutaric acid. The carboxyl carbon atom of acetyl-CoA is starred. Because the carboxyl carbon is found only in the γ-carboxyl group of α-ketoglutarate, citrate must yield α-ketoglutaric acid according to pathway A.

$CH_3\overset{*}{C}OOH$

+

$HOOCCOCH_2COOH$

Citric acid

cis-Aconitic acid

Isocitric acid

α-Keto-glutaric acid

Pathway A / Pathway B

Product found / Not found

Figure 17-22
The prochirality of the citric acid molecule. At the top is shown a projection structural formula for citric acid, which has no asymmetric carbon atom. Next is shown the order of the three substituent groups in the top half of the molecule, i.e., when bond (b) is perpendicular to and extending behind the plane of the page. At the bottom is shown the order of the substituent groups in the bottom half of the molecule, i.e., when (a) is perpendicular to and extending behind the plane of the page. The substituent groups in the two halves of the citric acid molecule are seen to be opposite in order, as shown by the direction of the arrows, and thus are mirror images of each other. (Modified from W. L. Alworth, Stereochemistry and Its Application in Biochemistry, p. 114, Interscience Publishers, a division of John Wiley & Sons, Inc., New York, 1973.

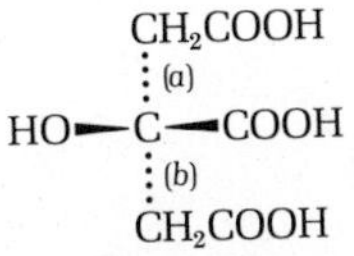

$$CH_2COOH$$
$$\vdots \,_{(a)}$$
$$HO\!-\!\!C\!-\!\!COOH$$
$$\vdots \,_{(b)}$$
$$CH_2COOH$$

Projection formula of citric acid

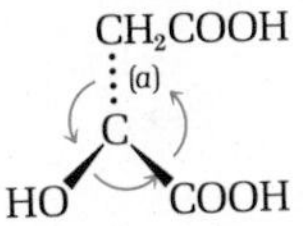

$$CH_2COOH$$
$$\vdots \,_{(a)}$$
$$C$$
$$HO \qquad COOH$$

View looking down on top

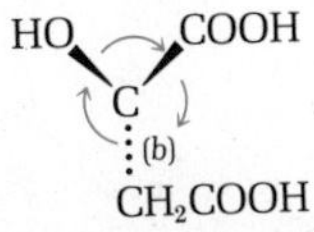

$$HO \qquad COOH$$
$$C$$
$$\vdots \,_{(b)}$$
$$CH_2COOH$$

View looking up at bottom half of
citric acid molecule

Aconitase is such an enzyme. It can combine with and transform only one of the two halves of the prochiral citrate molecule. This is why, when it acts on citrate formed from labeled acetate, aconitase yields isocitrate in which only one of the carboxyl groups is labeled. The nonequivalence of the two halves of the citrate molecule in the cycle has been confirmed unequivocally by several types of isotope tracer experiments. However, the precise manner and geometry by which the active site of aconitase recognizes and specifically binds one of the two halves of the citric acid molecule is not known and must await mapping of the active site (page 220). A three-point landing of the citrate molecule on the active site is not excluded but simply has not been proved.

The succinate arising from γ-carboxyl-labeled α-ketoglutarate ultimately yields labeled malate in which half of the molecules are labeled in the α-carboxyl group and half in the β-carboxyl group. In this reaction the succinate molecule itself is acting symmetrically. However, it will be recalled (page 459) that both succinate dehydrogenase and fumarase are stereospecific in their action.

Tests of the tricarboxylic acid cycle with other labeled precursors, such as $\overset{*}{C}H_3COOH$, $\overset{*}{C}H_3COCOOH$, and $H\overset{*}{C}O_3^-$ (see below), were found to yield labeled cycle intermediates with the isotopes in positions consistent not only with the postulated pathway of the cycle reactions but also with the asymmetrical reactions by which citrate is formed and converted to α-ketoglutarate.

When a molecule of acetic acid enters the cycle, the two molecules of carbon dioxide evolved during one revolution of the cycle are not the same two carbon atoms introduced as acetate; the latter actually remain in the four-carbon dicarboxylic acids (Figure 17-2).

Amphibolic Nature of the Cycle: Anaplerotic Reactions

We recall from Chapter 14 (page 371) that the tricarboxylic acid cycle is an amphibolic pathway and functions not only in catabolism but also to generate precursors for anabolic pathways. Certain intermediates of the cycle, particularly α-ketoglutarate and oxaloacetate, serve as precursors of amino acids, to which they are converted by enzymatic transamination reactions (page 562):

$$\alpha\text{-Ketoglutarate} + \text{alanine} \rightleftharpoons \text{glutamate} + \text{pyruvate}$$

$$\text{Oxaloacetate} + \text{alanine} \rightleftharpoons \text{aspartate} + \text{pyruvate}$$

Citrate can also be removed from the cycle to serve as a precursor of extramitochondrial acetyl-CoA for fatty acid biosynthesis (page 661), through the _ATP-citrate lyase_ reaction:

$$\text{Citrate} + \text{ATP} + \text{CoA} \longrightarrow \text{acetyl-CoA} + \text{oxaloacetate} + \text{ADP} + P_i$$

Moreover, succinyl-CoA can also be removed from the cycle for heme biosynthesis (page 719). Thus the tricarboxylic acid cycle can be drained of intermediates for biosynthetic reactions.

Tricarboxylic acid cycle intermediates can in turn be replenished by special enzymatic reactions called _anaplerotic_ ("filling up") reactions. The most important is the enzymatic carboxylation of pyruvate to form oxaloacetate, first discovered by H. G. Wood and C. Werkman in bacteria. The pathway of this carboxylation in animal tissues and the identity of the enzyme involved proved to be exceptionally difficult problems which required many years of research to clarify. Ultimately, M. F. Utter and his colleagues showed that in the liver oxaloacetate formation from pyruvate is catalyzed by _pyruvate carboxylase_, a mitochondrial enzyme:

$$\text{Pyruvate} + CO_2 + ATP + H_2O \underset{}{\overset{Mn^{2+}}{\rightleftharpoons}} \text{oxaloacetate} + ADP + P_i$$
$$\Delta G^{\circ\prime} = 0.5 \text{ kcal mol}^{-1}$$

In certain animal tissues, pyruvate may thus be converted to oxaloacetate whenever tricarboxylic acid cycle intermediates are deficient. Formation of pyruvate from oxaloacetate by reversal of this process in cells is unlikely since the intracellular concentration of oxaloacetate is very small. Instead, another route is followed (page 625).

Pyruvate carboxylase has a molecular weight of about 650,000. It is inactivated at 0°C, at which it dissociates into four subunits. The native enzyme contains four molecules of _biotin_ (page 345), covalently attached to the four subunits through amide linkages with the ε-amino group of specific lysine residues at the active site (page 224). Each subunit also binds one Mn^{2+} ion. The biotin prosthetic group on the enzyme active site serves as an intermediate carrier of the carboxyl group, which it transfers to pyruvate to form oxaloacetate. This reaction occurs in two steps:

$$\text{E–biotin} + ATP + CO_2 + H_2O \rightleftharpoons \text{E–carboxybiotin} + ADP + P_i$$

$$\text{E–carboxybiotin} + \text{pyruvate} \rightleftharpoons \text{E–biotin} + \text{oxaloacetate}$$

Sum:

$$ATP + CO_2 + \text{pyruvate} + H_2O \rightleftharpoons \text{oxaloacetate} + ADP + P_i$$

Pyruvate carboxylase is an allosteric enzyme. The rate of its forward reaction, leading to oxaloacetate formation, is negligible unless acetyl-CoA, its positive modulator, is present. Thus, whenever acetyl-CoA, the fuel of the tricarboxylic acid cycle, accumulates, it stimulates the pyruvate carboxylase reaction to produce more oxaloacetate, thus enabling the cycle to oxidize more acetyl-CoA.

Although the pyruvate carboxylase reaction is the most important anaplerotic reaction in the liver and kidney of higher animals, other reactions may also participate. One such reaction is that catalyzed by _malic enzyme_, the official name of which is _malate dehydrogenase (decarboxylating; NADP)_:

$$\text{Pyruvate} + CO_2 + NADPH + H^+ \rightleftharpoons \text{L-malate} + NADP^+$$
$$\Delta G^{\circ\prime} = -0.36 \text{ kcal mol}^{-1}$$

Intermediates of the tricarboxylic acid cycle may also be generated from aspartate and glutamate, which are converted into oxaloacetic and α-ketoglutaric acids, respectively, by transaminase reactions (page 562):

$$\text{Glutamate} + \text{pyruvate} \rightleftharpoons \alpha\text{-ketoglutarate} + \text{alanine}$$

$$\text{Aspartate} + \text{pyruvate} \rightleftharpoons \text{oxaloacetate} + \text{alanine}$$

In plants and many microorganisms the glyoxylate cycle (below) is also an important means of replenishing tricarboxylic acid cycle intermediates.

Regulation of the Tricarboxylic Acid Cycle

We briefly summarize the major regulatory reactions in the oxidation of pyruvate via the cycle (Figure 17-23). The activity of the pyruvate dehydrogenase complex, which furnishes a major portion of the acetyl-CoA input into the cycle, is diminished by the ATP-dependent phosphorylation of the dehydrogenase component and is activated by dephosphorylation of the phosphoenzyme. The condensation of acetyl-CoA with oxaloacetate to yield citrate is the primary control point of the tricarboxylic acid cycle in most tissues (page 453). However, there are other reactions in the cycle that are under allosteric regulation, at least in some tissues. The first of these is the NAD-linked isocitrate dehydrogenase reaction, which requires ADP as a positive or stimulatory allosteric modulator (page 234). In some tissues, e.g., insect flight muscle, Ca^{2+} also functions as a positive modulator of this reaction.

Figure 17-23
Regulation of pyruvate oxidation via the tricarboxylic acid cycle.

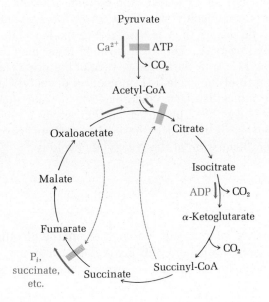

The other reaction that appears to be under regulation is succinate dehydrogenation, which is promoted by high concentrations of succinate, phosphate, ATP, and reduced ubiquinone (page 493) and is potently inhibited by oxaloacetate. Although this reaction is not usually the rate-setting step in the cycle, it competes with the NAD-linked reactions in donating electrons to the electron-transport chain (page 493) and thus may affect the integration of the dehydrogenation reactions of the cycle.

The tricarboxylic acid cycle also is regulated by the concentration of its various intermediates. Because some of the cycle reactions also function in biosynthesis (pages 625 and 719), a complex network of controls regulates the rate of the tricarboxylic acid cycle.

The Glyoxylate Cycle

The *glyoxylate cycle*, a modified form of the tricarboxylic acid cycle, takes place in most plants and microorganisms but not in higher animals. The primary purpose of the glyoxylate cycle, first delineated by Krebs and H. R. Kornberg, is to enable plants and microorganisms to utilize fatty acids or acetate, in the form of acetyl-CoA, as sole carbon source, particularly for the net biosynthesis of carbohydrate from fatty acids. Animals cannot bring about *net* synthesis of glucose from acetate or fatty acids, since the two carbon atoms are lost as CO_2 in the reactions of the tricarboxylic acid cycle leading from acetyl-CoA to oxaloacetate. The glyoxylate cycle bypasses the CO_2-evolving steps of the tricarboxylic acid cycle. The overall plan of the glyoxylate cycle is shown in Figure 17-24. Acetyl-CoA first condenses with oxaloacetate to form citrate, which is then converted by the action of aconitase into isocitrate, as in the tricarboxylic acid cycle. However, the breakdown of isocitrate occurs by a pathway in which three reactions of the tricarboxylic acid cycle are bypassed. Isocitrate is first cleaved by *isocitrate lyase* to form succinate and glyoxylate (Figure 17-25). *Malate synthase* (Figure 17-26) then catalyzes the condensation of glyoxylate with another molecule of acetyl-CoA to form malate (Figure 17-26). The malate then is oxidized to oxaloacetate by malate dehydrogenase. The oxaloacetate then may condense with acetyl-CoA, to start another turn of the glyoxylate cycle. In each turn of the cycle, two molecules of acetyl-CoA enter, and one molecule of succinate is formed. The succinate is used for biosynthetic purposes, particularly as a precursor in *gluconeogenesis*, the biosynthesis of "new" sugar (page 624). The overall equation of the glyoxylate cycle is

2 Acetyl-CoA + NAD$^+$ + 2H$_2$O $\longrightarrow$

succinate + 2CoA + NADH + H$^+$

In higher plants and microorganisms both the tricarboxylic acid cycle and the glyoxylate cycle may operate simultaneously, the former to provide energy needs via ox-

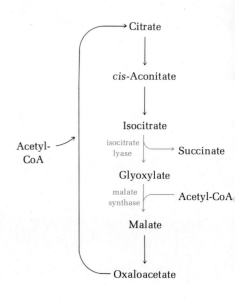

Figure 17-24
The glyoxylate cycle. The reactions shown by colored arrows are those catalyzed by isocitrate lyase and malate synthase; the others are reactions of the tricarboxylic acid cycle.

Figure 17-25
The isocitrate lyase reaction.

COOH
|
HOCH
|
HC—COOH   Isocitric acid
|
CH$_2$
|
COOH

$\Updownarrow$

COOH
|
CH$_2$
| Succinic acid
CH$_2$
|
COOH

+

CHO
| Glyoxylic acid
COOH

Figure 17-26
The malate synthase reaction.

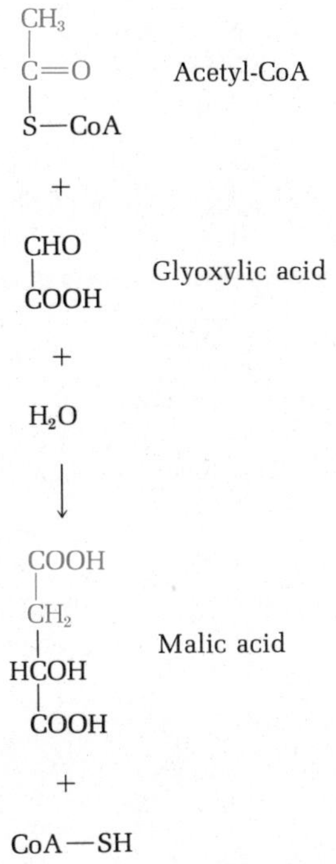

$$CH_3$$
$$|$$
$$C=O \qquad \text{Acetyl-CoA}$$
$$|$$
$$S—CoA$$

+

$$CHO$$
$$| \qquad \text{Glyoxylic acid}$$
$$COOH$$

+

$$H_2O$$

↓

$$COOH$$
$$|$$
$$CH_2$$
$$| \qquad \text{Malic acid}$$
$$HCOH$$
$$|$$
$$COOH$$

+

$$CoA—SH$$

idative phosphorylation (page 514) and the latter to provide succinate for the formation of new carbohydrate from fat.

Although the tricarboxylic acid cycle reactions in higher plants are localized in the mitochondria, the two characteristic enzymes of the glyoxylate cycle, *isocitrate lyase* and *malate synthase*, are localized in another class of cytoplasmic organelles, the *glyoxysomes*. These membrane-surrounded organelles lack most of the enzymes of the tricarboxylic acid cycle and have no cytochrome system. They are found only in plant cells capable of converting fatty acids into sugar. Since they are also rich in catalase, glyoxysomes may be related to the *microbodies* or *peroxisomes* or derived from them (page 33).

The glyoxylate cycle is under allosteric regulation. Isocitrate lyase is strongly inhibited by phosphoenolpyruvate, a key intermediate in the biosynthesis of glucose from non-sugar precursors (page 624). Isocitrate lyase and malate synthase are inducible enzymes (page 383); they are synthesized by plant cells only when they are needed.

The glyoxylate cycle is especially prominent in plant seeds, which can convert acetyl residues derived from the oxidation of fats into carbohydrate via succinic acid (page 628). Higher animals lack isocitrate lyase and malate synthase and are thus unable to convert acetyl-CoA into *new* glucose (page 630), although the carbon atoms of labeled acetyl groups will appear in glucose residues.

The Phosphogluconate Pathway

Many cells possess another pathway of glucose degradation whose first step is the enzymatic dehydrogenation of glucose 6-phosphate to 6-phosphogluconate. The *phosphogluconate pathway*, also known as the *pentose phosphate pathway* or the *hexose monophosphate shunt*, is not the main pathway for obtaining energy from the oxidation of glucose in animal tissues. Instead, it is a multifunctional pathway specialized to carry out four main activities, depending on the organism and its metabolic state. Its primary purpose in most cells is to generate reducing power in the extramitochondrial cytoplasm in the form of NADPH (pages 340 and 481). This function is especially prominent in tissues, e.g., the liver, mammary gland, and the adrenal cortex, that actively carry out the reductive synthesis of fatty acids and steroids from acetyl-CoA (page 661). Skeletal muscle, which is not active in synthesizing fatty acids, virtually lacks this pathway. The second special function of the phosphogluconate pathway is to convert hexoses into pentoses, particularly D-ribose 5-phosphate, required in the synthesis of nucleic acids (page 730). The third function is the complete oxidative degradation of pentoses by converting them into hexoses, which can then enter the glycolytic sequence. In the fourth function, to to be discussed in Chapter 23 (page 632), the phosphogluconate pathway is modified so as to participate in the formation of glucose from CO_2 in the dark reactions of photosynthesis (page 588).

The reactions of the phosphogluconate pathway take place in the soluble portion of the extramitochondrial cytoplasm of animal cells. All the enzymes required in this sequence have been highly purified and extensively studied, particularly by B. L. Horecker and E. Racker and their colleagues. The first reaction of the phosphogluconate pathway is the enzymatic dehydrogenation of glucose 6-phosphate to form 6-phosphogluconate (Figure 17-27) by _glucose 6-phosphate dehydrogenase_, also known as _Zwischenferment_:

Glucose 6-phosphate + NADP$^+$ $\rightleftharpoons$
$$6\text{-phosphogluconolactone} + NADPH + H^+$$
$$\Delta G^{\circ\prime} = -0.1 \text{ kcal mol}^{-1}$$

This enzyme, discovered by O. Warburg, was the first dehydrogenase found to be specific for NADP$^+$ as electron acceptor. It carries out dehydrogenation of carbon atom 1 of the pyranose form of glucose 6-phosphate to yield the corresponding 6-phosphoglucono-δ-lactone, which is hydrolyzed by a specific _lactonase_ (Figure 17-27):

6-Phosphogluconolactone + H$_2$O $\longrightarrow$ 6-phosphogluconate
$$\Delta G^{\circ\prime} = -5.0 \text{ kcal mol}^{-1}$$

The overall equilibrium of these two reactions lies far in the direction of formation of NADPH.

In the next step 6-phosphogluconate undergoes oxidation and decarboxylation by _6-phosphogluconate dehydrogenase_, a Mg^{2+}-dependent enzyme, to form D-ribulose 5-phosphate (Figure 17-27), a reaction that generates a second molecule of NADPH. Then, by the action of _ribosephosphate isomerase_, D-ribulose 5-phosphate is reversibly transformed into D-ribose 5-phosphate. Under some metabolic circumstances, the phosphogluconate pathway ends at this point, and its overall equation may then be written

Glucose 6-phosphate + 2NADP$^+$ + H$_2$O $\rightleftharpoons$
$$\text{D-ribose 5-phosphate} + CO_2 + 2NADPH + 2H^+$$

The net result is the production of NADPH for reductive biosynthetic reactions in the cytosol and the production of D-ribose 5-phosphate as a precursor for nucleotide synthesis.

Under other circumstances the phosphogluconate pathway may continue further, since the pentose 5-phosphates can undergo other transformations made possible by three additional enzymes, _ribulosephosphate 3-epimerase_, _transketolase_, and _transaldolase_. Ribulose phosphate 3-epimerase catalyzes conversion of D-ribulose 5-phosphate into its epimer at carbon atom 3, D-xylulose 5-phosphate (Figure 17-28). Transketolase, which contains tightly bound thiamin pyrophosphate (page 337) and Mg^{2+}, carries out the transfer of a glycolaldehyde group from D-xylulose 5-phosphate to D-ribose 5-phosphate to yield D-sedoheptulose 7-phosphate, a seven-carbon sugar phosphate, and D-glyceraldehyde 3-

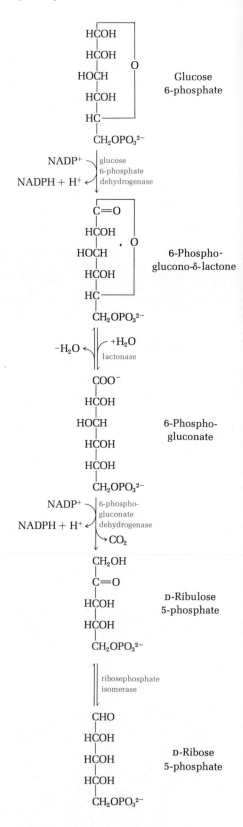

Figure 17-27
Early steps in the phosphogluconate pathway.

Figure 17-28
The epimerization of D-ribulose 5-phosphate.

$$
\begin{array}{l}
CH_2OH \\
| \\
C=O \\
| \\
HCOH \\
| \\
HCOH \\
| \\
CH_2OPO_3{}^{2-}
\end{array}
\qquad
\begin{array}{l}
\text{D-Ribulose} \\
\text{5-phosphate}
\end{array}
$$

ribulosephosphate
3-epimerase

$$
\begin{array}{l}
CH_2OH \\
| \\
C=O \\
| \\
HOCH \\
| \\
HCOH \\
| \\
CH_2OPO_3{}^{2-}
\end{array}
\qquad
\begin{array}{l}
\text{D-Xylulose} \\
\text{5-phosphate}
\end{array}
$$

phosphate, an intermediate of glycolysis (Figure 17-29). In this reaction the glycolaldehyde group (CH_2OH—CO—) is first transferred from D-xylulose 5-phosphate to enzyme-bound thiamin pyrophosphate to form the α,β-dihydroxyethyl derivative of the latter, which is analogous to the α-hydroxethyl derivative of thiamin pyrophosphate formed during the action of pyruvate dehydrogenase, described earlier (pages 338 and 450). The thiamin pyrophosphate acts as an intermediate carrier of this glycolaldehyde group, which is transferred to the acceptor molecule D-ribose 5-phosphate. Transketolase can also catalyze the transfer of a glycolaldehyde group from a number of other 2-keto sugar phosphates to carbon atom 1 of any of a number of different aldose phosphates. Free glycolaldehyde does not appear during such transformations.

Transaldolase is the third enzyme participating in further reactions of the phosphogluconate pathway. It acts on the products of the transketolase reaction, a seven- and a three-

Figure 17-29
The transketolase reaction.

$$
\begin{array}{l}
CH_2OH \\
| \\
C=O \\
| \\
HOCH \\
| \\
HCOH \\
| \\
CH_2OPO_3{}^{2-}
\end{array}
\qquad
\begin{array}{l}
\text{D-Xylulose} \\
\text{5-phosphate}
\end{array}
$$

$+$

$$
\begin{array}{l}
CHO \\
| \\
HCOH \\
| \\
HCOH \\
| \\
HCOH \\
| \\
CH_2OPO_3{}^{2-}
\end{array}
\qquad
\begin{array}{l}
\text{D-Ribose} \\
\text{5-phosphate}
\end{array}
$$

$$
\begin{array}{l}
CH_2OH \\
| \\
C=O \\
| \\
HOCH \\
| \\
HCOH \\
| \\
HCOH \\
| \\
HCOH \\
| \\
CH_2OPO_3{}^{2-}
\end{array}
\qquad
\begin{array}{l}
\text{D-Sedoheptulose} \\
\text{7-phosphate}
\end{array}
$$

$+$

$$
\begin{array}{l}
CHO \\
| \\
HCOH \\
| \\
CH_2OPO_3{}^{2-}
\end{array}
\qquad
\begin{array}{l}
\text{D-Glyceraldehyde} \\
\text{3-phosphate}
\end{array}
$$

Figure 17-30
The transaldolase reaction.

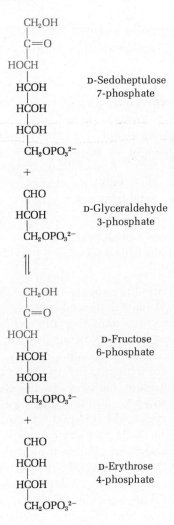

$$
\begin{array}{l}
CH_2OH \\
| \\
C=O \\
| \\
HOCH \\
| \\
HCOH \\
| \\
HCOH \\
| \\
HCOH \\
| \\
CH_2OPO_3{}^{2-}
\end{array}
\qquad
\begin{array}{l}
\text{D-Sedoheptulose} \\
\text{7-phosphate}
\end{array}
$$

$+$

$$
\begin{array}{l}
CHO \\
| \\
HCOH \\
| \\
CH_2OPO_3{}^{2-}
\end{array}
\qquad
\begin{array}{l}
\text{D-Glyceraldehyde} \\
\text{3-phosphate}
\end{array}
$$

$$
\begin{array}{l}
CH_2OH \\
| \\
C=O \\
| \\
HOCH \\
| \\
HCOH \\
| \\
HCOH \\
| \\
CH_2OPO_3{}^{2-}
\end{array}
\qquad
\begin{array}{l}
\text{D-Fructose} \\
\text{6-phosphate}
\end{array}
$$

$+$

$$
\begin{array}{l}
CHO \\
| \\
HCOH \\
| \\
HCOH \\
| \\
CH_2OPO_3{}^{2-}
\end{array}
\qquad
\begin{array}{l}
\text{D-Erythrose} \\
\text{4-phosphate}
\end{array}
$$

carbon sugar, to transform them into a six- and a four-carbon sugar:

D-Sedoheptulose 7-phosphate + D-glyceraldehyde 3-phosphate $\rightleftharpoons$
D-fructose 6-phosphate + D-erythrose 4-phosphate

The dihydroxyacetone group corresponding to carbon atoms 1, 2, and 3 of sedoheptulose 7-phosphate is transferred to D-glyceraldehyde 3-phosphate to form D-fructose 6-phosphate, leaving behind D-erythrose 4-phosphate (Figure 17-30). This reaction is formally similar to that catalyzed by fructose diphosphate aldolase and proceeds by a similar mechanism (page 210), but transaldolase cannot react with or form *free* dihydroxyacetone or its phosphate. The transketolase and transaldolase reactions provide not only for conversion of pentoses into hexoses for subsequent degradation by glycolysis and the tricarboxylic acid cycle, but also make possible, with the help of enzymes of the glycolytic sequence, the interconversion of three-, four-, five-, six-, and seven-carbon sugars, by reversible transfer of either two-carbon (glycolaldehyde) or three-carbon (dihydroxyacetone) moieties.

Another prominent reaction catalyzed by transketolase is

D-Xylulose 5-phosphate + D-erythrose 4-phosphate $\rightleftharpoons$
D-fructose 6-phosphate + D-glyceraldehyde 3-phosphate

in which two intermediates of the phosphogluconate pathway can be reversibly converted into two intermediates of the glycolytic pathway (Figure 17-31).

In some microorganisms, e.g., the lactobacilli, the phosphogluconate pathway is utilized in yet another way, to carry out fermentative degradation of pentoses. Horecker has found that this pathway begins with the enzyme _phosphoketolase_, catalyzing the reaction

Xylulose 5-phosphate + phosphate $\rightleftharpoons$
glyceraldehyde 3-phosphate + acetyl phosphate + H_2O

The glyceraldehyde 3-phosphate is then converted into lactate and the acetyl phosphate into acetate.

Thus a variety of nonoxidative and oxidative reactions of simple sugar phosphates may occur through the action of enzymes of the phosphogluconate pathway acting independently or in concert with enzymes of the glycolytic sequence. The phosphogluconate pathway is therefore not a well-defined route leading to a single end product but a set of diverging pathways capable of great metabolic flexibility.

The phosphogluconate pathway can also serve to carry out the *complete* oxidation of glucose 6-phosphate to CO_2, with simultaneous reduction of $NADP^+$ to NADPH, by a complex sequence of reactions (Figure 17-32) in which six molecules of glucose 6-phosphate are oxidized to six molecules each of ribulose 5-phosphate and CO_2; five molecules of glucose 6-

Figure 17-31
Another reaction catalyzed by transketolase.

CH_2OH
|
$C{=}O$
|
$HOCH$ D-Xylulose
| 5-phosphate
$HCOH$
|
$CH_2OPO_3{}^{2-}$

+

CHO
|
$HCOH$ D-Erythrose
| 4-phosphate
$HCOH$
|
$CH_2OPO_3{}^{2-}$

$\Updownarrow$ transketolase

CH_2OH
|
$C{=}O$
|
$HOCH$ D-Fructose
| 6-phosphate
$HCOH$
|
$HCOH$
|
$CH_2OPO_3{}^{2-}$

+

CHO
|
$CHOH$ D-Glyceraldehyde
| 3-phosphate
$CH_2OPO_3{}^{2-}$

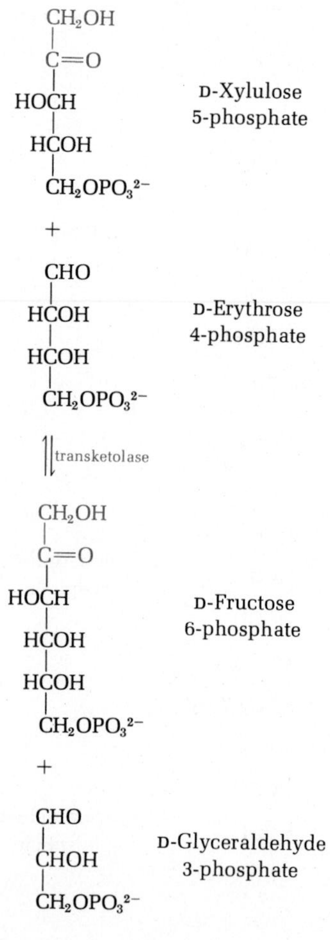

Figure 17-32
Schematic representation of the complete oxidation of one molecule of glucose 6-phosphate (G6P) to CO_2, with equivalent reduction of $NADP^+$, via the phosphogluconate pathway. The full equation is given in the text. End products are in color.

Key:

6PG = 6-phosphogluconate
Ru5P = ribulose 5-phosphate
R5P = ribose 5-phosphate
X5P = xylulose 5-phosphate
S7P = sedoheptulose 7-phosphate
G3P = glyceraldehyde 3-phosphate
G6P = glucose 6-phosphate
E4P = erythrose 4-phosphate
DHAP = dihydroxyacetone phosphate
FDP = fructose 1,6-diphosphate
F6P = fructose 6-phosphate

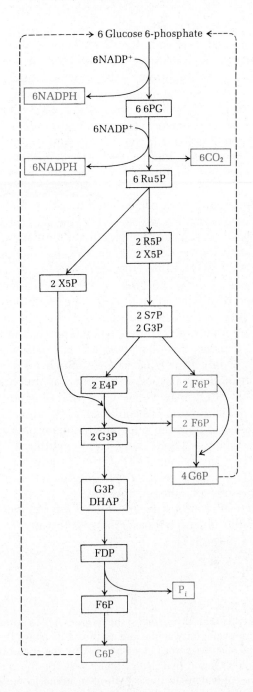

phosphate are then regenerated from the six molecules of ribulose 5-phosphate. The overall equation is

$$6 \text{ Glucose 6-phosphate} + 12NADP^+ + 7H_2O \longrightarrow$$
$$5 \text{ glucose 6-phosphate} + 6CO_2 + 12NADPH + 12H^+ + P_i$$

If we cancel out common terms, we get

$$\text{Glucose 6-phosphate} + 12NADP^+ + 7H_2O \longrightarrow$$
$$6CO_2 + 12NADPH + 12H^+ + P_i$$

The direction of flow and the path taken by glucose 6-phosphate after entry into the phosphogluconate pathway reactions are determined largely by the relative requirements

471

of the cell for NADPH and ribose 5-phosphate. If the requirement for NADPH exceeds that for ribose 5-phosphate, the excess pentose phosphate can be converted back into hexose phosphate. If the requirement for ribose 5-phosphate predominates, the flow through the transketolase and transaldolase reactions will be from fructose 6-phosphate to yield pentose phosphates. Only in certain cells and tissues, such as microorganisms active in biosynthesis of fat, or the lactating mammary gland, does the complete oxidative pathway leading exclusively to NADPH (Figure 17-32) prevail.

An isotopic approach can be used to assess the fraction of glucose catabolism in a given cell or tissue proceeding via the glycolytic vs. the phosphogluconate pathways. The cells are divided into two batches, one being incubated with glucose-1-^{14}C and the other with glucose-6-^{14}C. A comparison is then made of the initial rates at which ^{14}C appears in the CO_2 formed by oxidation of glucose. The combined action of the glycolytic sequence and the tricarboxylic acid cycle yields $^{14}CO_2$ from both types of labeled glucose at equal initial rates, whereas the phosphogluconate pathway initially yields $^{14}CO_2$ only from glucose-1-^{14}C. Such tests have been carried out on many tissues. In rat liver, as much as 20 percent of the $^{14}CO_2$ may come from the phosphogluconate pathway, depending on metabolic state. A much larger fraction of glucose enters the phosphogluconate pathway in the mammary gland, where fatty acid synthesis, requiring NADPH, is a major metabolic activity. However, in heart and skeletal muscle, relatively little glucose oxidation occurs via the phosphogluconate pathway.

An interesting although minor pathway of glucose degradation is the glucuronate-ascorbate pathway described in Chapter 23 (page 641).

Summary

Respiration occurs in three stages: the oxidative formation of acetyl-CoA from pyruvate, fatty acids, and amino acids (stage I); the degradation of the acetyl residues of acetyl-CoA by the tricarboxylic acid cycle to yield CO_2 and hydrogen atoms (stage II); and the transport of electrons equivalent to these hydrogen atoms to molecular oxygen (stage III), a process that is accompanied by coupled phosphorylation of ADP. Acetyl-CoA arises from carbohydrates via the oxidation of pyruvate. The tricarboxylic acid cycle, which takes place in the mitochondria, begins when citrate synthase catalyzes the condensation of acetyl-CoA with oxaloacetic acid to form citric acid. Aconitase then catalyzes the reversible formation of isocitrate from citrate, a reaction in which citrate, a prochiral molecule, is acted upon asymmetrically. Isocitrate is then oxidized to α-ketoglutarate and CO_2 by NAD-linked isocitrate dehydrogenase, an allosteric enzyme which is activated by ADP. The α-ketoglutarate then undergoes oxidation to succinyl-CoA in a sequence of reactions similar to the oxidation of pyruvate to acetyl-CoA. Succinyl-CoA reacts with GDP and phosphate to form free succinate and GTP; the terminal phosphate group of GTP is transferred to ADP. Succinate is then oxidized to fumarate by succinate dehydrogenase, a flavin enzyme. Fumarate is hydrated by fumarase to L-malate, which is oxidized by NAD-linked L-malate dehydrogenase to regenerate a mole-cule of oxaloacetate. The latter can then combine with another molecule of acetyl-CoA to start another revolution of the cycle. Iso-topic tracer tests with carbon-labeled intermediates have established that the tricarboxylic acid cycle is the major mechanism of oxida-tive degradation of carbohydrate and other fuels in intact cells. The overall rate of the cycle is controlled by the citrate synthase reaction.

Intermediates of the tricarboxylic acid cycle are also used as precursors in biosynthesis. Cycle intermediates are then replen-ished by anaplerotic reactions, the most important of which is the reversible carboxylation of pyruvate to oxaloacetate at the expense of ATP. In organisms able to live on acetate as sole carbon source, a variation of the citric acid cycle, the glyoxylate cycle, comes into play. It makes possible the net formation of succinate from acetate for net biosynthesis of glucose from fatty acids.

Glucose 6-phosphate can, alternatively, be catabolized via the phosphogluconate pathway to form NADPH, required in reductive biosynthesis of fatty acids and steroids, to form ribose 5-phosphate for nucleic acid biosynthesis, and to make possible photosynthetic formation of glucose from CO_2 in some plants. This pathway also makes possible interconversion of various three-, four-, five-, six-, and seven-carbon sugars and connects all such sugars metabolically with the glycolytic sequence.

References

Books

ALWORTH, W. L.: *Stereochemistry and Its Application to Biochemis-try*, Wiley-Interscience, New York, 1972. Examines the stereo-chemistry of many enzymatic reactions, including those of the citric acid cycle.

BOYER, P. D.: *The Enzymes*, 3d ed., Academic, New York, 1971, par-ticularly vol. 5, which contains reviews of aconitase, fu-marase, and enolase; vol. 6, which has extensive reviews on oxaloacetate formation; and vol. 7, with reviews on aldolases and transaldolases.

GOODWIN, T. W. (ed.): *The Metabolic Roles of Citrate*, Academic, New York, 1968. A Biochemical Society Symposium in honor of Sir Hans Krebs. An important collection of informative ar-ticles.

KREBS, H. A., and H. L. KORNBERG: *Energy Transformations in Living Matter*, Springer-Verlag, Berlin, 1957.

LOWENSTEIN, J. M. (ed.): *The Citric Acid Cycle*, vol. 13 of *Methods in Enzymology*, Academic, New York, 1969. Comprehensive col-lection of authoritative articles on experimental methods.

LOWENSTEIN, J. M. (ed.): *Citric Acid Cycle: Control and Compart-mentation*, Dekker, New York, 1969. Another collection of valuable articles.

MEHLMAN, M. A., and R. W. HANSON (ed.): *Energy Metabolism and the Regulation of Metabolic Processes in Mitochondria*, Aca-demic, New York, 1972. Articles on the regulation of the tricar-boxylic acid cycle.

WOOD, W. B., J. H. WILSON, R. M. BENBOW, and L. E. HOOD: *Biochemis-try: A Problems Approach*, Benjamin, Menlo Park, Calif., 1974. Chapters 9 and 10 constitute a study guide with problems.

Articles

AXELROD, B.: "Other Pathways of Carbohydrate Metabolism," pp. 272–308 in D. M. Greenberg (ed.), *Metabolic Pathways*, 3d ed., vol. 1, Academic, New York, 1967. The phosphogluconate and other pathways.

BARRERA, C. R., G. NAMIHIRA, L. HAMILTON, P. MUNK, M. H. ELEY, T. C. LINN, and L. J. REED: "α-Keto Acid Dehydrogenases; XVI: Studies on the Subunit Structure of the Pyruvate Dehydrogenase Complexes from Bovine Kidney and Heart," *Arch. Biochem. Biophys.*, 148: 343–358 (1972).

DAVIS, K. A., and Y. HATEFI: "Succinate Dehydrogenase; I: Purification, Molecular Properties, and Substructure," *Biochemistry*, 10: 2509–2516 (1971).

DENTON, R. M., P. J. RANDLE, and B. R. MARTIN: "Stimulation by Ca^{2+} of Pyruvate Dehydrogenase Phosphate Phosphatase," *Biochem. J.*, 128: 161–163 (1972).

GLUSKER, J. P.: "Mechanism of Aconitase Action Deduced from Crystallographic Studies of Its Substrates," *J. Mol. Biol.*, 38: 149–162 (1968).

HANSTEIN, W. G., K. H. DAVIS, M. A. GHALAMBOR, and Y. HATEFI: "Succinate Dehydrogenase; II: Enzymatic Properties," *Biochemistry*, 10: 2517–2524 (1971).

KREBS, H. A.: "The History of the Tricarboxylic Acid Cycle," *Perspect. Biol. Med.*, 14: 154–170 (1970). A personal account of the experimental basis and origin of the cycle.

LAI, C. Y., and B. L. HORECKER: "Aldolase: A Model for Enzyme Structure-Function Relationships," *Essays Biochem.*, 8: 149–178 (1972).

LA NOUE, K. F., J. BRYLA, and J. R. WILLIAMSON: "Feedback Interactions in the Control of Citric Acid Cycle Activity in Rat Heart Mitochondria," *J. Biol. Chem.*, 247: 667–679 (1972).

LINN, T. C., J. W. PELLEY, F. H. PETTIT, F. HUCHO, D. D. RANDALL, and L. J. REED: "Purification and Properties of the Component Enzymes of the Pyruvate Dehydrogenase Complexes from Bovine Kidney and Heart," *Arch. Biochem. Biophys.*, 148: 327–342 (1972).

LOWENSTEIN, J. M.: "The Tricarboxylic Acid Cycle," pp. 146–270 in D. M. Greenberg (ed.), *Metabolic Pathways*, 3d ed., vol. 1, Academic, New York, 1967. An excellent review, including stereochemical relationships.

MOSS, J., and M. D. LANE: "The Biotin-Dependent Enzymes," *Adv. Enzymol.*, 35: 321–442 (1971).

NISHIMURA, J., and F. GRINNELL: "Mechanism of Action and Other Properties of Succinyl Coenzyme A Synthetase," *Adv. Enzymol.*, 36: 183–202 (1972).

PLAUT, G. W. E.: "DPN-Linked Isocitrate Dehydrogenase of Animal Tissues," *Curr. Top. Cell Regul.*, 2: 1–27 (1970).

SINGER, T. P., E. B. KEARNEY, and W. C. KENNEY: "Succinate Dehydrogenase," *Adv. Enzymol.*, 37: 189–272 (1973).

VOGEL, O., B. HOEHN, and U. HENNING: "Subunit Structure of the Pyruvate Dehydrogenase Complex," *Eur. J. Biochem.*, 30: 354–360 (1972).

WALKER, W. H., T. P. SINGER, S. GHISLA, and P. HEMMERICK: "Studies on Succinate Dehydrogenase: 8-α-Histidyl–FAD as the Active Center," *Eur. J. Biochem.*, 26: 279–289 (1972).

Problems

1. How many turns of the citric acid cycle are implicated in accounting for the net oxidation of each of the following compounds to carbon dioxide and water: (*a*) glucose, (*b*) glyceraldehyde, (*c*) citric acid, (*d*) succinic acid?

2. Suppose that acetate labeled at the methyl position with ^{14}C is added to each of the following respiring cell preparations: (*a*) liver cells, (*b*) liver cells to which malonate has been added, (*c*) bean seedlings. Subsequently, succinic acid is isolated from each of the preparations. Which carbon atoms of succinic acid would be expected to be labeled?

3. Isotopically labeled substrates were incubated with respiring muscle suspensions poisoned with malonate, and the tricarboxylic acid cycle intermediates listed were isolated. In each case, predict which carbon atoms of the isolated products would be expected to be labeled.
   (*a*) Substrate is $^{14}CH_3COCOOH$ and excess unlabeled oxaloacetate. Product isolated is (1) isocitric acid, (2) α-ketoglutaric acid, (3) succinic acid.
   (*b*) Substrate is $^{14}CH_3COCOOH$. Product is (1) isocitric acid, (2) α-ketoglutaric acid, (3) succinic acid.
   (*c*) Substrate is $H^{14}CO_3^-$. Product is (1) isocitric acid, (2) α-ketoglutaric acid, (3) succinic acid, (4) malic acid.
   (*d*) Substrate is 5-^{14}C-fructose. Product is (1) isocitric acid, (2) succinic acid.

4. Pyruvate isotopically labeled in the 2-carbon atom $(CH_3{}^{14}COCOO^-)$ is fed into the tricarboxylic acid cycle of liver. Assuming that all the labeled pyruvate is immediately converted to citrate via acetyl-CoA, determine the fraction of the initial radioactivity lost as $^{14}CO_2$ during (*a*) the first turn of the cycle, (*b*) the second turn, (*c*) the third turn.

5. Suppose that under the conditions of Problem 4, the pyruvate were labeled instead in the 3-carbon atom $(^{14}CH_3COCOO^-)$. Determine the fraction of the initial radioactivity lost as $^{14}CO_2$ during (*a*) the first turn of the cycle, (*b*) the second turn, (*c*) the third turn, (*d*) the fourth turn, (*e*) the fifth turn.

6. The substrates indicated below were added to suspensions of minced rat liver in which succinic dehydrogenase was completely inhibited by the addition of malonate. Assuming that the glycogen supply is plentiful, write balanced equations for the conversion to succinic acid of (*a*) citric acid, (*b*) pyruvic acid, and (*c*) fumaric acid. (Reference to page 551 may be helpful in solving this problem and subsequent problems of the same type.)

7. Repeat Problem 6 under conditions giving negligible production of pyruvate in the cytosol.

8. Write a balanced equation for the conversion of 3-^{14}C-glucose into D-ribose 5-phosphate in the liver by a pathway involving the use of pyridine nucleotides. What carbon atom(s) of the product would be expected to be labeled?

9. Write a balanced equation for the conversion of glucose 6-

phosphate to ribose 5-phosphate by a pathway that does not involve participation of pyridine nucleotides.

10. Write a balanced equation for the conversion of D-glucose into pyruvic acid by a pathway that does not require phosphofructokinase as an essential enzyme.

11. Write a balanced equation for the conversion of D-ribose 5-phosphate into lactic acid in the liver. If ribose 5-phosphate is labeled with ^{14}C in carbon atom 1, which carbon atom(s) in lactic acid will become labeled?

12. From data in the text calculate the equilibrium constant and standard-free-energy change for the conversion of malate to citrate according to the equation

Malate + NAD$^+$ + acetyl-CoA + H$_2$O $\longrightarrow$

citrate + CoA + NADH + H$^+$

13. The intramitochondrial concentration of malate under some conditions is 0.22 mM, and the NAD/NADH ratio is 20. Calculate the maximum intramitochondrial concentration of oxaloacetate under these conditions. (Assume a temperature of 25°C.)

We shall now see how pairs of electrons derived from intermediates of the tricarboxylic acid cycle and other substrates
flow down the respiratory chain to molecular oxygen, the ultimate electron acceptor in respiration. This process, called
electron transport, is the mainspring of cell activities, since
it releases a large amount of free energy, much of which is
conserved in the form of the phosphate-bond energy of ATP,
in the process called *oxidative phosphorylation* (see Figure
17-1, page 445).

Oxidation-reduction enzymes, particularly those participating in electron transport, generally are more complex in
structure and mechanism and less well understood than
other classes of enzymes. Most of the members of the
electron-transport chain are embedded in the mitochondrial
inner membrane and exceedingly difficult to extract in soluble form and purify. Moreover, we do not yet fully understand how the free-energy release occurring during electron
transport is conserved and transformed into phosphate-bond
energy during oxidative phosphorylation. For these reasons
electron transport and oxidative phosphorylation are still
challenging frontier areas of biochemical research.

In this chapter we shall first consider the major classes of
oxidation-reduction enzymes and then see how they participate in the mainstream of electron transport in aerobic cells.
We shall also examine briefly the properties of other types of
oxidative reactions specialized for different purposes. Then,
in the following chapter, we shall see how the energy delivered by electron transport is utilized to generate ATP in the
process of oxidative phosphorylation.

Oxidation-Reduction Reactions

Although several enzymatic oxidation-reduction reactions
have already been described (pages 427 and 450), they will
now be treated in a more systematic manner so that their
mechanisms, equilibria, and energetic relationships can be
compared. Oxidation-reduction reactions (also called oxidoreductions or redox reactions) are those in which there is
transfer of electrons from an electron donor (the reducing
agent or reductant) to an electron acceptor (the oxidizing

agent or oxidant). In some oxidation-reduction reactions the transfer of one or more electrons is made via the transfer of hydrogen; dehydrogenation is thus equivalent to oxidation. Often the terms reducing equivalents or electron equivalents are used to refer to electrons and/or hydrogen atoms participating in oxidoreductions.

Oxidizing and reducing agents function as conjugate redox pairs or couples, consisting of an electron donor and its conjugate electron acceptor, just as Brönsted acids and bases function as conjugate acid-base pairs:

Acid-base reactions:

$$\text{Proton donor} \rightleftharpoons H^+ + \text{proton acceptor}$$

Oxidation-reduction reactions:

$$\text{Electron donor} \rightleftharpoons e^- + \text{electron acceptor}$$

As different acids differ in their tendency to dissociate their protons (page 47), so different reducing agents differ in their tendency to lose electrons. The tendency of a reducing agent to lose electrons (or an oxidizing agent to gain electrons) is given by the standard oxidation-reduction potential, defined as the electromotive force (emf) in volts given by a half-cell in which the reductant and oxidant are both present at 1.0 M concentration, at 25°C and pH 7.0, in equilibrium with an electrode which can reversibly accept electrons from the reductant species (Figure 18-1). By convention the electrode equation is written in the direction

$$\text{Oxidant} + ne^- \rightleftharpoons \text{reductant}$$

where n is the number of electrons transferred. Also by convention, the standard oxidation-reduction potential of the hydrogen-electrode reaction

$$2H^+ + 2e^- \rightleftharpoons H_2$$

is used as the reference potential. It is set at 0.0 V when the pressure of H_2 gas is 1.0 atm, $[H^+]$ is 1.0 M (that is, pH = 0.0), and the temperature is 25°C. When this value is corrected to pH 7.0 ($[H^+] = 1 \times 10^{-7}$ M), the reference pH assumed in all biochemical calculations, the standard oxidation-reduction potential of the hydrogen electrode becomes -0.42 V.

The standard oxidation-reduction potentials (also called standard redox potentials) of a number of biologically important redox pairs are given in Table 18-1. In redox pairs having a more negative standard potential than the $2H^+$–H_2 couple the reductant has a greater tendency to lose electrons than molecular hydrogen; conversely, in pairs with a more positive potential, the reductant has a lesser tendency to lose electrons than hydrogen. Note that the oxygen-water couple has a strongly positive standard potential, $+0.82$ V. Molecular oxygen thus has a very high affinity for electrons and is a very good oxidizing agent. Conversely, water has very little

Figure 18-1
Measurement of the oxidation-reduction potential. The solution containing the mixture of the oxidized and reduced forms of the compound to be examined is placed in the left vessel; they equilibrate with the inert electrode. A salt bridge containing saturated KCl solution provides electrical connection to a reference cell of known potential (right). From the total observed emf and the known emf of the reference cell, the emf of the left cell containing the mixture of oxidant and reductant is determined. When the oxidant and reductant are equal in concentration, the potential of the left cell is the standard or midpoint oxidation-reduction potential.

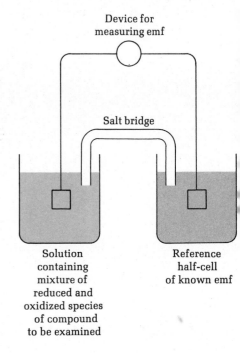

Device for measuring emf

Salt bridge

Solution containing mixture of reduced and oxidized species of compound to be examined

Reference half-cell of known emf

Table 18-1 Standard oxidation-reduction potentials of some conjugate redox pairs expressed on the basis of two-electron transfers at pH near 7.0 and temperature 25 to 30°C

The standard potentials of the cytochromes and of ubiquinone vary somewhat with their state, i.e., whether isolated or present in the mitochondrial membrane; values given are for the latter case

Electrode equation	E_0', V
Acetate + $2H^+$ + $2e^-$ ⇌ acetaldehyde	−0.58
$2H^+$ + $2e^-$ ⇌ H_2	−0.421
α-Ketoglutarate + CO_2 + $2H^+$ + $2e^-$ ⇌ isocitrate	−0.38
Acetoacetate + $2H^+$ + $2e^-$ ⇌ β-hydroxybutyrate	−0.346
NAD^+ + $2H^+$ + $2e^-$ ⇌ NADH + H^+	−0.320
$NADP^+$ + $2H^+$ + $2e^-$ ⇌ NADPH + H^+	−0.324
Acetaldehyde + $2H^+$ + $2e^-$ ⇌ ethanol	−0.197
Pyruvate + $2H^+$ + $2e^-$ ⇌ lactate	−0.185
Oxaloacetate + $2H^+$ + $2e^-$ ⇌ malate	−0.166
Fumarate + $2H^+$ + $2e^-$ ⇌ succinate	−0.031
Ubiquinone + $2H^+$ + $2e^-$ ⇌ ubiquinol	+0.10
2 cytochrome $b_{K(ox)}$ + $2e^-$ ⇌ 2 cytochrome $b_{K(red)}$	+0.030
2 cytochrome c_{ox} + $2e^-$ ⇌ 2 cytochrome c_{red}	+0.254
2 cytochrome $a_{3(ox)}$ + $2e^-$ ⇌ 2 cytochrome $a_{3(red)}$	+0.385
$\frac{1}{2}O_2$ + $2H^+$ + $2e^-$ ⇌ H_2O	+0.816

tendency to lose electrons and is thus a very weak reducing agent.

Just as the Henderson-Hasselbalch equation (page 50) expresses the quantitative relationship between the dissociation constant of an acid, its pH, and the concentration of its proton donor and acceptor species, a formally similar relationship, the *Nernst equation*, expresses the relationship between the standard redox potential of a given redox couple, its observed potential, and the concentration ratio of its electron-donor and electron-acceptor species. The Nernst equation is

$$E_h = E_0' + \frac{2.303RT}{n\mathscr{F}} \log \frac{[\text{electron acceptor}]}{[\text{electron donor}]} \tag{1}$$

in which E_0' is the standard redox potential (pH = 7.0, $T = 25°C$ or 298 K, all concentrations at 1.0 M), E_h the observed electrode potential, R the gas constant (8.31 J deg^{-1} mol^{-1}), T the temperature in Kelvins, n the number of electrons being transferred, and $\mathscr{F}$ the faraday (23,062 cal V^{-1} = 96,406 J V^{-1}). At 25°C (298 K) the term 2.303 $RT/n\mathscr{F}$ has the value 0.059 when $n = 1$ and 0.03 when $n = 2$. Since it is customary to calculate equilibria of biological redox couples in terms of two-electron transfers, the Nernst equation simplifies to

$$E_h = E_0' + 0.03 \log \frac{[\text{electron acceptor}]}{[\text{electron donor}]} \tag{2}$$

The Nernst equation expresses mathematically the shape of a titration curve of a given electron donor with a strong oxidizing agent (Figure 18-2). As the titration proceeds, an increasing fraction of the electron donor is converted into its oxidized form, resulting in an increase in the ratio [electron acceptor]/[electron donor] until a midpoint is reached at which [electron acceptor] = [electron donor]. At this point it is clear that the term 0.03 log ([electron acceptor]/[electron donor]) is zero, and the Nernst equation simplifies to

$$E_h = E_0'$$

The standard redox potential is therefore the emf in volts of the half-cell at the precise midpoint of the titration curve of a given reductant at pH 7.0, 25°C, and 1.0 atm, just as the pK' of an acid equals the pH at the midpoint of an acid-base titration curve. E_0', the standard redox potential, is often called the _midpoint potential_. Unlike acid-base titrations, which are fast ionic reactions requiring no catalyst, oxidation-reduction reactions of organic compounds are rather slow and ordinarily require a catalyst or enzyme to ensure that equilibrium will be attained quickly.

The standard redox potentials of various biological oxidation-reduction systems allow us to predict the direction in which electrons will tend to flow from one redox pair to another under standard conditions, just as the phosphate-group transfer potential (page 403) allows us to predict the direction in which phosphate groups will be enzymatically transferred. For example, from the oxidation-reduction potentials in Table 18-1 we can conclude that the NAD^+–NADH pair will tend to lose electrons to the oxaloacetate-malate pair when all four components are present at their standard concentrations of 1 M, since the NAD^+–NADH couple has a more negative potential and thus a greater electron "pressure" than the oxaloacetate-malate couple, which has a higher affinity for electrons. Thus, if we start with 1 M concentrations of all components, the point of equilibrium of the reaction

$$NADH + H^+ + \text{oxaloacetate} \rightleftharpoons NAD^+ + \text{malate}$$

will be to the right. Note that this reaction will not occur at a measurable rate unless we add a catalyst, e.g., the enzyme malate dehydrogenase (page 460). Although this reaction will tend to go to the right under standard conditions, it can also be made to go to the left if we adjust the concentrations of all components appropriately, i.e., if we reduce the concentrations of the reactants and/or increase the concentrations of the products. Indeed, in the mitochondrion this reaction, which is a step in the tricarboxylic acid cycle (page 446), normally does go to the left because oxaloacetate is very rapidly removed, so that its concentration is normally very low in comparison with that of the other components of the reaction.

The equilibrium concentrations of all four components in an oxidoreduction reaction can be calculated from the standard potentials of the interacting systems and the total concentrations of all components.

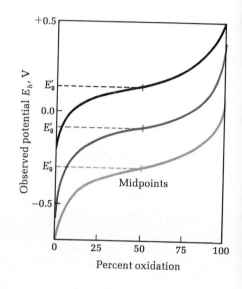

Figure 18-2
Oxidation-reduction titration curves of three conjugate redox pairs of differing standard or midpoint potentials, which can be extrapolated from perpendiculars drawn from the midpoint to the ordinate.

Classes of Electron-Transferring Enzymes

Four types of oxidation-reduction enzymes or electron-transferring proteins participate in the mainstream of electron transport from organic substrates to molecular oxygen. They are (1) *pyridine-linked dehydrogenases,* which require either NAD or NADP as coenzyme, (2) *flavin-linked dehydrogenases,* which contain flavin adenine dinucleotide (FAD) or flavin mononucleotide (FMN) as prosthetic group, (3) *iron-sulfur proteins,* and (4) *cytochromes,* which contain an iron-porphyrin prosthetic group. In addition to these proteins, the lipid-soluble coenzyme *ubiquinone* or *coenzyme Q* also functions in electron transport. These components will now be described in turn.

Pyridine-Linked Dehydrogenases

Since members of this class of dehydrogenases require as coenzyme either NAD or NADP, which contain nicotinamide (page 340), a derivative of pyridine, they are generically called *pyridine-linked dehydrogenases.* Some 200 or more, functioning in different aspects of metabolism, are known; important examples are listed in Table 18-2. They catalyze the general reactions

$$\text{Reduced substrate} + \text{NAD}^+ \rightleftharpoons$$
$$\text{oxidized substrate} + \text{NADH} + \text{H}^+ \quad (3)$$

or

$$\text{Reduced substrate} + \text{NADP}^+ \rightleftharpoons$$
$$\text{oxidized substrate} + \text{NADPH} + \text{H}^+ \quad (4)$$

depending upon which pyridine nucleotide they require or prefer. These reactions involve reversible transfer of two reducing equivalents from the substrate, in the form of a hydride ion (H^-), to the 4 position of the nicotinamide ring (in oxidized form) of the pyridine nucleotide; the other hydrogen is removed from the substrate as a free H^+ ion (Figure 13-6, page 342). Direct transfer of a hydrogen atom from the substrate to NAD^+ has been proved by using substrate molecules in which the hydrogen atoms normally removed by the dehydrogenase were first labeled with deuterium or tritium. Following the enzymatic dehydrogenation of the labeled substrate, one of the labeled hydrogen atoms was recovered in the 4 position of the pyridine ring of the NADH formed, without dilution by the H^+ of the water medium.

The pyridine nucleotides are noncovalently and relatively loosely bound to the dehydrogenase protein. NAD and NADP are therefore to be regarded not as fixed prosthetic groups but as substrates, since in most cases they bind to, and dissociate from, the active site during the catalytic cycle. The pyridine nucleotides thus serve as dissociable carriers of electrons (page 427).

Many of the pyridine-linked dehydrogenases have been obtained in pure crystalline form, including glyceraldehyde

Table 18-2   Properties of pyridine-linked dehydrogenase systems

System	E_0' of substrate couple	Stereo-specificity for position 4
NAD-linked		
Isocitrate	−0.38	
D-β-Hydroxybutyrate	−0.346	
Glyceraldehyde 3-phosphate	−0.29	
Dihydrolipoyl	−0.29	A or B
L-β-Hydroxyacyl-CoA	−0.238	B
Ethanol	−0.197	A
Lactate	−0.185	A
Glycerol 3-phosphate	−0.19	B
L-Malate	−0.166	A
NADP-linked		
Isocitrate	−0.38	A
Glucose 6-phosphate	−0.32	B
NAD or NADP		
L-Glutamate	−0.14	B

3-phosphate dehydrogenase (page 427), lactate dehydrogenase (page 431), and malate dehydrogenase (page 460). Some pyridine-linked dehydrogenases, e.g., β-hydroxybutyrate dehydrogenase, are localized in the mitochondria, some, e.g., lactate dehydrogenase, are localized in the cytosol, and some, e.g., malate dehydrogenase, are found in both compartments. In animal cells NAD is usually present in much greater amounts than NADP. In the liver about 60 percent of the total NAD is present in the mitochondria, and the rest is in the extramitochondrial cytoplasm. The NADP content of cells is in proportion to their biosynthetic activity. NAD-linked dehydrogenases serve primarily in respiration, e.g., in the transfer of electrons from substrates toward oxygen, whereas NADP-linked dehydrogenases serve primarily to transfer electrons from intermediates of catabolism to the intermediates of biosynthesis (pages 375, 467, and 666). Free NAD^+ and $NADP^+$ occur in solution in two types of conformation, the *open,* or *extended, forms* and *closed,* or *stacked, forms,* in which the planar nicotinamide and adenine rings are parallel to each other. The stacked forms predominate in solution, but when NAD^+ binds to some dehydrogenases, it assumes the open conformation. Most of the pyridine-linked dehydrogenases are specific for either NAD or NADP, but a few, e.g., glutamate dehydrogenase, can react with both coenzymes (Table 18-2).

Many pyridine-linked dehydrogenases contain tightly bound divalent metal ions; alcohol dehydrogenase, for example, contains Zn^{2+}.

Measurement of Pyridine-Linked Dehydrogenases

The enzymatic reduction of NAD^+ and $NADP^+$ [equations (3) and (4)] is accompanied by three characteristic changes useful for measuring the activity of pyridine-linked dehydrogenases. Neither the oxidized nor the reduced form of these coenzymes absorbs light in the visible range, but they both absorb very strongly in the ultraviolet near 260 nm, the absorption maximum of the adenine ring (page 314). When NAD^+ and $NADP^+$ are reduced, a new absorption maximum at 340 nm appears (Figure 18-3). The increase at 340 nm reflects the reduction of the aromatic pyridine ring of the nicotinamide moiety. The appearance (or disappearance) of the 340-nm absorption is used to follow the course of reactions catalyzed by pyridine-linked dehydrogenases.

Pyridine-linked oxidoreductions can also be followed by measurements of changes in pH, since reduction of NAD^+ (or $NADP^+$) results in formation of an H^+ ion [equations (3) and (4)]. Another method is by measurement of fluorescence changes. However, although very sensitive, the fluorescence method is subject to quenching or interference by a number of factors.

NAD^+ and $NADP^+$ can be reduced nonenzymatically by reducing agents such as sodium dithionite or sodium borohydride. NADH and NADPH can in turn be nonenzymatically reoxidized with ferricyanide, but they are not oxidized directly by molecular oxygen at pH 7.0.

Figure 18-3
The absorption spectra of NAD^+ and NADH. A is the absorbance, $\log I_0/I$. The molar absorption coefficient (see Figure 12-7, page 314) of NADH at 340 nm is 6.23×10^3 cm^2 mol^{-1}.

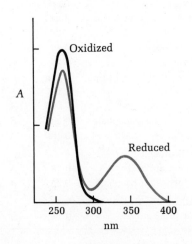

Stereospecificity of Pyridine-Linked Dehydrogenases

Pyridine-linked dehydrogenases may be stereospecific in two senses. Many of them are specific for one given stereoisomer of their substrates; e.g., lactate dehydrogenase is specific for L-lactate. However, most pyridine-linked dehydrogenases are stereospecific in another sense. When the pyridine ring of NAD^+ or $NADP^+$ is reduced by nonenzymatic means, the hydrogen atom that is transferred to carbon atom 4 of the nicotinamide ring may approach either side of the ring, to yield equal amounts of the two possible stereoisomeric forms (Figure 18-4). However, when the pyridine ring is reduced enzymatically by a dehydrogenase, the hydrogen atom from the substrate is transferred stereospecifically to only one side of the ring. Table 18-2 shows that some dehydrogenases transfer hydrogen to the A side of the ring and some to the B side. This type of stereospecificity for the pyridine ring has been established by using substrate molecules labeled with tritium or deuterium. Only one dehydrogenase is known to have no stereospecificity for the nicotinamide ring, namely, dihydrolipoyl dehydrogenase (page 450). These observations indicate that both the substrate and the NAD^+ molecules must possess a specific stereochemical orientation toward each other on the enzyme catalytic site.

Kinetics and Mechanism of Pyridine-Linked Dehydrogenases

Pyridine-linked dehydrogenases show characteristic Michaelis-Menten behavior with both the substrate and the pyridine nucleotide. By general methods described elsewhere (page 189) the Michaelis constants for the oxidized and reduced substrates and for the oxidized and reduced pyridine nucleotides have been established for a number of dehydrogenases; an example is shown in Table 18-3.

Nearly all pyridine-linked dehydrogenases that have been examined show the typical kinetic behavior of ordered bisubstrate reactions, in which there is a compulsory sequence for adding the substrate and coenzyme to the active site (page 202). The pyridine nucleotide is the leading substrate (page 204) and must bind to the active site first; then the substrate is bound. After the transfer of the reducing equivalents from substrate to coenzyme on the active site, the oxidized substrate departs first, followed by the reduced coenzyme. Because of this compulsory sequence of interactions, there are normally two binary complexes, E—NAD^+ and E—NADH, and two ternary complexes,

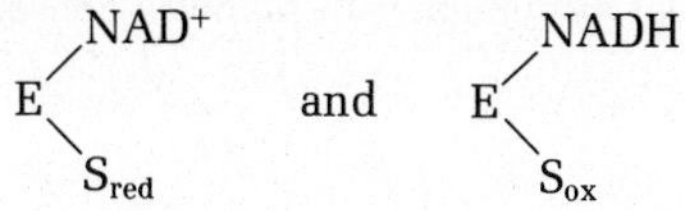

$$E\diagdown \genfrac{}{}{0pt}{}{NAD^+}{S_{red}} \quad \text{and} \quad E\diagdown \genfrac{}{}{0pt}{}{NADH}{S_{ox}}$$

participating in the catalytic cycle.

For lactate dehydrogenase important information, summarized in Figure 18-5, has been obtained on the geometry of the active site and the reaction mechanism by various experi-

---

Figure 18-4
Oxidized and reduced forms of NAD. In the oxidized form the hydrogen atom at position 4 is in the plane of the pyridine ring; in the reduced form the two hydrogens at position 4 are out of the plane of the ring. When NAD is reduced enzymatically with a deuterium-labeled substrate, the deuterium combines with carbon atom 4 from one side or the other, depending on the dehydrogenase, to yield one of two possible products, one with deuterium on the A side, the other with deuterium on the B side of the plane of the ring. Note the quaternary nitrogen in the oxidized form and the quinonoid bond system in the reduced form.

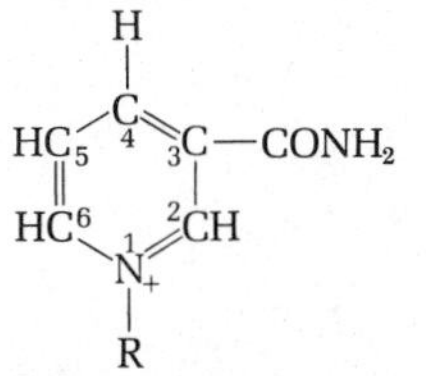

NAD^+, oxidized form

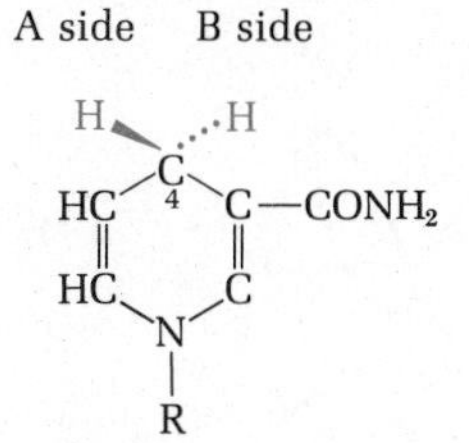

A side   B side

NADH, reduced form

Table 18-3 Approximate K_M values for bovine heart lactate dehydrogenase

Each was determined with the other reactants at saturating concentrations

Substrate	K_M, mM
Lactate	9.0
NAD^+	0.075
Pyruvate	0.14
NADH	0.001

Figure 18-5
The mechanism of the lactate dehydrogenase reaction. This enzyme (mol
wt 140,000) contains four identical subunits, each of which binds the coenzyme nicotinamide adenine dinucleotide (NAD) first (page 203), followed
by binding of the substrate. The structure of the NAD⁺ and substrate sites
has been deduced by x-ray analysis of the inactive NAD⁺-pyruvate complex
of the enzyme.

Schematic representation of the binding site for the open form of NAD⁺
showing the specific binding interactions with amino acid residues at the
active site. The second substrate, which is actually the enol form of pyruvate (page 405), is shown covalently linked via its methylene group to position 4 of the nicotinamide ring. His 195, which is essential for catalytic
activity, is shown hydrogen-bonded to the carbonyl oxygen of nicotinamide. [Adapted from M. J. Adams and 11 colleagues from the laboratories of M. G. Rossmann and N. O. Kaplan, "Structure-Function Relationships in Lactate Dehydrogenase," Proc. Natl. Acad. Sci. (U.S.), 70: 1970
(1973).]

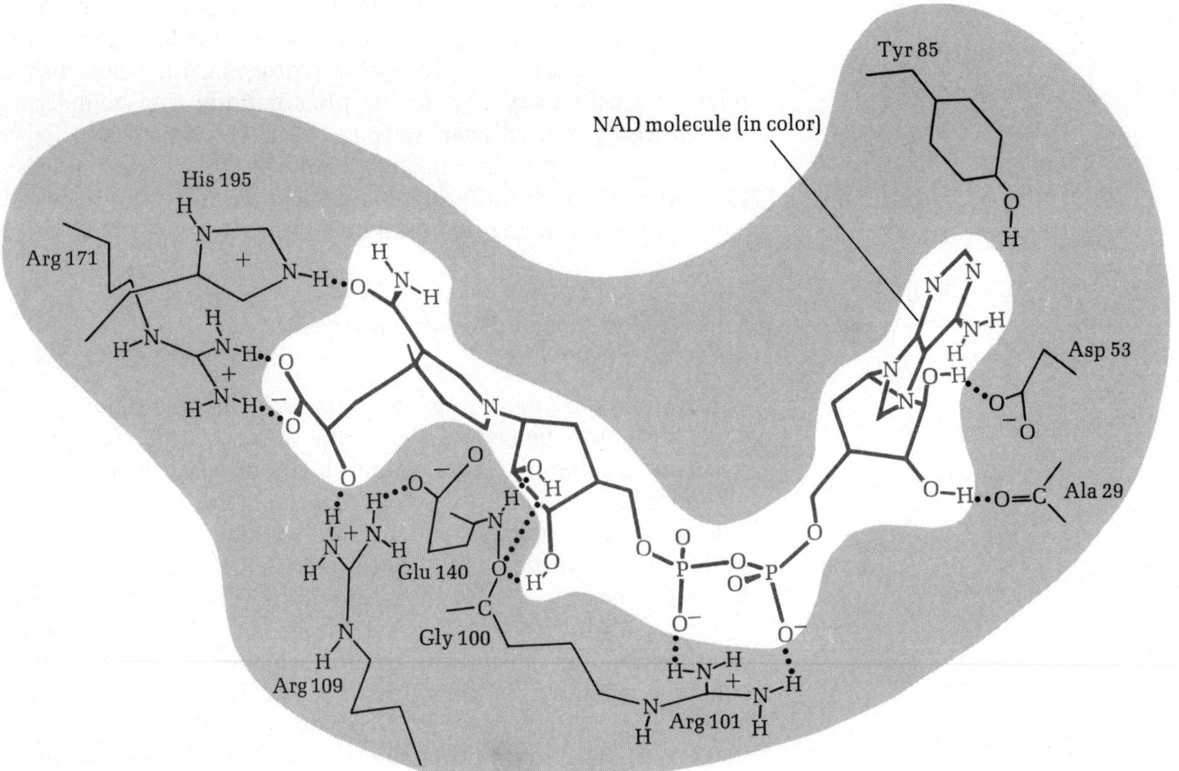

Possible reaction mechanism for the formation of the nicotinamide-
pyruvate intermediate showing the participation of His 195. Once His 195
has become protonated, it may shift position to form a hydrogen bond with
the carbonyl oxygen of nicotinamide as depicted above.

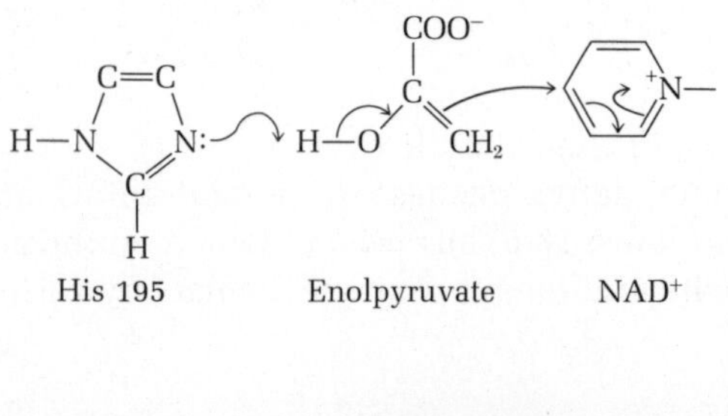

mental approaches, including x-ray analysis of the structure of the crystalline enzyme, the amino acid sequence, and nuclear magnetic resonance (page 151) studies of the conformational changes of the NAD^+ and NADH molecules as they bind to the dehydrogenase protein. Lactate dehydrogenase is the first pyridine-linked dehydrogenase for which the three-dimensional structure has been obtained.

Equilibria of Pyridine-Linked Dehydrogenases

The direction of reaction and the equilibrium composition of pyridine-linked oxidoreduction systems can be predicted from the standard oxidation-reduction potentials of the NADH–NAD^+ (NADPH–$NADP^+$) couple ($E_0' = -0.32$ V) and of the reduced substrate-oxidized substrate couple. Table 18-2 gives the standard oxidation-reduction potentials of some substrate redox pairs which react with pyridine-linked dehydrogenases. Substrate systems having a more negative standard potential than the NAD (or NADP) redox pair tend to lose electrons from the reduced substrate to the oxidized form of the coenzyme, and those having a more positive standard potential tend to accept electrons from NADH or NADPH, when tested under standard conditions of 1.0 M concentration of all reactants and products.

The NADH–NAD^+ system can also transfer electrons from one substrate couple to another by virtue of the capacity of NADH to act as a common intermediate shared by two pyridine-linked reactions, each catalyzed by a specific dehydrogenase. For example, in the glycolytic sequence (page 422) glyceraldehyde 3-phosphate is oxidized by pyruvate in the following reactions, which share NADH (in color) as common intermediate:

Glyceraldehyde 3-phosphate + P_i + NAD^+ $\rightleftharpoons$
$$1,3\text{-diphosphoglycerate} + NADH + H^+ \quad (5)$$

$$NADH + H^+ + \text{pyruvate} \rightleftharpoons NAD^+ + \text{lactate} \quad (6)$$

Sum:

Glyceraldehyde 3-phosphate + P_i + pyruvate $\rightleftharpoons$
$$1,3\text{-diphosphoglycerate} + \text{lactate} \quad (7)$$

If we start this set of reactions with equimolar concentrations of all components at pH 7.0, the direction of flow of electrons will be from the more electronegative glyceraldehyde 3-phosphate couple ($E_0' = -0.29$ V) to the more positive lactate-pyruvate couple ($E_0' = -0.19$ V) regardless of the standard potential of the intermediate electron carrier NAD (in thermodynamic calculations we are concerned only with the initial and final states, not the pathway). The overall equilibrium will therefore be far in the direction of lactate.

Since a proton is always formed or absorbed in pyridine-linked reactions, their equilibria will change with the pH of the system. The standard oxidation-reduction potential of the NADH–NAD^+ couple becomes 0.03 V more negative for each pH unit above pH 7.0 and 0.03 V more positive for each pH unit below 7.0.

Flavin-Linked Dehydrogenases and Oxidases

These enzymes contain as tightly bound prosthetic groups either *flavin mononucleotide* (FMN) or *flavin adenine dinucleotide* (FAD) (pages 339 and 340). The active portion of FMN or FAD that participates in the oxidoreduction is the *isoalloxazine* ring of the riboflavin moiety, which is reversibly reduced (Figure 18-6). The reaction is formally shown as a direct transfer of a pair of hydrogen atoms from the substrate to yield the reduced forms, designated $FMNH_2$ and $FADH_2$.

$$SH_2 + E\text{—}FMN \rightleftharpoons S + E\text{—}FMNH_2$$

$$SH_2 + E\text{—}FAD \rightleftharpoons S + E\text{—}FADH_2$$

The most important flavin-linked dehydrogenases in the mainstream of respiration and electron transport, all localized in the mitochondria, are (1) *NADH dehydrogenase,* which contains FMN and catalyzes transfer of electrons from NADH to the next member of the electron-transport chain, (2) *succinate dehydrogenase,* active in the tricarboxylic acid cycle (page 459), (3) *dihydrolipoyl dehydrogenase,* a component of the pyruvate (page 450) and α-ketoglutarate dehydrogenase (page 457) systems, and (4) *acyl-CoA dehydrogenase,* which catalyzes the first dehydrogenation step during fatty acid oxidation (page 548). There are many other flavin-containing enzymes, which catalyze specialized oxidoreductions not in the mainstream of electron transport, e.g., D-*amino acid oxidase* (page 566), *xanthine oxidase* (page 741), *orotate reductase* (page 736), and *aldehyde oxidase* (Table 18-4).

Flavin-linked dehydrogenases differ significantly from pyridine-linked dehydrogenases in that the flavin nucleotide is very tightly bound to the enzyme protein and thus func-

Figure 18-6
Reduction of the isoalloxazine ring of flavin nucleotides. R represents the remainder of the flavin nucleotide molecule. See Figure 13-3 (page 339) for complete structural formulas.

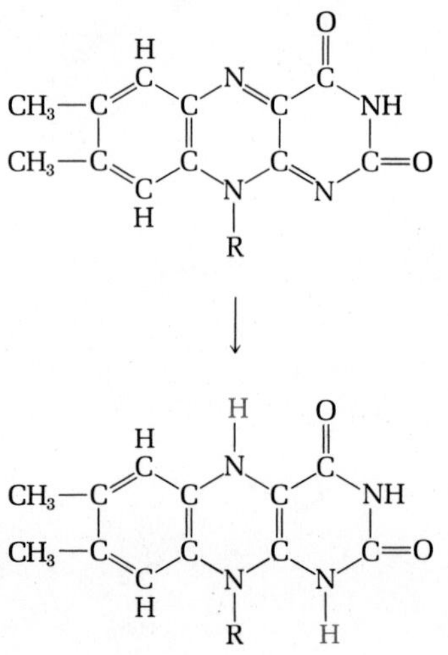

Table 18-4 Properties of some flavin-linked dehydrogenases and oxidases

Dehydrogenases	Mol wt	Flavin nucleotides	Metal
"Mainstream"			
NADH dehydrogenase	~ 300,000†	FMN	Fe
	78,000‡	FMN	Fe
Succinate dehydrogenase	100,000	FAD	Fe
Dihydrolipoyl dehydrogenase		FAD	None
Acyl-CoA dehydrogenase		FAD	None
Electron-transferring flavoprotein		FAD	None
Others			
Glycerol 3-phosphate dehydrogenase		FAD	Fe
D-Amino acid oxidase	100,000	FAD	None
Glucose oxidase	154,000	FAD	None
Xanthine oxidase	300,000	FAD	Fe, Mo
Aldehyde oxidase	280,000	FAD	Fe, Mo
Dihydroorotate oxidase	115,000	FMN, FAD	Fe

† Particulate.
‡ Soluble.

Figure 18-7
The covalent linkage of FAD to a histidyl residue in succinate dehydrogenase. R indicates the remainder of the FAD molecule.

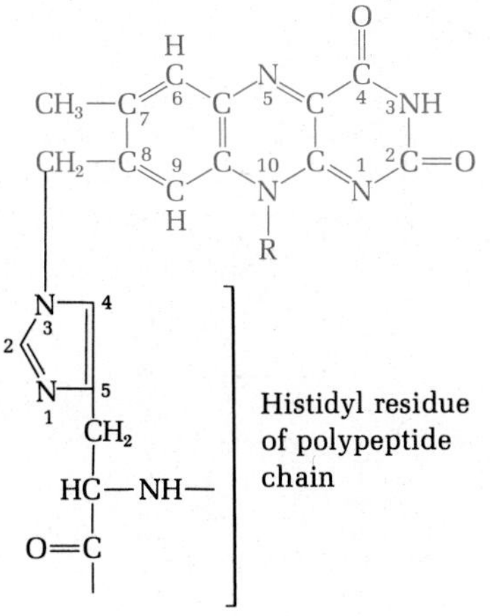

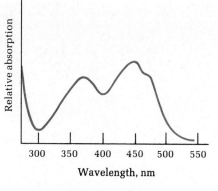

Histidyl residue of polypeptide chain

Figure 18-8
Two reducible dyes often used as artificial electron acceptors from flavin dehydrogenases.

Phenazine methosulfate

2,6-Dichlorophenolindophenol

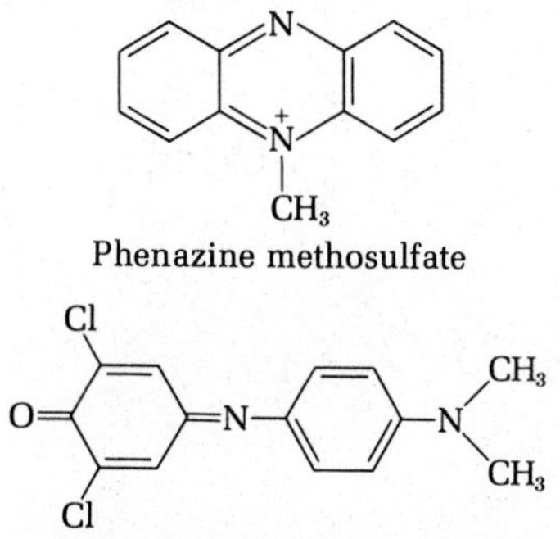

Figure 18-9
Absorption spectrum of the fully oxidized form of a flavin dehydrogenase. Reduction of the enzyme causes loss of the absorption peak near 450 nm.

tions as a prosthetic group rather than as a coenzyme (page 185); the flavin nucleotide does not leave the enzyme during or after the catalytic cycle. In most flavin dehydrogenases, including NADH dehydrogenase, the flavin nucleotide is noncovalently bound; dissociation constants are in the range 10^{-8} to 10^{-11} M. Such flavoenzymes can sometimes be dissociated by exposure to high ionic strength or low pH. However, in other flavin enzymes, such as succinate dehydrogenase, the flavin nucleotide is covalently linked by a bond between the isoalloxazine ring of FAD and an amino acid residue of the protein (Figure 18-7).

Flavin-linked oxidation-reduction enzymes may be placed into two classes, *dehydrogenases* and *oxidases,* according to their ability to react with electron acceptors. In the flavin dehydrogenases, such as NADH dehydrogenase and succinate dehydrogenase, there is little or no tendency of the reduced form of the flavin nucleotide to be reoxidized by molecular oxygen. Reduced flavin oxidases, in contrast, are reoxidized by oxygen to yield hydrogen peroxide. In members of this group, which includes D-amino acid oxidase and xanthine oxidase, the flavin nucleotide is bound to the protein in such a way that the reduced form of the nucleotide is available to react with oxygen. In the cell the immediate acceptor of electrons from the flavin-linked dehydrogenases appears to be ubiquinone (coenzyme Q) of the electron-transport chain, as we shall see. However, reduced flavin dehydrogenases are also reoxidized by certain artificial electron acceptors, such as ferricyanide, or the reducible dyes methylene blue (page 447), phenazine methosulfate, or 2,6-dichlorophenolindophenol (Figure 18-8). These dyes undergo changes in their absorption spectra when they are reduced by flavin dehydrogenases; an example is the reaction with methylene blue:

$$E\text{—}FADH_2 + \text{methylene blue}_{ox} \longrightarrow E\text{—}FAD + \text{methylene blue}_{red}$$
(Blue)   (Colorless)

Because of this property the artificial electron acceptors are often used in quantitative assay of flavin-linked dehydrogenases.

In their fully oxidized form different flavin dehydrogenases are yellow, red, brown, or green; they usually have broad absorption peaks near 370 and 450 nm (Figure 18-9). When they are fully reduced, enzymatically or chemically, they undergo bleaching with loss of the 450-nm absorption. Characteristic changes in fluorescence also occur during the oxidation-reduction cycle of some flavin enzymes.

Although some flavoproteins shuttle between fully oxidized and fully reduced forms by simultaneous two-electron transfers, others appear to transfer only one electron at a time in their normal catalytic cycle and thus can split electron pairs. Transfer of only one hydrogen atom or electron to a molecule of FAD or FMN (or loss of one hydrogen or electron from a molecule of FADH₂ or FMNH₂) leads to formation of the half-reduced, or *semiquinone,* form of the flavin nucleo-

Relative absorption

300   350   400   450   500   550

Wavelength, nm

tide. These forms, which are free radicals, can be detected either by their characteristic absorption spectra or by _electron spin resonance spectroscopy_, which signals the presence of free radicals, i.e., molecules with unpaired electron spins, through their characteristic behavior in a magnetic field.

Some flavoproteins contain metals in addition to the flavin nucleotide, particularly iron and molybdenum; these metal components are essential for catalytic activity (Table 18-4). The iron-containing flavoenzymes (examples are succinate dehydrogenase and NADH dehydrogenase) also contain especially reactive sulfur atoms. When _iron-sulfur flavoproteins_ are treated with acid, H_2S is evolved. Since there are no heme groups in such proteins, the iron is often termed _nonheme iron._ The iron atoms in iron-sulfur flavoproteins undergo Fe(II)–Fe(III) changes and appear to participate in electron transfers to or from the flavin prosthetic group. Other flavin-linked enzymes, such as xanthine oxidase and aldehyde oxidase, contain molybdenum as well as iron.

NADH dehydrogenase, which is an important member of the electron-transport chain, has been studied by many investigators, but much remains to be learned of its structure and mechanism. The enzyme has been isolated from the inner mitochondrial membrane, to which it is tightly bound, in two different forms. One is a high-molecular-weight particulate complex containing up to 18 atoms of iron and acid-labile sulfur, as well as lipids; this form can reduce coenzyme Q as acceptor and is inhibited by rotenone and amytal (page 497). The low-molecular-weight form (mol wt 78,000) contains only four atoms of iron and acid-labile sulfur per molecule, has an altered sensitivity to inhibitors, and also reacts with different specificity toward artificial electron acceptors than the high-molecular-weight form. Presumably the low-molecular-weight form is a subunit of the larger, more native form of the enzyme as it occurs in the inner membrane.

The four iron-sulfur centers of NADH dehydrogenase have different standard oxidation-reduction potentials. All are believed to participate in electron transport by undergoing Fe(II)–Fe(III) transitions. It appears possible from recent research that one of these centers serves to transfer electrons from NADH to the FMN prosthetic group; the remainder of the iron-sulfur centers appear to be concerned in the transfer of electrons from $FMNH_2$ to the ultimate acceptor ubiquinone.

Iron-Sulfur Proteins

The iron-sulfur proteins contain iron and acid-labile sulfur in equimolar amounts. The first to be discovered, _ferredoxin,_ was found in an anaerobic bacterium, _Clostridium pasteurianum,_ which is capable of fixing atmospheric nitrogen. Similar proteins were later isolated from higher plants, where they occur in the chloroplasts and participate in photosynthetic electron transport. Iron-sulfur proteins have also been found in other microorganisms and in animal tissues,

Table 18-5   Properties of some iron-sulfur proteins or centers

Type	Mol wt	Number of iron-sulfur centers per molecule	Standard oxidation-reduction potential, V
Mitochondrial			
NADH dehydrogenase	78,000	4	−0.30, +0.03
Succinate dehydrogenase	100,000	2	0.00
High-potential iron-sulfur protein (with cytochrome c_1)			+0.22
Other sources			
Ferredoxin (*Chromatium*)	10,000	4	−0.49
Ferredoxin (spinach)	12,000	2	−0.42
Adrenodoxin (adrenal cortex)	16,000	2	−0.27
Putidaredoxin (*Pseudomonas putida*)	12,000	2	−0.24
Ferredoxin (*Clostridium*)	40,000	2	−0.39

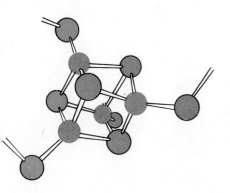

Figure 18-10
A postulated structure for the iron and acid-abile sulfur atoms in the iron-sulfur proteins. The iron atoms are in color and the sulfur atoms are gray. [From L. H. Jensen et al., Biochem. Soc. Trans., 1: 27 (1973).]

particularly in mitochondria. In addition, iron-sulfur groups also occur in certain flavoproteins (see above), where they are called *iron-sulfur centers*.

The molecular weights of the iron-sulfur proteins range from 6,000 to 100,000 or more (Table 18-5). These proteins appear to function as electron carriers by undergoing reversible Fe(II)–Fe(III) transitions. The Fe(III) forms are red or green; on reduction they undergo partial bleaching. The reduced or Fe(II) forms of most iron-sulfur proteins show electron spin resonance spectra at very low temperatures (4 to 100 K) with a characteristic signal lying between 1.90 and 2.10 gauss (G), indicative of the unpaired electrons of the Fe(II) form. Figure 18-10 shows one of the postulated structures for the iron-sulfur groups of these proteins.

The standard oxidation-reduction potentials of different iron-sulfur proteins vary over a wide range, from the very electronegative ferredoxin of the photosynthetic bacterium *Chromatium* ($E_0' = -0.49$ V) to a rather electropositive iron-sulfur center found in beef-heart mitochondria (see below), for which $E_0' = +0.22$ V.

In the mitochondrial chain from NADH to oxygen there appear to be at least seven different iron-sulfur centers detectable by electron spin resonance. Four are located in the NADH dehydrogenase complex, two are associated with cytochrome *b*, and one with cytochrome c_1 (page 492). Although they are undoubtedly very important in electron transport, their precise function is not yet known.

Cytochromes

The cytochromes are electron-transferring proteins containing iron-porphyrin groups; they are found only in aerobic cells. Some are located in the inner mitochondrial membrane, where they act sequentially to carry electrons originating from various dehydrogenase systems toward molecular oxygen. Other cytochromes are found in the endoplasmic reticulum, where they play a role in specialized hydroxylation reactions (page 500). All cytochromes undergo reversible Fe(II)–Fe(III) valence changes during their catalytic cycles. Their reduced forms cannot be oxidized by molecular oxygen, with the exception of the terminal cytochrome of mitochondrial respiration, namely, cytochrome a_3 or cytochrome c oxidase, which also contains tightly bound copper. In the mitochondria of higher animals, where the respiratory chain has been most thoroughly studied, at least five different cytochromes have been identified in the inner membrane: cytochromes b, c_1, c, a, and a_3. At least one of these, cytochrome b, occurs in two or more forms (page 498). In addition to cytochromes found only in the inner membrane of mitochondria, another type, cytochrome b_5, occurs in the endoplasmic reticulum (page 33). Animal and plant cells also contain other heme enzymes, such as peroxidase (page 556) and catalase (page 504).

The porphyrin ring is present not only in various heme proteins but also in the chlorophylls of green plant cells. In their structure porphyrins can be considered as derivatives of a parent tetrapyrrole compound, porphin (Figure 18-11). The porphyrins are named and classified on the basis of their side-chain substituents, e.g., etioporphyrins, mesoporphyrins, uroporphyrins, coproporphyrins, and protoporphyrins. Of these, protoporphyrins are by far the most abundant. Protoporphyrin contains four methyl groups, two vinyl groups, and two propionic acid groups. Fifteen different isomeric protoporphyrins differing in the sequence of substitution of the above groups in the eight available side-chain positions can be written. Of these many possible forms, one, protoporphyrin IX (Figure 18-11), is the only form in nature. It is found in hemoglobin, myoglobin, and most of the cytochromes.

Protoporphyrin forms quadridentate ("four teeth") chelate complexes with iron, magnesium, zinc, nickel, cobalt, and copper ions, in which the metal is held by four coordination bonds. Such a chelate complex of protoporphyrin with Fe(II) is called protoheme or more simply heme; a similar complex with Fe(III) is called hemin or hematin. In heme, the four ligand groups of the porphyrin form a square-planar complex with the iron; the remaining fifth and sixth coordination positions of iron are perpendicular to the plane of the porphin ring. When the fifth and sixth positions of iron are occupied, the resulting structure is a hemochrome or hemochromogen (Figure 18-11). In the heme proteins myoglobin and hemoglobin the fifth position is occupied by an imidazole group of a histidine residue and the sixth position is either unoccupied (deoxyhemoglobin and deoxymyoglobin) or occupied by oxygen (oxyhemoglobin and oxymyoglobin) or other ligands, such as carbon monoxide. In nearly all the cytochromes, on the other hand, both the fifth and sixth positons of the iron are occupied by the R groups of specific amino acid residues of the proteins. These cytochromes therefore cannot bind with ligands like oxygen, carbon monoxide, or cyanide; an important exception is cytochrome a_3, which normally binds oxygen in its biological function.

Figure 18-11
Structures of porphin, protoporphyrin, and hemes.

Porphin

Hemochrome

Protoporphyrin IX

Space-filling model of protoheme

Heme A (the prosthetic group of cytochrome aa_3)

In the normal function of hemoglobin and myoglobin the iron atom does not undergo change in valence as oxygen is bound and lost; it remains in the Fe(II) state. However, both hemoglobin and myoglobin can be oxidized to the Fe(III), or hemin, form by oxidizing agents such as ferricyanide, with a change in color from red to brown. The respective products, which do not function reversibly as oxygen carriers, are called *methemoglobin* and *metmyoglobin*. In the cytochromes, however, the iron atom undergoes reversible changes between Fe(II) and Fe(III) forms; cytochromes serve as electron carriers, whereas hemoglobin and myoglobin act as ligand (oxygen) carriers.

The cytochromes were first discovered and called *histohematins* in 1866 by C. MacMunn, but their significance in biological oxidations did not become clear until 1925, when they were rediscovered by D. Keilin in England. With a simple hand spectroscope he directly observed in intact insect muscles a number of absorption bands resembling those of reduced heme proteins. He showed that these bands appear and disappear in relationship to muscle activity. Keilin renamed these respiratory pigments cytochromes, postulated that they acted in a chain to carry electrons from nutrient molecules to oxygen, and grouped them into three major classes, a, b, and c, depending on the characteristic positions of their absorption bands in the reduced state. Each type of cytochrome in its reduced state has three distinctive absorption bands in the visible range, the α, β, and γ or Soret bands (Figure 18-12; Table 18-6).

With one exception the cytochromes are very tightly bound to the mitochondrial membrane and difficult to obtain in soluble and homogeneous form. The exception is cytochrome c, which is readily extracted from mitochondria by strong salt solutions. The cytochrome c's of many species have been obtained in crystalline form and their amino acid sequences determined (page 112). The iron protoporphyrin group of cytochrome c is covalently linked to the protein via thioether bridges between the porphyrin ring and two cysteine residues in the peptide chain. Cytochrome c is the only common heme protein in which the heme is bound to the protein by a covalent linkage. In hemoglobin and myoglobin, as in cytochromes b and a, the porphyrin ring is noncovalently bound and can be removed by extraction of the acidified protein with pyridine or other solvents. In cytochrome c the fifth and sixth coordination positions of iron are believed to be occupied by the side chains of a histidine residue and a methionine residue, which prevent cytochrome c from reacting with oxygen or carbon monoxide at pH 7.0. Some properties of various cytochromes are summarized in Table 18-6.

Cytochromes a and a_3, together called cytochrome c oxidase, the respiratory enzyme, or ferrocytochrome c–oxygen oxidoreductase, deserve special attention. Instead of protoheme they contain heme A, which differs from protoheme in having a formyl group instead of a methyl group at position 8, no methyl group at position 5, and a long hydrophobic 17-carbon isoprenoid side chain at position 2 instead of a vinyl group (Figure 18-11). Porphyrin A is structurally related to the porphyrin of chlorophyll (page 595), which also has a long isoprenoid side chain.

Figure 18-12
The absorption spectrum of cytochrome c (1.0-cm path; 10 μM solutions; pH 7.0). The molar absorption coefficient of the α peak (550 nm) of the reduced form is 27.7 cm^2 mol^{-1}.

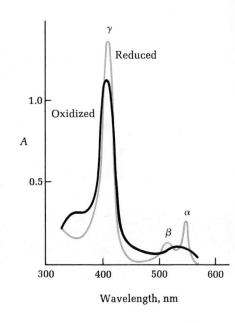

Table 18-6 Properties of mitochondrial cytochromes

Cytochrome	Mol wt	E_0', V	Absorption maxima in reduced form, nm		
			α	β	γ
b_K	25,000	+0.030	563	532	429
c_1	37,000	+0.225	554	524	418
c	12,500	+0.235	550	521	415
a	200,000	+0.210	600	—	439
a_3		+0.385	603.5	—	443

There is still some uncertainty about the structure and mechanism of action of cytochromes a and a_3. For many years they were thought to be two separate entities, since their hemes react differently with cyanide and carbon monoxide and have different spectra. However, it now appears certain that cytochromes a and a_3 are combined in the same large oligomeric protein molecule; the hemes are chemically identical but differ in reactivity toward certain ligands. This complex is referred to as cytochrome aa_3. It has a molecular weight of about 200,000 and contains a number of subunits of different molecular size. The enzyme contains two molecules of heme A and two atoms of copper. Electrons are received from cytochrome c by the a heme and are then transferred to the a_3 heme. The two copper atoms of cytochrome c oxidase, which give characteristic electron spin resonance signals and undergo Cu(II)–Cu(I) transitions during electron transport, are believed to catalyze transfer of electrons from the a_3 heme to oxygen. Cyanide inhibits the reoxidation of reduced cytochrome a_3 by oxygen; hydrogen sulfide has a similar action. Cytochrome a_3, like hemoglobin, combines with carbon monoxide, presumably at the site normally occupied by oxygen. The inhibition of cytochrome oxidase by carbon monoxide is reversed by illumination with visible light. The action of cytochrome aa_3 is discussed further below (page 495).

Ubiquinone (Coenzyme Q)

In addition to the electron-transferring proteins described above, a lipid-soluble electron-carrying coenzyme, ubiquinone or coenzyme Q, also participates in the transport of electrons from organic substrates to oxygen in the respiratory chain of mitochondria. This coenzyme, which is a reversibly reducible quinone with a long isoprenoid side chain (Figure 18-13), was first discovered by R. A. Morton, who named it ubiquinone because of its ubiquitous occurrence in animals, plants, and microorganisms. Later, F. L. Crane and his colleagues found a lipid-soluble quinone in mitochondria which they postulated to function in electron transport; they called it coenzyme Q (for quinone). It was later found to be identical with ubiquinone.

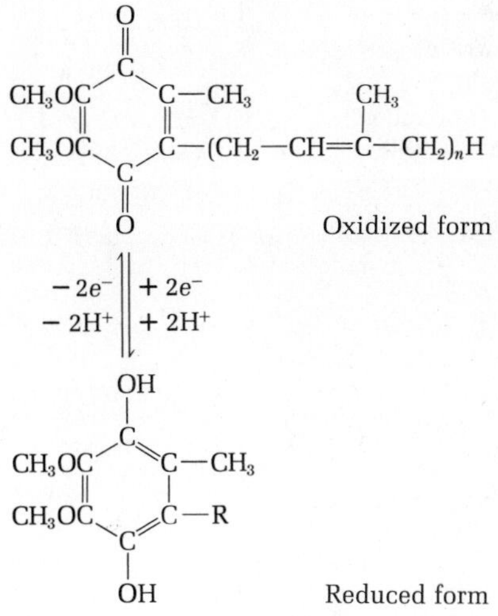

Actually, several ubiquinones are known, differing only in the length of the isoprenoid side chain, which has 6 isoprene units in some microorganisms and 10 in the mitochondria of animal tissues (Figure 18-13). In plant tissues the closely related plastoquinones (page 607) perform similar functions in photosynthetic electron transport.

Ubiquinone has a characteristic light-absorption band at 270 to 290 nm, which disappears when it is reduced to its quinol form; this spectral change is used to measure oxidation and reduction of ubiquinone.

The Pathway of Electron Transport: The Respiratory Chain

The concept that a chain of electron carriers is responsible for transferring electrons from substrate molecules to molec-

Figure 18-13
Ubiquinone or coenzyme Q. The oxidized or quinonoid form absorbs strongly at 290 nm. CoQ_6 is present in some microorganisms and CoQ_{10} in mitochondria of most mammals. For CoQ_6, $n = 6$; for CoQ_{10}, $n = 10$.

Oxidized form

$$-2e^- \;\|\; +2e^-$$
$$-2H^+ \;\|\; +2H^+$$

Reduced form

493

ular oxygen represents the confluence of two lines of investigation. Early investigators of biological oxidations in the period 1900–1920, particularly T. Thunberg, had discovered the dehydrogenases, which catalyze removal of hydrogen atoms from different metabolites in the complete absence of oxygen. From such experiments H. Wieland later postulated that activation of hydrogen atoms is the basic process involved in biological oxidation and that molecular oxygen does not need to be activated to react with the active hydrogen atoms yielded by dehydrogenases. However, in 1913 O. Warburg discovered that cyanide in very small concentrations almost completely inhibits the oxygen consumption of respiring cells and tissues. Since cyanide does not inhibit dehydrogenases but does form very stable complexes with iron (an example is ferricyanide), Warburg postulated that biological oxidation requires an iron-containing enzyme (the "respiratory enzyme"); in his view activation of oxygen by this enzyme is the basic mechanism involved in biological oxidation. The differing views of Wieland and Warburg, i.e., hydrogen activation vs. oxygen activation, were later brought together by A. Szent-Györgyi in Hungary, who postulated that both processes take place and that flavoproteins play the role of intermediate electron carriers between the dehydrogenases and the "respiratory enzyme." Keilin also provided important evidence that the cytochromes act as a consecutive series of electron carriers. He and other investigators, among them D. E. Green, K. Okunuki, T. Singer, T. King, and E. Racker, carried out in vitro reconstructions of segments of the electron-transport chain starting from purified components.

The sequence of electron-transfer reactions in the respiratory chain from NADH to oxygen shown in Figure 18-14 and Table 18-7 is now fairly well established. NADH is the form in which electrons are collected from many different substrates through the action of NAD-linked dehydrogenases. These electrons funnel into the chain via the flavoprotein NADH dehydrogenase. On the other hand, other respiratory

Figure 18-14
The respiratory chain in mammalian mitochondria (right). The points of entry of electrons from various substrates are shown, as well as the sites of inhibition of electron transport and the probable sites of energy conservation as ATP. The symbol FP designates flavoprotein; FP_1 is NADH dehydrogenase. Fe·S indicates iron-sulfur centers; their positions in the chain are still uncertain. Q is ubiquinone (coenzyme Q).

Table 18-7 The sequential reactions of electron transport from NADH to oxygen, written in balanced form

Not all the known carriers are included, such as cytochrome c_1 and several iron-sulfur centers; note particularly the formation and utilization of H^+ ions (in color) in some of the reactions

$NADH + H^+ + FMN \rightleftharpoons NAD^+ + FMNH_2$

$FMNH_2 + 2Fe·S(III) \longleftarrow FMN + 2Fe·S(II) + 2H^+$

$2Fe·S(II) + 2H^+ + Q \rightleftharpoons 2Fe·S(III) + QH_2$

$QH_2 + 2$ cytochrome $b(III) \rightleftharpoons Q + 2H^+ + 2$ cytochrome $b(II)$

2 cytochrome $b(II) + 2$ cytochrome $c(III) \rightleftharpoons 2$ cytochrome $b(III) + 2$ cytochrome $c(II)$

2 cytochrome $c(II) + 2$ cytochrome $a(III) \rightleftharpoons 2$ cytochrome $c(III) + 2$ cytochrome $a(II)$

2 cytochrome $a(II) + 2$ cytochrome $a_3(III) \rightleftharpoons 2$ cytochrome $a(III) + 2$ cytochrome $a_3(II)$

2 cytochrome $a_3(II) + \frac{1}{2}O_2 + 2H^+ \rightleftharpoons 2$ cytochrome $a_3(III) + H_2O$

Key: FMN = prosthetic group of NADH dehydrogenase, Fe·S = prosthetic group of an iron-sulfur protein or center, Q = uniquinone, and roman numerals II and III refer to the oxidation-reduction states of iron.

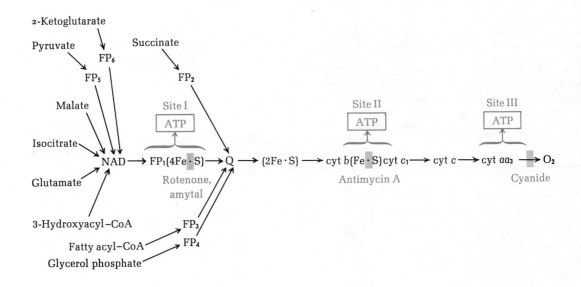

substrates are dehydrogenated by flavin-linked dehydrogenases, such as succinate dehydrogenase and acyl-CoA dehydrogenase (page 548), which funnel electrons into the chain via ubiquinone (Figure 18-14). NAD^+ and ubiquinone thus serve to collect reducing equivalents from respiratory substrates oxidized by pyridine-linked and flavin-linked dehydrogenases, respectively, again illustrating the principle of convergence in catabolic pathways (page 372).

The sequence from NAD to oxygen shown in Figure 18-14 is supported by several lines of evidence. First, it is consistent with the standard oxidation-reduction potentials of the different electron carriers, which become more positive as electrons pass from substrate to oxygen (Table 18-1). Second, in vitro reconstruction experiments with isolated electron carriers have shown that NADH can reduce NADH dehydrogenase but cannot directly reduce cytochromes *b*, *c*, or cytochrome aa_3. Similarly, reduced NADH dehydrogenase cannot react directly with cytochrome *c* but requires the presence of ubiquinone and cytochromes *b* and c_1. Third, complexes containing groups of functionally linked carriers have been isolated from mitochondria, e.g., a complex of cytochromes *b* and c_1 and an iron-sulfur protein, and a complex of NADH dehydrogenase and one or more iron-sulfur proteins.

Perhaps the most important and direct evidence has come from measurements of the oxidation-reduction state of the individual electron carriers as they function in *intact* mitochondria, by measurement of *difference spectra,* a procedure developed by B. Chance and G. R. Williams. Because mitochondrial suspensions are turbid and absorb and scatter much light, the absorption spectra of the electron carriers in intact mitochondria cannot be measured by direct spectrophotometry. However, it is possible to measure the amounts of the carriers in their reduced states in such a turbid suspension by reading its optical absorption in a sensitive spectrophotometer against a blank or control suspension of mitochondria in which the carriers are in their oxidized state, thus canceling out the large absorption due to

Figure 18-15
A difference spectrum of the fully reduced electron carriers in intact rat-liver mitochondria. The blank or control suspension is kept fully saturated with oxygen, maintaining the electron carriers in their oxidized states. Read against this control is a second mitochondrial suspension in which the carriers are kept in the fully reduced state, by making the suspension anaerobic in the presence of excess respiratory substrate. The peaks of reduced cytochrome c_1 are not shown; they can be differentiated from those of cytochrome c only at very low temperatures.

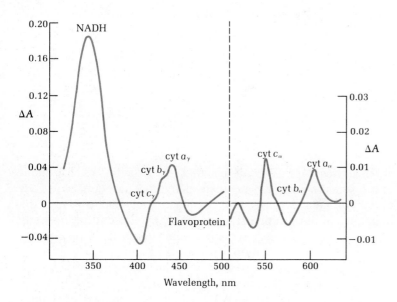

the turbidity of the suspension (Figure 18-15). When isolated mitochondria in the aerobic steady state are allowed to oxidize intermediates of the tricarboxylic acid cycle, difference spectra show that the electron carrier nearest the reducing or substrate end of the chain (NAD) is the most reduced member of the chain in the aerobic steady state whereas the electron carrier at the oxygen end (cytochrome aa_3) is almost entirely in the oxidized form. The intermediate electron carriers of the chain, in the direction from substrate to oxygen, are present in successively more oxidized form in the aerobic steady state, indicating that the electrons flow along a gradient from NAD to oxygen. In another spectrophotometric approach measurements have been made of the rate and sequence of reoxidation of the carriers when oxygen is admitted to an anaerobic suspension of mitochondria in the presence of excess substrate. Under anaerobic conditions all the carriers are fully reduced; when oxygen is admitted, reduced cytochrome aa_3 becomes oxidized first, followed by cytochrome c, and then cytochrome b, which is in turn followed by reoxidation of NADH. It is important to emphasize that a difference spectrum is not the spectrum of the reduced form of the electron carrier, but the *difference* between the spectra of the reduced and oxidized forms.

Other aspects of the dynamics of electron transport will be discussed in Chapter 19, after consideration of oxidative phosphorylation.

Figure 18-16
Inhibitors of electron transport.

Piericidin

Antimycin A₁; R = n-hexyl

Rotenone

Amytal

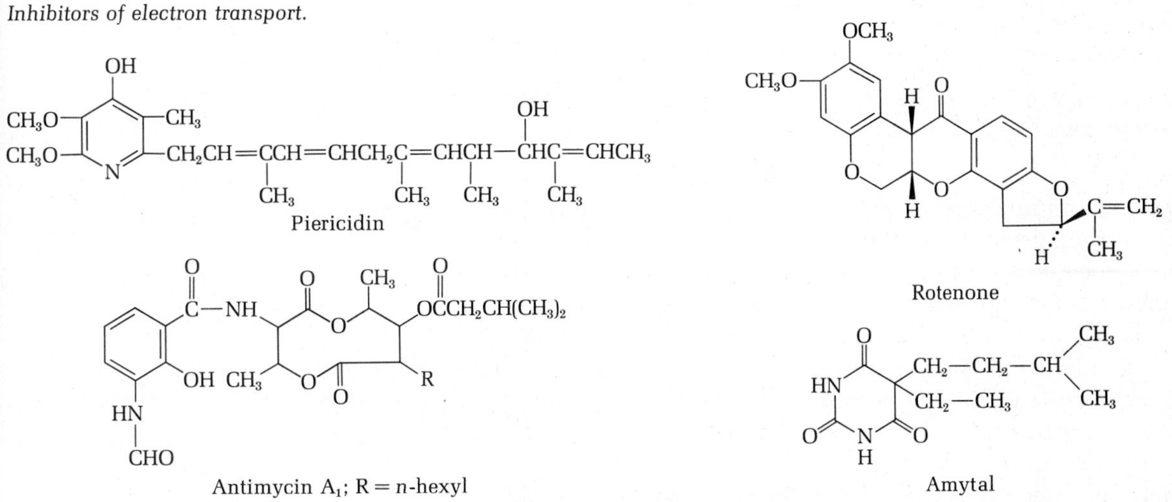

Figure 18-17
Hydraulic analogs of the state of reduction of the electron carriers in the aerobic steady state and following inhibition. The inhibition produces a crossover point; the carriers to the left of the crossover point become more reduced and those to the right become more oxidized. This crossover phenomenon is often used to identify points of regulation in multienzyme systems.

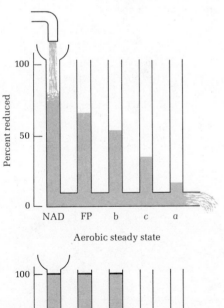

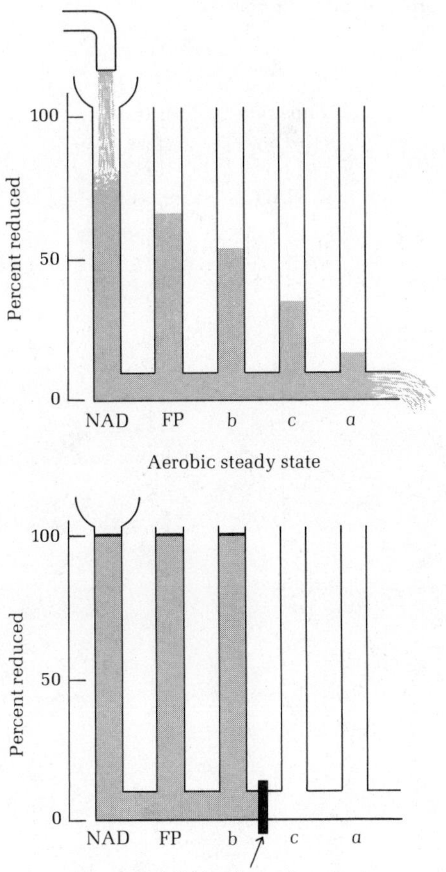

Aerobic steady state

Inhibition by antimycin A

Inhibitors of Electron Transport

Inhibitors that block specific carriers in the electron-transport chain have also yielded valuable information on the sequence of electron carriers in the respiratory chain. Three inhibitors have been found to block electron transport in the span between NADH and ubiquinone: *rotenone,* an extremely toxic plant substance used by South American Indians as a fish poison and now employed as an insecticide; *amytal,* a barbiturate drug; and *piericidin,* an antibiotic that resembles ubiquinone in structure and thus can compete with it. These compounds (Figure 18-16) are believed to act on NADH dehydrogenase. Another characteristic inhibitor is the antibiotic *antimycin A,* isolated from *Streptomyces griseus,* which blocks electron transport in the span from cytochrome *b* to *c*. A third class of inhibitor blocks electron transport from cytochrome *aa₃* to oxygen; it includes hydrogen cyanide, hydrogen sulfide, and carbon monoxide. The site of action of these inhibitors has been established by spectrophotometric measurement of the oxidation-reduction states of the different electron carriers before and after application of the inhibitor to actively respiring mitochondria, which produces a "crossover point" (Figure 18-17). Such crossover points are useful in identifying the rate-limiting step in a multienzyme sequence; another example of the use of crossover points is discussed later (page 537).

Proton Exchanges during Electron Transport

Another distinctive and important property of the electron-transport chain is that hydrogen ions are formed or utilized in certain of the sequential reactions of electron transport (Table 18-7). Some of the electron carriers, such as NADH and ubiquinone, carry electrons in the form of hydrogen atoms, whereas the cytochromes carry electrons as such and apparently do not take up or lose H^+ ions. Until a few years ago the possible significance of these proton exchanges had been completely overlooked. We shall see, however, that these proton-yielding and proton-absorbing reactions are now believed to serve an important role in the conservation of the energy of electron transport (page 527).

Some Uncertainties in the Electron-Transport Chain of Mitochondria

Although much evidence indicates that iron-sulfur proteins or centers are involved at several points in the chain of electron carriers from NADH to oxygen (Figure 18-14), their precise location and function are not yet known. It may also be pointed out that cytochrome b is now known to occur in two forms, cytochrome b_K and cytochrome b_T, differing in their standard oxidation-reduction potentials. The function of these two forms is not yet clear, although cytochrome b_T has been postulated to function in the mechanism of energy transduction during electron transport.

Another uncertainty involves the number of electrons transferred in each step of the respiratory chain. It is generally believed that electron transport occurs in two-electron steps between NAD and ubiquinone and in one-electron steps from cytochrome b to oxygen (Figure 18-18). On the other hand, the reduction of one molecule of oxygen to two water molecules requires a total of four electrons. How electron flow in the respiratory chain is coordinated to yield four electrons to ensure complete reduction of an O_2 molecule is not yet known. One suggestion is that the cytochromes may function in pairs. This is an especially important question since partial reduction of the oxygen molecule yields extremely toxic products. Reduction by a single electron yields the superoxide radical O_2^-, whereas reduction by two electrons yields hydrogen peroxide. Both products are highly reactive and potentially destructive to certain types of functional groups present in biomolecules. There is now increasing evidence that superoxide and hydrogen peroxide are in fact produced during reduction of oxygen in animal tissues (see below, page 503).

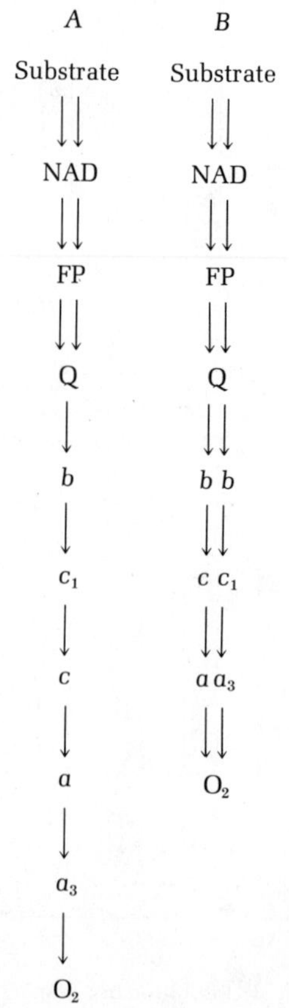

The Energetics of Electron Transport

The standard-free-energy change occurring when two conjugate redox pairs of known standard oxidation-reduction potentials react with each other is given by the equation

$$\Delta G^{\circ\prime} = -n\mathscr{F}\,\Delta E_0^\prime$$

where n is the number of electrons transferred, $\mathscr{F}$ is the faraday (23,062 cal), and $\Delta E_0^\prime$ is the $E_0^\prime$ of the electron-accepting couple minus the $E_0^\prime$ of the electron-donating couple. With this relationship we can calculate the standard-free-energy change as a pair of electron equivalents is transferred from NADH ($E_0^\prime = -0.32$ V) to molecular oxygen ($E_0^\prime = +0.82$ V), that is, along the entire length of the respiratory chain:

$$\Delta G^{\circ\prime} = -2 \times 23,062 \times [0.82 - (-0.32)]$$
$$= -52,700 \text{ cal mol}^{-1} = -52.7 \text{ kcal mol}^{-1}$$

Figure 18-18
Electron-transfer patterns. In (A) electron transfers are in two-electron steps to CoQ and one-electron steps thereafter. In (B) all are in two-electron steps.

This thermodynamic calculation is of course independent of the pathway taken by the electrons and requires no knowledge of the intervening electron-carrying molecules. A very large free-energy decrease, some 53 kcal, thus occurs during transport of a pair of electron equivalents from NADH to molecular oxygen via the respiratory chain. This value may be compared with the standard free energy of formation of ATP at pH 7.0 from ADP and phosphate:

$$P_i + ADP \rightleftharpoons ATP + H_2O$$
$$\Delta G^{\circ\prime} = +7.3 \text{ kcal mol}^{-1}$$

Clearly, the decline in free energy during the passage of one pair of electrons from NADH to molecular oxygen is sufficiently large to make possible the synthesis of not just one but several molecules of ATP from ADP and phosphate under standard conditions provided an energy-coupling mechanism is available. We know today that three molecules of ATP are generated during passage of a pair of electrons down the respiratory chain from NADH to oxygen in the process of oxidative, or respiratory-chain, phosphorylation, to be discussed in the following chapter (page 514). However, the precise mechanism of this energy conversion is not yet known with complete certainty. None of the electron-transferring components of the respiratory chain is known to occur in phosphorylated form. In fact, no evidence has been found that any of the electron carriers isolated to date contains a chemical grouping capable of acting as an energy donor for the synthesis of ATP. Nevertheless two striking characteristics of the electron-transport process are relevant to the mechanism of energy conservation during electron transport: (1) the fact that a large number of sequential electron-transferring steps is involved, which suggests stepwise release of energy, and (2) the fact that H^+ ions are absorbed and released at some of these steps, suggesting that proton exchanges are involved in energy conversion.

Pyridine-Nucleotide Transhydrogenase

Animal tissues and microorganisms contain enzymes that catalyze the general reaction

$$NADPH + NAD^+ \rightleftharpoons NADP^+ + NADH$$

This reaction permits the utilization of the reducing equivalents of NADPH by the respiratory chain, which normally accepts electrons from NADH as immediate donor. In the reverse direction it allows the reduction of $NADP^+$ for biosynthetic purposes. Such enzymes are called pyridine-nucleotide or $NAD(P)^+$ transhydrogenases.

In animal tissues the $NAD(P)^+$ transhydrogenase is located in the mitochondrial membrane, from which it can be extracted with detergents. The direction of the transhydrogenase reaction in intact mitochondria depends on the energy generated by electron transport and the presence of ATP. The rate of the reaction from NADPH to NAD^+ is usually greater than the rate of the reverse reaction. However, when energy is being generated by the electron-transport chain and excess ATP is present, the tendency is for electrons to pass from NADH to $NADP^+$, to yield NADPH in concentrations exceeding that predicted from the equilibrium constant. While the mechanism of this effect is still obscure, NADPH so generated plays an important role as a reductant in biosynthetic reactions.

Electron Transport in Other Membrane Systems

Electron transport takes place not only in mitochondria but also in other types of membrane systems. The heterotrophic bacteria carry out electron transport very similar in principle to that of animal mitochondria. In these prokaryotic organisms, which have no mitochondria, the electron carriers, including flavoproteins, iron-sulfur proteins, and cytochromes, are located in the cell membrane. Indeed, the cell membrane of heterotrophic bacteria and the inner membrane of mitochondria have homologous functions, a fact strongly supporting the hypothesis that mitochondria arose from bacteria during the course of evolution of eukaryotic cells (page 951). Most of the aerobic gram-negative bacteria utilize ubiquinone (page 493) as a "collector" of electrons from flavoproteins, whereas the gram-positive bacteria appear to employ vitamin K or one of its derivatives (page 358) for this purpose. E. coli contains both ubiquinone and vitamin K_3.

In photosynthetic organisms, both prokaryotic and eukaryotic, chains of electron carriers containing flavoproteins, iron-sulfur proteins, and cytochromes participate in light-induced photosynthetic electron transport (page 604).

Electron transport is also believed to take place in the nucleus of eukaryotic cells, although in only small amounts compared with mitochondrial respiration. ATP generated during nuclear electron transport may be utilized for nucleic acid biosynthesis.

Mitochondrial, bacterial, and photosynthetic electron-transport processes are accompanied by coupled phosphorylation of ADP, as a means of conserving the energy of oxidation. As we shall now see, however, there are other types of specialized electron-transport processes leading to the reduction of oxygen that are not accompanied by phosphorylation.

Utilization of Oxygen by Oxygenases

Although most of the molecular oxygen consumed by aerobic cells is reduced to water at the expense of electrons flowing down the respiratory chain of mitochondria, small amounts of oxygen are used in enzymatic reactions in which one or both atoms of the oxygen molecule are directly inserted into the organic substrate molecule to yield hydroxyl groups. Enzymes catalyzing such reactions are called oxygenases, of which there are two classes. The dioxygenases catalyze insertion of both atoms of the oxygen molecule into the organic substrate molecule, whereas the monooxygenases insert only one.

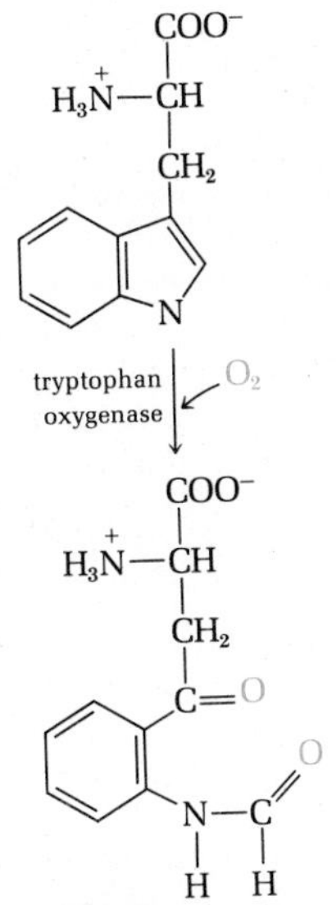

Figure 18-19
The tryptophan oxygenase reaction, showing the insertion of both atoms of the oxygen molecule into the substrate.

L-Tryptophan

tryptophan oxygenase

N-Formyl-L-kynurenine

Dioxygenases

The dioxygenases, also called oxygen transferases, catalyze reactions of the type

$$AH_2 + O_2 \longrightarrow A(OH)_2$$

in which AH_2 is the substrate molecule and $A(OH)_2$ is its dihydroxylated form; usually the two hydroxyl groups so introduced are adjacent or vicinal. In such reactions the product $A(OH)_2$ is often unstable and spontaneously undergoes cleavage of the carbon-carbon bond between the hydroxyl groups. That the two oxygen atoms of molecular oxygen are directly inserted into the substrate has been proved with the use of the ^{18}O isotope as tracer.

A number of dioxygenases have been obtained in highly purified form. All appear to contain iron, either in the labile iron-sulfur linkage (page 488) or as heme; a few also contain copper. An important example is tryptophan pyrrolase, whose newly recommended name is tryptophan 2,3-dioxygenase, a heme enzyme that catalyzes a step in the oxidative degradation of tryptophan (page 573). In this reaction two oxygen atoms are introduced into the six-membered ring of tryptophan to yield an unstable vicinal dihydroxyl product, which decomposes spontaneously with cleavage of the aromatic ring into L-formylkynurenine (Figure 18-19). Other examples of dioxygenases are described elsewhere (pages 570 and 573).

Monooxygenases.

The monooxygenases, or hydroxylases, catalyze insertion of one oxygen atom of molecular oxygen into the organic substrate; the other oxygen atom is reduced to water. Monooxygenases require a second substrate to donate electrons for the reduction of the second oxygen atom in the oxygen molecule, the one reduced to water. Hence the monooxygenases are also called mixed-function oxygenases. The general equation of such reactions is

$$AH + XH_2 + O_2 \longrightarrow AOH + H_2O + X$$

in which AH is the substrate undergoing hydroxylation, XH_2 is the electron donor, AOH is the hydroxylated substrate, and X the oxidized electron donor. One of the two oxygen atoms of the oxygen molecule (in color) is recovered in the hydroxylated product and the other in the water formed; this was established by use of ^{18}O as tracer.

In most monooxygenase reactions the second substrate that furnishes electrons to reduce one atom of oxygen to water is ultimately NADH or NADPH; however, different electron carriers are employed to transfer electrons from NADPH or NADH to the oxygen.

The simplest monooxygenases, those in bacteria, are flavo-proteins containing FAD, catalyzing the sequential reactions

$$NAD(P)H + E\text{—}FAD \longrightarrow NAD(P)^+ + E\text{—}FADH_2$$

$$E\text{—}FADH_2 + AH + O_2 \longrightarrow E\text{—}FAD + AOH + H_2O$$

where E represents the oxygenase protein, AH the substrate, usually an aromatic compound, and AOH its hydroxylated product.

A somewhat more complex type of monooxygenase reaction is that catalyzed by the liver enzyme _phenylalanine hydroxylase_, whose new recommended name is _phenylalanine 4-monooxygenase_. In this reaction the ultimate donor of the electrons required to reduce one of the oxygen atoms to water is NADPH, but the reducing equivalents are transferred via _tetrahydrobiopterin_ (page 570), a coenzyme whose structure resembles that of _tetrahydrofolate_ (page 345). The full sequence of reactions is described elsewhere (page 569).

Microsomal Electron Transport

Many hydroxylation reactions in animal tissues utilize more complex enzymes than the monooxygenases and dioxygenases described above. The endoplasmic reticulum, i.e., the microsome fraction, contains membrane-bound nonphosphorylating electron-transport systems that participate not only in hydroxylation but also in desaturation reactions. One of the microsomal electron-transport systems of liver (Figure 18-20) consists of a flavoprotein called _NADPH–cytochrome P450 reductase_ and a specialized microsomal cytochrome, _cytochrome P450,_ often simply referred to as P450; in some organisms an iron-sulfur protein also participates. In the first step an electron equivalent is transferred from NADPH to the semiquinone form (page 486) of the flavoprotein, reducing the latter completely. Electrons are transferred from the reduced flavoprotein to the oxidized, or Fe(III), form of P450, which is found only in the microsomes of liver cells and is absent from liver mitochondria; it has an α band at 420 nm and is a cytochrome of the B class. P450 (P for pigment) was first identified by the fact that the carbon monoxide derivative of its reduced form absorbs maximally at 450 nm. The reduced [Fe(II)] form of P450 reacts with molecular oxygen in such a way that one of the oxygen atoms is reduced to water and the other is introduced into the organic substrate (Figure 18-20).

Liver microsomes catalyze hydroxylation of many different kinds of substrates, including steroids (page 687), fatty acids (page 668), squalene (page 682), and certain amino acids. They also promote hydroxylation of various drugs, e.g., phenobarbital, morphine, codeine, amphetamines, and carcinogenic hydrocarbons like methylcholanthrene. Hydroxylation of these normally foreign substances is a step in their metabolism. The flavoprotein and P450 are inducible; i.e., they greatly increase in concentration in the liver of

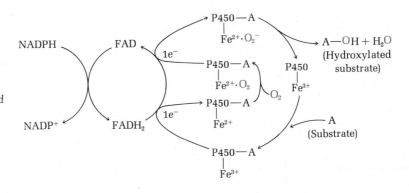

animals dosed with barbiturates and other drugs. In fact, these agents stimulate proliferation of the smooth portion (page 33) of the endoplasmic reticulum of liver, presumably a defensive or protective adaptation.

Liver microsomes have a second type of nonphosphorylating electron-transport chain, which includes the flavoprotein *cytochrome b_5 reductase* and *cytochrome b_5*, a cytochrome found in the endoplasmic reticulum but not in the inner mitochondrial membrane. These two proteins have been highly purified and studied in great detail. This type of microsomal electron-transport chain appears to be particularly active in the desaturation of fatty acids (page 668).

Superoxide Dismutase and Catalase

During electron transport to molecular oxygen via the mitochondrial respiratory chain, as well as in various hydroxylation and oxygenation reactions, toxic partial reduction products of oxygen may be formed, presumably as transient intermediates on the active sites of such enzymes. The most important are the superoxide anion O_2^- (see above, page 498) and hydrogen peroxide, which are extremely reactive and capable of irreversible damage to various biomolecules.

It has been found, largely through the research of I. Fridovich and his colleagues, that aerobic cells generally contain the enzyme *superoxide dismutase*, which converts superoxide into hydrogen peroxide and molecular oxygen:

$$2O_2^- + 2H^+ \longrightarrow H_2O_2 + O_2$$

Superoxide dismutase is found in two forms, one in the extramitochondrial cytosol and another in the mitochondria. The mitochondrial superoxide dismutase of eukaryotes is similar to the superoxide dismutase of many bacteria with respect to its characteristic content of Mn^{2+} and many homologies in amino acid sequence. The cytosol form of superoxide dismutase, on the other hand, has quite a different structure and contains Cu^{2+} and Zn^{2+}. These enzymes are present in high concentration and are extraordinarily active, suggesting that superoxide radicals are being continuously produced during the enzymatic reduction of oxygen by various enzymes and enzyme systems and quickly removed.

Hydrogen peroxide formed by superoxide dismutase and by the flavin-linked oxidases (page 486) is decomposed by the heme enzyme *catalase* in the reaction

$$H_2O_2 \longrightarrow H_2O + \tfrac{1}{2}O_2$$

Catalase is found in the microbodies (page 567) of animal cells, also called *peroxisomes* (page 467).

The protective action of superoxide dismutase and catalase is probably supported by ascorbic acid, glutathione, and vitamin E, which readily accept electrons and may serve a backup function by scavenging free radicals.

Transformation of Oxidation-Reduction Energy into Bioluminescence

Luminescent organisms include certain bacteria, protozoa, fungi, worms, and crustaceans, and particularly the firefly. In many of these organisms enzymatic oxidoreductions take place in which the free-energy change is utilized to excite a molecule to a high-energy state (page 594). This is followed by the return of the excited molecule to the *ground state*, a process accompanied by emission of visible light. This phenomenon is called *bioluminescence*.

The molecular components and the mechanism of firefly luminescence have been investigated by W. D. McElroy and his associates. The two necessary components, a heat-stable heterocyclic phenol, *luciferin* (Figure 18-21), and a heat-labile enzyme, *luciferase,* have been extracted from fireflies and crystallized. Luciferase (mol wt $\approx$ 100,000) appears to have no prosthetic group. In the first step, luciferin (LH$_2$) and ATP react to form *luciferyl adenylate* (LH$_2$—AMP), which remains tightly bound on the catalytic site of luciferase:

$$LH_2 + ATP + E \overset{Mg^{2+}}{\rightleftharpoons} E-LH_2-AMP + PP_i$$

Figure 18-21
Components of the luciferase reaction.

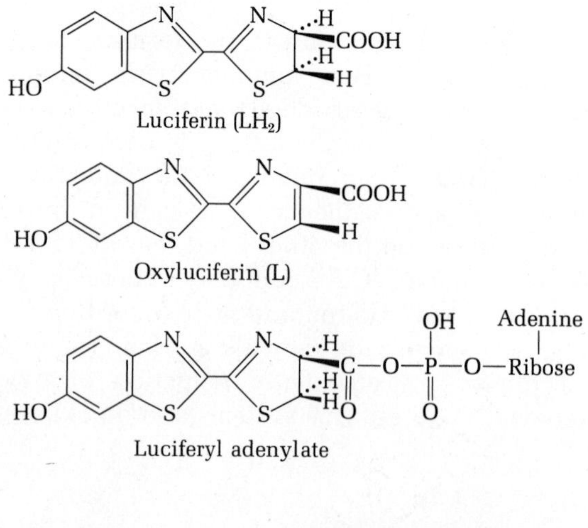

Luciferin (LH$_2$)

Oxyluciferin (L)

Luciferyl adenylate

When this form of the enzyme is exposed to molecular oxygen, the enzyme-bound luciferyl adenylate is oxidized to yield *oxyluciferin* (L), which emits light on returning to the ground state:

$$E—LH_2—AMP + O_2 \longrightarrow L + H_2O + bioluminescence$$

One quantum of light (page 594) is emitted for each molecule of luciferin oxidized. A peroxide is presumed to be formed as an intermediate. The color of the emitted light is determined by the enzyme protein, since different species of firefly have the same luciferin but emit light of different colors. This unusual type of energy transformation serves the firefly as a mating signal. The luciferin-luciferase reaction, followed in a recording photometer, is often employed as a highly sensitive quantitative assay for ATP.

Summary

In oxidation-reduction reactions electrons are transferred from the reductant, or electron donor, to the oxidant, or electron acceptor. The tendency of any biological reductant to donate electrons is given by the standard oxidation-reduction potential E'_0, the emf given by 1.0 M reductant in the presence of 1.0 M oxidant species at an inert electrode at 25°C and pH 7.0. The E'_0 value of any given conjugate redox pair allows prediction of the direction of net electron flow in its reaction with another redox pair.

There are four major classes of oxidation-reduction enzymes. (1) The pyridine-linked dehydrogenases catalyze reversible transfer of electrons from substrates to the loosely bound coenzymes NAD^+ or $NADP^+$, to form NADH and NADPH, respectively. The enzymatic reduction of the pyridine ring of NAD^+ or $NADP^+$ is accompanied by a spectral change and is stereospecific. (2) The flavin-linked dehydrogenases contain tightly bound FMN or FAD as prosthetic groups and often a metal. In their oxidized forms they are intensely colored, but on complete reduction they are bleached. The most important flavin dehydrogenases are succinate dehydrogenase and NADH dehydrogenase. (3) The iron-sulfur proteins contain two to eight atoms of iron and an equal number of acid-labile sulfur atoms; the iron atoms undergo Fe(II)–Fe(III) transitions. (4) The cytochromes, acting in series, transfer electrons from flavoproteins to oxygen. They contain iron-porphyrin prosthetic groups and undergo reversible Fe(II)–Fe(III) transitions, which can be followed spectrophotometrically.

The respiratory chain of mitochondria consists of the sequence NADH, NADH dehydrogenase, iron-sulfur protein, ubiquinone, and cytochromes b, c_1, c, a and a_3; several other iron-sulfur centers or proteins also participate. Electron transport is characteristically inhibited at specific points by rotenone, antimycin A, and cyanide. Spectrophotometric observations make study of the sequence of electron carriers possible. Certain reactions of electron transport are accompanied by uptake or formation of H^+ ions. Electron transport from NADH to oxygen proceeds with a very large decrease in free energy, which is in part conserved by the coupled phosphorylation of ADP to ATP.

Electron-transport chains also occur in the bacterial cell membrane, in chloroplasts, and in cell nuclei. Molecular oxygen is also utilized for hydroxylation of a variety of different organic metabolites. Dioxygenases insert both oxygen atoms of O_2 into the substrate, whereas monooxygenases insert only one. Specialized electron-transport chains in endoplasmic reticulum, which contain either cytochrome P450 or cytochrome b_5, are also active in a number of enzymatic hydroxylation reactions.

References

See also references to Chapter 19.

Books

ESTABROOK, R. W., and M. E. PULLMAN (eds.): *Oxidation and Phosphorylation*, vol. 10, *Methods in Enzymology*, Academic, New York, 1967. Experimental methods.

HAYAISHI, O. (ed.): *Oxygenases*, Academic, New York, 1972. A collection of reviews.

KEILIN, D.: *The History of Cell Respiration and Cytochromes*, Cambridge University Press, New York, 1966.

KING, T. E., and M. KLINGENBERG (eds.): *Electron and Coupled Energy Transfer in Biological System*, vols. I and II, Dekker, New York, 1971. Reviews on various aspects of electron transport and electron carriers.

LEHNINGER, A. L.: *The Mitochondrion: Molecular Basis of Structure and Function*, Benjamin, Menlo Park, Calif., 1964. Traces the development of current knowledge.

LEMBERG, R., and J. BARRETT: *Cytochromes*, Academic, New York, 1973. Comprehensive treatise with complete bibliography.

LOVENBERG, W. (ed.): *Iron-Sulfur Proteins*, vols. 1 and 2, Academic, New York, 1973. Reviews on the chemistry and biochemistry of these proteins, including the iron-sulfur flavoproteins.

SUND, H. (ed.): *Pyridine Nucleotide Dependent Dehydrogenases*, Springer-Verlag, New York, 1970. A collection of papers.

WAINIO, W. W.: *The Mammalian Respiratory Chain*, Academic, New York, 1970.

Articles

CHANCE, B., and G. R. WILLIAMS: "The Respiratory Chain and Oxidative Phosphorylation," *Adv. Enzymol.*, 17: 65 (1956). An early review of the dynamics of the electron-transport chain.

EVERSE, J., and N. O. KAPLAN: "Lactate Dehydrogenase: Structure and Function," *Adv. Enzymol.*, 37: 61–134 (1973).

FRIDOVICH, I.: "Superoxide Dismutases," *Adv. Enzymol.*, 41: 35–97 (1974).

HEMMERICH, P., G. NAGELSCHNEIDER, and C. VEEGER: "Chemistry and Biology of Flavins and Flavoproteins, *FEBS Lett.*, 8: 69–83 (1970).

KAPLAN, N. O.: "Pyridine Nucleotide Transhydrogenases," pp. 105–133, in *The Harvey Lectures*, ser. 66, Academic, New York, 1972.

KLINGENBERG, M.: "The Respiratory Chain," p. 3 in T. P. Singer (ed.), *Biological Oxidations*, Wiley-Interscience, New York, 1968.

KROGER, A., V. DIDAK, M. KLINGENBERG, and F. DIEMER: "On the Role of Quinones in Bacterial Electron Transport," *Eur. J. Biochem.,* 21: 322–333 (1971).

LEE, C.-Y., R. D. EICHNER, and N. O. KAPLAN: "Conformation of Diphosphopyridine Coenzymes on Binding to Dehydrogenases," *Proc. Natl. Acad. Sci. (U.S.),* 70: 1593–1597 (1973).

MCCAPRA, F.: "The Chemistry of Bioluminescence," *Endeavour,* 39: 139–145 (1973).

OHNISHI, T.: "Mechanisms of Electron Transport and Energy Conservation in the Site I Region," *Biochim. Biophys. Acta,* 301: 105–128 (1973).

ORME-JOHNSON, W. H.: "Iron-Sulfur Proteins," *Ann. Rev. Biochem.,* 42: 159–204 (1973).

SINGER, T. P., and M. GUTMAN: "The DPNH Dehydrogenase of the Mitochondrial Respiratory Chain," *Adv. Enzymol.,* 34: 79–153 (1971).

SINGER, T. P., E. B. KEARNEY, and W. C. KENNEY: "Succinate Dehydrogenase," *Adv. Enzymol.,* 37: 184–272 (1973).

WILSON, D. F., P. L. DUTTON, and M. WAGNER: "Energy-Transducing Components in Mitochondrial Respiration," p. 234 in D. R. Sanadi and L. Packer (eds.), *Current Topics in Bioenergetics,* vol. 5, Academic, New York, 1973. Detailed treatment of the oxidation-reduction potentials of the members of the electron-transport chain.

Problems

1. Calculate the electromotive force in volts registered by a responsive electrode immersed in a solution containing the following mixtures of NAD^+ and NADH at pH 7.0 and 25°C, with reference to a half-cell of 0.00 V.
   (a) 1.0 mM NAD^+ and 10 mM NADH
   (b) 1.0 mM NAD^+ and 1.0 mM NADH
   (c) 10 mM NAD^+ and 1.0 mM NADH

2. List the following substances in order of increasing tendency to donate electrons: (a) cyt c_{red}, (b) NADH, (c) H_2, (d) ubiquinol, (e) lactate.

3. List the following substances in order of increasing tendency to accept electrons: (a) α-ketoglutarate + CO_2, (b) oxaloacetate, (c) O_2, (d) $NADP^+$

4. Which of the following reactions would be expected to proceed in the direction shown under standard conditions, assuming that the appropriate enzymes are present to catalyze them?
   (a) Malate + $NAD^+ \longrightarrow$ oxaloacetate + NADH + H^+
   (b) Acetoacetate + NADH + $H^+ \longrightarrow$
   β-hydroxybutyrate + NAD^+
   (c) Pyruvate + NADH + $H^+ \longrightarrow$ lactate + NAD^+
   (d) Pyruvate + β-hydroxybutyrate $\longrightarrow$ lactate + acetoacetate
   (e) Malate + pyruvate $\longrightarrow$ oxaloacetate + lactate
   (f) Acetaldehyde + fumarate $\longrightarrow$ acetate + succinate

5. Calculate the standard-free-energy change when a pair of electron equivalents passes from (a) isocitrate to NAD^+, (b) succinate to cytochrome b, (c) malate to NAD^+, (d) NADH to cytochrome c. (Assume pH 7.0 and 25°C.)

6. Starting from standard oxidation-reduction potentials, calculate equilibrium constants at 25°C and pH 7.0 for the reactions
   (a) Malate + NAD$^+$ $\rightleftharpoons$ oxaloacetate + NADH + H$^+$
   (b) Acetoacetate + NADH + H$^+$ $\rightleftharpoons$
   $$\beta\text{-hydroxybutyrate} + NAD^+$$

7. Calculate the equilibrium concentrations of malate, oxaloacetate, NAD$^+$, and NADH in the reaction

   $$\text{Malate} + NAD^+ \rightleftharpoons \text{oxaloacetate} + NADH + H^+$$

   when the reaction is carried out at pH 7.0 and 25°C and the initial concentration of each component is 10 mM.

8. Calculate the equilibrium concentrations of malate and oxaloacetate in the overall reaction

   $$\beta\text{-Hydroxybutyrate} + \text{oxaloacetate} \rightleftharpoons \text{acetoacetate} + \text{malate}$$

   catalyzed by NAD-linked β-hydroxybutyrate dehydrogenase and malate dehydrogenase in the presence of NAD. Assume that the reaction takes place at 25°C and pH 7.0, with β-hydroxybutyrate and acetoacetate present initially at a concentration of 10 mM and oxaloacetate and malate at 20 mM.

9. Using standard potentials, calculate the standard-free-energy change for the oxidation of succinate to fumarate by (a) a flavoprotein, as in the cell, and (b) by NAD.$^+$

10. The conversion of 25-hydroxycholecalciferol (25-hydroxy vitamin D$_3$) to 1,25-dihydroxycholecalciferol in the kidney is catalyzed by a pyridine-nucleotide-requiring mixed-function oxidase. Write a balanced equation for the reaction.

11. Dideuteroethanol (CH$_3$CD$_2$OH) was oxidized in the presence of yeast alcohol dehydrogenase and NAD. The resulting reduced pyridine nucleotide was then incubated with substrate A in the presence of a specific NADH-requiring dehydrogenase. The reduced compound, AH$_2$, contained no deuterium, whereas NAD$^+$ contained one atom of deuterium per molecule. Is it more likely that the dehydrogenase for substrate A transfers hydrogen from the A side or the B side of NADH?

12. Consider a typical dehydrogenase reaction involving a pyridine nucleotide:

    $$AH_2 + NAD^+ \rightleftharpoons A + NADH + H^+$$

    (a) When plots of $1/v_0$ vs. $1/[AH_2]$ at different concentrations of NAD(P) are made, what pattern would be expected?
    (b) What kinetic relationship would be expected to prevail between NAD$^+$ and NADH? See Chapter 8.

13. Predict the oxidation-reduction states of NAD, NADH dehydrogenase, cytochrome b, cytochrome c, and cytochrome a in liver mitochondria amply supplied with isocitrate as substrate, P$_i$, ADP, and oxygen, but inhibited by (a) rotenone, (b) antimycin A, and (c) cyanide.

CHAPTER **19** **OXIDATIVE PHOSPHORYLATION,**
MITOCHONDRIAL STRUCTURE, AND THE
COMPARTMENTATION OF RESPIRATORY
METABOLISM

In the preceding chapter we examined the components of the
respiratory chain and the process of electron transport from
organic substrates to oxygen. We now consider oxidative
phosphorylation, the mechanism by which the free-energy
decrease accompanying the transfer of electrons along the
respiratory chain is coupled to the formation of the high-
energy phosphate groups of ATP. Oxidative phosphorylation
is fundamental to all aspects of cellular life in aerobic organisms since it is their main source of useful energy. It is also
a particularly challenging research problem, the ultimate solution of which will depend on a detailed understanding of
the molecular organization of the mitochondrial inner membrane, the site of this important energy-transducing system.
We shall therefore begin our discussion with a description of
the structure of mitochondria.

In this chapter we shall also see that the mitochondrial
membrane contains a number of transport systems that promote the movement of certain metabolites and mineral ions
between the mitochondrial matrix and the surrounding cytosol. These transport systems are elements in the compartmentation and regulation of energy metabolism; they also
participate in other important metabolic activities.

Structure of Mitochondria

The number of mitochondria per cell appears to be relatively
constant and characteristic for any given cell type. A rat-liver
cell, for example, contains about 800 mitochondria (see Figure 1-8, page 32). Mitochondria are often located near
structures that require ATP, the major product of their biochemical activity, or near a source of fuel, on which they
depend. For example, in the flight muscles of some insects
the mitochondria are regularly arranged along the myofibrils
(page 769). The ATP molecules formed by these mitochondria thus need diffuse only a short distance to the
ATP-requiring contractile elements. Mitochondria are also

Figure 19-1
Structure of mitochondria.

Scanning electron micrograph of intact liver
mitochondria in isolated state.

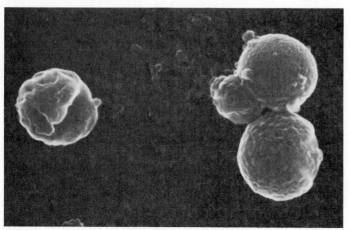

1.0 μm

R. H. Kirschner

Electron micrograph of a thin section of a
mitochondrion in a bat pancreas cell.

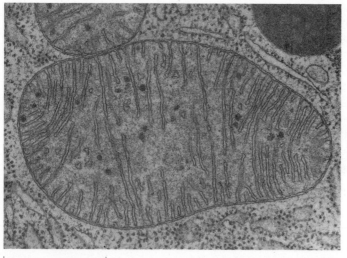

0.5 μm

K. R. Porter

Drawing of heart mitochondrion showing
three-dimensional arrangement of membranes. (*From Peter Raven and Helena
Curtis,* Biology of Plants, *Worth Publishers,
Inc., New York, 1970.*)

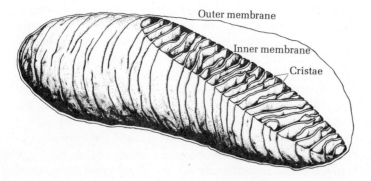

Outer membrane

Inner membrane

Cristae

Figure 19-2
*Variations in structure of the cristae in mi-
tochondria from different types of cells.*

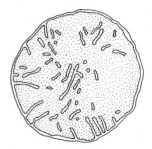

Rat liver (platelike)

Rat brown fat (septate)

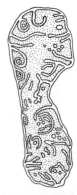

Paramecium (tubular)

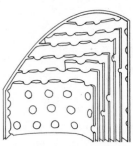

Blowfly flight muscle
(perforated leaflets)

Figure 19-3
Electron micrographs of inner-membrane spheres on cristae, following negative-contrast staining. In this procedure the ruptured mitochondria are mixed with phosphotungstate solution, dried, and examined under the electron microscope. Because the phosphotungstate is electron-opaque, it surrounds the membrane structure, outlining its profile.

Edge-on profile of a crista in blowfly flight muscle.

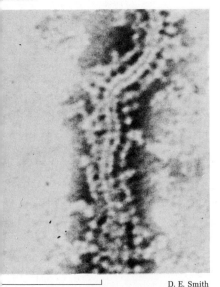

100 nm   D. E. Smith

Inner surface of inner membrane.

100 nm   A. Claude and V. Meneghelli

frequently located adjacent to cytoplasmic fat droplets, which serve as a source of fuel for oxidation. The mitochondria may make up a relatively large fraction of the total cytoplasmic volume; in the liver cell this is about 20 percent (page 33) and in a heart-muscle cell over 50 percent.

Mitochondria are spherical, or nearly so, in brown fat cells, football-shaped in liver cells, cylindrical in the kidney, and threadlike in fibroblasts. Sometimes they have a very complex irregular structure, with extended processes, as in yeast cells. The most intensively studied mitochondria are those of rat liver, which electron microscopy shows to be about 2 μm long and somewhat less than 1 μm wide in the intact cell (Figure 19-1). They are thus about the same size as bacteria (page 30).

Mitochondria have two membranes (Figure 19-1), an outer membrane that is smooth and somewhat elastic and an inner membrane that has inward folds, or invaginations, called *cristae*. The cristae vary in number and structure depending on the cell type (Figure 19-2). They appear to be devices for increasing the surface area of the inner membrane in relation to the mitochondrial volume.

Inside the inner compartment is the *matrix*. This gel-like phase contains about 50 percent protein, some of which is organized into a reticular network apparently attached to the inner surface of the inner membrane. The matrix undergoes dramatic changes in volume and state of organization during changes in respiratory activity (page 519). The matrix also contains DNA and ribosomes; the latter are often located near that portion of the inner membrane that is closely apposed to the outer membrane.

The outer and inner mitochondrial membranes differ in ultrastructure, as revealed by *negative-contrast staining* (Figure 19-3). With this method H. Fernández-Morán showed that the inner surface of the inner membrane is covered with regularly spaced spherical particles (diameter 8.0 to 9.0 nm) connected to the membrane by a narrow stalk (Figure 19-3). These knoblike structures, earlier called elementary particles but now known as *inner-membrane spheres* (page 522), are not present on the outer surface of the inner membrane or on either surface of the outer membrane.

Enzyme Localization in Mitochondria

The outer membrane can be removed from liver mitochondria. The remaining mitochondrial structure, consisting of the intact inner membrane plus matrix, is often called a *mitoplast* (Figure 19-4). The matrix contents can be extracted from such preparations with neutral detergents, e.g., Lubrol. The outer membrane, the inner membrane, and the matrix have been analyzed for their molecular composition and enzyme content (Table 19-1). The inner membrane contains cytochromes b, c, c_1, a, and a_3, the F_1 ATPase associated with the mechanism of oxidative phosphorylation (page 514), and certain dehydrogenases, particularly those for succinate (page 459) and NADH (page 486). The outer-

Table 19-1 Location of some enzymes in rat-liver mitochondria

Outer membrane
  Monoamine oxidase
  Kynurenine 3-monooxygenase
  NADH dehydrogenase (antimycin-insensitive)
  Acyl-CoA synthetases
  Phospholipase A_2
  Nucleoside diphosphate kinase

Space between the membranes
  Adenylate kinase

Inner membrane
  NADH dehydrogenase (antimycin-sensitive)
  Iron-sulfur proteins
  Cytochromes b, c, c_1, and aa_3
  F_1 ATPase
  Succinate dehydrogenase
  D-β-Hydroxybutyrate dehydrogenase
  Carnitine acyltransferase

Matrix
  Citrate synthase
  Isocitrate dehydrogenase
  Fumarase
  Malate dehydrogenase
  Glutamate dehydrogenase
  Aspartate transaminase
  Fatty acyl–CoA oxidation enzymes (Chapter 20)

membrane fraction does not contain any of these components but does have characteristic enzymes absent from the inner membrane, the most distinctive being monoamine oxidase, a flavoprotein that catalyzes the oxidation of various monoamines such as epinephrine (page 811). Monoamine oxidase is used as a "marker" enzyme to indicate the presence of outer membrane. Similarly, cytochromes a and a_3, which are found nowhere else in the cell, are used as markers for the inner membrane. The matrix characteristically contains most of the tricarboxylic acid cycle enzymes. Malate and glutamate dehydrogenases are frequently used as markers for the mitochondrial matrix.

The outer membrane of liver mitochondria contains nearly 50 percent lipids, whereas the inner membrane is distinctive in containing only about 20 percent lipids and nearly 80 percent proteins. The inner membrane is characteristically rich in cardiolipin (page 288), which makes up about 20 percent of the inner-membrane lipids.

The space between the membranes also contains specific enzymes, particularly adenylate kinase (page 411). Certain enzymes can be more precisely localized; e.g., the creatine kinase (page 767) of heart mitochondria has been located on the outer surface of the inner membrane.

Structural Organization of the Inner Membrane

The inner membrane of liver mitochondria contains at least 60 different biologically active proteins, embedded in a

Figure 19-4
Rat-liver mitochondria after removal of the outer membrane (mitoplasts). The protruding structures are presumably everted cristae.

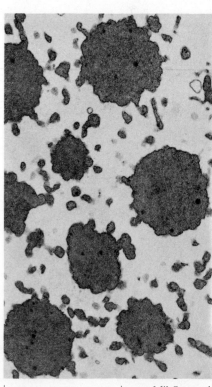

J. W. Greenawalt

1.0 μm

Figure 19-5

Electron micrograph of the concave fracture faces of the cores of the outer membrane (single arrow) and inner membrane (double arrow) of a rat-liver mitochondrion, following freeze etching. In this procedure the mitochondria are quick-frozen and then fractured or cleaved with a sharp blade, splitting each membrane into two leaflets. The molecular core of the outer membrane shows hexagonal arrays of small particles, presumably protein molecules. The larger particles in the core of the inner membrane are clusters of electron transport proteins and enzymes concerned in oxidation, phosphorylation, and transport. [From C. R. Hackenbrock, Ann. N.Y. Acad. Sci., 195: 499 (1972).]

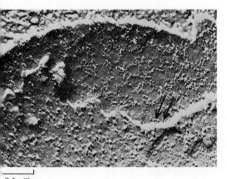

0.1 μm

Figure 19-6

Drawing of a model of the inner mitochondrial membrane (profile view). The tail-like structures represent phospholipid molecules; the globular structures are protein molecules drawn to scale. There is no regular repeating unit of structure. The inner membrane contains about 80 percent protein, including a great number of different enzymes and transport systems, and only 20 percent lipid. [From F. S. Sjöstrand and L. Barajas, J. Ultrastructure Res., 32: 298 (1970).]

phospholipid bilayer system (page 300). They include the electron-transferring enzymes and proteins, enzymes concerned in ATP synthesis, various dehydrogenases, and the protein components of transport systems for various metabolites. Most of these proteins are very difficult to extract from the membrane structure in intact, functional form. The close association of the proteins of the electron-transport and phosphorylation systems with the nonpolar lipid phase of the membrane has made it exceedingly difficult to study the enzymatic mechanisms of oxidative phosphorylation.

Valuable information about the organization of the inner membrane has been provided by the *freeze-etch* method of electron microscopy. With this method the faces of the inner membrane show many pits and mounds corresponding to the position of enzyme molecules (Figure 19-5). A model of the mitochondrial inner membrane structure is shown in Figure 19-6. As we shall see (page 527), many of the inner membrane enzymes and cytochromes have a fixed sidedness in the membrane and cannot rotate.

Spectrophotometric examination indicates that there is considerable variation in the molar ratios between the different cytochromes, flavoproteins, and iron-sulfur centers in mitochondria from different species. Table 19-2 shows the molar ratios of the electron carriers in rat-heart mitochondria. Some of the electron carriers occur in the form of organized supramolecular complexes which can be isolated as units from the membrane. One such complex contains NADH dehydrogenase, four or more iron-sulfur centers, and a number of lipids. Another complex contains cytochromes b and c_1. Cytochromes a and a_3 form yet another complex. The ratios of these complexes to each other may vary from one cell type to another.

The number of cytochrome a molecules per unit area of inner membrane (and thus the number of functional respiratory chains) appears to be constant from one species to another. Liver mitochondria, which have relatively sparse cristae, a relatively small inner membrane area, and a relatively low rate of respiration, contain about 17,000 molecules of cytochrome a, but heart mitochondria, which have very profuse cristae and a relatively high rate of respiration, contain 60,000 to 70,000 molecules of cytochrome a. The surface area of the inner membrane appears to bear a relationship to the intensity of respiration of the tissue. Most prodigious are the mitochondria of blowfly flight muscle, one of the most intensely respiring tissues known. These mitochondria have an enormous inner membrane surface, amounting to about 400 m² per gram of mitochondrial protein.

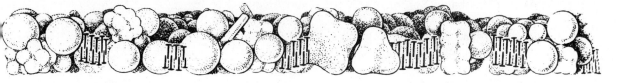

Oxidative Phosphorylation

We shall now examine the process of energy coupling in the respiratory chain which results in the formation of ATP from ADP and phosphate at the expense of the energy yielded by electron transport to oxygen.

That phosphorylation is coupled to the respiration of animal tissues was first postulated by V. A. Engelhardt in the Soviet Union in the early 1930s, but it was not until after the tricarboxylic acid cycle was formulated in 1937 that solid evidence appeared. H. Kalckar in Denmark and V. Belitser in the Soviet Union independently reported that when various intermediates of the cycle were oxidized by buffered suspensions of freshly minced liver, kidney, or muscle tissue, inorganic phosphate present in the medium disappeared. Concomitantly, there was an increase in the concentration of organic phosphate compounds, such as glucose 6-phosphate and fructose 6-phosphate, whose phosphate groups are derived from ATP. When the tissue suspensions were deprived of oxygen or poisoned with cyanide, uptake of inorganic phosphate did not take place. It was therefore concluded that phosphorylation of ADP is coupled to aerobic respiration as a mechanism for energy recovery.

A few years later a more quantitative picture emerged when it was found that *more* than one molecule of ATP was generated from ADP per atom of oxygen consumed. The P/O ratio (the number of molecules of inorganic phosphate taken up to phosphorylate ADP, per atom of oxygen consumed) was established to be 3, on the average, for each of the five oxidative steps involved in the oxidation of pyruvate via the tricarboxylic acid cycle (Table 19-3).

It was also recognized rather early that oxidative phosphorylation is an extremely labile process; it is most active in relatively fresh, unfractionated tissue suspensions and does not take place in aged preparations or in soluble extracts of tissues. Moreover, oxidative phosphorylation is sensitive to the osmotic pressure of the medium, suggesting that it occurs in a membrane-surrounded organelle. In 1948 E. P. Kennedy and A. L. Lehninger discovered the reason for this behavior: they showed that the process of oxidative phosphorylation takes place exclusively in the mitochondrial fraction isolated by differential centrifugation from homogenates of rat liver (page 381). In the years since, it has been found that in all cell types and species examined the mitochondria are the site of oxidative phosphorylation.

Coupling of Oxidative Phosphorylation to Electron Transport

In 1949 to 1951 Lehninger provided experimental proof that electron transport from NADH to oxygen is the direct source of the energy used for the coupled phosphorylation of ADP. Pure NADH was incubated aerobically with water-treated mitochondria, phosphate, and ADP in the absence of tricarboxylic acid cycle intermediates or any other added organic metabolite. (The hypotonic water treatment was necessary to

Table 19-2 Molar ratio of some electron-transferring components of the respiratory chain in beef-heart mitochondria

Component	Molar ratio relative to cytochrome c = 1.0
NADH + NAD$^+$	9
NADH dehydrogenase	0.14
Ubiquinone	7
Cytochrome b	1.1
Cytochromes $c_1 + c$	1.2
Cytochrome a	1.0
Cytochrome a_3	1.1

Table 19-3 P/O ratios of the oxidative steps in the tricarboxylic acid cycle

Step	P/O
Pyruvate ⟶ acetyl-CoA	3
Isocitrate ⟶ α-ketoglutarate	3
α-Ketoglutarate ⟶ succinyl-CoA	3
Succinate ⟶ fumarate	2
Malate ⟶ oxaloacetate	3

make the mitochondria permeable to NADH.) The NADH was rapidly oxidized to NAD$^+$ at the expense of molecular oxygen; simultaneously, up to three molecules of ATP were formed from ADP and phosphate. Such experiments indicated that at three points in the chain of electron carriers leading from NADH to oxygen, oxidation-reduction energy is transformed into phosphate-bond energy. For this reason, oxidative phosphorylation is more accurately called _respiratory-chain phosphorylation_.

The overall equation for the respiratory-chain phosphorylations could then be written as

$$\text{NADH} + \text{H}^+ + 3\text{ADP} + 3\text{P}_i + \tfrac{1}{2}\text{O}_2 \longrightarrow \text{NAD}^+ + 4\text{H}_2\text{O} + 3\text{ATP} \quad (1)$$

This reaction equation can be separated into an exergonic component,

$$\text{NADH} + \text{H}^+ + \tfrac{1}{2}\text{O}_2 \longrightarrow \text{NAD}^+ + \text{H}_2\text{O}$$
$$\Delta G^{\circ\prime} = -52.7 \text{ kcal mol}^{-1}$$

and an endergonic component,

$$3\text{ADP} + 3\text{P}_i \longrightarrow 3\text{ATP} + 3\text{H}_2\text{O}$$
$$\Delta G^{\circ\prime} = 3 \times 7.3 = +21.9 \text{ kcal mol}^{-1}$$

Coupled phosphorylation of three molecules of ATP thus conserves $21.9/52.7 \times 100$, or about 42 percent of the total free-energy decline during transport of a pair of electrons from NADH to oxygen under the usual standard conditions.

The approximate sites of the three energy-delivering segments of the respiratory chain have also been worked out. They have been predicted from calculations of the free-energy change occurring during the transfer of a pair of electron equivalents in each of the successive electron-transferring steps in the respiratory chain. Such calculations are carried out using the relationship

$$\Delta G = -n \mathscr{F} \, \Delta E_0^\prime$$

described in the preceding chapter (page 498), using the standard oxidation-reduction potentials listed in Tables 18-1 and 18-4. Figure 19-7 shows that there are three spans in the chain in which relatively large decreases in free energy occur, each sufficient to provide the energy (theoretically, at least) for the formation of ATP from ADP and phosphate. Such calculations of course assume thermodynamic equilibrium, which does not necessarily exist in any of these steps in the intact cell, and standard concentrations of 1.0 M of all components, also an unlikely condition. Nevertheless, direct experimental approaches, in which specific portions of the respiratory chain have been tested for their capacity to support ATP formation, have proved that these three spans are indeed the energy-coupling sites of the respiratory chain. They are designated _site I_ (the span between NADH and

Figure 19-7
The decline in free energy as electron pairs flow down the respiratory chain to oxygen. Each of the three segments denoted in color yields sufficient energy to generate a molecule of ATP from ADP and phosphate.

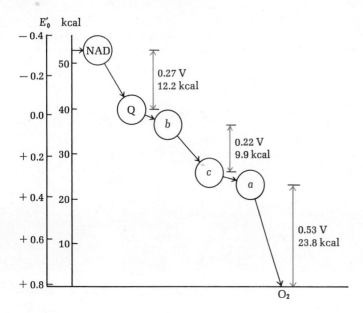

coenzyme Q), *site II* (the span between cytochrome *b* and cytochrome *c*), and *site III* (between cytochrome *a* and oxygen). Thus the multimembered respiratory chain may be regarded as a device for breaking up into a series of smaller energy drops the rather large decline in free energy occurring as a pair of electrons moves from NADH ($E_0' = -0.32$ V) to molecular oxygen ($E_0' = +0.82$ V), a process yielding 52.7 kcal of free energy per atom of oxygen that is reduced. Three of these spans deliver sufficient energy to generate a molecule of ATP from ADP and phosphate (page 398). The respiratory chain may therefore be regarded as an energy-transforming device.

From Figure 19-8 we see that the mitochondrial oxidation of *many* NAD-linked substrates, not only those of the tricarboxylic acid cycle, leads to the formation of three molecules of ATP per atom of oxygen reduced (P/O ratio = 3.0). However, some metabolites, e.g., succinate (page 459), fatty acyl–CoA's (page 548), and glycerol phosphate (page 534), are dehydrogenated by flavoproteins that bypass site I and feed electrons directly into ubiquinone (Figure 18-14), with the result that only two molecules of ATP are formed (at sites II and III) per atom of oxygen reduced (P/O ratio = 2.0).

The Energy Balance Sheet for Glucose Oxidation

From the data in Table 19-3 we see that for each molecule of pyruvate oxidized to completion, 12 molecules of ATP are formed in the four NAD-linked steps, i.e., the oxidation of pyruvate, isocitrate, α-ketoglutarate, and malate, 2 molecules

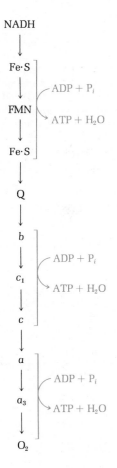

of ATP are formed during the flavin-linked oxidation of succinate (page 459), and 1 molecule of ATP is formed by the substrate-level phosphorylation at the expense of succinyl-CoA (page 458), to make a total of 15 ATPs formed per molecule of pyruvate oxidized. We can therefore write an equation for the complete oxidation of pyruvate by mitochondria, including the coupled phosphorylations:

$$\text{Pyruvate} + 2\tfrac{1}{2}O_2 + 15P_i + 15ADP \longrightarrow 3CO_2 + 15ATP + 17H_2O$$

We can now write a set of equations for the complete aerobic oxidation of glucose to CO_2 and water and the conservation of free energy as ATP. For the glycolytic sequence to pyruvate we have the reaction

$$\text{Glucose} + 2P_i + 2ADP + 2NAD^+ \longrightarrow$$
$$2 \text{ pyruvate } + 2NADH + 2H^+ + 2ATP + 2H_2O \quad (2)$$

and for the tricarboxylic acid cycle

$$2 \text{ Pyruvate} + 5O_2 + 30ADP + 30P_i \longrightarrow$$
$$6CO_2 + 30ATP + 34H_2O \quad (3)$$

To these we must add the equation for the oxidation of the two molecules of extramitochondrial NADH formed in the glycolytic conversion of glucose to pyruvate. Oxidation of extramitochondrial NADH may generate either two or three molecules of ATP per pair of electrons, depending upon how the electrons from extramitochondrial NADH enter the mitochondria (page 533). If we assume that two molecules of ATP are formed in this process, we have

$$2NADH + 2H^+ + O_2 + 4P_i + 4ADP \longrightarrow$$
$$2NAD^+ + 4ATP + 6H_2O \quad (4)$$

The sum of equations (2), (3), and (4) is therefore

$$\text{Glucose} + 6O_2 + 36P_i + 36ADP \longrightarrow 6CO_2 + 36ATP + 42H_2O \quad (5)$$

If we now dissociate this overall equation into its energetic components, we have:

Exergonic component:

$$\text{Glucose} + 6O_2 \longrightarrow 6CO_2 + 6H_2O$$
$$\Delta G^{\circ\prime} = -686 \text{ kcal mol}^{-1}$$

Endergonic component:

$$36P_i + 36ADP \longrightarrow 36ATP + 36H_2O$$
$$\Delta G^{\circ\prime} = +263 \text{ kcal mol}^{-1}$$

The overall efficiency of energy recovery in the complete oxidation of glucose is thus $263/686 \times 100 = 38$ percent under standard conditions. However, under intracellular conditions the true efficiency is much higher (page 432).

Acceptor Control of the Rate of Electron Transport

The overall equation for respiratory-chain phosphorylation [equation (1)] indicates that phosphate and ADP are necessary reactants for the process of electron transport from NADH to oxygen. Indeed, electron transport proceeds at a maximal rate in intact mitochondria only when phosphate and ADP are present in the suspending medium. When only ADP is lacking, the rate of respiration is very low and no phosphorylation occurs because there is no phosphate acceptor. This condition, known as *state 4 respiration*, is the idling or resting state of respiration (Figure 19-9). When a known amount of ADP is then added to such a system, the oxygen-uptake abruptly increases to a maximum; concomitantly, the added ADP becomes phosphorylated to yield ATP. This is called *state 3* or *active respiration*. When all the added ADP has been phosphorylated, the rate of oxygen consumption abruptly returns to the idling, or state 4, rate (Figure 19-9). This phenomenon, in which the rate of electron transport is controlled by the concentration of ADP, is called *acceptor control* or, less accurately, *respiratory control*. Intact mitochondria have a very high affinity for ADP and will continue to phosphorylate it, providing all other required components are present, until the ADP concentration is very low. The *acceptor-control ratio* or *index* is the ratio of the rate of respiration of mitochondria in the presence of ample ADP to the rate of respiration in the absence of ADP. This ratio is normally very high; it may be 10 or more in intact mitochondria and even higher in the intact cell. However, when the mitochondria are damaged or aged, they lose their ability to phosphorylate ADP and the ratio falls to 1.0. In such damaged mitochondria electron transport occurs at a maximal rate in the absence of ADP. The acceptor-control ratio is a useful measure of the integrity of isolated mitochondria: the higher the ratio, the more nearly intact the mitochondria. Moreover, as shown in Figure 19-9, the amount of extra oxygen consumption induced by a known amount of ADP can give us the ADP/O ratio, which is equal to the P/O ratio described above.

The ultrastructure of mitochondria undergoes striking reversible alterations as a result of transitions between respiratory states 4 (no ADP) and 3 (excess ADP) (Figure 19-10). This effect, first described by C. R. Hackenbrock and also studied by many others, results in a change in the volume and configuration of the inner-membrane–matrix compartment. In the absence of ADP the inner compartment of respiring mitochondria completely fills the space bounded by the outer membrane; this is called the *orthodox state*. When ADP is added to initiate state 3 (active) respiration, the matrix condenses to a volume that is only about 50 percent of that in the orthodox state and the inner membrane and cristae become more tightly folded and more contorted. This is called the *condensed state*. The orthodox and condensed configurations reflect the "off" and "on" states, respectively, of the mitochondrial ATP-generating system. In intact respiring liver cells the mitochondria are halfway between state 4 and state 3. The concentration of ADP in

Figure 19-9
Acceptor control of respiration. In state 4 respiration all available ADP has been phosphorylated to ATP, and the system is idling. When ADP is added (ΔADP), the rate of oxygen consumption abruptly increases to the state 3, or active, rate, during which the added ADP is phosphorylated to ATP. When nearly all the ADP has been phosphorylated, the mitochondria return to the state 4 rate. The ratio of the moles of ADP added to atom of extra oxygen consumed (ADP/O ratio) is equal to the P/O ratio. The ratio of the rate of state 3 respiration to the rate of state 4 respiration is the acceptor- or respiratory-control ratio.

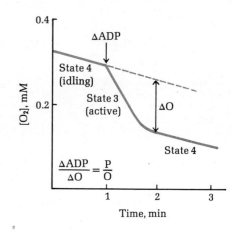

Figure 19-10
Electron micrographs showing the ultrastructural changes in mouse-liver mitochondria during the transition from resting (state 4) to active (state 3) respiration. This striking change in the structure and volume of the inner-membrane–matrix compartment appears to be caused by the binding of ADP to the ADP–ATP translocase molecules in the inner mitochondrial membrane.

Orthodox conformation

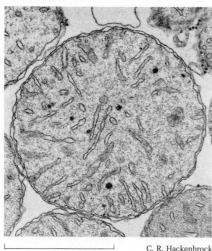

0.5 μm   C. R. Hackenbrock

Condensed conformation

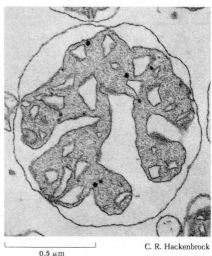

0.5 μm   C. R. Hackenbrock

normal intact cells is relatively low, insufficient to evoke maximal rates of respiration. When cells are stimulated to activity the rate of respiration increases and the mitochondria will then be in the condensed state, corresponding to state 3 respiration.

Uncoupling and Inhibition of Oxidative Phosphorylation

The molecular mechanism of oxidative phosphorylation has long been a challenging problem. Many experimental approaches have been taken to elucidate the intermediate steps. One important approach is the use of specific inhibitors to dissect the overall process into individual reactions. Oxidative phosphorylation is influenced by a number of chemical agents, which can be grouped into three major classes (Table 19-4). The <u>uncoupling agents</u> allow electron transport to continue but prevent the phosphorylation of ADP to ATP; i.e., they *uncouple* the energy-yielding from the energy-conserving reactions. Characteristically they stimulate the rate of oxygen uptake by intact mitochondria in the absence of ADP (Figure 19-11). Moreover, they evoke a large amount of ATP-hydrolyzing activity in mitochondria; in the absence of uncoupling agents mitochondria have very little ATPase activity. The first uncoupling agent to be described, by W. F. Loomis and F. Lipmann in 1948, is 2,4-<u>dinitrophenol</u>. Today many different uncoupling agents are known. Most are lipid-soluble substances containing an acidic group and usually an aromatic ring; representative uncouplers are shown in Figure

Table 19-4   Types of agents affecting oxidative phosphorylation

Uncoupling agents
  2,4-Dinitrophenol
  Dicumarol
  Carbonylcyanide phenylhydrazones
  Salicylanilides
  Arsenate

Inhibitors of ATP formation
  Oligomycin
  Rutamycin
  Aurovertin
  Triethyltin

Ionophores (carriers of cations)
  Valinomycin
  Gramicidin
  Nonactin
  Nigericin

19-12. These agents do not uncouple glycolytic phosphorylation or directly affect cellular reactions other than oxidative phosphorylation. Uncoupling agents function by breaking down or discharging a high-energy intermediate or state generated by electron transport (pages 493 and 527). We shall see later that uncoupling agents can promote the passage of H^+ ions through the mitochondrial membrane, which is normally impermeable to them.

The second class of agents, the _inhibitors_ of oxidative phosphorylation, differ from the uncoupling agents; they prevent both the stimulation of oxygen consumption by ADP and the phosphorylation of ADP to ATP. However, these agents do not directly inhibit any of the electron carriers of the respiratory chain. Instead they prevent the ATP-forming mechanism from utilizing the high-energy intermediate or state generated by electron transport. As a consequence, electron transport cannot continue unless the high-energy intermediate or state is used up. The antibiotic _oligomycin_, whose action was first described by H. A. Lardy and his colleagues, is the prototype of this class (Table 19-4). The inhibitory action of these agents on oxygen consumption is characteristically relieved by 2,4-dinitrophenol and other uncoupling agents, which can promote the breakdown of the high-energy intermediate or state generated by electron transport (Figure 19-11).

A third class of agents, the _ionophores_, causes the breakdown of the high-energy intermediate state only if certain monovalent cations are present. The prototype of this group is the antibiotic _valinomycin_ (Table 19-4), which requires K^+ for its inhibitory effect. Other agents of this class include the antibiotics _nigericin_ and _nonactin_, which also require K^+ for their action, and _gramicidin_, which functions in the presence of either K^+ or Na^+. Over 50 different antibiotics with similar activity have been described (page 802). These agents are called ionophores since they form lipid-soluble complexes with specific cations, which are thus carried through the mitochondrial membrane. The valinomycin–K^+ complex readily passes through the membrane; in the absence of valinomycin, K^+ passes only very slowly (page 529). Ionophores prevent oxidative phosphorylation because they force the mitochondiria to use the energy of respiration to pump cations, such as K^+, into the matrix, instead of using the energy to make ATP. The cations leak out as fast as they are pumped in.

The Partial Reactions of Oxidative Phosphorylation

Four reactions catalyzed by intact mitochondria are influenced in a characteristic way by 2,4-dinitrophenol and oligomycin (Table 19-5) and are therefore believed to represent individual steps or sequences of steps in the mechanism by which ATP is formed during oxidative phosphorylation. These reactions, often called _partial reactions_, appear to take place in the absence of a net flow of electrons down the chain. The first is the _ATPase activity_ of mitochondria, which is normally very low but which is greatly stimulated

Figure 19-11
Effect of uncoupling agents and oligomycin on the rate of oxygen consumption of mitochondria.

Action of uncoupling agents. Addition of ADP to mitochondria respiring in a buffered medium of substrate, phosphate, and Mg^{2+} causes the usual jump in the oxygen-uptake rate. However, addition of an uncoupling agent such as 2,4-dinitrophenol causes an indefinite stimulation of respiration, usually at a higher rate than given by ADP.

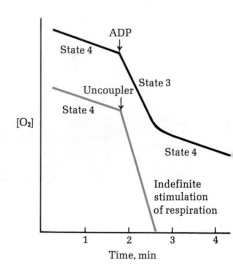

Action of oligomycin. Oligomycin does not inhibit state 4 respiration but prevents the usual stimulation given by ADP. Characteristically, uncoupling agents can still stimulate oxygen consumption in the presence of oligomycin.

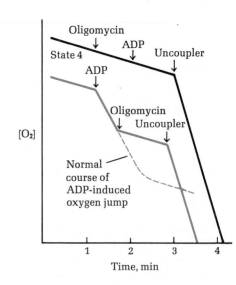

Figure 19-12
Some uncoupling agents.

Figure 19-12
Some uncoupling agents.

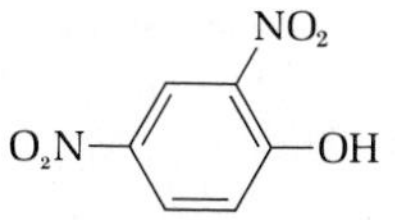

2,4-Dinitrophenol

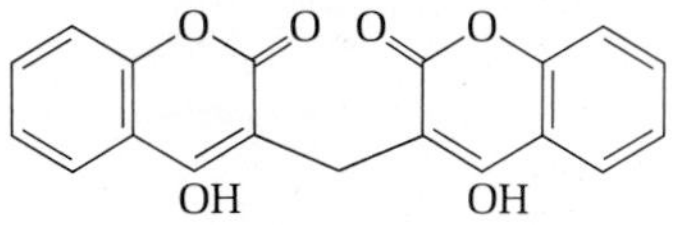

Dicumarol

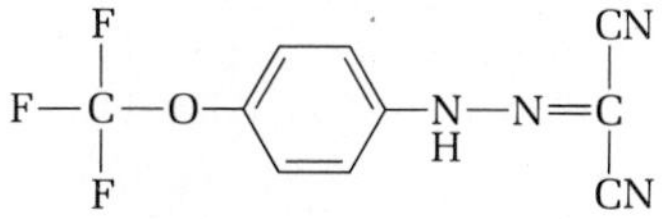

Carbonylcyanide p-trifluoro-
methoxyphenylhydrazone

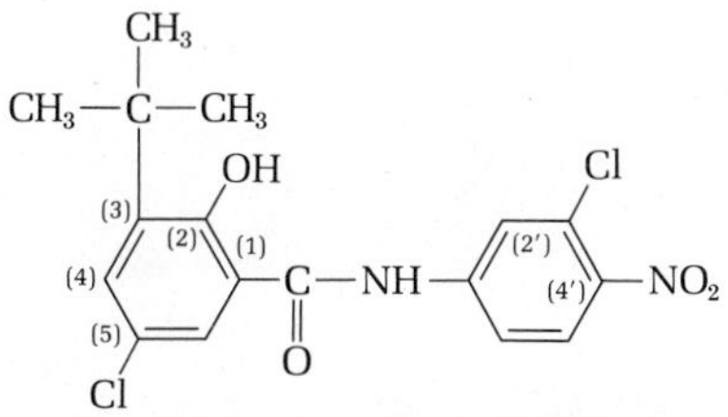

5-Chloro-3-*t*-butyl-2'-chloro-4'-
nitrosalicylanilide

by 2,4-dinitrophenol and other uncoupling agents. Characteristically, the stimulated ATPase activity is inhibited by oligomycin. The ATPase activity of mitochondria is believed to represent reversal of the reactions by which ATP is normally generated from ADP and phosphate. The second partial reaction is an isotopic exchange reaction in which inorganic phosphate labeled with ^{32}P exchanges rapidly with the terminal phosphate group of ATP in the absence of electron transport. This reaction, called the *phosphate-ATP exchange*, is completely inhibited by either 2,4-dinitrophenol or oligomycin. The third partial reaction is an exchange of the oxygen atoms of the water of the medium with those of inorganic phosphate, measured by labeling either the phosphate or the water with ^{18}O; during this exchange there is no net disappearance of phosphate. This reaction, the *phosphate-water exchange*, is also inhibited by 2,4-dinitrophenol and oligomycin. The fourth partial reaction is a rapid reversible transfer of the terminal phosphate group from ATP to ADP, called the *ADP–ATP exchange* reaction; it is also inhibited by 2,4-dinitrophenol and oligomycin.

These properties of the partial reactions strongly indicate that the sequence by which ATP is formed from ADP and P_i is reversible. Moreover, they provide an approach for experimental study of the individual steps in ATP formation.

The Reversibility of Phosphorylating Electron Transport

That the reactions of oxidative phosphorylation are indeed reversible has been shown in more direct ways. In one type of experiment mitochondria supplemented with succinate and oxaloacetate are incubated with cyanide to inhibit electron flow to oxygen. When ATP is added to such a system, electrons will flow from succinate to oxaloacetate via the electron-transport chain in the *reverse* direction, i.e.,

Succinate $\longrightarrow$ succinate dehydrogenase $\longrightarrow$ ubiquinone $\longrightarrow$
NADH dehydrogenase $\longrightarrow$ NAD$^+$ $\longrightarrow$ oxaloacetate

resulting in the reduction of oxaloacetate to malate. Concomitantly, ATP undergoes cleavage to ADP and P_i, by reversal of the energy-coupling reactions of site I. ATP energy

---

Table 19-5   Partial reactions of oxidative phosphorylation

The isotopically labeled component is in color; AMP~P~P represents ATP, and AMP~P represents ADP

ATPase activity, stimulated by uncoupling agents
  ATP + H_2O $\longrightarrow$ ADP + P_i

ATP-phosphate exchange
  AMP~P~P + P_i $\rightleftharpoons$ AMP~P~P + P_i

Phosphate-water exchange
  HPO_4^{2-} + H_2O $\rightleftharpoons$ HPO_4^{2-} + H_2O

ADP–ATP exchange
  AMP~P + AMP~P~P $\rightleftharpoons$ AMP~P~P + AMP~P

is thus required to reverse energy-coupled electron transport. The other two energy-conserving steps of the chain can also be reversed under suitable conditions.

Oxidative Phosphorylation in Submitochondrial Systems

Although oxidative phosphorylation is a very labile process and for a long time could be observed only in intact freshly prepared mitochondria, it has been found possible to prepare submitochondrial particles capable of electron transport and oxidative phosphorylation. These particles are obtained by treating mitochondria with membrane-dispersing agents, e.g., the nonionic detergents digitonin or Lubrol, or by high-frequency sonic irradiation. The particles consist of membranous vesicles resulting from the resealing of in-ner-membrane fragments. They catalyze oxidative phospho-rylation only if the membrane is completely sealed, indi-cating that an intact vesicular structure is required for the energy-conserving process to occur.

Significantly, such membranous vesicles contain inner-membrane spheres (page 511), usually on the outside surface of the vesicles, suggesting that the majority of the vesicles arise by pinching off of cristae (Figure 19-13). The submi-tochondrial vesicles are thus largely "inside out" compared with the intact inner mitochondrial membrane. Biochemical and electron-microscopic study of such vesicles has led to the isolation of some of the protein components of the ATP-forming system of the mitochondrial membrane.

Coupling Factors and the Reconstitution of Oxidative Phosphorylation

E. Racker and his colleagues have found that when phos-phorylating submitochondrial vesicles prepared from beef-heart mitochondria are subjected to mechanical shaking or to treatment with trypsin and urea, two fractions result: (1) membranous vesicles still capable of catalyzing electron transport but no longer able to phosphorylate ADP and (2) a soluble protein fraction that catalyzes the hydrolysis of ATP but not electron transport. The vesicles treated in this way no longer retain inner-membrane spheres on their outer sur-face, as observed by negative-contrast electron microscopy. This finding suggested that the spheres are involved in cou-pling phosphorylation to electron transport. When the solu-ble and membranous fractions were combined, a significant amount of oxidative-phosphorylation activity was restored (Figure 19-13); moreover, inner-membrane spheres were now observed on the surface of the vesicles. The phosphoryl-ation activity was uncoupled by 2,4-dinitrophenol and inhib-ited by oligomycin. From such reconstitution experiments Racker and his colleagues concluded that the mitochondrial membrane contributes the enzymes of electron transport, whereas the soluble, easily detached protein fraction, com-prising the inner-membrane spheres, represents the enzyme complex required to make ATP. Soluble proteins necessary for the restoration of energy-coupling activity to the mi-

Figure 19-13
(Right) *Schematic representation of the preparation of phosphorylating submi-tochondrial vesicles, their resolution into nonphosphorylating vesicles and* F_1 *ATPase and the reconstitution of phosphorylating vesicles. Most of the vesicles in the popula-tion are inside out compared with the mi-tochondria from which they are derived. (Adapted from E. Racker, Essays in Bio-chemistry, pp. 1–22, vol. 6, Academic Press, Inc., London, 1970.)*

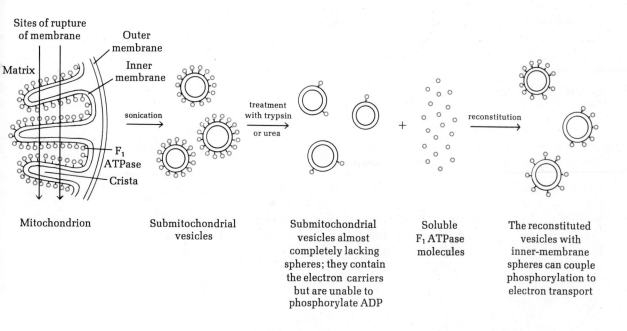

Mitochondrion | Submitochondrial vesicles | Submitochondrial vesicles almost completely lacking spheres; they contain the electron carriers but are unable to phosphorylate ADP | Soluble F_1 ATPase molecules | The reconstituted vesicles with inner-membrane spheres can couple phosphorylation to electron transport

Figure 19-14

Schematic representation of the mitochondrial ATPase complex. The F_1 portion contains five or six different kinds of subunits, including two sets of large subunits (shaded). The stalk and part of the membrane portion contain the protein(s) conferring oligomycin sensitivity. It is not at all certain that the subunits fit as shown; the drawing conforms to the sphere-and-stalk images seen after negative-contrast staining of cristae.

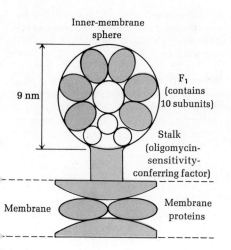

tochondrial membrane are called <u>coupling factors</u>. The soluble coupling factor of Racker is called F_1 (for factor 1); because it also catalyzes ATP hydrolysis it is often called <u>F_1 ATPase.</u>

Racker and his colleagues have succeeded in purifying the soluble F_1 ATPase of heart mitochondria; others have similarly purified that of rat-liver mitochondria. The F_1 ATPase is a large, complex molecule of molecular weight 360,000 to 380,000 and diameter 9 nm. In the presence of Mg^{2+} it catalyzes the slow hydrolysis of ATP to ADP and phosphate; this reaction is not inhibited by oligomycin. This ATPase activity is believed to represent reversal of the normal function of F_1 in intact mitochondria, namely, the synthesis of ATP from ADP and phosphate. F_1 is perhaps more appropriately named ATP synthase than ATPase. The F_1 factor is unstable at 0°C but relatively stable at room temperature, a curious property also shown by some allosteric enzymes (page 234). F_1 contains five or six kinds of protein subunits, having molecular weights of about 60,000, 57,000, 36,000, 12,500, and 7,500; the precise number of each type of subunit is not entirely clear. The F_1 molecule binds ADP very tightly but does not bind phosphate. Very significantly, the antibiotic <u>aurovertin,</u> which inhibits oxidative phosphorylation (Table 19-4), inhibits the binding of ADP by F_1.

Another specific protein functioning in the coupling of ATP synthesis to electron transport is the <u>oligomycin-sensitivity-conferring factor,</u> abbreviated OSCF or F_0. When the F_0 factor is added to the F_1 factor, the ATPase activity of the vesicles is inhibited by oligomycin. F_0 therefore confers oligomycin sensitivity on the F_1 factor. F_0 itself is a large molecule containing five or six specific subunits. A complex containing F_1 and F_0 can be isolated from mitochondria; it also contains a protein of molecular weight 10,000 that inhibits the ATPase activity of F_1. Figure 19-14 shows a schematic representation of the F_1–F_0 complex.

523

These important experiments with submitochondrial vesicles and their components have not only opened the door to molecular examination of the ATP-forming enzymes but also have shown that the complex assemblies of energy-coupling enzymes of the mitochondrial membrane can be dissociated or resolved and then reassembled or reconstituted.

The Mechanism of Oxidative Phosphorylation

Despite intensive investigations in many laboratories over the past quarter century we still lack a detailed molecular picture of the mechanism by which the oxidoreduction energy of electron transport is converted into the phosphate-bond energy of ATP. Three hypotheses have gained wide attention, the _chemical-coupling, conformational-coupling,_ and _chemiosmotic-coupling_ hypotheses. These hypotheses have broad significance in cell biology since they are applicable not only to oxidative phosphorylation in mitochondria and in bacteria but also, as we shall see, to photosynthetic phosphorylation (page 611).

The Chemical-Coupling Hypothesis

The earliest mechanism proposed for oxidative phosphorylation is the chemical-coupling hypothesis. This postulates that the energy-yielding electron-transfer reaction is coupled to the energy-requiring reaction by which ATP is formed from ADP and phosphate through a common chemical intermediate, a high-energy compound generated by electron transport and then utilized as the reactant in a second reaction to form ATP from ADP and phosphate. The principle of the common intermediate (page 403) in consecutive reactions is of course basic to virtually all metabolic sequences in the cell. A simple, well-known model or prototype of a chemical-coupling mechanism for the conversion of oxidation-reduction energy into ATP energy is the energy-conserving oxidoreduction catalyzed by 3-phosphoglyceraldehyde dehydrogenase of glycolysis (page 427). In this reaction (Figure 19-15) the energy released on oxidation of the aldehyde group to yield the carboxyl group is conserved as the common intermediate 3-phosphoglyceroyl 1-phosphate (page 427), which then donates its high-energy phosphate to ADP and thus couples these reactions chemically. Also shown in Figure 19-15 is an early form of the chemical-coupling hypothesis for oxidative phosphorylation, proposed by E. C. Slater in 1953. It is postulated that a high-energy chemical intermediate, i.e., a compound having a strongly negative standard free energy of hydrolysis, is generated by the transfer of electrons from one electron carrier of the respiratory chain to the next. This common intermediate then provides the energy for the formation of ATP from ADP and P_i. More elaborate chemical-coupling schemes, with additional steps, have been proposed by others.

Although the chemical-coupling hypothesis has guided

Figure 19-15
Chemical-coupling mechanisms for energy transduction between electron-transfer reactions and ATP synthesis via a common chemical intermediate.

Chemical coupling in glycolysis. The energy yielded on oxidation of the aldehyde group of 3-phosphoglyceraldehyde (RCHO) by NAD^+ _is conserved as 3-phosphoglyceroyl phosphate_ $(RCOOPO_3{}^{2-})$, _which serves as the common intermediate carrying energy to the subsequent reaction in which ADP is phosphorylated to ATP._

$$RCHO + P_i + NAD^+ \rightleftharpoons$$

$$\boxed{RCOOPO_3{}^{2-}} + NADH + H^+$$

$$\boxed{RCOOPO_3{}^{2-}} + ADP \rightleftharpoons RCOO^- + ATP$$

A simple chemical coupling hypothesis for the respiratory chain. AH_2 _is a reduced electron carrier, and B is the oxidized form of the next carrier in the chain. C is a third molecular component, a coupling factor. Transfer of electrons from_ AH_2 _to B is postulated to cause formation of a high-energy bond ($\sim$) between A and C, which is the common intermediate linking this reaction to the ATP-forming reaction:_

$$AH_2 + B + C \rightleftharpoons \boxed{A \sim C} + BH_2$$

$$\boxed{A \sim C} + P_i + ADP \rightleftharpoons A + C + ATP$$

much research, it has two serious shortcomings. One is that the postulated high-energy chemical intermediate required for energy coupling has never been detected in mitochondria after over 20 years of intensive search; many investigators therefore feel that such intermediates do not exist. The second is that the chemical-coupling hypothesis provides no satisfactory explanation for the fact that the inner mitochondrial membrane must be intact and continuous, as a completely closed vesicle, for oxidative phosphorylation to take place. Indeed, a membrane is not required for a chemical-coupling reaction to take place; we recall that the energy-conserving oxidation of 3-phosphoglyceraldehyde in glycolysis occurs in homogeneous aqueous solution (page 427). Although the chemical-coupling hypothesis now has fewer adherents, it is nevertheless still conceivable that the lipid bilayer of the membrane provides a nonpolar phase in which a labile coupling intermediate, possibly one easily hydrolyzed in an aqueous phase, may be generated and utilized.

The Conformational-Coupling Hypothesis

Several investigators, particularly P. D. Boyer, have postulated that the energy yielded by electron transport is conserved in the form of a conformational change in an electron carrier protein or in the coupling factor (F_1 ATPase) molecule. Such a high-energy conformational state would be the result of an energy-dependent shift in the number or location of the weak bonds (hydrogen bonds, hydrophobic interactions) maintaining the three-dimensional conformation of the protein. It is proposed that the energy inherent in this "energized" conformation is used to cause the formation of ATP from ADP and P_i; simultaneously, the energy-carrying protein undergoes reversion to its original low-energy conformational state. The conformational-coupling hypothesis is in principle a variant of the chemical-coupling hypothesis; it postulates that a large number of weak noncovalent bonds in a macromolecule serve as the common intermediate or carrier of energy for the formation of ATP, whereas the classical chemical-coupling hypothesis postulates that a single covalent bond of the common intermediate is the vehicle of energy transfer.

The conformational changes occurring in the actomyosin system of skeletal muscle as ATP is bound and hydrolyzed and the ADP released (page 761) have been taken as a prototype for conformational coupling. One piece of evidence favoring conformational coupling is the observation that the inner mitochondrial membrane undergoes very rapid physical changes as electrons pass along the respiratory chain. These are detected by fluorescence measurements on membrane-bound probe molecules, such as 1-anilinonaphthalene 8-sulfonic acid. Conformational coupling is also suggested by the dramatic ultrastructural changes that accompany ADP addition to respiring mitochondria (see page 519). However, to date there is little more than suggestive evidence for the conformational-coupling hypothesis.

Figure 19-16
*Simplified representation of the chemios-
motic-coupling hypothesis.*

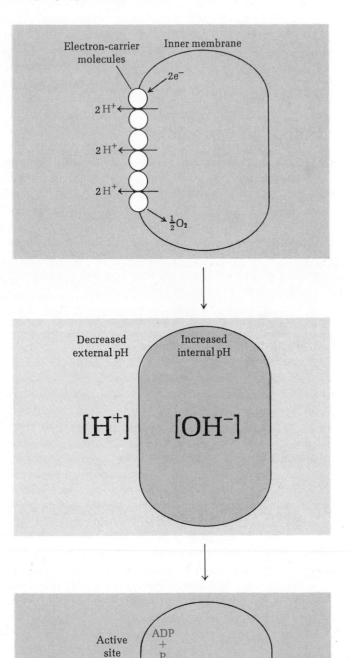

Electron transport causes H^+ ions to be pumped outward, across the inner membrane of the mitochondrion, to yield a gradient of H^+. This phase is shown in more detail in Figure 19-17.

The H^+ gradient is the energy-rich state into which electron-transport energy is transformed. The inner compartment becomes alkaline, the outer compartment more acid.

The H^+ gradient is the immediate driving force for the phosphorylation of ADP, which proceeds with the removal of HOH. The relatively high internal OH^- concentration pulls H^+ (color) from the active site of the F_1 ATPase and the relatively high external H^+ concentration pulls OH^- in the outward direction. Since the ion product of water ($K_w = [H^+] [OH^-]$) is very low (10^{-14}), the sinks of OH^- and H^+ generated by electron transport are very effective traps for H^+ and OH^-, respectively. ATP formation is shown in more detail in Figure 19-18.

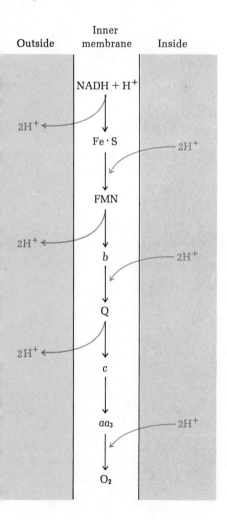

The Chemiosmotic-Coupling Hypothesis

This hypothesis, first proposed in 1961 and since championed vigorously by P. Mitchell, a British biochemist, denies that a common *chemical* intermediate couples the energy-yielding and energy-requiring reactions of oxidative phosphorylation. It proposes instead that these reactions are coupled by a high-energy intermediate *state*, rather than a compound. Mitchell proposes that an electrochemical gradient (page 782) of H^+ ions across the mitochondrial inner membrane serves as the means of coupling the energy flow from electron transport to the formation of ATP. According to the chemiosmotic hypothesis, the membrane is an integral part of the coupling mechanism and must be intact in the form of a continuous closed vesicle for oxidative phosphorylation to take place. Mitchell postulates that it is the function of the electron carriers of the respiratory chain to serve as an active-transport system or "pump" (page 779) to transport H^+ ions from the mitochondrial matrix across the inner membrane, thus generating a gradient of H^+ ions across the membrane, which he postulated to be impermeable to H^+ ions. The electrochemical gradient thus generated is then postulated to drive the synthesis of ATP by causing dehydration of ADP and P_i (Figure 19-16).

A special feature of the chemiosmotic-coupling hypothesis is that it requires the operation of *vectorial* chemical reactions across the mitochondrial membrane, i.e., reactions having a geometric direction, in contrast to the nondirectional, or *scalar,* chemical reactions occurring in a free, homogeneous solution. Electron transport is viewed as generating a gradient of H^+ ions across the membrane by virtue of a specific vectorial arrangement of the electron-carrying proteins in the inner mitochondrial membrane, so that the H^+-absorbing reactions (Table 18-7, page 494) take place on the inner face of the inner membrane and the H^+-yielding reactions on the outer face. The electron-transport chain is thus viewed as a device for converting the energy released by electron transport into the energy of an electrochemical gradient of H^+ ions. The free energy stored in such a gradient is a function of the relative concentration of H^+ ions across the membrane (page 780); it has been called the protonmotive force. It is believed that $2H^+$ ions are pumped out of the mitochondria per pair of electrons passing through each energy-conserving site (Figure 19-17).

The energy-rich gradient of H^+ ions is then used to cause formation of ATP from ADP and P_i by a vectorial reaction involving presumably the F_1–F_0 ATPase complex of the inner membrane. The removal of H_2O from ADP and phosphate, usually denoted by the overall equation

$$ADP + P_i \longrightarrow ATP + H_2O$$

$$\Delta G^{\circ\prime} = +7.3 \text{ kcal mol}^{-1}$$

may be most simply viewed as the asymmetric, or vectorial, removal of H^+ and OH^- from ATP on the active site of F_1 ATPase, in such a fashion that H^+ appears on the inside of the membrane and OH^- on the outside:

$$ADP + P_i \longrightarrow ATP + H^+_{\text{inside}} + OH^-_{\text{outside}}$$

The H^+ is thus delivered into the "sink" of alkalinity generated in the matrix by electron transport, and the OH^- is delivered into the sink of acidity generated on the outside of the membrane (Figure 19-18). Thus both the electron-carrier molecules and the ATP-forming enzyme must be so fixed in the membrane that their active sites have a specific directionality. Mitochondrial electron transport continuously generates the energy-rich H^+ gradient, and the phosphorylation of ADP continuously uses up the energy inherent in the gradient. The result is a steady state in which little or no *net* H^+ gradient can be observed.

Much evidence is consistent with the chemiosmotic-coupling hypothesis. An intact mitochondrial membrane is necessary for oxidative phosphorylation; the membrane has also been found to be impermeable to H^+ ions. Furthermore, it has been shown that the electron-transport chain can pump H^+ ions outward as postulated and that ATP formation is accompanied by inward H^+ movement. Moreover, in accordance with Mitchell's prediction, uncoupling agents like 2,4-dinitrophenol allow protons to cross the otherwise impermeable mitochondrial membrane, thus "collapsing" the proton gradient generated by electron transport. Submitochondrial membrane vesicles, which we have seen are inside out (page 522), *absorb* H^+ from the medium during electron transport to oxygen, whereas intact right-side-out mitochondria *eject* H^+ into the medium, showing that the mitochondrial membrane does indeed possess sidedness with respect to the direction of proton pumping.

An early objection to the chemiosmotic-coupling hypothesis was that it appeared to require a large gradient of H^+ across the membrane to generate ATP from ADP under standard thermodynamic conditions. It can be calculated (page 398) that in order to provide the 7.3 kcal of free energy required to synthesize ATP from ADP under standard conditions, an H^+ gradient across the membrane of about 3,000:1, or about 3.5 pH units, would be required. However, Mitchell has pointed out that not all of the required electrochemical gradient need be manifest as a large difference in pH across the membrane; some might be contributed by a difference of electric potential across the membrane. Actually a gradient of 3.5 pH units, acid outside, is energetically equivalent to a potential difference across the membrane of about 0.235 V, with the inside negative with respect to the outside. Mitchell has suggested that the actual electrochemical gradient or protonmotive force generated across the membrane of respiration-energized mitochondria may consist of a gradient of about 1.0 pH unit (acid outside) plus a transmembrane potential of about 0.150 V (negative inside). Indirect experiments strongly suggest that electron transport can generate a membrane potential of this magnitude with the proper polarity of charge.

Although there is now substantial agreement that mitochondria can eject H^+ ions during electron transport, it is not yet clear whether the proton gradient so generated is an *obligatory* step in oxidative phosphorylation, as proposed in the chemiosmotic-coupling hypothesis, or whether it is the

Figure 19-18
Schematic representation of the vectorial action of the F_1 ATPase during ATP formation at the expense of a gradient of H^+ across the membrane. The active site is assumed to have the binding sites for ADP and phosphate located so as to deliver H_2O derived from the dehydration as H^+ and OH^-, H^+ to the matrix side and OH^- to the outside.

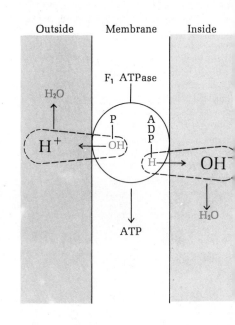

Figure 19-19
The role of the H⁺ gradient in ATP forma-
tion: the current dilemma.

Chemiosmotic coupling. The gradient of H⁺
across the membrane is postulated to be the
primary means of energy conservation and
to be an obligatory step in oxidative phos-
phorylation.

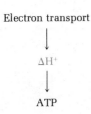

Electron transport

↓

ΔH^+

↓

ATP

Chemical coupling. A high-energy chemical
intermediate A~ C is postulated to be ob-
ligatory for oxidative phosphorylation. An
H⁺ gradient is considered to result from a
side reaction.

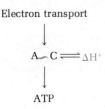

Electron transport

↓

$A{\sim} C \rightleftharpoons \Delta H^+$

↓

ATP

result of a side reaction (Figure 19-19) caused by vectorial hydrolysis of a high-energy chemical intermediate, $A{\sim}C$, generated by electron transport:

$$A{\sim}C + H^+_{inside} + OH^-_{outside} \longrightarrow H{-}A + C{-}OH$$

as proposed in the chemical-coupling hypothesis.

Although the chemical-coupling and conformational-coupling models cannot yet be ruled out, at present the chemiosmotic-coupling hypothesis appears to account most simply and directly for the available experimental evidence regarding the mechanism of energy transduction during electron transport in mitochondria and bacteria and during light-induced electron transport in photosynthetic organisms (page 611).

Metabolite Transport Systems of the Inner Membrane

Under some conditions mitochondria can utilize a significant fraction of the total available respiratory energy for purposes other than the phosphorylation of ADP. One of these activities is the transport of certain metabolites or mineral ions across the mitochondrial membrane against concentration gradients (page 785). This activity of mitochondria, which is concerned in the compartmentation and regulation of metabolism, involves specific transport systems (page 802) in the inner membrane.

The outer mitochondrial membrane is freely permeable to most solutes of low molecular weight, but the inner membrane is not: its permeability is highly selective. Intact nonrespiring liver mitochondria are not permeable to simple sugars such as glucose or sucrose, cations such as K^+ and Na^+, or anions such as Cl^- and Br^-. Moreover, the inner membrane is also impermeable to NAD^+, NADH, $NADP^+$, and NADPH; to other nucleotides, such as AMP, CTP, GTP, CDP, and GDP; and to CoA and acyl-CoA's. The inner compartment, or matrix, of mitochondria contains pools of these coenzymes and nucleotides that are thus physically separated from the extramitochondrial pools in the cytosol (page 381). On the other hand, the inner membrane of liver mitochondria does allow certain specific nucleotides and metabolites to pass quite readily. Among these are ATP and ADP, inorganic phosphate, pyruvate, citrate, succinate, α-ketoglutarate, and malate, as well as the amino acids glutamate and aspartate. The passage of this group of solutes through the mitochondrial membrane takes place by the action of specific membrane transport systems (Chapter 28, page 802).

The mitochondrial transport systems, which are probably specialized proteins or groups of proteins, are specific for certain metabolites and will not transport even closely similar molecules. For example, the ATP transport system will transport only ADP, ATP, dADP, and dATP but not AMP or other closely related nucleotides such as GTP, GDP, CTP, and CDP. Some of the mitochondrial transport systems can be quite specifically inhibited. For example, the transport system for ATP and ADP is inhibited by very low concentrations of atractyloside (see below).

Figure 19-20 shows the major transport systems of rat-liver mitochondria and the type of transport processes they promote. These systems are also called carriers, translocases, or porters. The ATP–ADP carrier, which has been most intensively studied, normally promotes the reversible equimolar exchange of a molecule of external ADP for a molecule of internal ATP formed by oxidative phosphorylation in the matrix. This carrier has a very high affinity for ADP; it is nearly saturated at less than 10 μM ADP. It is strongly and specifically inhibited by the toxic plant compound atractyloside, and by the antibiotic bongkrekic acid (Figure 19-21). The phosphate carrier promotes an exchange of $H_2PO_4^-$ ion and OH^- ion or, its equivalent, the cotransport of both $H_2PO_4^-$ and H^+ in the same direction. The phosphate carrier is inhibited by certain sulfhydryl reagents.

The dicarboxylate carrier (Figure 19-20) promotes equimolar exchange of malate, succinate, and fumarate with each other or with phosphate. The tricarboxylate carrier can promote equimolar exchange of citrate and isocitrate with each other, as well as equimolar exchange of a tricarboxylate with a dicarboxylate such as malate.

The inner-membrane transport systems of mitochondria are species-specific and genetically determined. Although mitochondria from all tissues appear to contain the transport systems for phosphate, ADP, and ATP, other transport systems vary in their species distribution.

These and related observations on mitochondria from various cell types have led to the general conclusion that fuel molecules such as pyruvate, fatty acids, and amino acids, as well as phosphate and ADP, must pass from the cytosol through the inner membrane via specific transport systems into the matrix compartment of mitochondria, where reactions of the tricarboxylic cycle, electron transport, and oxidative phosphorylation take place. The ATP formed during oxidative phosphorylation in the inner compartment must be transported across the membrane into the cytosol by the ADP–ATP carrier. The intramitochondrial pools of ATP and ADP are thus segregated from the cytosol pools of these nucleotides, but the two pools communicate via the atractyloside-sensitive carrier. Such compartmentation and specific transport of various metabolites across the mitochondrial membrane is important in the regulation of the glycolytic and respiratory pathways, as well as biosynthetic pathways in which the tricarboxylic acid cycle participates (pages 628 and 661).

The Coupling of Metabolite Transport to Electron Transport

The various metabolite-transport systems shown in Figure 19-20 are passive systems (defined in Chapter 28, page 788); in the absence of respiration they transport metabolites down concentration gradients, in the direction that will result in thermodynamic equilibrium across the membrane. However,

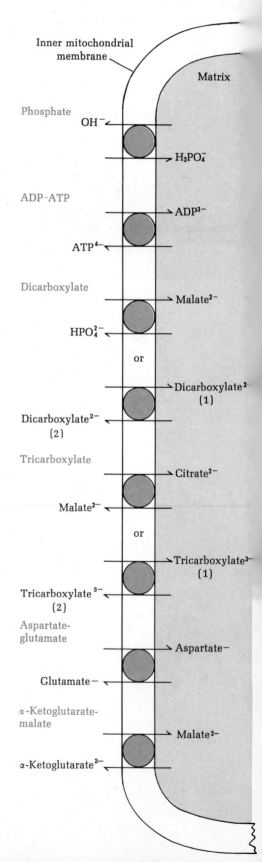

Figure 19-20

Important transport systems of the inner membrane in rat-liver mitochondria (colored circles). They can function in either direction in response to the concentration gradients of the transported metabolites.

Inner mitochondrial membrane

Matrix

Phosphate

OH^-

$H_2PO_4^-$

ADP-ATP

ADP^{3-}

ATP^{4-}

Dicarboxylate

$Malate^{2-}$

HPO_4^{2-}

or

$Dicarboxylate^{2-}$ (1)

$Dicarboxylate^{2-}$ (2)

Tricarboxylate

$Citrate^{2-}$

$Malate^{2-}$

or

$Tricarboxylate^{3-}$ (1)

$Tricarboxylate^{3-}$ (2)

Aspartate-glutamate

$Aspartate^-$

$Glutamate^-$

α-Ketoglutarate-malate

$Malate^{2-}$

α-Ketoglutarate^{2-}

Figure 19-21
Structures of atractyloside and bongkrekic
acid.

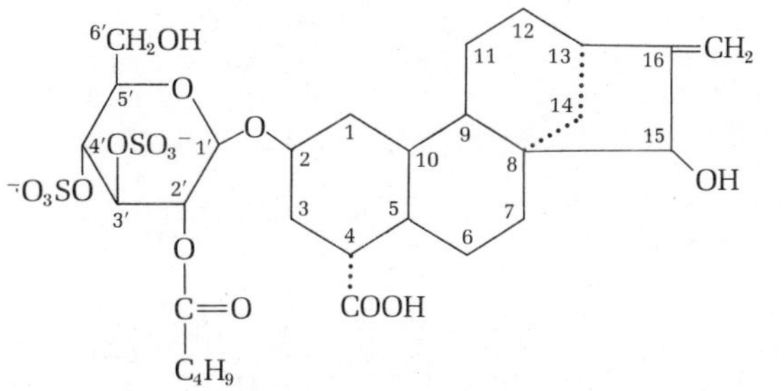

Atractyloside, a toxic glycoside from *Atractylis gummifera*,
the Mediterranean thistle

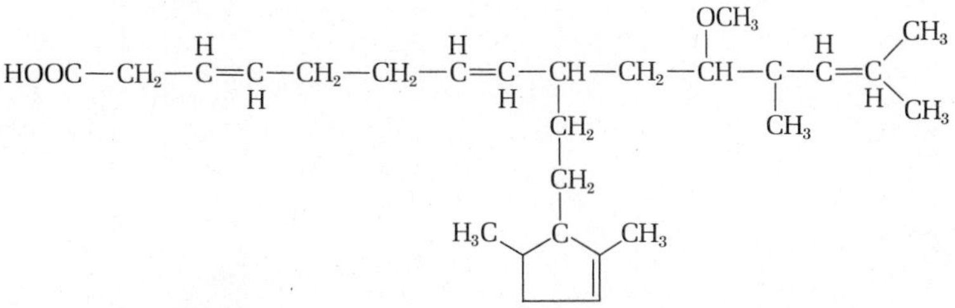

Bongkrekic acid, an antibiotic formed by a mold in decaying
bongkrek, a coconut meal used as food in Indonesia

these systems can also transport their substrates *against*
gradients of concentration when they are coupled to electron
transport as a source of energy. We have already seen that
the energy derived from electron transport can drive the formation of an H^+-ion gradient across the mitochondrial membrane. This gradient in turn may be utilized to transport
certain metabolites *against* gradients, into or out of mitochondria. For example, phosphate can be accumulated
from the surrounding medium by respiring mitochondria,
against a concentration gradient, through the action of the
phosphate carrier (Figure 19-22). The phosphate gradient so
generated can then cause the accumulation of malate against
its gradient via the dicarboxylate carrier. The malate gradient
can in turn be used to generate a gradient of citrate via the
tricarboxylate carrier (Figure 19-22). In this way, through
shared substrate specificities, the various transport systems
of the mitochondria can be linked to the electron-transport
chain as a source of energy.

Figure 19-22 also shows that the phosphate carrier
makes possible the equimolar exchange of external ADP^{3-} for
internal ATP^{4-} by balancing the charge distribution across
the membrane. The phosphate carrier thus is an essential
link in coupling energy delivered by the respiratory chain to
the transport of various metabolites.

Respiration-Dependent Calcium Transport
by Mitochondria

Mitochondria of animal tissues can also accumulate certain
cations, particularly Ca^{2+}, against a gradient, in a process
energetically coupled to electron transport. Accumulation of

Figure 19-22
Central role of the phosphate carrier in energy coupling between electron transport and transmembrane transport. The phosphate-hydroxide carrier appears to be the vehicle for converting the primary H^+ gradient into an internal accumulation of $H_2PO_4^-$ and thus a negative inside-membrane potential, which can be used to make the transport of various anions and cations possible, as shown.

Inward transport of dicarboxylates and tricarboxylates coupled to electron transport

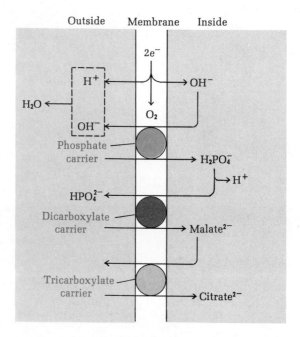

Entry of ADP^{3-} and exit of ATP^{4-} via the ADP–ATP carrier

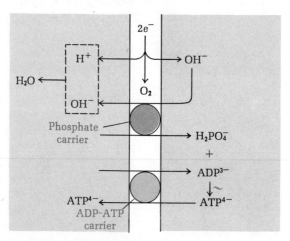

Inward transport of Ca^{2+}; other cations, such as Mn^{2+}, Fe^{2+}, and K^+, may also enter in response to the negative inside potential

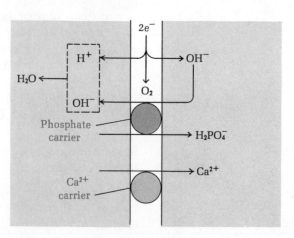

Figure 19-23
Phosphorylation of ADP and accumulation of cations are alternative processes coupled to electron transport.

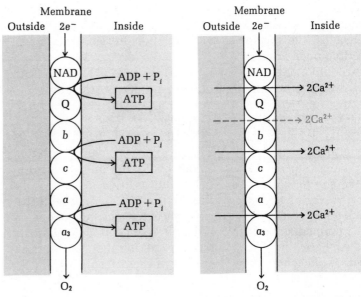

Oxidative phosphorylation of ADP

Oxidative accumulation of Ca^{2+}, which enters via a specific carrier

Ca^{2+} is accompanied by uptake of an equivalent amount of phosphate. For every pair of electrons passing from NADH to oxygen, about six Ca^{2+} ions are accumulated from the medium, two for each energy-conserving site (Figure 19-23). When Ca^{2+} is accumulated by mitochondria in this fashion, no oxidative phosphorylation of ADP occurs. Thus the energy delivered by electron transport can be used to carry out either Ca^{2+} accumulation or ATP formation but not both simultaneously. Mn^{2+} and Fe^{2+} ions are also accumulated in a similar manner by liver mitochondria, as is Sr^{2+}, but Mg^{2+} is not. Ca^{2+} transport by mitochondria is inhibited by the metal complex ruthenium red and by La^{3+} and other rare-earth cations.

Mitochondria of animal tissues can accumulate very large amounts of Ca^{2+} and phosphate, resulting in the deposition of electron-dense granules of calcium phosphate in the matrix (Figure 19-24), a process believed to be concerned in biological calcification. This property, as well as the very high affinity for Ca^{2+} shown by mitochondria, strongly indicates that mitochondrial Ca^{2+} transport plays an important biological role in vertebrate cells (page 796).

The inward transport of Ca^{2+} against a gradient of concentration is coupled to electron transport by the action of the phosphate carrier (Figure 19-22).

Figure 19-24
Electron micrograph of rat-liver mitochondria after respiration-coupled accumulation of Ca^{2+} and phosphate. The dark electron-dense spots correspond to granular deposits of amorphous calcium phosphate.

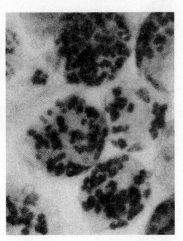

1.0 μm

J. W. Greenawalt

Shuttle Systems for Entry of Electrons from External (Cytosol) NADH

When NADH is added to suspensions of mitochondria from liver and other animal tissues, it is not oxidized, even though such mitochondria readily oxidize added NAD-linked substrates such as malate via *internal* NAD. From these and other observations it has been concluded that the mi-

533

tochondrial membrane in most animal tissues is impermeable to NADH. This permeability barrier effectively segregates the cytoplasmic from the intramitochondrial pyridine-nucleotide pools.

An important question now arises. Many NAD$^+$-linked dehydrogenases in the cytosol can reduce NAD$^+$. How can NADH in the cytosol be reoxidized to NAD$^+$ by the electron-transport chain in the mitochondria?

While extramitochondrial NADH cannot itself penetrate the mitochondrial inner membrane, electrons derived from it can enter the electron-transport chain by indirect routes called _shuttles_. The first NADH shuttle to be discovered is the _glycerol phosphate shuttle_ (Figure 19-25). NADH of the cytosol first reacts with dihydroxyacetone phosphate, one of the intermediates of glycolysis, to reduce it to L-glycerol 3-phosphate, in a reaction catalyzed by the NAD-linked _glycerol phosphate dehydrogenase_ of the cytosol:

Dihydroxyacetone phosphate + NADH + H$^+$ $\rightleftharpoons$

L-glycerol 3-phosphate + NAD$^+$

Figure 19-25

The glycerol phosphate shuttle for transfer of reducing equivalents from cytosol NADH to the mitochondrial electron-transport chain. The mitochondrial glycerol phosphate dehydrogenase is a flavoprotein, tightly bound to the inner membrane. It delivers its electrons to ubiquinone (Q), thus bypassing site I. Hence only two ATPs are generated per pair of electrons fed into the chain. The outer membrane, which is freely permeable to dihydroxyacetone phosphate and glycerol phosphate, is not shown.

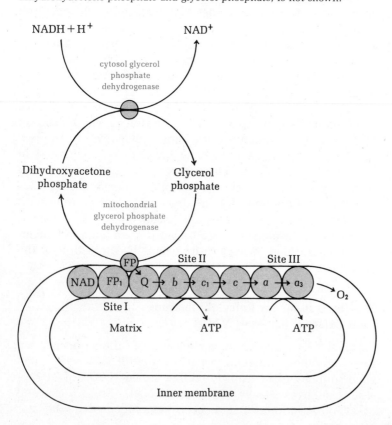

The glycerol 3-phosphate so formed passes readily through the outer membrane and is then oxidized by a second type of glycerol 3-phosphate dehydrogenase located on the outer surface of the inner mitochondrial membrane; the glycerol 3-phosphate need not pass through the inner membrane to be oxidized. The mitochondrial glycerol 3-phosphate dehydrogenase is not NAD-linked but is a flavin-containing dehydrogenase, designated FP, whose prosthetic group becomes reduced:

$$\text{Glycerol 3-phosphate} + \text{FP} \longrightarrow$$
$$\text{dihydroxyacetone phosphate} + \text{FPH}_2$$

The reducing equivalents of the reduced enzyme are then transferred to ubiquinone, from which they pass to oxygen via the cytochrome system in the inner membrane. The dihydroxyacetone phosphate formed in this reaction now can return to the cytosol and accept a pair of electrons from another molecule of extramitochondrial NADH. The glycerol phosphate–dihydroxyacetone phosphate redox couple thus acts as a shuttle of reducing equivalents from extramitochondrial NADH to the intramitochondrial respiratory chain. The pair of electrons that enters the respiratory chain via this shuttle causes the oxidative phosphorylation of only two molecules of ADP as it passes to oxygen, since the electrons from L-glycerol 3-phosphate enter the chain *after* the first energy-conserving site (see page 516).

The glycerol phosphate shuttle is unidirectional; it transports reducing equivalents into mitochondria in certain muscles and nerve cells. In other tissues, particularly the liver and heart, another type of shuttle operates, the malate-aspartate shuttle (Figure 19-26). This complex shuttle

Figure 19-26
The malate-aspartate shuttle for introducing reducing equivalents from NADH in the cytosol to the electron-transport chain of mitochondria. This shuttle involves the participation of cytosol and mitochondrial forms of malate dehydrogenase and of aspartate transaminase (page 562), as well as two membrane carriers. This shuttle, which predominates in the liver and heart, results in the phosphorylation of three molecules of ADP per molecule of cytosol NADH oxidized.

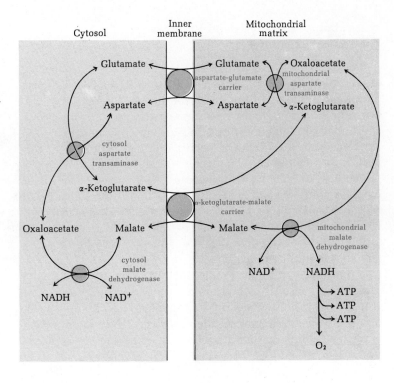

system is bidirectional; it can transport electrons from extramitochondrial NADH into the mitochondria or from intramitochondrial NADH to the cytosol. The malate-aspartate shuttle requires the cooperation, in a cyclic manner, of cytoplasmic and mitochondrial isozymes of malate dehydrogenase and aspartate-glutamate transaminase (page 562), as well as the mitochondrial membrane transport systems for malate, α-ketoglutarate, aspartate, and glutamate. The malate-aspartate shuttle yields three ATPs per molecule of cytosol NADH oxidized, whereas the glycerol phosphate shuttle yields only two.

Movement of Reducing Power from the Mitochondria to the Cytosol

As will be seen in Chapter 23, the biosynthesis of glucose from pyruvate requires NADH in the cytosol to reduce 1,3-diphosphoglycerate to glyceraldehyde 3-phosphate. Although mitochondria continuously generate internal NADH from the oxidation of respiratory substrates, NADH itself cannot pass into the cytosol directly, but its reducing equivalents may be transferred to extramitochondrial NAD^+ via the malate-aspartate shuttle described above.

Similarly, intramitochondrial reducing power can be used to form extramitochondrial NADPH. Intramitochondrial isocitrate may pass through the mitochondrial membrane on the tricarboxylate carrier (page 530) to the cytosol, where it may donate electrons to $NADP^+$ through the action of the cytosol form of NADP-linked isocitrate dehydrogenase, thus generating NADPH in the cytosol. The operation of the shuttle systems for NADH and the tricarboxylate carrier for transfer of the reducing equivalents of NADPH effectively regulates the ratio of the intramitochondrial and extramitochondrial $NADH-NAD^+$ and $NADPH-NADP^+$ couples.

Integration of Glycolysis and Respiration: The Pasteur Effect

A facultative cell can utilize glucose under either anaerobic or aerobic conditions. When such a cell utilizes glucose under anaerobic conditions to form lactate, the rate of breakdown of glucose is many times greater than it is in the same cell under aerobic conditions, in order to compensate for the smaller yield of ATP per molecule of glucose during glycolysis. Recall that anaerobic glycolysis delivers a net of only 2 molecules of ATP per molecule of glucose (page 432) whereas complete aerobic oxidation delivers 36 molecules of ATP per molecule of glucose (page 516). The anaerobic cell would therefore require $36/2 = 18$ times as much glucose per unit weight per unit time if it is to generate ATP at the same rate as it would aerobically.

If oxygen is admitted to an anaerobic suspension of cells utilizing glucose at a high rate by glycolysis, the rate of glucose consumption declines dramatically to a small fraction of its anaerobic rate. At the same time, the accumulation of lactate, which is quantitative under anaerobic conditions,

Figure 19-27
Schematic representation of the crossover point in glycolysis. It shows the relative concentrations of some intermediates of the glycolytic sequence (above) in anaerobic cells when phosphofructokinase is allosterically activated and (below) in aerobic cells when phosphofructokinase is allosterically inhibited. The crossover point (page 497) is that point in the sequence where removal of the inhibition causes the preceding intermediates to decline in their steady-state concentration and those beyond to increase in concentration.

Key:
G6P = glucose 6-phosphate
F6P = fructose 6-phosphate
FDP = fructose diphosphate
 TP = triose phosphates

Anaerobic cells; low ATP/ADP ratio

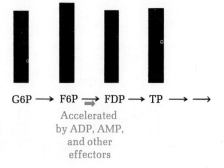

$$G6P \longrightarrow F6P \Longrightarrow FDP \longrightarrow TP \longrightarrow \longrightarrow$$
Accelerated
by ADP, AMP,
and other
effectors

Aerobic cells; high ATP/ADP ratio

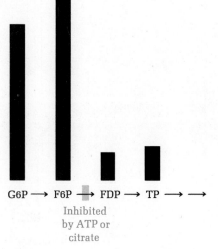

$$G6P \longrightarrow F6P \Longrightarrow FDP \longrightarrow TP \longrightarrow \longrightarrow$$
Inhibited
by ATP or
citrate

Relative concentrations of intermediates

is decreased to near zero in the presence of oxygen, since respiration leads to quantitative formation of CO_2 and H_2O from glucose without accumulation of lactate. This phenomenon, the inhibition of glucose consumption and the cessation of lactate accumulation with the onset of oxygen consumption, is called the _Pasteur effect_. Pasteur discovered it over a century ago during his important investigations of fermentation processes in winemaking, but it is a general property of all facultative cells, including those of higher animals. It is a spectacular instance of the regulation and integration of two multienzyme systems.

The first question raised by the Pasteur effect is: How does the onset of oxygen consumption slow down the rate of glucose consumption? We have seen (page 530) that mitochondria have a high affinity for ADP and that when oxygen is available, they will phosphorylate ADP until high [ATP]/[ADP] ratios are achieved. We have also seen (page 424) that phosphofructokinase, an allosteric enzyme, catalyzes a regulatory reaction in the glycolytic sequence, namely, the phosphorylation of fructose 6-phosphate to fructose 1,6-diphosphate. Phosphofructokinase activity is stimulated by ADP and inhibited by excess ATP. When the [ATP]/[ADP] ratio is high, phosphofructokinase is severely inhibited; conversely, when this ratio is low, the activity of the enzyme is increased. The high [ATP]/[ADP] ratio that results from oxidative phosphorylation therefore causes the rate of the phosphofructokinase reaction to be greatly decreased and thus slows down the rate of glycolysis. That this mechanism actually occurs in intact cells has been demonstrated by analysis of the concentrations of glucose, glucose 6-phosphate, fructose 6-phosphate, and other glycolytic intermediates in intact cells before and after the transition between anaerobic and aerobic conditions. When aerobic cells are deprived of oxygen, the concentrations of glucose 6-phosphate and fructose 6-phosphate decrease and the concentration of fructose 1,6-diphosphate increases; i.e., we have a _crossover point_ (page 497). Conversely, when oxygen is supplied to anaerobic cells, the concentrations of glucose 6-phosphate and fructose 6-phosphate in the cell increase but the concentration of fructose 1,6-diphosphate becomes very low (Figure 19-27). These observations show that the phosphofructokinase step is indeed crucial in modulating the rate of glycolysis under aerobic conditions.

In addition to ATP, other products of mitochondrial respiration can serve as inhibitory modulators of glycolysis. Citrate and isocitrate produced by the tricarboxylic acid cycle can leave the mitochondria via the tricarboxylate carrier and enter the cytosol, where they can act as inhibitory modulators of phosphofructokinase. Overproduction of citrate during respiration then leads to a decrease in the rate of glycolysis and thus in the supply of pyruvate to the mitochondria.

Although allosteric inhibition of the phosphofructokinase reaction by ATP and citrate appears to be the major regulatory mechanism responsible for the Pasteur effect, other mechanisms may participate or even predominate in some types of cells. In any case, the coordination of the rates of glycolysis and respiration involves a close interplay between the mitochondrial and the cytosol compartments of cells, in which the specific mitochondrial membrane transport systems play a central role. The major regulatory mechanisms involved are summarized in Figure 19-28.

The Energy Charge of the ATP System

We have seen that the concentrations of ATP and ADP are important elements in allosteric regulation of glycolysis and respiration; later we shall see numerous other instances in which these nucleotides participate in metabolic regulation. Moreover, AMP is also an important modulator of certain allosteric enzymes. AMP is formed during pyrophosphate cleavage of ATP (page 410) in certain biosynthetic reactions; it can be rephosphorylated to ADP again by the action of adenylate kinase:

$$\text{ATP} + \text{AMP} \rightleftharpoons 2\text{ADP}$$

In view of the many metabolic reactions in which ATP, ADP, and AMP participate as modulators, D. E. Atkinson has pointed out that the energy status of the cell and the "poise" of its allosteric modulators may be expressed by what he calls the *energy charge* of the cell, i.e., the extent to which the ATP–ADP–AMP system is "filled" with high-energy phosphate groups. If all the adenine nucleotide in the cell is ATP, the adenylate system is completely filled and is considered to have an energy charge of 1.0. At the other extreme, if all the adenine nucleotide is present as AMP, the system is empty of high-energy phosphate groups and has an energy charge of 0. If all the adenine nucleotide is present as ADP or as an equimolar mixture of ATP and AMP, it is only half-full of high-energy groups and has an energy charge of 0.5. The energy charge of the ATP–ADP–AMP system can easily be calculated for any given set of concentrations of ATP, ADP, and AMP by the equation

$$\text{Energy charge} = \frac{1}{2}\left(\frac{[\text{ADP}] + 2[\text{ATP}]}{[\text{AMP}] + [\text{ADP}] + [\text{ATP}]}\right)$$

Atkinson has suggested that the energy charge is a major factor in the regulation of pathways that produce and utilize high-energy phosphate groups. The curves in Figure 19-29 show the relationship of the rates of ATP-generating and ATP-utilizing metabolic processes as a function of the energy charge. The metabolic steady state in which ATP production is equal to ATP utilization is given by the intersection of the two curves, corresponding to an energy charge of about 0.85. This value happens to be the actual energy charge of the adenylate system of different types of cells under normal and fasting conditions (Table 19-6). If the energy charge decreases below 0.85, the ATP-generating sequences are accelerated through the response of their regulatory enzymes to the relative concentrations of ATP, ADP, and AMP; con-

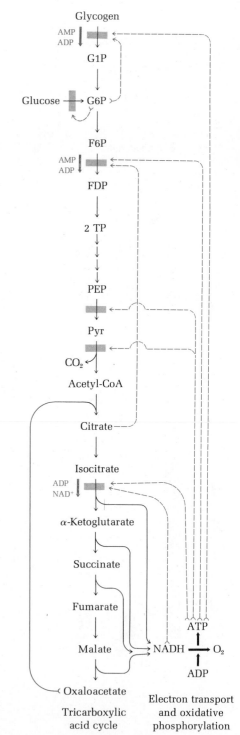

Figure 19-28
Summary of the important points of regulation of the rates of glycolysis and respiration. The dashed arrows in color show the origin of the feedback inhibitors (ATP, NADH, citrate, glucose 6-phosphate) and the reactions where they exert inhibition (colored bars across the reaction arrows). The reactions that are promoted by positive modulators (AMP, ADP, NAD+) are designated by colored arrows parallel to the reaction arrows.

Electron transport and oxidative phosphorylation

Tricarboxylic acid cycle

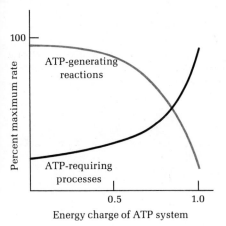

Figure 19-29
Response of ATP-generating and ATP-utilizing reactions to the energy charge of the ATP system.

Table 19-6   The adenine nucleotide content of some tissues in micromoles per gram wet weight

Tissue	ATP	ADP	AMP
Guinea pig brain	2.98	0.70	0.13
Rat heart	13.3	2.6	0.43
Blowfly muscle	14.0	3.0	0.25

versely, the ATP-requiring sequences slow down. If the energy charge increases above the normal level, reverse changes take place. The ATP–ADP–AMP system thus is poised to run optimally in a steady state in which the energy charge is about 0.85 and strongly resists any deviations from it.

The Phosphorylation Potential

Another way of expressing the energy status of cells is in terms of the *phosphorylation potential*, the ratio $[ATP]/[ADP][P_i]$. This ratio is equivalent to the expression for the mass-action constant, Γ, of the reaction

$$P_i + ADP \rightleftharpoons ATP + H_2O$$

$$\Gamma = \frac{[ATP]}{[ADP][P_i]}$$

which in the absence of a source of energy input has the very low value of about 5 μM^{-1} at 25°C. The extent to which this value is exceeded by the observed $[ATP]/[ADP][P_i]$ ratio in the cell is a measure of the potential of the cell for carrying out ATP-dependent processes. The phosphorylation potential is more satisfactory thermodynamically than the energy charge, which does not take into consideration the concentration of inorganic phosphate, an essential reactant in oxidative and glycolytic phosphorylations.

The phosphorylation potential is a more sensitive indicator of the energy status of a cell than the energy charge. Its value, computed from the actual concentrations of ATP, ADP, and phosphate in cells, varies from 200 to 800 M^{-1}, depending on the metabolic state. The higher the phosphorylation potential, the more highly "energized" the cell. The energy charge, on the other hand, tends to vary little from about 0.85.

Summary

Mitochondria, present in all aerobic eukaryotic cells, are the sites of the tricarboxylic acid cycle reactions, electron transport, and oxidative phosphorylation. They contain a smooth outer membrane, an inner membrane invaginated to form folds called cristae, and an internal matrix rich in protein. The two membranes differ in chemical composition, permeability, and enzyme content. The enzymes of electron transport and oxidative phosphorylation are located in the inner membrane; those of the tricarboxylic acid cycle are largely in the matrix. Oxidative phosphorylation of ADP occurs in respiring intact mitochondria supplemented with a substrate, phosphate, ADP, Mg^{2+}, and oxygen. The P/O ratio (number of molecules of phosphate taken up per atom of oxygen consumed) is 3.0 for NAD-linked electron transport and 2.0 for flavin-linked electron transport. The energy for the phosphorylation of ADP arises from the decrease in free energy as electrons flow from the primary electron donor (NADH or reduced flavin nucleotide) to molecular oxygen. Three sites in the chain have been identified as furnishing the energy for the phosphorylations: site I, between NAD and ubiquinone, site II, between cytochrome b and cytochrome c, and site III, between cytochrome a and oxygen. Phosphate and ADP are required components for maximal rates of electron transport in intact mitochondria; in the absence of ADP the rate of oxygen consumption is very low. The acceptor-control ratio is the ratio of the respiration in the presence of ADP to the rate in the absence of ADP.

Phosphorylation is inhibited without affecting electron transport by compounds called uncoupling agents; an example is 2,4-dinitrophenol. A second class of agents inhibits both electron transport and phosphorylation; an example is oligomycin. Ionophores (antibiotics that complex with K^+, Na^+, or other cations) also can inhibit oxidative phosphorylation by utilizing the energy of electron transport for the cation transport. Oxidative phosphorylation appears to be a reversible process; ATP can stimulate reverse electron flow in the respiratory chain.

The three main hypotheses for the mechanism of oxidative phosphorylation differ in the nature of the intermediate postulated to transfer energy from the oxidation-reduction reactions of electron transport to the synthesis of ATP. The chemical-coupling hypothesis postulates formation of a high-energy covalently bonded intermediate, the conformational-coupling hypothesis postulates an intermediate high-energy conformational state, and the chemiosmotic-coupling hypothesis, which is most consistent with known data, postulates that an electrochemical gradient of H^+ ions across the inner membrane is the coupling vehicle.

The outer mitochondrial membrane is freely permeable to most solutes. The inner mitochondrial membrane is not; however, it contains specific transport systems for phosphate, ADP, ATP, dicarboxylic acids, tricarboxylic acids, certain amino acids, and Ca^{2+}. These solutes can be transported against gradients of concentration when coupled to an energy-rich gradient of H^+ generated by electron transport.

The inner mitochondrial membrane contains components of shuttle systems that can transfer electrons from cytosol NADH to the mitochondrial electron-transport chain or from mitochondrial NADH to the cytosol. The rates of glycolysis in the cytoplasm and respiration in the mitochondria are integrated by the concentration of ATP, ADP, and phosphate in the cytosol and the mitochondria. The inhibition of glycolysis by respiration is due largely to allosteric inhibition of phosphofructokinase. The energy charge of the ATP–ADP–AMP system, i.e., the extent to which it is filled with high-energy phosphate groups, and the phosphorylation potential, the ratio $[ATP]/[ADP][P_i]$, express the energy states of the cells.

References

See also references for Chapter 18 (page 506). This field has a particularly extensive bibliography, and only some of the most important contributions can be given.

Books

KALCKAR, H.: *Biological Phosphorylations: Development of Concepts*, Prentice-Hall, Englewood Cliffs, N.J., 1969. Reprinted classical papers with comments by the editor.

KAPLAN, N. O., and E. P. KENNEDY: *Current Aspects of Biochemical Energetics, Fritz Lipmann Dedicatory Volume*, Academic, New York, 1966.

LEHNINGER, A. L.: *The Mitochondrion: Molecular Basis of Structure and Function*, Benjamin, Menlo Park, Calif., 1965.

MITCHELL, P.: *Chemiosmotic Coupling and Energy Transduction*, Glynn Research, Bodmin, England, 1968.

RACKER, E. (ed.): *Membranes of Mitochondria and Chloroplasts,* American Chemical Society Monograph 165, Van Nostrand Reinhold, New York, 1970.

Articles

BOYER, P. D., R. L. CROSS, and W. MOMSEN: "A New Concept for Energy Coupling in Oxidative Phosphorylation Based on a Molecular Explanation of the Oxygen Exchange Reactions," *Proc. Natl. Acad. Sci. (U.S.),* 70: 2837–2839 (1973).

CHANCE, B.: "The Nature of Electron Transfer and Energy-Coupling Reactions," *FEBS Lett.,* 23: 3–19 (1972).

CHAPPELL, J. B.: "Systems Used for the Transport of Substances into Mitochondria," *Br. Med. Bull.,* 24: 150–157 (1968).

GREVILLE, G. D.: "A Scrutiny of Mitchell's Chemiosmotic Hypothesis," *Curr. Top. Bioenerg.,* 3: 1–78 (1969).

HACKENBROCK, C. R.: "Ultrastructural Bases for Metabolically Linked Mechanical Activity in Mitochondria; I: Reversible Ultrastructural Changes with Change in Metabolic Steady State in Isolated Liver Mitochondria," *J. Cell Biol.,* 30: 269–297 (1966).

HACKENBROCK, C. R.: "Ultrastructural Bases for Metabolically Linked Mechanical Activity in Mitochondria; II: Electron Transport-Linked Ultrastructural Transformations in Mitochondria," *J. Cell Biol.,* 37: 345–369 (1968).

HAROLD, F. M.: "Conservation and Transformation of Energy by Bacterial Membranes," *Bact. Rev.,* 36: 172 (1972).

KLINGENBERG, M.: "Metabolite Transport in Mitochondria: An Example for Intracellular Membrane Function," *Essays Biochem.,* 6: 119–159 (1970).

KREBS, H. A.: "The Pasteur Effect and the Relations between Respiration and Fermentation." *Essays Biochem.,* 8: 2–34 (1972).

LEHNINGER, A. L.: "Mitochondria and Ca^{2+} Transport," *Biochem. J.,* 119: 129–138 (1970).

MITCHELL, P.: "Cation-Translocating ATPase models," *FEBS Lett.,* 33: 267–274 (1973).

MITCHELL, P.: "Chemiosmotic Coupling in Oxidative and Photosynthetic Phosphorylation," *Biol. Rev. Camb. Phil. Soc.,* 41: 445–502 (1966).

PALMER, J. M., and D. O. HALL: "The Mitochondrial Membrane System," *Prog. Biophys. Biophys. Chem.,* 24: 125–176 (1972).

RACKER, E.: "The Two Faces of the Inner Mitochondrial Membrane," *Essays Biochem.,* 6: 1–22 (1970).

SENIOR, A. E.: "The Structure of Mitochondrial ATPase," *Biochim. Biophys. Acta,* 301: 249–277 (1973).

SKULACHEV, V.: "Energy Transformations in the Respiratory Chain," *Curr. Top. Bioenerg.,* 4: 127–191 (1972). An important analysis of energy coupling.

SLATER, E. C.: "The Coupling between Energy-Yielding and Energy-Utilizing Reactions in Mitochondria," *Q. Rev. Biophys.,* 4: 35–71 (1971).

WRIGGLESWORTH, J. M., L. PACKER, and D. BRANTON: "Organization of Mitochondrial Structures as Revealed by Freeze-Etching," *Biochim. Biophys. Acta,* 205: 125–135 (1970).

Attention is also drawn to many excellent articles in the annual series *Current Topics in Bioenergetics* (Academic, New York) and the series of annual proceedings of conferences on mitochondria and bioenergetics held in Bari and in Bressanone, Italy, published by Editrice Adriatica and Academic Press. Also recommended are the series of articles in a special issue of the *Journal of Bioenergetics* on "Mechanism of Biological Energy Transformation," *J. Bioenerg.*, 3: 1–202 (1972).

Problems

1. Write the overall equation for the oxidation of extramitochondrial NADH to NAD$^+$ by oxygen via the respiratory chain, assuming that the glycerol phosphate shuttle is operating.

2. Write a similar equation for the oxidation of extramitochondrial NADH, assuming the malate-oxaloacetate shuttle is operating.

3. Predict theoretical P/O ratios for the following processes, considering GTP to be the equivalent of ATP:
   (a) Isocitrate $\longrightarrow$ succinate
   (b) α-Ketoglutarate $\longrightarrow$ succinate, in the presence of dinitrophenol
   (c) Succinate $\longrightarrow$ oxaloacetate
   (d) Glyceraldehyde 3-phosphate $\longrightarrow$ pyruvate, in the presence of arsenate

4. Write balanced equations for the complete combustion of the following substrates to carbon dioxide and water in a typical liver cell, including all associated phosphorylation steps; assume operation of the glycerophosphate–dihydroxyacetone phosphate shuttle: (a) Glyceraldehyde, (b) pyruvic acid, (c) citric acid, (d) fumaric acid, (e) oxaloacetic acid.

5. Write a balanced equation, including all associated phosphorylation steps, for the complete combustion of fructose to CO_2 and H_2O as it would occur in a typical cell preparation to which dinitrophenol has been added.

6. Calculate the percentage efficiency of energy conservation in ATP as succinate is oxidized to fumarate by molecular oxygen in intact mitochondria.

7. Calculate the percentage efficiency of energy conservation in ATP as glucose is oxidized to CO_2 and H_2O by molecular oxygen, assuming the operation of the malate-oxaloacetate shuttle.

8. Inorganic phosphate and adenine nucleotide concentrations in isolated, perfused rat hearts under both aerobic and anaerobic conditions were determined with the following results:

Condition	Concentration, mM			
	ATP	ADP	AMP	P_i
Aerobic	4.2	0.8	0.06	1.5
Anaerobic	3.4	1.2	0.2	5.2

Calculate the energy charge and the phosphorylation potential under each of these conditions.

CHAPTER **20** OXIDATION OF FATTY ACIDS

Fatty acids play an extremely important part as an energy-rich fuel in higher animals and plants since large amounts can be stored in cells in the form of triacylglycerols. Triacylglycerols are especially well adapted for this role because they have a high energy content (about 9 kcal g^{-1}) and can be accumulated in nearly anhydrous form as intracellular fat droplets. In contrast, glycogen and starch can yield only about 4 kcal g^{-1}; moreover, since they are highly hydrated, they cannot be stored in such concentrated form. Fatty acids provide up to 40 percent of the total fuel requirement in man on a normal diet. In fasting or hibernating animals and in migrating birds, fatty acids become virtually the sole source of energy.

In this chapter we shall examine the enzymatic pathway for the oxidation of fatty acids to yield acetyl-CoA, which then enters the tricarboxylic acid cycle. We shall also trace the formation and utilization of the ketone bodies, important intermediates in fatty acid oxidation.

Sources of Fatty Acids

Mammalian tissues normally contain only vanishingly small amounts of free fatty acids, which are in fact somewhat toxic. By the action of hormonally controlled lipases (page 816) free fatty acids are formed from triacylglycerols in fat or adipose tissue. The free fatty acids are then released from the tissue, become tightly bound to serum albumin, and in this form are carried via the blood to other tissues for oxidation. Fatty acids delivered in this manner are first enzymatically "activated" in the cytoplasm and then enter the mitochondria for oxidation.

Long-chain fatty acids are oxidized to CO_2 and H_2O in nearly all tissues of vertebrates except the brain. However, under certain conditions the brain can oxidize β-hydroxybutyrate, an intermediate of fatty acid catabolism. Some tissues, such as heart muscle, obtain most of their energy from the oxidation of fatty acids. The mobilization, distribution, and oxidation of fatty acids are integrated with the utilization of carbohydrate fuels; both are under complex endocrine regulation (page 839).

The Pathway of Fatty Acid Oxidation

Even before the end of the nineteenth century it was suspected that fatty acids are synthesized and degraded in the cell by addition or subtraction of two-carbon fragments, since most natural fatty acids have an even number of carbon atoms. This idea was also supported by the early observation that in fasting mammals or in people with diabetes mellitus fatty acids are incompletely oxidized, leading to the accumulation of the four-carbon acids *acetoacetic acid* and D-*β-hydroxybutyric acid* (Figure 20–1) in the blood and urine. Another line of evidence came from classical experiments by the German biochemist F. Knoop in 1904, who observed that when rabbits were fed an even-carbon fatty acid labeled with a phenyl group at the methyl-terminal (ω-carbon) atom, *phenylacetic acid* was excreted in the urine, regardless of the length of the carbon chain of the acid fed. On the other hand, when ω-phenyl derivatives of odd-carbon fatty acids were fed, *benzoic acid* was excreted (Figure 20–2). (Phenylacetic and benzoic acid appear in the urine as esters of D-glucuronic acid.) These results suggested that ω-phenyl fatty acids are degraded by oxidative removal of successive two-carbon fragments starting from the carboxyl end.

From these experiments, perhaps the first metabolic tracer experiments reported, Knoop postulated that fatty acids are oxidized by *β oxidation*, i.e., oxidation at the *β* carbon to yield a *β*-keto acid, which was assumed to undergo cleavage to form acetic acid and a fatty acid shorter by two carbon atoms.

For many years, however, efforts to demonstrate fatty acid oxidation in cell-free extracts of animal tissues were unsuccessful; the enzymes involved were believed to require intact cell structure. In 1943 oxidation of fatty acids in a cell-free system was demonstrated by L. F. Leloir and J. M. Munoz, in Argentina, who found the system to be extremely labile. Shortly later, A. L. Lehninger, in the United States, showed that ATP is required for the oxidation of fatty acids and gave evidence that the fatty acid is enzymatically activated at the carboxyl group. He also found that fatty acids are oxidized to yield two-carbon units that can enter the tricarboxylic acid cycle. In 1948, E. P. Kennedy and Lehninger showed that the oxidation of fatty acids occurs exclusively in the mitochondria.

The next important clue, which quickly led to recognition of the nature of the enzymatic steps of fatty acid oxidation, came from work of F. Lynen and his colleagues in Germany. They found that the ATP-dependent activation of fatty acids involves their esterification with the thiol group of coenzyme A (page 342), to yield an acyl-CoA derivative, and that all the subsequent oxidative steps take place with the fatty acid in the form of its CoA thioester. The various enzymes cata-

Figure 20-1
The ketone bodies.

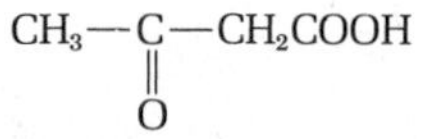

$$CH_3-C-CH_2COOH$$
$$\underset{O}{\|}$$

Acetoacetic acid

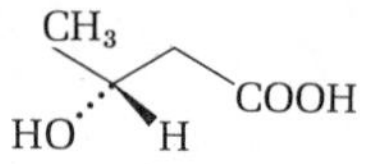

D-β-Hydroxybutyric acid

Figure 20-2
Knoop's experiments on oxidation of ω-phenyl fatty acids in rabbits. Even-carbon fatty acids always yield phenylacetic acid; odd-carbon acids always yield benzoic acid as the excreted end product. From these results he concluded that oxidative attack begins at the β-carbon atom followed by cleavage of successive two-carbon fragments from the chain at the arrows shown, presumably as acetic acid.

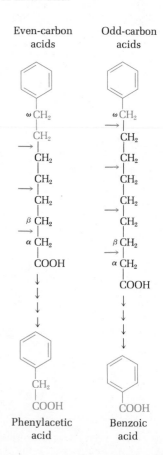

Figure 20-3
The fatty acid oxidation spiral. One acetyl-CoA is removed during each pass through the sequence. One molecule of palmitic acid (C_{16}), *after activation to palmitoyl-CoA, yields eight molecules of acetyl-CoA.*

$$R—CH_2—CH_2—CH_2—COCoA \quad (C_{16})$$

FAD

FADH$_2$

$$R—CH_2—CH{=}CH—COCoA$$

H$_2$O

$$R \diagdown \hspace{-1em} \underset{H \quad OH}{\cdots} \hspace{-1em} \diagup COCoA$$

NAD$^+$

NADH

$$R—CH_2—\underset{\underset{O}{\|}}{C}—CH_2—COCoA$$

CoA

$$R—CH_2—COCoA \ +$$

→ acetyl-CoA

→ acetyl-CoA

→ acetyl-CoA

→ acetyl-CoA

→ acetyl-CoA

→ acetyl-CoA

acetyl-CoA

Acetyl-CoA

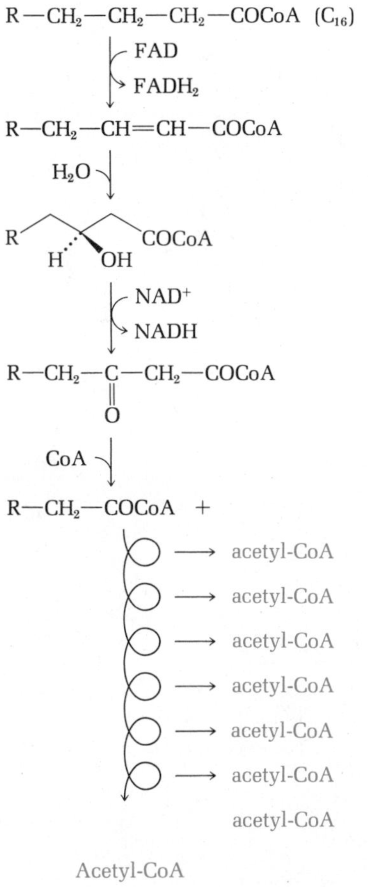

lyzing the successive steps in fatty acid oxidation have since been isolated in highly purified form in the laboratories of Lynen, D. E. Green, S. Ochoa, and other investigators.

Outline of the Fatty Acid Oxidation Cycle

Before oxidation, long-chain fatty acids from the cytosol must undergo a rather complex enzymatic activation, followed by transport across the mitochondrial membranes into the major compartment. There the fatty acyl group is transferred to intramitochondrial coenzyme A, yielding a fatty acyl–CoA thioester. The subsequent oxidation of the fatty acyl–CoA, outlined in Figure 20–3, takes place entirely in the mitochondrial matrix (page 511). The fatty acyl–CoA is dehydrogenated by removal of a pair of hydrogen atoms from the α and β carbon atoms (atoms 2 and 3) to yield the α,β- or Δ^2-unsaturated acyl-CoA. This is then enzymatically hydrated to form a β-hydroxyacyl-CoA, which in turn is dehydrogenated in the next step to yield the β-ketoacyl-CoA. It then undergoes enzymatic cleavage by reaction with a second molecule of CoA. One product is acetyl-CoA, derived from carbon atoms 1 and 2 of the original fatty acid chain. The other product, a long-chain saturated fatty acyl–CoA having two fewer carbon atoms than the original fatty acid, now becomes the substrate for another round of reactions, beginning at the first dehydrogenation step and ending with the removal of a second two-carbon fragment as acetyl-CoA (Figure 20–3). At each passage through this spiral the fatty acid chain loses a two-carbon fragment as acetyl-CoA. The 16-carbon palmitic acid thus undergoes a total of seven such cycles, to yield altogether 8 molecules of acetyl-CoA and 14 pairs of hydrogen atoms. The palmitate must be primed or activated only once, since at the end of each round the shortened fatty acid appears as its CoA thioester.

The hydrogen atoms removed during the dehydrogenation of the fatty acid enter the respiratory chain; as electrons pass to molecular oxygen via the cytochrome system, oxidative phosphorylation of ADP to ATP occurs. The acetyl-CoA formed as product of the fatty acid oxidation system enters the tricarboxylic acid cycle.

The individual enzymatic steps in fatty acid activation and oxidation will now be examined in detail.

Activation and Entry of Fatty Acids into Mitochondria

There are three stages in the entry of fatty acids into mitochondria from the extramitochondrial cytoplasm: (1) the enzymatic ATP-driven esterification of the free fatty acid with extramitochondrial CoA to yield fatty acyl–CoA, a step often referred to as the *activation* of the fatty acid, (2) the transfer of the acyl group from the fatty acyl–CoA to the carrier molecule *carnitine,* followed by the transport of the acyl carnitine across the inner membrane, and (3) the transfer of the acyl group from fatty acyl carnitine to intramitochondrial CoA, which occurs on the inner surface of the inner membrane.

Activation of Fatty Acids

At least three different enzymes catalyze formation of acyl-CoA thioesters, each being specific for a given range of fatty acid chain length. These enzymes are called _acyl-CoA synthetases_. _Acetyl-CoA synthetase_ activates acetic, propionic, and acrylic acids, _medium-chain acyl-CoA synthetase_ activates fatty acids with 4 to 12 carbon atoms, and _long-chain acyl-CoA synthetase_ activates fatty acids with 12 to 22 or more carbon atoms. The last two enzymes activate both saturated and unsaturated fatty acids, as well as 2- and 3-hydroxy acids. Otherwise the properties and mechanisms of all three synthetases, which have been isolated in highly purified form, are nearly identical. The overall reaction catalyzed by the ATP-linked acyl-CoA synthetases is

$$\underset{\substack{\text{Fatty}\\\text{acid}}}{\text{RCOOH}} + \text{ATP} + \text{CoA–SH} \rightleftharpoons \underset{\substack{\text{Fatty}\\\text{acyl–CoA}}}{\text{RCO–S–CoA}} + \text{AMP} + \text{PP}_i$$

In this reaction a thioester linkage is formed between the fatty acid carboxyl group and the thiol group of CoA; the ATP undergoes _pyrophosphate cleavage_ (page 410) to yield AMP and inorganic pyrophosphate (PP_i). The equilibrium favors the formation of the acyl-CoA significantly, since the standard free energy of the hydrolysis of ATP to AMP and PP_i is about -10.0 kcal and that of acyl-CoA about -7.52 kcal.

An enzyme-bound intermediate has been implicated in the action of acyl-CoA synthetases. It is a mixed anhydride of the fatty acid and the phosphate group of AMP, an _acyl adenylate_ (Figure 20–4). It is formed on the active site, with discharge of pyrophosphate to the medium. The bound acyl adenylate then reacts with CoA to yield acyl-CoA and free AMP as products. The acyl-CoA synthetases are found in the outer mitochondrial membrane and in the endoplasmic reticulum.

The pyrophosphate formed in the activation reaction may be hydrolyzed to inorganic orthophosphate by _inorganic pyrophosphatase_:

$$\text{PP}_i + \text{H}_2\text{O} \longrightarrow 2\text{P}_i$$

The net effect is the utilization of two high-energy bonds of ATP to activate one molecule of fatty acid. The AMP formed as the other product of the acyl-CoA synthetase reaction is rephosphorylated to ADP by _adenylate kinase_ (page 411):

$$\text{ATP} + \text{AMP} \rightleftharpoons 2\text{ADP}$$

Transfer to Carnitine

Long-chain saturated fatty acids have only a limited ability to cross the inner membrane as CoA thioesters, but their entry is greatly stimulated by _carnitine_. This substance was long known to be present in animal tissues, but its impor-

Figure 20-4
Structure of the fatty acyl adenylate intermediate. The acyl group is in color; R is an alkyl chain.

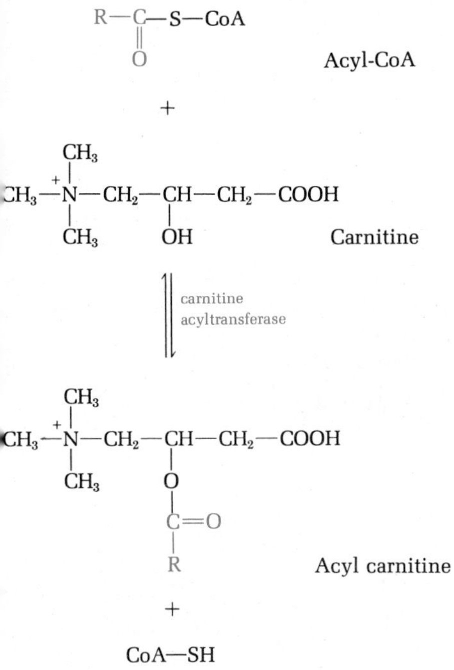

Figure 20-5
Transfer of the acyl group to carnitine.

$$R-\overset{\displaystyle O}{\underset{\displaystyle \|}{C}}-S-CoA \qquad \text{Acyl-CoA}$$

+

$$\underset{\displaystyle CH_3}{\overset{\displaystyle CH_3}{CH_3-\overset{+}{N}-CH_2-CH-CH_2-COOH}} \qquad \text{Carnitine}$$
$$\qquad\qquad\quad OH$$

carnitine
acyltransferase

$$\underset{\displaystyle CH_3}{\overset{\displaystyle CH_3}{CH_3-\overset{+}{N}-CH_2-CH-CH_2-COOH}}$$
$$\qquad\qquad\quad O$$
$$\qquad\qquad\quad |$$
$$\qquad\qquad\quad C=O$$
$$\qquad\qquad\quad |$$
$$\qquad\qquad\quad R \qquad \text{Acyl carnitine}$$

+

CoA—SH

tance went unrecognized until it was found to be an essential growth factor for the mealworm, *Tenebrio molitor*.

I. B. Fritz and others showed that the stimulation of fatty acid oxidation by carnitine is due to the action of an enzyme, *carnitine acyltransferase,* which catalyzes transfer of the fatty acyl group from its thioester linkage with CoA to an oxygen-ester linkage with the hydroxyl group of carnitine (Figure 20-5). Since the standard-free-energy change of this reaction is quite small, the acyl carnitine linkage evidently represents a high-energy bond. The O-acyl carnitine ester so formed then passes through the inner membrane into the matrix, presumably via a specific transport system.

Transfer to Intramitochondrial CoA

In the last stage of the entry process the acyl group is transferred from carnitine to intramitochondrial CoA by the action of a second type of carnitine acyltransferase located on the inner surface of the inner membrane:

$$\text{Acyl carnitine} + \text{CoA} \rightleftharpoons \text{acyl-CoA} + \text{carnitine}$$

This complex entry mechanism, often called the fatty acid shuttle (page 529), has the effect of keeping the extramitochondrial and intramitochondrial pools of CoA and of fatty acids separated. The intramitochondrial fatty acyl–CoA now becomes the substrate of the fatty acid oxidation system, which is situated in the inner matrix compartment.

Activation of Fatty Acids by Other Mechanisms

Mitochondria contain a second type of acyl-CoA synthetase which participates in fatty acid activation and which is located in the mitochondrial matrix. This enzyme requires GTP instead of ATP and causes an orthophosphate cleavage of the GTP. The overall reaction is catalyzed by the enzyme *acyl-CoA synthetase (GDP-forming):*

$$\text{Fatty acid} + \text{GTP} + \text{CoA} \rightleftharpoons \text{acyl-CoA} + \text{GDP} + P_i$$

This enzyme therefore can utilize GTP generated in the mitochondrial matrix during the substrate-level phosphorylation associated with oxidation of α-ketoglutarate (page 457) to activate free fatty acids formed internally in mitochondria.

In some microorganisms that ferment short-chain fatty acids, e.g., the strict anaerobe *Clostridium kluyveri*, the formation of acyl-CoA thioesters occurs by an indirect route. In the first step a high-energy acyl phosphate of acetate or butyrate is formed by direct phosphorylation of the carboxyl group by ATP, catalyzed by *acetate kinase* or *butyrate kinase,* respectively:

$$\text{ATP} + \text{acetate} \rightleftharpoons \text{ADP} + \text{acetyl phosphate}$$

$$\text{ATP} + \text{butyrate} \rightleftharpoons \text{ADP} + \text{butyryl phosphate}$$

The acyl phosphates so formed, mixed anhydrides of a carboxylic acid and phosphoric acid (Figure 20–6), are analogous in structure to 3-phosphoglyceroyl phosphate, formed during glycolysis (pages 404 and 427). The reactions written above are endergonic but are pulled to the right by the subsequent exergonic reaction with coenzyme A, catalyzed by *phosphate acyltransferase*:

$$\text{Acyl phosphate} + \text{CoA} \rightleftharpoons \text{acyl-CoA} + \text{phosphate}$$

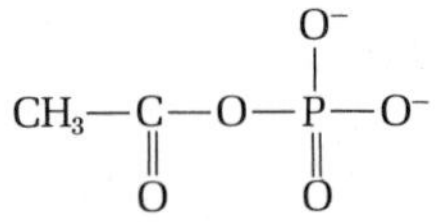

The First Dehydrogenation Step in Fatty Acid Oxidation

Following the formation of intramitochondrial acyl-CoA, all subsequent reactions of the fatty acid oxidation cycle take place in the inner compartment. In the first step the fatty acyl–CoA thioester undergoes enzymatic dehydrogenation by *acyl-CoA dehydrogenase* at the α and β carbon atoms (carbons 2 and 3) to form Δ^2-enoyl-CoA (Figure 20–7) as product; the position of the double bond is designated by Δ (page 280). The Δ^2 double bond formed in this reaction has the trans geometrical configuration. Recall, however, that the double bonds of the unsaturated fatty acids of natural fats nearly always have the cis configuration (page 281). We shall come back to this interesting point later.

There are four different *acyl-CoA dehydrogenases*, each specific for a given range of fatty acid chain lengths. All contain tightly bound flavin adenine dinucleotide (FAD) as prosthetic groups. The FAD becomes reduced at the expense of the substrate, a process that probably occurs through distinct one-electron steps (page 486).

The $FADH_2$ of the reduced acyl-CoA dehydrogenase cannot react directly with oxygen but donates its electrons to the respiratory chain (Figure 20–8) via a second flavoprotein, *electron-transferring flavoprotein*, which in turn passes the electrons to some carrier of the respiratory chain that has not yet been identified with certainty but is probably coenzyme Q.

The Hydration Step

The double bond of the Δ^2-*trans*-enoyl-CoA ester is then hydrated to form 3-hydroxyacyl-CoA by the enzyme *enoyl-CoA hydratase*, which has been isolated in crystalline form. The reaction catalyzed is shown in Figure 20–9. The addition of water across the Δ^2-trans double bond is stereospecific and results in the formation of the L-stereoisomer of the 3-hydroxyacyl-CoA. Enoyl-CoA hydratase also hydrates Δ^3-trans-unsaturated acyl-CoA's (see below), as well as Δ^2-cis-unsaturated acyl-CoA thioesters, in the latter case yielding the corresponding D stereoisomer of 3-hydroxyacyl-CoA (Figure 20–9). This apparent nonspecificity may appear puzzling, since the next enzyme in the fatty acid oxidation cycle is absolutely specific for the L stereoisomer of 3-hydroxyacyl-CoA. We shall see later, however, that both the L and D stereoisomers of 3-hydroxyacyl-CoA play a role in fatty acid metabolism but differ in origin and function (page 666).

Figure 20-6
Acetyl phosphate.

$$CH_3-\underset{\underset{O}{\|}}{C}-O-\underset{\underset{O}{\|}}{\overset{\overset{O^-}{|}}{P}}-O^-$$

Figure 20-7
The acyl-CoA dehydrogenase reaction. The enzyme protein (E) and its tightly bound flavin prosthetic group function as a unit.

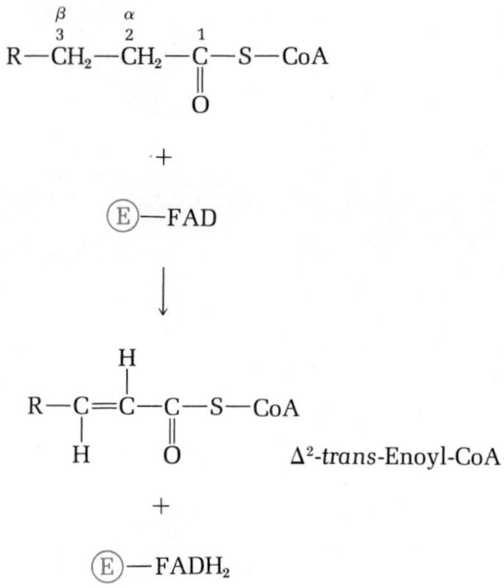

$$\overset{\beta}{\underset{3}{R}}-CH_2-\overset{\alpha}{\underset{2}{C}}H_2-\overset{1}{\underset{\underset{O}{\|}}{C}}-S-CoA$$

$+$

$\text{E}-FAD$

$\downarrow$

$$R-\overset{H}{\underset{H}{C}}=\overset{}{C}-\overset{}{\underset{\underset{O}{\|}}{C}}-S-CoA$$

Δ^2-*trans*-Enoyl-CoA

$+$

$\text{E}-FADH_2$

Figure 20-8
Electron transport from fatty acyl–CoA to oxygen (see also Figure 18-14, page 495).

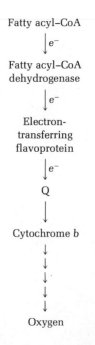

Fatty acyl–CoA

$\downarrow e^-$

Fatty acyl–CoA
dehydrogenase

$\downarrow e^-$

Electron-
transferring
flavoprotein

$\downarrow e^-$

Q

$\downarrow$

Cytochrome b

$\downarrow$
$\downarrow$
$\downarrow$
$\downarrow$

Oxygen

Figure 20-9
The action of Δ²-enoyl-CoA hydratase.

A. On trans substrates.

$$R—CH_2—\overset{H}{\underset{H}{C}}=\overset{}{C}—\overset{}{\underset{O}{C}}—S—CoA \qquad Δ^2\text{-}trans\text{-Enoyl-CoA}$$

$$-H_2O \parallel +H_2O \quad \text{enoyl-CoA hydratase}$$

$$R\underset{H \quad OH}{\wedge\wedge}\overset{}{\underset{O}{C}}—S—CoA \qquad \text{L-3-Hydroxyacyl-CoA}$$

B. On cis substrates.

$$R—CH_2—\overset{H\;H}{C}=\overset{}{C}—\overset{}{\underset{O}{C}}—S—CoA \qquad Δ^2\text{-}cis\text{-Enoyl-CoA}$$

$$-H_2O \parallel +H_2O \quad \text{enoyl-CoA hydratase}$$

$$R\underset{HO \quad H}{\wedge\wedge}\overset{}{\underset{O}{C}}—S—CoA \qquad \text{D-3-Hydroxyacyl-CoA}$$

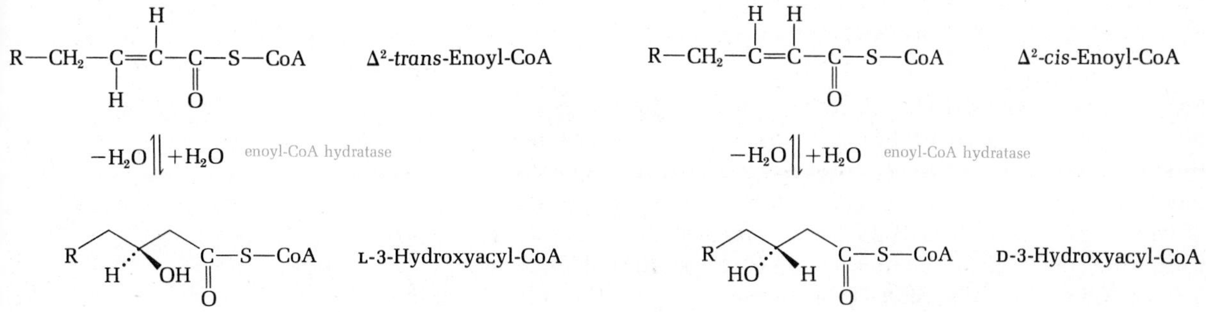

Figure 20-10
3-Ketoacyl-CoA.

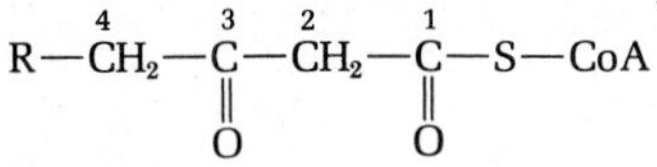

$$R—\overset{4}{C}H_2—\overset{3}{\underset{O}{C}}—\overset{2}{C}H_2—\overset{1}{\underset{O}{C}}—S—CoA$$

Figure 20-11
Thiolytic cleavage of 3-ketoacyl-CoA.

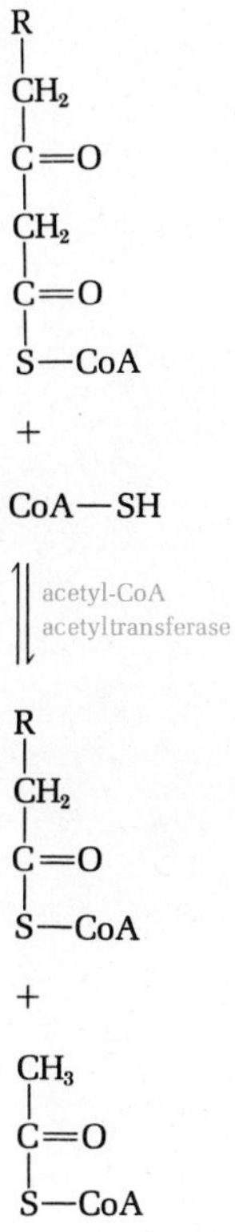

$$\begin{array}{c} R \\ | \\ CH_2 \\ | \\ C=O \\ | \\ CH_2 \\ | \\ C=O \\ | \\ S—CoA \\ + \\ CoA—SH \end{array}$$

acetyl-CoA
acetyltransferase

$$\begin{array}{c} R \\ | \\ CH_2 \\ | \\ C=O \\ | \\ S—CoA \\ + \\ CH_3 \\ | \\ C=O \\ | \\ S—CoA \end{array}$$

The Second Dehydrogenation Step

In the next step of the fatty acid oxidation cycle, the L-3-hydroxyacyl-CoA is dehydrogenated to form 3-ketoacyl-CoA (see Figure 20–10) by *3-hydroxyacyl-CoA dehydrogenase.* NAD⁺ is the specific electron acceptor. The reaction is

L-3-Hydroxyacyl-CoA + NAD⁺ ⇌ 3-ketoacyl-CoA + NADH + H⁺

This enzyme is relatively nonspecific with respect to the length of the fatty acid chain but is absolutely specific for the L stereoisomer. The NADH formed in the reaction donates its electron equivalents to the NADH dehydrogenase of the mitochondrial respiratory chain (page 486).

The Cleavage Step

In the last step of the fatty acid oxidation cycle, which is catalyzed by *acetyl-CoA acetyltransferase,* more commonly known as *thiolase,* the 3-ketoacyl-CoA undergoes cleavage by interaction with a molecule of free CoA to yield the carboxyl-terminal two-carbon fragment of the fatty acid as acetyl-CoA. The remaining fatty acid, now shorter by two carbon atoms, appears as its coenzyme A thioester (Figure 20–11).

This cleavage reaction, also called a *thiolysis* or a thiolytic cleavage, is analogous to hydrolysis. Since the reaction is highly exergonic, cleavage is favored. There appear to be two (perhaps three) forms of the enzyme, each specific for different fatty acid chain lengths.

The Balance Sheet

We have described one turn of the fatty acid oxidation cycle, in which one molecule of acetyl-CoA and two pairs of hydrogen atoms have been removed from the starting long-chain fatty acyl–CoA. The overall equation for one turn of the cycle, starting from palmitoyl-CoA, is

Palmitoyl-CoA + CoA + FAD + NAD⁺ + H₂O ⟶
   myristoyl-CoA + acetyl-CoA + FADH₂ + NADH + H⁺

We can now write the equation for the seven turns of the cycle required to convert one molecule of palmitoyl-CoA into eight molecules of acetyl-CoA:

Palmitoyl-CoA + 7CoA + 7FAD + 7NAD⁺ + 7H₂O ⇌
$$\text{8 acetyl-CoA} + 7\text{FADH}_2 + 7\text{NADH} + 7\text{H}^+$$

Each molecule of $FADH_2$ donates a pair of electron equivalents to the respiratory chain at the level of coenzyme Q; thus two molecules of ATP are generated during the ensuing electron transport to oxygen (page 495). Similarly, oxidation of each molecule of NADH by the respiratory chain results in formation of three molecules of ATP (page 495). Hence, a total of five molecules of ATP is formed by oxidative phosphorylation per molecule of acetyl-CoA cleaved. We can therefore write the following equation, which includes the oxidative phosphorylations:

Palmitoyl-CoA + 7CoA + 7O₂ + 35P$_i$ + 35ADP ⟶
$$\text{8 acetyl-CoA} + 35\text{ATP} + 42\text{H}_2\text{O} \quad (1)$$

The eight molecules of acetyl-CoA formed in the fatty acid cycle may now enter the tricarboxylic acid cycle. The following equation represents the balance sheet for their oxidation and the coupled phosphorylations (see page 516):

8 Acetyl-CoA + 16O₂ + 96P$_i$ + 96ADP ⟶
$$8\text{CoA} + 96\text{ATP} + 104\text{H}_2\text{O} + 16\text{CO}_2 \quad (2)$$

Combining equations (1) and (2), we get the overall equation

Palmitoyl-CoA + 23O₂ + 131P$_i$ + 131ADP ⟶
$$\text{CoA} + 16\text{CO}_2 + 146\text{H}_2\text{O} + 131\text{ATP} \quad (3)$$

Since two molecules of ATP are in effect utilized to form palmitoyl-CoA from palmitate (see page 546), the *net* yield of ATP per molecule of palmitate is 129. The overall equation for palmitate oxidation is therefore

Palmitate + 23O₂ + 129P$_i$ + 129ADP ⟶
$$16\text{CO}_2 + 145\text{H}_2\text{O} + 129\text{ATP} \quad (4)$$

The partial equations for the exergonic and endergonic processes are

$$\text{Palmitic acid} + 23\text{O}_2 \longrightarrow 16\text{CO}_2 + 16\text{H}_2\text{O}$$
$$\Delta G^{\circ\prime} = -2340 \text{ kcal mol}^{-1}$$

$$129\text{ADP} + 129\text{P}_i \longrightarrow 129\text{ATP} + 129\text{H}_2\text{O}$$
$$\Delta G^{\circ\prime} = 129 \times 7.30 = +942 \text{ kcal mol}^{-1}$$

Thus some $942/2340 \times 100 = 40$ percent of the standard free energy of oxidation of palmitic acid is recovered as high-energy phosphate.

Balancing Overall Metabolic Equations

A great deal of biochemistry is summarized in an overall balanced equation for a metabolic process. Summation of the balanced equations for each component step in a metabolic sequence can be tedious and time-consuming. Another method, illustrated here, provides a convenient shortcut in many cases.

Suppose we wish to write the overall equation, including the associated phosphorylation reactions, for the complete combustion of lauric acid, a C_{12} saturated fatty acid, to CO_2 and H_2O, as it occurs in a typical cell. We start by writing and balancing the chemical equation for the combustion of lauric acid as it would occur, for example, in a bomb calorimeter in the absence of coupled phosphorylation reactions:

$$C_{11}H_{23}COOH + 17O_2 \longrightarrow 12CO_2 + 12H_2O \qquad (5)$$
$$\text{Lauric acid}$$

Now we add to this equation the reactions involving ATP formation. For the cleavage of lauric acid into 6 acetyl-CoA's there are 5 $FADH_2$-generating steps, each giving 2 ATPs per step, or 10 ATPs, and 5 NADH-generating steps, each giving 3 ATPs per step, or 15 ATPs. The sum is 25 ATPs. Now, for the complete combustion of the acetyl portions of the 6 acetyl-CoA's produced, 6 turns of the tricarboxylic acid cycle are involved, yielding 6 × 12 ATPs = 72 ATPs. Thus we have a total of 97 ATPs produced. We can write a balanced equation for their formation:

$$97ADP + 97P_i \longrightarrow 97ATP + 97H_2O \qquad (6)$$

Adding together (5) and (6), we then have

$$C_{11}H_{23}COOH + 17O_2 + 97ADP + 97P_i \longrightarrow$$
$$12CO_2 + 97ATP + 109H_2O \quad (7)$$

Now, when lauric acid is converted to lauroyl-CoA, one ATP is converted to AMP and PP_i:

$$ATP + H_2O \rightleftharpoons AMP + PP_i \qquad (8)$$

The PP_i produced in equation (8) is converted to two P_i by pyrophosphatase:

$$PP_i + H_2O \longrightarrow 2P_i \qquad (9)$$

Summing equations (8) and (9), we have

$$ATP + 2H_2O \longrightarrow AMP + 2P_i \qquad (10)$$

When equations (7) and (10) are added together, we obtain

$$C_{11}H_{23}COOH + 17O_2 + 95P_i + 97ADP \longrightarrow$$
$$12CO_2 + 96ATP + 107H_2O + AMP \quad (11)$$

AMP is now phosphorylated to ADP at the expense of ATP

by adenylate kinase:

$$AMP + ATP \rightleftharpoons 2ADP \qquad (12)$$

We now add equations (11) and (12) to yield the overall equation for the combustion of lauric acid:

$$C_{11}H_{23}COOH + 17O_2 + 95P_i + 95ADP \longrightarrow$$
$$12CO_2 + 95ATP + 107H_2O \quad (13)$$

A total of 34 atoms of oxygen are consumed. We can account for these as follows. The reoxidation of the 5 reduced flavoproteins produced during cleavage of lauroyl-CoA to 6 acetyl-CoA's requires 5 atoms of oxygen. Reoxidation of the 5 NADHs produced during cleavage of lauroyl-CoA to 6 acetyl-CoA's requires 5 atoms of oxygen. Reoxidation of the reduced electron carriers generated by the 6 turns of the tricarboxylic acid cycle during oxidation of the 6 acetyl-CoA's requires 24 atoms of oxygen. The total is 34 atoms, or 17 molecules, of oxygen.

We note that a great deal of water is produced in the complete combustion of a fatty acid, most of which derives from the associated phosphorylation reactions. This fact is of some biological significance, since it explains how some animals can obtain both fuel and water from fat stores. In this way water is "stored" in the form of fat in the camel's hump.

Oxidation of Unsaturated Fatty Acids

Unsaturated fatty acids, such as oleic acid, are oxidized by the same general pathway as saturated fatty acids, but two special problems arise. The double bonds of naturally occurring unsaturated fatty acids are in the cis configuration (page 281), whereas the Δ^2-unsaturated acyl-CoA intermediates in the oxidation of saturated fatty acids are trans, as we have seen. Moreover, the double bonds of most unsaturated fatty acids occur at such positions in the carbon chain that successive removal of two-carbon fragments from the carboxyl end yields a Δ^3-unsaturated fatty acyl–CoA rather than the Δ^2 fatty acyl–CoA serving as the normal intermediate in the fatty acid cycle. This is illustrated for oleic acid in Figure 20–12.

These problems have been resolved with the discovery of an auxiliary enzyme, _enoyl-CoA isomerase_, which catalyzes a reversible shift of the double bond from the Δ^3-cis to the Δ^2-trans configuration (Figure 20–13). The resulting Δ^2-trans-unsaturated fatty acyl-CoA is the normal substrate for the next enzyme of the fatty acid oxidation sequence, _enoyl-CoA hydratase_, which hydrates it to form L-3-hydroxyacyl-CoA (see above). The complete oxidation of oleyl-CoA to nine acetyl-CoA units by the fatty acid oxidation cycle thus requires an extra enzymatic step catalyzed by the enoyl-CoA isomerase, in addition to those steps required in the oxidation of saturated fatty acids.

Polyunsaturated fatty acids, such as linoleic acid (page 281), require a second auxiliary enzyme to complete their ox-

Figure 20-12
Oxidative removal of 3 acetyl-CoA units from oleic acid to yield a Δ^3-cis-enoyl-CoA.

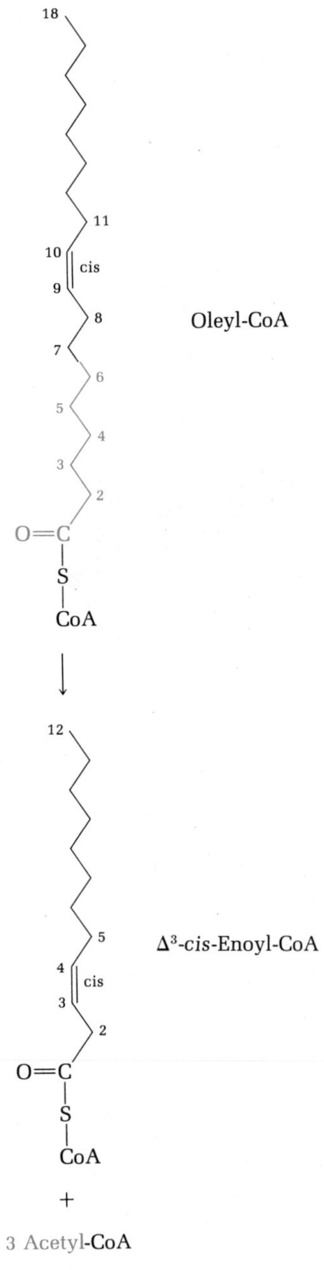

Figure 20-13
The enoyl-CoA-isomerase reaction.

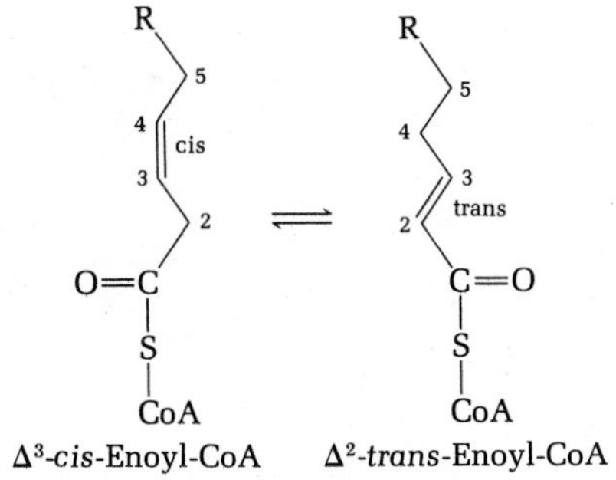

Δ³-cis-Enoyl-CoA   Δ²-trans-Enoyl-CoA

idation, since they contain two or more cis double bonds. When three successive acetyl-CoA units are removed from linoleyl-CoA, a Δ³-cis double bond remains, as in the case of oleyl-CoA (Figure 20–14). This is then transformed by the enoyl-CoA isomerase described above to the Δ²-trans isomer. This undergoes the usual reactions, with loss of two acetyl-CoA's, leaving an eight-carbon Δ²-unsaturated acid (Figure 20–14). Note, however that the double bond of the latter is in the cis configuration. Although the Δ²-cis double bond can be hydrated by enoyl-CoA hydratase, the product is the D stereoisomer of a 3-hydroxyacyl-CoA, not the L stereoisomer normally formed during oxidation of saturated fatty acids. Utilization of the D stereoisomer requires a second auxiliary enzyme, 3-hydroxyacyl-CoA epimerase, which catalyzes epimerization at carbon atom 3 to yield the L isomer (Figure 20–14). The product of this reversible reaction is then oxidized by the L-specific 3-hydroxyacyl-CoA dehydrogenase and cleaved by thiolase to complete the oxidation cycle. The remaining six-carbon saturated fatty acyl–CoA derived from linoleic acid can now be oxidized to three molecules of acetyl-CoA. These two auxiliary enzymes of the fatty acid oxidation cycle make possible the complete oxidation of all the common unsaturated fatty acids found in naturally occurring lipids. The number of ATP molecules yielded during the complete oxidation of an unsaturated fatty acid is somewhat lower than for the corresponding saturated fatty acid since unsaturated fatty acids have fewer hydrogen atoms and thus fewer electrons to be transferred via the respiratory chain to oxygen.

Ketone Bodies and Their Oxidation

In many vertebrates the liver has the enzymatic capacity to divert some of the acetyl-CoA derived from fatty acid or pyruvate oxidation, presumably during periods of excess formation, into free acetoacetate and D-β-hydroxybutyrate, which are transported via the blood to the peripheral tissues, where they may be oxidized via the tricarboxylic acid cycle.

Figure 20-14
Pathway of oxidation of linoleic acid,
showing action of auxiliary enzymes.

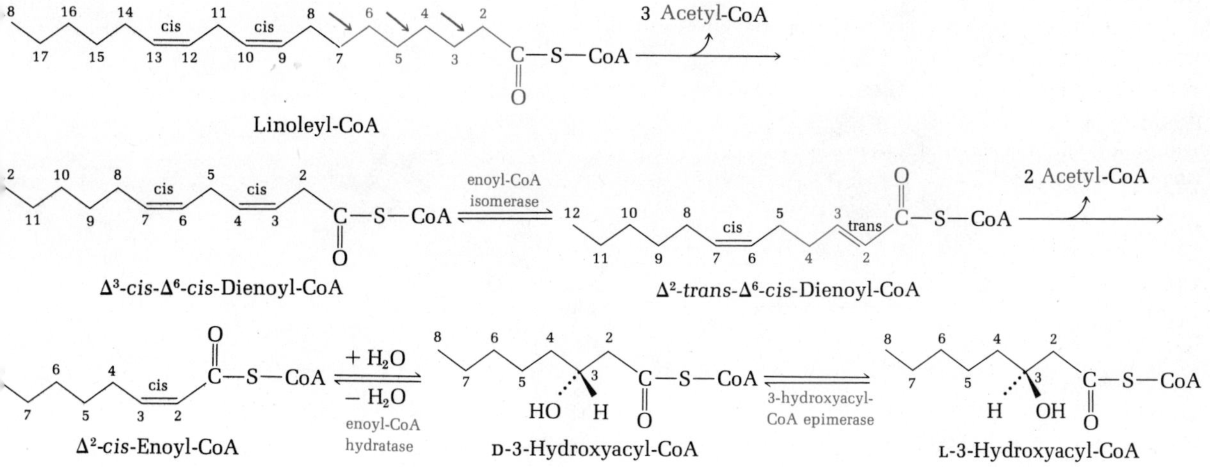

Linoleyl-CoA

Δ³-cis-Δ⁶-cis-Dienoyl-CoA   Δ²-trans-Δ⁶-cis-Dienoyl-CoA

Δ²-cis-Enoyl-CoA   D-3-Hydroxyacyl-CoA   L-3-Hydroxyacyl-CoA

These compounds (Figure 20–1), together with acetone, are collectively called the *ketone bodies.* Free acetoacetate, which is the primary source of the other ketone bodies, is formed from acetoacetyl-CoA. Some of the acetoacetyl-CoA arises from the last four carbon atoms of a long-chain fatty acid after oxidative removal of successive acetyl-CoA residues in the mitochondrial matrix. However, most of the acetoacetyl-CoA formed in the liver arises from the head-to-tail condensation of two molecules of acetyl-CoA derived from fatty acid oxidation by the action of acetyl-CoA acetyltransferase:

$$\text{Acetyl-CoA} + \text{acetyl-CoA} \rightleftharpoons \text{acetoacetyl-CoA} + \text{CoA}$$

The acetoacetyl-CoA formed in these reactions then undergoes loss of CoA, a process called *deacylation,* to yield free acetoacetate in a special pathway taking place in the mitochondrial matrix. It involves the enzymatic formation and cleavage of *β-hydroxy-β-methylglutaryl-CoA* (Figure 20–15), an intermediate which also serves as a precursor of sterols. The sum of the two reactions in Figure 20–15 is

$$\text{Acetoacetyl-CoA} + \text{H}_2\text{O} \rightleftharpoons \text{acetoacetate} + \text{CoA}$$

The free acetoacetate so produced is enzymatically reduced to D-β-hydroxybutyrate by the NAD-linked D-*β-hydroxybutyrate dehydrogenase,* which is located in the inner mitochondrial membrane. This enzyme is specific for the free D stereoisomer; it does not oxidize D-β-hydroxybutyryl-CoA:

$$\text{Acetoacetate} + \text{NADH} + \text{H}^+ \rightleftharpoons \text{D-}β\text{-hydroxybutyrate} + \text{NAD}^+$$

The mixture of free acetoacetate and β-hydroxybutyrate resulting from these reactions may diffuse out of the liver cells into the bloodstream, to be transported to the peripheral tissues. Normally the concentration of ketone bodies in the blood is rather low, but in fasting or in the disease diabetes mellitus, it may reach very high levels. This condition, known as *ketosis,* arises when the rate of formation of the ketone bodies by the liver exceeds the capacity of the peripheral tissues to utilize them, with a resulting accumulation in the blood.

Figure 20-15
Deacylation of acetoacetyl-CoA.

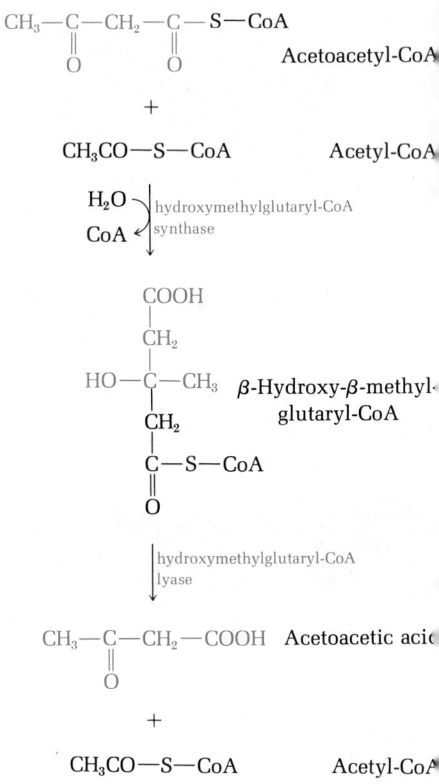

Figure 20-16
Transfer of CoA from succinyl-CoA to acetoacetic acid.

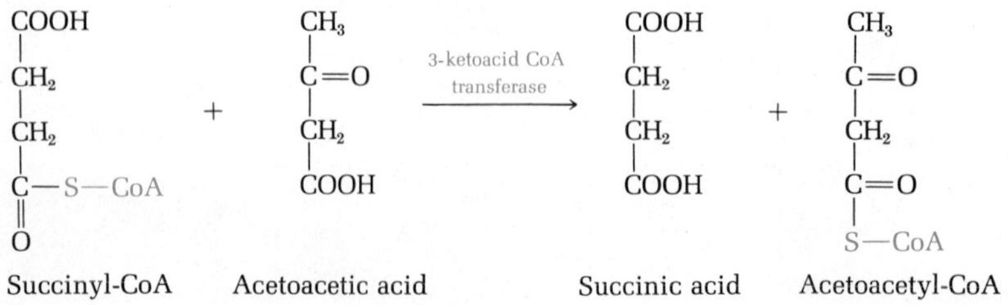

Figure 20-17
Carboxylation of propionyl-CoA.

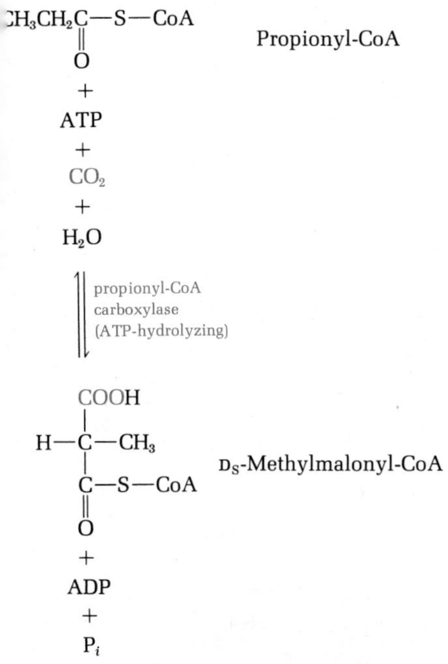

$$CH_3CH_2C-S-CoA$$
$$\overset{\|}{O}$$

Propionyl-CoA

+

ATP

+

CO_2

+

H_2O

‖ propionyl-CoA
‖ carboxylase
‖ (ATP-hydrolyzing)

COOH
|
H—C—CH₃
|
C—S—CoA
‖
O

D$_S$-Methylmalonyl-CoA

+

ADP

+

P$_i$

Figure 20-18
The racemization of methylmalonyl-CoA
and its conversion into succinyl-CoA. In the
latter reaction note than the entire
—C—S—CoA group is transferred from
‖
O
carbon atom 2 to the methyl carbon atom.

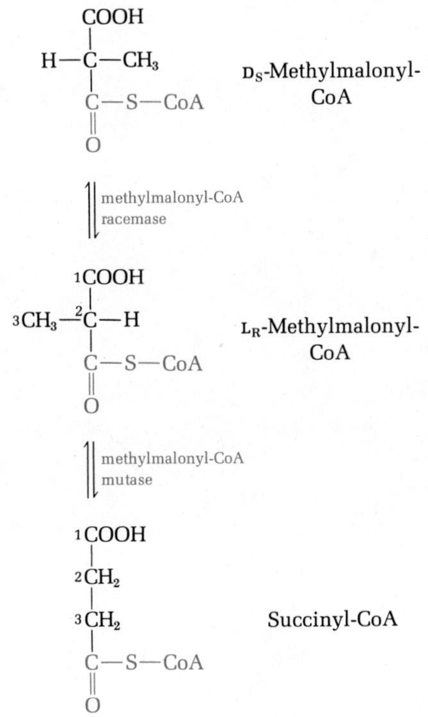

COOH
|
H—C—CH₃
|
C—S—CoA
‖
O

D$_S$-Methylmalonyl-
CoA

‖ methylmalonyl-CoA
‖ racemase

¹COOH
|
³CH₃—²C—H
|
C—S—CoA
‖
O

L$_R$-Methylmalonyl-
CoA

‖ methylmalonyl-CoA
‖ mutase

¹COOH
|
²CH₂
|
³CH₂
|
C—S—CoA
‖
O

Succinyl-CoA

In the peripheral tissues the D-β-hydroxybutyrate is oxidized to acetoacetate, which is then activated by transfer of CoA from succinyl-CoA (Figure 20–16). The succinyl-CoA required arises from the oxidation of α-ketoglutarate (page 457). The acetoacetyl-CoA formed in the peripheral tissues by these reactions then undergoes thiolytic cleavage to two molecules of acetyl-CoA, which then may enter the tricarboxylic acid cycle.

Although β-hydroxybutyrate can be utilized via the tricarboxylic acid cycle in all tissues, its utilization by the brain under some conditions is especially noteworthy. Normally the brain uses glucose almost exclusively as its fuel. However, in prolonged fasting, when the supply of glucose is limited, the brain may utilize β-hydroxybutyrate generated from fatty acids in the liver as its major oxidative fuel. By the reactions shown above it is converted into acetyl-CoA, which then enters the tricarboxylic cycle.

Oxidation of Odd-Carbon Fatty Acids and the Fate of Propionyl-CoA

Odd-carbon fatty acids, which are rare but do occur in some marine organisms (page 281), can also be oxidized in the fatty acid oxidation cycle. Successive acetyl-CoA residues are removed until the terminal three-carbon residue propionyl-CoA is reached. This compound is also formed in the oxidative degradation of the amino acids valine and isoleucine (page 577). Propionyl-CoA undergoes enzymatic carboxylation in an ATP-dependent process to form D$_S$-methylmalonyl-CoA (Figure 20–17), a reaction catalyzed by propionyl-CoA carboxylase. This enzyme contains biotin as its prosthetic group (page 345). In the next step D$_S$-methylmalonyl-CoA undergoes enzymatic epimerization to L$_R$-methylmalonyl-CoA, by action of methylmalonyl-CoA racemase (Figure 20–18). In the next reaction step, catalyzed by methylmalonyl-CoA mutase, L$_R$-methylmalonyl-CoA is isomerized to succinyl-CoA, which may then undergo deacylation by reversal of the succinyl-CoA synthetase reaction (page 458) to yield free succinate, an intermediate of the tricarboxylic acid cycle.

Methylmalonyl-CoA mutase requires as cofactor coenzyme B$_{12}$ (page 348). Study of this intramolecular reaction with isotope tracers has revealed that it takes place by the migration of the entire —CO—S—CoA group from carbon atom 2 of methylmalonyl-CoA to the methyl carbon atom in exchange for a hydrogen atom.

Patients suffering from pernicious anemia, who are deficient in vitamin B$_{12}$ because of their lack of intrinsic factor (page 349), excrete large amounts of methylmalonic acid and its precursor propionic acid in the urine, showing that in such patients the coenzyme B$_{12}$–dependent methylmalonyl-CoA mutase reaction is defective.

Minor Pathways of Fatty Acid Oxidation

The α-hydroxy fatty acids found in some cerebrosides and gangliosides (Chapter 11, pages 292 and 294) of animal tissues are formed by enzymatic hydroxylation of saturated fatty acids through the action of a _monooxygenase_ (page 501):

$$RCH_2CH_2COOH + \text{reduced cofactor} + O_2 \longrightarrow$$

$$RCH_2CHOHCOOH + H_2O + \text{oxidized cofactor}$$

α-Hydroxy
fatty acid

The reduced cofactor may be ascorbic acid or tetrahydrobiopterin (page 570).

Fatty acids also undergo oxidation in the liver at the ω carbon atom to form ultimately α,ω-dicarboxylic acids. This pathway is called _ω oxidation_; its function in mammals is not known.

In germinating plant seeds a special pathway of fatty acid oxidation occurs, called _α oxidation_ (Figure 20–19), in which the carboxyl carbon of the fatty acid is lost as CO_2 and the α carbon atom is oxidized to an aldehyde group at the expense of hydrogen peroxide. This reaction is catalyzed by _fatty acid peroxidase_. The hydrogen peroxide required is furnished by the direct oxidation of reduced flavoproteins by molecular oxygen. The fatty aldehyde formed is oxidized to the corresponding carboxylic acid. This two-enzyme reaction sequence is then repeated on the shortened free fatty acid. Since fatty acid peroxidase attacks only fatty acids having from 13 to 18 carbon atoms, this pathway cannot lead to complete oxidation of long-chain fatty acids. The aldehydes produced by α oxidation may alternatively undergo reduction to yield long-chain fatty alcohols, which occur in large amounts in plant waxes.

Figure 20-19
The α-oxidation pathway.

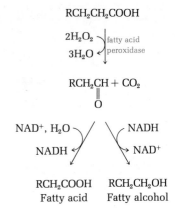

Summary

In animal tissues free fatty acids are first activated by esterification with CoA to form acyl-CoA thioesters at the outer mitochondrial membrane and are then converted into O-fatty acyl carnitine esters, which can cross the inner mitochondrial membrane into the matrix, where fatty acyl–CoA thioesters are reformed. All subsequent steps in the oxidation of fatty acids take place in the form of CoA esters within the mitochondrial matrix. The successive oxidative removal of acetyl-CoA units from long-chain saturated fatty acyl-CoA is called β oxidation. Four reaction steps are required to remove each acetyl-CoA residue: (1) the dehydrogenation of carbon atoms 2 and 3 by FAD-linked fatty acyl–CoA dehydrogenases, (2) hydration of the resulting Δ^2-trans double bond by enoylhydratase, (3) dehydrogenation of the resulting L-β-hydroxy fatty acyl–CoA by an NAD$^+$-linked dehydrogenase, and (4) a CoA-requiring cleavage (thiolysis) of the resulting β-keto fatty acyl–CoA, to form acetyl-CoA and the CoA thioester of a fatty acid shortened by two carbons. The shortened fatty acyl–CoA can then reenter the sequence. For complete oxidation of the 16-carbon palmitic acid, seven cycles through the system are required, yielding altogether eight molecules of acetyl-CoA. Electrons removed in the two dehydrogenation steps flow to oxygen via the respiratory chain, accompanied by oxidative phosphorylation of ADP. The acetyl-CoA formed during fatty acid oxidation is then oxidized to CO_2 and H_2O via the tricarboxylic acid cycle. The overall equation for oxidation of palmitic acid is

$$\text{Palmitic acid} + 23O_2 + 129P_i + 129ADP \longrightarrow$$
$$16CO_2 + 145H_2O + 129ATP$$

In this process about 40 percent of the standard free energy of oxidation of palmitic acid is recovered as high-energy phosphate.

Unsaturated fatty acids require additional enzymatic steps in order (1) to shift their double bonds into the proper position for the hydration step and (2) to yield the L stereoisomer of the 3-hydroxy intermediate. The ketone bodies acetoacetate and β-hydroxybutyrate are formed in the liver following deacylation of acetoacetyl-CoA. They are carried to other tissues, where they are oxidized via acetyl-CoA and the tricarboxylic acid cycle. Odd-carbon fatty acids are oxidized to acetyl-CoA and propionyl-CoA. Carboxylation of propionyl-CoA yields methylmalonyl-CoA, which undergoes isomerization to succinyl-CoA. α Hydroxylation and ω oxidation represent minor pathways of fatty acid metabolism in animals. In plants α oxidation of long-chain fatty acids leads to formation of fatty aldehydes, which are then reduced to fatty alcohols, components of plant waxes.

References

Books

DAWSON, R. M. C., and D. N. RHODES (eds.): *Metabolism and Physiological Significance of Lipids*, Wiley, New York, 1964.

FLORKIN, M., and E. H. STOTZ (eds.): *Lipid Metabolism*, vol. 18 of *Comprehensive Biochemistry*, American Elsevier, New York, 1967.

HOFMANN, K.: *Fatty Acid Metabolism in Microorganisms*, Wiley, New York, 1963. The occurrence and metabolism of branched and cyclic acids.

STANBURY, J. B., J. B. WYNGAARDEN, and D. S. FREDRICKSON (ed.): *The Metabolic Basis of Inherited Disease*, 3d ed., McGraw-Hill, New York, 1972. Chapter 22 has an excellent account of disorders of propionate metabolism.

Articles

BARNESS, L. A.: "Methylmalonic Acid," *Pediatrics*, 51: 1012–1018 (1973).

FRITZ, I. B.: "Carnitine and Its Role in Fatty Acid Metabolism," *Adv. Lipid Res.*, 1: 285–334 (1963).

FULCO, A. J.: "Metabolic Alterations of Fatty Acids," *Ann. Rev. Biochem.*, 43: 147–168 (1974).

GARLAND, P. B.: "Control of Citrate Synthesis in Mitochondria," in T. W. Goodwin (ed.), *Metabolic Roles of Citrate*, Academic, New York, 1968. Short review of relationships between fatty acid oxidation and the tricarboxylic acid cycle.

GREVILLE, G. D., and P. K. TUBBS: "The Catabolism of Long-Chain Fatty Acids in Mammalian Tissues," *Essays Biochem.*, 4: 155–212 (1968). Excellent and readable review.

KREBS, H. A.: "The Regulation of Release of Ketone Bodies by the Liver," *Adv. Enzyme Regul.*, 4: 339–354 (1966).

KUSUNOSE, M., E. KUSUNOSE, and M. J. COON: "Enzymatic ω-Oxidation of Fatty Acids," *J. Biol Chem.*, 239: 1374–1380 (1964).

SOKOLOFF, L.: "Metabolism of Ketone Bodies by the Brain," *Ann. Rev. Med.*, 24: 271–288 (1973).

STUMPF, P. K.: "Metabolism of Fatty Acids," *Ann. Rev. Biochem.*, 38: 159–212 (1969).

Problems

1. Write balanced equations for the following metabolic processes, including all required activation steps and all oxidative phosphorylations. Assume that the reactions are taking place in the liver, kidney, or heart and that oxaloacetate is readily available.
    (a) Oxidation of myristic acid to acetyl-CoA
    (b) Oxidation of arachidonic acid to CO_2 and H_2O
    (c) Oxidation of palmitic acid to acetoacetic acid
    (d) Oxidation of monooleyldipalmitoylglycerol to CO_2 and H_2O
    (e) Oxidation of n-nonanoic acid to CO_2 and H_2O

2. If n-nonanoic acid labeled with ^{14}C in carbon atom 7 is oxidized under conditions in which the tricarboxylic acid cycle is operating, which carbon atoms of the following intermediates will become labeled most rapidly: (a) succinate, (b) oxaloacetate, (c) α-ketoglutarate?

3. Pyruvate labeled with ^{14}C in carbon atom 2 is incubated with liver tissue. Which carbon atoms of β-hydroxy-β-methylglutaryl-CoA will become labeled rapidly?

CHAPTER 21 OXIDATIVE DEGRADATION OF AMINO ACIDS

In higher animals amino acids serve as building blocks of proteins and as precursors of many other important biomolecules, such as hormones, purines, pyrimidines, porphyrins, and some vitamins. However, they also serve as a source of energy, particularly when they are ingested in excess of the amounts required to replace body proteins. When they are used as fuel, amino acids undergo loss of their amino groups; their remaining carbon skeletons then have two major fates: (1) conversion into glucose in the process of gluconeogenesis (page 629) or (2) oxidation to CO_2 via the tricarboxylic acid cycle. Amino acids converted into glucose by mammals must first undergo enzymatic transformation to intermediates on the direct pathway to glucose synthesis, such as pyruvate or the dicarboxylic intermediates of the tricarboxylic acid cycle. Of course, amino acids that are converted into glucose will also ultimately be oxidized to completion via the tricarboxylic acid cycle.

Vertebrates actively oxidize both exogenous amino acids (from ingested proteins) and endogenous amino acids (from the metabolic turnover of body proteins). It has been estimated from isotopic tracer experiments that in a 70-kg man on an average diet about 400 g of protein turns over each day. Up to one-fourth of this amount undergoes oxidative degradation or conversion into glucose and is replaced daily from the exogenous intake; the remaining three-fourths is recycled. Anywhere from 6 to 20 g of nitrogen, derived from the amino groups of amino acids, is daily excreted in the urine in the form of nitrogenous compounds, principally urea. Even when no protein is being ingested, a person may excrete up to 5 g of nitrogen per day, corresponding to the daily loss of over 30 g of endogenous protein.

In most bacteria, on the other hand, oxidative degradation of amino acids is not a prominent process. Bacteria are ordinarily much more active in carrying out the biosynthesis of amino acids than degradation, particularly during rapid growth. However, some bacteria can degrade amino acids, particularly if they represent the sole source of carbon. In higher plants the net direction of amino acid metabolism is also generally toward synthesis rather than oxidative degradation, since plants tend to grow continuously.

This chapter describes the oxidative degradation by vertebrates of the 20 building-block amino acids into acetyl-CoA, the fuel of the tricarboxylic acid cycle, and into pyruvate and other precursors of glucose. We shall also consider the formation of the characteristic nitrogenous waste products that arise from the amino groups of amino acids, namely, urea, ammonia, and uric acid.

The rate of utilization and the proportion of the different amino acids utilized by animals depend on many factors, including (1) the availability of other fuels, (2) the availability of exogenous amino acids, (3) the need of the organism for amino acids in protein synthesis, (4) the nutritional dependence of the organism on essential amino acids (page 693), and (5) the need for specific amino acids as precursors of other important biomolecules.

Proteolysis

The amino acids ingested by vertebrates are largely in the form of proteins. Since amino acids can enter the metabolic pathways only in free form, proteins and peptides are first hydrolyzed by proteolytic enzymes in the gastrointestinal tract. These enzymes are secreted by the stomach, pancreas, and small intestine.

Digestion of proteins begins in the stomach. Here the major proteolytic enzyme, _pepsin_ (mol wt 33,000), is secreted as its zymogen _pepsinogen_ (mol wt 40,000) by the chief cells of the gastric mucosa. Pepsinogen is converted into active pepsin by pepsin itself at the acid pH of gastric juice, with liberation of 42 amino acid residues from the N-terminal end of its single polypeptide chain. Pepsin has very broad specificity but preferentially attacks peptide bonds involving residues of aromatic amino acids, as well as methionine and leucine, to yield peptides but few free amino acids (page 106).

Pancreatic juice, which is secreted into the small intestine, contributes the zymogens _chymotrypsinogen, trypsinogen, procarboxypeptidases A and B,_ and _proelastase._ Chymotrypsinogen (mol wt 24,000) is converted into active chymotrypsin by removal of two dipeptides by the action of free trypsin and chymotrypsin (page 227). Chymotrypsin hydrolyzes peptide bonds containing carboxyl groups of aromatic amino acids. Trypsinogen (mol wt 24,000) is converted into active trypsin by removal of the N-terminal hexapeptide through the action of the proteolytic enzyme _enterokinase._ Trypsin hydrolyzes peptide linkages involving the carboxyl groups of arginine and lysine. Carboxypeptidase A (mol wt 34,000), a Zn^{2+}-containing enzyme (page 233), hydrolyzes nearly all types of carboxyl-terminal peptide bonds, while carboxypeptidase B cleaves C-terminal arginine or lysine residues.

As a result of the action of pepsin in the stomach, followed by the action of the pancreatic proteases, proteins are converted into short peptides and free amino acids. The remaining short peptides are then degraded completely to yield free amino acids by peptidases found in, and secreted by, the intestinal mucosa, particularly *leucine aminopeptidase,* another Zn^{2+} enzyme, which removes N-terminal residues of peptides. The resulting free amino acids are absorbed into the blood and reach the liver, where much of the further metabolism of the amino acids, including their degradation, takes place.

Endogenous proteins must also undergo degradation to amino acids before they are used as fuel. However, very little is known about the mechanism of intracellular proteolysis, which in some tissues, particularly the liver, may take place at a high rate, in association with the metabolic turnover of body proteins.

The Flow Sheet of Amino Acid Oxidation

In vertebrates there are 20 different multienzyme sequences for the oxidative degradation of the 20 different amino acids found in proteins. These sequences ultimately converge into a few terminal pathways leading to pyruvate, acetyl-CoA, or to intermediates of the tricarboxylic acid cycle. As is shown in Figure 21-1, the carbon skeletons of eleven of the amino acids ultimately yield acetyl-CoA, either directly or via pyruvate or acetoacetyl-CoA, five are converted into α-ketoglutarate, three yield succinyl-CoA, and two yield oxaloacetate. Two amino acids, phenylalanine and tyrosine, are so degraded that one portion of the carbon skeleton enters the tricarboxylic acid cycle as acetyl-CoA and the other as fumarate. However, not all the carbon atoms of each of the 20

Figure 21-1
Pathways by which the carbon skeletons of amino acids enter into the tricarboxylic acid cycle. Some of the amino acids (leucine, tryphophan, and isoleucine) undergo fragmentation in such a way that the products enter the cycle by two different routes.

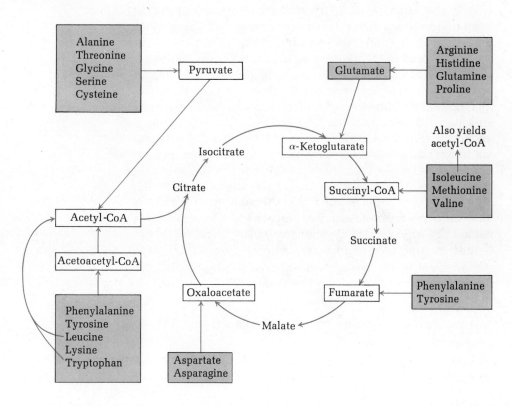

amino acids enter the tricarboxylic acid cycle, since some are lost en route by decarboxylation reactions. It is especially important to note that the catabolic pathways are not the reverse of the biosynthetic pathways. However, there are often common steps in the two routes, as will be developed more fully elsewhere (page 693).

The catabolic pathways of amino acids are often long and complex, with many intermediates; it may indeed appear that they are unnecessarily long for the end achieved. However, many of the intermediates of amino acid oxidation have additional functions in the cell, particularly as essential precursors of other cell components. The catabolic pathway of a given amino acid may thus lead to branches and collateral routes. In vertebrates amino acid catabolism takes place largely in the liver, although the kidney is also significantly active. Skeletal muscle is relatively inactive.

The α-amino nitrogen atoms removed from amino acids during their oxidative degradation are ultimately excreted in the urine of vertebrates as urea, ammonia, or uric acid, depending on the species. However, the enzymatic mechanisms by which the α-amino groups are removed from various amino acids and collected for their ultimate conversion to one of these three end products are similar in all vertebrates. Since removal of the α-amino group constitutes the first stage in catabolism of most of the amino acids, we shall first examine the two major enzymatic pathways involved, transamination and oxidative deamination.

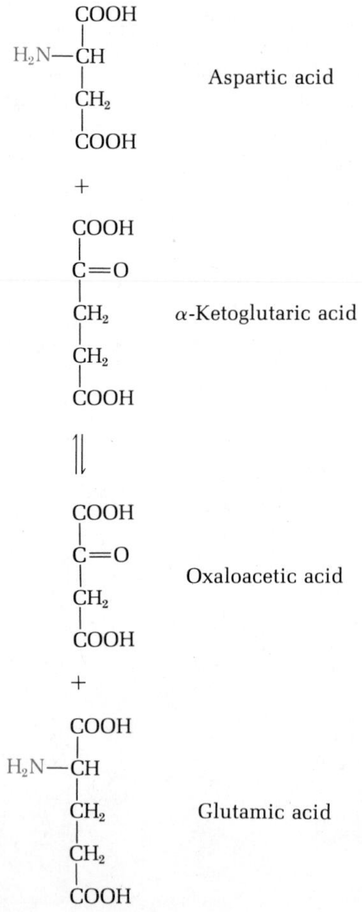

Transamination

In the catabolism of at least 12 of the amino acids (alanine, arginine, asparagine, aspartic acid, cysteine, isoleucine, leucine, lysine, phenylalanine, tryptophan, tyrosine, and valine) the α-amino group is enzymatically removed by transamination. In such reactions the α-amino group is transferred to the α carbon atom of an α-keto acid, in most cases α-ketoglutarate, leaving behind the corresponding α-keto acid analog of the amino acid and causing the amination of the α-ketoglutarate to L-glutamic acid (Figure 21-2). Such reactions are catalyzed by enzymes known generically as aminotransferases or transaminases. A large number of transaminases are now known. Most require α-ketoglutarate as one amino-group acceptor; they are therefore specific for the substrate couple α-ketoglutarate–L-glutamate. The specificity for the other substrate couple is less rigid, although usually there is one showing greatest activity, for which the enzyme is named. As an example, a prominent transaminase in animal tissues is aspartate aminotransferase (Figure 21-2), more commonly called aspartate transaminase, which catalyzes the reversible reaction

L-Aspartate + α-ketoglutarate $\rightleftharpoons$ oxaloacetate + L-glutamate (1)

Although the enzyme is most active with aspartate as amino-group donor in the forward direction, it will accept a number

Figure 21-2
The aspartate transaminase reaction. The specific amino-group acceptor of this and most other transaminases of animal tissues is α-ketoglutarate, thus yielding L-glutamate as the product in which α-amino groups from other amino acids are collected.

$$
\begin{array}{ll}
\text{COOH} & \\
| & \\
\text{H}_2\text{N}-\text{CH} & \\
| & \text{Aspartic acid} \\
\text{CH}_2 & \\
| & \\
\text{COOH} & \\
\end{array}
$$

+

$$
\begin{array}{ll}
\text{COOH} & \\
| & \\
\text{C=O} & \\
| & \\
\text{CH}_2 & \alpha\text{-Ketoglutaric acid} \\
| & \\
\text{CH}_2 & \\
| & \\
\text{COOH} & \\
\end{array}
$$

$$
\begin{array}{ll}
\text{COOH} & \\
| & \\
\text{C=O} & \\
| & \text{Oxaloacetic acid} \\
\text{CH}_2 & \\
| & \\
\text{COOH} & \\
\end{array}
$$

+

$$
\begin{array}{ll}
\text{COOH} & \\
| & \\
\text{H}_2\text{N}-\text{CH} & \\
| & \text{Glutamic acid} \\
\text{CH}_2 & \\
| & \\
\text{CH}_2 & \\
| & \\
\text{COOH} & \\
\end{array}
$$

of other α-amino acids as donors. In addition to aspartate transaminase, animal tissues contain other transaminases requiring α-ketoglutarate as amino-group acceptor, such as *alanine transaminase*, *leucine transaminase*, and *tyrosine transaminase*, catalyzing, respectively,

$$\text{L-Alanine} + \alpha\text{-ketoglutarate} \rightleftharpoons \text{pyruvate} + \text{L-glutamate} \quad (2)$$

$$\text{L-Leucine} + \alpha\text{-ketoglutarate} \rightleftharpoons$$
$$\alpha\text{-ketoisocaproate} + \text{L-glutamate} \quad (3)$$

$$\text{L-Tyrosine} + \alpha\text{-ketoglutarate} \rightleftharpoons$$
$$\text{p-hydroxyphenylpyruvate} + \text{L-glutamate} \quad (4)$$

The reactions catalyzed by the transaminases are freely reversible and have an equilibrium constant of about 1.0. Although they can proceed in either direction, it is appropriate, here, where the context is the oxidative degradation of amino acids, to write the transaminase reactions with α-ketoglutarate as the amino-group acceptor, as in reactions (1) to (4). By the action of the transaminases the amino groups of various amino acids are collected in the form of only one α-amino acid, usually glutamic acid. The collection of amino groups in this manner is another example of the convergence of catabolic pathways (page 323). Glutamate, the end product of most transaminations, then serves as the specific amino-group donor in a final series of reactions in which its amino group is converted into nitrogenous excretory products.

Transaminases are found both in the mitochondria and in the cytosol of eukaryotic cells. The mitochondrial and extramitochondrial forms of pig-heart aspartate transaminase differ in their isoelectric pH and amino acid composition. Both forms have a molecular weight of about 90,000, and both contain two subunits of about equal size. In mammals the collection of amino groups from other amino acids takes place in the cytosol, catalyzed by the cytosol form of aspartate transaminase, with formation of glutamate. The glutamate so formed then enters the mitochondrial matrix via a specific membrane transport system (page 530). In the mitochondrial matrix glutamate is either directly deaminated or becomes the amino-group donor to oxaloacetate by the action of mitochondrial aspartate transaminase to yield aspartate, one of the immediate amino-group donors in the formation of urea (see below, page 581).

All the transaminases appear to have the same prosthetic group, pyridoxal phosphate, and to share a common reaction mechanism (page 344). The tightly bound pyridoxal phosphate, which is noncovalently linked to the enzyme protein, presumably through the charged ring nitrogen atom, functions as a carrier of amino groups. During its catalytic cycle it undergoes reversible transitions between its free aldehyde form, pyridoxal phosphate, and its aminated form, pyridoxamine phosphate (Figure 21-3). In this cycle the unprotonated α-amino group of the amino donor is covalently bound to the carbon atom of the aldehyde group of the enzyme-bound pyridoxal phosphate with elimination of water to form an aldimine, which tautomerizes to the corre-

Figure 21-3
Intermediate steps in the transaminase reaction. The pyridoxal phosphate–enzyme complex is symbolized H—C—$\widehat{E}$; *the pyridoxamine phosphate–en-*
$$\|$$
$$O$$
zyme complex as H_2N—CH_2—$\widehat{E}$. *The structures of pyridoxal phosphate and pyridoxamine phosphate are given on page 344. The pyridoxal phosphate prosthetic group is the intermediate amino-group carrier between the amino acid and the keto acid.*

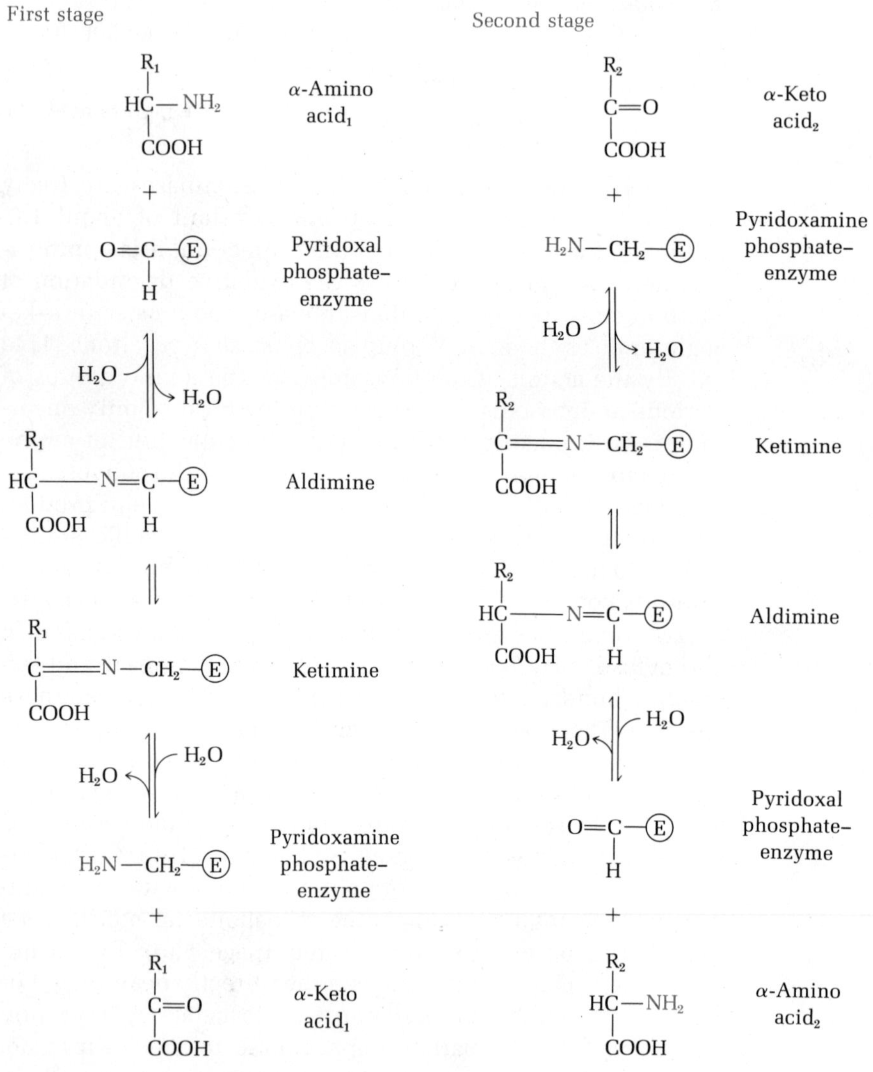

First stage   Second stage

sponding ketimine. Both the aldimine and the ketimine contain the $>$C$=$N— structure characteristic of a Schiff's base. Addition of water leads to formation of free α-keto acid and the aminated form of the prosthetic group, pyridoxamine phosphate. The pyridoxamine phosphate–enzyme complex then forms a Schiff's base with the incoming α-keto acid, to which the amino group is donated by reversal of the steps described above, with regeneration of pyridoxal phosphate (Figure 21-3). By oscillating between the aldehyde and amino forms, the prosthetic group acts as a carrier of amino groups from an amino acid to a keto acid. The transamina-

Figure 21-4
Aldimine linkage between the pyridoxal
phosphate prosthetic group and the ε-amino
group of a specific lysyl residue (color) of
the transaminase protein (shaded). The enzyme reacts with the incoming α-amino
acid, which displaces the lysyl residue from
its aldimine linkage with pyridoxal phosphate.

Figure 21-4
Aldimine linkage between the pyridoxal
phosphate prosthetic group and the ε-amino
group of a specific lysyl residue (color) of
the transaminase protein (shaded). The enzyme reacts with the incoming α-amino
acid, which displaces the lysyl residue from
its aldimine linkage with pyridoxal phosphate.

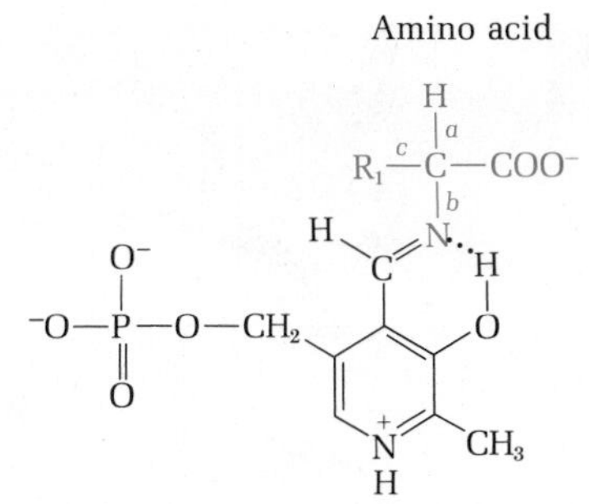

Schiff's base of pyridoxal phosphate
and an amino acid

Figure 21-5
Postulated intermediate at the catalytic site
of pyridoxal phosphate enzymes. The amino
acid substrate molecule is in color. When an
amino acid reacts with pyridoxal phosphate,
an electron pair moves from the amino acid
toward the positively charged pyridine ring.
In the second stage these electrons then flow
away from pyridoxamine phosphate toward
the acceptor α-keto acid. Thus the coenzyme
serves as a carrier not only of an amino
group but of an electron pair as well, oxidizing the α-carbon atom of the donor
amino acid and reducing the α-carbon atom
of the acceptor keto acid. Different pyridoxal
phosphate enzymes promote different kinds
of reactions of α-amino acids, including
transamination, decarboxylation, dehydration of β-hydroxyamino acids, racemization
of α-amino acids, and removal of hydrogen
sulfide from cysteine.

tion reaction is an example of a double-displacement reaction and shows the corresponding ping-pong kinetic pattern
(page 204).

In the free enzyme the pyridoxal phosphate is bound not
only via the ring nitrogen but also through the formation of a
Schiff's base with the ε-amino group of a specific lysine residue of the enzyme protein. The incoming amino acid substrate displaces the lysyl ε-amino group of the enzyme from
the pyridoxal phosphate to form the substrate–pyridoxal
phosphate aldimine (Figure 21-4). Pyridoxal phosphate has
a characteristic absorption spectrum which is useful in following the course of the formation and disappearance of
intermediate enzyme-substrate complexes.

The formulation in Figure 21-5 suggests a mechanistic
basis for the fact that pyridoxal phosphate is a prosthetic
group not only for transamination reactions but also for enzymatic decarboxylation of α-amino acids, dehydration of
serine, removal of sulfur from cysteine, and enzymatic racemization reactions in which L- and D-amino acids are interconverted. The formation of an intermediate complex
between the amino acid and the prosthetic group, together
with an electron-withdrawing structure contributed by the
enzyme, makes possible labilization of bonds a, b, and c
(Figure 21-5), rendering the amino acid molecule susceptible
to several types of transformation.

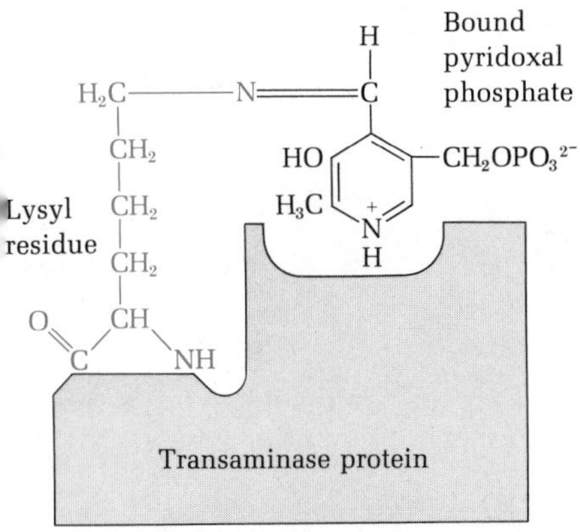

Oxidative Deamination

Glutamate formed by the action of the transaminases may
undergo rapid oxidative deamination catalyzed by the
pyridine-linked glutamate dehydrogenase, which is present
in both the cytosol and mitochondria of the liver,

$$\text{L-Glutamate} + \text{NAD}^+ (\text{NADP}^+) + \text{H}_2\text{O} \rightleftharpoons$$
$$\alpha\text{-ketoglutarate} + \text{NH}_4^+ + \text{NADH (NADPH)}$$

thus discharging as NH_4^+ ions the amino groups collected
from the other amino acids. Note that a dehydrogenation is
necessary to yield the α-keto acid; simple hydrolytic removal
of the amino group would yield the α-hydroxy acid, not the
α-keto acid. It is thought that the first step in the glutamate
dehydrogenase reaction is the formation of α-iminoglutarate
by dehydrogenation, followed by hydrolytic decomposition
of the imino acid to the keto acid, according to the reactions

$$\underset{\underset{\text{NH}_2}{|}}{\text{COOHCH}_2\text{CH}_2\text{CHCOOH}} + \text{NAD}^+ \rightleftharpoons$$
$$\underset{\underset{\text{NH}}{\|}}{\text{COOHCH}_2\text{CH}_2\text{CCOOH}} + \text{NADH} + \text{H}^+$$

α-Iminoglutaric acid

$$\underset{\underset{\text{NH}}{\|}}{\text{COOHCH}_2\text{CH}_2\text{CCOOH}} + \text{H}_2\text{O} \rightleftharpoons \underset{\underset{\text{O}}{\|}}{\text{COOHCH}_2\text{CH}_2\text{CCOOH}} + \text{NH}_3$$

L-Glutamate dehydrogenase can use either NAD^+ or NADP^+
as electron acceptor, but NAD^+ is preferred. The NADH

formed is ultimately oxidized by the electron-transport chain (page 495). L-Glutamate dehydrogenase plays a central role in amino acid deamination because in most organisms glutamate is the only amino acid that has such an active dehydrogenase. Glutamate dehydrogenase of beef liver has a molecular weight of 336,000 and contains six identical subunits, whose amino acid sequence has been established. It is an allosteric enzyme inhibited by the specific modulators ATP, GTP, and NADH and stimulated by ADP, GDP, and certain amino acids. Its activity is also influenced by the thyroid hormone *thyroxine* and certain steroid hormones. Glutamate dehydrogenase tends to associate reversibly into rod-shaped polymers of 2 to 3 megadaltons.

Many organisms contain flavin-linked *amino acid oxidases*, which also catalyze oxidative deamination of amino acids, but they play a relatively minor role and are not considered part of the mainstream of amino-group metabolism. One, L-*amino acid oxidase*, is specific for deamination of L-amino acids and promotes the reaction

$$\text{L-Amino acid} + H_2O + E\text{—FMN} \longrightarrow$$
$$\alpha\text{-keto acid} + NH_3 + E\text{—FMNH}_2$$

L-Amino acid oxidase contains tightly bound FMN as the prosthetic group and is present in the endoplasmic reticulum of the liver and kidneys, where it probably functions in the deamination of lysine (page 572). L-Amino acid oxidase is present in large amounts in some snake venoms.

The other flavoenzyme functioning in oxidative deamination is D-*amino acid oxidase*, present in the liver and kidneys, which catalyzes the oxidation of D-amino acids, again to the corresponding α-keto acids:

$$\text{D-Amino acid} + H_2O + E\text{—FAD} \longrightarrow$$
$$\alpha\text{-keto acid} + NH_3 + E\text{—FADH}_2$$

D-Amino acid oxidase contains FAD as prosthetic group. Presumably its function is to initiate the degradation of D-amino acids arising from the enzymatic breakdown of the cell-wall peptidoglycans of intestinal bacteria, which contain D-glutamic and other D-amino acids (page 653).

The reduced forms of the L- and D-amino acid oxidases can react directly with molecular oxygen to form hydrogen peroxide and regenerate the oxidized forms of the enzymes:

$$E\text{—FMNH}_2 + O_2 \longrightarrow E\text{—FMN} + H_2O_2$$
$$E\text{—FADH}_2 + O_2 \longrightarrow E\text{—FAD} + H_2O_2$$

These and other properties of flavin-linked oxidases are given elsewhere (page 486). Hydrogen peroxide is decomposed to water and oxygen by *catalase*, a heme-containing enzyme (page 504):

$$H_2O_2 \longrightarrow H_2O + \tfrac{1}{2}O_2$$

In eukaryotic cells both L- and D-amino acid oxidases, as

Figure 21-6
Microbody (peroxisome) in a rat-liver cell.
Note the single surrounding membrane. The
regular latticelike structure in the center of
the microbody consists of crystalline urate
oxidase.

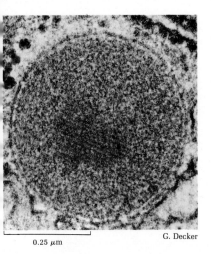

0.25 μm   G. Decker

Figure 21-7
Pathways to acetyl-CoA via pyruvic acid.

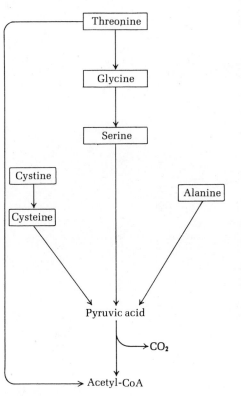

well as <u>urate oxidase</u> (page 741), are localized in the <u>mi-crobodies</u> (page 33), which may be regarded as specialized oxidative organelles. They also contain the catalase required to catalyze the decomposition of the hydrogen peroxide generated by these oxidases. For this reason microbodies are also called <u>peroxisomes.</u> An electron micrograph of a liver microbody is shown in Figure 21-6.

With this survey of the important transamination and oxidative deamination processes as background, we shall now examine the pathways of degradation of the 20 common amino acids.

Pathways Leading to Acetyl-CoA

Figure 21-1 shows the portals through which the carbon skeletons of the 20 different amino acids gain entry into the tricarboxylate cycle. In subsequent illustrations the pathways for all the amino acids are separated according to point of entry and given as a series of metabolic maps. Not all the enzymes and intermediates in these 20 pathways will be discussed in detail; however, reactions noteworthy for their mechanisms or biological properties will be emphasized.

The major point of entry into the tricarboxylate cycle is via acetyl-CoA; 11 amino acids enter by this route. Of these, five (alanine, glycine, serine, threonine, and cysteine) are degraded to acetyl-CoA via pyruvate, five (phenylalanine, tyrosine, leucine, lysine, and tryptophan) are degraded via acetoacetyl-CoA, and two (threonine and leucine) yield acetyl-CoA directly. (Threonine yields two molecules of acetyl-CoA, one via pyruvate and the other directly.) Leucine and tryptophan yield both acetoacetyl-CoA and acetyl-CoA as end products.

Pathways to Pyruvate (Alanine, Glycine, Serine, and Cysteine)

The four amino acids that lead to formation of pyruvate may either be converted into glucose and thus are <u>glycogenic</u> (page 629) or be oxidized via the tricarboxylic acid cycle (Figure 21-7).

<u>Alanine</u> yields pyruvate directly on transamination with α-ketoglutarate:

$$\text{Alanine} + \alpha\text{-ketoglutarate} \rightleftharpoons \text{pyruvate} + \text{glutamate}$$

<u>Glycine</u> has two pathways of degradation. The major route, which does not lead to acetyl-CoA, is by reversible oxidative cleavage to form CO_2, ammonia, and N^5,N^{10}-methylenetetrahydrofolate (page 345), by the action of <u>glycine synthase:</u>

$$\text{Glycine} + FH_4 + NAD^+ \rightleftharpoons$$
$$N^5,N^{10}\text{-methylene-}FH_4 + CO_2 + NH_3 + NADH + H^+$$

where FH_4 denotes tetrahydrofolate. Glycine may also be converted to serine by serine hydroxymethyltransferase, a pyridoxal phosphate enzyme. Serine in turn is dehydrated and deaminated to pyruvate by <u>serine dehydratase,</u> also a pyridoxal phosphate enzyme (Figure 21-8). Note that serine may also be degraded to glycine and N^5,N^{10}-methylene-tetrahydrofolate through the reverse action of serine hydroxymethyltransferase, which is reversible under intracellular conditions.

Cysteine degradation may proceed by at least three different pathways, as shown in Figure 21-9, depending on the nature of the last sulfur-containing product. The conversion of _cysteine_ to pyruvate (Figure 21-9) via cysteine sulfinic acid is catalyzed by an enzyme containing pyridoxal phosphate.

Pathways to Acetoacetyl-CoA (Phenylalanine, Tyrosine, Leucine, Lysine, and Tryptophan)

The pathway to acetoacetyl-CoA and thence to acetyl-CoA is taken by the amino acids _phenylalanine, tyrosine, leucine, lysine,_ and _tryptophan_ (Figure 21-10). The steps in the degradation of _phenylalanine_ and _tyrosine_ are shown in Figure 21-11. Both are _ketogenic_ amino acids since they yield free acetoacetate. However, the latter may be activated at the expense of succinyl-CoA to form acetoacetyl-CoA (page 550). One of the five remaining carbon atoms of these two amino acids appears as CO_2, following oxidative decarboxylation of the intermediate _p-hydroxyphenylpyruvic acid_. The other four carbon atoms of tyrosine and phenylalanine are recovered as fumaric acid, an intermediate of the tricarboxylic

Figure 21-8
Conversion of glycine to pyruvate via serine, a minor pathway. Serine hydroxymethyltransferase also acts on threonine (page 574).

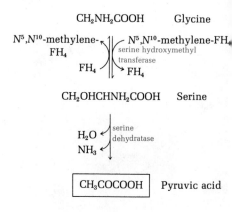

Figure 21-9
Conversion of cystine and cysteine to pyruvic acid. The route taken varies in different organisms.

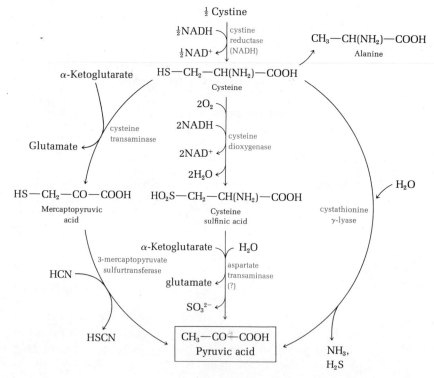

Figure 21-10
Pathways to acetyl-CoA via acetoacetyl-CoA.

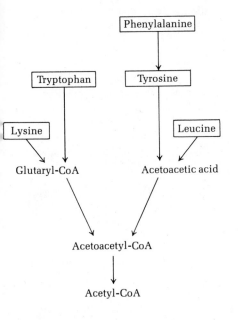

acid cycle. These two amino acids are thus *both* ketogenic and glycogenic.

Several enzymes in the phenylalanine-tyrosine pathway deserve special mention. *Phenylalanine 4-monooxygenase*, also called *phenylalanine hydroxylase*, is representative of the *monooxygenases* or *mixed-function oxygenases* (page 501). It incorporates one oxygen atom from molecular oxygen into phenylalanine to yield the p-hydroxyl group; the other oxygen atom is reduced to water. The reducing agent is NADPH, and the reaction takes place in two steps:

$$NADPH + H^+ + dihydrobiopterin \longrightarrow$$
$$NADP^+ + tetrahydrobiopterin$$

$$\text{L-Phenylalanine} + tetrahydrobiopterin + O_2 \longrightarrow$$
$$\text{L-tyrosine} + dihydrobiopterin + H_2O$$

The electron donor in this reaction is *tetrahydrobiopterin*, the reduced form of *dihydrobiopterin* (Figure 21-12), a reduced pteridine derivative related to folic acid (page 345). Tetrahydrobiopterin functions as a coenzyme in carrying reducing equivalents from NADPH, the ultimate electron donor, to the electron acceptor, one of the oxygen atoms of O_2.

Phenylalanine 4-monooxygenase is absent in about 1 in every 10,000 human beings owing to a recessive mutant gene. In the absence of this enzyme, a secondary pathway of phenylalanine metabolism that is normally little used comes

Figure 21-11
Conversion of phenylalanine and tyrosine to acetoacetic and fumaric acids.

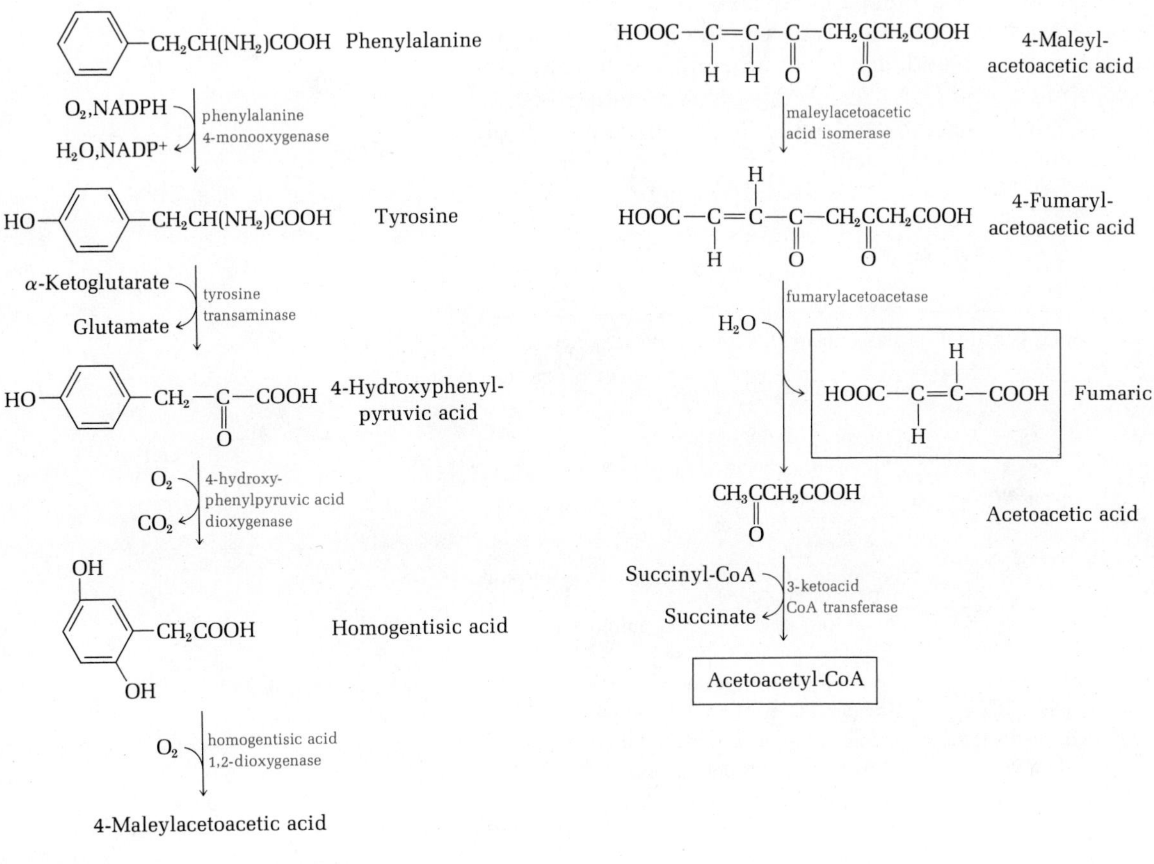

into play. In this minor pathway, phenylalanine undergoes transamination with α-ketoglutarate to yield phenylpyruvic acid, which accumulates in the blood and is excreted in the urine. In childhood excess circulating phenylpyruvate impairs normal brain development, causing severe mental retardation. This condition, _phenylketonuria_, often referred to as PKU, was among the first genetic defects of metabolism recognized in man. Restriction of dietary phenylalanine during childhood prevents the mental retardation. Many other genetic defects in amino acid metabolism have been found in human beings (Table 21-1).

4-Hydroxyphenylpyruvic acid dioxygenase, which catalyzes oxidation of hydroxyphenylpyruvic acid to _homogentisic acid_ (Figure 21-11), contains copper (page 493). This oxidation step is very complex and involves hydroxylation of the phenyl ring and decarboxylation, oxidation, and migration of the side chain. The next step is catalyzed by _homogentisic acid 1,2-dioxygenase_, an Fe(II) enzyme that requires reduced glutathione (page 96); its reaction mechanism is also complex. Both these reactions require vitamin C or ascorbic acid (Chapter 13) for maximum activity in vivo. Dietary deficiency of this vitamin in guinea pigs causes urinary excretion of phenylpyruvic and homogentisic acids. The urine of people genetically defective in homogentisic acid 1,2-dioxygenase contains homogentisic acid, which when made alkaline and exposed to oxygen, turns dark because it is oxidized and polymerized to a black melanin pigment. This condition is known as _alkaptonuria_. Patients with this condition have abnormal pigmentation of connective tissue.

Figure 21-13 shows the pathway by which four carbon atoms of _leucine_ are converted into acetoacetyl-CoA; leucine is thus ketogenic. The other two carbon atoms of leucine are converted into acetyl-CoA. Following transamination of the α-amino group of leucine and oxidative decarboxylation of the corresponding α-keto acid, _isovaleryl-CoA_ is formed.

Figure 21-12
Structure of tetrahydrobiopterin and dihydrobiopterin. Dihydrobiopterin is shown as the p-quinonoid tautomer, the form most probably involved in the phenylalanine 4-monooxygenase reaction.

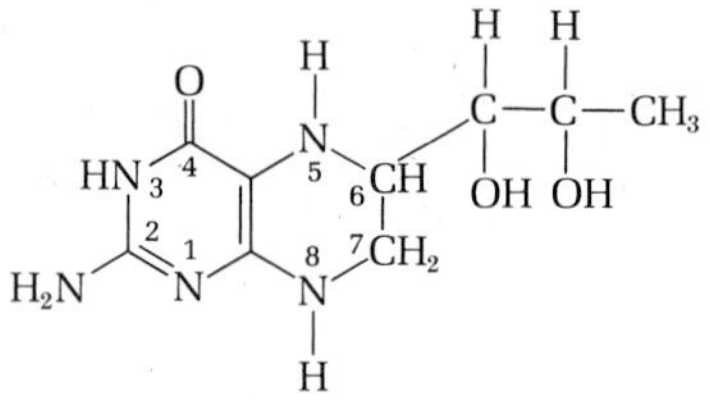

Tetrahydrobiopterin

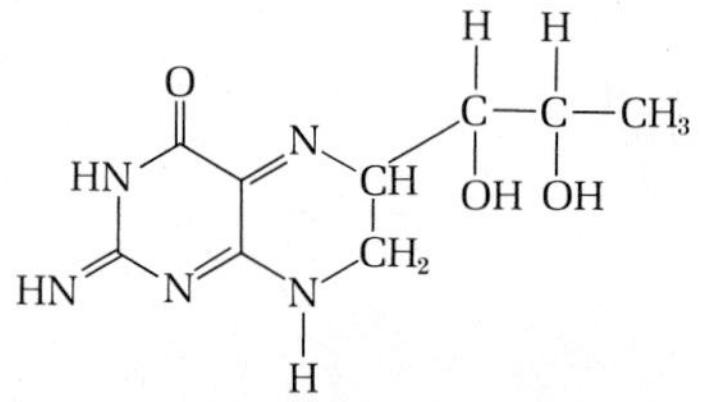

p-Quinonoid dihydrobiopterin

Table 21-1 Some genetic disorders in man affecting amino acid metabolism

Name	Defective enzyme or process
Albinism	Tyrosine 3-monooxygenase
Alkaptonuria	Homogentisic acid 1,2-dioxygenase
Argininosuccinic acidemia	Argininosuccinate lyase
Cystinosis	Storage and/or release of cystine from lysosomes
Cystinuria	Renal and intestinal transport of cystine and certain other amino acids
Hartnup's disease	Renal transport of neutral amino acids
Histidinemia	Histidine ammonia-lyase
Homocystinuria	Cystathionine β-synthase
Isovaleric acidemia	Isovaleryl-CoA dehydrogenation
Maple syrup urine disease	Branched-chain α-keto acid dehydrogenases
Phenylketonuria	Phenylalanine 4-monooxygenase
Hypervalinemia	Valine transaminase

Figure 21-13
Conversion of leucine to acetyl-CoA and acetoacetic acid.

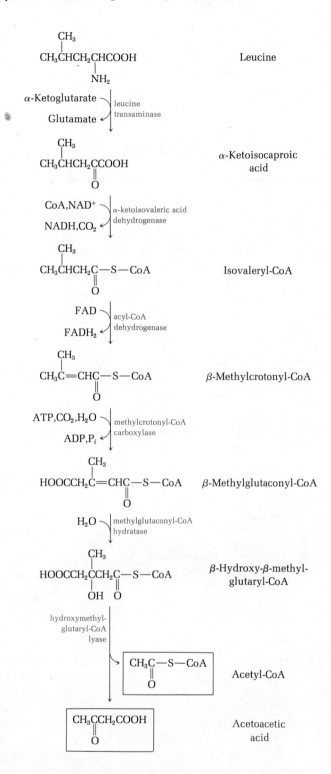

After its dehydrogenation, addition of a carboxyl group, and hydration, the resulting six-carbon *β-hydroxy-β-methylglutaryl-CoA* (page 554) is cleaved to yield one molecule of acetyl-CoA and one of acetoacetate, which is in turn converted into acetoacetyl-CoA by reaction with succinyl-CoA (page 554). The intermediate β-hydroxy-β-methylglutaryl-CoA formed during leucine degradation is also an important precursor in the biosynthesis of cholesterol (page 681).

Figure 21-14
Conversion of lysine to acetoacetyl-CoA. There are two alternate routes from lysine to α-aminoadipic semialdehyde. The route via the intermediate saccharopine predominates in the liver.

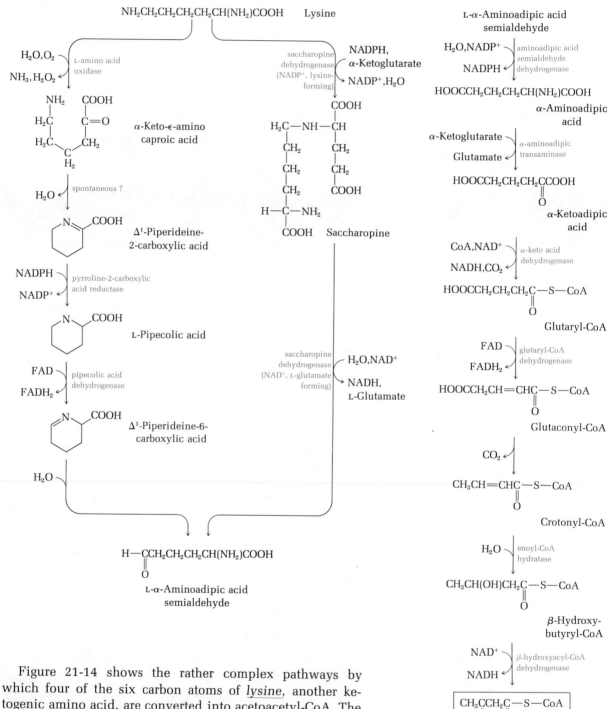

Figure 21-14 shows the rather complex pathways by which four of the six carbon atoms of *lysine*, another ketogenic amino acid, are converted into acetoacetyl-CoA. The other two carbon atoms are lost in decarboxylation reactions. Lysine does not undergo transamination. In one pathway lysine first condenses with α-ketoglutarate to yield *saccharopine*, which is ultimately converted into acetoacetyl-CoA. In the other pathway the α-amino group of

Figure 21-15
Conversion of tryptophan to acetyl-CoA and acetoacetyl-CoA.

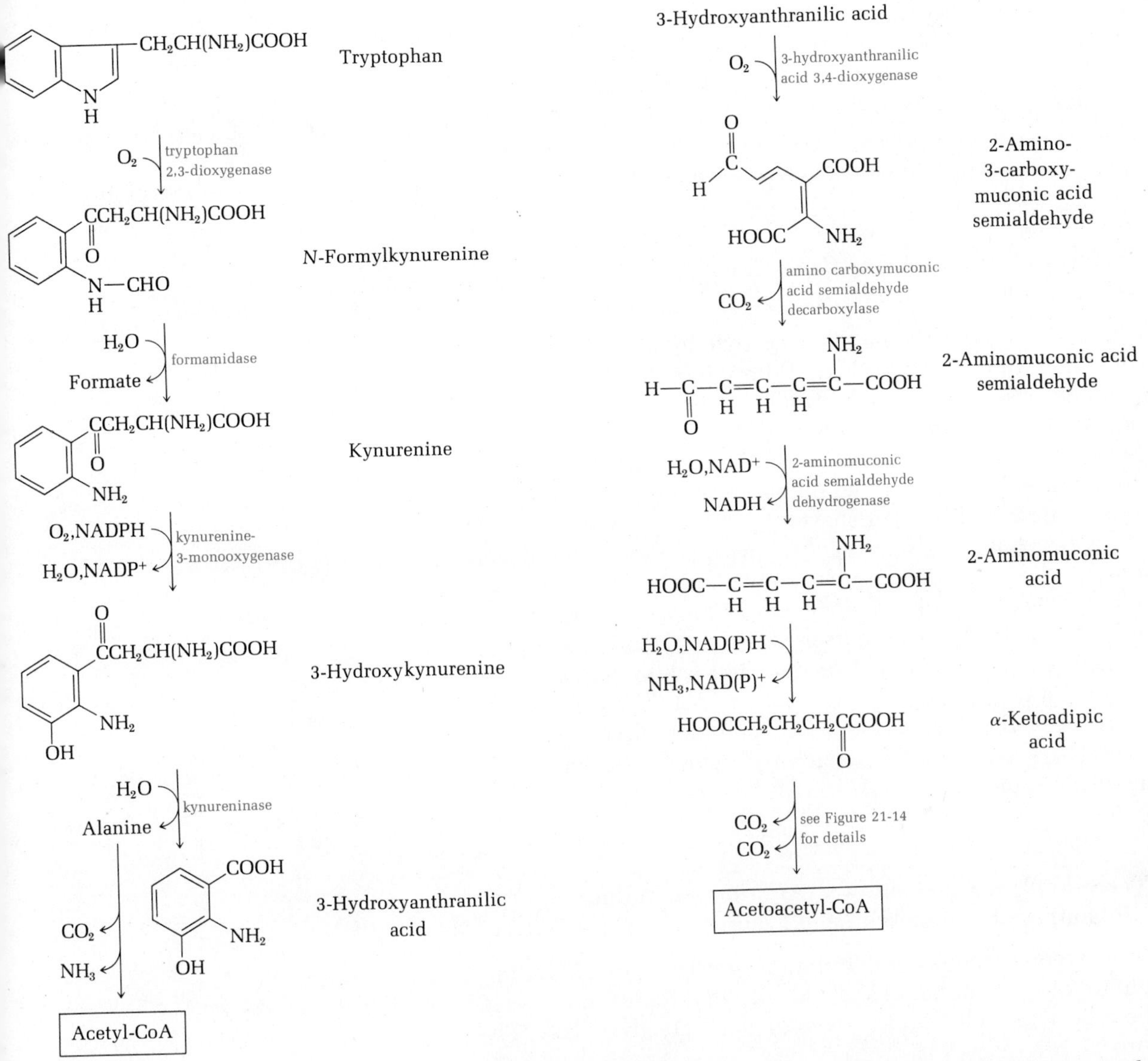

lysine is oxidized, presumably by L-amino acid oxidase. Both pathways converge into formation of α-aminoadipic semialdehyde.

Figure 21-15 shows the important steps in the pathway by which four of the eleven carbon atoms of *tryptophan* are converted into acetoacetyl-CoA and two into acetyl-CoA; the remainder appear as four molecules of CO_2 and one of formate. The first step is catalyzed by *tryptophan 2,3-dioxygenase*, also called *tryptophan pyrrolase*. This enzyme, which contains both copper and heme groups, employs molecular oxygen (page 501) to oxidize tryptophan to *N-formyl-L-kynurenine*. The enzyme is sometimes genetically defective in man, giving rise to mental retardation. The later intermediate 3-hydroxykynurenine is utilized by some insects as a precursor of the pigments known as *ommochromes*. The enzyme *kynureninase*, which catalyzes the cleavage of 3-hydroxykynurenine to yield alanine and 3-

hydroxyanthranilic acid, contains pyridoxal phosphate. In deficiency of vitamin B_6 in mammals, large amounts of L-kynurenine are excreted in the urine. The intermediate 3-hydroxyanthranilic acid (Figure 21-15) also serves as a precursor in the biosynthesis of the vitamin nicotinic acid (page 340). Intermediates in tryptophan catabolism are precursors for the biosynthesis of other important substances, including serotonin (5-hydroxytryptamine), a neurotransmitter substance and vasoconstrictor, and the plant hormone indoleacetic acid (Figure 21-16, see also page 717).

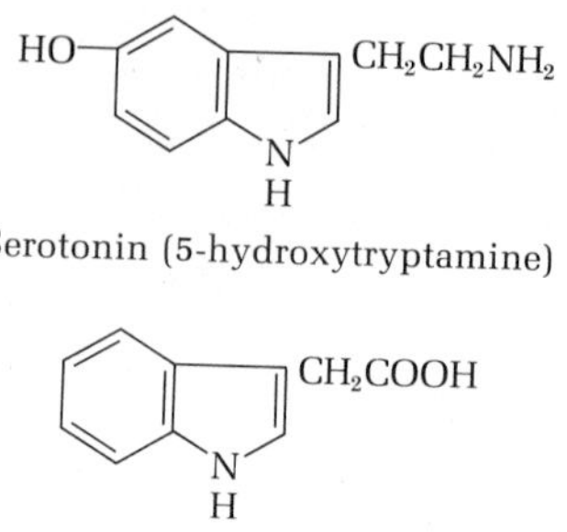

Pathways Direct to Acetyl-CoA (Threonine, Leucine, Tryptophan, and Isoleucine)

Threonine (four carbon atoms) has two possible pathways. In the first, threonine is cleaved into two two-carbon compounds, acetaldehyde and glycine, by serine hydroxymethyltransferase, which is capable of aldol cleavage of either serine or threonine (Figure 21-17). The acetaldehyde formed from threonine is converted into acetyl-CoA; the degradation of glycine was discussed above.

In a second, less important pathway for threonine degradation, threonine dehydratase converts threonine into α-ketobutyric acid, which undergoes oxidative decarboxylation to propionyl-CoA, a precursor of succinyl-CoA (page 555).

It was noted above that leucine and tryptophan fragment in such a way during their catabolism that they yield acetyl-CoA as well as acetoacetyl-CoA. Another direct source of acetyl-CoA is isoleucine, which is fragmented to yield succinyl-CoA and acetyl-CoA, as will be seen below.

The α-Ketoglutarate Pathway (Arginine, Histidine, Glutamine, Glutamic Acid, and Proline)

The carbon skeletons of five amino acids (arginine, histidine, glutamine, glutamic acid, and proline) enter the tricarboxylic acid cycle via α-ketoglutarate; all are glycogenic (Figure 21-18).

The pathway for arginine shown in Figure 21-19 is that occurring in the liver of mammals. Arginine is converted into ornithine by the action of arginase; this step is also employed in the synthesis of urea via the urea cycle (below). Ornithine is then converted into glutamic acid semialdehyde, which is also an intermediate in the oxidation of proline (see below).

The pathway for oxidation of histidine to glutamic acid (Figure 21-20) is noteworthy because the imidazole ring is opened to yield N-formiminoglutamic acid, from which the formimino group (HN=CH—) is removed by an enzyme using tetrahydrofolate as a one-carbon group acceptor (page 345).

Glutamine is hydrolyzed to glutamic acid by the enzyme glutaminase, particularly in the kidney (page 583):

$$\text{Glutamine} + H_2O \longrightarrow \text{glutamic acid} + NH_3$$

Figure 21-16
Two biologically active derivatives of tryptophan (see page 716).

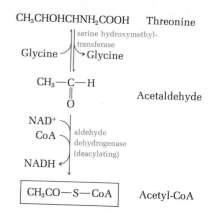

Serotonin (5-hydroxytryptamine)

Indoleacetic acid

Figure 21-17
Pathway leading from threonine to acetyl-CoA.

CH₃CHOHCHNH₂COOH   Threonine

Glycine — serine hydroxymethyl-transferase → Glycine

$$CH_3-\overset{\displaystyle O}{\underset{\displaystyle \|}{C}}-H$$ Acetaldehyde

NAD⁺
CoA — aldehyde dehydrogenase (deacylating)
NADH

CH₃CO—S—CoA   Acetyl-CoA

Figure 21-18
Pathways to α-ketoglutarate.

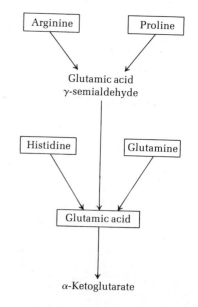

Figure 21-19
Major mammalian pathway for the conversion of arginine to glutamic acid.

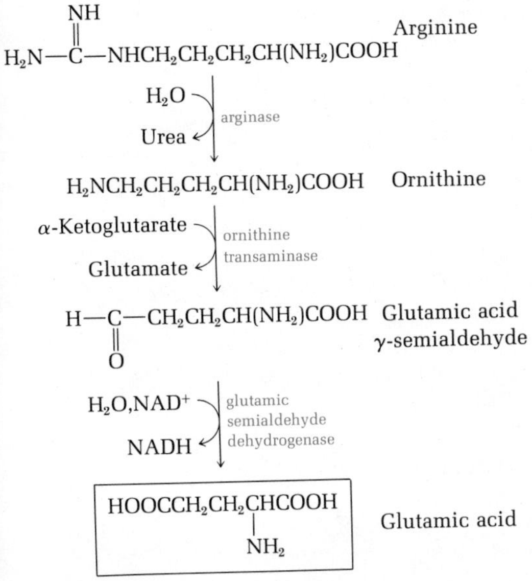

$$H_2N-\overset{\overset{\displaystyle NH}{\|}}{C}-NHCH_2CH_2CH_2CH(NH_2)COOH \quad \text{Arginine}$$

H₂O ⤵
        arginase
Urea ⤶

$$H_2NCH_2CH_2CH_2CH(NH_2)COOH \quad \text{Ornithine}$$

α-Ketoglutarate ⤵
                ornithine
                transaminase
Glutamate ⤶

$$H-\overset{\overset{\displaystyle}{}}{\underset{\underset{\displaystyle O}{\|}}{C}}-CH_2CH_2CH(NH_2)COOH \quad \begin{array}{l}\text{Glutamic acid}\\ \gamma\text{-semialdehyde}\end{array}$$

H₂O, NAD⁺ ⤵ glutamic
              semialdehyde
NADH ⤶   dehydrogenase

$$\boxed{HOOCCH_2CH_2CHCOOH \atop \qquad\quad NH_2} \quad \text{Glutamic acid}$$

Figure 21-20
The major mammalian pathway for the conversion of histidine to glutamic acid. FH₄ symbolizes tetrahydrofolate, which serves as acceptor for the formimino group.

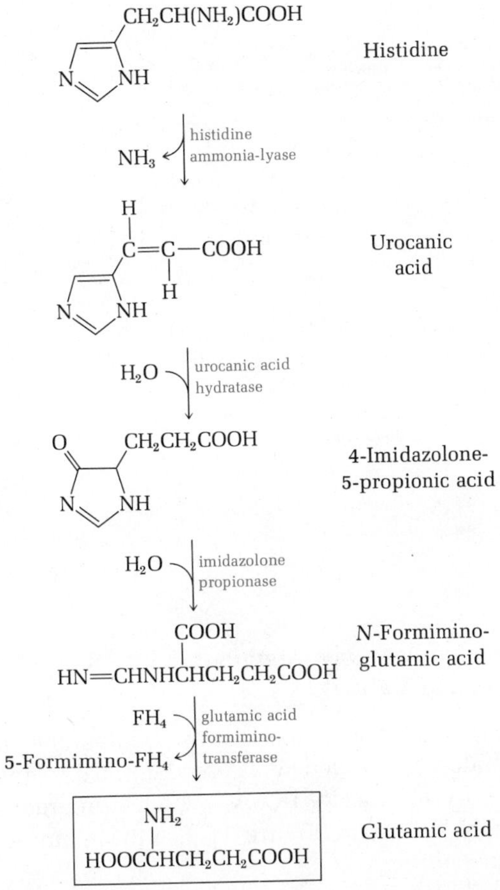

$$\text{Histidine}$$

CH₂CH(NH₂)COOH on imidazole ring

histidine
ammonia-lyase
NH₃ ⤶

$$\text{Urocanic acid}$$

H₂O ⤵ urocanic acid
        hydratase

$$\text{4-Imidazolone-5-propionic acid}$$

H₂O ⤵ imidazolone
        propionase

$$HN{=}CHNH\overset{\overset{\displaystyle COOH}{|}}{C}HCH_2CH_2COOH \quad \begin{array}{l}\text{N-Formimino-}\\ \text{glutamic acid}\end{array}$$

FH₄ ⤵ glutamic acid
        formimino-
5-Formimino-FH₄ ⤶ transferase

$$\boxed{\overset{\overset{\displaystyle NH_2}{|}}{HOOCCHCH_2CH_2COOH}} \quad \text{Glutamic acid}$$

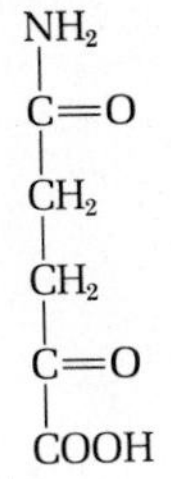

Glutamine is also converted into glutamic acid by *glutamate synthase*:

$$\text{Glutamine} + \alpha\text{-ketoglutarate} + NADPH + H^+ \longrightarrow$$
$$2 \text{ glutamate } + NADP^+$$

In a third pathway glutamine may undergo transamination with α-ketoglutaric acid to yield *α-ketoglutaramic acid* (Figure 21-21), which in turn either hydrolyzes to α-ketoglutarate and ammonia or cyclizes to form a lactam, 2-hydroxy-5-oxoproline (page 795).

L-*Proline*, after dehydrogenation, undergoes ring opening to yield L-glutamic acid semialdehyde, which is reduced to L-glutamic acid (Figure 21-22). 4-*Hydroxyproline*, a component of collagen, takes a different pathway via γ-hydroxyglutamic acid to alanine and glycine, which are degraded by pathways described earlier.

Figure 21-21
α-Ketoglutaramic acid.

$$\begin{array}{c} NH_2 \\ | \\ C{=}O \\ | \\ CH_2 \\ | \\ CH_2 \\ | \\ C{=}O \\ | \\ COOH \end{array}$$

Figure 21-22
Catabolism of proline and hydroxyproline.

Conversion of proline to glutamic acid.

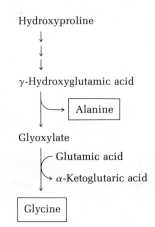

Proline

Δ¹-Pyrroline 5-carboxylic acid

Glutamic acid γ-semialdehyde

HOOCCH₂CH₂CH(NH₂)COOH   Glutamic acid

Catabolism of hydroxyproline. The reactions from hydroxyproline to γ-hydroxyglutamic acid are analogous to those of proline catabolism (left). Subsequent metabolism of γ-hydroxyglutamic acid takes a somewhat different course, to yield alanine and glycine.

Hydroxyproline
↓
↓
↓
γ-Hydroxyglutamic acid
↘ → Alanine

Glyoxylate
⌐ Glutamic acid
⌊→ α-Ketoglutaric acid

Glycine

The Succinate Pathway (Methionine, Isoleucine, and Valine)

The carbon skeletons of <u>methionine, isoleucine,</u> and <u>valine</u> are ultimately degraded <u>via propionyl-CoA</u> and <u>methylmalonyl-CoA</u> to succinyl-CoA, which undergoes deacylation to yield succinate (Figure 21-23); these amino acids are thus glycogenic.

<u>Methionine</u> (Figure 21-24) loses its methyl group to yield <u>homocysteine</u> in a sequence of three important reactions, involving <u>S-adenosylmethionine</u> as intermediate, to be described in Chapter 25 (page 713). Homocysteine combines with serine to yield <u>cystathionine,</u> which undergoes breakdown to cysteine, ammonia, and α-ketobutyrate. The latter undergoes oxidative decarboxylation to become propionyl-CoA. Carboxylation of propionyl-CoA yields D-methylmalonyl-CoA, which is converted to L-methylmalonyl-CoA by methylmalonyl-CoA racemase. Rearrangement of the resulting L form to succinyl-CoA then follows in a process involving the coenzyme B₁₂-dependent methylmalonyl-CoA mutase (page 555). In this way three carbon atoms of methionine are converted into succinate.

<u>Isoleucine</u> and <u>valine</u> have rather similar patterns of degradation (Figure 21-25). Both undergo transamination followed by oxidative decarboxylation of the resulting α-keto acids. The branched chains of the latter are then degraded in a parallel manner. Propionyl-CoA is formed from both valine and isoleucine. After carboxylation of propionyl-CoA the methylmalonyl-CoA formed is converted into succinyl-CoA, as described above. Thus, three of the carbon atoms of isoleucine and of valine are converted into succinate.

Figure 21-23
Pathways from methionine, isoleucine, and valine to succinyl-CoA.

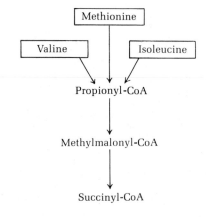

Figure 21-24
Conversion of methionine to succinyl-CoA.

Figure 21-25
Conversions of valine to succinyl-CoA and isoleucine to succinyl-CoA and acetyl-CoA. Note the similarities between these two pathways and that for the catabolism of leucine (Figure 21-13).

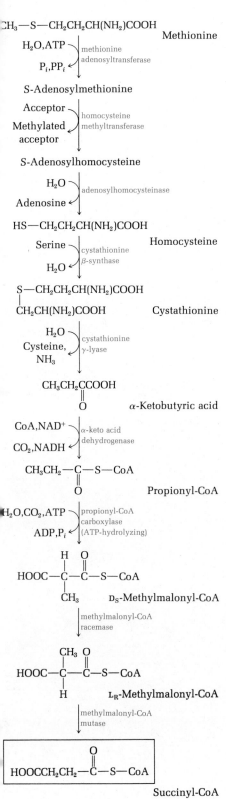

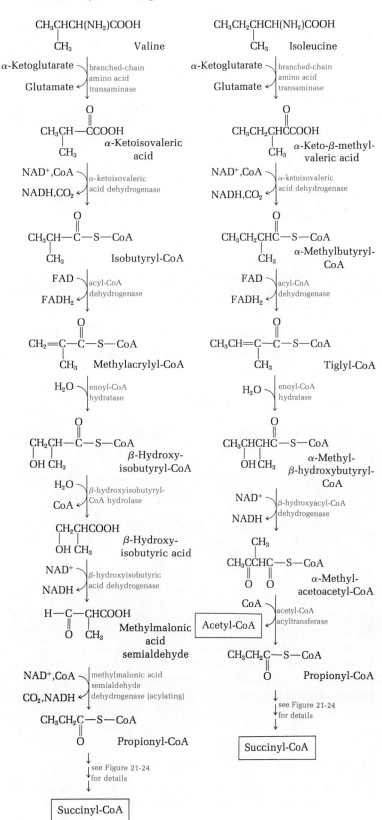

It is noteworthy that the oxidative decarboxylations of α-ketoisovalerate, α-keto-β-methylvalerate, and α-keto-iso-caproate (the deamination products of valine, isoleucine, and leucine, respectively) are catalyzed by the same enzyme, *α-ketoisovaleric acid dehydrogenase*. This enzyme is genetically defective in some people, leading to the excretion of these α-keto acids in the urine. This rare condition, which causes severe mental retardation, is called *maple syrup urine disease*, because of the characteristic odor imparted to the urine by these keto acids. It is remarkable that several of the heritable genetic defects involving amino acid metabolism in man lead to mental retardation, which appears in some cases to be caused by failure of certain nerve bundles to become myelinated.

The Fumarate Pathway (Phenylalanine and Tyrosine)

This pathway is taken by four of the nine carbon atoms of *phenylalanine* and *tyrosine*. As pointed out above, four carbon atoms of these amino acids enter the tricarboxylic acid cycle via acetoacetyl-CoA and acetyl-CoA. Four of the remaining five carbon atoms are converted to fumarate by the pathway shown in Figure 21-11.

The Oxaloacetate Pathway (Asparagine and Aspartic Acid)

Asparagine is first hydrolyzed to aspartic acid and ammonia by *asparaginase*:

$$\text{Asparagine} + H_2O \longrightarrow \text{aspartic acid} + NH_3$$

Asparaginase has wide distribution in animal and plant tissues. Injection of asparaginase into the blood has been found to be effective in controlling leukemia in some patients; presumably the enzyme functions by limiting the availability of asparagine to the malignant white blood cells (page 193).

Aspartic acid undergoes transamination with α-ketoglutaric acid to form oxaloacetic acid:

$$\text{Aspartic acid} + \alpha\text{-ketoglutaric acid} \rightleftharpoons$$
$$\text{oxaloacetic acid} + \text{glutamic acid}$$

In this way all four carbon atoms of these two amino acids can enter the tricarboxylic acid cycle; they are glycogenic amino acids. In plants and some microorganisms aspartate undergoes direct elimination of NH_3 to yield fumarate, catalyzed by *aspartate ammonia-lyase* (page 218), also called aspartase, which is not present in animal tissues:

$$\text{Aspartate} \longrightarrow \text{fumarate} + NH_3$$

Decarboxylation of Amino Acids

A few of the amino acids undergo decarboxylation in animal tissues to yield the corresponding primary amines, which have special biological functions. For example, histidine is decarboxylated by *histidine decarboxylase*, a pyridoxal phosphate enzyme, to yield *histamine,* a potent vasodilator which is released in certain tissues as a result of allergic hypersensitivity or inflammation:

$$\text{Histidine} \longrightarrow \text{histamine} + CO_2$$

Tryptophan is decarboxylated in a similar manner to yield *tryptamine,* a precursor of the plant-growth hormone *indoleacetic acid* (page 717).

Arginine is decarboxylated to yield *agmatine,* $NH_2—C—NH—(CH_2)_4NH_2$, a precursor of *spermine* and *spermidine* (page 716).

$\overset{||}{\underset{NH}{}}$

Formation of Nitrogenous Excretion Products

Most higher organisms tend to salvage and reuse ammonia derived from the catabolism of amino acids by reversing the glutamate dehydrogenase reaction

$$\alpha\text{-Ketoglutarate} + NH_3 + NADH(NADPH) + H^+ \rightleftharpoons$$
$$\text{glutamate} + NAD^+(NADP^+) + H_2O$$

However, a certain fraction of the ammonia formed from amino acids is ultimately excreted by vertebrates in one of three forms: urea, ammonia itself, or uric acid. Amino nitrogen is excreted by most terrestrial vertebrates as urea; these organisms are termed *ureotelic.* Most aquatic animals, e.g., the teleost fishes, excrete amino nitrogen as ammonia; they are termed *ammonotelic.* Ammonia is a rather toxic compound, as we shall see (page 583). Fishes can readily excrete ammonia into their aqueous environment, but terrestrial vertebrates have evolved the ability to excrete nitrogen as the nontoxic compound urea. Birds and land-dwelling reptiles, whose water intake is very limited, excrete amino nitrogen in a semisolid form as suspensions of solid uric acid; these organisms are termed *uricotelic.* The amphibia occupy a midposition. The tadpole, which is aquatic, excretes ammonia. After metamorphosis, during which the liver acquires the necessary enzymes, the adult frog forms and excretes urea.

The Urea Cycle

Urea formation, which takes place in the liver of ureotelic organisms, is brought about by the *urea cycle,* a cyclic pathway first postulated by H. A. Krebs and K. Henseleit in 1932 (Figure 21-26). They deduced the outlines of the urea cycle starting from their observation that addition of small amounts of ornithine and arginine catalytically stimulated the production of urea from ammonia by liver slices. Subsequent research in the laboratories of S. Ratner and P. P. Cohen established the details of the enzymatic steps in urea synthesis. In this sequence (Figure 21-26) two amino groups, originally derived from α-amino acids, and one molecule of carbon dioxide enter and, through a cyclic process requiring input of ATP, give rise to one molecule of urea, a neutral

Figure 21-26
The urea cycle, showing the compartmentation of its steps in the cytosol and mitochondrion.

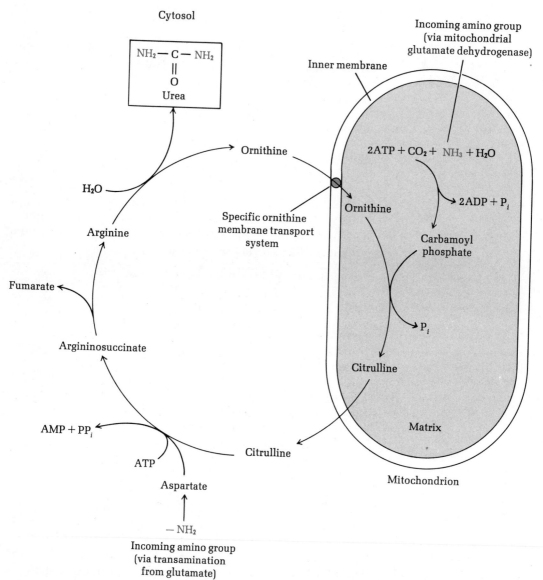

nontoxic compound which is transported via the blood to the kidneys and excreted in the urine.

The first amino group entering the urea cycle arises as free ammonia following the oxidative deamination of glutamate in liver mitochondria:

Glutamic acid + NAD$^+$ (NADP$^+$) + H$_2$O $\rightleftharpoons$
$\quad$ α-ketoglutaric acid + NH$_3$ + NADH (NADPH) + H$^+$

The free ammonia so formed is then utilized, together with carbon dioxide, to form *carbamoyl phosphate* (Figure 21-27), a very unstable compound, in a complex reaction catalyzed by *carbamoyl phosphate synthetase (ammonia)*, present in the mitochondrial matrix:

2ATP + CO$_2$ + NH$_3$ + H$_2$O $\longrightarrow$ carbamoyl phosphate + 2ADP + P$_i$

Figure 21-27
Structures of components of the carbamoyl phosphate synthetase (ammonia) reaction.

Carbamoyl phosphoric acid

N-Acetylglutamate

Figure 21-28
Conversion of ornithine to citrulline. The amino group introduced by carbamoyl phosphate is in color.

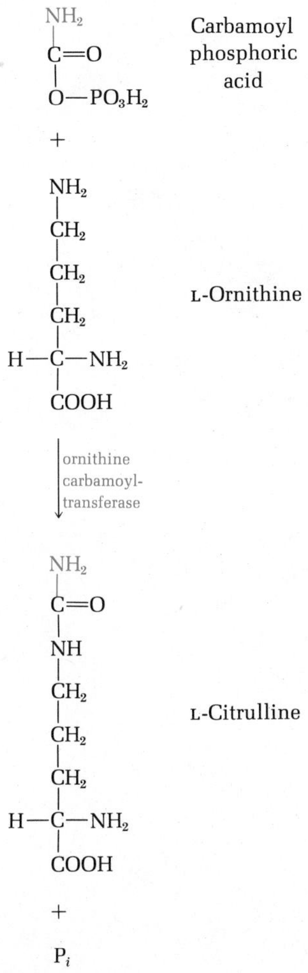

NH₂
|
C=O   Carbamoyl
| phosphoric
O—PO₃H₂   acid

+

NH₂
|
CH₂
|
CH₂
| L-Ornithine
CH₂
|
H—C—NH₂
|
COOH

ornithine
carbamoyl-
transferase

NH₂
|
C=O
|
NH
|
CH₂
| L-Citrulline
CH₂
|
CH₂
|
H—C—NH₂
|
COOH

+

Pᵢ

Two molecules of ATP are required to form each molecule of carbamoyl phosphate in this reaction, which is essentially irreversible. This complex reaction, which occurs in at least two steps, requires *N-acetylglutamate* (Figure 21-27) as a stimulatory allosteric activator. The formation of carbamoyl phosphate by this pathway in the mitochondria is specialized for urea synthesis. However, in the cytosol of some tissues, as well as in bacteria and fungi, carbamoyl phosphate is made by a different reaction catalyzed by a different enzyme, *carbamoyl phosphate synthetase (glutamine)* (page 735).

The carbamoyl phosphate generated in the mitochondria now donates its carbamoyl group to ornithine, which is formed in the cytosol but enters the mitochondrion via a specific inner-membrane transport system (page 529). The product is *citrulline*:

$$\text{Carbamoyl phosphate} + \text{ornithine} \longrightarrow \text{citrulline} + P_i$$

This reaction (Figure 21-28) is catalyzed by *ornithine carbamoyltransferase* of the mitochondrial matrix. The citrulline formed now leaves the mitochondrial matrix and passes to the cytosol, where the remaining reactions of the urea cycle take place.

The second amino group required for urea synthesis now arrives in the form of aspartate, which in turn acquired it from glutamate by the action of aspartate transaminase in the cytosol. The amino group of aspartate condenses reversibly with the carbamoyl carbon atom of citrulline in the presence of ATP to form *argininosuccinate* (Figure 21-29); this reaction is catalyzed by *argininosuccinate synthetase*:

$$\text{Citrulline} + \text{aspartate} + \text{ATP} \rightleftharpoons \text{argininosuccinate} + \text{AMP} + PP_i$$

The pyrophosphate formed in this reaction is hydrolyzed by pyrophosphatase to inorganic phosphate, thus pulling the overall reaction to the right. In the next reaction argininosuccinate undergoes a β elimination reaction by the action of *argininosuccinate lyase* (Figure 21-30) to form free arginine and fumarate:

$$\text{Argininosuccinate} \longrightarrow \text{arginine} + \text{fumarate}$$

The arginine formed in this reaction becomes the immediate precursor of urea, whereas the fumarate returns to the pool of tricarboxylic acid cycle intermediates.

Up to this point the reaction sequence is that employed by all organisms capable of the biosynthesis of arginine. However, only ureotelic animals possess large amounts of *arginase*, which cleaves urea from arginine and regenerates ornithine, a reaction taking place in the cytosol:

$$\text{Arginine} + H_2O \longrightarrow \text{ornithine} + \text{urea}$$

Arginase has a molecular weight of 120,000 and contains four subunits, each of which has a tightly bound Mn^{2+} ion.

Figure 21-29
Formation of argininosuccinic acid. The amino groups that end up in urea are in color.

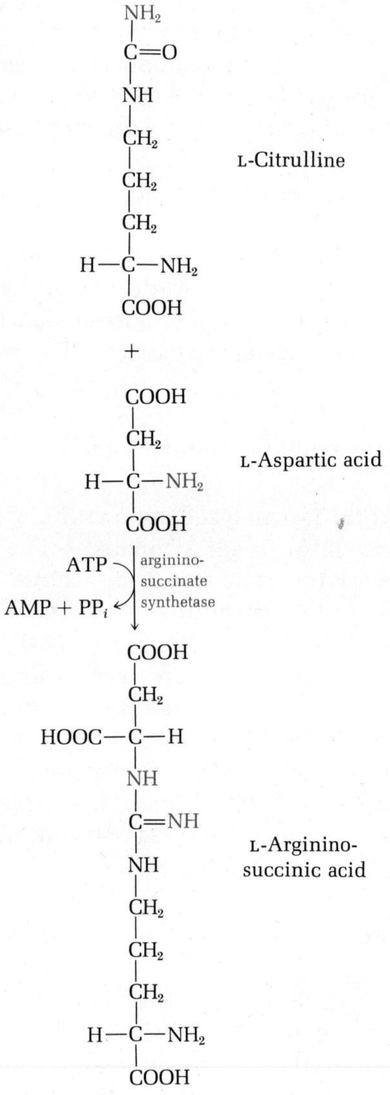

Figure 21-30
Formation and hydrolysis of arginine. The amino groups that end up in urea are in color.

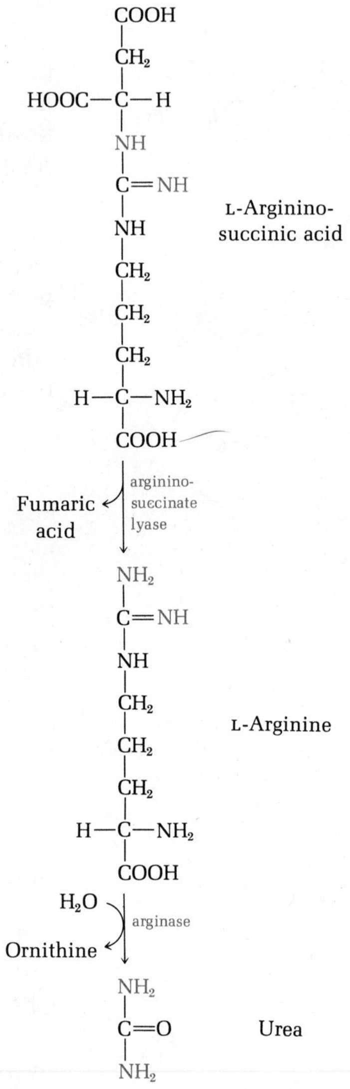

The overall equation of the urea cycle is

$$2NH_3 + CO_2 + 3ATP + 3H_2O \longrightarrow urea + 2ADP + AMP + 4P_i$$

The formation of one molecule of urea therefore requires the hydrolysis of four high-energy phosphate groups provided by ATP.

Genetic defects of each of the various enzymes of the urea cycle have been observed in man. Such patients are intolerant to ingestion of proteins; they show mental deficiency, retarded development of the nervous system, and excessive amounts of free ammonia in the blood. Some of these patients can be treated by replacing dietary protein with a mixture of α-keto acid analogs of the essential amino acids. The α-keto acids are converted into α-amino acids required for tissue protein synthesis, with removal of ammonia from the blood.

As pointed out above, some of the enzymes catalyzing reactions that feed amino groups into the urea cycle, particularly aspartate transaminase, glutamate dehydrogenase, carbamoyl phosphate synthetase (ammonia), and ornithine carbamoyltransferase, are localized in the mitochondria of the liver cell, thus providing a rather complex compartmentation of the reactions of amino acid catabolism and urea synthesis between the cytosol and the mitochondria (Figure 21-26). This separation appears to be necessary to prevent accumulation in the blood of free ammonia, which is exceedingly toxic to ureotelic vertebrates, particularly to the central nervous system. Ammonia is toxic because it leads to the reductive amination of α-ketoglutarate in mitochondria, catalyzed by glutamate dehydrogenase:

$$NH_3 + \alpha\text{-ketoglutarate} + NADH + H^+ \longrightarrow$$
$$\text{glutamate} + NAD^+ + H_2O$$

Since the equilibrium of this reaction is far to the right (page 481), ammonia effectively removes α-ketoglutarate from the tricarboxylic acid cycle and can cause severe inhibition of respiration in the brain, as well as excess ketone body formation from acetyl-CoA in the liver. The concentration of free ammonia in the liver is thus carefully regulated.

Ammonia Excretion

In ammonotelic animals, the amino groups derived from various α-amino acids are transaminated to form glutamate, which then undergoes oxidative deamination via glutamate dehydrogenase. The ammonia so formed is then converted into the amide group of *glutamine*, the transport form of ammonia in such organisms. Glutamine is formed from glutamate and ammonia at the expense of ATP energy by *glutamine synthetase*:

$$ATP + \text{glutamate} + NH_3 \longrightarrow \text{glutamine} + ADP + P_i$$

This enzyme plays an extremely important role in the intermediary metabolism of amino acids, since glutamine is not only a nontoxic transport form of ammonia but also functions as an amino-group donor in many biosynthetic reactions (pages 575, 712, and 731). The glutamine synthetases of animal tissues and bacteria, which differ significantly from each other, are described further in Chapter 25 (page 695). The enzymatic formation of glutamine apparently takes place via the intermediate formation of γ-*glutamyl phosphate* (Figure 21-31), which remains enzyme-bound before undergoing reaction with ammonia. Glutamine is the source of free ammonia in the urine formed in the kidney tubules of most vertebrates, but this reaction is especially prominent in ammonotelic animals. The reaction is a hydrolysis, catalyzed by *glutaminase*:

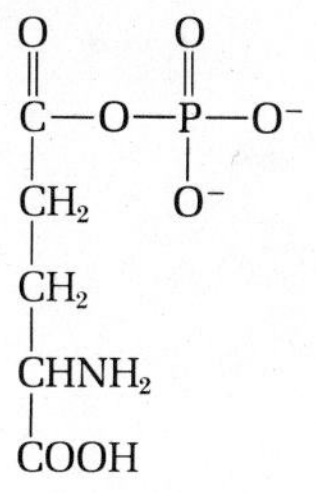

$$\text{Glutamine} + H_2O \longrightarrow \text{glutamate} + NH_3$$

The ammonia so formed enters the urine directly.

Figure 21-31
γ-Glutamyl phosphate.

Formation of Uric Acid

In uricotelic organisms (terrestrial reptiles and birds) uric acid is the chief form in which the amino groups of α-amino acids are excreted. Uric acid also happens to be the end product of purine metabolism in primates, birds, and terrestrial reptiles. The pathway of formation of uric acid from ammonia is complex, since the purine ring must first be formed from smaller precursors; it will be discussed in detail in Chapter 26 (page 740). The structure of uric acid is shown in Figure 21-32.

Urea, ammonia, and uric acid are not the only forms in which different species excrete nitrogen arising from amino acid catabolism. Spiders excrete amino nitrogen as guanine (page 741) instead of uric acid, and many fishes excrete nitrogen as trimethylamine oxide (Figure 21-33). In higher plants, glutamine and asparagine function in the transport and storage of amino groups.

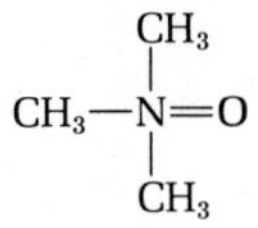

Summary

In the digestive tract of vertebrates, the proteolytic enzymes pepsin, trypsin, chymotrypsin, carboxypeptidase, and leucine aminopeptidase function to carry out complete hydrolysis of ingested proteins, yielding free amino acids, which are then absorbed into the blood for transport to the liver, where the major portion of amino acid catabolism takes place.

The carbon skeletons of amino acids undergo oxidative degradation to compounds that can enter the tricarboxylic acid cycle for oxidation. The amino groups of most of the L-amino acids are removed by transamination to α-ketoglutarate, yielding glutamate, which undergoes oxidative deamination by glutamate dehydrogenase. The amino groups of D-amino acids are removed by D-amino acid oxidase. There are five pathways by which carbon atoms of L-amino acids enter the tricarboxylate cycle: via (1) acetyl-CoA, (2) α-ketoglutarate, (3) succinate, (4) fumarate, and (5) oxaloacetate. The amino acids entering via acetyl-CoA are divided into two groups. The first, which includes alanine, threonine, glycine, serine, and cysteine, yields pyruvate (and is thus glycogenic) en route to acetyl-CoA, and the second group (phenylalanine, tyrosine, leucine, lysine, and tryptophan) yields acetoacetyl-CoA en route to acetyl-CoA. The amino acids arginine, histidine, glutamine, glutamic acid, and proline enter via α-ketoglutarate; methionine, isoleucine, and valine enter via succinate; four carbon atoms of phenylalanine and tyrosine enter via fumarate; and asparagine and aspartic acid enter via oxaloacetate. In man, a number of genetic defects in enzymes of amino acid catabolism occur. Many of these give rise to serious pathological disturbances. The pathways of amino acid catabolism are complex and have many intermediates which often serve as precursors of other important cell components.

In ureotelic animals (terrestrial mammals and adult amphibia), urea, formed by the urea cycle, is the final excretion product of amino nitrogen. Urea results from the action of arginase on arginine, the other cleavage product being ornithine. Arginine is resynthesized from ornithine by carbamoylation of the latter to citrulline at the expense of carbamoyl phosphate, followed by addition of an imino group to citrulline at the expense of aspartic acid. The urea cycle takes place in the liver. Ammonotelic animals (most fishes) excrete amino nitrogen as ammonia, which derives from the hydrolysis of glutamine. Glutamine is formed by the action of glutamine

Figure 21-32
Excretion of amino nitrogen as uric acid by birds and reptiles. The nitrogen atoms (color) of uric acid are derived from α-amino groups of amino acids. The pathway of formation of the purines and uric acid is given elsewhere (page 741).

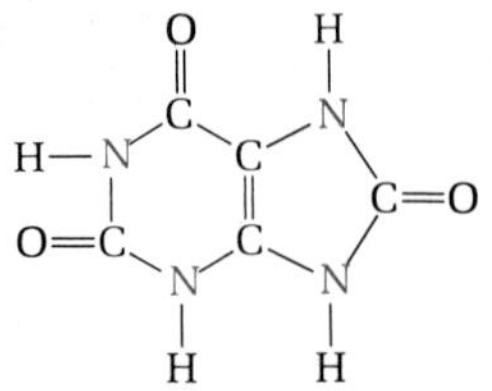

Uric acid (keto form)

Figure 21-33
Trimethylamine oxide.

synthetase from glutamic acid and ammonia derived from α-amino groups. Uricotelic animals (birds and terrestrial reptiles) excrete amino nitrogen as uric acid, a purine derivative.

References

Books

DAGLEY, S., and D. E. NICHOLSON: *Metabolic Pathways*, Wiley, New York, 1970.

GREENBERG, D. M. (ed.): *Metabolic Pathways*, vol. 3, Academic, New York, 1969.

MEISTER, A.: *Biochemistry of the Amino Acids*, 2d ed., vols. 1 and 2, Academic, New York, 1965. Comprehensive and detailed treatise.

STANBURY, J. O., J. B. WYNGAARDEN, and D. S. FREDRICKSON (eds.): *The Metabolic Basis of Inherited Disease*, 3d ed., McGraw-Hill, New York, 1972. Excellent comprehensive text on human genetic defects affecting metabolism.

TABOR, H., and C. W. TABOR (eds.): *Metabolism of Amino Acids and Amines*, vol. 17, pts. A and B of *Methods in Enzymology*, Academic, New York, 1970–1971.

Articles

ADAMS, E.: "The Metabolism of Hydroxyproline," *Mol. Cell. Biochem.*, 2: 109 (1973).

BRAUNSTEIN, A. E.: "Amino Group Transfer," pp. 379–482 in P. D. Boyer (ed.), *The Enzymes*, 3d ed., vol. 9, pt. B, Academic, New York, 1973. Definitive review of transamination reactions.

COHEN, P. P., and G. W. BROWN, JR.: "Ammonia Metabolism and Urea Biosynthesis," pp. 161–294 in M. Florkin and H. S. Mason (eds.), *Comparative Biochemistry*, vol. 11, Academic, New York, 1961. Comparative and developmental aspects of urea excretion.

CUNNINGHAM, L.: "The Structure and Mechanism of Action of Proteolytic Enzymes," pp. 85–188 in M. Florkin and E. H. Stotz (eds.), *Comprehensive Biochemistry*, vol. 16, Academic, New York, 1965. Review of proteolytic reactions.

EISENBERG, H.: "Glutamate Dehydrogenase: Anatomy of a Regulatory Enzyme," *Acc. Chem. Res.*, 4: 379–385 (1971).

KIKUCHI, G.: "The Glycine Cleavage System: Composition, Reaction Mechanism, and Biological Significance," *Mol. Cell. Biochem.*, 1: 169–187 (1973).

KOBERSTEIN, R., and H. SUND: "Studies of Glutamate Dehydrogenase," *Eur. J. Biochem.*, 36: 545–552 (1973).

RATNER, S.: "Enzymes of Arginine and Urea Synthesis," *Adv. Enzymol.*, 39: 1 (1973).

Problems

1. In an inherited human disease abnormally high levels of isovaleric acid in the blood plasma were found.

   (a) The metabolism of which amino acid is likely implicated?

   (b) Assuming that the blood levels of this amino acid and its

α-keto analog are normal, predict which enzyme is likely to be defective.

2. The E_0' value for the alanine–pyruvate + NH_4^+ substrate couple is -0.13 V. Calculate the change in standard free energy, $\Delta G^{\circ\prime}$, for the oxidation of alanine to pyruvate assuming the oxidizing agent is (a) a pyridine nucleotide, (b) a flavoprotein ($E_0' = 0.00$ V).

3. Write a balanced equation for the complete oxidation of phenylalanine, including all activation and energy-conserving steps for the process as it occurs in (a) an ammonotelic, (b) a ureotelic animal.

4. Calculate the number of ATP molecules generated during the oxidation of (a) valine to CO_2, H_2O, and NH_3, (b) threonine to CO_2, H_2O, and urea.

5. During catabolism of histidine, in which positions of glutamic acid would the following numbered atoms be expected to appear?

$$\underset{3\ N}{\overset{5}{HC}}=\overset{6}{C}-\overset{}{CH_2}-\underset{NH_2}{\overset{8}{CH}}-COOH$$

6. Write an overall balanced equation for the conversion of alanine to acetoacetate plus urea.

7. Which carbon atoms of α-ketoglutaric acid would be expected to become labeled after oxidation of the following labeled compounds in animal tissues?

   a. $CH_3\overset{*}{C}HNH_2COOH$

   b. ⬡—$CH_2\overset{*}{C}HNH_2COOH$

   c. $\overset{*}{}$⬡—CH_2CHNH_2COOH

8. On a given diet yielding 3,000 kcal per day, a 70-kg man excretes 27.0 g of urea daily. What percentage of his daily energy requirement is met by protein? Assume that 1.0 g of protein yields 4.0 kcal and 0.16 g of nitrogen as urea.

9. Compare the net energy yield in terms of ATP produced per carbon atom upon complete combustion of glucose, butyric acid, and alanine in the human liver.

10. On the basis of the description of a typical transamination reaction (pages 562 through 565) and the summary of bisubstrate reaction kinetics (page 204), predict what patterns would be expected to be found in kinetic experiments on alanine aminotransferase if (a) $1/v_0$ is plotted against $1/[\text{pyridoxal phosphate}]$ at various concentrations of the amino acid substrate and a fixed high concentration of the keto acid substrate, (b) $1/v_0$ is plotted against $1/[\text{amino acid substrate}]$ at various concentrations of the keto acid substrate and a high fixed concentration of pyridoxal phosphate.

CHAPTER 22 PHOTOSYNTHETIC ELECTRON TRANSPORT AND PHOSPHORYLATION

Photosynthesis, which we may define for the moment as the utilization of solar light energy by plant cells for the biosynthesis of cell components, is a metabolic process that is fundamental to all living organisms. Solar energy is not only the immediate source of energy for green plants and other photosynthetic autotrophs but also the ultimate source of energy for nearly all heterotrophic organisms, through the operation of food chains in the biosphere. Moreover, solar energy captured by the process of photosynthesis is the source of well over 90 percent of all the energy used by man for heat, light, and power, since coal, petroleum, and natural gas, the fuels for most man-made machines, are all decomposition products of biological material generated millions of years ago by photosynthetic organisms.

It is convenient and useful to consider photosynthesis as occurring in two major phases. The biochemical nature of these two phases is most simply illustrated by the case of photosynthesis in higher plants, which is usually represented by the equation

$$6CO_2 + 6H_2O \xrightarrow{\text{light}} C_6H_{12}O_6 + 6O_2$$

Glucose

We know today that this overall process can be resolved into two phases. The first is the capture of light energy by light-absorbing pigments and its conversion into the chemical energy of ATP and certain reducing agents, particularly NADPH. In this process hydrogen atoms are removed from water molecules and used to reduce $NADP^+$, leaving behind molecular oxygen, a by-product of plant photosynthesis; simultaneously, ADP is phosphorylated to ATP. The general equation for the first phase of photosynthesis, which we shall not write in balanced form at this point, is

$$\text{Water} + NADP^+ + P_i + ADP \xrightarrow{\text{light}} \text{oxygen} + NADPH + H^+ + ATP$$

In the second phase of photosynthesis the energy-rich products of the first phase, NADPH and ATP, are used as the sources of energy to bring about the reduction of carbon dioxide to yield glucose; simultaneously, NADPH is reoxidized to $NADP^+$, and the ATP is broken down again into ADP and phosphate. This second phase of photosynthesis, which may be represented in general terms as

$$CO_2 + NADPH + H^+ + ATP \longrightarrow glucose + NADP^+ + ADP + P_i$$

is brought about by conventional enzyme-catalyzed reactions which do not require light. Indeed, many of the enzymes catalyzing the conversion of carbon dioxide into glucose in the second phase of photosynthesis are also found in animal tissues.

The set of reactions in the first phase of photosynthesis, involving the conversion of light energy into the chemical energy of NADPH and ATP, is somewhat loosely referred to as the *light reactions* or the *light phase* of photosynthesis. The second stage, in which glucose and other reduced products are formed from CO_2 in plants, is referred to as the *dark reactions* or *dark phase*. Although there are other types of photosynthesis in addition to that taking place in higher green plants, we can in all cases distinguish a light phase and a dark phase.

In this chapter we focus our attention primarily on the light reactions of photosynthesis; the dark reactions will be described in the following chapter, which deals with the biosynthesis of carbohydrates.

Some Historical Landmarks

Joseph Priestley, one of the discoverers of oxygen, carried out some of the first important experiments on the exchange of matter during photosynthesis in the period 1770–1777. He established that an enclosed volume of air "depleted" by a burning candle, and thus unable to support the oxygen needs of a mouse, could be "restored" by a sprig of mint over a period of some weeks, so that it could again support the respiration of a mouse or the burning of a candle. He concluded that green plants evolve oxygen, a process that appeared to be the reverse of respiration in animals, which results in consumption of oxygen. Curiously, Priestley failed to note that this action of a green plant requires light, a fact recognized only some years later by J. Ingenhousz, a Dutch physician, who also first established that only the green part of a plant carries out the formation of oxygen in the light. Later, in the early nineteenth century, quantitative balance studies were performed on carbon dioxide assimilation, oxygen evolution, and plant matter produced. Such studies, together with Robert Mayer's recognition that sunlight contributes the energy for the formation of the photosynthetic products, led in the mid-nineteenth century to the general equation

$$CO_2 + H_2O \xrightarrow{\text{light}} O_2 + \text{organic matter}$$

Another landmark was G. Engelmann's discovery in 1880 that chloroplasts are responsible for oxygen evolution. He observed under the microscope that when small oxygen-requiring motile bacteria are added to a suspension of the eukaryotic alga *Spirogyra*, they migrate toward the portions of the algal cell surface in the neighborhood of its single large spiral chloroplast, but only when the cell is illuminated. Since these bacteria seek out zones rich in oxygen, Engelmann concluded that the chloroplast is the site of oxygen production.

Biological Occurrence of Photosynthesis

The capacity to carry out photosynthesis is found in a wide range of organisms in both the prokaryotic and eukaryotic domains. The photosynthetic eukaryotes include not only the familiar higher green plants but also unicellular and multicellular green, brown, and red algae, as well as euglenoids, dinoflagellates, and diatoms.

The photosynthetic prokaryotes are less familiar to us and require some comment, since they play an important role in the biosphere and are intensively used in research on photosynthesis. They are a very ancient class of organisms, descendants of the first photosynthetic cells. They include the blue-green algae, the green sulfur bacteria, and the purple bacteria. The blue-green algae, which may live as single cells or in colonies, are abundant in soil, fresh water, and the oceans. They can live on carbon dioxide as sole carbon source; some blue-green algae also can fix atmospheric nitrogen (page 366). The green sulfur bacteria are strict anaerobes that live in ponds and lakes rich in sulfur-containing organic matter. The green color of some mountain lakes is due to the green sulfur bacterium *Chlorobium*. The purple sulfur bacteria, such as *Chromatium*, are strict anaerobes that require hydrogen sulfide, sulfur, or thiosulfate; they are found in ponds and sulfur springs. The nonsulfur purple bacteria, e.g., *Rhodospirillum rubrum*, require organic molecules such as ethanol, acetate, β-hydroxybutyrate, or isopropanol.

It is a common misconception that photosynthesis largely takes place in higher plants. Actually, more than half of all the photosynthesis on the surface of the earth is carried out in the oceans by microscopic algae, diatoms, and dinoflagellates, which together constitute the phytoplankton.

Intracellular Organization of Photosynthetic Systems

In eukaryotic cells the photosynthetic apparatus is localized in the *chloroplasts*, one of several different kinds of plastids, membrane-surrounded organelles peculiar to plant cells (page 35). Chloroplasts are self-replicating organelles; like mitochondria, they contain DNA (page 870). An electron micrograph of a photosynthetic cell of a higher plant is shown in Figure 1-9 (page 34). Chloroplasts are generally much larger than mitochondria, but they have a wide range of size, from 1 to 10 μm in diameter. They are usually globu-

lar or discoid but sometimes assume exotic forms, as in
Spirogyra, in which they are ribbonlike spiral structures. In
some eukaryotic algae there is only one chloroplast per cell;
however, a higher plant cell may contain as many as 40
chloroplasts.

Chloroplasts are readily isolated from some green-plant
tissues; spinach leaves are the most common source. The
leaves are gently homogenized in a blender in a medium of
0.35 M NaCl or 0.3 to 0.4 M sucrose, buffered at pH 8.0. The
leaf suspension is filtered through muslin and centrifuged
lightly to remove cell debris. The chloroplasts are then sedimented by centrifugation for some minutes at 1,000g. They
are washed one or more times by resuspending them in
0.35 M NaCl and centrifuging. When carefully isolated,
chloroplasts remain relatively intact and retain the capacity to carry out all the reactions involved in the light-
induced reduction of carbon dioxide to hexose.

Chloroplasts are surrounded by a single continuous outer
membrane, which is rather fragile (Figure 22-1). The inner-
membrane system, which is continuous but arranged in
paired folds called *lamellae*, encloses a compartment containing the *stroma*, comparable to the mitochondrial matrix.
At regular intervals, the lamellae widen to form flattened
membrane sacs, or vesicles, called *thylakoids* (Greek,
"baggy trousers"), which occur in stacked arrangements
called *grana*. The paired membranes between the grana are

Figure 22-1
*Electron micrograph of a chloroplast of let-
tuce, Lactuca sativa; G = a granum; I = intergranal lamella; S = stroma; E = envelope
or outer membrane.*

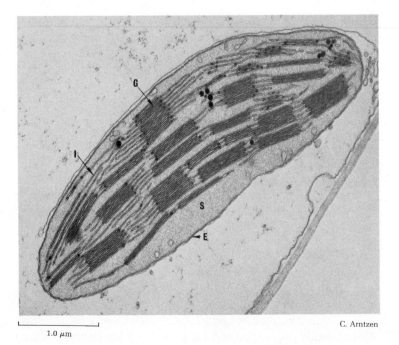

1.0 μm

C. Arntzen

*Schematic drawing of a chloroplast,
showing the arrangement of the membranes
in the thylakoids, which are actually vesicular, although they are sometimes referred
to as disks. Studies of developing chloroplasts indicate that the thylakoid vesicles
are derived from the inner membrane.*

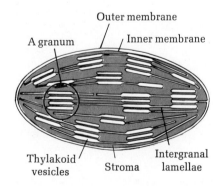

Outer membrane

A granum — Inner membrane

Thylakoid vesicles — Stroma — Intergranal lamellae

Figure 22-2
Electron micrographs of photosynthetic prokaryotes.

The blue-green alga Oscillatoria rubescens. The cell membrane infolds to form the photosynthetic lamellae, which run across the entire cell.

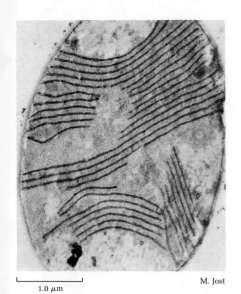

1.0 μm   M. Jost

Electron micrograph of a thin section of a photosynthetic bacterium, Rhodospirillum molischianum, showing the cytoplasmic lamellae which contain the photosynthetic pigments. The insert at the upper right shows isolated chromatophores derived from the lamellae; such preparations catalyze light-induced electron flow and phosphorylation.

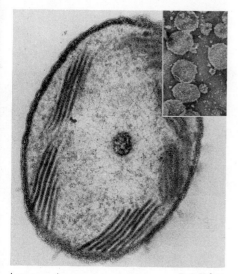

0.2 μm   R. A. Nelson

called *intergranal lamellae*. The membranes of the thylakoids and the intergranal lamellae contain the photosynthetic pigments of the chloroplast, as well as the enzymes required for the primary light-dependent reactions; they are thus functionally comparable to the cristae of mitochondria (page 511). The inner-membrane system of chloroplasts can be fragmented and washed free of stroma to yield resealed thylakoids and grana, which can be recovered by differential centrifugation. When suitably supplemented, such preparations carry out the light-dependent phase of photosynthesis.

Photosynthetic bacteria and blue-green algae, which are prokaryotes, lack chloroplasts. The molecular components of their light-receptor systems are located either in the cell membrane or in vesicular structures called *chromatophores*, which are probably derived from the cell membrane. Figure 22-2 shows the structure of a photosynthetic blue-green alga and of a photosynthetic bacterium.

Fundamental Processes in Photosynthesis

All photosynthetic organisms *except* bacteria use water as electron or hydrogen donor to reduce various electron acceptors, and from the water they evolve molecular oxygen. The overall equation of photosynthesis for this group of organisms is

$$H_2O + CO_2 \xrightarrow{\text{light}} (CH_2O) + O_2 \tag{1}$$

in which (CH_2O) designates the carbohydrate formed as the end product of photosynthesis. However, this equation does not apply to all photosynthetic organisms. The photosynthetic bacteria normally neither produce nor use molecular oxygen; in fact, most of them are strict anaerobes and are poisoned by oxygen. Instead of water these organisms use other compounds as electron donors. The green and purple sulfur bacteria use hydrogen sulfide, according to the equation

$$2H_2S + CO_2 \xrightarrow{\text{light}} (CH_2O) + H_2O + 2S$$

The elemental sulfur formed is deposited as globules which either accumulate in the cell or are extruded from it (Figure 22-3). Some nonsulfur purple bacteria use an organic hydrogen donor, such as isopropanol, which is oxidized to acetone:

$$2CH_3CHOHCH_3 + CO_2 \xrightarrow{\text{light}} (CH_2O) + 2CH_3COCH_3 + H_2O$$

Isopropanol   Acetone

Despite these differences, C. Van Niel, a pioneer in the study of the comparative aspects of metabolism and photosynthesis, postulated that plant and bacterial photosynthesis are fundamentally similar, as is evident if the equation of photosynthesis is written in a more general form,

$$2H_2D + CO_2 \xrightarrow{\text{light}} (CH_2O) + H_2O + 2D$$

in which H_2D symbolizes a hydrogen donor and D its oxidized form. Thus H_2D may be water, hydrogen sulfide, isopropanol, or any one of a number of different hydrogen donors. The nature of the hydrogen donor that can be used depends on the species of photosynthetic organism. Van Niel also predicted that the molecular oxygen formed during plant photosynthesis is derived exclusively from the oxygen atoms of water and not from the carbon dioxide, as indicated in the following version of equation (1):

$$2H_2O + CO_2 \xrightarrow{\text{light}} (CH_2O) + H_2O + O_2$$

Photosynthesis taking place in water labeled with an oxygen isotope does in fact yield labeled O_2.

Study of the comparative aspects of photosynthesis has also revealed that CO_2 is not the universal electron or hydrogen acceptor in all photosynthetic cells. Carbon dioxide is of course the *major* electron acceptor in all photosynthetic autotrophs (page 363), such as higher plants, which must manufacture all their organic biomolecules from carbon dioxide. However, most higher plants can also use as electron acceptor nitrate, which they reduce to ammonia. In nitrogen-fixing photosynthetic organisms, molecular nitrogen as well as carbon dioxide may be used as electron acceptors during photosynthesis; the nitrogen is reduced to ammonia. Moreover, many photosynthetic organisms can use hydrogen ions as the ultimate electron acceptor, from which they form molecular hydrogen; still others can use sulfate as electron acceptor. Typical equations for photosynthesis with different electron acceptors follow:

Electron donor	Electron acceptor		
$2H_2D +$	CO_2	$\longrightarrow$	$(CH_2O) + H_2O + 2D$
$9H_2D +$	$2NO_3^-$	$\longrightarrow$	$2NH_3 + 6H_2O + 9D$
$3H_2D +$	N_2	$\longrightarrow$	$2NH_3 + 3D$
$H_2D +$	$2H^+$	$\longrightarrow$	$2H_2 + D$

From these considerations, it is clear that photosynthesis may involve different electron donors and different electron acceptors, depending on the species of photosynthetic organism. Thus we can write a completely general equation for photosynthesis,

$$H_2D + A \xrightarrow{\text{light}} H_2A + D$$

in which H_2D is the electron or hydrogen donor and A is the electron or hydrogen acceptor. Therefore we should not look upon photosynthesis exclusively as a mechanism for the synthesis of carbohydrates from carbon dioxide. Indeed, even in higher plants the products of the light reactions (ATP and NADPH) are used to carry out the biosynthesis of many cell components other than carbohydrates.

We now come to a highly important characteristic of the light reactions of photosynthesis. In all photosynthetic organisms, regardless of the electron donor and electron acceptor, the light-induced flow of electrons from electron

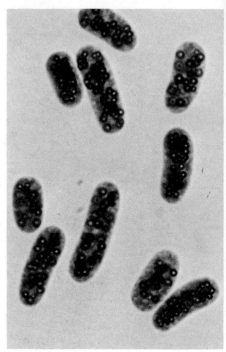

Figure 22-3
Purple sulfur bacteria. In these cells H_2S is the electron donor. The sulfur released accumulates as globules visible within the cells.

10 μm   N. Pfennig

donor to electron acceptor is *against* the normal gradient of the standard oxidation-reduction potentials of the electron-donor and electron-acceptor systems; i.e., the net flow of electrons is in the direction of the system having the lower or more electronegative standard potential (page 478). The direction of flow of electrons in photosynthesis does not violate the thermodynamic laws, since it is the energy of the absorbed light that causes the electrons to flow in reverse, in the direction of a more negative or more energy-rich state, opposite to the direction of flow of electrons in respiration.

The Light and Dark Reactions

We have seen that photosynthesis has two phases, the *light reactions*, which are directly dependent on light energy, and the *dark reactions*, which can occur in the absence of light. This division was first suggested by observations that the rate-limiting step in plant photosynthesis is some step that can take place in the dark. When photosynthetic organisms are subjected to intermittent illumination with very short flashes of light (milliseconds or less) followed by dark intervals of varying duration, the maximum O_2 evolution after a single light flash of 10^{-5} s can be realized only if it is followed by a much longer dark period, about 0.06 s or more. This difference in rate is greatly accentuated when the temperature of the cells is lowered.

A more direct experiment proving that there are light and dark phases of photosynthesis was carried out by D. I. Arnon and his colleagues in 1958. They showed that the light and dark phases could be separated temporally. First, they illuminated chloroplasts in the absence of carbon dioxide, which resulted in trapping some of the light energy in a chemical form. They then disrupted the chloroplasts and removed the grana, in which the light-trapping reaction takes place, and added radioactive carbon dioxide to the remaining stroma. They found that the carbon dioxide was converted in the dark into radioactive hexoses at the expense of the chemical energy generated in the preceding light period. These experiments also showed that chloroplasts are capable of the entire photosynthetic process leading to hexose formation; i.e., they are complete photosynthetic units, just as the mitochondria are complete respiratory units (page 509).

Today we know that the light reactions of photosynthesis are primarily responsible for converting light energy into chemical energy in the form of ATP and NADPH, whereas the dark reactions involve the utilization of the chemical energy of ATP and NADPH to bring about the reduction of carbon dioxide to hexose and other products. The term dark reactions should not be taken to mean that they take place only in the dark or at night; in living plants they take place, together with the light reactions, in the daytime. At night green leaf cells respire, utilizing oxygen and consuming glucose and other organic fuels generated by photosynthesis in daylight.

We shall now examine the individual steps in the reactions by which plants convert light energy into the chemical energy of ATP and NADPH.

Excitation of Molecules by Light

Visible light is a form of electromagnetic radiation of wavelength 400 to 700 nm (Figure 22-4). Light has properties suggesting that it is propagated in a discontinuous or corpuscular manner, in the form of _photons_ or _quanta_. The total amount of solar energy falling on the surface of the earth in the form of photons is immense; it is estimated to exceed 2×10^{25} cal/year. Only 12 percent of this energy is actually available to plant life; the rest is either outside the visible range or is absorbed by the atmosphere or the nonliving portion of the earth's surface.

The energy content of a photon is represented by $h\nu$, in which h is Planck's constant (1.58×10^{-34} cal-s) and ν is the frequency of the radiation. E, the energy in kilocalories of 1.0 einstein, that is, 1.0 mol of light, containing 6.023×10^{23} (Avogadro's number) quanta, is most simply given by the formula

$$E = \frac{28,600}{\text{wavelength (nm)}}$$

In the visible range 1 einstein carries from 40 to 72 kcal of energy, depending on the wavelength of the light (Figure 22-4). Photons of short wavelength, at the violet end of the spectrum, have the greatest energy content. Note that the energy content of a "mole" of photons, regardless of wavelength, is considerably greater than the amount of energy required to synthesize 1 mol of ATP from ADP and phosphate, which is 7.3 kcal under standard thermodynamic conditions (page 398).

The ability of a compound to absorb photons depends on its atomic structure, particularly on the arrangement of electrons surrounding its atomic nuclei. The _absorption spectrum_ of a compound indicates its capacity to absorb light as a function of wavelength. When a photon strikes an atom or molecule capable of absorbing light at a given wavelength, energy is absorbed by some of the electrons, which are thus boosted to higher energy levels; the atom or molecule is then in an energy-rich _excited state_. Only photons of certain wavelengths can excite a given atom or molecule because the excitation of molecules is not continuous but quantized; i.e., light energy is absorbed only in discrete packets on an all-or-none basis, leading to the term _quantum_. Excitation of a molecule by light is very rapid, taking less than 10^{-5} s. The excited molecule has two possible fates. It may return to its original low-energy state, the _ground state_, with simultaneous emission of the energy originally absorbed during excitation, which reappears either as light or as heat or both. The emission of light by excited molecules is called _fluorescence._ Fluorescent decay of excited molecules, which is complete in less than 10^{-8} s, occurs at a longer wavelength

Figure 22-4
Energy equivalent of 1 einstein at different wavelengths.

Wavelength, nm	Color	Kilo-calories	Kilo-joules
700	Far red	40.9	171
	Red		
600	Orange	47.7	199
	Yellow		
500	Green	57.2	239
	Blue		
400	Violet	71.5	299

than the exciting wavelength. However, an excited molecule has another possible fate; it may, thanks to its energy-rich condition, react more readily with some other molecule. In such a photochemical reaction the excited molecule may lose an electron to the other reacting molecule.

With this background we shall now examine the characteristic light-absorbing pigments of photosynthetic cells. It is these molecules that become excited and thus energy-rich when they are illuminated.

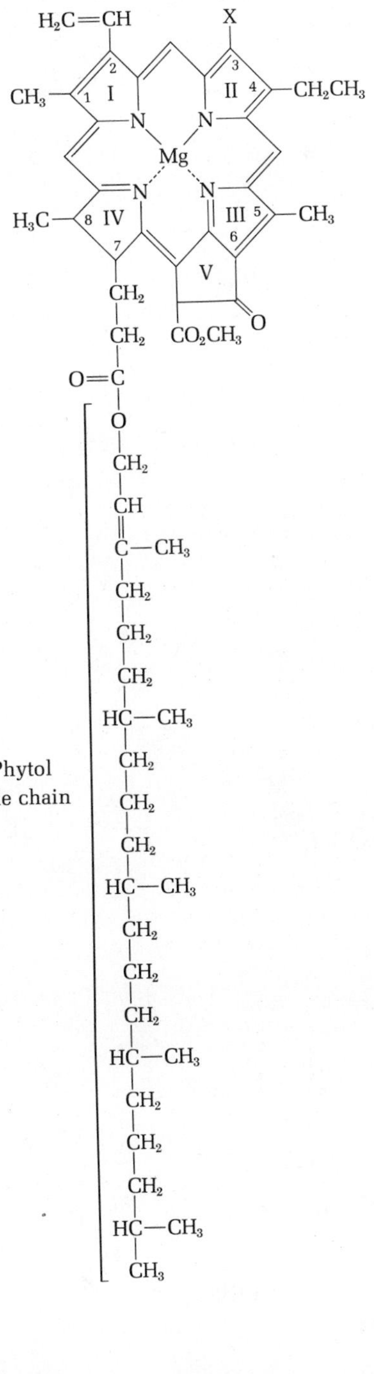

The Photosynthetic Pigments

All photosynthetic cells contain one or more of the pigments known as *chlorophylls*, most of which are green. Not all photosynthetic cells are green, however; for example, photosynthetic algae and bacteria may be green, brown, red, or purple, depending on the species. This variety of colors results because, besides chlorophyll, most photosynthetic cells contain one or more *accessory pigments*, which include the yellow, red, or purple *carotenoids* (page 296) and the blue or red *phycobilins*.

Chlorophylls

Chlorophylls, which are normally associated with specific proteins, can be extracted from leaves with alcohol or acetone and purified by chromatography. In fact, the term chromatography (Greek, "color writing") was first used by M. Tswett, a Russian botanist who described the separation of pigments extracted from plants with acetone by their adsorption, development, and elution on columns of inert adsorbents. Today the term "chromatography" is used to describe the separation of any substances, colored or not, by partition between two phases (page 87). Higher plants contain two forms of chlorophyll, designated chlorophyll *a* and chlorophyll *b*. The structure of chlorophyll *a* was established from degradation studies by H. Fischer in Germany in 1940 and unequivocally proved by the total synthesis carried out by R. B. Woodward in 1960. Chlorophyll *a* (Figure 22-5) is a magnesium-porphyrin complex; its porphyrin differs from protoporphyrin (page 490) in the nature and position of the substituents of the pyrrole rings and the fact that it contains a fused cyclopentanone ring. The four central nitrogen atoms are coordinated with a Mg^{2+} ion to form an extremely stable, essentially planar complex. Chlorophyll also has a long, hydrophobic terpenoid side chain, consisting of the alcohol *phytol* (page 296) esterified to a propionic acid substituent in ring IV (Figure 22-5). When the phytol is removed from chlorophyll *a* by the hydrolytic enzyme *chlorophyllase*, the remaining structure is called *chlorophyllide a*, an intermediate in the biosynthesis of chlorophyll.

Pure chlorophyll *a* in acetone has absorption maxima at 663 and 420 nm. In intact cells, however, chlorophyll *a* shows several different absorption maxima at the longer wavelength, typically at 660, 670, 678, and 685 nm. These maxima are not due to the existence of different molecular

Figure 22-5
The structure of chlorophylls. In chlorophyll a, X = —CH₃; in chlorophyll b, X = —CHO. Note that there is a fused cyclopentanone ring (V) in addition to the four pyrrole rings. In bacteriochlorophyll, pyrrole ring II is reduced. When the ester linkage with phytol is hydrolyzed, a chlorophyllide results.

forms of chlorophyll *a* but represent spectral shifts caused by different states of aggregation or binding of chlorophyll *a* molecules with specific proteins in the plant cell.

Oxygen-producing photosynthetic cells contain two kinds of chlorophyll, of which one is always chlorophyll *a*. While in green plants the second chlorophyll is chlorophyll *b*, in brown algae, diatoms, and dinoflagellates it is chlorophyll *c*. Prokaryotic photosynthetic cells that produce no oxygen do not contain chlorophyll *a*. They contain *bacteriochlorophyll a* or *bacteriochlorophyll b*; in addition, green bacteria contain *Chlorobium* chlorophyll. Bacteriochlorophyll *a* differs from chlorophyll *a* of higher plants in that ring I contains an acetyl group and ring II is reduced. All the chlorophylls absorb visible light efficiently because of their many conjugated double bonds. Moreover, the light energy of the photons absorbed by a chlorophyll molecule may become delocalized and spread throughout the entire electronic structure of the excited molecule (see below).

That chlorophyll is the major light-absorbing pigment in most green cells has been established by measurements of the *photochemical action spectrum* of photosynthesis, a plot of the efficiency of different wavelengths of visible light in supporting oxygen evolution. Figure 22-6 shows the action spectrum of photosynthesis in a green plant, as well as the light-absorption spectra of its photosynthetic pigments. The

Figure 22-6
The action spectrum of photosynthesis in a green-plant cell, compared to the absorption spectra of the chlorophylls and carotenoids present. From 550 to 680 nm, the action spectrum reflects the absorption spectra of the chlorophylls alone. (Prepared by Govindjee.)

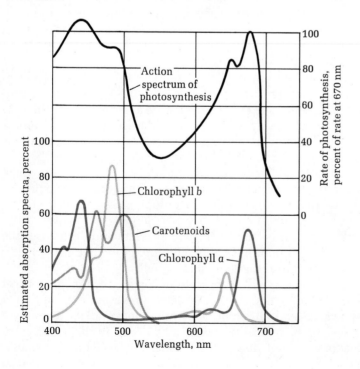

Figure 22-7
Structure of β-carotene and spirilloxanthin.

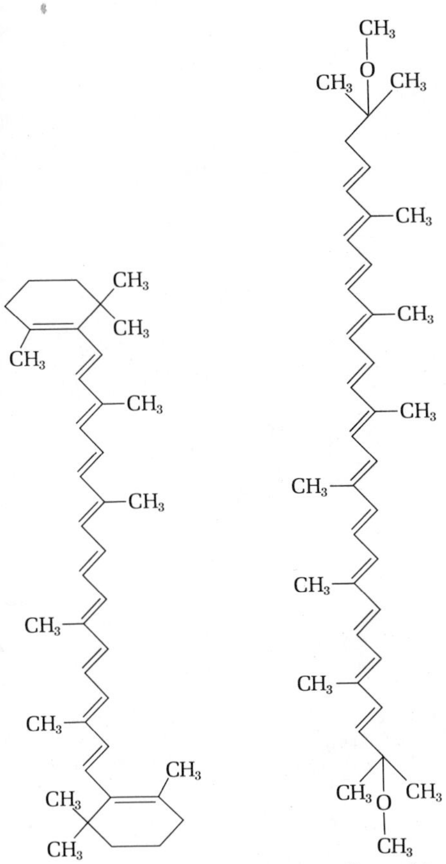

β-Carotene

Spirilloxanthin

Figure 22-8
Phycoerythrobilin, a red photosynthetic pigment. In phycocyanobilin, a blue pigment, the $CH_2 = CH-$ group on the fourth pyrrole ring is replaced by CH_3CH_2-.

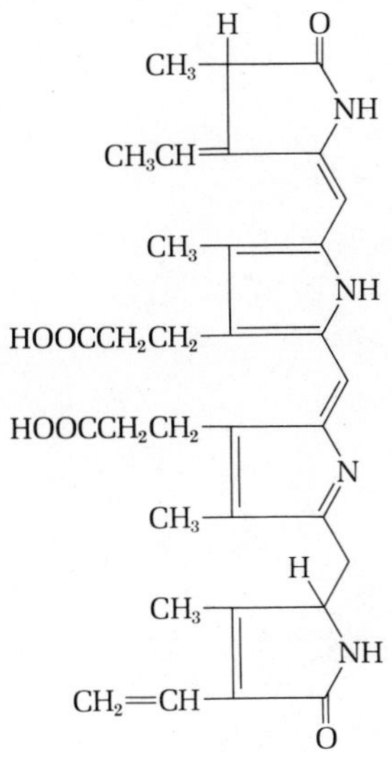

action spectrum closely coincides with the sum of the absorption spectra of the chlorophylls and carotenoids (see below). However, over much of the visible range the action spectrum mirrors the absorption spectra of chlorophylls a and b. From such evidence it has been concluded that chlorophyll is the predominant light-trapping molecule in green-plant cells. In many plant cells the action spectrum in the far-red region of the spectrum, above 680 nm, drops sharply compared with the absorption spectrum. As we shall see, this phenomenon was an important clue to identification of different kinds of photosystems in green plants (page 600).

The Accessory Pigments: Carotenoids and Phycobilins

Like chlorophyll, the *accessory pigments* also can serve as receptors of light energy. They include the *carotenoids* and the *phycobilins*. For example, red algae contain relatively little chlorophyll but considerable amounts of *phycoerythrobilin*, a red phycobilin. The photochemical action spectrum of red algae largely coincides with the absorption spectrum of its red phycobilin–protein conjugate, indicating that this conjugate is the major light-absorbing pigment in this organism. Hence these accessory pigments, which have absorption maxima at wavelengths other than those of the chlorophylls, serve as supplementary light receptors for portions of the visible spectrum not completely covered by chlorophyll. However, whenever light energy is absorbed by such accessory pigments, it must be transferred as excitation energy to chlorophyll molecules before it can be used for photosynthesis. Chlorophyll is thus an indispensable component of the photosynthetic apparatus.

The *carotenoids* (page 296) are long polyisoprenoid molecules having conjugated double bonds; each end of the molecule contains an unsaturated substituted cyclohexene ring. There are two major classes of carotenoid pigments in chloroplasts, the *carotenes*, which are isoprenoid hydrocarbons (page 293) and contain no oxygen, and the *xanthophylls*, which are very similar in structure but contain oxygen atoms in their terminal rings. Figure 22-7 shows the structure of β-carotene, the most abundant carotene, and *spirilloxanthin*, a xanthophyll.

The phycobilin pigments occur in red and blue-green algae but not in higher plants. They are linear tetrapyrroles (Figure 22-8), in contrast to chlorophyll, which is a cyclic tetrapyrrole; phycobilins differ also in lacking bound Mg^{2+}. Phycobilins are conjugated to specific proteins. The protein conjugate of phycoerythrobilin is *phycoerythrin*, the major red pigment of red algae. The blue *phycocyanin* is the analogous conjugate in the blue-green algae.

The photosynthetic pigments in the chloroplasts of plants are organized into two functional sets, or assemblies, which in turn are connected with characteristic electron-transport chains. These functional units, *photosystems I and II*, will be described later.

P700 and P680

With refined spectroscopic methods B. Kok discovered that oxygen-producing photosynthetic cells contain a very small amount of a pigment which has a light-absorption maximum at 700 nm and which undergoes bleaching when the cell is illuminated. Since this bleaching effect can also be produced by the oxidizing agent ferricyanide, it is believed to correspond to loss of an electron from the pigment, which is designated P700 (P stands for pigment). P700 is believed to consist of chlorophyll *a* associated with a specific type of protein; however, P700 constitutes only about one-four-hundredth of the total chlorophyll in chloroplasts. Much evidence now indicates that P700 is a trap for collecting *excitons*, quanta of excitation energy, from the other chlorophyll molecules in the thylakoid membrane. When an exciton is trapped, a high-energy electron is lost from P700.

Similar experimental approaches have revealed the presence of a second exciton-collecting pigment in green-plant cells, also a specialized chlorophyll-protein complex, designated P680. We shall see that P700 and P680 are associated with two different photosystems of green-plant cells.

The Hill Reaction and Light-Induced Electron Transport

We now examine the process of light-induced electron transport, by which the light energy trapped by the pigment system is converted into chemical energy. Modern research on photosynthetic electron transport began with the discovery of R. L. Hill, at Cambridge University, that light-induced oxygen evolution can be observed in cell-free granular preparations extracted from green leaves. Illumination of such preparations in the presence of aritificial electron acceptors, such as ferricyanide or reducible dyes (page 487), caused evolution of oxygen and simultaneous reduction of the electron acceptor, according to the general equation

$$H_2O + A \xrightarrow{\text{light}} AH_2 + \tfrac{1}{2}O_2 \qquad (2)$$

in which A is the hydrogen (electron) acceptor and AH_2 its reduced form. However, carbon dioxide was apparently not required, nor was it reduced to a stable form that accumulated, suggesting that the photoreduction of carbon dioxide to hexose is a *later* step in photosynthesis. The reduction of acceptor A did not take place unless the leaf extracts were illuminated.

The reaction summarized in equation (2) is known today as the *Hill reaction* and acceptor A as a *Hill reagent*. The Hill reagent thus functions as an artificial acceptor for electrons arising from water, just as methylene blue may act as an artificial acceptor for electrons removed from a substrate by dehydrogenase action (page 487). An important feature of the Hill reaction is that electrons are induced to flow *away* from water molecules to acceptor A, thus yielding molecular oxygen from the water. Yet in animal tissues electrons arising from organic substrates flow *toward* molecular oxygen, which is reduced to water. Clearly, the direction of electron flow in the Hill reaction is opposite to that in respiration. The energy for this reversed electron flow, which takes place only on illumination, comes from the absorbed light.

In normal photosynthesis carbon dioxide is the ultimate acceptor of the high-energy electrons generated by the absorbed light, but in the Hill reaction an artificial acceptor intercepts the electrons before they reach carbon dioxide. This consideration implies that chloroplasts must contain electron carriers capable of leading electrons from water to carbon dioxide. In 1951 three groups of investigators reported that $NADP^+$, a normal component of chloroplasts, could replace artificial Hill reagents as an electron acceptor. When incubated with chloroplasts in light, $NADP^+$ became reduced to NADPH and oxygen was evolved stoichiometrically, according to the equation

$$H_2O + NADP^+ \xrightarrow{\text{light}} NADPH + H^+ + \tfrac{1}{2}O_2$$

In the dark no reduction of $NADP^+$ took place. The NADPH formed in the light reaction could then be used to reduce NADP-linked substrates via their specific dehydrogenases (page 481). From these and other observations $NADP^+$ was identified as a common carrier of electrons from water to the various terminal electron acceptors of photosynthesis, which in green plants is carbon dioxide. The light-induced movement of electrons via electron-carrying molecules from water or other electron donors to various electron acceptors is called *photosynthetic electron transport*.

The standard oxidation-reduction potential (page 478) of the NADPH–$NADP^+$ couple has the value −0.32 V (page 479), which we may compare with +0.82 V, the standard potential of the water-oxygen couple (page 479). Normally, electrons would of course tend to flow from NADPH to oxygen, i.e., in the direction of the more positive system, but in photosynthetic electron transport electrons are induced to flow in the reverse direction, from water to $NADP^+$.

Several other electron carriers participate in photosynthetic electron transport between water and $NADP^+$. They will be described after we consider the formation of the other major energy-rich product of the light reactions of photosynthesis, namely, ATP.

Photosynthetic Phosphorylation

In 1954 D. I. Arnon, M. B. Allen, and F. R. Whatley discovered that when isolated spinach chloroplasts are illuminated in the presence of ADP and phosphate, ATP is formed. Since the amount of ATP yielded in such experiments was rather high and comparable to the magnitude of the light-induced electron flow, it was concluded that the formation of ATP must represent a major mechanism for the conservation of the absorbed light energy, just as ATP formation is the major process by which the energy of respiration is conserved during electron transport in mitochondria. This process was called *photosynthetic phosphorylation* or simply *pho-*

tophosphorylation. Independently and nearly simultaneously A. W. Frenkel discovered that a very similar phosphorylation process occurs on illumination of membrane vesicles from photosynthetic bacteria. These observations also indicated that the formation of ATP from ADP and phosphate results from the energetic coupling of the phosphorylation to the process of photoinduced electron transport, in much the same way that oxidative phosphorylation is coupled to electron transport in mitochondria.

We shall return to a more detailed description of the electron carriers participating in photoinduced electron flow and the mechanism of photosynthetic phosphorylation after we have examined a fundamental set of relationships between the function of the light-absorbing pigment systems of chloroplasts and the pathway of electron flow from water to $NADP^+$.

The Two Light Reactions of Plant Photosynthesis

We now know that there are two sets of light reactions in oxygen-evolving plant photosynthesis. This conclusion had its origin in studies of the photochemical action spectrum (page 596) of photosynthesis in certain plants. We have seen (Figure 22-6) that the efficiency of absorbed light in promoting photosynthesis coincides with the absorption spectrum of the cell pigments over most of the visible spectrum in some plants. However, in most plant cells the efficiency of light in the far red, above 680 nm, has been found to drop sharply compared with the absorption spectrum of the cells. This phenomenon is called _red drop_ (Figure 22-9). R. Emerson showed that if light at 710 nm, which is not very effective in promoting photosynthesis alone, is supplemented with some light at 670 nm, the efficiency of photosynthesis at 710 nm can be greatly enhanced. These findings suggested that two different light reactions, each having a distinctive optimum wavelength, are required for maximal photosynthetic efficiency. To account for this and other observations, L. N. M. Duysens has postulated that the long-wavelength system with a maximum near 710 nm, which he designated _photosystem I_, is associated with those forms of chlorophyll _a_ absorbing at longer wavelengths (page 595); it is not responsible for oxygen evolution. _Photosystem II_, on the other hand, is activated by shorter wavelengths, 670 nm and below; it is required for oxygen evolution. All oxygen-evolving photosynthetic cells contain both photosystems I and II, whereas the photosynthetic bacteria, which do not evolve oxygen, contain only a single photosystem resembling plant photosystem I. It was also postulated that photosystem I arose first during biological evolution; the capacity of plants to use water as reductant and thus to cause oxygen evolution, which is conferred by photosystem II, arose later. The relationship between the two photosystems and other evidence for a functional difference between them is developed further below.

We shall now examine the somewhat different composition of these two photosystems. Each has its own character-

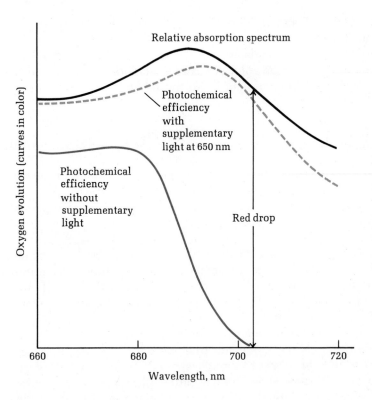

Figure 22-9
Red drop in Chlorella, a green alga. Above about 680 nm, the red region of the spectrum, the efficiency of monochromatic light in supporting oxygen evolution decreases greatly in relation to the absorption spectrum, which is largely due to chlorophyll. This deficit is called red drop. However, if supplementary light at 650 nm is added, the efficiency of the longer wavelengths is restored. Therefore two light-absorbing systems, one absorbing in the region 680 to 720 nm and the other at shorter wavelengths must cooperate to yield maximal rates of photosynthesis.

istic set or assembly of light-absorbing pigment molecules, which functions like an antenna to absorb and transmit light energy. Although both photosystems contain chlorophyll *a* and chlorophyll *b*, the ratio of chlorophyll *a* to *b* is higher in photosystem I than in photosystem II. The more significant difference between the photosystems, however, is the presence of large amounts of chlorophyll *a*–protein complexes absorbing at long wavelengths in photosystem I and their absence from photosystem II. This difference is responsible for red drop. Both photosystems must function for high rates of oxygen-evolving photosynthesis to occur. The pigment assembly or *photosynthetic unit* of photosystem I in higher green plants contains some 200 molecules of chlorophyll *a*, particularly of the long-wavelength type, perhaps 50 molecules of chlorophyll *b*, 50 to 200 molecules of carotenoid pigments, depending on the species, and a single molecule of P700. A quantum of light energy absorbed anywhere in this photosynthetic unit, whether by a carotenoid or a chlorophyll molecule, migrates through the set of pigment molecules, a process called *exciton transfer*, until it reaches the single molecule of P700, which accepts the exciton and, as a consequence, loses an electron having a large amount of energy. Photon capture and exciton transfer within the photosystem are extremely rapid processes and (like all photochemical processes) are temperature-insensitive. A similar process of photon capture and exciton transfer occurs in the other photosystem, until the exciton reaches P680, the reactive center of photosystem II, which then loses an electron.

The pigment assemblies of photosystems I and II are embedded in the membranes of the thylakoid vesicles. Photosystems I and II appear to be physically distinct in the membrane structure, since treatment of chloroplasts with digitonin or other detergents, followed by density-gradient fractionation (page 158), yields two membrane fractions, one enriched in photosystem I and the other in photosystem II.

Noncyclic Electron Flow and Noncyclic Photophosphorylation

We must now consider some important questions raised by the preceding discussion. What is the function of the two photosystems? How are photosystems I and II of green plants related? Do they function independently, or are they connected?

To answer these questions it will be simplest to examine the main outlines of the process of photosynthetic electron transport as we know it today, resulting from several different lines of research. In the simplest terms, photosystems I and II are the energy-delivering components in a continuous electron-transport chain extending from water, the electron donor, to NADP⁺, the electron acceptor, shown in linear form in Figure 22-10 to emphasize points of similarity with the electron-transport chain of mitochondria (page 495). First we note, as pointed out before, that electron flow along the photosynthetic chain from water to NADP⁺ is opposite in direction to mitochondrial electron transport. Second, the photosynthetic electron-transport chain contains a rather large number of electron carriers, as is also true for mitochondrial electron transport. The major difference, however, is the presence of the two pigment systems, photosystems I and II, which function as electron boosters utilizing light energy to push electrons via a series of carriers from H_2O to NADP⁺. The photosynthetic electron-transport chain has three functional segments, a rather short segment from H_2O to photosystem II, a central chain from photosystem II to photosystem I, and a segment from photosystem I to NADP⁺ (Figure 22-10).

We may now trace the flow of electrons along this chain from water to NADP⁺. When photons are absorbed by the pigment molecules of photosystem I, resulting in their excitation, the energy-rich excitons formed are trapped by P700; as a result P700 loses electrons, which are transferred to its primary electron acceptor, a pigment designated as P430. These electrons flow via a chain of electron carriers (whose identity is described later) to NADP⁺ causing its reduction to yield NADPH. Of course, two electrons are required to reduce each molecule of NADP⁺. However, the loss of an electron from P700 leaves it in its oxidized form, P700⁺, which is deficient in an electron. To use an apt term, P700 is left with an electron hole. The electron required to fill this hole is provided by the central chain of electron carriers extending from photosystem II to photosystem I. However, electrons are available for reducing P700⁺ only after photosystem II is in turn illuminated, causing it to become excited, with the result that the excitation energy is trapped by P680, which

Figure 22-10

(Right) *The photosynthetic electron-transport chain, showing the flow of electrons from H_2O to NADP⁺ via a sequence of electron carriers (see key). The energy required comes from the light quanta (hν) absorbed by the two photosystems, which push electrons toward NADP⁺. The photosynthetic electron-transport chain occurs in three segments, set apart by the two photosystems. In the central segment electron flow from P680 to P700 provides energy for the formation of ATP.*

Key:

c_{550} = *cytochrome* c_{550}
b_{559} = *cytochrome* b_{559}
PQ = *plastoquinone*
f = *cytochrome f*
PC = *plastocyanin*
P430 = *pigment 430*
FRS = *ferredoxin-reducing substance*
Fd = *ferredoxin*
FP = *ferredoxin—NADP⁺ reductase*

$$NADP^+ \longleftarrow FP \longleftarrow Fd \longleftarrow FRS \longleftarrow P430 \longleftarrow \boxed{P700} \longleftarrow PC \longleftarrow f \longleftarrow PQ \longleftarrow b_{559} \longleftarrow C550 \longleftarrow \boxed{P680} \longleftarrow ? \overset{O_2}{\underset{H_2O}{\nwarrow}}$$

Photosystem I — $h\nu$

Photosystem II — $h\nu$

ATP

Shown for comparison is the respiratory chain of mitochondria, in which electron flow proceeds in the direction NADH to oxygen (see also Figure 15-14, page 495).

$$NADH \longrightarrow Fe \cdot S \longrightarrow FP \longrightarrow Fe \cdot S \longrightarrow Q \longrightarrow b \longrightarrow c_1 \longrightarrow c \longrightarrow a \longrightarrow a_3 \longrightarrow \overset{O_2}{\underset{H_2O}{\big(}}$$

ATP   ATP   ATP

loses an electron to the primary acceptor designated as C550, thus leaving behind the oxidized form of P680 or $P680^+$. The electron lost from P680 flows along the central chain of electron carriers to the electron hole in $P700^+$ of photosystem I, restoring $P700^+$ to its reduced state. But now, $P680^+$ of photosystem II must in turn be restored to its reduced state. The electron required comes from a water molecule via another chain of electron carriers whose nature is still unknown. The removal of electrons from water results in the evolution of molecular oxygen. This unidirectional light-induced electron flow from water to $NADP^+$ is called *noncyclic photosynthetic electron transport*. The two electrons required to reduce each molecule of $NADP^+$ are furnished by one molecule of water.

Figure 22-10 also shows that the phosphorylation of ADP to ATP is coupled to the flow of electrons from water to $NADP^+$; it is localized in the central chain leading from photosystem II to photosystem I. It is not known with certainty whether one or two ATPs are formed per pair of electrons traveling from photosystem II to photosystem I. In any case, if we assume that only one phosphorylation takes place, we can represent the overall equation for photosynthetic electron flow from H_2O to $NADP^+$, with its coupled phosphorylation, as

$$H_2O + NADP^+ + P_i + ADP \xrightarrow{\text{light}} \tfrac{1}{2}O_2 + NADPH + H^+ + ATP + H_2O$$

The water molecule on the left-hand side of the equation is the donor of the two electrons required to reduce $NADP^+$ and of the oxygen atom that is released as $\tfrac{1}{2}O_2$. The water molecule on the right-hand side of the equation arises from the formation of ATP from ADP and phosphate (page 398). Although these water molecules can be canceled out, they are left in the equation to show all the inputs and outputs of photosynthetic electron transport.

Energy Relationships in Photosynthetic
Electron Transport

To clarify how light energy is utilized to bring about the flow of electrons from H_2O to $NADP^+$, as well as the photophosphorylation of ADP, we now show the photosynthetic electron-transport chain, which was represented in a linear form in Figure 22-10, in a way that indicates the energy relationships. Figure 22-11 is an energy diagram of photosynthetic electron transport from H_2O to NADPH, in terms of the standard oxidation-reduction potentials of the initial reactants, the final products, and the various electron carriers of the photosynthetic electron-transport chain.

First it should be noted that electrons are raised in energy level from the electropositive standard oxidation-reduction potential of the water-oxygen system, $+0.82$ V, to the electronegative potential of the $NADPH–NADP^+$ system, -0.32 V, in a pathway featuring two large boosts furnished by the input of light energy into photosystems I and II.

P700 in its ground state has a relatively positive standard oxidation-reduction potential (page 478), believed to be somewhere between $+0.40$ and $+0.50$ V; it therefore has little tendency to lose an electron (see page 480). However, after it is excited by absorption of light energy, the standard potential of P700 becomes much more negative, i.e., more energy-rich. Absorption of an exciton by P700 thus serves to boost an electron from a low-energy to a high-energy form, so that it has more than sufficient "pressure" or energy to reduce $NADP^+$ ($E_0' = -0.32$ V) in a "downhill" flow via the chain of electron carriers extending from excited P700 to $NADP^+$ (see Figure 22-11). Similarly, absorption of light quanta by photosystem II boosts electrons from a low energy level in P680 to a high energy level, so that they can now flow downhill from excited P680 to deexcited $P700^+$ via the central electron-transport chain. To complete the process, electrons flow downhill from water to the very electropositive deexcited $P680^+$. During the downhill passage of electrons via the excited photosystem II to photosystem I, coupled photophosphorylation of ADP to ATP occurs. Ample energy is available on passage of a pair of electrons from the primary electron acceptor of photosystem II (E_0' about -0.05 V) to deexcited P 700 (E_0' about $+0.4$ V) for the coupled synthesis of at least one molecule of ATP, which requires 7.3 kcal, equivalent to a difference in standard potential of about 0.16 V.

In each photosystem there is only one truly light-dependent reaction, namely, the absorption of a light quantum and transfer of its energy to P700 (or P680), with displacement of high-energy electrons to the first acceptor. The ensuing passage of these electrons along the electron-transport chain can take place in the dark. For this reason the term "light reactions," traditionally used in reference to the whole set of reactions shown in Figures 22-10 and 22-11, is actually inaccurate, since the only reactions that require light are the steps in which P700 and P680 lose electrons; the rest of the electron-transferring steps can occur in the absence of light.

Figure 22-11
Energy diagram of photosynthetic electron transport via photosystems I and II, plotted in terms of the standard oxidation-reduction potentials E_0' of the interacting redox couples. The symbols for the electron carriers are defined in Figure 22-10. This representation is often called the Z scheme. The two systems are connected by a central electron-transport chain between the primary electron acceptor of photosystem II and P700 of photosystem I. Noncyclic electron flow employs both systems, starting from water and ending in NADPH. Cyclic electron flow requires only photosystem I; electrons boosted to the primary electron acceptor (P430) in photosystem I can return to $P700^+$ via a shunt provided by cytochrome b_6 or by an artificial electron carrier such as phenazine methosulfate. The point of entry of the shunt is not known precisely. The boosting of electrons to higher energy levels by the absorption of light energy ($h\nu$), is shown by the colored arrows. Phosphorylation of ADP to ATP is coupled to electron flow in the central chain. There is increasing evidence that two molecules of ATP are formed per pair of electrons.

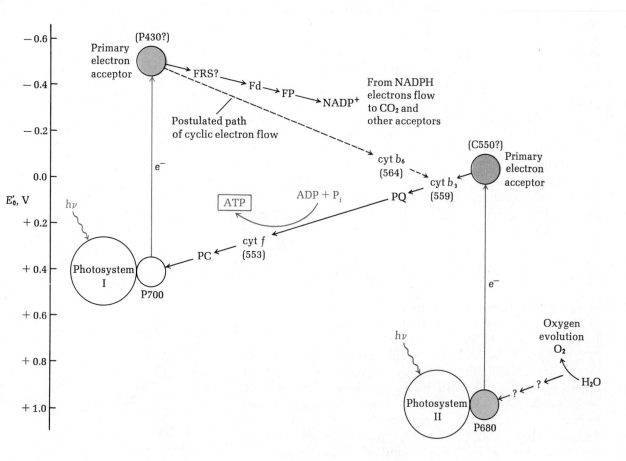

We shall now examine the identity and sequence of the electron carriers participating in photosynthetic electron transport.

Electron Transport from Photosystem I to NADP$^+$

In 1958 A. San Pietro and H. M. Lang discovered that the reduction of NADP$^+$ and the evolution of oxygen by illuminated spinach-chloroplast suspensions could be greatly accelerated by addition of a soluble protein isolated from spinach. They showed this factor to be an enzyme that promotes the transfer of electrons to NADP$^+$ and called it _photosynthetic pyridine nucleotide reductase_. Highly purified preparations were found to contain iron and sulfur in a labile form. Later Arnon and his colleagues found that a crystalline preparation of the iron-sulfur protein _ferredoxin_ (page 488) from the nonphotosynthetic anaerobic bacterium *Clostridium pasteurianum* is also highly active in promoting photoreduction of NADP⁺ in spinach chloroplasts. This observation suggested that an iron-sulfur protein resembling bacterial ferredoxin (see page 488) is one of the components carrying electrons from photosystem I to NADP⁺. Arnon and his colleagues ultimately isolated a crystalline ferredoxin from spinach leaves. It has a molecular weight of about 11,600 and contains two iron atoms bound to two sulfur atoms. Like other iron-sulfur proteins, spinach ferredoxin occurs in Fe(II) and Fe(III) forms. Its oxidized, or Fe(III), form shows two strong absorption bands in the visible spectrum, at 463 and 420 nm, which decrease in intensity on reduction. It has a standard oxidation-reduction potential of -0.42 V at pH 7.0, which is thus about 0.1 V more negative than the standard potential of the NADPH–NADP⁺ couple (page 479).

In its reduced state ferredoxin cannot pass its electrons directly to NADP⁺, but an enzyme capable of catalyzing this reaction in the dark, *ferredoxin-NADP oxidoreductase*, has been isolated from spinach chloroplasts and crystallized. It is a flavoprotein that can utilize either NAD⁺ or NADP⁺ as electron acceptor, although its affinity for NADP⁺ is 400 times greater than for NAD⁺; in plant photosynthesis NADP⁺ is its normal acceptor. The equation for this reaction may be written

$$2Fd(II) + 2H^+ + NADP^+ \longrightarrow 2Fd(III) + NADPH + H^+$$

in which Fd designates ferredoxin.

There is still some uncertainty about the identity of the primary acceptor of electrons from P700 (Figures 22-10 and 22-11). Since it shows characteristic light-absorption changes at 430 nm on reduction, the pigment responsible is designated P430. Electron spin resonance measurements (page 489) suggest that P430 is an iron-sulfur protein distinct from ferredoxin; in fact, such a membrane-bound iron-sulfur protein has been isolated from chloroplasts. Recent measurements indicate that P430 has a standard oxidation-reduction potential of about -0.4 V. It has been proposed that another factor, *ferredoxin-reducing substance* (FRS), is required for transfer of electrons to ferredoxin. However, FRS may be identical with P430.

The NADPH formed in this segment of the photosynthetic electron-transport chain is ultimately used to reduce CO_2, the major electron acceptor in plant photosynthesis, to form carbohydrate. NADPH can also be used to reduce other electron acceptors via various NADP-linked dehydrogenases.

Electron Transport from Photosystem II to Photosystem I

Chloroplasts from higher plants contain a number of other electron carriers, which function in the transfer of electrons from photosystem II to photosystem I (Figure 22-11). Among these are at least three distinctive plant cytochromes. The first, found in 1940 by R. L. Hill and H. Davenport, is *cytochrome f* (Latin *frons,* "leaf"); it is tightly bound to the chloroplast structure and can be released only by using alkaline nonpolar solvents. It has a molecular weight of about 100,000 and contains two hemes per molecule. The γ band of its reduced form is at 553 nm and it is therefore also called cytochrome b_{553}. Its standard oxidation-reduction potential is about +0.365 V, compared with the value of +0.265 V for mitochondrial cytochrome *c*. The second cytochrome is *cytochrome b_3*, which has an absorption maximum at 559 nm and is also called b_{559}. The third is *cytochrome b_6*, whose function is not well understood. Cytochrome b_6 has not been obtained in water-soluble form because it is very tightly bound to the chloroplast membrane structure. Its γ band is at 564 nm, and its standard potential is −0.06 V. Chloroplasts also contain a blue copper-protein, called *plastocyanin*, which has a standard oxidation-reduction potential of about +0.32 V. Also present in the chain are two fat-soluble quinones, *plastoquinone* (Figure 22-12), an analog of ubiquinone (page 493), and *vitamin K₁* (page 358).

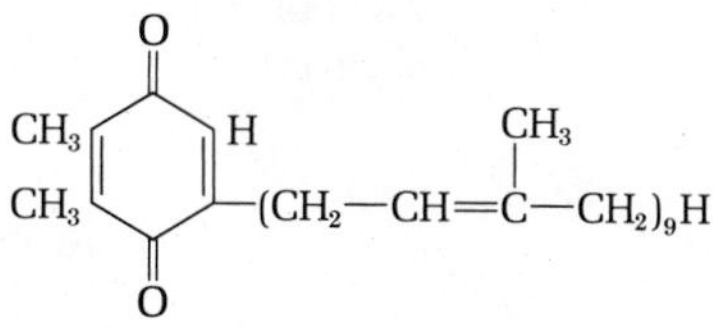

Elucidation of the sequence of these electron carriers in the photosynthetic electron-transport chain between photosystems II and I has been approached by many of the methods used to analyze mitochondrial electron transport. The sequence of carriers shown in Figure 22-11 has been deduced from information on the standard potentials of the carriers, the kinetics and sequence of their interaction as observed spectroscopically, and the action of inhibitors and artificial electron carriers. The chain leading from photosystem II to P700 begins with the unidentified primary electron acceptor, shown by spectral measurements to undergo a characteristic change at 550 nm on photoreduction; it is often designated C550. The next carrier is believed to be cytochrome b_{559} ($E_0' \approx$ +0.080V), followed by plastoquinone (E_0' = +0.100 V). Cytochrome f(553) and plastocyanin, which have more positive standard potentials, carry electrons further; it has been proposed that plastocyanin is the direct electron donor to P700⁺. Another experimental approach to analyzing the sequence of electron carriers in this chain is provided by genetic mutants of algae defective in one or another of the carriers. Information from such mutants also supports the formulation shown in Figure 22-10.

Electron Transport from Water to Photosystem II

The pathway and mechanism of electron transfer from the electron donor water to P680⁺ of photosystem II in plants is not well understood. Four electrons must be removed from two water molecules to bring about formation of one molecule of oxygen (O_2). It is probable that the dehydrogenation of water occurs in two (or more) sequential steps, the nature of which is unknown. This process is known to require Mn^{2+} and the chloride ion. Various artificial electron donors such as hydroxylamine, benzidine, and semicarbazide may replace water as the electron donor to photosystem II.

Figure 22-12
Plastoquinone A. This is the most abundant plastoquinone in plants and algae. Other plastoquinones differ in the length of the side chain and the nature of the substituents in the quinone ring.

Cyclic Photosynthetic Electron Transport and Cyclic Photophosphorylation

Another type of light-induced electron flow in green-plant cells, *cyclic electron flow*, cannot be detected by the methods used for measuring noncyclic electron flow. It can be recognized only by an effect produced by the flow, namely, the phosphorylation of ADP to ATP. Isolated chloroplasts can cause phosphorylation of ADP to ATP when they are illuminated in the absence of any added electron donor or electron acceptor and without accumulation of a reduced substance. Since there is no *net* transfer of electrons and thus no accumulation of a reduced product, it has been concluded that absorption of light energy causes a flow of electrons from P700 of excited photosystem I around a circular chain of electron carriers, in such a way that the electrons ultimately return to reduce deexcited P700$^+$ of photosystem I, without involving photosystem II. Thus this process is called cyclic electron flow. Although it is not possible to measure the rate of cyclic electron flow directly, since the electrons do not accumulate in any product, it is possible to measure the ATP produced by cyclic electron flow, a process called *cyclic photophosphorylation*. Cyclic electron flow is greatly stimulated by various oxidation-reduction carriers, such as the dye pyocyanin or by flavin nucleotides.

Cyclic electron flow is thought to occur by a shunt or bypass (Figures 22-11 and 22-13), which comes into play when NADP$^+$ is not available to act as electron acceptor. Electrons ejected from P700 of photosystem I, instead of passing to NADP$^+$, are shunted into the central electron-transport chain, probably via cytochrome b_6, and pass back to the electron hole in photosystem I. Coupled to this electron flow is the phosphorylation of ADP to ATP. It is also possible that cyclic electron flow and phosphorylation are promoted by a separate photosystem that is entirely independent of the noncyclic system.

To summarize, noncyclic electron flow, the normal process occurring in plant photosynthesis, involves both photosystems I and II and is accompanied by net formation of both ATP and NADPH and by evolution of oxygen. Cyclic electron flow, which requires the operation of only photosystem I, involves no oxygen evolution or net reduction of an electron acceptor, only the coupled synthesis of ATP. It is not known whether cyclic electron flow normally occurs in living plants.

Figure 22-13
Cyclic electron flow and photophosphorylation (see also Figure 22-11).

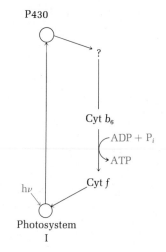

The Energetics of Photosynthesis

The maximum thermodynamic efficiency of photosynthesis, i.e., its *quantum efficiency*, has been a celebrated and hotly debated problem. In principle, the problem reduces to the experimental determination of the value of n, the number of light quanta required in the overall equation of photosynthesis. The standard-free-energy change $\Delta G^{\circ\prime}$ for the synthesis of hexose from CO_2 and H_2O is $+686$ kcal. If we divide this value by 6, we obtain the energy input, namely,

+114 kcal, required to reduce one molecule of CO_2 to (CH_2O), as called for in the equation

$$H_2O + CO_2 \xrightarrow{\text{light}} (CH_2O) + O_2 \qquad (3)$$

We have seen that the caloric value of a light quantum depends on its wavelength and ranges from 72 kcal einstein^{-1} at 400 nm to about 41 kcal at 700 nm (Figure 22-4). If we take the smaller of these values, 41 kcal, then the minimum number of quanta n required for reaction (3) would be 114/41, or about 2.7. Since light quanta are indivisible, the quantum requirement must be an integral number, which is therefore at least 3.0 if *all* the energy required for hexose formation comes from light.

The actual quantum requirement of photosynthesis is determined by experimental measurement of the amount of light absorbed by a suspension of photosynthesizing cells in relation to the amount of carbon dioxide reduced and/or oxygen evolved. The first systematic measurements of n were carried out by O. Warburg in Germany, in the 1920s. At first he found values of about 5, but many years later, with different techniques, he observed values as low as 3. He postulated that the true value for the quantum yield is actually 1 per molecule of CO_2 reduced. Although 3 quanta are theoretically required, as we have seen, Warburg proposed that the balance of the energy required comes from respiration, a dark process.

Warburg's hypothesis has since been found to be inconsistent with a number of other experimental observations. For one thing, most available evidence indicates that *all* of the energy required for reduction of carbon dioxide to hexose and for oxygen evolution normally comes from light. Moreover, most other investigators, particularly Emerson, have found that in photosynthesis fully supported by light energy the minimum number of quanta required is 8. This value corresponds to a thermodynamic efficiency nearly the same as that of the overall process of respiration, about 38 percent (page 517).

To understand the significance of the quantum yield of 8 we must anticipate a point more fully developed in Chapter 23 and give the overall equation for the reduction of CO_2 to (CH_2O) during the dark phase of photosynthesis in most green plants, which is

$$CO_2 + 3ATP + 2NADPH + 2H^+ + 2H_2O \longrightarrow$$
$$(CH_2O) + 3ADP + 3P_i + 2NADP^+$$

To provide the two molecules of NADPH required per molecule of CO_2 reduced in the dark reactions, four electrons are ejected by photosystem I. Correspondingly, four electrons are also ejected by photosystem II, thus yielding one molecule of oxygen (O_2) from water. Since 1 light quantum is required to eject each electron, we require 4 quanta for photosystem I and 4 for photosystem II. Now, if only one photophosphorylation per pair of electrons occurs during the

overall process, as many investigators believe, only two molecules of ATP will be generated by the two pairs of electrons
flowing from water to NAD^+, according to the equation

$$2H_2O + 2NADP^+ + 2ADP + 2P_i \xrightarrow{\text{light}}$$
$$O_2 + 2NADPH + 2H^+ + 2ATP + 2H_2O$$

in which colored type is used to show the formation of ATP.
But three molecules of ATP are required to reduce one molecule of CO_2 in the dark phase, as we have seen. However, if
two ATPs are formed during light-induced transport of a pair
of electrons from water to $NADP^+$, as some investigators have
postulated, then four ATPs will be formed per 8 light quanta,
which is more than enough to reduce 1 mol of CO_2 to carbohydrate.

There is another way in which extra ATP required for
dark biosynthetic reactions may be generated, i.e., cyclic
photophosphorylation, which may proceed independently of
noncyclic electron flow to generate extra ATP from ADP.

The efficiency of photosynthesis in nature is much lower
than the 38 percent calculated for the basic molecular
process. From the amount of carbon fixed by a field of corn
in one growing season it has been found that only about 1 to
2 percent of the solar energy falling on the field is recovered
in the form of new photosynthetic products; uncultivated
plant life yields far less, perhaps only 0.2 percent. Cultivated
sugarcane is much more efficient; it can yield up to 8 percent
of the captured light energy in the form of organic products.
As will be seen elsewhere (page 639), the process of photorespiration tends to lower the net efficiency of photosynthesis in some plants.

Properties of Photosynthetic Phosphorylation; Coupling Factors

The phosphorylation of ADP accompanying photoinduced
electron transport in chloroplasts strongly resembles the
analogous process of oxidative phosphorylation coupled to
electron transport in mitochondria (page 514). For example,
light-induced noncyclic electron transport is stimulated by
addition of the phosphate acceptor ADP to chloroplasts,
which thus shows the phenomenon of acceptor control of
photoinduced electron transport, similar to the requirement
of ADP for maximal rates of electron transport in mitochondria (page 518).

Like mitochondrial phosphorylation, photosynthetic
phosphorylation also can be uncoupled by certain chemical
agents, so that electron flow continues but no phosphorylation takes place. Among the uncoupling agents effective on
photosynthetic phosphorylation are NH_4^+ ions and carbonylcyanide phenylhydrazone; the latter is effective against
either oxidative or photosynthetic phosphorylation (page
519). Ionophores (page 520), such as gramicidin, also prevent
photophosphorylation. Phloridzin, a toxic glycoside from the
bark of pear trees and a synthetic compound called Dio-9

inhibit photophosphorylation by an action resembling that of oligomycin on mitochondrial phosphorylation.

Yet another line of evidence showing the similarity between mitochondrial and photosynthetic phosphorylation is provided by studies of ATPase activity. Chloroplasts normally show no ATPase activity, but after treatment with certain sulfhydryl compounds, such as dithiothreitol, a light-dependent ATPase activity is evoked, similar to the ATPase activity of mitochondria evoked by uncoupling agents such as 2,4-dinitrophenol (page 519). From corn chloroplasts E. Racker and his colleagues have extracted and purified a protein fraction that acquires the ability to hydrolyze ATP in the presence of Ca^{2+} when exposed to trypsin. The chloroplast ATPase has properties that resemble those of mitochondrial F_1 ATPase functioning in the coupled synthesis of ATP during oxidative phosphorylation (page 523). It has about the same molecular weight (380,000) and contains five different kinds of subunits (page 523). The chloroplast ATPase, which is designated CF_1, acts as a coupling factor and restores photosynthetic phosphorylation in chloroplasts depleted of CF_1.

A further similarity between chloroplast ATPase and mitochondrial ATPase has been revealed by electron microscopy of the lamellar membranes of corn chloroplasts. These membranes show the presence of 9-nm spheres protruding from the outer surface of the thylakoid membrane; in mitochondria such spheres protrude from the inner surface of the inner membrane (page 522). These spheres are lost from chloroplasts when the CF_1 ATPase is extracted from them; the spheres are restored when the CF_1 ATPase is returned to the depleted membranes under appropriate conditions, with restoration of photosynthetic phosphorylation.

Figure 22-14 shows an electron micrograph of the membrane surfaces of a thylakoid. The granular bodies are believed to be enzyme complexes, among which are molecules of CF_1 ATPase.

Figure 22-14
Ultrastructure of thylakoids. Electron micrograph of the membrane surfaces of a thylakoid disk of a spinach chloroplast, following freeze-fracturing. In this technique the sample is frozen and then fractured with a knife edge to yield cleavage surfaces, which are replicated by metal-casting. The image is that of the metal replica.

100 nm

D. Branton

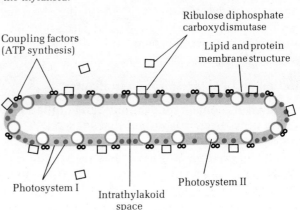

A schematic representation of the ultrastructure of the thylakoid.

Coupling factors
(ATP synthesis)

Ribulose diphosphate
carboxydismutase

Lipid and protein
membrane structure

Photosystem I

Intrathylakoid
space

Photosystem II

The Mechanism of Photosynthetic Phosphorylation

We have seen that currently there are three hypotheses for the mechanism of oxidative phosphorylation, chemical coupling, conformational coupling, and chemiosmotic coupling (page 524). Because of the striking similarity between oxidative phosphorylation in mitochondria and photophosphorylation in chloroplasts, these three hypotheses for mitochondrial phosphorylation are also applicable to the mechanism of photophosphorylation.

Much evidence is consistent with the chemiosmotic-coupling hypothesis as the basic mechanism of photosynthetic phosphorylation. When chloroplasts are illuminated under conditions in which they exhibit cyclic electron flow, they absorb H^+ ions from the suspending medium, which becomes more alkaline. When the light is turned off and the photoinduced electron flow stops, H^+ ions slowly return from the chloroplasts to the medium. These movements of H^+ ions are remarkably similar to those occurring during electron transport in isolated mitochondria, with one important difference. In mitochondria, H^+ ions are *ejected* into the medium during electron transport, whereas in chloroplasts H^+ ions are *absorbed* during electron transport. The sidedness of the chloroplast membrane thus is the reverse of that of the mitochondrial membrane (page 527).

Another important piece of evidence favoring the chemiosmotic hypothesis is the discovery by A. Jagendorf and his colleagues that a pH gradient artificially imposed across the chloroplast membrane can drive the phosphorylation of ADP in the dark, without the input of light energy (Figure 22-15). They lowered the internal pH of chloroplasts artificially by soaking them in an acid bath, a medium buffered at pH 4.0. The chloroplasts were then quickly mixed in the dark with an alkaline buffer (pH 8.5), in order to impose a momentary pH gradient across the membrane; the alkaline buffer also contained phosphate and ADP. Mixing was followed by a burst of ATP formation; simultaneously, the pH gradient disappeared.

When chloroplasts are illuminated with a very short light flash given by a single laser pulse (20 ns), only a single "turnover" of the electron carriers ensues. This is accompanied by absorption of H^+ ions by the chloroplasts. H. Witt and his colleagues in Berlin have used such biophysical methods to establish the quantitative relationships between electron flow, H^+ movements, and ATP formation. Their observations also support the chemiosmotic hypothesis.

Respiration and Photorespiration in Plants

Green-plant cells contain mitochondria in addition to chloroplasts, and it has been established that such cells exhibit mitochondrial respiration and oxidative phosphorylation in the dark, at the expense of substrates generated by photosynthesis in earlier light periods. The question arises whether green-plant cells also respire in the light, during active photosynthesis, or whether respiration is turned off. From careful measurements of the rates of oxygen and

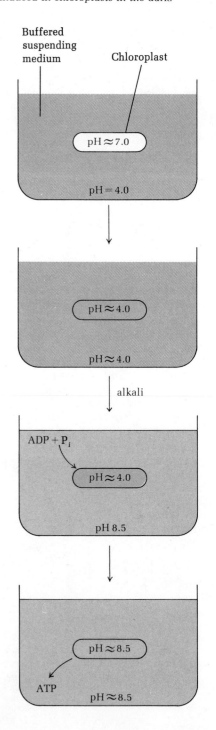

Figure 22-15
Light-independent phosphorylation of ADP at the expense of an artificial pH gradient induced in chloroplasts in the dark.

carbon dioxide exchanges in illuminated plants, particularly with the use of isotopic oxygen, it has been found that plants do in fact respire in the light while they are carrying out photosynthesis. However, the type of respiration that occurs in illuminated green plants is not mitochondrial, since it is not sensitive to characteristic inhibitors of mitochondrial electron transport. Such "light" respiration in green plants, which is called *photorespiration*, short-circuits photosynthesis. We shall consider the mechanism of photorespiration in the next chapter (page 639), since it involves intermediates in hexose synthesis that are generated during the dark reactions of photosynthesis.

Summary

Water is used as electron donor, and from it oxygen is evolved in all photosynthetic organisms except the photosynthetic bacteria, which employ H_2, H_2S, or certain organic compounds as electron donors. Although CO_2 is the major electron acceptor in higher plants, NO_3^-, N_2, and H^+ may also serve as acceptors. The direction of net light-induced electron flow in photosynthesis is toward the more electronegative system, i.e., against the normal gradient of the standard oxidation-reduction potentials. The energy for such reverse electron flow is provided by the absorbed light. The first stage of photosynthesis (the light reactions) results in the reduction of $NADP^+$ and the phosphorylation of ADP; in the second stage (the dark reactions), NADPH and ATP are used to reduce CO_2 to hexose. In eukaryotic cells photosynthesis takes place in the thylakoids, flattened membranous vesicles within chloroplasts; they are stacked into structures called grana. Photosynthetic cells contain three types of light-capturing pigments, chlorophylls, carotenoids, and phycobilins, arranged into two sets called photosystems I and II. Photosystem I contains chlorophyll *a* and β-carotene, as well as a single molecule of P700, a specialized chlorophyll *a* which serves as an energy trap. When a pigment molecule in photosystem I is excited, its energy level is raised. The excitation energy is then transferred to P700, from which electrons are expelled for the reduction of $NADP^+$ to yield NADPH. Photosystem II has its own set of pigments and a characteristic reactive center, namely P680, a specialized chlorophyll-protein complex. Photosystem II is excited by shorter wavelengths and is responsible for oxygen evolution. Organisms that do not evolve O_2 lack photosystem II. In green plants photosystems I and II are linked in series. Boosting an electron to a highly reducing potential by excitation of photosystem I leads to reduction of $NADP^+$, via a chain of carriers that includes ferredoxin and ferredoxin-NADP oxidoreductase. The electrons required to fill the electron holes left in photosystem I come from excited photosystem II via a central electron-transport chain, which includes a plastoquinone, cytochrome b_{559}, cytochrome *f*, and plastocyanin. The electrons required to fill the electron holes in photosystem II come from H_2O, which is dehydrogenated and yields oxygen by mechanisms still unknown.

Phosphorylation of ADP is coupled to photoinduced electron transport in the chain between photosystems I and II. Noncyclic photoinduced electron transport is that occurring during the net transfer of electrons from H_2O to $NADP^+$ via photosystems I and II. At least one molecule of ATP is generated per pair of electrons passing from H_2O to $NADP^+$. Cyclic photoinduced electron flow and phosphorylation occur in photosystem I alone, through a shunt mechanism in which no net reduction of an electron acceptor takes

Spinach chloroplasts are exposed to an acid bath at pH 4 for a prolonged period.

As a result the internal pH of the chloroplast approaches that of the medium.

ADP and P_i are added in the dark, together with sufficient alkali to bring the external pH to 8.5, thus creating a momentary pH gradient across the membrane.

ATP is formed from ADP and P_i in the dark at the expense of the pH gradient.

place. In noncyclic electron transport, apparently 8 light quanta, 4 absorbed by photosystem I and 4 by photosystem II, are required to evolve each molecule of oxygen and to reduce one molecule of CO_2, for which two molecules of NADPH and three molecules of ATP are necessary.

Photosynthetic phosphorylation appears to have many of the same properties as oxidative phosphorylation; it may be uncoupled or inhibited by specific agents. A coupling factor for photosynthetic phosphorylation has been isolated; it has ATPase activity. Photophosphorylation is reconstituted in depleted chloroplasts by addition of the chloroplast ATPase. Chloroplasts absorb H^+ during illumination. Considerable evidence suggests that chemiosmotic coupling occurs during photosynthetic electron flow and phosphorylation.

References

Books

CLAYTON, R. K.: *Molecular Physics in Photosynthesis*, Blaisdell, New York, 1965.

GOVINDJEE (ed.): *Bioenergetics of Photosynthesis*, Academic, New York, 1974.

GREGORY, R. P. F.: *Biochemistry of Photosynthesis*, Wiley-Interscience, London, 1971.

HALL, D. O., and K. K. RAO: *Photosynthesis*, Arnold, London, 1972. A useful primer (67 pages).

KROGMANN, D. W.: *The Biochemistry of Green Plants*, Prentice-Hall, Englewood Cliffs, N.J., 1973.

RABINOWITCH, E., and GOVINDJEE: *Photosynthesis*, Wiley, New York, 1969. A more comprehensive survey of photosynthesis.

TRIBE, M., and P. WHITTAKER: *Chloroplasts and Mitochondria*, Arnold, London, 1972. Elementary review of chloroplast structure and function.

VERNON, L. P., and G. R. SEELEY (eds): *The Chlorophylls*, Academic, New York, 1966.

Articles

AMESZ, J.: "The Function of Plastoquinone in Photosynthetic Electron Transport," *Biochim. Biophys. Acta*, 301: 35–51 (1973).

BOARDMAN, N. K.: "The Photochemical System of Photosynthesis," *Adv. Enzymol.*, 30: 1–80 (1968). A review of earlier work.

FRENKEL, A. W.: "Multiplicity of Electron Transport Reactions in Bacterial Photosynthesis," *Biol. Rev.*, 45: 569–616 (1970).

HALL, D. O., and M. C. W. EVANS: "Photosynthetic Phosphorylation in Chloroplasts," *Sub-Cell. Biochem.*, 1: 197–206 (1972).

HILL, R.: "The Biochemists' Green Mansions: The Photosynthetic Electron Transport Chain in Plants," pp. 121–152 in P. N. Campbell and G. D. Greville (eds.), *Essays in Biochemistry*, vol. 1, Academic, New York, 1965. An interesting account written by a pioneer.

KE, B.: "The Primary Electron Acceptor of Photosystem I," *Biochim. Biophys. Acta*, 301: 1–33 (1973).

LEMON, E., D. W. STEWART, and R. W. SHAWCROFT: "The Sun's Work in a Cornfield," *Science*, 174: 371–378 (1971). The energetics of photosynthesis in a cultivated plant.

OLSON, J. M.: "The Evolution of Photosynthesis," *Science*, 168: 438–446 (1970).

REEVES, S. G., and D. O. HALL: "The Stoichiometry (ATP/2e Ratio) of Non-Cyclic Photophosphorylation in Isolated Spinach Chloroplasts," *Biochim. Biophys. Acta*, 314: 66–78 (1973).

TREBST, A.: "Energy Conservation in Photosynthetic Electron Transport of Chloroplasts," *Ann. Rev. Plant Physiol.*, 25: 423–458 (1974).

WITT, H. T.: "Coupling of Quanta, Electrons, Fields, Ions, and Phosphorylation in the Functional Membrane of Photosynthesis," *Q. Rev. Biophys.*, 4, 365–477 (1971).

Problems

1. Using the data given on pages 604 and 605, calculate the standard-free-energy change accompanying the transfer of an electron pair from the primary electron acceptor of photosystem II to deexcited P700.

2. Assuming that the formation of 1 ATP accompanies the process described in Problem 1, calculate the efficiency of energy conservation at that step.

3. Assuming a quantum requirement for photosynthesis of 8, calculate a thermodynamic efficiency under standard conditions for the synthesis of 1 mol of glucose if the wavelength of light used is (a) 400 nm, (b) 750 nm.

PART 3 BIOSYNTHESIS AND THE UTILIZATION OF PHOSPHATE-BOND ENERGY

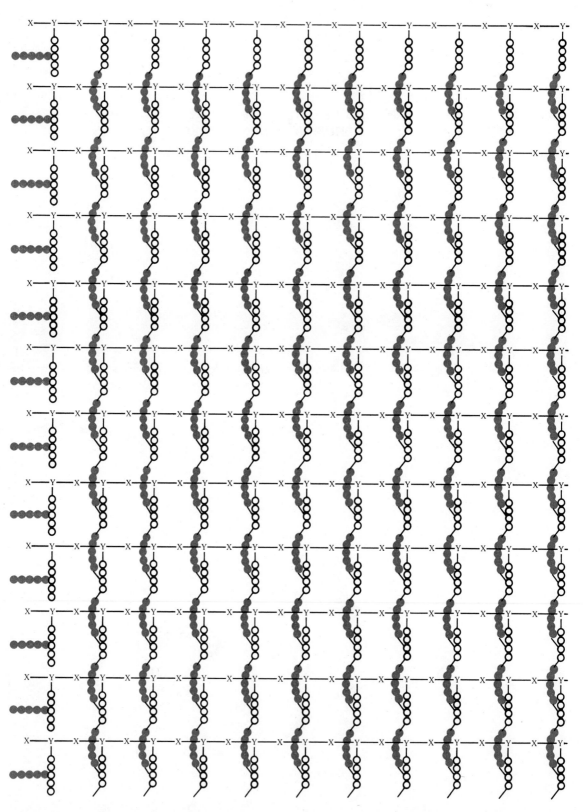

Schematic representation of the peptidoglycan backbone of the cell wall of *Staphylococcus aureus*. Its enzymatic assembly outside the cell membrane from precursors made in the cytoplasm is among the more complex biosynthetic activities known. [From J. L. Strominger, K. Izaki, M. Matsuhashi, and D. J. Tipper, Fed. Proc., 26:9 (1967).]

Key:

    X = *N-acetylglucosamine residues*

    Y = *N-acetylmuramic acid residues*

○○○○ = *side chains of* L-*alanyl-*D-*isoglutamyl-*L-*lysyl-*D-*alanine*

●●●●● = *cross bridges of pentaglycine*

PART 3 BIOSYNTHESIS AND THE UTILIZATION OF PHOSPHATE-BOND ENERGY

Cellular processes that require free energy can be placed in one of three primary categories: (1) *biosynthesis*, in which chemical work is carried out, (2) *contraction* and *motility*, which are forms of mechanical work, and (3) *active transport*, which is a reflection of osmotic or concentration work. The free energy required for these activities is largely furnished by the phosphate-bond energy of ATP and by the reducing power of NADPH and other reduced coenzymes.

Foremost in complexity among the energy-requiring activities is the *biosynthesis* of the major molecular components of cells from simple precursors, the principal process involved in the formation and maintenance of the intricate orderliness of living cells at the expense of their environment. In Part 3 we shall discuss the biosynthesis of what we have called noninformational biomolecules (page 20), namely, carbohydrates, lipids, amino acids, and nucleotides. In Part 4 we shall consider biosynthesis of nucleic acids and proteins, the informational macromolecules of cells, with particular reference to the mechanisms by which genetic information is preserved and transferred.

As we now turn from the biochemistry of catabolism to the biochemistry of anabolism, some of the organizing principles of biosynthetic pathways require reiteration and emphasis. The first principle, mentioned in Chapter 14 (page 373), is that the chemical pathway for the biosynthesis of a biomolecule is not usually identical to the pathway for its degradation. The two pathways may contain one or even several identical steps, but there is nearly always at least one enzymatic step that is dissimilar in the anabolic and catabolic pathways. Thus the mechanisms of biosynthesis and degradation of cell components are not simply the reverse of each other. This fact has profound biological significance. If the reactions of catabolism and anabolism were catalyzed by the same set of enzymes acting reversibly, no stable biological structure of any complexity could exist, since the various equilibria between precursors and products mediated by freely reversible reaction systems would, through mass-action effects, undergo fluctuation with changes in the concentration of the precursors.

The second organizing principle underlying biosynthetic reactions involves the energetics of these processes. We have seen that ATP serves as a common intermediate between those chemical processes in the cell which yield free energy, such as glycolysis, respiration, and the light reactions of photosynthesis, and those cellular processes which consume free energy, such as the biosynthetic pathways. The conservation in the form of ATP of much of the free energy released in a catabolic reaction sequence is made possible because the overall coupled reaction is usually exergonic and hence can proceed in the direction of ATP formation. For example, the overall reaction of glycolysis, which produces ATP, is highly exergonic, as we have seen (page 419):

$$\text{Glucose} + 2P_i + 2ADP \longrightarrow 2 \text{ lactate} + 2ATP$$
$$\Delta G^{\circ\prime} = -32.4 \text{ kcal mol}^{-1}$$

In a similar way, energy-requiring biosynthetic processes are obligatorily coupled to the energy-yielding breakdown of ATP, so that the overall coupled reaction is also exergonic and thus proceeds in the direction of biosynthesis. For example, the synthesis of sucrose from glucose and fructose under standard thermodynamic conditions is strongly endergonic and requires input of energy:

$$\text{Glucose} + \text{fructose} \rightleftharpoons \text{sucrose} + H_2O$$
$$\Delta G^{\circ\prime} = +5.5 \text{ kcal mol}^{-1}$$

However, the overall equation for the biosynthesis of sucrose in sugarcane, which involves coupling to the breakdown of ATP, is actually exergonic:

$$\text{Glucose} + \text{fructose} + 2ATP \longrightarrow \text{sucrose} + 2ADP + 2P_i$$
$$\Delta G^{\circ\prime} = -8.5 \text{ kcal mol}^{-1}$$

By coupling the breakdown of two molecules of ATP to the synthesis of the covalent bond between glucose and fructose the overall biological reaction becomes exergonic and thus can proceed in the direction of sucrose formation.

Another way in which reactions can be "pulled" in the direction of biosynthesis is _pyrophosphate cleavage_ of ATP, the release of pyrophosphate (pages 397 and 410) rather than orthophosphate during a biosynthetic reaction coupled to ATP breakdown. The decrease in standard free energy of such reactions is substantially greater than that of reactions in which ATP undergoes orthophosphate cleavage (page 410). Moreover, pyrophosphate undergoes subsequent enzymatic hydrolysis to orthophosphate by the action of pyrophosphatase, so that in biosynthetic reactions involving pyrophosphate cleavage two high-energy phosphate bonds may ultimately be expended to create only one new covalent linkage in the biosynthetic product.

The third important principle underlying biosynthetic reactions is that their regulation is independent of that of the corresponding catabolic reactions. Such independent control is made possible by the circumstance that the catabolic and anabolic pathways are not identical. Usually the regulatory enzyme controlling the rate of the catabolic pathway does not participate in the anabolic pathway, and, conversely, the regulatory enzyme controlling the biosynthetic pathway is usually not shared by the catabolic pathway.

We have also seen that most catabolic pathways appear to be regulated by the energy charge or the phosphorylation potential of the cell (pages 538 and 539). AMP and ADP often serve as stimulating modulators and ATP as an inhibiting modulator of the regulatory enzymes controlling the rate of catabolism. Certain other catabolic products, such as citrate or NADH, also serve to inhibit catabolic pathways. Biosynthetic pathways are also regulated by ATP, ADP, or AMP, particularly the pathways leading to fuel storage in the form of glycogen or fat. However, the biosynthesis of amino acids and nucleotides is primarily regulated by the concentration of the end product of the biosynthetic process, in such a way that the cell synthesizes only enough of a given biomolecule to meet its immediate needs.

In biosynthetic pathways the regulatory enzyme that is under allosteric control is almost always the first enzyme in the sequence, starting from some key precursor, or from a branch point in a metabolic chain. This arrangement has a biological advantage: allosteric inhibition of the *first* step in a biosynthetic sequence avoids wasting precursors to make unused intermediates. Moreover, the first reaction in a biosynthetic sequence is usually irreversible. This first reaction is often spoken of as the *committing step*, since once it occurs, the remainder of the biosynthetic process nearly always proceeds to completion.

One final point needs to be raised. Does catabolism, which produces ATP, "drive" anabolism, which requires ATP? Or does anabolism "drive" catabolism? The answer is neither. The correct statement is that biosynthesis and all other ATP-requiring activities of the cell *pull* the process of ATP-yielding catabolism. In general, the rate of utilization of the phosphate-bond energy of ATP determines the rate at which ATP is regenerated from ADP at the expense of energy from the environment. This relationship is but another aspect of the principle of maximum economy in the molecular logic of living cells: ATP is generated only as fast as it is needed.

CHAPTER 23 THE BIOSYNTHESIS OF CARBOHYDRATES

From the standpoint of sheer mass, the biosynthesis of glucose and other carbohydrates from simpler precursors is the most prominent biosynthetic process carried out in the biosphere. In the domain of photosynthetic organisms, hexoses generated from carbon dioxide and water are converted into starch, cellulose, and other polysaccharides, which are found in enormous amounts in the plant world. In the domain of heterotrophic organisms, the conversion of pyruvate, lactate, amino acids, and other simple precursors into glucose and then glycogen is also a central biosynthetic route.

We shall now trace the enzymatic pathways involved in the biosynthesis of glucose and other hexoses from simpler precursors, the conversion of monosaccharides into disaccharides, and the biosynthesis of the various storage and structural polysaccharides. We shall also examine the regulatory mechanisms for controlling the rates of key enzymatic steps in these processes.

Major Pathways in Carbohydrate Synthesis

In most cells, the conversion of glucose or glucose 6-phosphate to pyruvate, catalyzed by the glycolytic enzymes, is the central pathway of carbohydrate catabolism, under either anaerobic or aerobic conditions. In a comparable manner, the reverse process, the conversion of pyruvate to glucose 6-phosphate, is the central pathway in the biosynthesis of carbohydrates by many different organisms (Figure 23-1). Converging into this central pathway are various feeder pathways leading from noncarbohydrate precursors (Figure 23-1). One such feeder pathway consists of the reaction sequences by which intermediates of the tricarboxylic acid cycle are transformed into precursors of glucose; this pathway is also employed when the carbon chains of certain amino acids are converted into glucose. Another major feeder pathway consists of the reactions bringing about the net reduction of CO_2 to form precursors of glucose in photosynthetic cells.

Although nearly all organisms, including both heterotrophs and autotrophs, employ the central biosynthetic pathway leading from pyruvate to glucose 6-phosphate, different types of cells utilize the feeder pathways to different degrees. For example, only photosynthetic and chemosynthetic autotrophs (page 363) can cause the *net* reduction of CO_2 to form new glucose. On the other hand, nearly all organisms can convert the glycogenic amino acids into certain tricarboxylic acid cycle intermediates and thus into glucose. In vertebrates the net synthesis of glucose from blood lactate, which occurs largely in the liver, is a particularly active process during recovery from intense muscular activity.

Starting from the glucose 6-phosphate formed in the central pathway, several diverging biosynthetic pathways (Figure 23-1) then lead to formation of (1) other monosaccharides and their derivatives, (2) various disaccharides, (3) fuel-storage polysaccharides such as starch and glycogen, and (4) cell-wall and cell-coat components such as cellulose, xylans, peptidoglycans, and the acid mucopolysaccharides and glycoproteins. The diverging pathways leading from glucose 6-phosphate to these products also differ widely among different organisms. For example, the capacity to form the disaccharide sucrose is present in many plants but absent in the tissues of mammals. The various biosynthetic pathways leading to the extracellular polymers are also species-specific. For example, cellulose is manufactured in large amounts by plants but not by mammals; conversely, various mucopolysaccharides are produced by animals but not by plants. On the other hand, the pathways to starch (or glycogen) appear to be nearly universal in occurrence but are used to widely varying degrees, depending on metabolic demands and nutritional supply.

We shall begin by examining the central biosynthetic pathway leading from pyruvate to glucose 6-phosphate, which is utilized in the process of _gluconeogenesis_, the synthesis of "new" glucose from such precursors as pyruvate, lactate, certain amino acids, and intermediates of the tricarboxylic acid cycle.

The Biosynthetic Pathway from Pyruvate to Glucose 6-Phosphate

Most of the reaction steps in the heavily traveled central pathway from pyruvate to glucose 6-phosphate are catalyzed by enzymes of the glycolytic sequence and thus proceed by reversal of steps employed in glycolysis. However, there are two irreversible steps in the normal "downhill" glycolytic pathway (pages 424 and 431) which cannot be utilized in the "uphill" conversion of pyruvate to glucose 6-phosphate. In the biosynthetic direction these steps are bypassed by alternative reactions (Figure 23-2), which are thermodynamically favorable in the direction of synthesis.

Conversion of Pyruvate to Phosphoenolpyruvate

The first of these bypass steps is the phosphorylation of pyruvate to phosphoenolpyruvate, which does not occur at any significant rate by direct reversal of the pyruvate kinase reac-

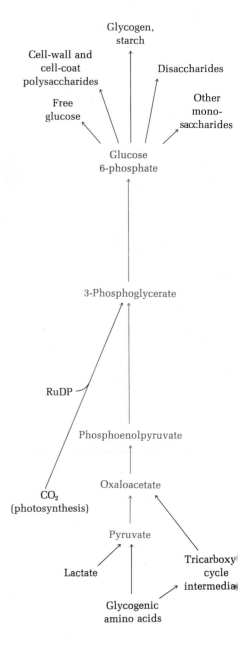

Figure 23-1
The central pathway of hexose biosynthesis (color). The auxiliary pathways feeding into the central pathway from CO_2, lactate, and amino acids are in black, as are the diverging pathways leading from glucose 6-phosphate to other carbohydrates. RuDP is ribulose 1,5-diphosphate, an important intermediate in photosynthetic formation of hexose.

Figure 23-2
Bypasses in the pathway of gluconeogenesis.
The enzymatic steps that bypass the corre-
sponding steps of glycolysis are in color.

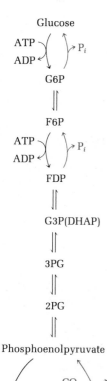

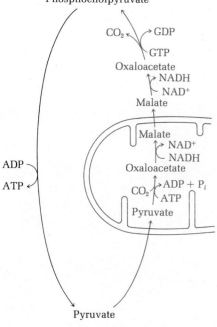

tion, presumably because of the large positive standard-free-energy change:

$$\text{Pyruvate} + \text{ATP} \rightleftharpoons \text{phosphoenolpyruvate} + \text{ADP}$$
$$\Delta G^{\circ\prime} = +7.5 \text{ kcal mol}^{-1}$$

During gluconeogenesis the phosphorylation of pyruvate is achieved by an alternate pathway, through a somewhat roundabout sequence of reactions that requires the cooperation of enzymes in both the cytosol and mitochondrial compartments in the liver of the rat and certain other species (Figure 23-2).

The first step is catalyzed by *pyruvate carboxylase* of mitochondria, which we have seen (page 464) catalyzes the major anaplerotic reaction by which tricarboxylic acid cycle intermediates are generated from pyruvate. This reaction is likely irreversible in cells, as discussed earlier (page 464).

$$\text{Pyruvate} + \text{CO}_2 + \text{ATP} \xrightarrow{\text{acetyl-CoA}} \text{oxaloacetate} + \text{ADP} + \text{P}_i \quad (1)$$
$$\Delta G^{\circ\prime} = -0.5 \text{ kcal mol}^{-1}$$

However, this reaction is tightly regulated since pyruvate carboxylase, an allosteric enzyme, is completely inactive in the absence of its specific positive modulator acetyl-CoA. Mn^{2+} is required for the reaction. The oxaloacetate formed in this mitochondrial reaction is then reduced to malate at the expense of NADH by the mitochondrial form of malate dehydrogenase:

$$\text{NADH} + \text{H}^+ + \text{oxaloacetate} \rightleftharpoons \text{NAD}^+ + \text{malate} \quad (2)$$
$$\Delta G^{\circ\prime} = -6.7 \text{ kcal mol}^{-1}$$

The malate so formed may then leave the mitochondria via the dicarboxylate transport system of the inner mitochondrial membrane, in exchange for phosphate or some other dicarboxylate (page 530). In the cytosol the malate is then reoxidized by the cytoplasmic form of NAD-linked malate dehydrogenase (page 461) to form extramitochondrial oxaloacetate:

$$\text{Malate} + \text{NAD}^+ \longrightarrow \text{oxaloacetate} + \text{NADH} + \text{H}^+ \quad (3)$$
$$\Delta G^{\circ\prime} = +6.7 \text{ kcal mol}^{-1}$$

Although this reaction is strongly endergonic as written, it proceeds to the right because the end products are quickly removed. In the last step of the bypass, oxaloacetate is acted upon by *phosphoenolpyruvate carboxykinase (GTP)* to yield phosphoenolpyruvate and CO₂, a reaction in which GTP (or ITP) serves as the phosphate donor:

$$\text{Oxaloacetate} + \text{GTP} \longrightarrow \text{phosphoenolpyruvate} + \text{CO}_2 + \text{GDP} \quad (4)$$
$$\Delta G^{\circ\prime} = +1.0 \text{ kcal mol}^{-1}$$

Phosphoenolpyruvate carboxykinase (GTP) has a molecular weight of about 75,000. Since it has a very low affinity for CO₂, the reactive chemical species, the enzyme is biologically active only in the direction of phosphoenolpyruvate formation.

We can now write the overall equation [sum of equations (1) to (4)] for the formation of phosphoenolpyruvate, together with a summation of the free-energy changes:

Pyruvate + ATP + GTP $\rightleftharpoons$

$$\text{phosphoenolpyruvate} + \text{ADP} + \text{GDP} + P_i$$
$$\Delta G^{\circ\prime} = +0.5 \text{ kcal mol}^{-1}$$

The overall reaction is thermodynamically reversible because of its small standard-free-energy change. It will tend to go to the right whenever the ATP/ADP ratio is high and excess pyruvate is present. The overall equation can be dissected into its endergonic and exergonic components. The energy-requiring component is

$$\text{Pyruvate} + \text{GTP} \rightleftharpoons \text{phosphoenolpyruvate} + \text{GDP}$$
$$\Delta G^{\circ\prime} = +7.8 \text{ kcal mol}^{-1}$$

The energy-yielding process is

$$\text{ATP} + H_2O \rightleftharpoons \text{ADP} + P_i$$
$$\Delta G^{\circ\prime} = +7.3 \text{ kcal mol}^{-1}$$

Two high-energy phosphate bonds, one from ATP and one from GTP and each providing -7.3 kcal mol^{-1}, are ultimately expended to phosphorylate one molecule of pyruvate. We shall see other instances (pages 682 and 949) in which two or more high-energy phosphate bonds of ATP are ultimately consumed to bring about the formation of but a single covalent bond during a biosynthetic reaction. This bypass pathway from pyruvate to phosphoenolpyruvate in rat liver capitalizes on the fact that the NADH/NAD$^+$ ratio is relatively high in the mitochondria, a circumstance that would cause intramitochondrial oxaloacetate to be readily reduced to malate, whereas the NADH/NAD$^+$ ratio is very low in the cytoplasm, which would favor the reoxidation of extramitochondrial malate to oxaloacetate (page 535). Since oxaloacetate is unable to pass as such through the mitochondrial membrane in rat liver, the mitochondrial reduction of oxaloacetate to malate, the transport of malate to the cytosol via a specific membrane transport system (page 529), and the subsequent reoxidation of malate to oxaloacetate in the cytosol may be seen as an indirect means of transferring oxaloacetate from the mitochondrial matrix to the cytosol. In species other than the rat, particularly the guinea pig and the rabbit, phosphoenolpyruvate carboxykinase (GTP) is found in both the cytosol and the mitochondria; apparently both forms of the enzyme participate in phosphoenolpyruvate formation.

In some microorganisms and plants the phosphorylation of pyruvate to phosphoenolpyruvate is carried out by an entirely different reaction, catalyzed by the enzyme *pyruvate orthophosphate dikinase*,

$$\text{Pyruvate} + \text{ATP} + P_i \rightleftharpoons \text{phosphoenolpyruvate} + PP_i + \text{AMP}$$

followed by the enzymatic hydrolysis of the pyrophosphate to phosphate.

Conversion of Phosphoenolpyruvate to Fructose 1,6-Diphosphate

Phosphoenolpyruvate generated from pyruvate by the above reactions is now easily converted into fructose 1,6-diphosphate by reversal of the glycolytic reactions, beginning with that catalyzed by enolase and ending with that catalyzed by fructosediphosphate aldolase (pages 424 to 430):

$$\text{Phosphoenolpyruvate} + H_2O \xrightleftharpoons{\text{enolase}} \text{2-phosphoglycerate}$$

$$\text{2-Phosphoglycerate} \rightleftharpoons \text{3-phosphoglycerate}$$

$$\text{ATP} + \text{3-phosphoglycerate} \rightleftharpoons$$
$$\text{ADP} + \text{3-phosphoglyceroyl phosphate}$$

$$\text{3-Phosphoglyceroyl phosphate} + \text{NADH} + H^+ \rightleftharpoons$$
$$\text{3-phosphoglyceraldehyde} + \text{NAD}^+ + P_i$$

$$\text{3-Phosphoglyceraldehyde} \rightleftharpoons \text{dihydroxyacetone phosphate}$$

$$\text{3-Phosphoglyceraldehyde} + \text{dihydroxyacetone phosphate} \xrightleftharpoons{\text{aldolase}}$$
$$\text{fructose 1,6-diphosphate}$$

Conversion of Fructose 1,6-Diphosphate into Fructose 6-Phosphate

We now come to the second crucial point in gluconeogenesis in which a reaction of the downhill glycolytic sequence is bypassed by an enzyme functioning primarily in the direction of synthesis. The downhill reaction of glycolysis at this point is that catalyzed by phosphofructokinase:

$$\text{Fructose 6-phosphate} + \text{ATP} \rightleftharpoons \text{fructose 1,6-diphosphate} + \text{ADP}$$
$$\Delta G^{\circ\prime} = -3.4 \text{ kcal mol}^{-1}$$

It does not function in the reverse direction biologically, in part because of the unfavorable $\Delta G^{\circ\prime}$. During gluconeogenesis this reaction is bypassed (Figure 23-2) by the cytosol enzyme hexosediphosphatase, more commonly known as *fructose diphosphatase*, which carries out the essentially irreversible hydrolytic removal of the 1-phosphate group:

$$\text{Fructose 1,6-diphosphate} + H_2O \rightleftharpoons \text{fructose 6-phosphate} + P_i$$
$$\Delta G^{\circ\prime} = -4.0 \text{ kcal mol}^{-1}$$

Hexosediphosphatase is an allosteric enzyme; it is strongly inhibited by the negative modulator AMP and stimulated by 3-phosphoglycerate and citrate. The enzyme has at least three binding sites for AMP, which are distinct from the substrate binding site(s). It contains four or more subunits. The enzyme is maximally active and thus favors formation of glucose when the concentration of certain glucose precursors is high and the AMP concentration is low, i.e., when the energy charge is high. Hexosediphosphatase of the liver has other properties related to the regulation of gluconeogenesis. It is converted by lysosomal proteases into a form having a more alkaline optimum pH; this change appears to be the result of endocrine regulation.

In the last step, fructose 6-phosphate is reversibly converted into glucose 6-phosphate by _glucosephosphate isomerase,_ which functions reversibly in both glycolysis (page 424) and gluconeogenesis:

$$\text{Fructose 6-phosphate} \rightleftharpoons \text{glucose 6-phosphate}$$

We can now sum up the reactions in gluconeogenesis leading from pyruvate to glucose 6-phosphate:

$$\text{2 Pyruvate} + \text{4ATP} + \text{2GTP} + \text{2NADH} + \text{2H}^+ + \text{6H}_2\text{O} \longrightarrow$$
$$\text{glucose 6-phosphate} + \text{4ADP} + \text{2GDP} + \text{2NAD}^+ + \text{5P}_i$$

For each molecule of glucose 6-phosphate formed, six high-energy phosphate bonds are consumed and two molecules of NADH are required as reductant; the overall reaction is exergonic. This equation is clearly very different from that for the downhill conversion of glucose 6-phosphate into pyruvate, which generates three molecules of ATP:

$$\text{Glucose 6-phosphate} + \text{3ADP} + \text{2P}_i + \text{2NAD}^+ \longrightarrow$$
$$\text{2 pyruvate} + \text{3ATP} + \text{2NADH} + \text{2H}^+ + \text{3H}_2\text{O}$$

In some animal tissues, particularly the liver, kidney, and intestinal epithelium, glucose 6-phosphate may be dephosphorylated to form free glucose; the liver is the major site of formation of blood glucose. The hydrolytic cleavage of glucose 6-phosphate does not occur by reversal of the hexokinase reaction (page 424) but is brought about by _glucose-6-phosphatase,_ which catalyzes the exergonic hydrolytic reaction

$$\text{Glucose 6-phosphate} + \text{H}_2\text{O} \longrightarrow \text{glucose} + \text{P}_i$$
$$\Delta G^{\circ\prime} = -3.3 \text{ kcal mol}^{-1}$$

This Mg^{2+}-dependent enzyme is characteristically found in the endoplasmic reticulum of the liver of vertebrates. Its activity is dependent on lipids and on the intactness of the membrane. Glucose-6-phosphatase is not present in muscles or in the brain, which thus cannot donate free glucose to the blood.

Gluconeogenesis from Tricarboxylic Acid Cycle Intermediates

The pathway from pyruvate to glucose described above allows the net synthesis of glucose from various precursors of pyruvate or phosphoenolpyruvate (Figure 23-1). Chief among them are the tricarboxylic acid cycle intermediates, which may undergo oxidation to oxaloacetate. The latter is

Figure 23-3
Oxaloacetic acid, showing the carbon atoms (color) that are the direct precursors of the carbon atoms of phosphoenolpyruvate.

$$COOH$$
$$\beta \ CH_2$$
$$\alpha \ C{=}O$$
$$COOH$$

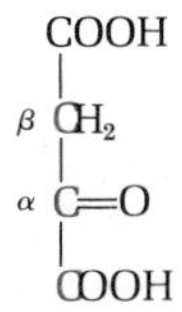

Figure 23-4
Phloridzin, a toxic glycoside from the bark of the pear tree; it causes excretion of glucose in the urine.

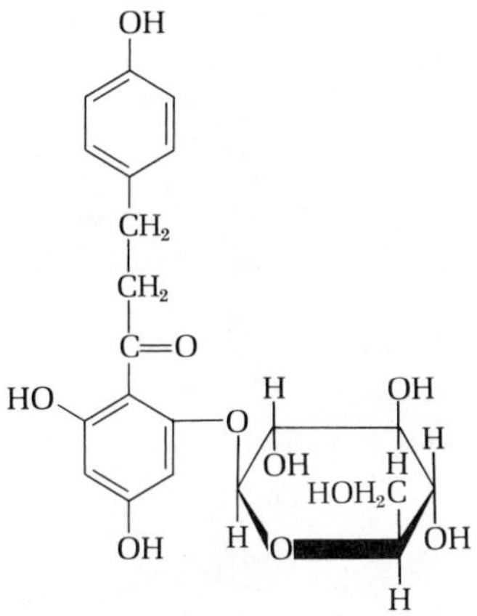

Table 23-1 Fate of amino acids in mammals

Glucogenic	
Alanine	Histidine
Arginine	Methionine
Aspartic acid	Proline
Asparagine	Serine
Cysteine	Threonine
Glutamic acid	Tryptophan
Glutamine	Valine
Glycine	
Ketogenic	
Leucine	
Glucogenic and ketogenic	
Isoleucine	
Lysine	
Phenylalanine	
Tyrosine	

then converted into phosphoenolpyruvate by the action of phosphoenolpyruvate carboxykinase. By this pathway three carbon atoms of the various tricarboxylic acid cycle intermediates are ultimately convertible into the three carbon atoms of phosphoenolpyruvate. These carbon atoms arise from the α carboxyl, α (carbonyl), and β carbon atoms of oxaloacetate (Figure 23-3). This pathway has been amply verified by many isotopic tracer experiments carried out on intact animals, as well as on tissue slices or extracts.

It has also been established by various experimental approaches that the tricarboxylic acid cycle intermediates give rise to net synthesis of new glucose in the whole animal. In one type of experiment, rats are fasted for 24 h or longer, a treatment which reduces the glycogen level in the liver from about 7 percent of the wet weight to 1 percent or less. When succinate or other tricarboxylic acid cycle intermediates are fed, they cause a net increase in the total amount of glycogen in the fasted animal, largely because of an increase in the liver glycogen. Such a net conversion of tricarboxylic cycle intermediates into glucose is also observed in animals treated with the toxic glycoside phloridzin (Figure 23-4). This poison blocks reabsorption of glucose from the kidney tubule into the blood and thus causes blood glucose to be excreted nearly quantitatively into the urine (page 844). Feeding of succinate or other tricarboxylic acid cycle intermediates to phloridzin-poisoned animals causes excretion of an amount of glucose nearly equivalent to three of the carbon atoms of the intermediate fed.

Gluconeogenesis from Amino Acids

As shown in Chapter 21 (page 567), some or all of the carbon atoms of certain amino acids are ultimately convertible by vertebrates either into pyruvate or into intermediates of the tricarboxylic acid cycle, which in turn are precursors of phosphoenolpyruvate. Such amino acids, which are thus also precursors of glucose, are called glucogenic amino acids (page 559). Two examples are glutamic and aspartic acids, which are directly convertible by transamination into tricarboxylic acid cycle intermediates, α-ketoglutarate and oxaloacetate. Table 23-1 lists the amino acids that are glucogenic in mammals. Also shown are the amino acids that are both glucogenic and ketogenic, such as phenylalanine and tyrosine, which on degradation (Chapter 21) are cleaved to form fumaric acid, which is glucogenic, and acetoacetate, one of the ketone bodies.

The amino acid leucine yields neither pyruvate nor a tricarboxylic acid cycle intermediate during its oxidative degradation. Since it does yield acetyl-CoA, which can be converted into ketone bodies but not into pyruvate (see below), it is a ketogenic amino acid.

In plants and many microorganisms no distinction can be made between glucogenic and nonglucogenic amino acids because all the amino acids may ultimately contribute to the net formation of glucose through the combined reactions of the tricarboxylic acid and glyoxylate (page 466) cycles.

Gluconeogenesis from Acetyl-CoA in Plants and Microorganisms

It is most important to distinguish between *net* synthesis of glucose from smaller precursors and mere incorporation of an isotopic carbon atom from a labeled metabolite into glucose without *net* synthesis of new glucose. For example, when carbon-labeled acetic acid is fed to an animal, its isotopic carbon atoms will be incorporated into the glucose residues of liver glycogen, specifically into carbon atoms 1, 2, 5, and 6, as some pencil-and-paper work will show (pages 419, 461, and 624). However, there is no *net* formation of new glucose from the two carbon atoms of the acetyl group of acetyl-CoA. Citrate, the six-carbon condensation product of acetyl-CoA and oxaloacetate, ultimately undergoes loss of two carbon atoms as CO_2 during its oxidation to oxalacetate; thus no more glucose can be formed from citrate than from oxaloacetate. Moreover, in animal tissues acetyl-CoA cannot be directly converted into pyruvate by reversal of the reactions catalyzed by the pyruvate dehydrogenase complex (page 450) or into succinate via the glyoxylate cycle, which is lacking in mammals (page 466). In higher animals there is *no* metabolic pathway by which the carbon atoms of fatty acids or acetyl-CoA can be used to form *new* glucose.

On the other hand, plants and many microorganisms do carry out the net synthesis of carbohydrate from fatty acids by way of acetyl-CoA, a process made possible by the reactions of the glyoxylate cycle (page 466), which permits the *net* conversion of acetyl-CoA into succinate according to the overall reaction

2 Acetyl-CoA + NAD⁺ + 2H₂O ⟶

$$\text{2 Acetyl-CoA} + NAD^+ + 2H_2O \longrightarrow$$
$$\text{succinate} + 2CoA + NADH + H^+$$

Two specific enzymes are required for this pathway, *isocitrate lyase* and *malate synthase* (page 466); they are completely lacking in higher animals. The succinate formed in the glyoxylate cycle yields oxaloacetate, which in turn is the precursor of phosphoenolpyruvate. By this pathway stored fat can be converted into glucose by germinating seeds.

Regulation of Gluconeogenesis and Glycolysis

Figure 23-5 summarizes the control points in the pathways between pyruvate and glucose in animal tissues. The first reaction in the uphill pathway from pyruvate to glucose is catalyzed by a regulatory enzyme, pyruvate carboxylase, which exerts primary control. This reaction is promoted by the allosteric modulator acetyl-CoA. As a consequence, whenever excess mitochondrial acetyl-CoA builds up beyond the immediate needs of the cell for fuel, glucose synthesis is promoted. The secondary control point of this pathway is the reaction catalyzed by hexosediphosphatase, which is stimulated by the glucose precursors citrate and 3-phosphoglycerate and inhibited by AMP. Thus the pathway from pyruvate to glucose is regulated both by the level of respiratory fuels such as acetyl-CoA and citrate and by the energy charge (page 538) of the ATP system.

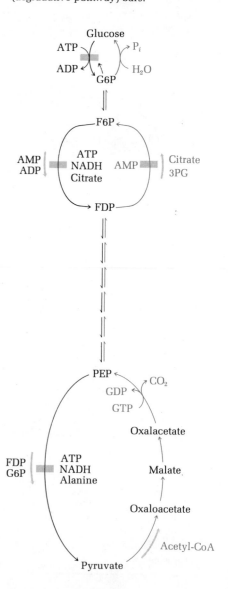

Figure 23-5
Allosteric regulation of opposing pathways from pyruvate to glucose (reaction arrows in color) and from glucose to pyruvate (black arrows). Points of action of positive or stimulatory modulators are indicated by wide colored (biosynthetic-pathway) or wide gray (degradation-pathway) arrows; points of action of negative modulators are shown by colored (biosynthetic-pathway) or gray (degradative pathway) bars.

In contrast, the downhill pathway from glucose to pyruvate is primarily regulated by the allosteric enzyme phosphofructokinase, which is stimulated by AMP and ADP but inhibited by ATP, citrate, and NADH (page 424). A secondary point of regulation of glycolysis is provided by hexokinase, which is inhibited by glucose 6-phosphate and possibly acetyl-CoA and phosphoenolpyruvate. Another regulator of glycolysis is pyruvate kinase, which is inhibited by ATP, NADH, and alanine but stimulated by fructose 1,6-diphosphate and glucose 6-phosphate.

We therefore see that the important regulatory enzymes of the opposing pathways of glycolysis and gluconeogenesis are located at those points where the two pathways are not identical, thus providing independent regulation of each. To summarize, whenever the cell has an ample level of ATP, and whenever respiratory fuels such as acetyl-CoA, citrate, or NADH are readily available, glycolysis is inhibited and gluconeogenesis promoted. On the contrary, when the energy charge is low or respiratory fuels are not available, glycolysis is accelerated and gluconeogenesis is inhibited.

Superimposed on the allosteric regulation of the reactions of glycolysis and gluconeogenesis is the regulation of the biosynthesis of certain key enzymes involved in these pathways. Administration of excess glucose, fructose, or glycerol to fasted rats depresses the phosphoenolpyruvate carboxykinase activity of the liver; it will be recalled that this enzyme catalyzes the formation of phosphoenolpyruvate from oxaloacetate. This effect is due to suppression of the biosynthesis of the enzyme, presumably because it is not needed when the liver is provided with ample supplies of glucose, a more direct precursor of glycogen than phosphoenolpyruvate. On the other hand, feeding excess glycogenic amino acids enhances the biosynthesis of phosphoenolpyruvate carboxykinase, required for the conversion of these amino acids into glucose.

Futile Cycles

Inspection of the reactions shown in Figure 23-5 will indicate that there are two points in the opposing pathways of glycolysis and gluconeogenesis where *futile cycles* may operate. Such cycles are best defined by description. The two opposing reactions between glucose and glucose 6-phosphate (the hexokinase and glucose-6-phosphatase reactions) would appear to cancel each other and result in the net dephosphorylation of ATP:

Glucose + ATP $\xrightarrow{\text{hexokinase}}$ glucose 6-phosphate + ADP

Glucose 6-phosphate + H_2O $\xrightarrow{\text{glucose-6-phosphatase}}$ glucose + phosphate

Sum:

ATP + $H_2O \longrightarrow$ ADP + phosphate

Such a cycle is futile because it accomplishes nothing but the wasteful hydrolysis of ATP.

Another possible futile cycle is that given by the opposing reactions catalyzed by phosphofructokinase and hexose-diphosphatase:

$$\text{ATP} + \text{fructose 6-phosphate} \xrightarrow{\text{phosphofructokinase}} \text{ADP} + \text{fructose 1,6-diphosphate}$$

$$\text{Fructose 1,6-diphosphate} + H_2O \xrightarrow{\text{hexosediphosphatase}} \text{fructose 6-phosphate} + \text{phosphate}$$

Sum:

$$\text{ATP} + H_2O \longrightarrow \text{ADP} + \text{phosphate}$$

Such futile cycles represent an interesting metabolic puzzle, for which a biologically meaningful answer must be found. Traditional belief held that futile cycles probably do not occur in intact cells because under any given metabolic circumstances one reaction of each pair would be relatively very slow and the other rather fast, depending on the *net* direction of the two opposing reactions, whether toward lactate formation or toward glucose formation. However, isotope tracer experiments have revealed that glycolysis and gluconeogenesis may go on simultaneously, sometimes at quite comparable rates, and that under some conditions futile cycles may in fact operate in normal liver cells. Although some investigators prefer to think of futile cycles as the result of an imperfection in the biological design of metabolism, others feel that they have a biological significance. One idea is that they are elements in the regulation of opposing metabolic pathways, supplementing the usual allosteric controls in such a way as to hydrolyze excess ATP, which may under some circumstances be necessary. Another idea is that futile cycles leading to ATP hydrolysis may be a device to produce heat. Indeed, exposure of mammals to low temperatures appears to increase the activity of certain enzymes capable of participating in futile cycles, such as hexosediphosphatase. The net hydrolysis of ATP in the absence of energy-conserving mechanisms would of course yield much of the free-energy decline as heat.

Photosynthetic Formation of Glucose by the Calvin Pathway

Part of the central biosynthetic pathway leading from pyruvate to glucose 6-phosphate is also utilized in the formation of glucose from CO_2 during photosynthesis. The enzymatic reactions by which light energy is conserved as the phosphate-bond energy of ATP and as reducing power in the form of NADPH were described in Chapter 22 (page 605). The ATP and NADPH generated in the light reactions are then utilized by green-plant cells to bring about the reduction of CO_2 to form carbohydrates and other reduced products. The major end products of plant photosynthesis are starch, cellulose, and other polysaccharides; free glucose per se is not present in significant quantities in most higher-plant tissues. We shall use the general term "hexose" to designate all free and combined six-carbon sugar residues formed in photosynthesis.

Although higher animals can fix carbon dioxide as the carboxyl carbon of oxaloacetate and other compounds in enzymatic reactions described elsewhere (pages 464 and 555), these reactions do not result in net synthesis of new hexose from carbon dioxide. For example, in the pyruvate carboxylase reaction

$$\text{Pyruvic acid} + CO_2 + \text{ATP} \rightleftharpoons \text{oxaloacetic acid} + \text{ADP} + P_i$$

the CO_2 incorporated into oxaloacetate is ultimately lost again, as CO_2, in the subsequent reactions by which three of the carbon atoms of oxaloacetate are converted into phosphoenolpyruvate and then glucose (page 625). Moreover, the oxidative decarboxylation of pyruvate to acetyl-CoA and of α-ketoglutarate to succinyl-CoA are irreversible reactions in animal tissues and therefore cannot lead to net formation of glucose from CO_2 (page 450). Thus the biosynthetic pathways in green-plant cells that lead to net hexose formation from CO_2 differ qualitatively from the carboxylation reactions taking place in animal tissues.

An important clue to the nature of the pathway from CO_2 to hexose in photosynthetic organisms first came from the work of M. Calvin and his associates. They illuminated green algae in the presence of radioactive carbon dioxide ($^{14}CO_2$) for very short intervals (only a few seconds) and then quickly killed the cells, extracted them, and with the aid of chromatographic methods searched for those metabolites in which the labeled carbon was incorporated earliest. One of the compounds that became labeled very early was 3-phosphoglyceric acid, a known intermediate of glycolysis; the carbon isotope was found predominantly in the carboxyl carbon atom. This carbon atom, which corresponds to the carboxyl carbon atom of pyruvate (page 431), is not labeled rapidly in animal tissues incubated with radioactive CO_2.

These findings strongly suggested that the labeled 3-phosphoglycerate is an early intermediate in photosynthesis in algae, particularly since 3-phosphoglycerate is readily converted into glucose by reversal of the steps of glycolysis. However, no enzymatic reactions capable of incorporating CO_2 into the carboxyl group of 3-phosphoglycerate were then known. After an intensive search, such an enzyme was found in extracts from spinach leaves. This enzyme, *ribulose-diphosphate carboxylase*, catalyzes the carboxylation and cleavage of ribulose 1,5-diphosphate:

$$\text{Ribulose 1,5-diphosphate} + CO_2 \rightleftharpoons \text{2 3-phosphoglycerate}$$

Two molecules of 3-phosphoglycerate are formed, one of which bears the carbon atom introduced as CO_2 (Figure 23-6). The enzyme has a molecular weight of 550,000 and contains 16 subunits. 2-Carboxy-3-ketoribitol-1,5-diphosphate is an intermediate in the reaction, presumably enzyme-bound. The true substrate in the reaction is CO_2 rather than HCO_3^-. The affinity of the isolated enzyme for CO_2 is very low, in contrast with the relatively high affinity

Figure 23-6
The ribulosediphosphate carboxylase reaction. Also shown is the structure of the proposed enzyme-bound intermediate (in brackets). The carbon atom introduced as CO_2 is in color.

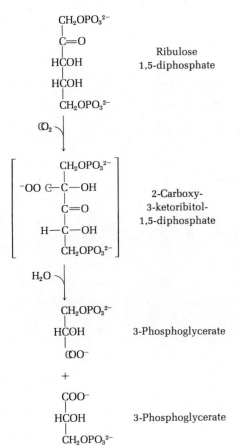

of most photosynthetic plants for CO_2. The enzyme appears to have allosteric properties; it shows a sigmoid activity-concentration curve for CO_2 and is reported to be modulated by fructose 1,6-diphosphate. It is inhibited by molecular oxygen; the significance of this effect is considered below (page 639). Ribulosediphosphate carboxylase is extremely abundant in the cell, making up about 15 percent of the total protein of chloroplasts. Electron-microscopic observations on spinach chloroplasts have suggested that individual molecules of the enzyme, which have a diameter of about 20 nm, are located in or on the thylakoid membranes.

The 3-phosphoglycerate formed by ribulosediphosphate carboxylase can be converted into glucose 6-phosphate by the pathway already described, i.e., reversal of the glycolytic reactions and the hexosediphosphatase reaction.

Although this sequence of reactions shows how one of the glucose carbons arises from CO_2, it does not account for the fact that all six carbon atoms of hexose are ultimately formed from CO_2 during photosynthesis. To provide such a pathway, Calvin and his colleagues proposed a cyclic mechanism for hexose synthesis. In the Calvin cycle, a modification and extension of the phosphogluconate and glycolytic sequences (page 467), one molecule of ribulose 1,5-diphosphate is regenerated for each molecule of CO_2 reduced. Figure 23-7 shows the schematic outline of this pathway; its individual steps are shown in the following sequence of reactions:

Figure 23-7

(Right) The photosynthetic formation of glucose from CO_2 via the Calvin cycle in spinach leaves. The inputs are shaded in gray, and the products are in color.

Key:

3PG = 3-phosphoglyceric acid
G3P = glyceraldehyde 3-phosphate
DHAP = dihydroxyacetone phosphate
FDP = fructose 1,6-diphosphate
F6P = fructose 6-phosphate
G6P = glucose 6-phosphate
E4P = erythrose 4-phosphate
X5P = xylulose 5-phosphate
SDP = sedoheptulose 1,7-diphosphate
S7P = sedoheptulose 7-phosphate
R5P = ribose 5-phosphate
Ru5P = ribulose 5-phosphate
RuDP = ribulose 1,5-diphosphate

$$6CO_2 + 6 \text{ ribulose 1,5-diphosphate} \xrightarrow{\substack{\text{ribulosediphosphate} \\ \text{carboxylase}}}$$
$$12 \text{ 3-phosphoglycerate} \qquad (5)$$

$$12 \text{ 3-Phosphoglycerate} + 12ATP \xrightarrow{\substack{\text{phosphoglycerate} \\ \text{kinase}}}$$
$$12 \text{ 3-phosphoglyceroyl phosphate} + 12ADP \qquad (6)$$

$$12 \text{ 3-Phosphoglyceroyl phosphate}$$
$$+ 12NADPH + 12H^+ \xrightarrow{\substack{\text{glyceraldehyde-phosphate} \\ \text{dehydrogenase (NADP)}}}$$
$$12 \text{ glyceraldehyde 3-phosphate} + 12NADP^+ + 12 \text{ } P_i \qquad (7)$$

$$5 \text{ Glyceraldehyde 3-phosphate} \xrightarrow{\substack{\text{triosephosphate} \\ \text{isomerase}}}$$
$$5 \text{ dihydroxyacetone phosphate} \qquad (8)$$

$$3 \text{ Glyceraldehyde 3-phosphate}$$
$$+ 3 \text{ dihydroxyacetone phosphate} \xrightarrow{\substack{\text{fructosediphosphate} \\ \text{aldolase}}}$$
$$3 \text{ fructose 1,6-diphosphate} \qquad (9)$$

$$3 \text{ Fructose 1,6-diphosphate} \xrightarrow{\text{hexosediphosphatase}}$$
$$3 \text{ fructose 6-phosphate} + 3P_i \quad (10)$$

$$\text{Fructose 6-phosphate} \xrightarrow{\substack{\text{glucosephosphate} \\ \text{isomerase}}} \text{glucose 6-phosphate} \qquad (11)$$

$$\text{Glucose 6-phosphate} \xrightarrow{\substack{\text{glucose-} \\ \text{6-phosphatase}}} \boxed{\text{glucose}} + P_i \qquad (12)$$

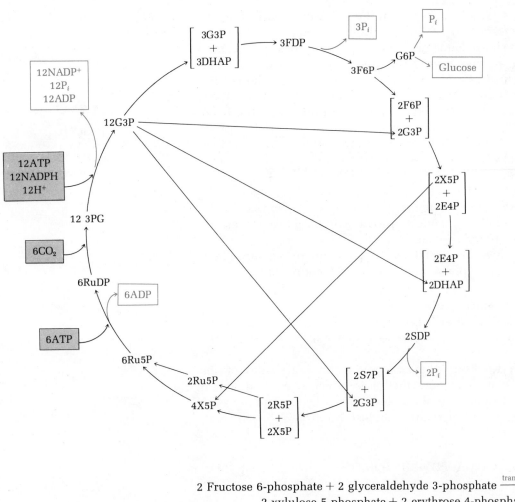

2 Fructose 6-phosphate + 2 glyceraldehyde 3-phosphate $\xrightarrow{\text{transketolase}}$
$$\text{2 xylulose 5-phosphate} + \text{2 erythrose 4-phosphate} \quad (13)$$

2 Erythrose 4-phosphate

$+$ 2 dihydroxyacetone phosphate $\xrightarrow[\text{aldolase}]{\text{fructosediphosphate}}$
$$\text{2 sedoheptulose 1,7-diphosphate} \quad (14)$$

2 Sedoheptulose 1,7-diphosphate $\xrightarrow{\text{hexosediphosphatase}}$
$$\text{2 sedoheptulose 7-phosphate} + 2P_i \quad (15)$$

2 Sedoheptulose 7-phosphate

$+$ 2 glyceraldehyde 3-phosphate $\xrightarrow{\text{transketolase}}$
$$\text{2 ribose 5-phosphate} + \text{2 xylulose 5-phosphate} \quad (16)$$

2 Ribose 5-phosphate $\xrightarrow[\text{isomerase}]{\text{ribosephosphate}}$ 2 ribulose 5-phosphate $\quad (17)$

4 Xylulose 5-phosphate $\xrightarrow[\text{3-epimerase}]{\text{ribulosephosphate}}$ 4 ribulose 5-phosphate $\quad (18)$

6 Ribulose 5-phosphate + 6ATP $\xrightarrow{\text{phosphoribulokinase}}$
$$\text{6 ribulose 1,5-diphosphate} + \text{6ADP} \quad (19)$$

Reactions (5) to (12) represent the pathway described above for the formation of glucose from CO_2 and ribulose 1,5-diphosphate, whereas reactions (13) to (19) are concerned with the regeneration of ribulose 1,5-diphosphate. The glucose formed as end product in reaction (12) is shown in a box. Reaction (15) is catalyzed by hexosediphosphatase (page 627); it is exergonic and is thus a "pulling" reaction. Reactions (17) and (18) are catalyzed by the _ribosephosphate isomerase_ and _ribulosephosphate 3-epimerase_ described earlier (page 468), and reaction (19) by _phosphoribulokinase_, an enzyme similar in many respects to phosphofructokinase (page 424). The intermediate seven-carbon sugar _sedoheptulose 1,7-diphosphate_ (Figure 23-8) is similar in its chemical and biochemical properties to its six-carbon analog fructose 1,6-diphosphate.

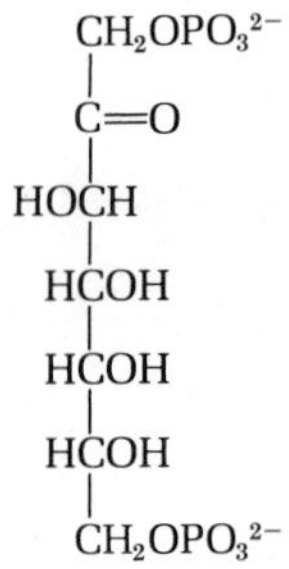

One way of writing the overall equation for one turn of the Calvin cycle, expressed in terms of CO_2 equivalents, is

Ribulose 1,5-diphosphate + CO_2
    + 3ATP + 2NADPH + 2H$^+$ + 2H$_2$O $\longrightarrow$
        ribulose 1,5-diphosphate + (CH$_2$O) + 3P$_i$ + 3ADP + 2NADP$^+$

Ribulose 1,5-diphosphate is written on both sides of the equation only to show that it is a necessary component which is regenerated at the end of each cycle. The _net_ reaction, after canceling out the ribulose 1,5-diphosphate on both sides, is

CO_2 + 3ATP + 2NADPH + 2H$^+$ + 2H$_2$O $\longrightarrow$
        (CH$_2$O) + 3P$_i$ + 3ADP + 2NADP$^+$

If we sum up the reactions through six turns of the Calvin cycle (see Figure 23-7), we have the following overall equation, again showing the input and regeneration of ribulose 1,5-diphosphate:

6 Ribulose 1,5-diphosphate + 6CO$_2$
    + 18ATP + 12NADPH + 12H$^+$ + 12H$_2$O $\longrightarrow$
        6-ribulose 1,5-diphosphate + hexose + 18P$_i$ + 18ADP + 12NADP$^+$

Subtracting the ribulose diphosphate from both sides, we obtain for the net equation:

6CO$_2$ + 18ATP + 12NADPH + 12H$^+$ + 12H$_2$O $\longrightarrow$
        hexose + 18P$_i$ + 18ADP + 12NADP$^+$

Note that three molecules of ATP are required for each molecule of CO_2 reduced.

The C₄ or Hatch-Slack Pathway of Formation of Glucose

In the mid-1960s evidence came to light that in some green plants 3-phosphoglycerate is not the earliest intermediate into which radioactive CO_2 is incorporated. H. Kortschak,

Figure 23-8
Sedoheptulose 1,7-diphosphate.

$$CH_2OPO_3^{2-}$$
$$|$$
$$C{=}O$$
$$|$$
$$HOCH$$
$$|$$
$$HCOH$$
$$|$$
$$HCOH$$
$$|$$
$$HCOH$$
$$|$$
$$CH_2OPO_3^{2-}$$

and later M. D. Hatch and C. R. Slack, found that in such plants, which include sugarcane and maize, four-carbon dicarboxylic acids (oxaloacetic, malic, and aspartic acids) appear to be the earliest products of CO_2 fixation. In plants of this class, now called C_4 plants, CO_2 is first fixed in a reaction catalyzed by phosphoenolpyruvate carboxylase:

$$\text{Phosphoenolpyruvate} + CO_2 \longrightarrow \text{oxaloacetate} + P_i$$
$$\Delta G^{\circ\prime} = -6.5 \text{ kcal mol}^{-1}$$

[This enzyme is not to be confused with phosphoenolpyruvate carboxykinase (GTP) (page 625).] The oxaloacetate so formed in C_4 plants may then be reduced to malate by NADP-linked malate dehydrogenase:

$$\text{NADPH} + H^+ + \text{oxaloacetate} \rightleftharpoons \text{NADP}^+ + \text{L-malate}$$

or converted into aspartate by transamination:

$$\text{Oxaloacetate} + \text{glutamate} \rightleftharpoons \text{aspartate} + \alpha\text{-ketoglutarate}$$

The next stages in the assimilation of CO_2 by the C_4 plants are interesting because they involve the cooperation of two different cell types. The leaves of C_4 plants contain two types of photosynthetic cells, which have quite different biochemical and structural organization: the bundle-sheath cells, which surround the veins, and the mesophyll cells, which are loosely arranged around the bundle-sheath cells. Recent research on isolated bundle-sheath and mesophyll cells indicates that the fixation of CO_2 by phosphoenolpyruvate carboxylase as described above occurs in the mesophyll cells. In some C_4 plants the malate formed by reduction of the oxaloacetate is transported to the bundle-sheath cells, where it is decarboxylated by malic enzyme [malate dehydrogenase (decarboxylating) (NADP$^+$)]:

$$\text{Malate} + \text{NADP}^+ \rightleftharpoons \text{pyruvate} + CO_2 + \text{NADPH} + H^+$$

In other C_4 plants aspartate formed from oxaloacetate by transamination is also ultimately decarboxylated in the bundle-sheath cells. The CO_2 released in these two reactions then reacts with ribulose 1,5-diphosphate to yield 3-phosphoglycerate, which is converted into hexose by the Calvin cycle. The pyruvate formed in the bundle-sheath cells returns to the mesophyll cells and is converted into phosphoenolpyruvate at the expense of two high-energy phosphate groups. Another cycle may then begin. The C_4 cycle in these tropical plants is called the Hatch-Slack pathway (Figure 23-9).

It may seem pointless for the C_4 plants to fix CO_2 as oxaloacetate in one type of cell and then decarboxylate and refix it again in another type of cell. However, it has been established that phosphoenolpyruvate carboxylase of the C_4 pathway in the mesophyll cells has a very high affinity for CO_2, whereas we have seen that the ribulosediphosphate carboxylase of the Calvin pathway has a low affinity for CO_2. The mesophyll cells serve to collect CO_2 with high effi-

Figure 23-9
The Hatch-Slack pathway for CO_2 fixation in C_4 plants. CO_2 is first fixed in mesophyll cells as oxaloacetate, which is then reduced to malate (or transaminated to form aspartate). Malate is then transported to cells of the bundle sheath, where it is decarboxylated by malate dehydrogenase (decarboxylating). The CO_2 formed enters the Calvin cycle and is converted into hexose. The Hatch-Slack cycle is then completed by transport of pyruvate to the mesophyll cell, where it is converted into oxaloacetate via phosphoenolpyruvate.

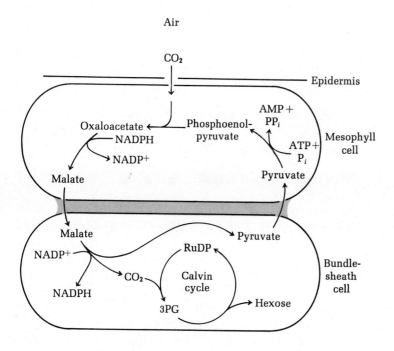

ciency and then transfer it into the bundle-sheath cells, apparently to yield a very high local concentration of CO_2 within the latter, thus enabling the ribulosediphosphate carboxylase of the bundle-sheath cells to fix CO_2 very effectively. Note, however, that two additional high-energy phosphate groups are required for each molecule of CO_2 fixed.

We therefore have two large classes of photosynthetic plants, the C_3 plants, largely of temperate zones, which fix CO_2 directly into the three-carbon 3-phosphoglycerate and the C_4 plants, largely tropical, which fix CO_2 first into oxaloacetate in the mesophyll and then "refix" CO_2 into 3-phosphoglycerate in bundle-sheath cells. As will be seen below, C_3 and C_4 plants also differ in other important respects. The C_4 plants include over 100 species. Among them are not only sugarcane and corn but also desert plants, crabgrass, Bermuda grass, and sorghum.

We can now compare the overall equations for the photosynthetic generation of hexose in C_3 vs. C_4 plants:

C_3 plants:
$$6CO_2 + 18ATP + 12NADPH + 12H^+ + 12H_2O \longrightarrow$$
$$\text{hexose} + 18P_i + 18ADP + 12NADP^+$$

C_4 plants:

$$6CO_2 + 30ATP + 12NADPH + 12H^+ + 24H_2O \longrightarrow$$
$$hexose + 30ADP + 30P_i + 12NADP^+$$

The C_4 plants therefore require considerably more ATP to synthesize a hexose unit than the C_3 plants. Despite the extra ATP they require, the C_4 plants can synthesize hexose much faster per unit leaf area, grow much faster, and function efficiently at much higher light intensities than C_3 plants, as we shall now see.

Photorespiration

Closely related to the mechanism and regulation of hexose biosynthesis in photosynthetic plants is the phenomenon of _photorespiration_. Green-plant cells contain mitochondria in addition to chloroplasts; such cells exhibit mitochondrial respiration and oxidative phosphorylation in the dark at the expense of substrates generated by photosynthesis in earlier light periods. The question arises whether green-plant cells also respire in the light, during active photosynthesis, or whether respiration is turned off. Careful measurements of the rates of oxygen and carbon dioxide exchanges in illuminated plants, particularly with the use of isotopic oxygen, show that most plants do indeed respire in the light while they are carrying out photosynthesis. However, the type of respiration that occurs in the light is not mitochondrial respiration, since it is not sensitive to characteristic inhibitors of that process. Such "light" respiration, which is called _photorespiration_, diverts light-induced reducing power from the biosynthesis of glucose into the reduction of oxygen. Moreover, photorespiration is not accompanied by oxidative phosphorylation of ADP and thus appears to be wasteful of energy-rich reducing power generated by the light reactions. Photorespiration is very active in C_3 plants of temperate zones but nearly or completely absent in most C_4 plants of tropical origin (see above).

A number of observations indicate that the major substrate for photorespiration in C_3 plants is _glycolic acid_ (Figure 23-10). Leaves of such plants contain not only glycolic acid but also a very active enzyme for its oxidation by molecular oxygen to _glyoxylic acid_, namely _glycolate oxidase_, a flavin oxidase. This enzyme is present in neither the mitochondria nor the chloroplasts but is localized in the microbodies or peroxisomes. The oxidation of glycolate yields hydrogen peroxide, which is decomposed by the catalase present in the peroxisomes. Glyoxylate is metabolized further to other products, such as glycine, oxalate, formate, or CO_2, depending on the species.

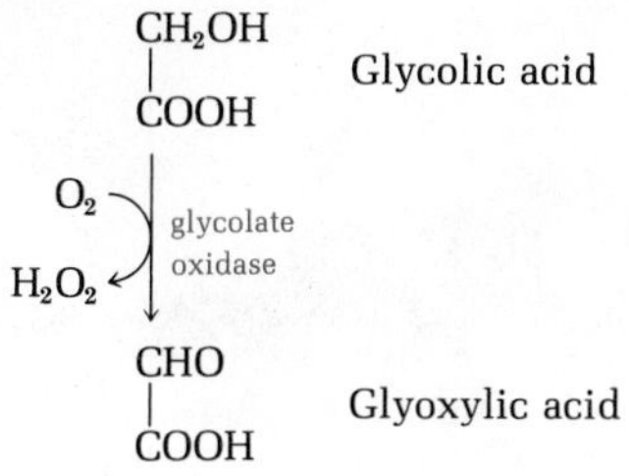

Although this much has been known for some time, only recently has the source of glycolic acid been identified. It will be recalled (see page 634) that ribulosediphosphate carboxylase is inhibited by molecular oxygen. This inhibition appears to be a mechanism for the regulation of CO_2 fixation when oxygen partial pressure is high and CO_2 pressure is low. Subsequent research revealed, moreover, that molecular

Figure 23-10
Oxidation of glycolic acid during photorespiration.

$$\begin{array}{l} CH_2OH \\ | \\ COOH \end{array} \quad \text{Glycolic acid}$$

$$\begin{array}{l} CHO \\ | \\ COOH \end{array} \quad \text{Glyoxylic acid}$$

oxygen can actually replace CO_2 in the ribulosediphosphate carboxylase reaction, resulting in the *oxygenation* of ribulose diphosphate, instead of its carboxylation, to yield *phosphoglycolic acid* and 3-*phosphoglyceric acid* (Figure 23-11). The resulting phosphoglycolate then undergoes enzymatic hydrolysis to yield free glycolate, the substrate of photorespiration.

The rate of photorespiration is rather high in C_3 plants; it is about 5 times the rate of respiration in the dark. Since it results in the oxidation of a reduction product of CO_2, one whose synthesis has required the expenditure of both NADPH and ATP, it is evidently a wasteful process. In C_3 plants the net rate of photosynthesis under normal atmospheric conditions is much less than maximal, limited by the relatively high concentration of oxygen and low concentration of CO_2 in the air. C_4 plants, which show little or no photorespiration, are considerably more efficient because they can carry out photosynthesis at much lower concentrations of CO_2 and at higher oxygen tensions via the Hatch-Slack pathway, which is not diverted by oxygen. The precise function of photorespiration in C_3 plants is still a riddle. Much effort is being invested in improving the efficiency of C_3 crop plants by reducing the rate of photorespiration, by use of inhibitors or by crossbreeding of C_3 and C_4 plants.

Biosynthetic Pathways Leading from Glucose 6-Phosphate: Nucleoside Diphosphate Sugars

Now that we have seen how glucose or glucose 6-phosphate is formed from simpler precursors in both animals and plants, we shall trace the biosynthetic pathways leading from glucose 6-phosphate to (1) other hexoses and hexose derivatives, (2) disaccharides, (3) storage polysaccharides, and (4) the complex structural polysaccharides of cell walls, cell coats, and intercellular spaces. In these pathways hexose residues must often be transformed into hexose derivatives or transferred to other monosaccharides or to the ends of polysaccharide chains. Such reactions show a common pattern in that they employ as the energized glycosyl donor a *nucleoside diphosphate sugar* (NDP-sugar). An example of an NDP-sugar is *uridine diphosphate glucose* (Figure 23-12).

The role of the nucleoside diphosphate sugars was first discovered by L. F. Leloir and his colleagues in Buenos Aires. They showed that a nucleoside triphosphate, such as UTP, reacts with a hexose 1-phosphate to yield a nucleoside diphosphate sugar (Figure 23-12), by the action of enzymes generically called *glycosyl-1-phosphate nucleotidyltransferases* or by their older, less descriptive name, *pyrophosphorylases*:

$$NTP + \text{sugar 1-phosphate} \rightleftharpoons \text{NDP-sugar} + PP_i$$

The pyrophosphate formed derives from the two terminal phosphate groups of the NTP. In the NDP-sugar there are two bonds with a large negative $\Delta G^{\circ\prime}$ of hydrolysis, that between the two phosphate groups and that between the terminal

Figure 23-11
Formation of phosphoglycolate from ribulose 1,5-diphosphate.

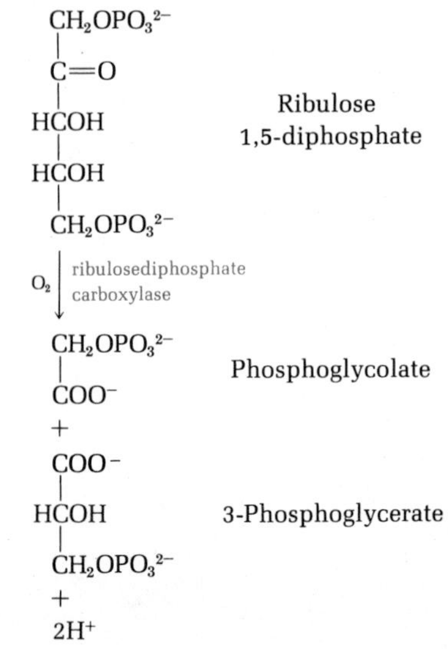

Figure 23-12
Structure and space-filling model of uridine diphosphate glucose. The glucose molecule carried by UDP is in color.

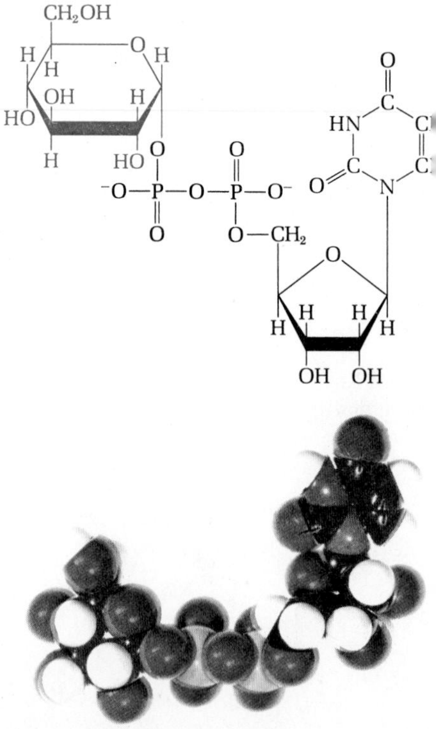

phosphate of the NDP moiety and the glycosyl group. There is only a small standard-free-energy change in the nucleotidyl transferase reaction, but the subsequent hydrolysis of the inorganic pyrophosphate by pyrophosphatase pulls the reaction further in the direction of formation of the NDP-sugar. Although uridine diphosphate usually serves as the specific glycosyl carrier in higher animals, ADP, CDP, and GDP also function as specific sugar carriers in various enzymatic sugar-transfer reactions in different plants, microorganisms, and animal tissues. We thus see another important biochemical function served by nucleotides.

Once an NDP-sugar has been formed, its sugar residue may undergo a variety of enzymatic reactions, including oxidation, reduction, and epimerization, as well as transfer to other sugars or sugar polymers.

Formation of Monosaccharide Derivatives from UDP-Glucose

We have earlier seen that D-glucose 6-phosphate can be converted into D-fructose 6-phosphate and then into D-mannose 6-phosphate (page 437); these reactions do not require NDP-sugars as intermediates. However, nearly all other pathways by which D-glucose is converted into other hexoses or hexose derivatives in animal tissues proceed via UDP-glucose.

UDP–D-glucose is the precursor of *glucuronic acid*, an important building block in some polysaccharides and also a precursor of ascorbic acid or vitamin C. UDP–D-glucose undergoes oxidation to *UDP–D-glucuronic acid* (Figure 23-13) by the action of *UDP-glucose dehydrogenase*:

$$\text{UDP–D-glucose} + 2\text{NAD}^+ + \text{H}_2\text{O} \rightleftharpoons$$
$$\text{UDP–D-glucuronic acid} + 2\text{NADH} + 2\text{H}^+$$

In this reaction, two oxidation steps occur, one oxidizing the hydroxyl group at carbon atom 6 to the aldehyde and the second oxidizing the aldehyde to a carboxyl group. UDP–D-glucuronate is the donor of a glucuronic acid residue to various acceptor molecules in animal tissues, such as foreign phenols and amines, by the action of UDP-glucuronate transferases in the liver. The general reaction catalyzed is

$$\text{UDP–glucuronic acid} + \text{ROH} \longrightarrow \text{R—O—glucosiduronide} + \text{UDP}$$

where ROH is the foreign phenol. This reaction serves to detoxify and/or promote the excretion of foreign phenols and amines, including various drugs. Phenol is excreted by some animals as *phenol glucosiduronide* (Figure 23-13).

Free D-glucuronic acid, which is formed from UDP–D-glucuronic acid by enzymatic hydrolysis, is a precursor in the biosynthesis of L-ascorbic acid (vitamin C), which takes place in plants and in the liver of all vertebrates except man, monkeys, guinea pig, and the Indian fruit bat (page 350). D-Glucuronic acid is first reduced to L-gulonic acid, which lactonizes in the presence of a lactonase to L-gulonolactone. The latter is in turn oxidized to L-ascorbic acid (Figure 23-14).

Figure 23-13
UDP–D-glucuronic acid, the donor of D-glucuronyl residues in the formation of polysaccharides and of glucosiduronides, excretory forms of many foreign alcohols, phenols, and amines. Shown below is a typical glucosiduronide.

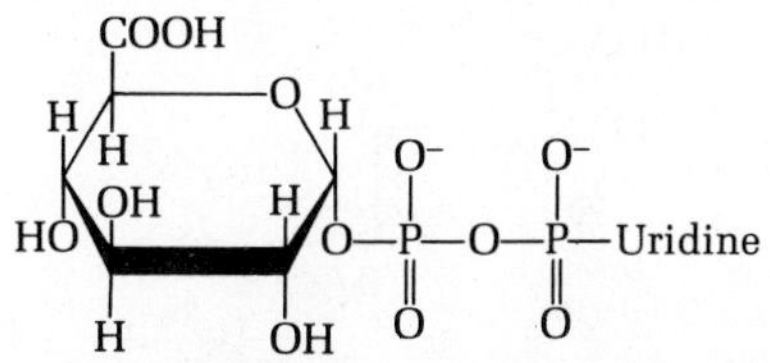

UDP–D-glucuronic acid

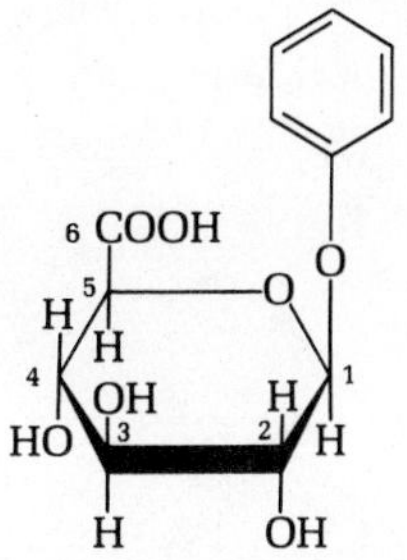

Phenol glucosiduronide

Nucleoside diphosphate sugars also may undergo reduction to form nucleoside diphosphate derivatives of deoxy sugars, such as L-fucose and L-rhamnose, which are important components of lipopolysaccharides of bacterial cell walls (page 268). One such reaction is

$$\text{GDP–D-mannose} + \text{NADPH} + \text{H}^+ \longrightarrow$$
$$\text{GDP–L-fucose} + \text{NADP}^+ + \text{H}_2\text{O}$$

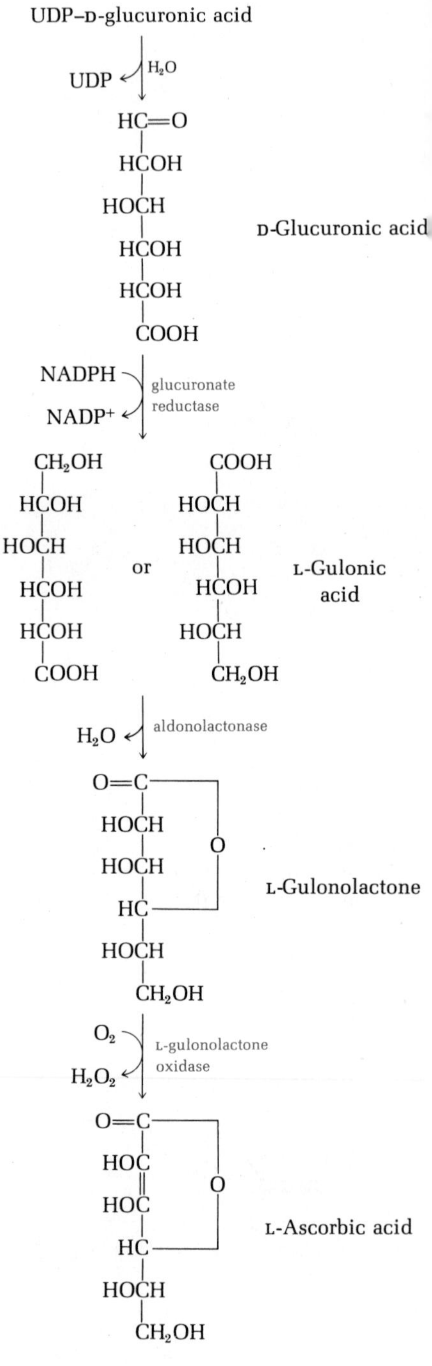

Galactose Metabolism: Galactosemia

A very important reaction in animals is the reversible conversion of D-glucosyl to D-galactosyl residues by enzymatic epimerization of UDP-D-glucose at carbon atom 4 of the glucose residue to form uridine diphosphate D-galactose (UDP-galactose):

$$\text{UDP-D-glucose} \rightleftharpoons \text{UDP-D-galactose}$$

Since the enzyme catalyzing this reaction, *UDP-glucose 4-epimerase*, has an absolute requirement for NAD, it is believed that epimerization occurs in two steps:

$$\text{UDP-D-glucose} + \text{NAD}^+ \rightleftharpoons \text{UDP–4-keto-D-glucose} + \text{NADH} + \text{H}^+$$

$$\text{UDP–4-keto-D-glucose} + \text{NADH} + \text{H}^+ \rightleftharpoons$$
$$\text{UDP–D-galactose} + \text{NAD}^+$$

The postulated 4-ketoglucose intermediate (Figure 23-15), which can accept a pair of hydrogens at the keto group to form either epimer, remains tightly bound to the enzyme active site during the catalytic cycle. UDP-galactose formed in this reaction is a required precursor in the biosynthesis of the disaccharide lactose in the mammary gland, as well as the galactose-containing oligosaccharide side chains of glycoproteins (below).

UDP-galactose is also an important intermediate in the metabolism of free D-galactose, formed by the enzymatic hydrolysis of lactose or milk sugar in the intestinal tract. D-Galactose is converted into D-glucose in the liver by two series of reactions which have attracted much attention because they are subject to genetic defects in man, resulting in different forms of the hereditary disease *galactosemia*, whose biochemistry was analyzed in detail by H. Kalckar and other investigators. In the liver free D-galactose is first phosphorylated at carbon atom 1 by *galactokinase*, to yield D-galactose 1-phosphate:

$$\text{ATP} + \text{D-galactose} \longrightarrow \text{ADP} + \text{D-galactose 1-phosphate}$$

The resulting galactose 1-phosphate can be converted into UDP-galactose by one of two possible reactions. The minor route is catalyzed by *galactose-1-phosphate uridylyltransferase*:

$$\text{D-Galactose 1-phosphate} + \text{UTP} \rightleftharpoons \text{UDP–D-galactose} + \text{PP}_i$$

Figure 23-14
Enzymatic synthesis of L-ascorbic acid.

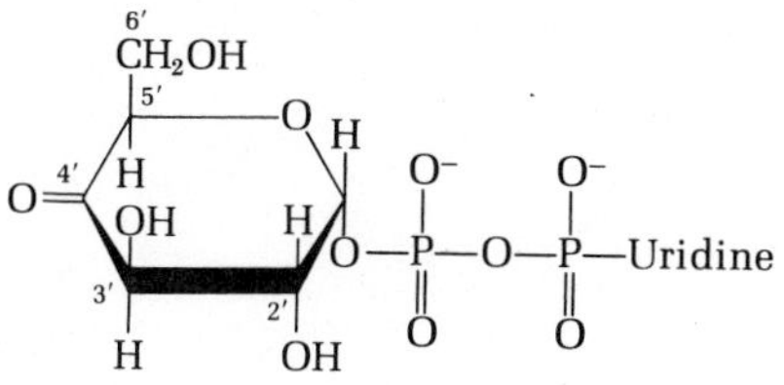

Figure 23-15
UDP–4-ketoglucose.

The UDP-galactose so formed is normally converted into UDP-glucose by UDP-glucose 4-epimerase (above). These reactions make it possible for the galactose residue to enter into the main pathways of glucose metabolism, since UDP-glucose, as we shall see, can donate its glucose residue to glycogen. This uridylyltransferase is present in high amounts in the liver of adults but lacking in infants.

The second pathway for utilization of free galactose also begins with galactose 1-phosphate. It is catalyzed by *hexose-1-phosphate uridylyltransferase*, which catalyzes the reaction

$$\text{UDP–D-glucose} + \text{D-galactose 1-phosphate} \rightleftharpoons$$
$$\text{UDP–D-galactose} + \text{D-glucose 1-phosphate}$$

This particular enzyme is present in normal fetal liver but is lacking in infants with one form of galactosemia. Such infants are thus unable to utilize galactose by either pathway. Galactosemic infants have an excessively high concentration of D-galactose in the blood and suffer from cataract (opacity) of the lens of the eye as well as mental disorders. This condition can be successfully treated by withholding milk and other sources of galactose from the diet during infancy and childhood.

A second type of galactosemia, milder in nature but also characterized by the development of cataracts and blindness, is caused by a genetic defect in the biosynthesis of *galactokinase*, which is necessary to phosphorylate galactose to enable it to enter either galactose pathway. In both the hexose-1-phosphate uridylyltransferase and galactokinase types of galactosemia, the cataracts appear to be caused by an accumulation of the sugar alcohol *galactitol*, formed in the lens from D-galactose by *aldose reductase*:

$$\text{D-Galactose} + \text{NADPH} + \text{H}^+ \rightleftharpoons \text{D-galactitol} + \text{NADP}^+$$

Biosynthesis of Disaccharides and Other Glycosides

In some plants, such as the sugarcane, sucrose is formed from glucose and fructose by the following series of reactions, which employ UDP–D-glucose as an intermediate:

$$\text{ATP} + \text{glucose} \xrightarrow{\text{hexokinase}} \text{glucose 6-phosphate} + \text{ADP}$$

$$\text{Glucose 6-phosphate} \xrightleftharpoons{\text{phosphoglucomutase}} \text{glucose 1-phosphate}$$

$$\text{UTP} + \text{glucose 1-phosphate} \xrightarrow[\text{uridylyltransferase}]{\text{glucose-1-phosphate}} \text{UDP-glucose} + \text{PP}_i$$

$$\text{ATP} + \text{fructose} \xrightarrow{\text{fructokinase}} \text{fructose 6-phosphate} + \text{ADP}$$

$$\text{UDP-glucose} + \text{fructose 6-phosphate} \xrightarrow[\text{synthase}]{\text{sucrose-phosphate}}$$
$$\text{UDP} + \text{sucrose 6}'\text{-phosphate}$$

$$\text{Sucrose 6}'\text{-phosphate} + \text{H}_2\text{O} \xrightarrow{\text{sucrose-phosphatase}} \text{sucrose} + \text{P}_i$$

Sum:

2ATP + UTP + glucose + fructose $\longrightarrow$
$$\text{sucrose} + 2\text{ADP} + \text{UDP} + \text{PP}_i + \text{P}_i$$

If the pyrophosphate formed is hydrolyzed to P_i, the overall reaction is

2ATP + UTP + glucose + fructose $\longrightarrow$
$$\text{sucrose} + 2\text{ADP} + \text{UDP} + 3\text{P}_i$$

Thus three high-energy phosphate bonds are required to form the single glycosidic bond of sucrose, whose $\Delta G^{\circ\prime}$ for hydrolysis is about -5.5 kcal mol^{-1}. The overall reaction of sucrose synthesis from glucose and fructose is therefore quite irreversible. Presumably the strongly exergonic nature of this set of reactions enables sugarcane to form a juice containing a high concentration of sucrose from very dilute monosaccharide precursors. In other plants sucrose is formed by a reaction employing fructose as glucosyl acceptor, rather than fructose 6-phosphate:

$$\text{UDP-glucose} + \text{fructose} \xrightarrow{\substack{\text{sucrose} \\ \text{synthase}}} \text{UDP} + \text{sucrose}$$

Some bacteria contain the enzyme *sucrose phosphorylase* (page 205), which catalyzes the reversible reaction

$$\text{Sucrose} + \text{P}_i \rightleftharpoons \text{glucose 1-phosphate} + \text{fructose}$$

This reaction normally catalyzes the breakdown of sucrose under intracellular conditions.

The disaccharide lactose is formed in the mammary gland from D-glucose and UDP-galactose by the action of two enzyme proteins, which together constitute the *lactose synthase* system. The first, *protein A*, found in the mammary gland and also in the liver and small intestine, catalyzes a reaction between UDP-galactose and various acceptors, particularly N-acetyl-D-glucosamine:

UDP-galactose + N-acetyl-D-glucosamine $\longrightarrow$
$$\text{UDP} + \text{N-acetyllactosamine}$$

Protein A by itself is only weakly active with D-glucose as acceptor because the K_M for D-glucose is very high. The second component, *protein B*, has been long known as the *α-lactalbumin* of milk; it has no catalytic activity of its own. This protein has been found to decrease greatly the K_M of protein A for D-glucose, so that the enzyme will utilize D-glucose in preference to N-acetyl-D-glucosamine as galactose acceptor, causing it to make lactose instead of N-acetyllactosamine:

$$\text{UDP-galactose} + \text{D-glucose} \longrightarrow \text{UDP} + \text{lactose}$$

α-Lactalbumin is remarkable in another respect: it has a striking homology to the enzyme lysozyme (page 115) in its amino acid sequence, discovered by R. L. Hill and his colleagues.

Synthesis of Glycogen and Starch and the Role of Nucleoside Diphosphate Sugars

A very heavily traveled biosynthetic pathway starting from glucose 6-phosphate leads to formation of the storage polymers glycogen in animals and starch in higher plants. This path begins with the conversion of glucose 6-phosphate to glucose 1-phosphate, catalyzed by phosphoglucomutase (page 435):

$$\text{Glucose 6-phosphate} \rightleftharpoons \text{glucose 1-phosphate}$$

We have seen that glucose 1-phosphate is a product of the action of the glycogen-degrading enzyme phosphorylase (page 433). It was once thought that glycogen phosphorylase catalyzes both the synthesis and degradation of glycogen since the phosphorylase reaction is easily reversed in vitro. However, under intracellular conditions glycogen phosphorylase catalyzes only the breakdown of glycogen. For the conversion of glucose 1-phosphate to glycogen, a different enzyme is involved, which employs UDP-glucose as glucosyl donor. The first step in glycogen synthesis in animals is catalyzed by *glucose-1-phosphate uridylyltransferase*

$$\alpha\text{-D-Glucose 1-phosphate} + \text{UTP} \rightleftharpoons \text{UDP-D-glucose} + \text{PP}_i$$

The structure of UDP–D-glucose was given in Figure 23-12. In the second step the glucosyl group of UDP–D-glucose is transferred to the terminal glucose residue at the nonreducing end of an amylose chain (page 264) to form an $\alpha(1 \rightarrow 4)$ glycosidic linkage between carbon atom 1 of the added glucosyl residue and the 4-hydroxyl of the terminal glucose residue of the chain. This reaction is catalyzed by *glycogen synthase*:

$$\text{UDP-D-glucose} + (\text{glucose})_n \longrightarrow \text{UDP} + (\text{glucose})_{n+1}$$
$$\text{Glycogen}$$

The $\Delta G^{\circ\prime}$ of this reaction step is about -3.2 kcal mol^{-1}. The overall $\Delta G^{\circ\prime}$ for insertion of one glucosyl residue, starting from glucose 1-phosphate and assuming that inorganic pyrophosphate is completely hydrolyzed, is approximately -10 kcal mol^{-1}. The overall equilibrium thus greatly favors synthesis of glycogen.

Glycogen synthase requires as a primer an $\alpha(1 \rightarrow 4)$ polyglucose chain having at least four glucose residues, to which it adds successive glucosyl groups; however, it is far more active with long-chain glucose polymers, such as amylose. In higher animals UDP-glucose is the most active glucosyl donor; ADP-glucose is only about 50 percent as active. In some lower organisms, however, ADP-glucose is the preferred substrate for glycogen synthase. Glycogen synthase of mammalian liver and muscle contains four subunits and has a total molecular weight of about 360,000. The enzyme is usually associated with glycogen granules in the cell.

Glycogen synthase cannot make the $\alpha(1 \rightarrow 6)$ bonds found in the branch points of glycogen (page 266). However, 1,4-α-

glucan branching enzyme, which is present in many animal tissues, catalyzes transfer of a terminal oligosaccharide fragment of six or seven glucosyl residues from the end of the main glycogen chain to the 6-hydroxyl group of a glucose residue of the same or of another glycogen chain, in such manner as to form an $\alpha(1 \rightarrow 6)$ linkage and thus create a new branch point (Figure 23-16).

In human beings several hereditary disorders of glycogen metabolism have been recognized in which abnormally large amounts of glycogen are accumulated, particularly in the liver. The genetically defective enzymes in such patients have been detected by assay of small samples of tissue obtained by surgical biopsy. In one of these disorders, von Gierke's disease, glucose-6-phosphatase activity is defective. In Andersen's disease the branching enzyme is defective, and the glycogen has abnormally long, unbranched chains. In McArdle's disease muscle glycogen phosphorylase is defective, leading to excessive glycogen deposition in muscles. In Cori's disease there is a deficiency in the debranching enzyme (page 434).

In plants and some bacteria starch synthesis occurs by a pathway similar to that in glycogen synthesis, catalyzed by starch synthase. However, ADP-glucose rather than UDP-glucose is the more active glucose donor in many plants. The sequence is

$$\text{ATP} + \alpha\text{-D-glucose 1-phosphate} \rightleftharpoons \text{ADP-glucose} + \text{PP}_i$$

$$\text{ADP-glucose} + \underset{\text{starch}}{(\text{glucose})_n} \longrightarrow \text{ADP} + (\text{glucose})_{n+1}$$

The first of these reactions is catalyzed by glucose-1-phosphate adenylyltransferase, an allosteric enzyme that is stimulated by 3-phosphoglycerate and fructose 1,6-diphosphate, early products of the photosynthetic reduction of CO_2 (page 634).

Regulation of Glycogen Synthesis and Breakdown

Glycogen synthase, like glycogen phosphorylase, is subject to regulation through both allosteric (page 234) and covalent (page 242) modification. Glycogen synthase occurs in two forms. The phosphorylated form is inactive by itself. This form is, however, stimulated by the presence of the allosteric modulator glucose 6-phosphate and is therefore called the D or dependent form. The phosphate groups in the D form of the enzyme are acted upon by a phosphoprotein phosphatase to yield the dephosphorylated or active form of glycogen synthase, which does not require glucose 6-phosphate and is thus called the I or independent form. The dephosphorylation of glycogen synthase is itself inhibited by glycogen, a negative allosteric modulator. The I form of glycogen synthase can be inactivated by protein kinase (page 812), an enzyme that can phosphorylate several different proteins at the expense of ATP. An example seen earlier is the activation of the inactive form of phosphorylase kinase (page 434); other examples are described elsewhere (page

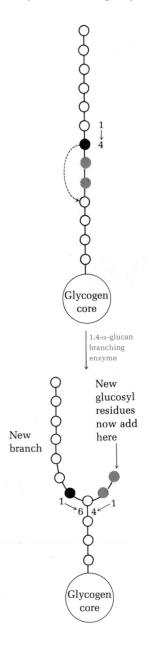

Figure 23-16
Formation of a branch in glycogen by the action of 1,4-α-glucan branching enzyme.

Figure 23-17
Summary of allosteric and covalent regulation of glycogen synthesis and breakdown in mammalian liver and muscle. The roles of adenylate cyclase, cyclic AMP, protein kinase, and hormones and hormone receptors are described elsewhere (Chapter 29, page 812).

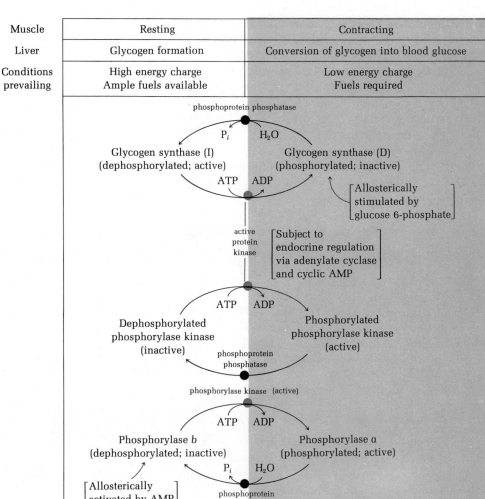

Muscle	Resting	Contracting
Liver	Glycogen formation	Conversion of glycogen into blood glucose
Conditions prevailing	High energy charge Ample fuels available	Low energy charge Fuels required

812). Protein kinase in turn is allosterically activated by cyclic AMP (page 812), formed as a result of the binding of certain hormones, such as epinephrine (page 813) and glucagon (page 816), to specific receptor sites on cell membranes.

Although the interconversion of glycogen synthase between active and inactive forms by the loss or addition of phosphate groups resembles that of glycogen phosphorylase (page 812), the net effect of the glycogen synthase regulatory cycle is opposite to that of glycogen phosphorylase, as seen in Figure 23-17. Recall that phosphorylase *a*, the phosphorylated form of the enzyme, is fully active, whereas the dephosphorylated form, phosphorylase *b*, is inactive unless AMP, which is a positive allosteric modulator, is present. Phosphorylase *a* is dephosphorylated by *phosphorylase phosphatase,* and the resulting phosphorylase *b* is rephosphorylated to the active form by *phosphorylase kinase*. This enzyme also exists in active and inactive, or phosphorylated, or dephosphorylated, forms. The inactive form is activated by ATP through the action of protein kinase, which also phosphorylates (and thus inactivates) glycogen synthase directly (see above).

The regulation of glycogen synthesis and utilization is locked into the regulatory mechanisms of glycolysis and of the tricarboxylic acid cycle. An excess of glucose, reflected as a high concentration of glucose 6-phosphate, or an ample supply of other fuels, reflected as a high energy charge (page 538), tends to turn on glycogen synthase and turn off glycogen phosphorylase, resulting in storage of glucose as glycogen in the liver and muscles. Conversely, when fuels are required because the energy charge is low, as in muscular work, or when the blood glucose is low, glycogen phosphorylase is stimulated and glycogen synthase inhibited, thus causing liver glycogen to be broken down to yield blood glucose and muscle glycogen to yield glucose 6-phosphate as fuel for glycolysis.

Superimposed on these controls over glycogen synthase and glycogen phosphorylase, which reflect the level of glucose 6-phosphate and the energy charge, is another set of controls involving regulation by hormones, particularly epinephrine, glucagon, and insulin, whose effects are mediated by cyclic AMP, or cyclic GMP, and other factors. The action of these hormones on their specific receptors in their target tissues and the amplification cascades for endocrine regulation of glycogen metabolism are summarized elsewhere (page 813).

Turnover of Glucose Residues in Glycogen

Glycogen occurs in particulate granules in such tissues as the liver and skeletal muscle; these granules also contain tightly bound glycogen phosphorylase and glycogen synthase. The glucose residues of liver glycogen undergo constant, very rapid turnover because of the large amounts of glucose and other hexoses reaching the liver from the intestine and leaving again as blood sugar, after being temporarily stored as glycogen. Not all parts of the highly branched structure of glycogen turn over at the same rate. Most of the turnover of glucose residues occurs at the outer branches (page 266), whereas the inner-core structure of glycogen is metabolically more stable, with a much lower rate of turnover.

Structural Polysaccharides of Cell Walls and Coats

In the biosynthesis of the structural polysaccharides of cell walls and coats, much more complex pathways operate than for the storage polysaccharides. Some cell-wall and cell-coat polysaccharides are hetero- rather than homopolysaccharides. Moreover, the precursors or intermediates must be extruded through the cell membrane and assembled outside the cell.

As in glycogen synthesis, nucleoside diphosphate sugars function as glycosyl donors in biosynthesis of cell walls and coats: glycosyl units are added to one end of a preexisting polysaccharide molecule, to lengthen it by one unit, with liberation of the free nucleoside diphosphate. The extrusion of precursors from the cell and the assembly of the wall outside the membrane is effected by the participation of *lipid intermediates*, phosphate esters of long, isoprenoid alcohols in the membrane. These act as arms to carry hydrophilic sugar residues across the membrane. This area of biosynthesis is still very much a frontier and provides many challenges in analysis of the strategies cells have perfected to build their external walls and coats from internally generated precursors.

Cell Walls of Plants; Insect Exoskeletons

Cellulose, the major structural polysaccharide in plant cell walls, is a $\beta(1 \to 4)$ glucan (page 267). It is synthesized in some plants from GDP-glucose and in others from UDP-glucose by *cellulose synthase*:

$$\text{NDP-glucose} + (\text{glucose})_n \longrightarrow \text{NDP} + (\text{glucose})_{n+1}$$

Preexisting   Lengthened
cellulose chain   cellulose chain

In similar reactions plant *xylans*, which contain $\beta(1 \to 4)$ linkages between D-xylose residues, are formed from UDP–D-xylose. Plant cell walls also contain other complex materials, such as *hemicellulose*, *pectin*, *lignin*, and the glycoprotein *extensin* (page 267). Relatively little is known of the biosynthesis of these accessory substances. In some plant cells short segments of the wall are prefabricated within the cell body, inside membranous vesicles generated by the Golgi apparatus. The finished segment is then extruded from the cell and attached to the wall structure outside the membrane (Figure 23-18).

Chitin, a $\beta(1 \to 4)$ homopolymer of N-acetylglucosamine found in insect exoskeletons, is formed from UDP–N-acetylglucosamine in a reaction catalyzed by *chitin synthase*:

$$\text{UDP–N-acetylglucosamine} + (\text{N-acetylglucosamine})_n \longrightarrow$$

$$\text{UDP} + [\beta(1 \to 4)\text{-N-acetylglucosamine}]_{n+1}$$

Lengthened chitin chain

Animals

The different types of polysaccharides found in the intercellular space of animal tissues and in the flexible coats of animal cells are described elsewhere (page 271). *Hyaluronic acid*, the simplest acid mucopolysaccharide, which is a major component of the ground substance or intercellular "filler" between cells, consists of alternating D-glucuronic acid and N-acetylglucosamine residues (page 273). It is formed by successive alternating reactions of UDP-glucuronic acid and UDP-N-acetylglucosamine at the grow-

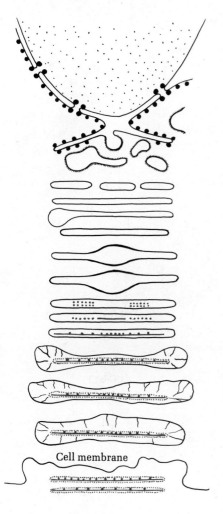

Figure 23-18
Role of Golgi vesicle in transport of prefabricated section of cell wall to exterior of plant cell. [Redrawn from R. M. Brown, Jr., et al., J. Cell Biol., 45:267 (1970)]

Cell membrane

ing end of the chain, in a manner similar to the formation of homopolysaccharides. Chondroitin, dermatan, and
keratan sulfates (page 273) are formed in similar reactions.
However, these acid mucopolysaccharides normally occur
covalently attached to polypeptide chains, as glycoproteins
or protein-polysaccharides (page 273). Although the structure and biosynthesis of such glycoproteins are exceedingly
complex, some of the major enzymatic reaction steps have
been worked out in the laboratories of A. Dorfman, S.
Roseman, R. Jeanloz, E. C. Heath, and other investigators.

The glycoproteins of animal-cell coats contain many oligosaccharide groups covalently attached to asparagine, serine,
threonine, or hydroxylysine residues of their polypeptide
chains (page 274). The oligosaccharide groups are believed to
be built up in part inside the cell, in the Golgi apparatus, and
in part outside the cell membrane. The synthesis of the
oligosaccharide side chains begins at the Golgi membrane
with the glycosylation of an amino acid residue of the protein in reactions with NDP derivatives of D-galactose, D-
glucose, or D-xylose. Then different monosaccharide residues
are added in succession to extend the oligosaccharide side
chain in a specific sequence, which is determined by the
specificity of the active sites of the glycosyl transferases.
Although nucleoside diphosphate sugars are the ultimate
donors of the monosaccharide units, some of the monosaccharide units are directly added to the growing oligosaccharide side chain through the intervention of *lipid intermediates*. Such lipid intermediates of animal tissues were identified by Leloir and his colleagues as long-chain unsaturated
isoprenoid alcohols (isoprenols) called *dolichols*, which contain 16 to 21 isoprenoid units and are thus up to 84 carbon
atoms long (Figure 23-19). Dolichols are found either in free
or phosphorylated form (phosphoryldolichol) in membranes
of the endoplasmic reticulum and Golgi apparatus but not in
the mitochondrial or plasma membranes. Phosphoryldolichol
is glycosylated at the expense of nucleoside diphosphate
sugars in reactions such as

Phosphoryldolichol + GDP-mannose ⟶
                mannosyl phosphoryldolichol + GDP

The mannose residue is then transferred to the end of the
growing oligosaccharide side chain of the glycoprotein:

Mannosyl phosphoryldolichol + oligosaccharide ⟶
         phosphoryldolichol + lengthened oligosaccharide

Vitamin A, or retinol (page 352), as its phosphate or pyrophosphate, also appears to function as a sugar-carrying lipid
intermediate in similar reactions.

The *sialic acids* or *N-acylneuraminic acids* (page 274),
which serve as the terminal residues in the oligosaccharide
side chains of gangliosides and glycoproteins of animal cell
coats, are derived from glucosamine. The following sequence
of reactions shows the formation of N-acetylneuraminic acid:

Figure 23-19
A phosphoryldolichol.

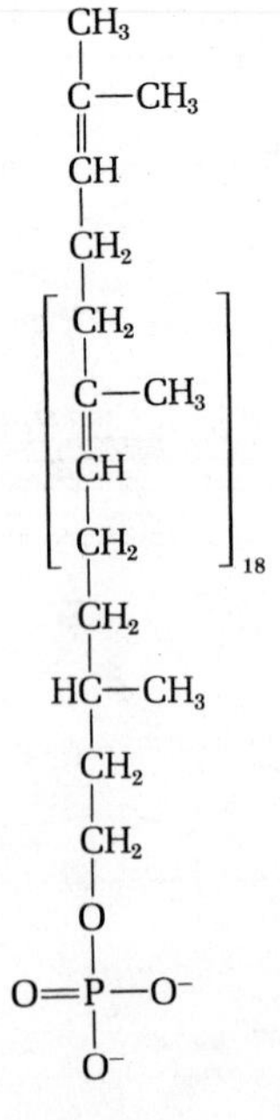

$$\text{UDP–N-acetylglucosamine} + H_2O \xrightarrow{\text{acylglucosamine 2-epimerase}}$$
$$\text{N-acetylmannosamine} + \text{UDP}$$

$$\text{N-Acetylmannosamine} + \text{ATP} \xrightarrow{\text{N-acetylmannosamine kinase}}$$
$$\text{N-acetylmannosamine 6-phosphate} + \text{ADP}$$

$$\text{N-Acetylmannosamine 6-phosphate}$$
$$+ \text{ phosphoenolpyruvate} \xrightarrow[\text{9-phosphate synthase}]{\text{N-acylneuraminate-}}$$
$$\text{N-acetylneuraminic acid 9-phosphate} + P_i$$

$$\text{N-Acetylneuraminic acid 9-phosphate} + H_2O \xrightarrow[\text{9-phosphatase}]{\text{N-acylneuraminate-}}$$
$$\text{N-acetylneuraminic acid} + P_i$$

N-Acetylneuraminic acid (NANA) is prepared for reaction with oligosaccharides and polysaccharides by formation of a CMP derivative:

$$\text{CTP} + \text{N-acetylneuraminic acid} \xrightarrow[\text{transferase}]{\text{acylneuraminate cytidylyl-}}$$
$$\text{CMP–N-acetylneuraminic acid} + PP_i$$

The product of this reaction, abbreviated CMP–NANA, then becomes the donor of its N-acetylneuraminic acid residue to the terminal sugar residue of glycoprotein side chains by the action of *sialytransferase*, with release of free CMP:

$$\text{CMP–NANA} + \text{—X—X—X—galactose} \xrightarrow{\text{sialyltransferase}}$$
$$\text{CMP} + \text{—X—X—X—galactose—NANA}$$
Sialyloligosaccharide

Mucopolysaccharidoses

There are a number of genetic diseases in man, called *mucopolysaccharidoses*, in which excessive accumulations of certain mucopolysaccharides occur in the tissues. Skeletal defects and severe mental retardation are the major symptoms. The question arises as to whether the excessive deposition of these mucopolysaccharides is the result of abnormally high rates of synthesis or abnormally low rates of breakdown. Intensive study of the complex biosynthetic and degradative pathways of mucopolysaccharide metabolism has revealed that these disorders are probably the result of genetic deficiencies in specific hydrolytic enzymes responsible for the degradation of different mucopolysaccharides. In *Hurler's disease*, also called *gargoylism*, the most striking abnormalities are a grotesque facial appearance, bone defects, mental retardation, a large abdomen, and corneal clouding. Excessive amounts of dermatan sulfate and heparan sulfate are deposited in the connective tissue cells and in the liver, where they accumulate in Golgi vesicles and lysosomes. It has been found that the hydrolytic enzyme α-L-iduronidase, which catalyzes an essential step in degradation of these mucopolysaccharides, is lacking.

In Hunter's disease, which resembles Hurler's disease but is genetically distinct, dermatan and heparan sulfates also accumulate, apparently owing to a genetic deficiency of a

Figure 23-20
Steps in the biosynthesis of the pep-
tidoglycan of a bacterial cell wall. The three
stages are described in the text.

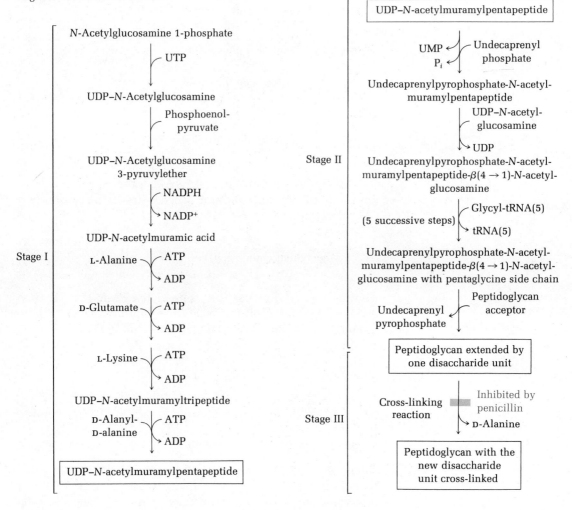

specific sulfatase. Another type of sulfatase is deficient in Sanfilippo syndrome A, in which heparan sulfate accumulates. β-Galactosidase appears to be low or lacking in all three diseases.

Bacterial Cell Walls

The biosynthesis of the cell-wall polymers of bacteria has attracted many investigators, not only because of the complex structure of these polymers, but also because cell-wall biosynthesis in bacteria is inhibited by penicillin and other antibiotics.

The cell-wall peptidoglycan, or murein, of bacteria, described in Chapter 10 and also on page 618, is synthesized in three major stages, as revealed by research of J. L. Strominger and other investigators. This process will be briefly outlined (Figure 23-20). In stage I, N-acetylmuramylpentapeptide, a recurring unit of the backbone structure, is synthesized in the form of its UDP derivative (Figure 23-21) from its precursors N-acetylglucosamine, phosphoenolpyruvate, and five

Figure 23-21
UDP–N-acetylmuramylpentapeptide. Note
the γ peptide linkage in the D-glutamic acid
residue.

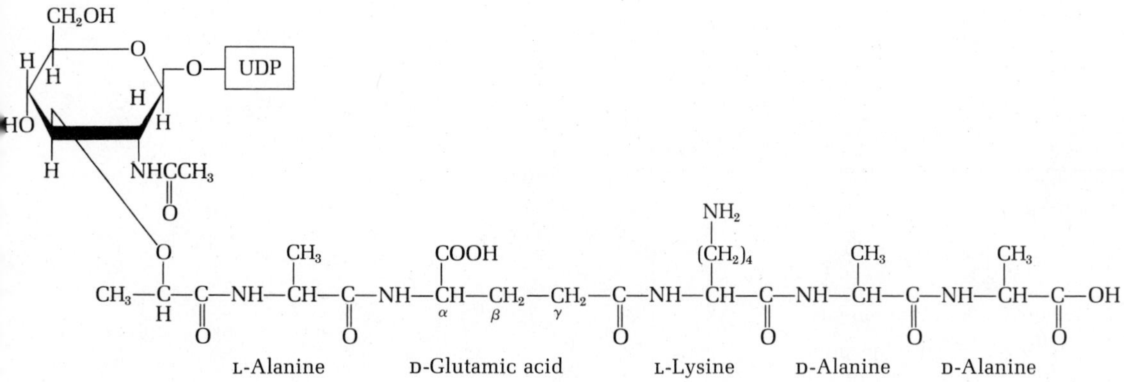

L-Alanine   D-Glutamic acid   L-Lysine   D-Alanine   D-Alanine

Figure 23-22
D-Cycloserine, an antibiotic inhibiting the
formation of D-alanyl-D-alanine. The similar structures in the antibiotic and D-alanine
are in color.

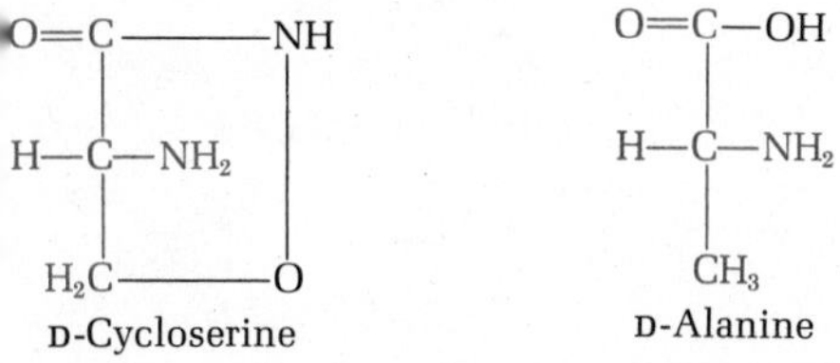

D-Cycloserine

D-Alanine

Figure 23-23
Undecaprenyl phosphate. Undecaprenol has
11 isoprene units, whereas dolichol (Figure
23-18) has 20.

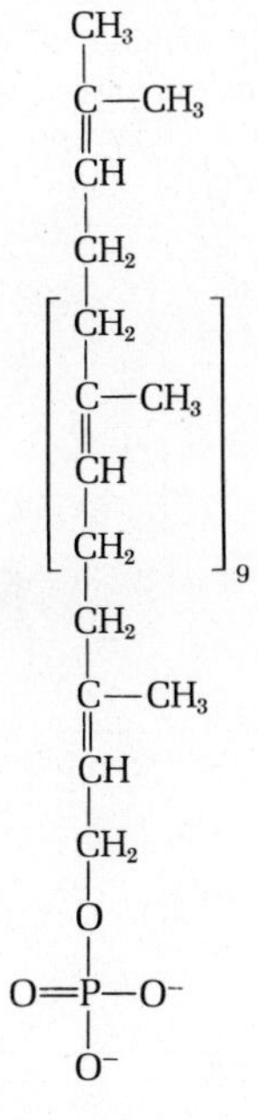

amino acid units. The last step in the first stage is the attachment of the dipeptide D-alanyl-D-alanine to the tripeptide
side chain of the recurring sugar unit. This dipeptide is
formed by the reaction

$$2 \text{ D-Alanine} + \text{ATP} \longrightarrow \text{D-alanyl-D-alanine} + \text{ADP} + P_i$$

This reaction is noteworthy since it is inhibited by the antibiotic cycloserine (Figure 23-22), a structural analog that presumably competes with D-alanine in the enzymatic reaction.

Up to this point, all the reactions have taken place within
the cell. In stage II, which takes place in the membrane, the
completed N-acetylmuramylpentapeptide is first transferred
enzymatically from UDP to a membrane-bound lipid intermediate, undecaprenyl phosphate (Figure 23-23). The nonpolar end of its hydrocarbon chain is buried within the lipid
portion of the cell membrane, whereas the polar end carries
the monosaccharide unit. To the N-acetylmuramylpentapeptide attached to the undecaprenyl arm is now added the
other monosaccharide unit, N-acetylglucosamine, donated by
UDP–N-acetylglucosamine, to yield the recurring disaccharide of the peptidoglycan backbone, N-acetylmuramyl-
pentapeptide-β-(4 → 1)-N-acetylglucosamine, attached to the
undecaprenyl phosphate. Short cross-linking peptide chains
are then added to the disaccharide. The length of the cross-
linking peptides and their amino acid components vary with
the species; in Staphylococcus aureus the chain is pentaglycine. The entire disaccharide unit, with its cross-linking
chain, is now transferred to the growing end of the peptidoglycan polymer. Then undecaprenol is left as its pyrophosphate, from which the orthophosphate ester is regenerated by enzymatic hydrolysis. The latter reaction is inhibited
by the cyclic peptide bacitracin, an antibiotic from Bacillus
licheniformis.

In the last stage (III) of cell-wall synthesis, the cross-
linkage between parallel peptidoglycan chains is established
through a transpeptidation reaction in which the amino ter-

653

Figure 23-24
Completion of a cross-link between two ad-
jacent peptidoglycan chains in the bacterial
cell wall. This reaction is blocked by peni-
cillin.

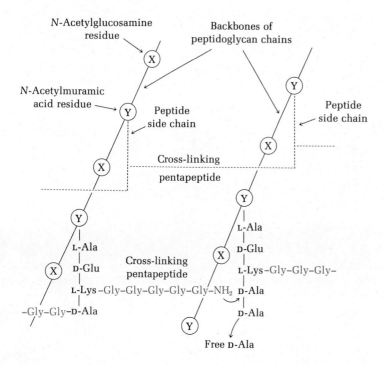

minal glycine residue of the cross-linking chain displaces the carboxyl-terminal D-alanine from the end of the pentapeptide side chain of an N-acetylmuramylpentapeptide residue in the adjacent peptidoglycan chain (Figure 23-24); it is this reaction that is inhibited by _penicillin_ (Figure 23-25). Penicillin thus prevents completion of the synthesis of the cell wall at its last stage, after the bacterial cell has already made a very heavy biosynthetic investment.

By repetitive additions of the disaccharide units with their pentapeptide side chains and cross-linking chains, followed by cross-linking, the entire bacterial cell becomes surrounded by a single, gigantic saccular molecule of peptidoglycan. In some bacteria several layers of peptidoglycan are built in this manner.

In some gram-positive bacteria the rigid peptidoglycan framework of the cell wall is covered with a layer of _teichoic acid_ (page 271), whose backbone structure is synthesized from CDP-ribitol or CDP-glycerol precursors by the action of _teichoic-acid synthase_. Sugar side chains are then introduced into the teichoic acid chain from UDP derivatives.

In gram-negative organisms, such as _E. coli_, the peptidoglycan framework is surrounded by a complex polymeric lipopolysaccharide (page 271). Although the details of the biosynthesis of this very complex polymer cannot be given here, the major stages are now understood from recent research in a number of laboratories.

Figure 23-25
Benzylpenicillin (penicillin G). Many dif-
ferent penicillins are known. They differ in
the structure of the R group (color).
Penicillin G is the most widely used form in
medicine.

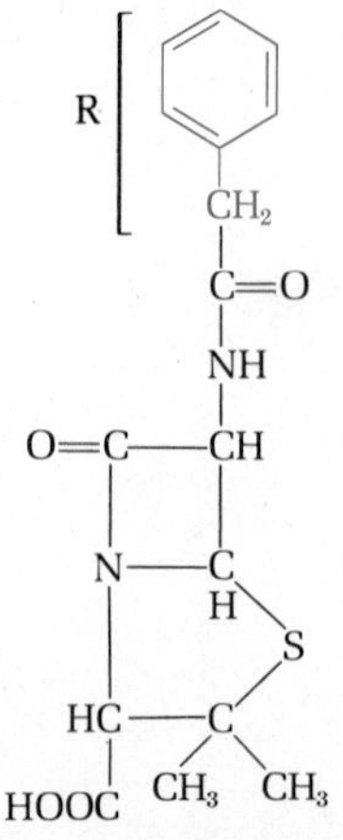

Summary

The central common pathway in the biosynthesis of most carbohydrates from noncarbohydrate precursors is the route from pyruvate to glucose 6-phosphate. Most of the reversible reactions of glycolysis are utilized in the biosynthetic direction. However, two irreversible reactions of glycolysis are replaced by bypass reactions that are energetically favorable for synthesis. In the first, pyruvate is converted into phosphoenolpyruvate by the mitochondrial sequence

$$\text{Pyruvate} + CO_2 \xrightarrow{\text{ATP}} \text{oxaloacetate} \longrightarrow \text{malate}$$

followed by the cytoplasmic sequence

$$\text{Malate} \longrightarrow \text{oxaloacetate} \xrightarrow{\text{GTP}} \text{phosphoenolpyruvate} + CO_2$$

The second bypass is the hydrolysis of fructose 1,6-diphosphate to fructose 6-phosphate by hexosediphosphatase. Glucose 6-phosphate formed in this central pathway from pyruvate may be dephosphorylated to form free glucose by glucose-6-phosphatase. The rate of gluconeogenesis is primarily regulated by the first reaction of the sequence (pyruvate carboxylase) and secondarily by hexosediphosphatase. Glycolysis and gluconeogenesis are independently regulated. They may occur simultaneously and thus may give rise to futile cycles, in which ATP energy is lost.

Lactate and the intermediates of the tricarboxylic acid cycle can undergo net conversion into glucose, as can the glycogenic amino acids. On the other hand, neither acetyl-CoA nor CO_2 can undergo net conversion into glucose in animal tissues. However, in plants and microorganisms, acetyl-CoA is converted into glucose by operation of the glyoxylate cycle.

In photosynthesis in most plants of temperate zones, CO_2 gains entry into the carbon backbone of glucose by a dark reaction with ribulose 1,5-diphosphate to yield 3-phosphoglycerate; this is called the C_3 pathway. At the expense of ATP and NADPH generated in the light reactions, six molecules of CO_2 can ultimately be converted into glucose by operation of the Calvin cycle, which includes reactions of the phosphogluconate and glycolytic pathways. In some tropical plants (C_4 plants) CO_2 is first fixed in mesophyll cells into C_4 acids, which are then transported into bundle-sheath cells and decarboxylated, and the CO_2 is refixed by the C_3 pathway. C_3 plants show considerable photorespiration, the oxidation of glycolic acid derived from a reaction of oxygen rather than CO_2 with ribulose diphosphate. C_4 plants are much less active in photorespiration; they are far more efficient in net photosynthetic activity than C_3 plants.

Nucleoside diphosphate sugars, particularly uridine diphosphate derivatives, are precursors of other monosaccharides such as D-galactose, of disaccharides such as sucrose and lactose, and of various polysaccharides. Formation of glycogen by glycogen synthase requires uridine diphosphate glucose as the glucose donor. Glycogen synthase occurs in a phosphorylated, or D, form, which is relatively inactive but is stimulated somewhat by glucose 6-phosphate, and a dephosphorylated, or I, form, which is maximally active and is independent of glucose 6-phosphate as modulator. The phosphate group of the D-glycogen synthase is removed by a phosphoprotein phosphatase. The I form can be rephosphorylated by protein kinase, whose inactive form is converted into the active enzyme by cyclic adenylate. Glycogen phosphorylase and glycogen synthase activities are independently controlled in muscles and the liver. Nucleoside diphosphate sugars are also glycosyl donors in the biosynthesis of extracellular structural polysaccharides such as cellulose and xylans found in plant-cell walls, hyaluronic acid and the oligosaccharide side chains of glycoproteins in animal tissues, and the peptidoglycans of the cell walls of bacteria. Certain steps in the biosynthesis of bacterial cell walls are inhibited by specific antibiotics. For example, penicillin inhibits the final cross-linking reaction in the biosynthesis of the peptidoglycan framework of the cell wall.

References

Books

MEHLMAN, M. A., and R. W. HANSON (eds.): *Energy Metabolism and the Regulation of Metabolic Processes in Mitochondria*, Academic, New York, 1972.

STANBURY, J. B., J. B. WYNGAARDEN, and D. S. FREDRICKSON (eds.): *The Metabolic Basis of Inherited Disease*, McGraw-Hill, New York, 1971. Comprehensive account of genetic disorders of carbohydrate metabolism.

ZELITCH, I.: *Photosynthesis, Photorespiration, and Plant Productivity*, Academic, New York, 1971.

Articles

ATKINSON, D. E.: "Adenine Nucleotides as Coupling Agents in Metabolism and as Regulatory Modifiers: The Adenylate Energy Charge," pp. 1–21 in H. J. Vogel (ed.), *Metabolic Regulation*, vol. 5, Academic, New York, 1971.

BARTHOLOMEW, B. A., G. W. JOURDIAN, and S. ROSEMAN: "The Sialic Acids: Transfer of Sialic Acid to Glycoproteins by a Sialyltransferase from Colostrum," *J. Biol. Chem.*, 248: 5751–5762 (1973).

BASSHAM, J. A.: "The Control of Photosynthetic Carbon Metabolism," *Science*, 172: 526–534 (1971).

BAYNES, J. W., A-F. HSU, and E. C. HEATH: "The Role of Mannosyl-Phosphoryl-Dihydropolyisoprenol in the Synthesis of Mammalian Glycoproteins," *J. Biol. Chem.*, 248: 5693–5704 (1973).

BEHRENS, N. H., and L. F. LELOIR: "Dolichol Monophosphate Glucose: An Intermediate in Glucose Transfer in Liver," *Proc. Natl. Acad. Sci. (U.S.)*, 66: 153–159 (1970).

BJORKMAN, O., and J. BERRY: "High-Efficiency Photosynthesis," *Sci. Am.*, 229: 80–93 (1973).

CLARK, D. G., R. ROGNSTAD, and J. KATZ: "Isotopic Evidence for Futile Cycles in Liver Cells," *Biochem. Biophys. Res. Commun.*, 54: 1141–1148 (1973).

FISCHER, E. H., L. M. G. HEILMEYER, JR., and R. H. HASCHKE: "Phosphorylase and the Control of Glycogen Degradation," *Curr. Top. Cell Regul.*, 4: 211–251 (1971).

GOLDSWORTHY, A.: "The Riddle of Photorespiration," *Nature*, 224: 501–502 (1969).

HASSID, W. Z.: "Biosynthesis of Oligosaccharides and Polysaccharides in Plants," *Science*, 165: 137–144 (1969).

LARNER, J., and C. VILLAR-PALASI: "Glycogen Synthase and Its Control," *Curr. Top. Cell Regul.*, 3: 195–236 (1971).

LENNARZ, W. J.: "Lipid-linked Sugars in Glycoprotein Synthesis," *Science*, 188: 986–991 (1975).

MCCLURE, W. R., and H. A. LARDY: "Rat Liver Pyruvate Carboxylase: Factors Affecting the Regulation *in Vivo*," *J. Biol Chem.*, 246: 3591–3596 (1971).

SOLS, A., and R. MARCO: "Concentrations of Metabolites and Binding Sites: Implications in Metabolic Regulation," *Curr. Top. Cell. Regul.*, 2: 227–273 (1970).

STROMINGER, J. L.: "The Actions of Penicillin and Other Antibiotics on Bacterial Cell Wall Synthesis," *Johns Hopkins Med. J.*, 133: 63–81 (1973).

Problems

1. Write balanced equations for the following processes: (a) the net formation of D-glucose from citric acid in the liver of the rat, (b) the net formation of D-glucose from phenylalanine in the liver of the rat, (c) the net conversion of palmitic acid into D-glucose in the seeds of a higher plant, (d) the net conversion of L-leucine into D-glucose in a higher plant.

2. In which carbon atoms of glucosyl residues in liver glycogen and palmitic acid in liver phosphatidyl choline will the isotopic carbon atom in $3\text{-}^{14}C\text{-}L$-leucine be found?

3. Write a balanced equation for the synthesis of one D-glucosyl residue of glycogen starting from pyruvic acid. How many high-energy phosphate bonds are required?

4. Write a balanced equation for the biosynthesis of L-ascorbic acid from pyruvic acid.

5. If a green leaf is illuminated in the presence of radioactive CO_2 for a short period and then extracted, in which positions of the following compounds found in the extract would you expect to find isotopic carbon? (a) D-Glucose 6-phosphate, (b) sedoheptulose 7-phosphate, (c) ribose 5-phosphate. Assume we are dealing with a C_3 plant.

CHAPTER 24 THE BIOSYNTHESIS OF LIPIDS

The biosynthesis of lipids is a prominent metabolic process in most organisms. Because of the limited capacity of higher animals to store polysaccharides, glucose ingested in excess of immediate energy needs and storage capacity is converted by glycolysis into pyruvate and then acetyl-CoA, from which fatty acids are synthesized. These in turn are converted into triacylglycerols, which have a much higher energy content than polysaccharides (page 840) and may be stored in very large amounts in adipose or fat tissues. Triacylglycerols are also stored in the seeds and fruits of many plants.

The formation of the various phospholipids and sphingolipids of cell membranes (page 287) is also an important biosynthetic process. These complex lipids undergo continuous metabolic turnover in most cells.

In this chapter we shall consider the biosynthesis of fatty acids, the formation of triacylglycerols and various other lipids containing fatty acids as building blocks, as well as some of the heritable lipid-storage diseases. We shall also examine the biosynthesis of cholesterol and other steroids and of the prostaglandins. Although the formation of this last group of compounds accounts for only a small fraction of the total biosynthetic capacity of most cells, the biological activity of steroid hormones and prostaglandins in the regulation of many different cell functions makes their biosynthesis an important metabolic activity.

Biosynthesis of Saturated Fatty Acids

The biosynthesis of saturated fatty acids from their ultimate precursor acetyl-CoA occurs in all organisms but is particularly prominent in the liver, adipose tissues, and mammary glands of higher animals. It is brought about by a process that differs significantly from the opposed process of fatty acid oxidation (page 545). In the first place total biosynthesis of fatty acids occurs in the cytosol, whereas fatty acid oxidation occurs in the mitochondria (pages 374 and 544). Second, the presence of citrate is necessary for maximal rates of synthesis of fatty acids, whereas it is not required in fatty acid oxidation. Perhaps the most unexpected difference is that CO_2 is essential for fatty acid synthesis in cell extracts, although isotopic CO_2 is not itself incorporated into the newly synthesized fatty acids. These and many other observations have revealed that fatty acid synthesis from acetyl-CoA takes place with an entirely different set of enzymes from those employed in fatty acid oxidation.

In the overall reaction of fatty acid synthesis, which is catalyzed by a cluster of seven proteins in the cytosol, the *fatty-acid synthetase complex*, acetyl-CoA derived from carbohydrate or amino acid sources is the ultimate precursor of all the carbon atoms of the fatty acid chain. However, of the eight acetyl units required for biosynthesis of palmitic acid, only one is provided by acetyl-CoA; the other seven arrive in the form of malonyl-CoA, formed from acetyl-CoA and HCO_3^- in a carboxylation reaction. One acetyl residue and seven malonyl residues undergo successive condensation steps, with release of seven molecules of CO_2, to form palmitic acid; the reducing power is furnished by NADPH:

$$\text{Acetyl-CoA} + 7 \text{ malonyl-CoA} + 14\text{NADPH} + 14\text{H}^+ \longrightarrow$$
$$\underset{\text{Palmitic acid}}{CH_3(CH_2)_{14}COOH} + 7CO_2 + 8CoA + 14NADP^+ + 6H_2O$$

The single molecule of acetyl-CoA required in the process serves as a *primer*, or starter; the two carbon atoms of its acetyl group become the two terminal carbon atoms (15 and 16) of the palmitic acid formed. Chain growth during fatty acid synthesis thus starts at the carboxyl group of acetyl-CoA and proceeds by successive addition of acetyl residues at the carboxyl end of the growing chain. Each successive acetyl residue is derived from two of the three carbon atoms of a malonic acid residue entering the system in the form of malonyl-CoA (Figure 24-1); the third carbon atom of malonic acid, i.e., that of the unesterified carboxyl group, is lost as CO_2. The final product is a molecule of palmitic acid. This sequence of reactions was largely elucidated in the laboratories of F. Lynen, of S. J. Wakil, and of P. R. Vagelos.

A distinctive feature of the mechanism of fatty acid biosynthesis is that the acyl intermediates in the process of chain lengthening are thio esters, not of CoA, as in fatty acid oxidation (page 545), but of a low-molecular-weight conjugated protein called *acyl carrier protein* (ACP). This protein can form a complex or complexes with the six other enzyme proteins required for the complete synthesis of palmitic acid. In most eukaryotic cells all seven proteins of the fatty-acid synthetase complex are associated in a multienzyme cluster. The fatty-acid synthetase complex of yeast cells has been isolated in pure form by Lynen and his colleagues (Figure 24-2). Its particle weight is 2.3 million, and it cannot be dissociated into its component enzymes without loss of activity. On the other hand, the fatty-acid synthetase complex of pigeon liver can be dissociated into two major components without losing activity. Most easily dissociated is the synthetase complex from *E. coli*; each of its seven en-

Figure 24-1
Malonyl-CoA.

COOH
|
CH$_2$
|
C—S—CoA
‖
O

Figure 24-2
The yeast fatty-acid synthetase complex. The electron micrograph (above) shows individual molecules of the synthetase complex outlined by negative-contrast staining with phosphotungstate. Each molecule contains the seven enzyme proteins, arranged in an apparently regular three-dimensional manner. The yeast fatty-acid synthetase has also been crystallized (below); each crystal of course contains many molecules of the complex.

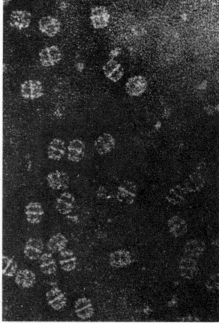

50 nm

F. Lynen

0.2 mm

F. Lynen

Figure 24-3
The carboxylation of acetyl-CoA.

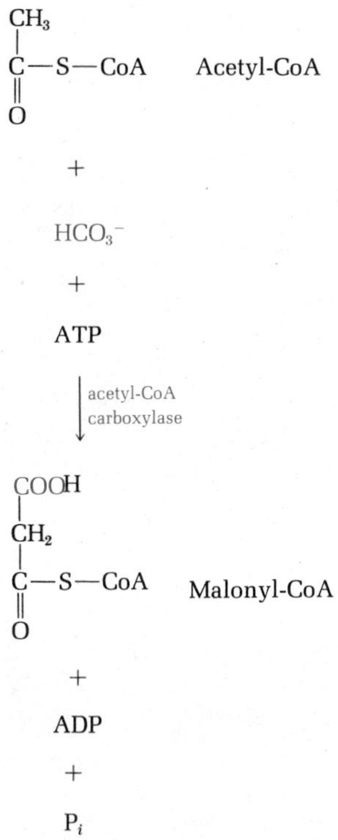

CH₃
|
C—S—CoA   Acetyl-CoA
‖
O

+

HCO₃⁻

+

ATP

| acetyl-CoA
| carboxylase

COOH
|
CH₂
|
C—S—CoA   Malonyl-CoA
‖
O

+

ADP

+

P_i

zymes can be separated in active form and studied individually. Most of our information on the sequential steps of fatty acid synthesis comes from study of the enzymes from *E. coli* by Vagelos and his colleagues. In most organisms the end product of the fatty-acid synthetase system is palmitic acid, the precursor of all other higher saturated fatty acids and of all unsaturated fatty acids.

We shall now examine in detail the enzymatic reactions responsible for the biosynthesis of fatty acids.

The Carbon Source for Fatty Acid Synthesis

The ultimate source of all the carbon atoms of fatty acids is acetyl-CoA, formed in the mitochondria by the oxidative decarboxylation of pyruvate (page 450), the oxidative degradation of some of the amino acids (pages 567 to 574), or by the β oxidation of long-chain fatty acids (page 545).

Acetyl-CoA itself cannot pass out of the mitochondria into the cytosol; however, its acetyl group is transferred through the membrane in other chemical forms. As pointed out earlier (page 529), citrate, formed in mitochondria from acetyl-CoA and oxaloacetate, may pass through the mitochondrial membrane to the cytoplasm via the tricarboxylate transport system. In the cytosol acetyl-CoA is regenerated from citrate by *ATP-citrate lyase*, also called *citrate cleavage enzyme*, which catalyzes the reaction:

$$\text{Citrate} + \text{ATP} + \text{CoA} \longrightarrow \text{acetyl-CoA} + \text{ADP} + \text{P}_i + \text{oxaloacetate}$$

In a second pathway the acetyl group of acetyl-CoA is enzymatically transferred to carnitine, which, as we have already seen, acts as a carrier of fatty acids into mitochondria preparatory to their oxidation (page 546). Acetylcarnitine passes from the mitochondrial matrix through the mitochondrial membrane into the cytosol; acetyl-CoA is then regenerated by transfer of the acetyl group from acetylcarnitine to cytosol CoA.

Formation of Malonyl-CoA: Acetyl-CoA Carboxylase

Before the acetyl groups of acetyl-CoA can be utilized by the fatty-acid synthetase complex, an important preparatory reaction must take place to convert acetyl-CoA into malonyl-CoA, the immediate precursor of 14 of the 16 carbon atoms of palmitic acid. Malonyl-CoA is formed from acetyl-CoA and bicarbonate in the cytosol by the action of *acetyl-CoA carboxylase*, a complex enzyme that catalyzes the reaction:

$$\text{Acetyl-CoA} + \text{HCO}_3^- + \text{H}^+ + \text{ATP} \rightleftharpoons \text{malonyl-CoA} + \text{ADP} + \text{P}_i$$

The carbon atom of the HCO_3^-, which is the active species rather than CO_2, becomes the distal or free carboxyl carbon of malonyl-CoA, as shown in Figure 24-3. However, the above equation and Figure 24-3 give only the overall reaction, the sum of at least three intermediate reactions.

Acetyl-CoA carboxylase contains biotin (page 345) as its prosthetic group. The carboxyl group of biotin is bound in amide linkage to the ϵ-amino group of a specific lysine residue of a subunit of the enzyme. The covalently bound biotin serves as an intermediate carrier of a molecule of CO_2 (page 345) in a two-step reaction cycle:

$$HCO_3^- + H^+ + ATP \diagup \text{biotin-enzyme} \diagup \text{malonyl-CoA}$$
$$ADP + P_i \diagdown \text{carboxybiotin-} \diagdown \text{acetyl-CoA}$$
$$\text{enzyme}$$

The reaction catalyzed by acetyl-CoA carboxylase, an allosteric enzyme, is the primary regulatory, or rate-limiting, step in the biosynthesis of fatty acids. Acetyl-CoA carboxylase is virtually inactive in the absence of its positive modulators citrate or isocitrate. The striking allosteric stimulation of this enzyme by citrate accounts for the fact that citrate is required for fatty acid synthesis in cell extracts without being used as a precursor.

Acetyl-CoA carboxylase occurs in both an inactive monomeric form and an active polymeric form. As it occurs in the avian liver, the inactive enzyme monomer has a molecular weight of 410,000 and contains one binding site for HCO_3^- (that is, one biotin prosthetic group), one binding site for acetyl-CoA, and one for citrate. Citrate shifts the equilibrium between the inactive monomer and the active polymer, to favor the latter:

$$\text{Monomer} \underset{-\text{citrate}}{\overset{+\text{citrate}}{\rightleftarrows}} \text{polymer}$$
$$\text{(inactive)} \qquad \text{(active)}$$

Polymeric acetyl-CoA carboxylase of animal tissues consists of long filaments of enzyme monomers; each monomer unit contains a molecule of bound citrate. The length of the polymeric form varies, but on the average each filament contains about 20 monomer units, has a particle weight of some 8 megadaltons, and is about 400 nm long. Such filaments have been studied in the electron microscope (Figure 24-4) and have actually been observed in the cytoplasm of adipose cells.

The acetyl-CoA carboxylase reaction is actually more complex than described thus far. In fact, the monomeric unit of the enzyme contains four different subunits. The sequence of reactions in the formation of malonyl-CoA has been deduced from study of the four subunits of the monomer. One of these subunits, _biotin carboxylase_ (BC), catalyzes the first step of the overall reaction, namely, the carboxylation of the biotin residue covalently bound to the second subunit, which is called _biotin carboxyl-carrier protein_ (BCCP):

$$\text{Biotin-BCCP} + HCO_3^- + H^+ + ATP \overset{\substack{BC \\ \text{subunit}}}{\rightleftharpoons}$$
$$\text{carboxybiotin-BCCP} + ADP + P_i$$

The second step in the overall reaction is catalyzed by the

Figure 24-4
Electron micrograph of the enzymatically active polymeric form of chicken-liver acetyl-CoA carboxylase. A schematic interpretation of the relationship of the monomeric units is shown at the right.

50 nm   A. Kleinschmidt and M. D. Lane

third type of subunit, called *carboxyl transferase* (CT):

$$\text{Carboxybiotin-BCCP} + \text{acetyl-CoA} \xrightleftharpoons[\text{subunit}]{\text{CT}}$$

$$\text{biotin-BCCP} + \text{malonyl-CoA}$$

In these reactions the biotin residue of the carboxyl carrier protein serves as a swinging arm to transfer the bicarbonate ion from the biotin carboxylase subunit to the acetyl-CoA bound to the active site of the carboxyltransferase subunit. The change from the inactive monomeric form of acetyl-CoA carboxylase to the polymeric, active form of the enzyme occurs when citrate is bound to the fourth subunit of each monomeric unit. Binding of the allosteric modulator citrate greatly increases V_{max} for the transfer of the carboxyl group of carboxybiotin to acetyl-CoA, with little or no change in K_M for the binding of acetyl-CoA.

The Reactions of the Fatty-Acid Synthetase System

Once malonyl-CoA has been generated from acetyl-CoA by the action of the rather complex acetyl-CoA carboxylase reaction, the ensuing reactions of fatty acid synthesis occur in a sequence of six successive steps catalyzed by the six enzymes of the fatty-acid synthetase system. The seventh protein of this system, which has no enzymatic activity itself, is acyl carrier protein, to which the growing fatty acid chain is covalently attached.

Acyl Carrier Protein (ACP)

Acyl carrier protein, universally symbolized as ACP, was first isolated in pure form from E. coli by Vagelos and his colleagues and has since been studied from many other sources. The E. coli ACP is a relatively small (mol wt 10,000), heat-stable protein containing 77 amino acid residues, whose sequence has been established, and a covalently attached prosthetic group. Vagelos and his colleagues have synthesized a number of analogs of the ACP polypeptide utilizing the solid-phase procedure of polypeptide synthesis (page 118). Study of these analogs has revealed which portions of the polypeptide chain of ACP are responsible for binding the six enzyme molecules of the fatty-acid synthetase complex.

The single sulfhydryl group of ACP, to which the acyl intermediates are esterified, is contributed by its prosthetic group, a molecule of *4'-phosphopantetheine,* which is covalently linked to the hydroxyl group of serine residue 36 of the protein (Figure 24-5). The 4'-phosphopantetheine moiety is identical with that of coenzyme A (page 342), from which it is derived. The function of ACP in fatty acid synthesis is analogous to that of CoA in fatty acid oxidation: it serves as an anchor to which the acyl intermediates are esterified (Figure 24-6).

The sequential reactions involved in fatty acid synthesis may now be examined.

Figure 24-5
Structure of the prosthetic group of acyl carrier protein (ACP). The sulfhydryl group (color) forms a thio ester with fatty acids.

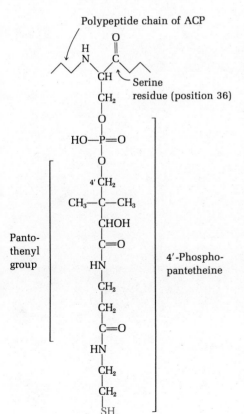

Figure 24-6

A schematic representation of the fatty-acid synthetase complex. The central protein molecule is ACP. Its long phosphopantetheine side chain apparently serves as a swinging arm to carry acyl groups from one enzyme molecule to the next, to accomplish the six steps needed for addition of each two-carbon unit.

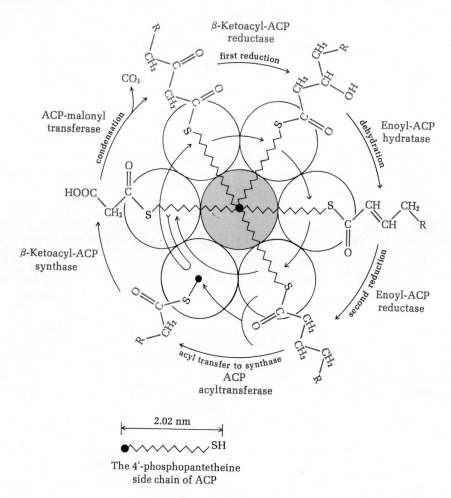

The 4′-phosphopantetheine
side chain of ACP

The Priming Reaction

To prime the fatty-acid synthetase system, acetyl-CoA first reacts with the sulfhydryl group of ACP by the action of one of the six enzymes of the synthetase system, *ACP-acyltransferase*, which catalyzes the reaction:

Acetyl—S—CoA + ACP—SH $\rightleftharpoons$ acetyl—S—ACP + CoA—SH

This acetyl group does not, however, remain on the ACP but is transferred to a specific cysteine residue of another enzyme of the fatty-acid synthetase complex, *β-ketoacyl-ACP synthase*, in the reaction

Acetyl—S—ACP + synthase—SH $\rightleftharpoons$

ACP-SH + acetyl—S—synthase

Figure 24-7
The condensation reaction in fatty acid synthesis.

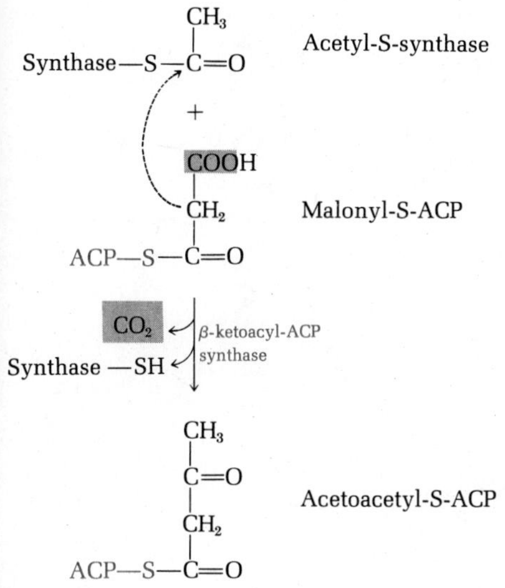

in which β-ketoacyl-ACP synthase is simply designated as synthase. With this reaction the fatty-acid synthetase complex is now primed and ready to carry out the sequence of reactions required to add a two-carbon unit in the chain-lengthening process.

The Malonyl Transfer Step

In the next reaction, catalyzed by *ACP malonyltransferase,* malonyl-S-CoA formed in the acetyl-CoA carboxylase reaction reacts with the —SH group of the 4'-phosphopantetheine arm of ACP, with loss of free CoA, to form *malonyl-S-ACP:*

$$Malonyl—S—CoA + ACP—SH \rightleftharpoons$$
$$malonyl—S—ACP + CoA—SH$$

As a result of this step and of the preceding priming reaction, a malonyl group is now esterified to ACP and an acetyl group is esterified to an —SH group on the β-ketoacyl-ACP synthase molecule.

The Condensation Reaction

In the next reaction of the sequence, catalyzed by *β-ketoacyl-ACP synthase,* the acetyl group esterified to the cysteine residue of this enzyme is transferred to carbon atom 2 of the malonyl group on ACP, with release of the free carboxyl group of the malonyl residue as CO_2 (Figure 24-7):

$$Acetyl—S—synthase + malonyl—S—ACP \longrightarrow$$
$$acetoacetyl—S—ACP + CO_2 + synthase—SH$$

The molecule of CO_2 so released contains the same carbon atom that was introduced as HCO_3^- in the acetyl-CoA carboxylase reaction. In essence, HCO_3^- plays a catalytic role in fatty acid synthesis since it is regenerated as CO_2 as each two-carbon unit is inserted into the growing fatty acid chain.

Study of the reaction equilibrium has revealed the probable basis for the biological selection of malonyl-CoA as the precursor of two-carbon residues for fatty acid synthesis. If acetoacetyl-CoA were to be formed from two molecules of acetyl-CoA by the action of *acetyl-CoA acetyltransferase* (page 554),

$$Acetyl—S—CoA + acetyl—S—CoA \rightleftharpoons$$
$$acetoacetyl—S—CoA + CoA—SH$$

the reaction would be endergonic, with its equilibrium lying to the left. On the other hand, the formation of acetoacetyl-S-ACP by the reaction

$$Acetyl—S—synthase + malonyl—S—ACP \longrightarrow$$
$$acetoacetyl—S—ACP + synthase—SH + CO_2$$

is strongly exergonic. Its equilibrium lies far to the right, in the direction of synthesis, because the decarboxylation of the malonyl residue, which is strongly exergonic, provides a strong thermodynamic pull toward completion.

The First Reduction Reaction

The acetoacetyl-S-ACP now undergoes reduction by NADPH to form β-hydroxybutyryl-S-ACP. This reaction is catalyzed by *β-ketoacyl-ACP reductase:*

$$\text{Acetoacetyl—S—ACP} + \text{NADPH} + \text{H}^+ \rightleftharpoons$$
$$\text{D-}\beta\text{-hydroxybutyryl—S—ACP} + \text{NADP}^+$$

It is noteworthy that the product is the D stereoisomer of the β-hydroxy acid, whereas the isomer formed during fatty acid oxidation has the L-configuration (page 548).

The Dehydration Step

D-β-Hydroxybutyryl-S-ACP is next dehydrated to the corresponding *trans-α,β-* or Δ^2-unsaturated acyl-S-ACP, namely, *crotonyl-S-ACP*, by *enoyl-ACP hydratase:*

$$\text{CH}_3\text{CHOHCH}_2\text{CO—S—ACP} \rightleftharpoons \overset{\text{H}}{\underset{\text{H}}{\text{CH}_3\text{C}=\text{C}}}\text{—CO—S—ACP} + \text{H}_2\text{O}$$

Crotonyl-S-ACP

The Second Reduction Step

Crotonyl-S-ACP is now reduced to butyryl-S-ACP by *enoyl-ACP reductase (NADPH)*; the electron donor is NADPH in *E. coli* and in animal tissues:

$$\text{CH}_3\overset{\text{H}}{\underset{\text{H}}{—\text{C}=\text{C}}}\text{—CO—S—ACP} + \text{NADPH} + \text{H}^+ \rightleftharpoons$$

Crotonyl-S-ACP

$$\text{CH}_3\text{CH}_2\text{CH}_2\text{CO—S—ACP} + \text{NADP}^+$$

Butyryl-S-ACP

This reaction also differs from the corresponding reaction of fatty acid oxidation in mitochondria in that a pyridine nucleotide rather than a flavoprotein is involved (page 548). Since the NADPH–NAD$^+$ couple has a more negative standard potential than the fatty acid oxidizing flavoprotein (page 495), NADPH favors reductive formation of the saturated fatty acid.

The formation of butyryl-ACP completes the first of seven cycles en route to palmitoyl-S-ACP. To start the next cycle the butyryl group is transferred from ACP to the —SH group of the β-ketoacyl-ACP synthase molecule, thus allowing ACP to accept a malonyl group from another molecule of malonyl-CoA. Then the cycle repeats, the next step being the condensation of malonyl-S-ACP with butyryl-S-β-ketoacyl-ACP synthase to yield β-ketohexanoyl-S-ACP and CO_2.

After seven complete cycles, palmitoyl-ACP is the end product. The palmitoyl group may be removed to yield free palmitic acid by the action of a thioesterase, or it may be transferred from ACP to CoA, or it may be incorporated directly into phosphatidic acid (page 670) in the pathway to phospholipids and triacylglycerols.

It is remarkable that in most organisms the fatty-acid synthetase system stops with the production of palmitic acid and does not yield stearic acid, which has only two more carbon atoms than palmitic acid and thus does not differ greatly in physical properties. This chain-length specificity is conferred on the system by β-ketoacyl-ACP synthase, which is very active in accepting the tetradecanoyl group (14 carbons) from ACP but does not accept the hexadecanoyl group (16 carbons), presumably because the binding site of the enzyme for the saturated acyl group can accommodate only a specific range of chain lengths. In addition, palmitoyl-CoA functions as a feedback inhibitor of the fatty-acid synthetase system.

Saturated fatty acids having an odd number of carbon atoms, which are found in many marine organisms, are also made by the fatty-acid synthetase complex. In this case the synthesis is primed by a starter molecule of propionyl-S-ACP (instead of acetyl-S-ACP), to which are added successive two-carbon units via condensations with malonyl-S-ACP.

We can now write the overall equation for palmitic acid biosynthesis starting from acetyl-S-CoA:

$$8 \text{ Acetyl—S—CoA} + 14\text{NADPH} + 14\text{H}^+ + 7\text{ATP} + \text{H}_2\text{O} \longrightarrow$$
$$\text{palmitic acid} + 8\text{CoA} + 14\text{NADP}^+ + 7\text{ADP} + 7\text{P}_i$$

The 14 molecules of NADPH required for the reductive steps in the synthesis of palmitic acid arise largely from the NADP-dependent oxidation of glucose 6-phosphate via the phosphogluconate pathway (page 467). Liver, mammary gland, and adipose tissue of vertebrates, which have a rather high rate of fatty acid biosynthesis, also have a very active 6-phosphogluconate pathway (page 467). The NADPH for fatty acid biosynthesis in plants arises through photoreduction of $NADP^+$ (page 605).

The enzymatic steps leading to the biosynthesis of palmitic acid differ from those involved in oxidation of palmitic acid in the following respects:

1 Their intracellular location
2 The type of acyl-group carrier
3 The form in which the two-carbon units are added or removed
4 The pyridine nucleotide specificity of the β-ketoacyl–β-hydroxyacyl reaction
5 The stereoisomeric configuration of the β-hydroxyacyl intermediate
6 The electron donor-acceptor system for the crotonyl-butyryl step
7 The response to citrate and HCO_3^-

These differences illustrate how two opposing metabolic processes may proceed independently of each other in the cell.

Elongation of Saturated Fatty Acids in Mitochondria and Microsomes

Palmitic acid, the normal end product of the fatty-acid synthetase system, is the precursor of the other long-chain saturated and unsaturated fatty acids in most organisms. Elongation of palmitic acid to longer-chain saturated fatty acids, of which stearic acid is most abundant, occurs by the action of two different types of enzyme systems, one in the mitochondria and the other in the endoplasmic reticulum.

In mitochondria palmitic and other saturated fatty acids are lengthened by successive additions to the carboxyl-terminal end of acetyl units in the form of acetyl-CoA; malonyl-ACP cannot replace acetyl-CoA. The mitochondrial elongation pathway occurs by reactions similar to those in fatty acid oxidation. Condensation of palmityl-CoA with acetyl-CoA yields β-ketostearyl-CoA, which is reduced by NADPH to β-hydroxystearyl-CoA. The latter is dehydrated to the Δ^2-unsaturated stearyl-CoA, which is then reduced to yield stearyl-CoA at the expense of NADPH. This system will also elongate unsaturated fatty acids.

Microsome preparations can elongate both saturated and unsaturated fatty acyl–CoA esters, but in this case malonyl-CoA rather than acetyl-CoA serves as source of the acetyl groups. The reaction sequence is identical to that in the fatty-acid synthetase system except that the microsomal system employs CoA and not ACP as acyl carrier.

Formation of Monoenoic Acids

Palmitic and stearic acids serve as precursors of the two common monoenoic (monounsaturated) fatty acids of animal tissues, namely, palmitoleic and oleic acids, both of which possess a cis double bond in the Δ^9 position. Although most organisms can form palmitoleic and oleic acids, the pathway and enzymes employed differ between aerobic and anaerobic organisms. In vertebrates (and most other aerobic organisms) the Δ^9 double bond is introduced by a specific monooxygenase system (page 501); it is located in the endoplasmic reticulum of liver and adipose tissue. One molecule of molecular oxygen (O_2) is used as the acceptor for two pairs of electrons, one pair derived from the palmitoyl-CoA or stearyl-CoA substrate and the other from NADPH, which is a required coreductant in the reaction. The transfer of electrons in this complex reaction involves a microsomal electron-transport chain which carries electrons from NADPH (or NADH) to microsomal cytochrome b_5 via cytochrome b_5 reductase, a flavoprotein (page 486). A terminal cyanide-sensitive factor (CSF), a protein, is required to activate the acyl-CoA and the oxygen (Figure 24-8). In some plants and some lower aerobic organisms cytochrome b_5 does

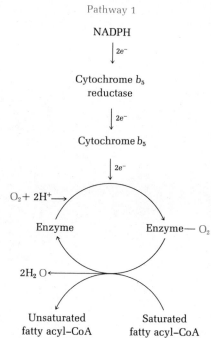

Figure 24-8
Pathways of electron transport in the desaturation of fatty acids. Pathway 1 represents the mechanism in animal tissues, whereas pathway 2 occurs in some plants and microorganisms. The reaction with oxygen in pathway 1 is inhibited by cyanide.

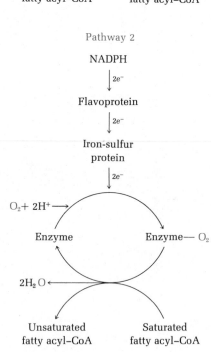

not participate but is replaced by an iron-sulfur protein (page 488). The overall reaction for palmitoyl-CoA is

$$\text{Palmitoyl—CoA} + \text{NADPH} + \text{H}^+ + \text{O}_2 \longrightarrow$$
$$\text{palmitoleyl—CoA} + \text{NADP}^+ + 2\text{H}_2\text{O}$$

In many bacteria an entirely different mechanism, one not employing molecular oxygen, comes into play. It involves _dehydration_ of a specific saturated intermediate-length β-hydroxyacyl-ACP, rather than oxidative desaturation of a fatty acyl–CoA. In _E. coli_, the biosynthesis of palmitoleic acid starts with β-hydroxydecanoyl-ACP, formed by the fatty-acid synthetase system. The enzyme _β-hydroxydecanoyl-ACP dehydratase_ dehydrates β-hydroxydecanoyl-ACP to yield the β,γ- or Δ³-decenoyl-ACP, presumably in the cis form. Three more two-carbon units, brought in the form of malonyl-ACP, are then added to the carboxyl end of the ten-carbon unsaturated acyl-ACP to yield the ACP ester of palmitoleic acid.

Formation of Polyenoic Acids

Bacteria do not contain polyenoic acids; however, these acids are abundant both in higher plants and in animals. Mammals contain four distinct families of polyenoic acids, which differ in the number of carbon atoms between the terminal methyl group and the nearest double bond. These families are named from their precursor fatty acids, namely, palmitoleic, oleic, linoleic, and linolenic acids (Table 24-1). All polyenoic acids found in mammals are formed from these four precursors by further elongation and/or desaturation reactions. Two of these precursor fatty acids, linoleic and linolenic acids, cannot be synthesized by mammals and must be obtained from plant sources; they are therefore called _essential fatty acids_ (page 281).

The important animal polyenoic acids derived from palmitoleic and oleic acids are shown in Figure 24-9, together with the pathway of their formation. The elongation of chains occurs at the carboxyl end by the mitochondrial or microsomal systems described above. The desaturation steps occur by the action of the cytochrome b_5–oxygenase system with NADPH as coreductant of oxygen, like the steps in the formation of palmitoleic and oleic acids, also described above.

Figure 24-10 shows the interrelationships between some of the polyenoic acids derived from linoleic and linolenic acids. Arachidonic acid is the most abundant polyenoic acid. When young rats are placed on diets deficient in essential fatty acids, they grow slowly and develop a scaly dermatitis and thickening of the skin (page 281). This condition can be relieved by administration not only of linoleic or linolenic acid but also of arachidonic acid. The essential fatty acids and some of their derivatives (Figure 24-10) serve as precursors of the _prostaglandins_ (page 686), as we shall see below.

Table 24-1 Families of polyenoic acids, named after the parent unsaturated fatty acid

Each family includes the polyenoic acids derived from the parent acid by C_2 elongation at the carboxyl end and/or by desaturation at positions between the carboxyl group and the parent structures indicated below; thus all members of each family will have the methyl-terminal chain structure indicated

Family	Chain structure
Palmitoleic	CH_3—$(CH_2)_5$—$CH{=}CH$—
Oleic	CH_3—$(CH_2)_7$—$CH{=}CH$—
Linoleic	CH_3—$(CH_2)_4$—$CH{=}CH$—
Linolenic	CH_3—CH_2—$CH{=}CH$—

Figure 24-9
Formation of unsaturated fatty acids from palmitic acid. These transformations (as well as those in Figure 24-10) take place with the CoA esters of the indicated fatty acids. In the symbols used for the fatty acids, the prefixed number indicates the double-bond position, the first subscript the number of carbon atoms in the chain, and the second subscript the number of double bonds.

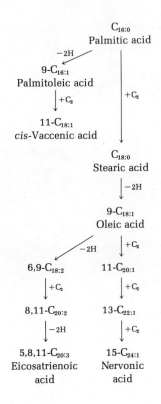

$C_{16:0}$
Palmitic acid

−2H

9-$C_{16:1}$
Palmitoleic acid

+C_2 +C_2

11-$C_{18:1}$
cis-Vaccenic acid

$C_{18:0}$
Stearic acid

−2H

9-$C_{18:1}$
Oleic acid

−2H   +C_2

6,9-$C_{18:2}$ 11-$C_{20:1}$

+C_2 +C_2

8,11-$C_{20:2}$ 13-$C_{22:1}$

−2H   +C_2

5,8,11-$C_{20:3}$ 15-$C_{24:1}$
Eicosatrienoic   Nervonic
acid   acid

In plants linoleic and linolenic acids are synthesized from oleic acid via aerobic desaturation reactions catalyzed by specific oxygenase systems requiring NADPH as coreductant.

The double bonds of naturally occurring fatty acids do not in general undergo hydrogenation to yield more completely saturated fatty acids; only a few organisms appear to carry out this process. Unsaturated fatty acids, however, are completely oxidized by the fatty acid oxidation system (page 552).

In most organisms the conversion of saturated to unsaturated fatty acids is promoted by low environmental temperatures. This is an adaptation to maintain the melting point of the total cell lipids below the ambient temperature; unsaturated fatty acids have lower melting points than saturated (page 280). In some organisms the enzymes involved in fatty acid desaturation increase in concentration in response to low temperatures; in others the unsaturated fatty acids are inserted into lipids at increased rates.

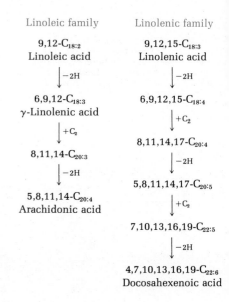

Figure 24-10
Pathways of polyenoic acid synthesis from the essential fatty acids.

Linoleic family

9,12-$C_{18:2}$
Linoleic acid
$\downarrow$ −2H
6,9,12-$C_{18:3}$
γ-Linolenic acid
$\downarrow$ +C_2
8,11,14-$C_{20:3}$
$\downarrow$ −2H
5,8,11,14-$C_{20:4}$
Arachidonic acid

Linolenic family

9,12,15-$C_{18:3}$
Linolenic acid
$\downarrow$ −2H
6,9,12,15-$C_{18:4}$
$\downarrow$ +C_2
8,11,14,17-$C_{20:4}$
$\downarrow$ −2H
5,8,11,14,17-$C_{20:5}$
$\downarrow$ +C_2
7,10,13,16,19-$C_{22:5}$
$\downarrow$ −2H
4,7,10,13,16,19-$C_{22:6}$
Docosahexenoic acid

Biosynthesis of Triacylglycerols

The triacylglycerols, which function as *depot*, or *storage, lipids*, are actively synthesized in the cells of vertebrates, particularly liver and fat cells, as well as those of higher plants. Bacteria in general contain relatively small amounts of triacyglycerols.

In higher animals and plants two major precursors are required for the synthesis of triacyglycerols: L-glycerol 3-phosphate (sn-glycerol 3-phosphate) and fatty acyl–CoA's. L-Glycerol 3-phosphate is derived from two different sources. Its normal precursor is dihydroxyacetone phosphate, the product of the aldolase reaction of glycolysis (page 425). Dihydroxyacetone phosphate is reduced to L-glycerol 3-phosphate by the NAD-linked *glycerol-3-phosphate dehydrogenase* of the cytosol:

$$\text{Dihydroxyacetone phosphate} + \text{NADH} + \text{H}^+ \rightleftharpoons$$
$$\text{L-glycerol 3-phosphate} + \text{NAD}^+$$

It may also be formed from free glycerol arising from degradation of triacylglycerols, through the action of *glycerol kinase*:

$$\text{ATP} + \text{glycerol} \xrightarrow{\text{Mg}^{2+}} \text{L-glycerol 3-phosphate} + \text{ADP}$$

The first stage in triacyglycerol formation is the acylation of the free hydroxyl groups of glycerol phosphate by two molecules of fatty acyl–CoA to yield first a *lysophosphatidic acid* and then a *phosphatidic acid* (Figure 24-11). Free glycerol is not acylated. These reactions occur preferentially with 16- and 18-carbon saturated and unsaturated acyl-CoA's. In some microorganisms, such as E. coli, the fatty acyl donor is fatty acyl–ACP rather than fatty acyl–CoA. An alternate pathway for the formation of phosphatidic acid involves es-

Figure 24-11
The formation of a triacylglycerol. The stereospecific numbering system (page 287) has been followed in naming compounds prefixed by sn. All naturally occurring acylglycerols and phosphoglycerides derive from sn-glycerol 3-phosphoric acid. Note that the projection formulas given here become identical with those of Figure 11-10 upon rotation of 180° in the plane of the paper.

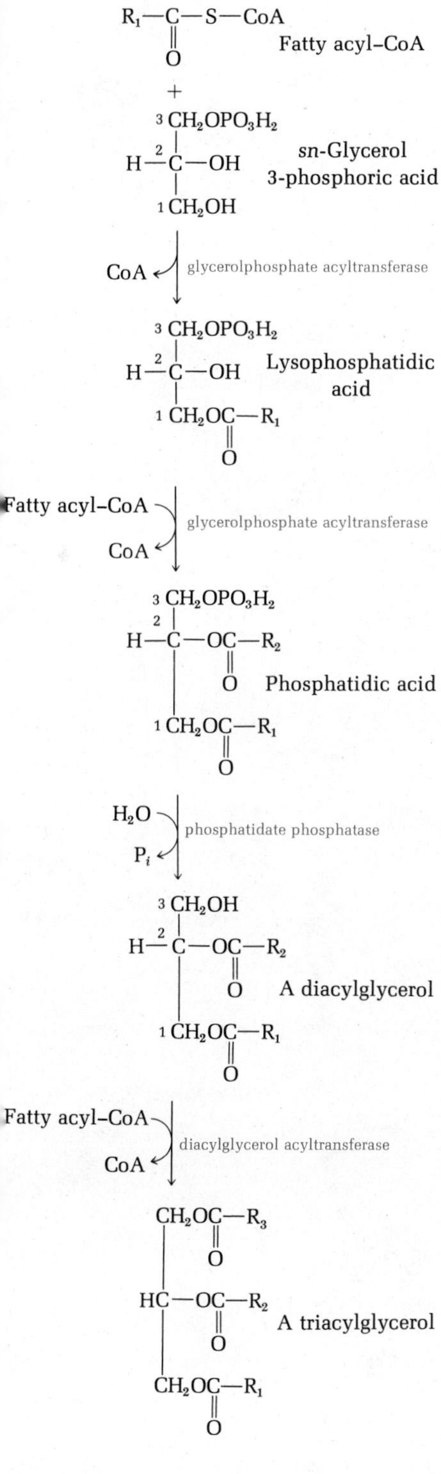

$$R_1-\overset{\overset{\displaystyle O}{\|}}{C}-S-CoA \qquad \text{Fatty acyl–CoA}$$

+

$$
\begin{array}{l}
3\ CH_2OPO_3H_2 \\
2 \\
H-C-OH \qquad \text{sn-Glycerol} \\
\qquad\qquad\qquad \text{3-phosphoric acid} \\
1\ CH_2OH
\end{array}
$$

CoA ⤆ glycerolphosphate acyltransferase

$$
\begin{array}{l}
3\ CH_2OPO_3H_2 \\
2 \\
H-C-OH \qquad \text{Lysophosphatidic} \\
\qquad\qquad\qquad\quad \text{acid} \\
1\ CH_2O\overset{\overset{\displaystyle O}{}}{C}-R_1
\end{array}
$$

Fatty acyl–CoA ⤹
    glycerolphosphate acyltransferase
CoA ⤆

$$
\begin{array}{l}
3\ CH_2OPO_3H_2 \\
2 \\
H-C-OC-R_2 \\
\qquad\quad \| \\
\qquad\quad O \qquad \text{Phosphatidic acid} \\
1\ CH_2OC-R_1 \\
\qquad\quad \| \\
\qquad\quad O
\end{array}
$$

H₂O ⤹
    phosphatidate phosphatase
P_i ⤆

$$
\begin{array}{l}
3\ CH_2OH \\
2 \\
H-C-OC-R_2 \\
\qquad\quad \| \\
\qquad\quad O \qquad \text{A diacylglycerol} \\
1\ CH_2OC-R_1 \\
\qquad\quad \| \\
\qquad\quad O
\end{array}
$$

Fatty acyl–CoA ⤹
    diacylglycerol acyltransferase
CoA ⤆

$$
\begin{array}{l}
CH_2OC-R_3 \\
\qquad \| \\
\qquad O \\
HC-OC-R_2 \\
\qquad \| \\
\qquad O \qquad \text{A triacylglycerol} \\
CH_2OC-R_1 \\
\qquad \| \\
\qquad O
\end{array}
$$

terification of the fatty acyl group with dihydroxyacetone phosphate. This pathway consists of the reactions

Dihydroxyacetone phosphate + fatty acyl–CoA $\xrightarrow{\substack{\text{dihydroxyacetone-}\\\text{phosphate}\\\text{acyltransferase}}}$
fatty acyl dihydroxyacetone phosphate + CoA

Fatty acyl dihydroxyacetone phosphate + NADPH + H⁺ $\longrightarrow$
lysophosphatidic acid + NADP⁺

Lysophosphatidic acid + fatty acyl–CoA $\longrightarrow$
phosphatidic acid + CoA

Phosphatidic acids occur only in trace amounts in cells, but they are important intermediates in the biosynthesis of triacylglycerols and phosphoglycerides.

In the pathway to triacylglycerols, phosphatidic acid undergoes hydrolysis by *phosphatidate phosphatase* to form a *diacylglycerol* (Figure 24-11):

$$\text{L-Phosphatidic acid} + H_2O \longrightarrow \text{diacylglycerol} + P_i$$

The diacylglycerol then reacts with a third molecule of a fatty acyl–CoA to yield a triacylglycerol by the action of *diacylglycerol acyltransferase* (Figure 24-11):

Fatty acyl–CoA + diacylglycerol $\longrightarrow$ triacylglycerol + CoA

In the intestinal mucosa of higher animals, which actively synthesizes triacylglycerols during absorption of fatty acids from the intestine, another type of acylation reaction comes into play. Monoacylglycerols formed during intestinal digestion may be acylated directly by *acylglycerol palmitoyltransferase* and thus phosphatidic acid is not an intermediate:

Monoacylglycerol + palmitoyl-CoA $\rightleftharpoons$ diacylglycerol + CoA

In storage fats of animal and plant tissues the triacylglycerols are usually mixed, i.e., contain two or more different fatty acids (page 284). However, little is known of the enzymatic mechanisms by which the identity and position of the different fatty acyl groups in triacylglycerols are specified.

Biosynthesis of Phosphoglycerides

The major phosphoglycerides, which serve as components of membranes and of transport lipoproteins, are formed from phosphatidic acid in two different pathways, one prevailing in higher animals and plants and the other in some bacteria. In both pathways cytidine nucleotides serve as carriers, either of the "head" alcohol groups (page 289) or of phosphatidic acid. The central role of cytidine nucleotides in lipid biosynthesis was discovered by E. P. Kennedy and his colleagues. We shall first consider the biosynthesis of the major phosphoglycerides in animal tissues.

Phosphatidylethanolamine

The biosynthesis of phosphatidylethanolamine in animal tissues begins with the phosphorylation of ethanolamine by the action of *ethanolamine kinase*:

$$\text{Ethanolamine} + \text{ATP} \longrightarrow \text{phosphoethanolamine} + \text{ADP}$$

In the next step phosphoethanolamine reacts with cytidine triphosphate (CTP) to yield *cytidine diphosphoethanolamine*, an activated form of phosphoethanolamine (Figure 24–12):

$$\text{CTP} + \text{phosphoethanolamine} \underset{\xrightarrow{\hspace{2cm}}}{\overset{\substack{\text{phosphoethanolamine} \\ \text{cytidylyltransferase}}}{}}$$
$$\text{CDP-ethanolamine} + \text{PP}_i$$

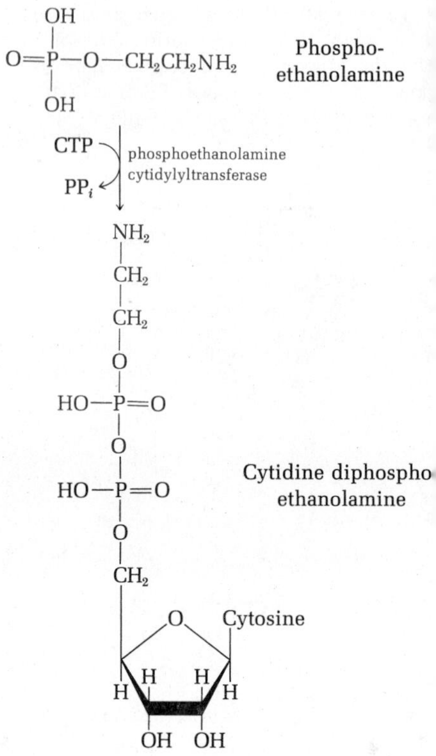

In the last step the CMP part of CDP-ethanolamine is cleaved and the phosphoethanolamine portion is transferred to diacylglycerol to yield phosphatidylethanolamine:

$$\text{CDP-ethanolamine} + \text{diacylglycerol} \xrightarrow{\substack{\text{phosphoethanolamine} \\ \text{transferase}}}$$
$$\text{phosphatidylethanolamine} + \text{CMP}$$

The enzymes catalyzing these reactions occur in the endoplasmic reticulum membrane, to which they are very tightly bound.

Phosphatidylcholine

In animal tissues phosphatidylcholine can be formed by two different pathways. One pathway is via direct methylation of the amino group of phosphatidylethanolamine by the methyl-group donor (page 697) *S-adenosylmethionine* (Figure 24-13), a three-step process. The overall equation is

$$\text{Phosphatidylethanolamine} + 3 \text{ S-adenosylmethionine} \xrightarrow{\substack{\text{phosphatidyl-} \\ \text{ethanolamine} \\ \text{methyltransferase}}}$$
$$\text{phosphatidylcholine} + 3 \text{ S-adenosylhomocysteine}$$

The second pathway of synthesis of phosphatidylcholine in animal tissues begins with dietary choline or choline salvaged from the enzymatic degradation of phosphatidylcholine. It proceeds by a sequence of reactions analogous to those involved in formation of phosphatidylethanolamine:

$$\text{Choline} + \text{ATP} \xrightarrow{\text{choline kinase}} \text{phosphocholine} + \text{ADP}$$

$$\text{Phosphocholine} + \text{CTP} \xrightarrow{\substack{\text{phosphocholine} \\ \text{cytidylyltransferase}}} \text{CDP-choline} + \text{PP}_i$$

$$\text{CDP-choline} + \text{diacylglycerol} \xrightarrow{\substack{\text{phosphocholine} \\ \text{transferase}}}$$
$$\text{phosphatidylcholine} + \text{CMP}$$

Figure 24-12
The formation of cytidine diphosphoethanolamine. The portion arising from phosphoethanolamine is in color.

Figure 24-13
Conversion of phosphatidylethanolamine to
phosphatidylcholine.

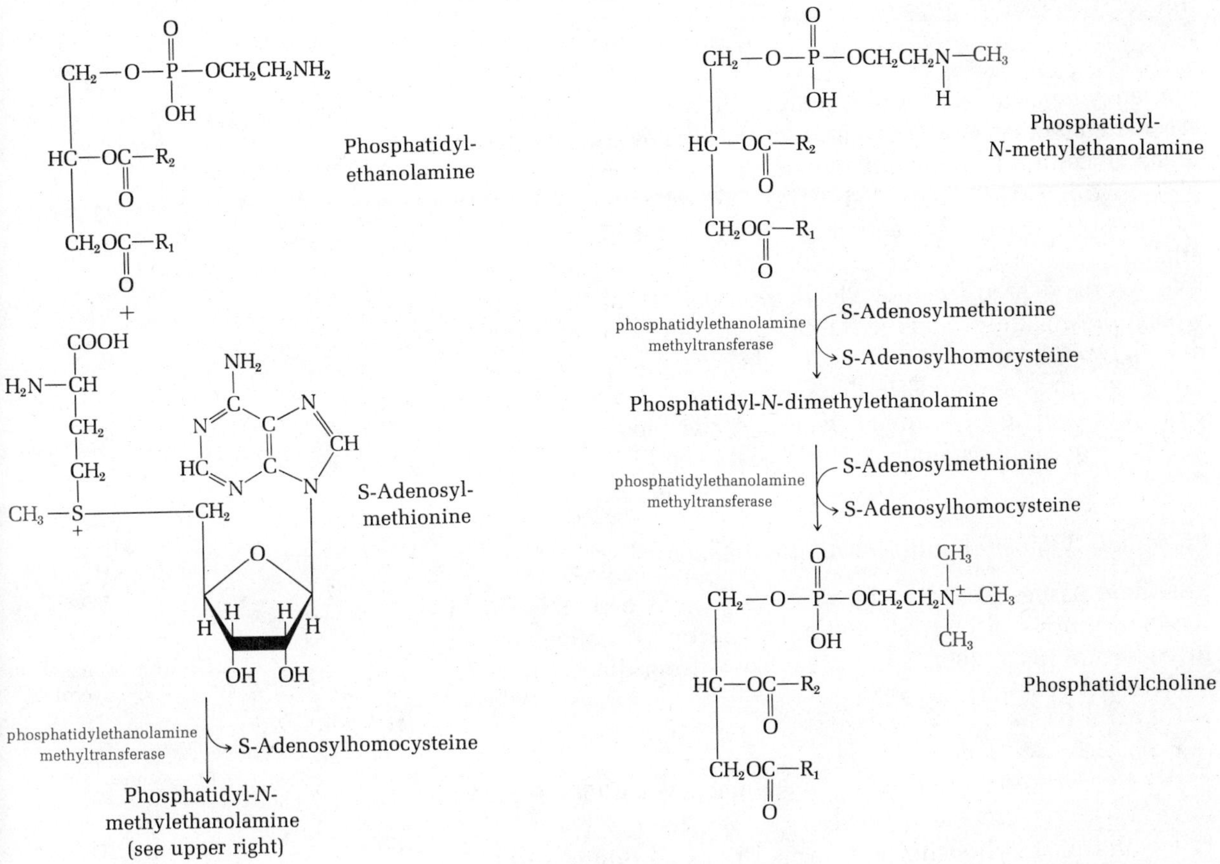

Phosphatidylserine

In animal tissues phosphatidylserine (page 289) is formed by the enzymatic exchange of the "head" alcohol of phosphatidylethanolamine, namely, ethanolamine, with another alcohol group, that of L-serine:

Phosphatidylethanolamine + serine $\rightleftharpoons$
phosphatidylserine + ethanolamine

In animal tissues and in E. coli phosphatidylserine may undergo decarboxylation to yield phosphatidylethanolamine:

Phosphatidylserine $\longrightarrow$ phosphatidylethanolamine + CO_2

In some bacteria, such as E. coli, phosphatidylserine is formed by a different pathway, in which the phosphate group of phosphatidic acid is activated by reaction with CTP, resulting in the formation of *cytidine diphosphate diacyglycerol* (Figure 24-14):

L-Phosphatidate + CTP $\xrightarrow[\text{cytidylyltransferase}]{\text{phosphatidate}}$ CDP-diacylglycerol + PP_i

CDP-diacylglycerol then reacts with L-serine to form phosphatidylserine:

$$\text{CDP-diacylglycerol + serine} \underset{\substack{\text{CDPdiacylglycerol serine} \\ \textit{O}\text{-phosphatidyltransferase}}}{\rightleftharpoons} \text{phosphatidylserine + CMP}$$

The enzyme catalyzing this reaction in *E. coli* is apparently associated with the ribosomes. The CMP moiety of CDP-diacylglycerol may be regarded as a carrier of a molecule of phosphatidic acid. Note that when phosphoethanolamine or phosphocholine is activated in the form of CDP-ethanolamine or CDP-choline, esterification occurs with the free hydroxyl group of diacylglycerol. On the other hand, when phosphatidic acid is activated in the form of CDP-diacylglycerol, esterification takes place with the free hydroxyl group of serine. CDP-diacylglycerol is also an important intermediate in animal tissues for the biosynthesis of other classes of phosphoglycerides, as we shall now see.

Phosphatidylinositol and Phosphatidylglycerol

In animal tissues CDP-diacylglycerol is used as precursor in the biosynthesis of both phosphatidylinositol and phosphatidylglycerol (page 289). The former arises from the reaction

$$\text{CDP-diacylglycerol + inositol} \xrightarrow{\substack{\text{CDPdiacylglycerol inositol} \\ \text{phosphatidyltransferase}}} \text{phosphatidylinositol + CMP}$$

From phosphatidylinositol are formed *phosphatidylinositol monophosphate* and *phosphatidylinositol diphosphate* by two successive phosphorylations of free hydroxyl groups in the inositol residue by ATP. These phosphorylated forms of phosphatidylinositol are present in the brain; their function is not entirely clear.

Phosphatidylglycerol is formed from CDP-diacylglycerol in the reactions

$$\text{CDP-diacylglycerol + } \textit{sn}\text{-glycerol 3-phosphate} \xrightarrow{\substack{\text{glycerolphosphate} \\ \text{phosphatidyltransferase}}}$$
$$\text{3-}\textit{sn}\text{-phosphatidyl-1'-}\textit{sn}\text{-glycerol 3'-phosphate + CMP}$$

$$\text{3-}\textit{sn}\text{-phosphatidyl-1'-}\textit{sn}\text{-glycerol 3'-phosphate + H}_2\text{O} \xrightarrow{\substack{\text{phosphatidylglycero-} \\ \text{phosphatase}}}$$
$$\text{3-}\textit{sn}\text{-phosphatidyl-1'-}\textit{sn}\text{-glycerol + P}_i$$

The resulting phosphatidylglycerol is the precursor of *diphosphatidylglycerol*, more commonly called *cardiolipin* (page 288), which characteristically makes up 20 percent of the lipids of the inner mitochondrial membrane and a significant fraction of bacterial membrane lipids. The reaction in animal tissues appears to be (Figure 24-15):

$$\text{Phosphatidylglycerol + CDP-diacylglycerol} \longrightarrow \text{cardiolipin + CMP}$$

Figure 24-14
Formation of phosphatidylserine in some bacteria. Phosphatidylinositol is formed in a similar reaction.

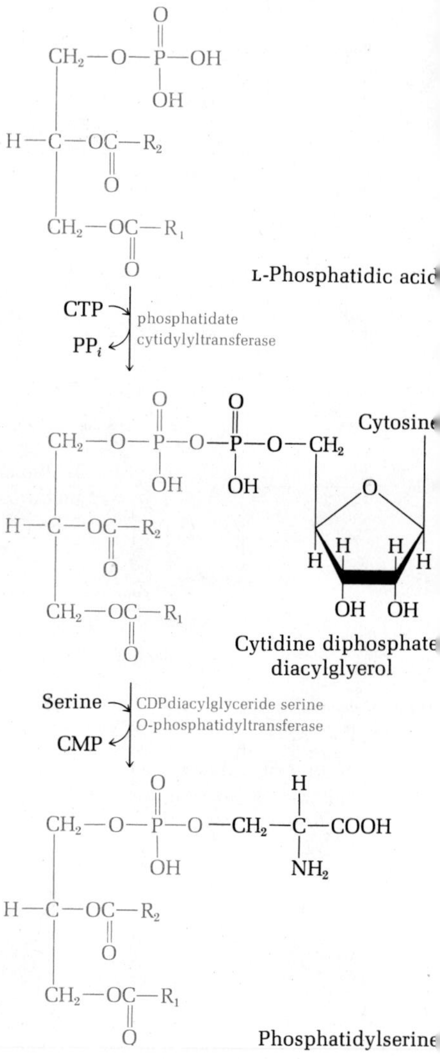

Figure 24-15
Pathway of cardiolipin biosynthesis in animal tissues.

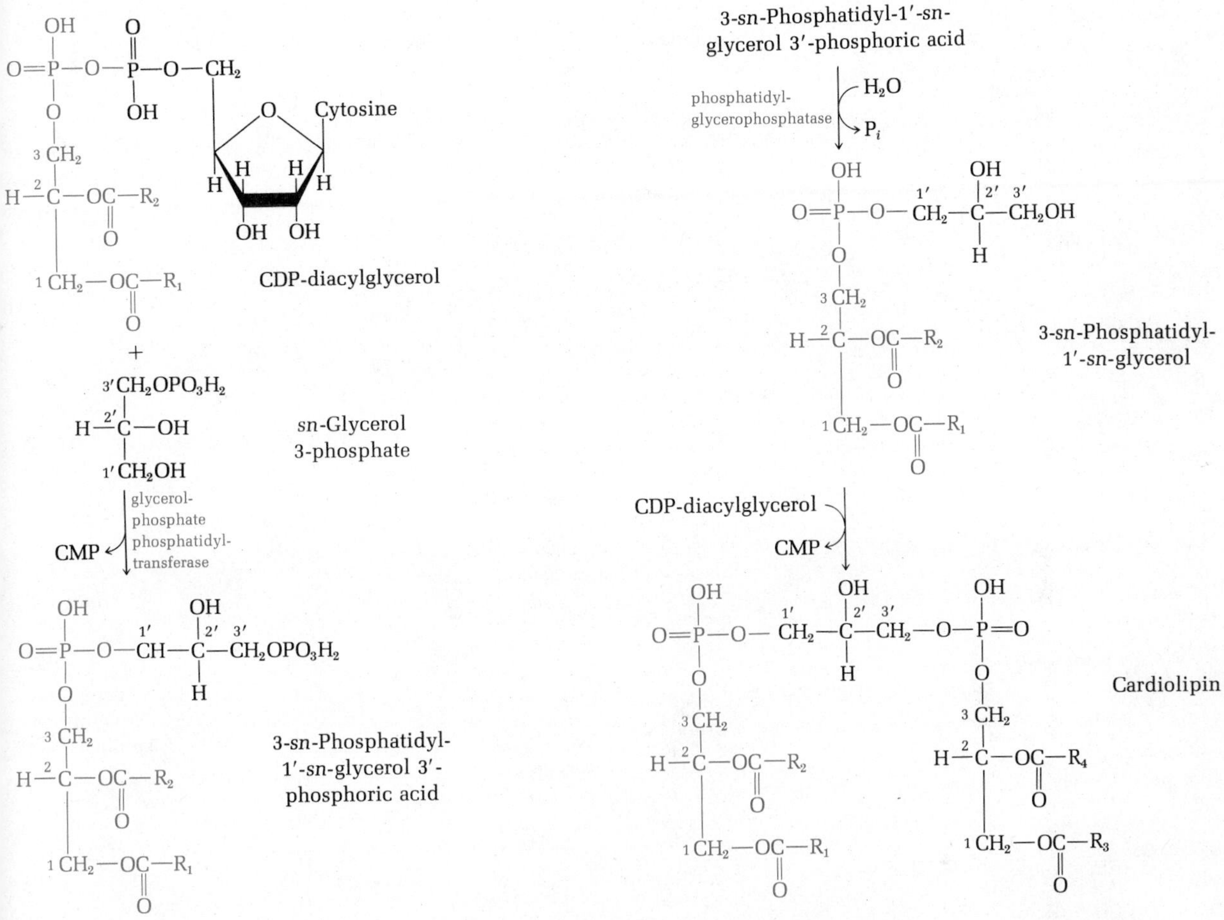

CDP-diacylglycerol

+

sn-Glycerol
3-phosphate

glycerol-
phosphate
phosphatidyl-
transferase

CMP

3-sn-Phosphatidyl-
1'-sn-glycerol 3'-
phosphoric acid

3-sn-Phosphatidyl-1'-sn-
glycerol 3'-phosphoric acid

phosphatidyl-
glycerophosphatase

H_2O

P_i

3-sn-Phosphatidyl-
1'-sn-glycerol

CDP-diacylglycerol

CMP

Cardiolipin

In bacteria cardiolipin is formed by the reaction

2 Phosphatidylglycerol $\longrightarrow$ cardiolipin + glycerol

Biosynthesis of Sphingomyelin and Other Sphingolipids

The long-chain aliphatic amine sphingosine (4-sphingenine, page 291), which is the characteristic building block of sphingolipids, is formed from palmitoyl-CoA by the following sequence of enzymatic reactions (Figure 24-16):

Palmitoyl-CoA + L-serine $\longrightarrow$ 3-dehydrosphinganine + CoA + CO_2

3-Dehydrosphinganine + NADPH + H^+ $\longrightarrow$ NADP + sphinganine

Sphinganine + FAD $\longrightarrow$ $FADH_2$ + sphingosine

In the next step of sphingolipid synthesis the amino group of sphingosine is acylated by a long-chain fatty acyl–CoA to yield N-acylsphingosine, or ceramide (Figure 24-17):

Sphingosine + fatty acyl–CoA $\xrightarrow[\text{acyltransferase}]{\text{sphingosine}}$ ceramide + CoA

Figure 24-16
The formation of sphingosine. The portion
furnished by serine is indicated in color.

Figure 24-17
The formation of a ceramide. The N-palmi-
toyl group arising from palmitoyl-CoA is in
color.

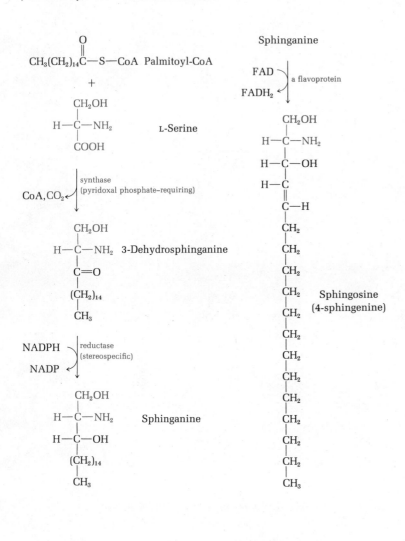

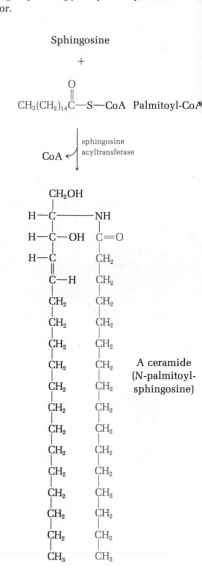

Sphingomyelin (page 292) is formed by reaction of a ceramide with CDP-choline (Figure 24-18):

$$\text{Ceramide} + \text{CDP-choline} \xrightarrow[\text{cholinephosphotransferase}]{\text{ceramide}} \text{sphingomyelin} + \text{CMP}$$

The _cerebrosides_, hexose derivatives of ceramides (page
292), are formed by a similar reaction between a ceramide
and UDP–D-glucose or UDP–D-galactose (Figure 24-19):

$$\text{Ceramide} + \text{UDP–D-glucose} \longrightarrow \text{glucocerebroside} + \text{UDP}$$

$$\text{Ceramide} + \text{UDP–D-galactose} \longrightarrow \text{galactocerebroside} + \text{UDP}$$

The cerebrosides may also be made by an alternative pathway, in which the order of substitution of the groups into
sphingosine is different. In this pathway, a galactose-substi-
tuted sphingosine, called _psychosine_ (page 292), is formed

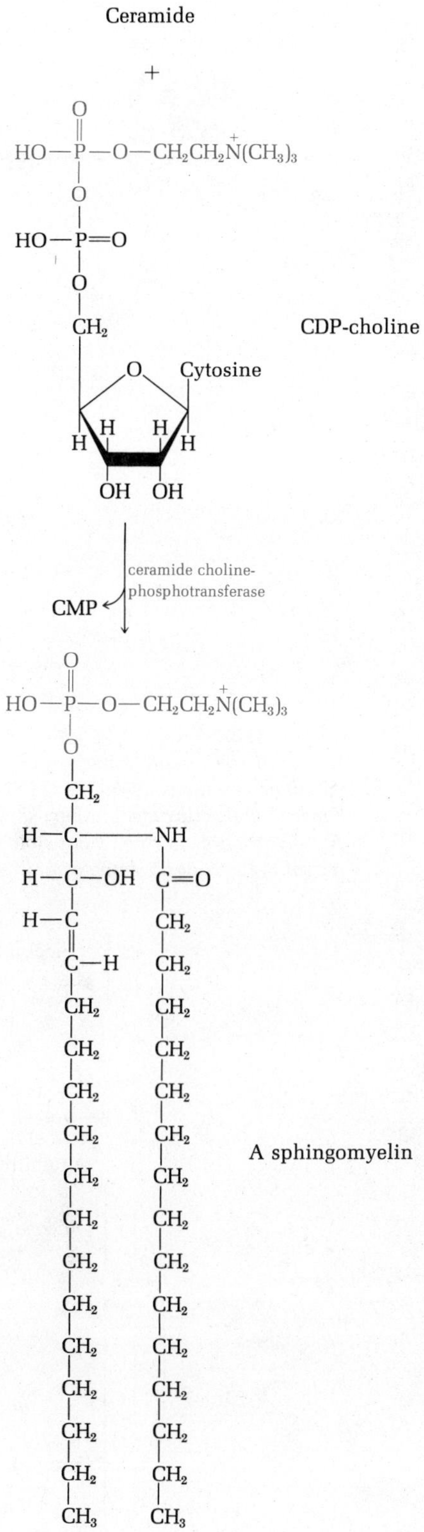

Figure 24-18
The formation of sphingomyelin. The portion arising from CDP-choline is in color.

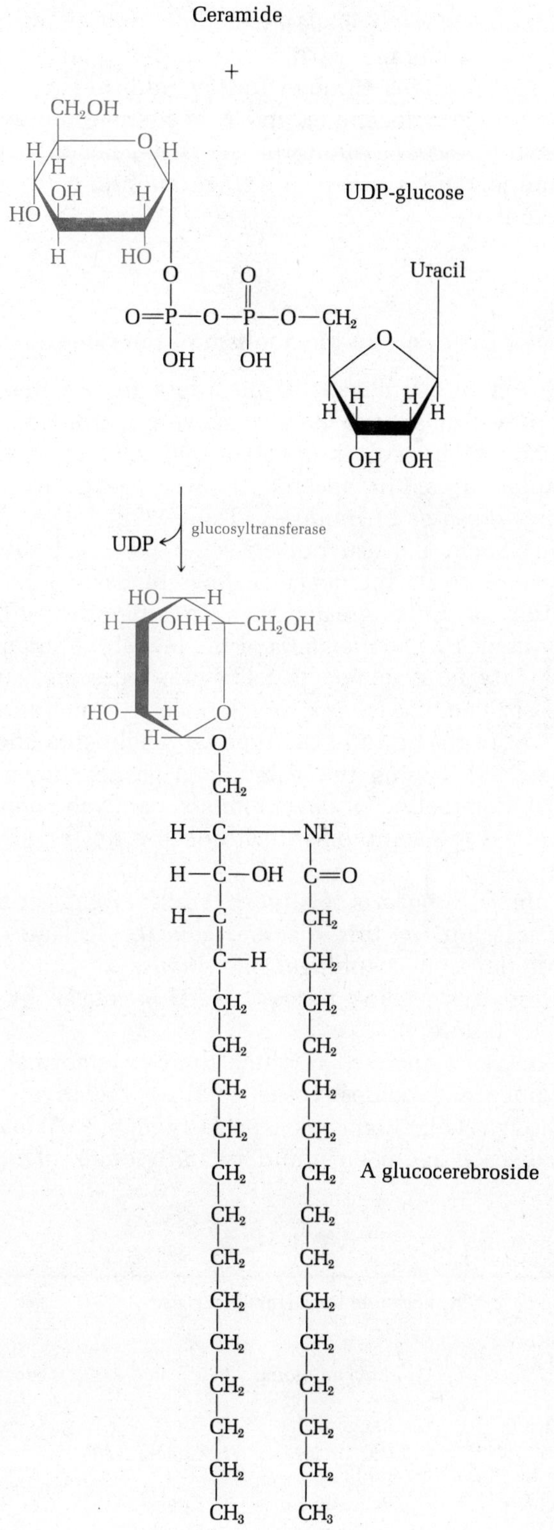

Figure 24-19
The biosynthesis of a glucocerebroside. The
portion arising from UDP-glucose is shown
in color. Note the inversion of configuration
about carbon atom 1 of the glucosyl residue.

Ceramide

+

$$HO-\overset{\overset{\displaystyle O}{\|}}{P}-O-CH_2CH_2\overset{+}{N}(CH_3)_3$$

$$HO-\underset{\underset{\displaystyle O}{|}}{\overset{\|}{P}}{=}O$$

CH₂ — Cytosine   CDP-choline

OH   OH

ceramide choline-
phosphotransferase

CMP ↙

$$HO-\overset{\overset{\displaystyle O}{\|}}{P}-O-CH_2CH_2\overset{+}{N}(CH_3)_3$$

O

CH₂
H—C——NH
H—C—OH C=O
H—C   CH₂
C—H   CH₂
CH₂   CH₂
CH₂   CH₂
CH₂   CH₂
CH₂   CH₂   A sphingomyelin
CH₂   CH₂
CH₂   CH₂
CH₂   CH₂
CH₂   CH₂
CH₂   CH₂
CH₂   CH₂
CH₂   CH₂
CH₃   CH₃

Ceramide

+

CH₂OH

UDP-glucose

HO   OH

$$O{=}\overset{\underset{\displaystyle OH}{|}}{P}-O-\overset{\underset{\displaystyle OH}{|}}{P}{=}O-CH_2$$

Uracil

OH   OH

UDP ↙ glucosyltransferase

HO—H
H—OHH—CH₂OH
HO—H   O
H—O

CH₂
H—C——NH
H—C—OH C=O
H—C   CH₂
C—H   CH₂
CH₂   CH₂
CH₂   CH₂
CH₂   CH₂
CH₂   CH₂   A glucocerebroside
CH₂   CH₂
CH₂   CH₂
CH₂   CH₂
CH₂   CH₂
CH₂   CH₂
CH₂   CH₂
CH₃   CH₃

first; it is then acylated to form the cerebroside:

$$\text{Sphingosine} + \text{UDP-galactose} \rightleftharpoons \text{psychosine} + \text{UDP}$$

$$\text{Psychosine} + \text{fatty acyl–CoA} \rightleftharpoons \text{cerebroside} + \text{CoA}$$

Gangliosides (page 294), which are important lipids in the neuronal membrane, particularly at the synapses, are formed from cerebrosides through further addition of a number of galactose, galactosamine, and N-acetylneuraminic acid residues donated in the form of UDP-galactose, UDP-galactosamine, and CMP–N-acetylneuraminic acid (page 651), respectively.

Genetic Disorders in Metabolism of the Complex Lipids

A dozen or more genetic disorders in the metabolism of complex lipids have been discovered in man. These disorders, which result in the abnormal accumulation of certain complex lipids in specific tissues, are also called *lipid-storage diseases* or *lipidoses* (Table 24-2).

Tay-Sachs disease has attracted particular public attention because its incidence in Jews of Eastern European extraction is much greater than in any other ethnic group examined; 1 in 30 members of the Jewish population in New York City is estimated to carry the recessive gene for this disease. The disease is characterized by an abnormally high content in the brain of the type of ganglioside known as G_{M2} (page 295), having the polar head group shown in Figure 24-20. Retardation of development, paralysis, dementia, and blindness are symptoms; death occurs by the age of two to four years.

Fabry's disease is relatively rare; it is subject to maternal transmission. In this disease, galactosylgalactosylglucosylceramide, a glycosphingolipid (Figure 24-21), accumulates abnormally in many tissues; death is caused by cardiac or kidney failure.

Gaucher's disease, in which there is abnormal deposition of glucocerebrosides (page 292), is relatively common. Although its incidence is especially high in Ashkenazi Jews, the disease has been found in individuals of many ethnic

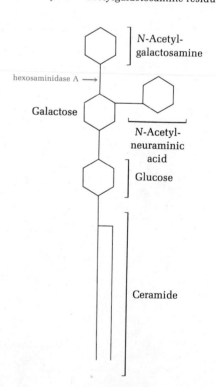

Figure 24-20
Schematic representation of ganglioside G_{M2}, which accumulates in Tay-Sachs disease because of a genetic defect in hexosaminidase A, a hydrolytic enzyme responsible for removal of the N-acetylgalactosamine residue.

Table 24-2   The heritable lipid-storage diseases

Disease	Lipid accumulated	Organ affected	Defective enzyme
Tay-Sachs	G_{M2} gangliosides	Brain	Hexosaminidase A
Fabry's	Gal-Gal-glucosyl-ceramide	Generalized	α-Galactosidase
Gaucher's	Glucocerebrosides	Spleen, liver	Glucocerebrosidase
Niemann-Pick	Sphingomyelin	Generalized, but particularly liver and spleen	Sphingomyelinase

groups. In *Niemann-Pick disease* sphingomyelin deposits accumulate. In nearly all the lipid-storage diseases mental retardation and neurological dysfunction are major symptoms, apparently because of abnormal deposition of the lipids in the brain and nerve tissue; there are heavy lipid deposits in other tissues as well.

After the chemical nature of the abnormal lipids was identified in these diseases, a paramount question arose: Are the lipid-storage diseases the result of abnormally high rates of synthesis of these lipids, with otherwise normal rates of breakdown? Or are they the result of abnormally low rates of breakdown, with essentially normal rates of synthesis? This question could not be approached until recently, after the enzymatic mechanism of the synthesis and degradation of these lipids was established. Small tissue specimens were obtained from patients by surgical biopsy (or from cadavers at autopsy), assayed for the relevant enzyme activities, and examined microscopically. Alternatively, fibroblasts from such biopsies have been cultured in vitro in a growth medium and then analyzed.

As a result of such investigations in a number of laboratories, particularly those of R. O. Brady, J. S. O'Brien, and L. Svennerholm, we now understand the enzymatic basis of the major lipid-storage diseases. All are disorders in which one or more enzymes of the *degradative* pathway are genetically defective, whereas the biosynthetic pathways and rates are essentially normal (page 373). In Tay-Sachs disease the defective enzyme is *hexosaminidase A,* an enzyme that normally causes hydrolytic cleavage of the N-acetylgalactosamine residue of G_{M2} ganglioside (Figure 24-20).

In Fabry's disease the defective enzyme is *α-galactosidase,* which normally causes hydrolytic cleavage of the galactose residues of galactosylgalactosylglucosylceramide (Figure 24-21). In Gaucher's disease, *glucocerebrosidase* activity is deficient, whereas in Niemann-Pick disease the defective enzyme is *sphingomyelinase,* which catalyzes hydrolysis of sphingomyelin to phosphorylcholine and ceramide.

In normal individuals these enzymes are present in the *lysosomes* (page 382), which are known to contain some 40 different hydrolytic enzymes, including ribonuclease and several proteases. The lysosomes of affected cells in patients with the lipid-storage diseases not only lack one or more of these degradative enzymes, but also show pronounced abnormalities of structure. For this reason the lipid-storage diseases are often called *lysosomal diseases.*

Early diagnosis of the lipid-storage diseases and genetic counseling of parents represent the only constructive means of control.

The Pathway of Cholesterol Biosynthesis

Most of the steps in the enzymatic biosynthesis of cholesterol are now known in some detail, as a result of the important research of K. Bloch in the United States, F. Lynen in Germany, and G. Popjak and J. Cornforth in Great Britain.

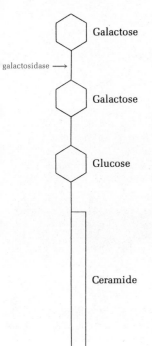

Figure 24-21
The trihexosylceramide accumulating in Fabry's disease. The accumulation is the result of a genetic defect in an α-galactosidase normally responsible for removal of the terminal galactose residue.

galactosidase →

Galactose

Galactose

Glucose

Ceramide

Figure 24-22
Incorporation of the carbon atoms of acetate
into cholesterol. All the carbon atoms of
cholesterol derive from acetate, in the
manner shown.

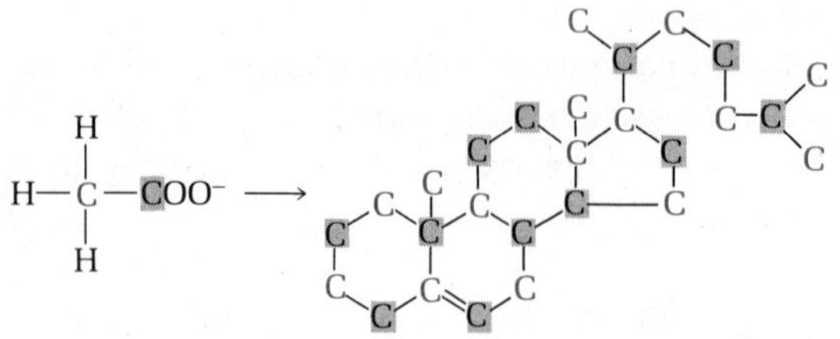

Acetate   Cholesterol carbon skeleton

Their interesting analyses of this complex process, which requires altogether some 25 reaction steps, revealed numerous new metabolic intermediates and greatly illuminated the pattern of biosynthesis of many other complex natural products, particularly the terpenes. The story began in the early 1940s, when Bloch and his colleagues demonstrated that the carbon atoms of carbon-labeled acetate fed to rats are incorporated into the cholesterol of the liver. Both the steroid nucleus and the eight-carbon side chain of cholesterol were found to be labeled. Experiments with methyl-labeled and carboxyl-labeled acetate revealed that both carbon atoms of acetic acid were incorporated into cholesterol and in approximately equal amounts; in fact, all the carbon atoms of cholesterol derive from acetate. Systematic degradation of biologically labeled cholesterol revealed the origin of each of its carbon atoms, as shown in Figure 24-22. This labeling pattern then became the guide for the elucidation of the pathway leading from acetate to cholesterol.

An important clue to the nature of this pathway came from the discovery of R. G. Langdon and K. Bloch that *squalene*, an open-chain isoprenoid hydrocarbon (more specifically, a dihydrotriterpene) (Figure 24-23), is an intermediate in cholesterol biosynthesis. This hydrocarbon, first found in the liver of sharks, is present in small amounts in the liver of most higher animals. When carbon-labeled acetate is fed to animals, the squalene of the liver becomes extensively labeled. When such biologically labeled squalene was fed to animals or incubated with liver slices, much of the isotope was incorporated into cholesterol. This incorporation occurred with such a high yield that squalene was concluded to be a rather direct precursor of cholesterol. These observations therefore strongly suggested that the steroid nucleus of cholesterol, which contains four condensed rings, is formed by cyclization of the open-chain hydrocarbon squalene (Figure 24-23).

The mechanism by which the isoprene units of squalene are formed from acetate remained a mystery, however, until work in another area led to the discovery of *mevalonic acid*

Figure 24-23
Squalene, shown in a form suggesting how it
might be folded to serve as the precursor of
cholesterol. The dashed lines (color) show
the isoprene units.

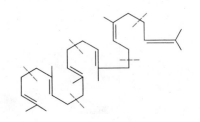

CO—S—CoA   Acetyl-CoA
|
CH₃

+

CH₃
|
C=O
| Acetoacetyl-CoA
CH₂
|
CO—S—CoA

    ↓ hydroxymethyl-
CoA—SH ↙ glutaryl-CoA
       synthase

COOH
|
CH₂
| Hydroxymethyl-
HO—C—CH₃   glutaryl-CoA
|
CH₂
|
CO—S—CoA

2NADPH ↘
       hydroxymethyl-
2NADP⁺ ↙ glutaryl-CoA
       reductase
CoA—SH ↙

COOH
|
CH₂
| L-Mevalonic
HO—C—CH₃   acid
|
CH₂
|
CH₂OH

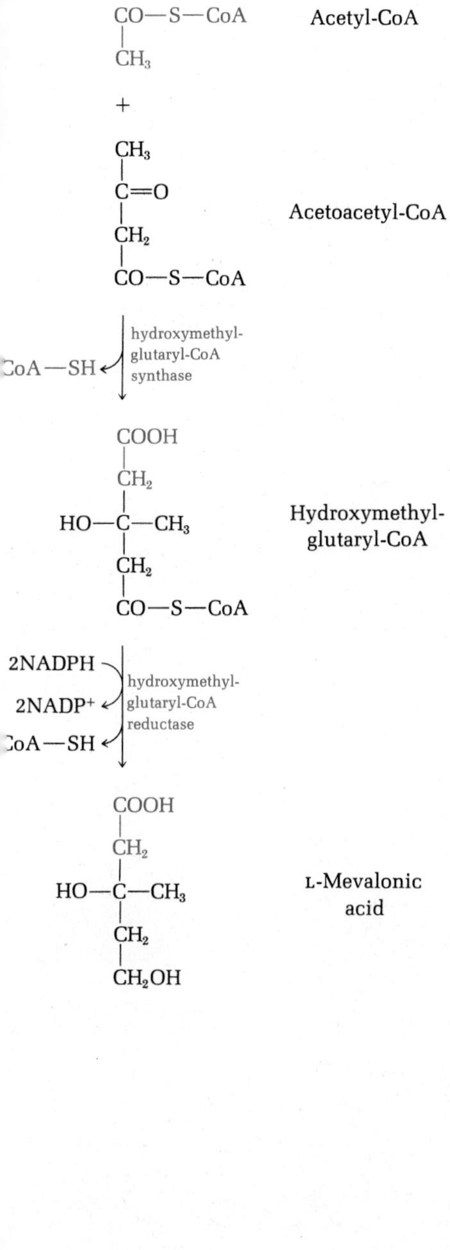

(Figure 24-24), a metabolite formed from acetic acid. Certain bacteria require acetate as a growth factor. Detailed study of the nutritional requirements of such bacteria revealed that acetate could be replaced as a growth factory by an unidentified organic acid found in many different natural sources, including the liver of mammals. The acetate-replacing growth factor was soon isolated and identified as mevalonic acid (Figure 24-24). It was at once obvious that this six-carbon acid could theoretically give rise to a five-carbon isoprene unit by decarboxylation. When isotopically labeled mevalonic acid was incubated with liver slices, it was incorporated into squalene and cholesterol with a very high yield. Furthermore, incubation of labeled acetate with liver slices showed that acetate carbon was an immediate precursor of mevalonic acid. These experiments set the stage for more detailed research on the enzymatic mechanisms by which (1) acetic acid is converted to mevalonic acid, (2) mevalonic acid is converted into squalene, and (3) squalene is converted into cholesterol.

Conversion of Acetic Acid into Mevalonic Acid

Mevalonic acid is formed by condensation of three molecules of acetyl-CoA. The key intermediate in this process is *β-hydroxy-β-methylglutaryl-CoA* (HMG-CoA), which is formed as follows:

$$\text{Acetyl-CoA + acetoacetyl-CoA} \xrightarrow{\text{hydroxymethylglutaryl-CoA synthase}} \text{β-hydroxy-β-methylglutaryl-CoA + CoA}$$

The β-hydroxy-β-methylglutaryl-CoA can undergo two reactions. It may be cleaved to form acetoacetate plus acetyl-CoA, as described elsewhere (page 554), and it may also undergo an irreversible two-step reduction of one of its carboxyl groups to an alcohol group, with concomitant loss of CoA, by the action of *hydroxymethylglutaryl-CoA reductase*, to yield mevalonate (Figure 24-24):

$$\text{β-Hydroxy-β-methylglutaryl-CoA + 2NADPH + 2H⁺} \longrightarrow \text{mevalonate + 2NADP⁺ + CoA}$$

This complex reaction is an important control point in cholesterol biosynthesis (see below).

There is a second route for formation of mevalonic acid. In principle it is identical to that shown above but in it acetyl-CoA reacts with acetoacetyl-ACP to form β-hydroxy-β-methylglutaryl-ACP, which is then reduced to mevalonic acid. This pathway takes place in the cytosol.

Conversion of Mevalonate into Squalene

This sequence of reactions begins with the phosphorylation of mevalonate by ATP, first to the 5-monophosphate ester and then to the 5-pyrophosphate (Figure 24-25). A third

Figure 24-25
The pathway from mevalonic acid to the
isomeric pentenyl pyrophosphates.

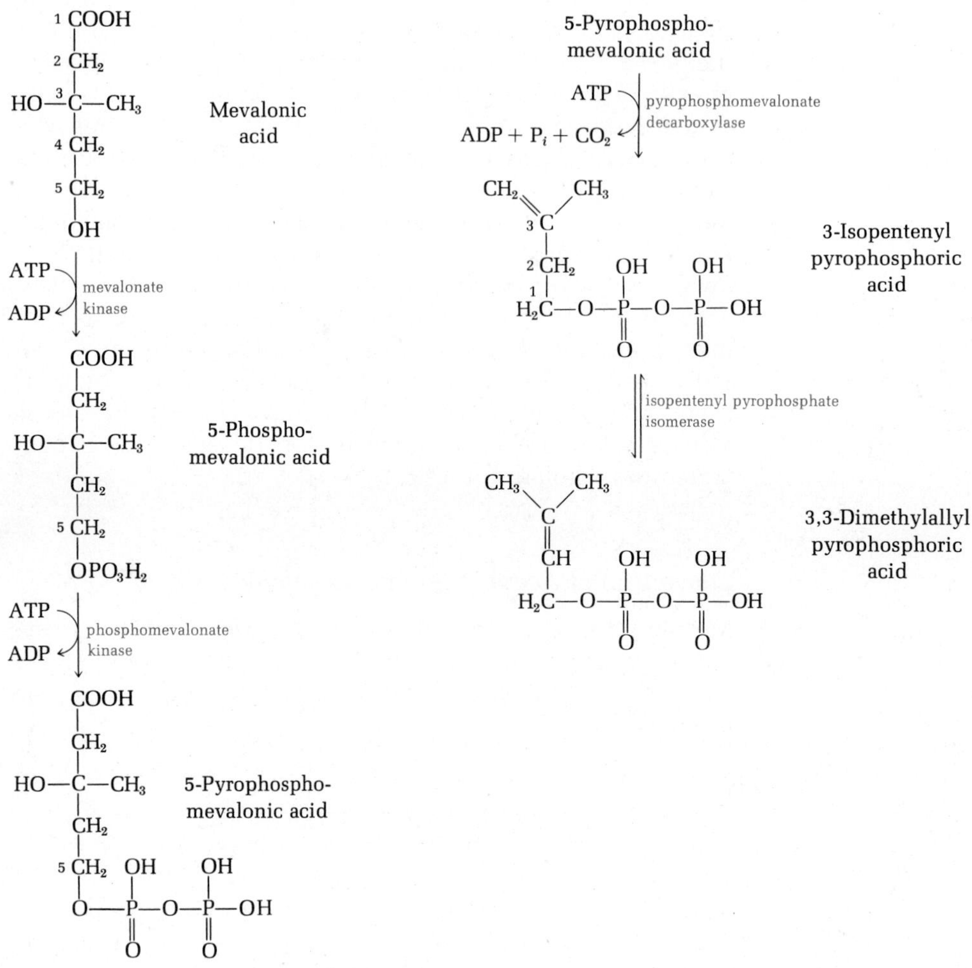

phosphorylation, at carbon atom 3, yields a very unstable intermediate which loses phosphoric acid and decarboxylates to form 3-*isopentenyl pyrophosphate*, which isomerizes to 3,3-*dimethylallyl pyrophosphate*. These two isomeric isoprenyl pyrophosphates then undergo condensation with elimination of pyrophosphoric acid to form the monoterpene derivative *geranyl pyrophosphate*. Another molecule of isopentenyl pyrophosphate then reacts, again with elimination of pyrophosphate, to yield the sesquiterpene derivative *farnesyl pyrophosphate* (Figure 24-26). Two molecules of this compound undergo condensation by a microsomal enzyme to yield *presqualene pyrophosphate* (Figure 24-27), which is reduced by NADPH to yield squalene and pyrophosphate.

Conversion of Squalene into Cholesterol

In the last stage of cholesterol biosynthesis, squalene undergoes attack by molecular oxygen to form *squalene 2,3-*

Figure 24-26
Formation of farnesyl pyrophosphate.

Isopentenyl
pyrophosphate

Dimethylallyl
pyrophosphate
PP$_i$ — dimethylallyl-
transferase

CH_3 CH_3
C
‖
CH
CH$_2$
CH$_2$ CH$_3$
C
‖
CH
CH$_2$—O—P—O—P—OH
OH   OH

Geranyl
pyrophosphoric
acid

Isopentenyl
pyrophosphate
PP$_i$ — dimethylallyl-
transferase

CH_3 CH_3
C
‖
CH
CH$_2$
CH$_2$ CH$_3$
C
‖
CH
CH$_2$
CH$_2$ CH$_3$
C
‖
CH
CH$_2$—O—P—O—P—OH
OH   OH

Farnesyl
pyrophosphoric
acid

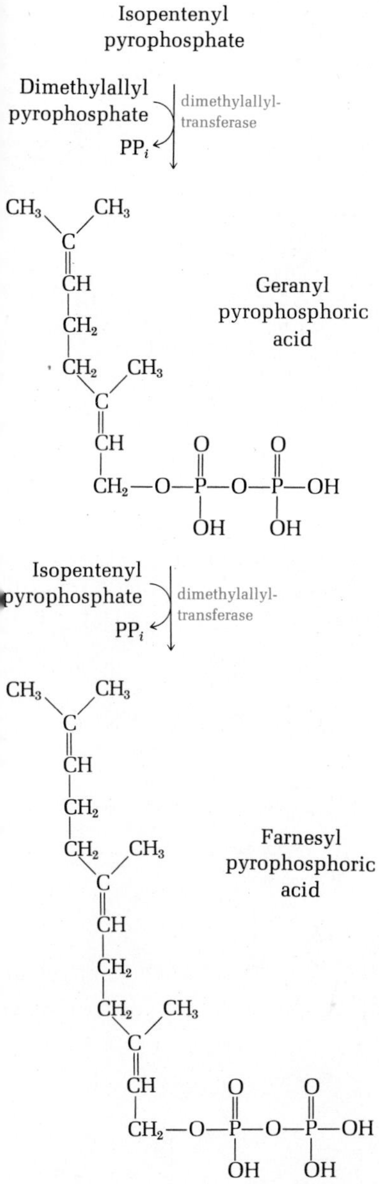

Figure 24-27
Formation of squalene and its conversion
into lanosterol.

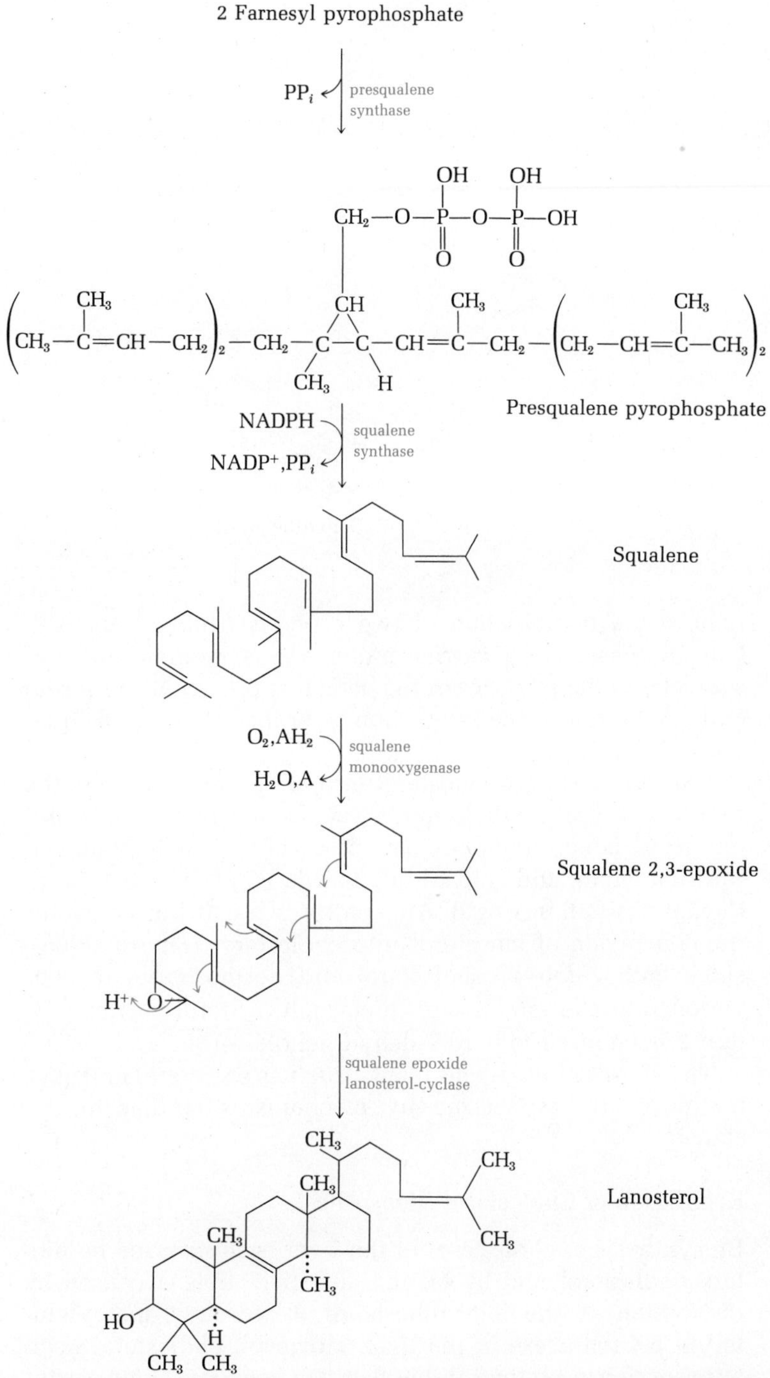

2 Farnesyl pyrophosphate

PP$_i$ — presqualene
synthase

$$\left(CH_3-\overset{CH_3}{\underset{}{C}}=CH-CH_2\right)_2-CH_2-\overset{CH_2-O-P-O-P-OH}{\underset{CH_3}{\overset{}{C}}}-\overset{CH_3}{\underset{H}{C}}-CH=\overset{CH_3}{\underset{}{C}}-CH_2-\left(CH_2-CH=\overset{CH_3}{\underset{}{C}}-CH_3\right)_2$$

Presqualene pyrophosphate

NADPH — squalene
NADP$^+$,PP$_i$ — synthase

Squalene

O$_2$,AH$_2$ — squalene
H$_2$O,A — monooxygenase

Squalene 2,3-epoxide

H$^+$ O

squalene epoxide
lanosterol-cyclase

Lanosterol

epoxide, a reaction catalyzed by squalene monooxygenase 2,3-epoxidizing). The squalene 2,3-epoxide then undergoes cyclization to lanosterol, the first sterol to be formed (Figure 24-27). In this extraordinary reaction, a series of concerted 1,2 methyl-group and hydride shifts along the squalene chain brings about closure of the four rings; the methyl-group shifts have been established by appropriate isotopic experiments. These reactions occur in the microsomes but

Figure 24-28
Conversion of lanosterol into cholesterol.
Two pathways are possible; that via 7-
dehydrocholesterol is more important.

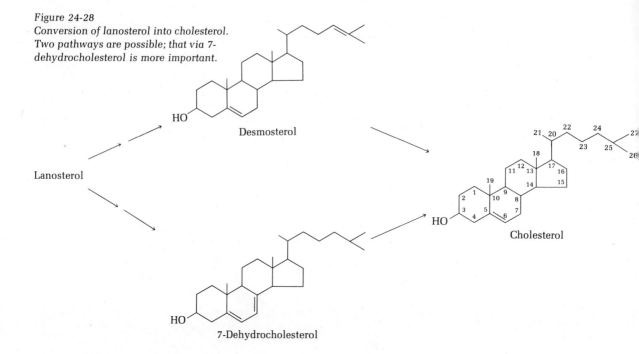

require the participation of two proteins from the cytosol. One of these, *sterol carrier protein 1*, is required to bind squalene in the presence of the cofactors phosphatidyl serine and FAD, prior to the interaction with the microsomal squalene monooxygenase.

The conversion of lanosterol into cholesterol involves the removal of three methyl groups (two from carbon atom 4 and one from carbon atom 14), saturation of the double bond in the side chain, and shift of the double bond from the 8,9 to the 5,6 position in ring B. Apparently two pathways exist for the conversion of lanosterol into cholesterol (Figure 24-28), since both 7-dehydrocholesterol and desmosterol are converted into cholesterol with high yields. *Sterol carrier protein 2* is required to bind 7-dehydrocholesterol.

The sterols of fungi and plants, such as ergosterol and stigmasterol, are synthesized by mechanisms similar to the above.

Regulation of Cholesterol Biosynthesis

Biosynthesis of cholesterol in the liver is suppressed by dietary cholesterol and by fasting, an effect that is caused by depression of the biosynthesis of β-hydroxy-β-methylglutaryl-CoA reductase in the liver. However, cholesterol itself appears not to be the inhibitor. It has been variously postulated that a cholesterol-containing lipoprotein, a bile acid, or a specific protein found in bile is the true inhibitor. Fasting also inhibits cholesterol biosynthesis, but high-fat diets accelerate the process.

The inhibition of cholesterol biosynthesis by the feeding of cholesterol has been observed in nearly all animal species and tissues tested. It is remarkable, however, that in cancer cells this type of feedback inhibition of cholesterol biosynthesis is completely absent. While the significance of this fact in the metabolic economy of cancer cells (page 853) is not known, it is noteworthy that it is one of several regulatory mechanisms that are defective or modified in malignant cells.

The transport and deposition of cholesterol in mammals are also subject to a number of regulatory mechanisms which are still unknown in detail; defects in these processes lead to pathological abnormalities. Cholesterol is often deposited on the inner walls of blood vessels, together with other lipids, a condition known as *atherosclerosis*, which often leads to occlusion of blood vessels in the heart and the brain, resulting in heart attacks and strokes, respectively. For this reason much current research is devoted to a better understanding of the regulation and control of the biosynthesis, transport, degradation, and excretion of cholesterol and other blood lipids.

The concentration of cholesterol in the blood is abnormally high in the genetic disorders *familial hypercholesterolemia* and *xanthomatosis*; in the latter, lipid deposits rich in cholesterol also form in the skin.

Synthesis of Cholesterol Esters

Most of the cholesterol in the tissues of higher organisms is esterified at the 3-hydroxyl group with long-chain fatty acids. The liver contains an enzyme that forms cholesterol esters by the reaction

$$\text{Cholesterol} + \text{fatty acyl--CoA} \xrightarrow[\text{acyltransferase}]{\text{cholesterol}} \text{cholesterol ester} + \text{CoA}$$

Cholesterol esters are also formed by an enzyme in blood plasma, *phosphatidyl-cholesterol acyltransferase*, which catalyzes transfer of a fatty acyl group from the 2 position of phosphatidylcholine to cholesterol:

$$\text{Phosphatidylcholine} + \text{cholesterol} \rightleftharpoons$$
$$\text{lysophosphatidylcholine} + \text{cholesterol ester}$$

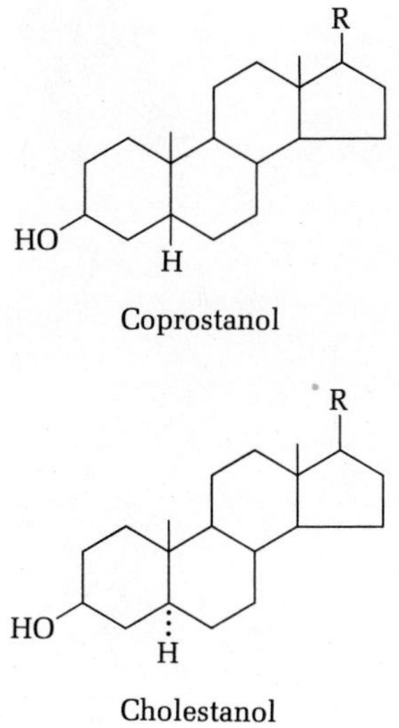

Formation of Other Steroids

Cholesterol is the precursor of fecal sterols, bile acids, and steroid hormones in animals. Cholesterol, coprostanol, and cholestanol are the major excretory sterols in mammals; coprostanol and cholestanol, which are stereoisomers (Figure 24-29), are formed from cholesterol by microbial action in the intestine.

The major pathway of degradation of cholesterol in animals is conversion to the bile acids, whose structure varies characteristically with the species. Bile acids contain a carboxyl group in their side chain, which is often bound in amide linkage with glycine or taurine (ethanolaminesulfonic acid). The bile acids are formed in the liver and secreted into the small intestine, where they aid in absorption of lipids by forming *chylomicrons*, droplets of triacylglycerols stabilized by a small amount of protein (page 838). The bile acids are largely reabsorbed again during lipid absorption. The secre-

Figure 24-29
Two fecal sterols. R represents the eight-carbon side chain of cholesterol.

Coprostanol

Cholestanol

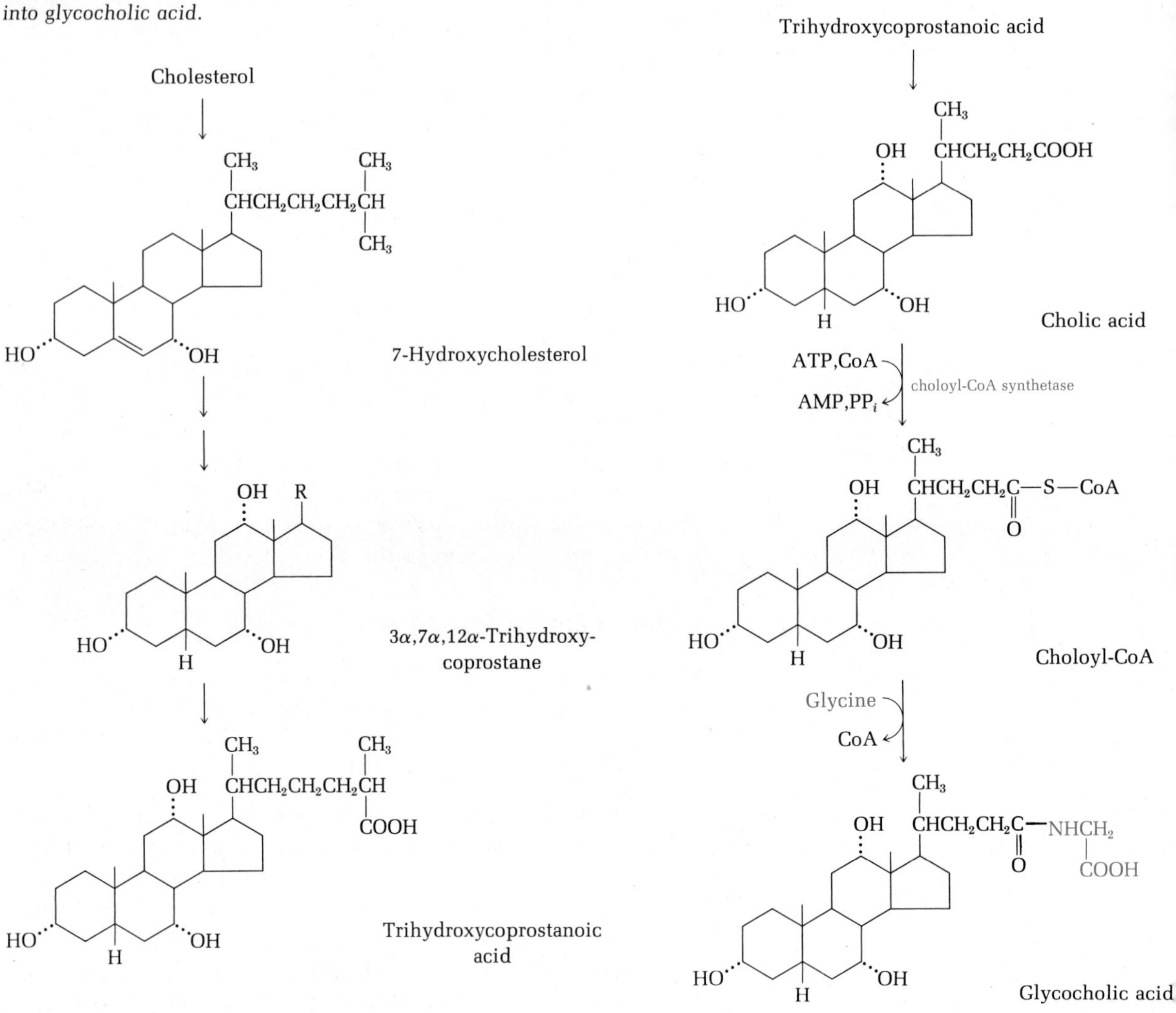

Figure 24-30
Formation of cholic acid and its conversion
into glycocholic acid.

Cholesterol

7-Hydroxycholesterol

$3\alpha,7\alpha,12\alpha$-Trihydroxy-coprostane

Trihydroxycoprostanoic
acid

Trihydroxycoprostanoic acid

Cholic acid

choloyl-CoA synthetase

Choloyl-CoA

Glycocholic acid

tion and reabsorption cycle of the bile acids is called the *enterohepatic circulation* (page 838). The major steps in the formation of one of the bile acids, *cholic acid*, as well as its substituted amides *glycocholic* and *taurocholic acids*, are shown in Figure 24-30.

The formation of steroid hormones from cholesterol occurs through intermediary formation of *pregnenolone* (Figure 24-31), which contains the cholesterol nucleus but has only a two-carbon side chain. Pregnenolone is the precursor of *progesterone*, the progestational hormone of the placenta and corpus luteum, and this in turn is the precursor of androgens or male sex hormones, such as *testosterone* and *androsterone*, of the estrogens or female sex hormones, such as *estrone* and *estradiol*, and of the adrenal corticosteroids, such as *corticosterone* and *aldosterone*.

Biosynthesis of Prostaglandins

The prostaglandins are cyclic derivatives of certain unsaturated fatty acids having 20 carbon atoms. They are found in

Figure 24-31
Formation of some steroid hormones. The multiple arrows do not designate the number of steps but only show that several steps are involved.

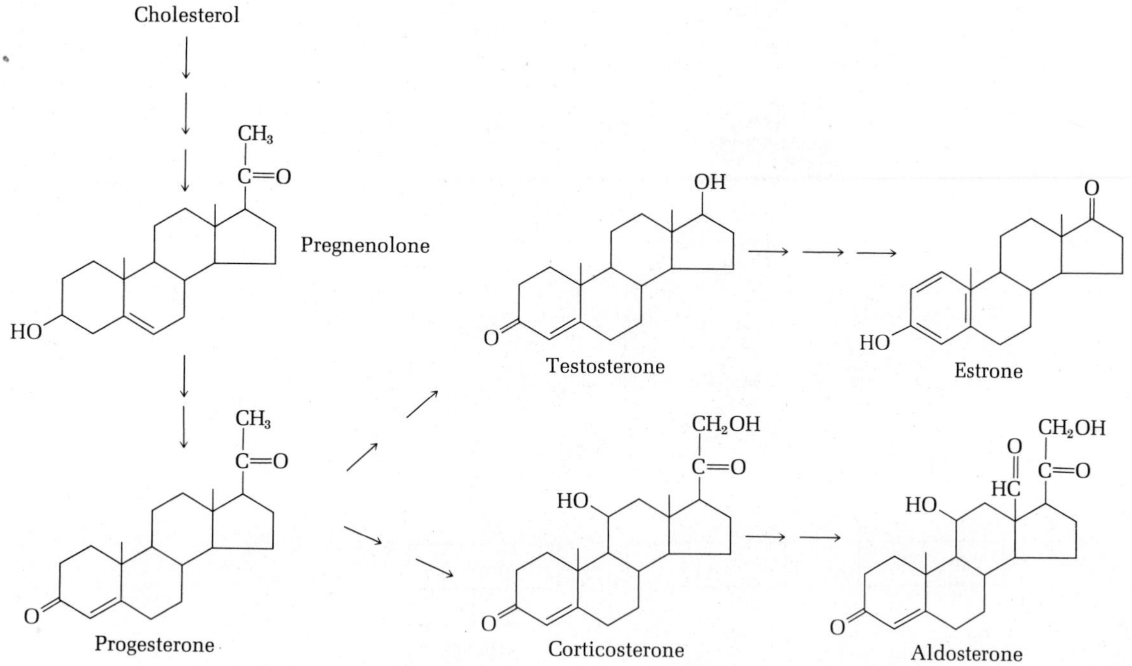

Cholesterol

Pregnenolone

HO

Progesterone

Testosterone

Estrone

HO

CH₂OH
C=O
HO

Corticosterone

CH₂OH
HC
C=O
HO

Aldosterone

Figure 24-32
Formation of prostaglandin PGE₂ from arachidonic acid. In addition to oxygen, an electron donor, GSH, is required. GSH represents reduced glutathione (page 96).

$CH_3(CH_2)_4(CH=CHCH_2)_3CH=CH(CH_2)_3COOH$
Arachidonic acid

O_2

COOH

O_2

COOH

O—O

2GSH
H_2O, GSSG

COOH

OH

OH

Prostaglandin PGE₂

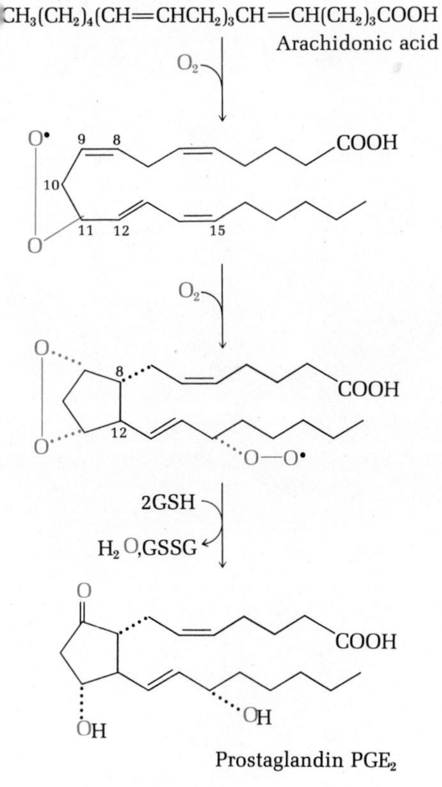

many animal tissues and have a variety of profound hormonelike physiological and pharmacological effects (page 300). Many natural and synthetic prostaglandins are known. They are classified according to their structural derivation from various unsaturated fatty acids.

The biosynthesis of prostaglandins begins with the essential fatty acids. In a typical pathway the essential fatty acid linoleic acid is first converted to the 20-carbon arachidonic acid (Figure 24-32), which is then acted upon by *prostaglandin synthase*, a dioxygenase (page 501) that adds one oxygen molecule to carbon atom 9 of arachidonate and a second molecule of oxygen at carbon 15. The product then undergoes cyclization by formation of a bond between carbon atoms 8 and 12 (Figure 24-32). In the presence of reduced glutathione (page 96) the cyclized product undergoes conversion into prostaglandin PGE₂. Other types of prostaglandins are derived from other polyunsaturated C₂₀ acids.

The physiological effects of the various prostaglandins are summarized elsewhere (page 300). Prostaglandins undergo oxidative degradation to inactive excretory products.

The Central Role of Acetic Acid as a Biosynthetic Precursor of Lipids

Acetic acid, in the form of acetyl-CoA, is the precursor not only of fatty acids, steroids, and prostaglandins but also of many other naturally occurring compounds, particularly the terpenes (page 296) and the acetogenins, substances built of recurring two-carbon units, in which one of the two carbon

Figure 24-33
Acetyl-CoA as the key precursor in biosynthesis of many different lipids.

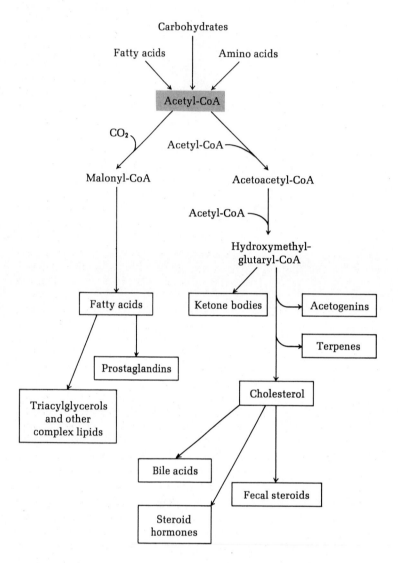

atoms, that corresponding to the carboxyl carbon of acetic
acid, is frequently oxidized. The scheme in Figure 24-33
shows the important role of acetyl-CoA as a precursor of
these products.

Summary

The long-chain saturated fatty acid palmitic acid is synthesized by
a complex of enzymes, the fatty-acid synthetase system of the cytosol, which employs a pantetheine-containing protein, acyl carrier
protein (ACP), as an acyl-group carrier. Malonyl-CoA, formed from
acetyl-CoA and HCO_3^- by acetyl-CoA carboxylase, is the direct precursor of seven of the eight two-carbon units of palmitic acid.
Acetyl-ACP, formed from acetyl-CoA, reacts with malonyl-ACP,
derived from malonyl-CoA, to yield acetoacetyl-ACP and CO_2. Reduction of acetoacetyl-ACP to the β-hydroxy derivative and the

dehydration of the latter to the Δ^2-unsaturated compound is followed by reduction to butyryl-ACP at the expense of NADPH. Six more molecules of malonyl-ACP react successively at the carboxyl end of the growing fatty acid chain to form ultimately palmitoyl-ACP, the usual end product. Palmitic acid is the precursor of all other long-chain saturated and unsaturated fatty acids. Palmitic acid can be elongated by reaction with successive residues of acetyl-CoA in mitochondria or of malonyl-CoA in microsomes. Palmitoleic and oleic acids are formed in animals from palmitic and stearic acids, respectively, by action of an oxygenase system containing cytochrome b_5, which requires NADPH as coreductant. In bacteria monounsaturated fatty acids are formed by Δ^2 dehydration of β-hydroxydecanoyl-ACP, followed by chain elongation. Polyenoic acids are formed from oleic and palmitoleic acids by further action of desaturating oxygenases. Linoleic and linolenic acids, essential fatty acids, are readily formed by plants but not by mammals, which require them in the diet.

Triacylglycerols are formed in a sequence of reactions in which two molecules of fatty acyl–CoA react with glycerol 3-phosphate to form L-phosphatidic acid, which is dephosphorylated and then acylated by a third molecule of fatty acyl–CoA. Phosphatidic acid is also the major precursor of phosphoglycerides. In animals it reacts with cytidine diphosphoethanolamine to form phosphatidylethanolamine. The latter may be methylated at the expense of S-adenosylmethionine to yield phosphatidylcholine, which can also be formed from CDP-choline and phosphatidic acid. Phosphatidic acid may react with CTP to form cytidine diphosphate diacylglycerol, a precursor of phosphatidylserine, phosphatidylinositol, and cardiolipin. Phosphatidylserine, in turn, may be decarboxylated to yield phosphatidylethanolamine.

Sphingosine, the characteristic base of sphingolipids, is formed by reaction of palmitoyl-CoA and serine. Sphingosine is acylated by long-chain fatty acyl–CoA's to yield a ceramide, which reacts with CDP-choline to yield sphingomyelin. Cerebrosides are also formed from ceramide by reaction of the latter with UDP-glucose or UDP-galactose or by an alternative pathway starting from psychosine, a hexose-substituted sphingosine.

Cholesterol is formed from acetic acid in three major stages. In the first, mevalonic acid is formed from acetyl-CoA via β-hydroxy-β-methylglutaryl-CoA. In the second stage mevalonate is converted into the isomers 3-isopentenyl pyrophosphate and dimethylallyl pyrophosphate, which condense to form geranyl pyrophosphate. The last in turn leads to farnesyl pyrophosphate, the direct precursor of squalene. In the third stage squalene undergoes cyclization to the steroid lanosterol and finally yields cholesterol, the precursor of bile acids, fecal sterols, and steroid hormones.

Prostaglandins are synthesized by oxygen-requiring ring-closure reactions from essential fatty acids and their derivatives.

References

Books

ANSELL, G. M., J. N. HAWTHORNE, and R. M. C. DAWSON: *Form and Function of Phospholipids,* 2d ed., Elsevier, New York, 1973. A comprehensive and valuable treatise.

DORFMAN, R. I., and F. UNGAR: *Metabolism of Sex Hormones,* Academic, New York, 1965.

GREENBERG, D. M. (ed.): *Metabolic Pathways,* 3d ed., vol. II, Academic, New York, 1968. Contains reviews of fatty acid and lipid biosynthesis.

HASLEWOOD, G. A. D.: *Bile Salts*, Methuen, London, 1967. The evolution and comparative biochemistry of bile salts, reviewed in brief, readable form.

RICHARDS, J. H., and J. B. HENDRICKSON: *Biosynthesis of Steroids. Terpenes, and Acetogenins*, Benjamin, Menlo Park, Calif., 1964. The organic approach to problems of biosynthesis.

STANBURY, J. B., J. B. WYNGAARDEN, and D. S. FREDRICKSON (eds.): *The Metabolic Basis of Inherited Disease*, 3d ed., McGraw-Hill, New York, 1972. Genetic disorders of lipid metabolism are treated in detail.

SWEELEY, C. C. (ed.): *Chemistry and Metabolism of Sphingolipids*, North-Holland, Amsterdam, 1970.

Articles

BEYTIA, E., A. A. QURESHI, and J. PORTER: "Squalene Synthetase: Mechanism of the Reaction," *J. Biol. Chem.*, 248: 1856–1867 (1973).

BLOCH, K.: "The Biological Synthesis of Cholesterol," *Science*, 150: 19–28 (1965). Nobel award address; a historical recapitulation and experimental narrative.

BOSCH, H. VAN DEN: "Phosphoglyceride Metabolism," *Ann. Rev. Biochem.*, 43: 243–278 (1974).

BRADY, R. O.: "Inborn Errors of Lipid Metabolism," *Adv. Enzymol.*, 38: 293–316 (1973).

DEMPSEY, M. E.: "Regulation of Steroid Biosynthesis," *Ann. Rev. Biochem.*, 43: 967–990 (1974).

FULCO, A. J.: "Metabolic Alterations of Fatty Acids," *Ann. Rev. Biochem.*, 43: 215–241 (1974).

GATT, S., and Y. BARENHOLZ: "Enzymes of Complex Lipid Metabolism," *Ann. Rev. Biochem.*, 42: 61–90 (1973).

KENNEDY, E.P.: "The Metabolism and Function of Complex Lipids," *Harvey Lect.*, 57: 143–171 (1962). Short review of early research.

LANE, M. D., J. MOSS, and S. E. POLAKIS: "Acetyl Coenzyme A Carboxylase," *Curr. Top. Cell. Regul.*, 8: 139–195 (1974).

LOWENSTEIN, J. M.: "Citrate and the Conversion of Carbohydrate into Fat," p. 61 in T. W. Goodwin (ed.), *Metabolic Roles of Citrate*, Academic, New York, 1968. An excellent and clearly organized review.

LYNEN, F.: "The Role of Biotin-Dependent Carboxylations in Biosynthetic Reactions," *Biochem. J.*, 102: 381–400 (1967). Includes discussion of the yeast fatty-acid synthetase complex.

MILUNSKY, A., J. W. LITTLEFIELD, J. N. KANFER, E. H. KOLODNY, V. E. SHIH, and L. ATKINS: "Prenatal Genetic Diagnosis (First of Three Parts)," *N. Engl. J. Med.*, 283: 1370–1381 (1970).

MUSCIO, F., J. P. CARLSON, L. KUEHL, and H. C. RILLING: "Presqualene Pyrophosphate: A Normal Intermediate in Squalene Biosynthesis," *J. Biol. Chem.*, 249: 3746–3749 (1974).

OESTERHELT, D., H. BAUER, and F. LYNEN: "Crystallization of a Multienzyme Complex: Fatty Acid Synthetase from Yeast," *Proc. Natl. Acad. Sci. (U.S.)*, 63: 1377–1382 (1969).

SIPERSTEIN, M. D.: "Regulation of Cholesterol Biosynthesis in Normal and Malignant Tissues," *Curr. Top. Cell. Regul.*, 2: 65–100 (1970).

STOFFEL, W.: "Sphingolipids," *Ann. Rev. Biochem.*, 40: 57–82 (1971). A comprehensive review of the subject.

VOLPE, J. J., and P. R. VAGELOS: "Saturated Fatty Acid Biosynthesis and Its Regulation," *Ann. Rev. Biochem.*, 42: 21–60 (1973).

WLODAWER, P., and B. SAMUELSSON: "On the Organization and Mechanism of Prostaglandin Synthetase," *J. Biol. Chem.*, 248: 5673–5678 (1973).

Problems

1. Write a balanced equation for the biosynthesis of stearic acid starting from acetyl-CoA, NADPH, and ATP.

2. Write a balanced equation for the biosynthesis of palmitoleic acid starting from pyruvate, NADPH, and ATP.

3. Write a balanced equation for the formation of stearic acid from lauric acid and NADPH, as it occurs in microsomes.

4. Write a balanced equation for the biosynthesis of tripalmitin in an aerobic liver cell starting from glucose. Include changes in the ATP system.

5. How many high-energy phosphate bonds are required for the synthesis of a molecule of dipalmityl phosphatidylserine starting from palmitic acid, glucose, and serine?

6. How many high-energy phosphate bonds are required in the biosynthesis of one molecule of dipalmityl phosphatidylcholine starting from palmitic acid, serine, glycerol, and methionine? (See Chapter 25.)

7. Write a balanced equation for the synthesis of cholesterol from free acetic acid. How many high-energy phosphate bonds are required per molecule of cholesterol formed?

8. Contrast the number of high-energy phosphate bonds required per carbon atom for an albino rat to store the carbon of ingested D-glucose as glycogen and as the palmitoyl groups of tripalmitoylglycerol, respectively.

CHAPTER 25 THE BIOSYNTHESIS OF AMINO ACIDS AND SOME DERIVATIVES; METABOLISM OF INORGANIC NITROGEN

Although amino acids are of central importance in the metabolism of all organisms, principally because they are the precursors of proteins, different organisms vary considerably in their ability to synthesize amino acids and in the forms of nitrogen utilized for this purpose. Vertebrates are not able to synthesize all the common amino acids; for example, man and the albino rat can make only 10 of the 20 amino acids required as building blocks of proteins. The remainder, the essential, or nutritionally indispensable, amino acids, must be obtained from plants or bacteria. Higher animals can utilize ammonium ions as the nitrogen source for the synthesis of nonessential amino acids, but they are unable to use nitrite, nitrate, or atmospheric nitrogen. Higher plants are more versatile; they can make all the amino acids required for protein synthesis and can utilize ammonia, nitrite, or nitrate as nitrogen source. The leguminous plants, which harbor symbiotic nitrogen-fixing bacteria in their root nodules, are even able to use atmospheric nitrogen, which they convert into ammonia.

Microorganisms differ widely in their capacity to synthesize amino acids. For example, the bacterium *Leuconostoc mesenteroides* cannot grow unless it is supplied with a total of 16 different amino acids. Such bacteria can survive only in environments rich in preformed amino acids from decaying biological matter. Other bacteria, such as *E. coli*, can manufacture all their amino acids starting from ammonia. Although most microorganisms require a reduced form of nitrogen, such as ammonia, numerous bacteria and fungi, as well as the higher plants, can utilize nitrite or nitrate.

The 20 different amino acids are synthesized by 20 different multienzyme sequences, some of which are exceedingly complex. As with most biosynthetic routes, the pathways of amino acid synthesis are for the most part different from those employed in their degradation (pages 373 and 619). The complexity of the pathways of amino acid synthesis and degradation may at first be discouraging to contemplate; however, some exceedingly interesting biochemical relationships are involved in this area of metabolism.

Frequently the pathways of synthesis and degradation of amino acids have branches or extensions leading to the formation of various specialized biomolecules. Moreover, the amino acids themselves also serve as precursors of other biomolecules having specialized functions.

Biologically available forms of nitrogen are relatively scarce in the nonliving environment (page 365). For this reason, most living organisms tend to be economical in their metabolic use of reduced forms of nitrogen. Often amino groups or nitrogenous intermediates resulting from catabolic processes are salvaged and reutilized for amino acid synthesis. We shall also see that the biosynthesis of most of the amino acids is under feedback control, through the action of regulatory enzymes. Moreover, the biosynthesis of the enzymes catalyzing the formation of the amino acids is also under regulation; their synthesis is often repressed if the cell is amply supplied with amino acids from exogenous sources.

In this chapter a series of metabolic maps will show the biosynthetic pathways for the different amino acids. In these maps the major regulatory enzymes and the characteristic action of their inhibitory or stimulatory allosteric modulators will be indicated using the conventions developed earlier (pages 234 and 538).

Biosynthesis of the Nonessential Amino Acids

The nonessential, or dispensable, amino acids are those which can be synthesized by man and the albino rat, organisms which have identical amino acid requirements. It is convenient to consider this group of amino acids first because in most organisms their biosynthesis can be brought about from other precursors. Moreover, these amino acids are distinctive because they have relatively short biosynthetic pathways.

Glutamic Acid, Glutamine, and Proline

The main pathway for glutamate formation in most organisms is the amination of α-ketoglutarate by the action of L-*glutamate dehydrogenase*, which in animal tissues can utilize either NAD or NADP. In plants the enzyme is often specific for NADP. The reaction is

$$NH_3 + \alpha\text{-ketoglutarate} + NAD(P)H + H^+ \rightleftharpoons$$
$$\text{L-glutamate} + NAD(P)^+ + H_2O$$

This reaction is of fundamental importance in the biosynthesis of all amino acids, since in many species it is the major pathway for the formation of α-amino groups directly from ammonia. Subsequently the amino group of glutamate is transferred to various α-keto acids to yield the corresponding amino acids.

Glutamate dehydrogenase is a regulatory enzyme. As isolated from beef liver the enzyme has a molecular weight of about 320,000 and contains six identical subunits, each with two coenzyme binding sites. The enzyme self-associates re-

Figure 25-1
γ-Glutamyl phosphate.

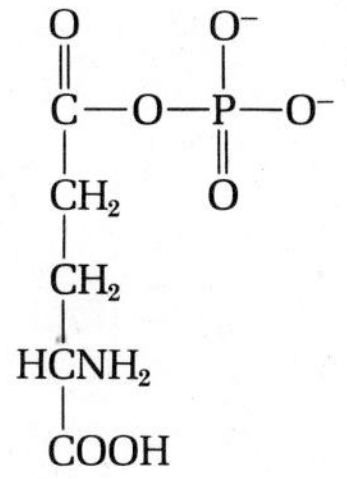

$$
\begin{array}{c}
\text{O} \qquad \text{O}^- \\
\| \qquad | \\
\text{C--O--P--O}^- \\
| \qquad | \\
\text{CH}_2 \quad \text{O} \\
| \\
\text{CH}_2 \\
| \\
\text{HCNH}_2 \\
| \\
\text{COOH}
\end{array}
$$

Figure 25-2
Biosynthesis of proline. All five carbon atoms arise from glutamic acid. The end product proline is an allosteric inhibitor of the first reaction step.

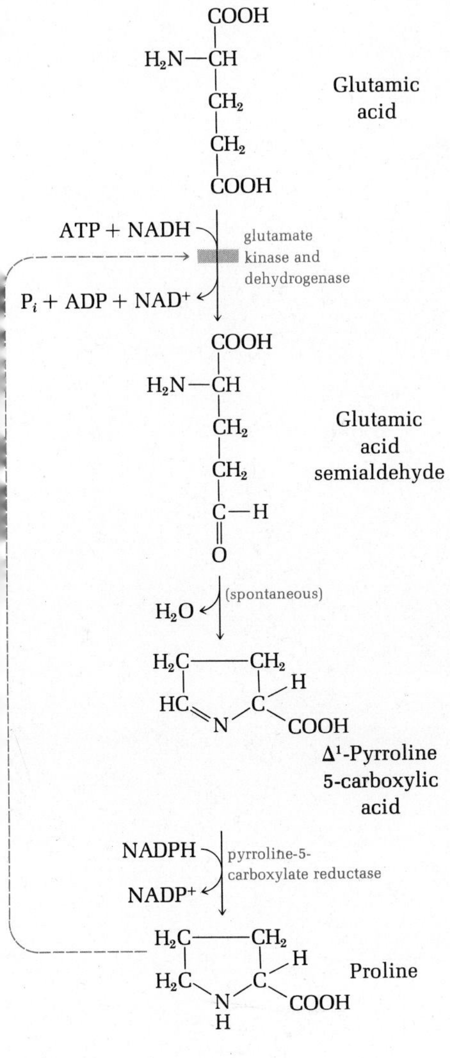

versibly to yield long linear polymers. Such polymers are dissociated into monomers by NADH or by GTP. Glutamate dehydrogenase has a number of allosteric effectors whose complex effects depend on the concentrations of the enzyme, of the NAD$^+$ or NADH, and of the substrate glutamate. ADP is either stimulatory or inhibitory, depending on the NAD$^+$ concentration, whereas GTP is strongly inhibitory to the reaction. The reaction rate is also influenced by Zn^{2+}, the thyroid hormone thyroxine, and the synthetic estrogen (female sex hormone) diethylstilbestrol.

Glutamine is formed from glutamic acid by the action of *glutamine synthetase,* which is also discussed elsewhere (pages 243 and 583):

$$\text{NH}_3 + \text{glutamate} + \text{ATP} \rightleftharpoons \text{glutamine} + \text{ADP} + \text{P}_i$$

This reaction is rather complex and involves two or more intermediate steps. It has been shown that *γ-glutamyl phosphate* (Figure 25-1) functions as an enzyme-bound intermediate:

$$\text{ATP} + \text{glutamic acid} \rightleftharpoons \text{ADP} + [\text{γ-glutamyl phosphate}]$$

$$[\text{γ-Glutamyl phosphate}] + \text{NH}_3 \rightleftharpoons \text{glutamine} + \text{P}_i$$

The specificity, kinetics, and mechanism of action of glutamine synthetase have been studied in great detail by A. Meister and his colleagues. Extensive substrate-specificity studies have permitted mapping of the active site of the enzyme, which has been shown to bind the extended conformation of the glutamic acid molecule at three sites, namely, at the two carboxyl groups and the amino group. Curiously, the enzyme will accept either the D or the L stereoisomer of glutamic acid as substrate.

Glutamine synthetase is a structurally complex enzyme of molecular weight 350,000 to 600,000, depending on the species. In some cells it is a regulatory enzyme, but its allosteric characteristics, discussed later in this chapter, vary widely with the cell type.

The glutamine synthetase reaction also participates in the conversion of free ammonia into the α-amino groups of amino acids, since it may cooperate with the reaction catalyzed by the flavin enzyme *glutamate synthase* to convert free ammonia into the α-amino group of glutamate:

$$\text{Glutamate} + \text{ATP} + \text{NH}_3 \xrightarrow{\substack{\text{glutamine} \\ \text{synthetase}}} \text{glutamine} + \text{ADP} + \text{P}_i$$

$$\text{Glutamine} + \text{α-ketoglutarate} + \text{NADPH} + \text{H}^+ \xrightarrow{\substack{\text{glutamate} \\ \text{synthase}}}$$
$$\text{2 glutamate} + \text{NADP}^+$$

Proline is synthesized from glutamic acid by the pathway shown in Figure 25-2. The γ-carboxyl group of glutamate presumably is first phosphorylated by ATP prior to its reduction by NADH or NADPH, by analogy with the reduction of 3-phosphoglycerate to glyceraldehyde 3-phosphate in the re-

695

versal of glycolysis (pages 427 and 627). The end product of this sequence, L-proline, is an allosteric inhibitor of the first reaction step.

The 4-*hydroxyproline* residues present in collagen and a few other fibrous proteins are formed from certain protein-bound proline residues through *proline 4-monooxygenase*, which acts on proline residues only after they have been incorporated into the polypeptide chain. This mixed-function oxygenase (page 501) utilizes α-ketoglutarate as coreductant:

$$\text{Proline residue} + O_2 + \alpha\text{-ketoglutarate} + \text{CoA} \longrightarrow$$
$$\text{4-hydroxyproline residue} + \text{succinyl-CoA} + CO_2 + H_2O$$

Fe^{3+} and ascorbic acid are required in this reaction as cofactors.

Alanine, Aspartic Acid, and Asparagine

In most organisms alanine and aspartate arise from pyruvate and oxaloacetate by transamination from L-glutamate:

$$\text{Glutamate} + \text{pyruvate} \rightleftharpoons \alpha\text{-ketoglutarate} + \text{alanine}$$

$$\text{Glutamate} + \text{oxaloacetate} \rightleftharpoons \alpha\text{-ketoglutarate} + \text{aspartate}$$

The properties of transaminases involving glutamate as amino-group donor are described elsewhere (page 562). In some plants, however, alanine and aspartate may be formed by reductive amination with ammonia in reactions analogous to the glutamate dehydrogenase reaction.

Aspartic acid is the direct precursor of *asparagine*, in a reaction catalyzed by *asparagine synthetase*, which is analogous in its mechanism to glutamine synthetase:

$$NH_3 + \text{aspartate} + \text{ATP} \longrightarrow \text{asparagine} + \text{ADP} + P_i$$

In some organisms an alternative pathway may occur in which the amide amino group of glutamine is transferred to the β-carboxyl group of aspartic acid, by the action of *asparagine synthetase (glutamine-hydrolyzing)*:

$$\text{ATP} + \text{glutamine} + \text{aspartate} \rightleftharpoons$$
$$\text{glutamate} + \text{asparagine} + \text{AMP} + PP_i$$

Tyrosine

Tyrosine is formed from the essential amino acid phenylalanine in a hydroxylation reaction catalyzed by *phenylalanine 4-monooxygenase*, which, as we have seen, also participates in the degradation of phenylalanine (page 569). This mixed-function oxygenase requires NADPH as coreductant and dihydrobiopterin as cofactor, as outlined elsewhere (page 570). The overall reaction is

$$\text{Phenylalanine} + \text{NADPH} + H^+ + O_2 \longrightarrow \text{tyrosine} + \text{NADP}^+ + H_2O$$

The biosynthesis of phenylalanine will be considered below.

Figure 25-3
Biosynthesis of L-serine from 3-phosphoglycerate.

OH
|
HO—P—O—CH$_2$—CH—COOH
‖ |
O   OH   3-Phospho-D-glyceric acid

NAD$^+$ ⌐
        phosphoglycerate
        dehydrogenase
NADH ↙

OH
|
HO—P—O—CH$_2$—C—COOH
‖   ‖
O   O   3-Phosphohydroxy-pyruvic acid

Glutamate ⌐
        phosphoserine
        transaminase
α-Ketoglutarate ↙

OH
|
HO—P—O—CH$_2$—CH—COOH
‖ |
O   NH$_2$ 3-Phosphoserine

H$_2$O ⌐
      phosphoserine
      phosphatase
P$_i$ ↙

HO—CH$_2$—CH—COOH
         |
         NH$_2$ Serine

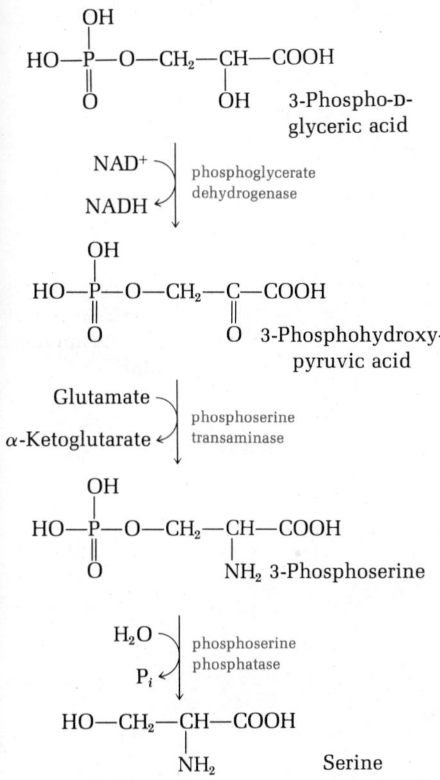

Serine and Glycine

L-*Serine*, which also serves as a precursor of glycine and cysteine, is formed from 3-phosphoglycerate, its major source, in the pathway shown in Figure 25-3. 3-Phospho-D-glycerate, an intermediate of glycolysis, is oxidized by *phosphoglycerate dehydrogenase* at the expense of NAD to yield 3-*phosphohydroxypyruvate*. Transamination from glutamate yields 3-*phosphoserine*, which forms free L-serine on hydrolysis by *phosphoserine phosphatase*.

In an alternate route to serine, hydroxypyruvate formed from D-glycerate undergoes transamination with glycine or alanine to yield serine and either glyoxylate or pyruvate, respectively:

$$\text{Hydroxypyruvate + glycine} \underset{\text{aminotransferase}}{\overset{\text{serine-glyoxylate}}{\rightleftharpoons}} \text{serine + glyoxylate}$$

$$\text{Hydroxypyruvate + alanine} \underset{\text{aminotransferase}}{\overset{\text{serine-pyruvate}}{\rightleftharpoons}} \text{serine + pyruvate}$$

Glycine is formed from carbon dioxide and ammonia by the action of *glycine synthase* (page 567), a pyridoxal phosphate enzyme, which catalyzes the reversible reaction:

$$CO_2 + NH_3 + N^5,N^{10}\text{-methylenetetrahydrofolate} + NADH + H^+ \rightleftharpoons$$
$$\text{glycine + tetrahydrofolate} + NAD^+$$

This appears to be the major pathway in the liver of vertebrates. Glycine may also be formed from L-serine by the action of *serine hydroxymethyltransferase*, which catalyzes transfer of the β carbon atom of serine to tetrahydrofolate to yield N^5,N^{10}-methylenetetrahydrofolate (page 568):

$$\text{L-Serine + tetrahydrofolate} \rightleftharpoons$$
$$\text{glycine} + N^5,N^{10}\text{-methylenetetrahydrofolate}$$

Since this reaction is reversible, it also provides a pathway for the biosynthesis of serine. Pyridoxal phosphate is the prosthetic group of the enzyme. The structure and various functions of the folic acid coenzymes are described elsewhere (page 345).

Cysteine

Cysteine is not an essential amino acid, but it arises in mammals from methionine, which is essential, and serine, which is not. In this pathway the sulfur atom of methionine is transferred to replace the hydroxyl oxygen atom of serine, thus converting serine into cysteine. This process is often called *transsulfuration*. In the first step of this sequence (Figure 25-4), methionine loses the methyl group from its sulfur atom to become homocysteine. This conversion, which takes place in three or more steps, first requires ATP to convert methionine into an activated form, *S-adenosylmethionine*, in a very unusual reaction first described by G. Cantoni and his colleagues, in which the adenosyl group of ATP is transferred to methionine:

$$\text{L-Methionine} + \text{ATP} \longrightarrow \text{S-adenosylmethionine} + PP_i + P_i$$

S-Adenosylmethionine is an important biological methylating agent. Its methyl group, which is attached in a sulfonium linkage having high-energy characteristics, may be donated to any of a large number of different methyl-group acceptors in the presence of the appropriate enzyme, leaving _S-adenosylhomocysteine_ as the demethylated product (pages 577 and 672). S-Adenosylhomocysteine then undergoes hydrolysis to yield free homocysteine.

In the second stage of cysteine synthesis, homocysteine reacts with serine in a reaction catalyzed by _cystathionine β-synthase_ to yield _cystathionine_ (Figure 25-4):

$$\text{Homocysteine} + \text{serine} \longrightarrow \text{cystathionine} + H_2O$$

In the last step, _cystathionine γ-lyase_, a pyridoxal phosphate enzyme, catalyzes the cleavage of cystathionine to yield free cysteine, with α-ketobutyrate and NH_3 as the other products:

$$\text{Cystathionine} \longrightarrow \alpha\text{-ketobutyrate} + NH_3 + \text{cysteine}$$

Cysteine is an allosteric inhibitor of cystathionine γ-lyase. The overall equation of cysteine synthesis is thus

$$\text{Methionine} + \text{ATP} + \text{methyl-group acceptor} + H_2O + \text{serine} \longrightarrow$$
$$\text{cysteine} + \text{methylated acceptor} + \text{adenosine}$$
$$+ \alpha\text{-ketobutyrate} + NH_3 + PP_i + P_i$$

Cystathionine, the key intermediate in these reactions, has two asymmetric centers and thus occurs in four stereoisomeric forms, L, D, L-allo, and D-allo; only the L form is biologically active. In mammals its only function is to serve as an intermediate in the transfer of sulfur from methionine to form cysteine. Cystathionine is present in rather high concentrations in the brain of man, but lesser amounts are found in lower vertebrates. People with genetic defects in cystathionine synthase have mental deficiencies; in one such genetic disorder the unused homocysteine is excreted as homocystine in the condition known as _homocystinuria_. In another genetic disease involving defective cystathionine γ-lyase, cystathionine is excreted in the urine (_cystathionuria_).

In some microorganisms cysteine is made from serine by a different pathway, catalyzed by _cysteine synthase_, a pyridoxal phosphate enzyme:

$$\text{L-Serine} + H_2S \rightleftharpoons \text{L-cysteine} + H_2O$$

Sources of Organic Sulfur for Cysteine and Methionine

The organic sulfur of cysteine and methionine originates biologically from various inorganic forms, such as SO_4^{2-}, $S_2O_3^{2-}$,

Figure 25-4
The pathway from methionine to cysteine.

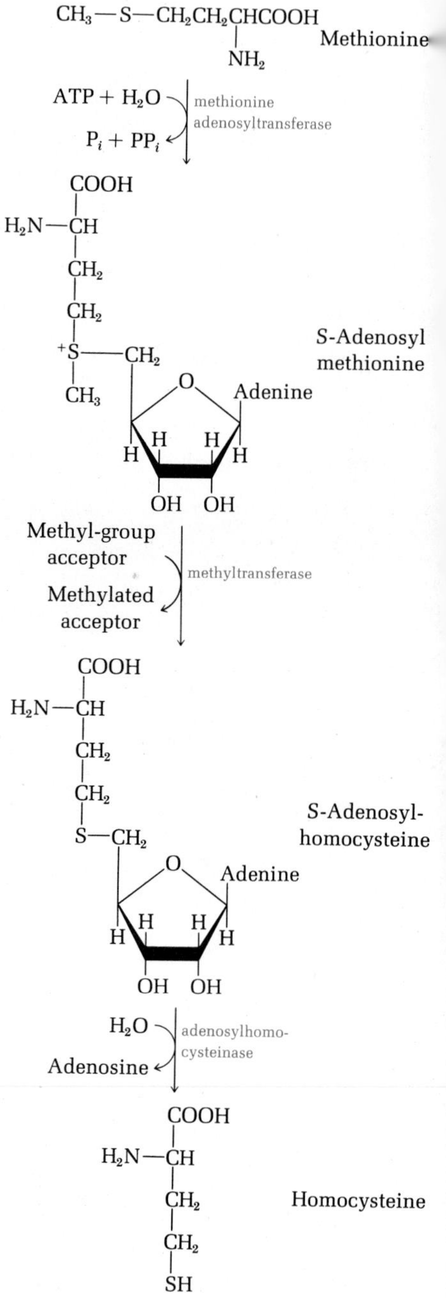

698

H$_2$S, and even elemental sulfur, which some bacteria can oxidize to sulfate. Various microorganisms and plants can bring about the enzymatic reduction of sulfate and thiosulfate to yield H$_2$S, which may then be converted into the thiol group of cysteine by the reactions

$$\text{L-Serine} + \text{acetyl-CoA} \rightleftharpoons \text{O-acetylserine} + \text{CoA}$$

$$\text{O-Acetylserine} + \text{H}_2\text{S} \rightleftharpoons \text{cysteine} + \text{acetate} + \text{H}_2\text{O}$$

catalyzed by *serine acetyltransferase* and *cysteine synthase,* respectively. Yet another reaction for utilization of H$_2$S is

$$\text{Pyruvate} + \text{NH}_3 + \text{H}_2\text{S} \rightleftharpoons \text{cysteine} + \text{H}_2\text{O}$$

catalyzed by the pyridoxal phosphate enzyme *cystathionine γ-lyase* (page 568), which can act on (or form) either cystathionine or cysteine.

Biosynthesis of the Essential Amino Acids

The pathways for the synthesis of the amino acids essential in the nutrition of the rat and man have been largely deduced from biochemical and genetic studies on bacteria. In general, the pathways for this group of amino acids are similar, if not identical, in most species of bacteria and higher plants. Sometimes, however, there are species differences in certain individual steps. The biosynthetic pathways to the essential amino acids are more complex and longer (five to fifteen steps) than those leading to the nonessential amino acids, most of which have fewer than five steps.

Threonine and Methionine

These two essential amino acids have a common denominator; their four-carbon skeletons arise from *homoserine,* a four-carbon analog of serine. The carbon chain of homoserine is in turn derived from aspartic acid in a series of reactions (Figure 25-5) that does not take place in mammals. The reaction pathway for the reduction of the β-carboxyl group of aspartic acid to the aldehyde, which takes place via an acyl phosphate (intermediate), resembles the reduction of 3-phosphoglyceroyl phosphate to glyceraldehyde 3-phosphate catalyzed by glyceraldehyde-3-phosphate dehydrogenase (page 427). Homoserine formed by the two reduction steps is then phosphorylated to *homoserine phosphate,* in an ATP-requiring reaction (Figure 25-5). Homoserine phosphate is now converted into threonine by *threonine synthase,* a pyridoxal phosphate enzyme (Figure 25-6). This complex reaction, which proceeds in several steps, is believed to take place with the α-amino group of the substrate linked as a Schiff's base to the aldehyde group of the pyridoxal phosphate of the enzyme; in this complex the α hydrogen atom is labile (Figure 25-6). Threonine, the end product of the sequence, is an inhibitory modulator of one of the three isozyme forms of *aspartate kinase* (Figure 25-5).

Homocysteine

Serine

cystathionine
β-synthase

H$_2$O

COOH
|
H$_2$N—C—H
|
CH$_2$
|
CH$_2$
|
S   L-Cystathionine
|
CH$_2$
|
H—C—NH$_2$
|
COOH

H$_2$O

cystathionine γ-lyase

NH$_3$

α-Keto-
butyrate

SH
|
CH$_2$ Cysteine
|
HCNH$_2$
|
COOH

Figure 25-5
Biosynthesis of threonine from aspartate.

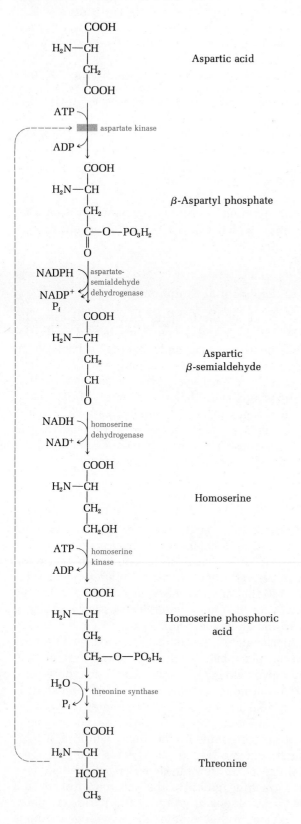

Figure 25-6
The threonine synthase reaction. The enzyme and its prosthetic group pyridoxal phosphate are in color.

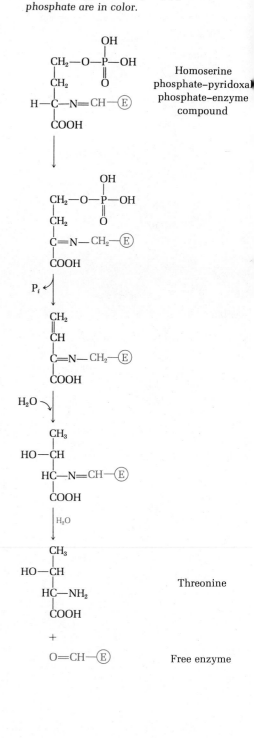

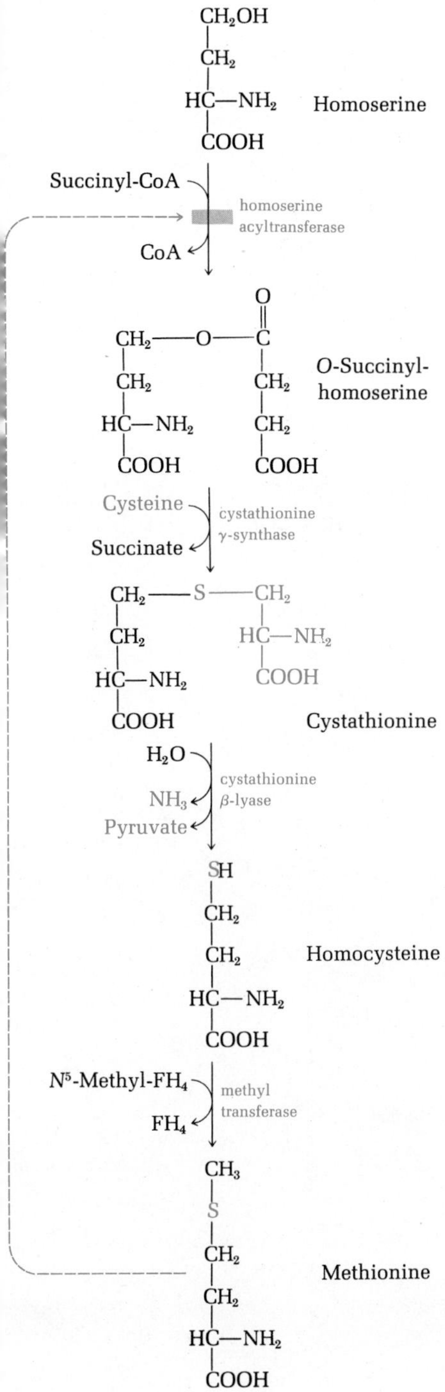

An alternate pathway to threonine is provided by the pyridoxal phosphate enzyme *serine hydroxymethyltransferase* (page 574), which can catalyze the reaction

$$\text{Acetaldehyde} + \text{glycine} \longrightarrow \text{L-threonine}$$

The conversion of homoserine into *methionine* (Figure 25-7) begins with the enzymatic formation of O-succinylhomoserine by transfer of the succinyl group of succinyl-CoA to homoserine. In the next reaction cysteine displaces succinate from O-succinylhomoserine, resulting in the formation of cystathionine. Cystathionine in turn is cleaved hydrolytically to yield homocysteine, pyruvic acid, and NH_3 by the action of *cystathionine β-lyase*. [Recall that cystathionine can also undergo cleavage on the other side of the sulfur atom to yield cysteine by the action of cystathionine γ-lyase (page 698).] Since homocysteine can be methylated to form methionine, it is apparent that cystathionine can serve as an intermediate in the conversion of cysteine to methionine in plants and bacteria and in the conversion of methionine to cysteine in mammals.

The methylation of homocysteine to methionine in *E. coli* takes place by transfer of the methyl group from N^5-methyltetrahydrofolate. This is probably the only reaction in which N^5-methyltetrahydrofolate donates a methyl group. Under some special circumstances methylcobalamin, a form of vitamin B_{12} in which the methyl carbon is directly bonded to the cobalt (page 348), is required as an intermediate in methyl-group transfer. Other direct methyl-group donors for methionine biosynthesis are *betaine* and *dimethylthetin* (Figure 25-8):

$$\text{Betaine} + \text{homocysteine} \xrightarrow{\substack{\text{betaine-homocysteine} \\ \text{methyltransferase}}}$$
$$\text{dimethylglycine} + \text{L-methionine}$$

$$\text{Dimethylthetin} + \text{homocysteine} \xrightarrow{\substack{\text{dimethylthetin-homocysteine} \\ \text{methyltransferase}}}$$
$$\text{S-methylthioglycolate} + \text{L-methionine}$$

Lysine

There are two major routes of synthesis of lysine, one proceeding via *diaminopimelic acid*, which is the major route in

Figure 25-7
Conversion of homoserine into methionine. The sulfur atom of methionine arises from cysteine and the carbon chain from homoserine. The atoms contributed by cysteine are in color.

Figure 25-8
Other methyl-group donors in methionine biosynthesis.

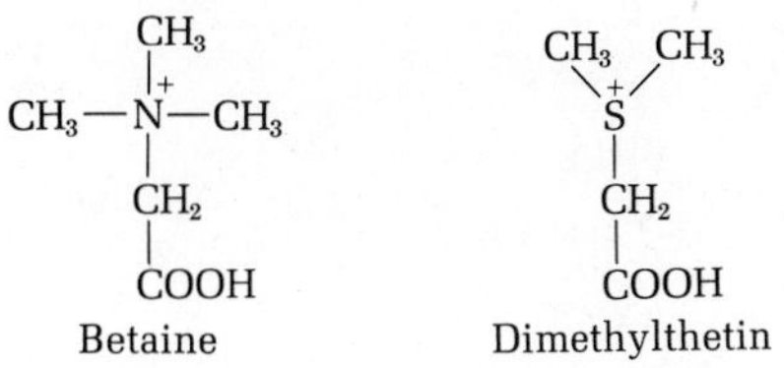

Figure 25-9
Two pathways leading to lysine (below and opposite).

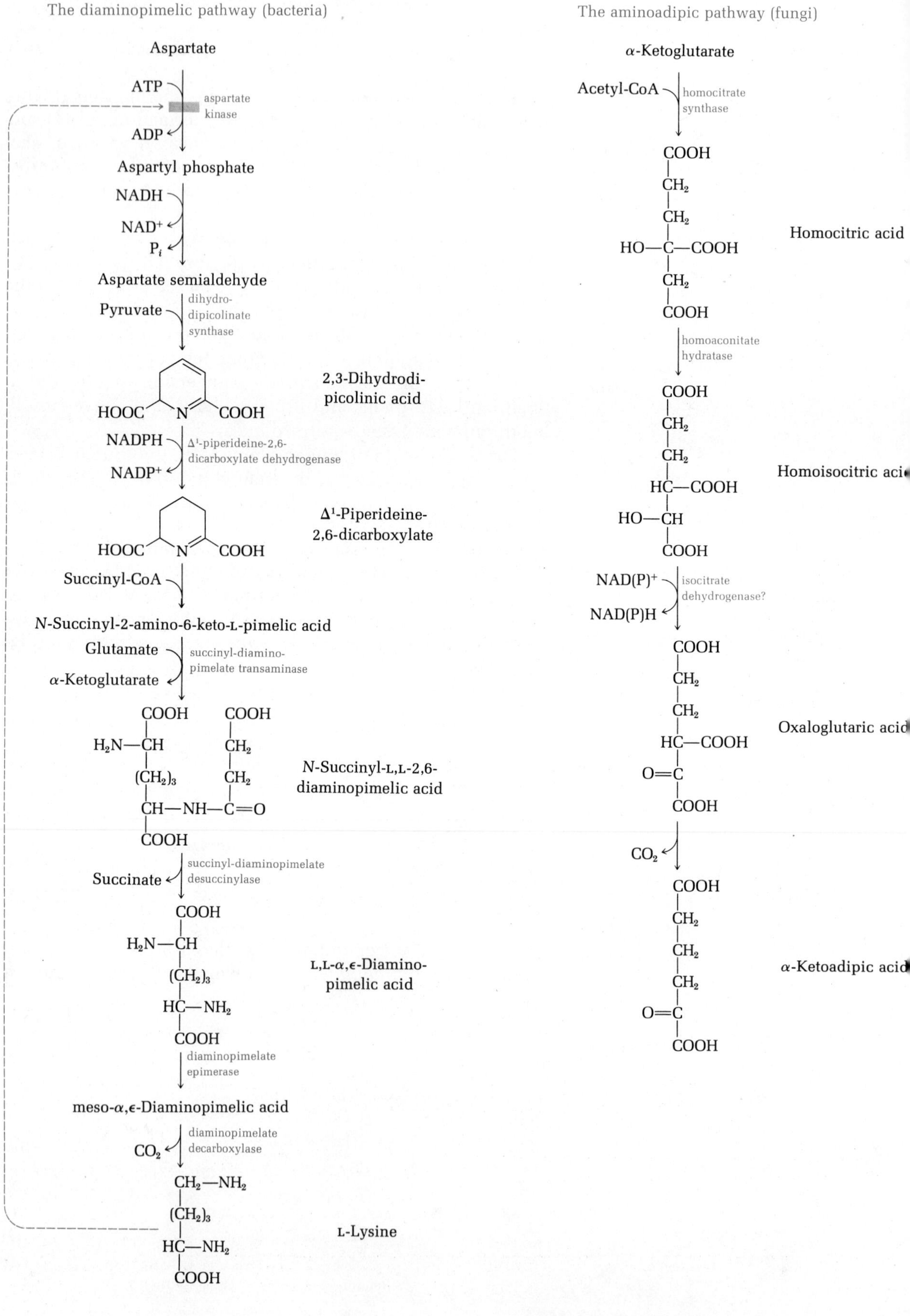

The diaminopimelic pathway (bacteria)

Aspartate

ATP ⟍
aspartate kinase
ADP ⟋

Aspartyl phosphate

NADH ⟍
NAD⁺ ⟵
P_i ⟋

Aspartate semialdehyde

Pyruvate ⟍ dihydro-dipicolinate synthase

2,3-Dihydrodi-picolinic acid

HOOC⁀N⁀COOH

NADPH ⟍ Δ¹-piperideine-2,6-dicarboxylate dehydrogenase
NADP⁺ ⟋

Δ¹-Piperideine-2,6-dicarboxylate

HOOC⁀N⁀COOH

Succinyl-CoA ⟍

N-Succinyl-2-amino-6-keto-L-pimelic acid

Glutamate ⟍ succinyl-diamino-pimelate transaminase
α-Ketoglutarate ⟋

N-Succinyl-L,L-2,6-diaminopimelic acid

COOH   COOH
| |
H₂N—CH CH₂
| |
(CH₂)₃ CH₂
| |
CH—NH—C=O
|
COOH

Succinate ⟍ succinyl-diaminopimelate desuccinylase

L,L-α,ε-Diamino-pimelic acid

COOH
|
H₂N—CH
|
(CH₂)₃
|
HC—NH₂
|
COOH

diaminopimelate epimerase

meso-α,ε-Diaminopimelic acid

CO₂ ⟍ diaminopimelate decarboxylase

L-Lysine

CH₂—NH₂
|
(CH₂)₃
|
HC—NH₂
|
COOH

The aminoadipic pathway (fungi)

α-Ketoglutarate

Acetyl-CoA ⟍ homocitrate synthase

COOH
|
CH₂
|
CH₂
|
HO—C—COOH
|
CH₂
|
COOH

Homocitric acid

homoaconitate hydratase

COOH
|
CH₂
|
CH₂
|
HC—COOH
|
HO—CH
|
COOH

Homoisocitric acid

NAD(P)⁺ ⟍ isocitrate dehydrogenase?
NAD(P)H ⟋

COOH
|
CH₂
|
CH₂
|
HC—COOH
|
O=C
|
COOH

Oxaloglutaric acid

CO₂ ⟍

COOH
|
CH₂
|
CH₂
|
CH₂
|
O=C
|
COOH

α-Ketoadipic acid

702

α-Ketoadipic acid

Glutamate ⟍
         ⟩ aminoadipate
α-Ketoglutarate ⟋ transaminase

COOH
|
CH₂
|
CH₂   α-Aminoadipic
| acid
CH₂
|
CHNH₂
|
COOH

ATP ⟍
NADPH ⟍ aminoadipate
         ⟩ semialdehyde
NADP⁺ ⟋ dehydrogenase
P$_i$ + ADP ⟋

CHO
|
CH₂
|
CH₂   α-Aminoadipic
| semialdehyde
CH₂
|
CHNH₂
|
COOH

Glutamate ⟍
NADPH ⟍ saccharopine
         ⟩ dehydrogenase
NADP⁺ ⟋ (glutamate-
H₂O ⟋   forming)

        H
        |
H₂C—NH—C—COOH
| |
CH₂   CH₂
| |
CH₂   CH₂   Saccharopine
| |
CH₂   COOH
|
HCNH₂
|
COOH

NADP⁺ ⟍
         ⟩ saccharopine
NADPH ⟋ dehydrogenase
α-Ketoglutarate ⟋ (lysine-forming)

CH₂—NH₂
|
CH₂
|
CH₂   L-Lysine
|
CH₂
|
HC—NH₂
|
COOH

bacteria and higher plants, and the other via α-*aminoadipic acid,* the route in most fungi (Figure 25-9).

The diaminopimelic route begins with aspartic semialdehyde and pyruvate, which undergo an aldol condensation and lose water to yield a cyclic intermediate, 2,3-*dihydrodipicolinic acid.* At a later stage, L,L-α,ε-*diaminopimelic acid* is formed, which is converted to its meso form and then decarboxylated to yield L-lysine. Lysine biosynthesis by this pathway is subject to feedback inhibition of one of the isozymes of aspartate kinase. Recall that aspartyl phosphate is also a precursor of threonine and that threonine inhibits *its* isozyme of aspartate kinase. The synthesis of the entire set of enzymes leading from aspartate semialdehyde to lysine is coordinately repressed by lysine.

The *aminoadipic acid pathway* begins with acetyl-CoA and α-ketoglutarate and proceeds via *homoisocitric acid* to α-*ketoadipic acid,* in reactions analogous to the tricarboxylic acid cycle reactions from citric to α-ketoglutaric acids (page 454). The α-ketoadipic acid is aminated to α-aminoadipic acid, which is ultimately converted into L-lysine (Figure 25-9).

Valine, Isoleucine, and Leucine

These three amino acids, which have branched aliphatic R groups, are made by similar pathways. The pathways to valine and isoleucine, which are catalyzed by the same set of enzymes, begin with the formation of the α-keto acids pyruvate and α-ketobutyrate, respectively, to which an active acetaldehyde group, derived from pyruvate in the form of α-hydroxyethylthiamin pyrophosphate (page 338), is added. The products are the corresponding α-aceto-α-hydroxy acids (Figure 25-10). These undergo reduction with simultaneous migration of a methyl or ethyl group, a reaction formally similar to the ketol rearrangement. The products are then dehydrated to yield the α-keto analogs of isoleucine and valine, which are aminated by transaminases (Figure 25-10).

The formation of leucine (Figure 25-11) begins by condensation of α-ketoisovaleric acid (which is also the precursor of valine) with acetyl-CoA to yield α-*isopropylmalic acid.* The subsequent steps are similar to those leading from citric acid to α-ketoglutaric acid in the tricarboxylic acid cycle (page 454).

In bacteria the biosynthesis of valine, isoleucine, and leucine is subject to feedback inhibition at the first step by the end products.

Ornithine and Arginine

Ornithine (page 77) is the precursor of arginine in many species. Although ornithine can be converted into arginine by mammals during the operation of the Krebs urea cycle (page 579), arginine is so rapidly broken down by arginase to form urea and ornithine that insufficient amounts are available for protein synthesis. For this reason arginine is an essential amino acid in mammals. Ornithine is formed in bac-

Figure 25-10
Pathways to valine and isoleucine.

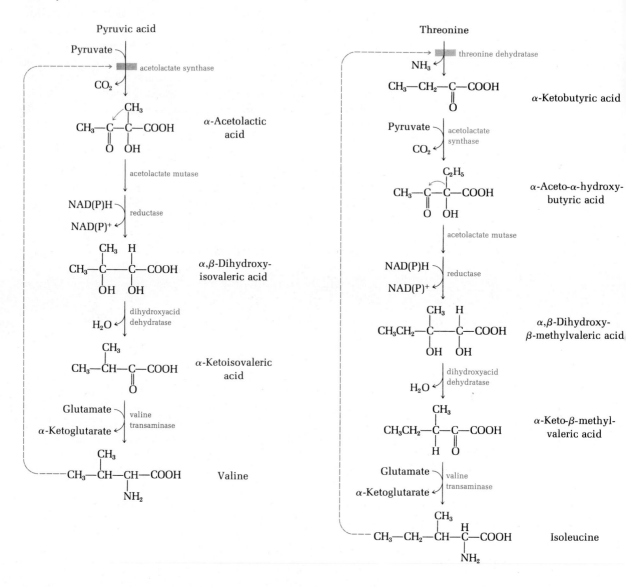

teria and plants from glutamic acid by two general routes; the major pathway is that shown in Figure 25-12.

The final step in this route depends on the species. In *E. coli*, N-acetylornithine is hydrolyzed to yield ornithine and free acetic acid, whereas in other microorganisms and plants, N-acetylornithine donates its acetyl group to glutamic acid to yield free ornithine and ultimately, N-acetylglutamic γ-semialdehyde. The N-acetyl group appears to prevent glutamic acid semialdehyde from spontaneous cyclization. This reaction sequence constitutes the *N-acetylornithine cycle*. In some organisms ornithine is made by transamination from different amino acids to L-glutamate γ-semialdehyde:

$$\text{Glutamate semialdehyde} + \text{aspartate} \xrightleftharpoons[\text{transaminase}]{\text{ornithine}} \text{ornithine} + \text{oxaloacetate}$$

Figure 25-11
The biosynthetic pathway to leucine.

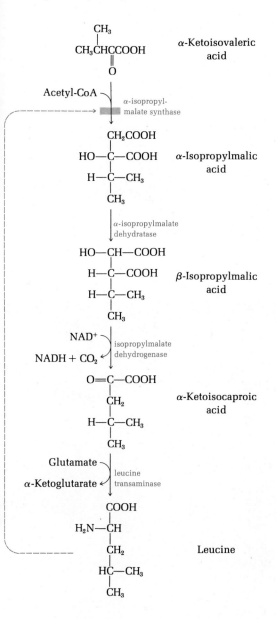

Figure 25-12
The pathway to ornithine and arginine. The conversion of N-acetylornithine to ornithine may occur by two different reactions. Displacement of the N-acetyl group with glutamic acid yields N-acetylglutamic acid and institutes the N-acetylornithine cycle. The first step in ornithine synthesis is inhibited by the final end product arginine (see text).

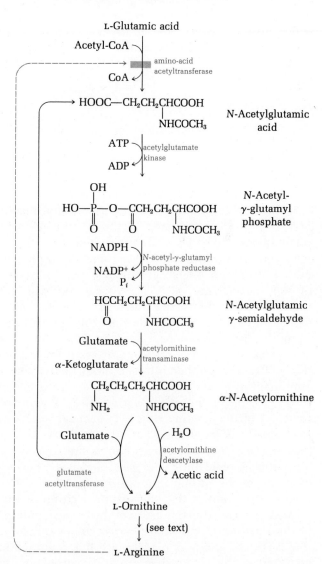

The ornithine formed by these pathways is converted into arginine by the following reactions of the urea cycle, described more fully in Chapter 21 (page 581):

Ornithine + carbamoyl phosphate $\longrightarrow$ citrulline + P_i

Citrulline + aspartate + ATP $\longrightarrow$ argininosuccinate + AMP + PP_i

Argininosuccinate $\longrightarrow$ arginine + fumarate

Since arginase is lacking in many bacteria, this pathway yields net synthesis of arginine.

Histidine

The pathway of biosynthesis of histidine, a difficult biochemical problem, was solved in an outstanding series of investigations by B. Ames and others, in which mutants of *Salmonella typhimurium* and *E. coli* were employed. The pathway of histidine formation is given in Figure 25-13; it contains a number of unusual and complex reactions. The first step is extraordinary; 5-phosphoribosyl 1-pyrophosphate reacts with ATP in such a manner that the pyrophosphate group of the former is lost and the 5-phosphoribosyl moiety forms an N-glycosyl linkage with nitrogen atom 1 of the purine ring of ATP. As can be seen by tracing through the reactions, the three-carbon side chain and two carbon atoms of the imidazole ring of histidine arise from the 5-phosphoribosyl moiety. One of the —N=C— structures of the imidazole ring arises from the adenine of the ATP and the other nitrogen atom of the ring from the amide nitrogen of glutamine, after an unusual enzymatic fragmentation of the adenine ring of ATP. The remaining fragment of the adenine ring is in fact salvaged and reused as a precursor of purines, as we shall see in Chapter 26 (page 731); evidently no by-products are wasted during the biosynthesis of histidine. In fact, the biosynthesis of histidine and of the purine ring are linked (page 731). Another noteworthy point is that the carboxyl group of histidine is formed by a two-step oxidation of the corresponding α-amino alcohol *histidinol,* apparently by a single enzyme, whereas the carboxyl group of nearly all the other amino acids arises from the carboxyl group of a corresponding α-keto acid.

The first step in the reaction sequence leading to histidine is catalyzed by an allosteric enzyme, *ATP phosphoribosyltransferase,* which is inhibited by histidine, the end product of the sequence. Moreover, the entire enzyme system undergoes *coordinate repression* (pages 383 and 980); in the presence of excess histidine in the culture medium the biosynthesis of the entire sequence of histidine-forming enzymes is suppressed. We shall see later that genetic studies of the *his* operon, the segment of DNA coding this group of nine enzymes, have provided extremely important information about the regulation of the biosynthesis of enzyme systems (page 984).

Figure 25-13
The biosynthetic pathway to histidine (right). ⓟ *denotes the phosphate group. Atoms in color arose from 5-phosphoribosyl 1-pyrophosphate.*

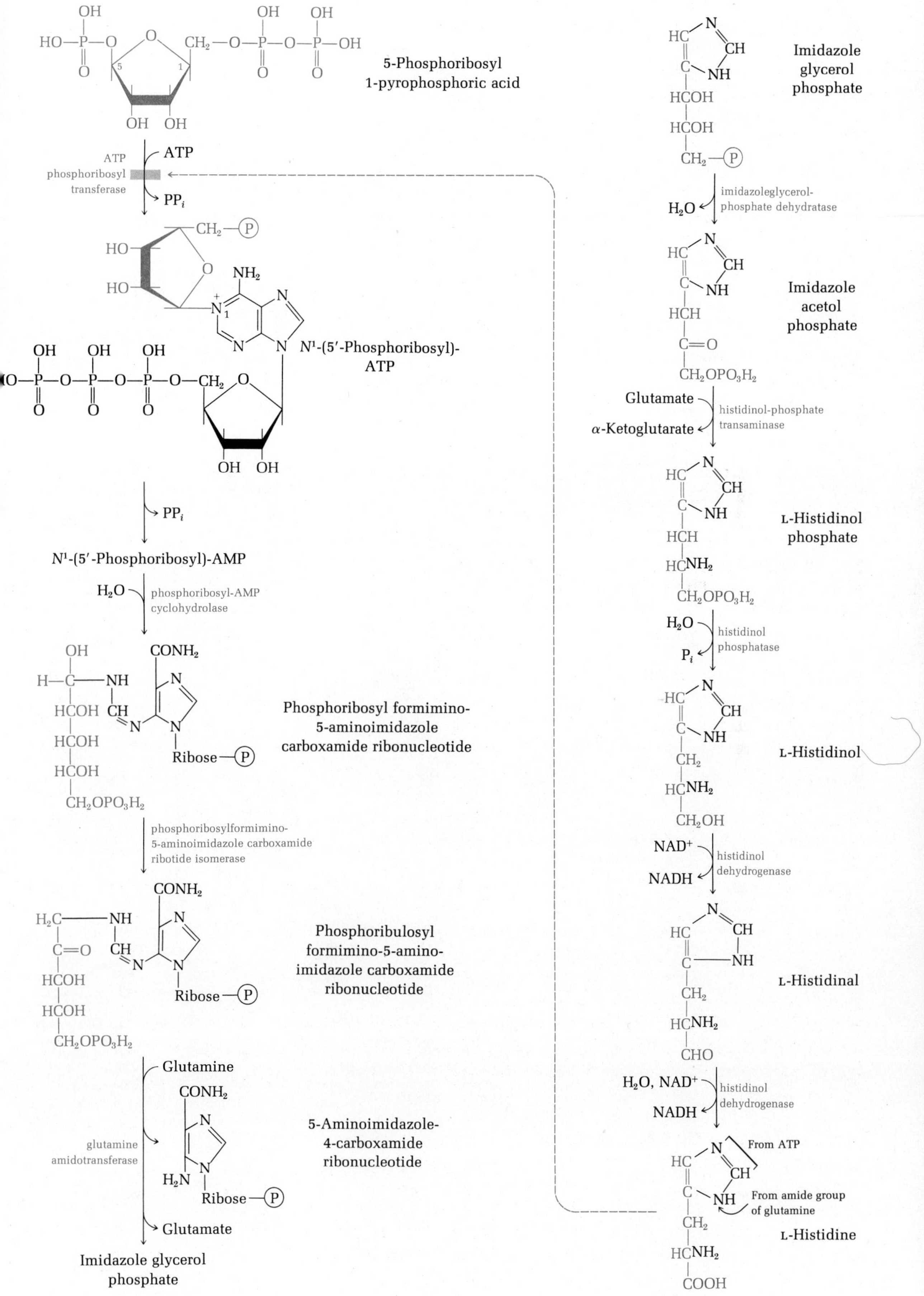

5-Phosphoribosyl
1-pyrophosphoric acid

ATP
phosphoribosyl
transferase

ATP

PP$_i$

N^1-(5'-Phosphoribosyl)-
ATP

PP$_i$

N^1-(5'-Phosphoribosyl)-AMP

H$_2$O

phosphoribosyl-AMP
cyclohydrolase

CONH$_2$

Ribose—(P)

Phosphoribosyl formimino-
5-aminoimidazole
carboxamide ribonucleotide

phosphoribosylformimino-
5-aminoimidazole carboxamide
ribotide isomerase

CONH$_2$

Ribose—(P)

Phosphoribulosyl
formimino-5-amino-
imidazole carboxamide
ribonucleotide

Glutamine

CONH$_2$

glutamine
amidotransferase

H$_2$N

Ribose—(P)

5-Aminoimidazole-
4-carboxamide
ribonucleotide

Glutamate

Imidazole glycerol
phosphate

Imidazole
glycerol
phosphate

H$_2$O

imidazoleglycerol-
phosphate dehydratase

Imidazole
acetol
phosphate

CH$_2$OPO$_3$H$_2$

Glutamate

α-Ketoglutarate

histidinol-phosphate
transaminase

L-Histidinol
phosphate

CH$_2$OPO$_3$H$_2$

H$_2$O

P$_i$

histidinol
phosphatase

L-Histidinol

CH$_2$OH

NAD$^+$

NADH

histidinol
dehydrogenase

L-Histidinal

CHO

H$_2$O, NAD$^+$

NADH

histidinol
dehydrogenase

From ATP

From amide group
of glutamine

L-Histidine

COOH

Figure 25-14
The biosynthetic pathway to tyrosine, phenylalanine, and tryptophan.

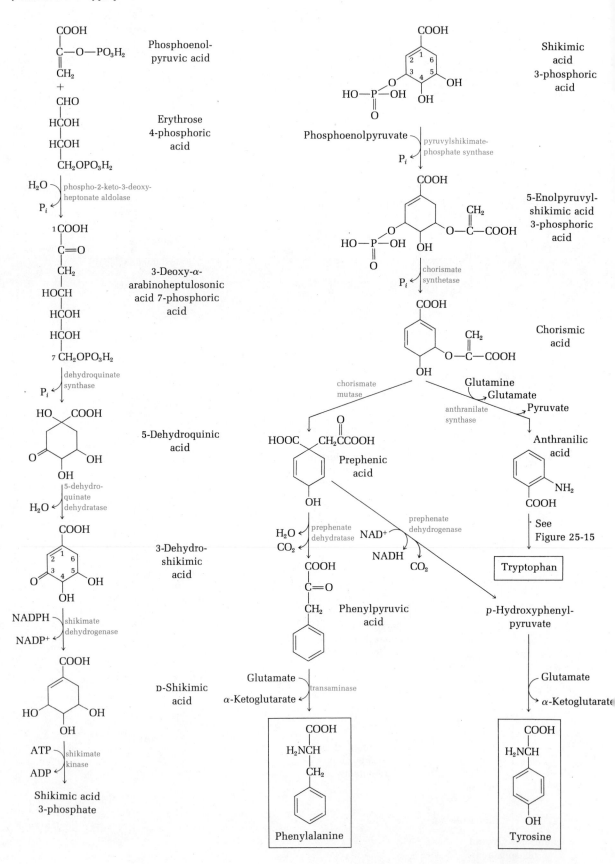

Biosynthesis of the Aromatic Amino Acids

Phenylalanine and Tryptophan

The most noteworthy aspect of the biosynthesis of these amino acids is the mechanism by which their aromatic rings are formed from purely aliphatic precursors. The pathway of these reactions (Figure 25-14) was deduced from experiments on auxotrophic mutants (page 377) of *E. coli* and *Aerobacter aerogenes* that required phenylalanine, tyrosine, and tryptophan for growth. B. D. Davis made the important observation that the compound *shikimic acid* (Figure 25-14), a hydroaromatic acid abundant in some higher plants, can replace the aromatic amino acids in supporting growth of these mutants. Studies to determine the capacity of compounds related to shikimic acid to support the growth of these mutants, as well as isotopic experiments carried out by D. B. Sprinson, ultimately revealed the pathway of biosynthesis of the aromatic amino acids. The mechanism of aromatization employed in this pathway, in which shikimic acid is a key intermediate, is of the widest biological significance, since it extends well beyond the formation of the aromatic amino acids. For example, enormous quantities of *lignin*, a polymerized aromatic substance making up a substantial portion of the woody portions of plant tissues, are synthesized via shikimic acid, as are many other aromatic biomolecules such as ubiquinone (page 493) and plastoquinone (page 607).

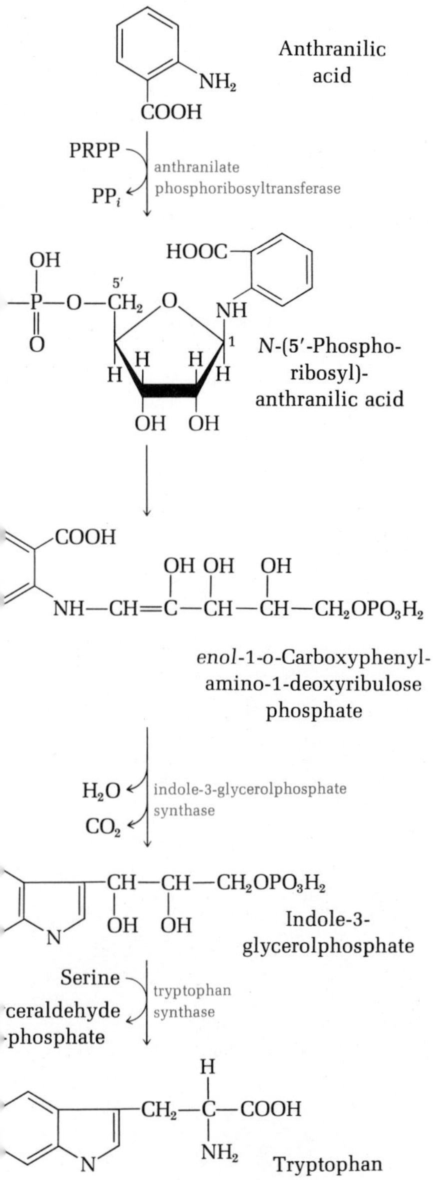

In the pathway to shikimic acid, a four-carbon sugar phosphate, D-erythrose 4-phosphate, reacts with phosphoenolpyruvate to yield a phosphorylated seven-carbon keto sugar acid; it in turn cyclizes to 5-*dehydroquinic acid*, which has a six-carbon aliphatic ring. This intermediate is then converted into shikimic acid, which leads via phosphorylated intermediates to *chorismic acid* (Greek, "fork"), at which there is an important metabolic branch point. One branch leads to *anthranilic acid* and thence to tryptophan, whereas the other leads to the quinonoid compound *prephenic acid*, the last nonaromatic compound in the sequence. Prephenic acid can be aromatized in two ways: (1) by dehydration and simultaneous decarboxylation to yield phenylpyruvic acid, the precursor of phenylalanine, and (2) by dehydrogenation and decarboxylation to yield p-hydroxyphenylpyruvic acid, the precursor of tyrosine. Some of the enzymes catalyzing this sequence have not yet been well characterized.

Details of the formation of tryptophan from anthranilic acid are shown in Figure 25-15. The terminal step in this sequence is catalyzed by *tryptophan synthase*, an enzyme that has received considerable attention, not only because it catalyzes an interesting reaction but especially because its enzymatically active subunits have been utilized as elements in the experimental proof of one of the basic principles of molecular genetics, namely, the *colinearity* of the nucleotide triplet sequences in DNA and the amino acid sequence of its gene product (page 69). Tryptophan synthase is a pyridoxal phosphate enzyme having a molecular weight of about

Figure 25-15
Details of the pathway from anthranilic acid to tryptophan. PRPP denotes phosphoribosyl pyrophosphate.

135,000; it has been isolated in crystalline form. It catalyzes the overall reaction

Indole-3-glycerolphosphate + serine $\longrightarrow$
$$\text{tryptophan + glyceraldehyde 3-phosphate}$$

which takes place in two steps, with the intermediate indole remaining bound to the active site:

Indole-3-glycerolphosphate $\longrightarrow$
$$[\text{indole}] + \text{glyceraldehyde 3-phosphate} \quad (1)$$

$$[\text{Indole}] + \text{serine} \longrightarrow \text{tryptophan} + H_2O \quad\quad\quad\quad (2)$$

The enzyme of *E. coli* contains four polypeptide chains, made up of two α chains and two β chains, which can be separated. The α subunit alone can catalyze reaction (1), but at a low rate, and the β chains, associated as the dimer β_2, can catalyze reaction (2), also at a low rate. When the α and β chains are mixed, both reaction rates are greatly increased. The amino acid sequence of both the α and β chains, as well as of many of their mutants, has been determined. Tryptophan synthase of *Neurospora* differs in some respects from the *E. coli* enzyme.

The allosteric regulation of the biosynthesis of the aromatic amino acids is described below.

Regulation of Amino Acid Biosynthesis

Two general mechanisms for regulation of amino acid biosynthesis have been noted in the preceding discussion: (1) *allosteric,* or *feedback, inhibition* (page 234) and (2) *repression* of the biosynthesis of one or more of the enzymes of the pathway, thus lowering their concentration in the cell. *Feedback inhibition* of the first reaction in the biosynthetic sequence by its end product yields fine control over biosynthesis, since it is capable of second-by-second adjustment of the rate of biosynthesis of the amino acid to the steady-state level of the biosynthetic end product. Allosteric control of amino acid biosynthesis is especially conspicuous in bacteria because of their basic need to conserve nitrogen sources. In mammalian tissues allosteric regulation of the biosynthesis of amino acids does not appear to be quite so conspicuous.

The second type of control mechanism, repression of enzyme synthesis, is often spoken of as yielding coarse control. Such control is afforded by changes in the rate of DNA transcription or of messenger RNA translation, thus changing the rate of synthesis of the enzymes. Repression and derepression of enzyme synthesis are slower to respond to changing metabolic conditions than feedback inhibition of regulatory enzymes. Repression of enzyme synthesis has another purpose: to exert maximum economy in the use of amino acids and energy by preventing the synthesis of unused enzymes. Since in many biosynthetic pathways the entire multienzyme system can be repressed in coordinate fashion, the end result is a large saving in all amino acids.

Figure 25-16
Feedback controls in the regulation of the biosynthesis of the aromatic amino acids. The first step, which is shared in the synthesis of all three amino acids, is catalyzed by three different isozymes of phospho-2-keto-3-deoxyheptonate aldolase (Figure 25-14), each inhibited by one specific end product. In this way the metabolic flow through this step can be adjusted if only one or two of the aromatic amino acids must be made. Each of these end products also causes repression of the formation of its corresponding isozyme. Two regulatory isozymes also are involved in the conversion of chorismate to prephenate, each specific for an end product of this branching point.

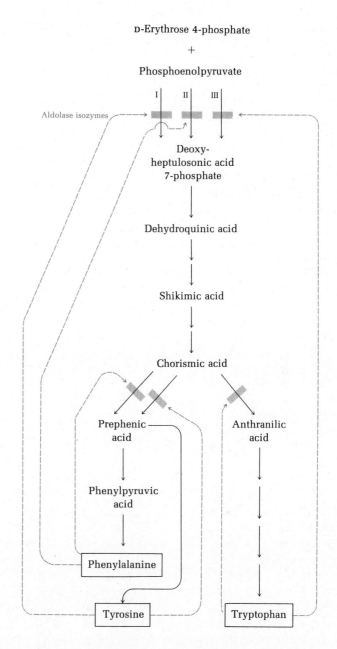

Most of the biosynthetic pathways leading to formation of amino acids in bacteria are regulated by a combination of allosteric and repression mechanisms, which are often complex and which vary from one species to another. When biosynthetic pathways involve branches leading to two (or more) different amino acids, the feedback and repression mechanisms become rather complex. For example, the two end products may show *concerted inhibition* of the first enzyme in the sequence, so that separately they have no effect, but together are very inhibitory. In other cases they may show *cumulative inhibition*, in which each end product inhibits by a given percentage, which is additive. Another type of control is given by a regulatory enzyme that occurs in multiple forms, i.e., isozymes (page 244). Figure 25-16 shows that in *E. coli* the first enzyme in the biosynthesis of the three aromatic amino acids, namely, the aldolase responsible for condensation of erythrose 4-phosphate and phosphoenol-

Figure 25-17
Multivalent allosteric inhibition of glutamine synthetase. Although glycine and
alanine are not direct products of glutamine
metabolism, they are very strong inhibitors,
suggesting that the steady-state levels of glycine and alanine in the cell are critically
related to glutamine synthesis and metabolism.

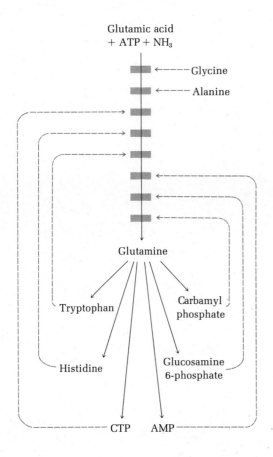

pyruvate, occurs in the form of three different isozymes, each
of which is inhibited by one of the three end products,
phenylalanine, tyrosine, or tryptophan, the latter being the
least effective inhibitor. In this way the flow through this
enzymatic step can be adjusted if only one or two of the
aromatic amino acids needs to be made. Moreover, the
biosynthesis of each of the three aldolase isozymes is also
specifically repressed by its corresponding end-product inhibitor. The isoenzymes catalyzing the conversion of
chorismic acid to prephenic acid are also independently inhibited and/or repressed (Figure 25-16). The pattern of
feedback inhibition and repression of a given pathway
varies from one species to another.

   Especially noteworthy is the remarkable set of multivalent
allosteric controls exerted on the activity of glutamine synthetase of E. coli. Glutamine is a precursor or amino-group
donor of many metabolic products (Figure 25-17). As many
as eight products of glutamine metabolism in E. coli are

Figure 25-18
Electron micrographs of glutamine synthetase molecules from E. coli, showing the arrangement of the 12 subunits. (Photograph by the late Robin Valentine, courtesy of E. Stadtman.)

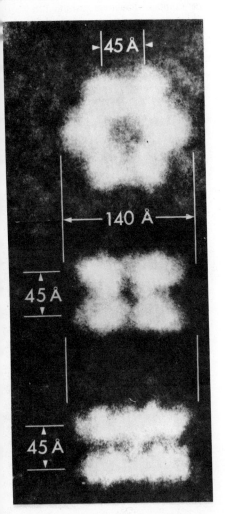

believed to serve as separate and independent negative-feedback inhibitors of the activity of this glutamine synthetase, perhaps the most complex regulatory enzyme now known. Work of E. R. Stadtman and his colleagues has shown that glutamine synthetase of E. coli contains 12 identical subunits, each of molecular weight 50,000, arranged in two hexagonal layers (Figure 25-18). The enzyme occurs in two forms, the active, or free, form and a covalently modified inactive form. The active form of the enzyme is subject to allosteric feedback inhibition by AMP and tryptophan, among other metabolites. Each of the 12 subunits has a binding site for the substrate ATP and binding sites for each of the feedgack inhibitors. Glutamine synthetase may be covalently modified by an enzymatic reaction with ATP to yield the relatively inactive adenylylated form, which contains 12 molecules of adenylic acid linked via their 5'-phosphate groups to the hydroxyl groups of specific tyrosine residues in each of the 12 subunits. The adenylylated form of the enzyme may be deadenylylated by a specific enzyme.

The glutamine synthetases of animal tissues, which are smaller proteins with fewer subunits, do not show the complex allosteric behavior of the enzyme in E. coli. The enzyme of rat brain shows little or no regulatory activity, whereas the liver enzyme is affected by only a few modulators. Since glutamine plays different metabolic roles in different types of cells, the allosteric behavior of glutamine synthetase appears to have evolved in different ways.

Precursor Functions of Amino Acids

In addition to their role as building blocks of proteins, the amino acids are precursors of many other important biomolecules, including various hormones, vitamins, coenzymes, alkaloids, porphyrins, antibiotics, pigments, and neurotransmitter substances. Table 25-1 shows some examples of biomolecules that arise from some of the amino acids. The aromatic amino acids are particularly versatile as precursors; from them are made many alkaloids, such as morphine, codeine, and papaverine, and a number of hormones, such as the thyroid hormone thyroxine, the plant hormone indoleacetic acid, and an adrenal hormone, epinephrine. The following will briefly outline a few of the important pathways from various amino acid precursors to other products.

Biosynthesis of Methylated Compounds

The methyl group of methionine is utilized as precursor for the formation of a large number of methylated biomolecules. Methionine is first activated by reaction with ATP to yield S-adenosylmethionine, which contains the high-energy methylsulfonium group:

$$ATP + \text{L-methionine} \longrightarrow \text{S-adenosylmethionine} + PP_i + P_i$$

Table 25-1 Precursor functions of some amino acids

Arginine	Serine
Spermine	Sphingosine
Spermidine	Tyrosine
Putrescine	Epinephrine
Aspartic acid	Norepinephrine
Pyrimidines	Melanin
Glutamic acid	Thyroxine
Glutathione	Mescaline
Glycine	Tyramine
Purines	Morphine
Glutathione	Codeine
Creatine	Papaverine
Phosphocreatine	Tryptophan
Tetrapyrroles	Nicotinic acid
Histidine	Serotonin
Histamine	Kynurenic acid
Ergothioneine	Indole
Lysine	Skatole
Cadaverine	Indoleacetic acid
Anabasine	Ommochrome
Coniine	Valine
Ornithine	Pantothenic acid
Hyoscyamine	Penicillin

Figure 25-19
The origin and pathway of transfer of methyl groups. Nearly all biological methylation processes in higher animals (and probably most of those in plants and bacteria) involve methionine as the key intermediate. Plants and bacteria manufacture methionine from homocysteine and serine, the major source of the methyl group. S-Adenosylmethionine is the major methyl-group donor to a large number of methyl acceptors. Betaine and dimethylthetin are active donors of methyl groups to homocysteine in some microorganisms.

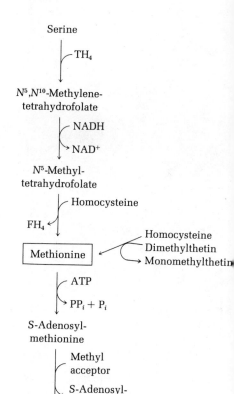

Methylated acceptors:

Creatine
Phosphatidylcholine
Epinephrine
Norepinephrine
Cyclic fatty acids
Ergosterol
Vitamin B_{12}
N-Methylhistamine
N-Methylnicotinamide
Choline
Sarcosine
Anserine

S-Adenosylmethionine is the direct methyl-group donor to some 40 different methyl-group acceptors, of which a partial list is given in Figure 25-19. Such transfers are catalyzed by *methyltransferases*, which yield S-adenosylhomocysteine as the demethylated product.

Although methionine serves as a general methyl-group donor to many methyl-group acceptors, the formation of the methyl group of methionine occurs by only a very few reactions. The major pathway is by transfer of a methyl group from N^5-methyltetrahydrofolate to homocysteine (Figure 25-19). The methyl group of N^5-methyltetrahydrofolate in turn derives from a limited number of metabolites capable of donating one-carbon functional groups to tetrahydrofolate, particularly serine. The sources of one-carbon groups and other aspects of their metabolism are discussed elsewhere (pages 345, 567, and 731).

Peptides

Amino acids are biological precursors of a number of peptides, many of which, such as the hormones oxytocin, vasopressin, and bradykinin (page 96), have intense biological activity.

The simple tripeptide *glutathione* of animal tissues (Figure 25-20), which serves as a component of an amino acid transport system (page 795), an activator of certain enzymes, and in the protection of lipids against autoxidation, is syn-

Figure 25-20
Glutathione (γ-L-glutamyl-L-cys-
teinylglycine). Note that the glutamic acid
residue is joined by a peptide linkage
through the γ- rather than the usual α-carboxyl group.

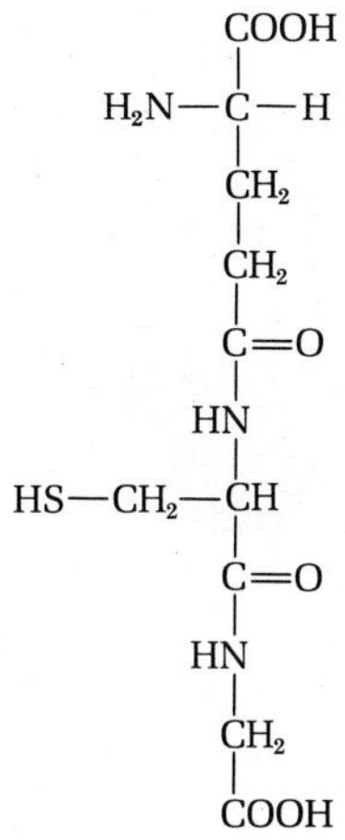

```
          COOH
           |
  H₂N — C — H
           |
          CH₂
           |
          CH₂
           |
          C=O
           |
          HN
           |
HS—CH₂—CH
           |
          C=O
           |
          HN
           |
          CH₂
           |
          COOH
```

thesized in two steps from L-glutamate, L-cysteine, and glycine. One molecule of ATP is broken down to ADP and
phosphate for each peptide bond generated:

$$\text{Glutamate} + \text{cysteine} + \text{ATP} \xrightarrow{\overset{\text{γ-glutamylcysteine}}{\text{synthetase}}}$$

$$\text{γ-glutamylcysteine} + \text{ADP} + \text{P}_i$$

$$\text{γ-Glutamylcysteine} + \text{glycine} + \text{ATP} \xrightarrow{\overset{\text{glutathione}}{\text{synthetase}}}$$

$$\text{glutathione} + \text{ADP} + \text{P}_i$$

In bacteria even rather long peptides, having as many as
20 peptide bonds, are also enzymatically synthesized in a
stepwise manner, without the participation of mRNA templates or ribosomes (page 929). Especially noteworthy is the
enzymatic synthesis by cells of *Bacillus brevis* of the cyclic
peptide antibiotic gramicidin S (Figure 25-21), which has 10
amino acid residues. The steps in its biosynthesis have been
worked out by F. Lipmann and his colleagues. This peptide
contains residues of L-ornithine and D-phenylalanine, which
are not found in proteins (page 71). Its synthesis is brought
about by a soluble complex of two extraordinary enzymes,
together called gramicidin synthetase. Enzyme I (mol wt
280,000) catalyzes the formation from the incoming amino
acid of an enzyme-bound acyl-AMP derivative (Figure
25-22). The aminoacyl group is then transferred to an —SH
group of the enzyme:

$$\text{RCH(NH}_2\text{)COOH} + \text{ATP} \longrightarrow [\text{RCH(NH}_2\text{)CO—AMP}] + \text{PP}_i$$

$$[\text{RCH(NH}_2\text{)CO—AMP}] + \text{HS—}\boxed{\text{E-I}} \longrightarrow$$

$$\text{RCH(NH}_2\text{)CO—S—}\boxed{\text{E-I}} + \text{AMP}$$

The —SH group involved in this reaction is furnished by
4′-phosphopantetheine, the covalently bound prosthetic
group of enzyme I, which resembles the acyl carrier protein
functioning in fatty acid biosynthesis (page 660). Enzyme I
functions in this way to activate, in the correct sequence,
the four L-amino acids required for the synthesis of gramicidin S.

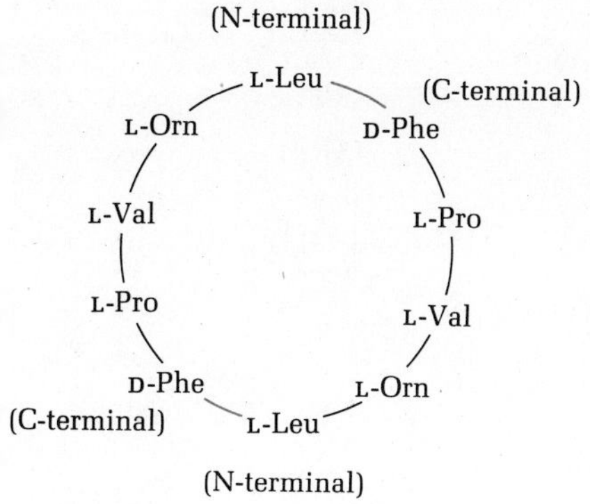

Enzyme II of the gramicidin synthetase complex activates
L-phenylalanine in a similar manner and then promotes the
inversion of the α carbon atom to yield the bound D stereoisomer. The amino acid unit esterified to enzyme I is then
transferred to the free amino group of the D-phenylalanine attached to enzyme II to form a dipeptide. Successive activated
L-amino acid residues are transferred to the growing peptide
chain on enzyme II until the pentapeptidyl derivative Leu–
Orn–Val–Pro–D-Phe–S–E-II results. Two such pentapeptide
chains are then joined to yield the complete cyclic decapeptide shown in Figure 25-21. What is remarkable about this
and similar enzyme complexes which synthesize such peptides is their ability to select the proper amino acid for each
successive step. This property appears to be conferred by in-

Figure 25-21
The structure of gramicidin S. The pathway
of its biosynthesis is described in the text.
First the two pentapeptide chains are built,
starting from D-phenylalanine as the C-terminal residue. The two chains are then
joined at the peptide bonds indicated in
color.

```
                (N-terminal)
                 L-Leu          (C-terminal)
        L-Orn            D-Phe
       /                      \
   L-Val                      L-Pro
     |                          |
   L-Pro                      L-Val
      \                      /
      D-Phe            L-Orn
   (C-terminal)    L-Leu
                (N-terminal)
```

duced conformational changes in the peptidyl–E-II complex at each step, which are transmitted to enzyme I, resulting in alterations in the active site of the latter that render it specific for the next amino acid to be added. Since some peptide antibiotics have considerably longer chains, the enzymes synthesizing them have a most remarkable ability to store and transfer information, presumably conferred by quite specific conformational changes. Gramicidin synthetase carries out 19 different reactions. In animal tissues, however, the synthesis of peptide hormones appears to require the participation of mRNA and ribosomes.

Creatine and Phosphocreatine

As discussed elsewhere, creatine and phosphocreatine play important roles in the storage and transmission of phosphate-bond energy (pages 404 and 767). Creatine (α-methyl-guanidoacetic acid) is synthesized from arginine, glycine, and methionine in the following reaction steps, in which guanidinoacetic acid (Figure 25-23) is an intermediate and S-adenosylmethionine is the methyl-group donor:

$$\text{Arginine} + \text{glycine} \xrightarrow[\text{amidinotransferase}]{\text{glycine}} \text{ornithine} + \text{guanidinoacetic acid}$$

$$\text{Guanidinoacetic acid} + \text{S-adenosylmethionine} \xrightarrow[\text{methyltransferase}]{\text{guanidinoacetate}} \text{creatine} + \text{S-adenosylhomocysteine}$$

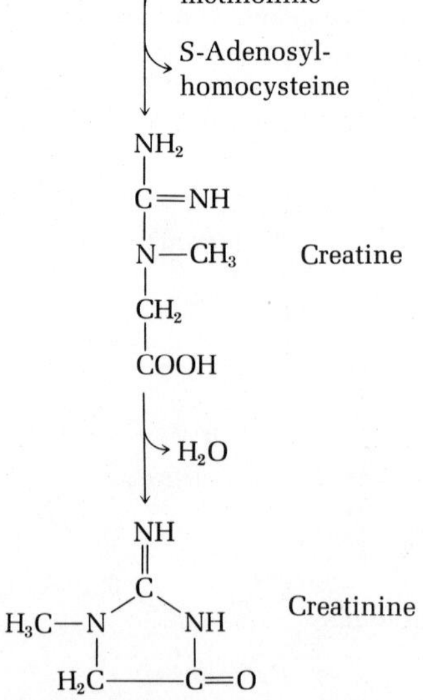

Serotonin and Indoleacetic Acid

Serotonin, a neurohumor and vasoconstrictor in vertebrates, and indoleacetic acid, a plant growth hormone, are formed from tryptophan by the pathways shown in Figure 25-24.

Epinephrine and Norepinephrine

These hormones, secreted by the adrenal medulla, function physiologically in regulation of heart rate and blood pressure. Epinephrine is an activator of glycogen breakdown in liver and muscle via its stimulation of adenylate cyclase (page 813). These hormones are formed from tyrosine in the reactions shown in Figure 25-25. Histamine, an important vasodilator secreted by mast cells of the lungs, liver, and gastric mucosa, is formed from histidine by decarboxylation:

$$\text{Histidine} + \text{H}_2\text{O} \xrightarrow[\text{decarboxylase}]{\text{histidine}} \text{histamine} + \text{CO}_2$$

Polyamines

The polyamines spermine and spermidine (Figure 25-26) are found in all bacteria and most animal cells. They are growth factors for some microorganisms and serve to stabilize membrane structures of bacteria, as well as the structure of ribosomes, some viruses, and the DNA of many organisms. They

Figure 25-22
General structure of aminoacyl adenylic acids (see also page 934).

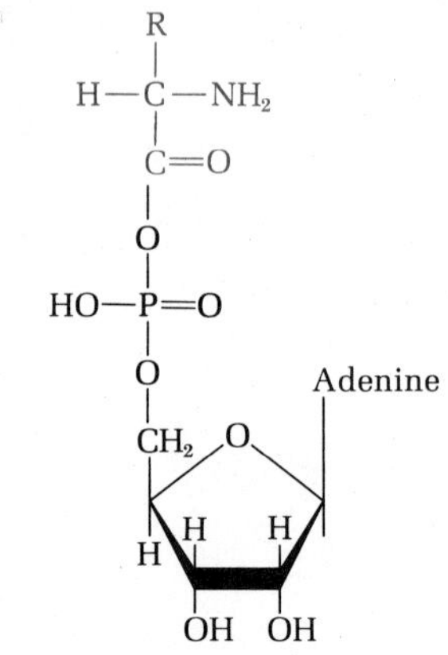

Figure 25-23
Conversion of guanidinoacetic acid to creatine; the methyl group is donated by S-adenosylmethionine (see text). Creatinine, the anhydride of creatine, is excreted in the urine of mammals as a degradation product.

Figure 25-24
Formation of serotonin and indoleacetic acid from tryptophan.

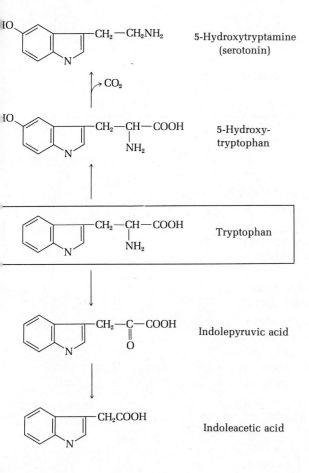

5-Hydroxytryptamine (serotonin)

CO_2

5-Hydroxy-tryptophan

Tryptophan

Indolepyruvic acid

Indoleacetic acid

Figure 25-25
Conversion of L-tyrosine into norepinephrine and epinephrine.

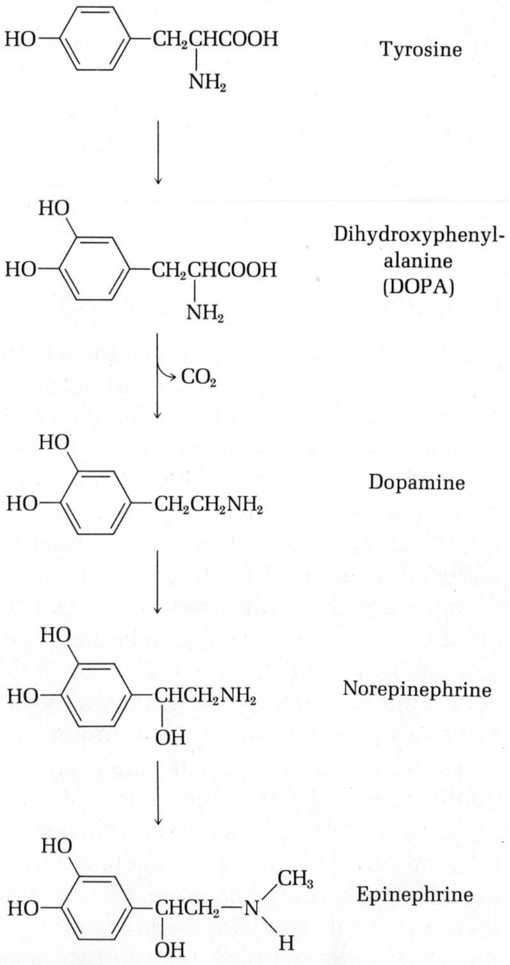

Tyrosine

Dihydroxyphenyl-alanine (DOPA)

CO_2

Dopamine

Norepinephrine

Epinephrine

Figure 25-26
Spermidine and spermine, polyamines derived from arginine and methionine. At cell pH they occur as polycations.

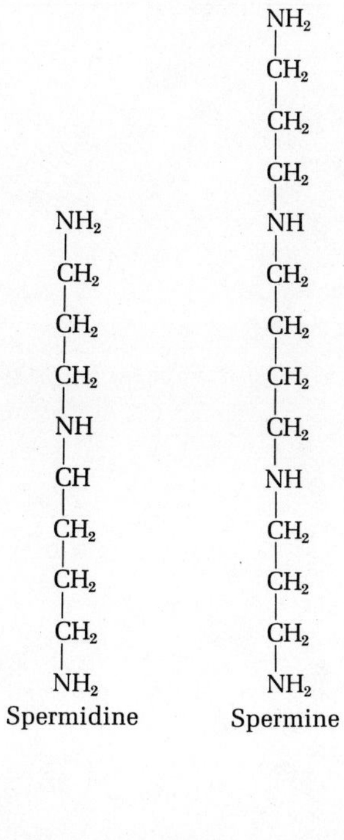

Spermidine

Spermine

are formed from _putrescine_ and S-adenosylmethionine. In mammals putrescine is formed by decarboxylation of ornithine by the action of _ornithine decarboxylase_:

$$\text{Ornithine} \longrightarrow H_2NCH_2CH_2CH_2CH_2NH_2 + CO_2$$
Putrescine

In _E. coli_ putrescine is formed from arginine:

$$\text{Arginine} \xrightarrow[\text{decarboxylase}]{\text{arginine}} H_2N\overset{\overset{\displaystyle NH}{\|}}{C}NH(CH_2)_4NH_2 + CO_2$$
Agmatine

$$\text{Agmatine} + H_2O \xrightarrow{\text{agmatinase}} H_2N(CH_2)_4NH_2 + \text{urea}$$
Putrescine

S-Adenosylmethionine is first decarboxylated before reacting with putrescine to form spermidine:

717

$$\text{S-Adenosylmethionine} \xrightarrow{\substack{\text{S-adenosylmethionine} \\ \text{decarboxylase}}}$$

$$\text{CO}_2 + \text{S-(5'-adenosyl)-3-methylmercaptopropylamine}$$

S-(5'-Adenosyl)-3-methylmercaptopropylamine

$$+ \text{ putrescine} \xrightarrow{\substack{\text{aminopropyl-} \\ \text{transferase}}} \text{5'-methylthioadenosine} + \text{spermidine}$$

Spermidine is converted into spermine by a repetition of the last two reactions.

Porphyrins

The biosynthesis of porphyrins, for which glycine is a major precursor, is an especially important pathway because of the central role of the porphyrin nucleus in hemoglobin, in the cytochromes, and in chlorophyll. The tetrapyrroles are constructed from four molecules of the monopyrrole derivative *porphobilinogen,* which is synthesized in the steps shown in Figure 25-27. The pathway in animal tissues was largely deduced from isotopic-tracer and enzyme studies by D. Shemin and his colleagues. Glycine first reacts with succinyl-CoA to yield enzyme-bound *α-amino-β-ketoadipic acid,* which then decarboxylates to yield *δ-aminolevulinic acid.* This reaction is catalyzed by a pyridoxal phosphate enzyme in the endoplasmic reticulum of liver cells. Two molecules of δ-aminolevulinate then condense to form *porphobilinogen,* through the action of *δ-aminolevulinate dehydratase.* The mechanism of this unusual and complex reaction involves the intermediate formation of a Schiff's base between the keto group of one molecule of δ-aminolevulinic acid and an ε-amino group of a lysine residue of the enzyme. Both the synthetase and dehydratase are regulatory enzymes; they are inhibited by heme, hemoglobin, and other heme proteins, which are the ultimate end products of this biosynthetic pathway. Four molecules of porphobilinogen now serve as the precursors of the cyclic tetrapyrrole protoporphyrin, through a series of complex reactions. Iron is not incorporated until the protoporphyrin molecule is completed. The incorporation of iron requires an enzyme, *ferrochelatase,* localized in the mitochondria. Another enzyme, *heme synthetase,* puts together Fe^{2+}, protoporphyrin, and the globin protein to form hemoglobin. Figure 25-27 also shows the origin of the specific carbon and nitrogen atoms of protoporphyrin IX from glycine and succinyl-CoA.

The porphyrinlike corrin ring of vitamin B_{12} (page 348) is synthesized from δ-aminolevulinic acid in a pathway related to that of the cyclic tetrapyrroles.

Nitrogen-Fixing Organisms

The fixation of molecular nitrogen of air, i.e., its reduction to yield ammonia and other biologically useful forms of nitrogen, is an extremely important process in the biosphere (page 365). However, it can be carried out by only a limited

Figure 25-27
Biosynthesis of porphobilinogen and the conversion of porphobilinogen into protoporphyrin IX. The carbon and nitrogen atoms originating from glycine are in color; the carbon atoms originating from the carboxyl carbon atoms of succinyl-CoA are marked with •. Bilirubin is one of the degradation products of protoporphyrin.

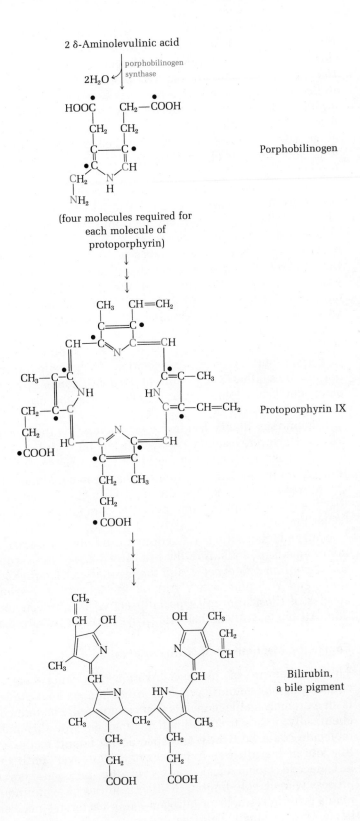

number of organisms. Most leguminous plants can fix atmospheric N_2, as can some 250 or more species of nonleguminous plants. The fixation of nitrogen by legumes requires the cooperative action of the host plant and bacteria present in the root nodules; it is referred to as *symbiotic nitrogen fixation*. Representative nitrogen-fixing plants include peas, beans, clover, alfalfa, and soybeans among the legumes, and alder, sea buckthorn, and wax myrtle among nonlegumes. Enormous amounts of atmospheric nitrogen are fixed by crop legumes.

The microorganisms that invade the roots of leguminous plants are largely species of the bacterial genus *Rhizobium*. The infecting bacteria gain entrance into the cortical parenchyma of the root and give rise to a _nodule_, a highly organized structure with membranous sacs containing colonies of the bacteria. The nodule is in direct connection with the vascular system of the plant. Curiously, such nodules contain considerable amounts of hemoglobin, which is otherwise absent in the plant kingdom. The hemoglobin appears to aid nitrogen fixation indirectly, by enhancing the transport of oxygen at a low partial pressure to the nodule; free oxygen is an inhibitor of nitrogen fixation. The genetic information for biosynthesis of nodule hemoglobin comes from the plant, but hemoglobin is not formed in the absence of the bacteria.

Leguminous plants lacking such bacteria cannot fix nitrogen. However, bacteria removed from the nodules of the host plant can do so when properly fortified, whereas the bacteria grown in free culture cannot. It has been concluded that the nitrogen-fixing enzymes are located in the bacteria and that the plant supplies some essential components that the bacterium lacks.

Nonsymbiotic nitrogen fixation occurs in a number of microorganisms, including the blue-green algae, the aerobic soil bacterium *Azotobacter,* and the facultative bacteria *Klebsiella* and *Achromobacter.* Among the anaerobes, various species of *Clostridia* are especially active in nitrogen fixation. All photosynthetic bacteria (page 589) can fix nitrogen.

Enzymatic Mechanism of Nitrogen Fixation

The mechanism of nitrogen fixation has long been a challenging biochemical problem. Since molecular nitrogen is an extremely stable molecule that cannot easily be reduced chemically, it has been difficult to imagine how this inert molecule could be tightly and specifically bound to the active site of the nitrogen-fixing enzyme. Moreover, until the last decade biochemical studies of the nitrogen-fixation process were largely limited to intact nodules of leguminous plants or to intact cells of nitrogen-fixing bacteria, since earlier efforts to obtain cell-free extracts which could fix nitrogen had been unsuccessful. Nevertheless, a number of significant properties of the nitrogen-fixing mechanism were deduced from early studies. The K_M for nitrogen in the fixation process in soybean nodules or blue-green algae was found to be quite low, about 0.02 atm of nitrogen. Clearly, the nitrogen-fixing system is fully saturated at the normal partial pressure of nitrogen in air (~ 0.8 atm). Another important observation is that nitrogen fixation is competitively inhibited by carbon monoxide (CO). Later studies showed that, in addition to nitrogen, other triply bonded compounds, such as acetylene (CH≡CH), cyanide (H—C≡N), and azide, may also be reduced. In fact, the reduction of acetylene to ethylene (CH$_2$=CH$_2$) is convenient for measurement of nitrogen-fixing activity since acetylene (substrate) and ethylene (product) can be separated and measured easily by gas chromatography. Other observations were concerned with the identification of the electron or hydrogen donors for the reduction of molecular nitrogen to ammonia. Many nitrogen-fixing organisms can use molecular hydrogen for this purpose. However, other reductants, such as pyruvate or formate, can also provide hydrogen or electrons for the reduction of nitrogen to ammonia.

In 1960 L. E. Mortenson and his colleagues finally succeeded in reproducibly obtaining cell-free extracts from the anaerobic bacterium *Clostridium pasteurianum* that could fix nitrogen when supplemented with a suitable electron donor and ATP; the enzyme system in these extracts is called the *nitrogenase system*. The earliest stable end product of the fixation reaction is ammonia. The nitrogenase system from this and other nitrogen-fixing bacteria was found to be exceedingly sensitive to molecular oxygen, which inactivates it. It is therefore necessary to prepare and purify the system in an oxygen-free atmosphere.

In 1964 J. E. Carnahan, L. E. Mortenson, and their colleagues isolated a protein from extracts of *Cl. pasteurianum* which they showed to be essential for nitrogen fixation by serving as the electron carrier to nitrogenase. The purified factor was found to have a molecular weight of 6,000 and to contain seven atoms of iron and seven atoms of acid-labile sulfur. No heme was present. They called this protein *ferredoxin* (derived from iron and redox); it was the first of many iron-sulfur proteins to be isolated from natural sources. As pointed out elsewhere, similar proteins participate in both mitochondrial (page 488) and photosynthetic (page 606) electron transport. The ferredoxin of *Cl. pasteurianum* has a very negative standard oxidation-reduction potential, −0.43 V. Reduced ferredoxin serves as the electron donor in the nitrogenase system of *Cl. pasteurianum*. Some nitrogen-fixing bacteria also contain *flavodoxin*, which is functionally similar to ferredoxin; it contains FMN but no iron.

The nitrogenase system is a complex of two proteins, neither of which has any demonstrable activity by itself. One is a molybdenum-iron-sulfur protein (symbolized Mo-Fe protein) with a molecular weight near 220,000; this protein, which has two types of subunits, has been crystallized. It contains 2 atoms of molybdenum, 32 of iron, and 25 to 30 of acid-labile sulfur. The other component is an iron protein (Fe protein) with a molecular weight near 60,000. These components are present in the ratio of one Mo-Fe protein to two Fe proteins. At very low temperatures both proteins show electron-spin-resonance signals characteristic of iron-sulfur centers (page 489).

Figure 25-28 shows one hypothesis for the mechanism of nitrogen fixation. Reduced ferredoxin serves as the immediate electron donor. Six electrons are required to reduce one molecule of N_2 to two NH_3. At least one molecule of ATP is required for each electron transferred; in fact, much evidence suggests that two ATPs are required. The overall equation is therefore

$$N_2 + 6H^+ + 6e^- + 12ATP + 12H_2O \longrightarrow 2NH_3 + 12ADP + 12P_i$$

Regeneration of reduced ferredoxin occurs through the enzyme *NADH-ferredoxin reductase*:

$$NADH + Fd_{ox} \longrightarrow NAD^+ + Fd_{red} + H^+$$

or by the pyruvate dehydrogenase system. The precise function of ATP is not known, but it is believed that the binding of MgATP to the reduced Fe protein converts it into a conformation capable of reacting with the oxidized form of the Mo-Fe protein (Figure 25-28) and passing its electrons to the N_2 bound at the active site of the latter.

The biosynthesis of the nitrogenase system is regulated by genetic repression. If the cells are provided with ample NH_3, they stop making the enzyme. Thus nitrogen fixation occurs only when its product is needed.

Other Steps in the Nitrogen Cycle

Although reduced nitrogen in the form of ammonia or amino acids is the form utilized in most living organisms, some soil bacteria derive their energy by oxidizing ammonia with the formation of nitrite and, from it, nitrate. Because these organisms are extremely abundant and active, nearly all ammonia reaching the soil ultimately becomes oxidized to nitrate, a process known as *nitrification*.

Nitrification occurs in two steps. In the first ammonia is oxidized to nitrite, almost exclusively by the aerobic chemolithotroph *Nitrosomonas*. In effect, ammonia plays the role of energy-yielding fuel, since it is the main electron donor in the respiration of *Nitrosomonas*. The electrons flow from the primary ammonia dehydrogenase system along a cytochrome-containing respiratory chain to oxygen; oxidative phosphorylation is coupled to this electron transport. Similarly, nitrite is then oxidized to nitrate by *Nitrobacter*, which also obtains nearly all its energy from the oxidation of the nitrite. *Nitrosomonas* and *Nitrobacter* obtain their carbon for growth from CO_2.

Nitrate is the principal form of nitrogen available to higher plants from the soil. Some nitrate is lost from the soil as molecular nitrogen by the action of bacteria that use nitrate as terminal electron acceptor; this process is called *denitrification*. However, much nitrate is absorbed from the soil by higher plants, in which its metabolic assimilation into the form of ammonia proceeds in two major steps: (1) reduction of nitrate to nitrite and (2) reduction of nitrite to ammonia.

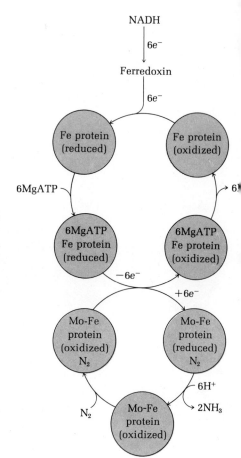

Figure 25-28
Schematic representation of one hypothesis for the action of the nitrogenase system.

The first reaction in nitrate utilization is catalyzed by *nitrate reductase,* which is widely distributed in plants and fungi; it has been most intensively studied in *Neurospora* by A. Nason and his colleagues. This enzyme is a flavoprotein containing both molybdenum and cytochrome *b*; it employs NADPH as electron donor. The overall process of electron flow to nitrate can be summarized in the scheme

$$\text{NADPH} \xrightarrow{e^-} \text{FAD} \xrightarrow{e^-} \text{Mo} \xrightarrow{e^-} \Bigg\langle \begin{array}{l} \text{NO}_2^- \\ \text{NO}_3^- \end{array}$$

The molybdenum appears to undergo cyclic valence changes between Mo(V) and Mo(VI) during the reduction of nitrate.

Nitrate may also be used by certain microorganisms as a terminal electron acceptor in place of oxygen, usually under anaerobic or partially anaerobic conditions. Because of its physiological and enzymological similarity to aerobic respiration the process is usually spoken of as *nitrate respiration.* Respiratory nitrate reductase, which has been extensively studied in *E. coli,* employs formate as electron donor and utilizes cytochrome *b,* molybdenum, and nonheme iron as components. Formation of ATP from ADP and phosphate is coupled to nitrate respiration.

The reduction of nitrite to ammonia by *nitrite reductase* of plants requires a highly electronegative reductant. In green plants ferredoxin reduced during the light reactions of photosynthesis may serve as the reductant of nitrite in the following chain of reactions:

$$\text{Ferredoxin} \xrightarrow{e^-} \text{NADP} \xrightarrow{e^-} \text{FAD} \xrightarrow{e^-} \Bigg\langle \begin{array}{l} \text{NH}_3 \\ \text{NO}_2^- \end{array}$$

The reduction of nitrite to ammonia requires the input of six electrons. It will be recalled that respiration, which requires four electrons to reduce a molecule of O_2, and nitrogen fixation, which requires six electrons per molecule of N_2, are other examples of multielectron reduction, which offer a puzzle with respect to reaction mechanism. Recent research indicates that nitrite reductase and sulfite reductase, which also catalyzes six-electron steps, contain a special prosthetic group called *siroheme,* an iron-porphyrin in which two of the pyrrole rings are reduced. Free ammonia, the end product of nitrate reduction, is then utilized to aminate α-ketoglutarate, thus providing amino groups for transaminations.

Summary

The albino rat and man can synthesize 10 of the 20 amino acids required as building blocks for protein biosynthesis. The remainder of the amino acids are nutritionally indispensable and must be obtained from other sources. Higher plants and many microorganisms can synthesize all the amino acids starting from ammonia as nitrogen source. In the nonessential, or nutritionally dispensable, group, glutamic acid is formed by reductive amination of α-ketoglutarate and is the direct precursor of glutamine and proline. Alanine and aspartic acid are formed by transamination to pyruvate and oxaloacetate, respectively. Tyrosine is formed by hydroxylation of phenylalanine. Cysteine is formed from methionine by a complex series of reactions in which S-adenosylmethionine, S-adenosylhomocysteine, and cystathionine are the most significant intermediates. The carbon chain of cysteine derives from serine and the sulfur atom from methionine. Serine is synthesized from 3-phosphoglycerate. Serine is also a precursor of glycine; its β carbon atom is transferred to tetrahydrofolate.

The pathways of biosynthesis of the essential amino acids have been established largely from studies on bacteria. The carbon skeletons of methionine and threonine arise from aspartic acid, whereas the methyl group of methionine arises from N^5-methyltetrahydrofolate. Lysine is made by two routes, the aminoadipic (bacteria) and the aminopimelic acid (fungi) pathways. Isoleucine, valine, and leucine synthesis begin with α-keto acids, involve unusual migrations of alkyl groups, and utilize some common steps. Arginine is formed from ornithine, which in turn is derived from glutamic acid. The precursors of the aromatic amino acids are aliphatic compounds which are cyclized to shikimic acid, an essential precursor of many aromatic biomolecules. Shikimic acid yields phenylalanine and tyrosine via prephenic acid and tryptophan via anthranilic acid. The pathway to histidine is most complex and unusual and involves the carbon chain of a pentose and the fragmentation of the purine ring of ATP.

Most of the biosynthetic pathways leading to the amino acids are subject to allosteric, or end-product, inhibition; the regulatory enzyme is usually the first in the sequence. Some of the regulatory enzymes in the branched pathways occur as isozymes and thus respond to more than one modulator. Amino acids are precursors of many other important biomolecules, including biologically active peptides such as glutathione and the antibiotic gramicidin, creatine (from glycine and methionine), the hormones epinephrine and norepinephrine (from tyrosine), serotonin and indoleacetic acid (from tryptophan), and the polyamines spermine and spermidine from lysine and methionine. The porphyrin ring of heme and chlorophyll is derived from glycine and succinyl-CoA.

Fixation of molecular nitrogen by bacteria in the root nodules of legumes is catalyzed by the nitrogenase system, composed of an iron-protein and molybdenum-iron protein. Electrons for the reduction are furnished by reduced ferredoxin. For each electron used to reduce nitrogen, at least one ATP is required. The nitrification of ammonia by soil organisms to form nitrate and the denitrification of nitrate by higher plants to yield NH_3 complete the nitrogen cycle.

References

Books

DAGLEY, S., and D. E. NICHOLSON: *An Introduction to Metabolic Pathways*, Wiley, New York, 1970. The "bible" of metabolic pathways. A comprehensive bibliography is included.

GREENBERG, D. M. (ed.): *Metabolic Pathways*, vols. I to III, 3d ed., Academic, New York, 1967–1969. Excellent papers in review form.

KUN, E., and S. GRISOLIA (eds.): *Biochemical Regulatory Mechanisms in Eukaryotic Cells*, Interscience, New York, 1972.

MANDELSTAM, J., and K. MCQUILLEN: *Biochemistry of Bacterial Growth*, 2d ed., Wiley, New York, 1973. Excellent and complete account of amino acid biosynthesis and its regulation in bacteria.

MEISTER, A.: *Biochemistry of the Amino Acids,* vols. I and II, 2d ed., Academic, New York, 1965. Comprehensive review of amino acid biosynthesis and precursor functions, particularly in vol. II.

PRUSINER, S., and E. R. STADTMAN (eds.): *The Enzymes of Glutamine Metabolism,* Academic, New York, 1973.

STANBURY, J. B., J. B. WYNGAARDEN, and D. B. FREDRICKSON (eds.): *The Metabolic Basis of Inherited Disease,* 3d ed., McGraw-Hill, New York, 1972. Comprehensive review of amino acid metabolism in man and its genetic defects.

TABOR, H., and C. W. TABOR (eds.): *Metabolism of Amino Acids and Amines,* vols. XVIIA and B of S. P. Colowick and N. O. Kaplan (eds.), *Methods in Enzymology,* Academic, New York, 1970 and 1971.

Articles

BURNHAM, B. F.: "Metabolism of Porphyrins and Corrinoids," pp. 403–537 in D. M. Greenberg (ed.), *Metabolic Pathways,* 3d ed., vol. III, Academic, New York, 1969.

BURNS, R. C., and R. W. F. HARDY: "Purification of Nitrogenase and Crystallization of Its Mo-Fe Protein," *Methods Enzymol.,* 24B: 480–496 (1972).

BURRIS, R. H.: "Fixation by Free-Living Microorganisms: Enzymology," pp. 105–160 in J. R. Postgate (ed.), *Chemistry and Biochemistry of Nitrogen Fixation,* Plenum, London, 1971.

DATTA, P.: "Regulation of Branched Biosynthetic Pathways in Bacteria," *Science,* 165: 556–562 (1969). Multiple feedback control patterns in amino acid biosynthesis.

DAVIES, R. C., A. GORCHEIN, A. NEUBERGER, J. D. SANDY, and G. H. TAIT: "Biosynthesis of Chlorophyll," *Nature,* 245: 15–19 (1973).

EADY, R. R., and J. R. POSTGATE: "Nitrogenase," *Nature,* 249: 805–810 (1974). An excellent review of recent advances.

GIBSON, F., and J. PITTARD: "Pathways of Biosynthesis of Aromatic Amino Acids and Vitamins and Their Control in Microorganisms," *Bacteriol. Rev.,* 32: 465 (1968).

LIPMANN, F. "Attempts to Map a Process Evolution of Peptide Biosynthesis," *Science,* 173: 875–884 (1971).

MEISTER, A.: "Specificity of Glutamine Synthetase," *Adv. Enzymol.,* 31: 183–218 (1968). An interesting review, illustrated with color stereophotographs of molecular models of substrates of the enzyme.

MEISTER, A.: "On the Synthesis and Utilization of Glutamine," *Harvey Lect.,* 63: 137–168 (1969).

MURPHY, M. J., L. M. SIEGAL, S. R. TOVE, and H. KAMIN: "Siroheme: A New Prosthetic Group Participating in 6-Electron Reduction Reactions," *Proc. Natl. Acad. Sci. (U.S.),* 71: 612–616 (1974).

ROBINSON, T.: "Metabolism and Function of Alkaloids in Plants," *Science,* 184: 430–435 (1974).

SANWAL, B. D., M. KAPOOR, and H. W. DUCKWORTH: "Regulation of Branched and Converging Pathways," *Curr. Top. Cell Regul.,* 3: 1–115 (1971).

SCHIMKE, R. T.: "On the Roles of Synthesis and Degradation in Regulation of Enzyme Levels in Mammalian Tissues," *Curr. Top. Cell Regul.*, 1: 77–124 (1969).

STADTMAN, E. R.: "Allosteric Regulation of Enzyme Activity," *Adv. Enzymol.*, 28: 41–154 (1966). A broad survey of the regulation of metabolic pathways.

STADTMAN, E. R., B. M. SHAPIRO, H. S. KINGDON, C. A. WOOLFOLK, and J. S. HUBBARD: "Cellular Regulation of Glutamine Synthetase Activity in *Escherichia coli*," *Adv. Enzyme Regul.*, 6: 257–289 (1968).

TAYLOR, R. T., and H. WEISSBACH: "N⁵-Methyltetrahydrofolate–Homocysteine Methyltransferases," pp. 121–165 in P. D. Boyer (ed.), *The Enzymes*, vol. 9, 3d ed., Academic, New York, 1973.

Problems

1. Write a balanced equation for the synthesis of glycine starting from succinate as sole carbon source.

2. Write a balanced equation for the biosynthesis of threonine starting from isocitric acid as sole carbon source.

3. From isotopic tracer experiments on intact albino rats it was established that the labeled atoms of the metabolites shown were precursors of the specified atoms in the creatine molecule (margin). From these observations construct a possible pathway of biosynthesis of creatine.

4. How many high-energy phosphate bonds are required for the biosynthesis of one molecule of arginine from glutamate, ammonia, and carbon dioxide? Assume that acetyl-CoA is available.

5. Administration of serine labeled with ^{14}C in the α carbon atom to rats was found to yield choline labeled in the 2-carbon atom, that is, $(CH_3)_3N-^{14}CH_2CH_2OH$. In a parallel experiment, dimethylethanolamine deuterated in the methyl groups gave methyl-labeled choline when fed with methionine but gave no labeled choline when fed homocysteine. Suggest a mechanism for choline synthesis.

6. The dipeptide anserine (margin) is found in skeletal muscle. In which carbon atoms of anserine would you expect to find the label from $1-^{14}C$-D-ribose? How would you determine the source of the methyl group of anserine?

7. Write a balanced equation for the formation of glutamate from nitrate, NADPH, NAD⁺, ATP, and CO_2 in a photosynthetic cell of a higher plant.

8. Write a balanced equation for the synthesis of proline from α-ketoglutarate. If the α-carbonyl carbon atom of α-ketoglutarate is labeled, where would the label appear in the proline?

9. In a bacterial mutant unable to synthesize panthothenic acid, the pantothenic acid requirement can be abolished by giving pantoic acid, α-ketopantoic acid, or α-ketoisovaleric acid in the presence of p-aminobenzoic acid. Labeled p-aminobenzoic acid does not give rise to labeled pantothenic acid in this mutant whereas $2-^{14}C$-α-ketoisovaleric acid gives rise to pantothenic acid labeled as shown in the margin. In wild-type bacteria of the same species, $\alpha-^{14}C$-aspartate was found to give rise to pantothenic acid labeled in the β carbon atom of the β-alanine

Problem 3

NH
‖
$NH_2CNHCH_2CH_2CH_2CHCOOH$ NH_2 Arginine

NH_2
|
C=NH   Creatine
|
CH_3-N-CH_2COOH

NH_2CH_2COOH
Glycine

CH_3
|
S
|
CH_2
|
CH_2 Methionine
|
$CHNH_2$
|
COOH

Problem 6

COOH
|
CH=C—CH_2CHNHC=O
| | |
N   N—CH_3 CH_2
\C/ |
H CH_2
|
NH_2

Anserine
(β-alanyl-1-methyl-L-histidine)

Problem 9

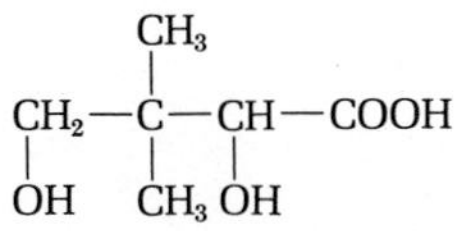

CH$_3$
|
CH$_2$—C—CH—COOH
| |
OH   CH$_3$ OH

Pantoic acid

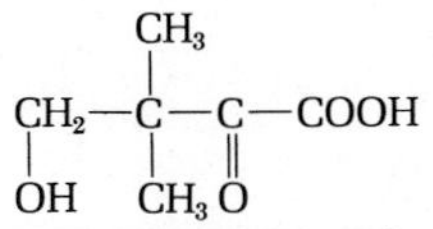

CH$_3$
|
CH$_2$—C—C—COOH
| | ‖
OH   CH$_3$ O

α-Ketopantoic acid

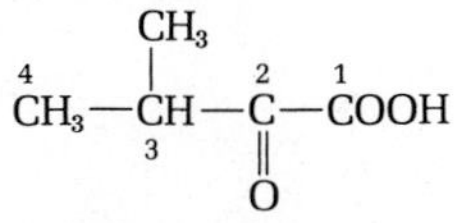

CH$_3$
|
$\overset{4}{CH_3}$—$\overset{}{CH}$—$\overset{2}{C}$—$\overset{1}{COOH}$
      $\underset{3}{}$ ‖
                       O

α-Ketoisovaleric acid

CH$_2$OH
|
CH$_3$—C—CH$_3$
|
^{14}CHOH
|
C=O
|
HN
|
CH$_2$
|
CH$_2$
|
COOH

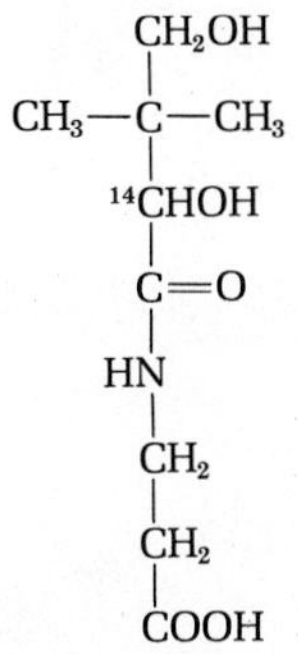

Labeled pantothenic acid

moiety. From these observations deduce the pathway for pantothenic acid synthesis.

10. How many high-energy phosphate bonds are utilized in the conversion of serine and methionine to cysteine? If (a) 2-^{14}C-labeled methionine or (b) 2-^{14}C-labeled serine is utilized as the precursor, where will the label appear in the cysteine formed?

The biosynthesis of the deoxyribonucleotides and ribonucleotides is a central process for all cells, since the nucleotides are direct precursors of DNA and RNA and many also participate in metabolism as coenzymes. An important aspect of the biosynthesis of the nucleotides is the pathway of formation of their bases, the pyrimidines and purines. Nearly all living organisms, except for some bacteria, appear able to synthesize these bases from simpler precursors.

The biosynthetic pathways leading to the nucleotides are under strict regulation. Since the four major deoxyribonucleotides and the four major ribonucleotides are inserted into DNA and RNA of cells in specific molar ratios (Chapter 31), these regulatory mechanisms are geared to yield the proper "mix" of nucleotides appropriate for each type of nucleic acid and for each type of cell. Nucleotides and their nitrogenous bases are used with economy; in fact, they are not employed as energy-yielding fuels in the great majority of organisms. Indeed, many cells possess salvage mechanisms for recovery of free purines and pyrimidines resulting from the hydrolytic breakdown of nucleotides.

Also to be discussed in this chapter are the biosynthetic pathways to the nucleotide coenzymes.

Biosynthesis of the Purine Ribonucleotides

The first important clues to the biosynthetic origin of the purine bases came from the experiments of J. M. Buchanan and his colleagues. They fed various possible isotopic precursors to birds and determined the sites of incorporation of the labeled atoms into the purine ring. Birds were chosen for such experiments since they excrete nitrogen largely in the form of uric acid, a purine derivative that is easily isolated in pure form (page 584). Chemical degradation of the uric acid revealed the origin of the atoms in the purine ring system. As seen in Figure 26-1, nitrogen atoms 3 and 9 arise from the amide group of glutamine, nitrogen 1 from aspartate, and nitrogen 7 from glycine. Carbon atoms 4 and 5 also derive from glycine, indicating that the backbone of the glycine molecule is directly incorporated into the purine

Figure 26-1
Precursors of the purine ring.

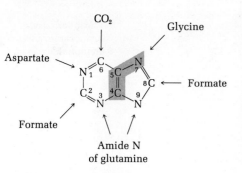

Figure 26-2
Biosynthesis of inosinic acid (IMP) (below
and opposite). Enzyme names are in color.
Note the inversion from the α to the β configuration of carbon atom 1 of D-ribose in the
second reaction.

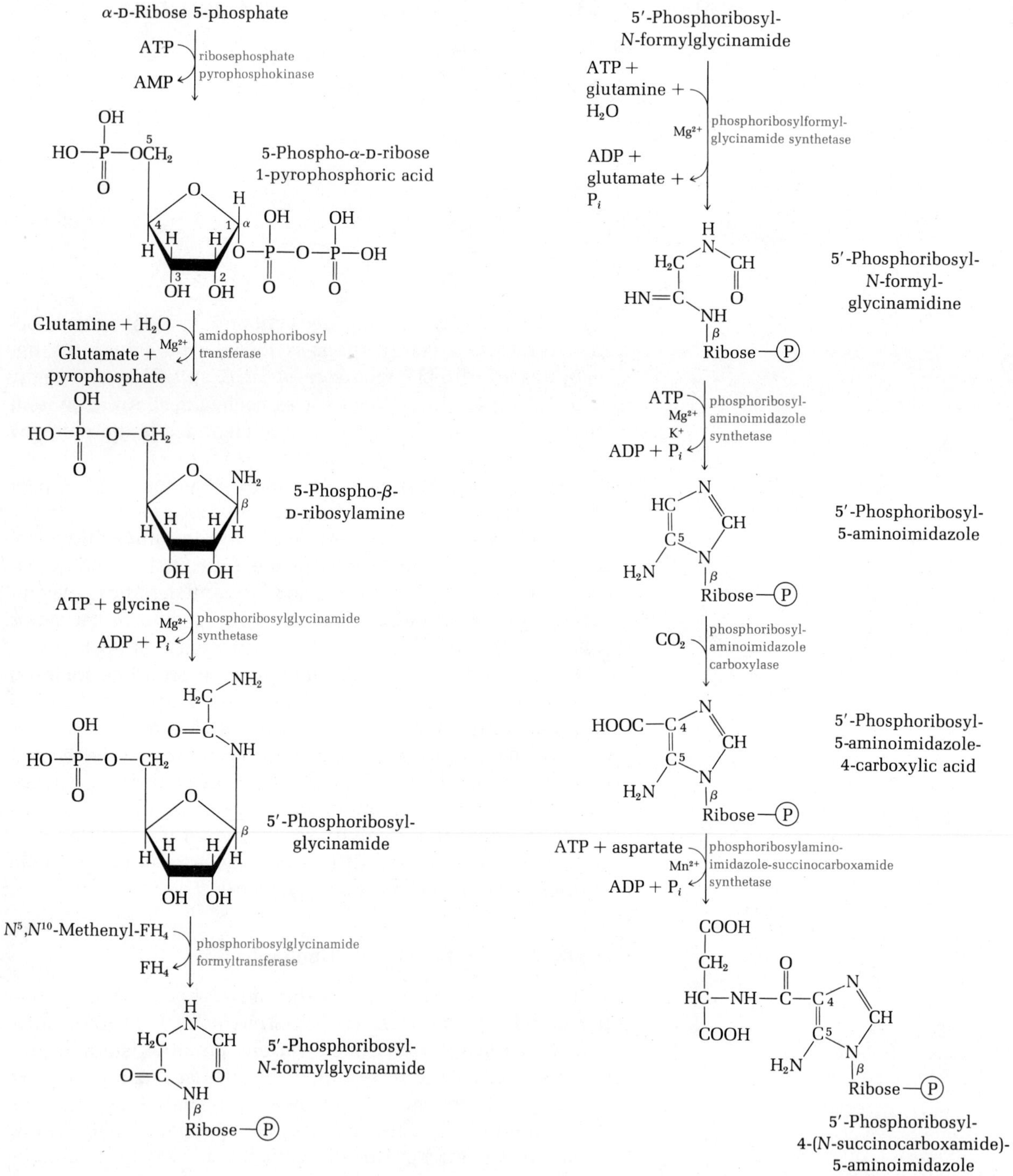

α-D-Ribose 5-phosphate

ATP ⌐ ribosephosphate
AMP ⌐ pyrophosphokinase

5-Phospho-α-D-ribose
1-pyrophosphoric acid

Glutamine + H₂O ⌐ amidophosphoribosyl
Glutamate + ⌐ Mg²⁺ transferase
pyrophosphate

5-Phospho-β-
D-ribosylamine

ATP + glycine ⌐ phosphoribosylglycinamide
ADP + P_i ⌐ Mg²⁺ synthetase

5'-Phosphoribosyl-
glycinamide

N⁵,N¹⁰-Methenyl-FH₄ ⌐ phosphoribosylglycinamide
FH₄ ⌐ formyltransferase

5'-Phosphoribosyl-
N-formylglycinamide

5'-Phosphoribosyl-
N-formylglycinamide

ATP +
glutamine +
H₂O ⌐ phosphoribosylformyl-
⌐ Mg²⁺ glycinamide synthetase
ADP +
glutamate +
P_i

5'-Phosphoribosyl-
N-formyl-
glycinamidine

ATP ⌐ phosphoribosyl-
⌐ Mg²⁺ aminoimidazole
⌐ K⁺ synthetase
ADP + P_i

5'-Phosphoribosyl-
5-aminoimidazole

CO₂ ⌐ phosphoribosyl-
⌐ aminoimidazole
⌐ carboxylase

5'-Phosphoribosyl-
5-aminoimidazole-
4-carboxylic acid

ATP + aspartate ⌐ phosphoribosylamino-
⌐ Mn²⁺ imidazole-succinocarboxamide
ADP + P_i ⌐ synthetase

5'-Phosphoribosyl-
4-(N-succinocarboxamide)-
5-aminoimidazole

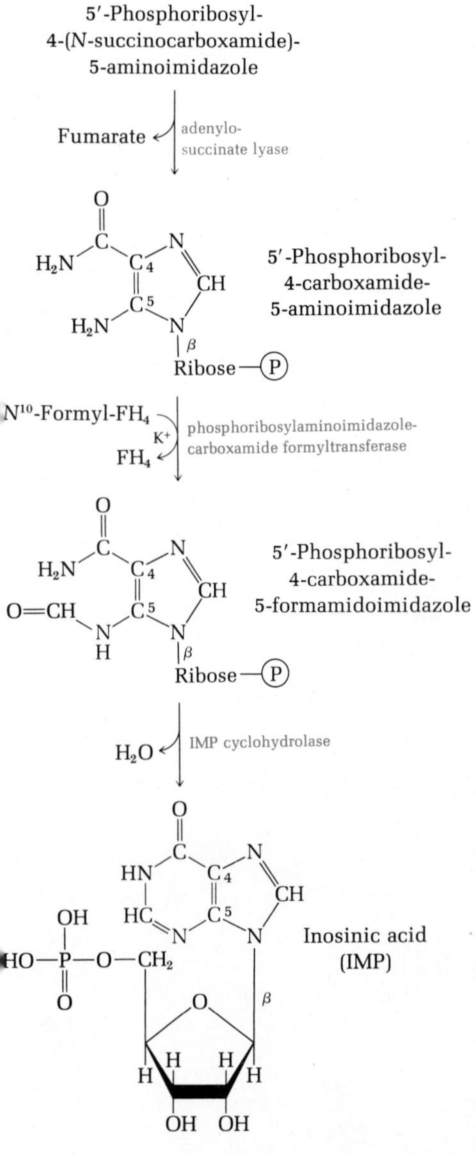

5'-Phosphoribosyl-
4-(N-succinocarboxamide)-
5-aminoimidazole

Fumarate ⟵ adenylo-
              succinate lyase

5'-Phosphoribosyl-
4-carboxamide-
5-aminoimidazole

H₂N—C
      C₄   N
H₂N—C₅   CH
            N
            | β
          Ribose—(P)

N^{10}-Formyl-FH₄ ⟍
            K⁺   phosphoribosylaminoimidazole-
            FH₄ ⟋ carboxamide formyltransferase

5'-Phosphoribosyl-
4-carboxamide-
5-formamidoimidazole

H₂N—C
      C₄   N
O=CH—C₅   CH
      N   N
      H | β
          Ribose—(P)

H₂O ⟵ IMP cyclohydrolase

Inosinic acid
(IMP)

HN—C
     C₄   N
HC—C₅   CH
   N   N

OH
|
HO—P—O—CH₂
||
O   O   β

H   H
H   H
OH OH

ring. Carbon atoms 2 and 8 are furnished by formate and carbon 6 by CO_2. However, several years of research were required to elucidate the complex enzymatic steps involved in the biosynthesis of the purine nucleotides; Buchanan's laboratory and that of G. R. Greenberg made important contributions.

Contrary to the early expectation that the purine ring is formed first, followed by attachment of the D-ribose phosphate side chain, it was ultimately found that the starting material is an activated form of D-ribose 5-phosphate, on which a purine ring is built step by step, resulting directly in the formation of a nucleotide. The biosynthetic pathway to adenylic and guanylic acids begins with the activation of α-D-ribose 5-phosphate by enzymatic pyrophosphorylation at the expense of ATP to form 5-phospho-α-D-ribose 1-pyrophosphate (PRPP, Figure 26-2). This is an unusual reaction in that the pyrophosphate group of ATP is transferred intact. We shall see later that PRPP is also a precursor of the pyrimidine nucleotides. It is therefore significant for the regulation of nucleotide biosynthesis that PRPP formation is catalyzed by an allosteric enzyme, which is inhibited by both ADP and GDP.

In the next step (Figure 26-2) the PRPP reacts with glutamine so that the amino group from the amide portion of glutamine displaces the pyrophosphate group from the 1 position in the pentose to yield *5-phospho-β-D-ribosyl-1-amine*. The amide nitrogen atom of the glutamine is the first atom of the purine ring to be introduced; it corresponds to nitrogen atom 9 of the finished purine ring. In this step, catalyzed by an iron-containing enzyme, the anomeric carbon atom 1 of the D-ribose undergoes inversion from the α to the β configuration. The β configuration so introduced is retained in the final purine nucleotide products. This reaction is a secondary control point for purine biosynthesis; it is inhibited by purine nucleotides. As shown below, this step is also inhibited by the antibiotic *azaserine*.

In the third step, the carboxyl group of glycine reacts with the amino group of 5-phosphoribosylamine to form an amide linkage between glycine and the amino sugar. ATP is required as energy source. The product is *5'-phospho-β-D-ribosyl-1-glycinamide*, together with ADP and phosphate.

The purine ring is now built up around the glycinamide moiety, which already contains atoms 4, 5, 7, and 9 of the purine ring (Figure 26-2). The remaining atoms are introduced one by one. The next step is the addition of the one-carbon formyl group to the free α-amino group of 5'-phosphoribosylglycinamide to yield what will become carbon atom 8 of the purine ring (Figure 26-2). The formyl group is donated by the formyl-group carrier (page 347) N^5,N^{10}-methenyltetrahydrofolate, which is derived from N^5,N^{10}-methylenetetrahydrofolate (page 347). The normal biological source of the formyl carbon atom is the β carbon atom of serine (page 568), but the formyl group may also arise from endogenous or administered formate, via the formation of N^{10}-formyltetrahydrofolate (page 347).

The next step in purine nucleotide biosynthesis involves the introduction of nitrogen atom 3, which is derived from the amide group of glutamine (Figure 26-2), resulting in the formation of 5′-*phosphoribosyl-N-formylglycinamidine*. The transfer of the amino group from glutamine requires input of energy from ATP, which is cleaved to ADP and phosphate. Azaserine blocks this step also.

The product of this reaction is seen to possess the five atoms of the imidazole-ring portion of purine. In the next reaction the imidazole ring is closed by elimination of a molecule of water to yield 5′-*phosphoribosyl-5-aminoimidazole*, also an ATP-requiring reaction.

Carbon atom 6, which ultimately arises from CO_2 (Figure 26-1), is now introduced in a carboxylation reaction. Only one nitrogen and one carbon remain to be introduced. The nitrogen atom is introduced first into what will become position 1 of the purine ring (Figure 26-1). This nitrogen is derived from aspartic acid. The entire aspartic acid molecule is incorporated to yield 5′-*phosphoribosyl-4-(N-succinocarboxamide)-5-aminoimidazole*, from which free fumaric acid is subsequently eliminated. The enzyme catalyzing the latter reaction is probably identical with *adenylosuccinate lyase*, which catalyzes a later reaction in the biosynthesis of adenylic acid (see below).

The remaining carbon atom of the purine ring (number 2) is now introduced by transfer of the formyl group of N^{10}-formyltetrahydrofolate to the 5-amino group of the almost completed ribonucleotide, 5′-*phosphoribosyl-4-carboxamide-5-formamidoimidazole*. The pyrimidine portion of the purine ring system is then closed by elimination of water to form the ribonucleotide *inosinic acid* (IMP), the first product in this biosynthetic pathway to possess a complete purine ring system. This ring closure, unlike that of the imidazole ring, does not require ATP. Altogether, six high-energy phosphate groups of ATP are utilized for the formation of inosinic acid from D-ribose 5-phosphate if it is assumed that the pyrophosphate group displaced from 5-phosphoribosyl-1-pyrophosphate is ultimately hydrolyzed to orthophosphate by pyrophosphatase.

In the reaction sequence leading from D-ribose 5-phosphate to inosinic acid, there are three steps that are characteristically inhibited by specific antibacterial agents. As shown above, the antibiotic *azaserine* (Figure 26-3), formed by a species of *Streptomyces*, blocks the two enzymatic transfers of the amide group of glutamine. This antibiotic is a structural analog of glutamine and competes with it (Figure 26-3). Azaserine apparently blocks the essential —SH groups in the transferases catalyzing these reactions.

Antibacterial agents of the *sulfonamide* class (Figure 26-4), which are widely used in medicine because they inhibit growth of many bacteria, prevent the formation of folic acid (page 345) in susceptible microorganisms and hence, by indirectly inhibiting the last step, the formylation of 5′-phosphoribosyl-4-carboxamide-5-aminoimidazole, prevent purine biosynthesis. Sulfonamides are structural analogs of *p-aminobenzoic acid* (Figure 26-4), a building block of folic acid. Through competitive inhibition *sulfanilamide* and other sulfonamides prevent the incorporation of

Figure 26-3
Azaserine, an antibiotic analog of glutamine.

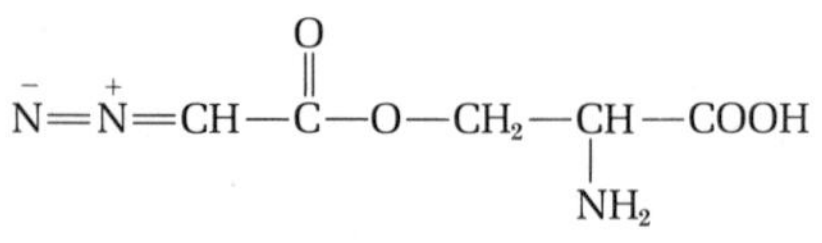

Azaserine

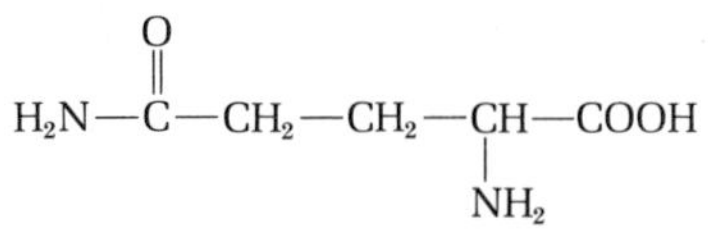

Glutamine

Figure 26-4
Sulfanilamide, a competitor for p-aminobenzoic acid in the biosynthesis of folic acid.

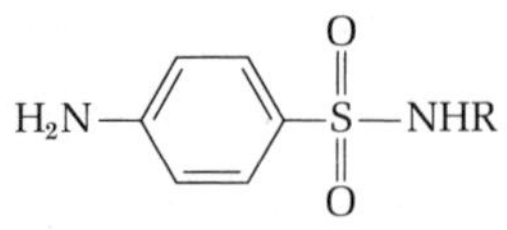

General structure of sulfonamides

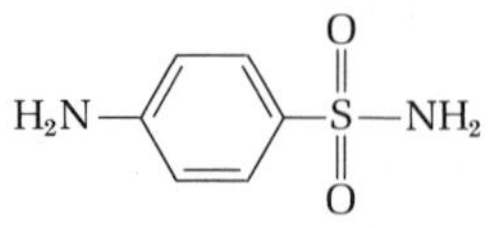

Sulfanilamide

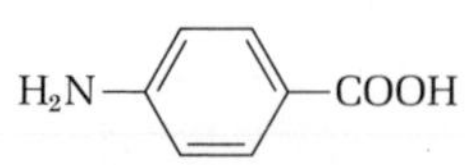

p-Aminobenzoic acid

Figure 26-5
The pathway from inosinic acid to adenylic
acid.

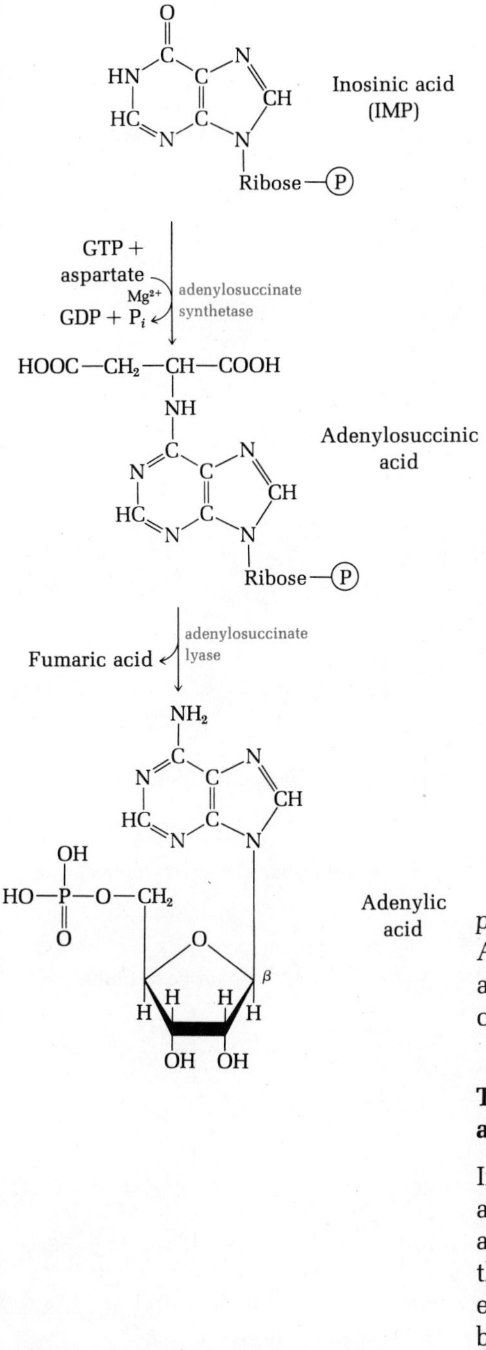

Figure 26-6
The pathway from inosinic acid to guanylic
acid.

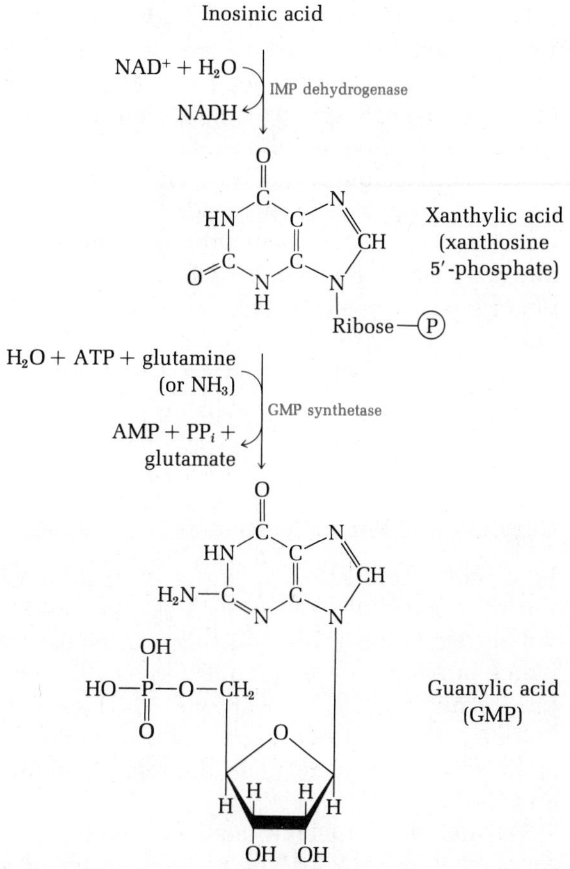

p-aminobenzoic acid during the biosynthesis of folic acid.
As a consequence, free 5'-phosphoribosyl-4-carboxamide-5-
aminoimidazole accumulates in the culture medium of susceptible microorganisms.

The Pathways from Inosinic Acid to Adenylic
and Guanylic Acid

Inosinic acid (IMP) is the precursor of adenylic and guanylic
acids (AMP and GMP). The conversion of inosinic acid to
adenylic acid (Figure 26-5) requires only the replacement of
the hydroxyl group at position 6 by an amino group. However, this exchange is brought about in a circuitous fashion
by the reaction of inosinic acid with aspartic acid to yield
adenylosuccinic acid, which is accompanied by the cleavage
of GTP to GDP and P_i. Fumaric acid is then eliminated from
adenylosuccinate to yield adenylic acid, by the action of
adenylosuccinate lyase, the same enzyme responsible for
eliminating fumaric acid from 5'-phosphoribosyl-4-(N-succinocarboxamide)-5-aminoimidazole in the pathway to inosinic acid. A total of seven high-energy phosphate groups is
required to make adenylic acid from D-ribose 5-phosphate.

In the formation of guanylic acid (Figure 26-6) inosinic
acid is first dehydrogenated to *xanthylic acid*, a nucleotide

containing the base *xanthine*, in an NAD-linked reaction. Xanthylic acid is then aminated to form guanylic acid in a reaction that requires glutamine as amino donor, as well as ATP. Since this step involves pyrophosphate cleavage of ATP (page 410) and ultimate hydrolysis of pyrophosphate, two high-energy phosphate groups of ATP are sacrificed to convert inosinic acid to guanylic acid. A total of eight high-energy phosphate groups is thus required to convert D-ribose 5-phosphate into guanylic acid.

The conversion of adenylic and guanylic acids into ATP and GTP, respectively, proceeds by two successive transfers of high-energy phosphate groups from ATP, catalyzed by *nucleosidemonophosphate kinase* and *nucleosidediphosphate kinase*, respectively:

$$GMP + ATP \rightleftharpoons GDP + ADP$$

$$GDP + ATP \rightleftharpoons GTP + ADP$$

Regulation of Purine Nucleotide Biosynthesis

In *E. coli* there are two levels of regulatory control over purine nucleotide biosynthesis (Figure 26-7). The first involves regulation of the pathway leading to inosinic acid and hence provides common control over the biosynthesis of all purine nucleotides. The second involves regulation of the branch pathways from inosinic acid to adenylic and guanylic acids and serves to regulate the relative abundance of AMP and GMP.

Regulation of inosinic acid production is exerted at the early reaction step leading to the transfer of an amino acid group to 5-phosphoribose 1-pyrophosphate. The amidotransferase catalyzing this reaction is a bivalent regulatory enzyme (page 234), which is inhibited by ATP, ADP, or AMP, or by GTP, GDP, or GMP, each of the two types of nucleotides apparently binding to a separate allosteric site on the enzyme. The inhibition is cumulative, so that when both types of purine nucleotides are present at a high level, the enzyme is inhibited to a greater degree than by either type alone.

Regulation of the branching pathways to AMP and GMP is achieved in two ways. The first of these is unusual. Note that the reaction pathway leading from xanthylic acid to guanylic acid (GMP) requires ATP as a reactant, whereas the pathway leading from inosinic acid to adenylic acid requires GTP. Thus an excess of ATP accelerates the pathway leading to guanylic acid; similarly, an excess of GTP accelerates the synthesis of adenylic acid. This reciprocal arrangement helps to balance the rates of production of AMP and GMP.

A second type of coordination between AMP and GMP levels is provided by two feedback inhibitions, that of the conversion of inosinic acid to adenylosuccinic acid by AMP and that of the conversion of inosinic acid to xanthylic acid by GMP. The first of the processes regulating AMP vs. GMP levels is a hurry-up process, in which GMP synthesis is accelerated when AMP levels are high, whereas the second is

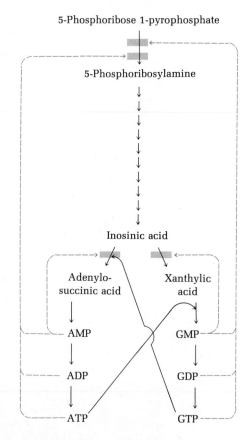

Figure 26-7
Regulation of the pathways to the purine nucleotides.

a slow-down process, in which AMP and GMP retard their own synthesis.

Biosynthesis of Pyrimidine Nucleotides

The biosynthetic pathway leading to the pyrimidine nucleotides is much simpler than the pathway to the purine nucleotides. It differs from the purine pathway in that the D-ribose 5-phosphate moiety (also derived from PRPP) is attached *after* the pyrimidine ring has been formed from its open-chain precursors.

Although early isotopic experiments revealed that CO_2 and ammonia are precursors of the pyrimidine ring, the first important advance in our knowledge of pyrimidine biosynthesis came from studies of mutants of *Neurospora crassa* deficient in the ability to make pyrimidines and thus unable to grow unless cytosine or uracil is present in the medium. Such *pyrimidineless* mutants can, however, grow normally when presented with the pyrimidine derivative orotic acid (Figure 26-8), which was first found in cow's milk (Greek, *oros*, "whey"). Tests with isotopic orotic acid showed that it is an immediate pyrimidine precursor in *Neurospora* and a number of bacteria.

The pathway to the pyrimidine nucleotides via orotic acid, mapped by A. Kornberg and his colleagues, first leads to uridylic acid (UMP), from which both cytidylic (CMP) and deoxythymidylic (dTMP) acid are derived (Figure 26-9). The pyrimidine ring is formed by the condensation of aspartic acid, which provides three carbons and a nitrogen, and carbamoyl phosphate, which provides the remaining carbon and nitrogen atoms. In this reaction, catalyzed by the allosteric enzyme *aspartate carbamoyltransferase*, more commonly known by its older name *aspartate transcarbamoylase*, the carbamoyl group of carbamoyl phosphate is transferred to aspartate to yield N-carbamoylaspartate. This reaction is the committing step in pyrimidine biosynthesis. The carbamoyl phosphate required is made in the reaction

$$2ATP + glutamine + CO_2 + H_2O \longrightarrow$$
$$2ADP + P_i + glutamate + carbamoyl\ phosphate$$

catalyzed by *carbamoyl-phosphate synthase (glutamine)*. In eukaryotes this enzyme is found in the cytosol, in contrast to *carbamoyl-phosphate synthase (ammonia)* (page 580), which is a mitochondrial enzyme primarily involved in urea production (page 579). Carbamoyl-phosphate synthase (glutamine) functions primarily to make carbamoyl phosphate for pyrimidine biosynthesis.

In the second step of pyrimidine biosynthesis the ring is closed by removal of water from N-carbamoylaspartate by the action of *dihydroorotase* to yield L-*dihydroorotic acid*. This compound is now oxidized by the flavoprotein *orotate reductase* to yield orotic acid, in a rather unusual reaction utilizing NAD^+ as the ultimate electron acceptor. This very complex dehydrogenase contains both FAD and FMN, as well as iron-sulfur centers.

Figure 26-8
Orotic acid (6-carboxyuracil).

735

At this point the D-ribose 5-phosphate side chain, provided by 5-phosphoribose 1-pyrophosphate, is attached to orotic acid by the action of *orotate phosphoribosyltransferase* to yield *orotidine 5'-phosphoric acid* or *orotidylic acid*. Orotidylic acid is now decarboxylated to yield uridylic acid (Figure 26-9). The last two steps are believed to be catalyzed by the same enzyme protein in some organisms.

Uridylic acid (UMP) is the precursor of the cytidine nucleotides. This conversion requires that uridylic acid first be phosphorylated to uridine triphosphate (UTP) by the successive reactions

$$\text{UMP} + \text{ATP} \underset{\text{nucleosidemonophosphate kinase}}{\rightleftharpoons} \text{UDP} + \text{ADP}$$

$$\text{UDP} + \text{ATP} \underset{\text{nucleosidediphosphate kinase}}{\rightleftharpoons} \text{UTP} + \text{ADP}$$

UTP then undergoes amination by ammonia (bacteria) or glutamine (animals) at the 4 position of the pyrimidine ring to yield cytidine 5'-triphosphate or CTP (Figure 26-10). ATP is cleaved to ADP and P_i in this reaction.

Orotic aciduria is a genetic disorder of pyrimidine biosynthesis in man in which orotic acid accumulates in the blood and is excreted in the urine. Children in whom this defect is detected do not grow normally and suffer a failure in the normal formation of red blood cells. The disorder is alleviated by feeding of uridine or cytidine.

Regulation of Pyrimidine Nucleotide Biosynthesis

The regulation of the rate of pyrimidine nucleotide synthesis has been the subject of a number of penetrating investigations, particularly those initiated by A. B. Pardee and his colleagues. They found that E. coli aspartate transcarbamoylase, which catalyzes the first reaction in the sequence (the condensation of carbamoyl phosphate with aspartic acid to yield carbamoylaspartic acid), is inhibited by CTP, the end product of this sequence of reactions (Figure 26-11). This inhibition is prevented by ATP (Figure 26-12). Aspartate transcarbamoylase of E. coli is perhaps the most thoroughly studied allosteric enzyme; its kinetics and molecular structure (pages 237 to 240) are known in considerable detail.

Figure 26-9
Biosynthesis of uridylic acid (below and opposite).

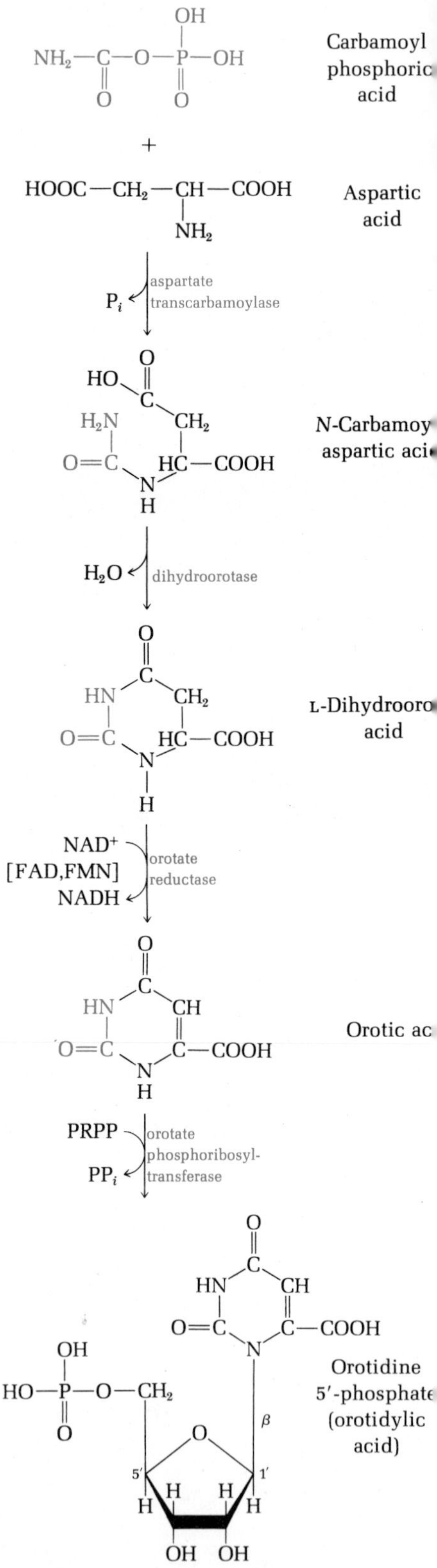

Figure 26-10
Amination of UTP to CTP by CTP synthetase of bacteria.

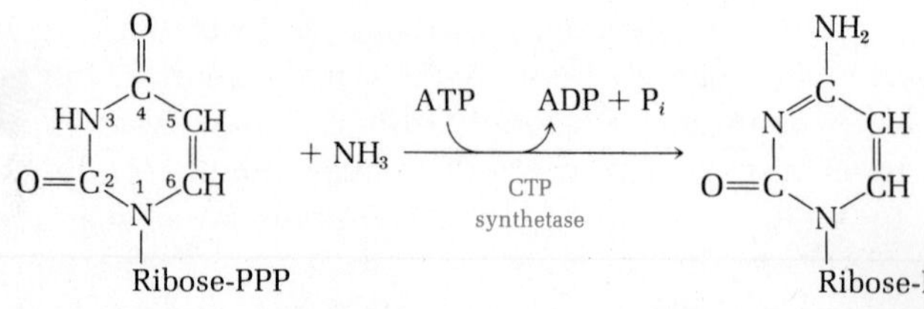

Uridine triphosphate (UTP)　　　　Cytidine triphosphate (CTP)

Orotidine 5′-phosphate
(orotidylic acid)

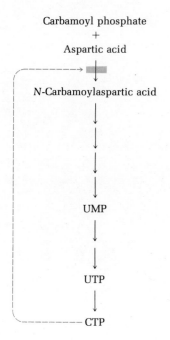

Uridylic acid

Comparative studies of the behavior of aspartate transcarbamoylase have revealed that its regulatory properties differ widely in different organisms. The regulatory pattern described above, in which the inhibitory modulator is CTP, occurs in the bacteria *E. coli* and *Aerobacter aerogenes.* However, in the bacterium *Pseudomonas fluorescens* the major inhibitory modulator is UTP; in some higher plants, it is UMP. Moreover, in some microorganisms, such as *Bacillus subtilis,* aspartate transcarbamoylase is not inhibited by any of these compounds, presumably because it lacks the regulatory subunits.

It is emphasized again that 5-phospho-α-D-ribose 1-pyrophosphate is a common precursor for both the purine and pyrimidine nucleotides and that the enzyme catalyzing its formation, *ribosephosphate pyrophosphokinase* (Figure 26-2), is an allosteric enzyme, whose inhibitory modulators are ADP and GDP. In addition, the synthesis of this enzyme is repressed by pyrimidine nucleotides.

Biosynthesis of Deoxyribonucleotides

Since the deoxyribonucleotides differ from the ribonucleotides only in containing 2-deoxyribose instead of ribose as the pentose unit, it was assumed that they would be formed by the pathway described above for the ribonucleotides, with the phosphorylated ribose precursor replaced by an analogous 2-deoxyribose derivative. Much early research therefore centered on the mechanism of biosynthesis of 2-deoxy-D-ribose. Although an enzymatic pathway for the formation of

Figure 26-11
Regulation of the biosynthetic pathway to CTP by end-product inhibition of aspartate transcarbamoylase. The inhibitory effect of CTP is prevented by ATP.

Figure 26-12
Effect of CTP and ATP on the aspartate transcarbamoylase reaction. Note that increasing concentrations of CTP, the end product of the overall pathway, increases the substrate concentration giving half-maximal velocity from about 10 mM to about 23 mM, an effect that is reversed by ATP.

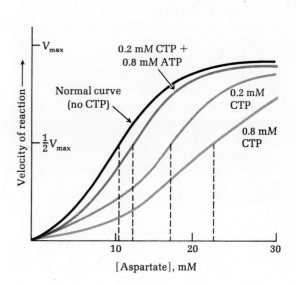

free 2-deoxyribose 5-phosphate from simpler precursors does in fact exist in some cells, it is now known that the deoxyribonucleotides are not normally synthesized starting from deoxyribose as a building block but are formed by direct reduction of the 2' carbon atom of the corresponding ribonucleotides. An important experiment pointing to this conclusion was carried out by I. A. Rose and B. S. Schweigert with cytidine labeled with ^{14}C in both the pyrimidine ring and the ribose. When presented to animal or bacterial cells, this compound was incorporated into the cytidylic acid residues of RNA and into the deoxycytidylic acid residues of DNA. The ratios of the isotopes in the pyrimidine and ribose (or deoxyribose) portions were identical in the cytidylic and deoxycytidylic acid residues. This finding indicated that the ribose moiety was not detached from the cytosine ring of the labeled cytidine prior to the formation of the labeled deoxycytidylic residues of DNA. Had ribose been detached first, it would have become diluted by nonisotopic ribose or deoxyribose prior to formation of the deoxyribonucleotide, with a corresponding change in the ratio of ^{14}C in the pyrimidine and pentose residues. Direct isotopic tests on cell-free bacterial extracts subsequently confirmed that labeled ribonucleotides are in fact directly converted into the corresponding deoxyribonucleotides.

There appear to be two pathways for the direct reduction of ribonucleotides, depending on the species. P. Reichard and his colleagues have shown that in *E. coli* all four ribonucleoside diphosphates (ADP, GDP, UDP, and CDP) are directly reduced to the corresponding deoxy analogs dADP, dGDP, dUDP, and dCDP, respectively, by a multienzyme system. In the overall process the reduction of the ribose moiety to 2-deoxyribose requires a pair of hydrogen atoms, which are ultimately donated by NADPH and H^+. However, the immediate electron donor is not NADPH but the reduced form of a small heat-stable protein (108 amino acid residues) called *thioredoxin*, which has two free —SH groups contributed by two cysteine residues. Thioredoxin may be reversibly oxidized and reduced. Its oxidized, or disulfide, form, which contains a residue of cystine, is reduced to the dithiol form by NADPH + H^+ in a reaction catalyzed by *thioredoxin reductase*:

$$\text{Thioredoxin (—S—S—)} + \text{NADPH} + H^+ \xrightleftharpoons{\overset{\text{thioredoxin}}{\text{reductase}}}$$
$$\text{thioredoxin (—SH)}_2 + \text{NADP}^+$$

Thioredoxin reductase is a flavoprotein of molecular weight 68,000; it contains two molecules of bound FAD. The reducing equivalents of the reduced thioredoxin formed are then transferred to the ribonucleoside 5'-diphosphate acceptor (NDP) by *ribonucleoside diphosphate reductase*:

$$\text{Thioredoxin (—SH)}_2 + \text{NDP} \xrightarrow{\text{Mg}^{2+}}$$
$$\text{thioredoxin (—S—S—)} + \text{dNDP} + H_2O$$

Figure 26-13
Methylation of dUMP to deoxythymidylate.

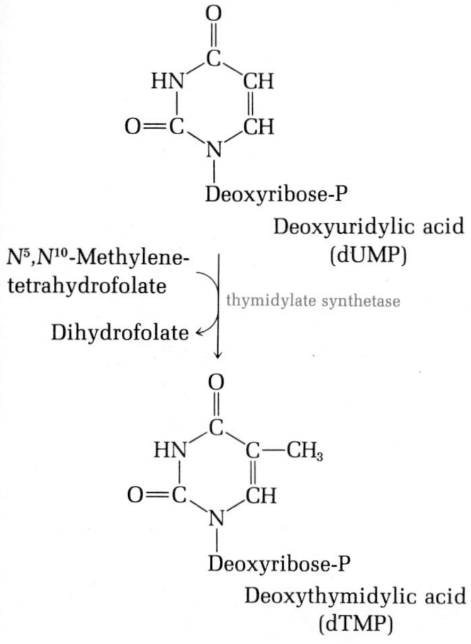

Deoxyribose-P

Deoxyuridylic acid
(dUMP)

N^5,N^{10}-Methylene-
tetrahydrofolate

thymidylate synthetase

Dihydrofolate

Deoxyribose-P

Deoxythymidylic acid
(dTMP)

*Below are shown the structures of dihydro-
folate and the drugs aminopterin and
amethopterin, which inhibit thymidylate
formation through their action on dihydrofolate reductase. Glu symbolizes the
glutamic acid residue.*

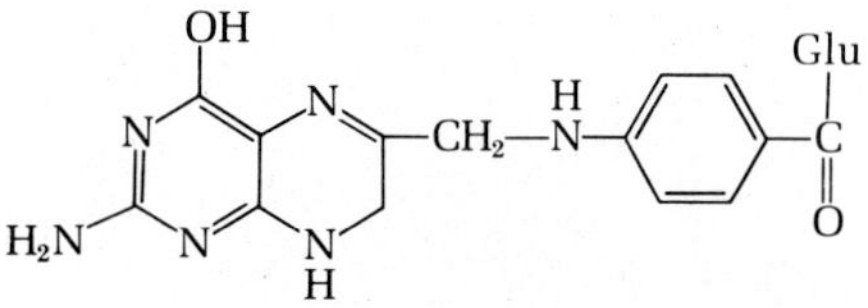

Dihydrofolic acid
(normal substrate)

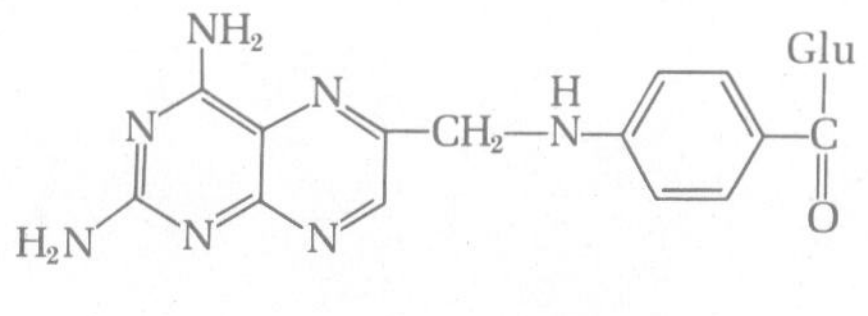

Aminopterin

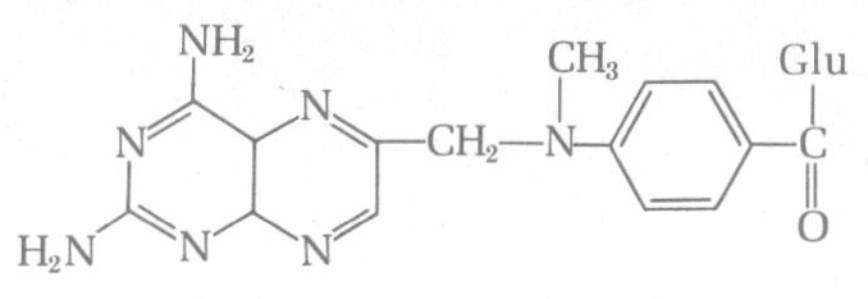

Amethopterin

This reaction is rather complex and requires two proteins or subunits, as well as Mg^{2+}. The complete system can reduce any of the four ribonucleoside 5′-diphosphates to yield the corresponding 2-deoxy-D-ribose analogs.

In various other microorganisms, including some species of *Lactobacillus, Rhizobium, Euglena,* and *Clostridium,* however, there is a different pathway, which in most cases requires the ribonucleoside 5′-triphosphates rather than the diphosphates. Moreover, it requires coenzyme B_{12} and can use either thioredoxin or dihydrolipoic acid (page 347) as reducing agent. Since coenzyme B_{12} is capable of acting as an intermediate carrier for the exchange of a hydrogen atom on one carbon atom for a substituent group on a vicinal carbon atom (page 350), the mechanism of reduction of the ribose moiety to 2-deoxyribose in *Lactobacillus* may possibly involve not only carbon atom 2 but also an adjacent carbon atom.

Formation of Deoxythymidylic Acid

DNA contains thymine (5-methyluracil) instead of the uracil present in RNA. Deoxythymidylic acid (dTMP) is formed from deoxyuridylic acid (dUMP) by *thymidylate synthetase,* which catalyzes the methylation of the uracil moiety in a reaction requiring the participation of a folic acid coenzyme, N^5,N^{10}-methylenetetrahydrofolate (Figure 26-13) as methyl donor. The formation of thymidylate and thus of DNA is greatly retarded by the drugs *aminopterin* and *amethopterin* (Figure 26-13). These agents, which are called *anti-folic* drugs, slow the growth of some types of cancer, particularly leukemias. Because of their structural resemblance to dihydrofolate, they inhibit competitively the conversion of dihydrofolate into tetrahydrofolate by *dihydrofolate reductase.*

$$NADPH + H^+ + dihydrofolate \longrightarrow NADP^+ + tetrahydrofolate$$

The formation of thymidylate is particularly sensitive to depressed levels of tetrahydrofolate. In this way DNA synthesis in the rapidly dividing malignant leukocytes is inhibited.

Regulation of Deoxyribonucleotide Biosynthesis

In *E. coli* one set of allosteric mechanisms for regulation of deoxyribonucleotide biosynthesis (Figure 26-14) involves the reaction between reduced thioredoxin and the NDPs, catalyzed by the ribonucleoside-diphosphate reductase system described above. The reduction of CDP to dCDP and of UDP to dUDP by this enzyme is strongly accelerated by ATP, whereas the reduction of ADP to dADP and of GDP to dGDP is stimulated by dGTP and dTTP. On the other hand, dATP acts as a feedback inhibitor for the reduction of all the ribonucleoside 5′-diphosphates.

Since 5-phosphoribose 1-pyrophosphate is the ultimate source of the pentose units of both the ribonucleotides and

Figure 26-14
Allosteric control mechanisms in the bio-synthesis of deoxyribonucleoside 5'-triphosphates. ATP, dGTP, and dTTP are important stimulatory modulators and dATP is a general inhibitor.

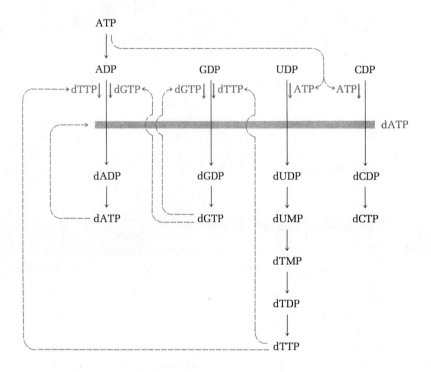

deoxyribonucleotides, regulation of ribosephosphate pyro-phosphokinase by the allosteric inhibitors ADP and GDP (see above) is also important in the biosynthesis of deoxy-ribonucleotides.

Degradation of Purines

In higher animals nucleotides resulting from degradation of the nucleic acids by the action of nucleases (page 323) usually undergo enzymatic hydrolysis to yield, ultimately, the free purine and pyrimidine bases. If not salvaged and reused (see below), the free bases are degraded further and the end products excreted. In some vertebrates, including the primates, the Dalmatian dog, birds, and some reptiles, the end product of purine degradation is *uric acid*, whereas in other mammals and reptiles, and also in mollusks, the end product is *allantoin*. In fishes, allantoin is broken down to *allantoic acid* and urea. In aquatic invertebrates ammonia is the major nitrogenous end product of purine catabolism. Curiously, guanine is the excretory form of purines in the spider and the pig. The reactions shown in Figure 26-15 outline the major steps in the degradation of purines.

The degradation of purines to the end product uric acid in man has been intensively studied, since genetic aberrations of this pathway are known. The major purines adenine and guanine are first converted into *xanthine* (Figure 26-15),

Figure 26-15
Degradation of purines to nitrogenous
excretory products.

AMP

H_2O

P_i

H_2O

AMP
deaminase

adenosine deaminase

NH_3

Adenosine

H_2O

NH_3

IMP   Inosine

PP_i

hypoxanthine
phosphoribosyl
transferase

H_2O

hydrolase

PRPP

Ribose

OH

Hypoxanthine

NH_3 H_2O

Adenine

H_2O, O_2

xanthine oxidase

H^+, O_2^-

OH

Xanthine

NH_3 H_2O

Guanine

H_2O, O_2

xanthine oxidase

H^+, O_2^-

Uric acid
(see above, right)

OH

Uric acid (enol form),
excreted by primates,
Dalmatian dogs, etc.

$\frac{1}{2}O_2 + H_2O$

urate oxidase

CO_2

Allantoin, excreted by
other mammals, turtles,
mollusks

H_2O

allantoinase

Allantoic acid,
excreted by
some fishes

H_2O

allantoicase

Urea, excreted by most
fishes and amphibia

$2H_2N-C-NH_2$

+

COOH

Glyoxylic acid C—H

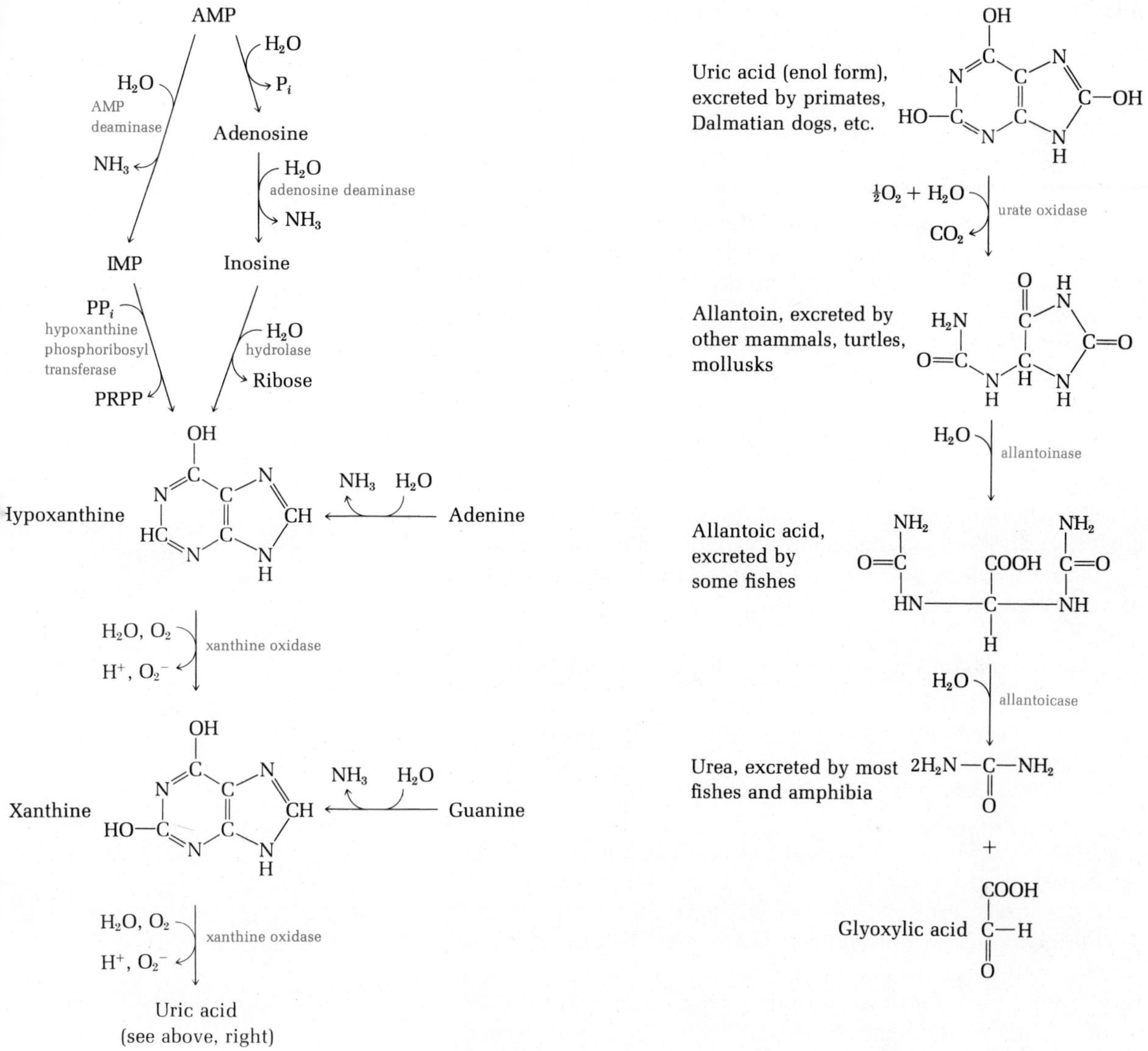

which is then oxidized by the complex flavoprotein *xanthine
oxidase* to *uric acid:*

$$\text{Xanthine} + H_2O + O_2 \longrightarrow \text{uric acid} + O_2^-$$

Superoxide

The superoxide radical (page 503) undergoes conversion to
hydrogen peroxide by the action of superoxide dismutase
(page 503). Isotopic studies on vertebrates that excrete uric
acid have shown it to derive from both exogenous and endogenous nucleic acids. Only about 0.5 g of uric acid is
excreted daily by the normal person, although up to 5 g of
free purines are formed daily; evidently the greater part of
the free purines is salvaged or recycled, by the pathway
described below. Uric acid is present in blood largely as
monosodium urate; however, both the free acid and the urate
salts are relatively insoluble in water, with the result that in
some individuals uric acid precipitates and crystallizes in

741

the urine, forming kidney stones and causing damage to this organ. Uric acid deposits are also formed in cartilaginous tissues, to produce the disease gout, which apparently results from overproduction of uric acid. This disease can be alleviated by treatment with the drug allopurinol (Figure 26-16), an analog of hypoxanthine. Allopurinol inhibits xanthine oxidase and thus decreases the formation of uric acid.

Figure 26-16
Allopurinol, an inhibitor of xanthine oxidase. It is an analog of hypoxanthine in which the positions of nitrogen atom 7 and carbon atom 8 are reversed.

Salvage of Purines

Free purines formed from nucleotides (Figure 26-15) are salvaged in vertebrates for reuse in nucleotide and nucleic acid biosynthesis. The major mechanism is by the action of adenine phosphoribosyltransferase and guanine (or hypoxanthine) phosphoribosyltransferase:

Adenine + 5-phosphoribose 1-pyrophosphate $\rightleftharpoons$ AMP + PP$_i$

Guanine + 5-phosphoribose 1-pyrophosphate $\rightleftharpoons$ GMP + PP$_i$

These important enzymes convert the free purines into the corresponding purine nucleoside 5'-phosphates for reuse.

Another salvage pathway is promoted by sequential action of purine nucleoside phosphorylase:

Purine + ribose 1-phosphate $\rightleftharpoons$ purine nucleoside + P$_i$

and nucleoside kinases, such as adenosine kinase:

Adenosine + ATP $\longrightarrow$ AMP + ADP

Since up to 90 percent of the free purines formed by man are salvaged and recycled, these salvage pathways, particularly the first described above, are very important in the purine economy of vertebrates. Deficiency in one of these salvage pathways results in the Lesch-Nyhan syndrome, a rare genetic disorder in which there is a deficiency in the enzyme guanine (hypoxanthine) phosphoribosyltransferase. This enzyme catalyzes the transfer of a ribose phosphate group from PRPP to either guanine (see above) or hypoxanthine:

Hypoxanthine + 5-phosphoribose 1-pyrophosphate $\longrightarrow$

inosinic acid + PP$_i$

When this enzyme is deficient, guanine and hypoxanthine are not salvaged and hence are degraded further to uric acid. Patients with the Lesch-Nyhan syndrome show mental symptoms, including very aggressive behavior and a bizarre tendency to self-mutilation, as well as massive deposition of uric acid in the kidneys with resulting renal failure.

Pyrimidine Degradation

In most species pyrimidines are degraded to urea and ammonia. They may also be utilized as precursors in the bio-

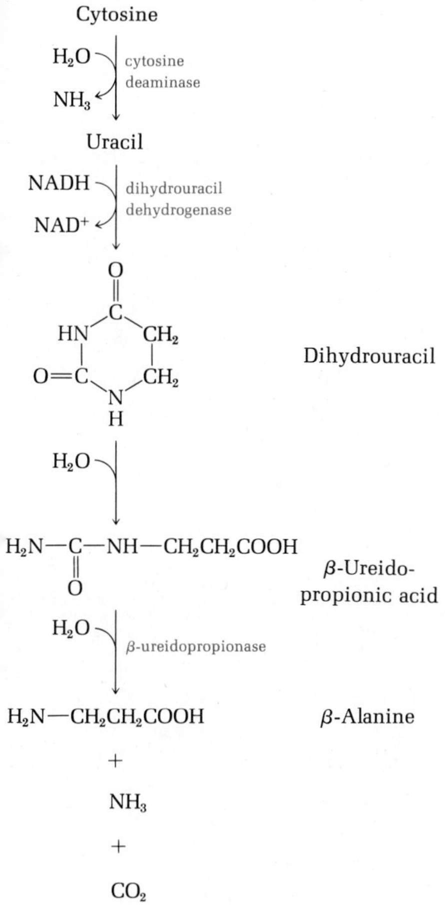

Figure 26-17
Degradation of pyrimidine bases.

Cytosine

H_2O ⤿ cytosine
          deaminase
NH_3 ⤾

Uracil

NADH ⤿ dihydrouracil
         dehydrogenase
NAD^+ ⤾

Dihydrouracil

H_2O ⤿

$H_2N—C—NH—CH_2CH_2COOH$
      ‖
      O

β-Ureido-
propionic acid

H_2O ⤿ β-ureidopropionase

$H_2N—CH_2CH_2COOH$ β-Alanine

+

NH_3

+

CO_2

synthesis of β-alanine and thus of coenzyme A. The major degradative pathway for cytosine and uracil is shown in Figure 26-17. A similar pathway leads to the formation of β-aminoisobutyric acid from thymine.

The Purine Nucleotide Cycle

Two of the reactions involved in the biosynthesis of adenylic acid (AMP) appear to participate in a cyclic mechanism described by J. Lowenstein for the deamination of aspartate in muscle tissue; this cycle also serves important regulatory functions. This pathway, the *purine nucleotide cycle*, consists of the following sequence of reactions:

$$AMP + H_2O \xrightarrow[\text{deaminase}]{\text{adenylate}} IMP + NH_3$$

$$IMP + \text{aspartate} + GTP \xrightarrow[\text{synthetase}]{\text{adenylosuccinate}} \text{adenylosuccinate} + GDP + P_i$$

$$\text{Adenylosuccinate} \xrightarrow{\text{adenylosuccinate lyase}} AMP + \text{fumarate}$$

Sum:

$$\text{Aspartate} + GTP + H_2O \longrightarrow \text{fumarate} + NH_3 + GDP + P_i$$

The net effect of this cycle is the deamination of aspartate to yield ammonia in skeletal muscle, which does not contain significant glutamate dehydrogenase, the key enzyme in amino acid deamination in liver and other tissues (page 565). The purine nucleotide cycle also serves to generate fumarate and thus other tricarboxylic acid cycle intermediates in muscle, which does not contain other enzymes capable of anaplerotic reactions (page 464).

Biosynthesis of Nucleotide Coenzymes

Flavin adenine dinucleotide, the pyridine nucleotides, and coenzyme A are derivatives of adenylic acid. Flavin mononucleotide (FMN), which is not a true nucleotide and is more accurately called riboflavin phosphate (page 339), is synthesized by the enzyme *riboflavin kinase* from free riboflavin (vitamin B$_2$) and ATP in the reaction

$$\text{Riboflavin} + ATP \longrightarrow \text{riboflavin 5'-phosphate} + ADP$$

FAD is then formed from FMN by the action of *FMN adenylyltransferase:*

$$\text{Riboflavin 5'-phosphate} + ATP \longrightarrow$$
$$\text{flavin adenine dinucleotide} + PP_i$$

NAD is synthesized from free nicotinic acid in bacteria in the following series of reactions:

$$\text{Nicotinic acid} + \text{5-phospho-}\alpha\text{-D-ribose 1-pyrophosphate} \rightleftharpoons$$
$$\text{nicotinic acid mononucleotide} + PP_i$$

Figure 26-18
Biosynthesis of coenzyme A.

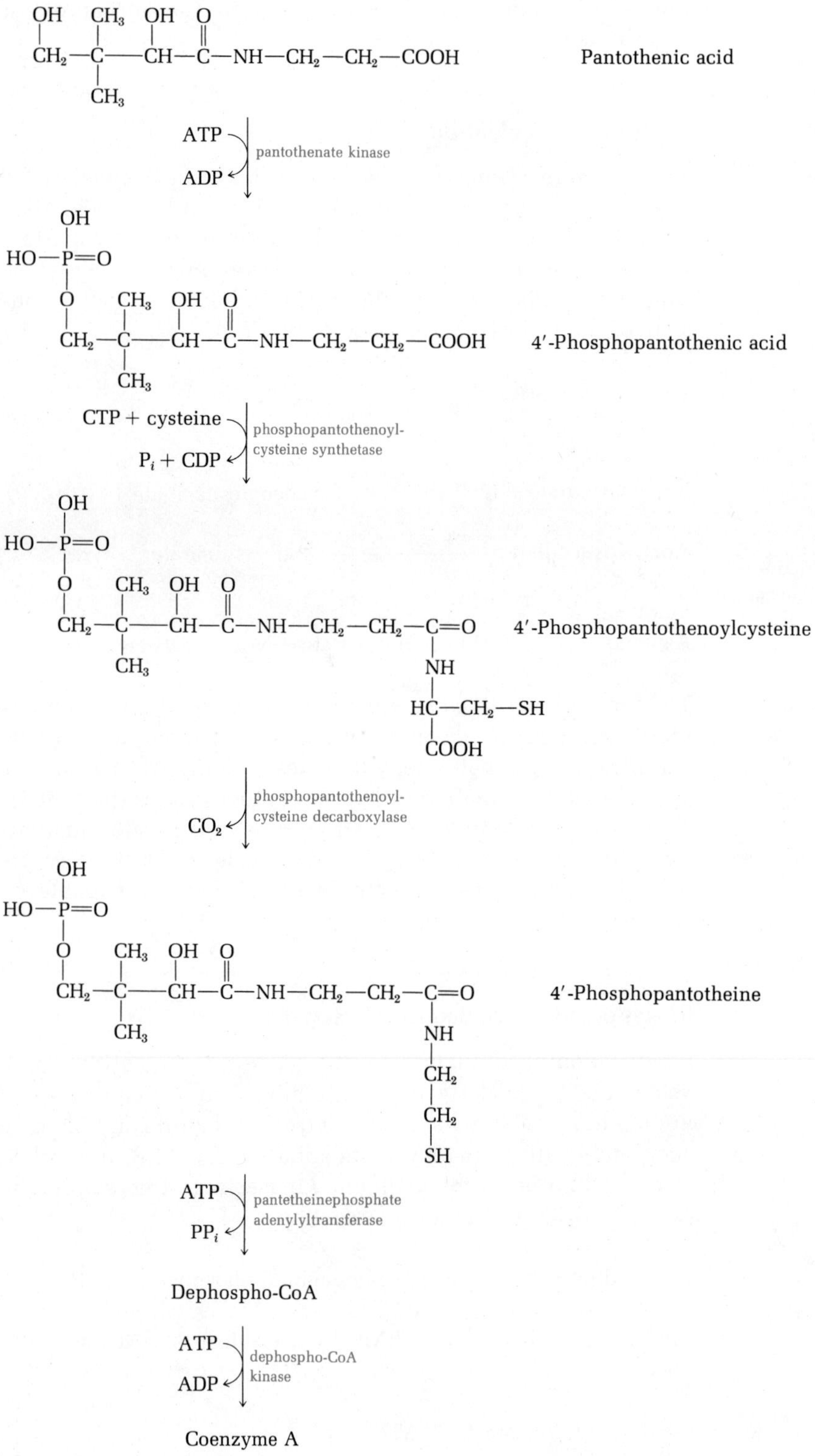

$$\underset{\underset{CH_3}{|}}{\overset{OH}{\underset{|}{CH_2}}-\overset{CH_3}{\underset{|}{C}}-\overset{OH}{\underset{|}{CH}}-\overset{O}{\overset{||}{C}}-NH-CH_2-CH_2-COOH}$$ Pantothenic acid

ATP ⎤
    ⎬ pantothenate kinase
ADP ⎦

$$HO-\overset{OH}{\underset{||}{P}}=O$$
$$\underset{O}{|}$$
$$\underset{\underset{CH_3}{|}}{\overset{}{CH_2}-\overset{CH_3}{\underset{|}{C}}-\overset{OH}{\underset{|}{CH}}-\overset{O}{\overset{||}{C}}-NH-CH_2-CH_2-COOH}$$ 4'-Phosphopantothenic acid

CTP + cysteine ⎤
              ⎬ phosphopantothenoyl-cysteine synthetase
P_i + CDP ⎦

4'-Phosphopantothenoylcysteine

$$HO-\overset{OH}{\underset{||}{P}}=O$$

$$CH_2-C(CH_3)_2-CH-C-NH-CH_2-CH_2-C=O$$
$$NH$$
$$HC-CH_2-SH$$
$$COOH$$

    ⎤
    ⎬ phosphopantothenoyl-cysteine decarboxylase
CO_2 ⎦

4'-Phosphopantotheine

$$HO-\overset{OH}{\underset{||}{P}}=O$$

$$CH_2-C(CH_3)_2-CH-C-NH-CH_2-CH_2-C=O$$
$$NH$$
$$CH_2$$
$$CH_2$$
$$SH$$

ATP ⎤
    ⎬ pantetheinephosphate adenylyltransferase
PP_i ⎦

Dephospho-CoA

ATP ⎤
    ⎬ dephospho-CoA kinase
ADP ⎦

Coenzyme A

$$\text{Nicotinic acid mononucleotide} + \text{ATP} \rightleftharpoons \text{desamido-NAD}^+ + \text{PP}_i$$

$$\text{Desamido-NAD}^+ + \text{glutamine} + \text{ATP} + \text{H}_2\text{O} \rightleftharpoons$$
$$\text{NAD}^+ + \text{glutamic acid} + \text{ADP} + \text{P}_i$$

In bacteria the amide group of NAD is inserted *after* construction of the nicotinic acid analog of NAD, desamido-NAD. In mammals, however, free nicotinamide can be directly utilized via nicotinamide mononucleotide instead of free nicotinic acid and nicotinic acid mononucleotide. NADP$^+$ is formed from NAD$^+$ by the reaction

$$\text{NAD}^+ + \text{ATP} \rightleftharpoons \text{NADP}^+ + \text{ADP}$$

Coenzyme A is assembled, starting from the free vitamin *pantothenic acid*, in the reaction sequence shown in Figure 26-18. In the last reaction, a phosphate group is introduced into the 3'-hydroxyl group of the adenosine portion of dephospho-CoA to yield the complete coenzyme A molecule.

Summary

The biosynthesis of the purine ribonucleotides adenylic acid and guanylic acid begins with the introduction of an amino group at the 1 position of 5-phospho-D-ribose 1-pyrophosphate, with loss of pyrophosphate. Glycine is then attached to the amino function to form a glycinamide derivative. Successive additions of a formyl group and an amino group to the glycinamide moiety are followed by closure of the imidazole ring. In subsequent reactions, the second ring of the purine nucleus is assembled and closed, yielding as the first purine-containing product the nucleotide inosinic acid. Inosinic acid is the precursor of adenylic acid via adenylosuccinic acid and of guanylic acid via xanthylic acid. The biosynthesis of AMP and GMP is allosterically inhibited at the first step in the sequence by adenine and guanine nucleotides.

Pyrimidine ribonucleotides are formed in a somewhat different pattern. The pyrimidine ring is synthesized in the form of orotic acid, before the attachment of ribose 5-phosphate. Uridylic acid is then formed and phosphorylated to UTP. UTP is the direct precursor of CTP. The first step in pyrimidine biosynthesis, which is catalyzed by the allosteric enzyme aspartate transcarbamoylase, is inhibited by CTP in some organisms.

Ribonucleoside 5'-diphosphates are directly reduced to the corresponding 2-deoxyribonucleoside 5'-diphosphates. In *E. coli*, electrons are transferred from NADPH + H$^+$ to the oxidized form of the protein thioredoxin, which contains a cystine residue, converting it into its dithiol form, reduced thioredoxin, the immediate reductant for conversion of the NDPs into the corresponding dNDPs. In some species, the 5'-triphosphates or diphosphates of ribonucleosides are reduced to the corresponding deoxy compounds in a reaction requiring coenzyme B$_{12}$. Deoxythymidylic acid is formed by methylation of deoxyuridylic acid.

Free purines and pyrimidines formed on degradation of nucleic acids are salvaged and reused for the synthesis of nucleotides and nucleic acids or may undergo degradation to yield nitrogenous end products that are excreted. The purines are ultimately degraded to uric acid by man and to allantoin and other end products by certain other vertebrates. In man overproduction of uric acid and its deposition in cartilage results in the disease gout. Genetic failure of the salvage of guanine and hypoxanthine results in the Lesch-Nyhan syndrome.

The biosynthesis of the nucleotide coenzymes FAD, NAD, NADP, and coenzyme A requires the participation of ATP, in addition to their corresponding precursors.

References

Books

DAGLEY, S., and D. E. NICHOLSON: *An Introduction to Metabolic Pathways*, Wiley, New York, 1970.

GREENBERG, D. M. (ed.): *Metabolic Pathways*, vol. IV, Academic, New York, 1970.

HENDERSON, J. F.: Regulation of Purine Biosynthesis, *Am. Chem. Soc. Monog.* 170, Washington, D.C., 1972.

HENDERSON, J. F., and A. R. P. PATERSON: *Nucleotide Metabolism*, Academic, New York, 1972.

REICHARD, P.: *Biosynthesis of Deoxyribose*, Wiley, New York, 1967.

STANBURY, J. B., J. B. WYNGAARDEN, and D. S. FREDRICKSON: *The Metabolic Basis of Inherited Disease*, 3d ed., McGraw-Hill, New York, 1972. Genetic disorders of purine and pyrimidine metabolism are described in detail.

Articles

BUCHANAN, J. M.: "The Amidotransferases," *Adv. Enzymol.*, 39: 91–183 (1973). These enzymes participate in purine and pyrimidine biosynthesis.

FRIEDKIN, M.: "Thymidylate Synthetase," *Adv. Enzymol.*, 38: 235–292 (1973). An important review of this important enzyme and its relationship to tetrahydrofolate coenzymes and cancer chemotherapy.

LARSSON, A., and P. REICHARD: "Enzymatic Reduction of Ribonucleotides," *Prog. Nucleic Acid Res. Mol. Biol.*, 7: 303–347 (1967).

MURRAY, A. W.: "The Biological Significance of Purine Salvage," *Ann. Rev. Biochem.*, 50: 811–826 (1971).

O'DONOVAN, G. A., and J. NEUHARD: "Pyrimidine Metabolism in Microorganisms," *Bacteriol. Rev.*, 34: 278–343 (1970).

Problems

1. In which atoms of adenylic acid isolated from a hydrolyzate of liver RNA would you expect to find the designated atoms of the following precursors: (a) the carboxyl carbon atom of acetyl-CoA, (b) carbon atom 3 of D-glucose, (c) the α carbon atom of serine, (d) the amino nitrogen atom of aspartic acid, and (e) the amide nitrogen atom of glutamine?

2. In which atoms of cytidylic acid of rat liver RNA would you expect to find the designated atoms of the following precursors: (a) the β carboxyl carbon atom of oxaloacetic acid, (b) the nitrogen atom of ammonia, and (c) the nitrogen atom of glutamic acid?

3. Calculate the number of high-energy phosphate bonds "saved" by a cell if it salvages one molecule of free adenine and reuses it for nucleic acid synthesis.

4. Write a balanced equation for the formation of CTP from CO_2, NH_3, ATP, ribose 5-phosphate, and oxaloacetic acid in *E. coli*. How many high-energy bonds are required per CTP synthesized? How many moles of D-glucose must the cell oxidize to completion in order to obtain enough ATP to synthesize 1 mol of CTP from the precursors indicated?

5. Write a balanced equation for the formation of NADP from nicotinic acid, ATP, glucose, and ammonia.

6. What labeled end products would be expected from the breakdown of adenine labeled in the following positions: (*a*) nitrogen atom 3, (*b*) carbon atom 5, and (*c*) the nitrogen atom of the 6-amino group in man? In the turtle? In amphibia?

CHAPTER 27 THE BIOCHEMISTRY OF MUSCLE AND MOTILE SYSTEMS

Figure 27-1
A notice of a lecture presented by Professor D. R. Wilkie to the Institution of Electrical Engineers in London. The subject is muscle.

Available *now*. LINEAR MOTOR. Rugged and dependable: design optimized by world-wide field testing over an extended period. All models offer the economy of "fuel-cell" type energy conversion and will run on a wide range of commonly available fuels. Low stand-by power, but can be switched within msecs to as much as 1 KW mech/Kg (peak, dry). Modular construction, and wide range of available subunits, permit tailor-made solutions to otherwise intractable mechanical problems.

Choice of two control systems:

(1) *Externally triggered mode.* Versatile, general-purpose units. Digitally controlled by picojoule pulses. Despite low input energy level, very high signal-to-noise ratio. Energy amplification 10^6 approx. Mechanical characteristics: (1 cm modules) max. speed; optional between 0·1 and 100 mm/sec. Stress generated; 2 to 5×10^{-5} newtons m^{-2}.

(2) *Autonomous mode with integral oscillators.* Especially suitable for pumping applications. Modules available with frequency and mechanical impedance appropriate for
   (a) Solids and slurries (0·01–1·0 Hz).
   (b) Liquids (0·5–5 Hz): lifetime $2 \cdot 6 \times 10^9$ operations (typ.) $3 \cdot 6 \times 10^9$ (max.)—independent of frequency.
   (c) Gases (50–1.000 Hz).

Many *optional extras* e.g. built-in servo (length and velocity) where fine control is required. Direct piping of oxygen. Thermal generation. Etc.

Good to eat.

In the preceding chapters of Part 3 we have seen how the chemical energy of ATP is utilized to make the biosynthesis of major cell components possible through coupled enzymatic reactions. We now turn to another mode of utilization of energy generated by catabolic reactions, the transformation of the chemical energy of ATP into the mechanical energy of contraction and motion, a major route of energy utilization in animals. For example, the musculature of an adult man in the resting state utilizes some 30 percent of the total ATP energy generated by respiration. During very intense muscular activity, as in a sprint, the muscles consume 85 percent or more of the total ATP generated.

The performance of mechanical work is by no means limited to a few specialized tissues such as muscle. Actually, nearly all eukaryotic cells contain contractile filaments and microtubules, which participate in several important cell functions: (1) the organization of the cell contents, (2) spindle formation and cell division, (3) the activity of cilia and flagella, (4) the transport of materials within cells, and in other cell activities involving mechanical stress, torsion, or translocation.

The conversion of the chemical energy of ATP into the mechanical energy required to bring about contraction, translocation, and propulsion is one of the most challenging problems in biochemistry. One reason is that we lack simple conceptual models for this type of energy conversion. In fact, there are no common man-made machines that convert chemical energy directly into mechanical energy.

Another reason is that the mechanism of contractile activity is more than a biochemical problem. Most of the recent advances in our knowledge have come from the correlation of biochemical, biophysical, and electron-microscopic approaches. We shall therefore examine muscular contraction and other mechanochemical activities from this broader point of view. Indeed, muscles should be looked upon as highly efficient and versatile energy-converting devices, with remarkable engineering and performance characteristics when compared to man-made machines (Figure 27-1). More-

Figure 27-2
Electron micrographs of skeletal muscle.

Longitudinal section of frog muscle in contracted state, showing the characteristic cross striations in two parallel myofibrils

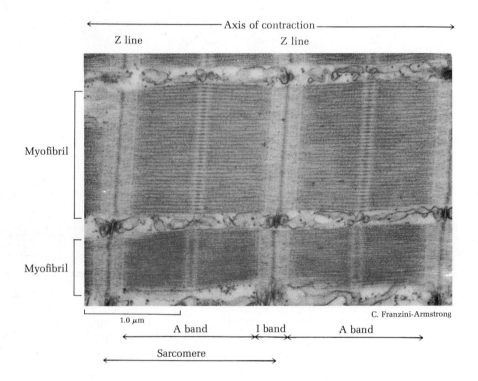

C. Franzini-Armstrong

Transverse section of myofibril of blowfly flight muscle, showing hexagonal array of myofilaments

over, they are capable of developing enormous tensions, in the neighborhood of 3.5 kg per square centimeter of cross section. In fact, the total tension that can be achieved by all the muscles of the human body is some 30 tons.

Ultrastructural Organization of Skeletal Muscle

The contractile cells of skeletal muscle are extremely long and multinucleated. Their plasma membrane is known as the *sarcolemma*. A large part of the volume of muscle cells is occupied by their contractile elements, the *myofibrils*, which are arranged in parallel bundles in the axis of contraction; each myofibril contains many myofilaments (Figure 27-2). The myofibrils in turn are surrounded and bathed by the *sarcoplasm*, the intracellular fluid of muscle, which contains glycogen, glycolytic enzymes and intermediates, ATP, ADP, AMP, phosphate, phosphocreatine, creatine, and inorganic electrolytes, as well as a number of amino acids and peptides. In continuously active, aerobic muscles, such as the flight muscles of birds and insects, the mitochondria are profuse and are regularly arranged along the myofibrils, with their positions relatively fixed. Such muscles are often called red muscles because of their high content of myoglobin and cytochromes. In white skeletal muscles, such as the little-used breast muscles of the chicken and turkey, the mi-

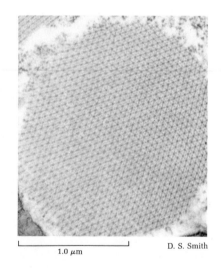

D. S. Smith

Figure 27-3
Schematic representation of the structure of
myofibrils. (Redrawn from S. C. Keene, in W.
Bloom and D. W. Fawcett, Textbook of His-
tology, W. B. Saunders Company, Philadelphia, 1968.)

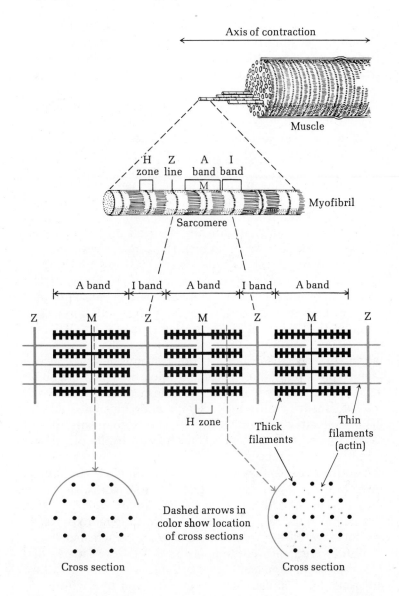

Figure 27-3
Schematic representation of the structure of
myofibrils. (Redrawn from S. C. Keene, in W.
Bloom and D. W. Fawcett, Textbook of His-
tology, W. B. Saunders Company, Philadelphia, 1968.)

tochondria are sparse and irregularly spaced. Muscle cells
also contain highly differentiated endoplasmic reticulum,
referred to as the *sarcoplasmic reticulum*; its structure and
function are described later.

Myofibrils, which are long, thin bundles of myofilaments,
manifest along their length a structural pattern that repeats
every 2.5 μm, more or less (Figures 27-2 and 27-3). These
repeated units, called *sarcomeres*, are the contractile units of
the myofibrils. In skeletal or striated muscle, the sarcomeres
of many parallel myofibrils are in transverse register,
yielding characteristic cross striations across the muscle cell,
the result of the alternation of light and dark transverse
bands. The light bands are called *isotropic* or *I bands*; the
dark bands are *anisotropic* or *A bands*. Isotropic structures
are those having uniform physical properties regardless of
the direction in which they are measured, whereas in anisotropic structures the physical properties depend on the
direction of measurement. The A bands of muscle are optically anisotropic; i.e., their index of refraction is not uni-

form in all directions, giving them the property of _double refraction_, or _birefringence_. In general, birefringence is shown by solids with asymmetric molecular components oriented in one direction.

In resting muscle, the dark A bands extend along the axis of contraction and are about 1.6 μm long; the I bands are about 1.0 μm long. The I cross striations are bisected by a dense transverse line about 80 nm thick, the Z line. The central portion ($\sim$0.5 μm) of the A band, the _H zone_, is less dense than the rest of the band; it is also bisected by a dense transverse line, the _M line_. The sarcomere, the complete longitudinal repeat unit, extends from one Z line to the next (Figures 27-2 and 27-3).

Detailed examination with the electron microscope, particularly by H. E. Huxley and by J. Hanson, has revealed that there are two types of myofilaments, thick and thin (Figure 27-3). In the transverse I bands, only thin filaments are present; they are about 6 nm in diameter. The dense portions of the transverse A bands contain not only the thin filaments found in the I band but also a regular array of thick filaments (diameter 15 to 17 nm), which give the A bands their characteristic birefringence. Cross sections of muscle made at different points have shown that the thick filaments are about 45 nm apart and are arranged in a hexagonal pattern. Each thick filament is surrounded by six thin filaments, also in hexagonal array (Figure 27-4). The thick filaments extend continuously from one end of the A band to the other. The thin filaments, on the other hand, are not continuous through the entire A band: they begin at the Z line, are continuous through the I bands, and extend into the A bands. However, they terminate at the edge of the central H zone of the A bands. Since the thin filaments are also in lengthwise register, the edges of the H zone are distinctly demarcated.

Within the dense portions of the A bands, regularly disposed projections extending from the thick filaments toward adjacent thin filaments have been observed by high-resolution electron microscopy; these projections form cross bridges. In skeletal muscle there are six such projections in a period of about 43 nm. They are arranged in pairs, each pair being rotated by about 120° from the preceding pair (Figure 27-4). The cross bridges represent the only structural connection between the thick and thin filaments.

Changes in the Sarcomere during Contraction

When they are maximally contracted, the sarcomeres, which are the contractile units of myofibrils, shorten by anywhere from 20 to 50 percent. When muscle is passively stretched, the sarcomeres may extend to about 120 percent of their resting length. From careful microscopic measurements of the length of the A bands and I bands in intact muscle in the relaxed, contracted, and stretched states, it has been demonstrated that the A bands, and thus the thick filaments responsible for their anisotropy, always remain constant in length. The thin filaments likewise undergo no change in length. It has therefore been concluded that changes in muscle length

Figure 27-4
The arrangement of cross bridges on the thick filaments in frog sartorius muscle, deduced from low-angle x-ray diagrams.
[From H. E. Huxley and W. Brown, J. Mol. Biol., 30: 383 (1967).]

The 6:2 helical arrangement of cross bridges on a thick filament

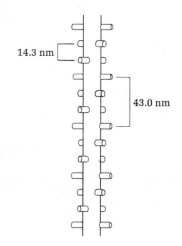

Cross section showing superlattice of thick and thin filaments and the arrangement of cross bridges

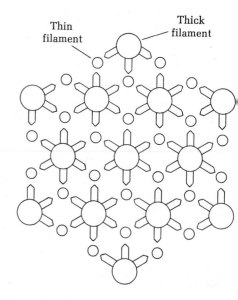

Figure 27-5
Schematic representation of sliding filaments during contraction.

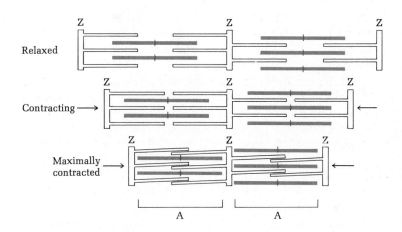

must be due to sliding of the thick and thin filaments along each other, so that the degree of penetration of the thin filaments into the zones of thick filaments varies (Figure 27-5). When muscle undergoes maximal shortening, the thin filaments may even slide past each other to form a new, dense central band within the H zone. These observations suggest that the cross bridges between the thick and thin filaments in the dense portions of the A band are rapidly formed and broken as the filaments slide along each other.

The Protein Components of Muscle Cells

The intracellular proteins of skeletal muscle cells consist of the water-soluble proteins of the sarcoplasm, which make up about 20 to 25 percent of the total muscle protein, and the water-insoluble filamentous proteins of the myofibrils. The soluble sarcoplasmic proteins are easily extracted from muscle with cold water; this fraction is often called myogen. The myogen fraction is rich in glycolytic enzymes, and from it the enzymes aldolase, glyceraldehyde-phosphate dehydrogenase, glycerol-phosphate dehydrogenase, and phosphorylase, among others, may be isolated in large yields.

The water-insoluble proteins of the myofilaments have proved to be much more difficult to separate and to obtain in pure form. The modern era of biochemical research on the contractile elements of muscle began with the classic investigations of J. T. Edsall and A. von Muralt in the early 1930s on the muscle protein myosin. After first removing the soluble myogen fraction of minced rabbit skeletal muscle by extraction with cold water, they found that much of the protein of the remaining insoluble muscle residue can be removed in soluble form by prolonged extraction with cold 0.6 M KCl solutions. The protein so extracted can then be precipitated by diluting or dialyzing the extract to reduce the KCl concentration; the precipitated protein can be redissolved, i.e., salted in (page 162), in 0.6 M KCl. This protein fraction, insoluble in water but soluble in 0.6 M KCl, was called myosin; it is now known that this fraction contains considerable actin, as we shall see.

Two lines of evidence suggested that the fraction called myosin is a major component of the anisotropic bands of skeletal muscle. First, Edsall and von Muralt showed that solutions of myosin possess the property of *double refraction of flow* or *flow birefringence.* This phenomenon is shown by solutions of long, thin, highly asymmetric molecules. When such solutions are unstirred, the solute molecules are randomly oriented; consequently, the refractive index is identical in all directions. However, when such a solution is forced to flow through a narrow capillary, the long solute molecules orient themselves in the direction of flow, causing the solution to become optically anisotropic. A second line of evidence suggesting that myosin is the major component of the A bands came later, from microscopic examination of the appearance of intact muscle fibers before and after the extraction of myosin with 0.6 M KCl. After extraction of the myosin, the A bands became less dense and lost most of their anisotropic character.

In 1939, V. A. Engelhardt and his wife M. N. Lyubimova, in Moscow, found that myosin prepared by the method of Edsall possesses enzymatic activity; in the presence of Ca^{2+} ions, myosin catalyzes hydrolysis of the terminal phosphate group of ATP to yield ADP and inorganic phosphate. Shortly thereafter, they showed that if solutions of myosin in 0.6 M KCl are squirted through a fine hypodermic needle into cold distilled water, the myosin is precipitated in the form of fragile threads. When ATP and Ca^{2+} ions are added to the medium bathing such artificial fibers of myosin, their length and extensibility undergo marked alteration; simultaneously, the ATP undergoes hydrolysis. These observations strongly supported the view that myosin is a major component of the A bands and that it is involved in the energy-dependent contractile process.

During World War II A. Szent-Györgyi and his colleagues in Hungary made a number of new discoveries regarding the molecular components of the myofibrils which greatly clarified the relationship of myosin to the contractile process. In

Table 27-1 Protein components of the myofilaments of rabbit skeletal muscle

Location	Subunits	Mol. wt.
Thick filaments		
Myosin	Two heavy chains, four light chains	460,000
C-protein		140,000
M-line protein I		155,000
M-line protein II		88,000
Thin filaments		
F-actin	nG-actin	$(46,000)_n$
Tropomyosin	Two chains	70,000
Troponin	Three subunits	80,000
α-Actinin		100,000
β-Actinin		70,000

Figure 27-6
Schematic representation of the myosin molecule, showing its globular head and long tail and the points of enzymatic fragmentation. The polypeptide chains in color are the light chains.

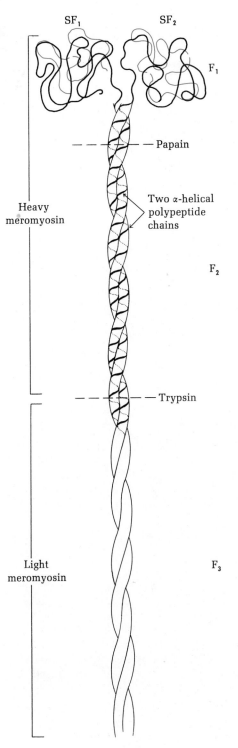

SF$_1$

SF$_2$

F$_1$

Papain

Heavy meromyosin

Two α-helical polypeptide chains

F$_2$

Trypsin

Light meromyosin

F$_3$

particular, they found that skeletal muscle contains another major protein component, which they called _actin_.

Today it is recognized that there are two major protein components in the water-insoluble, filamentous elements of skeletal muscle, _myosin_ and _actin_. They make up about 80 percent of the proteins of the contractile apparatus (Table 27-1). In addition, there are at least six other protein components, _tropomyosin_, _troponin_, _C-protein_, _M-line protein_, and _α- and β-actinin_. The two major proteins found in myofilaments can be selectively extracted from muscle after removal of the myogen fraction. Myosin can be extracted from the myogen-free residue of muscle with cold alkaline 0.6 M KCl. The muscle residue is then dried with acetone and the actin extracted with cold, slightly alkaline water; it is obtained in a globular form called _G-actin_. Further extraction of the residue at higher temperatures yields tropomyosin.

Myosin

Myosin, the major component of the thick muscle filaments, can be purified by cycles of solution in 0.6 M KCl and precipitation by dilution or dialysis. Myosin has been subjected to intensive physicochemical study by many investigators, but for a very long time there was still considerable ambiguity and disagreement about the size, subunit composition, and shape of the myosin molecule. These difficulties arose in part from the highly asymmetrical nature of myosin, which causes it to show substantial deviations from ideal behavior in hydrodynamic measurements, such as sedimentation and diffusion. Moreover, myosin molecules tend to associate tightly with each other and with other myofilament proteins, a property that can give an illusion of homogeneity.

Recent studies on highly purified myosin indicate conclusively that it has a molecular weight of about 460,000. It is a long (~160 nm) asymmetric molecule containing a globular head, as schematically represented in Figure 27-6. It contains two identical long polypeptide chains of ~200,000 daltons, called _heavy chains_, which can be separated on treating myosin with concentrated urea or guanidine solutions. They are the longest single polypeptide chains known to occur biologically, each having about 1,800 amino acid residues. Their complete amino acid sequence has not yet been established, and its determination remains a huge task. Skeletal-muscle myosin contains three unusual amino acids, 3-methylhistidine, ε-N-monomethyllysine, and ε-N-trimethyllysine (Figure 27-7), which are located in the head of the myosin molecule. Only 4 of the 170 lysine residues and only 1 of the 35 histidine residues of myosin are methylated; presumably they play some specialized role in the function of the myosin head.

Through most of the length of the myosin molecule, each heavy chain is in α-helical conformation and the two chains are wound around each other (Figure 27-6). Both the heavy polypeptide chains are folded into globular structures to form the head. In addition, the head contains four smaller, or _light_, polypeptide chains. Two of the light chains are identical and have a molecular weight of about 18,000; depending on how they are extracted, they may contain phosphorylated serine residues. The two other light chains have molecular weights of 16,000 and 21,000. Myosin thus contains six polypeptide chains, two heavy and four light.

Highly purified myosin preparations can hydrolyze the terminal phosphate group of ATP; they also attack GTP, ITP, and CTP. The ATPase activity of myosin is distinctive in that it (1) is stimulated by Ca^{2+}, (2) is inhibited by Mg^{2+}, (3) is profoundly influenced by KCl concentration, and (4) has two pH optima, one at pH 6.0 and the other at pH 9.5. In addition, myosin ATPase activity is characteristically dependent on two different sets of sulfhydryl groups which differ in their susceptibility to alkylation or to mercaptide formation, although they appear to be close together in the myosin molecule. When the more susceptible —SH groups are blocked, the ATPase activity of the myosin molecule is increased, suggesting that these sulfhydryl groups are normally inhibitory to the enzyme. When the second class of sulfhydryl groups is then blocked, the ATPase activity is completely abolished, suggesting that these —SH groups are required for the hydrolytic process.

Much important information regarding the structure of myosin and its ATPase activity has come from study of myosin fragments. When myosin is briefly exposed to trypsin or chymotrypsin, specific peptide linkages near the center of the tail are cleaved to yield a heavy and a light fragment, called *heavy meromyosin* and *light meromyosin*, respectively (Figure 27-6). These fragments have been isolated in purified form. They are also designated as $F_1 + F_2$ and F_3, respectively. On the other hand, when myosin is treated with the proteolytic enzyme papain, a different type of fragmentation results. It yields two fragments from the head, called SF_1 *fragments*, and a long fragment ($F_2 + F_3$), corresponding to the entire tail. If heavy meromyosin ($F_1 + F_2$) is treated further with trypsin, the F_1 portion is recovered in the form of two SF_1 fragments, identical to those obtained with papain. All these fragments have been recovered and purified; their characteristics are summarized in Table 27-2. Of particular interest is the finding that the ATPase activity of myosin survives these fragmentation procedures; it is located in the SF_1 fragments. It has therefore been concluded that the ATPase activity of the myosin molecule resides entirely in its head and that there are two catalytic sites, one in each of the two SF_1 fragments. Each contains an inhibitory and a catalytic sulfhydryl group. The light chains in the myosin heads are concerned in the binding of ATP and in the ATPase activity. From the head end of the myosin molecule an octapeptide capable of binding ATP has been isolated. It has been postulated to represent part of the active site of the ATPase. The amino acid sequence of this peptide has been deduced; it contains a cysteine residue, which is believed to represent that required for ATPase activity. Figure 27-8 shows a postulated model for the binding of ATP to this peptide, showing multiple sites of attachment between the ATP molecule and the octapeptide.

Figure 27-7
Unusual amino acids found in myosin.

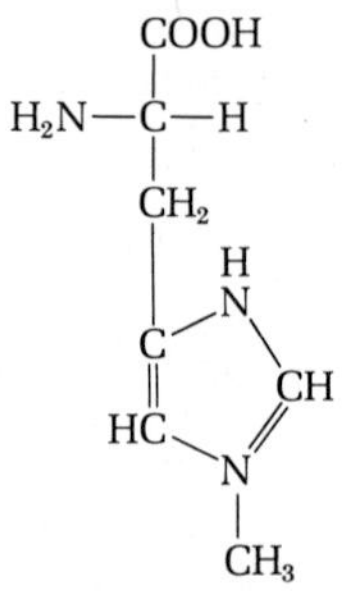

3-Methylhistidine

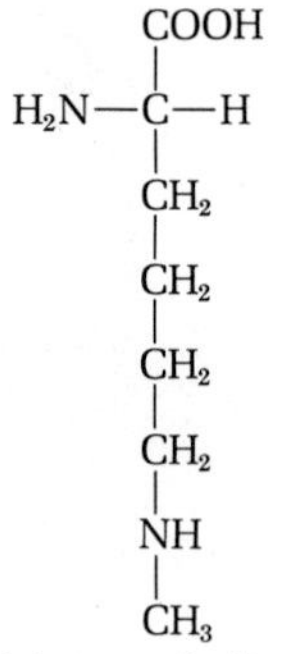

ε-N-Monomethyllysine

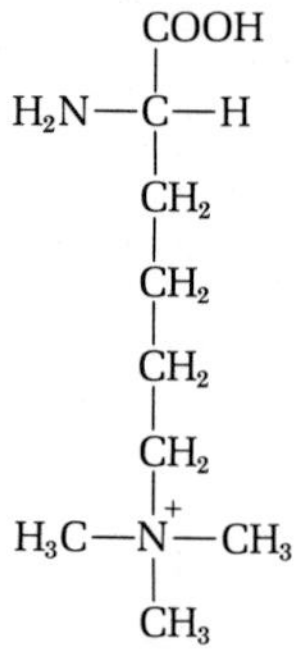

ε-N-Trimethyllysine

Table 27-2 Fragments resulting from enzymatic cleavage of myosin (see Figure 27-6)

	Particle weight, daltons	Length, nm
Myosin	460,000	150–160
$F_1 + F_2$	350,000	75
$F_2 + F_3$	210,000	135
F_2	61,000	46
F_3	150,000	86
SF_1	115,000	150

Figure 27-8
Structure of an ATP-binding peptide iso-
lated from myosin, showing how it binds
MgATP^{2-} (color). [Redrawn from G. Barany
and R. B. Merrifield, Cold Spring Harbor
Symp. Quant. Biol., 37: 122 (1972).]

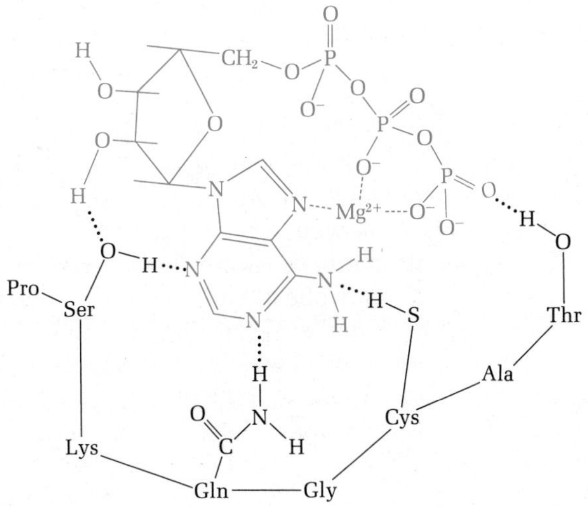

A second important conclusion has been drawn from studies of myosin fragments. Purified, intact myosin binds actin at two specific sites to form *actomyosin*, a phenomenon which represents a very crucial step in the contractile mechanism, as we shall see. When actin is bound to myosin, an interesting and significant change in the ATPase activity occurs. Whereas the ATPase activity of pure myosin requires Ca^{2+} and is inhibited by Mg^{2+}, the ATPase activity of actomyosin is stimulated by Mg^{2+}. Moreover, it has been found that the entire actin-binding activity of myosin resides in the SF_1 fragments bearing the ATPase activity. From such experiments it has been concluded that the two catalytic ATPase sites on the head of the myosin molecule are located at or near the actin binding sites.

The Mechanism of Myosin ATPase Activity

Because of the obvious possibility that the hydrolysis of ATP by myosin is related to the generation of force, the ATPase activity of myosin has been intensively studied. Of special interest is the nature of the enzyme-substrate complex(es) formed during myosin ATPase activity. Although it has often been proposed that the myosin molecule becomes phosphorylated on some specific amino acid residue on interaction with ATP, no conclusive evidence for the formation of a covalent phosphoenzyme intermediate exists. However, kinetic data have suggested that perhaps three different enzyme-substrate complexes, noncovalent in nature, are formed during the ATPase activity of myosin. The basic observation is that when equimolar amounts of ATP and myosin are mixed, there is a very rapid appearance of free H^+ in the medium but a much slower appearance of free phosphate and ADP. The H^+ arises as one of the products of hydrolysis of ATP at pH near 7.0:

$$ATP^{4-} + H_2O \longrightarrow ADP^{3-} + HPO_4^{2-} + H^+$$

Since free ADP and phosphate do not appear in the medium during the early burst of H^+ formation, it has been concluded that they remain bound to the enzyme during this period, presumably as a complex of myosin with its products ADP and phosphate, similar to the EP (P for product) complex formed during the action of all enzymes. Breakdown of this complex could presumably be the rate-limiting step in ATP hydrolysis by myosin. However, it has been found that such a slowly decomposing complex cannot be formed simply by mixing myosin, ADP, and phosphate, suggesting that at least two enzyme-product complexes are formed, which are not rapidly interconvertible. This and other evidence has led to the hypothesis that these two enzyme-product complexes differ in their conformation. Various hypotheses have been proposed for the mechanism of myosin ATPase activity. A simple version is given in the following equations (M is myosin and M* is an energized conformation of the myosin molecule):

$$M + ATP \rightleftharpoons M \cdot ATP \tag{1}$$

$$M \cdot ATP + H_2O \rightleftharpoons M^* \cdot ADP \cdot P + H^+ \tag{2}$$

$$M^* \cdot ADP \cdot P \longrightarrow M \cdot ADP \cdot P \tag{3}$$

$$M \cdot ADP \cdot P \rightleftharpoons M + ADP + P \tag{4}$$

Reaction (2) is that giving the early rapid burst of H^+ formation. The complex $M^* \cdot ADP \cdot P$ is postulated to be a high-energy complex, in which the free energy of hydrolysis of ATP is conserved in the form of an energized conformation of the myosin molecule. Reaction (3) is the slow, rate-limiting reaction, in which the energized complex breaks down, with release of the energy conserved in the conformation of $M^* \cdot ADP \cdot P$, to yield the low-energy complex $M \cdot ADP \cdot P$. The latter then rapidly dissociates to yield free myosin in its ground-state conformation. As we shall see below, the energy-releasing reaction (3) has been proposed to be responsible for the power stroke in contraction, in which the conformational change in the myosin head is used to slide the thin actin filaments along the thick myosin filaments.

Molecular Organization of the Thick Myofilaments

From detailed electron-microscopic investigations, particularly by H. E. Huxley and his colleagues, and from physico-chemical studies of myosin and its self-association, we now have a detailed picture of the molecular organization of the thick filaments of the A bands of rabbit skeletal muscle. Each thick filament is about 1,500 nm long and 10 nm thick. It consists of longitudinally bundled myosin molecules, each 160 nm long. The myosin molecules are oriented with their heads away from the midpoint of the thick filament. The heads project laterally out of the bundle in a regular, helical

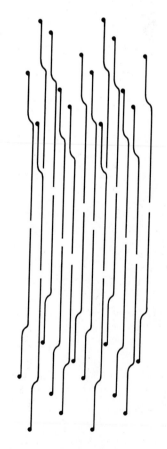

Figure 27-9
Exploded representation of the packing of myosin molecules in the center portion of a thick filament of skeletal muscle. The myosin molecules are arranged in a staggered fashion, with the heads projecting in a regular arrangement around and along the length of the filament (see Figure 27-4). The heads of the myosin molecules are oriented away from the midpoint of the thick filament. [Adapted from F. Pepe, Cold Spring Harbor Symp. Quant. Biol., 37: 106 (1972).]

Figure 27-10

Formation of cross bridges between actin and myosin filaments. The myosin molecule is postulated to have flexible joints at the points where it is susceptible to proteolytic cleavage.

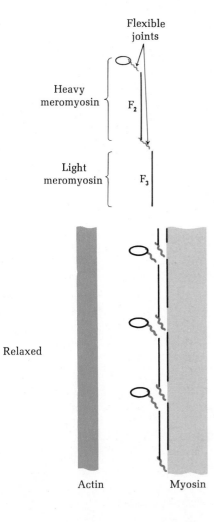

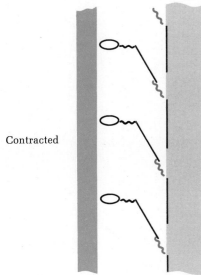

fashion, so that there are six cross bridges for each 43.0 nm, which represents one turn or repeat unit (Figure 27-4). In a cross section of the thick filament at any given point there are 18 myosin molecules in a regular packing arrangement (Figure 27-9). Altogether there are about 400 myosin molecules per complete thick filament. The myosin heads, which resemble barbs, vary in the distance they project from the thick filament, depending on the state of relaxation or contraction; they are shortest in relaxed muscle, in which they are detached from the adjacent thin myofilaments.

The schematic drawing in Figure 27-10 shows the arrangement of the myosin molecules in the thick filaments; it is particularly significant that each myosin head is located next to the hingelike trypsin-sensitive point separating the F_2 and F_3 fragments of an adjacent myosin molecule.

Recent research has revealed that the thick filaments contain two other proteins in addition to myosin. One is _C-protein_ (Table 27-1), which has been isolated from myosin preparations by careful fractionation. C-protein (mol wt 140,000) makes up about 3.5 percent of the total protein of the thick filaments. It is a highly elongated molecule and binds very strongly to the myosin tail. Immunochemical and microscopic approaches have revealed that C-protein molecules, which are about 35 nm long, are wound around the thick filaments at regular intervals. The function of C-protein is not known; one idea is that it serves as a clamp to hold the bundle of myosin molecules together. Another protein found in the thick filaments is the _M-line protein_, which may also serve to hold the bundle of myosin molecules together at the M line (Figure 27-3).

Actin

This protein occurs in two forms, _G-actin_ (globular actin) and _F-actin_ (fibrous actin), a polymer of G-actin. The monomeric G-actin is stable in distilled water, but under certain conditions it polymerizes to form F-actin. Rabbit muscle G-actin has been highly purified. It has a molecular weight of 46,000, is globular, and consists of a single polypeptide chain, much of whose amino acid sequence has been identified. G-actin contains one residue of the unusual amino acid _3-methylhistidine_, which is also present in myosin (Figure 27-7); G-actin also contains a large number of proline and cysteine residues.

Each molecule of G-actin binds one Ca^{2+} ion very tightly. It also binds one molecule of ATP or ADP with high affinity. Binding of ATP by G-actin is usually followed by its polymerization to form F-actin. For each molecule of G-actin added to the F-actin chain, one molecule of ATP is split to ADP and phosphate. The ADP so formed remains bound to the G-actin subunits of the F-actin chain. The polymerization is described by the equation

$$n(\text{G-actin–ATP}) \longrightarrow (\text{G-actin–ADP})_n + nP_i$$
F-actin

The Molecular Organization of the Thin Filaments

Two strands of F-actin, each composed of G-actin monomers, are coiled about each other in the thin filaments (Figure 27-11). Such a two-strand coil would have an average diameter of 6 nm, in agreement with the diameter of the thin I filaments of muscle as observed in electron micrographs. The thin filaments also contain two major accessory proteins, which serve a regulatory function in controlling the making and breaking of the cross bridges between the thick and thin myofilaments, as well as the generation of mechanical force. The first is _tropomyosin_, which makes up about 10 to 11 percent of the total contractile protein of muscle; it was first isolated in crystalline form by K. Bailey in the 1940s. The tropomyosin of rabbit muscle has a molecular weight of about 70,000 and contains two different kinds of α-helical polypeptide chains of molecular weight 33,000 and 37,000, which form a two-chain twisted coil about 40 nm long. Most of the amino acid sequence of the tropomyosin chains has been worked out. Tropomyosin molecules have been found to undergo end-to-end associations and, when crystallized, form square-net lattices.

The long, thin tropomyosin molecules are arranged end to end in the shallow grooves of the coiled F-actin filaments, in such a way that each tropomyosin molecule makes contact with only one of the two F-actin filaments (Figure 27-12). Each tropomyosin molecule extends along seven G-actin monomers. As we shall see, the tropomyosin molecules are not fixed in position but apparently can move along the grooves between the F-actin strands.

The second major regulatory protein present in the thin filaments is _troponin_, a large globular protein discovered by the Japanese biochemist S. Ebashi and his colleagues in 1965. Troponin contains three polypeptide subunits, each of which has a specific function; they have been given various symbols and names. The _Ca²⁺ binding subunit_ of troponin, symbolized TN-C, but also called troponin A, has a molecular weight of 18,000. Each molecule of TN-C binds two Ca^{2+} ions tightly; simultaneously, the TN-C molecule undergoes a change in conformation. The _inhibitory subunit_ of troponin, symbolized TN-I, has a molecular weight of 23,000. It has a binding site specific for actin but does not bind Ca^{2+}. Its function is to inhibit the interaction of actin with the myosin head cross bridges. It is of special interest that the TN-I subunit is phosporylated by phosphorylase kinase, which normally activates phosphorylase b to the active phosphorylase a (page 435). The third component of troponin, TN-T, is the _tropomyosin-binding subunit_. The complete troponin molecule contains one each of the TN-C, TN-I and TN-T subunits and has a globular shape.

Each troponin molecule is attached to the thin filament by two binding sites, one specific for an actin strand and the other for a tropomyosin strand. The binding to tropomyosin is apparently at a fixed site, whereas the binding to the actin filament undergoes making and breaking depending on the binding of Ca^{2+} (see below). One troponin molecule occurs every 40 nm along the thin filament; this periodicity is about

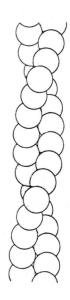

Figure 27-11
F-actin strands in the thin filaments, each consisting of G-actin monomers. The two strands are coiled around each other.

Figure 27-12
Arrangement of tropomyosin and troponin molecules in the thin filaments and their relationship to the myosin heads. In the end view (below) a cross section of a thin filament is shown, with two actin strands and the attached tropomyosin strands. The heads of two myosin molecules from thick filaments adjacent to the thin filament are also shown. [From H. E. Huxley, Cold Spring Harbor Symp. Quant. Biol., 37: 361–376 (1972).] In the side view (right) a thin filament is shown. There is one troponin molecule at or near every seventh G-actin monomer. When it binds Ca^{2+} it apparently activates a segment of seven actin monomers, through a conformational change of the tropomyosin, so that ATP competes successfully with the actin monomers for the binding site on the myosin head.

Thick filament   Thin filament

Actin monomers

Troponin, containing binding site for Ca^{2+}

Myosin head

Tropomyosin

Hinge

ATP-binding site

Myosin tail

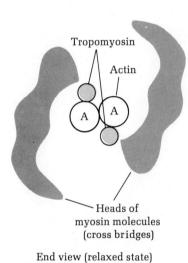

Tropomyosin

Actin

A   A

Heads of myosin molecules (cross bridges)

End view (relaxed state)

the same as that of the recurring tropomyosin molecules in the grooves. Thus, for every seven G-actin monomers there is one molecule of tropomyosin and one of troponin (Figure 27-12).

Actomyosin Complexes

When pure myosin and actin are mixed, actomyosin complexes are formed. Their formation is accompanied by a large increase in viscosity and flow birefringence. The ratio of myosin to actin and the particle weight of such actomyosin complexes depend on the experimental conditions, such as the pH, the KCl and $MgCl_2$ concentrations, as well as the concentration of protein. Actomyosin complexes can also be extracted from muscle by prolonged exposure to 0.6 M KCl. The molar ratio of G-actin monomers to myosin in actomyosin complexes may be as high as 1.7. Since an F-actin chain contains many G-actin monomers, each F-actin filament can bind many myosin molecules. Electron microscopy has not only confirmed this stoichiometry but has also revealed that only the heads of myosin molecules bind to the actin filaments, to yield structures resembling barbs (Figure 27-13).

A significant property of actomyosin complexes is that they undergo dissociation in the presence of ATP and Mg^{2+}; the dissociation is accompanied by a large and rapid decrease in the viscosity of the actomyosin solution. Following dissociation, hydrolysis of ATP occurs (Figure 27-14). When the ATP is completely hydrolyzed to ADP, the actin and myosin reaggregate. As we shall see, the dissociation and reassociation of actin and myosin are steps in the breaking and making of the cross-linkages between the thin and thick filaments of the myofibril.

The Triggering of the Thick-Filament–Thin-Filament Interaction by Ca²⁺ and the Generation of Force

We shall now put together the information developed above and examine current views of the molecular mechanism of the cyclic making and breaking of the cross bridges between the thick and thin filaments, the power stroke, and the regulation of contraction by the Ca^{2+} concentration in the sarcoplasm.

In skeletal muscle in the relaxed state, the sarcoplasm has a high $MgATP^{2-}$ concentration, but the concentration of Ca^{2+} is below the threshold required for initiation of contraction. No cross bridges are evident between the thick and thin filaments; the myosin heads are in a retracted position. Each myosin head in this state contains two molecules of tightly bound ATP. However, as indicated earlier, the ATP may not be present as such but as ADP and phosphate, both tightly bound to the energized conformation of the myosin head. The myosin head in this state of the muscle is unable to react with actin of the thin filaments because in the absence of Ca^{2+} the tropomyosin molecule masks the myosin binding site on the G-actin monomer or holds it in a conformation that is unreactive, through the action of the TN-I subunit of troponin. One tropomyosin molecule inhibits the myosin-binding activity of seven G-actin monomers.

When free Ca^{2+} is now released into the sarcoplasm by the incoming nerve signal (see below), Ca^{2+} is immediately bound to the Ca^{2+}-binding sites of troponin. Through a conformational change of the troponin molecule, the myosin-binding site on the G-actin monomer now becomes exposed and combines with the energized myosin head, with its bound ADP and phosphate, to form the _force-generating complex_, in which the myosin head is attached to a G-actin monomer.

The myosin head is now believed to undergo an energy-yielding conformational change, so that the cross bridge changes its angular relationship to the axis of the heavy filament, causing the thin filament to be moved along the thick filament; this is the _power stroke_ (Figure 27-15). Simultaneously, the myosin head reverts from its energized to its deenergized conformation and the bound ADP and phosphate leave their binding sites on the myosin. Two ATP molecules are ultimately hydrolyzed per cross bridge.

An alternative hypothesis has been proposed for the mechanism of the power stroke. It will be noted (Figure 27-10) that each myosin head is very close to the flexible hinge of a neighboring myosin molecule in the thick-filament structure. It has been postulated that the hydrolysis of ATP by the myosin head, which at pH 7 causes formation of H^+, results in a local acidification (page 527) in the environment of the hinge, causing it to undergo a conformational change, possibly a pH-dependent transition from an α-helical structure to a random conformation. In this way the energy of hydrolysis of ATP is suggested to be transmitted to the hinge, thus generating the power stroke. Very little else is known about the utilization of ATP energy for the generation of force. Whatever the mechanism, there appear to be many

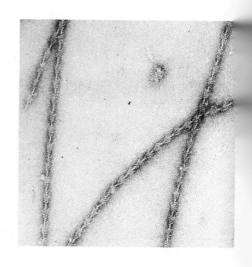

Figure 27-13

Actin filaments with attached barblike heads of myosin molecules. [From J. A. Spudich, H. E. Huxley, and J. T. Finch, J. Mol. Biol., 72: plate X (c) (1972).]

Figure 27-14

Dissociation of actomyosin by ATP in the presence of Mg^{2+}. A rapid drop of viscosity of the actomyosin solution accompanies dissociation. As the ATP is dephosphorylated, the actin and myosin reassociate, with increase in viscosity.

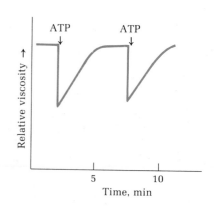

Figure 27-15
The power stroke of the myosin cross bridges. The energized myosin head, containing tightly bound ADP and P_i (step 1), makes contact with an actin monomer on activation of troponin by Ca^{2+} (step 2). This is followed by a conformational change so that the myosin head assumes a 45° angle with the tail (step 3). The bound ADP and P_i are released and replaced by ATP (step 4). Cleavage of ATP to yield bound ADP and P_i causes the head to revert to the 90° configuration, prepared to interact with the next actin monomer. [Adapted from H. G. Mannherz, J. B. Leigh, K. C. Holmes, and G. Rosenbaum, Nature New Biol., 241: 226 (1973).]

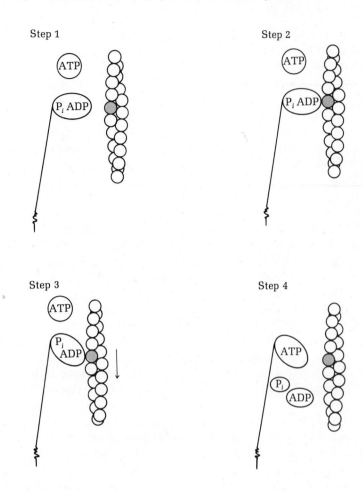

complex cooperative interactions between the various protein components of both the thick and thin filaments.

Relaxation of the muscle is brought about by resegregation of ionic Ca^{2+} from the sarcoplasm into the sarcoplasmic reticulum.

Excitation-Contraction Coupling

How is contraction triggered or set off by the incoming nerve impulse? How does the contracted muscle relax again? The stimulus for contraction of muscle is an electrical impulse arriving from the motor nerve at the motor end plate or neuromuscular junction; this impulse rapidly spreads over the sarcolemma. In resting muscle there is a potential difference across the sarcolemma such that the outside is about 60 mV more positive than the inside. As the excitation impulse spreads over the sarcolemma, this potential difference disappears, a phenomenon called _depolarization_. The depolarization is believed to be due to a sudden increase in permeability to K^+, Na^+, and Ca^{2+}, which flow in such a direction as to result in discharge of the transmembrane potential.

The mechanism by which the rapidly spreading impulse in the sarcolemma is instantly communicated to the interior of the muscle cell, some 50 μm deep, so that all filaments of the muscle fiber contract simultaneously, was one of the classical problems in muscle physiology. Simple diffusion of a chemical agent or messenger from the sarcolemma to the interior myofilaments did not seem likely as the means of communication, on the grounds that diffusion is much too slow to account for the great speed of the excitation process. The answer to this problem ultimately came from investigations with the electron microscope. Close scrutiny of the ultrastructure of muscle revealed that there are many repeating tubular invaginations of the sarcolemma or muscle-cell membrane. These tubular structures run across the muscle cell near the Z lines of the myofibrils in some muscles and near the A-I junctions in others. This complex system of transverse tubules is called the _T system_ (Figure 27-16). When the sarcolemma is excited by an incoming impulse and undergoes depolarization, the T system communicates the electrical impulse almost simultaneously to all the sarcomeres of the muscle fiber.

The mechanism by which the depolarization of the T system triggers the molecular events of contraction has been revealed by correlated electron-microscopic and biochemical investigations. Surrounding each set of adjacent sarcomeres in a myofibril is a flattened double-membrane vesicle with many perforations, resembling a lace cuff or sleeve. In the muscles of the frog and of mammals each vesicle extends from the A-I junction of one sarcomere to the A-I junction of the next (Figure 27-16). The entire sleevelike system of vesicles is called the _sarcoplasmic reticulum_; it is a specialization of the endoplasmic reticulum (page 382). The internal compartment, or cisterna, of each vesicle is connected to all others in the same cross striation by transversely oriented connections called _terminal cisternae._ Pairs of parallel terminal cisternae run across the myofibrils in close contact with the transverse tubules of the T system (Figure 27-16). When the sarcolemma is excited and the T system becomes depolarized, this change is transmitted to the membranes of the closely apposed sarcoplasmic reticulum, causing it to increase in permeability. As a result, Ca^{2+} ions escape from the cisternae of the reticulum, in which Ca^{2+} is normally segregated in resting muscle. The extremely rapid discharge of Ca^{2+} from the reticulum into the sarcoplasm triggers the interaction between the thick and thin myofilaments, as described above, and causes the hydrolysis of ATP. In the resting muscle cell ATP undergoes only very slow hydrolysis by myosin, because of inhibitory concentrations of Mg^{2+}. This inhibition is relieved when the cross bridge reacts with actin in the presence of activating concentrations of Ca^{2+} (about 10 μM). The Ca^{2+} concentration in the sarcoplasm of relaxed skeletal muscle is believed to be less than 0.2 μM. As long as motor-nerve impulses continue to arrive at the sarcolemma, Ca^{2+} remains in the sarcoplasm and keeps the muscle in the contracted state.

Relaxation

When the motor-nerve impulses cease, the sarcolemma regains its original permeability pattern and returns to its

Figure 27-16
The sarcoplasmic reticulum (color). The terminal cisternae are indicated by arrows.
[*Redrawn from L. D. Peachey, J. Cell. Biol.,*
25 (3,II): 222 (1965).]

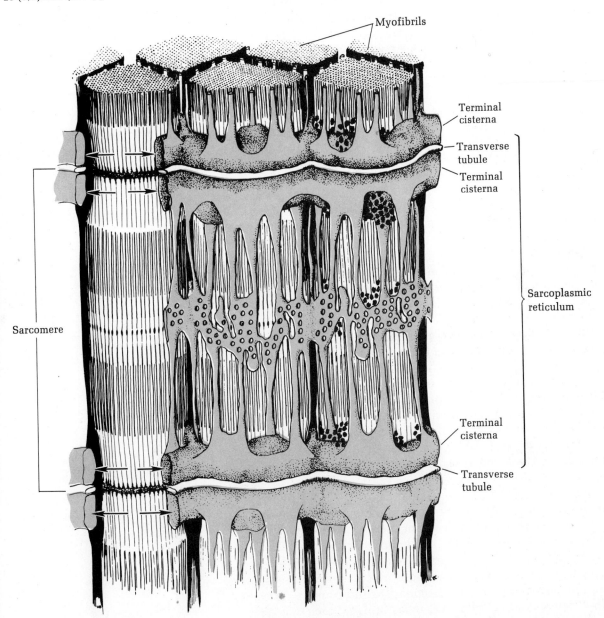

Myofibrils

Terminal
cisterna

Transverse
tubule

Terminal
cisterna

Sarcoplasmic
reticulum

Terminal
cisterna

Transverse
tubule

Sarcomere

original polarized state, in which the outside of the cell
membrane is some 60 mV more positive than the inside.
Similarly, the membrane of the sarcoplasmic reticulum returns to its original state. The Ca^{2+} present in the sarcoplasm
is now transported, in an energy-requiring process, across
the membrane into the cisternae of the sarcoplasmic reticulum, so that the concentration of Ca^{2+} in the sarcoplasm
is brought to its low resting concentration. The energy-
requiring active transport of Ca^{2+} from the sarcoplasm into
the sarcoplasmic reticulum has been studied directly by
biochemical methods. When skeletal muscle is homogenized, the sarcoplasmic reticulum becomes fragmented into

small closed vesicles, which can easily be isolated by differential centrifugation as the microsome fraction (page 381). Suspensions of such vesicles have been found to accumulate added Ca^{2+} ions very rapidly in the presence of excess ATP and a matching anion such as phosphate, by the action of an ATP-dependent Ca^{2+} pump or active-transport system (page 796) in the membrane. This system transports Ca^{2+} from the medium into the vesicle against a gradient of Ca^{2+} concentration, at the expense of the free energy of hydrolysis of ATP. For each molecule of ATP hydrolyzed to ADP and phosphate, two Ca^{2+} ions are accumulated in the vesicle, together with phosphate. Such vesicle preparations are capable of removing Ca^{2+} from the surrounding medium down to concentrations of less than 0.1 μM. The Ca^{2+}-stimulated transport ATPase has been extracted from the sarcoplasmic reticulum and obtained in soluble purified form. It has a molecular weight of about 100,000 and constitutes a large fraction of the total protein of the membrane. The ATPase molecule becomes phosphorylated during Ca^{2+}-stimulated hydrolysis of ATP.

In addition to the Ca^{2+}-pumping ATPase, the sarcoplasmic reticulum membrane also contains two or more Ca^{2+}-binding proteins; one of these, called _calsequestrin_, has been purified and has a molecular weight of 44,000. Some of the Ca^{2+} segregated by the sarcoplasmic reticulum is apparently bound to these membrane proteins, ready to be released again when the membrane is depolarized.

Recent research suggests that in some muscles rich in mitochondria, such as heart muscle and flight muscles of some birds, segregation of Ca^{2+} during relaxation may also take place in the mitochondria, which can accumulate Ca^{2+} at the expense of energy yielded by electron transport or ATP hydrolysis (page 531). In such muscles the sarcoplasmic reticulum and mitochondria may cooperate in segregation of Ca^{2+}.

It follows from the above considerations that ATP energy is required for both the contraction and relaxation of muscle. We have seen that at least two and perhaps as many as four ATP molecules are bound and hydrolyzed for each cross bridge activated during contraction. Moreover, much evidence suggests that two Ca^{2+} ions released from the sarcoplasmic reticulum must be bound to the troponin of the thin filaments to activate each cross bridge. Since one ATP molecule is required to segregate two Ca^{2+} ions in the cisternae of the sarcoplasmic reticulum to relax muscle, the data suggest that at least two-thirds of the ATP used in a single contraction-relaxation cycle is required for the contraction and up to one-third for the relaxation.

The Source of Energy for Muscular Contraction

Isolated vertebrate muscles can be electrically stimulated to contract repeatedly. Simultaneously, large amounts of lactic acid are formed, and the level of glycogen in the muscle decreases. These observations indicate that glycolysis can provide the energy for muscular contraction. However, classical experiments by E. Lundsgaard in the 1930s revealed that if a muscle is first treated with certain inhibitors of glycolysis, such as iodoacetate, which inhibits glyceraldehyde-phosphate dehydrogenase, it will still contract when stimulated, but without lactate formation, indicating that glycolysis per se is not essential for contraction. Similarly, muscles may also be poisoned with cyanide to inhibit respiration without blocking their ability to contract when stimulated. From such experiments it was concluded that muscle contains energy-rich substances that can supply the energy for contraction, at least for short periods, in the absence of energy-yielding glycolysis or respiration.

At first it was assumed that ATP, discovered in muscle extracts several years earlier, provides the energy for contraction when glycolysis is inhibited. However, for two reasons this explanation appeared to be inadequate: (1) the amounts of ATP found in muscle can provide the energy for only about one-fifth of the contractions that can take place after glycolysis is inhibited by iodoacetate, and (2) careful analysis of the ATP content of muscles before and after single contractions showed essentially no decrease in ATP content or increase in ADP. These experiments thus suggested that ATP may not be the immediate energy source for contraction. However, the answer to this dilemma ultimately came from consideration of the role of the high-energy compound *phosphocreatine* (page 406), discovered by C. Fiske and Y. Subbarow in 1927, which is present in muscle in about 5 times the concentration of ATP. Its high-energy phosphate group is very rapidly transferred to ADP by the action of *creatine kinase* (page 406), present in the sarcoplasm. At the pH of the sarcoplasm, which is about 6.0, the creatine kinase equilibrium,

$$\text{Phosphocreatine} + \text{ADP} \rightleftharpoons \text{creatine} + \text{ATP}$$

lies far to the right, so that ATP formation is favored at the expense of phosphocreatine. This fact therefore explains why the ATP concentration of muscle does not decline during a single contraction: the terminal phosphate group lost from ATP during the contraction is instantly replenished at the expense of phosphocreatine. If the muscle is stimulated long enough in the absence of glycolysis or respiration, the phosphocreatine supply will eventually become depleted. Only then does the ATP concentration decline. However, direct proof that ATP is the immediate energy source in intact muscles, and not phosphocreatine, did not come for many years. It was finally provided by R. E. Davies, who showed that the creatine kinase activity of intact muscles can be completely inhibited by treatment with 2,4-dinitrofluorobenzene, the Sanger reagent (page 102). Presumably an essential functional group of the enzyme undergoes substitution by the 2,4-dinitrophenyl group, rendering the enzyme inactive. In muscles poisoned in this manner, Davies found that the ATP quickly declines on stimulation but phosphocreatine remains unchanged, as expected if ATP is the direct energy source and creatine kinase is inhibited.

The ultimate source of metabolic energy for rephosphorylation of ADP, and thus of phosphocreatine, is not identical in all muscles. Although all muscles of vertebrates show both glycolytic and respiratory activity, the relative contribution of glycolysis and respiration to the regeneration of ATP from ADP may vary considerably. There are two major types of skeletal muscle fibers, red, or slow, fibers and white, or fast, fibers. In red muscles, which owe their color to their high content of myoglobin and cytochromes, respiration serves as the chief source of energy for rephosphorylation of ADP via oxidative phosphorylation. Such muscles, which have profuse mitochondria, include heart muscle and the active flight muscles of birds. Red muscles contract more slowly than white muscles and normally function in regular periodic cycles. White muscles, on the other hand, contain little myoglobin and few mitochondria; in such muscles glycolysis is the chief source of energy for rephosphorylation of ADP. The jumping muscles of the frog are an example of quickly contracting white muscles. Some muscles contain both red and white fibers. In general, red muscles use fatty acids as their major fuel, which they oxidize via the fatty acid oxidation cycle (page 545) to acetyl-CoA; the latter is oxidized to CO_2 via the tricarboxylic acid cycle. White muscles, on the other hand, use glucose as major fuel.

The rates of glycolysis and respiration, and thus of ATP production, are adjusted to the rate of ATP consumption in muscle by a series of feedback controls (page 536). In resting muscle the $[ATP]/[ADP][P_i]$ ratio is high; under these conditions the rates of glycolysis and the tricarboxylic acid cycle are low, because of their allosteric inhibition by the negative modulator ATP (pages 537 and 538). When the muscles are stimulated to maximal activity, the $[ATP]/[ADP][P_i]$ ratio is at first greatly decreased, with the consequence that the rate of consumption of fuels and oxygen is greatly increased. The oxygen uptake of mammalian skeletal muscle may increase twentyfold or more in the transition from rest to full activity; the oxygen consumption of the housefly may increase a hundredfold when it takes off. The sudden increase in respiration is the result of the abrupt breakdown of ATP to ADP and P_i as the muscle contracts, which stimulates both glycolysis and the tricarboxylic acid cycle, for which ADP is a positive modulator. Some of the ADP formed on contraction also undergoes conversion to AMP by the adenylate kinase reaction (page 411):

$$2ADP \rightleftharpoons ATP + AMP$$

The AMP so formed also stimulates glycolysis, since it is a potent positive modulator for phosphofructokinase.

After a period of maximal muscular exertion, such as a sprint, during which lactate appears in the blood in large amounts, an animal will continue to breathe in excess of the normal resting rate and consume considerable extra oxygen. The extra oxygen so consumed during the recovery period, called the <u>oxygen debt</u>, corresponds to the oxidation of some or all of the excess lactic acid formed during maximal muscular activity. Some of the excess lactic acid accumulating in the blood may be converted into glycogen in the liver by pathways described in Chapter 23 (page 624).

Specialized Muscles: Asynchronous and "Catch" Muscles

The three major types of muscle in vertebrates, namely, skeletal, cardiac, and smooth muscle, all contain actin and myosin filaments and appear to employ the same basic set of molecular interactions described earlier in this chapter, although there are some differences in the composition and dimensions of their molecular components and in their ultrastructural arrangement in the muscle fibers. However, there are two highly specialized types of muscle in invertebrates that deserve special comment because of their distinctive biochemical specialization. Most muscles yield a single contraction per nerve impulse received; they are called *synchronous* muscles. However, in insects having a high frequency of wing beat, such as the housefly, wasp, and mosquito, the flight muscles are *asynchronous,* since a single nerve impulse sets off a whole burst, or train, of contractions, the frequency of which may exceed 1,000 per second in the midge. The asynchronous muscles of such insects possess little or no sarcoplasmic reticulum, although they do contain a T system. These facts suggest that Ca^{2+} is segregated and released through means other than the active transport into, and release from, sarcoplasmic reticulum, such as occurs in vertebrate skeletal muscle. The flight muscles of these insects consist of two sets of opposed fibers which contract alternately, one set for the downbeat and the other for the upbeat of the wing. Recent research suggests that contraction of one set of fibers stretches the other set, which is thereby stimulated to contract. Once the asynchronous muscle is set off by a single nerve impulse, its activity is self-sustaining over many contractions. Asynchronous flight muscles of insects are also remarkable in their high respiratory activity; they are perhaps the most intensely respiring eukaryotic cells known. In these muscles the mitochondria are profuse and are regularly arranged along the myofibrils, so that there are either one or two mitochondria per sarcomere, depending on the species (Figure 27-17). The inner membranes of these mitochondria possess an enormous surface area (page 512).

The "catch," or adductor, muscles of mollusks like the clam or scallop are also of special interest, since they can remain locked in the contracted state for long periods after stimulation has ended with no more consumption of metabolic energy than in the relaxed state. Catch muscles contain both actin and myosin arranged in the form of thin and thick filaments, although the latter are usually thicker than those in vertebrate skeletal muscle. The thick filaments of catch muscles distinctively contain from 20 to 50 percent of the water-insoluble filamentous protein, *paramyosin* (mol wt 117,000), which forms the core of the thick filaments, around which myosin molecules are layered. However, it is not yet known how the cross bridges in catch muscles become

Figure 27-17

Longitudinal section of asynchronous flight muscle of wasp (Polistes), showing regular arrangement of mitochondria along the sarcomeres.

D. S. Smith

1.0 µm

769

locked into position; the available evidence suggests that Ca^{2+} and the troponin system are involved in regulation of the catch mechanism.

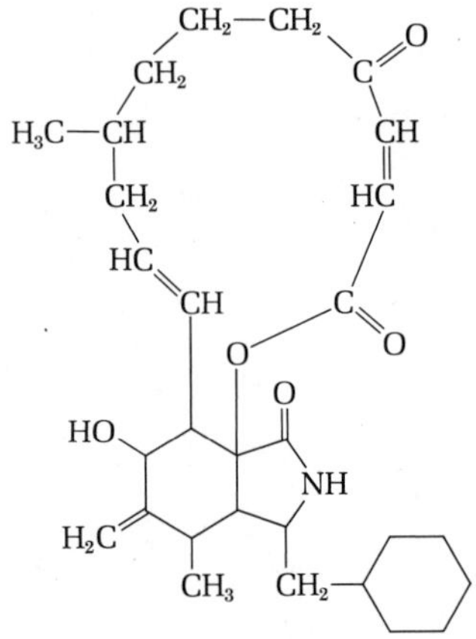

Contractile Proteins in Cells Other Than Muscle Fibers

Recent research has revealed that actin and myosin are found in many cells and tissues other than muscle. In such cells actin and myosin participate in localized contractile events in the cytoplasm, in motile activity, and in cell-aggregation phenomena. An especially interesting case is the mechanism of locomotion in the amoeba. E. D. Korn and his colleagues have isolated an actinlike protein from *Acanthamoeba cas-tellani*. This protein is remarkably similar to the actin of skeletal muscle; both contain a residue of 3-methylhistidine. Amoeba actin forms double-helical filaments resembling the F-actin or thin filaments of mammalian skeletal muscle (page 759); moreover, it will also combine with rabbit-muscle myosin to form hybrid actomyosin complexes. Perhaps the most remarkable finding from the evolutionary point of view is that the actin from *Acanthamoeba* has an amino acid composition very similar to that of rabbit-muscle actin, indicating that the actins of these widely different species have a common ancestor and that they have undergone very little evolutionary change in their amino acid composition, even less than is apparent in the cytochrome *c* of different species (page 113). This striking finding suggests that actin may be present in all eukaryotic animal cells.

Acanthamoeba also contains a protein resembling myosin; it has ATPase activity and actin-binding activity. However, it has a lower molecular weight, about 200,000, and is unable to associate to form thick filaments. Actin apparently functions in cell motility and pseudopod formation. One end of the actin filaments of *A. castellani* is attached to the inner surface of the cell membrane, whereas the actin-binding ATPase extends into the cytoplasm. Presumably their interaction generates a shear force that induces streaming of the cytoplasm and thus the amoeboid movement.

The slime mold *Physarum polycephalum* also contains actin closely similar to that in skeletal muscle, as well as a soluble, myosinlike protein with ATPase activity, which can combine with either slime mold actin or muscle actin. Another striking case is given by blood platelets, very small cells that aggregate into contractile clusters during blood clotting. Platelets also undergo changes in shape and show pseudopod formation, events that are responsible for the retraction of blood clots. These activities have been traced to the presence of actin and myosin which are very similar to muscle actin and myosin in molecular weight, polymerization properties, and ATPase activity.

Actomyosinlike protein complexes have also been isolated from mammalian brain, where they appear to be associated with the synaptic vesicles from which neurotransmitter substances are secreted during synaptic transmission of the nerve impulses.

Figure 27-18
The structure of cytochalasin B.

Figure 27-19
Proposed role of microtubules in conferring shape to animal cells. They often radiate from the centriole. [From J. Freed and M. M. Lebowitz, J. Cell Biol., 45: 344 (1970).]

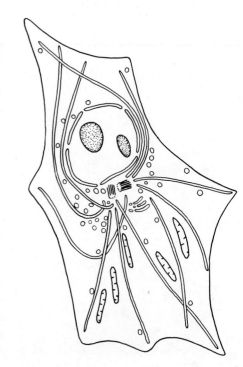

Microfilaments and Cytochalasin B

Some cell processes are dependent on microfilaments that are related to actin, but whose exact molecular nature is not yet known. These cell activities are inhibited or otherwise influenced by low concentrations of *cytochalasin B* (Figure 27-18), a metabolite from the mold *Helminthosporium dematoiderum*. Among the processes sensitive to this agent are *phagocytosis* and *pinocytosis* (engulfment of external solid and liquid matter, respectively, by the cell membrane), *cytokinesis* (the formation of a contractile ring around a dividing cell), *exocytosis* (the extrusion of materials from the cell), various types of cell and organelle movements associated with changes in membrane properties, as well as some membrane transport processes (page 779). Because cytochalasin B at relatively high concentrations has been found to combine with actin from muscle, one view is that cell processes inhibited by cytochalasin B are mediated by microfilaments of actin or an actinlike protein, which may be located in the cytoplasm or in the cell membrane. Many types of motile, contractile, and membrane phenomena in cells thus appear to be the result of basically similar molecular interactions, all sensitive to the single specific inhibitor, cytochalasin B.

Microtubules and Tubulin

Hollow, tubelike filaments of varying diameters have been detected in many plant and animal cells with the electron microscope; they are generically called *microtubules*, although they are found in a number of different cellular arrangements and associated in different ways. Microtubules are found in mitotic spindles, as supporting elements giving shape to some cells (Figure 27-19), in the axons of nerve fibers, in the cilia and flagella of eukaryotic cells, in the tails of spermatazoa, and in fibroblasts, the connective tissue cells that extrude collagen by the mechanical action of microtubules. Microtubules are constructed of recurring molecules of *tubulin*, a globular protein containing two subunits. Tubulin molecules are arranged in helical fashion to form the wall of a tube, the diameter of which is determined by the number of tubulin molecules in its circumference (Figure 27-20).

The formation of microtubules in cells is characteristically inhibited by certain agents long known to be inhibitors of cell division in eukaryotes, namely, the alkaloid *colchicine* and the mold products *vinblastine* and *vincristine* (Figure 27-21). Because these compounds interfere with the formation of mitotic spindles and thus block cell division, they are sometimes used in the treatment of rapidly growing cancer. They are also important tools in the study of the structure, formation, and function of microtubules.

Tubulin, the subunit protein of microtubules, has been isolated in unpolymerized form from a number of cell types, particularly from the brain, in which microtubules are constantly formed and dissociated; tubulin can also be obtained

Figure 27-20

Model of microtubule. The tubule is built of alternating helixes of α and β tubulin subunits as shown. The number of subunits in the circumference is usually 13 but may vary from 10 to 14, depending on the organism. [From J. Bryan, Fed. Proc., 33: 156 (1974).]

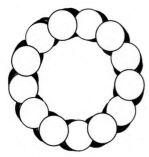

Protomer, 35 × 4.0 nm

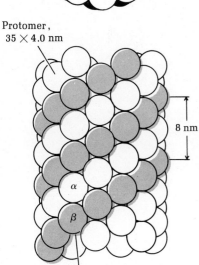

8 nm

α

β

Heterodimer

from preformed microtubules. It has a molecular weight of 120,000 and consists of two apparently nonidentical polypeptide chains of the same size, designated α and β. Significantly, tubulin contains tightly bound GDP; this fact suggests a similarity to muscle actin, which contains tightly bound ADP (page 759).

Assembly of microtubules begins with "activated" tubulin molecules, containing two tightly bound GTP molecules, one on each subunit. Assembly into a microtubule proceeds through both lateral and longitudinal association of tubulin molecules. As each tubulin subunit is attached, a molecule of GTP appears to undergo hydrolysis to form tightly bound GDP and free phosphate. Vincristine interferes with this assembly process by causing the precipitation of tubulin in a three-dimensional polymeric array, very different from the normal microtubule structure. Colchicine interferes with microtubule formation in a different way; it binds to tubulin with displacement of bound guanine nucleotide, thus preventing polymerization.

The apparently similar roles of ATP in actin polymerization and GTP in tubulin polymerization argue for a single basic type of molecular logic in the biological assembly of thin muscle filaments and of microtubules.

Flagella and Cilia of Eukaryotic Cells

Flagella and cilia occur in certain cells of both the animal and plant worlds. They range in length from about 2 μm in some protozoa to several millimeters in spermatozoa of some species; their diameter is about 0.2 μm (Figure 27-22). Flagella are usually very long, and there are usually only one or two per cell; they function to propel the cell. The tails of sperm cells are examples of flagella. Cilia are shorter, occur in larger numbers per cell, and usually function to move extracellular material along the cell surface. These differences are a matter of definition, since both types of appendage are remarkably similar in structural and molecular organization, although flagella are longer.

Flagella and cilia of eukaryotic cells consist of a bundle of paired microtubules, called an _axoneme_, embedded in a matrix and surrounded by a membrane, the _ciliary_ or _flagellar membrane_, which is continuous with the cell membrane (Figures 27-22 and 27-23). The microtubules are characteristically arranged in a 9 + 2 pattern; 9 double tubules are arranged in an outer ring surrounding 2 central, separated tubules. At the base of the cilium or flagellum, and within the cytoplasm, is a cylindrical structure called a _basal body_ or _kinetosome_, which is about 0.5 μm long and 0.15 μm in diameter. The wall of the kinetosome is made up of nine triple fibers, which have cross bridges. The kinetosome serves to transfer ATP and possibly other materials to the shaft of the cilium or flagellum.

The 9 + 2 structure of the shaft of cilia and flagella has been subjected to detailed study with the electron microscope (Figures 27-22 and 27-23). The two central tubules, which are surrounded by a central sheath, are circular in

Figure 27-21
Specific inhibitors of microtubule formation.

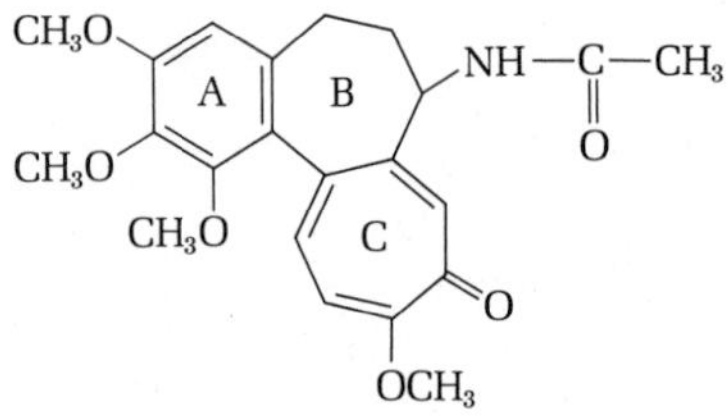

Colchicine

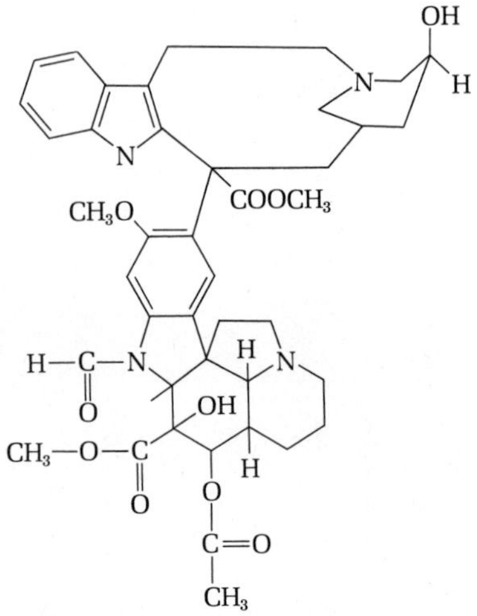

Vincristine (in the related compound vinblastine the formyl group of vincristine is replaced by a methyl group)

igure 27-22
lectron micrographs of cross sections of cilia
f Halteria.

Longitudinal section

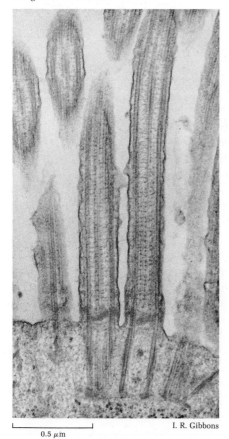

0.5 μm

I. R. Gibbons

Transverse section

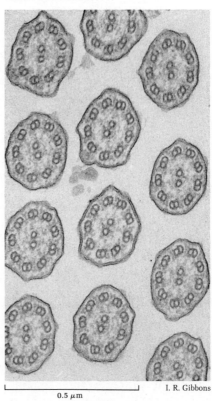

0.5 μm

I. R. Gibbons

igure 27-23
tructure of cilia and flagella of eukaryotic
rganisms, showing the 9 + 2 arrangement
f fibers. [From I. R. Gibbons and A. V.
rimstone, J. Biophys. Biochem. Cytol., 7:
97 (1960).]

Cross section

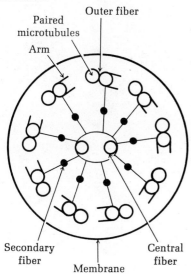

Paired
microtubules

Outer fiber

Arm

Secondary
fiber

Membrane

Central
fiber

Schematic representation of the structure of
the paired microtubules. One usually has 13
subunits in the circumference, the other 11
or 12.

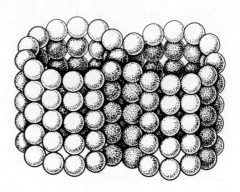

cross section and are about 24 nm in diameter; their walls are about 4.5 nm thick. Each of the nine outer double tubules consists of one complete tubule, to which a somewhat smaller tubule is fused. From the smaller tubule, short arms project. The outer and central microtubules are connected in some flagella by spokelike cross-linkages. The wall of each microtubule consists of eleven or more parallel protofilaments about 4.5 to 5.0 nm in diameter, arranged along the long axis; the protofilaments consist of strings of globular subunits 4.5 nm in diameter.

Flagella and cilia can be detached from protozoa and other unicellular eukaryotes and then isolated by differential centrifugation. Addition of ATP and Mg^{2+} to suspensions of isolated flagella and cilia causes them to undergo localized, wavelike oscillations resembling those of flagella in the intact cell; simultaneously, the ATP is hydrolyzed to ADP and phosphate. Such experiments prove that the shaft of eukaryotic cilia and flagella possesses intrinsic mechanochemical activity (Figure 27-24), in contrast to bacterial flagella, which do not (see below).

The proteins making up the outer tubules of the cilia of *Tetrahymena* and certain other ciliated organisms have been extracted in soluble form and fractionated. Three major fractions have been identified. Two consist of rather large proteins possessing ATPase activity in the presence of Mg^{2+}; they are called _dyneins_. They sediment at different rates and are called 14S dynein (mol wt ≈ 600,000) and 30S dynein. The third component, which is much smaller (4S) and much more abundant, has no ATPase activity; it is apparently identical with tubulin.

The 14S and 30S dyneins have been obtained in highly purified form. When 30S dynein in solution is mixed with a suspension of outer tubules from which the arms have been removed, it becomes bound to the tubules to reconstitute the arms. From such experiments it has also been concluded that the arms possess ATPase activity. The mechanism by which ATP hydrolysis produces wavelike undulations of the cilia or flagella is a matter of great current interest. High-resolution electron microscopy indicates that the pairs of tubules do not undergo a change in length during the sinusoidal undulations of the flagellum but slide along each other; they are attached to each other at the base. Here again we have evidence for the sliding-filament model of mechanochemical energy conversion.

Bacterial Flagella

Some bacteria possess but a single flagellum, others have a tuft of flagella, and still others have flagella distributed over the entire cell surface. Bacterial flagella are between 10 and 35 nm in diameter and may sometimes exceed 10 to 15 μm in length, or many times the diameter of the cell. Most bacterial flagella show a regular and uniform curl (Figure 27-25), with a wavelength of about 2.5 μm.

Bacterial flagella differ from the cilia and flagella of eukaryotes in that they are much thinner, are not surrounded

Figure 27-24
Six stages in the beat of a cilium in the gills of the marine worm Sabellaria; these cilia are about 30 μm long. These characteristic motions are apparently imparted by the ATP-dependent sliding of tubular filaments along each other. [From M. A. Sleigh, Endeavour, 30: 12 (1971).]

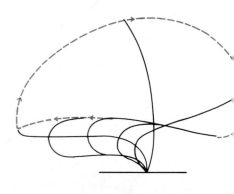

Figure 27-25
Electron micrographs of bacteria showing flagella (below and opposite).

Two cells of *Pseudomonas aeruginosa*, each with a single polar flagellum (negative contrast)

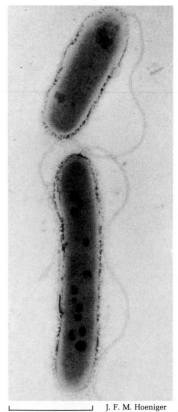

1.0 μm

J. F. M. Hoeniger
and H. D. Tauschel

by a membrane, and have no intrinsic mechanochemical activity. Their helical or gyratory motion is imparted to them at their base within the cell, by conversion of the chemical energy of ATP into mechanical energy in the *basal granules*, to which the flagella are attached. Flagellar motions may exceed 40 cycles per second and they may propel bacteria at velocities up to 50 μm s^{-1}. Many bacterial cells "swim" toward sources of nutrients, a process called *chemotaxis* (page 798). Such movements consist of a series of straight-line swims interrupted by turns and tumbles.

When bacterial flagella, which are protein in nature, are acidified to pH 3, they dissociate into identical monomeric subunits called *flagellin*, which has a molecular weight of 40,000 in most species. The flagellin of *Proteus vulgaris* has been crystallized. X-ray evidence indicates that bacterial flagella consist of three intertwined strands of flagellin monomers.

The flagellin from *Salmonella typhimurium* and other bacteria contains many residues of ϵ-N-methyllysine, which also occurs in muscle actin. It is remarkable that under appropriate conditions of pH and salt concentration, flagellin monomers will spontaneously reaggregate to form structures that appear to be identical with intact flagella, possessing periodic curls of the same wavelength as the native flagella.

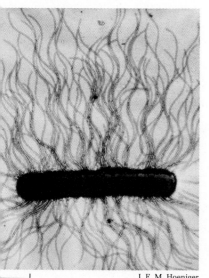

roteus mirabilis cell with profuse flagella. Note regular curl. The wavelength of the curl an be altered by mutation. [*From J. F. M. Hoeniger, J. Gen. Microbiol., 40: 29 (1965).*]

1.0 μm   J. F. M. Hoeniger

Summary

The myofibrils of skeletal-muscle cells contain functional segments, sarcomeres, each consisting of two types of interpenetrating parallel filaments. The anisotropic, or A, band of the sarcomere contains a set of parallel thick filaments, and the isotropic, or I, band a set of thin filaments. During contraction and relaxation, the thin filaments slide along the thick filaments, without change in the length of either. Cross bridges between the thick and thin filaments are rapidly made and broken during muscle contraction.

The thick filaments consist of parallel bundles of myosin molecules, which are long rods with projecting heads. Myosin contains two heavy polypeptide chains and four light chains. Myosin has Ca^{2+}-stimulated ATPase activity, which is kept in an inhibited state by interaction of MgATP with its sulfhydryl groups.

Myosin can be cleaved by trypsin to yield two types of fragments, light meromyosin and heavy meromyosin. The latter contains the head of the myosin molecule, in which the ATPase activity is located. Myosin also can bind actin, the major protein component of the thin filaments. Binding of actin involves the same —SH groups of the head that are involved in the ATPase activity. Hydrolysis of bound ATP by myosin leads to an energized form of myosin containing tightly bound ADP and phosphate.

The thin filaments of the sarcomere consist of two supercoiled F-actin strands. F-actin is a polymer of the globular monomer G-actin. Polymerization of G-actin is promoted by ATP, which undergoes simultaneous dephosphorylation to ADP. Actin also binds Ca^{2+}. Actin forms a complex with myosin, actomyosin, which undergoes dissociation in the presence of ATP, a phenomenon which may be basic to the release of the cross-linkages between actin and myosin filaments in contraction. The thin filaments also contain tropomyosin, which consists of two helical polypeptide chains; tropomyosin molecules are located in the grooves of the paired F-actin strands. At every seventh G-actin monomer there is a troponin molecule, which has three subunits, one for binding Ca^{2+}, one for binding to tropomyosin, and the third for binding to actin. Troponin is a regulatory molecule; when it binds Ca^{2+}, the energized myosin head is released from binding to actin and undergoes the power stroke.

Contraction is initiated by release of Ca^{2+} from vesicles of the sarcoplasmic reticulum following excitation of the muscle cell membrane. Relaxation is produced by removal of Ca^{2+} from the sarcoplasm and its segregation in the sarcoplasmic reticulum by an ATP-dependent transport process. ATP generated by glycolysis or respiration is the immediate energy source for contraction and relaxation; phosphocreatine serves as a reservoir of high-energy phosphate groups for rephosphorylation of ADP. Catch muscles of mollusks contain paramyosin.

Actinlike and myosinlike filaments are found in many kinds of cells and participate in motility, pseudopod formation, phagocytosis, pinocytosis, exocytosis, and cytokinesis. Cytochalasin B inhibits such functions; it also binds to actin. Microtubules consist of tubulin subunits, polymerized by interaction with GTP. Colchicine, vinblastine, and vincristine interfere with microtubule formation or function. Cilia and flagella of eukaryotes consist of nine double tubules and a central tubular system. They contain ATP-hydrolyzing proteins called dyneins, which give rise to a sliding-filament type of interaction. Bacterial flagella, which consist of recurring flagellin monomers, are involved in locomotion and chemotaxis.

References

Books

The Mechanism of Muscle Contraction, Cold Spring Harbor Symp. Quant. Biol., vol. 37, 1972. An important and comprehensive collection of over 70 papers on many different aspects of muscle, presented at an international symposium. The concluding article by H. E. Huxley summarizes the symposium.

NEEDHAM, D. M.: *Machina Carnis: The Biochemistry of Muscular Contraction in Its Historical Development*, Cambridge University Press, London, 1971.

TONOMURA, Y.: *Muscle Proteins, Muscle Contraction, and Cation Transport*, University of Tokyo Press, Tokyo, 1972. A comprehensive survey of the field (433 pages; in English).

Articles

BAGSHAW, C. R., and D. R. TRENTHAM: "The Characterization of Myosin-Product Complexes and of Product-Release Steps during the Mg^{2+}-dependent ATPase Reaction," *Biochem. J.*, 141: 331–349 (1974).

BROKAW, C. J.: "Flagellar Movement: A Sliding Filament Model," *Science*, 178: 455–462 (1972).

CARTER, S. B.: "The Cytochalasins as Research Tools in Cytology," *Endeavour*, 31: 77–82 (1972).

DAVIES, R. E.: "On the Mechanism of Muscular Contraction," pp. 29–56 in P. N. Campbell and C. D. Greville (eds.), *Essays in Biochemistry*, vol. 1, Academic, New York, (1965).

GIBBONS, I. R., and E. FRONK: "Some Properties of Bound and Soluble Dynein from Sea Urchin Sperm Flagella," *J. Cell Biol.*, 54: 365–381 (1972).

HARRINGTON, W. F.: "The Molecular Basis of Muscle Contraction," *PAABS Rev.*, 1: 615–664 (1972). A short, critical review and commentary on important articles from five different laboratories, which are included in reprinted form.

HUXLEY, H. E.: "The Mechanism of Muscular Contraction," *Science*, 164: 1356–1366 (1969). A model of cross-bridge function.

HUXLEY, H. E.: "Muscular Contraction and Cell Motility," *Nature*, 243: 445–449 (1973). Shows the wide distribution of actin and myosin filaments in many types of cells and the general applicability of the sliding-filament concept.

KIRSCHNER, M. W., R. C. WILLIAMS, M. WEINGAARTEN, and J. C. GERHART: "Microtubules from Mammalian Brain: Some Properties of Their Depolymerization Products and a Mechanism of Assembly and Disassembly," *Proc. Natl. Acad. Sci. (U.S.)*, 71: 1159–1163 (1974).

MACNAB, R., and D. E. KOSHLAND, JR.: "Bacterial Motility and Chemotaxis: Light-Induced Tumbling Response and Visualization of Individual Flagella," *J. Mol. Biol.*, 84: 399–406 (1974).

OLMSTED, J. B., and G. G. BORISY, "Microtubules," *Ann. Rev. Biochem.*, 42: 507–540 (1973).

PEACHEY, L. D.: "Sarcoplasmic Reticulum and Transverse Tubules of Frog Muscle," *J. Cell Biol.*, 25: 209–231 (1965). Excellent electron micrographs and drawings.

"Pharmacological and Biochemical Properties of Microtubule Proteins," *Fed. Proc.*, 33: 151–174 (1974). A symposium of articles.

POLLARD, T. D., and R. R. WEIHING: "Actin and Myosin and Cell Movement," *CRC Crit. Rev. Biochem.* 2: 1–65 (1974). An important survey and analysis of the role of actin and myosin in all aspects of cell motility.

SATIR, P.: "How Cilia Move," *Sci. Am.*, 231: 45–52 (1974). Excellent illustrations of molecular structure and motion of cilia.

SPUDICH, J. A., H. E. HUXLEY, and J. T. FINCH: "Regulation of Skeletal Muscle Contraction; II: Structural Studies of the Interaction of the Tropomyosin-Troponin Complex with Actin," *J. Mol. Biol.*, 72: 619–632 (1972).

WEBER, A., and J. M. MURRAY: "Molecular Control Mechanism in Muscle Contraction," *Physiol. Rev.*, 53: 612–673 (1973). Excellent comprehensive review of the regulation of contraction and relaxation.

Problems

1. Skeletal muscle contains about 5×10^{-6} mol of ATP and 30×10^{-6} mol of phosphocreatine per gram wet weight. Calculate how much work (in calories) a 400-g muscle can theoretically carry out at the expense of its high-energy phosphate compounds alone, assuming that both glycolysis and respiration are inhibited.

2. Calculate how long the supply of muscle glycogen (0.8 percent of wet weight of muscle) would last during full activity of skeletal muscle if it is the only fuel and if all of it is available. Recall (see text) that the normal content of ATP (5×10^{-6} mol gram^{-1}) is sufficient for only 0.5 s of maximal activity. Assume that the muscle remains aerobic throughout.

3. During a 100-yd dash an athlete consumes about 1.2 l of pure oxygen at STP, whereas during rest for an equal period he would consume only 40 ml. During the recovery period he consumes an additional 4.9 l of extra oxygen above the resting level. Assuming that glucose was the sole fuel calculate (a) the oxygen debt, (b) the amount of extra glucose oxidized during the 100-yd dash, (c) the amount of extra glucose oxidized during recovery, and (d) the amount of glucose converted into lactic acid during the sprint, above the basal resting level.

4. If the 100-yd dash in Problem 3 was completed in 10 s, calculate the rate of total glucose consumption (moles glucose per gram muscle per minute) by the muscles of a 70-kg athlete. Assume that muscle constitutes 40 percent of the total body weight.

5. From data in Problems 3 and 4 calculate the rate of ATP utilization (moles per gram muscle per minute) in the muscles of the athlete during a 100-yd dash.

6. Now calculate the rate of ATP utilization in the flight muscle of the blowfly, which consumes about 1,000 mm³ of oxygen per minute per gram body weight at 25°C. A blowfly weighs 50 mg of which about 10 mg is flight muscle.

7. Calculate the equilibrium constant for the phosphocreatine kinase reaction at 25°C and pH 7.0 from data in Table 15-4 (page 398).

8. In the table below, approximate concentrations for the substrates of the creatine kinase reaction in skeletal muscle tissue are given for conditions in which the high-energy phosphate supply is (a) plentiful and (b) diminished. Calculate the free-energy change for the creatine kinase reaction under these conditions.

	Phosphocreatine	ADP	ATP	Creatine
(a)	26.7	0.072	7.93	13.3
(b)	0.507	3.14	2.17	39.5

CHAPTER 28 ACTIVE TRANSPORT ACROSS MEMBRANES

We come now to the third major energy-requiring activity of cells, _active transport,_ which we define for the moment as the energy-requiring movement of a metabolite or an inorganic ion across a membrane against a gradient of concentration. This is by no means the least of the major energy-requiring activities of cells, since in some specialized animal tissues, such as the kidney, up to 70 percent of the available metabolic energy is utilized for active-transport processes.

Because they contain a continuous nonpolar hydrocarbon core, contributed by the phospholipid bilayer, biological membranes are intrinsically impermeable to most polar molecules. This property has a biological advantage, since it prevents the internal metabolites of the cell, most of which are ionized, from diffusing out of the cell. However, living cells must obtain certain polar nutrients such as glucose and amino acids from the environment, where they may be very dilute, and must also eject or secrete various ions or polar molecules. For this purpose they have perfected specific membrane transport systems which not only carry polar nutrient molecules across the membrane but may do so against a gradient of concentration. The concentration gradients generated by active-transport systems can be very high.

Active-transport systems have many important functions in addition to the transport of fuels and essential nutrients from the surrounding medium. They participate in the maintenance of metabolic steady states by their ability to maintain the internal concentration of organic nutrients and metabolites relatively constant in the face of wide fluctuations in the composition of the external environment. Active-transport systems also maintain constant and optimal internal concentrations of inorganic electrolytes, in particular K^+ and Ca^{2+}, which are essential for regulation of many important intracellular activities. Moreover, they also aid in maintaining the osmotic relationships between cells and surrounding media, and, as a consequence, maintaining the volume of cells. Active-transport systems are involved in the transmission of information by the nervous system and in the excitation and relaxation cycle of muscle tissue; they also function in the absorptive activity of intestinal epithelia and the secretory function of the kidneys. Another extremely important role of active transport has been suggested: the conversion of the energy of electron transport into the chemical energy of ATP during oxidative and photosynthetic phosphorylation, since in these processes "pumping" of H^+ across mitochondrial and chloroplast membranes may be an essential intermediate step.

Until recent years little biochemical information was available concerning the molecular basis of active transport. Most investigations were limited to the rates and the specificity of active-transport processes in intact cells and tissues. However, as a result of important new experimental approaches, a number of membrane proteins participating in active transport have been isolated, a development that has opened this important energy-requiring cell function to direct biochemical study. Today the molecular basis of membrane transport is a large and rapidly growing area of biochemical research.

The Energetics of Active Transport

When a solution is diluted by addition of water, its free energy decreases and its entropy increases; the solute molecules are now farther apart and thus more randomly disposed. Conversely, when a dilute solution is made more concentrated, its free energy is increased. Thus, input of free energy is required to transfer a solute at a given concentration into a compartment in which its concentration is higher; conversely, there is a decrease in free energy when a solute passes into a compartment where its concentration is lower (Figure 28-1). *Active transport* is thus rigorously defined in thermodynamic terms as a transport process in which the system gains free energy, and *passive transport* is defined as one in which the system loses free energy.

The free-energy change $\Delta G'$ occurring when 1.0 mol of an uncharged solute moves from a compartment in which it is present at concentration C_1 to a compartment in which it is present at concentration C_2, as in Figure 28-1, is given by

$$\Delta G' = RT \ln \frac{C_2}{C_1} \tag{1}$$

in which R and T have their usual meanings (page 391). This equation assumes that the transported solute carries no net electric charge and that it occurs as the same molecular species in both compartments. From this equation we can calculate the free-energy change that occurs at 20°C when 1.0 mol of a solute such as glucose is transported from a compartment in which its concentration is 0.01 M into a compartment in which its concentration is tenfold higher, namely, 0.1 M:

$$\Delta G' = RT \ln \frac{0.1}{0.01}$$
$$= 1.98 \times 293 \times 2.303 \times \log 10$$
$$= 1{,}340 \text{ cal} = 1.34 \text{ kcal}$$

Figure 28-1
Schematic representation of active and passive transport.

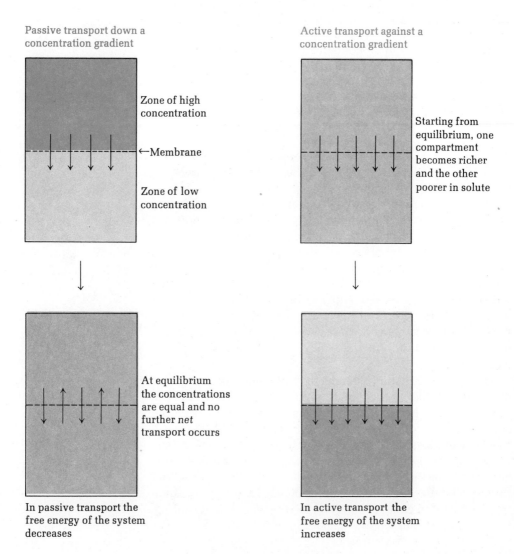

Passive transport down a
concentration gradient

Zone of high
concentration

←Membrane

Zone of low
concentration

At equilibrium
the concentrations
are equal and no
further *net*
transport occurs

In passive transport the
free energy of the system
decreases

Active transport against a
concentration gradient

Starting from
equilibrium, one
compartment
becomes richer
and the other
poorer in solute

In active transport the
free energy of the system
increases

Since the calculated free-energy change is positive in sign, it follows that transport of glucose against a 10:1 concentration gradient will not occur spontaneously. It can occur, however, if it is coupled to some other process proceeding with a decline in free energy, for example, the hydrolysis of ATP. Since the standard free energy of hydrolysis of ATP is -7.3 kcal mol^{-1}, the hydrolysis of 1 mol of ATP could theoretically yield more than sufficient energy to transport 1.0 mol of glucose (or any other uncharged solute) from one compartment to another against a tenfold gradient of concentration. However, the energy of ATP is not available for active transport unless a mechanism exists for the coupling of ATP hydrolysis to the process by which glucose is transported.

If the solute under consideration is electrically charged, the equation for the free-energy change of an active transport process is more complex, since *two* gradients then exist: (1) a gradient of mass and (2) a gradient of electric charge or potential. The sum of these two gradients is called the _electrochemical gradient_. The free-energy change during transport of an electrically charged ion is given by a modification of equation (1), which contains terms for both the mass gradient and the charge gradient:

$$\Delta G' = RT \ln \frac{C_2}{C_1} + Z\mathscr{F}\,\Delta\Psi \qquad (2)$$

The term $Z\mathscr{F}\,\Delta\Psi$ represents the free-energy contribution due to transport of electric charge. Z is the number of charges per solute molecule, $\mathscr{F}$ is the faraday (96,493 C g equiv^{-1}), and $\Delta\Psi$ the difference in electric potential between the two compartments in volts. The electrical component of a gradient across a biological membrane, i.e., the _membrane potential_, can in some cases be measured directly, as in large nerve or muscle cells. In other cases, it can only be approximated by indirect methods. The measured membrane potential is often quite small because of compensating movements of other ions. Membrane potential plays a major role in the function of nerve and muscle cells and very likely also in the membranes of mitochondria (page 527) and chloroplasts (page 612).

Most active-transport systems in cell membranes can drive transport against rather high gradients of concentration. Sugars and amino acids can be transported into bacterial cells against a concentration gradient of over 100:1. One of the most impressive active-transport gradients known occurs in the secretion of H^+ into the gastric juice of mammals (pH about 1) from the blood plasma (pH = 7.4). The H^+ gradient thus exceeds $10^{-1}/10^{-7} = 10^6$, or 1,000,000:1. By use of equation (1), the free-energy change occurring when 1.0 g equiv of H^+ is secreted from blood plasma into gastric juice, assuming that both are electroneutral, is

$$\Delta G' = RT \ln \frac{10^{-1}}{10^{-7}} = 8{,}040 \text{ cal} = 8.04 \text{ kcal}$$

Figure 28-2 gives the approximate composition of some body fluids in man, showing the magnitude of some of the solute gradients. Using equation (1) or (2), we can calculate the minimum amount of work required to produce a given volume of a body fluid, such as the urine, from blood plasma; it is the algebraic sum of the free-energy changes for transport of each of the solutes between the blood and the urine.

Calculations of this sort do not take account of the fact that cells are open systems and are not in thermodynamic equilibrium. A concentration gradient of any given solute across a cell membrane is a reflection of a steady state, the resultant of two opposing processes, the active transport of the solute into (or out of) the cell and a passive backleakage process. The work required to maintain a given concentra-

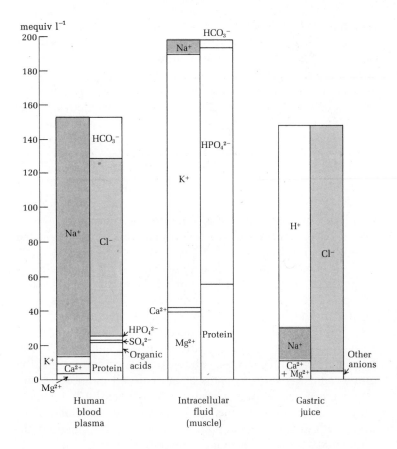

Figure 28-2
The ionic composition of human blood plasma, intracellular fluid, and gastric juice. Note the particularly large difference in Na^+ and K^+ concentrations between blood plasma and intracellular fluid and the very large difference in H^+ concentration between blood plasma (pH 7.4) and gastric juice (pH ≈ 1.0).

tion gradient across a cell membrane must take into account not only the magnitude of the gradient but also the rate of back-leakage.

Identifying Characteristics of Mediated Membrane Transport

Solutes may undergo a net movement or flux across a membrane by either _nonmediated_ or _mediated processes_; the latter are also called _facilitated processes_. These processes can be distinguished by measurements of the rate of transport and how it is influenced by certain conditions. The rate of a nonmediated-transport process is at all times directly dependent on the concentration of the solute, and its temperature coefficient is usually that of physical diffusion, namely, about 1.4 per 10°C rise in temperature. Nonmediated transport of a solute is entirely the result of simple physical diffusion of the solute alone, in response to its concentration or electrochemical gradient. In nonmediated flux the solute molecule is neither chemically modified nor associated with another molecular species in its passage through the membrane.

On the other hand, _mediated_, or _facilitated_, membrane transport processes, which may be either passive or active, show different and more complex behavior. The first identifying characteristic of mediated transport is that it exhibits _saturation kinetics_; i.e., the transport system can become saturated with the substance transported, just as enzymes can become saturated with their substrates (page 189). Plots of the initial rate of a mediated-transport process against the substrate concentration usually show a hyperbolic curve approaching a maximum at which the rate is zero order with respect to substrate concentration (Figure 28-3), similar to the Michaelis-Menten relationship of enzyme kinetics (page 189). Such behavior suggests that the membrane transport system contains a specific site to which the substrate must bind reversibly in order to be transported across the membrane. It also suggests that the rate of binding of the substrate on one side of the membrane, its transport across the membrane, or its release on the other side can set an upper limit to the rate of transport, just as the rate of formation (or breakdown) of an enzyme-substrate complex can limit the rate of an enzymatic reaction.

The second criterion of mediated transport is _specificity_ for the substance transported. For example, the erythrocytes of some vertebrates have a membrane transport system which promotes the influx of D-glucose and a number of structurally related monosaccharides but has little or no activity toward D-fructose or disaccharides such as lactose. Mediated transport also may show stereospecificity; thus, the amino acid transport systems of animal cell membranes are much more active with L-amino acids than the D isomers. From such findings it has been postulated that mediated-transport systems in membranes contain a binding site complementary to the substance transported, resembling in its specificity the active site of enzyme molecules.

The third criterion of mediated transport is that it can frequently be inhibited quite specifically. Some biological transport systems are inhibited competitively by substances structurally related to the substrate, which compete with the substrate for its specific binding site. Other transport systems may be inhibited noncompetitively by reagents capable of blocking or altering specific functional groups of proteins, such as N-ethylmaleimide, a sulfhydryl-group blocking agent, or 2,4-dinitrofluorobenzene, which blocks free amino groups (page 102).

These properties of mediated transport strongly suggest that biological membranes contain protein molecules capable of reversibly binding specific substrates and of transporting them across the membrane. Such proteins or systems of proteins are variously called _transport systems, carriers, porters,_ or _translocases._ Carrier molecules may carry their ligands across the membrane by diffusion, they may rotate within the membrane, or they may undergo conformational changes to create a "hole" in the membrane for the specific substrate transported.

Mediated transport is sometimes called _facilitated diffusion,_ in recognition of the fact that all transport processes, whether mediated or not, are ultimately the result of diffusion. In mediated transport it may be the substrate-carrier complex that undergoes diffusion, either translational or rotational, in response to a gradient of the complex within the membrane.

Figure 28-3
Saturation kinetics of a membrane transport process. Mediated transport has a maximum rate V_{max} at which the carrier is saturated, whereas unmediated transport by simple physical diffusion is usually proportional to the concentration of solute and does not show saturation. Small, lipid-soluble molecules, such as undissociated acetic acid or β-hydroxybutyric acid, can readily pass through biological membranes. At pH 7.0 sufficient concentrations of these weak acids exist to allow rapid, unmediated transport.

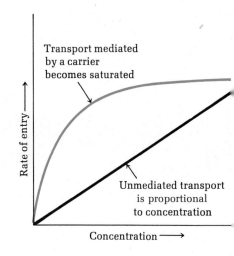

Transport mediated by a carrier becomes saturated

Rate of entry →

Unmediated transport is proportional to concentration

Concentration →

Active and Passive Mediated Transport

All mediated-transport processes across biological membranes share the three characteristic properties of saturability, substrate specificity, and specific inhibition described above. However, they can be divided into two types, *active* and *passive*, by application of other criteria.

The first criterion is whether a given transport process occurs up or down a gradient (page 781). This requires precise knowledge of the concentration of the transported substance in the two compartments concerned, as well as assurance that the substance exists in the same molecular or ionic species in the two compartments. In the case of the secretion of gastric juice from blood plasma, pH measurements on both fluids with a glass electrode that senses only *free* H^+ give considerable assurance that a large gradient of H^+ is generated by the gastric epithelium.

However, the precise magnitude of the concentration gradient of a solute across a biological membrane is not always easy to measure, since the available analytical methods may not be able to distinguish between free and bound solute. For example, Ca^{2+} is strongly bound to other intracellular solutes, particularly proteins. Since the bound Ca^{2+} does not contribute to the thermodynamic gradient of the free Ca^{2+}, an analytical method that measures only the total Ca^{2+} content of cells or body fluids cannot give an accurate measure of the true Ca^{2+} gradient across a cell membrane. Special analytical procedures are often required to determine the concentration of the thermodynamically free species.

The second criterion of an active-transport process is its dependence on metabolic energy. If the transport process under question depends upon glycolysis or respiration or ATP hydrolysis, then it may be considered to be an active-transport process. For example, the erythrocytes of some species normally maintain a high internal concentration of K^+ and extrude Na^+. If fluoride, an inhibitor of glycolysis, is added to the erythrocyte, then the internal K^+ concentration declines and the Na^+ concentration increases, to approach the state in which their concentrations are equal on both sides of the membrane. The maintenance of concentration gradients of Na^+ and K^+ by erythrocytes is thus an active-transport process, since it is dependent on glycolysis, an ATP-yielding process. On the other hand, other mediated-transport processes, such as the movement of glucose into human erythrocytes, are not inhibited when metabolic energy sources are blocked; such systems may be concluded to be carrying out *passive* mediated transport.

Active-transport systems are unidirectional in their normal energy-requiring function; i.e., they will move the substrate across the membrane in only one direction. For example, the active-transport system of erythrocytes moves Na^+ only in the outward direction and K^+ in the inward direction. This behavior is in sharp contrast to the passive mediated transport of glucose, which may be in either direction, depending on the relative concentration of glucose in the two compartments.

It is important to note that in the absence of a supply of metabolic energy, some active-transport systems of cells may also engage in passive transport of their substrates across the membrane, in either direction, depending on the relative concentration of the substrate in the two compartments.

Models of Mediated Transport

From the characteristic properties described above, generalized models of transport systems have been proposed (Figure 28-4). Such models postulate that a specific protein containing a binding site complementary to the substrate transported serves as a carrier of the substrate. By diffusing across the membrane, by rotating, or by a conformational change, it carries the substrate molecule across the membrane so that the binding site now faces the other compartment, into which the bound substrate is discharged. In passive mediated-transport systems, the process is completely reversible, and the net movement of solute molecules may be in either direction, depending on the relative concentration of the substrate in the two compartments. In models of active-transport systems, the carrier molecule is "energized" in some manner to achieve unidirectional transport against a gradient, for example, by an energy-dependent alteration of the substrate-binding affinity of the transport system on one side of the membrane (Figure 28-4).

Since some active-transport systems can function as passive-transport systems if their energy supply is cut off, they have been postulated to contain at least two components. One functions to recognize, bind, and carry the substrate reversibly, in either direction, whereas the other is required to couple the action of the carrier molecule to an energy source, such as electron transport or ATP hydrolysis, in such a way that the active process occurs in only one direction.

Group Translocation across Membranes

In active transport a concentration gradient of the transported substrate is generated, without net chemical modification of the substrate. But there is another type of membrane transport process in which the transported substrate is obligatorily transformed into another chemical species. For example, in many bacteria (see below) glucose is transported across the membrane into the cell in a process which simultaneously phosphorylates the glucose to glucose 6-phosphate, the form in which the transported glucose appears inside the cell. Such a process clearly requires metabolic energy to carry out the phosphorylation, but it cannot be called active transport, since the transported substrate has been chemically modified into another species. Such a membrane transport process which results in the translocation of a given chemical group rather than an unaltered molecule or ion is called group translocation. In a sense, group translocation achieves the same end as a true active-transport process, since it can bring about the concentration of a chemical

Figure 28-4
Models of membrane transport mechanisms.

A passive-transport mechanism. The substance transported (S) combines with a specific transport protein (T) to form a TS complex, which carries S across the membrane. In this case transport is from left to right, in the direction of the lower concentration of S. Passive transport can occur in either direction depending on the direction of the concentration gradient.

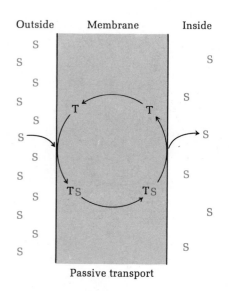

Passive transport

An active-transport mechanism. In this model the energy of ATP is applied to "unload" the substrate from the carrier by decreasing the affinity of the carrier on the inside surface, possibly by inducing a conformational change in the carrier protein.

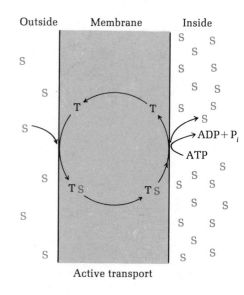

Active transport

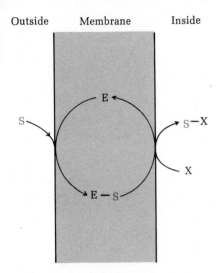

group in the cell from a dilute external medium at the expense of metabolic energy (Figure 28-5). Moreover, it may also have an advantage in that the transported substance may be simultaneously transformed into a derivative that is more directly utilized by the cell. For example, group translocation of free glucose into bacterial cells in the form of glucose 6-phosphate enables it to enter the glycolytic sequence directly. Group-translocation processes may also occur during amino acid transport (see below).

Genetic Evidence for Transport Systems

Genetic studies have yielded extremely important information on the function and biology of transport systems. An example may be described. When certain bacteria are grown on glucose as the sole carbon source, they are unable to utilize external citrate added to the medium, although they oxidize internally formed citrate at high rates via the tricarboxylic acid cycle. However, if such cells are removed from the glucose medium and placed in a medium where citrate is the sole carbon source, they quickly adapt and acquire the ability to utilize exogenous citrate as carbon source. This effect is not due to a generalized increase in permeability of the cell membrane, since the citrate-grown cells do not acquire the ability to utilize substrates other than citrate. Such findings led to the conclusion that these bacteria contain an inducible citrate-transport system. When the bacteria have an ample supply of glucose from the medium, particularly if glucose is the sole carbon source, the citrate-transport system is not needed and its biosynthesis is genetically repressed (page 979). Mutants of such bacteria have been found in which the ability to form the citrate-transport system in response to exogenous citrate has been lost. Such mutants, generically termed *transport-negative*, are believed to be metabolically normal in all other respects.

From such experiments it has been concluded that the molecular components of membrane transport systems are genetically determined, just like the protein components of a multienzyme sequence catalyzing a metabolic pathway (page 370).

The Organization of Membrane Transport Systems in Animal Tissues

The membrane transport systems of animal tissues are organized into three different morphological types, which have been termed *homocellular*, *transcellular*, and *intracellular* (Figure 28-6). Homocellular transport, the most common and general membrane transport process, which occurs in both prokaryotic and eukaryotic cells, is that taking place across the cell membrane, so that a given organic substrate or mineral ion is transported into or out of the cell. All cells possess such homocellular systems for inward transport of their nutrients, particularly sugars and amino acids. Moreover, most types of animal cells contain homocellular transport systems for extruding Na^+ into the medium and maintaining a high intracellular K^+ concentration.

Transcellular transport is a special case of homocellular transport in which the transport system is localized on a specific portion of the cell surface in such a way as to make transport of a substrate through or across the entire cell possible. This arrangement occurs in epithelial cell layers, such as the gastric and intestinal mucosa and the epithelial cells lining kidney tubules. These sheets of epithelial cells promote transport across the cell layer, for example, from the blood to the gastric juice. The transport systems in epithelial cell layers are so organized in the cell membrane that a directional transport *across* the cell occurs, presumably because the active-transport systems are located on only one side of the cell barrier (Figure 28-6).

The third type of membrane transport in eukaryotic cells is *intracellular transport*, in which the transport process occurs across the membrane of an intracellular organelle, e.g., mitochondrion, chloroplast, or sarcoplasmic reticulum. Such intracellular transport systems serve to transport mineral ions or metabolites between the cytosol and the internal medium of the organelle. This type of membrane transport process is involved in the compartmentation of ATP and many metabolites in mitochondria (page 529), in the regulation of the contraction-relaxation cycle of muscle by transport of Ca^{2+} between cytosol and sarcoplasmic reticulum (page 762), and in the transport of Ca^{2+} from the cytosol into the mitochondria (see below).

Passive-Transport Systems in Animal Tissues

The biochemical characteristics of some representative transport systems in cells of higher animals will now be described. We shall consider passive-transport systems first.

The Glucose Carrier of Human Erythrocytes

One of the best known passive-transport systems is the glucose carrier of the human erythrocyte, which promotes *net* transport of glucose down a concentration gradient of glucose. The rate of entry of D-glucose into human erythrocytes increases with substrate concentration but ultimately approaches a maximum rate, at which the transport system is saturated. However, the rate of entry of glucose into erythrocytes of certain other vertebrate species, such as the cow, does not show a saturation effect; apparently a similar glucose carrier is genetically lacking in the latter species. The glucose carrier in human erythrocytes has a rather broad substrate specificity and is able to transport many sugars other than D-glucose, including D-mannose, D-galactose, D-xylose, D-arabinose, and D-ribose, as well as the unnatural and nonmetabolized derivatives 2-deoxy-D-glucose and 3-O-methyl-D-glucose (Figure 28-7), which are often used as test substrates. Comparable L sugars are not transported. For maximal activity the transported sugar must be a pyranose in the chair conformation, i.e., with the hydroxyl groups in equatorial positions. It has been postulated that the glucose

Figure 28-6
Organization of membrane transport systems in cells and tissues.

Homocellular and intracellular transport

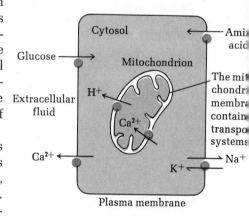

Plasma membrane

Transcellular transport across an epithelial cell layer of the gastric mucosa. The direction of net movement is from the blood to the gastric juice.

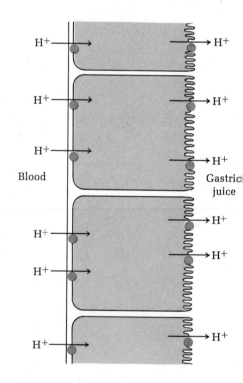

Figure 28-7
Two derivatives of α-D-glucose that are transported but not metabolized.

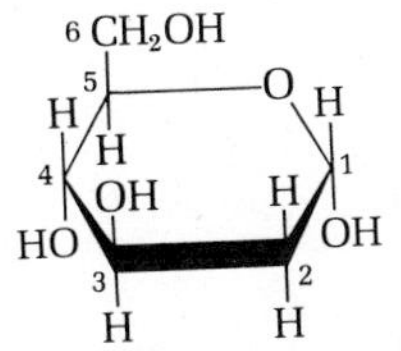

2-Deoxy-α-D-glucose

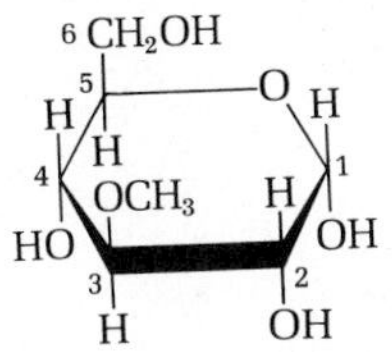

3-O-Methyl-α-D-glucose

Figure 28-8
Inhibitors of glucose transport by the human erythrocyte.

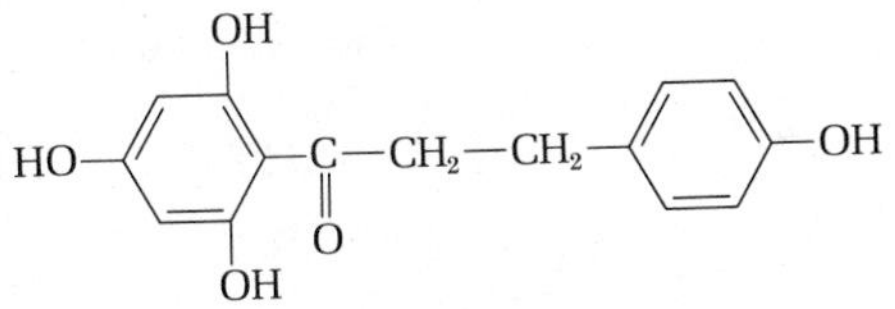

Phloretin

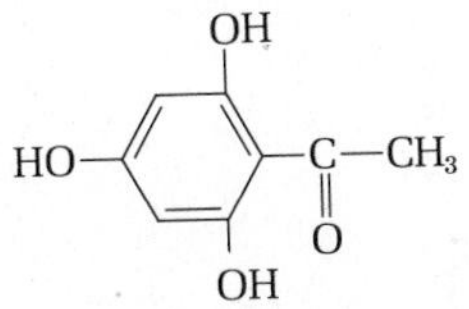

2,4,6-Trihydroxyacetophenone

carrier of human erythrocytes binds the sugar at three equatorial hydroxyl groups.

The erythrocyte sugar carrier has a characteristic affinity for each sugar. K_M, defined as the concentration of substrate giving a half-maximal rate of transport, is about 6.2 mM for D-glucose and about 18.5 mM for D-mannose. The glucose concentration in human blood is about 5.0 mM. The concentration of D-fructose required for half-maximal activity ($K_M > 2000$ mM) is so high that the glucose carrier does not function to transport fructose biologically. The compounds *phloretin* and *trihydroxyacetophenone* (Figure 28-8) are competitive inhibitors of glucose transport in erythrocytes. The glucose carrier of human erythrocytes definitely appears to be a protein, since it is inactivated by certain reagents capable of reacting with specific functional groups of amino acid residues. A sulfhydryl group, presumably of a cysteine residue, and an amino group are required for activity. Moreover, glucose transport is irreversibly inhibited by D-glucosyl isothiocyanate, which is an *affinity label* for the binding site (page 221). Because of its glucoselike structure this reagent binds to the glucose binding site; its highly reactive isothiocyanate group then reacts covalently with an amino group on or near the glucose binding site. From experiments with this and other reagents it has been deduced that each erythrocyte contains about 300,000 glucose binding sites which function in passive transport of glucose across the membrane.

The ATP–ADP Carrier of Mitochondria: Exchange-Diffusion

Several passive exchange-transport or exchange-diffusion systems occur in the mitochondrial inner membrane, such as the dicarboxylate and tricarboxylate carriers described earlier (page 530). The best-studied example is the ATP–ADP carrier of the mitochondrial membrane, which normally functions to transport one molecule of ADP into the mitochondrial matrix, where oxidative phosphorylation takes place, in exchange for one molecule of ATP returning to the cytosol from the matrix (page 530). The ATP–ADP carrier promotes only an exchange across the membrane; it does not bring about a *net* transport of nucleotides. Transport systems that promote such exchanges are called *antiport systems*.

The ATP–ADP carrier is specific for ATP and ADP and for dATP and dADP; it will not transport AMP or any other NDPs or NTPs. It is saturable and shows a very high affinity for ADP and ATP. Much evidence indicates that it has two substrate binding sites, one on the inside face and the other on the outside face of the inner membrane. As has been pointed out (page 530), the ATP–ADP carrier is strongly and specifically inhibited by atractyloside and bongkrekic acid. The carrier will also exchange ^{14}C-ADP for unlabeled ADP and ^{14}C-ATP for ATP, in either direction.

Active-Transport Systems of Animal Tissues

There are three major types of active-transport systems in animal tissues: (1) the Na$^+$ and K$^+$ "pump," which utilizes metabolic energy to transport Na^+ out of the cell and K^+ into it, (2) active-transport systems for glucose and other sugars, and (3) active-transport systems for amino acids. Of these the Na^+ and K^+ pump appears to be most active and of universal occurrence; it is present in the membrane of nearly all animal cells. Moreover, much evidence suggests that the Na^+ and K^+ pump is also essential for the operation of the pumps for glucose and for amino acids. We shall therefore examine the active transport of Na^+ and K^+ in some detail, particularly since it has been very intensively studied from the biochemical point of view.

The Na^+- and K^+-Transporting ATPase System

Most animal cells maintain intracellular K^+ at a relatively high and constant concentration, between 120 and 160 mM, whereas the intracellular Na^+ concentration is usually much lower, less than 10 mM. A substantial gradient of K^+ and Na^+ exists across the cell membrane, since the extracellular fluid of mammals contains a relatively high concentration of Na^+, about 150 mM, and a very low concentration of K^+, usually less than 4 mM. The constancy of the high internal K^+ concentration is maintained by the energy-requiring extrusion of Na^+ out of the cell and its replacement by K^+, promoted by an active-transport system called the Na^+K^+–ATPase.

Relatively high concentrations of internal K^+ are required for several processes vital in the internal economy or function of animals cells. One is protein biosynthesis by ribosomes (page 1025). Second, a number of enzymes require K^+ for maximum activity. For example, in the glycolytic sequence K^+ is required for maximal activity of pyruvate kinase (page 431). Third, metabolically supported gradients of Na^+ and K^+ across the cell membrane are involved in the maintenance of the membrane potential of excitable tissues, which is the vehicle for transmission of impulses in the form of an *action potential*, i.e., a transient discharge or collapse of the membrane potential caused by an abrupt increase in the permeability of the membrane to Na^+ and K^+ when it is stimulated or excited.

In 1957 an important breakthrough opened the molecular basis of Na^+ extrusion and K^+ accumulation to direct biochemical study. In Denmark, J. C. Skou discovered that the hydrolysis of ATP to ADP and phosphate by a fraction of homogenized crab nerve now known to contain cell-membrane fragments requires *both* K^+ and Na^+ for maximum activity, as well as Mg^{2+} (which is required for most enzymatic reactions involving ATP as substrate; page 402). Addition of either Na^+ or K^+ alone produced very little stimulation. This finding was unusual, since most K^+-requiring enzymes are inhibited by Na^+. Most significant, however, was the later observation that the stimulation of the ATPase activity of such cell-membrane preparations by Na^+ and K^+ is inhibited by the glycoside *ouabain*, long known to inhibit the transport of Na^+ and K^+ by intact cell membranes. Ouabain (Figure 28-9) is one member of a group of related compounds called *cardiac glycosides*, which are capable of al-

Figure 28-9
Ouabain (strophanthin G), an inhibitor of the Na^+K^+–ATPase

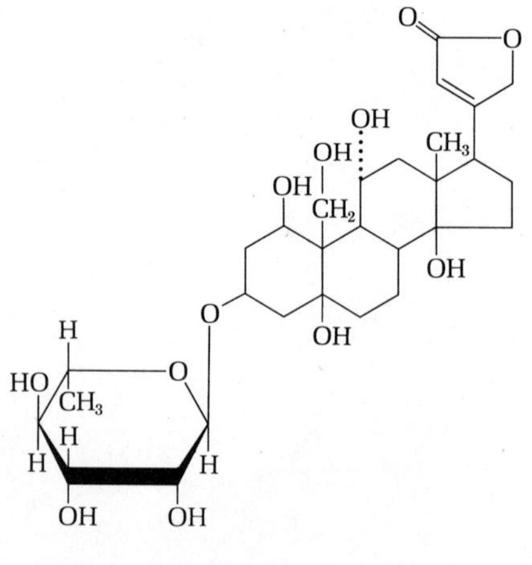

tering the excitability of heart muscle, a function dependent on the balance of Na^+ and K^+ across the membrane. Skou postulated that this Na^+- and K^+-stimulated ATPase activity of membrane fragments represents the Na^+ and K^+ pump of the nerve cell membrane, which he further postulated requires Na^+ from the inside and K^+ from the outside surface of the membrane for maximal activity and thus maximal rates of ATP hydrolysis. Cell-membrane fractions from many different animal tissues have since been found to contain such a Na^+- and K^+-stimulated ATPase activity. Excitable cells, such as brain, nerve, and muscle, and the electric organ of *Electrophorus electricus*, the electric eel, are particularly rich in the enzyme. Also very active are Na^+-transporting epithelial tissues such as the kidney cortex, the salivary gland, and the salt gland of the seagull.

The Na^+- and K^+-stimulated ATPase of cell membranes transports Na^+ and K^+ in a vectorial or directional manner. This was convincingly shown by R. Whittam and his colleagues in classical experiments on the Na^+K^+-ATPase of erythrocytes. When erythrocytes are exposed to distilled water under controlled conditions, they swell and their membranes increase in permeability. As a result, their internal electrolytes, as well as hemoglobin and other cytoplasmic proteins, leak out into the surrounding hypotonic medium. These preparations of erythrocytes are called *ghosts*. Erythrocyte ghosts can now be "loaded" with different kinds of salts. For example, when isotonic NaCl solution is added to a suspension of ghosts, they shrink to normal size and their membranes return to their usual relatively impermeable state. These preparations are called *reconstituted* or *resealed erythrocytes*. During the resealing process the ghosts entrap NaCl in a concentration equivalent to that in the suspending medium. On the other hand, if the reconstitution is carried out in an isotonic KCl solution, they will trap KCl. In this manner, reconstituted erythrocytes can be loaded with varying internal concentrations of KCl or NaCl, or various other salts, such as LiCl. Moreover, such erythrocytes can also be loaded with ATP if it is added to the salt medium in which the ghosts are reconstituted.

Whittam prepared reconstituted erythrocytes containing internal ATP and various concentrations of internal NaCl or KCl. He then examined the effect of various internal and external concentrations of Na^+ and K^+ on the rate of the enzymatic hydrolysis of the internal ATP and on the transport of Na^+ and K^+. When the external medium contained a high Na^+ concentration and the internal medium a high K^+ concentration, conditions resembling those existing in normal intact erythrocytes in the blood, the rate of hydrolysis of internal ATP was relatively low. Conversely, when the external medium contained a high concentration of K^+ and the internal medium a high concentration of Na^+, the rate of hydrolysis of internal ATP was high. If both the external and internal media contained only Na^+ (or only K^+), little ATP hydrolysis ensued.

The second important observation was that Na^+ and K^+ move across the erythrocyte membrane in the anticipated directions during hydrolysis of internal ATP, as Skou had postulated. If the Na^+ concentration was high on the inside and the K^+ concentration was high on the outside, conditions which evoked maximal ATP hydrolysis, then Na^+ moved out of the cell and K^+ moved in as the internal ATP was hydrolyzed. Moreover, it was also found that the reconstituted erythrocyte ghosts utilize only internal ATP; external ATP is not attacked.

Experiments on erythrocyte ghosts have also made it possible to examine the specificity and stoichiometry of the Na^+K^+–ATPase activity. For example, external K^+ can be replaced by Rb^+, NH_4^+, or certain other monovalent cations, but internal Na^+ is an absolute requirement for the ATPase, which is thus regarded as serving primarily as a Na^+ pump. ATP may be replaced by dATP, CTP, and ITP, but these nucleotides are far less active than ATP. It has also been possible to measure the amounts of K^+ and Na^+ transported per molecule of ATP hydrolyzed. In erythrocytes three molecules of Na^+ are extruded and two molecules of K^+ are accumulated per molecule of ATP hydrolyzed. These ratios are relatively constant and independent of the existing ion-concentration gradient. ATP and its hydrolysis products ADP and P_i remain inside the cell, as does the Mg^{2+}. The Mg^{2+} appears to be required only because it is part of the true substrate, which is the $MgATP^{2-}$ complex. We can then write a vectorial equation for the Na^+K^+–ATPase reaction:

$$3Na^+_{inside} + 2K^+_{outside} + ATP^{4-} + H_2O \longrightarrow$$
$$3Na^+_{outside} + 2K^+_{inside} + ADP^{3-} + P_i^{2-} + H^+$$

Experiments carried out in the laboratories of L. E. Hokin and of R. L. Post on cell-membrane vesicles or "solubilized" preparations of the Na^+K^+–ATPase have shown that a transient phosphoenzyme intermediate is formed by transfer of the ^{32}P-labeled terminal phosphate group of ATP to the ATPase molecule. Transfer of ^{32}P to the enzyme requires the presence of Mg^{2+} and Na^+, but K^+ must be absent. When K^+ is then added to the ^{32}P-labeled ATPase preparation, the isotope is discharged and appears as inorganic phosphate. From these and other findings a two-step mechanism for the transport reaction promoted by the Na^+K^+–ATPase has been postulated (Figure 28-10).

Figure 28-10
A postulated mechanism for transport of Na^+ and K^+ by the Na^+K^+–ATPase. The ATPase molecule is believed to undergo conformational changes between two forms designated ○ and □. Ouabain inhibits the K^+-dependent hydrolysis of Na—$\boxed{E}$—P.

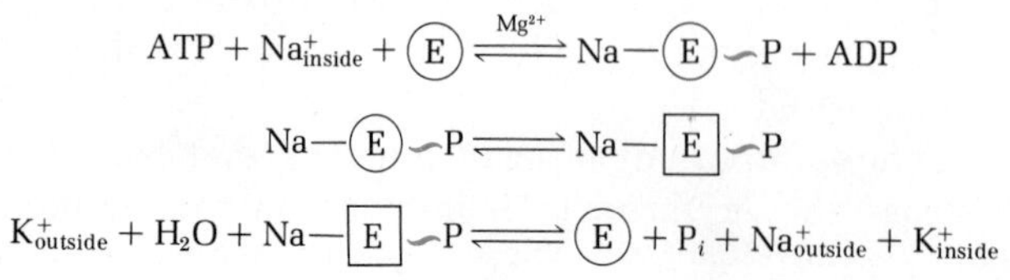

$$ATP + Na^+_{inside} + \bigcirc{E} \underset{}{\overset{Mg^{2+}}{\rightleftharpoons}} Na—\bigcirc{E}—P + ADP$$

$$Na—\bigcirc{E}—P \rightleftharpoons Na—\boxed{E}—P$$

$$K^+_{outside} + H_2O + Na—\boxed{E}—P \rightleftharpoons \bigcirc{E} + P_i + Na^+_{outside} + K^+_{inside}$$

Figure 28-11
The aspartyl phosphate residue at the active
site of the Na⁺K⁺–ATPase. The phosphate
group, donated by ATP, is in color.

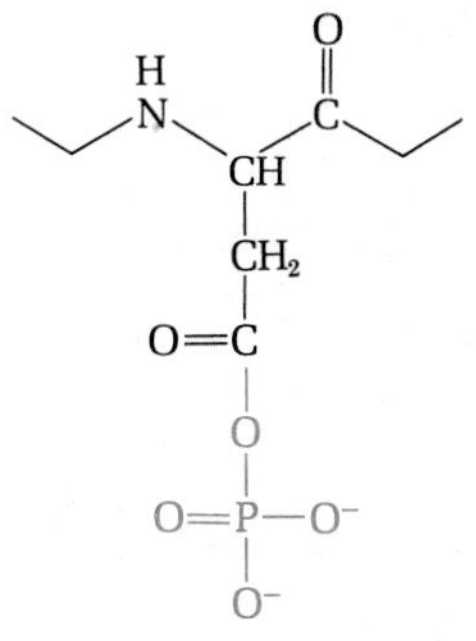

The properties of the phosphoenzyme intermediate resemble those of known acyl phosphates like 3-phosphoglyceroyl phosphate (page 427). On enzymatic degradation of the ^{32}P-labeled enzyme, Post and his colleagues found that an aspartyl side chain of the enzyme is phosphorylated at its free β-carboxyl group (Figure 28-11).

It is of particular interest that the overall Na⁺K⁺–ATPase activity is reversible. I. M. Glynn has shown that when erythrocyte ghosts are loaded with ADP and phosphate, rather than with ATP, and, in addition, a high concentration of K⁺ is imposed inside and a high concentration of Na⁺ outside the membrane, then K⁺ flows out of the cell, Na⁺ flows in, and ATP is generated from the ADP and phosphate. Thus an electrochemical gradient of Na⁺ and K⁺ across the erythrocyte membrane can drive the synthesis of ATP, just as an electrochemical gradient of H⁺ across the mitochondrial (page 527) or chloroplast (page 612) membrane appears to drive the synthesis of ATP.

The Na⁺K⁺–ATPase of various tissues, particularly the kidney medulla, brain, and electric organ, has been solubilized with detergents and purified over a hundredfold. It has a molecular weight of about 250,000 to 300,000 and contains two different types of subunit, the larger with a molecular weight of about 100,000 to 130,000 and the smaller about 50,000. The large subunit is the portion of the molecule that is phosphorylated as ATP is hydrolyzed. Since it has binding sites for Na⁺, K⁺, ATP, and ouabain, the large subunit appears to extend through the entire thickness of the cell membrane with the K⁺ and ouabain sites on its outer surface and the Na⁺ and ATP sites on its inner surface. The smaller subunit of the Na⁺K⁺–ATPase is a glycoprotein and contains sialic acid, as well as glucose, galactose, and other hexose residues. In some tissues the ATPase contains two large and one small subunit. The Na⁺K⁺–ATPase also appears to require or contain certain lipids.

The Na⁺K⁺–ATPase in some cells utilizes a very large fraction of their respiratory energy. For example, it has been found that ouabain inhibits the respiration of kidney cells or brain cells by some 70 percent. Since cells consume oxygen only as fast as needed to regenerate ATP utilized in energy-requiring functions, it has been concluded that kidney and brain cells use some 70 percent of their ATP output for the purpose of pumping Na⁺ and K⁺.

The Active Transport of Glucose and Amino Acids in Vertebrates

Glucose is transported against concentration gradients across the epithelial cell layer of the small intestine, as part of the process of absorption of sugars into the blood from the intestine. Glucose is also transported against a gradient by the epithelial cell layer of the kidney tubules, from the glomerular filtrate into the blood; thus glucose is not normally excreted in the urine but is salvaged and kept in the bloodstream. Although the substrate specificity of the intestinal and renal glucose-transport systems has been studied, little is known of the molecular mechanism of glucose transport.

One significant finding, however, is that glucose transport appears to be coupled to the transport of Na^+; inward glucose transport appears to be optimal when there is a large inward gradient of Na^+ into the cell.

An interesting model for the relationship of glucose transport to Na^+ movements has been postulated (Figure 28-12). It proposes that glucose enters the cell by binding to a specific carrier molecule, which can also bind Na^+, presumably at a second site on the carrier. The carrier molecule thus facilitates the simultaneous, compulsory transport of both Na^+ and glucose into the cell; such a process is designated *cotransport* or *symport*. If the external Na^+ concentration is much higher than the internal concentration, Na^+ bound to this carrier will tend to move inward, down the Na^+ gradient. Since the carrier must also bind glucose in order to function, the inward transport of Na^+ down a gradient can "drag" glucose into the cell. In this manner glucose can be accumulated against a glucose gradient, so long as the inward gradient of Na^+ generated by the Na^+K^+–ATPase exceeds the outward gradient of glucose. The energy required to transport glucose into the cell is furnished by the inward gradient of Na^+ generated by the Na^+K^+–ATPase. Much evidence supports this model of glucose transport in the intestine.

The active transport of amino acids shows a number of similarities with the active transport of glucose. Amino acid transport is particularly pronounced in the intestinal epithelium, where it functions in the intestinal absorption of amino acids into the bloodstream. It also occurs in the kidney tubules, where it functions to salvage amino acids from tubular urine and to retain them in the bloodstream. From the kinetics of transport of the various amino acids and from studies of competition between different amino acids for transport it has been deduced that there are five or more different amino acid transport systems, each of which can transport a group of closely related amino acids. Specific transport systems have been found for (1) small neutral amino acids, (2) large neutral amino acids, (3) basic amino acids, (4) acidic amino acids, and (5) the imino acid proline. Each of these transport systems is characterized by an optimum pH and by specific K_M values for its substrates; they can be inhibited competitively by certain amino acid analogs.

As with glucose transport, external Na^+ is required for inward transport of amino acids in some types of cells. The higher the external Na^+ concentration, the greater the capacity of the amino acid transport system to transport amino acids into the cell. These observations have led to the hypothesis that the energy inherent in a downward gradient of Na^+ into the cell is the immediate driving force for inward transport of amino acids against a gradient, as described above for glucose.

The γ-Glutamyl Cycle for Amino Acid Transport

An interesting and novel hypothesis, the *γ-glutamyl cycle*, has been developed by A. Meister and his colleagues for the mechanism of transport of amino acids into cells of certain

Figure 28-12
Postulated mechanism for the coupling of glucose transport to a Na^+ gradient.

The Na^+K^+-ATPase pumps Na^+ out of the cell to create an inward-directed Na^+ gradient at the expense of ATP.

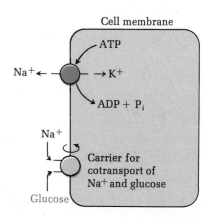

The inward-directed Na^+ gradient so generated "pulls" glucose into the cell by means of a passive carrier which has two binding sites, one for Na^+ and the other for glucose.

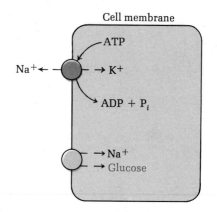

Figure 28-13
The γ-glutamyl cycle, a postulated group-translocation mechanism for the transport of amino acids.

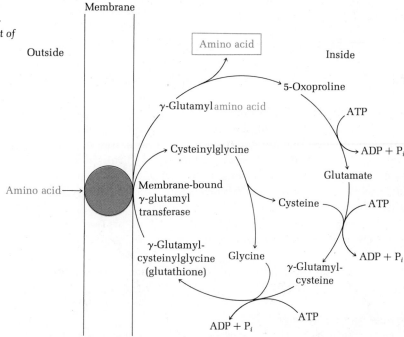

animal tissues. It is an example of a *group translocation* in which the substrate transported appears in a different chemical form inside the cell. The γ-glutamyl cycle involves a sequence of six enzymes, one of which is membrane-bound; the remainder are present in the cytosol. Also involved is the tripeptide *glutathione* (γ-glutamylcysteinylglycine), which is present in high concentrations (~5 m*M*) in all animal tissues but whose major biological function has been an unsolved question for many years (pages 96 and 714).

In the operation of this cycle (Figure 28-13) the membrane-bound enzyme *γ-glutamyltransferase* plays a key role. It catalyzes the reaction

$$\text{Amino acid} + \underset{\text{glutathione}}{\gamma\text{-glutamylcysteinylglycine}} \longrightarrow \gamma\text{-glutamyl amino acid} + \text{cysteinylglycine}$$

which results in the transfer of the glutamyl residue of glutathione to the incoming amino acid. In this reaction not only glutathione but certain other γ-glutamylpeptides can serve as the donor of the γ-glutamyl group; moreover, all the common amino acids except proline can accept the γ-glutamyl group. The free amino acid participating in this reaction comes from outside the cell and the glutathione from inside. Following this reaction, the γ-glutamylamino acid is discharged into the cytosol, as is the cysteinylglycine. In the next step the γ-glutamylamino acid undergoes cleavage to yield the free amino acid and 5-oxoproline, in a reaction catalyzed by *γ-glutamylcyclotransferase*, a cytosol enzyme:

$$\gamma\text{-Glutamylamino acid} \longrightarrow \text{amino acid} + \text{5-oxoproline}$$

The cysteinylglycine then is postulated to undergo hydrolysis by the action of a peptidase:

$$\text{Cysteinylglycine} + H_2O \longrightarrow \text{cysteine} + \text{glycine}$$

Thus one molecule of an amino acid has been transported into the cell, at the expense of the energy of hydrolysis of the peptide bonds of glutathione.

If this process is to be continuous, the glutathione required must be regenerated from free cysteine, free glycine, and 5-oxoproline. Formation of glutathione is brought about by a sequence of three reactions, first discovered by K. Bloch. In the first 5-oxoproline is converted into L-glutamate by the action of 5-*oxoprolinase*:

$$\text{5-Oxoproline} + \text{ATP} + 2H_2O \longrightarrow \text{L-glutamate} + \text{ADP} + P_i$$

In the next step the L-glutamate so formed reacts with L-cysteine by the action of γ-*glutamylcysteine synthetase*:

$$\text{L-Glutamate} + \text{L-cysteine} + \text{ATP} \longrightarrow$$
$$\text{γ-glutamylcysteine} + \text{ADP} + P_i$$

In the last step glycine is attached to γ-glutamylcysteine to yield glutathione by the action of *glutathione synthetase* (page 714):

$$\text{γ-Glutamylcysteine} + \text{glycine} + \text{ATP} \longrightarrow \text{glutathione} + \text{ADP} + P_i$$

The glutathione so formed is now ready to participate in another round of the cycle. Three terminal phosphate bonds of ATP are thus utilized to bring each molecule of an amino acid into the cell by this mechanism.

All the enzymes of this cycle are present in high concentrations in a number of tissues active in amino acid transport. Some people have a genetic defect, presumably in the enzyme 5-oxoprolinase, resulting in the excretion of 5-oxoproline in the urine. Such patients also show aberrations in amino acid transport and metabolism. It is of some interest that Na^+, which is required for amino acid transport (see above), is also necessary for optimal activity of γ-glutamyltransferase.

Intracellular Transport of Ca^{2+} in Animal Cells

An important example of an intracellular active-transport process is the ATP-dependent transport of Ca^{2+} from the sarcoplasm into the sarcoplasmic reticulum, which is responsible for the relaxation of muscle (page 762). This transport process occurs by the action of a Ca^{2+}-transporting ATPase system found in the sarcoplasmic reticulum membrane. This system has been intensively studied by W. Hasselbach, as well as other investigators, in resealed membrane vesicles obtained from the microsome fraction of skeletal muscle. Such vesicles transport Ca^{2+} inward at the expense of the energy yielded on the hydrolysis of ATP to ADP and phosphate. The Ca^{2+}–ATPase system has an extremely high affinity for Ca^{2+} and is capable of transporting Ca^{2+} against very high gradients. Two Ca^{2+} ions are transported into the vesicles for each ATP molecule hydrolyzed.

The mitochondria of animal cells also carry out active intracellular transport of Ca^{2+}. Mitochondria can accumulate Ca^{2+} against steep gradients and with high affinity, the energy being provided by electron transport (page 531). The balance between energy-dependent uptake of Ca^{2+} by mitochondria and its energy-independent release is believed to be an important factor in regulating the normally very low concentration of Ca^{2+} (about 0.1 to 1.0 μM) in the cytosol. The cytosol Ca^{2+} concentration is an important element in the regulation of a large number of cell activities.

Active-Transport Systems in Bacteria

Much progress has been made in the study of active-transport systems in bacteria. Bacterial transport systems are very powerful, in consonance with the very high rate of metabolism of bacteria compared with eukaryotic cells. Bacteria also offer the advantage that genetic analysis of transport systems can easily be carried out. Moreover, simple methods have been devised to obtain membrane vesicles from bacteria. Particular progress has been made in isolating specific protein components of bacterial systems for group translocation and for active transport of sugars and amino acids.

The first important success in separating transport proteins from bacteria was the extraction by C. F. Fox and E. P. Kennedy of "M-protein" from the membrane of *E. coli* cells adapted to transport β-galactosides. They showed that this protein (although isolated in inactive form as an N-ethylmaleimide derivative) contained the recognition and binding site for β-galactosides.

Another important advance came with the discovery of *periplasmic binding proteins* of bacteria. Bacterial cells characteristically contain in the *periplasmic space* one or more specific proteins capable of binding some specific amino acid or sugar. The periplasmic space is the space just outside and surrounding the cell membrane but inside the cell wall. *Osmotic shock* of certain bacteria placed in a hypotonic medium causes release of the periplasmic proteins. From the released material a large number of specific proteins capable of binding certain amino acids, sugars, and inorganic ions have been isolated. The first of the periplasmic binding proteins to be studied, the sulfate-binding protein, has been highly purified and crystallized by A. B. Pardee and his colleagues. It has a molecular weight of 32,000 and contains one binding site for sulfate per molecule.

Specific periplasmic binding proteins have been isolated for most of the common amino acids and for glucose, galactose, arabinose, ribose, and other sugars. They have molecular weights from 22,000 to 42,000 and are readily soluble in aqueous systems. Some of them change in conformation on binding their specific ligands.

Much evidence suggests that the periplasmic binding proteins are involved in active-transport mechanisms. Release of the binding protein from bacterial cells is usually accompanied by decrease of transport activity. The substrate specificity of the binding protein is often the same as that of the transport process of the intact cells. Moreover, some mutants that are transport-negative for a given substrate contain a defective binding protein for that substrate. Such evidence has suggested that the periplasmic binding proteins are dissociable components of membrane-bound transport systems that contain the specific substrate-recognition sites for their substrates. They are also believed to be receptor sites in chemotaxis, by which bacteria recognize nutrients and propel themselves by flagellar motion in the direction of increasing nutrient concentration.

Group Translocation of Sugars into Bacteria

Two general types of sugar-transport systems occur in bacteria, sometimes within the same species. One promotes a group translocation of the sugar across the membrane so that it enters the cytoplasm in a different chemical form, whereas the other type promotes active transport of the sugar into the cell, apparently without chemical modification of the sugar molecule.

In a series of important investigations S. Roseman and W. Kundig carried out a biochemical and genetic analysis of sugar transport into certain bacteria via a group-translocation system called the *phosphotransferase system* (PTS). The system has been studied for the most part in E. coli, Salmonella typhimurium, and Staphylococcus aureus. It is found in many facultative and photosynthetic bacteria but appears to be absent in strictly aerobic bacteria. Its properties vary somewhat with the species. All the basic components of PTS have been isolated and purified. The phosphotransferase system consists of three proteins or protein complexes. Two of these, called enzyme I and HPr, are normally present in the cytoplasm, and the third, enzyme II, is located in the cell membrane. Enzyme II contains two separable components, which are possibly different subunits.

The reactions taking place via the action of the phosphotransferase system can be summarized by two equations. In the first, the cytoplasmic protein HPr is phosphorylated at the expense of phosphoenolpyruvate, a reaction that is catalyzed by enzyme I:

$$\text{Phosphoenolpyruvate} + \text{HPr} \xrightarrow[\text{Mg}^{2+}]{\text{E}_1} \text{pyruvate} + \text{phospho–HPr}$$

This reaction is unusual in two respects. (1) It employs phosphoenolpyruvate as the immediate phosphate donor, whereas most biological phosphorylations require ATP or some other nucleoside triphosphate as phosphate donor. (2) The HPr protein molecule becomes phosphorylated on nitrogen atom 1 of the imidazole ring of one of its two histidine residues (Figure 28-14). HPr has been highly purified and has a molecular weight of only 9,400.

Figure 28-14
The 1-phosphohistidyl residue of HPr.

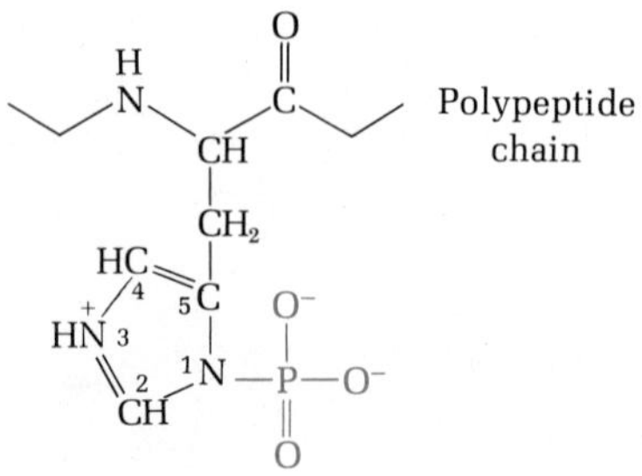

Figure 28-15
Energy-coupled group translocation of glucose across the bacterial membrane and accumulation of glucose 6-pyruvate (PEP =
phosphoenolpyruvate).

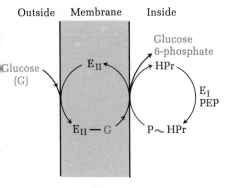

Outside Membrane Inside

In a second enzymatic reaction, which takes place in the membrane, the phosphorylated HPr donates its phosphate group to the sugar that is being transported, in a reaction catalyzed by the membrane-bound enzyme II. With glucose the product of the reaction is glucose 6-phosphate:

$$\text{Phospho–HPr} + \text{glucose} \xrightarrow{\text{E}_{\text{II}}} \text{HPr} + \text{glucose 6-phosphate}$$

In gram-negative (page 268) bacteria, such as *E. coli*, enzyme II is a complex of two proteins or subunits, designated IIA and IIB. Protein IIA carries the specificity for the sugar; several different IIA proteins have been isolated, each specific for a given sugar, such as glucose, fructose, or mannose. Protein IIB appears to be involved in the catalysis of phosphate transfer from the phosphorylated HPr to the sugar bound to IIA. Protein IIB has a molecular weight of 36,000 and makes up nearly 10 percent of the protein of the bacterial membrane. Many different sugars — including D-glucose and various glucosides, D-galactose and various galactosides, D-fructose, pentoses, pentitols, and hexitols — have been identified as phosphate-group acceptors in the phosphotransferase system. The postulated mechanism for sugar transport by the phosphotransferase system is shown in Figure 28-15.

Much genetic evidence supports this hypothesis. Bacterial mutants that lack enzyme I or HPr cannot accumulate any of the transportable sugars. On the other hand, mutants defective in only one species of enzyme IIA are defective in transporting only that particular sugar.

The phosphotransferase system has been reconstituted from purified components. Membrane vesicles isolated from bacterial cells in such a manner that they lack HPr and enzyme I are unable to accumulate sugars as their phosphates. When purified HPr and enzyme I are then added to the vesicles, they became bound to them and the vesicles then gain the ability to accumulate sugar phosphates.

E. coli cells contain not only the PTS system for group translocation of sugars, but also, as we shall now see, a system for respiration-coupled active transport of sugars, in which phosphorylation of the sugar does not appear to be involved.

Sugar and Amino Acid Transport Coupled to Electron Transport in Bacteria

H. R. Kaback and his colleagues have made substantial progress in the analysis of the coupling between the energy-requiring active transport of glucose and other sugars into bacteria and the energy-yielding transport of electrons from substrates to oxygen. They have developed a procedure for obtaining membrane vesicles from *E. coli* cells that are nearly devoid of cytoplasmic and cell-wall components. Such vesicles are empty sacs, 0.5 to 1.5 μm in diameter and right side out. They appear to retain full activity of the major active-transport systems and also contain the dehydrogenases and electron carriers of the respiratory chain, all bound to the membrane structure. Such bacterial membrane vesicles promote the active transport of a number of different sugars, including lactose and other β-galactosides, glucose, galactose, arabinose, and glucuronic acid, providing the vesicles are oxidizing certain substrates, such as D-lactate, β-hydroxybutyrate, or succinate, to provide the necessary energy. The concentration of sugars attained within such respiring vesicles may exceed by a hundredfold the external concentration of the sugar in the medium. The accumulated sugar is present in free form and does not appear to be phosphorylated during or after transport into the membrane vesicle. Thus this respiration-dependent transport of sugars differs from the group translocation promoted by the phosphotransferase system.

Because oligomycin, an inhibitor of oxidative phosphorylation (page 520), does not prevent sugar transport coupled to D-lactate oxidation, the intermediate formation of ATP is apparently not required for the coupling of the energy of electron transport to the sugar-transport mechanism. This conclusion is supported by the finding that membrane vesicles from certain mutants capable of electron transport but unable to carry out phosphorylation of ADP can still transport sugars. Membrane vesicles from bacteria are capable of accumulating most of the common amino acids in a very similar manner, coupled to electron transport from D-lactate or some other substrate via the electron-transport chain to oxygen.

The key question now concerns the mechanism by which the energy of electron transport is utilized to drive the active transport of sugars and amino acids into the vesicles against (in some cases) rather steep concentration gradients. Here two schools of thought have developed. One hypothesis, vigorously championed by Kaback, is that one or more of the electron carriers in the respiratory chain in the bacterial membrane functions not only to transfer electrons but also to transport the sugar or amino acid across the membrane (Figure 28-16). Kaback has postulated that one of the specific electron carriers in this chain occurs in disulfide (oxidized) and disulfhydryl (reduced) forms. These two forms also are capable of binding the substance being transported. The disulfide or oxidized form of this electron carrier is regarded as the high-affinity form capable of binding the transported substrate only on the outside surface of the membrane. The disulfhydryl or reduced form of this electron carrier is believed to bind the transported substrate on the inside surface of the membrane with much lower affinity. Thus the oxidized form of the carrier traps the substrate molecule from the external medium with high affinity, and the carrier molecule, as it is reduced by electrons coming from the electron donor D-lactate, undergoes a conformational change resulting in a large decrease in affinity and consequent unloading of the transported substrate at the inside surface of the vesicle. A specific role of sulfhydryl groups in sugar and amino acid transport is suggested by the fact that reagents capable of blocking —SH groups also block active transport.

The Kaback hypothesis suggests that one substrate molecule should be transported into the vesicle for each pair of

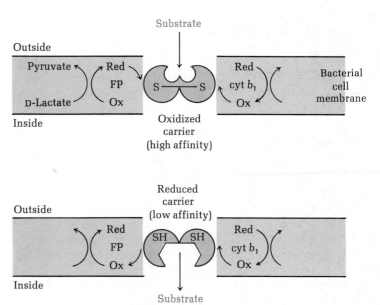

Figure 28-16
Coupling of the active transport of an external substrate to electron transfer. The respiratory chain of bacteria is in the cell membrane. One of the electron carriers (color) is postulated to transfer electrons, via its —SH and —S—S— groups, and to use the energy of electron transfer to carry the substrate across the membrane. The transport system has a high affinity on its outside face and a much lower affinity on its inside face; interconversion of the two forms is assumed to be responsible for transport against a gradient of concentration.

Figure 28-17
Active transport of galactosides into bacterial cells by obligatory cotransport of H^+ and the galactoside. The electron-transport chain in the cell membrane continuously pumps H^+ outward coupled to electron flow from substrate (SH_2) to oxygen. The galactoside is pulled into the cell by the inward gradient of H^+ via the carrier, which binds both H^+ and galactoside.

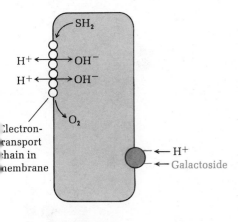

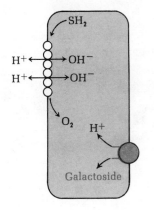

electrons being transported from the electron donor to oxygen. However, it has been pointed out that the rate of lactate oxidation is anywhere from 10 to several hundred times greater than the rate of sugar or amino acid transport on a molar basis.

An alternative hypothesis for the mechanism of energy coupling between electron transport and the inward transport of sugars and amino acids in bacteria has been proposed by P. Mitchell, F. M. Harold, and other investigators, who postulate that bacterial sugar transport is brought about by chemiosmotic coupling, as originally developed for oxidative (page 527) and photosynthetic (page 612) phosphorylation.

The chemiosmotic hypothesis proposes that electron transport generates an electrochemical gradient of H^+ across the membrane, composed of an outside H^+ gradient and an inside negative membrane potential. This gradient, called a proton-motive gradient, has been postulated as the immediate driving force for ATP formation and for other energy-dependent functions of mitochondria and chloroplasts, including ion transport (page 527). Several items of evidence support the chemiosmotic hypothesis as the principal means of energy coupling in bacterial membrane vesicles. Electron transport in such vesicles has been shown to generate a membrane potential and to cause extrusion of protons. Moreover, the membrane carrier for galactosides has been found to carry both H^+ and galactosides across the membrane into the cell, apparently in a 1:1 relationship (Figure 28-17). When membrane vesicles not capable of electron transport are placed in a medium containing a labeled galactoside and then acidified to pH 6.0 to create an inward gradient of H^+, the galactoside is transported into the vesicle, together with H^+. T. H. Wilson and his colleagues have shown ATP hydrolysis also can support galactoside transport, through its capacity to generate a H^+ gradient.

Ionophores

A large number of antibiotic substances capable of inducing passage of specific ions across biological membranes have been isolated and identified. Among the first in which this property was recognized were _gramicidin_, which induces the transport of K^+, Na^+, and other monovalent cations across the membranes of mitochondria, erythrocytes, and synthetic phospholipid bilayers, and _valinomycin_, which has a high degree of specificity for inducing transport of K^+ across mitochondrial and other membranes. The mechanism of action of these antibiotics has been intensively investigated by B. C. Pressman, H. A. Lardy, P. Mueller, and many other investigators. Valinomycin, like many of the other ionophores, is an annular or ringlike compound with a hydrophobic exterior and hydrophilic interior, within which the K^+ ion precisely fits through coordination bonds. The hydrophobic exterior of valinomycin allows the valinomycin-K^+ complex, which has a single positive charge, to dissolve in, and thus pass through, the nonpolar hydrocarbon layer in the membrane structure (Figure 28-18). Hence, the generic name _ionophore_ for these antibiotics. Another ionophore antibiotic specific for K^+ is _nigericin_, which differs from valinomycin in having a single negative charge; the nigericin-K^+ complex thus has no net electric charge. Nigericin promotes exchange of K^+ with H^+, whereas valinomycin promotes passage of K^+ alone. Certain other ionophores, such as _gramicidin_, a polypeptide, are linear molecules; however, they are believed to wrap around the monovalent cation to form an annular structure resembling that of valinomycin. Some ionophores are very active in carrying divalent cations, particularly Ca^{2+}. Synthetic ionophores have also been prepared.

Ionophores have received intensive study, not only for their characteristic effects on mitochondrial ion transport and oxidative phosphorylation but also as tools in experimental analysis of ion-transport functions of the heart and kidney.

Ionophores enable cations to be transported across synthetic phospholipid bilayers (page 301). Study of such artificial transport systems has given important information on the basic physical principles involved in biological transport systems, for which they are important models. For example, the antibiotic _alamethicin_, when combined with a phospholipid bilayer, enables the system to mimic an action potential when K^+ and Na^+ gradients are imposed and an electric stimulus is applied. Indeed, study of ionophores is giving important clues to the nature of the Na^+ _gates_ of nerve fibers, passive Na^+-transport systems that allow Na^+ to leak out transiently as the action potential passes. Such Na^+ gates, which are few in number but highly active, are vital to nerve activity; they are blocked by _tetrodotoxin_, the intensely toxic poison of the Japanese puffer fish.

Figure 28-18
Model structures for two K^+-ionophore complexes. The structure of the valinomycin-K^+ complex. A = L-lactate, B = L-valine, C = D-hydroxyisovalerate, and D = D-valine.

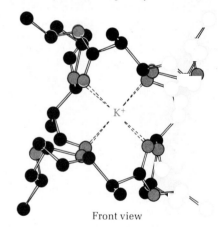

Below are shown front and side views of a model of the K^+ complex of nonactin.

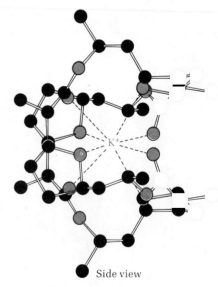

Front view

Side view

Summary

In active transport solutes move against gradients of concentration, a process resulting in an increase of free energy. In passive transport, or movement down a gradient, there is a decrease in free energy. The free-energy change in a transport process is a function of the relative concentrations of the solute across the membrane. If the solute has an electric charge, the gradient is spoken of as an electrochemical gradient, since it is a gradient of both mass and charge. Cell membranes contain transport systems capable of promoting the passage of biologically important solutes. Such systems are substrate-specific, show saturability with substrate, may be inhibited either competitively or noncompetitively, and are genetically determined. The substrate forms a complex with a specific carrier protein in the membrane, analogous to an enzyme-substrate complex. An example of a passive carrier is the glucose carrier of the human erythrocyte; it is competitively inhibited by phloretin and other analogs of glucose. Other membrane transport systems in animal tissues promote exchange diffusion, i.e., equimolar exchange of solute molecules across the membrane, such as the ADP–ATP carrier of the mitochondrial membrane.

The most widely distributed active-transport system in higher animals is the Na^+K^+–ATPase of the plasma membrane, which extrudes Na^+ from cells and allows K^+ to enter. High internal K^+ is required for ribosome function, glycolysis, nerve action, and many other functions. The Na^+K^+–ATPase acts on internal MgATP; it is stimulated by Na^+ from the inside and by K^+ from the outside. Three Na^+ are extruded and two K^+ accumulated per molecule of ATP hydrolyzed to ADP and phosphate. The transport reaction proceeds in two steps. The first requires Na^+; it leads to phosphorylation of the enzyme by ATP and binding of internal Na^+. In the second step, which is stimulated by K^+, the phosphoenzyme, which contains an aspartyl phosphate residue, is hydrolyzed and Na^+ is released to the outside. The Na^+K^+–ATPase contains two large and possibly one small subunits.

The active-transport systems for glucose and amino acids usually require an inward gradient of Na^+ as the energy source to transport amino acids into the cell, presumably by cotransport of both Na^+ and the substrate on the same carrier molecule. Amino acids may also be carried across some animal membranes by group translocation in the γ-glutamyl cycle.

In the periplasmic space bacteria contain a number of binding proteins that selectively bind different sugars, amino acids, or inorganic ions. These proteins function in transport and chemotaxis. Sugars are transported by two types of systems in bacteria. The phosphotransferase system transports simple sugars by group translocation to yield internal sugar phosphates. Three protein components of this system have been isolated: enzyme I, enzyme II, which carries the sugar specificity, and a high-energy phosphorylated protein HPr, formed by transfer of phosphate from phosphoenolpyruvate.

Aerobic bacteria contain a second system for transport of sugars and amino acids. It depends on electron transport to oxygen for energy, without the intermediate function of ATP. It is a true active-transport process; the substrate is transported without chemical change.

Ionophores, antibiotics capable of binding certain cations such as K^+, Na^+, and Ca^{2+}, among them valinomycin, gramicidin, and nigericin, are useful tools in the study of transport and serve as important models of biological transport systems.

References

Books

CHRISTENSEN, H. N.: *Biological Transport*, Benjamin, Menlo Park, Calif., 1974.

Current Topics in Membranes and Transport, Academic, New York. An annual volume of reviews, initiated in 1970.

GREENBERG, D. R. (ed.): *Metabolic Transport*, vol. 6 of *Metabolic Pathways*, 3d ed., Academic, New York, 1972.

HARRIS, E. J.: *Transport and Accumulation in Biological Systems*, 3d ed., Butterworth, London, 1972.

STEIN, W. D.: *Movement of Molecules across Cell Membranes*, Academic, New York, 1967. An excellent account of the theory and function of membrane transport systems.

Articles

BOOS, W.: "Bacterial Transport," *Ann. Rev. Biochem.*, 43: 123–146 (1974).

CHRISTENSEN, H. N.: "Linked Ion and Amino Acid Transport," pp. 393–422 in E. E. Bittar (ed.), *Membrane and Ion Transport*, Interscience, New York, 1970.

DAHL, J. L., and L. E. HOKIN, "The Sodium-Potassium ATPase," *Ann. Rev. Biochem.*, 43: 327–356 (1974).

EPSTEIN, E.: "Mechanisms of Ion Transport through Plant Cell Membranes," *Int. Rev. Cytol.*, 34: 123–168 (1973).

FICKEL, T. E., and G. C. GILVARG: "Transport of Impermeant Substances in *E. coli* by Way of Oligopeptide Permease," *Proc. Natl. Acad. Sci. (U.S.)*, 70: 2866–2869 (1973).

FOX, C. F., and E. P. KENNEDY: "Specific Labeling and Partial Purification of the M Protein, a Component of the β-Galactoside Transport System of *E. coli*," *Proc. Natl. Acad. Sci. (U.S.)*, 54: 891–899 (1965).

HAROLD, F. M.: "Conservation and Transformation of Energy by Bacterial Membranes," *Bacteriol. Rev.*, 36: 172–230 (1972).

HAYDEN, D. A., and S. B. HLADKY: "Ion Transport across Thin Lipid Membranes," *Q. Rev. Biophys.*, 5: 187–282 (1972). This review also contains much information on ionophore antibiotics.

HAZELBAUR, G. L., and J. ADLER: "Role of the Galactose Binding Protein in Chemotaxis of *E. coli* toward Galactose," *Nature New Biol.*, 230: 101–104 (1971).

KABACK, H. R.: "Transport across Isolated Bacterial Cytoplasmic Membranes," *Biochim. Biophys. Acta*, 265: 367–416 (1972).

KASHKET, E. R., and T. H. WILSON: "Proton-Coupled Accumulation of Galactoside in *S. lactis* 7962," *Proc. Natl. Acad. Sci. (U.S.)*, 70: 2866–2869 (1973).

KYTE, J.: "Properties of the Two Polypeptides of Na^+ and K^+ Dependent Adenosine Triphosphatase," *J. Biol. Chem.*, 247: 7642–7649 (1972).

MEISTER, A.: "On the Enzymology of Amino Acid Transport," *Science*, 180: 33–39 (1973). Proposal of the γ-glutamyl cycle for amino acid transport.

MITCHELL, P.: "Translocations through Natural Membranes," *Adv. Enzymol.*, 29: 33–88 (1967). A review of fundamental processes and concepts of membrane transport.

MITCHELL, P.: "Reversible Coupling between Transport and Chemical Reactions," pp. 192–256 in E. E. Bittar (ed.), *Membranes and Ion Transport*, Interscience, New York, 1970.

PARDEE, A. B.: "Membrane Transport Proteins," *Science*, 162: 632–637 (1968). A brief review article.

POST, R. L., and S. KUME: "Evidence for an Aspartyl Phosphate Residue at the Active Site of Na^+ and K^+ Adenosine Triphosphatase," *J. Biol. Chem.*, 248: 6993–7000 (1973).

ROSEMAN, S.: "Carbohydrate Transport in Bacterial Cells," pp. 41–88 in D. M. Greenberg (ed.), *Metabolic Pathways*, 3d ed., vol. 6, Academic, New York, 1972.

SCHULTZ, S. G., and P. F. CURRAN: "Coupled Transport of Sodium and Organic Solutes," *Physiol. Rev.*, 50: 637–718 (1970).

WEST, I., and P. MITCHELL: "Stoichiometry of Lactose-Proton Symport across the Plasma Membrane of *E. coli*," *Biochem. J.*, 132: 587–592 (1973).

Problems

1. Calculate the amount of free energy required to transport 1 gram ion of K^+ from blood plasma to intracellular fluid of muscle at 37°C, using data given in Figure 28-2.

2. Calculate the maximum gradient of Na^+-ion concentration that can theoretically be achieved by the hydrolysis of ATP to ADP and phosphate by a Na^+-transporting ATPase at equilibrium. Assume standard thermodynamic conditions.

3. (a) Calculate the maximum rate of K^+ uptake by an aerobic cell oxidizing glucose if its rate of respiration is 10.2 μg atoms of oxygen per milligram protein per hour at 37°C. Assume that 80 percent of the respiratory energy is utilized for K^+ transport via the Na^+K^+–ATPase. Use data furnished in the text.
    (b) How many K^+ ions would be extruded from this cell per molecule of citric acid undergoing complete combustion to CO_2 and H_2O?

4. If 75 percent of the glycolytic energy of adult human erythrocytes, available as ATP, is devoted to extrusion of Na^+, how many Na^+ ions could be extruded by the Na^+K^+–ATPase per molecule of glucose undergoing glycolysis to lactic acid?

5. If a single erythrocyte contains 5,000 molecules of the Na^+K^+–ATPase, what fraction of the cell membrane surface is occupied by the enzyme molecules? Assume that the erythrocyte is a disk 7 μm in diameter and 1 μm thick, and that the ATPase molecule is spherical and has a density of 1.2. Other necessary data are given in the text.

6. Calculate the minimum pH difference across a membrane that could, theoretically at least, support the formation of 1.0 mM ATP from 1.0 mM P_i and 1.0 mM ADP at pH 7.0 at 25°C. Now calculate the minimum pH gradient required to generate one molecule of ATP in a system in which the equilibrium concentrations are 3.0 mM ATP, 0.5 mM ADP, and 3.0 mM phosphate.

Hormones, trace substances produced by various endocrine glands, serve as chemical messengers carried by the blood to various target organs, where they regulate a variety of physiological and metabolic activities in vertebrates. Endocrinology, the study of hormones and their action, has long been an important field of vertebrate physiology, but apart from biochemical studies of the molecular structure of some of the hormones, until the late 1960s little if anything was known of the biochemical mechanisms of hormone action. In the last few years some important advances have been made in the molecular analysis of hormone function, which have transformed the study of endocrine regulation from an essentially descriptive area of biology into an important and productive branch of biochemical research.

In this chapter we shall not attempt to describe the physiological and anatomical consequences of hormone secretion and function. Instead we shall examine in detail a few of the hormones that are involved in regulating the central metabolic pathways in vertebrates, particularly those hormones whose biochemical function is becoming understood.

Although we have biochemical knowledge of the action of only two or three hormones, new insights have been gained into the kinds of biochemical processes that are regulated and how their regulation is effected. These events engender much optimism that we may soon learn, for example, how insulin and thyroxine, two hormones long used in medicine, exert their characteristic and often lifesaving effects.

The Organization of the Mammalian Endocrine System

Figure 29-1 outlines the organization of the endocrine system of mammals. The synthesis and/or release of several of the hormones is controlled in a hierarchical manner, involving three successive stages of hormone–target cell interactions. When the _hypothalamus_, at the base of the brain, receives specific neural messages, it secretes minute quantities of hormones called _releasing factors_, which are passed down nerve

Figure 29-1
*The hierarchical organization of endo-
crine regulation under the control of
the hypothalamus.*

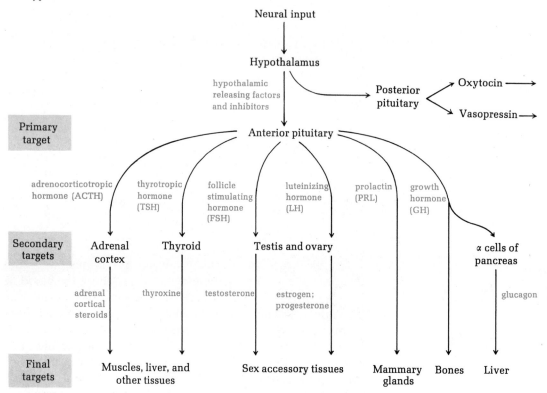

fibers to the anterior pituitary gland. There each of the releasing factors can trigger the release of a specific hormone
from the anterior pituitary gland. For example, *thyrotropin-
releasing factor* (TRF) triggers the release of *thyrotropic hormone* (TH), and *corticotropin-releasing factor* (CRF) causes
release of the *adrenocorticotropic hormone* (ACTH). The
hypothalamus also secretes hormonelike substances called
inhibitory factors, which can inhibit release of some of the
pituitary hormones.

The various hormones released from the anterior pituitary
(Figure 29-1) pass via the systemic blood to specific target
glands. The target of ACTH is the adrenal cortex, and that of
thyrotropic hormone is the thyroid gland. These glands, as
well as the target glands of other anterior-pituitary hormones, particularly the gonads, are in turn stimulated to
produce *their* characteristic hormones, which act on a
number of final target tissues, as shown in Figure 29-1. Besides secreting various releasing and inhibitory factors acting
on the anterior pituitary, the hypothalamus also generates at
least two other hormones, *oxytocin* and *vasopressin* or an-
tidiuretic hormone (structures on page 97), important in the
regulation of lactation and of water balance, respectively.
These hormones, bound to small proteins called *neuro-
physins,* pass to the posterior pituitary, from which they
are released into the systemic blood.

Hormones whose release is under less direct pituitary control include the polypeptides *calcitonin,* formed by certain types of cells in the thyroid and parathyroid glands, and *parathyroid hormone,* which regulates Ca^{2+} and phosphorus metabolism. Also in this group are *insulin* and *glucagon,* polypeptides formed by the β cells and α cells, respectively, in the *islets of Langerhans,* specialized endocrine regions of the pancreas, as well as *epinephrine* and *norepinephrine,* formed by the adrenal medulla. In addition to the hormones listed here and in Figure 29-1, several others controlling a variety of physiological activities have been found.

The secretion of hormones is regulated by a complex network of controls. External stimuli transmitted by the nervous system modulate the activity of the hypothalamus, as well as the secretion of epinephrine by the adrenal medulla. The secretion of the tropic hormones by the anterior pituitary is in turn modulated in a feedback relationship by the characteristic secretions of their target glands. For example, a high blood concentration of glucocorticoid hormones secreted by the adrenal cortex induces feedback inhibition of the secretion of ACTH by the pituitary. The secretion of some hormones is modulated by the concentration of specific metabolites in the blood; for example, the release of insulin from the pancreas is stimulated when the concentration of glucose in the blood rises. Moreover, high blood concentrations of thyroxine inhibit the action of thyrotropin-releasing factor in the anterior pituitary.

Some of the hypothalamic releasing and inhibitory factors, which are short peptides, have been isolated and identified; examples are shown in Figure 29-2. They are produced in only minute amounts; for example, only 1 mg of thyrotropin-

Figure 29-2
Structures of two hypothalamic releasing factors and the inhibitory factor for growth-hormone release.

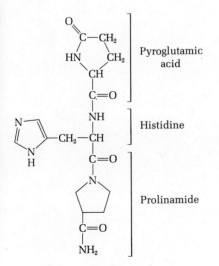

Pyroglutamic acid	
Histidine	
Prolinamide	

Thyrotropin-releasing factor

Pyroglutamic
|
His
|
Trp
|
Ser
|
Tyr
|
Gly
|
Leu
|
Arg
|
Pro
|
Gly
|
Glycinamide

LH-releasing factor

N-terminus   Ala
|
Gly
|
Cys
|
Lys
|
Asn
|
Phe
|
Phe
|
Trp   S
|
Lys   S
|
Thr
|
Phe
|
Thr
|
Ser
|
C-terminus   Cys

Growth-hormone
inhibitory factor

Table 29-1 Some characteristics of polypeptide hormones

	Approximate molecular weight	Approximate number of residues	Components other than amino acids
Anteriory pituitary			
Thyrotropin	28,300	220	Oligosaccharide
Adrenocorticotropin	4,700	39	None
Luteinizing hormone	28,500	200	Oligosaccharide
Follicle-stimulating hormone	34,000	200	Oligosaccharide
Prolactin	21,500	191	None
Growth hormone	21,000	191	None
Posterior pituitary			
Oxytocin	1,070	9	None
Vasopressin	1,070	9	None
Parathyroid and thyroid			
Parathyroid hormone	9,500	84	None
Calcitonin	4,500	32	None
Pancreas			
Insulin	5,500	51	None
Glucagon	3,500	29	None

releasing factor (TRF) was obtained from several tons of hypothalamic tissue obtained from slaughterhouses. Identification and synthesis of some of the releasing and inhibitory factors in the laboratories of R. Guillemin, A. V. Schally, and others has been an outstanding advance in biochemical endocrinology.

The anterior-pituitary hormones are large polypeptides (Table 29-1). The amino acid sequences of most of them have been established; some have been synthesized. It is of interest that TSH and LH contain an identical subunit. Analogs of some of these hormones have been synthesized in a search for drugs capable of inhibiting the action of specific hormones.

Hormone Receptors and Intracellular Messengers

Two basic principles of hormone action have emerged from research of the last decade. The first is that the responsive target cells for any given hormone contain specific hormone receptors, specialized proteins capable of binding the hormone molecule with very high specificity and affinity. Such hormone receptors occur in only very small amounts in the target cell. In the target cells of the water-soluble hormones, such as epinephrine, glucagon, and insulin, which do not pass through cell membranes readily, the hormone receptors are located on the cell surface. In target cells of the sex and adrenal-cortex hormones, which are lipid-soluble steroids and thus can pass through the cell membrane, the primary receptors are located within the cell in the cytosol.

The second principle is that binding of the hormone to its specific receptor causes the formation of an intracellular messenger molecule, which then stimulates (or depresses) some characteristic biochemical activity of the target tissue. For the polar, water-soluble hormones epinephrine and glucagon, the intracellular messenger is 3',5'-cyclic adenylic acid or cyclic AMP (page 316), often called the second messenger. With lipid-soluble steroids, the hormone-receptor complex itself becomes the intracellular messenger.

We shall now discuss the structure, biosynthesis, release, and mode of action of those hormones which exert major regulatory effects on the central metabolic pathways. We shall begin with *epinephrine*, secreted by the adrenal medulla, since studies of its action on the liver yielded the first important breakthroughs in our understanding of the molecular basis of hormone action.

Epinephrine and the Discovery of Cyclic AMP

Secretion of epinephrine (Figure 29-3) into the blood triggers a set of responses that prepares the vertebrate organism to "fight or flee." These include a rise in blood pressure, an increase in heart rate and output, and characteristic effects on smooth muscles, involving relaxation of some and contraction of others. The most prominent biochemical consequence is a greatly increased rate of breakdown of glycogen in the muscles, to yield lactate, and secondarily in the liver, to yield blood glucose. In the early 1950s the latter effect was found by E. W. Sutherland to occur in vitro; when epinephrine was added to slices of intact liver tissue, free glucose appeared in the suspending medium. Starting from this observation, Sutherland and his colleagues, principally T. W. Rall, carried out a classic series of investigations in which they determined the biochemical basis for the action of epinephrine. They found that incubation of epinephrine with intact liver slices caused a large increase in glycogen phosphorylase activity (page 433), which appeared to be rate-limiting in the formation of glucose from liver glycogen. However, the addition of epinephrine to isolated, highly purified phosphorylase preparations failed to elicit an increase in activity, indicating that the stimulating effect of epinephrine depends on factors contributed by the intact liver cell.

Further analysis of the action of epinephrine required a more complete understanding of the factors affecting the activity of phosphorylase. Research by C. F. Cori and G. T. Cori, by E. G. Krebs and E. H. Fischer, as well as by Sutherland, established that the active tetrameric form phosphorylase *a* is converted into inactive phosphorylase *b*, a dimeric molecule, by the action of *phosphorylase phosphatase* (page 434), which removes the phosphate groups from the four phosphoserine residues of the active enzyme. Phosphorylase *b* is converted back into the active phosphorylase *a* by *phosphorylase kinase*, a Ca^{2+}-dependent enzyme that catalyzes the phosphorylation of specific serine residues of phosphorylase *b* (page 435). However, direct tests showed that

Figure 29-3
Epinephrine and norepinephrine are formed in the adrenal medulla. Epinephrine is also known as adrenaline. Norepinephrine serves as a neurotransmitter in the nervous system.

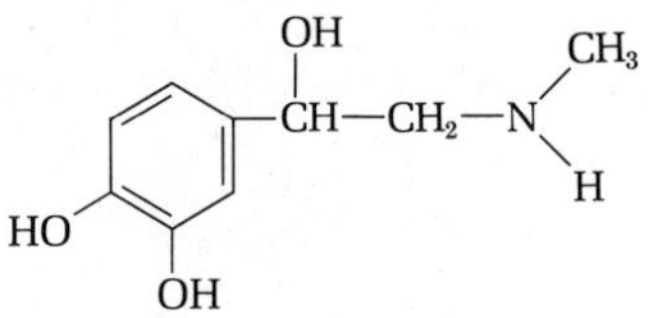

Epinephrine

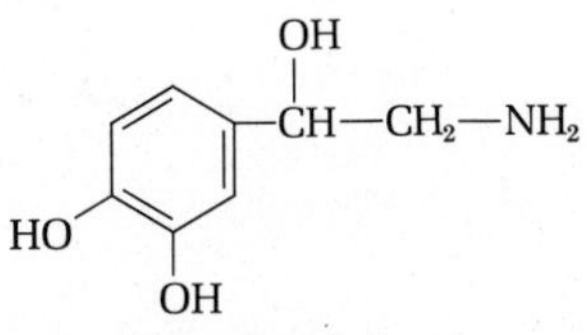

Norepinephrine

epinephrine has no effect on purified preparations of either phosphorylase phosphatase or phosphorylase kinase. Nevertheless, Sutherland and Rall did find that epinephrine promotes the conversion of phosphorylase b into phosphorylase a in crude, broken-cell preparations of liver supplemented with ATP and Mg^{2+}. This effect was ultimately found to be the result of two successive events. In the first, epinephrine, acting on the particulate, insoluble portion of the liver, causes the enzymatic formation of a heat-stable factor that can stimulate phosphorylase activity; this reaction requires ATP and Mg^{2+}. In a second stage, which also requires ATP, the heat-stable factor promotes the conversion of the inactive phosphorylase b in the soluble portion of the liver homogenate into active phosphorylase a. The active stimulatory factor, which normally occurs in only minute amounts in cells, was found to be a heat-stable, dialyzable substance of low molecular weight which could be purified by chromatography on ion-exchange resins. It was ultimately crystallized and found to contain adenine, ribose, and phosphate in the ratio 1:1:1, suggesting that it arose from the ATP required for its formation. On mild hydrolysis or by the action of a phosphatase it yielded adenylic acid (AMP). It was identified in 1960 as a quite unexpected derivative of adenylic acid, never before observed in biological material, namely, *3′,5′-cyclic adenylic acid* (page 316, see also Figure 29-4). Addition of very small amounts of cyclic AMP to soluble liver extracts promotes the formation of active phosphorylase a from phosphorylase b in the presence of ATP.

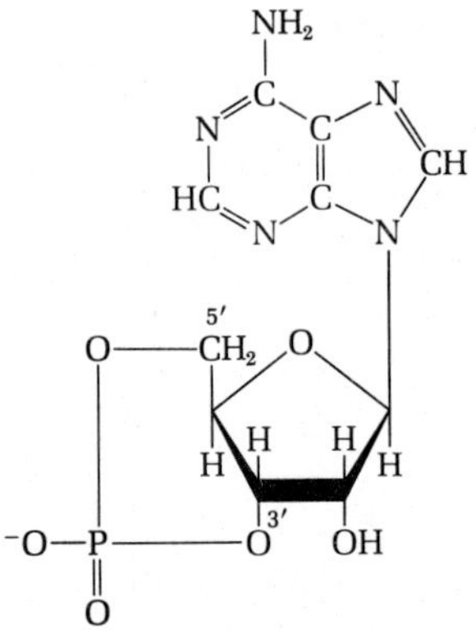

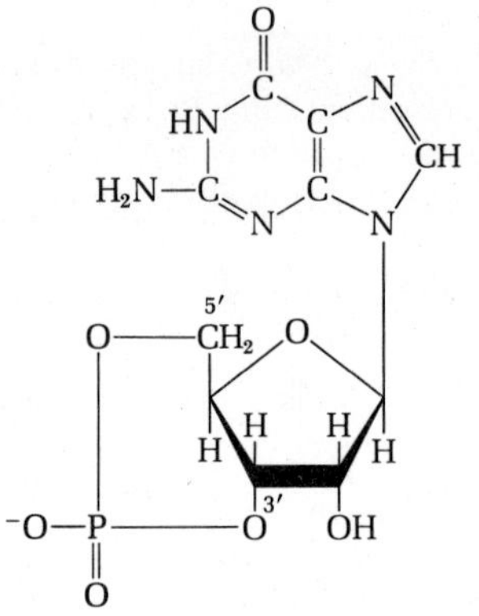

Adenylate Cyclase, Protein Kinase, and Phosphodiesterase

Further research by Sutherland and his colleagues revealed that epinephrine greatly stimulates a Mg^{2+}-dependent enzymatic reaction in the plasma-membrane fraction by which ATP is converted into cyclic AMP with loss of inorganic pyrophosphate:

$$\text{ATP} \xrightarrow{\text{Mg}^{2+}} 3',5'\text{-cyclic AMP} + \text{PP}_i$$

The enzyme catalyzing this reaction, called *adenylate cyclase*, is found in many animal tissues. The enzyme is tightly bound to the plasma membrane and has only recently been extracted in soluble form. Adenylate cyclase is specific for ATP as substrate; it will not form 3′,5′-cyclic nucleotides from ADP or from other NTPs such as GTP or CTP.

How does cyclic AMP stimulate the conversion of inactive phosphorylase b into phosphorylase a, a reaction catalyzed by phosphorylase kinase? The missing link has been provided by the discovery of an enzyme at first known as *phosphorylase kinase kinase* but now called simply *protein kinase*. This enzyme, studied in detail by Krebs and D. A. Walsh, occurs in inactive and active forms. Its active form catalyzes the phosphorylation of inactive phosphorylase kinase by ATP to yield the active phosphorylated form, by a

Figure 29-4
The structures of cyclic AMP and cyclic GMP (page 822).

Cyclic AMP

Cyclic GMP

Figure 29-5

The activation of protein kinase by cyclic AMP, which combines with the inhibitory subunit R, displacing it and releasing the catalytic subunit C in active form.

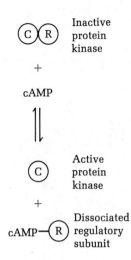

C R   Inactive protein kinase

+

cAMP

C   Active protein kinase

+

cAMP—R   Dissociated regulatory subunit

reaction in which ATP is the phosphate-group donor:

$$\text{Dephospho-phosphorylase kinase} + \text{ATP} \xrightarrow{\substack{\text{active} \\ \text{protein} \\ \text{kinase}}}$$

(Inactive)

$$\text{phospho-phosphorylase kinase} + \text{ADP}$$

(Active)

Phosphorylase kinase is a very large protein, over 1 million in molecular weight. It contains 16 subunits, each of which becomes phosphorylated by ATP at a serine residue through the action of protein kinase.

Protein kinase, the key enzyme in linking cyclic AMP to the phosphorylase system, is an allosteric enzyme. Its inactive form contains two types of subunit, a catalytic (C) subunit and a regulatory (R) subunit, which inhibits the catalytic subunit. The allosteric modulator of protein kinase is cyclic AMP, which binds to a specific site on the regulatory subunit, causing the inactive CR complex to dissociate, yielding an R—cAMP complex and the free C subunit, which is now catalytically active. Cyclic AMP therefore removes the inhibition of enzyme activity that is imposed by the binding of the regulatory subunit (Figure 29-5).

Protein kinase is not specific for phosphorylation of phosphorylase kinase; it can phosphorylate a number of other proteins, such as certain histones, ribosomal proteins, proteins of fat-cell membranes, as well as membrane proteins of mitochondria, microsomes, and lysosomes. In all cases protein kinase requires cyclic AMP for activity.

Although the concentration of cyclic AMP is greatly and quickly increased in the liver following binding of epinephrine, once the epinephrine is removed or destroyed, the cyclic AMP in the liver quickly falls to very low levels. The enzyme responsible for the destruction of cyclic AMP is *phosphodiesterase*, which catalyzes the hydrolytic reaction

$$\text{Cyclic AMP} + \text{H}_2\text{O} \longrightarrow \text{adenosine 5'-phosphate}$$

Phosphodiesterase activity, also dependent on Mg^{2+}, occurs in the soluble cytosol of most animal tissues. It is characteristically inhibited by *caffeine* and *theophylline*, alkaloids present in small amounts in coffee and tea, respectively. These drugs have long been known to prolong or intensify the activity of epinephrine; presumably this is due to increased persistence of cyclic AMP in cells stimulated by epinephrine. In some tissues, phosphodiesterase activity is modified by Ca^{2+}.

The Amplification Cascade in Epinephrine-Stimulated Glycogen Breakdown

We can now assemble the various elements of the amplification cascade involved when epinephrine stimulates the breakdown of glycogen to blood glucose in the liver (Figure

Figure 29-6
Amplification cascade in the stimulation of glycogenolysis by epinephrine in the liver cell to yield blood glucose. Epinephrine is released into the blood to give a concentration of about 10^{-9} M; as a result the increase in the blood glucose concentration is about 5 mM. The amplification produced is thus about 3-million-fold.

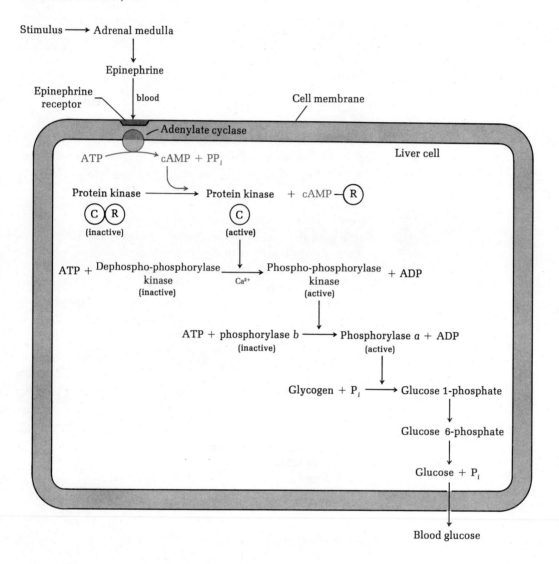

29-6). Epinephrine arriving at the surface of the liver cell in a concentration of 10^{-8} to 10^{-10} M becomes bound to the specific epinephrine receptor sites on the outer surface of the liver-cell membranes. Its binding is believed to cause a local conformational change in the membrane resulting in the activation of adenylate cyclase, located on the inner surface of the cell membrane. The active form of adenylate cyclase converts ATP into cyclic AMP, which may attain a peak concentration of about 10^{-6} M within the cell. The cyclic AMP so formed then binds to the regulatory subunit of protein kinase, releasing its catalytic subunit in an active form. The catalytic subunit then catalyzes the phosphorylation of the inactive form of phosphorylase kinase at the expense of ATP, to yield active phosphorylase kinase. This enzyme, which

Figure 29-7
Inhibition of glycogen synthesis in the liver
by epinephrine.

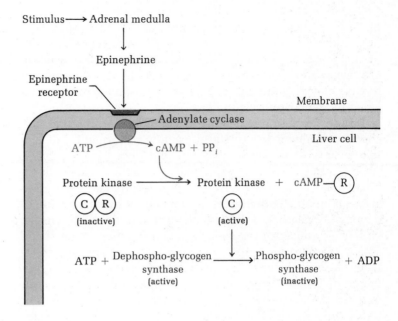

Figure 29-7
Inhibition of glycogen synthesis in the liver
by epinephrine.

requires Ca^{2+} for activity, now catalyzes the phosphorylation of inactive phosphorylase b at the expense of ATP to yield active phosphorylase a, which in turn catalyzes breakdown of glycogen to yield glucose 1-phosphate, from which glucose 6-phosphate and then free blood glucose are formed.

Each step in this cascade is catalytic and thus results in a large amplification of the incoming signal, namely, the binding of a relatively small number of epinephrine molecules to the liver-cell surface. Although there are many steps in this cascade of enzymes acting on enzymes, it can reach peak activity in a matter of only seconds.

Epinephrine not only stimulates glycogen breakdown but also inhibits glycogen synthesis in the liver, thus directing all available glucose residues and precursors into the production of free blood glucose. Figure 29-7 shows that the binding of epinephrine to the liver cell and the subsequent formation of cyclic AMP promotes the phosphorylation by protein kinase of the active, or dephospho, form of glycogen synthase to its phosphorylated, inactive form (page 646). Six serine residues of the glycogen synthase molecule are phosphorylated at the expense of ATP. The inhibition of glycogen synthase is thus brought about by a chain of events triggered by the same stimulus that causes acceleration of glycogen breakdown to yield blood glucose. Other aspects of the degradation and biosynthesis of glycogen are given elsewhere (pages 433 and 645).

So long as epinephrine is secreted into the blood by the adrenal medulla, the liver adenylate cyclase system remains activated, maintaining cyclic AMP at a high concentration. However, once epinephrine secretion stops, the bound epinephrine on the liver-cell membrane dissociates. Cyclic AMP is then no longer formed, and that remaining is destroyed by phosphodiesterase. The protein kinase subunits now reassociate into a complex which has no catalytic activity. The phosphorylated form of phosphorylase kinase then undergoes dephosphorylation, as does phosphorylase *a* itself, by the action of phosphorylase phosphatase. In this way the glycogenolytic system is brought back into the normal resting state; simultaneously, glycogen synthase is reactivated by dephosphorylation.

Besides its action on the liver, epinephrine promotes glycogen breakdown in skeletal muscle with formation of lactate, also by stimulating glycogen phosphorylase via cyclic AMP. Epinephrine also stimulates a lipase in fat cells to break down triacylglycerols to yield free fatty acids bound to serum albumin. This effect is brought about by stimulation of adenylate cyclase and the formation of active protein kinase, which is believed to phosphorylate an inactive precursor of active triacylglycerol lipase. Moreover, the characteristic effects of epinephrine on heart rate and output are also mediated by specific catecholamine receptors and the formation of cyclic AMP. However, heart glycogen is not converted into blood glucose but into lactate. Heart and skeletal muscles lack glucose 6-phosphatase. The epinephrine receptor of heart muscle has been isolated and highly purified.

Glucagon

In addition to epinephrine, a number of other hormones are capable of increasing the concentration of cyclic AMP in specific target cells. Among these is glucagon, also called the *hyperglycemic-glycogenolytic hormone*. Glucagon is a polypeptide hormone of the pancreas, secreted by the α cells of the islets of Langerhans into the blood whenever the blood glucose level drops below the normal value of about 80 mg per 100 ml. The glucagon so released causes the liver to break down glycogen to restore the blood glucose to its normal level. Glucagon thus counterbalances the action of insulin, which is secreted into the blood when the blood glucose level is high and subsequently causes glucose to be removed from the blood by the peripheral tissues. Glucagon and insulin are hormones of opposite effect on the blood sugar, secreted by two different cell types of the pancreas. Glucagon is a polypeptide having 29 amino acid residues (Figure 29-8). It is synthesized as an inactive precursor of higher molecular weight, *proglucagon*, which contains additional amino acid residues at the C-terminal end of the glucagon chain (Figure 29-8).

Glucagon acts primarily on the liver and does not affect glycogen breakdown in muscles, whereas epinephrine promotes glycogen breakdown in both. Like epinephrine, glucagon exerts its action by binding to specific glucagon-binding sites on the liver-cell membrane, resulting in stimulation of adenylate cyclase. The resulting increase in cyclic AMP causes an increase in the concentration of phosphorylase *a* in the liver, by the same sequence of amplification steps occurring in the action of epinephrine described above.

Figure 29-8
The amino acid sequence of bovine glucagon (29 residues; molecular weight 4,500). A precursor of glucagon, proglucagon, has eight more residues at the C-terminal end (color).

N-terminal
His
Ser
Glu
Gly
Thr
Phe
Thr
Ser
Asp
Tyr_{10}
Ser
Lys
Tyr
Leu
Asp
Ser
Arg
Arg
Ala
Glu_{20}
Asp
Phe
Val
Glu
Trp
Leu
Met
Asn
Thr
Lys_{30}
Arg
Asn
Asn
Lys
Asn
Ile
Ala
C-terminal

Other Endocrine and Regulatory Systems Mediated by Cyclic AMP

A number of other hormones are now known to increase the concentration of cyclic AMP in their target tissues, presumably because they bind to specific receptor sites on the surface of the target cells and stimulate membrane-bound adenylate cyclase. The anterior-pituitary hormones ACTH, LH, FSH, and TSH (Figure 29-1) are in this group, as are parathyroid hormone and calcitonin. The posterior-pituitary hormone vasopressin increases cyclic AMP in the kidney. Although many hormones act by stimulating formation of cyclic AMP, each hormone has a specific action because it stimulates cyclic AMP formation only in cells containing specific surface receptors for that hormone. Moreover, cyclic AMP remains within the stimulated cells and does not escape into the bloodstream to cause generalized stimulation of all cells. Although cyclic AMP appears to be the second messenger in the action of many hormones, only in the case of epinephrine and glucagon acting on the liver do we have a clear-cut biochemical picture of the complete chain of events in the regulatory process. As mentioned above, epinephrine stimulates triacylglycerol lipase in adipose tissue. This effect is also produced by glucagon, ACTH, and thyroid-stimulating hormone (TSH), all of which increase formation of cyclic AMP in adipose tissue and thus stimulate the lipase.

As we shall see, cyclic AMP is also a specific mediator, or messenger, in other types of regulatory systems in cells. It mediates the induced synthesis of enzymes (page 987), participates in synaptic transmission in the nervous system, functions in the regulation of cell division (page 996), and is a mediator in inflammatory and immune reactions of tissues, including allergies. The toxin of the cholera organism, *Vibrio cholerae*, which causes massive diarrhea and fluid loss from the intestine, binds to specific receptors in intestinal cells, causing them to form and maintain a high level of cyclic AMP, which in turn produces pathological acceleration of active-transport processes in intestinal cells.

The function of cyclic AMP in various cells may be profoundly influenced by two other cellular components, free Ca^{2+} and various prostaglandins (pages 300 and 686). In some cases Ca^{2+} enhances the action of cyclic AMP and in others inhibits it. Prostaglandin E_1 is thought to serve as intermediate messenger between the cell surface hormone receptor and the membrane-bound adenylate cyclase in some target cells. Prostaglandin A_1, on the other hand, inhibits adenylate cyclase of intestinal cells by binding to specific sites on the enzyme. Much is yet to be learned of the molecular steps involved in the complex regulatory events triggered by various hormones (Figure 29-9).

Insulin: Synthesis, Storage, and Secretion

Important information on the formation, storage, and release of insulin has come from biochemical studies of D. F. Steiner and his colleagues. Their original goal was to determine how

Figure 29-9
The second messenger about to be galvanized into action by a messenger from the outside. (From J. B. Finean, R. Coleman, and R. H. Michell, *Membranes and Their Cellular Functions,* Blackwell, Oxford, 1974.)

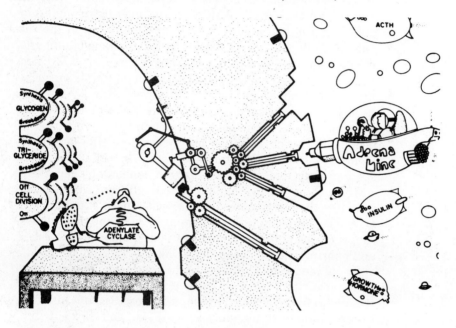

the A and B chains of insulin are synthesized and how their characteristic disulfide cross-linkages are constructed. They incubated radioactive leucine or phenylalanine with islet tissue of rat pancreas or human pancreatic tumors; tumors of this type may produce excessively large amounts of insulin. The pancreatic tumor produced two radioactive protein products capable of combining with a specific antibody to pure insulin. One was established to be insulin itself. However, the other, which obviously resembled insulin closely since it reacted with the anti-insulin antibody, was substantially larger in molecular weight than insulin. Treatment of this second product with trypsin and carboxypeptidase causes cleavage of a number of peptide bonds and the formation of a compound that was proved to be identical with native insulin.

Further chemical- and enzymatic-degradation studies showed that the large insulinlike molecule formed by the pancreas has the structure shown in Figure 29-10. This molecule, now called _proinsulin,_ consists of a single polypeptide chain having from 81 to 86 residues, depending on the species of origin. It contains both the A and B chains of insulin; the A chain constitutes the carboxyl-terminal end of the proinsulin chain and the B chain the amino-terminal end. Between the A and B chains is the connecting C chain. The A and B chains are separated from the C chain by two pairs of basic amino acids. The proinsulin molecule contains three disulfide cross-linkages in the same positions in its A and B segments as in native insulin. The amino acid sequence of the C chain of proinsulin of different species shows many replacements; its mutation rate (page 115) is

Figure 29-10
The structure of bovine proinsulin and insulin. Excision of the amino acids shown yields free insulin (black) and C-chain (color).

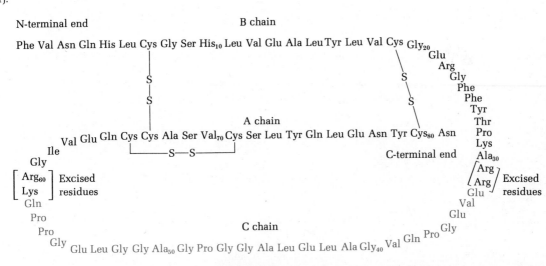

many times greater than the mutation rate of the A and B chains.

Proinsulin, which has only a small amount of insulinlike activity itself, is the biosynthetic precursor of insulin. Its transformation into insulin (Figure 29-10) is apparently accomplished by the action of peptidases in the islet tissue. The conversion of proinsulin into insulin is another example of the pattern of synthesis and activation of a number of proteins whose biological function is largely extracellular. Such proteins, which include trypsin and chymotrypsin, are synthesized in inactive single-chain forms, which are converted by action of proteases into active forms having two or more polypeptide chains held together by disulfide cross-linkages (pages 106 and 227).

Figure 29-11 outlines the stages and time relationships in the biosynthesis, secretion, and release of insulin from the β cells of the islet tissue. Insulin is first made on the ribosomes in the form of proinsulin, which is translocated via the cisternae of the endoplasmic reticulum to the Golgi apparatus. The proinsulin is cleaved to form insulin and C-peptide, which are packaged in Golgi vesicles, where the insulin and C-peptide crystallize with Zn^{2+} in an ordered array. Ultimately, on receipt of certain signals triggered by an increase in the blood glucose level, the contents of these vesicles are released by exocytosis through the plasma membrane into the blood. Ca^{2+} plays an important role in insulin release.

The concentration of insulin in the blood and tissues of man is so small that standard chemical methods or bioassay procedures are inadequate to carry out studies of its secretion and utilization. However, by an extremely sensitive radioimmunoassay procedure postabsorptive blood has been found to contain about 0.4 ng of insulin per milliliter, or about 1 nM. After a carbohydrate-rich meal it may rise to 3 or 4 times this level. The normal human pancreas contains

Figure 29-11
Schematic representation of the biosynthesis of insulin and its release from
β cells of islets of Langerhans, showing time relationships. Mature
granules are not released until blood glucose level is significantly elevated.
[From D. F. Steiner and colleagues, Fed. Proc., 33: 2107 (1974).]

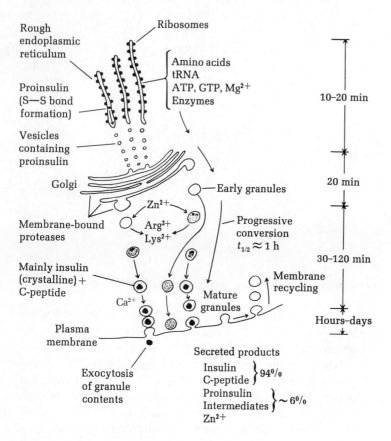

about 10 mg of insulin, but the amount secreted into the blood daily is only about 1 to 2 mg. Free C-peptide and a small amount of proinsulin also circulate in the blood; they are apparently released together with insulin.

The release of insulin from the pancreas depends on the glucose concentration in the blood and certain other factors (below). When the blood glucose increases significantly above its normal level of about 80 to 90 mg per 100 ml, after a meal, the contents of the secretion vesicles closest to the plasma membrane of the β cells are ejected into the blood. The insulin concentration then declines to the normal level in an hour or two after a meal. The half-life of insulin molecules in the blood is only about 3 to 4 min; the release of insulin from the pancreas therefore is very responsive to fluctuations in the blood glucose concentration. The release of insulin is also stimulated by increased levels of certain amino acids and by specific factors secreted by the stomach and intestine.

The Action of Insulin on Target Tissues

The most conspicuous effect of insulin administered to a mammal is a prompt reduction of the blood glucose level, which is believed to be due to enhancement of the transport of glucose from the blood across the plasma membrane of muscle and fat cells into the intracellular space. Insulin also has an immediate effect in promoting the conversion of the inactive to the active form of glycogen synthetase. Moreover, insulin inhibits lipolysis. As a consequence, in the peripheral tissues there is enhanced conversion of blood glucose into glycogen and lipids and an increased oxidation of glucose to carbon dioxide (page 851).

Insulin also promotes protein synthesis from amino acids, enhances the induction of glucokinase and phosphofructokinase, and suppresses the formation of certain enzymes in gluconeogenesis (page 624), such as pyruvate carboxylase and fructose diphosphatase. Insulin appears to have a generalized action on the plasma membrane of its target cells, causing it to undergo changes that lead to enhanced entry not only of glucose but also of amino acids, lipids, and K^+, followed by increased biosynthesis of protoplasm and storage products.

Insulin Receptors

The idea that insulin binds to specific receptors in or on its target cells was first proposed many years ago. However, experimental proof for the occurrence of such receptors did not come until the late 1960s, when it was found that radioactive insulin is bound with very high affinity to specific receptors of muscle and fat cells in a time- and temperature-dependent manner. The binding of insulin is accompanied by characteristic metabolic effects on fat cells, particularly increased synthesis of triacylglycerols from glucose and decreased enzymatic hydrolysis of lipids.

One approach to the conclusion that the binding of labeled insulin takes place on the outer surface of the cell was provided by novel experiments of P. Cuatrecasas. Insulin was first covalently attached to large beads of agarose, an inert polysaccharide material used in affinity chromatography (page 168). Such beads, which are much larger than the fat cells, cannot pass through the cell membrane (Figure 29-12). Mixing of the insulin-agarose beads with a suspension of fat cells caused the cells to be bound to the beads, presumably to the insulin side chains; simultaneously, the fat cells showed some of the characteristic metabolic effects of the hormone. Although it is difficult to prove that stimulation of the fat cells is caused by binding of the cells to the derivatized beads, and not by some free insulin adhering to the beads, this and other approaches strongly indicate that the insulin-binding sites are on the cell surface. No further insulin binding sites are exposed after the cell membrane is ruptured. Normal fat cells contain only about 10 insulin receptor sites per square micrometer of cell surface. Binding of insulin to as few as 100 receptor sites per cell suffices to give a measurable metabolic effect.

Cuatrecasas and his colleagues have succeeded in extracting insulin-receptor protein from fat cells with nonionic detergents and purifying it, by means of affinity chromatography (page 168) on insulin-agarose beads. In this way the high degree of specificity of the interaction between insulin

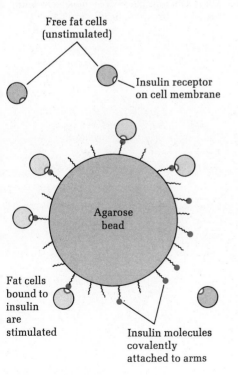

Figure 29-12
Binding of fat cells to insulin molecules covalently attached to a large agarose bead. Since the binding of the cells is specific for insulin, and since the bound cells are stimulated by the insulin, it is concluded that the insulin receptors are on the cell surface.

Free fat cells (unstimulated)

Insulin receptor on cell membrane

Agarose bead

Fat cells bound to insulin are stimulated

Insulin molecules covalently attached to arms

and the insulin receptor was utilized to retain only the solubilized insulin-receptor molecules, thus separating them
from most of the other proteins in the cell extract. A purification of over 250,000-fold has been achieved. The insulin-
receptor protein has a molecule weight of about 300,000. Its
affinity for insulin is very high; the dissociation constant of
the insulin-receptor complex is about 10^{-10} M.

The Effect of Insulin Binding on Cyclic Nucleoside Phosphates

Insulin differs strikingly from epinephrine and glucagon in
that its binding to the target-cell membrane does not cause
an increase in cyclic AMP; in fact, insulin usually causes a
decrease in cyclic AMP, as might be expected, since its action on the blood sugar is opposite to that of glucagon. However, Cuatrecasas and others have found that cyclic GMP
(page 316) *rises* in concentration when insulin is bound to its
receptors in the fat-cell membrane, simultaneous with the
decrease in cyclic AMP. The two regulatory nucleotides cyclic AMP and cyclic GMP (page 316, see also Figure 29-4)
often undergo inverse changes in concentration in cellular
events in which opposing or bidirectional processes are regulated. Such observations have suggested that cyclic AMP and
cyclic GMP function in a reciprocal relationship, which has
been called the *yin-yang hypothesis*. However, much research
is still required to clarify these relationships. Figure 29-13
summarizes the possible effects of the binding of insulin to
its receptor sites in a target cell.

The action of insulin is opposed in one way or another by
four other hormones: (1) glucagon, (2) pituitary growth hormone, (3) glucocorticoid hormones, and (4) epinephrine.
Growth hormone causes complex, generalized effects on
many tissues via a Ca^{2+}-dependent increase in cyclic AMP. It
also characteristically stimulates glucagon secretion by the α
cells, causing blood glucose to rise. One of the adrenal glucocorticoids, cortisol (Figure 29-14), causes both glucose release from the liver and increased glycogen deposition by
unknown mechanisms. The interplay between insulin, the
anterior pituitary, and the adrenal cortex with respect to
glucose metabolism is extremely complex, and its biochemical basis is not yet understood.

Steroid Hormones

Although most of the steroid hormones are highly specialized in their function and do not produce general or systemic
effects on metabolism, recent research has provided revealing insights into the biochemical basis of these and
perhaps other lipid-soluble hormones. The steroid hormones
include the estrogens, or female sex hormones, of which the
most important are 17(β) estradiol and estrone; the androgens, or male hormones (testosterone, dihydrotestosterone); the progestational hormone progesterone; and the
steroid hormones of the adrenal cortex (major forms, cortisol, aldosterone, and corticosterone). The structures of
these hormones are given in Figure 29-14 (see also page 299).

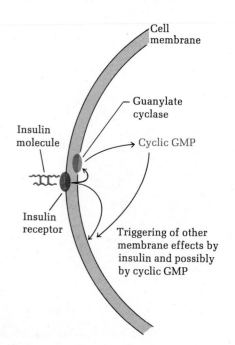

Figure 29-13
Triggering of cyclic GMP formation and specific membrane changes in fat cells by binding of insulin to the insulin receptor.

Cell membrane

Guanylate cyclase

Insulin molecule

Cyclic GMP

Insulin receptor

Triggering of other
membrane effects by
insulin and possibly
by cyclic GMP

Figure 29-14
The major steroid hormones.

The sex hormones.

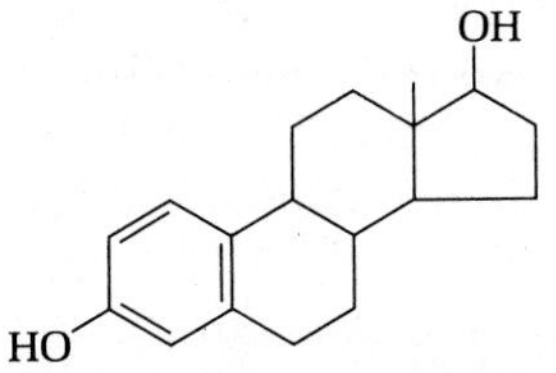

The major adrenal-cortical hormones. Over 30 steroids have been found in the adrenal cortex.

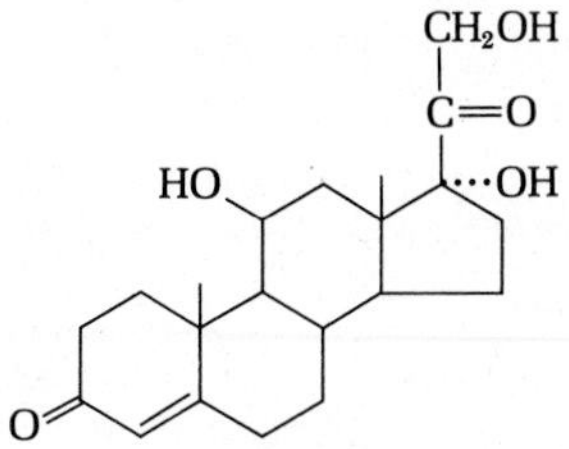

β-Estradiol, the major estrogen

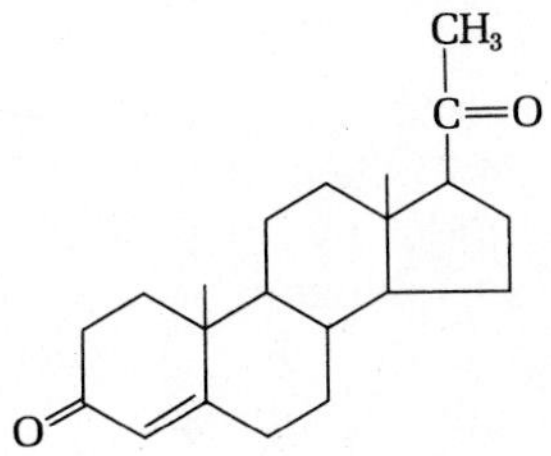

Progesterone

The major androgens

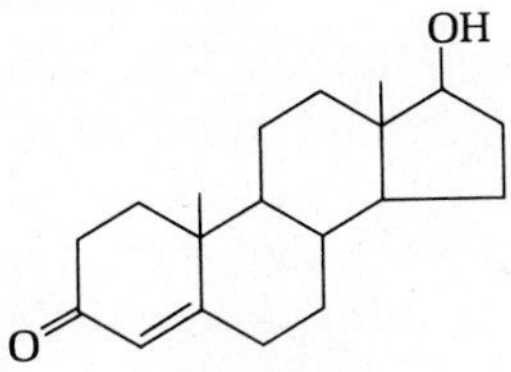

Testosterone

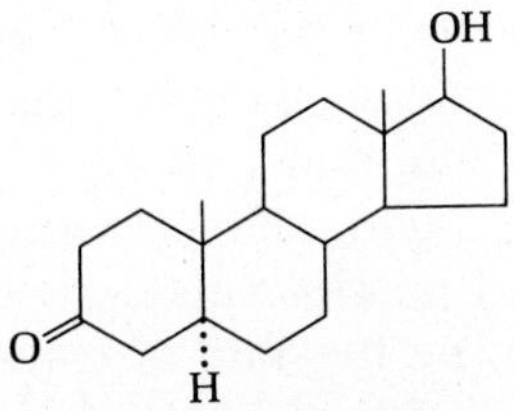

Dihydrotestosterone

Cortisol

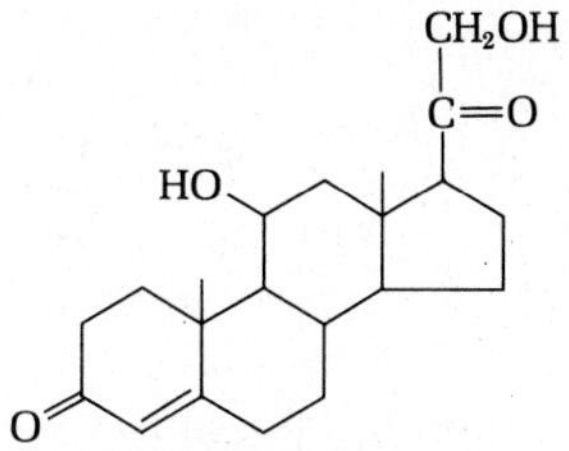

Corticosterone

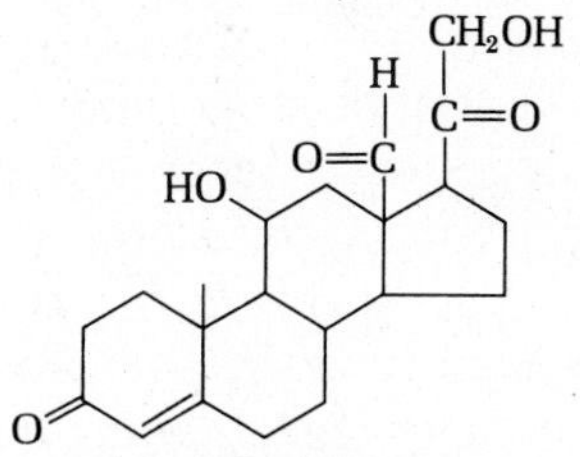

Aldosterone

Although the sex hormones act largely on the sex accessory organs, some of the steroids secreted by the adrenal cortex have profound effects on the metabolism of carbohydrate and protein in many tissues.

That the steroid hormones combine with specific receptors first emerged from early studies of E. V. Jensen and his colleagues on estrogen action. They studied the fate of very small amounts of highly radioactive estradiol injected into immature female animals. The labeled hormone became localized in the uterus and vagina, the targets of estrogen action, but underwent no chemical change. The uptake of labeled estrogen closely paralleled the timing of the earliest detectable hormonal effects of estrogen on these targets. Incubation of minute amounts of labeled estradiol with intact uterine tissue or homogenates of the uterus revealed that the hormone became bound in two cell fractions, the cytosol fraction and the nuclear fraction. The estradiol in the soluble cytosol fraction binds to a specific estrogen-binding protein sedimenting at 4S (page 175). The labeled estrogen found in the nuclear fraction has been localized in the chromatin

823

(page 869). No estrogen receptors are found in tissues unresponsive to the hormone.

The estrogen-receptor protein of the cytosol is present in only minute amounts in target cells. It has been highly purified (but only in microgram quantities) by means of affinity chromatography. Its molecular weight is about 200,000. The purified receptor has extraordinarily high affinity for estrogens, which it binds noncovalently; the dissociation constant for the estrogen-receptor complex is between 10^{-9} and 10^{-12} M.

Studies in the laboratories of Jensen and of J. Gorski have shown that the free estrogen-receptor protein normally occurs in the cytosol in a form sedimenting at 4S, which tends to form a dimeric molecule when isolated. Upon binding estrogen, the 4S receptor undergoes a change into a form sedimenting at 5S. The 5S estrogen-receptor complex then finds its way to the cell nucleus, where it binds to the chromatin (Figure 29-15).

Through similar approaches receptor proteins for other steroid hormones have been found, including receptors for androgens in prostate tissue and receptors for adrenal-cortex hormones. A specific receptor for progesterone has been found in chick-oviduct tissue by B. W. O'Malley and his colleagues, who have carried out a revealing study of its action. The binding of progesterone is followed by translocation of the progesterone-receptor complex from the cytosol to the nucleus. The progesterone-receptor complex binds to the chromatin, causing an increase in DNA-directed RNA polymerase activity (page 916) and an increase in the formation of messenger RNAs coding for several proteins made in large quantities by the oviduct, such as egg albumin and avidin (page 345). From these penetrating experiments (Figure 29-16) it appears that progesterone, and possibly also other steroid hormones, function by modulating the rate of transcription of certain specific genes in the nucleus of their target cells (page 993).

Thyroid Hormones and Basal Metabolic Rate

We do not yet understand the precise biochemical basis for the function of <u>thyroxine</u> and <u>3,5,3'-triiodothyronine</u> (Figure 29-17), characteristic hormones secreted by the thyroid gland, although their profound physiological effects on the basal metabolic rate of man and animals have long been known. Excessive secretion of thyroid hormone, called <u>hyperthyroidism,</u> is responsible for <u>Graves' disease,</u> or <u>exophthalmic goiter,</u> while deficiency in thyroid hormone, <u>hypothyroidism,</u> is characteristic of the disease <u>myxedema.</u>

The <u>basal metabolic rate</u> (BMR) is the rate of oxygen consumption measured in a subject lying at complete rest but not asleep; such measurements are made in the postabsorptive state, at least 12 h after a meal. The BMR is expressed either in terms of kilocalories of heat energy produced or volume of oxygen consumed per square meter per hour, since there is a direct relationship between the energy released from combustion of foods and the oxygen consumed. Table 29-2 shows the energy yield per gram of nutrient and per

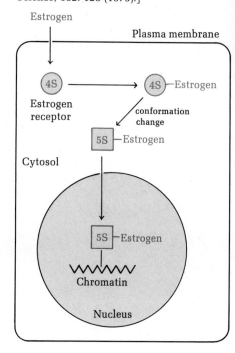

Figure 29-15
Schematic representation of interaction of estrogen with estrogen receptor in uterine cell. [After E. V. Jensen and E. R. DeSombre, Science, 182: 128 (1973).]

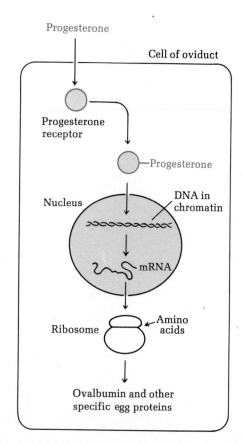

Figure 29-16
Schematic representation of the action of progesterone on an oviduct cell.

Figure 29-17
Thyroid hormones.

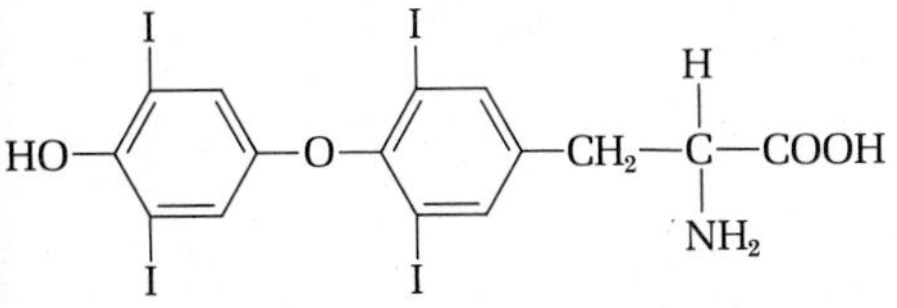

Thyroxine (L-3,5,3',5'-tetraiodothyronine)

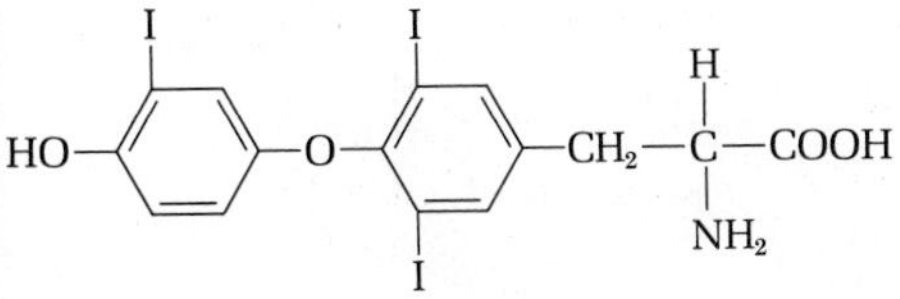

Triiodothyronine (L-3,5,3'-triiodothyronine)

liter of oxygen for carbohydrate, fat, and protein, as well as the *respiratory quotient* (RQ), the molar ratio of CO_2 produced to oxygen consumed. In clinical practice it is assumed that the respiratory quotient is 0.82, corresponding to the combustion of a normal fuel mixture (Table 29-3). The rate of oxygen consumption is a function of body surface area rather than weight.

In hyperthyroidism, the BMR rises anywhere from 20 to 80 percent above the normal level. Removal of part of the thyroid gland of hyperthyroid patients, by diminishing the total thyroid-hormone secretion, results in a decrease of the BMR toward the normal level. Conversely, administration of thyroid hormone to hypothyroid patients increases the basal metabolic rate. The oxygen consumption of nearly all the organs is influenced by the thyroid hormones. The major exception is the brain, whose rate of oxygen consumption is not affected by the thyroid state.

Thyroxine and *3,5,3'-triiodothyronine* (Figure 29-17) are iodinated derivatives of tyrosine. Although thyroxine is produced in much larger quantities than triiodothyronine, the latter is 5 to 10 times more active. The thyroid gland extracts and accumulates iodide ions from the blood, which normally contains only minute concentrations. The iodide is used to iodinate tyrosine through the action of a peroxidase to yield 3,5-diiodotyrosine, which is then converted into thyroxine and 3,5,3'-triiodothyrine. These hormones are synthesized from tyrosine residues of *thyroglobulin*, a glycoprotein of large molecular weight in the follicles of the thyroid gland. Thyroglobulin itself is not biologically active but on proteolysis releases the hormones into the blood, attached to a specific carrier glycoprotein, *thyroxine-binding globulin* (mol wt 55,000). The biosynthesis of thyroid hormones is inhibited by the drugs *thiourea* and *thiouracil* (Figure 29-18).

The precise mechanism by which thyroxine stimulates oxygen consumption of the tissues is not known, despite two generations of intensive biochemical studies. Thyroxine promotes an increased concentration of several enzymes concerned in respiration, particularly glycerol-3-phosphate dehydrogenase of mitochondria, a flavin dehydrogenase functioning in the glycerol phosphate shuttle for entry of the reducing equivalents of cytosol NADH into the mitochondrial respiratory chain (page 534). Thyroxine also increases the rate of respiration of liver mitochondria in the absence of ADP and causes mitochondrial swelling. Minute amounts of thyroxine induce metamorphosis in the tadpole, with an attendant switch from ammonotelic to ureotelic

Table 29-2 Respiratory quotient and energy yield of major fuels

		Energy produced, kcal	
	RQ	Per gram of fuel	Per liter of O_2
Carbohydrate	1.00	4.18	5.05
Triacylglycerols	0.71	9.46	4.69
Protein	0.80	4.32	4.46

Table 29-3 Normal fuel utilization (24 h) by adult male (basal conditions)

	Grams	Moles O_2 consumed	Moles CO_2 produced	Kilocalories produced	Average RQ
Carbohydrate	200	7.5	7.5	836	1.00
Protein	70	3.0	2.4	306	0.80
Fat	60	5.25	3.8	578	0.72
Total	330	15.75	13.7	1,720	0.82

(page 579) excretion of amino-group nitrogen, made possible by a large increase in the concentration of carbamoyl-phosphate synthase (ammonia) and other enzymes essential for the formation of urea (page 579). An important lead to the mode of action of thyroxine has been provided by the discovery that radioactive thyroxine is very tightly bound by the chromatin of liver cells of the tadpole, suggesting that thyroxine functions by derepressing specific genes.

Parathyroid Hormone, 1,25-Dihydroxycholecalciferol, and Calcitonin

Three hormones appear to play major roles in the regulation of Ca^{2+} and phosphate metabolism in mammals. Although their precise mode of action is not yet understood, much progress is being made toward a biochemical understanding of their activities. The parathyroid glands secrete _parathyroid hormone_ into the blood whenever the Ca^{2+} concentration of the blood falls below normal levels. This hormone is a polypeptide having 84 amino acid residues. It acts primarily on the kidneys, stimulating adenylate cyclase activity. The increase in cyclic AMP is reflected in the excretion of more phosphate into the urine, thus lowering the phosphate concentration in the blood. Parathyroid hormone also acts as a tropic hormone by stimulating the production of 1,25-dihydroxycholecalciferol from its precursor, 25-hydroxycholecalciferol, in the kidneys (page 356). 1,25-Dihydroxycholecalciferol is now regarded as a hormone that is secreted by the kidneys and acts on the small intestine and bones. It promotes the absorption of Ca^{2+} from the intestine by stimulating the biosynthesis of a Ca^{2+}-binding protein in the epithelial cells of the intestine, presumably by derepressing specific structural genes. Parathyroid hormone also stimulates release of Ca^{2+} from bone.

Calcitonin, a second polypeptide hormone secreted by the parathyroid gland (and also by the thyroid), has a molecular weight of about 4,500 and contains 32 amino acid residues. Its chief site of action is bone, with secondary effects on the kidney. The secretion of calcitonin increases when the Ca^{2+} level in the blood increases. Calcitonin inhibits loss of Ca^{2+} from bone to the blood and opposes the action of parathyroid hormone.

Figure 29-18
Inhibitors of thyroxine biosynthesis.

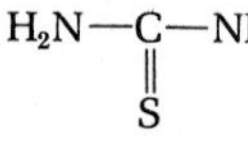

Thiourea

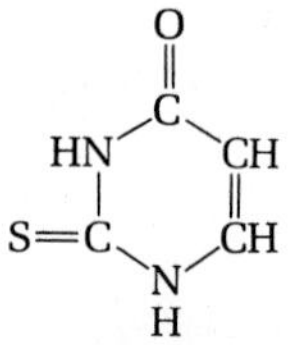

2-Thiouracil

Summary

There is a hierarchy in the organization of the mammalian endocrine system. The hypothalamus secretes peptides called releasing factors and inhibitory factors, each of which stimulates (or inhibits) the release of one of the anterior-pituitary hormones, such as adrenocorticotropic hormone, thyrotropic hormone, and the gonadotropic hormones. These in turn act on specific target tissues. Some (possibly all) of the hormones bind to specific hormone receptors, proteins located either on the cell surface or in the cytosol. Such binding causes formation of an intracellular messenger which stimulates or inhibits some basic activity of the cell.

Epinephrine, which stimulates glycogen breakdown in the liver and muscles, initiates a cascade of amplification reactions. It binds to a specific receptor site on the cell membrane, causing stimulation of the conversion of intracellular ATP to 3′,5′-cyclic adenylic acid by a membrane-bound adenylate cyclase. Cyclic AMP in turn activates protein kinase; this phosphorylates phosphorylase kinase at the expense of ATP, converting it into an active form which activates glycogen phosphorylase. This chain of events thus stimulates glycogen breakdown into blood glucose in the liver. Epinephrine also inhibits the biosynthesis of glycogen from glucose in the liver and stimulates formation of free fatty acids in adipose tissue, via formation of cyclic AMP. Glucagon, a polypeptide hormone, is secreted by the α cells of the islet tissue of the pancreas. Its glycogenolytic action on the liver is exactly like that of epinephrine; it does not act on muscle. Release of glucagon is stimulated by a lowered blood sugar and by growth hormone.

Many other hormones, including ACTH, the gonadotropic hormones, parathyroid hormone, calcitonin, and vasopressin, act by stimulating cyclic AMP production in specific target tissues, but the subsequent steps in the action of cyclic AMP are not known in these cases. Insulin is made in the β cells of the pancreas in the form of proinsulin, a single-chain polypeptide of 81 to 86 residues, which is cleaved to yield insulin. Insulin is released into the blood when the blood glucose level rises above normal; it promotes transport of glucose and other fuels into muscle and fat cells. It binds to specific receptors on the fat-cell membrane and causes the cyclic AMP level to fall and the cyclic GMP level to rise.

Steroid hormones, particularly estrogens and progesterone, bind to intracellular receptor proteins to form a hormone-receptor complex, which passes to the cell nucleus and binds to the chromatin. In the chicken oviduct the progesterone-receptor complex promotes the formation of messenger RNAs coding for the biosynthesis of characteristic egg proteins such as ovalbumin. Little is known of the specific biochemical action of the thyroid hormones, which stimulate basal oxygen consumption and heat production.

References

Books

LEFEBRE, P. J., and R. H. VAGER: *Glucagon: Molecular Physiology, Clinical and Therapeutic Implications*, Pergamon, New York, 1972. A collection of 22 review articles covering all aspects of glucagon chemistry and function.

NEWSHOLME, E. A., and C. START: *Regulation in Metabolism*, Wiley, New York, 1973. An extremely valuable book outlining the biochemistry and theory of metabolic regulation.

ROBISON, G. A., R. W. BUTCHER, and E. W. SUTHERLAND: *Cyclic AMP*, Academic, New York, 1971. A comprehensive review written by members of the laboratory in which cyclic AMP was discovered.

WHITE, A., P. HANDLER, and E. L. SMITH: *Principles of Biochemistry*, McGraw-Hill, New York, 5th ed., 1973. Includes a comprehensive treatment of biochemical endocrinology and the more physiological aspects of metabolism.

Articles

BROSTROM, C. O., J. D. CORBIN, C. A. KING, and E. G. KREBS: "Interaction of the Subunits of Cyclic AMP-Dependent Protein Kinase of Muscle," *Proc. Natl. Acad. Sci. (U.S.)*, 68: 2444–2447 (1971).

CUATRECASAS, P.: "Interaction of Insulin with the Cell Membrane: The Primary Action of Insulin," *Proc. Natl. Acad. Sci. (U.S.)* 63: 450–457 (1969).

CUATRÉCASAS, P.: "Insulin Receptor of Liver and Fat Cell Membranes," *Fed. Proc.*, 32: 1838–1846 (1973). A very readable review of the identification and isolation of insulin receptor.

CUATRECASAS, P.: "Membrane Receptors," *Ann. Rev. Biochem.*, 43: 169–214 (1974).

EHRLICHMAN, J., C. S. RUBIN, and O. M. ROSEN: "Physical Properties of a Purified Cyclic AMP-Dependent Protein Kinase from Bovine Heart Muscle," *J. Biol. Chem.*, 248: 7607–7609 (1973).

GREENGARD, P., and J. W. KEBABIAN: "Role of Cyclic AMP in Synaptic Transmission in the Peripheral Nervous System," *Fed. Proc.*, 33: 1054–1067 (1974).

GUILLEMIN, R.: "Hypothalamic Hormones: Releasing and Inhibiting Factors," *Hosp. Pract.*, November 1973. An excellent short review.

ILLIANO, G., G. P. E. TELL, M. I. SIEGEL, and P. CUATRECASAS: "Guanosine 3':5'-Cyclic Monophosphate and the Action of Insulin and Acetylcholine," *Proc. Natl. Acad. Sci. (U.S.)*, 70: 2443–2447 (1973).

JENSEN, E. V., and E. R. DESOMBRE: "Estrogen-Receptor Interactions," *Science*, 182: 126–134 (1973).

LEVINE, R.: "The Action of Insulin at the Cell Membrane," *Am. J. Med.*, 40: 691–694 (1966). An important early discussion of the cell membrane as the site of insulin action.

O'MALLEY, B. W., and A. R. MEANS: "Female Steroid Hormones and Target Cell Nuclei," *Science*, 183: 610–620 (1974). An important review of the interaction of steroid-receptor complexes with nuclear DNA and the regulation of transcription into mRNA.

PASTAN, I.: "Cyclic AMP," *Sci. Am.*, 227: 97–105 (1972).

RYAN, J., and D. R. STORM: "Solubilization of Glucagon and Epinephrine-Sensitive Adenylate Cyclase from Rat Liver Plasma Membranes," *Biochem. Biophys. Res. Commun.*, 60: 304–311 (1974).

SCHALLY, A. V., A. ARIMURA, and A. J. KASTIN: "Hypothalamic Regulatory Hormones," *Science*, 179: 341–350 (1973). A useful and comprehensive review.

SICA, V., I. PARIKH, E. NOLA, G. A. PUCA, and P. CUATRECASAS: "Affinity Chromatography and the Purification of Estrogen Receptors," *J. Biol. Chem.*, 248: 6543–6558 (1973).

STEINER, D. F., W. KEMMLER, H. S. TAGER, and J. D. PETERSON: "Proteolytic Processing in the Biosynthesis of Insulin and Other Proteins," *Fed. Proc.*, 33: 2105–2115 (1974).

TAGER, H. S., and D. F. STEINER: "Isolation of a Glucagon-Containing Peptide: Primary Structure of a Possible Fragment of Proglucagon," *Proc. Natl. Acad. Sci. (U.S.)*, 70: 2321–2325 (1973).

TANNER, J. M.: "Human Growth Hormone," *Nature*, 237: 433–439 (1972).

ZAHLTEN, R. N., A. A. HOCHBERG, F. W. STRATMAN, and H. A. LARDY: "Glucagon-Stimulated Phosphorylation of Mitochondrial and Lysosomal Membranes of Rat Liver *in Vivo*," *Proc. Natl. Acad. Sci. (U.S.)*, 69: 800–804 (1972).

CHAPTER 30 ORGAN INTERRELATIONSHIPS IN THE METABOLISM OF MAMMALS

The preceding chapters of Parts 2 and 3 have considered the chemical pathways, enzymatic catalysis, and regulation of the central pathways of metabolism, largely in terms of cells as metabolic units. We shall now integrate and summarize the organ interrelationships in the metabolism of vertebrates, particularly in man.

We shall first see how nutrients are transported to the different organs via the blood and how different metabolic activities are delegated to the various tissues and organs. We shall then examine in special detail the metabolic responses of the mammalian organism to the stresses of two conditions, starvation and diabetes mellitus. Some aberrations in the metabolism of cancer cells will also be noted.

Although biochemical principles have been utilized in the theory and practice of medicine for many years, our knowledge of human biochemistry is still quite fragmentary. The molecular basis of human disease is a particularly great challenge to biochemistry today and certainly one of its most important research areas.

Transport between Organs: The Blood

The blood is the liquid vehicle by which the major organic nutrients are transported from the intestine, where they are absorbed, to the liver, where they are processed, and thence to the other organs; and by which organic waste products and excess mineral ions are conveyed to the kidneys for excretion. The blood is also the vehicle for transport of oxygen from the lungs to the tissues and for the removal of CO_2 generated during respiratory metabolism. In addition, the blood also serves to transport hormones and other chemical messengers from various endocrine glands to their specific target organs.

The human vascular system contains approximately 5 to 6 l of blood. Nearly one-half its volume consists of cells: red blood cells (erythrocytes), which transport oxygen and also some carbon dioxide, much smaller numbers of white blood cells (leukocytes), and blood platelets. The noncellular por-

Table 30-1 Major protein components of human blood plasma

Some 50 minor protein components, including a number of enzymes, have been found in plasma

	Concentration, mg per 100 ml	Approximate molecular weight	Other components	Function
Serum albumin	3,000–4,500	68,000		Osmotic regulation; transport of fatty acids, bilirubin, etc.
α_1-Globulins	100	40,000–55,000	Carbohydrate	Not known
α_1-Lipoproteins†	350–450	200,000–400,000	40–70% lipid	Lipid transport
α_2-Globulins	400–900			
α_2-Glycoproteins		Up to 800,000	Carbohydrate	Unknown
Ceruloplasmin	30	150,000	Carbohydrate	Copper transport
Prothrombin		63,000	Carbohydrate	Blood clotting
β-Globulins	600–1,200			
β_1-Lipoproteins	350–450	3–20 million	80–90% lipid	Lipid transport
Transferrin	40	85,000	Carbohydrate	Iron transport
Plasminogen		90,000		Precursor of fibrinolysin
γ-Globulins‡	700–1,500	150,000	Carbohydrate	Antibodies
Fibrinogen	300	340,000	Carbohydrate	Blood clotting

† The distribution of plasma lipoproteins is shown on page 302.

‡ The γ-globulins occur in a very large number of different types (page 1002).

tion is the _blood plasma_, of which about 10 percent consists of various organic and inorganic solutes. The _plasma proteins_ make up three-fourths of the solutes. The different types of plasma proteins have a number of important functions (Table 30-1). Among these is the capacity to transport important nutrients, such as lipids and fatty acids, as well as some trace metals, vitamins, and hormones. The remainder of the dissolved solutes consists of organic nutrients and metabolites, waste products, and inorganic salts. Table 30-2 shows the major components of normal human blood plasma and their general function. The blood also contains a large number of minor components; undoubtedly many more are yet to be identified.

The components of blood plasma are maintained at characteristic steady-state concentrations by various regulatory devices, which smooth out fluctuations in composition, a necessity because of the intermittent feeding periods of vertebrates. The blood glucose level, as we shall see, is very tightly regulated by several different types of controls. The rate at which blood carries nutrients through the various organs is relatively high in relation to their rate of metabolism, so that the fraction of a given nutrient extracted from the blood in a single "pass" through most of the organs is relatively small. For example, less than 5 percent of the incoming glucose is removed from the blood on passing through the brain. The liver is an exception, since it can extract most of the incoming dietary hexoses arriving from the small intestine.

Measurements of the concentration of specific components of the blood plasma and the urine are extremely im-

Table 30-2   Major components of normal human blood plasma

	Approximate concentration, mg per 100 ml	Function
Proteins (total)	5,800–8,000	See Table 30-1
Serum albumin	3,000–4,500	
α, β, and γ-globulins	2,800–3,800	
Lipids (total)	400–700	
Triacylglycerols	100–250	Fuels en route to sites of storage or oxidation
Phospholipids	150–250	
Cholesterol and its esters	150–250	
Free fatty acids	8–20	Immediate fuel source for muscles, heart
Glucose	70–90	Transport form of carbohydrate
Amino acids	35–65	Precursors in protein synthesis
Urea	20–30	Excretion product of amino acid nitrogen
Uric acid	2–6	End product of purine catabolism
Creatinine	1–2	Excretory form of creatine
Lactate	8–17	Product of glycolysis in muscles, erythrocytes
Pyruvate	0.4–2.0	Product of glycolysis in muscles, erythrocytes
β-Hydroxybutyrate	3–6	The ketone bodies; transport forms of acetyl groups
Acetoacetate	0.8–2.0	
Bilirubin	0.2–1.4	Degradation product of protoporphyrin
	mequiv l^{-1}	
Inorganic components		
Na$^+$	140	
K$^+$	5	
Cl$^-$	100	
HCO$_3^-$	20	
Phosphate$^-$	4	

portant in medicine, since they can yield insight into the nature of metabolic disturbances and the effectiveness of therapeutic measures.

Organ Distribution of Major Metabolic Activities in the Mammal

Despite their high degree of specialization, all the major organs in vertebrates can carry out the reactions of the central metabolic pathways, such as glycolysis, the tricarboxylic acid cycle, and the biosynthesis of lipids and proteins. However, the "tuning" and apportioning of the central metabolic pathways differs from one organ to another; each has a characteristic pattern of fuel utilization and of biosynthetic

Table 30-3   Relative amounts of oxygen consumed by the major organs in the human body

	Rest	Light work	Heavy work
Skeletal muscles	0.30	2.05	6.95
Abdominal organs	0.25	0.24	0.24
Heart	0.11	0.23	0.40
Kidneys	0.07	0.06	0.07
Brain	0.20	0.20	0.20
Skin	0.02	0.06	0.08
Other	0.05	0.06	0.06
Total	1.00	2.90	8.00

activities in keeping with its specialized function in the organism.

We shall now summarize the special metabolic activities and functions of the major organs of mammals. For orientation, Table 30-3 shows the relative amounts of oxygen consumed and thus the relative amounts of ATP energy generated by the different organs of the human body at rest and during muscular activity.

The Liver and Its Function in Distribution of Nutrients

After being absorbed from the intestinal tract, most of the incoming nutrients pass directly to the liver, the major center for distribution of nutrients in the metabolism of mammals. The liver has perhaps the greatest metabolic flexibility of any organ, since it is the first to "see" the incoming nutrients from the intestine and must therefore adjust its metabolic activities to the highly variable composition of the nutrient mixture and to the discontinuous or intermittent nutrient intake. The liver functions to process and distribute the incoming nutrients and to maintain the constancy of their concentration in the systemic blood; it also has excretory functions.

The liver in the normal human adult weighs approximately 1.5 kg, which constitutes some 2 to 3 percent of the body weight. Its major cell type is the *hepatocyte*, a multipurpose epithelial cell. The liver undergoes rapid changes in size and in glycogen and protein content, depending on nutritional state. When a fasting subject resumes feeding, the liver may in a short time increase in weight, not only due to increased protein content but also to a very large increase in glycogen, which may attain 10 percent of the wet weight of the liver. Moreover, when a portion of the liver is removed surgically, new liver tissue may be regenerated.

The liver is the most flexible organ in the body with respect to the induction of specific enzymes in response to the nutrient mixture it receives. For example, when a mammal is placed on a high-protein diet, in a day or two the liver will show a large increase in the content of certain enzymes concerned in amino acid catabolism and in gluconeogenesis from the carbon skeletons of amino acids. If the animal should then be changed to a high-carbohydrate diet, the amino acid–catabolizing and gluconeogenic enzymes

Figure 30-1
Metabolic pathways of glucose 6-phosphate in the liver.

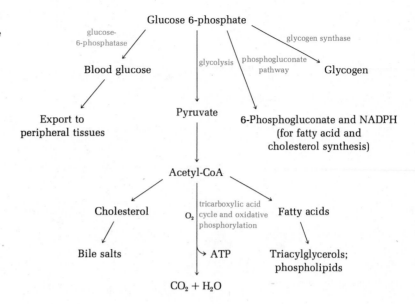

elicited by the previous amino acid feeding will almost completely disappear from the liver, some in a matter of hours and others in a day or two. The liver thus has the power to regulate the expression of the genes coding for the biosynthesis of the enzymes required to handle specific nutrients.

The liver receives a mixture of free monosaccharides resulting from the digestion of oligosaccharides and polysaccharides in the small intestine. About two-thirds of the free glucose coming to the liver enters its cells and is phosphorylated to glucose 6-phosphate by hexokinase; the remainder of the free glucose passes through the liver into the systemic blood. Monosaccharides other than glucose, such as D-fructose, D-galactose, and D-mannose, are phosphorylated in the liver and further transformed into glucose 6-phosphate by mechanisms described elsewhere (page 436). Systemic blood contains essentially no simple sugars other than D-glucose and minor amounts of D-fructose. D-Ribose, after phosphorylation, may be used in the biosynthesis of nucleotides (page 729) or may be transformed into intermediates of the glycolytic sequence via portions of the phosphogluconate pathway (page 467).

Figure 30-1 shows the major metabolic pathways glucose 6-phosphate can take in the liver. On a normal diet most of it is converted into glycogen, fatty acids, or blood glucose. Relatively little is oxidized to completion since oxidation of fatty acids and amino acids provides the bulk of the ATP required by the liver. About half of the glucose undergoing degradation in the liver enters the phosphogluconate pathway, which is responsible for generating NADPH required as reducing agent in the biosynthesis of fatty acids.

The amino acids entering the liver following absorption from the intestinal tract also have several possible metabolic routes (Figure 30-2). A fraction passes into the systemic blood directly for transport to peripheral tissues, where amino acids are utilized for protein biosynthesis. Some of

Figure 30-2
Amino acid metabolism in the liver.

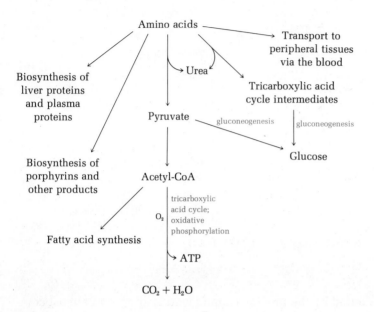

the amino acid influx is directly utilized by the liver for the biosynthesis of the intrinsic proteins of liver cells, particularly induced enzymes, and for the biosynthesis of the plasma proteins. When amino acids are available in excess, they are deaminated and degraded to yield pyruvate, acetoacetate, and intermediates of the tricarboxylic acid cycle (page 559). Some of these products may be oxidized to completion for ATP energy, and some may be used as precursors in gluconeogenesis. Certain of the amino acids may be converted to various specialized products such as porphyrins (page 718), polyamines (page 716), and purines (page 729). The amino groups of amino acids are converted into urea via the urea cycle (page 579); the liver is the only mammalian organ capable of urea synthesis. The human liver manufactures some 20 to 30 g of urea per day, a process which requires input of a substantial amount of ATP energy (page 582). When the liver is damaged, free NH_4^+ ions are formed instead of urea. Free ammonia is exceedingly toxic, especially to the brain (page 579).

Some of the lipids absorbed from the intestinal tract enter the liver directly as phospholipids via the portal vein; the remainder, largely triacylglycerols in the form of *chylomicrons* (page 302), enter the systemic blood through the thoracic duct, a large lymphatic vessel draining the intestine. The lipids entering the liver also have multiple metabolic options, as outlined in Figure 30-3. Some are used in the biosynthesis of plasma lipoproteins (page 301), in which form phospholipids are transported to the peripheral tissues. Some of the lipids, after hydrolysis, are oxidized to yield acetyl-CoA. This central intermediate may undergo complete oxidation via the tricarboxylic acid cycle for energy, or conversion into the blood ketone bodies, acetoacetic acid and D-β-hydroxybutyric acid, which are used by peripheral tis-

Figure 30-3
The fate of lipids in the liver.

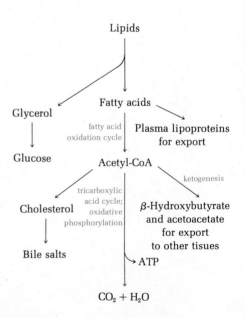

Figure 30-4
Bile salts. The biosynthesis of bile salts from cholesterol is given on page 686.

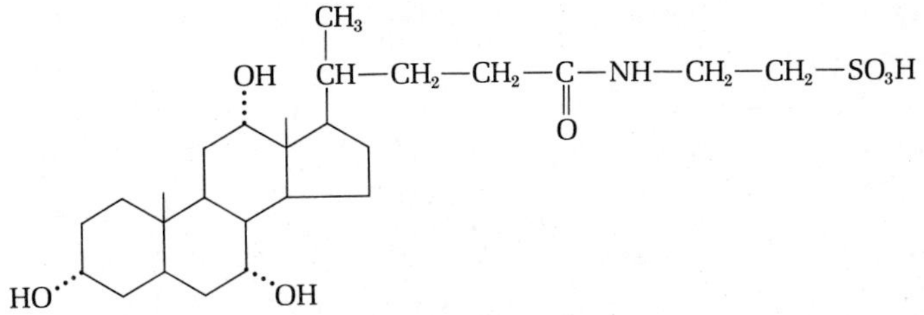

Taurocholic acid

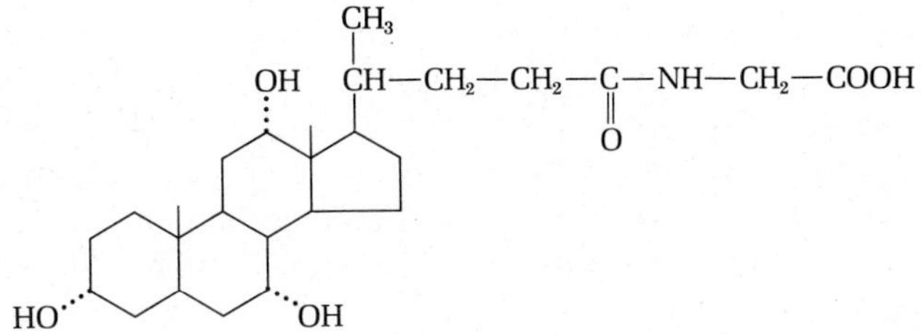

Glycocholic acid

Figure 30-5
Bile pigments.

Key:
M = methyl   P = propionyl
V = vinyl   E = ethyl

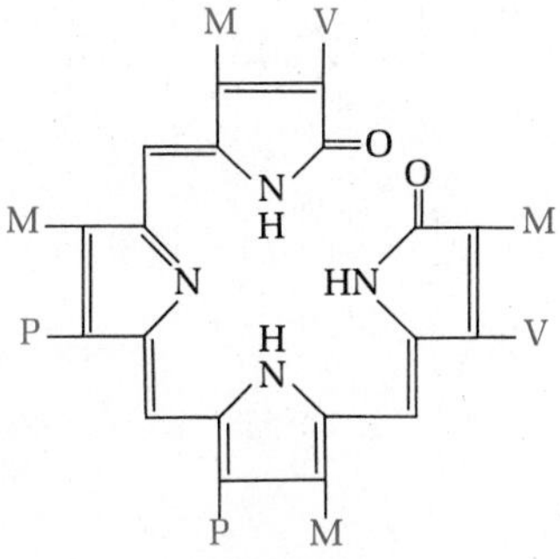

Biliverdin

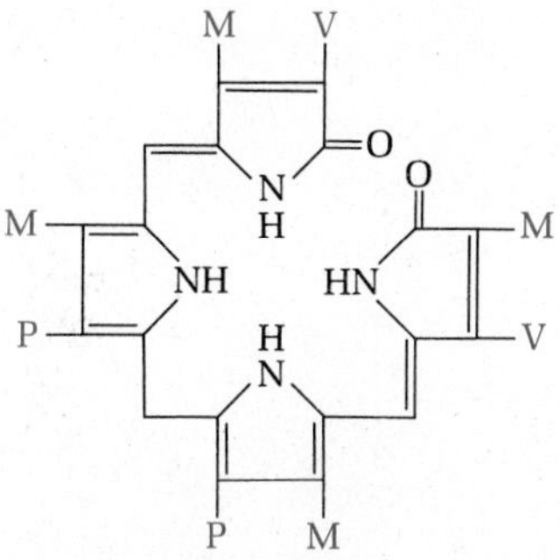

Bilirubin

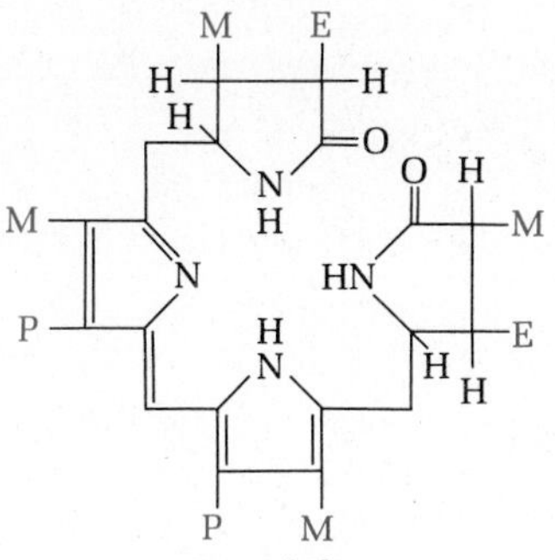

Stercobilin

sues as fuel (page 553). Acetyl-CoA may also undergo conversion into cholesterol and other steroids (page 679).

The liver also has an excretory function; it is responsible for the formation of bile, which drains into the small intestine. The liver converts cholesterol into *bile acids* and brings about their reaction with nitrogenous bases, such as glycine and taurine, to yield the corresponding *bile salts*; examples are *taurocholic* and *glycocholic acids* (Figure 30-4). The bile salts, after passing to the small intestine, are used as emulsifying agents to prepare fatty acids and acylglycerols for absorption. A large fraction of the bile salts is reabsorbed, together with lipids and fatty acids, and returned to the liver. The circulation of bile components between the liver and intestine in this manner is called the *enterohepatic circulation*.

Another excretory function of the liver is the degradation of protoporphyrin IX, derived from hemoglobin and other heme proteins, to the *bile pigments*, open-chain tetrapyrroles which contain no metal. These include *bilirubin* (red) and *biliverdin* (green) (Figure 30-5; page 718), which undergo further degradation through the action of intestinal microorganisms to yield the brown *stercobilin* (Figure 30-5), the chief pigment of feces. Another important component is *urobilinogen*, which is a reduced colorless reduction product of bilirubin; urobilinogen and bilirubin are in part reabsorbed into the bloodstream and returned to the liver. However, if the liver has undergone degeneration, if breakdown of hemoglobin is excessive, or if the bile duct is obstructed, bilirubin may accumulate to high levels in the blood, causing *jaundice*, in which the skin acquires a yellowish tinge.

The ATP energy generated by the liver is thus used for gluconeogenesis, glycogen synthesis, the biosynthesis of fatty acids and from them various complex lipids, in the biosynthesis of liver enzymes and the plasma proteins, and in the formation of urea, as well as other small biomolecules,

835

such as nonessential amino acids, purines, pyrimidines, and porphyrins.

Adipose Tissue

Second in importance to the liver with respect to the distribution and maintenance of fuel levels in the blood is *adipose tissue*. This consists of specialized fat cells, which contain large droplets or globules of triacylglycerols in nearly pure, unhydrated form; such droplets may make up 90 percent of the weight of the fat cell. Extensive formations of fat cells are present under the skin, around deep blood vessels, in the abdominal cavity, in skeletal muscles, and in mammary glands. Fat tissue is by no means merely an inert storage depot. It has a high rate of metabolism and is quickly responsive to the metabolic needs of the organism. Moreover, adipose tissue is very large in total weight and constitutes a major body tissue. The adipose tissue of normal adults contains some 15 kg of triacylglycerols; in obese individuals the mass of the stored triacylglycerols can be several or many times greater.

Fat cells utilize both glucose and fatty acids as fuels to generate energy, which is in turn employed largely in the biosynthesis of triacylglycerols for storage. Glucose is one of the two major precursors in triacylglycerol synthesis in fat tissue; it provides not only acetyl-CoA for the biosynthesis of fatty acids (page 659) but also the glycerol 3-phosphate required as precursor of the glycerol moiety of triacylglycerols (page 670). Fatty acids arriving from the blood in the form of lipoproteins are the other major precursors of triacylglycerols in adipose tissue. Fat cells are especially active in degrading glucose via the phosphogluconate pathway, to generate the NADPH required for biosynthesis and desaturation of fatty acids (page 667).

The most important function of adipose tissue is to serve as a readily mobilized storage depot of triacyglycerols. Triacylglycerols have important advantages over glycogen as a storage form of fuel and thus as a potential source of ATP. Gram for gram, pure triacylglycerols can yield on complete oxidation nearly 2.5 times as many molecules of ATP as pure glycogen. The greater energy content of triacylglycerols is due to the fact that their fatty acids are more highly reduced compounds and thus can yield a greater number of hydrogen atoms (electron equivalents) per unit weight than glycogen. Moreover, triacylglycerols can be stored in nearly pure form in fat cells whereas glycogen, a highly hydrated substance, binds its weight of water when it is deposited. It has been estimated that if the 15 kg of triacylglycerols in a normal adult male were replaced by glycogen in a quantity sufficient to yield the same amount of ATP on combustion, his total body weight would have to be increased by about 60 kg, or 130 lb.

Chylomicrons are rapidly removed from the blood by adipose tissue. Their triacylglycerols are in part hydrolyzed by *lipoprotein lipase* on the surface of fat cells to yield free fatty acids and glycerol, which are reconverted back into triacylglycerols within the fat cells. Chylomicrons have a very short half-life in the blood plasma, of the order of 10 min; storage of excess ingested triacylglycerols thus occurs rapidly. After a meal rich in triacylglycerols, the blood plasma is milky due to the high concentration of chylomicrons and lipoproteins. However, within 2 h the excess triacylglycerols are deposited in fat cells, and the plasma becomes clear.

The release of stored triacylglycerols into the blood from adipose tissue occurs by a different route. The triacylglycerols are first hydrolyzed by intracellular lipases, whose activity is controlled, via cyclic AMP, by various hormones, including epinephrine, adrenocorticotrophic hormone, and insulin (page 816). The fatty acids so released are transferred into the blood and are bound in unesterified form to serum albumin for transport, largely to the muscles, the heart, and the kidneys. At any one time there is only a small amount of such free fatty acid (FFA), also called unesterified fatty acid (UFA), in the blood. The turnover rate of the fatty acids bound to serum albumin is extremely high, however; their half-life is only about 3 min. Up to 300 g of fatty acids may be carried in this rapidly metabolized form per day.

Thus the liver and the adipose tissue constitute the major tissues for the storage, processing, and distribution of caloric fuels. Between them there is a massive and dynamic interplay, which is highly responsive to many factors, particularly to certain hormones; it enables mammals to maintain a constant flow of energy-rich fuels to the tissues despite their intermittent feeding periods.

Skeletal Muscle

In man, skeletal muscle utilizes some 30 percent of the resting oxygen consumption (Table 30-3); in people carrying out heavy muscular work the oxygen consumption of muscle may increase many times and then account for almost 90 percent of the total oxygen consumption of the organism. Most of the basal or resting oxygen requirement of muscle is for the oxidation of fatty acids brought by serum albumin. Only about 10 to 15 percent of the oxygen is utilized for oxidation of carbohydrate. However, during maximal muscular activity, as in a sprint, there is a large increase in glucose consumption by skeletal muscle. Most of the glucose consumed under these conditions comes from muscle glycogen, which is converted by glycolysis into lactate, causing the formation of an oxygen debt (page 768). The degradation of glucose to yield lactate provides a supplementary source of ATP, since under these conditions two molecules of ATP are generated per glucose residue (page 419). Most of the blood lactate formed during heavy muscular activity is salvaged and converted back into glucose in the liver, a process that requires the energy input of six high-energy phosphate groups per molecule of glucose formed (page 624). This is an example of metabolic interplay between organs; the liver indirectly helps the muscles in periods of extreme physical effort. The heart also consumes lactate from skeletal muscle.

Muscle contains about 1 percent of its wet weight as glycogen, much less than the glycogen content of the liver.

However, muscle glycogen cannot be converted into blood glucose, for lack of glucose 6-phosphatase.

The Brain

The brain of normal well-fed mammals uses glucose exclusively as its fuel, as indicated by its respiratory quotient (the molar ratio of CO_2 produced to O_2 consumed) of 1.0 (page 825). The brain is in fact dependent on a minute-to-minute supply of glucose, since it contains essentially no fuel reserves of either glycogen or triacylglycerols. Moreover, the metabolism of the brain is critically dependent on the concentration of glucose in the blood. If the blood glucose falls from its normal level of about 80 mg per 100 ml to about half this concentration, symptoms of brain dysfunction appear. If the blood glucose level falls to 20 mg percent or less, as may happen in a severe overdose of insulin (page 821), coma may result. Glucose is utilized in the brain entirely via the glycolytic sequence and the tricarboxylic acid cycle.

The human brain uses 20 percent or more of the total oxygen consumed by the adult in the resting state (Table 30-3); on a weight basis this is proportionately far more than most other organs. Most of this large energy demand, perhaps two-thirds, is used to maintain the characteristic transmembrane potential across the axonal membranes, by the action of the membrane Na^+K^+–ATPase. This conclusion follows from the fact that respiration of brain slices is inhibited well over 50 percent by ouabain, an inhibitor of Na^+K^+–ATPase (page 790). Although the brain is not capable of regeneration and its cells do not turn over, there is constant biosynthesis of certain brain proteins from amino acids, particularly the protein subunits of microtubules (page 771) and microfilaments (page 771), which together constitute more than half of the soluble proteins of developing brain and participate in axoplasmic flow. Energy is also required for the synthesis of acetylcholine and other neurotransmitter substances. The brain uses energy at a constant rate; active thought or intense concentration cause no significant increase in oxygen consumption.

The Heart

Many features of heart muscle, particularly the molecular design of the actin and myosin filaments and the nature of their interactions, are very similar to those of skeletal muscle. However, heart muscle, which is continuously although cyclically active, has a completely aerobic metabolism in the resting or moderately active subject and uses glycolysis as a source of extra energy only in an emergency. Heart-muscle cells are extremely rich in mitochondria, which make up 40 percent or more of the cytoplasmic space, whereas most skeletal muscles contain relatively few mitochondria. Free fatty acids carried by serum albumin from adipose tissue are the major fuel of the heart. Some ketone bodies, blood glucose, and lactate are also utilized; all these fuels are oxidized via the tricarboxylic acid cycle. Heart muscle also contains a small amount of glycogen; although it

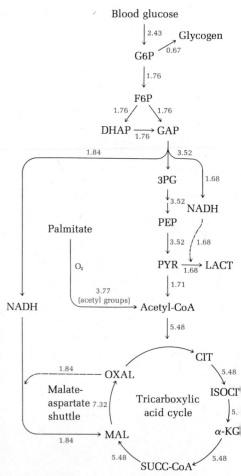

Figure 30-6
Metabolic flux maps for the rat heart before and after onset of a heavy work load (below and opposite). The rate of each reaction is given in micromoles per gram dry weight per minute. Note that the steady-state heart uses predominantly free fatty acids from blood as fuel but after the onset of the work load the predominant fuel is glucose, mostly from glycogen. However, most of the glucose residues yield lactate; the remainder are oxidized via the tricarboxylic acid cycle. Although the malate-aspartate shuttle for oxidation of cytosol NADH from the glycolytic sequence is activated, it does not reoxidize more than a small fraction of the total NADH. The overall rate of oxygen consumption increases about fourfold. (Provided by D. Garfinckel.)

Before onset of work

O_2 uptake = 16.1
Fatty acid contribution = 69%
Glucose contribution = 31%

Key:

G6P = *glucose 6-phosphate*
F6P = *fructose 6-phosphate*
DHAP = *dihydroxyacetone phosphate*
GAP = *glyceraldehyde 3-phosphate*
3PG = *3-phosphoglycerate*
PEP = *phosphoenolpyruvate*
PYR = *pyruvate*
LACT = *lactate*
CIT = *citrate*
ISOCIT = *isocitrate*
α-KG = *α-ketoglutarate*
SUCC-CoA = *succinyl-CoA*
MAL = *malate*
OXAL = *oxaloacetate*

12 s after onset of work

O_2 uptake = 69
Fatty acid contribution = 18%
Glucose + glycogen contribution = 82%

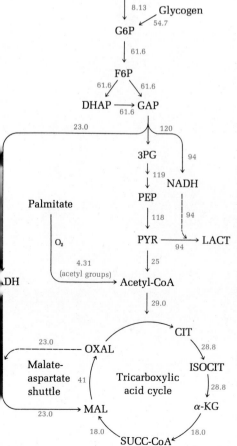

is insufficient to last for any great length of time, it is vitally important as an emergency fuel, as we shall see.

The dynamics of fuel utilization and ATP formation in the heart in relation to an imposed work load have been studied. The concentrations of various metabolic intermediates in the central pathways have been determined by direct analysis of quick-frozen heart tissue. Moreover, the Michaelis constants (K_M) of various enzymes in the central catabolic pathways of the heart have been determined, as well as the effects of allosteric modulators. From such data, the rates of glycolysis and the tricarboxylic acid cycle in heart muscle have been modeled with electronic computers. Two metabolic flux maps for perfused heart muscle are shown in Figure 30-6. That at the left represents a flux map for the resting steady state of the heart; the other represents a flux map taken 12 s after the onset of a heavy work load. The rate of oxygen consumption of the heart is increased approximately fourfold by the imposition of a heavy load; this is about the limit to which cardiac metabolism can be increased. In the resting heart fatty acids contribute 70 percent of the total fuel source and glucose the remainder. However, 12 s after the onset of a heavy work load the contribution of glucose from the blood and from glycogen rises to 82 percent; the rate of breakdown of glucose 6-phosphate is increased over thirtyfold. In the resting steady state very little lactate is formed from pyruvate in the heart; under these conditions one-half or more of the pyruvate is converted into acetyl-CoA, most of which is then oxidized via the tricarboxylic acid cycle. On the other hand, 12 s after sudden imposition of a heavy work load 80 percent of the pyruvate formed from glycogen is reduced to lactate, which appears in the blood. A fairly large amount of α-ketoglutarate is tapped away from the tricarboxylic acid cycle by transamination with aspartate to yield glutamate, which leaves the mitochondria to become a participant in the malate-aspartate shuttle for the introduction of reducing equivalents of cytosol NADH formed during glycolysis into the mitochondrial electron-transport chain (page 535). The supply of glycogen in the heart is limited and can suffice for only a short period of transition from the resting to the working state. Two minutes after imposition of the work load the heart has already readjusted its metabolism, with a large increase in the amount of fatty acid oxidized and a decrease in the rate of glucose utilization. The simulation experiments in Figure 30-6 did not include consideration of ketone-body utilization by the heart. These fuels are released from the liver by epinephrine and are important in heart metabolism because they can diffuse into the muscle cells from the blood faster than free fatty acids.

Unlike skeletal muscle, heart muscle has an active amino acid metabolism; the proteins of heart muscle have a relatively high turnover rate compared with skeletal muscle. Moreover, heart muscle has considerable capacity for biosynthesis of new contractile tissue, in response to sustained increases in the work load or in response to *myocardial infarction*, the irreversible degeneration of a portion of the contractile tissue after its blood supply has been blocked by the occlusion of a blood vessel.

Kidneys

The kidneys have an extremely high rate of respiratory metabolism and also show considerable metabolic flexibility. As sources of fuel they can utilize blood glucose, free fatty acids, ketone bodies, and also amino acids, degrading them ultimately via acetyl-CoA and the tricarboxylic acid cycle. From two-thirds to four-fifths of the total energy generated by the kidneys is utilized in the formation of urine, which takes place in a two-stage process. In the first stage the blood plasma is filtered through microscopic structures called *glomeruli*, in the cortex, or outer layer, of the kidney. These filters allow all the solutes of the blood plasma, except the proteins and lipoproteins, to pass into the renal tubules. As the glomerular filtrate passes down these tubules, Na^+, Cl^-, glucose, amino acids, and water are reabsorbed into the blood, which passes through fine capillaries surrounding the renal tubules. As a result, the glomerular filtrate becomes more concentrated as it proceeds down the renal tubules, so that each milliliter of final urine is formed by the concentration of some 50 to 100 ml of glomerular filtrate.

The solute composition of normal human urine is shown in Table 30-4. Some components, particularly glucose, are present in normal urine in much lower concentrations than in blood. Such solutes are reabsorbed from the glomerular filtrate into the blood against a concentration gradient. A second group of solutes, including ammonia and H^+, are present in relatively high concentration in urine compared with blood; they are actively secreted from the blood into the tubules. The ammonia is generated from the hydrolysis of glutamine by the enzyme glutaminase (page 574). A third group of substances, including urea and creatinine, the end product of phosphocreatine degradation, are neither absorbed nor secreted against gradients as the glomerular filtrate is concentrated to form urine. Na^+ is a special case; it is reabsorbed from the glomerular filtrate into the blood plasma by active transport in the first portion of the renal tubule, but some of it later passes back into the urine again by secondary exchange with other cations.

The transport of Na^+ and K^+ is an especially important process in the kidney, which must help regulate the concentration of these vital cations in the blood. Ouabain, a specific inhibitor of the Na^+K^+–ATPase (page 790), inhibits the rate of oxygen consumption of kidney-cell suspensions up to 65 to 70 percent, indicating that the active transport of Na^+ and K^+ by the Na^+K^+–ATPase requires most of the ATP energy generated by respiration in the kidney. The inward-directed Na^+ gradient so generated is used for the active transport of both glucose and amino acids, which are believed to enter epithelial cells by cotransport with Na^+ (page 794).

Adjustments of Metabolism to Stress

Starvation

With the preceding background on the metabolic relationships between the different mammalian organs, we can

Table 30-4 Major components of human urine

The volume and composition of urine vary widely depending on fluid intake and diet; these data are for an average 24-h specimen of total volume 1,200 ml

Component	Grams	Approximate urine/plasma concentration ratio
Glucose	<0.02	<0.02
Amino acids	0.5	1.0
Ammonia	0.8	100
Urea	25	70
Creatinine	1.5	70
Uric acid	0.7	20
H^+	pH 5–8	Up to 300
Na^+	3.0	1.0
K^+	1.7	15
Ca^{2+}	0.2	5
Mg^{2+}	0.15	2
Cl^-	6.3	1.5
Phosphate	1.2 (of P)	25
Sulfate	1.4 (of S)	50
Bicarbonate	0–3	0–2

Table 30-5 Available metabolic fuels in a normal 70-kg male and in an obese individual at the beginning of a fast

The survival time is calculated on the assumption of a basal energy expenditure of 1800 kcal day^{-1}

Type of fuel	Weight of fuel, kg	Caloric equivalent, kcal	Estimated survival time, months
Normal 70-kg man			
Triacylglycerols (adipose tissue)	15	141,000	
Proteins (mainly muscle)	6	24,000	
Glycogen (muscle and liver)	0.225	900	
Circulating fuels (glucose, fatty acids, triacylglycerols, etc.)	0.023	100	
	Total	166,000	~3
Obese individual			
Triacylglycerols (adipose tissue)	80	752,000	
Proteins (mainly muscle)	8	32,000	
Glycogen (muscle and liver)	0.23	920	
Circulating fuels	0.025	110	
	Total	785,030	~14

now examine two important situations in which dramatic metabolic adjustments take place, starvation and the disease diabetes mellitus.

Quantitative metabolic studies have been made on fasting human volunteers under controlled hospital conditions over long periods; particularly important are the quantitative investigations of G. F. Cahill, Jr., and his colleagues in Boston. Table 30-5 shows the amounts of potential fuels and their caloric equivalent in the body of a normal 70-kg adult of sedentary occupation and an obese individual weighing 135 kg. The first point made by these data is that the total amount of glucose and glycogen in the body are small and can provide basal energy requirements for less than a day. On the other hand, the amounts of triacylglycerols and protein are relatively very large, particularly the former. The stored triacylglycerols alone can theoretically provide the basal energy requirements for over 70 days of fasting in the normal adult and for well over a year in the very obese individual. If half of the body protein is also available, fasting can theoretically be prolonged further. In the first day or two of a fast the liver glycogen quickly falls to about 10 percent of its normal concentration and then stays approximately constant at this lower level for extended periods of fasting. Muscle glycogen also is decreased, but not so strikingly. The blood glucose level, however, remains relatively constant at about 80 mg per 100 ml, or 4.5 mM, for some 4 weeks or more during a total fast (Table 30-6). Once the supply of easily metabolized glycogen becomes exhausted, the rate of utilization of triacylglycerols from the fat depots in the abdominal and subcutaneous areas increases. The onset of

Table 30-6 Concentration of fuels in the blood, mM

	Glucose	FFA	Aceto-acetate	β-Hydroxy-butyrate	Amino acids
Normal (postabsorptive)	4.50	0.5	0.01	0.01	4.5
Fasting, 1 week	4.50	1.5	1.0	4.0	4.5
5 weeks	4.49	2.0	1.1	6.7	3.1

Figure 30-7

Utilization of body fuels in starvation (right). The upper diagram reflects the metabolic picture after 36 h of fasting; the lower, after 5 to 6 weeks, after adaptation of the brain for utilization of the ketone bodies. The figures given are for a 24-h period under basal conditions, with a daily energy output of 1800 kcal for the 36-h fast and an output of 1500 kcal for the 5- to 6-week fast. The abbreviations RBC and WBC are for red blood cells and white blood cells. [Adapted from G. F. Cahill, Jr., and O. E. Owen, p. 497 in F. Dickens, P. J. Randle, and W. J. Whelan (eds.), Carbohydrate Metabolism and Its Disorders, Academic Press, Inc., New York, 1968.]

accelerated triacylglycerol oxidation is accompanied by a rising blood ketone-body level and a fall in the respiratory quotient (page 825), from the normal RQ of about 0.82 to between 0.7 and 0.8. Within a few days after the beginning of a fast, the amount of nitrogen excreted in the urine begins to increase. Most of this increase is in the form of urea, the nitrogenous end product of amino acid catabolism (page 579), indicating that body proteins are beginning to undergo degradation. Figure 30-7 summarizes the utilization of fuels by the major organs during a period of fasting.

If triacylglycerols are available in large amounts, why are body proteins degraded during a fast? This question is particularly appropriate since the mammalian organism cannot store proteins. The utilization of body proteins as fuel can therefore be expected to impair some specific biological function dependent upon proteins.

The answer lies in the large amount of glucose required by the brain. We recall that the human brain has a very high metabolic rate, utilizing 20 percent or more of the total energy supply under basal conditions. Moreover, the brain normally utilizes only glucose as fuel and thus requires about 140 g of glucose per day. If the blood glucose level falls appreciably below the normal concentration of some 80 mg per 100 ml (about 4.5 mM), significant disturbances in the function of the central nervous system occur. Since the liver glycogen is nearly depleted after the first day of fasting, blood glucose must be manufactured from other body sources in order to meet the needs of the brain. Red blood cells also utilize glucose (Figure 30-7). However, red blood cells degrade glucose to lactate, which can be converted back into glucose by the liver.

Although some glucose can be formed from the glycerol portion of triacylglycerols, no glucose can be formed from fatty acids by mammals (page 630). The major source of blood glucose in the fasting individual must then be body protein. The carbon skeletons of most of the amino acids, i.e., those termed the glycogenic amino acids (page 629), can give rise to pyruvate or intermediates of the tricarboxylic acid cycle and thus yield blood glucose in the process of gluconeogenesis in the liver (page 624). The degradation of amino acids to form blood sugar results in the conversion of their α-amino groups into urea in the liver, to be excreted by the kidneys, thus accounting for the large increase in excretion of urinary nitrogen after the first few days of fasting. The yield of glucose from 100 g of body protein is about 57 g.

Fasting man (36 h)

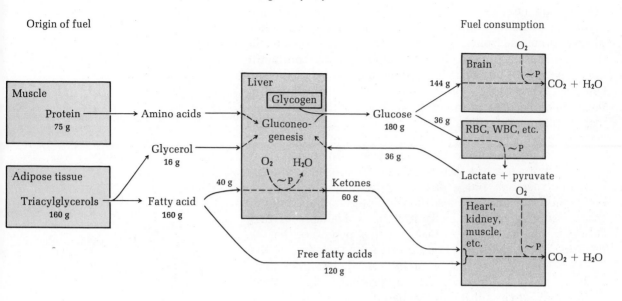

Fasting man, adapted (5–6 weeks)

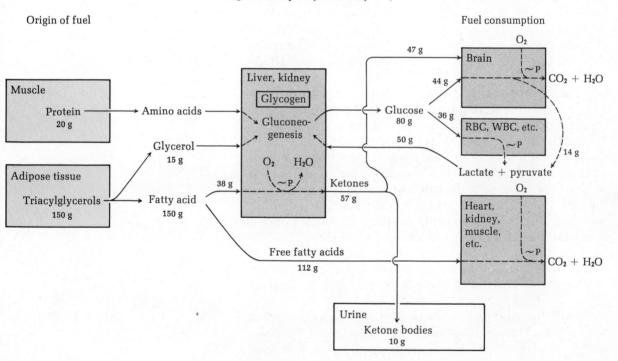

Because the metabolic demands of the brain are paramount and overriding, body proteins normally performing important biological functions are sacrificed in order to maintain the blood sugar level.

There is a definite sequence in which body proteins are lost during fasting in order to preserve the blood glucose level. Those lost first are the digestive enzymes secreted by the stomach, pancreas, and small intestine; they are no longer needed, nor are other enzymes and proteins that are required to make the digestive enzymes. Also lost very early are various enzymes of the liver that normally process incoming nutrients from the intestine and convert them into plasma proteins, lipids, and lipoproteins. Then begins the drain on muscle proteins, not only those of the contractile fibers (page 753) but also the glycolytic enzymes of the sarcoplasm. When muscle proteins begin to be utilized, fasting individuals become physically inactive, another physiological adaptation to fasting. Thus, the fasting organism makes a series of calculated choices in the utilization of body protein, in order to keep the central nervous system functional.

In the first week of a fast, body protein is used at a rather high rate, perhaps 100 g day^{-1}, a rate suggesting that the body could not sustain a fast for very long, perhaps only 30 days, assuming half the body protein is available as fuel. However, another major metabolic adjustment soon begins, resulting in the sparing of body protein. By the end of 4 to 6 weeks of fasting, very little protein is used, less than about 12 to 15 g day^{-1}. This metabolic adjustment, which allows body proteins to be substantially spared, is the acquisition by the brain of the capacity to utilize as fuel source, in addition to blood glucose, the blood ketone bodies, particularly D-β-hydroxybutyrate, which is produced during fatty acid oxidation in the liver. Concomitantly, the concentration of β-hydroxybutyrate in the blood rises (Table 30-6). The combination of increased blood ketones and the derepression of enzymes required for utilization of the ketone bodies as fuel [β-hydroxybutyrate dehydrogenase (page 554), β-ketoacid CoA transferase (page 554), and acetyl-CoA acetyltransferase (page 554)] enables the fasting brain to use β-hydroxybutyrate instead of (or in addition to) glucose. The brain now obtains a substantial fraction of its energy from the oxidation of β-hydroxybutyrate, which in turn arises from the relatively large stores of body fat. The organism is thus relieved in large part of the need to degrade body proteins to make blood glucose. Once this adjustment is made, a starving individual will continue to degrade body proteins, but at a minimal rate, until the fat supply has been totally utilized. No significant changes in mental capacity accompany the switch to ketone-body metabolism. Figure 30-7 shows a comparison of the metabolic fluxes in an obese individual after 1 week and after 6 weeks of fasting.

Finally the most severe phase of starvation begins. Once the supply of triacylglycerols has been consumed, the basal energy requirement of the entire body must be met from body protein. It is at this time that the muscle mass, the largest single source of protein, is very heavily drained. This third and last phase of starvation has an inevitable end.

The biochemistry of starvation in children is an especially important problem, since children have not built up the large masses of fat and muscle that allow adults to survive longer. Long-standing inadequate protein intake, which is akin to slow starvation, results in the condition known as *kwashiorkor*, an African word meaning displaced child (in the sense of having been weaned). Children with this condition often receive ample starchy vegetables, which supply metabolic energy and support brain function, but insufficient protein to maintain normal metabolic and physiological processes. Such children often have plump faces, but a very weak and poorly developed musculature, particularly of the legs, the result of inadequate intake of the indispensable amino acids. These children also have distended abdomens, due to *edema*, the abnormal balance of water between the blood and interstitial fluid, which in kwashiorkor is due to lack of sufficient serum albumin to maintain the osmotic pressure of the blood plasma. Lack of sufficient dietary protein also causes escape of K^+ from the tissues, the accumulation of fat in the liver, and abnormal deposits of iron in tissues, in the form of an iron protein called *hemosiderin*.

The biochemistry of starvation is not merely a research exercise on volunteers fasting in a well-controlled hospital laboratory. Starvation is an enormous problem in the real world, one threatening an ever-increasing fraction of the human race, as the disparity between increasing population and the world agricultural potential inevitably widens.

Diabetes Mellitus

We shall now outline the metabolic changes occurring in *diabetes mellitus,* one of the most intensively studied human diseases. Although some of the major symptoms of diabetes were first described in ancient times and insulin was first isolated over 50 years ago, we still do not have a clear picture of the basic biochemical defect(s) in this disease. Moreover, diabetes is still a serious and important health problem, involving 2 percent or more of the population of the United States.

Diabetes is a complex disease, exhibiting different kinds and degrees of pathology. It is a familial disease; susceptibility to diabetes has a large genetic component. Moreover, physical exercise and the quality of the diet have a profound effect on its incidence. In Europe there was a dramatic decline in mortality from diabetes during World Wars I and II but a return to a much higher incidence as the economy improved after each war.

The primary symptom of acute diabetes mellitus is hyperglycemia, often accompanied by glucosuria and polyuria, the excretion of large volumes of urine. The latter symptoms are at the origin of the names diabetes (Greek for siphon) and mellitus (sweet). Correspondingly, there is considerable hunger and thirst, weight loss, and in more severe cases, *ketonemia* (elevated blood ketone bodies), *ketonuria*, and *acidosis,* a decrease in the blood pH or in the acid-buffering capacity of the blood buffers. A secondary set of symptoms arises in chronic or long-standing diabetes. These include degeneration of the walls of blood vessels, particularly of fine capillaries and their basement membranes. Although many different organs are affected by these vascular changes, the eyes appear to be most susceptible; indeed, long-standing diabetes mellitus, even when treated with insulin, is a leading cause of blindness.

Two major types of diabetes are recognized: *juvenile-onset* and *adult-onset*. The juvenile form develops early in life, develops much more severe symptoms, and has a near-certain prospect of later vascular involvement. The adult type develops late in life, is milder, and more gradual in onset.

In addition to the overt forms of clinical diabetes, there are many otherwise normal individuals who show a functional predisposition toward the disease. Diabetes or a diabetic tendency is detected by the *glucose-tolerance test*. The patient is given an oral test dose of 1.0 g of glucose per kilogram of body weight, and his blood glucose level is followed for the next few hours. Sometimes the test dose is given intravenously. Normal subjects show a modest rise in the blood sugar level to about 120 to 140 mg per 100 ml at about 1 h, insufficient to exceed the *renal threshold* of about 160 to 180 mg per 100 ml, the concentration at which glucose begins to appear in the urine in most normal individuals. This rise in blood glucose is followed by a rapid decline to the normal level within a 2- to 3-h period (Figure 30-8). The rapid decline in blood glucose in normal individuals is due to an increased rate of glucose utilization by the tissues, evoked by increased secretion of insulin and a decreased secretion of glucagon from the pancreas, in response to the increased blood glucose level (Figure 30-9).

Diabetics, on the other hand, show a much larger increase in their blood glucose concentration, which is initially higher than normal. They also show a more prolonged period of hyperglycemia and a slower decline. They may also show (increased) glucosuria after the test dose of glucose (Figure 30-8).

More specific biochemical and enzymatic changes occurring in diabetes have been discovered by studying the metabolism of animals made diabetic by pancreatectomy or by destruction of the pancreatic islet tissue with toxic agents such as *alloxan* or the antibiotic *streptozotocin* (Figure 30-10).

Important early observations by S. Soskin and R. Levine showed that the peripheral tissues of diabetic animals are deficient in the removal of hexoses from the blood when tested at the normal levels of blood glucose. However, if the blood sugar is raised to the very high levels found in severe diabetes (up to 500 mg per 100 ml) the uptake of hexose is then substantially increased, almost to normal rates. This important finding indicated that the basic deficiency in glucose metabolism is more a matter of impaired transport of blood glucose across the membranes of skeletal muscles and other tissues than the lack of some intracellular enzymes required in the utilization of glucose. Although these and many similar experiments have shown that impaired glucose transport is certainly the major defect in glucose metabolism of diabetic animals, it has also been found that there is an enhanced rate of gluconeogenesis. This is indicated by the appearance of increased amounts of extra nitrogen in the urine (largely as urea) in diabetic animals, particularly after feeding amino acid mixtures. Comparison of the extra nitrogen excreted and the glucose appearing in the urine in-

Figure 30-8
Typical glucose tolerance curves for normal, mildly diabetic, and severely diabetic subjects. The test is described in the text.

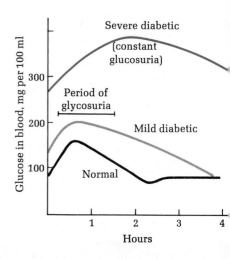

Figure 30-9
The increased release of insulin and the decreased release of glucagon during a glucose-tolerance test given to a normal individual.

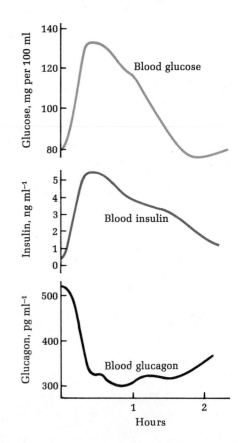

Figure 30-10
Diabetogenic agents.

Alloxan, a toxic agent that destroys the capacity to secrete insulin in some species.

Streptozotocin, an antibiotic from *S. achromogenes*. It has both diabetogenic and anti-tumor activity. It is a derivative of D-glucosamine.

dicates that the carbon skeletons of the glycogenic amino acids are almost entirely converted into glucose (pages 559 and 629). This conclusion is confirmed by experiments in which ^{14}C-labeled amino acids are fed to diabetic animals; much of the isotope is recovered in the excreted glucose.

Another major aberration of glucose metabolism in diabetes is the almost complete cessation of the conversion of glucose into fatty acids via acetyl-CoA (page 661). In normal animals perhaps as much as a third of the total ingested carbohydrate may be converted into fatty acids and thence into triacylglycerols in the liver and fat depots. The acceleration of gluconeogenesis from amino acids and the inhibition of fatty acid synthesis from glucose indicate that metabolism in the diabetic organism is tuned to maintain as high a concentration of glucose in the blood as possible, even though the blood level may already greatly exceed the renal threshold for glucose. The constant loss of glucose in the urine in diabetics, much of which occurs at the expense of ingested amino acids or body protein, accounts for the constant hunger and loss of body weight in severe diabetics, who literally "starve in the midst of plenty." Figure 30-11 summarizes the metabolism of the major nutrients in a severe diabetic.

In diabetic animals relatively little glucose is oxidized as fuel, except by the brain. The rest of the tissues burn a large amount of fat, particularly the liver, where the amount of acetyl-CoA formed from fatty acids exceeds the capacity of the tricarboxylic acid cycle to oxidize it. The excess acetyl-CoA is converted into ketone bodies, large amounts of which may appear in the blood (ketonemia) and urine (ketonuria), a condition known as <u>ketosis</u>.

All these metabolic aberrations can be reversed by administration of insulin, which increases the rate of removal of glucose from the blood at normal blood glucose concentrations, increases the amounts of glucokinase, phosphofructokinase, and pyruvate kinase in the liver, depresses the biosynthesis of enzymes specifically required in gluconeogenesis (and thus depresses the formation of glucose from amino acids), restores the normal rate of conversion of glucose into fatty acids, and inhibits the oxidative degradation of fatty acids. As a result, the blood sugar level becomes normal, glucosuria ceases, and the ketone bodies recede to their normal levels in the blood and urine. Administration of excessive amounts of insulin to normal animals or an overdose of insulin in a diabetic causes the blood sugar to fall to levels below 80 mg per 100 ml. Very large amounts of insulin cause <u>insulin shock,</u> convulsions and coma which occur when the blood sugar falls below 20 mg per 100 ml. It can be reversed by intravenous administration of glucose.

Deficiency of insulin in diabetics can theoretically be due to a number of different causes, such as a reduced amount of otherwise normal islet tissue, defective biosynthesis of proinsulin (page 817), defective conversion of proinsulin to insulin, deficient release of insulin into the blood in response to increased blood glucose (page 820), production of a genetically defective insulin, or an abnormally high rate

Figure 30-11
The source of endogenous fuels and the pattern of their distribution and
consumption in severe untreated diabetics. The figures are for a 24-h period
under basal conditions, assuming total energy output of 2400 kcal. Note the
heavy drain on muscle proteins for gluconeogenesis and the utilization of
body triacylglycerols as energy source for all organs except the brain and
blood cells, which require glucose. A large part of the glucose made from
muscle protein is lost in the urine. [Adapted from G. F. Cahill, Jr., and O. E.
Owen, in F. Dickens, P. J. Randle, and W. J. Whelan (eds.), Carbohydrate
Metabolism and Its Disorders, Academic Press, Inc., New York, 1968.]

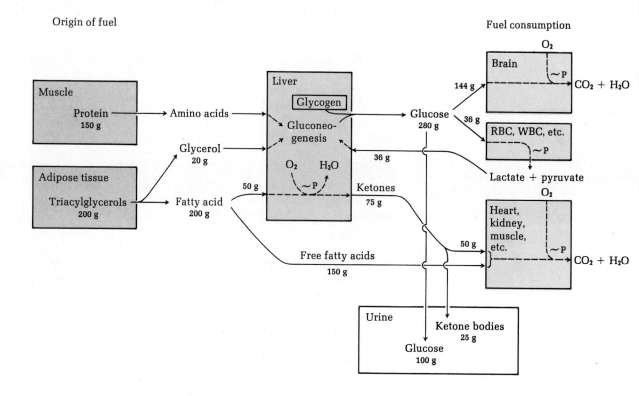

of destruction of insulin. In juvenile diabetes, the amount of
insulin found in the pancreas is less than 5 percent of the
normal amount; when the blood glucose level is increased,
no significant insulin release from the pancreas is elicited.
However, in adult diabetes the total amount of insulin in the
pancreas may be quite normal, although its release into the
blood in response to high blood glucose may be delayed.
Thus there are multiple ways in which the concentration of
active insulin in the blood in response to increased blood
glucose concentration may be abnormal or deficient. Some
diabetics are treated successfully with certain drugs, such as
tolbutamide (Figure 30-12), which is capable of stimulating
the release of preformed insulin from the islet cells. This effect suggests that such patients manufacture insulin normally but have a defect in the release mechanism.

Some diabetic patients are resistant or insensitive to insulin. In these cases insulin may not be deficient; instead,
there may be an overproduction of other hormones, such as
growth hormone or glucagon, that oppose the action of insulin in the checks and balances operating to maintain glucose homeostasis.

Figure 30-12
Tolbutamide (1-butyl-3-p-tolylsulfonylurea), a drug causing release of insulin from the pancreas.

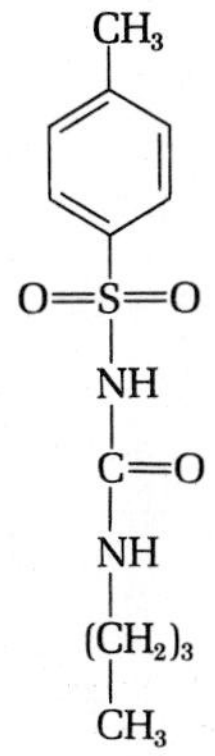

CH₃

O=S=O

NH

C=O

NH

(CH₂)₃

CH₃

Why is it that diabetics not only have a defect in the tissue utilization of glucose but also appear to be metabolically poised to produce maximal amounts of glucose from amino acids and to prevent glucose from being utilized to form fat? An old and possibly still correct hypothesis is that the enhanced gluconeogenesis and the diminished fat synthesis are biological compensations for the lack of insulin. In this view the concentration of glucose in the blood is increased so that the deficiency in glucose transport can be overcome, to permit the utilization of glucose by peripheral tissues in the absence of insulin. On the other hand, as pointed out earlier (page 821), insulin may be a generalized growth hormone, promoting entry of all nutrients into cells, as well as their storage and utilization.

Metabolic Interplay in Cancer

Besides diabetes there are other diseases in which alterations in the metabolic interplay between organs occur. One interesting case is provided by the growth of cancer cells. A cancer cell results from a permanent genetic change in an otherwise normal somatic cell. Such a change, called the *malignant transformation,* may be triggered by some external physical agency, such as x-rays or excess ultraviolet irradiation from sunlight, or by various carcinogenic (cancer-causing) chemical agents. Cancer cells tend to grow rather aggressively and do not obey normal patterns of tissue formation. Their abnormal growth pattern, which is not necessarily massive but is usually disruptive, tends to interfere with the normal activities of adjacent cells and tissues.

Cancer cells have a distinctive type of metabolism. Although they possess all the enzymes required for most of the central pathways of metabolism, cancer cells of nearly all types show an anomaly in the integration of the glycolytic sequence and the tricarboxylic acid cycle. The rate of oxygen consumption of cancer cells is somewhat below the values given by normal cells. However, malignant cells tend to utilize anywhere from 5 to 10 times as much glucose as normal tissues and convert most of it into lactate, even though they have a nearly normal rate of respiration. This phenomenon of aerobic lactate production is called *aerobic glycolysis.*

Aerobic formation of lactate occurs in only a few normal tissues, in erythrocytes, in skeletal and heart muscle under heavy work loads, and in the retina. The aerobic production of lactate by cancer cells is not merely an in vitro artifact but also occurs in vivo. The net effect of aerobic glycolysis on the bioenergetics of cancer cells is that in addition to the generation of ATP in the mitochondria from respiration and oxidative phosphorylation, there is a very large formation of ATP in the extramitochondrial compartment from glycolysis. It has been concluded that the normal allosteric factors regulating the rate of glycolysis to match the rate of utilization of pyruvate by the tricarboxylic acid cycle, i.e., through the Pasteur effect (page 536), are defective or altered in cancer cells.

The most important systemic effect of this metabolic im-balance in cancer cells is the utilization of a large amount of blood glucose and the release of correspondingly large amounts of lactate into the blood. The lactate so formed is then recycled in the liver to produce blood glucose again by the process of gluconeogenesis, exactly as occurs during heavy muscular work (page 769). Since the formation of a molecule of glucose from lactate requires input of six mole-cules of high-energy phosphate in the liver (page 624) whereas the cancer cell obtains only two molecules of ATP per molecule of glucose converted into lactate, the cancer cell may be looked upon as a metabolic parasite, dependent on the liver for a substantial part of its energy. Large masses of cancer cells thus can be a considerable metabolic drain on the host organism, in addition to causing other local and systemic effects.

Summary

The blood is the vehicle for transport of nutrients from the intestine to the liver and other tissues, for transport of organic waste prod-ucts to the kidneys, and for transport of the respiratory gases oxygen and carbon dioxide. The blood plasma contains about 7 percent of plasma proteins, which have manifold functions, particularly the transport of lipids, fatty acids, and other blood components. The remaining 3 percent of solids are metabolites, waste products, and inorganic ions.

The liver is the major organ for processing and distribution of nutrients. It is responsible for formation of blood glucose and plasma lipoproteins and for distribution of amino acids via the blood. It transforms other monosaccharides into glucose 6-phos-phate, which it converts into glycogen, into acetyl-CoA, and thence to fatty acids; glucose 6-phosphate is also used to generate NADPH for reductive synthesis of fatty acids and cholesterol. The liver also deaminates amino acids, carries out gluconeogenesis, and is the sole site of urea synthesis. It also forms and secretes bile salts and bile pigments into the intestine.

The adipose, or fat, tissue is the other major storage and fuel-dis-tribution center. It can store enormous amounts of triacylglycerols, which are received in the form of chylomicrons from the intestine or are generated from blood glucose under stimulation by insulin. Triacylglycerols have a higher energy content than glycogen, weight for weight, and are stored in anhydrous form. Free fatty acids are released from adipose tissue when it is stimulated by epinephrine and other hormones via the action of a cyclic AMP–dependent lipase. The free fatty acids are carried to other tissues bound to serum albumin. The brain utilizes 20 percent of the total oxygen consumed by the body, and it can utilize only glucose as fuel under normal conditions. The heart utilizes primarily fatty acids and ke-tone bodies as fuel but when a sudden work load is imposed, it can employ the glycolytic conversion of glycogen into lactate as ATP source. Skeletal muscles also utilize free fatty acids and ketone bodies as fuel but depend on conversion of glucose and glycogen to lactate for sudden bursts of energy. The kidneys employ most of their ATP production for the active transport of glucose, Na^+, and K^+ during the secretion of urine.

In starvation body stores of triacylglycerols are the major source of energy, supplemented by body protein, which, through gluconeogenesis from amino acids, furnishes blood glucose for the brain. The drain on body protein is minimized by a metabolic adaptation in which the brain acquires the ability to use blood ketone bodies from oxidation of fatty acids for energy instead of glucose. However, when the supplies of triacylglycerols are exhausted, body protein, especially that of the muscles, is heavily drained. Kwashiorkor is a condition of semistarvation in children due to a diet deficient in protein.

Diabetes mellitus is a complex disease, but its cardinal manifestations are hyperglycemia, glucosuria, polyuria, thirst, and hunger. In severe diabetes the body stores of protein are utilized to generate blood glucose, much of which is lost in the urine. Diabetics lack, for one reason or another, the capacity to release insulin into the blood in response to a rise in blood sugar, detected by the glucose-tolerance test. They show greatly increased gluconeogenesis from amino acids, nearly complete failure in conversion of glucose into fatty acids, and excessive production of ketone bodies from fatty acids. There is also diminished protein synthesis from amino acids. Although the major defect in diabetics appears to be the failure of glucose to be transported into muscles and fat tissue, the precise mode of action of insulin has not been identified.

Another type of metabolic aberration occurs in cancer-bearing animals. Cancer cells utilize abnormally large amounts of glucose, which they largely convert into blood lactate. The lactate is converted back into blood glucose in the liver at a large net cost in ATP. Cancer cells are thus metabolic parasites.

References

Books

HOFFMAN, W. S.: *The Biochemistry of Clinical Medicine*, 4th ed., Year Book Medical Publishers, Chicago, 1970.

NEWSHOLME, E. A., and C. START: *Regulation in Metabolism*, Wiley, New York, 1973. An extremely valuable book outlining the biochemistry and theory of metabolic regulation.

WHITE, A., P. HANDLER, and E. L. SMITH: *Principles of Biochemistry*, McGraw-Hill, New York, 5th ed., 1973. The biochemistry of body fluids, renal function, and other physiological activities is described in detail.

Articles

CAHILL, G. F., JR.: "Starvation in Man," *N. Engl. J. Med.*, 282: 668 (1970). A summary of classical investigations.

FERNSTROM, J. D., and R. J. WURTMAN: "Nutrition and the Brain," *Sci. Am.*, 230: 84–91 (1974).

GARFINCKEL, D.: "Simulation of the Krebs Cycle and Closely Related Metabolism in Perfused Rat Liver; 1: Construction of a Model; 2: Properties of the Model," and succeeding papers on gluconeogenesis, *Comput. Biomed. Res.*, 4: 1–125 (1971). Use of computer modeling to analyze the central pathways of glucose metabolism.

LEVINE, R., and D. E. HAFT: "Carbohydrate Homeostasis," *N. Engl. J. Med.*, 283: 175–183, 237–246 (1970). An important descriptive review, in two parts, of the regulation of carbohydrate metabolism.

OWEN, O. E., A. P. MORGAN, H. G. KENY, J. M. SULLIVAN, M. G. HER-RERA, and G. F. CAHILL, JR.: "Brain Metabolism in Fasting," *J. Clin. Invest.,* 46: 1589 (1967). A classic study showing the role of the ketone bodies.

RENOLD, A. E., W. STAUFFACHER, and G. F. CAHILL, JR.: "Diabetes Mellitus," in J. B. Stanbury, J. B. Wyngaarden, and D. S. Fred-rickson (eds.), *The Metabolic Basis of Inherited Disease,* McGraw-Hill, New York, 1972. An excellent short review of metabolism in diabetics.

Problems

1. Calculate theoretical RQ values (page 842) for (a) glucose, (b) citric acid, (c) phenylalanine, (d) palmitic acid, (e) glycerol, and (f) tripalmitin.

2. Alcohol dehydrogenase of human liver catalyzes the oxidation of ethanol to acetaldehyde; as does the enzyme from yeast (page 438). The acetaldehyde thus produced is converted to acetate in a pyridine nucleotide–requiring step, and the acetate is converted to acetyl-CoA by processes described earlier (page 546). Write a balanced equation for the complete combustion of ethanol as it occurs in the human liver, including all associated phosphorylation steps.

3. Calculate the number of ATP molecules produced per carbon atom upon complete combustion to carbon dioxide and water of the following compounds in a typical liver cell: (a) glucose, (b) ethanol, (c) palmitic acid, (d) tripalmitin, (e) leucine, and (f) phenylalanine.

4. Fatty acids cannot serve directly as an energy source for the brain when glucose is in short supply but are converted to β-hydroxybutyric acid, which gains rapid entry into brain cells. Write balanced equations for the following processes, including all associated phosphorylation steps:
   (a) Conversion of palmitic acid to β-hydroxybutyric acid
   (b) Complete combustion of β-hydroxybutyric acid to carbon dioxide and water

5. Compare the number of ATP molecules provided by direct combustion of palmitic acid to carbon dioxide and water with the number furnished when the combustion occurs via four molecules of β-hydroxybutyric acid.

6. Calculate the net ATP change occurring when a glucosyl unit of muscle glycogen is converted to two molecules of lactate and the resulting lactate is reconverted to glucose in the liver.

PART 4 REPLICATION, TRANSCRIPTION, AND
TRANSLATION OF GENETIC INFORMATION

MOLECULAR STRUCTURE OF NUCLEIC ACIDS

A Structure for Deoxyribose Nucleic Acid

WE wish to suggest a structure for the salt of deoxyribose nucleic acid (D.N.A.). This structure has novel features which are of considerable biological interest.

A structure for nucleic acid has already been proposed by Pauling and Corey[1]. They kindly made their manuscript available to us in advance of publication. Their model consists of three intertwined chains, with the phosphates near the fibre axis, and the bases on the outside. In our opinion, this structure is unsatisfactory for two reasons: (1) We believe that the material which gives the X-ray diagrams is the salt, not the free acid. Without the acidic hydrogen atoms it is not clear what forces would hold the structure together, especially as the negatively charged phosphates near the axis will repel each other. (2) Some of the van der Waals distances appear to be too small.

Another three-chain structure has also been suggested by Fraser (in the press). In his model the phosphates are on the outside and the bases on the inside, linked together by hydrogen bonds. This structure as described is rather ill-defined, and for this reason we shall not comment on it.

We wish to put forward a radically different structure for the salt of deoxyribose nucleic acid. This structure has two helical chains each coiled round the same axis (see diagram). We have made the usual chemical assumptions, namely, that each chain consists of phosphate di-ester groups joining β-D-deoxyribofuranose residues with 3′,5′ linkages. The two chains (but not their bases) are related by a dyad perpendicular to the fibre axis. Both chains follow right-handed helices, but owing to the dyad the sequences of the atoms in the two chains run in opposite directions. Each chain loosely resembles Furberg's[2] model No. 1; that is, the bases are on the inside of the helix and the phosphates on the outside. The configuration of the sugar and the atoms near it is close to Furberg's 'standard configuration', the sugar being roughly perpendicular to the attached base. There is a residue on each chain every 3·4 A. in the z-direction. We have assumed an angle of 36° between adjacent residues in the same chain, so that the structure repeats after 10 residues on each chain, that is, after 34 A. The distance of a phosphorus atom from the fibre axis is 10 A. As the phosphates are on the outside, cations have easy access to them.

The structure is an open one, and its water content is rather high. At lower water contents we would expect the bases to tilt so that the structure could become more compact.

The novel feature of the structure is the manner in which the two chains are held together by the purine and pyrimidine bases. The planes of the bases are perpendicular to the fibre axis. They are joined together in pairs, a single base from one chain being hydrogen-bonded to a single base from the other chain, so that the two lie side by side with identical z-co-ordinates. One of the pair must be a purine and the other a pyrimidine for bonding to occur. The hydrogen bonds are made as follows: purine position 1 to pyrimidine position 1; purine position 6 to pyrimidine position 6.

If it is assumed that the bases only occur in the structure in the most plausible tautomeric forms (that is, with the keto rather than the enol configurations) it is found that only specific pairs of bases can bond together. These pairs are: adenine (purine) with thymine (pyrimidine), and guanine (purine) with cytosine (pyrimidine).

In other words, if an adenine forms one member of a pair, on either chain, then on these assumptions the other member must be thymine; similarly for

This figure is purely diagrammatic. The two ribbons symbolize the two phosphate—sugar chains, and the horizontal rods the pairs of bases holding the chains together. The vertical line marks the fibre axis

guanine and cytosine. The sequence of bases on a single chain does not appear to be restricted in any way. However, if only specific pairs of bases can be formed, it follows that if the sequence of bases on one chain is given, then the sequence on the other chain is automatically determined.

It has been found experimentally[3,4] that the ratio of the amounts of adenine to thymine, and the ratio of guanine to cytosine, are always very close to unity for deoxyribose nucleic acid.

It is probably impossible to build this structure with a ribose sugar in place of the deoxyribose, as the extra oxygen atom would make too close a van der Waals contact.

The previously published X-ray data[5,6] on deoxyribose nucleic acid are insufficient for a rigorous test of our structure. So far as we can tell, it is roughly compatible with the experimental data, but it must be regarded as unproved until it has been checked against more exact results. Some of these are given in the following communications. We were not aware of the details of the results presented there when we devised our structure, which rests mainly though not entirely on published experimental data and stereochemical arguments.

It has not escaped our notice that the specific pairing we have postulated immediately suggests a possible copying mechanism for the genetic material.

Full details of the structure, including the conditions assumed in building it, together with a set of co-ordinates for the atoms, will be published elsewhere.

We are much indebted to Dr. Jerry Donohue for constant advice and criticism, especially on interatomic distances. We have also been stimulated by a knowledge of the general nature of the unpublished experimental results and ideas of Dr. M. H. F. Wilkins, Dr. R. E. Franklin and their co-workers at King's College, London. One of us (J. D. W.) has been aided by a fellowship from the National Foundation for Infantile Paralysis.

J. D. WATSON
F. H. C. CRICK

Medical Research Council Unit for the Study of the Molecular Structure of Biological Systems, Cavendish Laboratory, Cambridge. April 2.

[1] Pauling, L., and Corey, R. B., Nature, 171, 346 (1953); Proc. U.S. Nat. Acad. Sci., 39, 84 (1953).
[2] Furberg, S., Acta Chem. Scand., 6, 634 (1952).
[3] Chargaff, E., for references see Zamenhof, S., Brawerman, G., and Chargaff, E., Biochim. et Biophys. Acta, 9, 402 (1952).
[4] Wyatt, G. R., J. Gen. Physiol., 36, 201 (1952).
[5] Astbury, W. T., Symp. Soc. Exp. Biol. 1, Nucleic Acid, 66 (Camb. Univ. Press, 1947).
[6] Wilkins, M. H. F., and Randall, J. T., Biochim. et Biophys. Acta, 10, 192 (1953).

The above article by J. D. Watson and F. H. C. Crick is a historical landmark in modern biochemistry. (Reprinted in its entirety by special permission from Nature, April 25, 1953, p. 737.) It was followed some weeks later by a second article in which the replication process was more explicitly described.

PART 4 REPLICATION, TRANSCRIPTION, AND TRANSLATION OF GENETIC INFORMATION

In the last part of this book we shall consider the biochemical basis of some of the most fundamental and central questions that are posed by the genetic continuity and the evolution of living organisms. What is the molecular nature of the genetic material and its functional components, the chromosomes, genes, codons, and mutational units? How is genetic information replicated with such fidelity? How is genetic information ultimately transcribed and translated into the amino acid sequence of protein molecules, and thence into the characteristic three-dimensional structure of cells? How did living organisms develop the capacity for self-replication—indeed, how did life first arise?

The extraordinary advances made in our knowledge of the molecular basis of genetics have brought an intellectual revolution in biology that many consider comparable to that initiated by Darwin's theory of the origin of species. All fields of biology have been profoundly influenced by these developments. They have brought penetrating new insight into some of the most fundamental problems in cell structure and function and have led to a more comprehensive and a more widely applicable conceptual framework for the science of biochemistry. Today literally no biochemical problem can be studied in isolation from its genetic background. Moreover, these concepts have greatly increased the power of the molecular approach to biology and have encouraged biochemists to probe more confidently into such complex biological problems as cell differentiation, morphogenesis, immunity, and many others which, not too long ago, had been thought to be unapproachable by biochemistry except in a trivial way. Let us briefly examine the scientific and intellectual background which led to this new era in biochemistry.

Today's knowledge of the molecular basis of genetics arose from theoretical and experimental advances in three different fields of science: classical genetics, biochemistry, and molecular structure. In the field of genetics, experi-

mental progress was immensely accelerated by the use of
x-rays and other mutagenic agents, which greatly increase
the rate of mutation, and by the use of organisms with a very
short life cycle, such as molds, bacteria, and viruses. More
efficient methods for selecting rare mutants and genetic recombinants from huge populations have led to the construction of high-resolution genetic maps. Such maps have revealed the sequence of various genes in some chromosomes
and have shown that a given gene has a large number of sites
at which it may undergo mutation. But the most pervasive
developments in the field of genetics, which initiated the
confluence of genetics and biochemistry, were the emergence
of the one gene–one enzyme hypothesis from the work of
G. W. Beadle and E. L. Tatum beginnning in 1941 and the discovery by O. T. Avery and his colleagues in 1944 that genetic
information is contained in, and is transmitted by, DNA.

In the field of biochemistry important experimental advances were made possible by the refinement of chromatographic methods. These led to quantitative analysis of the
composition and the elucidation of the amino acid sequence
of many proteins. They also revealed that proteins of different species or of different functions vary in amino acid
sequence. New chromatographic methods were also in-
strumental in establishing the composition and covalent
structure of nucleic acids. One of the most significant developments was the discovery of the molar equivalence of certain bases in DNA by E. Chargaff. This equivalence later
became an important element in deducing the three-dimensional structure of DNA. The isotope tracer technique also
played a vital role, since it made possible direct experimental approaches to the enzymatic biosynthesis of nucleic
acids and proteins.

In the field of molecular structure the x-ray diffraction
analysis of the conformation of fibrous protein molecules by
W. Astbury and by L. Pauling and of globular proteins by
J. C. Kendrew and M. F. Perutz yielded the concept that
each type of protein molecule has a specific conformation of
precise dimensions, which determines its biological function. But the central event that triggered the development of molecular genetics was x-ray analysis of the three-
dimensional structure of DNA. In 1953 J. D. Watson and
F. H. C. Crick postulated the double-helical structure for
DNA, which not only accounted for the molar equivalence
of the bases and the characteristic x-ray diffraction pattern
of DNA, but also suggested a simple mechanism by which
genetic information can be precisely transferred from parent to daughter cells. This hypothesis quickly resulted in a
remarkable confluence of ideas and experimental approaches from genetics, biochemistry, and molecular physics.

The Watson-Crick hypothesis was soon built upon and extended to yield what Crick termed the _central dogma_ of
molecular genetics, which states that genetic information
flows from DNA to RNA to protein. Although this statement
has had to be modified more recently into the form

$$\text{DNA} \longrightarrow \text{RNA} \longrightarrow \text{protein}$$

the central dogma defined three major processes in the pres-
ervation and transmission of genetic information. The first
is *replication*, the copying of DNA to form identical daughter
molecules. The second is *transcription*, the process by which
the genetic message in DNA is transcribed into the form of
messenger RNA, to be carried to the ribosomes. The third is
translation, the process by which the genetic message is
decoded on the ribosomes, where RNA is used as a template
in directing the specific amino acid sequence during protein
biosythesis. The central dogma was supported not only by
the discovery of messenger RNA but also by the demonstration that the sequence of nucleotides in a gene bears a
linear correspondence to the sequence of amino acids in the
protein coded by the gene. Moreover, the "dictionary" of
triplet code words in messenger RNA for the various amino
acids has been deduced, in one of the greatest achievements
in modern science.

Now that the central dogma of molecular genetics has
been firmly substantiated, it may appear that most of the important work has been done; as one distinguished investigator put it, "that was the molecular biology that was." But
we have had only a glimpse. As in many fields of science,
the opening of one door leads to another, in an endless succession. Let us now see where we stand today and what
remains to be done.

Most of the illuminating research on the molecular basis
of genetic-information transfer over the last 20 years or so
has been carried out on prokaryotic organisms, particularly
the bacterium *Escherichia coli*, and on bacterial viruses. Prokaryotes contain only one chromosome, a single molecule of
double-stranded DNA, and they usually reproduce by an
asexual process of cell division, in which the daughter cells
normally receive a genome identical to that of the parent
cell. In the life cycle of prokaryotes the cell is usually in the
haploid state. Therefore the genetics of bacteria can be studied without the complications of dominance and reces-
siveness which are characteristic of diploid eukaryotic organisms and with which the mendelian principles are
concerned. Moreover, the problems of genetic information transfer are even simpler in the single "chromosomes"
of viruses, which contain only a few or some dozens of
genes, compared with several thousand genes in bacteria. Because of the relative simplicity of viruses and bacteria
and the speed and efficiency with which genetic experiments
can be carried out on them, we are now in command of solid
knowledge of many of the basic molecular processes in
genetic-information transfer, which we can with some
confidence regard as biologically universal. Nevertheless,
as we shall see, many details of the biochemistry of replication, transcription, and translation in prokaryotes are
still unknown and remain to be solved.

But today molecular genetics stands on a new threshold,
one that opens to still uncharted problems posed by the vast
gulf—in size, complexity, and the capacity for differentiation
and evolution—between prokaryotic and eukaryotic cells.
Thus there are new and difficult problems in defining the

molecular genetics of eukaryotic organisms and the regulation of gene expression during their differentiation and development. Although much evidence now supports the view that the basic molecular patterns of replication, transcription, and translation may be essentially identical in prokaryotic and eukaryotic organisms, in the latter many special complexities are superimposed. Eukaryotic organisms contain vastly more genetic information than prokaryotes, and it is divided among several or many chromosomes. The somatic cells of eukaryotic organisms are usually diploid and contain twice the number of chromosomes found in the germ cells, which are haploid. Each gene in a eukaryote thus occurs in two forms, or _alleles,_ one of which is genetically dominant, the other recessive. Eukaryotic organisms usually reproduce by sexual conjugation, during which genes from both parents are exchanged and incorporated into the genome of the progeny by the process of _recombination._ Although recombination occurs in prokaryotes during sexual conjugation or viral infection, it is not a normal process in their asexual division.

The molecular structure of the complex chromosomes of eukaryotic organisms, the manner in which various genes from different chromosomes interact together, the molecular mechanisms by which genes from the parents are recombined, how gene expression is controlled in eukaryotes — these are major unsolved problems of great complexity. Moreover, eukaryotic organisms contain DNA not only in the nucleus but also in the mitochondria and certain other organelles. The extranuclear DNA is involved in _cytoplasmic inheritance,_ a process expressed in a nonmendelian manner that is still little understood. It is therefore clear that the biochemical aspects of the genetics and development of higher organisms constitute enormous — indeed limitless — challenges for the biochemistry of today and tomorrow.

CHAPTER 31 DNA AND THE STRUCTURE OF THE GENETIC MATERIAL

We have already considered the nature of the nucleotide building blocks of DNA and the covalent structure of its backbone, matters that should first be reviewed (Chapter 12). In this chapter we shall try to answer three major questions: (1) What is the evidence that DNA is the usual form in which genetic information is stored? (2) What is the molecular basis for the ability of DNA molecules to store information? (3) What are the dimensions of some of the functional components of the genetic material: chromosomes, genes, and the units of mutation?

Transformation of Bacteria by DNA

Although DNA was first discovered in cell nuclei as long ago as 1869, by F. Miescher, it was not directly identified as bearing genetic information until 1944, when O. T. Avery and his colleagues discovered that some cells in a rapidly growing culture of a nonvirulent strain of the bacterium *Diplococcus pneumoniae*, usually called pneumococcus, are transformed in a heritable manner into a virulent or disease-producing strain by simple addition of DNA extracted from virulent pneumococci. Virulent pneumococci had long been known to form smooth colonies (designated S) when cultured on agar, because of the presence of a distinctive poly-saccharide capsule. Avirulent, or nonpathogenic, strains of pneumococci, on the other hand, form rough (R) colonies. Earlier investigators had found that injection into mice of heat-killed virulent S cells together with live nonvirulent R cells is lethal. From the infected mice, viable S cells could be isolated and cultured, indicating that the live R cells were transformed into S cells by a heat-stable component present in the killed cells. O. T. Avery, C. M. MacLeod, and M. Mc-Carty investigated this effect more closely. They found that DNA isolated in highly purified form from heat-killed S cells was capable of causing transformation when added to non-virulent R cells in vitro. However, DNA specimens from R cells or from species other than pneumococci were unable to transform R cells. Moreover, the DNA-induced transformation of R cells is a permanent and heritable characteristic, since all the progeny of the transformed cells, over many generations, were found to be virulent S cells. Avery and his colleagues therefore concluded that DNA can carry genetic information.

This conclusion was not immediately accepted, however. Some investigators suggested that the DNA preparations contained a mutagen which could promote mutation to the S form. Others suggested that the transformation of R into S cells was actually caused by traces of some specific protein remaining in the DNA samples used. Eventually, however, these objections were removed by further experiments. Moreover, many other instances of bacterial transformation by DNA alone have since been observed. Today the experiments of Avery and his colleagues are commonly regarded as a historical landmark. It is now known that the transforming DNA is covalently incorporated into the DNA of the recipient cell and is thus replicated when the host chromosome is replicated.

Although DNA is normally the primary form in which genetic information is stored and from which it is passed to daughter cells by replication, we know today that RNA can store genetic information in some viruses. Moreover, genetic information may sometimes be transferred from RNA to DNA, in the process of _reverse transcription,_ to be discussed elsewhere (page 915).

Biochemical Evidence for DNA as Genetic Material

Strong supporting evidence that DNA is the bearer of genetic information has come from other directions. First, the amount of DNA in any given species of cell or organism is remarkably constant and cannot be altered by environmental circumstances or by changes in the nutrition or metabolism of the cell, a property to be expected of genetic material.

Second, the amount of DNA per cell appears to be in proportion to the complexity of the cell and thus to the amount of genetic information it contains. Data in Table 31-1 show that the higher the organism in the evolutionary scale, the greater the content of DNA per cell. Bacteria contain about 0.01 pg (1 pg = 1×10^{-12} g) DNA per cell (about 1 percent of the total wet weight), whereas the tissues of higher animals contain about 6 pg DNA per cell. Moreover, the germ cells of higher animals, which are haploid and possess only one set of chromosomes, contain only one-half the amount of DNA found in somatic cells of the same species, which are diploid. Within any given species of higher organism, the amount of DNA per diploid cell is approximately constant from one cell type to another (Table 31-2). The amount of DNA present in DNA-containing animal and bacterial viruses, which have relatively few genes and thus relatively little genetic information, is correspondingly smaller.

Experiments on the replication of bacterial viruses also pointed to DNA as genetic material. Particularly significant were the experiments of A. D. Hershey and M. Chase, reported in 1952. They prepared virus particles that were isotopically labeled either in the protein or in the DNA by in-

Table 31-1 Approximate DNA content of some cells and viruses[†]

Species	DNA, pg per cell (or virion)	Number of nucleotide pairs, millions
Mammals	6	5,500
Amphibia	7	6,500
Fishes	2	2,000
Reptiles	5	4,500
Birds	2	2,000
Crustaceans	3	2,800
Mollusks	1.2	1,100
Sponges	0.1	100
Higher plants	2.5	2,300
Fungi	0.02–0.17	20
Bacteria	0.002–0.06	2
Bacteriophage T4	0.00024	0.17
Bacteriophage λ	0.00008	0.05

† The figures given for eukaryotic organisms are for somatic (diploid) cells.

Table 31-2 DNA content of cells of the chicken

Tissue	DNA, pg per cell
Heart	2.45
Kidney	2.20
Liver	2.66
Spleen	2.55
Pancreas	2.61
Erythrocyte	2.49
Sperm cells (haploid)	1.26

Table 31-3   Base equivalences in DNA (the deviations from precise equivalence are due to experimental error)

	Base composition, mole percent				Base ratio			Asymmetry ratio $\dfrac{A+T}{G+C}$
	A	G	C	T	A/T	G/C	Pu/Py	
Animals								
Man	30.9	19.9	19.8	29.4	1.05	1.00	1.04	1.52
Sheep	29.3	21.4	21.0	28.3	1.03	1.02	1.03	1.36
Hen	28.8	20.5	21.5	29.2	1.02	0.95	0.97	1.38
Turtle	29.7	22.0	21.3	27.9	1.05	1.03	1.00	1.31
Salmon	29.7	20.8	20.4	29.1	1.02	1.02	1.02	1.43
Sea urchin	32.8	17.7	17.3	32.1	1.02	1.02	1.02	1.58
Locust	29.3	20.5	20.7	29.3	1.00	1.00	1.00	1.41
Plants								
Wheat germ	27.3	22.7	22.8	27.1	1.01	1.00	1.00	1.19
Yeast	31.3	18.7	17.1	32.9	0.95	1.09	1.00	1.79
Aspergillus niger (mold)	25.0	25.1	25.0	24.9	1.00	1.00	1.00	1.00
Bacteria								
E. coli	24.7	26.0	25.7	23.6	1.04	1.01	1.03	0.93
Staphylococcus aureus	30.8	21.0	19.0	29.2	1.05	1.11	1.07	1.50
Clostridium perfringens	36.9	14.0	12.8	36.3	1.01	1.09	1.04	2.70
Brucella abortus	21.0	29.0	28.9	21.1	1.00	1.00	1.00	0.72
Sarcina lutea	13.4	37.1	37.1	12.4	1.08	1.00	1.04	0.35
Bacteriophages								
T7	26.0	24.0	24.0	26.0	1.00	1.00	1.00	1.08
λ	21.3	28.6	27.2	22.9	0.92	1.05	1.00	0.79
ϕX174, single strand	24.6	24.1	18.5	32.7	0.75	1.30	0.95	1.34
Replicative	26.3	22.3	22.3	26.4	1.00	1.00	1.00	1.18

cubating the host bacteria in media containing appropriate labeled precursors. With such labeled virus particles they showed conclusively that viral DNA rapidly enters the host cell whereas viral protein does not. These observations were later followed by the demonstration that the viral nucleic acid alone is infectious, in the absence of viral protein, and can lead to the formation of complete viral progeny in the host cell.

Base Equivalences in DNA

One of the most important pieces of biochemical evidence supporting the concept that DNA is the bearer of genetic information was the discovery that the base composition of DNA is related to the species of origin. Before reliable chromatographic methods became available, it was thought that the four major bases found in DNA—adenine, guanine, cytosine, and thymine—occur in equimolar ratio in all DNA molecules. Then, in the period 1949 to 1953, E. Chargaff and his colleagues applied quantitative chromatographic methods to the separation and quantitative analysis of the four bases in hydrolyzates of DNA specimens isolated from different organisms. Some of the data collected by them and by others are shown in Table 31-3. From such data the following important conclusions were drawn:

1 The base composition of DNA varies from one species to another.

2 DNA specimens isolated from different tissues of the same species have the same base composition.

3 The base composition of DNA in a given species does not change with age, nutritional state, or changes in environment.

4 In nearly all DNAs examined, the number of adenine residues is always equal to the number of thymine residues, that is, $A = T$, and the number of guanine residues is always equal to the number of cytosine residues ($G = C$). As a corollary, it is clear that the sum of the purine residues equals the sum of the pyrimidine residues; that is, $A + G = C + T$.

5 The DNAs extracted from closely related species have similar base composition, whereas those of widely different species are likely to have widely different base composition. The base composition of their DNAs can in fact be used to classify organisms.

The Watson-Crick Model of DNA Structure

The base equivalences observed in DNA from different species raised the intriguing possibility that there is a level of structural organization of DNA molecules which is compatible with certain base equivalences and incompatible with others. In fact, it had already been suspected that DNA has a specific three-dimensional conformation. Since solutions of native DNA are highly viscous, it appeared that DNA molecules are long and rigid rather than compact and folded. Moreover, heating freshly isolated DNA produces significant changes in its viscosity and other physical properties, without breaking the covalent bonds in the DNA backbone. Because of such observations and the illuminating discovery of the α-helical arrangement of the polypeptide chains in some fibrous proteins by L. Pauling and his colleagues, which had just been deduced by x-ray diffraction, by the early 1950s it was inevitable that the powerful x-ray method was to be applied to the problem of DNA structure. In fact, fibers of DNA had already been subjected to x-ray diffraction analysis as early as 1938, when W. Astbury and F. O. Bell first observed that fibrous DNA specimens show reflections corresponding to a regular spacing of 0.334 nm along the fiber axis, but the significance of this observation was not clear because the DNA was known to be impure. In the period 1950 to 1953 more refined x-ray data were obtained on fibers from highly purified DNA, particularly through the work of R. Franklin and M. H. F. Wilkins. DNA can be obtained in two forms, A and B, which differ in degree of hydration; the B form is the biologically important one. The B form of DNA was found to possess two periodicities, a major one of 0.34 nm and a secondary one of 3.4 nm (Figure 31-1), reminiscent of the major and minor periodicities observed in α-keratins (page 128). About this time important x-ray data on the dimensions of the purine and pyrimidine bases and of nucleosides were also being collected in a

Figure 31-1

X-ray diffraction photograph of DNA in the B form taken by Rosalind Franklin late in 1952. The reflections crossing in the middle are indicative of helical structure; they are often called the "helical cross." The heavy dark regions at the top and bottom are due to the closely stacked bases perpendicular to the fiber axis. (From James D. Watson, The Double Helix, *Atheneum Publishers, New York, 1968. Copyright © 1968 by James D. Watson.)*

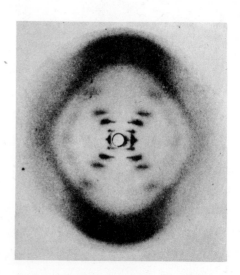

Figure 31-2
The Watson-Crick model for the structure of
DNA.

Dreiding model of the backbone. (W. Büchi.)

Space-filling model. (M. H. F. Wilkins, Medical Research Council, Biophysics Unit, King's College, London.)

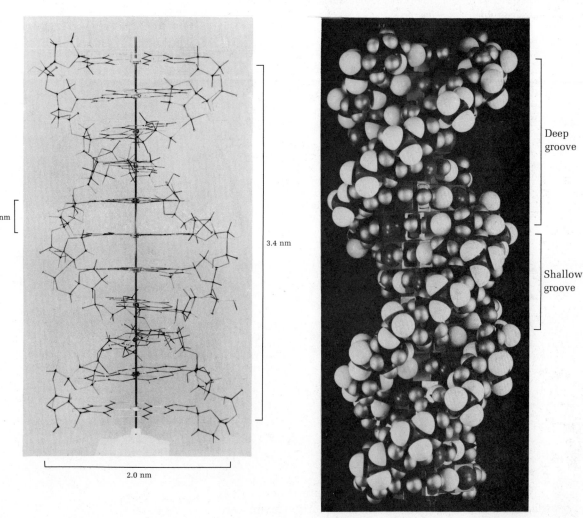

0.34 nm

3.4 nm

2.0 nm

Deep
groove

Shallow
groove

number of laboratories. The stage was now set for the formulation of a three-dimensional conformation of the DNA molecule which could account for the known dimensions of the bases, the observed periodicities in DNA structure, and the equivalence of certain bases. Interest in the problem was greatly heightened by the pervasive question of how the DNA molecule can undergo replication with such great fidelity. If the genetic information is encoded by a specific sequence of bases in DNA, as had been speculated, then how can this sequence be precisely replicated?

In 1953 J. D. Watson and F. H. C. Crick postulated a precise three-dimensional model of DNA structure (see page 854 for their announcement) based on the x-ray data of Franklin and Wilkins and the base equivalences observed by Chargaff. This model not only accounted for many of the observations on the chemical and physical properties of DNA but also suggested a mechanism by which genetic information can be accurately replicated.

Figure 31-3

Scale drawings and space-filling models of the hydrogen-bonded base pairs adenine-thymine and guanine-cytosine. The former has two hydrogen bonds, the latter three. Guanine-cytosine pairs are slightly closer together, more compact, and thus slightly denser than adenine-thymine pairs. [Redrawn from L. Pauling and R. B. Corey, Arch. Biochem. Biophys., 65:164 (1956), Academic Press, Inc., New York.]

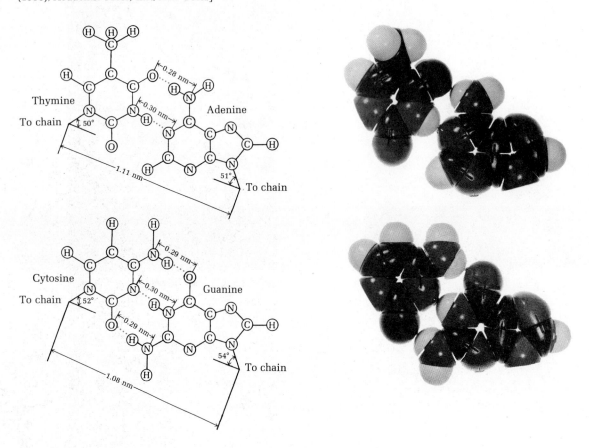

The Watson-Crick model of DNA structure (Figure 31-2) proposed that two right-handed polynucleotide chains are coiled in helical fashion around the same axis, thus forming a double helix. The two chains or strands are *antiparallel;* i.e., their 3′,5′-internucleotide phosphodiester bridges (page 318) run in opposite directions. The coiling of the two chains is such that they cannot be separated except by unwinding the coils; such coiling is termed *plectonemic.* The purine and pyrimidine bases of each strand are stacked on the inside of the double helix, with their planes parallel to each other and perpendicular to the long axis of the double helix. The bases of one strand are paired in the same planes with the bases of the other strand. The pairing of the bases contributed by the two strands is such that only certain base pairs fit inside this structure in such a manner that they can hydrogen-bond to each other. The allowed pairs are A-T and G-C, which are precisely the base pairs showing equivalence in DNA. How they form hydrogen bonds is shown in Figure 31-3. The disallowed purine pair A and G is rather large to fit inside a helix having a thickness of 2.0 nm and the disallowed pyrimidine pair C and T would appear to be too far apart within

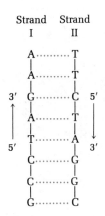

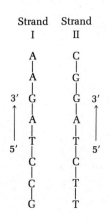

a 2.0-nm helix to form stable hydrogen bonds with each other. The allowed pairs A-T and G-C are not only about the same size, they are also more strongly hydrogen-bonded than A-G and C-T pairs. Thus, the Watson-Crick double helix involves not only the maximum possible number of hydrogen-bonded base pairs but also those pairs giving maximum fit and stability.

To account for the 0.34-nm periodicity observed by x-ray methods, Watson and Crick postulated that the bases are stacked at a center-to-center distance of 0.34 nm from each other. There are exactly 10 nucleotide residues in each complete turn of the double helix to account for the secondary repeat distance of 3.4 nm. DNA can theoretically exist in other helical forms, but they would not have the 0.34-nm repeat distance along the long axis observed in native DNA. The Watson-Crick double helix, which is about 2.0 nm in diameter, has one shallow and one deep groove. The relatively hydrophobic bases are within the double helix, shielded from water; the polar sugar residues and negatively charged phosphate groups are located on the periphery, exposed to water. The double helix is stabilized not only by the hydrogen bonding of complementary base pairs but also by electronic interactions between the stacked bases, as well as by hydrophobic interactions.

The two antiparallel chains of double-helical DNA are not identical in either base sequence or composition (Figure 31-4). Instead, the two chains are complementary to each other: wherever adenine appears in one chain thymine is found in the other and vice versa, and wherever guanine is found in one chain cytosine is found in the other, and vice versa.

The double-helical structure of DNA leads to the second element of the Watson-Crick hypothesis, namely, a mechanism by which genetic information can be accurately replicated. Since the two strands of double-helical DNA are structurally complementary to each other and thus contain complementary information, the replication of DNA during cell division was postulated to occur by replication of the two strands, so that each parent strand serves as the template specifying the base sequence of the new complementary strand. The end result of such a process is the formation of two daughter double-helical molecules of DNA, each identical to that of the parent DNA and each containing one strand from the parent (Figure 31-5).

Some years were to elapse before direct experimental proof emerged for the antiparallel polarity of the two strands of DNA and for the "semiconservative" nature of the replication process described above. Nevertheless, the Watson-Crick hypothesis had an immediate and profound impact on the nature of experimentation designed to unravel the molecular aspects of genetic information transfer. The double-helical, or *duplex*, structure of DNA quickly gained support from many other kinds of evidence. An example may be given here. The length of homogeneous native DNA molecules of known molecular weight, such as the relatively small DNAs of certain bacterial viruses, can be measured accurately and directly by electron microscopy (Figure 31-6). The observed lengths of viral DNA molecules of known molecular weight and their predicted lengths, calculated from the double-helical model, are in very good agreement.

Viral DNA: Rule of the Ring

For some years it was thought that DNA molecules have molecular weights of the order of 10 million on the basis of physical observations made on DNA specimens extracted from bacterial cells or animal tissues with aqueous phenol or neutral salt solutions. However, it is now known that most intact, native DNA molecules are far larger than these early estimates indicated. In fact, native DNA molecules of bacteria and animal cells are so large that they are not readily isolated in intact form, since they are easily broken by shear forces during extraction and handling. Even simple stirring or pipetting of DNA solutions can cause fragmentation of native DNA into pieces with molecular weight of about 10 million.

On the other hand, the DNA molecules of bacterial viruses can easily be obtained in native homogeneous form (Table 31-4); they have served as valuable prototypes of the much larger DNA molecules of bacteria and of eukaryotic cells. The DNAs of the small E. coli bacteriophages lambda (λ) and ϕX174 have received much attention. The former is a linear duplex DNA molecule with a molecular weight of 32 million. Its contour length, which can be measured from electron micrographs of specimens plated out on grids, is about 17.2 μm. The DNA of ϕX174 virus, which is one of the smallest DNA viruses known, is circular and single-stranded (mol wt 1.7 million); however, in the host E. coli cell it is converted into a double-strand replicative form (mol wt 3.4 million) (Figure 31-6). The contour length of the single-strand form is only 1.77 μm. The λ and ϕX174 viruses are of great experimental importance and are discussed in greater detail in Chapter 32. The DNA of the much larger bacteriophage T2 is also shown in Figure 31-6. From the data in Table 31-4 it is seen that the contour lengths of duplex DNA

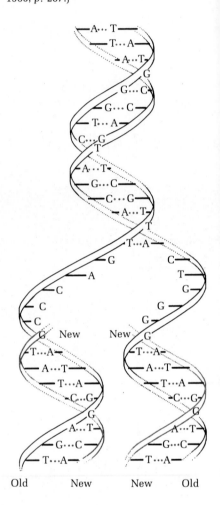

Figure 31-5
Replication of DNA as suggested by Watson and Crick. The complementary strands are separated, and each forms the template for synthesis of a complementary daughter strand (color). (Redrawn from James D. Watson, The Molecular Biology of the Gene, W. A. Benjamin Inc., Menlo Park, Calif., 1965, p. 267.)

Old   New   New   Old

Table 31-4 Dimensions of the DNAs of some viruses

	Contour length, μm	Weight, megadaltons	Percent G + C
E. coli bacteriophages			
ϕX174, single-strand	1.77	1.7	42
ϕX174, duplex	1.64	3.4	42
λ	17.2	32	49
T2	56.0	130	35
T7	12.5	25	48
Animal viruses			
Herpes simplex	35.0	68	68
Polyoma	1.56	3.1	48

Figure 31-6
Electron micrographs of DNA of bacteriophages φX174, λ, and T2.

Two molecules of circular φX174 DNA. The upper is double-stranded, the lower is a "kinky" single-stranded molecule.

T2 DNA. The field also shows an *E. coli* plasmid (page 869), at the left.

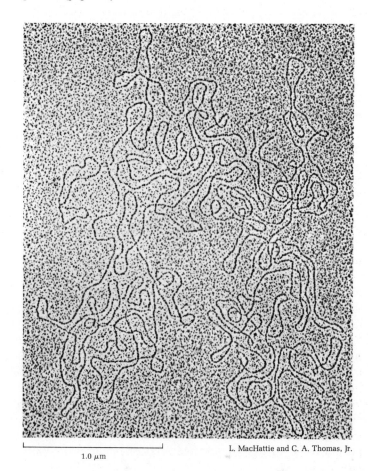

1.0 μm   L. MacHattie and C. A. Thomas, Jr.

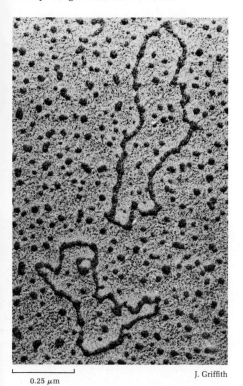

0.25 μm   J. Griffith

λ DNA (circular form)

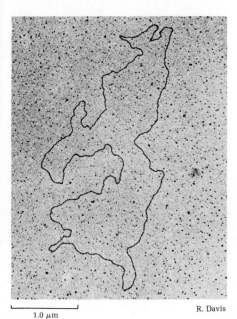

1.0 μm   R. Davis

molecules are related in a simple manner to their molecular weights; 1.0 μm in contour length is equivalent to about 2.0 megadaltons and about 3,000 base pairs.

Viral DNAs are in most cases linear duplexes, although φX174 DNA is a single-strand circle. However, the linear viral DNAs show some characteristic structural features indicating that they were originally formed as circles or that they are converted into circles before replication; this concept is the "rule of the ring." Some linear viral DNAs, such as that of phage λ, have *cohesive* ("sticky") *ends*. The 5′ ends of these duplex DNAs project as single strands beyond the 3′ ends; twelve nucleotides on these ends are complementary and thus pair readily to form circles, which can then be covalently sealed by DNA ligase (page 902). Moreover, some linear viral DNAs show *terminal repetition*, the presence of a base sequence at the 3′ end that repeats the sequence at the 5′ end of the strand. When such sequences are acted upon by a nuclease, they lead to the formation of complementary cohesive ends and thus to covalent circles. Another unique structural feature of several viral DNAs is that they do not have unique base sequences, consisting instead of a population of molecules whose sequences are <u>circular permutations</u> of each other. Circular permutations are illustrated by the following examples of three linear duplex DNAs, in which the bases are numbered:

$$1 \text{-} 2 \text{-} 3 \text{-} 4 \text{-} 5 \text{-} 6 \text{-} 7 \text{-} 8 \text{-} 9 \text{-} 10$$
$$1' \text{-} 2' \text{-} 3' \text{-} 4' \text{-} 5' \text{-} 6' \text{-} 7' \text{-} 8' \text{-} 9' \text{-} 10'$$

$$3 \text{-} 4 \text{-} 5 \text{-} 6 \text{-} 7 \text{-} 8 \text{-} 9 \text{-} 10 \text{-} 1 \text{-} 2$$
$$3' \text{-} 4' \text{-} 5' \text{-} 6' \text{-} 7' \text{-} 8' \text{-} 9' \text{-} 10' \text{-} 1' \text{-} 2'$$

$$7 \text{-} 8 \text{-} 9 \text{-} 10 \text{-} 1 \text{-} 2 \text{-} 3 \text{-} 4 \text{-} 5 \text{-} 6$$
$$7' \text{-} 8' \text{-} 9' \text{-} 10' \text{-} 1' \text{-} 2' \text{-} 3' \text{-} 4' \text{-} 5' \text{-} 6'$$

These circularly permuted linear DNAs will obviously form identical circular molecules, following exposure of complementary cohesive ends by exonuclease action and the action of DNA ligase (page 902).

Another characteristic of viral DNAs is that some of their bases are modified. The T-even phages contain hydroxymethylcytosine (page 312) instead of cytosine; moreover, the hydroxyl group of hydroxymethylcytosine is often glucosylated. These modifications protect the viral DNA from degradation by the host-cell endonucleases (page 881).

Although the term "chromosome" originally referred to the structures observed microscopically in eukaryotic nuclei following specific staining procedures, it is now more loosely used to refer to the genetic material of bacteria, as well as that of viruses.

Bacterial DNA

Intact, native DNA molecules of *E. coli* and other bacteria, because of their great size, are difficult to isolate and study by standard hydrodynamic methods, but some success has been achieved. Bacterial DNA molecules have been visualized not only by electron microscopy but also by autoradiography (Figure 31-7). The chromosome of *E. coli* appears to consist of a single, enormous double-strand DNA molecule with a molecular weight of about 2.8×10^9, a thickness of about 2.0 nm, and a contour length of about 1,360 μm (1.36 mm). *E. coli* DNA contains about 4 million nucleotide pairs (each pair has an average molecular weight of 670). This enormous DNA molecule is a closed loop, loosely called a circle, in accordance with the fact that the genetic map of the single chromosome of *E. coli* has been found to be circular (page 878). The bacterial chromosome is tightly and compactly looped in the nuclear zone, held together by some RNA, which forms the backbone or core of the condensed chromosome. The DNA molecule is gathered into 50 or more highly twisted or supercoiled loops (Figure 31-7).

Figure 31-7
The chromosome of E. coli *consists of a very large supercoiled (page 870) circular duplex, looped many times and held together by some RNA (color). Successive single-strand nicks (arrows) cause the supercoiled loops to relax, until the entire molecule is unlooped to yield a relaxed duplex circle, which is undergoing replication (page 871). [Adapted from A. Worcel and E. Burgi, J. Mol. Biol., 71: 143 (1972).]*

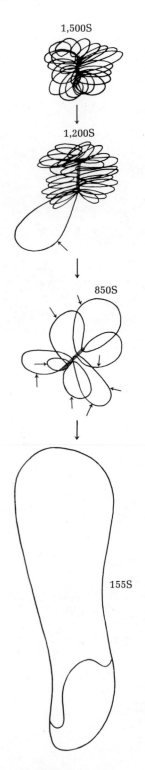

1,500S

1,200S

850S

155S

In addition to their single large circular chromosomes, most bacterial cells contain anywhere from 1 to 20 much smaller circular duplex DNA molecules called *plasmids*. They resemble viral DNAs in size, ranging from about 5 to 100 megadaltons. They are replicated in fixed numbers along with the chromosome. Their function is not clearly known, but some of them, called *episomes*, become incorporated into the host-cell chromosome. Plasmids also may be transferred from one bacterial cell to another during sexual conjugation and then confer new phenotypic characteristics on the recipient cell.

Eukaryotic Chromosomes

The nucleus of eukaryotic cells contains several or many nuclear chromosomes, depending on the species (Table 31-5). Each chromosome contains one very large DNA molecule. Very little information is available on the size of individual DNA molecules in cells of higher organisms; however, the largest chromosome of the fruit fly *Drosophila* has been found to be a linear molecule of double-strand DNA about 4.0 cm long, with a molecular weight of nearly 80×10^9, nearly 40 times larger than an *E. coli* chromosome. The total length of all the DNA present in a single mammalian cell has been calculated to be about 2 m, assuming that it is double-helical. This is equivalent to about 5.5×10^9 base pairs.

The ultrastructural organization of DNA within the nucleus of eukaryotic cells is highly complex and undergoes dramatic changes during the cell cycle (page 994). In *interphase*, the period between mitoses, the eukaryotic nucleus contains an irregular network of *chromatin*, so named because of its tendency to stain deeply with basic dyes (Figure 1-9). At this time the chromosomes are not demarcated. The chromosomes become most clearly visible during mitosis, at the end of prophase.

The chromatin fibers of somatic cells are about 20 nm in diameter. They contain two major components in about equal amounts, DNA and basic proteins called *histones*. In addition, chromatin contains some acidic nonhistone proteins, some enzyme proteins, such as DNA polymerase, as well as nuclear RNA and some lipids. Histones are relatively small, highly basic (i.e., at pH 7.0 positively charged) proteins with molecular weights in the range 10,000 to 20,000. There are five major classes, differentiated on the basis of their relative content of lysine and arginine (Table 31-6). The amino acid sequences of some histones are known. Some of the serine hydroxyl groups of histones are enzymatically phosphorylated through the action of protein kinase (page 812). Histone molecules are believed to be regularly arranged in the deep groove of the DNA double helix (page 996). The recurring positive charges of the histones form electrostatic associations with the negatively charged phosphate groups of DNA, which render the DNA more stable and flexible. The double helix of DNA, together with the associated histones, is supercoiled and folded back and forth many times in the chromatin fibers. These fibers in turn form a number of loops

Table 31-5   Normal chromosome number in different species

Prokaryotic organisms (haploid)	
Bacteria	1
Eukaryotic organisms (diploid)	
Drosophila	8
Red clover	14
Garden peas	14
Honeybee	16
Corn	20
Frog	26
Hydra	30
Fox	34
Cat	38
Mouse	40
Rat	42
Rabbit	44
Man	46
Chicken	78

Table 31-6   Classes of histones

Two different sets of designations, depending on the method of fractionation, are used to refer to the five groups of histones; their compositions
vary somewhat from one species to another

Group	Most abundant amino acids, %	Number of residues	Mol. wt.
I = f1	Lys 27	207	21,000
IIb2 = f2b	Lys 16	125	13,800
IIb1 = f2a2	Lys 11 Arg 9	129	15,000
III = f3	Arg 15 Lys 10	135	15,342
IV = f2a1	Gly 15 Arg 14	102	11,000

held together at the centromere. It has been calculated that
human chromosome 13 contains a DNA molecule about
32,000 μm (3.2 cm) long, looped to form a *chromatid* about 6
μm long and 0.8 μm in diameter. The histones appear to be
involved in the supercoiling of the DNA in eukaryotic chromosomes, either to serve as structural elements or to cover or
repress specific segments of DNA so that they are not transcribed (page 996). Such segments can be unwound and thus
exposed for transcription in response to certain molecular
signals (page 999). Figure 31-8 shows an electron micrograph
of one of the 46 chromosomes of a human cell. Important
clues to the molecular organization of the chromatin fibers
have come from examination of the unusual giant chromosomes of cells in the salivary gland of *Drosophila*. They contain extended segments of DNA-protein fibers separated by
bands in which the fibers are highly folded (Figure 31-9). It
has been suggested that the "puffs" in the folded regions are
loops of bare, unwound DNA exposed for transcription. The
enzymatic replication and transcription of eukaryotic chromosomes, which will be discussed elsewhere (pages 912 and
922), pose many topological and geometrical problems.

Mitochondria of eukaryotic cells contain DNA that is quite
different from that found in the nucleus. Mitochondrial DNA
is double-stranded and circular, but it has a molecular
weight of only about 10 million. In size it therefore resembles the DNA of the small bacteriophages. Mitochondrial
DNA also differs in base composition from nuclear DNA of
the same cell. There are about four to six molecules of DNA
per mitochondrion, but the total mitochondrial DNA is much
less than 1 percent of the total cellular DNA in somatic cells.
Chloroplasts of higher plants also contain DNA. A small fraction of the mitochondrial DNA occurs in the form of in-
terlocked circles called *catenated dimers* (Figure 31-10),
which are believed to be formed during replication of mitochondrial DNA.

Supercoiling of DNA Circles

Circular duplex DNAs carefully isolated in native form from
viruses, bacteria, and mitochondria are often *supercoiled* or

Figure 31-8
*Electron micrograph of human chromosome
12. Human chromosomes contain about 15
percent DNA, 10 percent RNA, and 75 percent protein. They consist of long chromatin
fibers about 20 to 30 nm in diameter and
about 700 to 800 μm long. Each chromatin
fiber is believed to contain one DNA duplex
about 4 cm long in supercoiled arrangement
around which the protein is wound
(33,750×). (From E. J. DuPraw, DNA and
Chromosomes. Copyright 1970 by Holt,
Rinehart and Winston, Inc.)*

Figure 31-9
Structural aspects of eukaryotic chromo-
somes. It is not yet possible to provide a
simple, satisfactory model for the structure
of chromosomes, which have been aptly de-
scribed as looking like a bad day in a maca-
roni factory. (Above) Each chromosomal
fiber (see Figure 31-8) appears to be made up
of a duplex DNA surrounded by histones
and is tightly coiled, as suggested in the
models by G. P. Georgiev (left) and B. M.
Richards and J. F. Pardon (right). (Below)
During replication such coiled fibers become
unwound, as suggested by Georgiev (left)
and D. E. Comings and T. A. Okada (right).

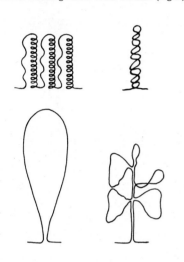

supertwisted (Figure 31-11). Supercoiling results when the DNA contains more base pairs per unit length than normally found in the double helix. As a consequence, the entire circular DNA structure compensates or relieves this torsional stress by twisting in the opposite or negative sense. Such supercoiled DNAs can be "relaxed" by a break in one strand, followed by resealing (page 902), by the insertion of certain chemical agents, such as *ethidium bromide* (page 920), between base pairs, or by binding certain proteins (page 909).

Properties of DNA in Solution

Solutions of small native DNA molecules or of fragments of large DNA molecules can be studied quite readily by physicochemical methods; they share a number of characteristic properties.

Acid-Base Properties

The recurring secondary phosphate groups of DNA, which constitute the bridges between adjacent mononucleosides, have a rather low pK' and are fully ionized at any pH above 4. DNA is thus strongly acidic. These phosphate groups, as we have seen, are located on the outer periphery of the double helix, exposed to water. They strongly bind divalent cations such as Mg^{2+} and Ca^{2+}, as well as the polycationic amines *spermine* and *spermidine* (Figure 31-12), which are found associated with the DNA in many viruses and bacteria. The binding of polyamines in the groove of double-helical DNA both stabilizes the DNA molecule and makes it more flexible.

Since the hydrogen-bonding properties of the different bases depend on their ionic form, which in turn depends on pH (see Chapter 12), the stability of the hydrogen-bonded base pairs of double-helical DNA is a function of pH. The base pairs are maximally stable between pH 4.0 and 11.0. Outside these unphysiological limits, double-helical DNA becomes unstable and unwinds.

Figure 31-10
Forms of circular mitochondrial DNA.

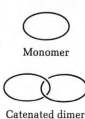

Monomer

Catenated dimer

Figure 31-11
Supercoiling of circular DNA.

A supercoiled duplex circle.

The result of relieving the superhelical twist by nicking one strand.

Single-strand nick

Viscosity

Because of the rigidity of the double helix and the immense length of DNA in relation to its small diameter, even very dilute DNA solutions are highly viscous. Viscosity measurements are often used to follow the course of unwinding or denaturation of duplex DNA molecules. There is another consequence of the immense length of DNA molecules. When they diffuse, they sweep with them a relatively enormous volume of solution, more than 10,000-fold greater than their own volume. For this reason DNA shows ideal behavior as a solute only in extremely dilute solutions.

Sedimentation Behavior

The sedimentation coefficient and the molecular weight of DNA specimens can be determined by the ultracentrifugal methods described earlier for the proteins (page 174). However, because of the extremely elongated nature of the DNA molecule and the high viscosity of DNA solutions, sedimentation measurements are carried out in a series of low concentrations of DNA and the sedimentation coefficient extrapolated to zero DNA concentration. The sedimenting boundary of DNA is usually detected by measuring the optical absorption at a wavelength of 260 nm, at which DNA absorbs strongly (page 314). Molecular weights of DNA molecules can also be obtained by comparing their rate of sedimentation in a sucrose density gradient (page 158) with the rate given by a DNA sample of known size and sedimentation coefficient.

Equilibrium sedimentation (page 176) in cesium chloride gradients is very widely used to determine the _buoyant density_ of DNA molecules. When a concentrated CsCl solution (8.0 M) is centrifuged to equilibrium in a high gravitational field (page 174), the CsCl becomes distributed in a linear gradient down the tube; at the top of a 1-cm column the density of the solution is about 1.55 g ml^{-1} and at the bottom about 1.80 g ml^{-1}. When DNA is present during formation of such a gradient, it concentrates into a stable band at that position in the tube at which its buoyant density is exactly equal to the density of the CsCl solution (Figure 31-13). The density of the DNA can be calculated directly or by comparison with the density of a known standard DNA specimen centrifuged in the same gradient. Single-strand DNA is denser in such a CsCl gradient than double-strand DNA, which in turn is denser than proteins in general. RNA can be distinguished from DNA since it is denser than either single- or double-strand DNA.

Most important, however, is the fact that buoyant-density measurements provide information on the base composition of the DNA specimen, because G-C base pairs, which are joined by three hydrogen bonds, are more compact and dense than A-T pairs, which are joined by only two hydrogen bonds. The buoyant density of DNA specimens in a CsCl gradient is a linear function of the ratio of G-C to A-T pairs (Figure 31-14). The CsCl gradient method has also revealed that the intact homogeneous DNAs of viruses give

Figure 31-12
Spermine and spermidine.

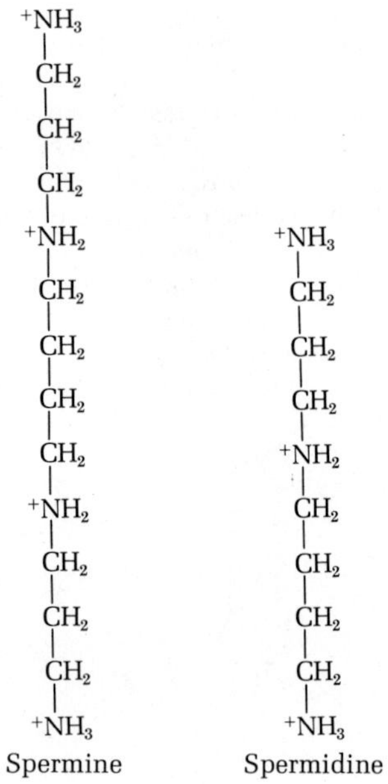

Figure 31-13
Determination of buoyant density.

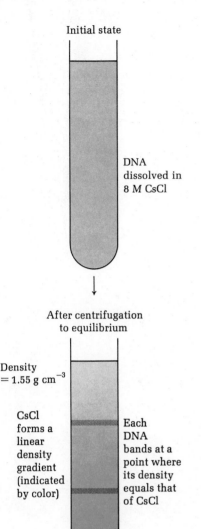

Initial state

DNA
dissolved in
8 M CsCl

After centrifugation
to equilibrium

Density
$= 1.55$ g cm^{-3}

CsCl
forms a
linear
density
gradient
(indicated
by color)

Each
DNA
bands at a
point where
its density
equals that
of CsCl

Density
$= 1.80$ g cm^{-3}

very sharp bands, whereas random heterogeneous DNA fragments from cells of higher animals give broad bands with a wide density range.

Denaturation of DNA

Solutions of the double-helical form of DNA undergo significant changes in a number of physical properties when subjected to (1) extremes of pH, (2) heat, (3) decrease in dielectric constant of the aqueous medium by alcohols, ketones, etc., and (4) exposure to urea, amides, and similar solutes. The viscosity of DNA solutions decreases, the light absorption at 260 nm increases, the optical rotation becomes more negative, and the buoyant density increases. This physical change in DNA is called *denaturation;* note that denaturing agents for DNA are identical to those producing denaturation or unfolding of globular proteins. During denaturation of DNA no covalent bonds are broken; instead the double-helical DNA structure unwinds and separates. Two sets of forces are responsible for maintaining the double-helical structure of DNA: (1) hydrogen bonding between base pairs and (2) stacking interactions between successive bases. When either or both sets of forces are interrupted, the native, double-helical structure undergoes transition into a randomly looped form, denoted as *single-stranded* or *denatured* DNA.

Figure 31-14
The relationship between buoyant density of
double-strand DNA and base composition.
[Data from C. L. Schildkraut, J. Marmur, and
P. Doty, J. Mol. Biol., 4: 430 (1962).]

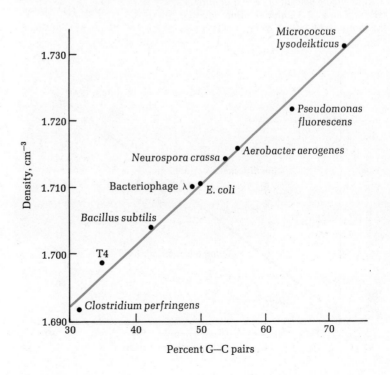

The Hyperchromic Effect

The physical change that is simplest to use as a measure of denaturation of DNA is the increase in light absorption at 260 nm, termed the <u>hyperchromic effect</u>. As pointed out in Chapter 12, the purine and pyrimidine bases and their corresponding nucleotides absorb ultraviolet light at 260 nm. However, native, double-strand DNA absorbs much less light at 260 nm than would be predicted from the summation of the light absorbed by its constituent mononucleotides. If native double-strand DNA is denatured by heating, there is a large increase in the light absorbed at 260 nm, up to 40 percent, depending on the DNA specimen. The total light absorption of fully denatured DNA is nearly equal to that of an equivalent number of the corresponding free mononucleotides. The percentage increase in light absorption at 260 nm produced by heating a native DNA sample is directly related to its content of A-T base pairs: the higher the proportion of A-T pairs, the greater the increase in light absorption (Figure 31-15).

The less-than-additive light absorption by double-strand DNA molecules, which is called <u>hypochromism</u>, is due to the electronic interactions between the stacked bases in the native double-helical structure, which diminish the amount of light each residue can absorb. When the double-helical structure is disordered, the bases "unstack" and in this less hindered form absorb nearly as much light as they would if they were present as free nucleotides.

Figure 31-15
Increase in absorption at 260 nm given by some DNA specimens on heating. The magnitude of the hyperchromic effect correlates with the mole percent of A-T base pairs.
[Data from H. R. Mahler, B. Kline, and B. D. Mehrotra, J. Mol. Biol., 9: 801 (1964).]

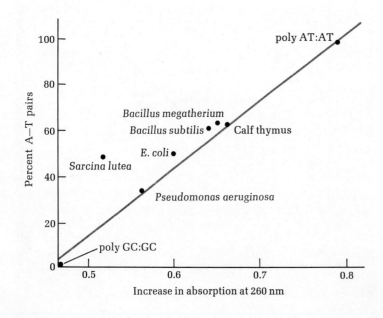

The Melting Point of DNA

Native DNA molecules usually denature within a very small increment of temperature (Figure 31-16); the sharpness of the transition is reminiscent of the melting point of a simple organic compound. In fact, thermal denaturation of DNA is often designated as melting. DNA specimens from different cell types have characteristically different melting points, defined as T_m, the temperature at the midpoint of the melting curve (Figure 31-16). T_m increases in a linear fashion with the content of G-C base pairs, which have three hydrogen bonds and are thus more stable than A-T pairs. The higher the content of G-C pairs, the more stable the structure and the more thermal energy required to disrupt it. Careful determination of the melting point of a DNA specimen, under fixed conditions of pH and ionic strength, can give a remarkably good estimate of its base composition.

Stages in Denaturation of DNA

There are two stages in the denaturation of duplex DNA molecules. In the first stage, the two strands are partially unwound but remain united by at least a short segment of double-helical structure, with some complementary bases still in register. The unwound segment of each strand assumes a random, changing conformation. In the second stage

Figure 31-16

Melting points (T_m) of DNA and their relationship to content of G-C pairs. [Redrawn from P. Doty in D. J. Bell and J. K. Grant (eds.), "The structure and biosynthesis of macromolecules," Biochem. Soc. Symp., 21: 8, Cambridge University Press, London, 1962.]

Characteristic melting curves for a viral DNA and a bacterial DNA. The former, being smaller, shows a sharper melting transition. The two samples differ in G-C content.

A plot of T_m for 40 different DNA specimens from plant, animal, and viral sources against G-C content. All samples were measured under identical conditions.

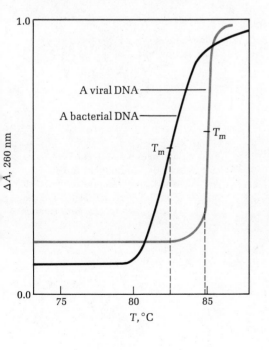

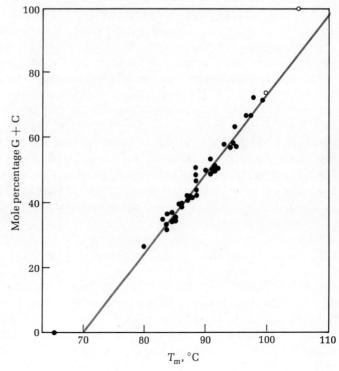

the two strands become completely separated from each other. When quickly cooled in this stage, each free strand doubles back on itself to form short, imperfect intrachain double-helical segments (Figure 31-17).

Denaturation of homogeneous DNA is easily reversible if the unwinding process has not gone past the first stage. As long as a double-helical segment unites the two strands, with base pairs in register, the unfolded segments of the two strands will spontaneously rewind to reform a complete duplex when the temperature is lowered. They literally snap back into their native conformation, which is the minimum-free-energy form. However, if the two strands have separated completely, renaturation is much slower.

The melting of homogeneous double-strand DNA shows a very sharp temperature transition because of the cooperative nature of the many successive interactions that occur, in which each interaction greatly increases the tendency of the next to take place (page 42). The rate of unwinding and rewinding of double-strand DNA has received much study, since these processes occur during the replication of DNA in the intact cell (Chapter 32).

When homogeneous DNA molecules are subjected to carefully controlled denaturing conditions, their less-stable segments, i.e., those richer in A-T pairs, open first. Electron microscopy of such partially denatured DNAs allows mapping of the segments rich in A-T pairs (Figure 31-18).

Mapping of Chromosome Structure

Genetic mapping of the relative positions of different genes in bacterial chromosomes is made possible by the fact that under special circumstances a segment of the chromosome of one type of cell, containing one or more genes, may become detached and transferred to the chromosome of a second type of cell, into which it becomes covalently incorporated. The identity of the recombined gene(s) can be deduced from changes in the phenotype of the acceptor cell and its progeny. Such recombination of genes is a common event in eukaryotic organisms, which usually reproduce by sexual conjugation, but it is relatively infrequent among prokaryotes. However, experimental conditions can be chosen that greatly increase the frequency of recombination processes in bacteria, thus yielding a powerful tool for mapping the relative positions of genes in bacterial chromosomes. Four types of genetic recombination process are useful in mapping bacterial and viral chromosomes: (1) *sexual conjugation,* (2) *transformation,* (3) *viral transduction,* and (4) *viral recombination.*

The first process, *sexual conjugation,* consists of the transfer of some or all of the chromosome of a "male," or (+), cell into a "female," or (−), cell, presumably through one of the pili or fimbriae (Chapter 1), which serves as a hollow connecting tube between the two cells (Figure 31-19). Conjugating E. coli cells require about 90 min under specified conditions for complete transfer of the (+) chromosome. The union between the cells and thus the conjugation process

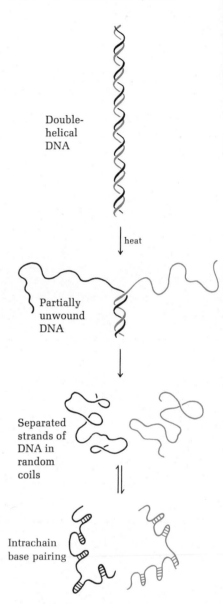

Figure 31-17
Stages in the denaturation of DNA.

Double-helical DNA

heat

Partially unwound DNA

Separated strands of DNA in random coils

Intrachain base pairing

Figure 31-18
Schematic diagram of a linear DNA molecule selectively denatured at A-T-rich regions, which are less stable than G-C-rich regions. The "bubbles" correspond to the A-T-rich regions, which can be correlated with the genetic map.

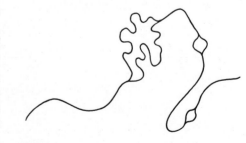

Figure 31-19

*Conjugation in E. coli. Schematic represen-
tation of the conjugation process, which
begins by incorporation of the sex factor F
into the chromosome to yield a (+) cell. The
cells come together, and one strand of the
chromosome of the (+) cell enters the (−) cell
through the conjugation bridge. When the
bridge is broken, the part of (+) DNA that
entered the (−) cell may become incorporated into the (−) chromosome, with loss of a
corresponding segment which is degraded
by nuclease action. From new phenotypes
introduced into the (−) cell after different
time periods of conjugation, mapping of the
chromosome is possible.*

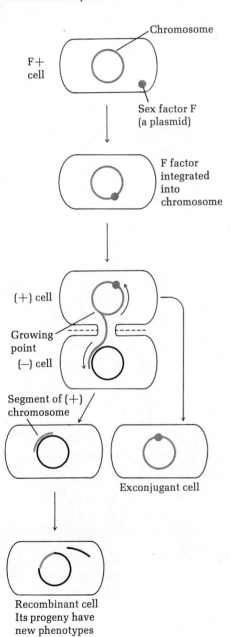

F+
cell

Chromosome

Sex factor F
(a plasmid)

F factor
integrated
into
chromosome

(+) cell

Growing
point

(−) cell

Segment of (+)
chromosome

Exconjugant cell

Recombinant cell
Its progeny have
new phenotypes

can be interrupted at any time by agitating the cell suspension in a blender. In this way (−) cells carrying varying
amounts of male chromosomal material can be recovered,
depending on the time of interruption. From the type and
number of phenotypic traits specified by the acquired segments of the male chromosome in the asexual progeny of the
(−) cells, as a function of the time of entry, the relative position of the genes specifying these traits in the inserted male
chromosome can be deduced. One of the most significant
conclusions derived from such sexual conjugation experiments is that the chromosome of *E. coli* must consist of an
endless, closed loop (Figure 31-20). The genetic map of bacterial chromosomes is thus circular.

Another approach to mapping genes in the chromosomes
of bacteria is made possible by the process of bacterial *transformation.* As we have seen (page 859), a recipient bacterial
cell can be genetically transformed by addition of exogenous
DNA derived from another strain. This effect is due to the
incorporation of one or more genes from the donor DNA into
the chromosome of the recipient cell. By determining the
frequency with which two or more phenotypes from the
donor cells appear together in the progeny of the recipient
cell it is possible to deduce the relative positions of the two
corresponding genes in the donor chromosome. Pairs of
genes incorporated into the recipient cells with high
frequency are closer together in the chromosome than pairs
of genes that enter together infrequently.

More useful for genetic mapping is the phenomenon of
transduction, in which a small portion of the chromosome of
a bacterial cell infected with a virus becomes incorporated
into the DNA molecules of the viral progeny. When the
progeny viral particles infect another cell, the genes derived
from the original host cell may be incorporated into the
chromosome of the new host cell. In effect, the transducing
viral particle carries a portion of the chromosome of one bacterial cell to the chromosome of another (Figure 31-21).
Transduction is of great importance in fine-structure genetic
mapping, since the transducing phage usually carries only a
very small portion of a bacterial chromosome, of the order of
1 percent. Thus transduction is particularly suitable for mapping very small segments of the chromosome rather than its
gross structure.

Fine-structure genetic mapping may be carried out on
viral chromosomes. Bacterial viruses undergo spontaneous
or induced mutations. When the chromosomes of two viral
particles (one normal and the other a mutant, or two different mutants) enter the same host cell, they can fragment
and then recombine in the course of viral replication to form
hybrid viral chromosomes. Such hybrid chromosomes can
then direct the formation of progeny which contain genes
derived from both parent viruses, recognizable by their phenotypic features. From such viral recombination experiments
it has been deduced that the genetic maps of T4 and other
bacteriophages of *E. coli* are also circular. We shall now see
that viral-recombination experiments have also permitted
some far-reaching conclusions concerning the size of genes.

Figure 31-20
The circular genetic map of E. coli K 12 *derived from sexual conjugation experiments. The scale is in terms of the time of entry of the male chromosome during conjugation; 89 min is required for the entire chromosome to enter. The positions of only a few genes are shown. The key indicates the effect of mutation of the gene. [Redrawn from A. L. Taylor and C. D. Trotter, Bacteriol. Rev., 31: 332 (1967).]*

Key

  Leu = *requires leucine for growth*
 ArgF = *requires arginine for growth*
  Lac = *unable to use lactose*
  Gal = *unable to use galactose*
  Trp = *requires tryptophan for growth*
  His = *requires histidine for growth*
PheA = *requires phenylalanine for*
      growth
LysA = *requires lysine for growth*
StrA = *resistant to streptomycin*
 Rha = *unable to use rhamnose*
 Thi = *requires thiamine*
MetA = *requires methionine*

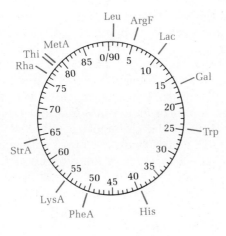

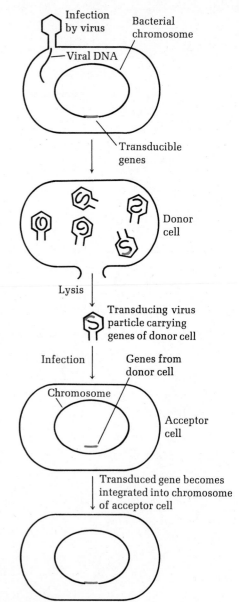

Figure 31-21
Transfer of genes from donor to acceptor bacterial cells by transduction via a carrier phage.

The Size of Genes and Mutational Units

In a classic genetic study carried out by S. Benzer, valuable information on the size of genes and mutational units was obtained. Benzer examined a small region (rII) of the T4 chromosome by the viral-recombination method. There is a class of viral mutants affecting this region which can be readily grown in one bacterial host strain (permissive) but not in others (nonpermissive). When nonpermissive cells were infected with two different mutants in the rII region, some recombinants between the two mutants were found which had regained the wild-type ability to grow in that host. That is, the "good" portions of the defective genes from two different T4 mutants came together to form a single "good" gene. Because many different pairs of T4 mutants were found to recombine to yield "good" rII genes, Benzer concluded that each gene must contain many mutable sites, which he was able to map in a linear sequence. This discovery was important because it had previously been thought that the gene itself was the unit of mutation, i.e., that each gene contained only one mutable site. From the known length of the DNA molecule in the T4 viral particle and the known small fraction of the genetic map occupied by the rII region, the rII A gene of the T4 chromosome was calculated to consist of about 2,500 nucleotide pairs (Figure 31-22). Benzer found nearly 500 different mutants of the rII A region, indicating that the unit of mutation is no greater than about five nucleotides. Later, through other genetic tests, Benzer and other investigators came to the conclusion that mutation may involve a chemical change in only a single nucleotide. Such mutations are called *point mutations*. Their molecular nature is discussed below.

These conclusions, which derived from genetic experimentation, were confirmed and much more precisely stated by the results of subsequent biochemical investigations. Most important was the discovery that each amino acid is coded for by a specific triplet of nucleosides (Chapter 33). Thus a gene specifying a single polypeptide chain of a protein may have anywhere from about 300 to 5,500 or more nucleotide pairs, corresponding to polypeptide chains having from 100 to 1,800 amino acid residues, the range of chain length of the polypeptides in most proteins. Since a given protein or enzyme may contain two or more polypeptide chains, each coded by a different gene, the original one gene–one enzyme hypothesis has been modified into the *one gene–one polypeptide chain* concept. The term *cistron* is also used to refer to genes.

The genes specifying transfer RNA molecules have fewer than 100 nucleotide units (page 935). Also among the smaller genes are those specifying the sequence of amino acids in short polypeptides, such as that of adrenocorticotropic hormone (ACTH), which has only 39 residues.

Molecular Nature of Mutation

Various types of chemical change in DNA that can lead to mutant gene products have been identified. Single-point mutations can be divided into four major classes on the basis of the change produced in the DNA by the mutagenic agent (Figure 31-23). In *transitional* mutants, one purine-pyrimidine base pair is replaced by another, that is, A-T for G-C or G-C for A-T, so that a purine in one chain is replaced by another purine and a pyrimidine in the other chain by another pyrimidine. Transitional mutations occur spontaneously but may also be induced by base analogs, in particular 5-*bromouracil* (BU) and 2-*aminopurine* (AP). Since 5-bromouracil has a close structural resemblance to thymine, it is readily inserted into DNA during its replication (as 5-bromouridylic acid) in the positions normally occupied by thymine. However, the keto form of 5-bromouracil, which pairs readily with A, may undergo tautomerization to the enol form, which more readily pairs with G. In this way replacement of T by BU in one chain may lead to incorporation of G rather than A in the complementary chain. When the latter chain in turn is replicated, the G now specifies C in the new complementary chain. The analog 2-aminopurine acts similarly; it can be read either as A or G. Transitional mutations are also produced by nitrous acid; it deaminates adenine, whose normal partner is T, to form *hypoxanthine* (page 741), which pairs with C.

In *transversional* mutations a purine-pyrimidine pair is replaced by a pyrimidine-purine pair (Figure 31-23). Transversions are common in spontaneous mutations; in fact,

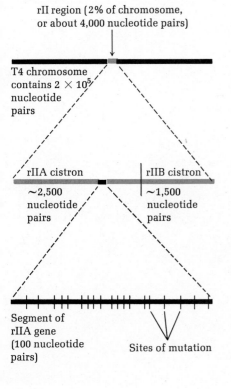

Figure 31-22
The fine structure of the rII region in the genetic map of bacteriophage T4. (Redrawn from J. D. Watson, The Molecular Biology of the Gene, W. A. Benjamin, Inc., Menlo Park, Calif., 1965.)

rII region (2% of chromosome, or about 4,000 nucleotide pairs)

T4 chromosome contains 2 × 10^5 nucleotide pairs

rIIA cistron

~2,500 nucleotide pairs

rIIB cistron

~1,500 nucleotide pairs

Segment of rIIA gene (100 nucleotide pairs)

Sites of mutation

Figure 31-23
Types of point mutations.

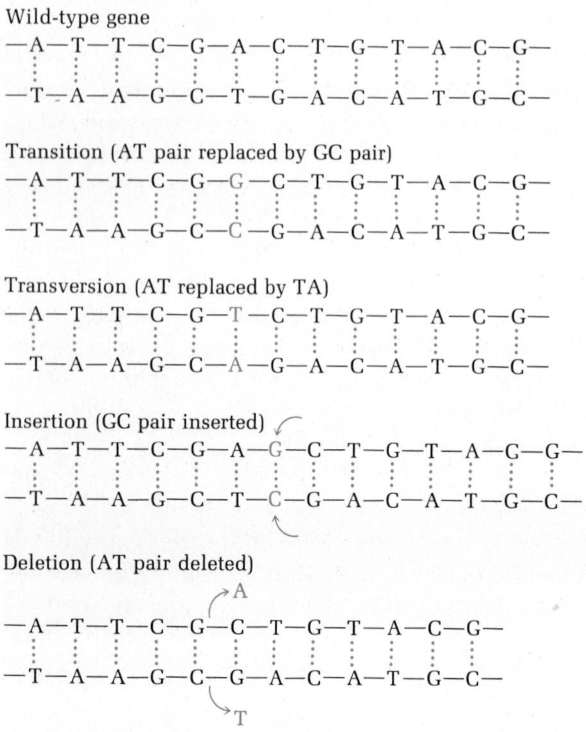

Wild-type gene

—A—T—T—C—G—A—C—T—G—T—A—C—G—

—T—A—A—G—C—T—G—A—C—A—T—G—C—

Transition (AT pair replaced by GC pair)

—A—T—T—C—G—G—C—T—G—T—A—C—G—

—T—A—A—G—C—C—G—A—C—A—T—G—C—

Transversion (AT replaced by TA)

—A—T—T—C—G—T—C—T—G—T—A—C—G—

—T—A—A—G—C—A—G—A—C—A—T—G—C—

Insertion (GC pair inserted)

—A—T—T—C—G—A—G—C—T—G—T—A—C—G—

—T—A—A—G—C—T—C—G—A—C—A—T—G—C—

Deletion (AT pair deleted)

A

—A—T—T—C—G—C—T—G—T—A—C—G—

—T—A—A—G—C—G—A—C—A—T—G—C—

T

about half of the mutations of the α and β chains of hemoglobin (Chapter 5) are transversions.

The third type of point mutation involves *insertion* of an extra base pair. The result is one type of *frame-shift mutation* (Figure 31-23), so called because the normal reading-frame relationship for readout of nucleotide triplets is put out of register by the mutation (Chapter 34). Insertions are readily induced by treating cells with acridine derivatives. *Acridine* (Figure 31-24) is a planar heterocyclic molecule. Much evidence suggests that, because of its flat, aromatic structure, acridine becomes noncovalently inserted, or *intercalated,* between two successive bases in DNA, separating them physically. When this chain is replicated, an extra base is covalently inserted into the complementary chain opposite the intercalated acridine. When this chain is then replicated in turn, the new complementary chain will also contain an extra base.

The fourth type of mutation results from a *deletion* of one or more bases from the DNA; it too is a frame-shift mutation (Figure 31-23). A deletion may result from hydrolytic loss of a purine base because of high pH or temperature, by action of covalent cross-linking reagents, or by alkylating or deaminating agents, which cause formation of bases that cannot pair. The compound *proflavin* also induces deletions.

Not all point mutations in DNA are necessarily lethal. Transitions and transversions are relatively benign, because they result in replacement of only one amino acid in the pep-

Figure 31-24
Acridine.

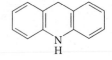

Figure 31-25
The structure of the thymine dimer. Ultraviolet irradiation forms a cyclobutane ring between successive thymine residues. The shaded portion is the plane of the cyclobutane ring.

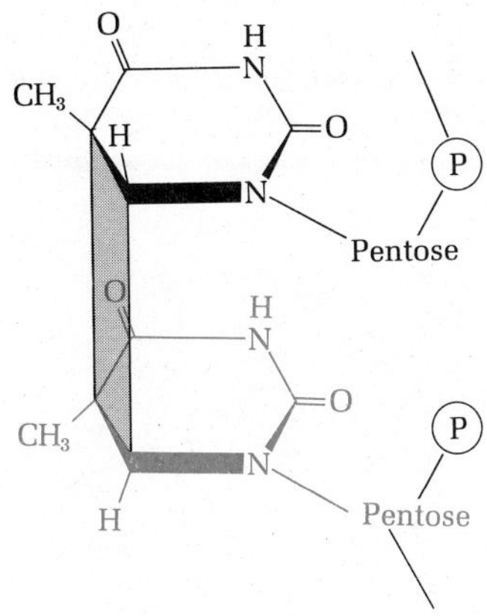

tide chain coded; very often the defective protein is still functional. Such a mutation is called a *silent mutation*. However, insertion or deletion mutants cause all the DNA beyond the point of the mutation to be misread; such mutations often are lethal. Another class of mutants results from deletion of segments of DNA which may be many nucleotides long. If the number of deleted nucleotides is 3 or some multiple of 3, there is no frame-shift error beyond the site of mutation, but the protein produced may be defective because one or more amino acids are deleted from its sequence.

Mutations may be caused by many other chemical agents. For example, lysergic acid diamide (LSD) and even caffeine have been reported to induce mutations. γ-Rays and x-rays are especially powerful mutagens. Mutations are also caused by ultraviolet light, a cause of skin cancer following exposure to excessive sunlight. The most common form of damage from ultraviolet irradiation is the formation of *thymine dimers*, in which two adjacent thymine residues become covalently joined by a cyclobutane ring (Figure 31-25). Thymine dimers stop the replication of the DNA. They can be enzymatically excised from the DNA and the resulting gap repaired (page 913).

Modification and Restriction of DNA

Bacteria can destroy an invading or foreign DNA from another species, thus preventing its replication, transcription, or incorporation into the host-cell genome. This is made possible by an ingenious combination of two enzymatic processes called modification and restriction. *Modification* is the enzymatic alteration of its own DNA by the cell, in a species-distinctive way, thus differentiating it from that of other species. *Restriction* is the action of specific enzymes which can degrade foreign invading DNAs but do not destroy the host-cell DNA, simply because the specific covalent alteration of the host-cell DNA produced by the modification enzymes does not allow the restriction enzymes to act on it.

The protective modification of the host cell DNA is brought about by *modification methylases*, which methylate certain adenine residues. Once the host-cell DNA is modified in this manner, it cannot be degraded by that cell's restriction enzymes. The restriction methylases transfer methyl groups from S-adenosylmethionine (page 698) to pairs of adenine residues in duplex DNA, one in each strand; the two adenines are on adjacent or nearby base pairs. The sequence of bases on the two strands between and near the methylated adenines is symmetrical on either side of a midpoint (Figure 31-26). Some viral DNAs are modified for protection against host-cell restriction enzymes by transfer of glucosyl residues from UDP-glucose to hydroxymethylcytosine residues.

The restriction enzymes are *endonucleases* (page 323) which recognize and cleave the foreign DNA at the same specific sequence of bases that in the host-cell DNA has been modified by methylation; the methyl groups of the host-cell DNA prevent the attack of the restriction endonuclease. The restriction endonucleases produce a double-strand break,

Figure 31-26
Specific sites in DNAs acted upon by modification methylases and restriction endonucleases. Examples from two species are given, showing the axis of symmetry of the restriction sites. The stars indicate the bases that are methylated in the host DNA and the arrows indicate the point of cleavage of the foreign DNA by the host's restriction endonucleases. Not all restriction endonucleases make staggered cuts.

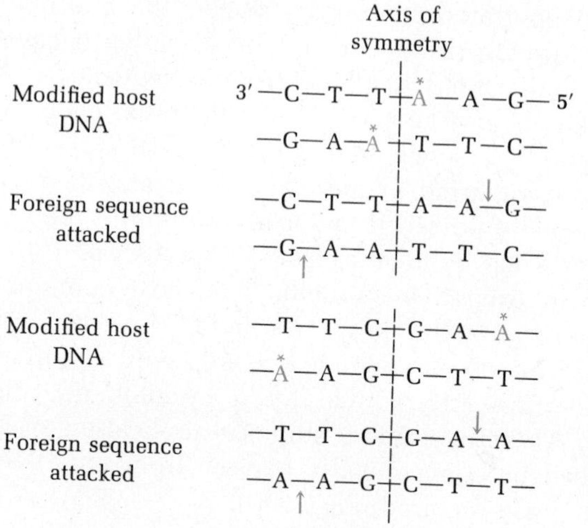

Figure 31-27
Hybridization tests with single-strand (denatured) reference DNA impregnated into an agar gel, which is then finely subdivided into particles and packed in a column (right). Radioactive DNA to be tested flows through the column. If it has nucleotide sequences complementary to the reference DNA, double-strand hybrids are formed, attached to the agar gel. The amount of labeled DNA retained by the column is a measure of the amount of complementarity.

Axis of
symmetry

Modified host DNA

$3'$ —C—T—T—$\overset{*}{A}$—A—G— $5'$

—G—A—$\overset{*}{A}$—T—T—C—

Foreign sequence attacked

—C—T—T—A—A—G—

—G—A—A—T—T—C—

Modified host DNA

—T—T—C—G—A—$\overset{*}{A}$—

—$\overset{*}{A}$—A—G—C—T—T—

Foreign sequence attacked

—T—T—C—G—A—A—

—A—A—G—C—T—T—

which cannot be repaired; as a result, the foreign DNA is degraded further by other nucleases in the cell. Another class of restriction endonucleases binds at specific points but cleaves at random points on DNA.

Restriction endonucleases have become exceedingly important tools for the selective, specific cleavage of native DNA molecules. They have permitted the controlled fragmentation and identification of the sites of specific genes in chromosomes. For example, considerable progress is being made in mapping the location of the "cancer genes" in the DNA of simian virus 40, an oncogenic (cancer-causing) virus.

The Hybridization Technique: Sequence Homologies in DNA of Different Species

Although the base composition of the DNA of many different bacteria and higher animals has been measured, very little information is available on the precise sequence of nucleotides in native intact DNA molecules. However, by means of hybridization tests it is possible to determine whether two given specimens of DNA have blocks or segments of complementary base sequences and what their extent may be. Such hybridization tests are based on the tendency of two DNA strands to pair and rewind at their complementary regions to form double helixes. The greater the tendency for a given pair of DNA strands to associate and to form duplexes, the greater the extent of complementarity of their sequences. To carry out such tests, the two DNA specimens are mixed and

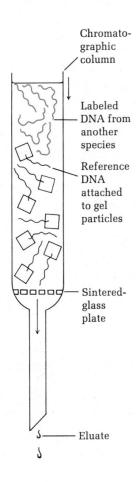

Chromato-
graphic
column

Labeled
DNA from
another
species

Reference
DNA
attached
to gel
particles

Sintered-
glass
plate

Eluate

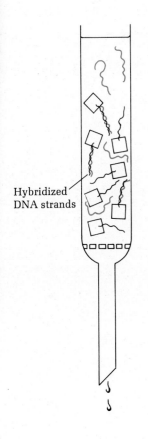

Hybridized
DNA strands

heated beyond the melting temperature of both and then cooled slowly. If they possess complementary base sequences, they will associate to form hybrid duplexes. The extent to which such hybrids form is easily determined if one of the DNA specimens is first labeled with ³²P. By means of density-gradient centrifugation the various single- and double-strand species in the mixture may be separated and the amount of radioactivity in the newly formed hybrid duplex fraction determined. From such experiments the important finding has been made that the tendency of the DNA from two given species to hybridize is a function of how close they stand in their taxonomic relationship. Thus, although the DNAs of various bacteria show very wide variations in base composition (Table 31-3), they may contain many zones of approximate complementarity of base sequence, possibly because the genes coding for homologous enzymes are similar in various bacterial species. On the other hand, the DNA of bacteria does not tend to hybridize readily with the DNA from higher animals or even with that of yeast cells. The tendency of DNA of various bacterial viruses to form hybrids has also been studied; DNA from various mutations of one virus tend to hybridize with each other much more readily than do DNAs from different types of viruses.

Hybridization between DNAs isolated from closely related higher animal species can be detected by means of a more sensitive technique. The denatured, single-strand DNA of the reference species is incorporated into an inert support, such as finely divided agar-gel particles, which is then packed into a chromatographic column. An isotopically labeled denatured DNA specimen from another species is then passed through the column; the extent of retention of the isotope on the column is an index of the tendency of the two DNAs to hybridize and thus of the extent of sequence homology (Figure 31-27).

Single-strand DNA will also hybridize with RNA derived from the same species, again through complementary base pairing (the uracil residues of RNA pair with adenine residues of DNA). This finding strongly suggests, as has been confirmed more directly, that RNA is synthesized on a DNA template. Hybridization between DNA and RNA can also be used to determine what fraction of the bacterial chromosome codes for RNA. For example, in some cells all the ribosomal RNA, which represents over half the total RNA, is coded by less than 2 percent of the bacterial chromosome.

Much more sensitive hybridization tests, based on chromatographic separations on hydroxyapatite columns, have been developed. Such tests are being used to determine whether DNA or RNA of normal and of cancer cells contains genes present in certain viruses known to cause cancer.

Palindromes in Eukaryotic DNA

The DNA of eukaryotic cells differs from that of prokaryotes not only in size and ultrastructural organization in the chromosomes but also with respect to some features of base sequence. One such special feature is the presence of numerous, perhaps thousands, of *palindromes* in eukaryotic DNA. A palindrome is a word or sentence that is spelled identically reading forward or backward, for example, "Able was I ere I saw Elba." The term is applied to specific regions in eukaryotic DNA in which there are *inverted repetitions*, symbolized as

$$5' \cdots A\text{-}B\text{-}C\text{-}D\text{-}t\text{-}D'\text{-}C'\text{-}B'\text{-}A' \cdots 3'$$
$$3' \cdots A'\text{-}B'\text{-}C'\text{-}D'\text{-}t\text{-}D\text{-}C\text{-}B\text{-}A \cdots 5'$$

where t stands for the "turnaround." Such inverted repetitions are quite long and involve hundreds or thousands of base pairs. They have been analyzed in some detail by C. A. Thomas, Jr., and his colleagues. These repetitions were detected by denaturing fragments of eukaryotic DNA with strong base and then following their renaturation to form duplexes by sedimentation and electron microscopic analysis. It was found that the renatured DNA showed extensive intrachain base pairing to form long "hairpins" with a short turnaround. These hairpins are formed by base pairing of the inverted repetitive base sequences. The possibility that these long inverted repetitions arise by chance appears remote. Thomas and his colleagues have postulated that their function is to form long duplex branches from the main duplex chromosome, perhaps to improve accessibility to the replicating enzymes (page 912).

Repetitive Sequences in the DNA of Eukaryotic Organisms

When the DNA of bacteria is fragmented by shearing forces into small pieces, all the fragments contain about the same content of bases and have the same density, as determined by density-gradient centrifugation. However, when DNA of higher organisms is fragmented and subjected to density gradient centrifugation, one obtains a large main peak and one or more small peaks of significantly different density and hence significantly different base composition. The DNA present in these small peaks is called *satellite DNA*. The satellite DNA of some species has been analyzed and found to have simple repetitive base sequences. The satellite DNA of the crab has the repetitive sequence -A-T-A-T-A-T-A-T-, that of the kangaroo rat has the repeated sequence -G-G-A-C-A-C-A-G-C-G-, and that of a fruit fly species is -A-C-A-A-A-T-T-. Such sequences occur in blocks of about 100 residues and may be repeated up to 10^7 times per cell. Satellite DNA may make up a large fraction of the total chromosomal DNA; that of the kangaroo rat makes up about 11 percent of the genome and that of the crab nearly 25 percent. Such repetitive sequences do not occur in viral or bacterial chromosomes and are thus believed to be related to the manner of organization of the DNA in eukaryotic chromosomes. Such short, highly repetitive sequences occur largely at the centromere region of the chromosome, suggesting that they may be informationless "spacers" or play a structural role in holding loops of informational DNA together.

In addition to satellite DNA, eukaryotic DNA has another class of repeated sequences of much larger size. They were detected by R. J. Britten and his colleagues by measurement of the rate of renaturation, i.e., base-pairing, of heat-denatured DNA. The rate of renaturation is given by the *cot value*, defined as the product of the molar concentration of nucleotide residues and the time of renaturation (in seconds). Satellite DNAs have very low cot values, since there are a relatively large number of short, nearly identical sequences. A homogeneous nonrepetitive DNA has a very high cot value. Surprisingly, about half of the DNA of eukaryotic cells has an intermediate cot value, interpreted to mean that there may be anywhere from about 10^2 to more than 10^4 repeated sequences of large size, containing several thousand residues.

The genes for the ribosomal RNAs in certain species occur in many repeated copies. For example, in the South African clawed toad *Xenopus laevis* there are more than 1,000 copies of these genes per cell, separated from each other by spacers rich in G-C pairs. However, apart from ribosomal RNA genes in certain species, no other structural genes coding for proteins or RNAs are known to occur in a large number of copies. It has therefore been postulated that the intermediate class of repeated copies of very similar segments of DNA, which may constitute more than half the total cell genome, has a regulatory function. This hypothesis is in agreement with estimates made from mutation rates, which suggest that structural genes make up only a small fraction of the total DNA in human cells, perhaps 10 percent, equivalent to one copy of each of about 100,000 genes.

Isolation of Specific Genes

In a few cases single genes or sets of genes have been isolated from DNA specimens by special procedures. The set of three genes and the regulatory regions of the DNA concerned in the transport and metabolism of β-galactosides in E. coli has been isolated by J. Beckwith and his colleagues (page 987). Moreover, genes coding for ribosomal RNA have been isolated by D. D. Brown. Such isolated DNA fragments are functional and capable of coding for characteristic gene products in vitro (page 915). The nucleotide sequence of isolated genes may be established indirectly (page 886) and used as a guide in working out direct methods for determining the nucleotide sequence of DNA.

Chemical Synthesis of Polydeoxyribonucleotides and Genes

The direct analysis of base sequence in a homogeneous DNA presents some formidable difficulties since even the smaller DNAs have a much larger number of residues than the RNAs and proteins that have been sequenced. Moreover, DNAs usually contain no minor bases to use as markers, and deoxyribonucleases fail to cleave DNA into pieces of manageable size. Despite these difficulties, base sequences of DNA fragments having 100 or more residues have been determined.

It is possible to deduce the base sequences in certain genes, particularly those specifying RNAs of known sequence, because the bases of the gene and those of the RNA it codes are complementary. The sequence of bases in the gene coding for yeast alanine tRNA, whose base sequence is known (page 324) is readily deduced. Yeast alanine tRNA has 77 ribonucleotide residues; its "bare" gene therefore has the same number of deoxyribonucleotide residues. The minor bases present in tRNA are formed from the corresponding parent bases by enzymatic reactions occurring after the tRNA backbone has been formed, i.e., by means of *post-transcriptional* reactions.

In a brilliant piece of work H. G. Khorana and his colleagues synthesized the gene for yeast alanine tRNA by a combination of chemical and enzymatic procedures, the first gene to be made by largely chemical methods. Basic to their approach is the use of protecting or blocking reagents for those reactive groups in nucleosides and nucleotides which are sensitive to the condensing agents required to form phosphodiester bridges. Such sensitive groups include the 3'- and 5'-hydroxyl groups of the deoxyribose moiety, the amino groups of the bases, and the phosphate group itself. The problem is thus similar to that faced in the chemical synthesis of polypeptides (page 117). Some of the protecting groups developed by Khorana's group are shown (Figure 31-28). After the appropriate groups have been blocked, the phosphodiester internucleotide linkage is formed by condensing the free 3'-hydroxyl group of one protected nucleotide with the free 5'-phosphate group on the next. The condensing agent employed in the earlier work was a carbodiimide derivative, such as dicyclohexylcarbodiimide, which is also used in peptide-bond synthesis (page 119), but more recently aromatic sulfonyl chlorides have been preferred. Usually short oligonucleotides of 5 to 10 residues are first built separately. They are then assembled to make longer chains. By such methods polydeoxyribonucleotide chains having up to 20 residues in known sequence have been synthesized; an example is shown in Figure 31-29. This chain corresponds to residues 21 through 40 of the gene coding for yeast alanine tRNA. By joining 15 such chemically synthesized double-strand polydeoxyribonucleotide chains in the correct sequence by means of the DNA ligase reaction (page 902), Khorana and his colleagues have achieved the total synthesis of both strands of the gene for yeast alanine tRNA. Moreover, they have also synthesized the gene for the precursor of tyrosine tRNA of *E. coli*, which includes two extra DNA segments, one at each end of the tyrosine tRNA gene. These terminal DNA segments are removed enzymatically during the conversion of the tRNA precursor into the finished tRNA (page 919). Altogether, the synthetic DNA of the tRNA precursor contains 126 residues.

Chemically synthesized oligodeoxyribonucleotides have also been used as templates for the enzymatic preparation of messenger RNAs of known sequence, which in turn have

Figure 31-28
Protection or blocking of sensitive groups of nucleotides. In this example the amino group of cytidylic acid is protected by the benzoyl group, the 3'-hydroxyl by an acetyl group, and the 5'-hydroxyl by a trityl group. The protecting groups are in color.

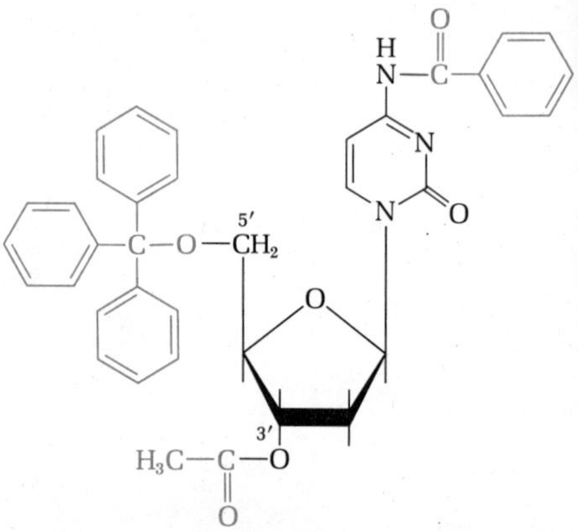

Figure 31-29

Chemical synthesis of an oligodeoxy-ribonucleotide chain of 20 residues, a segment of one strand of the gene coding for yeast alanine tRNA. The red bars are blocking groups for protection of bases; the black bars, for protection of 5'- or 3'-hydroxyl groups. Mononucleotide or oligonucleotide residues are added by the action of the condensing agent dicyclohexylcarbodiimide (DCCD). At the right is shown the working scheme for chemical synthesis of the 20-unit chain. The size of the added oligonucleotide blocks is shown by brackets. At the end of the synthesis the protecting groups are removed.

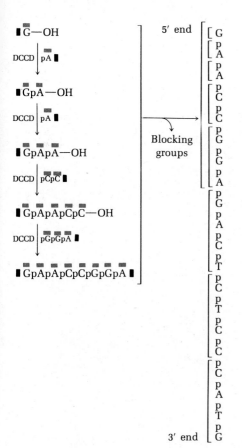

been used in deducing the nucleotide sequence of the genetic code words (page 958).

Summary

That DNA bears genetic information was first shown by experiments in which addition of DNA isolated from one strain of a bacterium transformed another strain in a heritable manner to the strain from which the added DNA was derived. The amount of DNA per cell increases with the cell's position on the evolutionary scale. All somatic (diploid) cells of a given species of higher organism contain the same amount of DNA, which is not modified by diet or environmental circumstances. The base composition of DNA specimens varies characteristically from one species to another. In double-stranded DNAs, the number of adenine residues equals the number of thymine residues; similarly, the number of guanine residues equals that of the cytosine residues.

From x-ray analysis of DNA fibers and from the base equivalences in DNA, Watson and Crick postulated that native DNA consists of two antiparallel chains in a double-helical arrangement, with the complementary bases A-T and G-C paired by hydrogen bonding within the helix and the covalent backbone on the outside. The base pairs are closely stacked perpendicular to the long axis, 0.34 nm apart. The double helix is about 2.0 nm in diameter. This structure provides an explanation for accurate replication of the two strands.

The molecular weight of native double-stranded DNA molecules in the smallest DNA viruses is about 1 million, in bacteria it is about 2 billion, and in eukaryotic chromosomes it is at least 80 billion. In most DNA viruses the DNA is linear but circularly permuted; in bacteria the chromosome is an endless, closed loop of DNA. The DNA of eukaryotic cells (somatic) is associated with histones and other proteins in the chromatin. There is apparently one very large DNA molecule in each chromosome, looped many times and bound together at the centromere.

Equilibrium centrifugation of DNA in cesium chloride gradients is a valuable analytical tool, since the buoyant density of DNA is proportional to its G-C base-pair content. DNA duplexes undergo unwinding (denaturation) on heating, during which light absorption at 260 nm increases (the hyperchromic effect). The hyperchromic effect increases with A-T content, and the "melting" temperature T_m increases with the G-C content.

Functional subdivisions and genetic maps of bacterial DNA have been deduced from experiments in which genes are transferred during sexual conjugation, transformation, or viral transduction. The genetic map of E. coli has been shown by such methods to be circular. From genetic-recombination experiments with mutant E. coli viruses and from biochemical approaches, genes (cistrons) have been found to have anywhere from about 75 to 5,500 or more nucleotide pairs. The smallest mutational unit is a single nucleotide pair. Point mutations are produced by substitution of one purine for the other, one pyrimidine for the other, or a purine for a pyrimidine or vice versa. Mutations also result when one or more base pairs are deleted or additional bases are inserted.

Bacterial cells can destroy foreign DNAs of other species by the action of restriction endonucleases. Their own DNA is protected by methylation of certain bases. Eukaryotic DNA contains many palindromes, inverted repetitive sequences.

From hybridization experiments and kinetic measurements on

renaturation rates the DNA of eukaryotic organisms appears to contain up to 1 million copies of very short repeated sequences (satellite DNA), 100 to 10,000 repeated copies of intermediate length, possibly functioning in regulation, and possibly single copies of most of the structural genes for proteins.

References

Books

BARRELL, B. G., and B. F. C. CLARK: *Handbook of Nucleic Acid Sequences*, Joynson-Bruvvers, Oxford, 1974.

CAIRNS, J., G. S. STENT, and J. D. WATSON (eds.): *Phage and the Origins of Molecular Biology*, Cold Spring Laboratory of Quantitative Biology, 1966. A collection of interesting articles written on the occasion of the sixtieth birthday of Max Delbrück, a pioneer in research on bacteriophages.

DAVIDSON, J. N.: *The Biochemistry of the Nucleic Acids*, 7th ed., Academic, New York, 1972. A popular, up-to-date survey.

DAVIDSON, J. N., and W. E. COHN (eds.): *Progress in Nucleic Acid Research and Molecular Biology*, vols. 1– , Academic, New York, 1963–present. Annual review volumes.

DRAKE, J.: *Molecular Basis of Mutation*, Holden-Day, San Francisco, 1970.

DUPRAW, E. J.: *DNA and Chromosomes*, Holt, New York, 1970. Superb electron micrographs of eukaryotic chromosomes.

GROSSMAN, L., and K. MOLDAVE (eds.): *Methods in Enzymology*, vols. 12, 20, and 21, Academic, New York, 1967–1971. Valuable compendium of experimental methods.

LEWIN, B. M.: *The Molecular Basis of Gene Expression*, Wiley-Interscience, New York, 1970. Excellent, well-documented, and well organized account of molecular aspects of replication, transcription, and translation.

LEWIN, B. M.: *Gene Expression*, vol. 2, *Eucaryotic Chromosomes*, Wiley, New York, 1974. Comprehensive review of present status, with extensive bibliography.

The Organization of Genetic Material in Eukaryotes, *Cold Spring Harbor Symp. Quant. Biol.*, vol. 38, 1973.

STENT, G. S.: *Molecular Genetics*, Freeman, San Francisco, 1971. Excellent textbook.

WATSON, J. D.: *The Double Helix*, Atheneum, New York, 1968. A personal narrative of the origins of the Watson-Crick hypothesis.

WATSON, J. D.: *Molecular Biology of the Gene*, 3d ed., Benjamin, Menlo Park, Calif., 1975. Well-illustrated elementary account of the principles of molecular genetics.

Articles

ARBER, W.: "DNA Modification and Restriction," *Prog. Nucleic Acid Res. Mol. Biol.*, 14: 1–38 (1974).

AVERY, O. T., C. M. MACLEOD, and M. MCCARTY: "Studies on the Chemical Nature of the Substance Inducing Transformation of Pneumococcal Types," *J. Exp. Med.*, 79: 137–158 (1944).

BRITTEN, R. J., and D. E. KOHNE: "Repeated Sequences in DNA," *Science,* 161: 529–540 (1968).

BROWN, D. D.: "The Isolation of Genes," *Sci. Am.,* 229: 20–29 (1973).

CHARGAFF, E.: "Preface to a Grammar of Biology: A Hundred Years of Nucleic Acid Research," *Science,* 172: 637–642 (1971). Personal philosophical commentary, in a paper given on the anniversary of the discovery of DNA by Friedrich Miescher; very interesting.

GALIBERT, F., J. SEDAT, and E. ZIFF: "Direct Determination of DNA Nucleotide Sequences: Structure of a Fragment of ϕX174 DNA," *J. Mol. Biol.,* 87: 377–407 (1974).

HESS, E. L.: "Origins of Molecular Biology," *Science,* 168: 664–669 (1970).

KHORANA, H. G., and colleagues: "Total Synthesis of the Gene for Alanine Transfer RNA from Yeast," *Nature,* 277: 27–34 (1970).

KHORANA, H. G., and colleagues: "Total Synthesis of the Structural Gene for an Alanine Transfer Ribonucleic Acid from Yeast," *J. Mol. Biol.,* 72: 209–522 (1972). A long series of papers describing the first synthesis of a gene.

LOEWEN, P. C., T. SEKIYA, and H. G. KHORANA: "The Nucleotide Sequence Adjoining the CCA End of an *Escherichia coli* Tyrosine Transfer Ribonucleic Acid Gene," *J. Biol. Chem.,* 249: 217–226 (1974).

MIRSKY, A. E.: "The Discovery of DNA," *Sci. Am.,* 218: 78–88 (1968). Early intimations in the nineteenth century that DNA is the genetic material.

STENT, G. S.: "That Was the Molecular Biology That Was," *Science,* 160: 390–393 (1968).

WILSON, D. A., and C. A. THOMAS, JR.: "Palindromes in Chromosomes," *J. Mol. Biol.,* 84: 115–144 (1974).

WORCEL, A., and E. BURGI: "On the Structure of the Folded Chromosome of *Escherichia coli,*" *J. Mol. Biol.,* 71: 127–147 (1972).

WYATT, H. V.: "When Does Information Become Knowledge?" *Nature,* 235: 85–88 (1972). An examination of the history of molecular biology from Avery (1944) to the Watson-Crick hypothesis (1953).

Problems

1. From data on the contour length of the DNAs of bacteriophages λ, ϕX174, and T2, calculate the average weight in daltons of double-helical DNA per micrometer.

2. Calculate the average number of nucleotide pairs per micrometer of DNA double helix according to the dimensions predicted by the Watson-Crick model.

3. Bacteriophage M13 DNA has the following base composition: A, 23 percent; T, 36 percent; G, 21 percent; C, 20 percent. What change in the absorbance at 260 nm would you expect upon heating the DNA from 83 to 88°C?

4. Bacteriophage T2 has a head about 95 nm long. Calculate the ratio of the length of its DNA to the length of the viral head.

5. Calculate the average length in nanometers and the average weight in daltons of the genes coding for (a) a tRNA (90 mono-

nucleotide residues), (b) ribonuclease (104 amino acid residues), and (c) myosin (1,800 amino acid residues).

6. From data in this chapter, predict (1) the buoyant density in grams per cubic centimeter, (2) the melting temperature T_m, and (3) the hyperchromic effect (in percent) of double-stranded DNA specimens having (a) 10 percent A-T pairs, (b) 50 percent A-T pairs, and (c) 90 percent A-T pairs.

7. Calculate how many genes can be contained in the DNA of a single human cell if each gene has 900 base pairs. Assume that one-third of the DNA is redundant and is not used for coding proteins.

8. (a) Calculate the total weight in grams and the length in miles of all the DNA in the human neonate; use data in Table 31-1 and assume the human neonate contains 2×10^{12} cells.
    (b) If the G + C base content is 40 percent, how many cubic centimeters does this DNA occupy.

9. The DNA isolated from an animal virus shows the usual base equivalences and is resistant to exonucleases.
    (a) What is its three-dimensional structure?
    (b) Cleavage with restriction enzyme EcoRI yields a single DNA of molecular weight 3.4×10^6. Total digestion with a restriction enzyme (HpaI) from *Haemophilus parainfluenzae* produces three fragments that are separable by gel electrophoresis and have molecular weights 1.4×10^6, 0.7×10^6, and 1.3×10^6. The shortest of the four fragments produced by combined cleavage by EcoRI and HpaI is 0.6×10^6; the longest is 1.3×10^6. The fragments are lettered by size, A being the largest. What are the molecular weights of the four fragments produced upon combined digestion by the two enzymes?
    (c) After partial digestion with both enzymes, the electrophoretic band corresponding to the fragment of molecular weight 1.3×10^6 was isolated and subjected to complete digestion by HpaI. This treatment produced three bands, one that migrated at a rate corresponding to a molecular weight of 1.3×10^6, and bands corresponding to two shorter fragments. Construct a possible genetic map of the fragments A to D of the viral DNA and indicate the cleavage sites.
    (d) The DNA from a mutant strain of the virus yielded only three fragments upon complete digestion by the two enzymes. The sizes were 2.0×10^6, 0.6×10^6, and 0.7×10^6. What is the simplest explanation of the lesion that results in this new pattern?

CHAPTER **32** **REPLICATION AND TRANSCRIPTION**
OF DNA

Now that we have surveyed the evidence that DNA is the genetic material and that its complementary double-helical structure furnishes a molecular basis for a template function, we shall proceed to examine the mechanisms by which DNA is replicated to yield identical daughter molecules and transcribed to yield complementary RNA molecules.

The enzymes and other proteins participating in replication and transcription of DNA are surely among the most remarkable biological catalysts known. A full molecular description of their action must account not only for the formation of the phosphodiester internucleotide linkages but also for the accurate transmission of genetic information from the template strand to the new strand in the form of a precisely complementary nucleotide sequence. Moreover, it must also account for the complex geometrical, mechanical, and kinetic problems posed by the unwinding of the parental duplex DNA, which is required to expose to the replicating and transcribing enzymes the information encoded in the sequence of bases within the duplex. These enzymes must also begin and end their action at specific points on the chromosome, in response to certain molecular signals. The polymerases and other proteins responsible for the replication and transcription of DNA thus have very complex functions, in contrast to most other enzymes.

Although a great deal of progress has been made in analysis of the molecular mechanisms involved in replication and transcription, many important problems remain to be solved, as we shall see.

Semiconservative Replication of DNA: The Meselson-Stahl Experiment

The most striking feature of the Watson-Crick hypothesis from the genetic point of view is its postulation that the two strands of double-helical DNA are complementary and that replication of each to form new complementary strands results in formation of two daughter duplex DNA molecules, each of which contains one strand from the parental DNA.

Figure 32-1
*Possible mechanisms of replication of DNA. Newly replicated strands
(or segments) are shown in color; parental strands (or segments) in black.*

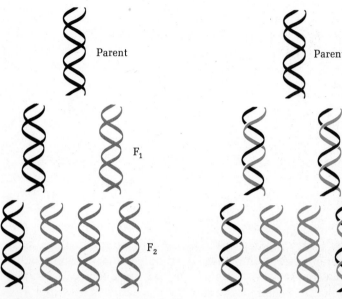
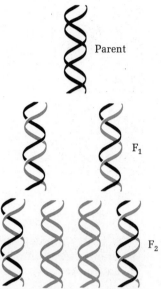
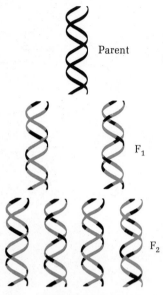

Conservative replication of DNA. Each
of the two strands of the parent DNA
is replicated, to yield the unchanged
parent DNA and a newly synthesized
DNA. The F_2 generation consists of
the parental DNA and three new
DNAs.

Semiconservative replication of DNA.
Each F_1 duplex contains one parent
strand. The F_2 generation consists of
two hybrid DNAs and two totally new
DNAs.

Dispersive replication. The parent
chains break at intervals, and the
parental segments combine with new
segments to form the daughter
strands.

This process, which is called *semiconservative replication,*
is not the only possible mechanism by which two complementary strands of the parental duplex DNA could yield
daughter DNA duplexes identical to those of the parent. Two
other schemes, *conservative* and *dispersive replication,* had
to be considered; these three alternatives are shown in Figure
32-1.

Ingeniously contrived experiments carried out by M. S.
Meselson and F. W. Stahl in 1957 conclusively proved that
in intact living *E. coli* cells DNA is replicated in the semiconservative manner postulated by Watson and Crick. They
grew *E. coli* cells for several generations in a medium in
which the sole nitrogen source was ammonium chloride,
whose nitrogen was nearly pure ^{15}N, the "heavy" nitrogen
isotope, instead of the normal ^{14}N. All nitrogenous components of these cells, including their DNA, thus contained ^{15}N
instead of ^{14}N. The ^{15}N-DNA of these cells (heavy DNA) has a
significantly greater density than normal ^{14}N-DNA; the heavy
and light forms of DNA can be separated by equilibrium centrifugation in a cesium chloride density gradient.

To carry out the crucial experiment on the mechanism of
DNA replication, the growth of the "heavy" ^{15}N-containing
E. coli cells was continued by placing them in a "light"
medium, which contained normal ^{14}N-NH_4Cl as sole nitrogen

Figure 32-2
Results of the Meselson-Stahl experiment. (Left) The positions of the DNA bands after density-gradient centrifugation to equilibrium. (Right) The forms of DNA in controls and after three cell divisions in the ^{14}N-containing medium. "Heavy" refers to ^{15}N-DNA; "light" refers to ^{14}N-DNA. The results conclusively support semiconservative replication.

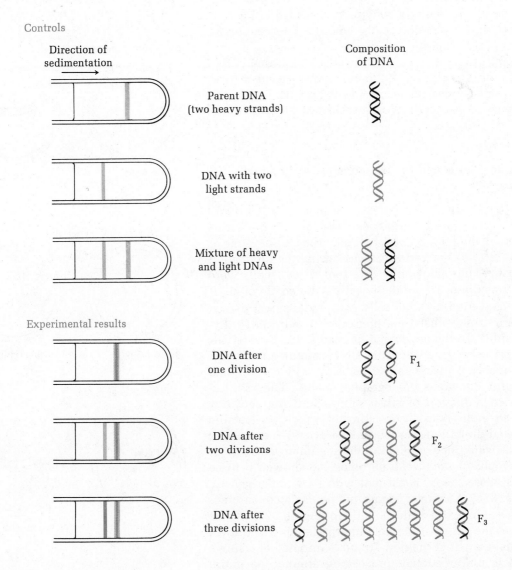

Controls

Direction of
sedimentation

Composition
of DNA

Parent DNA
(two heavy strands)

DNA with two
light strands

Mixture of heavy
and light DNAs

Experimental results

DNA after
one division

F_1

DNA after
two divisions

F_2

DNA after
three divisions

F_3

source. The cells were then allowed to grow for several generations and samples were harvested at intervals. From these samples the DNA was extracted and its buoyant density determined by centrifugation in CsCl density gradients. After exactly one generation time in the ^{14}N medium, which yielded doubling of the number of cells, the DNA isolated showed but a single band in the density gradient, midway in density between the normal or light ^{14}N-DNA and the heavy ^{15}N-DNA of cells grown exclusively on ^{15}N (Figure 32-2). This result is what might be expected if the duplex DNA of the daughter cells contained one ^{15}N strand and one newly synthesized ^{14}N strand. After two generations on ^{14}N-NH$_4$Cl, the DNA isolated exhibited two bands, one having a density equal to that of normal light DNA and the other equal to that of the hybrid DNA observed after one generation. These results, shown schematically in Figure 32-2, are exactly those expected from the hypothesis of semiconservative replication proposed by Watson and Crick, but they are not consistent with the alternative hypotheses of conservative or dispersive replication (Figure 32-1). The Meselson-Stahl experiment was particularly convincing because it was carried out on intact dividing cells without intervention of inhibitors or other injurious agents.

That semiconservative replication of chromosomal DNA also occurs during cell division in eukaryotic organisms was shown in similarly designed experiments by J. H. Taylor and his colleagues on root cells of bean seedlings grown in tissue culture.

The Initiation Points and the Direction of Replication of Chromosomes

Among the most significant early experiments on the direction of replication of chromosomes were those of J. Cairns, who examined the radioactive labeling of the chromosome of growing *E. coli* cells at various stages during the replication process, using autoradiography as a tool. *E. coli* cells were grown in a medium containing heavily tritiated thymine. It was first converted by the cells into radioactive deoxythymidine 5′-triphosphate, the precursor of radioactive thymidine residues in the newly synthesized DNA. At intervals the cells were harvested and the DNA carefully extracted. The DNA molecules were spread on grids, which were then covered with fine-grain photographic plates. The radioactivity caused reduction of silver chloride in the sensitive emulsion to yield tracks, consisting of grains of metallic silver located directly under the radioactive DNA molecule; these tracks were photographed under the microscope. Such autoradiographs of the *E. coli* chromosome showed it to be an endless, closed loop, consistent with its circular genetic map (page 878). Moreover, some of the isolated chromosomes showed an extra radioactive loop (Figure 32-3). From such images the simplest possible interpretation was drawn, namely, that growth of the new strand is initiated at a single point and is unidirectional, proceeding around the entire circle until the growing point returns to the original initiation point.

However, more complete genetic and biochemical experiments on *E. coli* have revealed that an alternative interpretation is more likely, namely, that replication from the single initiation point is *bidirectional* rather than unidirectional (Figure 32-3). Replication appears to begin at about 74 min on the genetic map of *E. coli* (Figure 31-20) and proceeds in both directions. Both strands of parental DNA are thus replicated simultaneously until the two growing points meet; at this point the two daughter DNA molecules separate. Moreover, eukaryotic chromosomes are also replicated bidirectionally, although we shall see that the replication process is much more complex in eukaryotes than in bacteria. On the other hand, the circular DNAs of mitochondria and a number

Figure 32-3
The bidirectional replication of the E. coli chromosome. Above are shown schematically the autoradiographic images; the newly formed portion is in color. Below are shown two possible interpretations. The new DNA strands are in color. The bidirectional model has been proved to be correct.

Autoradiographic images

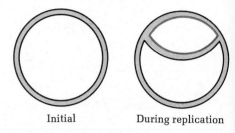

Initial   During replication

Interpretations
Unidirectional model

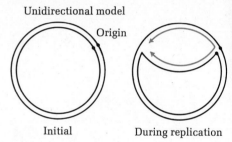

Initial   During replication

Bidirectional model

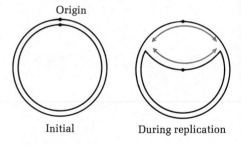

Initial   During replication

Figure 32-4
A rolling-circle model of DNA replication.
Some viral DNAs appear to be made by
variants of this model. A nick is formed in
one strand and the circle is rolled in the
direction shown by the inner arrows by a
pull on the end of one strand, which may be
fixed to some structure, possibly the cell
membrane. Daughter DNA strands (color)
are thus formed in the direction shown and
peeled from the circular strand; they are
then replicated in the linear portion. The
new strands are later cleaved to yield linear
viral duplexes.

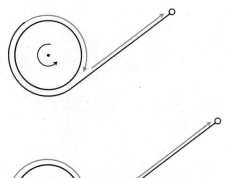
Membrane
attachment

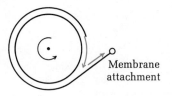

of viruses, such as φX174, appear to be replicated unidirectionally. Although the replicative form of viral DNAs is usually circular, the progeny viral DNA is linear. To account for this rolling circle, models have been proposed for viral DNA replication (Figure 32-4).

Bidirectional replication of circular DNAs poses many problems of mechanics and kinetics. The two chains must be partially unwound to allow replication to occur, since the bases, whose sequence carries the genetic information, are located within the parent double helix. Such unwinding would generate torque in the DNA. Moreover, unwinding of DNA at the replication forks must also be extremely rapid, since the entire chromosome of E. coli is replicated in a single division time of 30 min. Some 400,000 turns of the double-helical chromosome must be unwound, at a rate of at least 13,000 min^{-1}, comparable to the spin rate of a high-speed centrifuge. Another major problem is posed by the opposite or antiparallel polarity of the two strands of the parent DNA as proposed by the Watson-Crick hypothesis. As the replication fork moves along the duplex DNA, one of the strands must be replicated proceeding from its 5′ to its 3′ end, the other from its 3′ to its 5′ end.

DNA Polymerase I

The enzymatic mechanisms by which DNA is replicated were first opened to direct biochemical investigation by the important research of Arthur Kornberg and his colleagues beginning in 1956, on an enzyme today called DNA polymerase I. For some time it appeared that DNA replication is normally carried out by this enzyme. However, more recently two other enzymes having similar catalytic properties have been isolated, DNA polymerase II and DNA polymerase III. Today it appears that DNA polymerase III is the major enzyme concerned in the replication process, although DNA polymerase I also participates. DNA polymerase I has another role; it functions in the repair of DNA. Nevertheless, most of our present knowledge of the DNA polymerases has derived from the detailed studies of Kornberg on DNA polymerase I, which we shall examine first.

DNA polymerase I, isolated from E. coli, catalyzes the addition of deoxyribonucleotide residues to the end of a DNA strand, starting from a mixture of dATP, dGTP, dCTP, and dTTP. The overall reaction for the addition of a single residue is

$$dNTP + (dNMP)_n \underset{\text{Mg}^{2+}}{\overset{\text{Mg}^{2+}}{\rightleftharpoons}} (dNMP)_{n+1} + PP_i$$

DNA   Lengthened
                         DNA

This reaction requires Mg^{2+}. The reaction does not occur in the absence of preformed DNA, which has a more complex function, as we shall see, than implied by this equation. The enzyme is assayed by measuring the incorporation of ^{32}P in the α-phosphate group of one of the precursor nucleotides (Figure 32-5) into the 3',5'-phosphodiester internucleotide bridges of the lengthened DNA strand. Since DNA is insoluble in acid, it can thus be separated from the labeled precursor. The enzyme specifically requires the 5'-triphosphates of the deoxyribonucleosides as precursors; the 5'-diphosphates and the 5'-monophosphates are inactive. The 5'-triphosphates of ribonucleosides are also inactive in the presence of Mg^{2+}. Moreover, the 5'-triphosphates of all four deoxyribonucleosides normally found in DNA must be present for the reaction to occur at a significant rate.

DNA polymerase I catalyzes the addition of mononucleotide units to the free 3'-hydroxyl end of a DNA chain; the direction of DNA synthesis is thus 5'→3'. The reaction occurs by a nucleophilic displacement in which the 3'-hydroxyl group of the terminal mononucleotide residue at the growing end of the DNA chain attacks the electrophilic α-phosphorus atom of the incoming nucleoside 5'-triphosphate, causing displacement of pyrophosphate, the leaving group, and the formation of the phosphodiester linkage (Figure 32-6). The energy required to form the new phosphodiester linkage is provided by pyrophosphate cleavage (page 410) of the dNTP, which occurs with a large negative standard-free-energy change. However, under normal intracellular conditions the DNA polymerase reaction is pulled further to completion because the liberated pyrophosphate may be hydrolyzed to orthophosphate by inorganic pyrophosphatase. In the presence of a very large excess of pyrophosphate, the DNA polymerase reaction can be reversed, causing _pyrophosphorolysis_ of DNA, with formation of a mixture of deoxyribonucleoside 5'-triphosphates. Under intracellular conditions, however, the forward reaction is overwhelmingly favored.

DNA polymerase also catalyzes equilibrium exchange of ^{32}P-labeled pyrophosphate into the β- and γ-phosphate groups of dNTPs, a reaction which may be given as

$$^{32}\text{P}^{32}\text{P}_i + \text{dNPPP} \xrightleftharpoons{\text{DNA}} \text{PP}_i + \text{dNP}^{32}\text{P}^{32}\text{P}$$

in which dNPPP is a deoxyribonucleoside 5'-triphosphate. Both the pyrophosphorolysis of DNA and the pyrophosphate exchange require the presence of preformed DNA.

DNA polymerase I of _E. coli_ has a molecular weight of 109,000. It has a single polypeptide chain of about 1,000 amino acid residues. The enzyme possesses a sulfhydryl group, but this is not required for activity; it also has a single intrachain disulfide group. The pure enzyme contains one atom of zinc at the active site; it is essential for activity. The enzyme molecule appears to be roughly spherical and has a diameter of about 6.5 nm, in comparison with double-strand DNA, which has a diameter of about 2.0 nm. Pure DNA polymerase I can add about 1,000 nucleotide residues per minute per molecule of enzyme at 37°C.

Figure 32-5
A deoxyribonucleoside 5'-triphosphate labeled in the α-phosphate group.

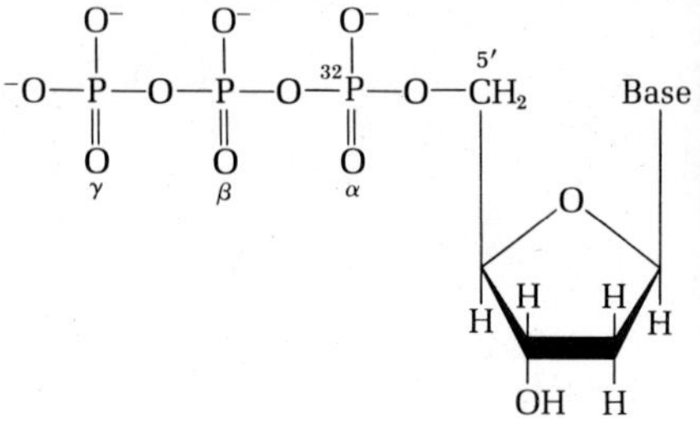

Figure 32-6
Mechanism of chain extension by DNA polymerase.

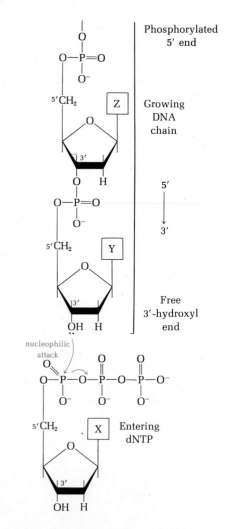

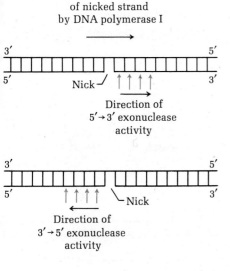

Figure 32-7
Action of the 5'→3' and 3'→5' exonuclease activities of DNA polymerase I. Successive residues are removed one at a time (colored vertical arrows) in the directions shown.

Exonuclease Activities of DNA Polymerase I, a Multifunctional Enzyme

DNA polymerase I also catalyzes two other reactions that were at first thought to be due to contaminating enzymes but have since been shown to be intrinsic in the DNA polymerase I molecule and important in its biological activity. One of these is 3'→5' *exonuclease* activity, i.e., the hydrolysis (*not* pyrophosphorolysis) of single nucleotide residues from the 3' end of a DNA chain (page 323). The other is 5'→3' *exonuclease* activity, namely, hydrolysis of mononucleotide residues from the 5' end of a DNA chain. These two exonuclease activities (Figure 32-7) play an important role in the function of DNA polymerase I, as we shall see later.

The single polypeptide chain of DNA polymerase I can be cleaved by the action of certain proteolytic enzymes into two fragments. The large fragment (76,000 daltons) contains the DNA polymerase and 3'→5' exonuclease activity; the small fragment (36,000 daltons) contains only the 5'→3' exonuclease activity. A mixture of the two fragments is similar to the intact enzyme in catalytic activity, and it must be concluded that the DNA polymerase I molecule in reality contains two distinct enzymes, one a DNA polymerase with 3'→5' exonuclease activity and the other a 5'→3' exonuclease, the two being joined by a peptide bond.

DNA polymerase activity has been found in many other bacteria. Through work in the laboratories of F. J. Bollum, J. N. Davidson, and many other investigators, it has also been found in animal and plant cells. Eukaryotic cells contain multiple forms of DNA polymerase, differing in molecular weight, which are located in different parts of the cell—in the nucleus, the cytoplasm, and in the mitochondria. In the last case the enzyme presumably functions to replicate mitochondrial DNA. DNA polymerase activity in bacterial cells increases greatly during replication of certain viruses, such as the *E. coli* phages T2 and T4. Viral DNA codes for the synthesis of a specific DNA polymerase that replicates viral DNA.

The Role of Preformed DNA in The Action of DNA Polymerase

The most striking and characteristic property of DNA polymerase is that it requires the presence of some preexisting DNA. In the absence of the latter, the purified enzyme will make no polymer at all, except under very special conditions.

At least two alternative explanations were possible for the role of the preformed DNA. It could function as a primer by providing an end, or growing point, to which additional nucleotide residues could be added by the polymerase, in much the same fashion that the end of a preformed glycogen chain is required as a primer to accept glucosyl residues in the action of glycogen synthase (page 645). Or the preformed DNA could serve as a template, on which the enzyme could make a parallel strand of DNA, complementary to the preformed DNA in base composition and sequence. Preformed DNA could, of course, be required for *both* template and primer functions.

Kornberg and his colleagues investigated the structural features of preformed DNA that are essential for the action of highly purified DNA polymerase I from *E. coli*. Significantly, native, intact, double-helical DNA, which one might expect to be active as a template for replication, does not support DNA synthesis. Melted or denatured DNAs, particularly those in which the two strands are largely separated, are active as templates. Moreover, denatured DNAs from many different sources (animal, plant, bacterial, and viral) can support the activity of *E. coli* DNA polymerase. Also active are duplex DNAs with a *single-strand nick* (defined as a break in which a free 3′-hydroxyl group and a free 5′-phosphate group are located in adjacent residues) and duplex DNAs that contain a gap in one strand. Single-strand DNA can support the action of DNA polymerase because it can form a hairpin with unequal ends, by intrachain duplex formation (Figure 32-8).

From such experiments it has been concluded that preformed DNA has two functions in the action of DNA polymerase I: it is required both as a template and as a primer. The enzyme requires the presence of a DNA strand having a free 3′-hydroxyl group, to which new nucleotide residues are added to extend the chain in the 5′→3′ direction. This is called the *primer strand,* and the terminal 3′ nucleotide is the *priming terminus.* The primer strand need not be very long, but it must be base-paired with a long DNA strand that serves as the *template.* Such a structure is now called a *template-primer.* Five different binding loci have been identified on the active site of DNA polymerase I (Figure 32-9):

1 A binding locus for the template
2 A binding locus for the primer strand, which must be antiparallel and base-paired to the template strand
3 A locus for the 3′-hydroxyl terminus of the primer (this locus can also bind monodeoxynucleotides having a 5′-phosphate group and a free 3′-hydroxyl group
4 A locus for the incoming deoxyribonucleoside 5′-triphosphate, which binds any of the four triphosphate precursors and requires Mg^{2+}
5 A locus of the 5′→3′ exonuclease activity, situated in the path of the growing chain (Figure 32-9)

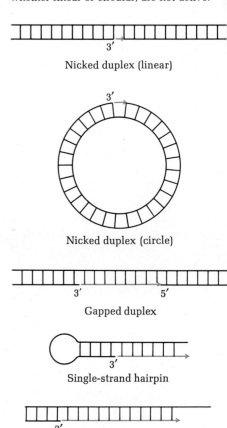

Figure 32-8
Template-primers for DNA polymerase I. The direction of polymerase action is shown by the arrows in color. Intact duplexes, whether linear or circular, are not active.

Nicked duplex (linear)

Nicked duplex (circle)

Gapped duplex

Single-strand hairpin

Primed single strand (linear)

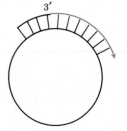

Primed single strand (circle)

Fidelity of Replication by DNA Polymerase I

The next major question can now be framed. Can DNA polymerase replicate its template strand with fidelity, so as to produce a perfectly complementary base sequence in the new DNA strand? Ideally, this question could be answered by determination and comparison of the base sequences of the template DNA and the new strand. However, this was not possible at that time and even today would be a most formidable undertaking. It was therefore necessary to utilize

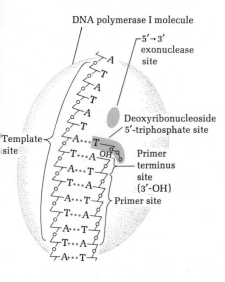

DNA polymerase I molecule

5′→3′ exonuclease site

Deoxyribonucleoside 5′-triphosphate site

Template site

Primer terminus site (3′-OH)

Primer site

less direct experimental approaches to compare the products of DNA polymerase with its template DNA.

In some conditions DNA polymerase can make many times the amount of template-primer added to the system. The molecular basis for this ability became clear when electron microscopy of the products showed that they contain many branches, which are not present in native DNA. Such branches result from the ability of DNA polymerase I to replicate the template starting at a single-strand nick in the primer strand, so that the segment beyond the nick peels off as it is replaced by the new strand. The DNA polymerase molecule may then switch to the peeled-off strand and replicate it to form a duplex branch (Figure 32-10). Such a duplex branch may migrate to another nick; moreover, one or both of its strands might then form hairpins that can serve as new template-primers, and so on. Although such branching is not a normal biological activity (the DNA so produced is called "junk" DNA), the large amount of new DNA formed made it possible to examine the base composition and fidelity of replication of the newly formed DNA from the template-primer DNA. Analysis of base composition showed that the newly formed DNA is identical to that of the template-primer within experimental error (Table 32-1), as would be expected if *both* strands of the denatured duplex DNAs were replicated. If only one DNA strand was replicated, the new DNA strand showed a base composition *complementary* to the template strand.

A more stringent test of the fidelity of replication of the template by DNA polymerase, developed by Kornberg and his colleagues, is called <u>nearest-neighbor base-frequency analysis</u>. This method determines the frequency with which any two bases occur as adjacent or nearest neighbors in a given DNA chain. Figure 32-11 shows that the four bases of DNA can occur in 16 different nearest-neighbor pairs. It can be expected that the DNA of each species of organism will have a distinctive and characteristic set of nearest-neighbor

Figure 32-10
Branching and formation of new single-strand hairpins by DNA polymerase I. (Adapted from A. Kornberg, DNA Synthesis, p. 111, W. H. Freeman and Company, San Francisco, 1974.)

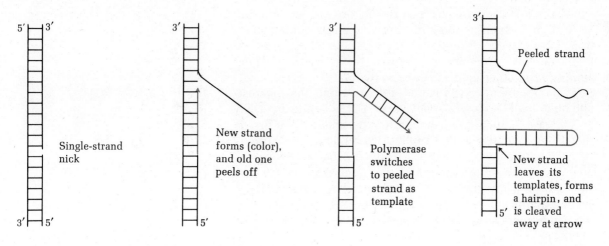

5′ 3′

Single-strand nick

3′ 5′

3′

New strand forms (color), and old one peels off

5′

3′

Polymerase switches to peeled strand as template

5′

3′

Peeled strand

New strand leaves its templates, forms a hairpin, and 5′ is cleaved away at arrow

Table 32-1 Base composition of templates and products of *E. coli*
DNA polymerase

DNA	A	T	G	C
Mycobacterium phlei	0.65	0.66	1.35	1.34
Product	0.65	0.65	1.34	1.37
E. coli	1.00	0.97	0.98	1.05
Product	1.04	1.00	0.97	0.98
Calf thymus	1.14	1.05	0.90	0.85
Product	1.12	1.08	0.85	0.85
T2 bacteriophage	1.31	1.32	0.67	0.70
Product	1.33	1.29	0.69	0.70
AT copolymer	2.00	2.00	0.00	0.00
Product	1.99	1.93	<0.05	<0.05

Figure 32-11
The 16 possible nearest-neighbor nucleotide pairs in a strand of DNA. The symbols are those of dinucleotides (page 322). For example, the symbol pGpA refers to the following dinucleotide:

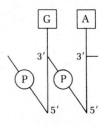

$5' \rightarrow 3'$

pApA
pGpA
pCpA
pTpA

pApG
pGpG
pCpG
pTpG

pApC
pGpC
pCpC
pTpC

pApT
pGpT
pCpT
pTpT

base frequencies. Figure 32-12 shows the principle of nearest-neighbor base analysis. As is seen, this method determines the frequency of occurrence of the four bases next to any given base introduced into a DNA specimen by the action of DNA polymerase.

Kornberg and his colleagues carried out such analyses on DNAs synthesized by DNA polymerase I in response to different template DNAs. An example is shown in Table 32-2, which gives the nearest-neighbor base frequencies for DNA synthesized by the polymerase using *Mycobacterium phlei* DNA as template-primer. All 16 possible nearest-neighbor pairs were found in the new DNA but in widely varying frequencies. Similar experiments with other natural DNAs used as templates indicated that each organism has a characteristic set of nearest-neighbor base frequencies.

Kornberg and his colleagues then compared the nearest-neighbor base frequencies of two generations of DNA products, the first-generation DNA being that synthesized by DNA polymerase in response to DNA isolated from a natural source, for example, calf-thymus DNA, and the second-generation DNA being that made by the polymerase in response to the DNA synthesized enzymatically from the calf-thymus DNA template. The results are shown in Table 32-3. Within experimental error, the first- and second-generation DNAs were essentially identical in their nearest-neighbor base frequencies. Similar data were obtained starting from other natural DNAs as template.

Two important conclusions follow from the nearest-neighbor base-frequency analysis of DNAs made by DNA polymerase in such two-generation experiments. The first is that DNA polymerase can replicate template DNA strands with a high degree of fidelity. However, because the experimental error in determination of the bases is at least 2 percent and probably higher, the nearest-neighbor type of analysis has significant limitations. The second conclusion is that the newly synthesized DNA chains run in the opposite direc-

Figure 32-12
Principle of nearest-neighbor base-frequency analysis, showing how the nearest-neighbor mononucleotide is labeled.

By the action of DNA polymerase, dATP (*color*), labeled with ^{32}P in the α position, reacts with the 3' end of DNA, thus inserting an A residue complementary to a T residue in the template strand.

The ^{32}P introduced as the 5'-phosphate group of the A residue forms a bridge to the 3' position of the preceding residue (G).

Further action of the polymerase adds a dTMP residue, for example, in response to a complementary A residue in the template chain, followed by subsequent template-complementary residues, until chain is complete.

Specific nucleases are then used to cleave the phosphodiester bridges at the points shown by the arrows, to yield 3'-mononucleotides. The ^{32}P originally introduced on the 5' position of the dAMP residue now appears as the 3'-phosphate group of the preceding dGMP residue, thus labeling the nearest neighbor of dAMP.

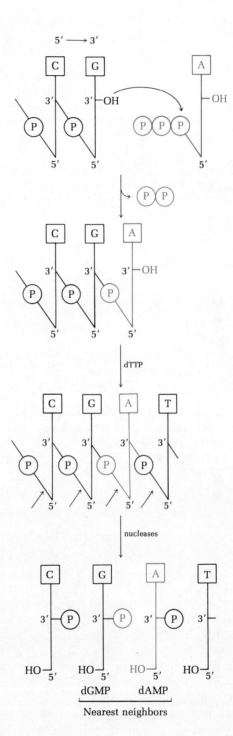

Table 32-2 Nearest-neighbor base frequencies in DNA synthesized from a template† of denatured DNA from *M. phlei*

The frequencies, expressed as fractions of all the nearest-neighbor pairs, are calculated as shown below; all dinucleotide sequences are read in the $5' \rightarrow 3'$ direction

Dinucleotide	Fraction of total ^{32}P	Frequency of occurrence
Labeled precursor = dATP		
T-A	0.075	0.012
A-A	0.146	0.024
C-A	0.378	0.063
G-A	0.401	0.065
Labeled precursor = dTTP		
T-T	0.157	0.026
A-T	0.194	0.031
C-T	0.279	0.045
G-T	0.370	0.060
Labeled precursor = dGTP		
T-G	0.187	0.063
A-G	0.134	0.045
C-G	0.414	0.139
G-G	0.265	0.090
Labeled precursor = dCTP		
T-C	0.182	0.061
A-C	0.189	0.064
C-C	0.268	0.090
G-C	0.361	0.122

† The mole fractions of the bases in the template were T = 0.165, A = 0.162, C = 0.335, and G = 0.338. The frequency of occurrence of C-A in the newly synthesized DNA is given by 0.378, the fraction of total A present as C-A, times the fraction of total bases represented by A, or 0.378 × 0.162 = 0.063.

tion from the corresponding template DNA chains, so that the template chain and the new chain have opposite, or *antiparallel*, polarity. This is clearly evident from data in Figure 32-13, which shows the observed nearest-neighbor frequencies of three specific pairs of residues, compared with the values expected on the basis (1) that the new strand has opposite polarity to the template and (2) that the two strands have the same polarity.

Although base composition and nearest-neighbor frequency measurements indicate that DNA polymerase replicates with considerable fidelity, there is a third and far more crucial test of the fidelity of replication, namely, whether the enzyme can make a biologically active DNA from the appropriate natural template. But first we must consider another important enzyme that functions in conjunction with DNA polymerase I, namely, DNA ligase.

DNA Ligase

Almost simultaneously several investigators reported the discovery in *E. coli* of an enzyme, *DNA ligase*, which can join

Table 32-3 Nearest-neighbor base frequencies of DNAs synthesized from templates of native and enzymatically formed calf-thymus DNA

Nearest neighbors $5' \rightarrow 3'$	Template	
	Native DNA	Enzymatically synthesized DNA
A-A	0.089	0.088
A-G	0.072	0.074
A-C	0.052	0.051
A-T	0.073	0.075
T-A	0.053	0.059
T-G	0.076	0.076
T-C	0.067	0.064
T-T	0.087	0.083
G-A	0.064	0.063
G-G	0.050	0.057
G-C	0.044	0.042
G-T	0.056	0.056
C-A	0.064	0.078
C-G	0.016	0.011
C-C	0.054	0.055
C-T	0.067	0.068

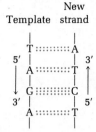

Assuming opposite polarity
(antiparallel chains)

New
Template   strand

| |
| T ::::::::: A |
5' | | 3'
| A ::::::::: T | ↑
| |
| ↓ G ::::::::: C |
3' | | 5'
| A ::::::::: T |
| |

Observed data

Template	New strand
T-A (0.012)	T-A (0.012)
A-G (0.045)	C-T (0.045)
G-A (0.065)	T-C (0.061)

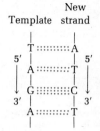

Assuming similar polarity
(parallel chains)

New
Template   strand

| |
| T ::::::::: A |
5' | | 5'
| A ::::::::: T |
| |
| ↓ G ::::::::: C | ↓
3' | | 3'
| A ::::::::: T |
| |

Observed data

Template	New strand
T-A (0.012)	A-T (0.031)
A-G (0.045)	T-C (0.061)
G-A (0.065)	C-T (0.045)

Figure 32-13

Comparison of matching nearest-neighbor frequencies in strands of opposite and similar polarity (opposite). The observed data (Table 32-2) for the template M. phlei strand and the newly synthesized strands are seen to correspond most closely when the strands have opposite polarity.

Figure 32-14

Three steps in the action of DNA ligase in E. coli (below). First the AMP portion of NAD⁺ forms an adenylyl-enzyme intermediate (Ⓟ—Ⓡ—Ⓐ designates phosphate-ribose-adenine of AMP). In Step 2 the AMP group is transferred to the 5'-phosphate group at the nick to yield a 5'-ADP derivative. In Step 3 nucleophilic attack of the 3'-hydroxyl group causes displacement of AMP and sealing of the nick. See Figure 32-15.

Step 1

$$E + NAD^+ \rightleftharpoons E - Ⓟ - Ⓡ - Ⓐ + NMN^+$$

(AMP)

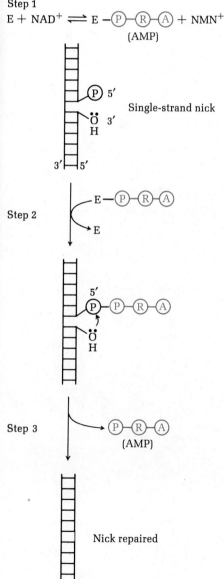

Step 2

Step 3

(AMP)

Single-strand nick

Nick repaired

the ends of two DNA chains by catalyzing the synthesis of a phosphodiester bond between a 3'-hydroxyl group at the end of one chain and a 5'-phosphate group at the end of the other. The formation of the new phosphodiester bond, which requires energy, is coupled to the cleavage of the pyrophosphate bond of NAD⁺ in the case of the E. coli enzyme or to the cleavage of the α,β-pyrophosphate bond of ATP in the case of the DNA ligase induced in E. coli by infection with bacteriophage T4. The eukaryotic enzyme also uses ATP. The enzyme requires that the chains to be joined be associated in a double helix with a complementary DNA strand and that the residues to be joined be paired to adjacent bases in the other strand. DNA ligase can form only a single phosphodiester bond between the ends of the chains; it cannot replicate even a short stretch of template.

DNA ligase from E. coli has been highly purified; it consists of a single polypeptide chain of molecular weight 77,000. About 200 to 400 copies of DNA ligase are present in each E. coli cell. The reaction mechanism is unusual. NAD⁺ is normally a coenzyme for oxidation-reduction reactions (page 481), but functions in the case of DNA ligase as a source of an adenylyl group. The DNA ligase reaction takes place in three steps (Figure 32-14), as shown by I. R. Lehman and his colleagues. In the first, a covalent adenylyl-enzyme compound (AMP—E) is formed on reaction of the enzyme with NAD⁺; nicotinamide mononucleotide (NMN⁺) is the other product of the cleavage of NAD⁺. The adenylyl group, which functions here in an energy-transfer capacity, is linked by a phosphoamide bond with the ϵ-amino group of a specific lysine residue of the enzyme (Figure 32-15). (In T4-induced DNA ligase a similar covalent AMP—E intermediate is formed by cleavage of ATP; the other product is pyrophosphate rather than NMN.) In the second step, the adenylyl group (and its chemical energy) is transferred from AMP—E to the 5'-phosphate end of one DNA strand to form a new pyrophosphate bond (Figure 32-14). In the third step the 3'-

hydroxyl terminus of the other DNA segment displaces the adenylyl group from the first DNA segment to form a new phosphodiester linkage, thus joining the 5'-terminal and 3'-terminal ends of the two segments of DNA.

DNA ligase has been found not only in bacteria but also in a number of eukaryotic cells. It has several functions: (1) to repair single-strand nicks in duplex DNA, (2) to link the ends of linear DNA duplexes to yield circles, a process that requires prior base pairing of unequal "sticky" ends (Figure 32-16), (3) to join segments of DNA in the process of recombination, which occurs during genetic transformation, transduction, and lysogenization in bacteria, and in meiosis in eukaryotes, and (4) to cooperate with DNA polymerases in the replication of DNA.

Enzymatic Synthesis of φX174 DNA by the Action of DNA Polymerase I and DNA Ligase

We have seen two types of evidence that DNA polymerase I replicates a template strand, namely, comparison of base composition and nearest-neighbor bases of first- and second-generation DNAs (page 900). However, these methods have a significantly large experimental error. The most stringent test of the fidelity of replication by DNA polymerase is whether it can catalyze the synthesis of a biologically active form of DNA. In 1968 M. Goulian, A. Kornberg, and R. Sinsheimer succeeded in demonstrating the in vitro enzymatic synthesis of the circular, biologically active double-strand, or replicative, form of the DNA of the small bacteriophage φX174 (page 866) by the combined action of highly purified DNA polymerase I and DNA ligase. Their procedure is outlined in Figure 32-17. In principle, they employed a circular strand of φX174 as template for DNA polymerase I to make a complementary strand; DNA ligase was then employed to convert the linear product into a circle. An interesting device was used to simplify separation of the template and product strands. DNA polymerase I had earlier been found to incorporate certain analogs of the normal nucleotides into the newly formed strand but only if the unnatural base could undergo hydrogen bonding with the corresponding coding base in the template strand. Thus, 5-bromodeoxyuridine triphosphate can replace dTTP in response to A residues in the template. Enzymatic incorporation of such bromouracil (BU) residues into the newly formed strand, instead of the normal T residues, increased the buoyant density of the new strand so that it could be separated from the template by density-gradient centrifugation. From the BU-containing strand a complementary strand with normal bases was then made by DNA polymerase and closed by DNA ligase. The newly synthesized circular DNA strand was found to be infectious. Figure 32-18 is an electron micrograph of partially synthetic φX174 DNA.

These elegant experiments established that DNA polymerase I and DNA ligase can synthesize a biologically active DNA molecule, presumably without error, the first ever

Figure 32-15
Structure of the covalent enzyme-AMP intermediate in DNA ligase action. The adenylyl group is in color.

Polypeptide chain of DNA ligase

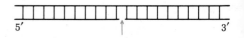

Figure 32-16
Action of DNA ligase. The colored arrows show the points where the enzyme can seal nicks or close circles.

Repair of nicks

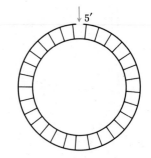

Formation of circles

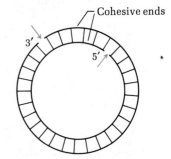

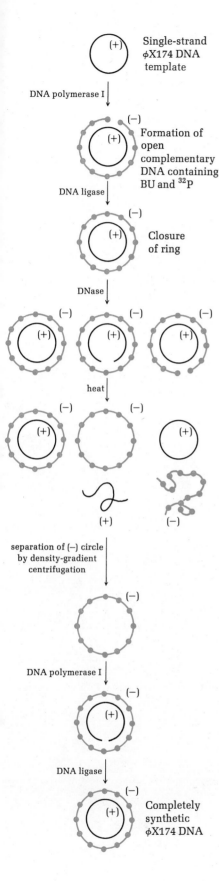

Figure 32-17
Enzymatic synthesis of both strands of circular φX174 DNA by DNA polymerase I (left). A single strand (+) of biologically labeled φX174 DNA (containing tritiated thymidine) was initially used as template to make a complementary linear DNA containing ^{32}P and 5-bromouracil (BU) instead of thymine [the (−) strand]. The ends were then closed by DNA ligase. Treatment with deoxyribonuclease followed by heat yielded some unchanged double circles, intact (+) circles, intact (−) circles, and linear (+) and (−) strands, which were subjected to density-gradient centrifugation. The enzymatically synthesized (−) circle, which has a greater density because of its Bu residues, was separated and used as a template to synthesize a complementary (+) strand in open form. Its ends were then closed by DNA ligase to yield a synthetic circular (+) strand of φX174 DNA, which was found to be infectious. [Redrawn from M. Goulian, A. Kornberg, and R. L. Sinsheimer, Proc. Natl. Acad. Sci. (U.S), 58: 2321 (1967).]

created in the test tube by the action of highly purified enzymes acting on highly purified substrates with a pure template of known biological activity. This experiment also proved that DNA polymerase can make both the (+) and (−) strands of a viral DNA, but it is important to note that the replication was carried out in two successive steps, rather than simultaneously, and thus cannot be taken as revealing the biological mechanism of DNA replication.

The Function of the Exonuclease Activities of DNA Polymerase I

The two exonuclease activities (3′→5′ and 5′→3′) of DNA polymerase I are essential auxiliary activities concerned in maintaining fidelity of replication. The 3′→5′ exonuclease activity appears to function by recognizing and hydrolytically removing (1) a primer terminus that is not base-paired with the template and (2) a frayed end of a primer strand, caused by partial melting. Thus the 3′→5′ exonuclease activity, which proceeds in the direction *opposite* to the direction of replication, ensures that the primer terminus is properly base-paired with the template before elongation begins. This activity has been aptly termed the *proofreading function* of DNA polymerase.

The 5′→3′ exonuclease activity of DNA polymerase I, on the other hand, literally clears the path for the new chain made by the polymerase. It hydrolyzes phosphodiester bonds starting from the 5′-phosphate terminus of a DNA strand, ahead of the polymerase activity, to excise a nonpaired stretch of DNA, or a segment containing chemically modified or mutated nucleotides, such as thymine dimers (page 881). As the 5′→3′ exonuclease excises the mismatched region, the polymerase replaces it, up to the point where a 3′-hydroxyl-5′-phosphate nick remains, which is covalently sealed by DNA ligase. The two exonuclease activities of DNA polymerase I thus equip it to remove various kinds of faults in the strand undergoing replication.

Figure 32-18
Electron micrograph of two duplex circular φX174 DNA molecules, in which one strand is natural and the other synthetic.

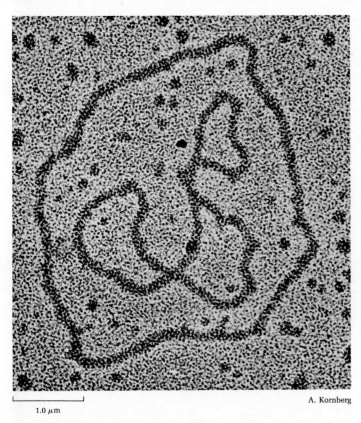

1.0 μm

A. Kornberg

DNA Polymerases II, III, and III*

However, we are still confronted with a fundamental question. Why is native double-strand DNA not used by DNA polymerase I as a template-primer, as might be expected to occur in intact cells? This question prompted the suggestion that DNA polymerase I is perhaps not the enzyme that normally replicates DNA but may function in some auxiliary capacity, possibly to repair DNA.

Crucial experiments to determine whether the original Kornberg DNA polymerase is responsible for the replication of DNA came from a search for mutants of *E. coli* deficient in the Kornberg DNA polymerase but still capable of replication. In 1969 P. De Lucia and J. Cairns reported finding such a mutant of *E. coli*, called *pol A1*, after examining over 3,000 mutants induced by nitrosoguanidine. This mutant, which apparently reproduces normally, was found to contain little or no Kornberg DNA polymerase when the cells were extracted under the same conditions used to obtain the enzyme from wild-type cells; however, the mutant cells did contain residual DNA polymerase activity having properties substantially different from those of the original Kornberg DNA polymerase. Although such mutants grew at a normal rate, they were abnormally sensitive to agents causing damage to DNA, such as ultraviolet and x-irradiation. They also showed an increased incidence of mutation by deletion and a deficiency in repair of single-strand nicks and gaps in DNA. These findings indeed suggested that the missing DNA polymerase is not responsible for DNA replication but is primarily concerned with repair. Attention was then focused on the residual DNA polymerase activity in these mutant cells. Two other DNA polymerases, DNA polymerase II and DNA polymerase III, have been isolated from such mutants, as well as from wild-type E. coli cells, which were subsequently found to contain all three polymerases.

DNA polymerase II, discovered in the laboratories of M. L. Gefter, C. C. Richardson, and R. Knippers almost simultaneously, is characteristically inhibited by sulfhydryl-group blocking agents, whereas DNA polymerase I is not, although, as pointed out earlier (page 896), the latter enzyme does contain an apparently nonessential —SH group. DNA polymerase II has been purified to homogeneity by methods resembling those used for DNA polymerase I. Its molecular weight is nearly the same, about 120,000, but its intrinsic molecular activity or turnover number is only about 5 percent of that of DNA polymerase I. Most important are two qualitative differences in catalytic activity: (1) DNA polymerase II possesses $3' \rightarrow 5'$ exonuclease activity but lacks the $5' \rightarrow 3'$ activity, whereas DNA polymerase I has both, and (2) DNA polymerase II has more stringent requirements for both template and primer, being effective only on duplex DNA with gaps or single-strand ends of less than 100 nucleotides, whereas DNA polymerase I functions with a broader range of templates and primers. In particular, DNA polymerase II cannot replicate long single strands with a short complementary primer, whereas DNA polymerase I is highly active with such a template-primer. The biological function of DNA polymerase II is not yet known.

DNA polymerase III was discovered in 1972 by T. Kornberg (son of A. Kornberg) and M. L. Gefter. This enzyme is rather unstable and was at first difficult to detect; however, it has been purified over 10,000-fold to near-homogeneity. It is by far the most active of the three polymerases, being 15 times more active than DNA polymerase I and 300 times more active than DNA polymerase II. DNA polymerase III is inhibited by sulfhydryl-group blocking reagents, has both exonuclease activities, has about the same template-primer dependence as DNA polymerase II, and has a rather high K_M for its dNTP substrates.

Later research by A. Kornberg and his colleagues revealed that DNA polymerase III occurs in a more complex form, designated as DNA polymerase III* (read "three star"). At this point we may as well begin to use the abbreviations pol I, pol II, pol III, and pol III* for the DNA polymerases of E. coli (Table 32-4). Pol III* is the biologically active form of pol III; it is believed to be a dimer of pol III. Pol III* requires an auxiliary protein, copolymerase III* (copol III*). A. Kornberg and his colleagues have provided some evidence that pol III* combines with copol III* to yield a pol III*–copol III* complex (Figure 32-19), which apparently

Figure 32-19
The pol III–copol III* complex. This complex, ATP, a DNA template, and an RNA primer are required to initiate replication. After the initiating residue has been added, copol III* dissociates and ADP and P_i are released.*

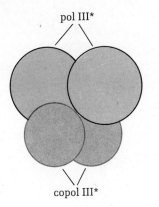

pol III*

copol III*

Table 32-4   Comparison of DNA polymerases I, II, and III of *E. coli*†

	pol I	pol II	pol III
DNA polymerization	+	+	+
5′ → 3′ exonuclease activity	+	−	−
3′ → 5′ exonuclease activity	+	+	+
Primers			
Intact duplex	−	−	−
Primed single strands	+	−	−
K_M for dNTPs	Low	Low	High
Inhibition by —SH reagents	−	+	+
Molecular weight	109,000	120,000	180,000
Molecules per cell	400	100	10
Relative molecular activity (page 208)	1	0.05	15

† From A. Kornberg, *DNA Synthesis*, p. 127, W. H. Freeman and Company, San Francisco, 1974.

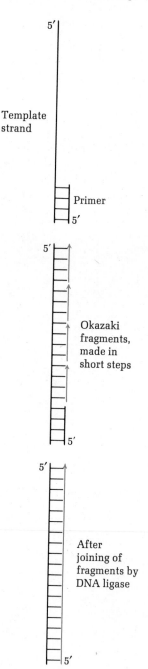

Figure 32-20
Synthesis of DNA in short steps. The nascent DNA, called Okazaki fragments, is about 1,000 to 2,000 residues long.

initiates the growth of a DNA chain. First, however, we must consider some other aspects of DNA replication.

Replication of DNA in Short Steps

R. Okazaki and his colleagues discovered that when viral DNA is replicated in *E. coli*, the most recently formed, or *nascent*, DNA occurs in short pieces, now called *Okazaki fragments*. This was established by following the incorporation of tritiated thymidine into DNA in infected cells grown at low temperatures to slow the rate of replication. Virtually all the labeled DNA formed at very short intervals after adding the labeled thymidine occurs as small fragments of 1,000 to 2,000 residues, with a sedimentation coefficient of only 8S to 10S. However, if the cells are exposed to labeled thymidine for long periods, the radioactivity is found only in high-molecular-weight DNA. These experiments thus indicate DNA to be replicated in short pieces that are quickly joined by covalent bonds (Figure 32-20).

Okazaki fragments have been found not only during viral replication but also in replication of bacteria and eukaryotic cells. They accumulate in particularly large amounts in *E. coli* mutants deficient in DNA ligase, which, as we have seen, can join the ends of single DNA strands. It therefore appears that DNA ligase functions in joining the Okazaki fragments. Further experiments have determined that *both* strands of duplex DNA in replicating *E. coli* or in replicating phages are made in short pieces. As we shall see, replication of DNA in short steps is a device that permits replication of both antiparallel strands of DNA by polymerases that replicate only in the 5′→3′ direction.

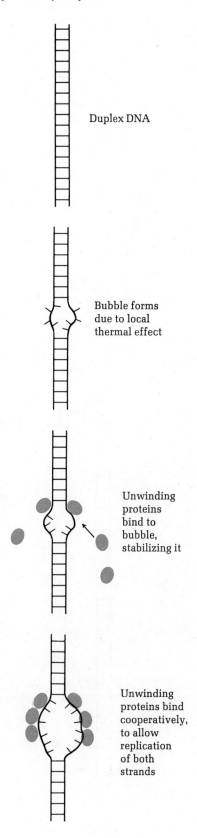

Duplex DNA

Bubble forms
due to local
thermal effect

Unwinding
proteins
bind to
bubble,
stabilizing it

Unwinding
proteins bind
cooperatively,
to allow
replication
of both
strands

DNA Unwinding and Untwisting Proteins

We have seen that unwinding the duplex DNA chromosome is a necessity if both strands are to be replicated. Some daylight on this puzzling problem has come with the discovery of "unwinding" proteins by B. Alberts and his colleagues, made possible by the recognition that gene 32 of phage T4 codes for a protein (neither DNA polymerase nor DNA ligase) which is essential for DNA replication. This protein was isolated and found to induce denaturation of duplex DNA at temperatures some 40°C lower than normally required (page 875). The gene 32 protein binds strongly to a single strand of DNA but weakly to intact double-strand DNA; moreover, it binds strongly to short regions of duplex DNA that happen to undergo transient opening. As soon as one protein molecule binds to the slightly open loop in the duplex DNA, the loop is opened further by binding of additional unwinding protein molecules. Such binding occurs in a cooperative manner, the binding of each protein molecule enhancing the binding of subsequent ones. Alberts and his colleagues have isolated unwinding proteins from bacterial cells infected with a number of different viruses and also from eukaryotic cells. Unwinding proteins vary in molecular weight from 10,000 to 75,000. They possess specific binding sites for short segments of about eight deoxyribonucleotide residues; a number of unwinding protein molecules bind in succession to one strand of duplex DNA in advance of the replicating fork (Figure 32-21).

As unwinding of a long duplex circular DNA takes place during replication, torque develops and results in supercoiling. An *untwisting protein,* also called a "swivelase," has been isolated. It apparently nicks one strand of supercoiled DNA, thus relieving the torsional stress by allowing some unwinding, and then reseals the nick.

The Role of RNA in the Initiation of DNA Chains

None of the purified DNA polymerases, including DNA polymerase III*, can utilize native double-strand DNA as primer-template. Yet in the intact cell DNA replication must surely take place on intact double-strand chromosomes. On the other hand, the enzyme *DNA-directed RNA polymerase* (discussed later), which participates in transcription of DNA to yield a complementary RNA, is known to transcribe from double-strand DNA and to recognize specific initiation points on chromosomes. This curious difference between DNA polymerase and RNA polymerase has been resolved by the discovery, made in the laboratories of A. Kornberg and R. Okazaki, that DNA replication is preceded by the formation of a short strand of RNA complementary to a section of double-strand DNA. This priming RNA is generated by a DNA-directed RNA polymerase from a mixture of the ribonucleoside 5'-triphosphates (page 916). Once the priming RNA strand has been made, DNA polymerase begins to add deoxyribonucleotide units, generated from deoxyribonucleoside 5′-triphosphates, to the 3′ end of the RNA priming strand. After a long stretch of DNA has been replicated in this manner, the short RNA priming strand is cut off by a nuclease (Figure 32-22). Moreover, Okazaki has found that the small DNA fragments generated during replication in *E. coli* contain short stretches of RNA at their 5′ ends, indicating that each short segment made in the stepwise synthesis of DNA begins with a run of RNA. Thus, initiation of DNA replication on RNA primers takes place not only at the start of a round of replication but also in the repetitive formation of the short Okazaki fragments.

A Hypothesis for DNA Replication

It is now clear from the preceding discussion that the replication of DNA requires the integrated action of several, perhaps many, enzymes or proteins, possibly functioning in a complex. The precise type and number of the enzymes and other proteins required will undoubtedly differ somewhat in the replication of viral, bacterial, and eukaryotic DNA.

Although significant and exciting progress is now being made, the information available does not yet permit construction of a detailed hypothesis of DNA replication. However, an outline of a reasonable hypothesis is sketched here and in Figure 32-23. This hypothesis may have to be modified as more evidence arises, but it will at least define some specific steps in the replication process in *E. coli*.

Recognition of the Origin

Replication of the bacterial chromosome begins at a specific *origin* or *initiation point* on the DNA. As will be seen later (page 917), one or more specific initiator proteins are required to recognize the initiation point and perhaps to provide the "starting mark" for a special DNA-directed RNA polymerase to generate the RNA primer of the new DNA chains. There is some evidence that the initiation point or some other specific point in the chromosome is anchored to the cell membrane. Such anchoring may be necessary to help overcome the torque forces when the chromosome is unwound.

Unwinding of DNA

Preceding the action of the replicating enzymes, successive molecules of unwinding protein bind to one strand of the chromosome and open up a loop, or bubble, in the duplex, making one or both strands accessible to replication. *E. coli* cells contain a large number of copies of unwinding proteins, perhaps 10,000 per cell, suggesting that they are used in large numbers at different stages in the replication process.

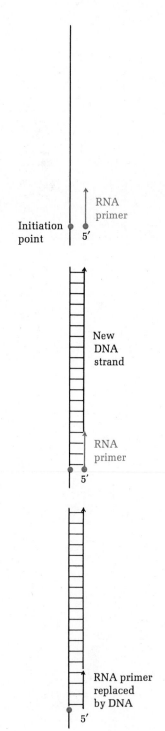

Figure 32-22
Initiation of replication by RNA polymerase, which recognizes the initiation point and makes a short complementary strand of 50 to 100 residues. DNA polymerase then makes a DNA strand on the RNA primer, which is excised and replaced with DNA inserted by DNA polymerase I and DNA ligase.

Figure 32-23
Hypothetical sequence of steps in the replication of DNA, proceeding in one direction, to the right, from the initiation point.
Simultaneously, by a similar process, replication also begins from the initiation
point and proceeds to the left.

Recognition of the origin by RNA polymerase

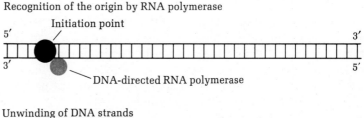

Unwinding of DNA strands

Formation of RNA primers for leading and following strands

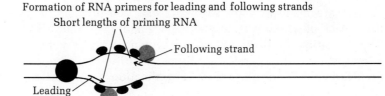

Formation of DNA on RNA primers by pol III*

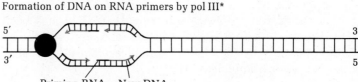

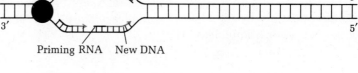

Excision of RNA primers by endonucleases to yield Okazaki fragments

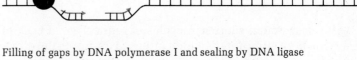

Filling of gaps by DNA polymerase I and sealing by DNA ligase

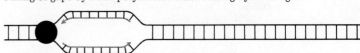

RNA Priming

Primer strands of RNA are then generated by a DNA-directed
RNA polymerase (page 916), which requires ribonucleoside
5′-triphosphates as precursors of the ribonucleotide residues.
The priming RNA strands are complementary to the two
strands of the chromosome. Such priming strands of complementary RNA are from 50 to 100 residues long.

Formation of DNA on the RNA Primers

The *leading strand* of the new DNA, which is that formed in
the 5′→3′ direction from the 3′→5′ template strand of the
chromosome, is believed to begin by addition of deoxy-

911

ribonucleotide residues to the 3' end of the priming "run" of RNA, by the action of the pol III*–copol III* complex in the presence of ATP. Once the DNA strand has been initiated, copol III* detaches and pol III* carries out replication of the template strand. Unwinding proteins separate the duplex strands of the parent DNA ahead of the replicating fork.

As the leading DNA strand is made in the 5'→3' direction, the *following strand* of the DNA is initiated and made, in the direction opposite to the direction of movement of the replicating fork. Presumably the following strand is started by a DNA-directed RNA polymerase to form the primer RNA, followed by the action of the pol III*–copol III* complex to replicate a DNA segment. Both the leading and following strands are made in short segments of 1,000 to 2,000 residues, each of which is begun with a segment of primer RNA. The initiation of these Okazaki fragments probably takes place with the aid of an RNA polymerase different from that involved in initiating a round of replication at the origin.

Excision of the RNA Primers

Once a short stretch (Okazaki fragment) of DNA has been made, the priming RNA is excised from its 5' end, residue by residue, possibly by the 5'→3' exonuclease activity of DNA polymerase I or ribonuclease H.

Joining of the Short DNA Fragments

The gaps left between the Okazaki fragments are then filled with complementary deoxyribonucleotide residues by DNA polymerase I, and the adjacent 5' and 3' termini are joined by DNA ligase.

After many such short-step cycles of replication at each of the two replicating forks, which proceed in opposite directions, the forks meet. Little is known of the events involved in completion of the daughter chromosomes and the release of the polymerases.

It is emphasized again that this is only one possible scenario for replication of DNA, which may differ in details between different organisms. For example, the enzymatic synthesis of φX174 by DNA polymerase I and DNA ligase described earlier did not appear to require RNA primers. Oligonucleotides, possibly present as impurities, may have provided primers for the φX174 DNA synthesis. Moreover, recent research indicates that replication of φX174 DNA requires a number of other specific proteins. Clearly, much further research will be required before we fully understand the mechanism of replication of viral and bacterial DNA.

Replication of DNA in Eukaryotic Cells

Although the replication of DNA in eukaryotic cells can be expected to be much more complex than in prokaryotes, significant information is now becoming available. Eukaryotic cells contain at least two different types of DNA polymerase, although it is not yet certain whether they correspond in function to DNA polymerase I, II, or III of prokaryotes. Eukaryotic cells also contain DNA ligase, in a form requiring ATP as adenylyl donor.

From autoradiographic approaches, as well as electron microscopy, it has been found that the DNA molecules in egg cells of the fruit fly Drosophila are replicated in a bidirectional fashion, thus resembling the replication process in E. coli described above. However, instead of beginning from a single point of origin, replication of eukaryotic DNA begins at multiple points of origin, thus forming a number of "eyes" as the parental strands of DNA are separated and replicated until the eyes meet. Multiple points of origin are a necessity because of the great length of eukaryotic DNA and the relatively slow movement of a replicating fork in eukaryotes, about 2,600 bases per min, compared to about 16,000 bases per min in E. coli. If there were only a single replicating fork in eukaryotes, complete replication of the nuclear DNA would require at least 2 weeks. However, a Drosophila egg cell completes replication in about 3 min and is believed to employ up to 6,000 replicating forks simultaneously.

As in prokaryotes, replication of eukaryotic DNA proceeds in short steps, in which each Okazaki fragment is made on a short RNA primer of about 10 nucleotides. However, the Okazaki fragments in eukaryotes are much shorter (about 100 to 150 nucleotides) than in prokaryotes (1,000 to 2,000 nucleotides). After excision of the RNA primers, the Okazaki fragments are linked together by DNA ligase. Replication of eukaryotic chromosomes requires soluble cytoplasmic factors, which apparently function in the initiation of replication, perhaps by dissociating histones and other proteins from the DNA.

Repair of DNA

DNA is a relatively fragile, easily damaged molecule. Nevertheless, bacterial genes can survive many millions of replication cycles without change. The integrity of the information content of DNA is protected in large measure by enzymes capable of accurately repairing certain kinds of damage or defects in one strand of DNA so long as the other strand, which contains the complementary information, is intact.

DNA can undergo several types of damage in vivo; (1) single-strand and double-strand breaks can result from bending or shear forces; (2) purine bases may be lost as a result of a local pH change (page 322); (3) chemical agents from the environment may modify one or more bases; (4) x-rays or ultraviolet light may bring about a chemical change in a base. Ultraviolet irradiation frequently produces dimerization of two adjacent thymine residues to produce a thymine dimer (page 881).

There are three principal types of repair mechanisms. The first, enzymatic photoreactivation, is still rather obscure. This process is brought about by illumination of the cell with visible blue light. Such photoreactivation requires a specific enzyme that binds to the defective site on the DNA. Illumination results in the absorption of light energy by the enzyme; the absorbed energy in some manner promotes cleavage of covalent bonds at the defect in a single strand. The most common photoreactivation process is the cleavage of the abnormal cyclobutane ring of a thymine dimer (Figure 32-24) to yield two free thymine residues. In this way visible light can be used to repair DNA damaged by ultraviolet light.

Another type of repair mechanism, *excision repair,* which is independent of light and is thus called a "dark" repair process, is brought about by the sequential action of four enzyme activities. If the defect is a thymine dimer, for example, it is detected by the 5'→3' nuclease activity of DNA polymerase I, which has a patrolling function and makes an incision at the 5' side of the defect. The defective segment is then replaced with the correct base pairs by the polymerase function of DNA polymerase I, which then cuts off, or excises, the defective segment. The new strand is then joined at its 3' end to the 5' end of the remaining intact strand, through the action of DNA ligase. This has been referred to as the "cut, patch, cut, and seal" process (Figure 32-25).

In the third type of repair process a large segment of one strand of damaged duplex DNA is replaced by the corresponding intact segment of another DNA molecule; in short, part of a gene is replaced by part of another molecule of DNA. This is known as *recombination repair;* it is a special case of a more general genetic process in which genes from one chromosome are combined covalently into another chromosome. Recombination repair occurs, for example, after damage to a viral chromosome in the host cell.

Xeroderma pigmentosum is a rare human genetic disease in which the skin is abnormally sensitive to the ultraviolet component of sunlight. Such patients have a tendency to develop skin cancer. J. E. Cleaver has discovered that this disease is due to a genetic deficiency of an enzyme normally participating in excision repair of ultraviolet-induced thymine dimers, probably the endonuclease required for the first incision, or cut. An assay has been developed for diagnosing xeroderma pigmentosum by in vitro analysis of the repair capacity of extracts of cultured skin fibroblasts. Several other human genetic disorders in which DNA repair is defective include *Bloom's disease, Fanconi's anemia,* and *progeria,* which causes early aging and death. A similar genetic disease of DNA repair in cattle, called "cattle eye," results in nearly 100 percent incidence of cancer.

Recombination of DNA: Directed Incorporation of Foreign Genes into Chromosomes

Recombination is the exchange of genes between chromosomes. It occurs frequently in the transformation of bacterial cells, in viral transduction, and in the exchange or crossing-over of genes during meiosis of eukaryotic germ cells. Many possible models for the mechanism of recombination have been proposed that employ enzymes already described, such as endonucleases, DNA ligase, and DNA polymerase I. Two

Figure 32-24
Repair of a thymine dimer by photoreactivation. [*After P. C. Hanawalt,* Endeavour, 31: 83 (1972).]

DNA duplex with thymine dimer in one strand

Thymine dimer

Photoreactivation enzyme (color) binds to dimer

Photoreactivation by blue light cleaves cyclobutane ring

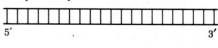

Completed repair

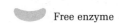

Free enzyme

Figure 32-25
Excision of a thymine dimer by enzymatic
processes. [After P. C. Hanawalt, Endeavour
31: 83 (1972).]

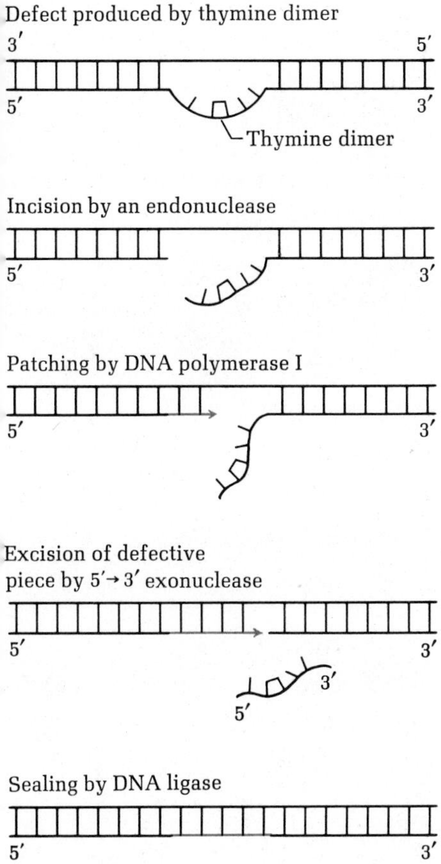

Defect produced by thymine dimer

3'   5'

5'   3'

Thymine dimer

Incision by an endonuclease

5'   3'

Patching by DNA polymerase I

5'   3'

Excision of defective
piece by 5'→3' exonuclease

5'   3'

3'

5'

Sealing by DNA ligase

5'   3'

Figure 32-26
Two types of genetic recombination.

possible models of recombination processes are shown in
Figure 32-26.

Such reactions have recently been employed for the in
vitro incorporation of certain genes from bacteriophage λ or
from the E. coli chromosome into the chromosome of an animal virus, the oncogenic simian virus 40, which can cause
transformation of normal animal cells into cancer cells.

Conversely, part of a eukaryotic chromosome (from
Xenopus laevis) coding for ribosomal RNA has been successfully incorporated into isolated plasmids (page 869) of
E. coli, which were then introduced into intact E. coli cells.
Such recipient cells replicated and transcribed the genes introduced from Xenopus, many times over. Thus genes can be
transferred experimentally between the chromosomes of animals and those of bacteria, a possibility that has raised some
social concerns and questions regarding human safety and
health. For example, laboratory incorporation of "cancer"
genes from animals or animal viruses into the common intestinal bacterium E. coli could conceivably result in wide-
spread transmission of cancer genes to human beings. On the
other hand, such experiments also raise hopes that gene
deficiencies in the chromosomes of higher animals and man
may one day be repaired by transplantation of "good" genes
from another organism.

RNA-Directed DNA Polymerases (Reverse Transcriptases)

A number of RNA viruses are known to cause cancer in susceptible animals. They include the Rous sarcoma virus
(RSV), which produces cancer in chickens, avian myeloblas-
tosis virus (AMV), which produces leukemia in birds, and a
virus producing mammary cancers in rodents. These viruses
are distinctive in being able to cause permanent, heritable
transformation of certain normal cells into malignant cancer
cells in vitro. It was postulated by H. M. Temin in the early
1960s that the RNA of these viruses, which contain cancer

Insertion of a gene into
duplex DNA.

Exchange of DNA segments. Endonuclease cleavage takes place at the
points shown, followed by filling of gaps
by DNA polymerase I and sealing by
DNA ligase.

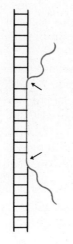

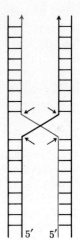

5'   5'

genes, must be transcribed into the form of DNA in order to become incorporated as a permanent, heritable part of the host-cell genome, possibly by the action of a DNA polymerase directed by the tumor-virus RNA. This view gained very little recognition for some years; it ran counter to the central dogma of molecular genetics, which held that information flows from DNA to RNA to protein (page 856). However, Temin's hypothesis was ultimately substantiated by his discovery, also made simultaneously by D. Baltimore, that RNA tumor-virus particles themselves contain *RNA-directed DNA polymerases*, also called *reverse transcriptases*. Such enzymes have been isolated and purified from several different RNA tumor viruses; they have molecular weights ranging from 70,000 to 160,000. The reverse transcriptases closely resemble the DNA-directed DNA polymerases and RNA polymerases in that they make DNA in the $5' \rightarrow 3'$ direction, utilize deoxyribonucleoside 5'-triphosphates as precursors, and require both a template and a primer strand, which must have a free 3'-hydroxyl terminus. The reverse transcriptases are very active on natural RNA templates, including the very large RNAs present in the viral particles. The DNAs produced hybridize with their RNA templates. The reverse transcriptases can also utilize DNA templates, although the latter are less effective than RNAs. Reverse transcriptases have been isolated from malignant cells of some animals and from human patients with leukemia; they closely resemble the reverse transcriptase of some RNA tumor viruses.

Reverse transcriptases, however, have also been found in cells of animals and people thought to be normal and not infected by tumor viruses; they have also been found in wild-type *E. coli*. Such transcriptases can utilize either RNA or DNA as templates. The function of RNA-directed DNA polymerase in normal cells is not understood; its presence suggests that the transcription of messages from RNA into DNA is a normal process, for example, in synthesis of multiple copies of certain genes (page 884). The recognition of the reverse transcriptases has thus opened some new avenues of research in biochemical genetics.

Biosynthesis of RNA: DNA-Directed RNA Polymerase

The discovery of DNA polymerase and its dependence on a template began an active search for the enzymes that participate in transcription, i.e., formation of RNA complementary to a strand of the chromosomal DNA. In 1959 three investigators, S. B. Weiss, J. Hurwitz, and A. Stevens, independently discovered a DNA-dependent enzyme capable of forming an RNA polymer from ribonucleoside 5'-triphosphates; the RNA formed is complementary to the DNA template. This enzyme, called *DNA-directed RNA polymerase*, is very similar in its action to DNA polymerase; in contrast, however, it needs no primer. The enzyme requires Mg^{2+} and yields pyrophosphate. RNA polymerase catalyzes the reaction

$$NTP + (NMP)_n \xrightleftharpoons{Mg^{2+}} (NMP)_{n+1} + PP_i$$

DNA is required for the reaction to take place. All four ribonucleoside 5'-triphosphates are required simultaneously. The RNA polymer formed has 3',5'-phosphodiester bridges and is hydrolyzed by ribonuclease and by spleen and venom phosphodiesterases (page 323). The polymerase adds mononucleotide units to the 3'-hydroxyl ends of the RNA chain and thus builds RNA in the 5'→3' direction, antiparallel to the DNA strand used as template. As with DNA polymerase, enzymatic hydrolysis of pyrophosphate helps ensure that the reaction goes in the direction of synthesis in the cell. RNA polymerase has been identified in many bacterial, plant, and animal cells. In the last there are at least three forms of the enzyme. RNA polymerases are also found in mitochondria and chloroplasts, where they participate in transcription of mitochondrial and chloroplast DNAs.

DNA-directed RNA polymerases from many different sources have been purified. The best known is that of *E. coli*, which has been obtained in homogeneous form by R. R. Burgess and his colleagues. It has a molecular weight of about 490,000 and contains four different kinds of subunits, designated α, β, β', and σ (sigma). The subunit ratio in the intact holoenzyme is given by $\alpha_2\beta\beta'\sigma$. Other proteins, designated ω (omega), ρ (rho), and κ (kappa), cooperate with RNA polymerase in transcription of chromosomal segments under intracellular conditions. The holoenzyme has been dissociated into subunits and successfully reconstituted in active form, by self-assembly in a specific sequence (page 1013). The σ subunit can be removed from RNA polymerase without loss of activity, to yield the core polymerase. This and other information indicates that the σ subunit does not function in catalysis but is a regulatory subunit.

Template Specificity and the Initiation of Transcription

RNA polymerase is most active with double-strand native DNAs from various sources and less active with single-strand or denatured DNAs. Synthetic DNA polymers, such as poly(dT) or the alternating copolymer poly(dTC), will also act as templates, although with some loss of fidelity.

The RNA polymerase molecule under intracellular conditions normally transcribes only one strand of double-strand DNA. The complete holoenzyme is able to recognize and bind to DNA at specific initiation sites, which appear to be pyrimidine-rich structures of some 10 or more residues. Such initiation sites are widely separated in the DNA template. The presence of the sigma subunit is necessary to keep the holoenzyme in the proper conformation to bind to the initiation site on the DNA template. Binding of the polymerase to the template is followed by strand separation in the template and a conformational change in the holoenzyme. Then the first or initiating nucleoside 5'-triphosphate, which is always ATP or GTP, is bound to the polymerase to form the *initiation complex*. The first internucleotide linkage is formed after binding of the second nucleoside 5'-triphosphate, which is usually a pyrimidine nucleoside (UTP or CTP). The first phosphodiester bond is then formed by nucleophilic attack of the 3'-hydroxyl group of the initiating purine nucleoside 5'-triphosphate on the α phosphorus atom of the second nucleoside triphosphate; inorganic pyrophosphate derived from the second NTP is the leaving group.

It will be noted that the first (purine) nucleoside 5'-triphosphate retains its 5'-triphosphate group after the second and the succeeding nucleotide residues have been added. This group thus permits recognition of the starting end, or 5' terminus, of RNA molecules.

Elongation and Termination of RNA Chains

Once the first two nucleotide residues have been joined, chain elongation proceeds rapidly, with transcription taking place in the 5'→3' direction, antiparallel to the 3'→5' strand of the template DNA. During elongation a segment of the newly formed RNA forms a transient complementary hybrid RNA–DNA duplex. However RNA–DNA hybrid duplexes are much less stable than DNA–DNA duplexes; the new RNA chain therefore tends to separate from the DNA at the transcription fork. Once about 10 ribonucleotide residues have been added, the σ factor dissociates from the holoenzyme to yield the core enzyme, which proceeds to replicate the DNA template further. The free σ factor then becomes available for initiating a new RNA chain with another molecule of core polymerase.

The elongation of RNA continues along the DNA template until the core polymerase molecule reaches a signal for chain termination. Specific termination signals are necessary because the DNA being transcribed contains many genes and often is circular; without a termination signal RNA would be made indefinitely. Moreover, the different kinds of RNA, such as tRNAs and rRNAs (pages 321 and 322), have quite specific chain lengths. The precise nature of the RNA termination signals is not known, but in the DNA of bacteriophage λ one such signal is a stretch of U residues. The ρ and κ proteins, which are auxiliary to the RNA polymerase but not part of it, appear to participate in terminating and releasing the RNA chain.

The Products of DNA-Directed RNA Polymerase

The RNA formed by RNA polymerase has a base composition complementary to that of the template DNA strand; in the RNA uracil residues are complementary to adenine residues of the DNA template. The complementarity of the RNA and the DNA template is shown by nearest-neighbor base-frequency analysis, which also demonstrates that the RNA chain has the opposite polarity of the template DNA chain. Complementariness has also been shown by the hybridization technique (page 882).

When highly purified RNA polymerase of E. coli is presented with a completely intact duplex DNA template, only one of the two strands is transcribed. However, not all RNAs in the cell are transcribed from the same specific strand; some RNAs are transcribed from one DNA strand, whereas other RNAs are transcribed from the other DNA strand. The selection of DNA strands for transcription occurs by an unknown mechanism.

Most available evidence indicates that in *E. coli* there is but a single DNA-directed RNA polymerase which makes mRNAs, tRNAs, and rRNAs. In eukaryotes, however, there are at least three different RNA polymerases, located in the nucleolus, in the nucleoplasm, and the mitochondria.

Posttranscriptional Processing of RNAs

RNAs synthesized by DNA-directed RNA polymerases require enzymatic processing or modification after transcription before the functional finished products, mRNAs, rRNAs, and tRNAs are released. This is particularly true in eukaryotic cells, in which complex RNA maturation processes take place.

One type of posttranscriptional processing in eukaryotes involves the attachment of blocks of poly A, containing 200 or more AMP residues, to the 3′ end of messenger RNAs, a reaction taking place in the nucleus. It is believed that the poly A "tail" aids in the transfer of mRNA from the nucleus to the cytoplasm; it is subsequently removed.

The 16S and 23S RNAs of *E. coli* appear to be formed as one single, long RNA strand, which is cleaved to yield two fragments, one containing 16S RNA and the other the 23S RNA. However, each of these fragments has extra nucleotide residues which are removed by nuclease action to yield the finished 16S rRNA and 23S rRNA, ready to be incorporated into the ribosomes (pages 322 and 1023). 5S rRNA is also made as a larger precursor.

In eukaryotic cells the ribosomal RNAs are made in the form of a single large 45S RNA strand bound to a protein in the nucleolus. The 45S precursor RNA is cleaved to yield 5S and 41S RNA; the latter is then cleaved to yield 20S RNA and 32S RNA. The latter are the precursors of the final 18S rRNA and 28S rRNA found in cytoplasmic ribosomes of eukaryotes (Figure 32-27).

Transfer RNAs also are formed from RNA precursors of higher molecular weight in both prokaryotic and eukaryotic cells. For example, the precursor of tyrosine tRNA has 41 additional nucleotide residues at the 5′ end and 3 residues at the 3′ end; in this form none of the bases are methylated. The extra nucleotide residues are removed enzymatically from the ends, and specific bases in the tRNA molecule are then modified enzymatically, by a variety of reactions, i.e., by methylation, acetylation, deamination, reduction, rearrangement, or attachment of isopentenyl or —SH groups (pages 321 and 326).

Inhibitors of RNA Synthesis

Many different compounds are known to inhibit transcription of DNA by RNA polymerase. Some have been very useful in identifying the different stages of transcription described above. One class interferes with transcription by

Figure 32-27
The maturation of ribosomal RNAs in eukaryotic cells.

Eukaryotic cell

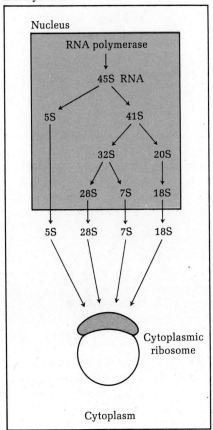

binding to the DNA template and modifying its structure; the other type binds to the RNA polymerase and interferes with its activity.

All the DNA-specific inhibitors form noncovalent complexes with DNA and impair its template function. The most important such agent is _actinomycin D,_ an antibiotic produced by _Streptomyces_ species (Figure 32-28), whose phenoxazone ring intercalates (inserts itself) between two G-C base pairs, while its pentapeptide side chains hydrogen-bond to G residues and project into the minor groove of duplex DNA. Actinomycin D has no effect on binding of RNA polymerase to DNA but preferentially blocks chain elongation. It penetrates into intact cells, in which it can inhibit transcription without affecting other aspects of cell metabolism. Replication of DNA is much less sensitive to actinomycin D than transcription.

Figure 32-28
Intercalating agents blocking template function of DNA. (Below) Actinomycin D. When it binds to duplex DNA, the flat phenoxazone ring system intercalates between two successive guanine-cytosine pairs while the two pentapeptide side chains lie in the minor groove of the double helix, each hydrogen-bonded to the opposite DNA chain. (Right) Three other planar condensed-ring compounds that intercalate between two base pairs.

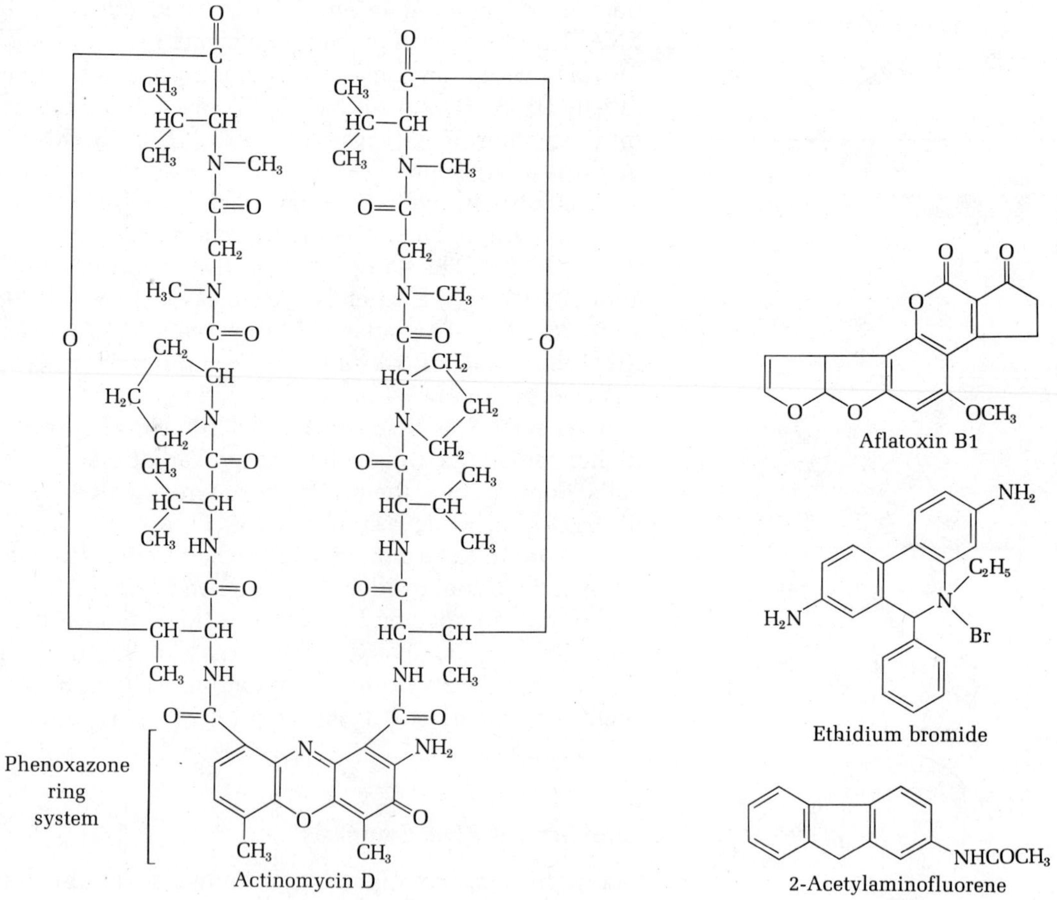

Actinomycin D

Aflatoxin B1

Ethidium bromide

2-Acetylaminofluorene

Another agent binding to DNA is *ethidium bromide* (Figure 32-28), which also is believed to intercalate between two successive base pairs. At low dye concentrations it is preferentially bound to supercoiled circular DNA, as in mitochondria (page 870). *Aflatoxin,* produced by the fungus *Aspergillus flavus,* which grows on peanuts, is an extremely potent carcinogen capable of inducing liver cancer; it inhibits both replication and transcription of DNA, as does the synthetic carcinogen 2-*acetylaminofluorene.*

Among the agents that act by inhibiting RNA polymerase itself (Figure 32-29) are a group of bacterial antibiotics called *rifamycins;* they are very potent inhibitors of bacterial RNA polymerases, but some viral and most eukaryotic nuclear RNA polymerases are unaffected. The most used of this class of compounds is *rifampicin,* a synthetic derivative of a natu-

Figure 32-29
Inhibitors of DNA-directed RNA polymerase.

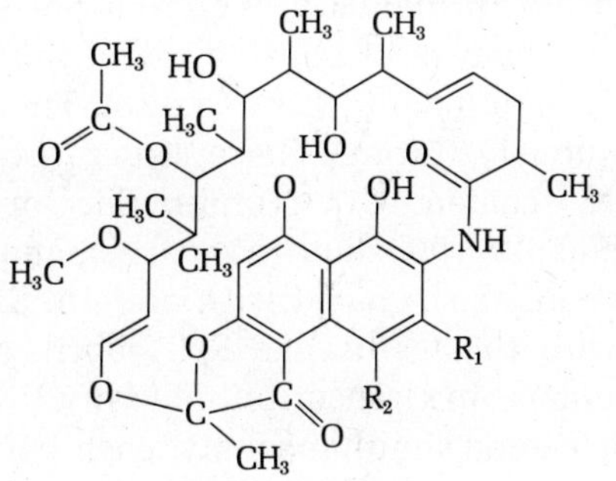

Rifamycin B ($R_1 = H$; $R_2 = O$—CH_2—COOH)

Rifampicin ($R_1 = CH=\overset{+}{N}$◯N—CH_3: $R_2 = OH$)

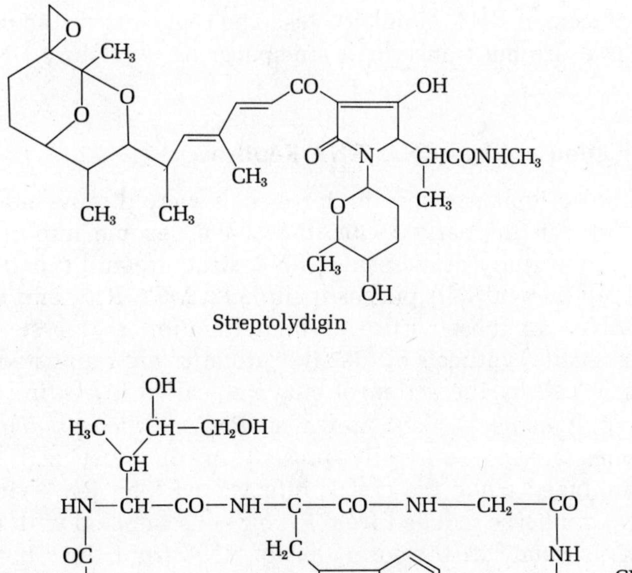

Streptolydigin

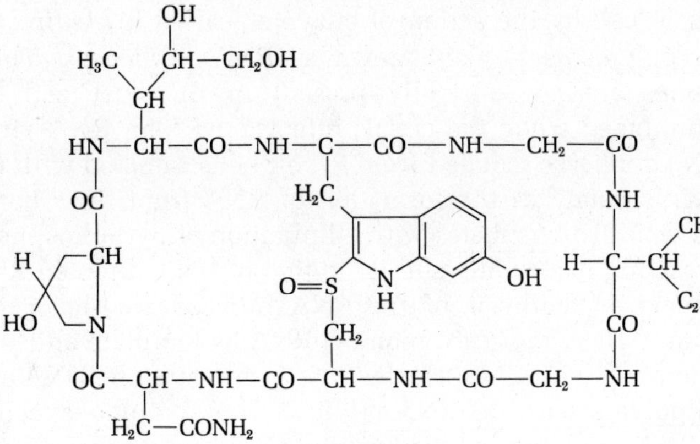

α-Amanitin

rally occurring rifamycin. Rifampicin binds noncovalently to the β subunit of RNA polymerase of some species and blocks the formation of the initiation complex, but it does not affect RNA elongation. _Streptolydigin,_ in contrast to rifampicin, blocks RNA chain elongation. _α-Amanitin,_ the toxic principle of the poisonous mushroom _Amanita phalloides_, blocks one of the nuclear RNA polymerases of eukaryotic cells but does not affect bacterial, mitochondrial, or chloroplast RNA polymerases.

Visualization of the Transcription Process

In the egg cells of amphibia, the genes coding for the synthesis of ribosomal RNA are located in the nucleoli. During early growth of the egg cell, ribosomal RNA is synthesized in large amounts. Electron microscopy of nucleolar material isolated from such cells shows it to contain long thin fibers about 10 to 20 nm in diameter, which at periodic intervals are coated with a matrix of hairlike radial fibrils; the fibrils gradually increase in length (Figure 32-30). The core fiber extending continuously through these structures consists of a filament of DNA coated with protein. The long hairlike fibrils, of which about 100 occur in each segment, are strands of RNA, presumably also coated with protein. Each DNA segment coated with the featherlike RNA fibrils is a gene for rRNA that is undergoing transcription. Many RNA chains are made on a single gene simultaneously, each being formed by an RNA polymerase molecule moving along the gene as the RNA chain grows. The length of the DNA segment coated with RNA fibrils is about 2 to 3 μm, which is about the length of DNA required to code for the precursor of 18S and 28S ribosomal RNA of eukaryotes. The segments of the core fiber that are not transcribed are spacer or regulatory DNA.

Replication of Viral RNA: RNA Replicases

RNA bacteriophages specific for E. coli were discovered by N. Zinder in the early 1960s; they have become important tools in the study of messenger RNA structure and function. The well-known RNA phages include f2, MS2, R17, and Qβ. The RNAs of these viruses, which function as messenger RNAs for the synthesis of the viral proteins, are replicated in the host cell by the action of enzymes called _RNA-directed RNA polymerases,_ also known as _RNA replicases_. These enzymes are not normally present in the host cell but are produced when the cell is infected with an RNA virus.

RNA replicase isolated from E. coli cells infected with the Qβ virus catalyzes the formation of RNA from the ribonucleoside 5'-triphosphates with elimination of pyrophosphate; the reaction is formally similar to that of DNA-directed RNA polymerase. Synthesis of the RNA proceeds in the $5' \rightarrow 3'$ direction. RNA replicase requires RNA as template and will not function with DNA. However, in contrast to the DNA and RNA polymerases, the RNA replicases are template-specific. The RNA replicase induced by the Qβ virus can employ as template only the RNA of the Qβ virus; the RNAs of the host cell are not replicated. This fact obviously explains how RNA viruses are preferentially replicated in the host cell, which contains many other types of RNA. Moreover, the Qβ replicase requires the intact Qβ RNA molecule and cannot utilize fragments of it as templates; thus it is prevented from making incomplete or defective viral RNAs.

Purified Qβ RNA replicase can bring about the net synthesis of new, biologically active Qβ RNA molecules. Starting with the Qβ (+) strand of RNA as template, the replicase can make complementary Qβ RNA (−) strands. These can then be used as templates in another incubation with the enzyme to yield Qβ RNA identical with the starting (+) strand.

The enzymatically synthesized (+) Qβ RNA was found to be infectious. S. Spiegelman and his colleagues repeated such cycles many times until the original natural template Qβ RNA was diluted to a negligible amount. Nevertheless, at each stage of replication the infectiousness of the RNA was

Figure 32-30
Electron micrograph of nucleolar genes of an amphibian egg cell. These genes, which code for ribosomal RNA, repeat along the DNA. They are identifiable because each is undergoing transcription into rRNA, about 100 strands of the latter being formed simultaneously as the RNA polymerase molecules move along the gene. The segments between the active genes are probably regulatory or "spacer" DNA.

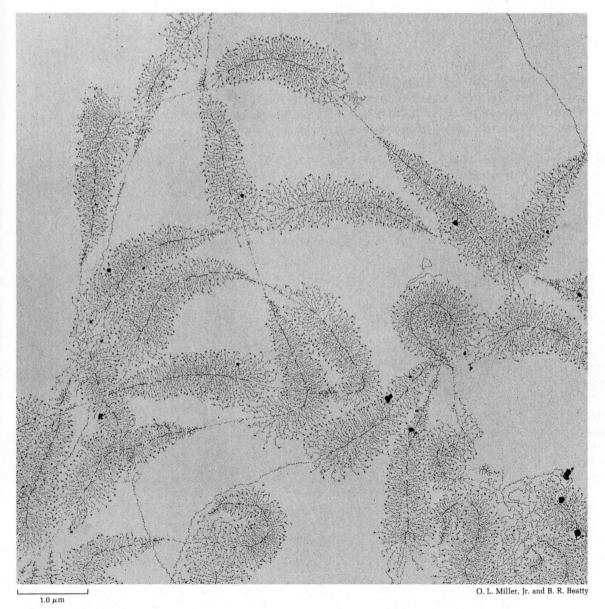

1.0 μm

O. L. Miller, Jr. and B. R. Beatty

in exact proportion to the new RNA formed by the enzyme. These experiments represented the first synthesis of a biologically active nucleic acid carried out with a highly purified enzyme.

Substantial portions of the amino acid sequence of the three protein gene products and much of the nucleotide sequence of the R17 RNA are known. They have given important information, to be discussed in Chapter 34 (page 966), on various structural features of mRNAs, on the genetic code, and on the use of specific codons for initiation and termination of translation.

Polynucleotide Phosphorylase

This enzyme, discovered by M. Grunberg-Manago and S. Ochoa in 1955, was the first enzyme found capable of synthesizing long-chain polynucleotides; it was a prototype or forerunner of the DNA and RNA polymerases. Polynucleotide phosphorylase, which is apparently found only in bacteria, catalyzes the reaction

$$NDP + (NMP)_n \rightleftharpoons (NMP)_{n+1} + P_i$$

It requires the 5'-diphosphates of ribonucleosides and cannot act on the homologous 5'-triphosphates or on deoxyribonucleoside 5'-diphosphates. Mg^{2+} is required for its action. The RNA-like polymer formed by polynucleotide phosphorylase contains 3',5'-phosphodiester linkages, which can be attacked by ribonuclease. The reaction is readily reversible and can be pushed in the direction of breakdown of the polyribonucleotide by increasing the phosphate concentration.

The polynucleotide phosphorylase reaction does not utilize a template and therefore does not form a polymer having a specific base sequence. It does require a priming strand of RNA, which merely furnishes a free 3'-hydroxyl terminus to which additional residues can be added. The reaction proceeds as well with one species of monomer as with all four. In general, the base composition of the polymer formed by the enzyme reflects the relative concentrations of the 5'-diphosphate precursors in the medium. For these reasons it is unlikely that polynucleotide phosphorylase normally functions to make RNA in the intact cell. It has been suggested to play a role in the degradation of messenger RNA in bacteria. However, polynucleotide phosphorylase has played a very important role in molecular biology, since it can be used for the laboratory preparation of many different kinds of RNA-like polymers, with differing sequences and frequencies of bases. Such synthetic RNA polymers were vital in deducing the genetic code words for the amino acids (page 958).

Summary

Double-strand DNA undergoes semiconservative replication in intact *E. coli* cells, shown in a classical experiment by Meselson and

Stahl. Autoradiographic, genetic, and biochemical experiments indicate that the circular bacterial chromosome is replicated bidirectionally; furthermore, both strands are replicated simultaneously.

DNA polymerase I of *E. coli* catalyzes synthesis of DNA polymers from the four deoxyribonucleoside 5′-triphosphates in the presence of Mg^{2+}, with elimination of pyrophosphate. The reaction requires the presence of some preexisting DNA as template-primer. Chain growth is in the 5′→3′ direction. Intact double-strand DNA does not prime the reaction unless it has single-strand nicks. DNA polymerase has specific binding sites for the template, for the primer, for the 3′ terminus of the primer chain, for the incoming deoxyribonucleoside 5′-triphosphate, and for its 5′→3′ exonuclease activity. The enzyme makes a strand of DNA complementary to the template strand, as shown by analysis of base composition. Nearest-neighbor base-frequency analysis proved that the polarity of the new strand is opposite to that of the template strand. The 3′→5′ exonuclease and 5′→3′ exonuclease activities of DNA polymerase I serve a proofreading function by eliminating unpaired nucleotides. DNA ligase can join the ends of two DNA chains if they are duplexed to a complementary strand. A. Kornberg and his colleagues have demonstrated that DNA polymerase I and DNA ligase can function with complete fidelity in the laboratory synthesis of a biologically active strand of circular DNA of phage ϕX174.

Mutants deficient in DNA polymerase I led to the discovery of DNA polymerases II, III, and III*. The latter, together with the protein copolymerase III*, appears to be the form responsible for replication of DNA, whereas DNA polymerase I functions in an auxiliary capacity and in DNA repair.

Okazaki has found that both strands of DNA are replicated in short pieces of up to 2,000 nucleotide residues, which are subsequently joined by DNA ligase. Replication of DNA requires the participation of unwinding and untwisting proteins. Moreover, DNA replication requires priming by DNA-directed RNA polymerase, which produces short priming lengths of RNA on which DNA is built; the RNA primers are then excised. DNA may also be synthesized by RNA-directed DNA polymerase or reverse transcriptase, found in some cancer-producing RNA viruses and also in normal cells. Single-strand nicks, defective bases, and thymine dimers in DNA can be repaired enzymatically.

RNAs are synthesized by DNA-directed RNA polymerase from ribonucleoside 5′-triphosphates, but only one strand of DNA is transcribed in vivo. The enzyme is very complex and requires special factors for recognizing the initiation point. Many RNA chains may be transcribed simultaneously from a single gene. Messenger RNAs are modified by attachment of a long poly-A tail at the 3′ end. Ribosomal RNAs are generated from a long precursor chain, which is cleaved to yield the finished rRNAs.

RNA-directed RNA replicases are formed in bacterial cells infected by RNA viruses; the replicases are template-specific and will accept as primer only intact homologous viral RNA. Polynucleotide phosphorylase can reversibly form RNA-like polymers by elimination of phosphate from ribonucleoside 5′-diphosphates; it adds or removes mononucleotides at the 3′-hydroxyl end of the primer.

References

Books

DAVIDSON, J. N.: *The Biochemistry of the Nucleic Acids,* 7th ed., Academic, New York, 1972.

KORNBERG, A.: *DNA Synthesis*, Freeman, San Francisco, 1974. A detailed, clarifying account of all aspects of DNA replication and transcription, illustrated with many helpful drawings. Indispensable.

LEWIN, B.: *Gene Expression*, vol. II, *Eukaryotic Chromosomes*, Wiley, New York, 1974. Replication and transcription of eukaryotic chromosomes.

STENT, G. S.: *Molecular Genetics: An Introductory Narrative*, Freeman, San Francisco, 1971.

STEWART, P. R., and D. S. LETHAM (eds.): *The Ribonucleic Acids*, Springer-Verlag, New York, 1973. Another extremely valuable book.

WOOD, W. B., J. H. WILSON, R. M. BENBOW, and L. E. HOOD: *Biochemistry: A Problems Approach*, Benjamin, Menlo Park, Calif., 1974. Short review and a number of excellent problems and questions.

ZUBAY, G. L., and J. MARMUR (eds.): *Papers in Biochemical Genetics*, 2d ed., Holt, New York, 1973. Reprinted classic papers on DNA replication and transcription.

Articles

Advances in Molecular Genetics, a special issue of the *Brit. Med. Bull.*, vol. 29 (Sept. 1973), with 16 short, authoritative reviews of the present status.

ALBERTS, B. and L. FREY: "T$_4$ Bacteriophage Gene 32: A Structural Protein in the Replication and Recombination of DNA," *Nature*, 227: 1313–1318 (1970).

BEERS, R. F., JR., R. M. HERRIOTT, and R. C. TILGHMAN (eds.): Molecular and Cellular Repair Processes, *Suppl. 7, Johns Hopkins Med. J.*, Johns Hopkins Press, Baltimore, 1972. Another important collection of reviews on an area of increasing importance.

CLEAVER, J.: "Xeroderma Pigmentosum: A Human Disease in Which an Initial Stage of DNA Repair Is Defective," *Proc. Natl. Acad. Sci. (U.S.)*, 63: 428–435 (1969).

DRESSLER, D.: "The Rolling Circle Model for ϕX174 DNA Replication; II: Synthesis of Single-Stranded Circles," *Proc. Natl. Acad. Sci. (U.S.)*, 67: 1934–1942 (1970).

GEFTER, M. L. (ed,): "DNA Synthesis in Vitro," *PAABS Rev.*, 1: 495–548 (1972). A short survey, followed by five reprinted original research papers on DNA polymerases II and III and other topics.

HAMKALA, B., and O. L. MILLER, JR.: "Electron Microscopy of Genetic Activity," *Ann. Rev. Biochem.*, 42: 379–396 (1973).

HANAWALT, P. C.: "Repair of Genetic Material in Living Cells," *Endeavour*, 31: 83–87 (1972).

KORNBERG, A.: "Active Center of DNA Polymerase," *Science*, 163: 1410–1418 (1969).

KORNBERG, T., A. LOCKWOOD, and A. WORCEL: "Replication of the *E. coli* Chromosome with a Soluble Enzyme System," *Proc. Natl. Acad. Sci. (U.S.)*, 71: 3189–3193 (1974).

LEHMAN, I. R.: "DNA Ligase: Structure, Mechanism, and Function," *Science*, 186: 790–797 (1974).

MCKENNA, W. G., and M. MASTERS: "Biochemical Evidence for the Bidirectional Replication of DNA in *E. coli*," *Nature*, 240: 536–539 (1972).

MILLER, O. L., JR.: "The Visualization of Genes in Action," *Sci. Am.*, March 1973, pp. 34–42.

Molecular Biology Comes of Age, *Nature*, 248: 765–788 (Apr. 26, 1974). A series of interesting essays, some by participants in the development of molecular biology, in a special issue devoted to the twenty-first anniversary of the Watson-Crick hypothesis.

MORGAN, R. (ed.): "DNA Replication in Bacteria and Bacteriophages," *PAABS Rev.*, 2: 323–386 (1973).

SALSER, W. A.: "DNA Sequencing Techniques," *Ann. Rev. Biochem.*, 43: 923–966 (1974).

SCHEKMAN, R., A. WEINER, and A. KORNBERG: "Multienzyme Systems of DNA Replication," *Science*, 186: 987–993 (1974). Identification of the components of systems replicating the DNA of small DNA phages of *E. coli*.

SOBELL, H. M.: "How Actinomycin Binds to DNA," *Sci. Am.*, August 1974, pp. 82–91.

SUGINO, A., S. HIROSE, and R. OKAZAKI: "RNA-linked Nascent DNA Fragments in *E. coli*," *Proc. Natl. Acad. Sci. (U.S.)*, 69: 1863–1867 (1972).

WEISSBACH, A. (ed.): "RNA-Dependent DNA Polymerases in Animal Cells: A Question," *PAABS Rev.*, 3: 167–208 (1974). A short survey of the field, followed by six reprinted research articles, important in this field, including papers of H. M. Temin and D. Baltimore.

Problems

1. Under conditions where the doubling time of a colony of *E. coli* cells is 30 min, (a) what is the turnover number (nucleotides polymerized per second) of the DNA polymerase in vivo if there are two enzyme molecules active at any one time? (b) How does this compare with the turnover number for pure pol I?

2. Calculate how long it takes for the gene for ribonuclease (104 residues) to be replicated during cell division in *E. coli*. Use data from Problem 1.

3. A single (+) strand of DNA (base composition: A, 21 percent; G, 29 percent; C, 29 percent; T, 21 percent) is replicated by DNA polymerase to yield a complementary (−) strand. The resulting duplex DNA is then used as a template by RNA polymerase, which transcribes the (−) strand. Indicate the base composition of the RNA formed.

4. Indicate the base composition of DNA synthesized by DNA polymerase on templates provided by an equimolar mixture of (+) and (−) single-stranded circular ϕX174 DNAs if the (+) strand base composition is A, 24.6 percent; G, 24.1 percent; C, 18.5 percent; T, 32.7 percent.

5. Ignoring the cost of RNA priming, (a) how many ATP molecules are utilized in replicating the entire E. coli genome, assuming the process begins with a mixture of ATP and deoxyribonucleoside 5'-monophosphates? (b) How many molecules of glucose would have to be degraded under aerobic conditions to supply this much energy? (c) Under anaerobic conditions?

6. A rapidly growing E. coli cell contains about 15,000 ribosomes. (a) Assuming the precursor rRNA genes total 5,000 nucleotide pairs, how many ATP molecules are expended in the transcription of rRNA if the process begins with ATP and ribonucleoside 5'-monophosphates? (b) How does this compare with the energy expended in replication?

7. How would you determine whether a short segment of double-helical DNA has parallel or antiparallel chains? The base sequence of one strand (5'→3' direction) is pApTpCpGpGpT-pCpApApCpCpTpGpTpApGpG.

8. (a) Sedimentation in alkaline sucrose density gradients, which causes DNA to denature, can be used to differentiate linear from circular DNA and to determine the relative size of DNA fragments. Replicative form II (RF II) of ϕX174 DNA is a double-stranded circle with a nick in one strand. What types and lengths of molecules should be distinguishable upon sedimentation of RF II DNA in alkaline sucrose?

(b) The in vivo replication of ϕX174 DNA is a complex process involving three separate stages. Replicative stage 2 involves a rolling circle mechanism as shown in Figure 32-4. After replication has proceeded to the point indicated in the third drawing in Figure 32-4, what types and lengths of DNA molecules should be distinguishable upon sedimentation in alkaline sucrose?

(c) Analysis of the unit length molecules of (b) showed that they had the base composition of the (−) strand ϕX174 DNA. Which of the fragments recovered in (b) would have eventually become (+) strands if replication had been allowed to proceed?

(d) Which fragments in (b) would have become (−) strands?

(e) When a short pulse of tritiated thymine is added to replicating ϕX174 DNA, what fragments(s) would not be labeled if replication were stopped and the DNA were sedimented as in (b)?

(f) When replication proceeds at low temperatures and the DNA, after exposure to a brief pulse of labeled thymine, is immediately extracted, it was found upon sedimentation in alkaline sucrose that radioactive labeled sequences complementary to (−) strands were present only in molecules greater than unit length. Although labeled sequences complementary to all segments of (+) strands were isolated, they were never found in fragments greater than 2,000 nucleotides in length (total length of ϕX174 DNA is 5,400 nucleotides). Interpret this information.

CHAPTER **33** **TRANSLATION: THE BIOSYNTHESIS OF PROTEINS**

We now consider the third major question posed by the genetic continuity of living organisms: How is the genetic information contained in the nucleotide sequence of messenger RNA translated into protein structure? In this chapter we shall deal with the mechanisms by which the polypeptide chain is built and by which ribosomes make possible the transfer of genetic information into the proper sequence of amino acids in the polypeptide chain. In following chapters we shall consider other elements of the translation process, namely, the identification of the coding triplets for the amino acids, the regulation of protein synthesis, and the assembly of three-dimensional biostructures from one-dimensional genetic information.

The enzymatic mechanisms by which the peptide bonds of protein molecules are formed have long attracted the attention of biochemists. One early view was that peptidases and proteolytic enzymes function not only in the breakdown but also in the synthesis of proteins, by reversal of their hydrolytic activity. However, as the central role of ATP in biosynthetic reactions emerged, as well as the concept that biosynthetic and catabolic reactions have different pathways, reversal of hydrolytic activity seemed an unlikely mechanism for protein biosynthesis. Indeed, early studies of reactions that could be regarded as models or prototypes of protein synthesis, such as the enzymatic synthesis of glutamine from glutamic acid (page 695) and the formation of glutathione (page 714) from its component amino acids, showed that ATP is required as energy donor to form amide or peptide bonds. Similarly, no information was available on how the polypeptide chain is built. One early view held that short peptides are made first and then assembled to form polypeptide chains.

It was not until the isotope-tracer technique was applied that the first significant progress was made in the study of protein biosynthesis. Studies on intact animals or intact cells with radioactive amino acids soon revealed that the proteins of animal tissues undergo metabolic turnover, that protein synthesis requires a source of metabolic energy, and that amino acids and not simple peptides are the immediate precursors of protein.

The modern era of research on the mechanism of protein biosynthesis began in the early 1950s with the important investigations of P. C. Zamecnik and his colleagues in Boston, who first developed cell-free systems for the study of protein synthesis, who identified ribonucleoprotein particles, the ribosomes, as the site of protein biosynthesis, and who discovered transfer tRNAs.

Ribosomes as the Site of Protein Synthesis

Zamecnik and his colleagues injected radioactive amino acids into rats. At successive time intervals the livers were removed, homogenized, and fractionated by differential centrifugation (page 380). The intracellular fractions were then examined for the presence of labeled amino acid incorporated into protein. When long periods, i.e., hours or days, were allowed to elapse between injection and fractionation of the liver, all the intracellular fractions contained labeled protein. However, if the liver was fractionated very shortly after the injection of labeled amino acid, only the microsome fraction (page 381) contained labeled protein. From such studies it was concluded that proteins are first synthesized in intracellular structures contained in the microsome fraction and that newly synthesized proteins are then transferred from these structures to the other cell components. Zamecnik and his colleagues soon found that incubation of labeled amino acids with freshly prepared cell-free homogenates of rat liver also resulted in the incorporation of radioactivity into the microsomal proteins. The incorporation of amino acids required the addition of ATP and respiratory substrates to provide energy. Thus it was demonstrated that intact cell structure is not required for protein synthesis.

Liver homogenates were then fractionated into the nuclear, mitochondrial, microsomal, and soluble fractions and various combinations of these fractions were tested for their ability to cause the incorporation of labeled amino acids into protein. A combination of two fractions was found to be necessary, the microsome fraction and the soluble, or cytosol, fraction. The microsome fraction served as the recipient of the newly incorporated amino acids, whereas the soluble fraction appeared to furnish essential cofactors. Mitochondria are required in such reconstruction experiments only insofar as they supply energy in the form of ATP.

When the labeled microsome fraction was further subfractionated, the incorporated radioactivity was largely recovered in the form of small ribonucleoprotein particles. Such particles had earlier been seen in the cytoplasm with the electron microscope by A. Claude and were later observed on the surface of the endoplasmic reticulum by K. R. Porter and G. E. Palade, but their function was then still unknown. These ribonucleoprotein particles, later named *ribosomes*, can be detached from the endoplasmic reticulum by treatment with a detergent and separated in homogeneous form by differential centrifugation. Ribosomes purified in this manner incorporate radioactive amino acids into peptide linkage on incubation with ATP, Mg²⁺, and the soluble or cytosol fraction of rat liver. Soon other workers found it possible to observe amino acid incorporation into ribosomes from many other cells, particularly from *reticulocytes,* immature red blood cells which are extremely active in hemoglobin synthesis, and from *E. coli* cells.

Essential Cofactors and Successive Stages in the Biosynthesis of Proteins

Zamecnik and his colleagues found that the soluble fraction of liver furnishes two categories of essential factors for protein biosynthesis. One type is heat-labile, and therefore presumably protein in nature, and the other is heat-stable. One of the heat-labile soluble factors in liver cytoplasm was found to be an enzyme catalyzing an ATP-dependent activation reaction by which free amino acids are esterified to a heat-stable component also present in the soluble fraction. The heat-stable amino acid acceptor is a low-molecular-weight type of RNA, which was first called "soluble RNA" or sRNA but now is called *transfer RNA* (tRNA) to designate its function more accurately. The aminoacyl-tRNA ester so formed in the activation reaction is then bound to the ribosome, where it serves as the donor of an aminoacyl residue in the ensuing reactions in which the peptide chain is elongated.

Subsequent research has revealed that the synthesis of polypeptide chains takes place in four major stages and that specific enzymes and cofactors are required at each stage. Table 33-1 provides some orientation for the more detailed discussion to follow. In the first stage, which is called the *activation* step and which takes place entirely in the soluble cytoplasm, amino acids are enzymatically esterified to their corresponding tRNAs, at the expense of ATP energy. In the second stage, called *initiation,* the messenger RNA, which carries the genetic message specifying the sequence of amino acids, and the first or initiating aminoacyl-tRNA are bound to the small subunit of the ribosome, a process which requires three specific proteins called *initiation factors* (IF-1, IF-2, IF-3), as well as GTP and Mg²⁺. The large ribosomal subunit then attaches to form a functional ribosome, ready for the next step.

In the third stage, called *elongation,* the polypeptide chain is lengthened by sequential addition of new aminoacyl residues enzymatically transferred from aminoacyl-tRNA esters, each of which is bound to the ribosome in response to a specific codon or base triplet in the mRNA. Two *elongation factors,* EF-T and EF-G, also proteins, are required in the lengthening of the chain. After formation of each new peptide bond, the ribosome moves along the mRNA to bring the next codon in position for alignment of the next aminoacyl-tRNA, a process which requires energy provided in the form of GTP.

In the last stage of protein synthesis, *termination,* the polypeptide chain is completed when appropriate termina-

Table 33-1 Components required in the four major steps in protein synthesis in E. coli

Stage	Necessary components
1. Activation of the amino acids	Amino acids
	tRNAs
	Aminoacyl-tRNA synthetases
	ATP
	Mg^{2+}
2. Initiation of the polypeptide chain	fMet-tRNA in prokaryotes (Met-tRNA in eukaryotes)
	Initiation codon in mRNA (AUG)
	mRNA
	GTP
	Mg^{2+}
	Initiation factors (IF-1, IF-2, IF-3)
	30S ribosomal subunit
	50S subunit
3. Elongation	Functional 70S ribosomes
	Aminoacyl-tRNAs specified by codons
	Mg^{2+}
	Elongation factors (EF-T and EF-G)
	GTP
4. Termination	Functional 70S ribosomes
	Termination codon in mRNA
	Polypeptide release factors (R_1, R_2, R_3)

tion signals in the mRNA are reached; the product is then released from the ribosome, a process that requires specific proteins called *releasing factors*.

Direction and Rate of Chain Growth during Polypeptide Synthesis

Before we examine the stages in protein synthesis more closely, we must consider the direction of growth of the polypeptide chain, whether it is built starting from the amino-terminal or the carboxyl-terminal residue. This information was vital in order to construct a chemically sound working hypothesis, since the amino acids are asymmetrical building blocks having a "head," the α-amino group, and a "tail," the α-carboxyl group. The answer to this basic question came from experiments of H. M. Dintzis. Tritium-labeled leucine was added to suspensions of reticulocytes actively synthesizing the α and β chains of hemoglobin, which contain many leucine residues and whose amino acid sequences were known. Incubation was carried out at a low temperature (15°C) to slow down the rate of protein synthesis. Samples of the reticulocytes were then taken at different intervals between 4 and 60 min. The labeled hemoglobin formed by the reticulocytes was isolated, the α and β chains separated and cleaved into peptide fragments with trypsin, and the resulting fragments separated by paper electro-

Figure 33-1
Addition of labeled leucine residues to the carboxyl-terminal end of the α chain of hemoglobin. The colored zone indicates the location of the labeled leucine residues along the chain following exposure of the reticulocytes to ³H-leucine for the periods shown at 15°C. After 4 min exposure only a few leucine residues at the carboxyl-terminal end were labeled, since these chains were already nearly finished at the time the labeled leucine was added.

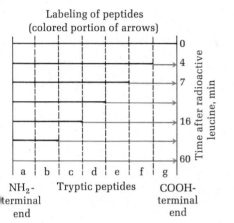

Labeling of peptides
(colored portion of arrows)

phoresis. The specific radioactivity of each of the leucine-containing tryptic peptides was measured. Since the positions of these peptides in the hemoglobin chains were already known (page 111), it was possible to deduce which end of the polypeptide chain is made first (Figure 33-1). We see that after 4 min incubation with ³H-leucine, only the leucine-containing peptide at the carboxyl-terminal end of the polypeptide chain was labeled. With longer exposure to ³H-leucine, there was an increase in the number of labeled peptides, all of which were grouped at the carboxyl-terminal end. In samples recovered after 60 min incubation with the labeled leucine, the leucine residues of all the tryptic peptides were labeled equally, indicating that the entire length of the polypeptide chain had been made within the 60-min period. Obviously the hemoglobin chains that became isotopically labeled at the 4-min interval were nearly complete at the time isotopic leucine was added. Therefore only a few labeled leucine residues could have been added at the end of such chains before they were completed and discharged from the ribosome. Conversely, in those chains which had just been started when isotopic leucine was added, all the leucine-containing peptides were labeled except that at the amino-terminal end. It was therefore concluded that polypeptide chains are constructed beginning with the amino-terminal amino acid, whose carboxyl group combines with the amino group of the incoming amino acid. The addition of successive residues at the free carboxyl-terminal end of the growing polypeptide continues until the chain is complete. This conclusion has now been amply confirmed by other types of experiments in both prokaryotic and eukaryotic species.

An interesting corollary of such experiments is that they give precise information on the time required to synthesize a complete polypeptide chain. At 37°C a single ribosome in the intact rabbit reticulocyte can complete an entire α chain of hemoglobin (146 residues) in about 3 min, a rate of a little less than one residue per second. In *E. coli* and other bacteria growing in an optimal medium, each functional ribosome can insert about 20 amino acids per second.

The Amino Acid Activation Reaction and Its Amino Acid Specificity

In the first stage of protein synthesis, which occurs in the cytosol, the 20 different amino acids functioning as building blocks of proteins are esterified to their corresponding tRNAs by the action of a class of enzymes known as amino acid–activating enzymes, or more specifically, *aminoacyl-tRNA synthetases*, each of which is highly specific for one amino acid and for its corresponding tRNA. The reaction catalyzed is

$$\text{Amino acid} + \text{tRNA} + \text{ATP} \xrightarrow{\text{Mg}^{2+}} \text{aminoacyl-tRNA} + \text{AMP} + \text{PP}_i \quad (1)$$

Nearly all the activating enzymes for the 20 different amino acids have been obtained in highly purified form, particularly from *E. coli;* several have been crystallized. They have molecular weights in the range 100,000 to 240,000. Some have only a single polypeptide chain of molecular weight 100,000 (for example, leucine), some have two or more identical subunits, and some have dissimilar subunits. Glycyl-tRNA synthetase has two subunits of molecular weight 35,000 and two of molecular weight 82,000. All have an absolute requirement for Mg^{2+}, and all contain one or more sulfhydryl groups that are essential for activity.

In the activation reaction pyrophosphate cleavage of ATP (page 410) takes place to yield AMP and pyrophosphate, as is true for the analogous activation reactions for fatty acids (page 546). The transfer of an amino acid to its tRNA occurs in two separate steps on the enzyme catalytic site. In the first step ATP reacts with the amino acid, with displacement of pyrophosphate, to form an enzyme-bound intermediate, an *aminoacyl adenylic acid* (Figure 33-2). In this intermediate the carboxyl group of the amino acid is linked by an acid anhydride bond with the 5′-phosphate group of the AMP; the high-energy anhydride bond activates the carboxyl group of the amino acid. In the second step the aminoacyl group is transferred from AMP to the 3′ end of its tRNA, followed by release of the products, an aminoacyl-tRNA and adenylic acid. The two steps are

ATP + amino acid $\rightleftharpoons$
$$[\text{aminoacyl adenylic acid}] + \text{pyrophosphate} \quad (2)$$

[Aminoacyl adenylic acid] + tRNA $\rightleftharpoons$
$$\text{aminoacyl-tRNA} + \text{adenylic acid} \quad (3)$$

The aminoacyl group becomes esterified by its activated carboxyl to the free 2′-hydroxyl group of the terminal AMP residue at the 3′ end of the tRNA molecule, that which bears the terminal C-C-A sequence (page 321), as shown in Figure 33-3. An equilibrium mixture of the aminoacyl esters of the terminal 2′- and 3′-hydroxyl groups of the tRNA subsequently forms.

Note that the overall activation reaction proceeds with a relatively small decrease in standard free energy. The new ester linkage between the amino acid and the tRNA, which is formed at the expense of the hydrolysis of a high-energy bond of ATP, is therefore a high-energy bond. Since the inorganic pyrophosphate formed in the activation reaction undergoes subsequent hydrolysis to orthophosphate by pyrophosphatase, two high-energy phosphate bonds may ultimately be split for each amino acid molecule activated. The overall reaction for amino acid activation in the cell is thus essentially irreversible. The activation enzymes are usually assayed by measuring the rate at which ^{32}P-labeled pyrophosphate is incorporated into added ATP in the first of the two reaction steps [reaction (2)]. This freely reversible exchange reaction requires the presence of the specific amino acid substrate but does not require its tRNA.

The aminoacyl-tRNA synthetases are highly specific for both the amino acid and the corresponding tRNA. This is a

Figure 33-2
General structure of aminoacyl adenylates, formed on the active site of aminoacyl-tRNA synthetases.

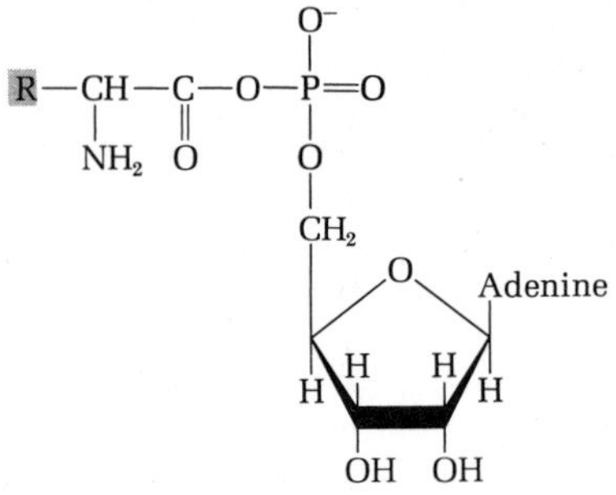

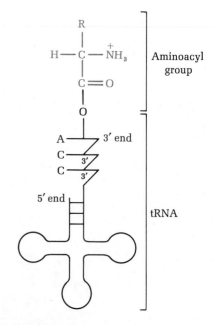

Figure 33-3
General structure of aminoacyl-tRNAs. See also Figure 33-5 for details of tRNA structure.

Figure 33-4
Synthetic amino acid analogs. They can be
incorporated into polypeptide chains in
place of the corresponding natural amino
acids.

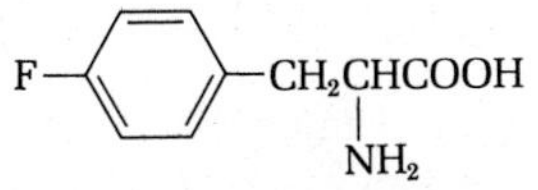

$$F-\hspace{-2mm}\bigcirc\hspace{-2mm}-CH_2\underset{\underset{NH_2}{|}}{C}HCOOH$$

p-Fluorophenylalanine

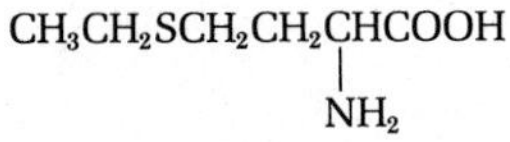

$$CH_3CH_2SCH_2CH_2\underset{\underset{NH_2}{|}}{C}HCOOH$$

Ethionine

$$CH_3CH_2CH_2CH_2\underset{\underset{NH_2}{|}}{C}HCOOH$$

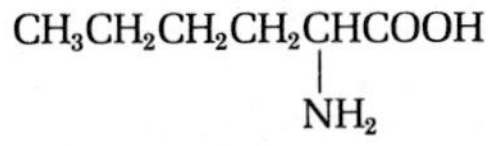

Norleucine

matter of the utmost importance, since once the aminoacyl-tRNA is formed, it is recognized only by the anticodon triplet in its tRNA moiety and not by the amino acid residue, as we shall see. Thus the aminoacyl-tRNA synthetases must possess two very specific binding sites, one for the amino acid substrate and the other for the corresponding tRNA; there is also a third site for binding ATP. The aminoacyl-tRNA synthetases are so specific in discriminating between the naturally occurring amino acids that there is less than 1 chance in 10,000 of an error under intracellular conditions. However, these enzymes can be fooled into accepting certain nonbiological amino acid analogs instead of their normal substrates. Among these are *p-fluorophenylalanine*, an analog of phenylalanine, and *ethionine* and *norleucine,* analogs of methionine (Figure 33-4). When these analogs are fed to animals, they are incorporated into proteins in positions normally occupied by phenylalanine or methionine, respectively.

Structure of tRNAs and the Specificity of the Activating Enzymes

The specificity of the aminoacyl-tRNA synthetases for their corresponding tRNAs is a complex and interesting matter, which first requires review of the covalent structure and the nucleotide components of tRNA molecules (see Chapter 12, page 321). The complete base sequences of over 50 tRNAs are now known, and important information is also available on their conformation, their specificity, and the function of different parts of their structure.

There is now substantial evidence that tRNA molecules, which have from 73 to 93 nucleotide residues and a molecular weight of about 25,000, possess a specific three-dimensional conformation. They show a substantial hyperchromic effect (page 874) at 260 nm when heated, suggesting that many of their bases are paired, presumably by looping of the single-strand tRNA chain. In fact, some 60 to 70 percent of the tRNA structure is in double-helix form. After tRNAs are melted by heating, they quickly resume their native configuration on cooling. Moreover, it has been demonstrated that the native conformation of tRNA molecules is essential for their biological activity; unfolded tRNA is not capable of acting as an amino acid acceptor. From the base sequences of tRNAs it has been possible to construct a number of conformations allowing intrachain base pairing. The surprising finding has been made that although the tRNAs for different amino acids have quite different base compositions, all tRNAs can be drawn two-dimensionally in the same four-arm cloverleaf arrangement (page 326) if the chains are arranged in such a way as to yield *maximal* intrachain base pairing (Figure 33-5). Some tRNAs, those with longer chains, are capable of forming a short fifth arm. Several tRNAs have been obtained in crystalline form. X-ray analysis of crystals of yeast phenylalanine tRNA shows that the molecule is asymmetrically folded to yield a compact structure about 9.0

Figure 33-5
General cloverleaf structure of tRNAs. The large dots on the backbone are nucleoside residues, and the crosslines represent base pairs. Characteristic and/or invariant residues common to all tRNAs are in color.

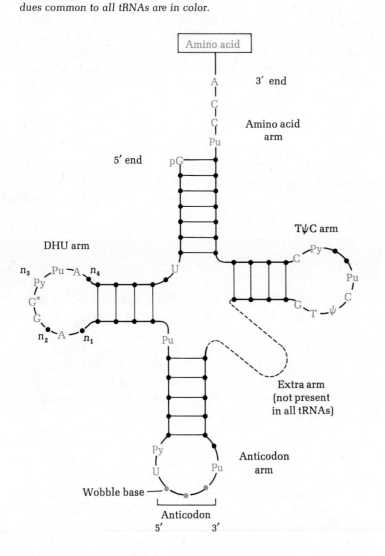

Key:
Pu = purine nucleoside
Py = pyrimidine nucleoside
ψ = pseudouridine
G* = guanosine or
    2'-O-methylguanosine
n_1 = 0 to 1
n_2 = 1 to 3
n_3 = 1 to 3
n_4 = 0 to 2 nucleoside
    residues in DHU
    arm, depending on
    the tRNA.

In some tRNAs the DHU stem has only three base pairs.

nm long and 2.5 nm thick, with the *anticodon arm* at one end and the *amino acid arm* at the other (Figure 33-6).

From base-sequence data some biologically important structural features of tRNAs have been identified (Figure 33-5). All tRNA molecules have the same terminal sequence Pu-C-C-A at the 3′ end of the *amino acid arm*. The last residue, adenylic acid, is the site to which the aminoacyl group is esterified by the aminoacyl synthetases (Figure 33-5). A pseudouridine residue (ψ) is characteristically present in the *TψC arm* of all tRNAs; similarly, a dihydrouridine residue characterizes the *dihydrouridine arm*. The *anticodon arm* contains a specific triplet of nucleosides that is different in the tRNAs for different amino acids. This triplet is the *anticodon*, which is complementary to the corresponding codon triplets in mRNA. The anticodon triplet of the amino-

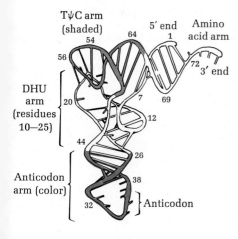

Figure 33-6
The three-dimensional conformation of yeast phenylalanine tRNA deduced from x-ray diffraction analysis at 0.3-nm resolution. [Adapted from S. H. Kim and colleagues, Science, 185: 436 (1974).]

Figure 33-7
Triacanthine [6-(3-methylbut-2-enylamino)purine], one of the minor purines found next to the anticodon. This compound is a cytokinin, a plant hormone promoting cell division.

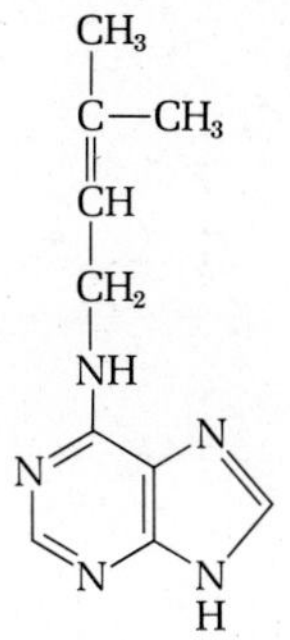

acyl-tRNA has the proper base sequence to form specific hydrogen bonds with the bases of the corresponding codon of the mRNA (page 969). In this manner the required amino acid is selected and also positioned correctly for transfer to the growing polypeptide chain. One of the bases in the anticodon is the wobble base (page 963), which is less specific in its interaction with its corresponding base in the codon than the other two anticodon bases. A uracil residue is found next to the 5′ end of the anticodon and a purine or purine derivative is found next to the 3′ end. In some tRNAs, particularly those of plants, an isopentenyl derivative of a purine (Figure 33-7) is found in the latter position. These characteristic bases apparently serve as "stoppers" to demarcate the anticodon (page 970).

tRNA molecules must contain at least two other specific sites, the *enzyme recognition site*, at which the tRNA is bound to the corresponding activating enzyme, and the *ribosome attachment site*. The variable nucleotide residues of tRNAs differ not only with the amino acid but also with the organ or species; an aminoacyl synthetase for any given amino acid works best with the corresponding tRNAs from the same species.

A striking feature of the molecular structure of tRNAs is the presence of many minor bases (page 321, Chapter 12) in addition to the normal bases A, U, G, and C. At least 40 different minor nucleosides have been discovered in tRNAs, most of which are methylated forms of the normal nucleosides. Enzymes catalyzing the methylation of specific bases in tRNAs were found by E. Borek; they require S-adenosylmethionine (page 698) as methyl-group donor. There are several possible functions of the minor bases. They may serve to disallow intrachain base pairing at certain points in order to yield the characteristic arm structure, to disallow unwanted base pairing with messenger RNA, and to render the tRNA less susceptible to the hydrolytic attack of nucleases.

There is more than one specific tRNA for each amino acid. Yeast cells contain five different homologous, or *isoaccepting*, tRNAs, which react specifically with leucine and with leucine aminoacyl synthetase, five for serine, four for glycine, and four for lysine. However, a given aminoacyl-tRNA synthetase is not equally reactive with all the homologous tRNAs. Moreover, in some cells there is more than one aminoacyl-tRNA synthetase for certain amino acids, which may differ in their relative specificity for the homologous tRNAs. The biological basis for the multiplicity of homologous tRNAs and aminoacyl-tRNA synthetases is not fully understood. *Neurospora crassa* cells have been found to contain at least two species of tRNA for many amino acids. One species is present only in the mitochondria and the other(s) only in the cytoplasm, suggesting that mitochondria and cytoplasmic protein synthesis employ different types of tRNA. Moreover, the mitochondrial synthetase is specific for the mitochondrial form of tRNA and the cytoplasmic synthetase is specific for the cytoplasmic tRNA. However, intracellular location is not the only consideration, since prokaryotic cells also contain multiple kinds of tRNAs for certain amino acids; altogether, E. coli cells contain some 80 different tRNAs for the 20 amino acids.

The Adapter Role of tRNA

Once an amino acid is esterified to its corresponding tRNA, it makes no contribution to the specificity of the aminoacyl-tRNA, since the aminoacyl group as such is not recognized by either the ribosome or the mRNA template. This was conclusively proved in clever experiments in which the amino acid residue of an aminoacyl-tRNA was chemically modified. For example, isotopically labeled cysteinyl-$tRNA_{Cys}$ was first formed by the action of the cysteine-activating enzyme on radioactive cysteine, ATP, and $tRNA_{Cys}$. It was then chemically converted to alanyl-$tRNA_{Cys}$ by reducing the cysteinyl residue with hydrogen and a catalyst, a procedure which does not cause any chemical change in the tRNA molecule itself. This hybrid aminoacyl-tRNA, which contains the anticodon for cysteine but actually carries alanine, was then incubated with a ribosomal system from reticulocytes containing all the other tRNAs, amino acids, and activating enzymes, as well as the mRNAs for hemoglobin chains. The newly synthesized polypeptide chains were examined to determine whether the hybrid alanyl-$tRNA_{Cys}$ transferred its anomalous alanine residue into those positions of the hemoglobin chain normally occupied by cysteine. The answer was positive: labeled alanine was incorporated into the cysteine positions in significantly large amounts. This experiment provided proof for the hypothesis, first postulated by F. H. C. Crick and by M. B. Hoagland, that tRNA is a molecular "adapter," into which the amino acid is plugged so that it can be adapted to the nucleotide triplet language of the genetic code. This experiment also proved that the activating enzymes must have extremely high specificity for both their amino acids and tRNAs, because any errors arising during the activation reaction cannot be corrected during the formation of the polypeptide chain.

Ribosomal Structure

The complex structure of ribosomes requires some review (page 327, also Chapter 36) before we consider the initiation stage in protein synthesis, since there is a specific spatial arrangement of binding sites on the ribosomes and a set of complex interactions between the bound components as the polypeptide chain is made. The 70S ribosomes of E. coli and other prokaryotic cells, which have a particle weight of about 2.7 megadaltons, have been most thoroughly studied. Under special conditions they dissociate into 50S and 30S subunits, each of which contains RNA and a number of proteins. The 50S subunit contains two RNA components, namely 23S and 5S RNAs, as well as 34 different kinds of proteins. The 30S subunits contain a molecule of 16S ribosomal RNA and 21 different kinds of proteins. The ribosomal proteins are insoluble in water at pH 7.0 under normal conditions but can be

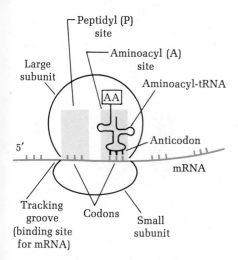

$$CH_3-S-CH_2-CH_2-\overset{\overset{\displaystyle H}{|}}{\underset{\underset{\displaystyle NH}{|}}{C}}-COOH$$

$$\underset{\underset{\displaystyle H}{|}}{C}=O$$

0.5 mm

A. Rich

selectively and sequentially extracted with concentrated salt solutions. Mild extraction of the 30S subunit with concentrated cesium chloride solutions removes several of the ribosomal proteins and leaves a stable inner core, from which additional proteins can be extracted by more drastic procedures. The many different proteins extracted from 30S and 50S subunits can then be separated by gel electrophoresis at pH 4.5. As will be seen (Chapter 36, page 1023), ribosomes can be reassembled from their component RNAs and proteins.

The ribosomes in the cytoplasm of eukaryotes are much larger than those of prokaryotes; they have a sedimentation coefficient of 73S to 80S, a diameter of about 22 nm, and a particle weight of about 4.0 megadaltons. Their subunits have sedimentation coefficients of about 60S and 40S. However, the size of the subunits varies somewhat between plants and animals, as does the size of the RNA. Plant ribosomes contain 25S and 16S or 18S rRNA and animal ribosomes 28S and 18S rRNA. The 28S rRNA of animal ribosomes varies somewhat in size in different species, depending on their phylogenetic position, and ranges in molecular weight from 1.4 million in sea urchins to 1.75 million in mammals.

Although at least some of the proteins of ribosomes have a catalytic function, the specific function of ribosomal RNA is still not clear, although it has binding sites for a number of ribosomal proteins (page 1026). It appears very likely that the rRNA and the different types of proteins clustered to form the 30S and 50S subunits of 70S ribosomes participate in conformational changes to move the ribosome, with its growing polypeptide chain, along the mRNA. Ribosomes thus may be mechanochemical systems, to be considered in the same class of biostructures as the actomyosin of muscle and the microtubules of eukaryotic flagella (page 1027).

Figure 33-8 is a schematic representation of ribosome structure, showing the peptidyl and aminoacyl sites on the 50S subunit. The mRNA tracks through a groove, presumably that between the large and small subunits. The sizes of the codons are not in proper proportion; about 8 codons are hidden in the groove.

Initiation of Polypeptide Chains

In *E. coli* and all other prokaryotes the synthesis of all proteins begins with the amino acid methionine, coded for by the initiation codon AUG. The initiating methionine residue does not enter as methionyl-tRNA but as *N-formylmethionyl-tRNA* (fMet-tRNA) (Figure 33-9), the product of an enzymatic transformylation in which a formyl group is transferred from N^{10}-formyltetrahydrofolate (page 347) to the α-amino group of methionyl-tRNA. There are two species of methionyl-tRNA$_{Met}$; only one, designated methionyl-tRNA$_F$, is capable of accepting the N-formyl group. The reaction is

N^{10}-Formyltetrahydrofolate + Met-tRNA$_F$ $\longrightarrow$

tetrahydrofolate + fMet-tRNA$_F$

The enzyme catalyzing this reaction does not formylate free methionine or a methionine residue attached to the other species of $tRNA_{Met}$. It now appears likely that $fMet-tRNA_F$ initiates peptide-chain synthesis not only in bacteria but also in the mitochondria of eukaryotic cells (see below). fMet-tRNA has been obtained in crystalline form (Figure 33-9). The initiating residue in cytoplasmic protein synthesis of eukaryotic cells has been found to be an unacylated methionine residue, esterified to one specific $tRNA_{Met}$, the _initiator tRNA._

During protein synthesis there is continuous dissociation of 70S ribosomes into their 50S and 30S subunits and continuous reassociation of the subunits to form intact ribosomes again. This was first shown by R. Kaempfer and his associates, who grew E. coli cells for several generations in a medium in which the carbon, nitrogen, and hydrogen sources were highly enriched with the heavy isotopes ^{13}C, ^{15}N, and ^2H. The ribosomes synthesized by these "heavy" cells have a greater density (page 872) than ribosomes from cells grown on normal sources of carbon, hydrogen, and nitrogen. The heavy E. coli cells were then incubated in a fresh medium containing normal sources of carbon, hydrogen, and nitrogen. The ribosomes isolated from such cells were subjected to density-gradient centrifugation and found to contain two species of hybrids. One species consisted of a heavy 50S and a light 30S subunit and the other, a light 50S and a heavy 30S subunit. From the rate of formation of such hybrids it was concluded that 70S ribosomes constantly undergo dissociation into subunits and that the subunits constantly reassociate. The significance of the dissociation-reassociation process became clear when M. Nomura and his colleagues found that the 70S ribosome must first dissociate into 50S and 30S subunits before polypeptide synthesis can begin; mRNA and the initiating aminoacyl-tRNA cannot bind directly to 70S ribosomes but are first bound to 30S subunits, which then combine with the 50S subunits. However, subsequent research in a number of laboratories showed that this is a rather complex multistep process (Figure 33-10), which requires participation of three specific proteins called _initiation factors_ (IF-1, IF-2, and IF-3). These proteins can be extracted in soluble form from ribosomes with strong salt solutions; they have molecular weights of about 9,000, 65,000, and 21,000, respectively. The initiation factors are not permanent components of the 30S subunits; as we shall see, they undergo cycles of attachment and release.

Although there is still some uncertainty about the details, the sequence of steps in the formation of a complete, functional 70S ribosome is as follows. The 30S subunit first forms a complex with IF-3, which then binds the mRNA; to this complex is added a molecule of IF-1. Then fMet-tRNA and GTP bind to IF-2 and the resulting complex combines with the complex of the 30S subunit, IF-3, mRNA, and IF-1 to yield what is termed the _initiation complex_ (Figure 33-10). The initiation complex then combines with the 50S subunit to form the complete, functional 70S ribosome; in this process GTP is hydrolyzed to GDP and P_i, and the three ini-

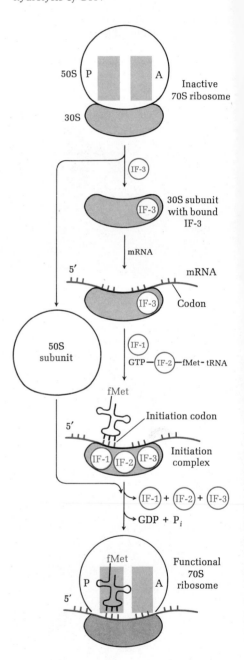

Figure 33-10
Formation of the initiation complex and the functional 70S ribosome (E. coli). IF-1, IF-2, and IF-3 are initiation factors. Binding of the 50S ribosome to the initiation complex results in release of the initiation factors and hydrolysis of GTP.

tiation factors IF-1, IF-2, and IF-3 dissociate from the ribosome.

Presumably this elaborate initiation process is required to ensure that the initiating aminoacyl-tRNA is bound at the peptidyl site and positioned at the initiation codon AUG, so that the ribosomes start translation at the correct point on the mRNA. The initiation factors are used over and over again to start new chains.

It has been established that translation of the codons of mRNA takes place in the $5' \rightarrow 3'$ direction; thus, the P site and A site on the ribosome recognize the polarity of the mRNA chain.

The Elongation Cycle

Elongation takes place when the initiating aminoacyl-tRNA (or the peptidyl-tRNA) is bound on the peptidyl site, positioned on its codon on the mRNA, with the A site free. Three major steps (Figure 33-11) occur in the elongation process.

Step 1   Binding of the Incoming Aminoacyl-tRNA

In the first step the incoming aminoacyl-tRNA is added to the aminoacyl site (A site) of the complete 70S ribosomal complex in a rather involved process. First the incoming aminoacyl-tRNA binds to a specific cytoplasmic protein; in prokaryotes it is called elongation factor T (EF-T) and has been obtained in crystalline form. EF-T contains two subunits, EF-T_S and EF-T_U. EF-T combines with GTP to form an EF-T_U–GTP complex, with release of EF-T_S. The EF-T_U–GTP complex now binds the aminoacyl-tRNA to form a ternary EF-T_U–GTP–aminoacyl-tRNA complex. This complex now binds to the ribosome in such a way that the aminoacyl-tRNA is positioned on the A site, with its anticodon hydrogen-bonded to the corresponding codon in the mRNA. Simultaneously, the bound GTP is hydrolyzed to GDP and P_i; the GDP leaves the ribosome in the form of an EF-T_U–GDP complex. The energy furnished by hydrolysis of GTP is apparently required to put the aminoacyl-tRNA in the correct position. EF-T does not readily accept the initiating fMet-tRNA. Binding of the aminoacyl-tRNA to the A site can be blocked by certain antibiotics, particularly tetracyclines (see below). In eukaryotic cells the elongation factors are designated EF-1 and EF-2; they function in a slightly different manner.

Step 2   Peptide-Bond Formation

With the fMet-tRNA on the P site and the newly bound aminoacyl-tRNA on the A site, the first peptide bond is now formed by nucleophilic attack of the amino group of the aminoacyl-tRNA on the esterified carboxyl carbon of the fMet-tRNA. This reaction is catalyzed by peptidyl transferase, which is present in the 50S subunit of the ribosome (Figures 33-11 and 33-12). The product is a dipeptidyl-tRNA bound to the A site; the discharged or empty tRNA remains bound to the P site. Neither ATP nor GTP is required for the

Figure 33-11
The elongation steps. EF-T and EF-G are
elongation factors.

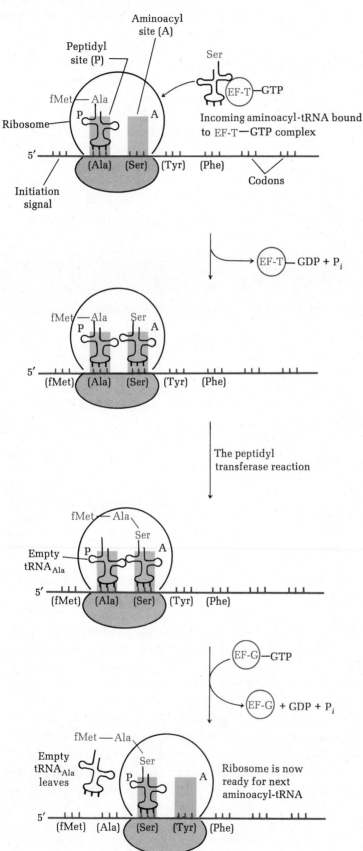

Figure 33-12
The peptidyl transferase reaction. The peptide linkage is formed by a nucleophilic displacement in which the amino group of the
new aminoacyl-tRNA displaces the tRNA of
the preceding amino acid from the carboxyl
carbon atom to yield a lengthened peptidyl-
tRNA.

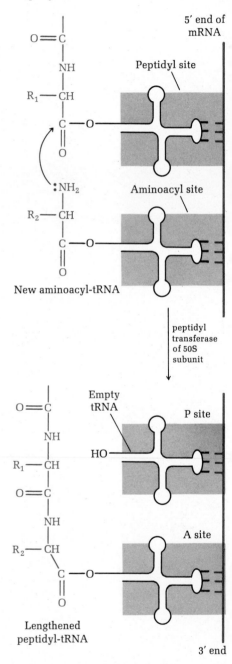

Figure 33-13
Mode of inhibition by puromycin.

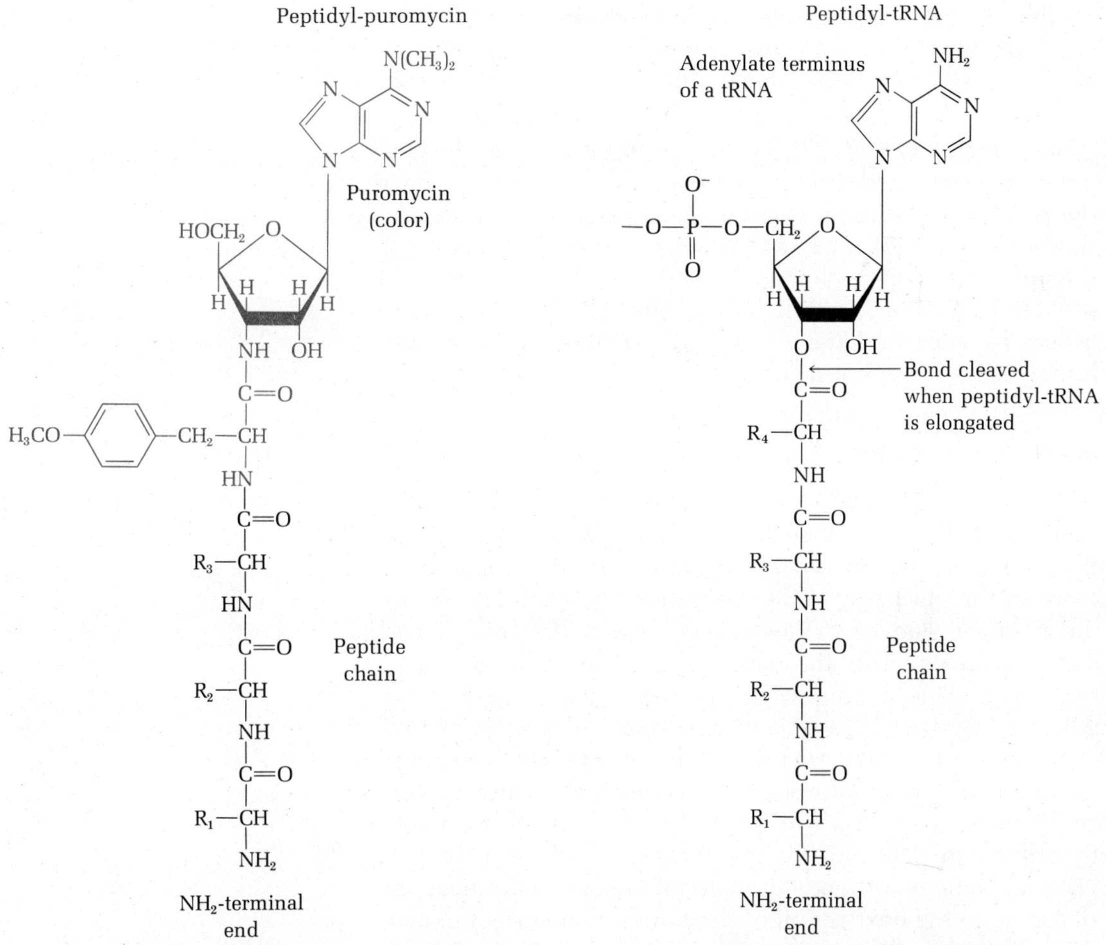

Peptidyl-puromycin

Peptidyl-tRNA

Puromycin resembles aminoacyl-tRNA and can react with a peptidyl-tRNA to yield peptidyl-puromycin. Since its polypeptide chain cannot be elongated further, the peptidyl-puromycin is discharged from the ribosome. A number of peptidyl-puromycins varying in chain length have been isolated.

The amino group of an incoming aminoacyl-tRNA can displace the tRNA from the COOH-terminal amino acid and thus elongate the chain.

formation of the new peptide bond, which is presumably made at the expense of the bond energy of the ester linkage between the N-formylmethionine and its tRNA. Although this description refers to the formation of the first peptide bond of a polypeptide chain, an identical process takes place at every elongation cycle; i.e., the free amino group of the newly bound aminoacyl-tRNA on the A site displaces the tRNA from the peptidyl-tRNA on the P site to form the new peptide bond.

This picture of the mechanism is supported by important experiments on the inhibition of protein synthesis by the antibiotic _puromycin_ (Figure 33-13), which has a structure very similar to that of an aminoacyl derivative of the terminal adenylic residue of a tRNA. However, instead of the normal ester linkage between the 2'- or 3'-hydroxyl group of the ribose moiety and the carboxyl group of the amino acid, puromycin possesses an amide linkage between its sugar component and the carboxyl group of 4-methoxyphenyl-alanine, i.e., the methyl ether of tyrosine. Puromycin interrupts peptide-chain elongation by virtue of its capacity to bind at the A site when the P site is occupied, with subsequent formation of a covalent *peptidyl-puromycin* derivative (Figure 33-13). The peptidyl-puromycin so formed then dissociates from the ribosome, since it does not have the precise structure for it to be recognized by the translocation apparatus and moved to the P site. These observations not only established the mode of action of this antibiotic but provided important confirmatory evidence that chain growth occurs by addition of aminoacyl-tRNAs at the carboxyl-terminal end of polypeptide chains.

Step 3 Translocation

After the peptide bond has been made, the peptidyl-tRNA which has just been elongated is still bound to the A site, positioned at the codon of the newly added amino acid. Now, in the third step of the elongation cycle (Figure 33-11), the ribosome moves to the next codon of the mRNA and simultaneously shifts the peptidyl-tRNA from the A site to the P site. This translocation is a rather complex process, which takes place in a sequence of steps. A specific protein called *elongation factor G* (EF-G), as well as GTP, is required for this step. The GTP first binds to EF-G, and the resulting complex binds to the ribosome. The GTP then undergoes hydrolysis to GDP and P_i. The hydrolysis of the GTP provides the energy for a conformational change that moves the ribosome to the next codon of the mRNA and shifts the peptidyl-tRNA from the A site to the P site. The A site is now open, with a new codon in place, ready to receive a new aminoacyl-tRNA and start a new elongation cycle. After the translocation step the EF-G dissociates from the ribosome.

The events involved in elongation require a highly specific topology of the sites for binding the peptidyl-tRNA, the aminoacyl-tRNA, the mRNA, the elongation factors, and the GTP (see also Chapter 36, page 1023). The ribosome probably undergoes complex and specific changes in shape during each bond-making cycle. The mRNA appears to "track" through a groove or tunnel in the ribosome, since in poly-ribosomes (see below) a substantial fraction of the mRNA chain is not susceptible to attack by ribonuclease. This fraction presumably consists of all the segments inside ribosomal tunnels at any instant (about 25 nucleotides per ribosome). A comparable segment of the polypeptide chain being formed is similarly protected from attack by proteolytic enzymes.

Termination of the Polypeptide Chain

The completion of a polypeptide chain and its detachment from the ribosome require special steps which are not yet fully understood. As we shall see in Chapter 34 (page 965),

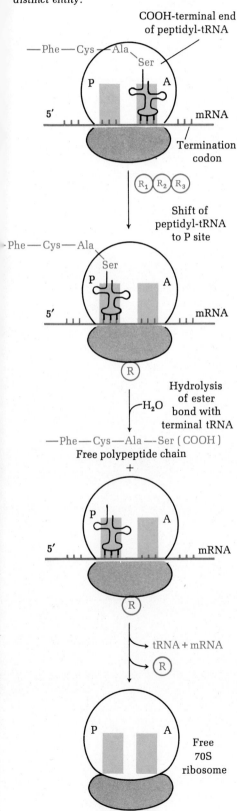

Figure 33-14
Steps in release of the completed polypeptide chain. R_1, R_2, and R_3 are release factors. Which one is used depends on the termination codon. It is not certain that R_3 is a distinct entity.

COOH-terminal end of peptidyl-tRNA

—Phe—Cys—Ala

Ser

P   A

5′   mRNA

Termination codon

R_1 R_2 R_3

Shift of peptidyl-tRNA to P site

·Phe—Cys—Ala

Ser

P   A

5′   mRNA

R

Hydrolysis of ester bond with terminal tRNA

—H_2O

—Phe—Cys—Ala—Ser (COOH)
Free polypeptide chain
+

P   A

5′   mRNA

R

tRNA + mRNA

R

P   A

Free 70S ribosome

the termination of polypeptide chains is signaled by one of three special termination codons in the mRNA. After the last or carboxyl-terminal amino acid residue has been added to a polypeptide chain on the ribosome, the polypeptide is still covalently attached by its terminal carboxyl group to the tRNA, which is on the A site of the ribosome. The release of the polypeptidyl-tRNA from the ribosome is promoted by three specific proteins called _release factors_, symbolized R_1, R_2, and R_3 (Figure 33-14). They bind to the ribosome to cause a shift of the polypeptidyl-tRNA from the A site to the P site. The ester bond between the polypeptide chain and the last tRNA is then hydrolyzed, apparently by the action of peptidyltransferase, whose specificity and catalytic action seem to be altered by the bound releasing factors. Once the polypeptide is released, the last tRNA and the mRNA leave. The free 70S ribosome then dissociates into its 50S and 30S subunits, a process that requires one of the specific initiation factors, and a new polypeptide chain may be started.

Very little is known about the state in which a completed polypeptide chain leaves the ribosome, whether as a randomly coiled chain or as a folded molecule in its native, tertiary conformation. The second alternative appears more likely since isolated ribosomes show a wide variety of enzymatic activities in addition to those concerned in the mechanism of peptide-bond synthesis. These activities are believed to be due to tightly bound, almost complete polypeptide chains of enzymes that are still undergoing synthesis.

Posttranslational Processing of Polypeptide Chains

Once a polypeptide has been made by a ribosome it may need to undergo covalent modification to yield its biologically active form.

The N-formyl group of the fMet residue does not appear in the finished protein of prokaryotes; it is removed by the action of a deformylase. Moreover, the initiating methionyl residue of some polypeptides may also be removed, by a methionine aminopeptidase. Some proteins are acetylated on the NH_2-terminal α-amino group, such as the coat protein of the tobacco mosaic virus (pages 104 and 1017). Such acetyl groups are introduced after protein synthesis is complete; they are not related to chain initiation during protein biosynthesis.

Intrachain disulfide cross-linkages are formed enzymatically after the polypeptide chain has been built. C. B. Anfinsen and other investigators have found that microsomes can catalyze the oxidation of the —SH groups of chemically reduced, inactive forms of ribonuclease, lysozyme, and other proteins, with formation of disulfide linkages. Because these oxidation products regain a large part of their native enzymatic activity, it appears likely that the reduced polypeptide chain folds itself spontaneously to approximately the native conformation, thus positioning the —SH groups properly to yield the correct cross-linkages when they are enzymatically oxidized (Chapter 6, page 141). The information necessary for establishing the correct cross-linkages of pro-

Figure 33-15
Other inhibitors of protein synthesis.

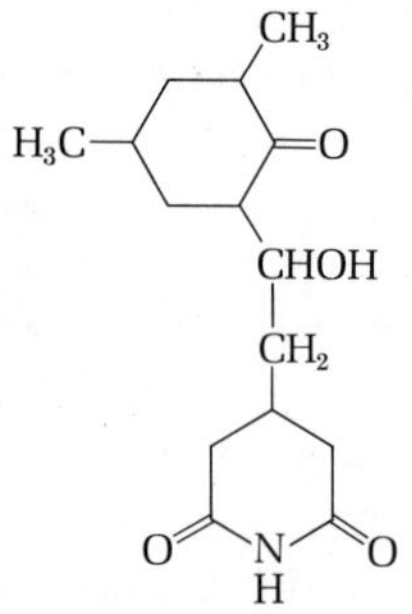

Cycloheximide (inhibits protein synthesis by 80S eukaryotic ribosomes).

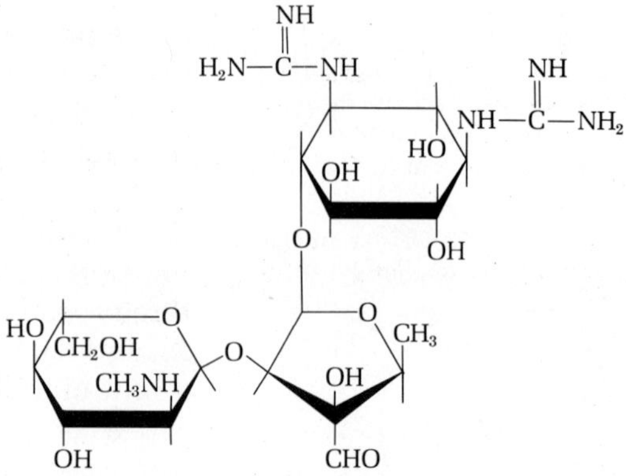

Streptomycin (binds to a specific protein of the 30S subunit of E. coli ribosomes and causes inhibition and misreading).

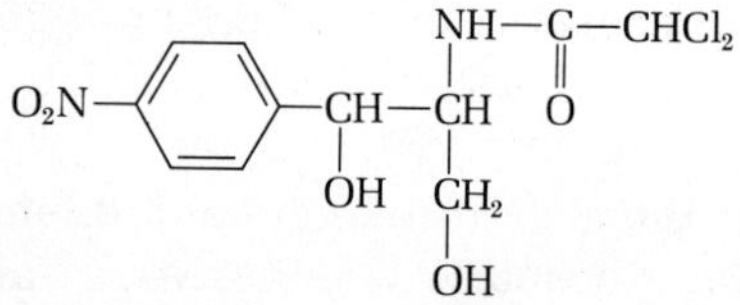

Chloramphenicol (inhibits protein synthesis by 70S ribosomes of bacteria by blocking peptidyl transfer).

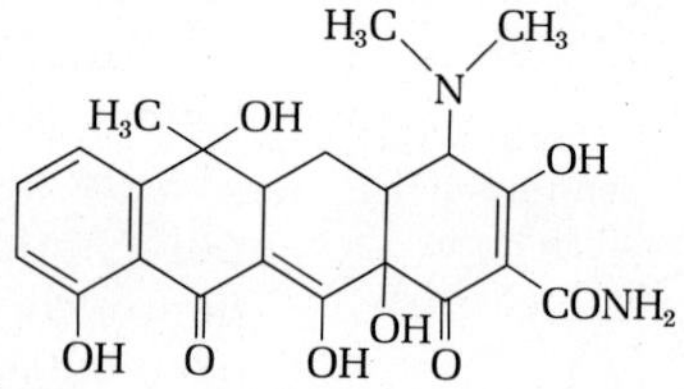

Tetracycline (inhibits the binding of amino-acyl-tRNA to the 30S subunit).

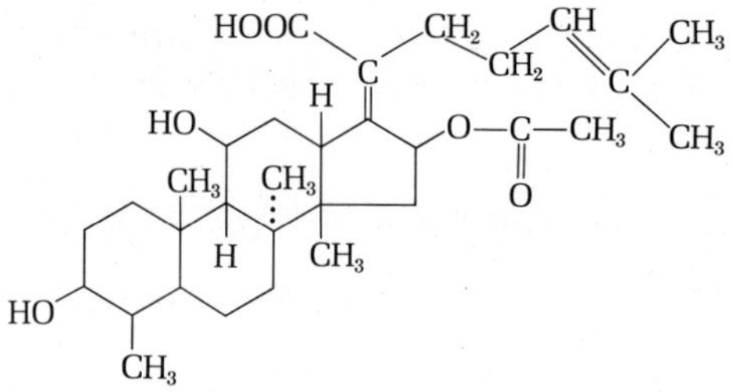

Fusidic acid (indirectly inhibits the translocation step).

teins is thus already inherent in the amino acid sequence of the polypeptide chain. Other aspects of the folding and assembly of polypeptide subunits of oligomeric proteins are described elsewhere (pages 144 and 1016).

Other covalent modifications occurring after translation include phosphorylation (page 434), methylation (page 755), and hydroxylation (page 696) reactions, as well as reactions involving the attachment of coenzymes or prosthetic groups.

Inhibitors of Protein Biosynthesis

In addition to the antibiotics already mentioned (tetracycline and puromycin) many others known to inhibit protein synthesis have become important tools in dissecting ribosome structure and function (Figure 33-15). _Chloramphenicol_ inhibits protein synthesis on the 70S ribosomes of prokaryotic cells (and of the mitochondria of eukaryotes) be-

Figure 33-16
Electron micrographs of polyribosomes of reticulocytes, functioning in synthesis of polypeptide chains of hemoglobin.

Shadowed preparation.

100 nm   A. Rich

Preparation stained with uranyl acetate to show the mRNA strand between the ribosomes.

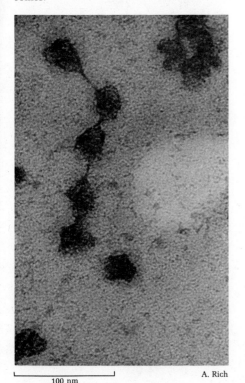

100 nm   A. Rich

cause it blocks the peptidyl transfer reaction. It does not inhibit protein synthesis by the 80S ribosomes of eukaryotic cells. Chloramphenicol has been used to treat certain infections in man, but it is relatively toxic, possibly because it inhibits protein synthesis in mitochondrial ribosomes. On the other hand, *cycloheximide* (also called *actidione*), inhibits protein synthesis in the 80S ribosomes of eukaryotes but does not inhibit the 70S ribosomes of prokaryotes. Cycloheximide blocks peptide-bond formation. Both chloramphenicol and cycloheximide bind to the large ribosomal subunits.

Streptomycin and related antibiotics (*neomycin, kanamycin*) bind to the small subunit of ribosomes of prokaryotic cells. They inhibit protein synthesis and cause misreading of the genetic code, presumably by altering the conformation of the ribosome so that the aminoacyl-tRNAs are less firmly bound to the codons of the mRNA. Mutant cells resistant to or dependent on streptomycin possess genetically altered 30S ribosomal subunits.

Inhibitors other than antibiotics include the alkaloid *emetine,* which inhibits binding of aminoacyl-tRNAs, and *diphtheria toxin,* a toxic protein secreted by the diphtheria organism. The diphtheria toxin is actually an enzyme that catalyzes a reaction between NAD^+ and elongation factor EF-2 in eukaryotic ribosomes to yield an inactive ADP-ribose–EF-2 complex, resulting in inhibition of translocation.

Abrin and *ricin,* toxic plant proteins, inhibit protein synthesis in eukaryotic ribosomes by inactivating the 60S subunits and thus blocking elongation. There is some evidence that these plant proteins are more toxic to cancer cells than normal cells. Both of these plant proteins, as well as diphtheria toxin, contain two polypeptide chains joined by disulfide cross-linkages.

Polyribosomes

Under careful conditions of isolation in which ribonuclease activity and mechanical shear are avoided, ribosomes are obtained in clusters containing from as few as 3 or 4 to as many as 100 individual ribosomes. Such clusters, which are called *polyribosomes* (also *polysomes*), can be fragmented into individual ribosomes (*monosomes*) by the action of ribonuclease, indicating that polyribosomes are held together by RNA. A connecting fiber between adjacent ribosomes can actually be seen in electron micrographs (Figure 33-16). The connecting strand of RNA is actually mRNA, which is being read by several ribosomes simultaneously, spaced some distance apart. Each individual ribosome of a polysome can make a complete polypeptide chain and does not require the presence of the other ribosomes. Such an arrangement therefore utilizes an mRNA template more efficiently, since several polypeptide chains can be made from it simultaneously (Figure 33-17). As pointed out above, translation of the mRNA is in the 5'→3' direction. Transcription and translation are usually coupled in bacteria, so that mRNA undergoes translation as it is being transcribed from DNA (Figure 33-18).

The polypeptide chains of hemoglobin, which contain about 150 amino acid residues, are coded for by mRNA molecules having about $3 \times 150 = 450$ nucleotide residues, which would be chains about 150 nm long. Since each 80S reticulocyte ribosome is 22 nm in diameter, there is obviously ample room along the mRNA for several ribosomes to read off the mRNA simultaneously. Polyribosomes isolated from reticulocytes have been found to contain five or six ribosomes. If six is the maximum number, the average clearance between successive 22-nm ribosomes along the mRNA is about 3.0 nm. Evidently individual ribosomes are able to work with relatively little elbow room along the mRNA chain; presumably their rates of movement are synchronized. Among the largest polyribosome systems are those responsible for the biosynthesis of the heavy chains of myosin, which have about 1,800 amino acid residues (page 755). Such polyribosomes contain 60 or more ribosomes.

High-resolution electron microscopy has shown that polyribosomes have a considerable degree of three-dimensional order. Polyribosomes with six or more ribosomes sometimes form regular, tight helical structures, in which the small subunits face inward (Figure 33-19). The extraordinary finding has been made that when certain tissues of higher organisms, particularly those of the chick embryo, are chilled for some time at 0°C prior to fixation and sectioning, the ribosomes form tetramers, which assume a very regular lattice within the cell and appear to organize into crystals (Figure 33-20).

Figure 33-18
Functional complex of DNA, mRNA, RNA polymerase and a ribosome in a bacterial cell. Ribosomes may begin to read an mRNA before it has been completed by RNA polymerase.

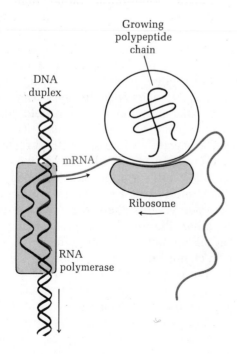

Figure 33-17
A polyribosome. Five ribosomes are reading the mRNA simultaneously, moving from the 5' end to the 3' end.

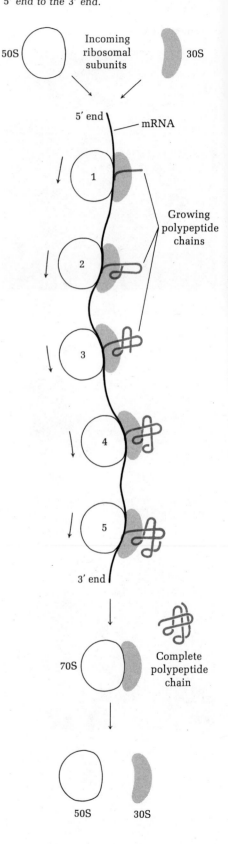

Figure 33-19
Tightly twisted helical arrangement of individual ribosomes in a free polyribosome. The ribosomes are closely spaced along the mRNA. [Redrawn from E. Shelton and E. L. Kuff, J. Mol. Biol., 22: 23 (1966).]

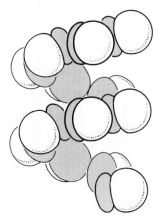

Figure 33-20
Electron micrographs showing ribosomes in regular crystalline array in liver of chicken embryo held at 0°C for some time before fixation.

Low magnification

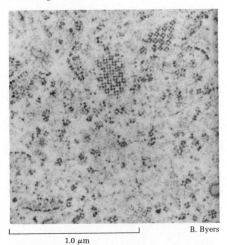

1.0 μm   B. Byers

High magnification

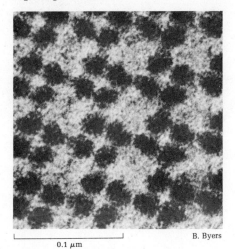

0.1 μm   B. Byers

The Energy Requirement in Protein Synthesis

We have seen that two high-energy phosphate bonds are probably utilized in the formation of each molecule of aminoacyl-tRNA during the amino acid activation reaction (page 934). In addition, one molecule of GTP is cleaved to GDP and P_i during the binding of each aminoacyl-tRNA to the A site of the ribosome, and another molecule of GTP is hydrolyzed in the translocation of the ribosome along the mRNA. Therefore a total of four high-energy phosphate groups is ultimately required for the synthesis of each peptide bond of the completed protein. This represents an exceedingly large thermodynamic pull in the direction of synthesis, since a total of some $7.3 \times 4 = 29.2$ kcal is invested to generate a peptide bond whose standard free energy of hydrolysis, although not known precisely, is at most -5.0 kcal. The net $\Delta G°'$ for peptide-bond synthesis is thus about -24.2 kcal, making peptide-bond synthesis at the expense of phosphate-bond energy overwhelmingly exergonic and essentially irreversible. Although this may appear to be exceedingly wasteful, it is the price the cell must pay to guarantee nearly perfect fidelity in the translation of the genetic message of mRNA into the amino acid sequence of proteins. Protein synthesis consumes more energy than any other biosynthetic process in most organisms; in E. coli cells up to 90 percent of biosynthetic energy goes into protein biosynthesis.

Protein Synthesis and Secretion in the Cytoplasm of Eukaryotic Cells

Biosynthesis of polypeptide chains on the large 80S cytoplasmic ribosomes of eukaryotic cells follows the same pattern as in the 70S ribosomes in bacteria, but there are some differences in details. Methionine rather than N-formyl-methionine is the initiating amino acid, in the form of a special type of methionyl-tRNA, which responds to the initiating AUG codon in the mRNA. Initiation factors and GTP are also required in the formation of the eukaryotic initiation complex, which must begin with the 40S ribosomal subunit. Elongation by eukaryotic ribosomes requires elongation factors EF-1 and EF-2; moreover, GTP is required for both the binding of the aminoacyl-tRNA and for the translocation step. The EF-2 of eukaryotic systems (but not the prokaryotic counterpart) is inactivated in a novel reaction promoted by diphtheria toxin, described above. The termination steps in eukaryotes have not been analyzed in detail.

In prokaryotic cells the ribosomes are largely free in the cytoplasm, mainly as polyribosomes, whereas in eukaryotic cells ribosomes may occur either free or bound to the endoplasmic reticulum, to constitute the rough-surfaced endoplasmic reticulum (page 33). It has been postulated that the free cytoplasmic ribosomes in eukaryotes are employed in the biosynthesis of proteins intended for internal use, as in the synthesis of hemoglobin in reticulocytes, whereas the ribosomes that are attached to the outer surface of the en-

Figure 33-21
Biosynthesis and secretion of zymogen proteins in an exocrine cell of the pancreas.

Electron micrograph of a section of an exocrine cell, showing the electron-dense zymogen granules.

Schematic representation of a pancreatic cell to show the steps in the secretion of newly synthesized zymogen proteins. [*Redrawn from D. Fawcett, "Structural and Functional Variations in the Membranes of the Cytoplasm," in S. Seno and E. V. Cowdry (eds.), Intracellular Membrane Structures, Japanese Society for Cell Biology, Okayama, Japan, 1963.*]

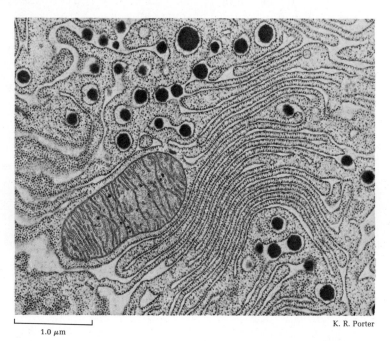

1.0 μm

K. R. Porter

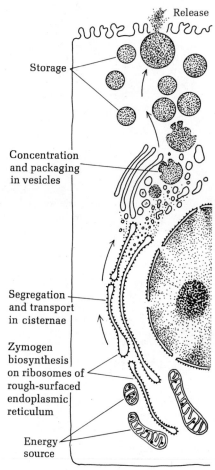

Release

Storage

Concentration and packaging in vesicles

Segregation and transport in cisternae

Zymogen biosynthesis on ribosomes of rough-surfaced endoplasmic reticulum

Energy source

doplasmic reticulum are believed to make proteins to be secreted outside the cell. Ribosomes bound to endoplasmic reticulum are particularly abundant in pancreatic exocrine cells, which synthesize and secrete the digestive enzymes, and in liver cells, which form and secrete proteins of the blood plasma.

The course of protein synthesis in pancreas exocrine cells has been studied in some detail by electron-microscopic and biochemical means (Figure 33-21). The newly formed proteins, such as the zymogens trypsinogen and chymotrypsinogen, leave the bound ribosomes and pass through the endoplasmic reticulum membrane, by a process that is not yet understood, and then enter the intracisternal space, which serves as a system of collecting channels. The contents of the intracisternal space are then concentrated and "packaged" by the Golgi apparatus into zymogen granules—large, dense, aggregates of zymogen molecules—which are surrounded by a single membrane. The zymogen molecules are stored in this form until they are secreted. They release their contents to the exterior of the cell after the membrane of the zymogen granule fuses with the plasma membrane and opens the lumen of the zymogen granule to the exterior (Figure 33-21).

Protein Synthesis in Mitochondria

Isolated mitochondria from many types of eukaryotic cells incorporate labeled amino acids into mitochondrial protein. Mitochondria contain all the essential components required for protein synthesis, including mRNA, ribosomes, tRNAs, and aminoacyl-tRNA synthetases, all of which are distinctive and different from those required in extramitochondrial protein synthesis in eukaryotes. The ribosomes of mitochondria vary considerably in size, depending on the position of the cell in the phylogenetic scale. Mitochondrial ribosomes from higher organisms have sedimentation values of 50S to 60S, whereas those from primitive eukaryotes, yeast and fungi, are 70S to 80S. Their subunits also vary correspondingly in size; in mitochondria from human cells the subunits are 45S and 35S, whereas from yeast they are 55S and 38S. The ribosomes are not always easily visible in electron micrographs of mitochondria in animal tissues; often they are located on the portion of the inner membrane between the cristae (page 511).

As pointed out above, mitochondria of eukaryotic cells contain distinctive and specific tRNAs and aminoacyl-tRNA synthetases, which differ from those found in the extramitochondrial cytoplasm. Protein synthesis by the ribosomes of mitochondria is inhibited by chloramphenicol, whereas extramitochondrial protein synthesis by 80S ribosomes of eukaryotic cells is not. Moreover, mitochondrial protein synthesis employs N-formylmethionine as initiating amino acid, whereas cytoplasmic 80S ribosomes employ methionine. Mitochondrial protein synthesis thus resembles that in bacteria. These observations support the hypothesis (page 1053) that mitochondria originated from parasitizing aerobic bacteria during the evolution of eukaryotic cells.

In addition to a specific circular mitochondrial DNA (Chapters 12 and 31), mitochondria also contain a DNA-dependent RNA polymerase that is inhibited by actinomycin D. Mitochondria thus are able to generate mRNA from their DNA. The observation that actinomycin D inhibits protein synthesis in isolated mitochondria indicates that mitochondrial mRNA is required for protein synthesis.

A point of major interest is the identity of the proteins synthesized by the mitochondria, since much evidence indicates that the great majority of mitochondrial enzymes and proteins are specified by genes in nuclear chromosomes and are synthesized on extramitochondrial ribosomes. Mitochondrial DNA codes for mitochondrial rRNAs and probably most of the mitochondrial tRNAs, leaving only a small amount of mitochondrial DNA with the function of coding protein synthesis. It is known that mitochondrial DNA specifies the synthesis of one or more hydrophobic protein subunits of F_1 ATPase (page 523), of a subunit of cytochrome aa_3 (page 493), and possibly a subunit of cytochrome b.

Chloroplasts also possess an independent protein synthetic machinery resembling that in mitochondria. One of the subunits of ribulose diphosphate carboxylase and some of the subunits of chloroplast ribosomes are synthesized in chloroplasts.

Aging and Errors in Protein Synthesis

Much interest attaches to the question of the accuracy of protein synthesis in animal tissues and to the consequences of errors. There is some evidence that structurally similar amino acids may replace a given amino acid no more than once in 3,000 residues; presumably the chance of errors in incorporation of dissimilar amino acids is much less. It has been postulated that the error frequency in protein biosynthesis rises with the age of the animal, to the point that the resulting protein molecules may be rendered less active or even inactive. This hypothesis has been confirmed by direct experiments in which the amount of catalytically active aldolase in skeletal muscles of experimental animals has been compared with the amount of aldolase protein detectable by titration with an antibody specific for aldolase, as a function of the age of the animal. Although the total amount of immunologically detectable aldolase remained unchanged in the muscles of aged animals, the isolated aldolase was only one-third as active catalytically as aldolase isolated from young animals. The implication is that in the old animals many errors were made in the biosynthesis of aldolase, rendering a large fraction of the molecules in the aldolase population inactive. These interesting observations, which deserve to be extended, imply that there may be a general deterioration in the accuracy of translation with age.

Summary

The synthesis of proteins from activated amino acids takes place on the surface of ribosomes. Amino acids are first activated in the cytoplasm by aminoacyl-tRNA synthetases, which catalyze the formation of the aminoacyl esters of homologous tRNAs; simultaneously, ATP is cleaved to AMP and pyrophosphate. The aminoacyl-tRNA synthetases are highly specific for both the amino acid and its corresponding tRNA. There is more than one type of tRNA for each amino acid. The aminoacyl-tRNA synthetases also have considerable species specificity. The amino acid moiety of aminoacyl-tRNA makes no contribution to the specificity of interaction of the aminoacyl-tRNA with the complementary codon of the mRNA; the tRNA thus is an "adapter."

Polypeptide chains are built starting from the N-terminal amino acid residue. New aminoacyl residues are added to the C-terminal carboxyl group of the peptidyl-tRNA by nucleophilic displacement of the tRNA from the carboxyl carbon atom. In E. coli the initiating amino acid is N-formylmethionine, which enters as N-formylmethionyl-tRNA (fMet-tRNA). fMet-tRNA and mRNA bind to the free 30S subunits of ribosomes to form the initiation complex, in a process requiring GTP and three initiation factors IF-1, IF-2, and IF-3. The initiation complex then combines with the 50S subunit to form a functional 70S ribosome. In the subsequent elongation process a cytoplasmic elongation factor EF-T and the hydrolysis of GTP are required for proper binding of the incoming aminoacyl-tRNA. The peptidyl group on the peptidyl site is transferred to the amino group of the new aminoacyl-tRNA at the aminoacyl site, by an enzyme activity present in the 50S subunit. A cytoplasmic elongation factor EF-G and the hydrolysis of a second molecule of GTP are required for translocation of the peptidyl-tRNA from the aminoacyl site to the peptidyl site. Four high-energy phosphate groups are required to insert each amino acid residue. Polypeptidyl-tRNA, on reaching a termination codon in the mRNA, is released from the tRNA and the ribosome by releasing factors R_1 and R_2. Many inhibitors of protein synthesis are known; most combine with either the small or large ribosomal subunits and interfere with a specific step. Puromycin inhibits peptide-chain synthesis by interacting with peptidyl-tRNA on the ribosome to form peptidyl-puromycin, which is discharged from the ribosome.

Polyribosomes, also called polysomes, consist of mRNA molecules to which are attached several or many ribosomes, each independently reading the mRNA and forming a polypeptide chain. In eukaryotic cells protein synthesis may occur on such free ribosomes or on ribosomes attached to the endoplasmic reticulum. In pancreatic cells the cisternae of the endoplasmic reticulum form channels for transport of the newly synthesized protein to be exported. Protein synthesis also takes place in mitochondria and chloroplasts, which possess mRNAs, specific tRNAs, aminoacyl-tRNA synthetases, and ribosomes; the latter resemble those of bacteria in many respects.

References

See also references at ends of Chapters 34 through 36.

Books

ARNSTEIN, H. R. V. (ed.): *Synthesis of Amino Acids and Proteins,* vol. 7 of *MTP International Review of Science,* Biochemistry Section, University Park Press, Baltimore, 1974. A collection of comprehensive reviews.

CAMPBELL, P. N., and J. R. SARGENT: *Techniques in Protein Biosynthesis,* 3 vols., Academic, New York, 1967.

FARBER, E. (ed.): *The Biochemistry of Disease,* vol. 2, *The Pathology of Transcription and Translation,* Dekker, New York, 1972.

Protein Synthesis, *Cold Spring Harbor Symp. Quant. Biol.,* vol. 34, 1969.

Articles

BYERS, B.: "Chick Embryo Ribosome Crystals: Analysis of Binding and Functional Activity in Vitro," *Proc. Natl. Acad. Sci. (U.S.),* 68: 440–444 (1971).

FORCHHAMMER, J., and L. LINDAHL: "Growth Rate of Polypeptide Chains as a Function of the Cell Growth Rate in a Mutant of E. coli," *J. Mol. Biol.,* 55: 563–568 (1971).

FULLER, W., and A. HODGSON: "Conformation of the Anticodon Loop in tRNA," *Nature,* 215: 817–821 (1967).

GRANT M. E., and D. J. PROCKOP: *New Engl. J. Med.,* 286: 194–199. 242–249, 291–300 (1972). A series of articles describing the biosynthesis of collagen, the most abundant protein in the body.

HASELKORN, R., and L. B. ROTHMAN-DENES: "Protein Synthesis," *Ann. Rev. Biochem.,* 42: 397–438 (1973). An important, clearly written review.

"Initiation of Protein Synthesis in Prokaryotic and Eukaryotic Systems," summary of an international workshop held in 1974, *FEBS Lett.*, 48: 1–6 (1974).

KIM, S. H., et al.: "Three-Dimensional Structure of Yeast Phenylalanine Transfer RNA," *Science*, 185: 435–440 (1974).

KISSELEV, L. L., and O. O. FAVOROVA: "Aminoacyl-tRNA Synthetases: Some Recent Results and Achievements," *Adv. Enzymol.*, 40: 141–238 (1974). Compilation of many data.

KURLAND, C. G.: "Ribosome Structure and Function Emergent," *Science*, 169: 1171–1177 (1970).

MORIMOTO, T., G. BLOBEL, and D. SABATINI: "Ribosome Crystallization in Chicken Embryos," *J. Cell. Biol.*, 52: 355–366 (1972).

NOMURA, M.: "Bacterial Ribosomes," *Bacteriol. Rev.*, 34: 228–277 (1970).

NONOMURA, Y., G. BLOBEL, and D. SABATINI: "Structure of Liver Ribosomes Studied by Negative Staining," *J. Mol. Biol.*, 60: 303–323 (1971).

OLSNES, S., K. REFSNES, and A. PIHL: "Mechanism of Action of the Toxic Lectins, Abrin and Ricin," *Nature*, 249: 627–631 (1974). These toxic plant proteins inhibit chain elongation in eukaryotic ribosomes.

PONGS, O., K. H. NIERHAUS, V. A. ERDMANN, and H. G. WITTMANN: "Active Sites in *E. coli* Ribosomes," *FEBS Lett.*, 40: S28–S35 (1974).

SCHATZ, G., and T. L. MASON: "The Biosynthesis of Mitochondrial Proteins," *Ann. Rev. Biochem.*, 43: 51–87 (1974).

SPRINZL, M., and F. CRAMER: "Accepting Site for Aminoacylation of tRNA$_{Phe}$ from Yeast," *Nature New Biol.*, 245: 3–5 (1973). Identification of the 2′ position of the terminal adenylic residue of tRNAs as the site of esterification.

VAZQUEZ, D.: "Inhibitors of Protein Synthesis," *FEBS Lett.*, 40: S63–S84 (1974). Comprehensive classification of antibiotics and other toxic agents according to site of action.

ZAMECNIK, P. C.: "Protein Synthesis," *Harvey Lect.*, 54: 256 (1960). A personal account of the early research which led to the identification of ribosomes, transfer RNA, and the activation reaction.

Problems

1. *E. coli* ribonuclease contains 104 amino acid residues. How long would translation of its mRNA take under intracellular conditions? Compare this with the time of synthesis of its gene. See Problem 2, Chapter 32.

2. Assuming the reaction begins with free amino acids, tRNAs, aminoacyl-tRNA synthetases, GDP, ATP, ribosomes, and translation factors, how many high-energy phosphate bonds are used to translate the mRNA of ribonuclease once?

3. The total mass of RNA in *E. coli* is 3.2×10^{10} daltons, of which 80 percent is rRNA, 15 percent is tRNA, and 5 percent is mRNA. One ribonucleotide has an average molecular weight of 340. (a) Assuming the molecular weight of RNA per ribosome is 1.7×10^6 and the average mRNA is 1,000 nucleotides long, how

many ribosomes are there per messenger? (*b*) How many tRNA molecules are there per ribosome if each adapter is 90 nucleotides long?

4. The total molecular weight of ribosomal proteins in *E. coli* is about 1 million per ribosome. Assuming that each protein is present in only one copy in each of the cell's 15,000 ribosomes and that each ribosomal protein mRNA molecule is translated 15 times, calculate the following:

   (*a*) How often must each ribosomal protein gene be transcribed?

   (*b*) How many nucleotides are contained in these genes if the average molecular weight of an amino acid residue in ribosomal proteins is 110?

   (*c*) In order to synthesize enough mRNA to code for all the proteins in the ribosomes of a single cell, how many high-energy phosphate bonds are expended in transcribing these genes, starting with ATP and ribonucleotide 5′-monophosphates?

   (*d*) How many high-energy phosphate bonds are expended in translating these mRNAs with an incubation mixture as in Problem 2? Make the approximation that the protein component of each ribosome is synthesized as a single peptide from a single mRNA.

   (*e*) How many high-energy phosphate bonds are expended in synthesizing the ribosomes, including the transcription of rRNA and mRNA and the translation of mRNA?

   (*f*) How does this compare with the energy expended in replication (Chapter 32, Problem 5)?

   (*g*) The cost of what other components must be considered in calculating the total energy cost of the protein synthetic apparatus?

5. The farther apart two mutations are, the more likely their recombination to form wild-type DNA. For the head-protein gene of bacteriophage T4 26 different mutants were isolated and lettered A to Z. The following relative recombination frequencies were observed in crosses of four representative mutants: $S \times D = 0.2$ percent, $Y \times D = 0.3$ percent, $E \times Y = 0.2$ percent, $Y \times S = 0.5$ percent. Transduction mapping localized mutant E to the 3′ end of the transcribed strand relative to the other mutations. (*a*) Construct a genetic map of the four mutations showing the direction of transcription. (*b*) All four mutations are of a type called nonsense mutations, which cause premature termination of mRNA translation. If the gene and the protein are colinear (as experiments such as these proved), arrange the mutant head proteins in order of decreasing size.

CHAPTER 34 THE GENETIC CODE

In the preceding chapter we saw how the ribosome functions in synthesizing the peptide bonds of proteins and how it aligns amino acids in the correct sequence specified by messenger RNA. We shall now consider how the 4-letter language of nucleic acids is translated into the 20-letter language of proteins. Our objectives are to examine the formal correspondence between nucleotide sequence and amino acid sequence, the genetic code words for the amino acids, the corresponding anticodons in the tRNAs, and the signals for the initiation and the termination of polypeptide chains.

Although early genetic experiments suggested that each amino acid is ultimately coded for by a small number of successive nucleotides in DNA, the identification of the sequence of nucleotides specifying each of the 20 amino acids had appeared to be a hopelessly difficult task. But a sudden breakthrough came, which made possible simple and direct biochemical approaches to the problem. As a result, the genetic dictionary of code words for the amino acids is now well known and is fully confirmed by independent genetic and biochemical evidence. The elucidation of the genetic code is one of the greatest scientific achievements in recent times. However, as we shall see, this new knowledge has opened a host of new and interesting problems. Indeed, the origin of the genetic code may have been the central event in the origin of life itself.

The Size of the Coding Units

From mathematical considerations alone, it had long appeared very likely that each amino acid residue is coded by only a small number of consecutive nucleotides in the DNA chain. Obviously, more than one nucleotide is required since there are only 4 bases in DNA and 20 amino acids in proteins. Furthermore, since the 4 code letters of DNA (A, G, C, and T) arranged in groups of 2 can yield only $4^2 = 16$ different pairs, the code words for amino acids must contain more than 2 letters. The 4 bases taken 3 at a time can theoretically specify for $4^3 = 64$ different amino acids. A triplet code is therefore feasible for coding amino acids.

That each amino acid residue is coded for by a relatively short nucleotide sequence also was evident from biochemical studies on viruses. For example, the tobacco necrosis satellite virus contains an RNA molecule having about 1,200 nucleotides, which codes for the synthesis of its coat protein subunits. Since each of the subunits has nearly 400 amino acid residues, it appeared likely that each amino acid requires $1,200/400 = 3$ nucleotides for coding, assuming that the code requires no punctuation, i.e., is commaless. Another approach came from genetic studies on frameshift mutants (page 880) of bacteriophage T4 in which one, two, or three nucleotide residues are deleted (or inserted). Single or double deletions resulted in synthesis of defective coat proteins, with many wrong amino acids, but triple deletions (or triple insertions) yielded coat proteins in which only one amino acid residue was lost or gained, all others being normal. These and other types of genetic experiments also supported a commaless triplet code. However, by 1960 it had become clear that a direct biochemical approach was required to identify precisely the nature of the code words for amino acids.

The Dependence of Ribosomal Systems on mRNA

In early experiments on the incorporation of labeled amino acids into protein by isolated ribosomes, supplemented with ATP, GTP, Mg^{2+}, amino acids, and appropriate fractions containing tRNAs and the aminoacyl synthetases, it was observed that amino acid incorporation often ceased after short periods of incubation, particularly in ribosomes from bacteria. Furthermore, a brief pretreatment of E. coli ribosomes with ribonuclease also caused loss of their activity in amino acid incorporation. Such observations suggested that freshly isolated ribosomes contain some form of RNA required for the synthesis of proteins, which is rapidly destroyed by action of endogenous ribonuclease. However, when RNA isolated from E. coli cells was added to such depleted ribosomes, their capacity for amino acid incorporation was restored. Most significant, however, was the finding that the activity of depleted ribosomes from E. coli could be restored by RNAs from other species. All these observations were consistent with the hypothesis that the labile RNA bears the message for the coding of amino acid sequence and that ribosomes of E. coli are able to "read" and translate RNAs from other species.

The stage was thus set for the dramatic experiments of M. Nirenberg and H. Matthaei, first reported in 1961, in which they added synthetically prepared polyribonucleotides of known base composition to isolated ribosomes in order to determine the relationship between the nucleotide composition of what was to be called messenger RNA and the amino acid composition of the polypeptide formed in response to it.

In their first experiments they incubated the RNA-like polymer polyuridylic acid (poly U), prepared by the action of polynucleotide phosphorylase (page 923), in a series of tubes containing messenger-depleted *E. coli* ribosomes supplemented with the usual ingredients required to support polypeptide synthesis, including all the amino acids. A different amino acid was labeled in each tube. Analysis of the acid-insoluble polypeptide fraction recovered from the tubes at the end of the incubation showed that radioactivity was present in the polypeptide fraction formed from only 1 amino acid, phenylalanine, of the 18 tested. Further analysis showed that the polypeptide chain formed in response to the synthetic poly-U messenger contained only phenylalanine residues. Nirenberg and Matthaei therefore suggested that the triplet UUU is the code word for phenylalanine. In similar experiments they found that the synthetic polyribonucleotide poly C codes for polyproline; later S. Ochoa and his colleagues found that poly A codes for polylysine. Thus it appeared that the triplet UUU codes for phenylalanine, CCC for proline, and AAA for lysine.

In order to determine the nucleotide composition of the coding triplets for other amino acids, various ribonucleotide copolymers were prepared by the action of polynucleotide phosphorylase on appropriate mixtures of nucleoside 5'-diphosphate substrates. It will be recalled (page 924) that the composition of such copolymers reflects the composition of the starting mixture of ribonucleoside 5'-diphosphates. By varying the ratio of two NDPs used as substrates, for example UDP and GDP, copolymers were prepared in which the ratio of U and G residues varied correspondingly. In such a UG copolymer the sequence of U and G residues is unknown. However, if the ratio of bases in the copolymer is known, the relative abundance of various possible triplets of U and G in poly UG, that is, UUU, UUG, UGU, GUU, UGG, GUG, GGU, and GGG, can be calculated (Table 34-1). By comparing the composition of the UG copolymer with the identity and amount of the radioactive amino acids found in the newly formed polypeptide, some conclusions could be drawn on the probable base composition of the code words specifying certain amino acids. By such experiments Nirenberg's and Ochoa's laboratories quickly succeeded in identifying the nucleotide composition of some 50 code words for the various amino acids.

Table 34-1   Amino acids coded by the random copolymer poly UG (U:G ratio 5:1)

Possible triplets	Calculated frequency of occurrence[†]	Relative amounts of amino acid incorporation observed	
UUU	100	Phe	100
UUG UGU GUU	20	Cys Val	20 20
UGG GUG GGU	4	Gly Trp	4 5
GGG	0.8		

† Assumes frequency of UUU = 100.

Sequence of Bases in the Code Words

Although the experimental approach described above gave the nucleotide *composition* of the code words, it yielded very little information on the *sequence* of nucleotides in the triplets, i.e., their spelling. But a direct approach to this problem soon came when Nirenberg and his colleague P. Leder described a new and more direct method for studying not only the composition but also the sequence of nucleotides in the code words. They found that isolated *E. coli* ribosomes incubated in 0.02 M MgCl$_2$ in the absence of GTP will bind synthetic messenger RNA preparations, such as poly U. Simultaneously, they will also bind a molecule of the aminoacyl-tRNA that is specified by the messenger. For example, ribosomes incubated with poly U simultaneously bind phenylalanyl-tRNA$_{Phe}$ but no other aminoacyl-tRNA. There is no formation of peptide bonds in such systems because GTP is absent. This test is called the *binding assay*. Simple trinucleotides were found to function as messengers in the binding assay; thus the trinucleotide UUU sufficed as a messenger to support the binding of phenylalanyl-tRNA$_{Phe}$. Similarly, the addition of AAA to ribosomes promoted the binding of lysyl-tRNA$_{Lys}$ but no other aminoacyl-tRNA.

Before the sequence of the code words could be established, it was necessary to determine conclusively the direction of readout of the messenger triplets, whether in the $5' \rightarrow 3'$ or the $3' \rightarrow 5'$ direction. (Remember that in the shorthand for polynucleotides, sequences are designated in the $5' \rightarrow 3'$ direction; reading from left to right, the residue on the left has a 5'-phosphate group and that on the right a free 3'-hydroxyl group.) In one approach Nirenberg and Leder found that the trinucleotide code words have a specific polarity, or direction. For example, GUU promotes binding of valyl-tRNA$_{Val}$, whereas UUG promotes the binding of leucyl-tRNA$_{Leu}$. Readout or decoding of the codons in mRNA thus occurs in a specific direction. When the oligonucleotide AAAUUU was tested as a messenger, it was found to code for Lys-Phe but not for Phe-Lys. Readout of this simple message therefore begins at the phosphorylated 5' end, since AAA codes for lysine. Later we shall see more direct biological evidence for the $5' \rightarrow 3'$ direction of readout in the case of natural viral RNAs (page 966). The $5' \rightarrow 3'$ direction of readout thus is the same as the direction of biosynthesis of DNA (page 896) and of RNA strands (page 917).

Once the direction of readout had been established, the binding assay was used to determine the sequence of bases in the code words. This was made possible because many trinucleotides of known base sequence could be made, either enzymatically or by synthetic methods. By determining which aminoacyl-tRNA was specifically bound by ribosomes in the presence of a given trinucleotide of known sequence. Nirenberg and Leder quickly established the spelling of the code words for many of the amino acids.

Another approach which proved to be especially valuable was the use of synthetic polyribonucleotides as messengers for the coding of polypeptides, particularly those having repeating dinucleotides or trinucleotides. A synthetic polynucleotide with repeating dinucleotides, such as UCUCU-CUCUCUC $\cdots$, contains two kinds of triplets in alternation, UCU and CUC, and should code for the synthesis of a polypeptide having two amino acids in alternating sequence. This was found to be the case; this messenger coded for the alternating polypeptide Ser-Leu-Ser-Leu-Ser-Leu etc. Repeating sequences of trinucleotides gave particularly interesting results. For example, the repeating polymer poly UUC when used as the messenger produced three different polypeptides, polyphenylalanine, polyserine, and polyleucine. This result follows from the fact that the poly

Table 34-2   Amino acid incorporation in cell-free ribosomal systems in response to synthetic polynucleotides containing repetitive sequences

Repeated sequence	Codons in polynucleotide	Amino acids incorporated
UC	UCU, CUC	Ser, Leu
AG	AGA, GAG	Arg, Glu
UG	UGU, GUG	Cys, Val
AC	ACA, CAC	Thr, His
UUC	UUC, UCU, CUU	Phe, Ser, Leu
AAG	AAG, AGA, GAA	Lys, Arg, Glu
GAU	GAU, AUG, UGA	Asp, Met
GUA	GUA, UAG, AGU	Val, Ser
UAC	UAC, ACU, CUA	Tyr, Thr, Leu
UAUC	UAU, CUA, UCU, AUC	Tyr, Leu, Ser, Ile
UUAC	UUA, CUU, ACU, UAC	Leu, Thr, Tyr

UUC can be read in three different ways, depending on the starting point:

$$\text{UUC-UUC-UUC-UUC-UUC}$$
$$\text{UCU-UCU-UCU-UCU-UCU}$$
$$\text{CUU-CUU-CUU-CUU-CUU}$$

The repeating polymer poly UAUC yielded a polypeptide in which the sequence Tyr-Leu-Ser-Ile is repeated. These and other results are shown in Table 34-2.

The data collected by H. G. Khorana and his colleagues with such repeating polynucleotide messengers provided independent evidence checking the results obtained by the binding assay, which tests codon specificity under somewhat artificial conditions and which left some ambiguities. Very rapid progress was made, and the combined results from the laboratories of Nirenberg, Khorana, and Ochoa led to the biochemical identification of all of the amino acid code words by 1965, a scant 4 years after Nirenberg reported his first experiments. Figure 34-1 shows the amino acid code-word dictionary.

The amino acid code words in Figure 34-1 have been extensively confirmed by other kinds of observations made on the correspondence between nucleotide sequence of natural messenger RNAs and the amino acid sequence of proteins formed biologically. This evidence will be reviewed later, after we examine the general features of the code words.

General Features of the Code-Word Dictionary

There are four particularly noteworthy features of the genetic code. First, and of course assumed throughout this discussion, is that no punctuation or signal is required to indicate the end of one codon and the beginning of the next, i.e., that the code is "commaless." The reading frame must therefore be correctly set at the beginning of the readout of an mRNA

Figure 34-1

The codon dictionary. The third nucleotide of each codon (color) is less specific than the first two. The codons read in the 5'→3' direction. For example, UUA = pUpUpA = leucine. The three termination (End) codons UAA, UAG, and UGA are in color. The initiation codon is AUG.

	U		C		A		G	
U	UUU	Phe	UCU	Ser	UAU	Tyr	UGU	Cys
	UUC	Phe	UCC	Ser	UAC	Tyr	UGC	Cys
	UUA	Leu	UCA	Ser	UAA	End	UGA	End
	UUG	Leu	UCG	Ser	UAG	End	UGG	Trp
C	CUU	Leu	CCU	Pro	CAU	His	CGU	Arg
	CUC	Leu	CCC	Pro	CAC	His	CGC	Arg
	CUA	Leu	CCA	Pro	CAA	Gln	CGA	Arg
	CUG	Leu	CCG	Pro	CAG	Gln	CGG	Arg
A	AUU	Ile	ACU	Thr	AAU	Asn	AGU	Ser
	AUC	Ile	ACC	Thr	AAC	Asn	AGC	Ser
	AUA	Ile	ACA	Thr	AAA	Lys	AGA	Arg
	AUG	Met	ACG	Thr	AAG	Lys	AGG	Arg
G	GUU	Val	GCU	Ala	GAU	Asp	GGU	Gly
	GUC	Val	GCC	Ala	GAC	Asp	GGC	Gly
	GUA	Val	GCA	Ala	GAA	Glu	GGA	Gly
	GUG	Val	GCG	Ala	GAG	Glu	GGG	Gly

molecule and then moved sequentially from one triplet to the next, or all codons will be out of register, resulting in formation of a protein with a garbled amino acid sequence. This fact emphasizes the need for some form of unique signal in the mRNA to indicate the correct point at which readout by the ribosome must begin; these initiation signals are discussed later.

The second major feature is that the amino acid code is *degenerate,* i.e., there is more than one code word for most of the amino acids. Degenerate does not mean imperfect, for there is no code word that specifies more than one amino acid. Moreover, the degeneracy of the amino acid code has some regularity and provides biological advantages, as we shall see.

It is striking that the degeneracy is not uniform. Thus, the code for arginine, leucine, and serine has sixfold degeneracy, i.e., there are six code words for each of these, whereas the code for other amino acids, such as glutamic acid, tyrosine, histidine, and several others, has only twofold degeneracy (Table 34-3). Of all the amino acids, only tryptophan and methionine have but a single code word.

A third important feature is that the third base in the codons is less specific than the first two. Note from Figure 34-1 that the degeneracy of the code involves only the third

Table 34-3 Degeneracy of the amino acid code

Amino acid	Number of codons
Ala	4
Arg	6
Asn	2
Asp	2
Cys	2
Gln	2
Glu	2
Gly	4
His	2
Ile	3
Leu	6
Lys	2
Met	1
Phe	2
Pro	4
Ser	6
Thr	4
Trp	1
Tyr	2
Val	4

base in the codon in most cases (exceptions are arginine, leucine, and serine). For example, alanine is coded by the triplets GCU, GCC, GCA, and GCG; the first two bases GC are common to all alanine codons, whereas the third may be any of the bases. Furthermore, when two different amino acids have code words in which the first two bases are identical, the third position of one is filled only by purines and the third position of the other only by pyrimidines. For example, for histidine (CAU, CAC) the third bases are the pyrimidines U and C, and for glutamine (CAA, CAG) the third positions are the purines A and G. Nearly all the amino acid codons occur in the code table in pairs that consist of the triplets XY^A_G or XY^U_C. Some amino acids are coded by two such pairs, others by one pair. Evidently the first two letters of each codon are the primary determinants of its specificity. The third position, i.e., the nucleotide at the 3'-hydroxyl end of the codon, is of less importance and does not fit as precisely; it is loose and tends to "wobble," to use F. H. C. Crick's phrase.

A plausible reason for the loose or wobbly codon-anticodon base pairing at position 3 of the codon comes from kinetic considerations. M. Eigen has pointed out that the anticodon must not stick too tightly to the codon, because the rate of dissociation of the hydrogen bonds holding together the codon-anticodon complex may then be too slow to be compatible with the known high rate of protein synthesis. The hydrogen bonds involved in base pairing of only one or two nucleotides would be relatively loose and thus have low specificity, whereas the hydrogen bonds formed when many successive nucleotide residues base-pair with a complementary sequence would have high specificity but would dissociate too slowly to be useful in coding for the synthesis of the proteins. Apparently a triplet of nucleotides optimizes both the specificity of hydrogen bonding and the rate of formation and dissociation of the codon-anticodon interaction. Possibly the wobble base provides fine tuning of the specificity and kinetics of the codon-anticodon interaction.

A fourth feature of the code is that 3 of the 64 triplets do not code for any amino acids: UAG, UAA, and UGA (Figure 34-1). As will be seen below, these triplets are signals for the termination of polypeptide chains.

Genetic Evidence for the Code-Word Assignments

Although the code words for the amino acids were originally deduced from in vitro biochemical experiments on isolated ribosomes, they have been fully confirmed by in vivo observations on the effects of mutations on amino acid sequence of specific proteins, not only induced mutations in bacteria but also spontaneous mutations in human beings. Single-point mutations, in which one base in a gene is replaced by another, often result in replacement of one amino acid by another. This can be expected, since each codon has three residues, any one of which may be replaced by mutation. In many cases the replacement of one nucleotide by another will result in a new code word for a different amino acid. We now know the complete amino acid sequence of human hemoglobin α and β chains, and we also have information on the amino acid replacements in different hemoglobin mutations (page 117). The identity of the amino acid replacements in mutant hemoglobins makes checking of the genetic code possible. Table 34-4 shows some amino acid replacements that have been observed in the α chain of some mutant human hemoglobins together with their code words, as well as the normal amino acids at these positions and their code words. All the amino acid replacements can be explained on the basis that only a single nucleotide in a coding triplet is altered, so that the codon for one amino acid is changed to a codon for another in a manner consistent with the code words elucidated by the in vitro biochemical methods. For example, in the case of the replacement of lysine by glutamic acid, a single base change in a codon for lysine (AAA or AAG) would produce a codon for glutamic acid (GAA or GAG).

The results of single-point mutations in a number of bacterial and viral proteins have also shown complete agreement with the code words established from in vitro tests. Data in Table 34-5 show some amino acid replacements observed in single-point mutations of the tobacco mosaic virus coat protein. In such analyses of the effect of mutations, nearly all the theoretically possible single-base changes in the codons have been found, and all agree with the code-word dictionary.

Frame-shift mutations have provided further confirmatory evidence. Such mutations are produced by deletion of a base or addition of an intercalated mutagen such as acridine (page 880). Such mutations shift the reading frame and produce a completely garbled product from the point of mutation to the carboxyl end of the chain. If such a mutation occurs early in the gene, so that most of the protein is defective, the mutation will in all probability be lethal. However, if it occurs late in the gene, i.e., near the point corresponding to the carboxyl-terminal end of the polypeptide chain, the mutation can often be tolerated. Comparison of the amino acid sequence of the defective "tail" of the protein with the normal carboxyl-terminal sequence in such cases also has provided confirmation of the composition and the sequence of the code words. More elaborate experiments with double frame-shift mutations (two bases deleted or added) have also verified the dictionary.

Further biological evidence for the code-word dictionary, and the most definitive, will be discussed later in this chapter.

Initiation Codons and Initiation Sites in RNAs

Although no punctuation is required between separate code words for the amino acids, specific signals are necessary to initiate the polypeptide chain at the proper point, so that the correct reading-frame relationship is established. We have seen that N-formylmethionine is the initiating amino acid in

Table 34-4 Some amino acid replacements in the α chain of mutant human hemoglobins

Bases undergoing mutation are in color

Normal		Mutant	Base replacement
Lys (AAA)	→	Glu (GAA)	A → G
Glu (GAA)	→	Gln (CAA)	G → C
Gly (GGU)	→	Asp (GAU)	G → A
His (CAU)	→	Tyr (UAU)	C → U
Asn (AAU)	→	Lys (AAA)	U → A
Glu (GAA)	→	Lys (AAA)	G → A
Glu (GAA)	→	Gly (GGA)	A → G

Table 34-5 Some amino acid replacements in the coat protein of tobacco mosaic virus mutants induced by nitrous acid

The bases undergoing mutation are in color

Wild type		Mutant	Base replacement
Pro (CCC)	→	Ser (UCC)	C → U
Pro (CCC)	→	Leu (CUC)	C → U
Ile (AUU)	→	Val (GUU)	A → G
Glu (GAA)	→	Gly (GGA)	A → G
Thr (ACA)	→	Ile (AUA)	C → U
Ser (UCU)	→	Phe (UUU)	C → U

prokaryotes and methionine in eukaryotes (page 937). AUG is the initiation codon in both prokaryotes and eukaryotes; GUG is an alternate initiation codon for methionine in prokaryotes. However, there are two transfer RNAs for methionine, but only one of them is commonly used as the initiation tRNA (page 937).

An important check on the identity of the initiating codon has come from another type of experiment. When ribosomes are incubated with a viral RNA, they do not necessarily begin translation from the 5' end of the RNA but often start at a site some distance from the 5' end, indicating that the initiation codon AUG may have an interior location in the RNA. This was ascertained by allowing ribosomes to bind to the RNA from the RNA phage f2 under appropriate conditions; only one ribosome is bound per RNA molecule. To determine the point of ribosome binding on the RNA, the exposed portion of the viral RNA was removed by digestion with pancreatic ribonuclease, and the bound RNA fragment, containing about 30 residues, was then released from the ribosome. This fragment was found to contain the initiation site for polypeptide synthesis, with the start codon AUG located approximately in the middle of the fragment. Similar tests of the RNA from the R17 virus, which codes for three proteins, identified three starting points to which ribosomes can bind; each contained the start codon AUG (see below).

Termination Codons

When protein synthesis is coded for by a natural messenger RNA, the completed polypeptide chain is released in free form after its hydrolytic removal from the carboxyl-terminal tRNA. However, when polypeptide synthesis is coded for by synthetic messengers, such as poly U, the newly formed chain remains attached to the ribosomes as polyphenylalanyl-tRNA$_{Phe}$, presumably because poly U contains no termination signal.

During the research on assignment of code words to the different amino acids three codons, UAG, UAA, and UGA, were found not to specify any of the 20 amino acids. They were at first called "nonsense" codons, but they were ultimately established as termination signals. This conclusion came in the first instance from genetic experiments to determine the consequences of mutation to or from these triplets (Figure 34-2). When a codon normally specifying a given amino acid undergoes mutation to form UAG, a polypeptide chain is synthesized which is terminated at the amino acid residue prior to that specified by the mutated codon. Among the amino acid codons that are mutational precursors of the codon UAG are serine (UCG), glutamine (CAG), and tyrosine (UAU or UAC). Each of these codons can be converted to UAG by a single-base change (Figure 34-2). An example of a chain-termination mutant (mutants to UAG are often called _amber_ mutants) is the _sus_ 13 mutant of the f2 RNA bacteriophage of E. coli. The wild-type f2 coat protein begins with the amino-terminal sequence fMet-Ala-Ser-Asn-Phe-Thr-Gln. In the _sus_ 13 amber mutant, only

Figure 34-2
Identification of the termination codon UAG from mutations to UAG and reversions from UAG. The bases undergoing mutation or reversion are in color.

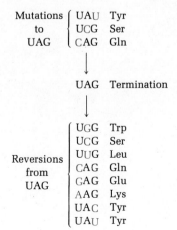

Mutations to UAG		
	UAU	Tyr
	UCG	Ser
	CAG	Gln

↓

UAG Termination

↓

Reversions from UAG		
	UGG	Trp
	UCG	Ser
	UUG	Leu
	CAG	Gln
	GAG	Glu
	AAG	Lys
	UAC	Tyr
	UAU	Tyr

the hexapeptide fMet-Ala-Ser-Asn-Phe-Thr is synthesized and released in free form, whereas in wild-type f2 the entire coat protein is synthesized. Since CAG, the codon for the next amino acid (glutamine) can undergo mutation to UAG, it was concluded that UAG signals the termination of the chain and its release in free form. In the absence of this mutation, the next amino acid, glutamine, would be introduced and the resulting heptapeptide would remain attached as heptapeptidyl-tRNA$_{Gln}$.

Other genetic experiments have shown that chain-termination mutants which make incomplete, prematurely discharged proteins, may revert on back mutation to a form which can now make a complete protein but in which one amino acid is not necessarily identical with the corresponding amino acid in the wild-type protein. In such cases, the chain-termination codon has been changed in a single-point mutation to a codon that specifies an amino acid, but not necessarily the same amino acid present in the wild-type protein, as is illustrated in Figure 34-2. From such chain-termination mutants and their revertants, UGA, UAG, and UAA have been identified as termination signals. The three termination codons are probably not used with equal frequency; indeed, recent research suggests that UAA is the most common termination signal.

Biochemical Mapping of Messenger RNAs

Much significant information is emerging regarding the coding sequences in natural mRNA molecules, among which we may include the RNAs of RNA bacteriophages, since they are directly utilized as messengers in the host cell. A number of pure RNAs from small RNA viruses of E. coli have been studied to determine the precision of correspondence between the codons in the RNA and the sequence of amino acid residues in the proteins they code. Among these viruses is R17 of E. coli. Its RNA has about 3,500 nucleotide residues and a molecular weight of about 1.1 million, and it codes for three viral proteins, namely, A protein, coat protein, and an RNA replicase. The nucleotide sequences of some large fragments of R17 RNA, the amino acid sequences of the amino-terminal ends of the three viral proteins, and a large portion of the sequence of the coat protein have been worked out in an important group effort by J. A. Jeppeson, J. A. Steitz, F. Sanger, and their colleagues in Cambridge. Figure 34-3 shows a genetic and biochemical map of the R17 RNA deduced from the combined nucleotide and amino acid sequence studies. The ribosome binding sites, where the polypeptide chains are initiated, were determined by allowing ribosomes to bind to the RNA and then subjecting the ribosome-RNA complex to ribonuclease action, which digested away all the RNA except that in the segments of 25 to 30 residues that were attached to the ribosomes.

From these and other experiments it was possible to deduce the location of the genes or cistrons for the three viral proteins and the approximate number of nucleotide residues in each (Figure 34-3). All three ribosome binding sites were

Figure 34-3
The genetic and biochemical map of the R17 genome (right). It shows the segments of the RNA coding for the three proteins, in the proper sequence. There are four untranslated regions, two at the ends of the RNA and two intercistronic regions. AUG is the initiation codon for A protein, coat protein, and the RNA replicase; AUG also appears, but does not initiate a chain, in the intercistronic region at R. Two successive termination codons UAA and UAG come at the end of the coat protein cistron; UGA also appears in the untranslated intercistronic region. [Data from P. G. N. Jeppesen and colleagues, Nature, 226: 230 (1970), and J. L. Nichols, Nature, 225: 147 (1970).]

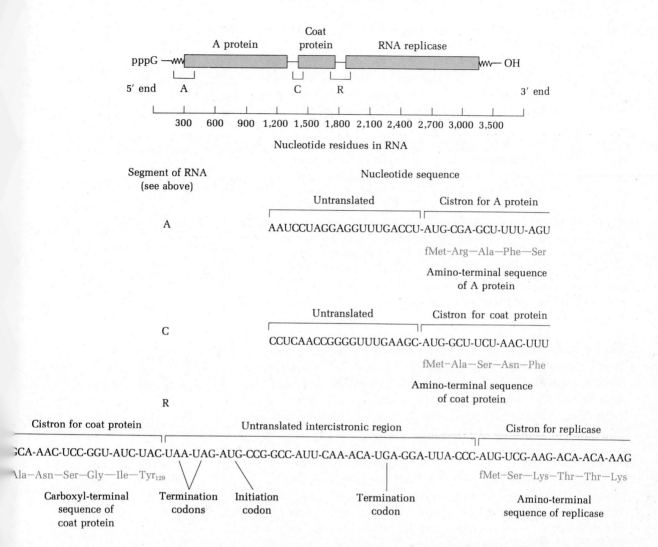

found to be internal. The first cistron begins at a point 300 nucleotide residues from the 5′ end of the RNA, where the first initiation codon (AUG) is located. Moreover, the 3′ end of the RNA has a stretch of over 300 residues following the last of the three cistrons. Neither of these long ends of the RNA is translated by *E. coli* ribosomes fully supplemented with all required cofactors; the ends therefore appear to have functions other than coding of the viral proteins. In addition, there are two other untranslated segments, one of 25 or more residues between the A-protein and the coat-protein cistrons, and another of 36 residues between the coat-protein and the replicase cistrons.

All three cistrons begin with the initiation codon AUG, followed by triplets coding for the amino-terminal sequences of the three viral proteins. A search for the termination codons has revealed that the coat-protein cistron is terminated by two successive termination codons, UAA and UAG, suggesting a fail-safe arrangement in case the first termination codon is bypassed or unrecognized by the ribosome. Examination of other oligonucleotide fragments showed that R17 RNA contains some extra termination codons, which are not at the end of the cistrons but lie in the untranslated intercistronic regions.

Most significant in this study of R17 RNA and its protein products is the fact that 77 of its codons could be matched with amino acid residues in the portions of the RNA and the polypeptides for which congruent sequence data were available. The 77 codons corresponded to 18 different amino acids. Moreover, 38 of the 61 known codons for amino acids were represented, including all the codons for serine, valine, threonine, phenylalanine, asparagine, and glutamic acid.

An even more extensive checking of codon–amino acid correspondence has been worked out by a group in the laboratory of W. Fiers, in Belgium, who have compared the nucleotide sequence in the RNA of the E. coli virus MS2 with the amino acid sequence of its coat protein (Figure 34-4). In this case, and in similar studies on the RNA virus FD2, there is again an exact correspondence with the code-word dictionary in Figure 34-1.

Very little information is available on nonviral messengers for either prokaryotic or eukaryotic organisms, since they are very difficult to isolate in pure form. However, the messenger RNAs for the α and β hemoglobin chains have been isolated in highly purified form. Each has long sequences of untranslated nucleotide residues on both ends of the molecule; the untranslated 3′ end contains blocks of poly A (page 321).

Universality of the Code

The genetic code words in Figure 34-1 are now regarded as biologically universal. They are certainly identical in E. coli, tobacco mosaic virus, man and other vertebrates, such as the rabbit, and in a large number of bacterial viruses, as indicated by the experiments described above. The question has also been tested in other ways. For example, when rabbit reticulocyte ribosomes, containing the mRNA for hemoglobin protein, are mixed with E. coli aminoacyl-tRNAs and other necessary factors, hemoglobin chains are formed, indicating that the anticodons in the E. coli tRNAs recognize the codons in the rabbit-hemoglobin mRNA. Such cross tests of tRNAs of one species vs. the codons of another have also been made between E. coli, the guinea pig, and the South African clawed toad, Xenopus laevis, and other species. In all species tested to date the code words are identical. However, the frequency of use of the different code words for a given amino acid may vary from one species to another. This fact has suggested that the multiplicity of code words for most of the amino acids is involved in evolutionary development or differentiation. The degeneracy of the genetic code is therefore a matter requiring much further study.

Formation of Uncoded Amino Acids in Proteins: Posttranslational Modification

Some proteins contain in peptide linkage amino acids that lack genetic code words. Thus, collagen contains hydroxyproline and hydroxylysine (page 76), some muscle pro-

Figure 34-4
The nucleotide sequence (5′→3′ direction) of the coat protein cistron of phage MS2, together with the amino acid sequence it codes (right). The cistron is preceded and followed by untranslated intercistronic regions. The numbers refer to the position of the amino acid residues in the coat protein (1 to 129) and the RNA polymerase (1 to 6). These data represent the longest RNA and corresponding amino acid sequence elucidated to date and provide valuable information on the occurrence of multiple codons for various amino acids. They provide yet another type of biological verification that the genetic code words worked out in in vitro reconstructed systems are correct. [Data from W. Min Jou, G. Haegeman, M. Ysebaert, and W. Fiers, Nature, 237: 83 (1972).]

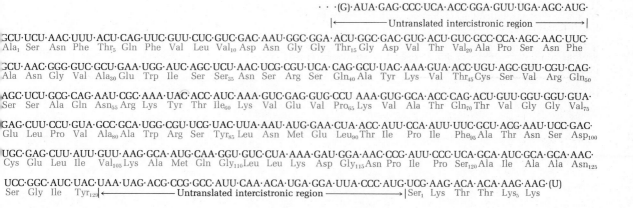

· · ·(G)·AUA·GAG·CCC·UCA·ACC·GGA·GUU·UGA·AGC·AUG·

|←——————Untranslated intercistronic region ——————→|

GCU·UCU·AAC·UUU·ACU·CAG·UUC·GUU·CUC·GUC·GAC·AAU·GGC·GGA·ACU·GGC·GAC·GUG·ACU·GUC·GCC·CCA·AGC·AAC·UUC·
Ala₁ Ser Asn Phe Thr₅ Gln Phe Val Leu Val₁₀ Asp Asn Gly Gly Thr₁₅ Gly Asp Val Thr Val₂₀ Ala Pro Ser Asn Phe

GCU·AAC·GGG·GUC·GCU·GAA·UGG·AUC·AGC·UCU·AAC·UCG·CGU·UCA·CAG·GCU·UAC·AAA·GUA·ACC·UGU·AGC·GUU·CGU·CAG·
Ala Asn Gly Val Ala₃₀ Glu Trp Ile Ser Ser₃₅ Asn Ser Arg Ser Gln₄₀ Ala Tyr Lys Val Thr₄₅ Cys Ser Val Arg Gln₅₀

AGC·UCU·GCG·CAG·AAU·CGC·AAA·UAC·ACC·AUC·AAA·GUC·GAG·GUG·CCU AAA·GUG·GCA·ACC·CAG·ACU·GUU·GGU·GGU·GUA·
Ser Ser Ala Gln Asn₅₅ Arg Lys Tyr Thr Ile₆₀ Lys Val Glu Val Pro₆₅ Lys Val Ala Thr Gln₇₀ Thr Val Gly Gly Val₇₅

GAG·CUU·CCU·GUA·GCC·GCA·UGG·CGU·UCG·UAC·UUA·AAU·AUG·GAA·CUA·ACC·AUU·CCA·AUU·UUC·GCU·ACG·AAU·UCC·GAC·
Glu Leu Pro Val Ala₈₀ Ala Trp Arg Ser Tyr₈₅ Leu Asn Met Glu Leu₉₀ Thr Ile Pro Ile Phe₉₅ Ala Thr Asn Ser Asp₁₀₀

UGC·GAG·CUU·AUU·GUU·AAG·GCA·AUG·CAA·GGU·GUC·CUA·AAA·GAU·GGA·AAC·CCG·AUU·CCC·UCA·GCA·AUC·GCA·GCA·AAC·
Cys Glu Leu Ile Val₁₀₅ Lys Ala Met Gln Gly₁₁₀ Leu Leu Lys Asp Gly₁₁₅ Asn Pro Ile Pro Ser₁₂₀ Ala Ile Ala Ala Asn₁₂₅

UCC·GGC·AUC·UAC·UAA·UAG·ACG·CCG·GCC·AUU·CAA·ACA·UGA·GGA·UUA·CCC·AUG·UCG·AAG·ACA·ACA·AAG·AAG·(U)
Ser Gly Ile Tyr₁₂₉ |←——————— Untranslated intercistronic region ———————→| Ser₁ Lys Thr Thr Lys₅ Lys

teins contain N-methyllysine (page 756), elastin contains desmosine and isodesmosine (page 76), glycogen phosphorylase contains phosphoserine (page 434), and prothrombin contains γ-carboxyglutamate. Such uncoded amino acids are always derivatives of amino acids that do possess code words. These uncommon amino acids are formed enzymatically from the parent amino acids *after* the latter have been inserted into the polypeptide chain in response to coding by their specific triplets. For example, hydroxyproline is formed by the enzymatic hydroxylation of prolyl residues by *proline monooxygenase* (page 696) in the formation of collagen.

Identification of Anticodons: Codon-Anticodon Pairing

With the identification of the codons for the different amino acids it was only natural that similar efforts would be made to identify their corresponding anticodons in their tRNAs. Many anticodons have been identified by sequence analysis of tRNAs (Table 34-6). Two major conclusions follow from these data. Codons and their corresponding anticodons pair in the *antiparallel* direction; we have already seen that the antiparallel direction of nucleotide strands is the rule in the replication of DNA (page 896) and its transcription to form RNA (page 917). Codons and anticodons are by convention written in the 5′→3′ direction, as are all polynucleotides, but we must be careful to remember that they pair in an antiparallel manner when aminoacyl-tRNAs bind to their codons in mRNA.

It is a second conclusion that many of the anticodons contain nucleoside residues other than the expected A, G, C, and U. In particular, inosinic acid residues (which are deaminated adenylic acid residues and designated I) are frequently found instead of the expected adenylic residues. Also found are pseudouridine (ψ), 2′-*O*-methylguanosine, 2-thiouridine, 4-acetylcytidine, and 2′-*O*-methylcytidine. These minor nucleosides are nearly always found at the 5′ end of the anticodon, i.e., that end which pairs with the nucleoside in the

Table 34-6 Some anticodons (color) found in tRNAs and their corresponding codons (black)

Note the degeneracy in the third, or wobble, position of most of the codons and the occurrence of minor bases in the anticodon paired with the wobble position

Amino acid	Anticodon-codon pairing†	Amino acid	Anticodon-codon pairing†

Ala	←5'
	CG I
	  U
	GCC
	  A
	5' →
		Glu	←
			CUUs²
			GAA
			→

Asp	←5' CUG GA C U 5' →	Lys	← UUC AAG → ← UUUs² AAA →

Arg	←
	GC I
	   A
	CGU
	   C
	→
		Phe	←
			AAG
			UU C
			   U
			→

	←
	GCG
	CGC
	→
		Ser	←
			AG I
			   A
			UCU
			   C
			→

	← CCG GG C U		
Gly	← CCC GGG →	Tyr	← A ψG UA C U →

	←
	CCU
	GG A
	   G
	→
		Val	←
			CAUoa⁵
			   A
			GUC
			   U
			→

† Us² = 2-thiouridine; ψ = pseudouridine; Uoa⁵ = uridine 5-oxyacetic acid; I = inosine.

3' end of the codon. The minor bases in the anticodons thus pair with the least specific of the three residues of the codon, i.e., that which is subject to wobble. Thus the pairing of the first two bases of the codon, reading from 5' to the 3' position, with the last two bases of the anticodon is "tight" and unambiguous, whereas the wobble base pair allows of some ambiguity, so that one anticodon may bind more than one codon. Adenosine is never found in the wobble position of an anticodon; it appears to be replaced by inosine.

From model-building experiments it has been deduced that the conformation in space of codon-anticodon pairs is nearly the same as that occurring in a segment of a double-helical RNA (Figure 34-5). Thus it appears that the codon and anticodon fit each other optimally in a steric sense.

Another interesting relationship between codons and anticodons relates to the residue in position 40 of tRNAs, that residue following the anticodon (Figure 33-5, page 936). This residue is always a purine. Moreover, in many tRNAs it has a long isopentenyl side chain and is methylated. These side chains appear to participate in the codon-anticodon interac-

Figure 34-5
Schematic diagram of the pairing between the codon bases (shaded gray) in the mRNA and the bases of the anticodon (solid color) in the tRNA. The anticodon stem, containing stacked base pairs, and the anticodon loop are shown in shaded color. The codon-anticodon pair are antiparallel in their direction and may be regarded as a short segment of a double helix. [From W. Fuller and A. Hodgson, Nature, 215: 817 (1967).]

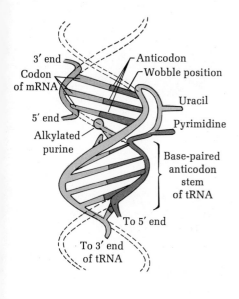

Table 34-7 A hypothetical primordial doublet code

The third position is assumed to be occupied by bases serving as commas; during the evolution from a doublet to a triplet code the commas may have been transformed into coding symbols to allow more than 15 amino acids to be coded; such a hypothesis is consistent with the wobble and degeneracy in the third position of the present triplet code

Primordial amino acid	Doublet codon	Additions in third position	"New" amino acids
Lys	AA	C or U	Asn
Thr	AC		
Ser	AG		
Ile	AU	G	Met
His	CA	A or G	Gln
Pro	CC		
Arg	CG		
Leu	CU		
Asp, Glu	GA		
Ala	GC		
Gly	GG		
Val	GU		
End	UA	C or U	Tyr
Ser	UC		
Cys	UG	G	Trp
Phe	UU		

tion and apparently provide extra binding power. Some purines with isopentenyl side chains have hormone activity in plants; such compounds, called *cytokinins* (Figure 33-7, page 937), stimulate cell division.

Evolution of the Amino Acid Code

A number of interesting ideas regarding the origin and evolution of the genetic code have emerged. First, the universality of the code suggests that it arose but once in the course of biological evolution. Since fossil evidence indicates that prokaryotes resembling present bacteria existed as early as 3,300 million years ago (page 1032), it appears possible that the present genetic code may have been functional at that time. The fact that the first two letters of the code words bear most of the specificity has been interpreted to mean that code words may once have been doublets coding for a group of 16 "primordial" amino acids or 15 amino acids plus a termination codon (Table 34-7). Such a doublet code may have employed every third nucleotide as a comma. It appears possible that a group of "new" amino acids may have arisen later, possibly including asparagine, glutamine, methionine, tyrosine, and tryptophan, which then required triplets for their coding. The third nucleotide for the new triplets may have derived from the symbols used for commas in the original doublet code. This idea receives some support from the fact, pointed out by T. H. Jukes, that ferredoxins (page 488) of the most ancient species of bacteria still living, the anaerobic *Clostridia*, contain only 13 different amino acids. The ferredoxins of these bacteria are very small proteins (55 residues) and consist of two homologous halves, apparently the result of duplication of a small gene corresponding to only 28 amino acid residues. Ferredoxins of higher species, however, are larger and contain more kinds of amino acids, suggesting that the acquisition of "new" amino acids occurred later in biological evolution. Jukes has also provided evidence that the primitive precursor of arginine is ornithine (pages 77 and 581), an amino acid not found in proteins today. Arginine, a derivative of ornithine, is not present in most ancient proteins, such as the clostridial ferredoxins, and may have taken over the codon of the simpler ornithine.

The degeneracy of the genetic code is a feature that gives it survival value. If there were only one code word for each amino acid, only 20 of the possible 64 codons would be used and most point mutations of coding triplets would result in triplets not coding for any amino acid. On the other hand, because of the degeneracy of the code, a mutation in most cases results in formation of a synonym triplet or replacement of one amino acid by another, often producing a silent mutation, i.e., a gene product with unaltered function. The degeneracy of the code thus allows formation of a mutant protein that is often still functional and, indeed, might even be a better protein than in the wild-type organism. The degeneracy of the code can thus allow for gradual "improvement" of the genome and its gene products; it permits a variety of different amino acid replacements to occur by mutation, from which the replacement with the best survival value can be selected. Since the most frequent mutations involve A-G or C-T exchanges, which usually have no effect on the amino acid coded when they occur in the third position, the degeneracy of the code may also be considered to have protective value in stabilizing the genome against mutation.

Another interesting and significant feature is the relationship between the code words of amino acids having similar R groups. For example, all codons having U as the second letter belong to amino acids having nonpolar or hydrophobic side chains (Phe, Leu, Ile, Met, Val). The codons of the very similar amino acids alanine and glycine contain only C and G as first or second letters. Moreover, the negatively charged amino acids aspartic and glutamic acid both have GA as their first two letters. Similar but less specific relationships may be observed between the structurally similar amino acids serine, threonine, and cysteine. It has been proposed that the specific code words arose because of structural or steric complementarity to their corresponding amino acids. However, the supporting evidence for this hypothesis is not very clear.

Despite the universality of the genetic code, the proportion of the four bases in DNA varies widely with the species (page 861). Thus, the DNA in various bacterial species contains anywhere from 30 to 70 percent G-C pairs. Those bacteria having a high G-C content may therefore use more frequently those codons in which the third position is either G or C, compared with the bacteria having a low G-C content. The reason for these striking variations is not clear, although DNA high in G-C content is more resistant to mutation by ultraviolet light. Other comments on the evolution of the genetic code, which may have been a central event in the origin of life, are made elsewhere (Chapter 37, page 1050).

Summary

From mathematical considerations, from structural studies on viruses, and from early genetic experiments the amino acid code words were considered to be triplets of successive nucleotides in messenger RNA with no commas. This conclusion was verified by direct biochemical experiments on ribosomes.

The nucleotide composition of the coding triplets for the amino acids was deduced by determining the identity of the labeled amino acids incorporated into polypeptides on incubation of E. coli ribosomes, previously depleted of adhering messenger RNA, with polyribonucleotides of known base composition.

The sequence of bases in each codon was then deduced (1) from experiments on the effect of nucleotide triplets of known base sequence on the binding of various labeled aminoacyl-tRNAs to ribosomes, in the absence of GTP, and (2) from the amino acid sequence of polypeptides synthesized from synthetic messenger templates consisting of repetitive doublets and triplets of bases. The codons identified by such in vitro biochemical experiments have been verified by the amino acid replacements observed in mutant human hemoglobins, single-point mutants of viral proteins, and frame-shift mutants.

The amino acid code is degenerate: it has multiple code words for all amino acids except tryptophan and methionine. The degeneracy is usually in the third position of the codon, i.e., at its 3' end; the nucleoside in this position is much less specific than the first and second. Chain initiation begins with the codon AUG, which codes for initiating N-formylmethionine in prokaryotes and for methionine in eukaryotes. The triplets UAA, UAG, and UGA code for no amino acids but are signals for chain termination; UAA is probably the normal chain terminator. The genetic code is almost certainly universal.

That multiple code words for most of the amino acids and the initiation and termination codons are employed biologically has been extensively verified by direct comparison of the nucleotide sequences of the RNAs of the R17, f2, and MS2 viruses with the amino acid sequences of their biologically formed protein products. Codons are read in the 5'→3' direction along the mRNA chain. The anticodons in most of the tRNAs have also been identified; many contain minor nucleosides. Anticodons are base-paired to their codons by hydrogen bonding in the antiparallel direction. The base pairing between position 3 of the codon and position 1 of the anticodon is relatively loose and wobbles. Uncoded amino acid derivatives in proteins, such as hydroxyproline and hydroxylysine, are formed from their parent coded amino acids by secondary enzymatic reactions after the chain is complete. The evolution of the genetic code is a central problem in biology.

References

Books

DAVIDSON, J. N.: *The Biochemistry of Nucleic Acids*, Academic, New York, 1972.

The Genetic Code, *Cold Spring Harbor Symp. Quant. Biol.*, vol. 31, 1966. A collection of reviews and articles on the experimental basis of the amino acid code words.

JUKES, T. H.: *Molecules and Evolution*, Columbia University Press, New York, 1966. A theory of the evolution of the genetic code.

STENT, G. S.: *Molecular Genetics*, Freeman, San Francisco, 1971. An excellent account of the genetic approaches to the amino acid code words.

STEWART, P. R., and D. S. LETHAM (eds.): *The Ribonucleic Acids*, Springer, New York, 1973.

WOESE, C. R.: *The Genetic Code*, Harper & Row, New York, 1967. The origin and evolution of the code.

YCAS, M.: *The Biological Code*, North-Holland, New York, 1969. The history of research on the genetic code, with complete references.

ZUBAY, G., and J. MARMUR: *Papers in Biochemical Genetics*, Holt, New York, 1968. Reprinted classic papers. A second edition has also appeared.

Articles

CRICK, F. H. C.: "Codon-Anticodon Pairing: The Wobble Hypothesis," *J. Mol. Biol.*, 19: 548–555 (1966). The structural explanation of the play, or wobble, in the fit of the third nucleotide of a codon.

CRICK, F. H. C.: "The Origin of the Genetic Code," *J. Mol. Biol.*, 38: 367–379 (1968). The biological evolution and logic of the code,

GHOSH, H. P., D. SOLL, and H. G. KHORANA: "Studies on Polynucleotides; LXVII: Initiation of Protein Synthesis *in Vitro* as Studied by Using Ribopolynucleotides with Repeating Nucleotide Sequences as Messengers," *J. Mol. Biol.,* 25: 275 (1967). Use of repeating or block polymers of ribonucleotides as synthetic messengers.

JEPPESEN, P. G. N., J. L. NICHOLS, F. SANGER, and B. G. BARRELL: *Cold Spring Harbor Symp. Quant. Biol.,* 35: 13 (1970). The biochemical map of the R17 genome.

LACEY, J. C., and K. M. PRUITT: "Origin of the Genetic Code," *Nature,* 223: 799–804 (1969). One view of the structural relationship of the code words to the amino acids.

MIN JOU, W., G. HAEGEMAN, M. YSEBAERT, and W. FIERS: "Nucleotide Sequence of the Gene Coding for the Bacteriophage MS2 Coat Protein," *Nature,* 237: 82–88 (1972). The correspondence between the code words in a viral RNA and the amino acid sequence of one of its translation products.

NIRENBERG, M. W., and P. LEDER: "RNA Codewords and Protein Synthesis," *Science,* 145: 1399–1407 (1964). The binding assay: the sequence of bases in amino acid codons.

NIRENBERG, M. W., and J. H. MATTHAEI: "The Dependence of Cell-Free Protein Synthesis in *E. coli* upon Naturally Occurring or Synthetic Polyribonucleotides," *Proc. Natl. Acad. Sci. (U.S.),* 47: 1588–1602 (1961). Classic paper showing dependence of ribosomes on RNA messengers and coding of polyphenylalanine by poly U.

STENFLO, J., P. FERNLUND, W. EGAN, and P. ROEPSTORFF: "Vitamin K-Dependent Modifications of Glutamic Acid Residues in Prothrombin," *Proc. Natl. Acad. Sci. (U.S.),* 71: 2730–2733 (1974). An account of the origin of γ-carboxyglutamic acid residues in prothrombin.

Problems

1. Predict the amino acid sequences of peptides formed by ribosomes in response to the following messengers, assuming each chain begins with the triplet on the left.
   (a) GGUCAGUCGCUCCUGAUU
   (b) UUGGAUGCGCCAUAAUUUGCU
   (c) CACGACGCUUGUUGCUAU
   (d) AUGGACGAA

2. Which of the following amino acid replacements in mutant proteins are consistent with the genetic code if they are due to single-point mutations? If any are inconsistent, state why.
   (a) Phe → Leu, (b) Ile → Leu, (c) Ala → Thr, (d) Pro → Ser, (e) Lys → Ala, (f) His → Glu, (g) Phe → Lys.

3. Using the information of Table 5-5 (page 117) and Figure 34-1 on the abnormal α chains of hemoglobin, state whether each mutation is a transition or transversion. Assume that all mutations are single-point base substitutions.

4. One strand of DNA contains the following sequence, reading from 5' to 3':

   T-C-G-T-C-G-A-C-G-A-T-G-A-T-C-A-T-C-G-G-C-T-A-C-T-C-G-A

   Write (a) the sequence of bases in the other strand of the DNA, (b) the sequence of bases in the mRNA transcribed from the first strand of DNA, (c) the amino acid sequence coded by the mRNA in (b), and (d) the amino acid sequence coded for if the second T from the 3' end of DNA is deleted.

5. A protein A of a wild-type bacterium (strain 1) has a tryptophan residue at position 26. Strain 2, derived from strain 1 by mutation, contained leucine at position 26. Mutation of strain 2 produced strain 3, in which no new mutant protein A was detected. Mutation of strain 3 produced strain 4, which contained proline at position 26. (a) Assuming that all mutations were base substitutions and that parent and daughter strain differ by no more than one base in the codon for residue 26 of protein A, are the observations in accord with the genetic code? (b) Give the progression of codon changes in this series of mutations.

6. A synthetic poly AC chain is synthesized from a mixture containing 20 times as many A residues as C residues. The chain is then cleaved with pancreatic ribonuclease and the resulting mixture of oligonucleotide fragments used as synthetic messenger RNA. (a) Assuming that the mRNA is always read in the 5'→3' direction and that polypeptide chains are always synthesized starting from the N-terminal end, after translation of the above mixture of oligonucleotide fragments, what amino acid is present at the N-terminal end in the great majority of the resulting peptides? (b) What amino acid will be present at the C-terminal end of the majority of the resulting peptides?

7. A wild-type bacterial protein has valine as its C-terminal amino acid. Two separate single-base mutants are isolated from the parent strain; their C-terminal amino acids are alanine and isoleucine, respectively. Another single-base mutation from the strain containing C-terminal alanine produces a protein that has a C-terminal glutamic acid residue. Deduce the codon coding for valine in the wild-type parent strain.

8. E. coli ribosomes depleted of their endogenous messenger RNA were incubated with the random polymer poly AG, in which the molar ratio A:G was 5:1. Predict which amino acids were incorporated by such a polymer and the relative frequency of their incorporation.

9. A segment of mRNA codes for the peptide Met-Thr-Phe-Ile-Trp. Proflavin induces the deletion of a single base from the gene coding for this RNA. The new peptide translated from the altered RNA is Met-Pro-Ser-Tyr-Gly. (a) In what codon did the deletion occur? (b) At what position in the original codon was the base that was deleted? (c) What was the base sequence in the wild-type mRNA? (d) What would the resulting amino acid sequence be if a G were inserted after the first three bases in the mutated RNA sequence?

CHAPTER 35 REGULATION OF GENE EXPRESSION

We come now to another basic question regarding the flow of genetic information from DNA to protein. How is the expression of genetic information regulated in the cell, so that the right kinds of proteins and the proper number of copies of each are synthesized?

That living cells possess accurately programmed mechanisms for regulating the relative amounts of different proteins synthesized is intuitively obvious. For example, the number of molecules of those enzymes catalyzing a mainstream metabolic pathway could be expected to be much greater than the number of molecules of the enzymes catalyzing the biosynthesis of the coenzymes, which are made in only trace amounts. But let us examine some actual data. *E. coli* cells contain the genes for perhaps 3,000 different proteins. If we assume that all the proteins have the same molecular weight and all are made in equal numbers, there would be about 3,000 copies of each protein in the *E. coli* cell. However, the number of copies of different types of proteins varies characteristically over a very wide range. A single wild-type *E. coli* cell contains about 15,000 ribosomes; therefore each of the 50 or more ribosomal proteins is present in about 15,000 copies. Some enzymes, particularly those of the glycolytic pathway, may be present in 100,000 or more copies. On the other hand, the enzyme β-galactosidase is normally present in only about 5 copies per cell. Moreover, some proteins are made in fixed, constant numbers, such as those of the glycolytic pathway, whereas others change dramatically in concentration in response to changes in the availability of certain nutrients from the environment. Thus regulation of the rate of enzyme synthesis provides each type of cell with the proper ensemble of enzymes required in its basic housekeeping activities and also makes it possible to economize in the use of amino acids for the synthesis of enzymes that are used only occasionally.

In eukaryotic organisms regulation of protein biosynthesis has another fundamental role, namely, as a vehicle for the differentiation of cells. Different cell types in higher organisms possess characteristically different ultrastructure, molecular composition, and biological functions. In part these differences are the result of the synthesis of distinctive and specific sets of enzymes, in addition to the basic ensemble all cells possess for catalysis of the central metabolic pathways. Tissues of higher organisms also have different sets and/or amounts of other types of specialized proteins, such as those found in collagen fibers, microtubules, actin and myosin filaments, and membrane transport systems. Moreover, the biosynthesis of different kinds of proteins during the differentiation of higher organisms must also be accurately programmed with respect to the time and sequence of their appearance.

Genetic and biochemical studies have yielded much important information and a set of principles for regulation of protein synthesis in prokaryotes. However, the regulation of gene expression in eukaryotic cells is much more complex than in bacteria and is not yet understood in detail; it affords some of the most interesting and important challenges in contemporary biochemical research.

There are at least two levels at which the biosynthesis of proteins is regulated. One is *transcriptional control*, the regulation of the transcription of DNA to yield the mRNA coding for a given protein or set of proteins. The other level is *translational control*, the regulation of the initiation and the rate of synthesis of polypeptide chains.

In this chapter we shall first examine the regulation of gene expression in bacteria in some detail and then outline some biochemical principles and problems involved in the domain of eukaryotic organisms.

Constitutive and Induced Enzymes in Bacteria

Early studies of the factors affecting enzyme synthesis in microorganisms led to the hypothesis that there are two classes of enzymes, differing in their occurrence and concentration under different metabolic conditions. The *constitutive* enzymes are those formed at constant rates and in constant amounts, regardless of the metabolic state of the organism. Constitutive enzymes were held to be part of the permanent and basic enzymatic machinery of the cell; examples are the enzymes of the glycolytic sequence, the most ancient energy-yielding catabolic pathway. The other class consists of *inducible* enzymes. An inducible enzyme is normally present only in trace amounts in a given species of bacterial cell, but its concentration can quickly increase a thousandfold or more when its substrate is present in the medium, particularly when this substrate is the only carbon source of the cell. Under these conditions the induced enzyme is required to transform the substrate into a metabolite that can be utilized by the cell directly. The classic example of an inducible enzyme is *β-galactosidase*. Wild-type *E. coli* cells do not utilize lactose (Figure 35-1) if glucose is available to them; as mentioned above, such cells contain only about five copies of *β*-galactosidase, which is required to hydrolyze lactose to glucose and galactose. However, if wild-type *E. coli* cells are placed in a culture medium containing lactose as the sole carbon source,

Figure 35-1
Substrates of *β*-galactosidase. Allolactose, which is formed from D-lactose, is the normal physiological inducer of *β*-galactosidase.

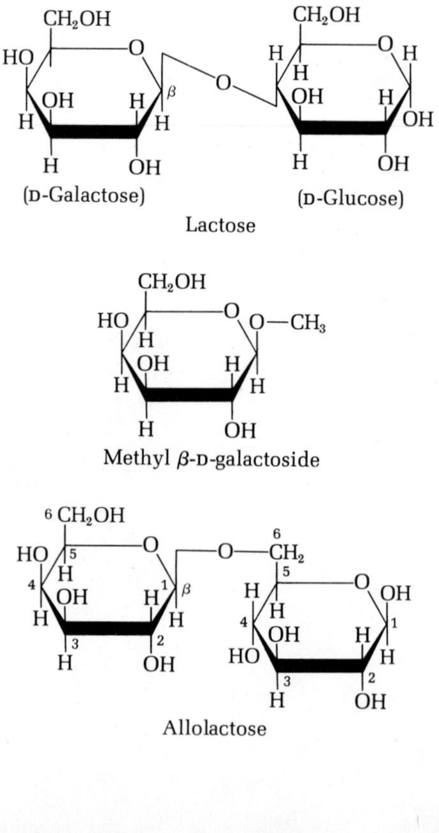

(D-Galactose)   (D-Glucose)

Lactose

Methyl *β*-D-galactoside

Allolactose

they are at first unable to utilize it, but within 1 or 2 min they respond by synthesizing β-galactosidase in large amounts, up to 5,000 copies per cell. The induced β-galactosidase hydrolyzes lactose to products that can be used directly as fuel and carbon source. If the induced cells are now transferred to a fresh medium containing glucose but no lactose, further synthesis of β-galactosidase ceases immediately. The previously induced galactosidase activity rapidly declines until it reaches its normally very low level. Enzyme induction thus makes economy in the use of amino acids and metabolic energy possible: inducible enzymes are made only when needed.

When a β-galactoside is added to E. coli cells, it induces the formation not only of β-galactosidase but also two functionally related enzymes, β-galactoside permease and β-thiogalactoside acetyltransferase. The former is a protein in the membrane of the E. coli cell which can promote the passage of β-galactosides into the cell against a concentration gradient; the function of the latter is not entirely clear, but presumably it is used in some step in the utilization of thiogalactosides in the cell. Induction of a group of related enzymes or proteins by a single inducing agent, as in this case, is known as coordinate induction. In an important series of genetic mapping experiments, J. Lederberg and his colleagues demonstrated that the genes for β-galactosidase, β-galactoside permease, and β-thiogalactoside acetyltransferase are probably contiguous to each other in the E. coli genome, in the sequence given.

Some species of bacteria show exceptional capacity for forming inducible enzymes in response to a great number of different substrates, many of which appear to be unphysiological. One species of Pseudomonas forms enzymes in response to hundreds of different organic compounds when presented as the sole carbon source. For this reason, early investigators suggested that the presence of the substrate molecule helped to "mold" or "adapt" the active site of an enzyme precursor, so that many different enzyme activities could be formed from a single protein precursor. Although this concept is now known to be incorrect, it is responsible for the fact that induced enzymes were at first called "adaptive" enzymes.

Enzyme Repression

Another important type of change in the concentration of an enzyme in the cell, seemingly opposite to that observed in enzyme induction, is enzyme repression. When E. coli cells grow on a medium containing an ammonium salt as sole nitrogen source, they must make all their nitrogenous components from the NH_4^+ ion and a carbon source. Such cells contain all the requisite enzymes for synthesis of the 20 different amino acids required in protein synthesis. However, if one of the amino acids, let us say histidine, is added to the culture medium, the enzymes required for making histidine from ammonia and a carbon source will quickly disappear from the cells. This effect is called enzyme repression.

Repression is also a reflection of the principle of cellular economy: when the enzymes required for histidine biosynthesis are no longer required, they are no longer made. Most cases of enzyme repression involve enzymes participating in biosynthetic reactions. In the case just described it is not merely the last enzyme in the sequence leading to histidine that is repressed when histidine is added to the medium. Actually, the entire sequence of enzymes required in histidine biosynthesis (page 706) is repressed under these circumstances. Such repression of the synthesis of a group of enzymes catalyzing a sequence of consecutive biosynthetic reactions is called *coordinate repression*. Coordinate repression is ordinarily evoked by the end product of the series of biosynthetic reactions catalyzed by the repressed enzymes; for this reason it is also called *end-product repression*.

Catabolite Repression

Yet another type of enzyme repression is common in bacteria; it functions in the repression of catabolic enzymes. It is a general observation that glucose, the most direct fuel source for most cells, can suppress the formation of inducible enzymes that make the use of some other fuel molecule possible. We have already seen that the presence of glucose in the culture medium represses the formation of β-galactosidase even when lactose is present; under these circumstances glucose is used preferentially as fuel, and the lactose is untouched. But glucose also represses some enzymes formerly thought to be constitutive. For example, the presence of glucose can repress certain of the enzymes of respiration and electron transport, possibly because the presence of a readily fermentable fuel supply makes it unnecessary for a facultative bacterial cell to use the more complex respiratory pathways. Thus when glucose is available as fuel, bacteria prefer to use the most primitive catabolic pathway, i.e., the glycolytic or fermentative pathway, and they turn all other energy-yielding catabolic pathways off. Since the signal for this effect is given by the elevation of the intracellular level of glucose or some catabolite derived from glucose, this type of enzyme repression is called *catabolite repression*. Its mechanism and regulation will be described below (page 987).

Regulatory Genes, Structural Genes, and Repressors

The molecular and genetic relationships between enzyme induction and enzyme repression, which at first were thought to be wholly different processes, were clarified by the genetic and biochemical studies of J. Monod and F. Jacob and their colleagues in the Pasteur Institute in Paris on the induction of β-galactosidase activity in *E. coli* cells. These studies, begun in the late 1940s, ultimately led some 15 years later to a general hypothesis for the genetic control of protein synthesis in prokaryotes, which has since been verified by direct biochemical experiments. Three important early discoveries set the stage. One was the finding that β-galactosidase can be

Figure 35-2
Gratuitous inducers of β-galactosidase.

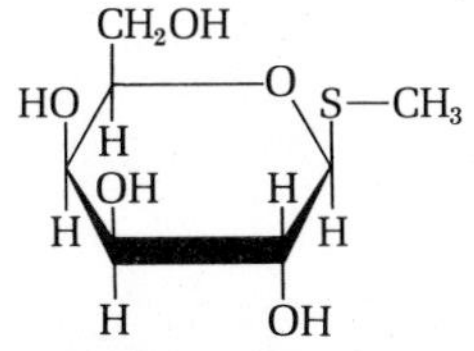

Methyl β-D-thiogalactoside

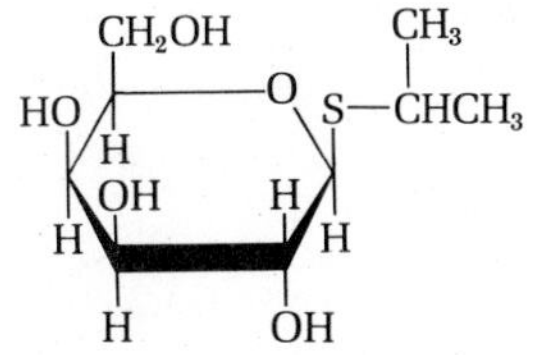

Isopropyl β-D-thiogalactoside

induced not only by those β-D-galactosides which are hydro-lyzed by the enzyme, such as lactose or methyl β-D-galac-toside, but also by some β-galactosides which are not, such as *methyl β-D-thiogalactoside*, the sulfur analog of methyl β-D-galactoside (Figure 35-2). In fact, methyl β-D-thiogalac-toside was found to induce formation of β-galactosidase even more effectively than the normal substrate of the enzyme. This discovery indicated that the process of enzyme induc-tion can be separated from those events concerned in the subsequent metabolism of the reaction products of the in-duced enzyme. The induction of enzyme activity by a spe-cific substance not acted upon by the enzyme is called *gratu-itous induction*.

The second important observation was that in certain mu-tants of *E. coli* the cells are unable to form β-galactosidase in response to the inducer but still retain the ability to form other kinds of induced enzymes in response to *their* charac-teristic inducers. Thus the capacity for forming any given in-ducible enzyme is dependent on a specific gene. This type of finding clearly laid to rest the older hypothesis of adaptation or "molding" of a common protein precursor capable of forming all kinds of inducible enzymes.

A third key observation was the discovery of another class of mutants of *E. coli* which contained large amounts of β-galactosidase activity, even though the cells had never been exposed to a β-galactoside as inducing agent. Such mutants are called *constitutive mutants* because they behave as though β-galactosidase were a constitutive rather than an in-ducible enzyme.

From these and certain other basic observations Jacob and Monod concluded that there are three different loci in the genetic map of *E. coli* that influence formation of β-galac-tosidase; these are designated z, i, and o (Figure 35-3). The z locus specifies the amino acid sequence of the β-galac-tosidase molecule; mutations in this locus lead to synthesis of an altered or inactive enzyme molecule possessing one or more amino acid replacements. This type of gene is called a *structural gene*, defined as a gene that carries the message coding for the amino acid sequence of a specific protein, in this case the β-galactosidase. The second, or i (for inhibi-tory), locus was postulated to have no influence on the struc-ture of β-galactosidase but to function as an inhibitory gene; it determines whether the structural gene for β-galactosidase

Figure 35-3
Schematic representation of the lac operon and its coordinate regulation. In the absence of inducer the repressor, coded by the i gene, binds to the operator (o) and prevents tran-scription of the structural genes z, y, and a for β-galactosidase, permease, and trans-acetylase, respectively. When the inducer molecule is present, it binds to the repressor, changing its conformation to an inactive form unable to bind to the operator. The structural genes are then free to be tran-scribed to yield a polycistronic messenger RNA, which is translated by ribosomes to yield the three proteins. The segment of the genome labeled p is the promoter (see page 987).

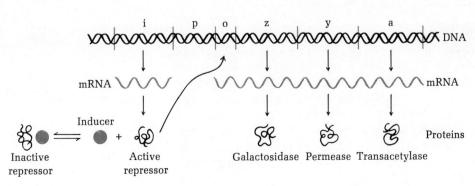

will be transcribed. Such a gene is called a *regulatory gene*. When the i gene undergoes mutation and becomes defective, it can no longer inhibit the transcription of the structural gene; β-galactosidase is then synthesized whether the inducer is present or not.

An important clue to the nature of the i-gene product came from experiments of A. B. Pardee, Jacob, and Monod. They found that when DNA containing intact i and z genes is introduced by sexual conjugation (page 876) into a mutant cell possessing a defective i gene (designated as an i^- cell), the intact i gene so introduced represses the transcription not only of the structural z gene in the DNA segment from the i^+ donor cell but also the structural gene in the i^- acceptor cell, which is thus transformed from a constitutive i^- mutant into an i^+ cell. This observation means that the i gene forms a product that is diffusible; it can reach its corresponding structural gene, wherever it is located in the cell, and inhibit its transcription.

From these and other experiments Jacob and Monod proposed that the regulatory gene codes for the amino acid sequence of a specific protein called the *repressor*. The repressor molecule was postulated to diffuse from the ribosomes, where it is formed, and to become physically bound to a specific site on the DNA of the cell next to or near the structural gene of the enzyme whose synthesis it controls. When the repressor is bound in this manner, it prevents the transcription of the structural gene coding for the synthesis of the enzyme.

The site in the DNA to which the repressor molecule was proposed to bind was named the *operator* or *o locus*. The existence of such a locus in DNA controlling the transcription of the structural gene for β-galactosidase, but distinct from the i gene, was proved by the discovery of a second class of constitutive mutants, different from those in which the i gene is defective. Genetic analysis indicated that this second group of constitutive mutants fails to respond to the repressor because the specific site on the chromosome to which the repressor normally binds, i.e., the operator locus, has undergone mutation and is defective; as a consequence, the z gene is transcribed and β-galactosidase is formed. Mutants of the operator locus are called *operator-constitutive* mutants. The independent existence of the operator, distinct from the structural genes and the regulatory gene, has since been firmly established by many genetic experiments. Presumably the operator is located close enough to the structural gene for β-galactosidase to be able to control its transcription.

Jacob and Monod assembled these facts into the following hypothesis. In the absence of the inducing agent the repressor molecule, presumably a protein, is made by the cell following transcription from the i gene. In the absence of inducer the repressor occurs in its active state and combines with the operator locus, thus preventing the transcription of the structural gene for β-galactosidase (Figure 35-3). When the inducer molecule is now added to these cells, it combines with the repressor protein, at a specific complementary binding site on the latter, to form a _repressor-inducer complex_. The binding of the inducer was proposed to render the repressor incapable of binding to the o locus, presumably because it elicits a conformational change in the repressor protein. The structural gene for β-galactosidase thus becomes free for transcription to yield mRNA, and the enzyme is then synthesized. The interaction between the inducing agent and the repressor molecule was postulated to be reversible, to account for the fact that when the inducing agent is removed from the culture medium or is used up by enzymatic action, the repressor reverts to its inhibitory form, binds to the o locus, and thus prevents further synthesis of the enzyme. The repressor molecule was thus conceived to be a protein having two specific binding sites, one for the inducer and one for the operator locus. When the inducer site is filled, the operator binding site is nonfunctional.

The Operon Model

Jacob and Monod extended their hypothesis for the regulation of protein synthesis to provide a mechanism for coordinate induction, in which a sequence of enzymes can be induced as a group by a single inducer. We have seen that β-galactosides coordinately induce a set of three proteins, β-galactosidase, β-galactoside permease, and β-thiogalactoside acetyltransferase, which are coded for by three genes, z, y, and a, respectively, grouped in that sequence in the E. coli genome. These three structural genes, regulated by the same i gene and the same operator o, were designated by Jacob and Monod the _lac operon_ (Figure 35-3). More generally, an operon is defined as a group of functionally related structural genes mapping close to each other in the chromosome, which can be turned on or off coordinately through the same regulatory loci. The operator locus was visualized as controlling the transcription of the entire group of coordinately induced genes, to form a single polycistronic messenger RNA. A number of operons have been identified in E. coli, Salmonella typhimurium, and other bacteria (Table 35-1). Of these, the _lac_ operon is by far the simplest in its organization and function. One of the most intensively studied and most

Table 35-1   Some operons in bacteria†

Operon	Number of enzymes	Function
lac	3	Hydrolysis and transport of β-galactosides
his	10	Synthesis of histidine
gal	3	Conversion of galactose to UDP-glucose
leu	4	Conversion of α-ketoisovalerate to leucine
trp	4	Conversion of chorismate to tryptophan
ara	4	Transport and utilization of arabinose
pyr	5	Conversion of aspartate to UMP

† The operons listed have been most thoroughly studied in E. coli, except the _his_ operon, which has been examined in detail in _Salmonella typhimurium_.

complex operons is that for histidine biosynthesis in *S. ty-phimurium*, which has nine structural genes coding for ten enzymatic activities (Figure 35-4).

Enzyme Repression

The hypothesis of Jacob and Monod also accounts for repression of an enzyme or enzyme system by a biosynthetic end product (Figure 35-5). For example, the presence of histidine in the culture medium of *E. coli* represses the formation of all the enzymes required to form histidine from its precursors. Constitutive mutants of *E. coli* have been found, however, which make these enzymes in maximal amounts whether histidine is present in the medium or not. These enzymes are therefore specified by a set of structural genes that can be repressed by the end product histidine. To account for such end-product repression, Jacob and Monod postulated that the repressor molecule in such cases is inactive by itself but binds the repressing metabolite, which is called the *corepressor*, to form a *repressor-corepressor complex*. This complex then combines with the operator locus for the structural genes of the operon and thus prevents transcription of the latter. Therefore there are two classes of repressor molecules, one operative in the induction of enzyme activity and the other in end-product repression. Both types of repressor contain at least two binding sites, one being specific for the operator locus, the other for either an inducer molecule or a repressing metabolite. Induction and repression are thus essentially similar in principle.

Although the biosynthesis of histidine is repressed when the histidine concentration exceeds a certain level, the true repressor of the *his* operon has been identified as histidinyl-tRNA$_{His}$.

The Isolation and Chemical Nature of Repressors

Although the hypothesis of Jacob and Monod brilliantly succeeded in explaining many aspects of enzyme induction and

Figure 35-4
The his operon. At the left is shown the genetic map of the operon; at the right the biosynthetic sequence catalyzed by its gene products. The full details of histidine bio-synthesis are given on page 706. Note that the sequence of structural genes in the his operon does not correspond to the sequence of action of their gene products.

Map of the nine genes (A to I) of the *his* operon.

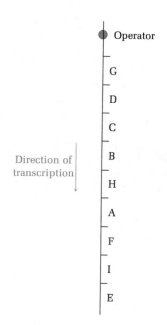

Figure 35-5
The Jacob-Monod model for enzyme repression. When a biosynthetic end product, such as histidine, is abundant, it acts as a corepressor. The core-pressor—repressor complex binds to the operator and prevents transcription of the structural genes.

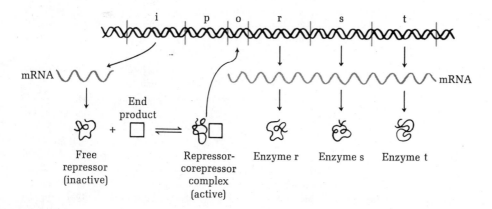

Steps in the enzymatic synthesis of histidine. The enzymes are lettered to correspond to the genetic map (left).

5-Phosphoribosyl
1-pyrophosphate

G ╱ ATP
   ATP phosphoribosyl
   transferase

5'-Phosphoribosyl-
ATP

E phosphoribosyl ATP
   pyrophosphohydrolase

[Phosphoribosyl
AMP]

I phosphoribosyl-AMP
   cyclohydrolase

Phosphoribosyl-
formimino-5-
aminoimidazole-4-
carboxamide ribonucleotide

A isomerase

Phosphoribulosyl-
formimino-5-
aminoimidazole-4-
carboxamide ribonucleotide

H ╱ Glutamine
   amidotransferase

[   ]

F cyclase

Imidazole glycerol
phosphate

B imidazole glycerol
   phosphate dehydratase

Imidazole acetol
phosphate

C histidinol
   phosphate transaminase

Histidinol phosphate

B histidinol phosphatase

Histidinol

D histidinol dehydrogenase

Histidine

repression and has withstood many stringent genetic tests, for a long time no direct chemical evidence for the existence of the hypothetical repressor molecules could be adduced. In fact, it appeared unlikely that direct chemical identification of such repressors would ever be feasible. For one thing, genetic calculations indicated that a single *E. coli* cell contains no more than about 10 of the hypothetical repressor molecules for the *lac* operon. Since the *E. coli* cell contains altogether many millions of protein molecules, it would appear to be a most formidable task to isolate the *lac* repressor in pure form. However, since the Jacob-Monod concept of repression is fundamental to so much genetic theory and experimentation, the task of isolating repressor molecules became an important goal.

In 1967 W. Gilbert and B. Müller-Hill succeeded in isolating the *lac* repressor in partially purified form. Because of the very small number of *lac* repressor molecules normally present in unadapted *E. coli* cells, Gilbert and Müller-Hill began their work by looking for mutant *E. coli* cells containing abnormally large amounts of *lac* repressor. They succeeded in finding a mutant containing 10 times as much *lac* repressor as normal *E. coli* cells. Extracts of such cells were made and tested for the presence of a specific protein that could bind a gratuitous inducer of β-galactosidase activity, namely, *isopropyl thiogalactoside* (IPTG). Analysis of binding activity was carried out by allowing the cell extract, placed within a dialysis bag, to come to equilibrium with an external solution of radioactive IPTG, a small molecule which freely passes through the membrane. The cell extracts bound significant amounts of IPTG, as shown by the fact that the equilibrium concentration of the IPTG within the bag containing the cell extract was higher than that in the external bathing medium. With this technique, called *equilibrium dialysis*, various nondiffusible fractions of the cell extract could be assayed for activity in binding IPTG. When ultimately the IPTG-binding material was purified some hundredfold, it proved to be heat-labile and showed other characteristic properties of a protein. In order to prove that the IPTG-binding protein is specific, similar experiments were carried out with cell extracts of constitutive mutants of *E. coli* having no *lac* repressor activity. No IPTG-binding protein could be found in such extracts.

Gilbert and Müller-Hill then grew *E. coli* cells on a medium containing radioactive sulfate and isolated the IPTG-binding repressor protein, which was labeled with ^{35}S in the sulfur atoms of its methionine and cysteine residues. The labeled repressor protein was then mixed with a fraction of DNA from *E. coli* known to contain the gene loci controlling the synthesis of β-galactosidase. The mixture was then subjected to density-gradient centrifugation. The DNA, which has a much higher sedimentation coefficient than proteins in general, sedimented first. It contained a significant amount of the bound ^{35}S-labeled repressor protein. However, when the inducer molecule IPTG was added, the labeled repressor protein failed to bind to the DNA. This finding verified a central feature of the Jacob-Monod hypothesis: that the repressor and inducer combine to yield an inactive complex unable to bind to the operator on the chromosome. The specificity of the isolated *lac* repressor protein was confirmed by other tests using DNA isolated from *E. coli* mutants defective in the operator region; as predicted from the Jacob-Monod hypothesis, no binding of the *lac* repressor to the DNA of such mutants was observed. It was also shown that the repressor molecule fails to bind to heat-denatured DNA from *E. coli*, whether it is obtained from cells possessing an intact or a defective structural gene for β-galactosidase. Moreover, Gilbert and Müller-Hill showed that the repressor molecule contains no RNA, since its binding activity is not influenced by treatment with ribonuclease.

More recent studies have resulted in much further purification of the *lac* repressor and ultimately its crystallization. Its molecular weight is about 150,000, and it contains four subunits, each capable of binding one molecule of IPTG. The isolated repressor molecule has an extraordinarily high affinity for its specific locus in *E. coli* DNA; half-maximal binding of repressor to operator occurs at a repressor concentration of only 10^{-13} M. The normal inducer capable of binding to the *lac* repressor in the cell is <u>allolactose</u> (Figure 35-1), an isomer of D-lactose in which the D-galactose and D-glucose residues are joined in a 1,6 linkage. Allolactose is formed from D-lactose in a transglycosylation reaction catalyzed by the small amounts of β-galactosidase present in uninduced *E. coli* cells. Figure 35-6 shows the *lac* repressor protein bound to the operator region of *E. coli* DNA.

Other repressors have since been isolated, including those of the gal and trp operons. Of special importance is the work of M. Ptashne, who has isolated the repressor for certain structural genes in DNA of bacteriophage λ. Phage λ DNA becomes incorporated into the *E. coli* genome, where it can remain dormant, largely untranscribed for generations, because it is kept repressed by a specific gene product of λ DNA, a repressor protein that prevents most of the viral DNA from being transcribed, thus preventing replication of the remainder of the viral genes. Another substantial step forward in the chemistry of operons was the isolation of a segment of *E. coli* DNA corresponding to the *lac* operon by J. Beckwith and his colleagues.

All these advances have provided convincing support for the Jacob-Monod theory of gene regulation and have opened the way for more penetrating investigation of other aspects of repression and derepression.

The Promoter

Yet another specific site in operons has been identified by genetic and biochemical approaches, the <u>promoter</u> site. This is the site which is recognized by DNA-directed RNA-polymerase as an initiation signal (page 917), to indicate where transcription to form mRNA begins. This site has been identified from genetic experiments as lying between the i, or regulatory gene and the operator locus of the *lac* operon (Fig-

Figure 35-6
Electron micrograph of a segment of E. coli
DNA, showing a lac *repressor molecule*
(arrow) bound to the lac *operator.*

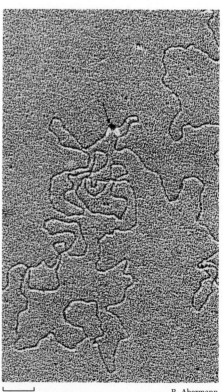

0.1 μm

R. Abermann

ure 35-3). The promoter represents only a relatively short segment of the DNA, less than 100 nucleotides. As we shall now see, the promoter is also the binding site for another specific type of protein required in transcription of certain operons, namely, the *cyclic AMP receptor protein*.

Cyclic AMP and the Transcription of Operons

In 1965 R. S. Makman and E. W. Sutherland found cyclic AMP in *E. coli* cells. This was somewhat surprising since this compound was at that time known largely for its function as the "second messenger" in the action of the animal hormones epinephrine and glucagon (page 811), which are not known to affect bacteria. But more surprising was their finding that glucose added to the medium greatly depressed the concentration of the cyclic AMP in the bacteria. At first the significance of this effect of glucose was obscure, but it was soon clarified by experiments of I. Pastan and R. Perlman and also of Monod. They showed that the addition of cyclic AMP to suspensions of *E. coli* overcame the normal catabolite repression of β-galactosidase synthesis by glucose. This effect was specific for cyclic AMP and was not given by other nucleotides or phosphate compounds. Moreover, cyclic AMP did not affect the synthesis of enzymes not subject to glucose repression. Pastan and Perlman surveyed many glucose-repressed enzymes in several species of bacteria; in all cases addition of cyclic AMP to the cell suspension overcame catabolite repression. They also found that cyclic AMP increases the amount of mRNA corresponding to the *lac* operon in glucose-repressed *E. coli* cells exposed to an inducing β-galactoside. They concluded that cyclic AMP promotes transcription of glucose-repressed operons, an effect which they ascribed to a direct or indirect interaction of cyclic AMP with the operator locus.

Pastan and Perlman and their colleagues have further found that the action of cyclic AMP on transcription is mediated by a specific protein capable of binding cyclic AMP. This protein, called *cyclic AMP receptor protein* (CRP) or *catabolite gene-activator protein* (CAP), has been isolated; it has a molecular weight of 45,000. Pastan and his colleagues have found that the regulation of transcription of isolated *lac* operon DNA can be observed in a cell-free system. When a mixture of the *lac* DNA, RNA polymerase, and the four NTP precursors is incubated, transcription of the structural genes of the operon takes place at only a very low rate. When both CAP and cyclic AMP are then added, there is a large increase in the rate of transcription of *lac* DNA by RNA polymerase. But when either cyclic AMP or CAP is omitted, transcription is reduced. To show that the system responds to regulation by *lac* repressor protein the addition of the latter to a mixture of *lac* DNA, RNA polymerase, NTPs, CAP, and cyclic AMP represses transcription, but when the inducer IPTG is added together with the *lac* repressor, repression is released and the structural genes are transcribed.

These and other experiments have shown that cyclic AMP

and CAP are components in initiating transcription of the *lac* operon, aiding in the recognition of the start signal, in or near the promoter, by RNA polymerase. Presumably cyclic AMP must bind first to CAP, which then interacts with the promoter (or possibly RNA polymerase) to enable the initiation process to take place. Cyclic AMP thus is not only a second messenger in hormone action in animal tissues but evidently also participates as a messenger in the regulation of transcription of glucose-repressible operons in bacteria. This type of regulation of gene expression, in which a cell component increases rather than inhibits transcription, is called *positive* transcriptional control. As long as glucose is available as fuel, the cyclic AMP level is low, thus preventing the synthesis of enzymes capable of utilizing other energy sources, such as lactose, galactose, arabinose, or maltose. When the glucose level in the cell goes down and an alternative carbon source is available, cyclic AMP rises and becomes bound to CAP. The cyclic AMP–CAP complex then binds to the promoter and stimulates transcription of its operon. How the presence of glucose in the medium depresses the cyclic AMP level in bacteria is still unknown.

The Identification of Specific Loci in the *lac* Control Region of DNA

A segment of bacterial DNA bearing the binding site for the *lac* repressor has been isolated by Gilbert and his colleagues. They first fragmented E. coli DNA into pieces about 1,000 base pairs long. The pieces containing intact *lac* operator regions were separated from the rest of the DNA by allowing them to bind to *lac* repressor protein, from which they could then be recovered on addition of the gratuitous inducer. The isolated DNA containing the *lac* promoter-operator region was then allowed to rebind to *lac* repressor, and the complex was exposed to pancreatic deoxyribonuclease to digest away all the DNA except that segment bound to the repressor molecule, which is about 27 base

Figure 35-7
Center of the binding site for the lac *repressor in the operator locus. This site shows twofold symmetry centered about base-pair 11. Presumably the tetrameric* lac *repressor molecule binds to two sites within this symmetrical region, in the major groove of the DNA.*

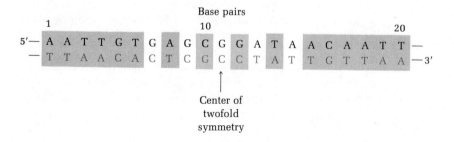

Base pairs

```
        1                    10                              20
5'—  A A T T G T G A G C G G A T A A C A A T T  —
  —  T T A A C A C T C G C C T A T T G T T A A  —3'
```

↑
Center of
twofold
symmetry

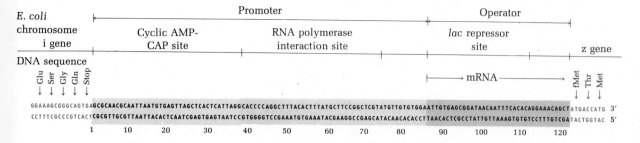

E. coli chromosome i gene

Promoter

Operator

| Cyclic AMP-CAP site | RNA polymerase interaction site | lac repressor site | z gene |

DNA sequence

Glu Ser Gly Gln Stop
↓ ↓ ↓ ↓ ↓

⟶ mRNA

fMet Thr Met
↓ ↓ ↓

GGAAAGCGGGCAGTGAGCGCAACGCAATTAATGTGAGTTAGCTCACTCATTAGGCACCCCAGGCTTTACACTTTATGCTTCCGGCTCGTATGTTGTGTGGAATTGTGAGCGGATAACAATTTCACACAGGAAACAGCTATGACCATG 3′
CCTTTCGCCCGTCACTCGCGTTGCGTTAATTACACTCAATCGAGTGAGTAATCCGTGGGGTCCGAAATGTGAAATACGAAGGCCGAGCATACAACACACCTTAACACTCGCCTATTGTTAAAGTGTGTCCTTTGTCGATACTGGTAC 5′

1   10   20   30   40   50   60   70   80   90   100   110   120

Figure 35-8
The molecular structure of the promoter-operator region of the lac operon. The base sequence of both DNA strands is shown, extending from the end of the i or regulatory gene through the first nine residues of the z gene. The binding site for the cyclic AMP–CAP complex is about 35 nucleotides long and the RNA polymerase site is about 40 nucleotides long. [R. C. Dickson, J. Abelson, W. M. Barnes, and W. S. Reznikoff, Science, 187: 32 (1975).]

pairs long. The base sequence of this segment was then determined (Figure 35-7). The binding site was shown to have twofold symmetry of its base sequence.

Further progress in identifying the regulatory regions of the lac operon has been made by R. C. Dickson and his colleagues, who succeeded in working out the complete nucleotide sequence of the mRNA complementary to the promoter and operator regions of the lac operon DNA. From this they deduced the sequence of the 122 base pairs in the regulatory region of lac DNA (Figure 35-8). Genetic and biochemical approaches allowed them to map the sites for the binding of cyclic AMP binding protein (CAP), the RNA polymerase binding site, the operator, and the beginning of the gene for β-galactosidase, and to describe a model for transcription of the lac operon (Figure 35-9).

Figure 35-9
A model for initiation of transcription of the lac operon. The model is drawn to scale. After the cyclic AMP–CAP complex binds to its specific site (a), which then destabilizes the DNA (b), the RNA polymerase is bound to its site (c) and "drifts" along a bubble to the initiation site (d). [From R. C. Dickson, J. Abelson, W. M. Barnes, and W. S. Reznikoff, Science, 187: 34 (1975).]

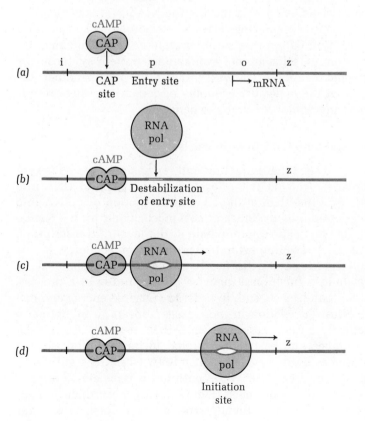

Stringent and Relaxed Control of Ribosomal RNA Synthesis

It has been known for some time, from experiments of O. Maaloe and his colleagues, that there is a maximum rate at which ribosomes can make polypeptide chains, about 15 to 20 amino acid residues per ribosome per second at 37°C. It is also known that the rate of protein synthesis in growing bacteria is limited by the number of ribosomes per cell; more specifically, it is the amount of ribosomal RNA per cell that is the critical factor. When *E. coli* cells are starved for amino acids and thus are unable to make proteins, the synthesis of ribosomal RNA stops and the number of ribosomes per cell may decrease to half the maximal value. Such control of RNA synthesis by lack of amino acids is called *stringent* control.

The capacity for stringent control is subject to mutation. In such mutants the synthesis of ribosomal RNA continues even though amino acids are lacking; the control of RNA synthesis is then said to be *relaxed*. In normal cells showing stringent control of RNA synthesis a hitherto unrecognized nucleotide was discovered as a spot appearing on paper electrophoresis of the cell extracts; however, this spot, called *magic spot I*, did not appear on analysis of extracts of relaxed mutants. The nucleotide in the spot was isolated and identified as guanosine 5′-diphosphate 2′(or 3′)-diphosphate, symbolized as ppGpp (Figure 35-10). This nucleotide, as well as another found in *magic spot II* and identified as guanosine 5′-triphosphate 2′(or 3′)-diphosphate (pppGpp), is enzymatically formed by ribosomes in the presence of ATP, GDP, mRNA, and uncharged tRNA. These and other observations of M. Cashel and J. Gallant and their colleagues established that ppGpp and pppGpp serve as nucleotide messengers that can turn off the synthesis of ribosomal RNA in some manner when amino acids are not available for protein synthesis. In stringent control of rRNA synthesis we have another example of the molecular economy of cells: ribosomes are not made unless they are going to be used.

Translational Control in Bacteria

We have seen that induction and repression of enzyme synthesis is regulated at the level of transcription of DNA into mRNA, rather than at the level of translation of mRNA into polypeptide chains. Indeed, most mechanisms for the regulation of gene expression operate at the level of transcription. This is in keeping with the principle of economy in the molecular logic of cells: nearly all regulatory mechanisms in metabolism function at or near the beginning of a process, rather than near its end, in order to conserve energy and metabolites. Translation in most cases appears to follow automatically once a messenger RNA is formed. In fact, translation appears to be tightly coupled to transcription, so that ribosomes begin translation of mRNA while it is still being transcribed from DNA and continue to translate as long as transcription is continued (Figure 35-11). Nevertheless, regulation at the translational level is also believed to occur

Figure 35-10
The magic-spot nucleotides responsible for stringent control of ribosomal RNA synthesis.

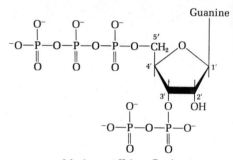

Magic spot I (ppGpp)
Guanosine 5′-diphosphate 2′(3′)-diphosphate (also called guanosine tetraphosphate)

Magic spot II (pppGpp)
Guanosine 5′-triphosphate 2′(3′)-diphosphate

Figure 35-11
Tight coupling of transcription and transla-
tion. The mRNA is translated by ribosomes
as it is being transcribed by RNA
polymerase from the DNA.

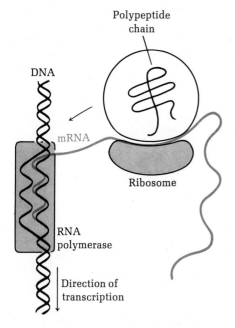

Polypeptide
chain

DNA

mRNA

Ribosome

RNA
polymerase

Direction of
transcription

under some special circumstances, either by regulation of the initiation of translation or by regulation of the drop-off of ribosomes from the polycistronic messenger RNA. For example, translational control has been invoked to explain the fact that in some operons the products of the different genes, i.e., the different enzymes coded by the operon, are not necessarily made in equimolar amounts. One gene product of an operon may be made in hundreds of copies while another, coded by the same polycistronic mRNA, may be made in only a few copies. In this case the ribosomes may drop off the mRNA before completing translation of all the structural genes in the operon. Such observations appear to be examples of translational control, for which a number of hypotheses have been proposed. However, transcriptional control, compared to translational control, appears to be a more direct and more economical means of regulating gene expression because regulation of transcription spares the cell the job of making an mRNA that is unused.

Gene Expression in Eukaryotic Cells

Although we now have a reasonably good understanding of the major molecular mechanisms involved in the regulation of gene expression in prokaryotes, as developed above, our present knowledge of gene expression in eukaryotic organisms is still quite meager, for many obvious reasons. The genome of the eukaryotic cells of vertebrates is very much larger than that of prokaryotes, by a factor of two or three orders of magnitude, it is divided among several or many chromosomes, and in somatic cells the genome is diploid. Moreover, the chromosomes have a very complex molecular composition and structure, and they are physically separated from the ribosomes in the cytoplasm by the perinuclear envelope. Although the basic enzymatic mechanisms of transcription in eukaryotes are probably very similar to those in prokaryotes, the complex physical geometry and supramolecular organization of eukaryotic chromosomes undoubtedly require the participation of regulatory factors not found in prokaryotes, to select and expose those DNA segments to be transcribed. In addition, eukaryotic organisms, due to their very much larger genomes, are capable of many kinds of cell differentiation, whereas prokaryotes show little or no ability to differentiate. In the highly differentiated eukaryotic cells of vertebrates a large fraction of the genome is permanently and perhaps irreversibly repressed; for example, a muscle cell cannot be transformed into a kidney cell, although each probably contains the structural genes characteristic of the other. Only a relatively small fraction of the total genome of eukaryotic cells can be induced or derepressed reversibly. In prokaryotes, on the other hand, there is considerable capacity for reversible repression, but little or no irreversible repression. There is the additional complexity afforded by the occurrence of a great deal of redundancy in the DNA of eukaryotes, in the form of short satellite sequences (page 884), as well as a large number of relatively long repeated sequences (page 885). For all these reasons, the molecular basis of the regulation of gene expression in eukaryotes can be expected to be much more complex than in prokaryotes.

One other principle of eukaryotic genetics must be stressed. It now appears certain that every cell of a highly differentiated eukaryotic organism contains the complete genome of that organism. Thus a muscle cell of a vertebrate not only contains the genes for its characteristic contractile system and all its characteristic enzymes, it also contains the genes required to specify the phenotype of all the other cell types in that organism. The genetic totipotence of the cells of eukaryotic organisms was first demonstrated by an example from the plant world: an entire carrot plant can be grown from a single somatic cell under appropriate conditions. Recently it has been shown that the nuclei of certain somatic cells contain all the necessary genetic information to specify the development of a mature animal organism.

As we consider various aspects of gene expression in eukaryotes, we must recognize that transcriptional control of gene expression, the major means employed by prokaryotes, is not the only way differential gene expression can be brought about. Gene expression may also be regulated by the processing of messenger RNA in the nucleus (page 919), by translational control at the level of ribosomes in the cytoplasm, or by posttranslational modification of the polypeptide product. Nevertheless, it appears probable that transcriptional control is the major regulatory process, on the grounds that this type of control would yield maximum economy of materials and energy.

In the following sections we shall consider enzyme induction by substrates and by hormones, as well as the regulation of gene expression during the cell cycle. Then we shall consider gene expression during differentiation of cells.

Enzyme Induction and Repression in Eukaryotic Organisms

We may now consider whether enzyme induction and repression take place in eukaryotes and, if so, whether the Jacob-Monod operon model developed for prokaryotes also applies.

Eukaryotic cells do in fact have the capacity for induction of some enzymes by their substrates. It is evident not only in lower eukaryotes, for example, in yeasts and in *Neurospora*, but also in vertebrates. The response of eukaryotic cells to inducing agents tends to be relatively slow and less dramatic than in prokaryotes. For example, induction of β-galactosidase activity by lactose in yeast or *Neurospora* cells takes many minutes, and the increase in activity may be only tenfold rather than the thousandfold increase seen in *E. coli*. Another important difference is that coordinate all-or-none induction or repression of a group of related enzymes does not appear to take place in eukaryotes. Indeed, there is little evidence that a sequence of enzymes catalyzing a given metabolic pathway is coded by adjacent genes in a single chromosome. Instead it appears that the structural genes required for such a sequence of related enzymes are scattered among several chromosomes in eukaryotic cells and that they are not regulated in the simple, coordinated manner implicit in the operon hypothesis.

Enzyme induction by substrates is especially evident in the liver of mammals. For example, the enzymes required to metabolize certain amino acids may vary considerably in concentration in the liver, as a reflection of the amount of the amino acid in the diet. Moreover, the enzymes concerned in the bypass portions of the gluconeogenesis pathway in the liver (page 624) also fluctuate in accordance with the nature of the diet. However, the full response of the liver to high levels of dietary amino acids may require hours or even days to develop. Return of the concentration of the enzyme to the normal or repressed level also may require hours or days after removal of the inducer from the diet. An induced liver enzyme that has received much study is *tryptophan 2,3-oxygenase*, responsible for the first stage in the degradation of tryptophan (page 573). This enzyme increases greatly in activity not only when tryptophan is present in large amounts in the diet but also when certain hormones are administered to the animal. That induction of tryptophan oxygenase in the liver is indeed due to net synthesis of new enzyme protein has been proved by the administration of actinomycin D, which inhibits the formation of the induced enzyme, presumably through inhibition of DNA transcription (page 920). Another enzyme readily induced in liver cells in vivo or in vitro is *tyrosine aminotransferase*.

Enzyme induction and repression in vertebrates occurs for the most part only in the liver and epithelial cells of the small intestine and not in muscle, brain, or other tissues. This fact is in keeping with the anatomical and physiological organization of vertebrates. The small intestine and the liver are the first tissues exposed to the incoming mixture of ingested nutrients, which may vary considerably in composition from hour to hour or day to day (page 836). The liver, which serves as the major nutrient-distributing center for the whole organism (page 836), thus requires induction and repression mechanisms to adjust to fluctuations in the nutritional intake of the animal. Most tissues are provided with a constant chemical environment and a constant supply of nutrients by way of the blood and interstitial fluid and thus have little need of quickly responding induction-repression systems.

Regulation of Gene Expression by Hormones

In addition to the induction of enzyme activity by nutrient substrates, eukaryotic cells of higher organisms have a highly developed capacity for enzyme induction and positive regulation of the synthesis of specific proteins in response to specific hormones. Of particular importance in this respect are the sex hormones, including androgens, estrogens, and progesterone (page 823), the *glucocorticoids*, a group of adrenal-cortical steroids regulating metabolism (page 823), and thyroxine (page 825). The steroid hormones bind to specific receptor proteins present in the cytoplasm of the hormone-sensitive cells; the hormone-receptor complex then becomes incorporated into the chromatin (page 824). For estrogen acting on the chick oviduct it has been demonstrated that binding of the hormone-receptor complex to chromatin causes greatly increased rates of transcription of certain genes to yield such characteristic egg proteins as ovalbumin and lysozyme, a process that has been shown to occur through the intermediate formation of mRNAs coding for these proteins. The hormone-receptor complex itself acts as the inducing agent, possibly binding to a specific locus in the appropriate chromosome, at a site comparable to the operator in prokaryotes.

Glucocorticoids appear to have a generalized effect on many if not all animal cells, which has been studied in some detail by G. M. Tomkins and his colleagues. This class of adrenal steroids, as well as a synthetic analog, dex-amethasone (Figure 35-12), not only promotes adaptation to stress but also promotes development of various organs of the mammal. Glucocorticoids are also highly active in induction of enzymes concerned in gluconeogenesis in the liver, such as phosphoenolpyruvate carboxykinase (page 625), as well as tyrosine aminotransferase and other enzymes required to convert the carbon skeletons of specific amino acids into glucose. Glucocorticoids also appear to be necessary factors for the growth of many types of mammalian cells in synthetic media in vitro. Nearly every tissue of the mammal has been shown to contain specific glucocorticoid receptors, and it now appears that glucocorticoids may be developmental hormones, capable of eliciting certain patterns of gene expression.

Thyroxine appears to have a similar ability to evoke the synthesis of specific enzymes. For example, administration of thyroxine to the tadpole causes not only its metamorphosis into the adult frog but also promotes the dramatic appearance of the liver enzymes required in the urea cycle (page 579), as this organism switches from an ammonotelic to a ureotelic habit of excreting amino nitrogen (page 579).

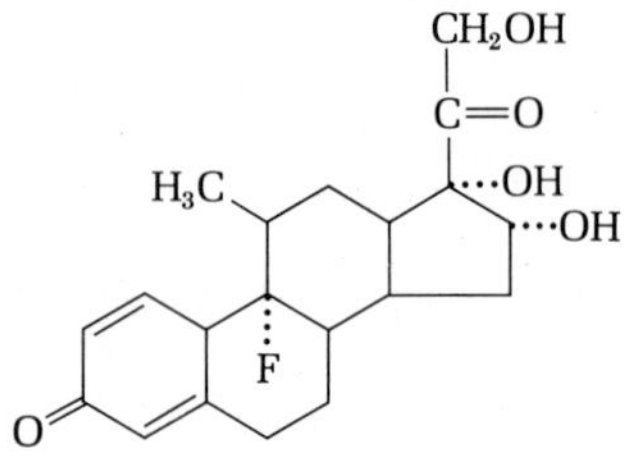

Gene Expression during the Cell Cycle

Another set of circumstances in which there are very dramatic changes in the expression of genes in eukaryotes occurs during the cell cycle, defined as the period from the beginning of one mitosis to the beginning of the next. All somatic cells undergo a characteristic cycle of morphological and biochemical changes in this period. Figure 35-13 shows the successive stages in the cell cycle of eukaryotes, which varies very considerably in its total duration, from minutes in some cells to weeks or months in others. The cell cycle has four distinct phases (Figure 35-13). The G_1 period is that beginning immediately after a cell division; during G_1 the cells are diploid. The next period is the S phase, during which DNA is synthesized and the chromatin replicates. At the end of the S phase the cells have twice the amount of chromatin and pass into the G_2 phase. From G_2 the cells pass into mitosis. G_1 and G_2 represent the gaps in the interphase

Figure 35-12
Dexamethasone (9α,16α-methylfluoroprednisolone), a synthetic glucocorticoid.

Figure 35-13
The cell cycle for a cell in tissue culture dividing once in 24 h.

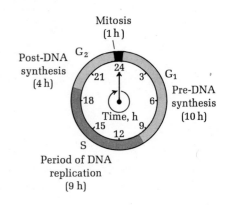

period of the cell cycle, when no DNA is synthesized. Changes in the total length of the cell cycle are largely the result of changes in the duration of G_1, which appears to be under specific control. Figure 35-13 shows the duration of the various phases of the cell cycle in a typical cell culture in which the doubling time is 24 h.

The synthesis of DNA, RNA, and proteins is very characteristically regulated during the cell cycle. DNA and the histone components of the chromatin are synthesized only during S phase, not during G_1 or G_2. In contrast, the cytoplasmic proteins and organelles are synthesized continuously during all of interphase, i.e., throughout G_1, S, and G_2. Moreover, RNA is also synthesized continuously throughout interphase. However, during mitosis the synthesis of all DNA and RNA is turned off; moreover, protein synthesis is greatly reduced.

From these properties it is clear that the DNA of the chromosomes is available for replication only during a specific portion of the cell cycle; in cells having a long cycle the S phase is but a relatively short period in interphase. At all other times the DNA templates must be restricted and unavailable for replication by DNA polymerases. On the other hand, RNA and most of the cell's proteins are synthesized continuously throughout interphase, indicating that many segments of the total genome, although restricted from replication, must be available for transcription. Clearly, all DNA is restricted from replication in G_1 and G_2 in such a way that selected, specific portions are still available for transcription. Simple masking or covering of DNA by proteins or other agents would not suffice to account for the differential availability of DNA to replication versus transcription.

There are two major experimental approaches currently being taken to study the biochemical factors controlling the different phases of the cell cycle, particularly G_1, and thus the rate of mitosis. In one approach, normal cells, such as fibroblasts, ovary cells, or kidney cells, are cultured in vitro in a medium capable of supporting cell growth. Such cells can then be transformed into malignant cancer cells by addition of oncogenic (cancer-causing) viruses, for example, the DNA-containing <u>simian virus 40</u> (SV40). When most normal cells attain a certain concentration or density in the culture medium, they abruptly stop dividing. However, after the transformation to malignancy induced by an oncogenic virus, the cells divide more rapidly and do not recognize the normal "stop" signals; they simply continue to proliferate to much higher cell densities. In some cases, the "stop" signal is simply the physical contact with neighboring cells of the same kind, suggesting that receptors on the cell surface are involved in receiving and transmitting the stop signal.

Another experimental situation often used to study regulation of the cell cycle is the addition of <u>mitogenic agents</u>, substances that stimulate mitosis, to cells growing in culture. The most widely used mitogenic agents are a group of plant proteins known as <u>lectins</u>. Two examples are <u>concanavalin A</u>, a protein from the jack bean, and <u>wheat phytohemagglu-</u>

tinin, so named because it agglutinates red blood cells, a rather general property of lectins. Lectins stimulate the division of many types of cells from animal tissues. The lectin molecule binds to specific receptors on the cell surface, analogous to, or possibly identical with, the cell-surface receptor sites normally functioning in binding certain hormones, such as insulin; in fact, lectins can mimic the action of insulin on some cells. Once the lectin is bound, a molecular signal is set off within the cell which profoundly affects the rate of division and the tendency to differentiate.

Much evidence suggests that cyclic AMP and cyclic GMP may be implicated in the regulation of gene expression during the cell cycle, although the specific manner in which they function is not yet clear. In rapidly dividing cells the cyclic AMP concentration is relatively low, whereas in cells undergoing cessation of proliferation, the cyclic AMP concentration increases. For example, in psoriasis, a skin disease in which epidermal cell division is accelerated manyfold, cyclic AMP and adenylate cyclase levels are much lower than in normal epidermal cells. Administration of a derivative of cyclic AMP capable of penetrating the cell membrane, dibutyryl cyclic AMP, greatly inhibits skin-cell division. When normal cells are transformed into rapidly growing cancer cells by oncogenic viruses or are stimulated to divide by addition of lectins, there is often a decrease in adenylate cyclase activity. Indeed, dibutyryl cyclic AMP can suppress the growth of some tumor cells. From these and other observations it has been suggested that cyclic AMP acts as an inhibitor of some phase of the cell cycle, thus decreasing the mitotic rate. Inhibition of mitosis is often accompanied by more extensive differentiation of cells.

On the other hand, cyclic GMP levels and guanylate cyclase activity appear to vary in the opposite direction in relation to cell-division rate. The concentration of cyclic GMP in cells stimulated to divide rises very sharply, as much as fiftyfold, directly after application of the stimulus; concurrently, there is either no change or an actual decrease in the concentration of cyclic AMP. From such observations it has been proposed that cyclic GMP is the intracellular messenger that induces cell division and limits differentiation and that cyclic AMP inhibits division and promotes differentiation, i.e., the yin-yang hypothesis (page 822).

It will be recalled (page 987) that cyclic AMP plays an important role as a messenger for turning off catabolite repression in prokaryotic cells, through its capacity to bind to CAP protein, which in turn binds to a specific locus in the operator region, thus initiating transcription and the synthesis of enzymes of catabolic pathways. It is therefore possible that cyclic AMP and cyclic GMP function in controlling the cell cycle in eukaryotes by binding to specific receptor proteins, which in turn modulate the transcription of certain genes in the chromatin. Alternatively, these nucleotides may function by activating a protein kinase, such as that regulating many enzymes (page 813), which in turn can bring about the phosphorylation of some protein controlling the rate of replication or transcription.

Regulation of Gene Expression in the Differentiation of Cells: Histones and Nonhistone Proteins

It has been pointed out above that all cells of higher plants and animals contain the complete genome of the organism, although in any given cell type only a small fraction of the total genome is actually expressed, all the rest of the structural genes being more or less permanently repressed. The molecular mechanisms responsible for this more permanent type of repression in eukaryotes must be rather different from the mechanisms employed in the easily reversible induction and repression processes taking place when the cells are exposed to specific substrates or hormones. These considerations have led to an intensive search for components of the chromatin which could conceivably act as permanent repressors of gene action. Chromatin contains only a few classes of components (Table 35-2), of which the histones are especially abundant. The general properties of histones are described elsewhere (page 869). It will suffice to recall here that they are small basic proteins of molecular weight 10,000 to 20,000, rich in lysine or arginine. The amino acid sequences of some histones have been established.

The first modern hypothesis for gene regulation in eukaryotes was proposed in 1943 by E. Stedman and E. Stedman, who believed that the amounts of histones in actively growing and nongrowing tissues vary characteristically. They postulated them to function as gene repressors during differentiation and development, long before the modern development of molecular genetics. This idea was taken up again in the early 1960s by James Bonner and his colleagues, as well as by V. Allfrey. They showed that preparations of histones could inhibit the ability of DNA to serve as a template for the synthesis of RNA by RNA polymerase.

Although this idea still remains very attractive, since histones are major components of chromatin, there are some major issues not readily explained by the histone hypothesis. In higher eukaryotic cells of mammals there are a great number of different structural genes, variously estimated from 40,000 to 100,000 or more. In any given cell type most of these are repressed. The large number of genes suggests that repressor molecules must also exist in large variety in the chromatin, in order to keep the great majority of the genes in any given cell from being expressed. Yet, as developed elsewhere (page 869), there are only five major classes of histones in eukaryotic cells. Moreover, they occur in nearly equimolar amounts and do not vary much between organs of a given animal, between different species, or in different metabolic or physiological states. In fact, the histones are among the most ancient proteins found in eukaryotic organisms, since there are many amino acid sequence homologies between species and relatively few variable residues. For example, in one of the histone classes, there are only two amino acid residues that differ in the preparations obtained from pea seedlings and from calf thymus. These facts suggest that histones play the same role in all organisms. Although the histones are present in sufficiently large amounts in

Table 35-2 Approximate composition of calf-thymus chromatin (relative to DNA = 100)

DNA	100
Histones	114
Nonhistone (acidic) proteins	33
RNA	7

chromatin to function as repressors, since there is 1 molecule of each type of histone for every 100 base pairs of DNA, there seems to be insufficient variety in the types of histones for them to serve as repressors, if each eukaryotic gene or set of genes is repressed by a specific protein.

However, the histones do have a property strongly suggestive of an important role in the cell cycle: they undergo a number of characteristic enzymatic modification reactions following their synthesis during S phase. The N-terminal amino group and the ϵ-amino group of certain lysine residues are enzymatically acetylated in certain histones. Moreover, the ϵ-amino groups of some lysine residues undergo methylation, to yield monomethyl, dimethyl, or trimethyl derivatives. Certain histidine and arginine residues also undergo methylation. Of especial interest is the fact that certain serine residues of histones are phosphorylated by protein kinase, following activation of the latter by cyclic AMP. This phosphorylation process begins at the end of G_1. Most of these modifying groups undergo constant metabolic turnover. The function of these modification reactions appears to be concerned with the assembly of histone molecules into the chromatin structure. Recent research indicates that specific sets of histones form pairs, tetramers, and oligomers, as they assemble with DNA to form chromatin. R. D. Kornberg has postulated that there is a repeat unit of 8 histone molecules and about 200 DNA base pairs in chromatin. Thus it appears that histones are of fundamental importance in the structural organization of chromatin, but whether histones function as specific repressors remains in question.

Chromatin also contains acidic nonhistone proteins, which have been separated into many bands by electrophoresis. In any given species the number of different nonhistone proteins is much larger than the number of histone types, but there simply are not enough of them to function as specific gene repressors. We are therefore left with a puzzling dilemma regarding the molecular identity of the repressor molecules in eukaryotic nuclei.

Heterogeneous Nuclear RNA

In the eukaryotic nucleus the three types of ribosomal RNA are made as a single, very long precursor strand that subsequently is cleaved and further processed to yield the final finished products, in a process called _RNA maturation_ (page 919). For some time it has been thought that messenger RNAs also are formed from larger precursors in the nucleus, from a pool of RNA called _heterogeneous nuclear RNA_ (hnRNA). HnRNA consists of a complex mixture of many RNAs that are very much longer than messenger RNAs. HnRNA also turns over very rapidly, as deduced from isotope tracer experiments; 90 percent of the hnRNA fraction has a half-life of only minutes. It has been estimated that most of the hnRNA is never transported into the cytoplasm but is degraded within the nucleus. Although some investigators have held that fragments arising from the 3′ end of the very long hnRNA chains become messenger RNA molecules, re-

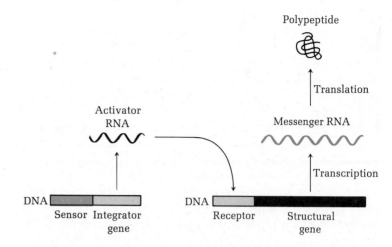

cent research appears to eliminate this possibility. Actually,
the great bulk of the total hnRNA hybridizes with nuclear
DNA but does not hybridize with mRNAs.

Thus a large fraction if not all of the hnRNA never leaves
the nucleus as part of mRNA molecules but contains presumably meaningful base sequences. The specific function of this
portion of hnRNA constitutes a major puzzle, particularly
because it is large in amount and turns over rapidly. One
idea is that the hnRNA may be involved in the regulation of
gene expression.

The Britten-Davidson Model of Gene Regulation in Eukaryotes

Several models have been proposed for the regulation of
gene expression in eukaryotes. We shall outline one of these,
postulated by R. J. Britten and E. H. Davidson. In some
respects their model borrows from the Jacob-Monod model
for prokaryotes, but in others it is significantly different. The
Britten-Davidson model takes account of the fact that the
eukaryotic genome does not contain simple operons consisting of adjacent genes, such as those in prokaryotic DNA;
it allows for the regulation of related genes in different chromosomes on an individual rather than a group basis. It also
accounts for the fact that the chromatin contains too few proteins to function as repressors; moreover, it accounts for the
great variety, size, and turnover of much of the hnRNA.

Britten and Davidson propose that the eukaryotic genome
contains a large number of *sensor* sites (Figure 35-14),
which recognize various molecular signaling agents, such as
substrates of specific enzymes, hormone-receptor complexes,
or regulatory nucleotides. Each sensor site is adjacent to an
integrator gene. When the sensor is activated on binding its
specific substrate or a specific hormone-receptor complex,
the integrator gene is transcribed to yield *activator RNA*. The
latter is in turn recognized by one or more *receptor sites*,
located elsewhere in the genome, on the same or some other
chromosome. When activator RNA becomes bound to the
receptor site, an adjacent structural gene is transcribed to

yield a messenger RNA, which is then translated by the ribosomes in the cytoplasm. This is the simplest form of the Britten-Davidson model, which is elaborated further below.

We see that this model employs as a diffusible regulator of transcription not a protein, as in prokaryotes, but a species of RNA. This unique feature gets around the fact that there are very few proteins in the eukaryotic nucleus in relation to the number of structural genes; presumably the many different activator RNAs of short lifetime are derived from the very labile hnRNA pool. The model also is logistically sound in that the activator RNA need not leave the nucleus. If a protein served as a repressor or derepressor, it would first have to be made on cytoplasmic ribosomes from mRNAs made in the nucleus; such a protein would then have to return to the nucleus to exert its derepressor function. Another important feature of the Britten-Davidson model is that it also accounts for the fact (page 884) that structural genes make up only a small part of the total genome of eukaryotes and that much of the total DNA apparently functions in regulation, in the form of sensor, integrator, and receptor loci.

Britten and Davidson have expanded this simple model to include a multiplicity of receptor genes for any given structural gene and a multiplicity of integrator genes, thus making coordinated regulation of many different combinations of genes possible (Figure 35-15). This provocative hypothesis allows for a much larger degree of flexibility in the regulation of gene transcription than the Jacob-Monod model. Several other interesting models of gene regulation in eukaryotes have been proposed, but because of the great complexity of eukaryotic genomes years of effort may be required before the problem is solved.

It appears very unlikely that gene expression in eukaryotes is regulated significantly by the life span of messenger RNAs. In eukaryotic cells, messenger RNAs appear to have a rather long half-life, in contrast to prokaryotes, in which mRNA has a half-life of only about 2 min. On the other hand, much evidence suggests that gene expression in eukaryotes may be regulated at the translational as well as at the transcriptional level. Moreover, the initiation and elongation phases of protein synthesis in eukaryotes are more complex than in prokaryotes and appear to involve a number of as yet unidentified regulatory factors.

The Immune Response

A particularly dramatic example of cell differentiation is the immune response, which affords unusual opportunities for the correlation of genetic and biochemical approaches to the solution of the problem of the regulation of gene expression in eukaryotes.

The basic elements of the immune response and the antigen-antibody reaction have been described in Chapter 3 (page 66). An antibody or immunoglobulin is produced by plasma cells of the lymphoid system of a given species of vertebrate in response to the presence of an antigen, a pro-

Figure 35-15
Elaboration of the Britten-Davidson model. In a regulatory network (top) multiple receptor sites allow expression of structural genes by several different activator RNAs. In another network (bottom) multiple integrators allow each sensor to control multiple structural genes. Through multiple sensors, integrators, and receptors, many different sets of structural genes can be regulated in response to many different sets of sensor stimuli.

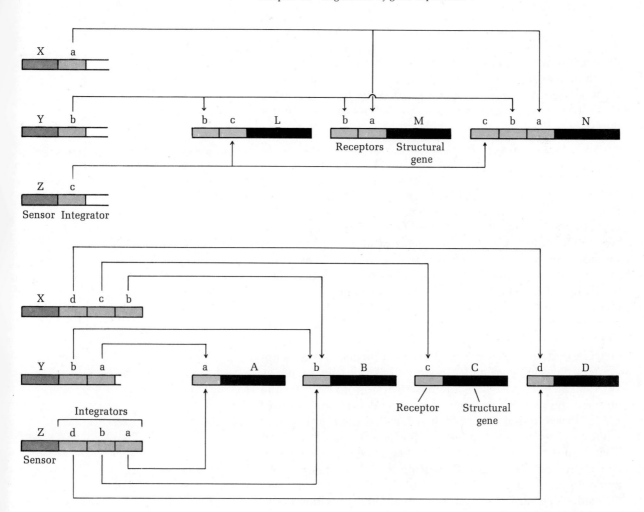

tein (or some other macromolecule) foreign to that species. Such an antibody is recognized by its ability to form a specific complex with only the antigen that evoked its formation; no other protein antigen will combine with this antibody with the same affinity. A remarkable feature of the immune response is that animals can be made immune to an extraordinarily large number of different macromolecules, including various synthetically prepared derivatives of proteins which animals would normally never encounter in their natural habitat. Moreover, a single animal can be made immune to many different antigens simultaneously. Indeed, it has been estimated that as many as a million different kinds of antibodies might be made by one species. As with enzyme induction, it was once thought that all antibody molecules were identical in amino acid composition but could be "molded" in the presence of different antigens to form complementary antibodies. However, this view had to be abandoned with the discovery that antibodies to different antigens do in fact differ in amino acid composition and sequence.

Because of the enormously great number of different antigens to which vertebrates can respond by synthesis of a specific antibody, the question has been raised whether the plasma cells that produce antibody molecules possess a sufficient number of genes to code for so many different antibodies. Moreover, the question arises about the mechanism by which a given type of antibody molecule is synthesized in response to a given antigen. Clues to these questions have come from analysis of the amino acid sequence of the polypeptide chains of immune globulins.

The Structure of Antibodies

There are five major classes of immunoglobulins in human blood plasma (IgG, IgA, IgM, IgD, and IgE), of which the IgG globulins are by far the most abundant and the best understood. They have a molecular weight of about 150,000 and possess four polypeptide chains: two identical "heavy" (H) chains, having about 430 amino acid residues, and two identical "light" (L) chains, having about 214 residues. These chains are linked together by disulfide bonds into the Y-shaped but flexible structure schematized in Figure 35-16. The heavy chains contain a covalently bound oligosaccharide component. Each chain has a region of constant amino acid sequence and a region in which the sequence varies. The antibody molecule has two binding sites for the antigen; the variable portions of the L and H chains contribute to these binding sites, which are further distinguished by having *hypervariable* regions in which there is an especially high frequency of amino acid replacement. The arms of the Y can be removed by the action of *papain*, a proteolytic enzyme, to yield fragments called Fab', which have been crystallized. Their structure has been elucidated by x-ray analysis by R. L. Poljak and his colleagues.

The blood serum of a normal person contains a mixture of many different IgG immunoglobulins, all having the basic structure shown in Figure 35-16. Each individual type of immunoglobulin is specific for combining with a certain antigen. From such a mixture it is very difficult to obtain a single molecular species of an immunoglobulin for analysis. However, in the disease *multiple myeloma*, certain cells of the bone marrow become cancerous and multiply, causing production of excessive amounts of a single type of immunoglobulin, which appears in the blood. In such patients, the urine contains large amounts of another type of protein, discovered by H. Bence-Jones in 1847 and since known as *Bence-Jones protein*. This abnormal protein is a dimer of light or L chains, which are evidently made in excess of the H chains and then excreted in very large amounts. In any given patient only a single type of Bence-Jones protein is synthesized; no two patients have identical Bence-Jones proteins. Amino acid–sequence determinations have been made on a number of Bence-Jones proteins by F. W. Putnam and other investigators. In 1968 G. Edelman and his colleagues succeeded in establishing the entire amino acid sequence of an IgG antibody from the blood of a multiple myeloma patient. Comparison of the sequences of the L and H chains obtained from a number of myeloma patients have revealed some extraordinary relationships (Figure 35-16). The car-

Figure 35-16
Structure of the IgG immunoglobulin. The gray segments of the H and L chains are the constant regions; the colored segments are the variable regions. The hypervariable portions of both chains (black bands) determine the antigen-binding specificity. The IgG immunoglobulin thus has two binding sites for antigens. The oligosaccharide groups are designated CHO *. The enzyme papain cleaves at the point shown by the colored arrows to yield Fab' fragments. Below is a drawing of a model of the constant regions of the light and heavy chains, deduced from x-ray analysis. The variable regions of both chains also have a similar conformation, with the addition of an extra loop indicated by the dotted lines. The numbers indicate the light-chain residues beginning at NH_3^+, which also corresponds to residue 110 of L chains. [R. J. Poljak et al., Proc. Natl. Acad. Sci. (U.S.), 70: 3308 (1973).]*

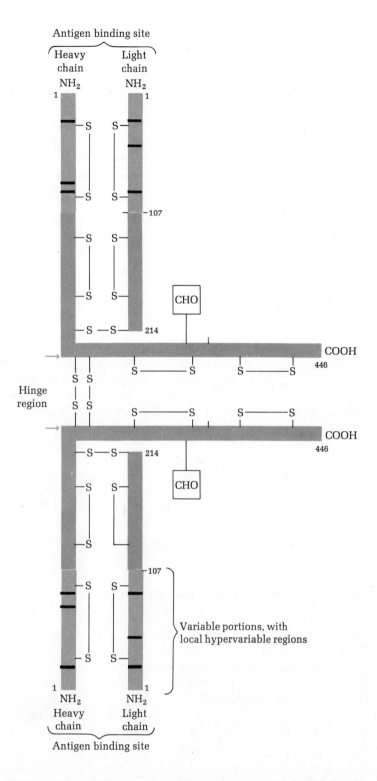

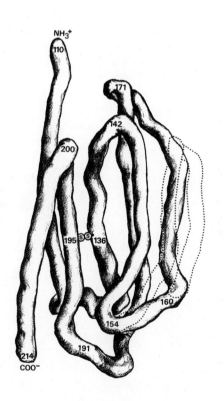

boxyl-terminal half of the Bence-Jones proteins, which contain about 214 residues, has essentially the same amino acid sequence in all patients, whereas the amino-terminal half differs in sequence from one patient to another, although there is some sequence homology. Moreover, there are some sequence homologies between the two halves of the L chains. There are two types of L chains, called *kappa* (κ) and *lambda* (λ). They have many sequence homologies but characteristically differ in their residues at position 191. Similarly, in the H chains from myeloma patients, which contain about 446 residues, the amino-terminal segment of about 108 residues varies in sequence from one individual to another, whereas the other 300 or more residues are essentially constant in sequence. The C (constant) and V (variable) regions of each chain are believed to be coded by different genes; this is the "two gene–one polypeptide chain" hypothesis.

The variable-sequence portion of each L and each H chain represent the binding site for the specific antigen. The immune globulin chains appear to be coded for by duplication or repetition of primordial precursor genes, because of the sequence homologies between the variable and constant halves of L chains and among the four segments of the H chains, each about 107 residues in length. However, there are no sequence homologies between the variable regions of L chains and the variable regions of H chains.

Of the 107 residues in the variable-sequence segment of the light chains, over 40 have been found to differ from one multiple myeloma patient to another. In a chain of 107 residues of which 40 can be occupied by two different amino acids, it may be calculated that 2^{40}, or more than 10 billion, different sequences are possible. If each of the 40 variable sites may be occupied by any one of the 20 amino acids, the number of different possible sequences is far greater. Since each antibody molecule also has a heavy chain, in which many residues may also vary, it is immediately evident that a fantastically large number of antibodies can be made by replacement of the variable amino acid residues.

These considerations now bring us to the question: How can such an immensely large number of different antibody molecules be genetically specified? Does each individual human being possess all the genes necessary to specify the million or more different antibodies which can theoretically be formed?

Cellular Mechanisms of Antibody Formation

The old view that the antigen helped to shape or mold the antibody from a single plastic precursor, the "instruction" hypothesis, has been abandoned in favor of the *clonal-selection hypothesis*. This postulates that all the cells potentially capable of forming antibodies (now known to be lymphocytes) have the same genes but they occur in many different cell lines, called *clones*, each of which makes only a single type of immunoglobulin with a characteristic amino acid sequence; the genes for all other antibodies are permanently repressed. This concept can most aptly be described as the one cell–one antibody hypothesis. In the absence of the antigen, each clone exists only in small numbers and makes only small amounts of its particular antibody; it is committed to making that type of antibody before it has ever been exposed to an antigen. When a specific antigen is introduced into a vertebrate, it stimulates one specific clone of lymphocytes to proliferate and increase in number; simultaneously, there is a large increase in the amount of specific antibody formed by that type of clone. The lymphocytes thus

stimulated are transformed into *plasma cells*, the actual pro-
ducers of the antibodies. Presumably the ability of the antigen to "select" and stimulate a single clone lies in its ability to attach to specific receptors on the cell surface and thus
to elicit cell division and production of its characteristic antibody. This idea would account for the fact that an animal
can be immunized against many antigens simultaneously,
each stimulating a different clone of lymphocytes to proliferate and make a specific antigen. The clonal-selection theory
is in fact supported by studies on single cells.

We now see why the immune response is such a useful
model for the study of cell differentiation. The different
clones of lymphocytes all possess the same genome, but they
differ from each other in a heritable manner in that only one
set of antibody genes is expressed, all others being silent and
presumably permanently repressed.

Two types of lymphocytes participate in the production of
antibodies, *B lymphocytes*, made in the bone marrow, and *T
lymphocytes*, derived from the thymus gland. The B cells are
the precursors of the many different clones of plasma cells,
each of which can make only one specific type of circulatory
(or "humoral") immunoglobulin. The T cells are primarily
concerned with production of cell or tissue immunity, but
are also required for the formation of circulating immunoglobulins by B cells. Antigens interact with specific
receptors on T cells and induce these cells to release a "cooperation" factor. This factor selectively interacts with and
activates those B cells that have also bound the antigen
through specific cell-surface receptors. The activated B cells
proliferate and some of their progeny become plasma cells
which synthesize and secrete specific antibody.

Three possible models have been proposed for the immense diversity of different antibodies that can be made by
an individual vertebrate organism. One is that each antibody-
producing cell contains an immensely large number of V and
C genes capable of specifying many different L and H chains.
However, each cell is committed in a heritable manner to
make only one specific antibody. In this model only two
pairs of genes in each clone would be transcribed: a C gene
and a V gene coding for a specific heavy chain and another C
and V pair for a light chain. All the other genes for antibodies are permanently repressed. A second model proposes that there are only a few C genes for the constant portions but an enormous number of V genes for the variable
portions of the two sets of chains.

However, both these models require that a significant fraction of the genome of vertebrates be used to code for antibodies alone. Since the immune response has great survival
value to higher animals, allowing them to resist bacterial and
viral infections, use of a rather large fraction of their
genomes to code for antibodies would not be surprising.
However, the evidence for the first two models has become
rather shaky with the finding that there are only a few C
genes and a relatively small number of V genes, certainly less
than 100. To account for the large number of possible antibodies, one model proposes that during the very rapid pro-

liferation of a clone of B cells induced by an antigen, many mutations occur in the V genes, particularly in the hypervariable portions, which can result in a very large diversity of antibodies, of which at least one will be relatively specific for binding the antigen. Another model suggests that the V genes undergo extensive exchange and reshuffling of gene segments by recombination processes. Whatever the explanation of antibody diversity, the solution of this problem should greatly clarify the genetic and biochemical basis of the differentiation of cells and may also lead to ways and means of altering immune reactions of the host which may be medically useful.

Summary

Constitutive enzymes are those made in constant amounts by cells, regardless of their metabolic state, whereas induced enzymes are made only when needed, in response to the presence of their substrates, particularly when the latter are the sole carbon source. Some enzymes are no longer made, i.e., they are repressed, when their products are no longer needed by the cell. Sometimes a whole sequence of enzymes is induced as a group (coordinate induction) or repressed as a group (coordinate repression). The sets of genes for such groups of enzymes are called operons. Studies of the inducible enzyme β-galactosidase in E. coli have revealed the mechanisms by which synthesis of induced enzymes is regulated. Each operon contains the structural genes, usually adjacent to each other, for each of the enzymes whose synthesis it controls, as well as a regulator gene and an operator. The regulator gene codes for the synthesis of a repressor protein, which normally is specifically bound to the operator and prevents transcription of the structural genes. If the substrate of the inducible enzyme is present, it binds to the repressor molecule, causing its release from the operator and freeing the structural genes for transcription. In enzyme repression, on the other hand, the end product of the enzyme sequence binds to the free repressor molecule; the resulting complex then binds to the operator and prevents transcription of the operon. The repressor of the *lac* operon, which controls synthesis of β-galactosidase and two other related enzymes, has been isolated and the nucleotide sequence of its binding site on the operator DNA identified. The controlling region of the *lac* operon also includes a binding site for a complex of cyclic AMP with a specific protein, as well as a binding site for RNA polymerase. The messenger RNA transcribed from bacterial DNA has a very short half-life of about 2 min. When amino acids are unavailable to a bacterial cell, it stops making ribosomal RNA; the signaling agent for such stringent control is guanosine 2'(3')-diphosphate 5'-diphosphate (ppGpp).

The regulation of gene expression in eukaryotes is much more complex. Eukaryotic cells of animals show enzyme induction in the presence of specific substrates and certain hormones. Some hormones bind to receptor proteins; the resulting hormone-receptor complex binds to specific sites in the chromatin and thus promotes transcription of certain genes. Specific enzyme activities are also turned on and off during the cell cycle; the mediators are cyclic AMP and cyclic GMP. During differentiation and development in eukaryotes some genes are transcribed and others repressed, the latter often permanently, in a complex pattern having a specific sequence and timing. Eukaryotic cells do not have simple operons; genes for enzyme sequences are scattered in different chromosomes. The histones and other proteins of the chromatin, which have been postulated to be repressors, are probably structural rather than regulatory elements. However, a portion of the heterogeneous nuclear RNA has been postulated to serve in the capacity of signaling molecules in the induction of gene transcription in eukaryotes. The formation of specific antibodies by vertebrates is a reflection of a specialized type of cell differentiation, in which each lymphoid cell type is capable of producing a unique kind of antibody. This is made possible by the combined action of a small number of genes, of which some portions are apparently highly mutable or undergo rapid genetic recombination.

References

Books

BAUTZ, E. K., P. KARLSON, and H. KERSTEN (eds.): *Regulation of Transcription and Translation in Eukaryotes*, Springer-Verlag, New York, 1973. A series of symposium articles.

LEWIN, B.: *Gene Expression*, vol. 1, *Bacterial Genomes*, Wiley, New York, 1974.

LEWIN, B.: *Gene Expression*, vol. 2, *Eukaryotic Chromosomes*, Wiley, New York, 1974. This pair of valuable books provides a comprehensive, balanced account of the experimental work in this area, together with a large bibliography.

Regulation of Gene Expression, *Cold Spring Harbor Symp. Quant. Biol.*, vol. 38, 1973. Many valuable articles.

Articles

BRITTEN, R. J., and E. H. DAVIDSON: "Gene Regulation for Higher Cells: A Theory," *Science*, 165: 349–357 (1969).

BURGER, M. M., B. M. BOMBIK, B. M. BRECKENRIDGE, and J. R. SHEPPARD: "Growth Control and Cyclic Alterations of Cyclic AMP in the Cell Cycle," *Nature New Biol.*, 239: 161–163 (1972).

DAVIDSON, E. H., and R. J. BRITTEN: "Organization, Transcription and Regulation in the Animal Genome," *Quart. Rev. Biol.*, 48: 565–613 (1973).

DE CROMBRUGGHE, B., B. CHEN, W. ANDERSON, P. NISSLEY, M. GOTTESMAN, I. PASTAN, and R. PERLMAN: "*lac* DNA, RNA Polymerase, Cyclic AMP Receptor Protein, Cyclic AMP, *lac* Repressor and Inducer Are Essential Elements for Controlled *lac* Transcription," *Nature New Biol.*, 231: 139–142 (1971).

DICKSON, R. C., J. ABELSON, W. M. BARNES, and W. S. REZNIKOFF: "Genetic Regulation: The *lac* Control Region," *Science*, 187: 32 (1975). The nucleotide sequence of the control DNA.

EDELMAN, G.: "Antibody Structure and Molecular Immunology," *Science*, 180: 830–840 (1973). Nobel prize lecture.

EDELMAN, G. M., B. A. CUNNINGHAM, W. E. GALL, P. D. GOTTLIEB, U. RUTTISHAUSER, and M. J. WAXDAL: "The Covalent Structure of an Entire γG Immunoglobulin Molecule," *Proc. Natl. Acad. Sci.* (U.S.), 63: 78–85 (1969).

GILBERT, W., and A. MAXAM: "The Nucleotide Sequence of the *lac* Operator," *Proc. Natl. Acad. Sci.* (U.S.), 70: 3581–3584 (1973).

GILBERT, W., and B. MÜLLER-HILL: "The *lac* Operator Is DNA," *Proc. Natl. Acad. Sci.* (U.S.), 58: 2415 (1967). The specific binding of the isolated *lac* repressor to the operator site of DNA.

GILBERT, W., N. MAIZELS, and A. MAXAM: "Sequences of Controlling Regions of the Lactose Operon," *Cold Spring Harbor Symp. Quant. Biol.*, 38: 845–855 (1973).

GROS, F.: "Control of Gene Expression in Prokaryotic Systems," *FEBS Lett.*, 40: S19–S27 (1974).

HOLLIDAY, R., and J. E. PUGH: "DNA Modification Mechanisms and Gene Activity during Development," *Science*, 187: 226–232 (1975).

HOOD, L. E.: "Two Genes, One Polypeptide Chain–Fact or Fiction?", *Fed. Proc.*, 31: 177–187 (1972).

JACOB, F., and J. MONOD: "Genetic Regulatory Mechanisms in the Synthesis of Proteins," *J. Mol. Biol.*, 3: 318 (1961). Classic paper postulating the function of repressors, operators, and structural genes, as well as the messenger concept.

KORNBERG, R. D.: "Chromatin Structure: A Repeating Unit of Histones and DNA," *Science*, 184: 686–872 (1974).

KORNBERG, R. D., and J. O. THOMAS: "Chromatin Structure: Oligomers of the Histones," *Science*, 184: 865–868 (1974). This and the preceding are two important articles on the molecular organization of chromatin.

PAIN, V. M., and M. J. CLEMENS: "The Role of Soluble Protein Factors in the Translational Control of Protein Synthesis in Eukaryotic Cells," *FEBS Lett.*, 32: 205–212 (1973). An excellent short review of a complex subject.

PASTAN, I., and R. PERLMAN: "Cyclic AMP in Bacteria," *Nature*, 169: 339–344 (1969).

PTASHNE, M.: "Isolation of the λ Phage Repressor," *Proc. Natl. Acad. Sci. (U.S.)*, 57: 306 (1967).

RIGGS, A. D., and S. BOURGEOIS: "On the Assay, Isolation and Characterization of the *lac* Repressor," *J. Mol. Biol.*, 34: 361–364 (1968).

RUTTER, W. J., R. L. PICTET, and P. E. MORRIS: "Toward Molecular Mechanisms of Developmental Processes," *Ann. Rev. Biochem.*, 42: 601–646 (1974). A particularly valuable review of biochemical approaches to differentiation and development.

SHAPIRO, J., L. MACHATTIE, L. ERON, G. IHLER, K. IPPEN, and J. BECKWITH: "Isolation of Pure *lac* Operon DNA," *Nature*, 224: 768–774 (1969).

SIMPSON, R. T.: "Structure and Function of Chromatin," *Adv. Enzymol.*, 38: 41–108 (1973).

STEIN, G. S., T. C. SPELSBERG, and L. J. KLEINSMITH: "Nonhistone Chromosomal Proteins and Gene Regulation," *Science*, 183: 817–823 (1974). An alternative to the Britten-Davidson model.

TOMKINS, G. M.: "Regulation of Gene Expression in Mammalian Cells," *Harvey Lect.* 68: 37–66 (1974). Role of glucocorticoids.

Problems

1. In a bacterium, the rate of biosynthesis of a protein containing 230 amino acid residues is subject to biological regulation. Assume that 10 mRNA molecules are transcribed per minute under normal conditions and that each mRNA molecule can be read

by 15 ribosomes. What is the cost per minute in ATP, in ribo-nucleotide bases, and in amino acids if protein synthesis is com-pletely shut off at (a) the transcriptional level, (b) the transla-tional level, (c) the posttranslational level, i.e., by degradation of the completed proteins after their synthesis?

2. The RNA transcribed from a region of bacterial DNA that par-ticipates in a site-specific reaction with a protein molecule yields the following oligonucleotide fragments upon cleavage with ribonuclease T_1: ACG, UCAG, pppCG, CUG. If the protein-binding site shows twofold (rotational) symmetry, deduce a possible DNA sequence of the protein-binding site.

3. Cis-dominant mutations are those in which the phenotypic defect cannot be alleviated by the proteins of a wild-type gene on a separate DNA molecule or on the same DNA molecule but in an anomalous position. Mutants in an operator gene are an ex-ample. Name two other sites in the *lac* operon that might be the site of cis-dominant mutations.

4. A mutant of *Salmonella typhimurium* is found to constitutively synthesize all the enzymes required for histidine biosynthesis. The mutated gene maps at a locus far from the structural genes of the *his* operon. During repression in wild-type S. *typhi-murium*, the product of another gene binds to the operator region. Addition of histidine to the culture medium of the con-stitutive mutant does not cause repression, but addition of his-tidinyl-tRNA$_{His}$ does. (a) What is the probable identity of the mutated gene in the constitutive mutant? (b) Antibody is pre-pared to the *his* repressor protein isolated from a wild-type strain. Competitive antibody binding assays show that several times more repressor is present in an extract prepared from the derepressed mutant described in part (a) than in an extract prepared from a repressed wild-type strain. What is a possible interpretation?

5. *E. coli* strain A is unable under any conditions to metabolize lactose, but it can metabolize galactose when it is the sole energy source. (a) List all the possible types of genetic defects that might cause this deficiency. (b) The information that the cell can metabolize galactose enables you to eliminate which other possible defects in the control of *lac* enzyme synthesis? (c) Which possible defects can you eliminate on the basis of experiments (1) to (4)?
(1) Hybridization studies with mRNA transcribed from func-tional *lac* genes show that essentially all of this RNA forms hybrids when mixed with excess DNA from strain A.
(2) A bacteriophage δ containing the *lac* regulatory region but only the first few bases of β-galactosidase gene is introduced into strain A. After recombination, strain A(δ) is now able to cleave lactose.
(3) The entire *lac* operon from strain A is placed on a plasmid that can autonomously replicate in E. *coli*. This plasmid is in-troduced into a bacterial strain that is recombination deficient (i.e., cannot recombine). The recipient strain has a deleted i gene and hence is synthesizing the *lac* operon structural pro-teins constitutively. After introduction of the plasmid, some progeny cells contain replicated versions of both the plasmid and the chromosome. In these cells, the *lac* structural proteins are synthesized in the presence of lactose but not in the absence of lactose.
(4) ^{35}S-labeled CAP isolated from a wild-type strain normally sediments at 5S in sucrose density gradients. When incubated

with purified *lac* DNA from strain A, the radioactivity's sedimentation behavior is unaffected. When incubated with cAMP in addition to *lac* DNA, the sedimentation coefficient of the radioactivity increases to 80S.

(*d*) Restriction endonuclease Hind II makes a single cut in DNA purified from phage δ or the plasmid constructed from strain A. When *E. coli* RNA polymerase holoenzyme is added to δ DNA, this protein-DNA complex is resistant to Hind II. Addition of RNA polymerase does not affect the action of Hind II on *lac* DNA isolated from the plasmid. What is the most likely nature of the original lesion in strain A?

6. DNA extracted from the exotic single-stranded phage FFC 88 contains nine types of modified bases. A 21-nucleotide-long region known to bind a repressor protein was isolated. Chromatographic analysis gave the following relative compositions: E and D (1.0 each); V, L, R, I, N, and A (0.5 each); Y (0.25). A trinucleotide obtained from this region was found to have the 5'→3' sequence · · ·DNA· · ·. If repressors bind to perfectly symmetrical sequences on single-stranded DNA, what is a possible base sequence for the entire repressor binding region?

CHAPTER 36 THE MOLECULAR BASIS OF MORPHOGENESIS

We come now to the third fundamental question implied by the discovery of DNA as the genetic material. How is the genetic information in DNA, which is encoded in one-dimensional form as a linear sequence of nucleotides, translated into the three-dimensional form characteristic of such supramolecular structures as enzyme complexes, viruses, ribosomes, and membranes and such higher-order biostructures as tissues, organs, and organisms?

Morphogenesis is the name applied to the set of processes by which the characteristic micro- and macrostructures of living organisms grow and develop in space and time, as a result of genetic programming. An important key to the molecular basis of morphogenesis is intrinsic in the properties of polypeptide chains, which we have seen possess the ability to convert the one-dimensional information inherent in their amino acid sequence into three-dimensional information by spontaneously folding into characteristic, biologically active globular conformations. The three-dimensional conformation of a native polypeptide chain is not imposed on it by external forces: it is the inevitable and automatic outcome of the tendency of the surrounding water molecules to seek the state of maximum entropy and the tendency of the polypeptide chain to seek its state of minimum free energy (page 143). In this chapter we shall see that much larger information-rich structures can assemble themselves spontaneously from their components. Such self-assembly processes do not violate thermodynamic principles. Rather, they are the result of the same processes governing the folding of polypeptide chains, the tendency of surrounding water molecules to seek the state of maximum entropy and the tendency of the component parts to seek their state of minimum free energy. Precision in the "fit" of the various components with each other is provided by their structural complementarity.

The study of the molecular basis of morphogenesis has become one of the frontier areas of biochemical research. In this chapter we shall survey some of the molecular principles involved in self-assembling systems and then examine some large supramolecular systems capable of self-assembly from purified components.

The Subunit Structure of Macromolecular and Supramolecular Systems

Nearly all large macromolecular biostructures in the size range between proteins and organelles contain a number of polypeptide subunits, which, alone or with other molecular components, are associated through noncovalent forces. Hemoglobin contains four polypeptide chains, tightly but noncovalently associated (page 145). The regulatory enzyme aspartate transcarbamoylase (page 237) contains twelve polypeptide chain subunits of two different kinds, six of each (page 239). Some proteins form polymers; acetyl-CoA carboxylase polymerizes in the cell to form long, filamentous structures (page 662). The simpler viruses, such as tobacco mosaic, are constructed from a molecule of a nucleic acid and many identical sheath- or coat-protein subunits. The more complex viruses may contain many different kinds of protein subunits. Ribosomes of bacteria contain 50 or more protein subunits and 3 RNAs (page 322). In all these cases the subunits of these supramolecular systems are held together almost entirely by noncovalent but highly specific interactions. Moreover, there is often a hierarchy of structural organization in such systems. For example, although aspartate transcarbamoylase has 12 polypeptide chains, they are associated in such a way that the enzyme has two catalytic subunits, each having 3 identical polypeptide chains, and three regulatory subunits, each with 2 identical polypeptide chains, different from those in the catalytic subunits (page 237). Thus we often find that a supramolecular structure contains subassemblies.

Advantages of Subunits

Why are supramolecular structures built out of many small subunits, associated in a noncovalent manner, rather than as a single, large "unitized" supermolecule, held together entirely by covalent bonds? There are perhaps four major biological advantages to the subunit pattern of construction. One is that use of a number of identical polypeptide subunits can minimize the effect of random errors that may occur in the biosynthesis of protein molecules. Let us consider two alternative ways of constructing a very large protein structure having 100,000 amino acid residues. In scheme 1, the structure is covalently built, residue by residue, into a single long polypeptide chain. In scheme 2, 1,000 polypeptide chains, each with 100 amino acid residues, are built separately and then noncovalently assembled to form an oligomeric protein. Let us assume that in the biosynthesis of polypeptide chains random mistakes happen, so that the wrong amino acid is inserted on the average of once every 100,000 times. Let us further assume that each mistake results in a biologically inactive or useless product. Under scheme 1, each complete molecule will contain on the average one wrong amino acid residue; thus most of the molecules produced would be inactive. Under scheme 2, mistakes will occur in the synthesis of only 1 of the 1,000 subunits, on the average.

There is a second major advantage to construction from subunits. If "bad" subunits can be rejected in some way, errors in synthesis can be further minimized. As an analogy, take the assembly lines on which television sets are constructed. At each stage of assembly, defective components are rejected and kept from being assembled into the finished product, at least in principle. Similarly, in the self-assembly of oligomeric protein molecules, only "good" subunits which fit their neighbors are likely to find their way into the final product; defective subunits that do not fit will be rejected. For example, there are some mutants of tobacco mosaic virus in which one specific amino acid residue in each subunit chain is replaced. In certain of these mutants the resulting polypeptide chains will not assemble into a stable sheath to yield a complete virus particle; moreover, when mixed with "good" subunits the mutant subunits are rejected and do not appear in the finished product.

There is a third advantage. The subunit structure of large oligomeric protein systems provides considerable economy of DNA. In our examples above, a single polypeptide chain of 100,000 residues would require for its coding a segment of DNA containing 300,000 base pairs (page 957). On the other hand, if this structure consisted of 1,000 identical polypeptides of 100 residues each, a segment of DNA containing only 300 base pairs would suffice. Thus if a DNA template can be used repetitively to code identical subunits, considerable economy of genetic material is achieved.

There is another important principle. Many supramolecular systems must participate in extremely rapid reactions, such as muscle contraction, flagellar motion, readout of mRNA during protein synthesis, conduction of nerve impulses, and DNA unwinding, all of which require rapid changes in conformation, or make-and-break interactions. Nearly all such very fast processes occur by noncovalent dissociation and reassociation of subunit components or by noncovalent changes in conformation of subunits. Hydrophobic interactions and the making and breaking of hydrogen bonds are intrinsically much faster (page 42) than covalent changes catalyzed by enzymes and, energetically speaking, much cheaper.

The Driving Forces for Self-Assembly of Subunits

We have earlier seen that two classes of noncovalent associations participate in the formation and the stabilization of the native structure of monomeric and oligomeric proteins: (1) hydrophobic interactions (pages 43 and 139) and (2) hydrogen bonds and ionic interactions (pages 40 and 139). We may add to these a third type of interaction in the special case of duplex nucleic acids, namely, the electronic interactions between stacked bases (page 873). Such noncovalent interactions also participate in the formation and stabilization of large multisubunit structures such as enzyme complexes, ribosomes, viruses, and membranes.

The first and second laws of thermodynamics (page 390) are combined in the expression

$$\Delta G = \Delta H - T\,\Delta S$$

where ΔG is the change in free energy, ΔH is the change in enthalpy, and ΔS is the change in entropy, or disorder, in the system (page 390). As supramolecular systems assemble themselves into a stable end product, the free energy of the system of subunits declines to a minimum. This decrease in free energy must come about as the result of a decrease in enthalpy, i.e., a negative value of ΔH, or an increase in entropy, i.e., a positive value for ΔS, or both. In most supramolecular systems the major driving force for self-assembly consists of hydrophobic interactions, which are driven by the tendency of the surrounding water molecules to seek their own state of maximum entropy. The self-assembly of most supramolecular systems is thus entropy-driven and proceeds with a large increase in ΔS of the aqueous surroundings. There is only one general type of assembly process driven largely by a decrease in ΔH, namely, the formation of duplex or double-helical nucleic acids from single-strand structures (page 873).

Although hydrophobic interactions provide most of the driving force and stability of oligomeric systems, they are relatively nonspecific, i.e., nondirectional, and thus do not contribute much to assuring the exact fit of subunits. On the other hand, hydrogen bonds and ionic interactions, although less important energetically in stabilization of supramolecular structures, are highly specific and directional; they confer both geometrical specificity and correct fit.

The Source of Information in Self-Assembly of Supramolecular Systems

Many supramolecular systems assemble themselves from their subunits without any external instruction. In such cases all the information required must be inherent in the structure of their subunits. Information for self-assembly may be provided by the protein subunits alone, as is obviously the case for the self-assembly of such oligomeric proteins as the regulatory enzyme aspartate transcarbamoylase, which contains no components other than polypeptide chains. In ribosomes both the protein subunits and the ribosomal RNA carry information specifying the assembly process; this is also true of tobacco mosaic virus. Moreover, in the self-assembly of membranes phospholipid molecules contribute information, since they inherently possess the capacity for self-assembly into bilayer systems, in the complete absence of other components (page 300).

Nevertheless, for the assembly of most supramolecular systems it is the protein subunits that are the major carriers of the required information. For example, the oligomeric proteins hemoglobin and aldolase, which contain no covalent linkages between their subunits, retain their molecular identity when mixed in solution; they show no measurable tendency to exchange subunits with each other. Normal adult hemoglobin contains two α and two β chains; its structure is symbolized $\alpha_2\beta_2$. Under no normal circumstances do hemoglobin molecules in solution reassemble to form α_4, β_4, $\alpha\beta_3$, or $\alpha_3\beta$ oligomers; hemoglobin is always present as the $\alpha_2\beta_2$ species. Thus the α and β subunits of hemoglobin contain quite specific information, which permits only one type of assembled product. Moreover, the fit of the subunits of hemoglobin molecules to each other is so specific that when native hemoglobins from two different mammalian species are mixed at normal pH and ionic strength, they will not significantly exchange subunits to form hybrids, even though the hemoglobins of different species are nearly identical in conformation and have many homologies in their amino acid sequences.

We may therefore conclude that the amino acid sequence of the polypeptide subunits of supramolecular systems carries two levels of information. At one level the sequence specifies the three-dimensional conformation of the individual polypeptide chains, i.e., their tertiary structure (page 141). At another level each polypeptide chain in its tertiary conformation must contain one or more recognition and binding sites, by which adjacent polypeptide subunits are bound in a specific geometrical relationship to form the characteristic quaternary structure of the system. The non-covalent interactions between the binding sites of two adjacent subunits must have a very high degree of specificity and precision in their fit to each other, through hydrogen bonds and ionic interactions, as well as a high degree of stability, conferred by hydrophobic interactions. The amino acid residues contributing to these binding sites are not necessarily confined to one short segment of the polypeptide chain but may be contributed from several segments because of the folding of the polypeptide chain in the subunit structure.

Sequence of Assembly of Subunits

Another characteristic of self-assembly processes is that they often follow a specific sequence in the assembly of the components; presumably this is also coded into the primary structure of the subunit polypeptide chains. Such specific sequences of assembly may be devices to minimize the number of "wrong" trial-and-error collisions and associations between subunits en route to the finished product, thus decreasing the time required for the system to assemble itself. We recall that polypeptide chains fold into their characteristic three-dimensional conformations very rapidly, in far less time than the chain would require to try out all possible conformations before attaining the correct minimum-free-energy form (page 144). For the same reason we may suppose that polypeptide subunits find their final positions in a supramolecular structure according to a specific pathway or sequence of events that leads quickly to the final, correct three-dimensional structure.

We shall see that the more complex supramolecular structures such as ribosomes and the larger bacteriophages undergo self-assembly according to a precoded sequence, presumably to assure the completion of assembly in a time span consistent with the kinetics of the biological process of which the assembly is a part. Thus ribosomes and viruses of *E. coli* cells must be made well within the division time of these cells, which may be as short as 20 min.

Cooperativity in Self-Assembling Systems

We have seen that the principle of cooperativity makes possible the formation of very stable macromolecular structures held together by nothing more than a large number of weak bonds or interactions (page 42). In cooperative systems the formation of one weak bond between two structures favors the formation of a second bond, which in turn increases the probability of another, and so on, so that both the stability and usually the rate of formation of the structure are greatly enhanced.

Cooperativity is also an important characteristic of self-assembling systems and is manifest at two levels. The binding of two subunits to each other is usually cooperative because there may be several or many weak bonds between the subunits. But cooperativity may also be manifest in the assembly of higher orders of structure; i.e., the association of two subunits enhances the addition of a third, which in turn enhances the addition of the fourth, and so on. The net result is that the subunits of macromolecular systems, after a relatively slow initiation or nucleation step, often assemble at ever-increasing rates to achieve structures of ever-increasing stability. As another consequence of cooperativity, intermediate stages in self-assembly of a supramolecular structure are not normally observed and can be seen only when the system is experimentally perturbed for this purpose. The principle of cooperatively thus gives an all-or-none quality to the self-assembly process, leaving no incomplete structures or spare parts behind.

Preferred Assembly Pathways in Oligomeric Proteins

The regulatory enzyme aspartate transcarbamoylase (ATCase) is one of the most thoroughly studied regulatory enzymes (pages 237 and 736). It contains six catalytic, or c, chains and six regulatory, or r, chains which are grouped into functional subunits. The catalytic subunits, designated C, contain three c chains and the regulatory subunits, designated R, contain two r chains. The complete functional ATCase molecule is designated C_2R_3 (Figure 36-1).

Even this relatively small oligomeric system has preferred pathways of self-assembly from its polypeptide chains, so that the complete molecule is assembled very rapidly from its 12 component polypeptide chains. These pathways have been studied by a combination of physicochemical methods

Figure 36-1
Schematic representation of the structure of aspartate transcarbamoylase and an assembly map showing the pathways taken in its formation from six c chains and six r chains (see text).

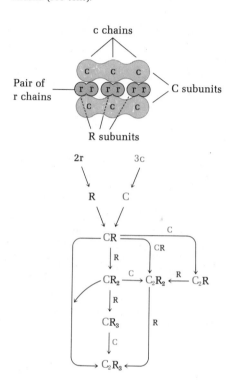

Figure 36-2
Electron micrograph of tobacco mosaic virus. The rods are 300 nm long and 18 nm in diameter.

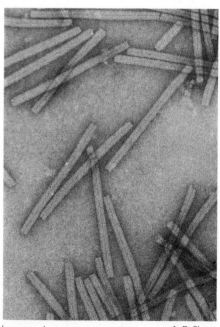

0.1 μ

L. D. Simon

Figure 36-3
An early conception of the assembly of tobacco mosaic virus, which postulated sequential addition of single protein subunits. (Redrawn from H. Fraenkel-Conrat, Design and Function at the Threshold of Life, Academic Press, Inc., New York, 1962.)

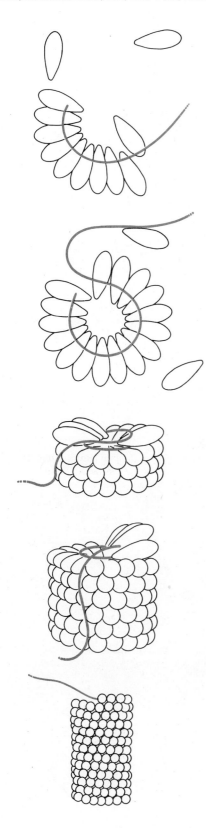

and by the use of isotopically labeled subunits, in the laboratory of H. K. Schachman. Figure 36-1 shows an assembly map indicating the probable pathways by which the complete C_2R_3 molecule is assembled. The C and R subunits are first assembled separately from the c and r polypeptide chains, respectively; only complete C and R subunits can enter into formation of the final product. C subunits do not interact with each other, so assembly pathways involving C_2 subassemblies are excluded; similarly, R subunits do not associate with each other alone. The first step appears to be the formation of CR. Which of the subsequent pathways (Figure 36-1) is taken depends in large measure on the relative concentrations of C and R in the system. When C is present in excess, formation of C_2R is the next step; when R is in excess, CR_2 formation is favored. The remaining steps are shown in Figure 36-1.

The entire ATCase molecule is assembled from C and R subunits in a few seconds, or much less than the time required to synthesize the polypeptide chains from the constituent amino acids. Once the complete C_2R_3 molecule is formed, its subunits do not exchange with free, isotopically labeled C or R subunits at any measurable rate.

Self-Assembly of Tobacco Mosaic Virus

The first large supramolecular structure found to assemble itself from its component parts in vitro was tobacco mosaic virus (TMV). In 1955 H. Fraenkel-Conrat and R. C. Williams reported that TMV can be separated into its component sheath proteins and RNA and then reconstituted, simply by mixing, to yield biologically active, infectious viral particles. The TMV particle (mol wt 40 million; length 300 nm) is composed of about 2,200 identical polypeptide chains, each having 158 amino acid residues, and a single RNA chain (mol wt 2.1 million), which makes up about 5 percent of the total weight. The viral particle is a long rod (Figure 36-2) in which a cylinder of globular protein subunits is arranged around a core furnished by the helically coiled RNA chain. The TMV particle can be dissociated into its polypeptide chains and RNA by exposure to detergents or weakly alkaline solutions. The polypeptide chains alone are noninfectious. The RNA is infectious alone, but very weakly so; only one molecule of many applied succeeds in entering the host cells in the tobacco leaf. Fraenkel-Conrat and Williams found that when the solubilized protein subunits of TMV are mixed with the pure viral RNA, the mixture becomes distinctly turbid, resembling the opalescence of a suspension of intact TMV virus particles. Application of this mixture to the surface of the tobacco leaves resulted in typical viral lesions, whereas none appeared on application of the protein fraction alone and very little with RNA alone. Very large yields of active reconstituted viral particles could be obtained, up to 50 percent. From electron-microscopic observations it was postulated that tobacco mosaic virus assembles itself according to the schematic representation in Figure 36-3. The protein subunits were thought to assemble serially into a helical

Figure 36-4
The ranges of pH and ionic strength in which different polymeric species of TMV protein predominate at 20°C. [From A. Klug, Fed. Proc., 31: 31 (1972).]

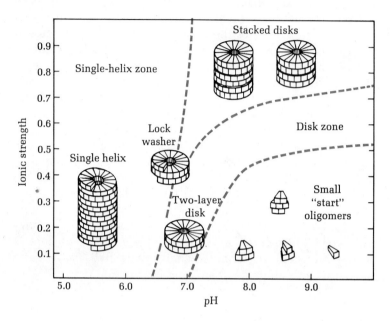

array in the shape of a hollow tube, bound to the coiled RNA molecule within. The assembly process stops when the rod reaches a length of 300 nm, apparently limited by the length of the RNA molecule. All the information required to assemble the protein subunits into the rod-shaped helical array was concluded to be present in the protein subunits, since the subunits alone spontaneously assemble themselves into hollow tubes in the absence of RNA. Such tubes are of indefinite length and are not infectious.

But more recent research by A. Klug and his colleagues in England has revealed that the assembly of TMV is much more complex and has a number of intermediate steps. From detailed physicochemical and electron-microscopic study of the effect of changing pH and ionic strength on the association of the protein subunits of TMV, Klug and his colleagues identified several different types of association products, which they were able to place into a specific sequence in the assembly process. Figure 36-4 shows the structure of the different polymeric species of the TMV protein (in the absence of RNA), as a function of pH and ionic strength (page 162). Two types of long tubelike structures were observed, one helical and the other consisting of stacks of flat disks. The basic unit of construction of the helical tubes appeared to be a "lock-washer" structure, consisting of two layers of subunits in helical arrangement. The unit of construction of the other tubelike structure is a flat, two-layer disk. The normal structure of intact TMV is helical.

At neutral pH, near that of the host-cell contents, the protein subunits form the two-layer flat disks predominantly, in

Figure 36-5
The assembly of the polypeptide subunits of TMV to form two-layer disks. [From A. Klug, Fed. Proc., 31: 35 (1972).]

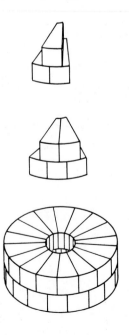

Figure 36-6
Initiation and growth of tobacco mosaic virus. To initiate assembly a complete two-layer disk (34 subunits) first binds to the initiation site on the RNA (color). A second disk then adds and both undergo dislocation to yield lock-washer structures. The RNA intercalates between the layers of subunits; successive subunits rise to allow the RNA to enter into the core. [From A. Klug, Fed. Proc., 31: 40 (1972).]

Initiation   Growth

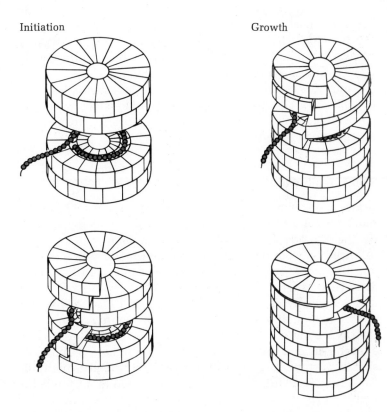

a process in which a two-layer trimer appears to be a necessary starter, as shown in Figure 36-5; each two-layer flat disk has 34 protein subunits. The viral RNA is not required for this step. Such a flat disk can be converted into a lock washer by lowering the pH. Klug and his colleagues then studied the interaction between viral RNA and the different polymeric forms of the protein subunits. They found that the RNA prefers to bind to the flat two-layer disks. Moreover, this binding is rather specific; it takes place at a stretch of about 50 nucleotide residues at the 5′ end of the RNA molecule. Evidently this end of the RNA contains specific nucleotide sequences capable of interacting with the flat disks to initiate the growth of the virus. Thus the RNA is not merely trapped within the core of the TMV particle; it is required to initiate the self-assembly process.

The next stage in the assembly of the TMV particle requires the addition of successive, complete two-layer disks, each containing 34 protein subunits, in contrast to the earlier hypothesis which proposed that the protein subunits were added singly in sequence. However, since the finished viral particles are helical and consist of the lock-washer arrangement of subunits, rather than the flat disks, the flat disks must evidently be converted into the lock-washer configuration *after* they are added to the growing TMV particle. This conversion occurs by a conformational change of the subunit molecules, coupled to a change in ionization state of two charged groups in each subunit. This conformational change accompanies the binding of the RNA to the subunits immediately beneath the two-layer disk. Thus the conformational change in each two-layer disk is triggered by the binding of the next two-layer disk. This series of changes causes the two-layer disks to assume the helical, or lock-washer, configuration and leads to the orderly helical coiling of the RNA within the core of the structure (Figure 36-6).

Thus the simple tobacco mosaic virus assembles itself through a specific sequence of intermediate steps which are observable only when the assembly process is perturbed by alterations in pH and ionic strength.

Self-Assembly of Bacteriophage T4

Another advance of major significance with respect to the molecular and genetic basis of morphogenesis is represented by the important work of R. S. Edgar and W. B. Wood, followed by other investigators, on the genetic determination and the assembly of the components of phage T4 of *E. coli*. This virus is far more complex than TMV, both structurally and genetically. It has a hollow polyhedral head, consisting of a number of different types of protein subunits, into which the DNA molecule is folded (Figure 36-7). The tail resembles a spring-loaded hypodermic syringe; it has a contractile sheath surrounding a central passage through which the DNA is injected into the host cell. Connected to the tail is the end plate, to which are attached six long, thin tail fibers and six short spikelike structures, both adapted to fasten to the cell wall. T4 is also genetically complex, since it contains well over 100 genes, compared with only a few in TMV. The T4 genes code not only for all the proteins of the virus but also for a number of enzymes required to synthesize viral DNA in the host cell and to cause lysis of the host-cell wall for release of the progeny virus particles.

Many mutations of the T4 bacteriophage have been observed. Some of them result in the biosynthesis of viral proteins that are defective owing to amino acid replacements. Others result in defective auxiliary enzymes, such as the viral deoxyribonuclease, which destroys the DNA of the host cell, or the viral lysozyme, which makes possible the rupture of the host-cell wall. When mutations occur in which the protein components of the viral structure are defective, the assembly of the progeny virus particles usually stops at the point where the defective protein molecule is inserted. Most such mutations are lethal, because the mutant virus cannot replicate or is unable to enter the host cell. However, some mutants of T4 defective in one or another structural component are *conditional* lethal mutants; i.e., the mutant protein may appear in either a functional or a defective form, depending on the conditions of culture and growth, particularly the temperature. In the latter case such mutants are called *temperature-sensitive mutants*. When infected *E. coli* cells are grown under *permissive conditions*, for instance at a temperature of 25°C, such mutant viruses will form intact, functional viral particles, whereas under *restrictive conditions*, i.e., at a high temperature, 37 to 40°C, defective viral particles will form.

From such experiments, Edgar mapped the various genes of the T4 virus that specify the biosynthesis of the various

Figure 36-7
Structure of bacteriophage T4. The head is 30-sided. The tail is hollow and surrounded by a contractile sheath. The six tail fibers are attached to the end plate, which has six short spikes. (From W. B. Wood and R. S. Edgar, "Building a Bacterial Virus." Copyright © July, 1967 by Scientific American, Inc. All rights reserved.)

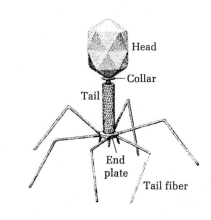

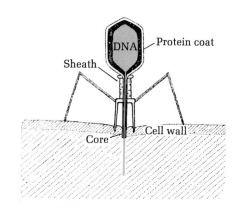

Figure 36-8
The genetic map of bacteriophage T4, showing the position of some of the
genes specifying its structure and assembly (black segments). The boxes
show the nature of the parts formed by the host cell in response to mutations at the site shown. When gene 11 or 12 undergoes mutation, a complete viral particle is formed but it is abnormally fragile. (From W. B. Wood
and R. S. Edgar, "Building a Bacterial Virus." Copyright © July, 1967 by
Scientific American, Inc. All rights reserved.)

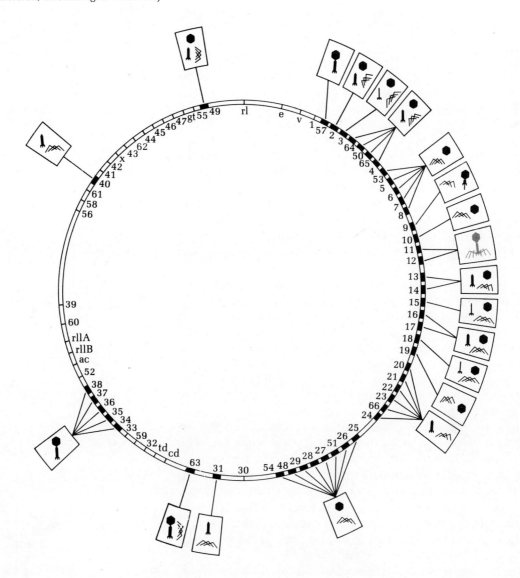

component parts of the T4 virus in the host *E. coli* cell. Figure 36-8 shows a genetic map of T4 (see also page 878). More
than half the T4 genes participate in the formation of the
complete virus particle, each specifying a distinct type of
protein subunit. In mutants in which one or another of these
genes are defective, the usually orderly process of viral morphogenesis is interrupted. This effect leads to the accumulation within the host cell of incomplete parts, such as heads,
tails, or tail fibers, depending on the site of mutation. These

parts can be visualized by electron microscopy inside the cells bearing the mutant virus. Moreover, such incomplete viral parts can be recovered from extracts of the mutant cells by differential centrifugation. The identity of the parts accumulating in various mutations of T4 is shown in the boxes (Figure 36-8).

Wood and Edgar prepared separate extracts of *E. coli* cultures infected with 40 different temperature-sensitive mutants of T4, each of which led to incomplete virus assembly. In no case were the extracts infectious for *E. coli* when tested singly. However, on mixing pairs of these extracts they found that some pairs produced active viral particles and others did not. For example, incomplete viral particles in extracts of mutants of genes 34, 35, 37, or 38, which lack tail fibers, were found to combine with the tail fibers in extracts of mutants of gene 23, which cannot form heads. The combination of the heads and the tails took place on simple mixing of the extracts, followed by incubation at 30°C. Intact, infectious T4 particles are formed, the number increasing with the time of incubation. Electron microscopy of the incubated mixture confirmed the net formation of new, intact viral particles (Figure 36-9). Such a recombination process also occurs if the heads and tails are first isolated from the extracts and then incubated together. The combination of the heads and tails to form complete viral particles requires no enzymes and is thus a self-assembly process. Such experiments, in which the gene products of one mutant can complement gene products of another, are called *complementation tests.*

From many such complementation tests, the various T4 mutants could be classified into 13 groups. A mutant from any given group is unable to complement with another of the same group, but it can complement with mutants of any of the other groups. On further analysis of the mutants in these groups, the sequence of the assembly reactions and the assignment of various genes to various critical steps could be established. Figure 36-10 shows the results. The first conclusion is that intact infectious T4 particles are formed by combination of intact heads, intact tails, and intact tail fibers, each previously formed in a separate assembly line. There is a specific sequence in the assembly process, since the viral particle cannot be built starting from the top of the head and finishing with the ends of the tail fibers.

For the formation of complete heads at least 16 genes participate, each presumably coding for synthesis of a specific type of structural protein. The assembly of the tail begins with the formation of the end plate, which requires participation of at least 15 genes. Four or more additional gene products are required to build the complete tail on the end plate. The heads and the tails then combine spontaneously; no enzyme-catalyzed reaction is required. Once they have combined, another gene product, that of gene 9, is required before the tail fibers can attach. The tail fibers require a total of five gene products to be assembled. However, the entire tail fiber must be built before it can be attached to the end plate. The latter step is of particular interest, since it takes place in what appears to be an enzyme-catalyzed reaction by

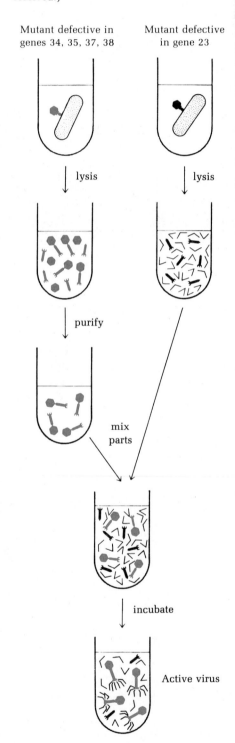

Figure 36-9
In vitro reconstitution of bacteriophage T4 by mixing mutant viral particles lacking tail fibers with an extract from cells infected with a virus defective in head structure. (From W. B. Wood and R. S. Edgar, "Building a Bacterial Virus." Copyright © July, 1967 by Scientific American, Inc. All rights reserved.)

Figure 36-10

Three separate assembly processes lead to heads, tails, and tail fibers, which are then combined in the sequence shown to form the complete particle. [Redrawn from W. B. Wood, R. S. Edgar, J. King, I. Lielausis and M. Henninger, Fed. Proc., 27: 1160 (1968).]

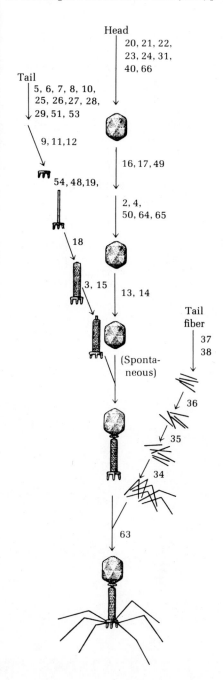

which two structural proteins are fused. It is the only step in the entire process of assembly of the T4 particle that requires a catalyst; all the other steps appear to be spontaneous, nonenzymatic processes.

Another significant finding is that in each of the three assembly lines there is a requirement for a given sequence of assembly. For example, assembly of the tail cannot begin from the middle. The obligatory sequence of assembly suggests that as each subunit is added and bound to the preceding one, it undergoes a conformational change which makes it more receptive to binding the next subunit, and so on. Each step then can promote the occurrence of the next step in a series of consecutive, cooperative interactions, so that the entire structure can be made only in one specific sequence. This conclusion is completely analogous to the observations on assembly of ATCase and tobacco mosaic virus.

Many essential details of T4 assembly are being filled in by other groups, particularly that of J. King, who has unraveled the identity of the genes specifying the tail proteins, the molecular weights of these proteins, and their sequence of assembly (Figure 36-11). Other investigators have identified the function of some of these proteins. For example, one of the end-plate proteins is the enzyme dihydrofolate reductase.

Encouraged by these successes in unraveling the self-assembly and morphogenesis of tobacco mosaic and T4 virions, many investigators have tackled other viruses, such as the RNA virus Qβ and the DNA virus T7. Some additional important principles of morphogenesis are emerging from these studies. One such principle has come from a study of the assembly of the bacteriophage P22 of Salmonella typhimurium. King and his colleagues have discovered that the construction of the viral head takes place in such a way that the 420 identical coat-protein subunits are assembled in the correct three-dimensional configuration to form the head of the virus, with the aid of a second protein, called the "scaffolding protein." This protein does not appear in the finished viral particle, but it aids in the proper assembly of the head subunits by forming a shell-like scaffold, which then detaches. The scaffolding protein, which occurs in about 250 copies per infected bacterial cell, is used over and over again in the assembly of P22 virions.

In these interesting sets of experiments we can see how three-dimensional structures of a high order of complexity can arise from the one-dimensional information programmed into the amino acid sequence of protein subunits. This sequence codes for the conformation of each subunit, how each subunit fits the next and how the assembly takes place in a specific sequence of steps.

Self-Assembly of Ribosomes

Ribosomes have extremely complex structures. The 70S ribosomes of E. coli have a particle weight of about 2.7 million and contain about 65 percent RNA and 35 percent proteins. The small 30S subunits contain one molecule of 16S RNA

Figure 36-11
Structure and assembly of the tail of bacteriophage T4. Genes listed above the arrows in
the assembly flow chart specify proteins that have been identified; those below the
arrows specify proteins that have not yet been identified. The columns under each struc-
ture list the approximate molecular weights of the polypeptides it contains; each is num-
bered according to the gene of which it is a product, the small arrows designating new
additions. [From J. King and N. Mykolajewycz, J. Mol. Biol., 75: 351 (1973).]

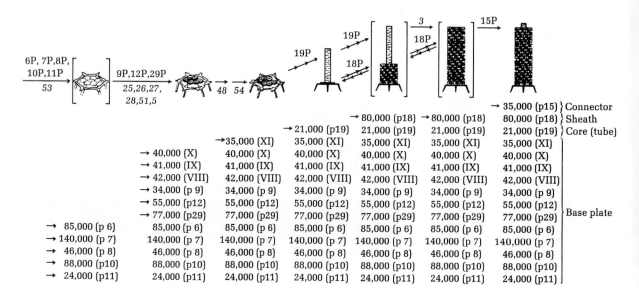

and 21 different subunits, which occur in no more than one copy per ribosome. The 50S subunit contains 5S RNA and 23S RNA, together with some 34 different kinds of proteins. Moreover, the biological function of ribosomes is exceedingly complex, involving the recognition and binding of mRNA, the binding of specific aminoacyl-tRNAs, the formation of peptide bonds, and the physical translocation of the ribosome along the mRNA through a GTP-dependent mechanochemical process. It would therefore appear to be a heroic task to dissociate all the components of ribosomes, recover them in pure form, and study their reassembly into a functional ribosome capable of synthesizing a polypeptide chain. Yet very interesting progress has been made in a number of laboratories. M. Nomura and his colleagues at the University of Wisconsin initiated this project and delineated many of the major features of the self-assembly of ribosomes. An important result of these studies is the recognition of the obligatory sequence in the self-assembly process and its cooperative nature.

These investigations began with the discovery that exposure of the 30S subunits of ribosomes to high salt concentrations, followed by cesium chloride density-gradient centrifugation (page 872), causes the loss of seven distinct ribosomal proteins that become soluble, leaving a 23S "core," consisting of RNA and protein. Such ribosome cores are not active in protein synthesis. Five of the detached proteins were separated and purified to homogeneity; the other two were obtained as a mixture. The five pure pro-

Table 36-1   Proteins of the 30S subunits of *E. coli* ribosomes as determined by SDS-gel electrophoresis

Not all ribosomes in a population have one copy of each of the 21 proteins

Protein	Molecular weight
S1	65,000
S2	28,300
S3	28,200
S4	26,700
S5	19,600
S6	15,600
S7	22,700
S8	15,500
S9	16,200
S10	12,400
S11	15,500
S12	17,200
S13	14,900
S14	14,000
S15	12,500
S16	11,700
S17	10,900
S18	12,200
S19	13,100
S20	12,000
S21	12,200

Figure 36-12
Gel electrophoresis of the proteins of native (left) and reconstituted (right) 30S subunits of bacterial ribosomes.

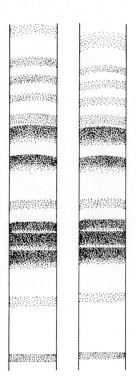

teins were found to have different amino acid compositions. In order to determine whether functionally active ribosomes could be reconstituted, the 23S core was mixed with all the solubilized proteins. Intact 50S ribosomal subunits were then added and the resulting mixture tested for its ability to synthesize a polypeptide when supplemented with a synthetic messenger such as poly U, labeled amino acids, ATP, GTP, tRNAs, and amino acid–activating enzymes. Such a reconstituted ribosomal system was found to be active in polypeptide synthesis. However, if any single one of the five purified proteins was omitted, activity of the product in protein synthesis was substantially less or totally lacking. These observations indicated that partially dissociated ribosomes could be reconstituted and pointed to the feasibility of attempting a total reconstitution of 30S subunits.

Nomura, and also C. G. Kurland and his colleagues, have succeeded in separating all the 21 different proteins found in the 30S subunits of *E. coli* ribosomes and obtaining them in pure form (Table 36-1). Their molecular weights, amino acid composition, and possible function are now known. All are immunologically distinct. Moreover, the number of copies of each protein in the 30S subunit has been ascertained. None occurs in more than one copy per 30S particle, and a few do not appear to be present in all the ribosomes of a given population. The 16S RNA of the 30S subunits has also been obtained in pure form by extraction from detergent-treated ribosomes with phenol.

Nomura and his colleagues mixed 16S RNA with the 21 different purified ribosomal proteins, incubated them in the cold, and then, after combining the mixture with intact 50S subunits, assayed the system for its ability to carry out polypeptide synthesis. No activity was observed. However, when the mixture of 16S RNA and the 21 ribosomal proteins was incubated at 40°C for 10 min in high KCl concentrations, they readily reassembled to give 30S subunits. The proteins could be added in any order so long as the reaction was carried out at 40°C. The reconstituted 30S subunits were found to contain the same proteins as native 30S subunits (Figure 36-12), indicating the accuracy with which the proper ribosomal proteins assembled. The reconstituted ribosomes were found to be capable of synthesizing labeled polyphenylalanine at about half the rate of normal ribosomes.

Nomura and his colleagues also found that the 30S ribosomal proteins isolated from *E. coli* may be mixed with the 16S rRNA isolated from the bacteria *Azotobacter vinelandii*, *Micrococcus luteus*, and *Bacillus stearothermophilus* to reconstitute hybrid 30S ribosomes, which were also found to possess activity in protein synthesis. Conversely, when the 16S RNA of *E. coli* was mixed with the 30S ribosomal proteins of two other bacterial species, it also yielded biologically active hybrids. On the other hand, there was no exchange of parts when intact 30S ribosomes of two different species were mixed.

Nomura and his colleagues then set about the task of determining the order of assembly of the 21 proteins and the

16S RNA. They first ascertained which of the ribosomal proteins binds specifically to the RNA. Three of the proteins (S4, S8, and S15) bound strongly, four others (S7, S13, S17, S20) bound weakly, but none of the rest were bound (Figure 36-13). They then tested the remaining proteins for their capacity for binding to the complex of 16S RNA with the three tightly bound proteins S4, S8, and S15. Three other proteins were found to bind more strongly than the rest. Then further tests were made to examine the sequence of binding of the rest of the proteins to the complex of 16S RNA with the six bound proteins. From a long series of such experiments, S. Mizushima and Nomura concluded that there is a specific sequence of self-assembly in the 30S ribosomal subunit and that the addition of the components is cooperative, i.e., that the addition of one component enhances the addition of the next. Figure 36-13 is an assembly map of the 30S subunits; it shows the sequence of addition of nearly all the 21 proteins. The actual three-dimensional structures of the 30S and 50S subunits are being mapped by immunological and electron-microscopic approaches; tests of the specific function of the different subunits have also been made (Table 36-2).

Nomura and others have gone ahead to dissociate the 50S subunit into its component parts. It contains 34 different proteins and two RNAs, 5S and 23S, all of which have been purified. All but two of the proteins of the 50S subunit are distinct from each other and distinct also from the proteins of the 30S subunit. The complete self-assembly of the 50S subunits of *B. stearothermophilus* ribosomes has been achieved in vitro. It is a slower process than 30S subunit assembly and requires elevated temperatures. After the reconstituted 50S subunits were combined with native 30S subunits, full activity in protein synthesis was observed. Again, a specific sequence and cooperativity of assembly appear to be involved.

The molecular details of the self-assembly of ribosomes are now being examined. Since the base sequences of both 5S and 16S rRNAs are known, as well as parts of the sequence of 23S rRNA, the specific binding sites for the different ribosomal proteins on the RNAs can be identified. Moreover, detailed study of the different ribosomal proteins

Figure 36-13
Assembly map for the 30S subunits of bacterial 70S ribosomes, showing the major pathways of assembly of the 21 proteins (numbered S1 to S21; S stands for small ribosomal subunit). The 16S RNA provides the framework. The thickness of the arrows is in proportion to the probability of each step; the arrowheads show the direction of binding. Thus the proteins S4 and S8 bind to 16S RNA most strongly, followed by S15. S7 and S20 are more weakly bound. S7 also binds to S4 and S8. Proteins S3 and S21 are the last to bind. [Adapted from M. Nomura, Science, 179: 869 (1973).]

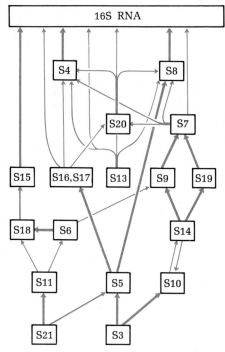

The sequences of binding of S1, S2, and S12 are not known

Table 36-2 Function of some proteins of the 30S ribosomal subunit (bacterial)

Function	Protein
mRNA binding	S1
fMet-tRNA binding	S2, S3, S10, S14, S19, S21
Codon recognition	S3, S4, S5, S11, S12
Function of A and P sites	S1, S2, S3, S10, S14, S19, S20, S21
Binding of aminoacyl tRNAs	S9, S11, S18
Near GTPase center	S2, S5, S9, S11

has revealed the function of some of them (Table 36-2). One of the proteins of the 50S subunit has been found to be responsible for the translocation step. This protein is involved in the hydrolysis of GTP and the accompanying translocation. It contains the rare amino acid ϵ-N-methyllysine, which is characteristic of muscle actin (page 756) and flagellin (page 775), an observation suggesting that these proteins, which have a mechanochemical function, may also have some common features of molecular structure.

Other Self-Assembly Processes: Membrane Systems

In addition to the examples described above many other self-assembly processes have been studied. Among them are the assembly of microtubules from tubulin subunits (page 771), bacterial flagella from their monomeric flagellin subunits (page 775), the partial assembly of thin and thick filaments of skeletal muscle (page 761), the self-association of glutamate dehydrogenase into filamentous and tubular structures, and the assembly of the fatty acid synthetase (page 660) and pyruvate dehydrogenase (page 450) complexes.

Some membrane systems may also spontaneously assemble themselves from their components. We have seen that the polar phosphoglycerides, in the absence of proteins, spontaneously form bilayers (page 300). The driving force for this assembly process is again the tendency of the surrounding aqueous phase to seek its state of maximum entropy, which forces the lipid molecules to arrange themselves in such a way as to minimize the exposure of their hydrocarbon chains to water. Because such self-assembled bilayers have about the same thickness and possess many of the properties of natural membranes, e.g., high electrical resistance and capacitance, high water permeability, and low permeability to electrolytes, they have been postulated to serve as the "core" of natural membranes (page 304). Various proteins have been found to attach themselves to such artificial lipid bilayer systems, which are often used to bring about the in vitro reconstitution of membrane-bound enzymes and transport systems.

Membrane-bound enzymes or enzyme complexes can in some cases be detached, purified, and reattached to the depleted membrane. For example, the F_1, or ATP-synthesizing, enzyme of oxidative phosphorylation (page 514) has been detached from the mitochondrial membrane and purified by E. Racker and his colleagues. The F_1 ATPase can be added back to the depleted mitochondrial membrane vesicles in a specific manner so that oxidative phosphorylation is restored (page 522). More recently, the Ca^{2+}-transporting ATPase of the sarcoplasmic reticulum (page 764) has been reconstituted from highly purified protein components and lipids, to yield vesicles capable of transporting Ca^{2+} across the membrane at the expense of the hydrolysis of ATP. Similarly, the Na^+K^+–ATPase (page 790) has been reconstituted from purified components.

Morphogenesis of Higher-Order Structures

Can self-assembly processes also account for the morphogenesis of more complex structures, such as mitochondria and chloroplasts? Indeed, can self-assembly operate in the morphogenesis of entire eukaryotic cells? In principle, there is no *a priori* reason why most of the ordered structure of cells cannot be formed by self-assembly processes, since much of it appears to be composed of noncovalently bonded subunits of proteins, lipids, and nucleic acids and at least some of its component parts show self-assembling characteristics.

The molecular aspects of the morphogenesis of mitochondria are just now coming under close scrutiny. Because yeast cells are facultative and can readily live by fermentation alone if their mitochondria are defective, the biosynthesis and assembly of certain mitochondrial components can be studied in mutations of yeast cells in which one or another specific mitochondrial protein is defective. Mitochondrial mutants have been isolated in which various specific proteins, such as cytochrome c, F_1 ATPase, or cytochrome oxidase are defective. For example, simple mixing of cytochrome c with mitochondria isolated from a mutant defective in this protein results in its binding to the genetically empty sites in the mitochondria and in the restoration of electron transport and oxidative phosphorylation. Important experiments on the morphogenesis of chloroplasts are also under way.

There is evidence that some large supramolecular structures may not be completely self-assembling and perpetuate themselves by a process of accretion of specific building blocks to a preformed structure serving as a template. For example, the gullet of certain *Paramecia* is built from small molecular parts furnished by the cytoplasm. However, assembly of the gullet does not occur *de novo*; some preformed gullet structure must be present. If a portion of an extra gullet is implanted into a *Paramecium*, which normally has only one gullet, a second complete gullet will grow. All progeny of such a *Paramecium* will then be found to have two gullets regardless of their genotype. *Paramecium* therefore has the genetic competence to make the building blocks for gullet morphogenesis, but ordinarily it makes only one gullet, evidently starting from some preformed growing point, which serves as a template for accretion of further subunits.

The molecular and genetic basis of the morphogenesis of eukaryotic cells, and of the formation of specific tissues and organs (for example, muscle, kidney, and the brain, or leaves and flowers of plants) offer many absorbing problems for the biochemistry and genetics of the future.

Summary

Most large macromolecular and supramolecular biostructures—regulatory enzymes, enzyme complexes, lipoproteins, ribosomes, viruses, and membranes—contain several or many polypeptide subunits, noncovalently associated through hydrophobic interactions and hydrogen bonds. This manner of construction has a number of biological advantages. It minimizes the effect of errors in protein synthesis, permits defective subunits to be rejected, is economical in use of DNA for coding, and allows very fast interactions between the subunits to take place. Supramolecular systems are stabilized largely by hydrophobic interactions between the subunits, the result of the tendency of the surrounding water to seek a maximum-entropy state. Specificity of interaction of the subunits is achieved by hydrogen bonds and ionic bonds. The information required for the self-assembly of oligomeric proteins and protein complexes is inherently present in the amino acid sequence of the polypeptide subunits. Self-assembling systems usually exhibit a specific sequence of steps in the assembly process. They also show cooperativity, so that addition of one subunit enhances the addition of the next.

Aspartate transcarbamoylase shows a specific sequence of assembly. First the catalytic subunits are formed, each from three identical polypeptide chains. In an independent process the regulatory subunits are formed, each from two identical chains. The final active product, which contains two catalytic and three regulatory subunits, may then assemble by different pathways, depending on the concentration of the two types of subunits. Tobacco mosaic virus, a long, cylindrical molecule which contains one long RNA chain and 2,200 identical protein subunits, is assembled from flat circular disks, each containing 34 protein subunits. The initiation of the assembly process begins with the combination of the RNA and the first disk. Successive disks are then added, each being converted into a lock-washer configuration as the RNA molecule winds around the inner core, until the complete viral particle is built.

Bacteriophage T4 of *E. coli* assembles itself from a very large number of different proteins and DNA. The head, the tail, and the tail fibers are made separately, each in a specific sequence, and then assembled to make a complete virion. All the many steps are noncovalent, with the exception of a single enzyme-catalyzed reaction.

The 30S and 50S subunits of bacterial ribosomes also assemble themselves under appropriate conditions. The 30S subunit is formed by mixing and warming the 16S RNA and the 21 different protein subunits. When combined with native 50S subunits, the resulting 70S ribosomes are active in protein synthesis. There is a specific sequence of assembly of the protein subunits. The 50S subunits have been reconstituted by self-assembly of the 5S RNA, 23S RNA, and 34 different protein subunits.

Microtubules, flagella, actomyosin, and various multienzyme complexes have been found to exhibit self-assembly, as have some membrane-bound transport systems.

References

Books

FRAENKEL-CONRAT, H.: *Design and Function at the Threshold of Life: The Viruses*, Academic, New York, 1962. Early experiments on the self-assembly of tobacco mosaic virus are described.

NOMURA, M., A. TISSIÈRES, and P. LENGYEL (eds.): *Ribosomes*, Cold Spring Harbor Laboratory of Quantitative Biology, Cold Spring Harbor, N.Y., 1974. An up-to-date compendium of reviews on all aspects of ribosome structure and function.

THOMPSON, D'ARCY: *On Growth and Form*, abridged edition, J. T. Bonner (ed.), Cambridge University Press, London, 1961. A classic analysis of biological form at the macroscopic level.

WOLSTENHOLME, G. E. W. (ed.): *Principles of Biomolecular Organiza-tion*, Ciba Foundation Symposium, Little, Brown, Boston, 1966. An interesting discussion of the molecular design of biostructures.

Articles

BOTHWELL, M., and H. K. SCHACHMAN: "Pathways of Assembly of Aspartate Transcarbamoylase from Catalytic and Regulatory Subunits," *Proc. Natl. Acad. Sci. (U.S.)*, 71: 3221–3225 (1974).

KING, J., and S. CASJENS: "Catalytic Head-Assembling Protein in Virus Morphogenesis," *Nature*, 251: 112–119 (1974).

KING, J., and U. K. LAEMMLI: "Bacteriophage T4 Tail Assembly: Structural Proteins and Their Genetic Identification," *J. Mol. Biol.*, 75: 315–337 (1973).

KING, J., and N. MYKOLAJEWYCZ: "Bacteriophage T4 Tail Assembly: Proteins of the Sheath, Core, and Baseplate," *J. Mol. Biol.*, 75: 339–358 (1973).

KURLAND, C. G.: "Ribosome Structure and Function Emergent," *Science*, 169: 1171–1177 (1970).

MIZUSHIMA, S., and M. NOMURA: "Assembly Mapping of 30S Ribosomal Proteins of E. coli," *Nature*, 226: 1214–1218 (1970).

NOMURA, M.: "Assembly of Bacterial Ribosomes," *Science*, 179: 864–873 (1973).

PONGS, O., K. H. NIERHAUS, V. A. ERDMANN, and H. G. WITTMANN: "Active Sites in E. coli Ribosomes," *FEBS Lett.*, 40: 528–535 (1974).

SPIRIN, A. S.: "Structural Transformations of Ribosomes (Dissociation, Unfolding, and Assembly)," *FEBS Lett.*, 40: S38–S47 (1974).

STÖFFLER, G., L. DAYA, K. H. RAK, and R. A. GARRETT: "The Number of Specific Binding Sites on 16S and 23S RNA of E. coli," *J. Mol. Biol.*, 62: 411–414 (1971).

STÖFFLER, G., and H. G. WITTMANN: "Immunological Studies of E. coli Ribosomal Proteins," *J. Mol. Biol.*, 62: 407–409 (1971).

STUDIER, F. W.: "Bacteriophage T7," *Science*, 176: 367–376 (1972). An analysis of the gene products and assembly of a small DNA virus.

WEISS, P.: "1 + 1 ≠ 2 (One Plus One Does Not Equal Two)," p. 801 in *The Neurosciences: A Study Program*, Rockefeller University Press, New York, 1967. An absorbing essay on biological form and its molecular basis.

WOOD, W. B., and R. S. EDGAR: "Building a Bacterial Virus," *Sci. Am.*, 217: 61–74 (1967). An account of the early experiments on self-assembly of bacteriophage T4.

WOOD, W. B., R. S. EDGAR, J. KING, I. LIELAUSIS and M. HENNINGER: "Bacteriophage Assembly," *Fed. Proc.*, 27: 1160–1166 (1968).

CHAPTER **37** **THE ORIGIN OF LIFE**

In our survey of biochemistry we have had many glimpses into the group of organizing principles which we have called the molecular logic of living matter. From the viewpoint of the physical scientist cells may be regarded as complex collections of organic molecules which are self-organizing and self-replicating, which are capable of exchanging energy and matter with their environment by means of consecutive enzyme-catalyzed organic reactions, and which appear to function on the basis of maximum economy of parts and processes. We have identified many of the molecular components of cells; we have seen how cells obtain and use energy and how they replicate themselves. We must now return to such fundamental questions as: How did biomolecules arise? How did organic biomolecules "learn" to interact with each other and organize themselves? How did the first cells—or the first "living" structure—arise from organic molecules? How did the first cells evolve into the remarkable panoply of living forms we know today?

Not too long ago inquiry into the origin of life was considered to be a matter of pure armchair speculation, with little hope of yielding conclusive information. But many scientific advances made in the last two decades support the view that valid answers to some of these fundamental questions can be deduced and that at least some of the steps in the origin of biomolecules and of living cells can be simulated in the laboratory. In this final chapter we shall survey some of the experimentation and the directions of thought in this increasingly active field. Even though this topic must always remain speculative, its study may one day reveal new insights into molecular biology and a clearer picture than we now have of the fundamental axioms in the molecular logic of living matter.

The Time Scale of Chemical and Biological Evolution

For orientation we shall first outline the geological and biological history of the earth (Figure 37-1). The earth is believed to have been formed about 4,800 million (4.8×10^9)

Figure 37-1
The time scale of chemical and biological
evolution. Although the evolution of higher
forms of life can be fairly accurately dated,
the times of origin of lower species indicated
are only approximations.

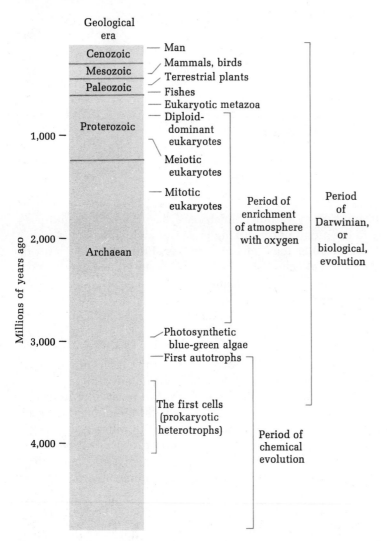

years ago, by condensation from a cloud of gases and dust. Because much of the dust consisted of radioactive elements, their decay in the interior of the earth generated much heat and volcanic action, so that the temperature of the surface of the earth was probably higher than it is now. There is still considerable controversy and uncertainty regarding the composition of the primitive atmosphere, which derived in part from the interior of the earth. The earliest atmosphere probably contained much nitrogen and was quite reducing—rich in hydrogen, methane, ammonia, and water—but hydrogen was soon lost and with time more oxidized components, such as carbon monoxide, increased in concentration. Most likely, the atmosphere at the time of the origin of life was still significantly reducing and no oxygen was present. Oxy-

Figure 37-2
Algal-like fossil from early Precambrian
South Africa. [From J. W. Schopf and E.
Barghoorn, Science 156: 508–512 (1967).]

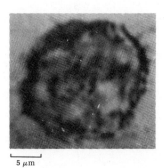

5 μm

gen did not appear until much later, largely as the product of photosynthesis.

Much evidence now suggests that early in the history of the earth organic compounds first arose by reactions between various inorganic components of the atmosphere and geosphere, activated by the energy of ultraviolet light, electric discharges, heat, or other forms, as outlined in Chapter 1 (page 23). These organic compounds gradually accumulated in the primitive sea. This early period, which has been called the period of _chemical evolution,_ may have lasted some 1,500 million years, or nearly one-third of the earth's history. In this period the primordial biomolecules, perhaps some 30 or 40 in number (page 20), were first generated, along with many other organic compounds. Later in this period these primordial building blocks are believed to have undergone abiotic condensation to form primitive polypeptides, polynucleotides, polysaccharides, and lipids. From this rich broth of organic molecules and polymers, the primordial organic "soup," the first living organisms are believed to have arisen. This is the sequence of events first postulated in the 1920s by A. I. Oparin in the Soviet Union and J. B. S. Haldane in Britain.

The oldest known organic material has been found in the oddly named Fig Tree shale deposits in South Africa, which contain many hydrocarbons, some of isoprenoid nature, as well as porphyrins, purines, and pyrimidines. Isotope dating methods indicate that these shales were deposited at least 3,100 million years ago. Recent investigations on the structure of the Fig Tree isoprenoid compounds show that some of them are related to those found in modern cells. If they had a biological origin, life may have originated much earlier than 3,100 million years ago. This view has in fact been given independent support by the discovery of microfossils by electron microscopy of the Fig Tree sediments (Figure 37-2). These are cell-like structures, about 0.6 μm in diameter, which strongly resemble present-day bacteria (page 30).

It is generally believed that the first living cells were anaerobic heterotrophs (page 363), which used organic compounds from the primitive sea as their building blocks and fuel source. With the growth and proliferation of these early cells, the sea gradually became depleted of organic compounds. Only those cells could survive which were able to use simpler carbon compounds, particularly carbon dioxide, as their carbon source and sunlight as energy source. Thus the first photosynthetic cells arose, perhaps 3,000 million years ago. At first they were probably not capable of evolving oxygen and may have utilized hydrogen sulfide instead of water as hydrogen donor, as purple sulfur bacteria do today. Later the first oxygen-producing photosynthetic cells arose, ancestors of modern blue-green algae. Until their appearance there was probably little or no free oxygen in the atmosphere. The first oxygen-utilizing aerobic heterotrophs arose much later, since the buildup of the atmospheric oxygen level as a result of photosynthesis was very slow. The concentration of oxygen in the atmosphere reached a level of 1 percent only about 600 to 1,000 million years ago and a level of 10 percent only about 400 million years ago. Thus, over 1,000 million years after oxygen first appeared in the atmosphere were required for the development of aerobic vertebrates and of vascular higher plants. The origin of *Homo sapiens* is a matter of only the last 2 million years, which in the history of the earth is comparable to the last 30 seconds of a 24-hour day.

Some Working Assumptions

From mathematical and philosophical considerations one cannot assume that life arose by a unique physical encounter among a large number of organic building-block molecules to yield, in one miraculous coalescence, a full-blown cell like those we know today. It is far more likely that the first "cells" or "living" structures were much simpler than present bacteria and arose as the result of a long sequence of single events, so that each stage in their evolution developed from the preceding one by only a very small change. Moreover, it is a corollary assumption that each single step in the evolution leading to the first cell line to survive must have had a reasonably high probability of happening, at least over a long span of time.

This assumption has a significant consequence. The sequential steps in prebiotic evolution may be sufficiently probable, in terms of the laws of chemistry and physics, to be observable in the laboratory within a reasonable period of time. If so, at least some of the early events leading to the origin of life can be studied by experimental methods.

Other working assumptions may be made. In the process of chemical evolution toward life, three stages may be postulated. In the first, certain types of organic molecules may have begun to accumulate in the primordial sea, because they were easily formed and had superior stability. The second stage in chemical evolution is that point where a specific set of organic molecules was better able to survive in the form of a cluster or complex than singly. Whatever the nature of the first cluster of organic molecules that "learned" to survive together, it is very likely that it formed because it was more stable thermodynamically than its precursor molecules. Put in another way, it appears that each successive step in chemical evolution took place because the system and its aqueous surroundings gained in entropy. Thus chemical evolution had a driving force, which in the most fundamental terms must have been the drive to maximize the entropy of the universe. In the search for the position of maximum entropy, groups of molecules may have tried out different "passes" over the activation-energy barrier (page 188) until the lowest and most probable pass was found. In this way chemical evolution may have taken a specific pathway, one that can be traversed quite reproducibly, given the same set of starting compounds.

In the third stage of chemical evolution some of the more complex clusters of organic molecules may have acquired a "function," such as a catalytic activity or the capacity for resisting changes in their environment or modifying it, and then "learned" to replicate themselves, through formation of some primitive molecular template. It is this stage in chemical evolution that is most difficult to understand, since there are few if any modern nonbiological prototypes of self-organizing, self-replicating molecular systems. Only recently have attempts been made to formulate a theory of the self-organization of matter, which invokes very special thermodynamic and kinetic conditions.

Finally, it can also be assumed that cells or living structures may have arisen from nonliving matter several or even many times, at different places or times, and possibly from different kinds of organic building blocks. However, it appears certain that only one line of cells survived, from which all organisms now on the earth descended. From studies on the evolution of species, it has been estimated that possibly 99 percent of all species of organisms that ever existed are now extinct and have no living descendants. By analogy it is probable that strong selection pressure was also exerted in prebiotic evolution, with the result that there may have been false starts in the origin of the first stable line of living cells.

Conditions Leading to the Abiotic Origin of Organic Compounds

We have seen (Chapter 1) how laboratory experiments under simulated primitive-earth conditions yielded amino acids, purines, pyrimidines, sugars, and many other organic compounds from inorganic precursors likely to have been present in the primitive environment, under activation by various forms of energy. The formation of amino acids and other nitrogenous organic compounds occurs most easily in a reducing gas mixture; amino acids are generally not produced from synthetic atmospheres containing oxygen. Such organic compounds would therefore not be produced in our contemporary atmosphere.

Central to most of the reaction pathways leading to abiotic formation of simple organic compounds containing nitrogen is hydrogen cyanide, which is readily formed by reactions such as

$$2CH_4 + N_2 \longrightarrow 2HCN + 3H_2$$
$$CO + NH_3 \longrightarrow HCN + H_2O$$

These reactions are promoted by heat (from volcanoes), ultraviolet light (from solar radiation), or electric discharges (lightning). Under simulated primitive-earth conditions hydrogen cyanide is in turn a precursor of other highly reactive compounds, such as *cyanoacetylene, cyanamide*, and *nitriles* (Figure 37-3). These in turn are precursors of a variety of amino acids, purines, and pyrimidines. Indeed, heating a simple mixture of hydrogen cyanide and ammonia yields a number of common amino acids, as well as a surprisingly large amount of the purine adenine (Figure 37-4). In fact, this reaction has been developed in Japan into a commercial method for the synthesis of adenine. Formaldehyde, also

Figure 37-3
Chemical evolution of biomolecules from HCN. From three known immediate products of HCN (cyanamide, nitriles, and cyanoacetylene) many different organic derivatives are formed under simulated primitive-earth conditions.

Figure 37-4
A possible pathway from HCN and NH_3 to adenine.

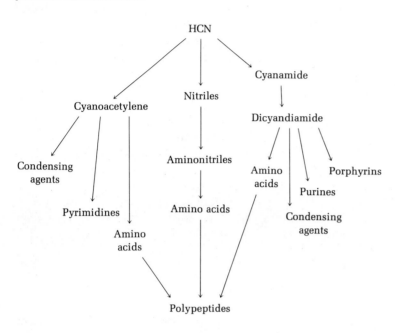

readily formed in simulated primitive-earth experiments, yields a variety of sugars when heated with limestone (calcium carbonate). As shown below, cyanoacetylene is an important condensing agent.

Several energy sources available on the primitive earth (Table 37-1) have been found to activate components of the primitive atmosphere sufficiently to generate organic molecules via the chemical routes sketched above. Ultraviolet light and lightning discharges are most likely to have served as major energy sources. Heat is also extremely effective in promoting abiotic formation of simple organic compounds; sufficiently high temperatures may have existed near the sites of volcanic action on the primitive earth. Shock waves, like those produced by implosion (thunder) after a lightning flash or by the impact of meteorites, are particularly efficient in activating chemical reactions.

In most primitive-earth simulation experiments the reacting systems are usually heated to temperatures considerably higher than may have existed on the primitive earth. The rationale is, of course, to accelerate the rate of reaction sufficiently to produce observable products in a period of days or weeks. Most chemical reactions in aqueous systems have about the same temperature coefficient, and it thus appears unlikely that reactions are evoked by heat that would not have occurred, however slowly, at a lower temperature.

Widely different estimates have been made of the amount of organic matter that may have been formed on the primitive earth as a result of the types of reactions described

Table 37-1 Sources of energy averaged over the surface of the earth

Source	Energy, cal cm^{-2} $year^{-1}$
Total solar radiation	260,000
Ultraviolet light	4,000
Electric discharges	4
Shock waves	1.1
Radioactivity	0.8
Volcanic action	0.13
Cosmic radiation	0.0015

above, given a period of 1,000 million years of a reducing or near-reducing atmosphere. Some calculations suggest that the maximum possible concentration of organic compounds in the primitive oceans was in the neighborhood of 10^{-12} M, but more recent calculations suggest that the primeval sea may once have contained as much as 10 percent of dissolved organic compounds, assuming that the energy of shock waves was utilized for activating chemical reactions. It is possible, of course, that a significant concentration of organic compounds arose only in small ponds, rather than in the open sea.

Extraterrestrial Formation of Organic Compounds

The proposition that organic compounds inevitably arise when the appropriate simple gases and sources of energy are available is greatly reinforced by the discovery of organic compounds in extraterrestrial bodies. One line of evidence is that carbonaceous meteorites reaching the earth contain traces of amino acids and other organic compounds, although there has been much controversy over whether they are the result of terrestrial contamination. The best case has been made from careful analysis of the Murchison meteorite, which fell near the town of Murchison in Australia in 1969. Fragments of this meteorite were collected and processed in such a way that contamination was minimized. Two laboratories, working independently, found a number of amino acids in the meteorite, including some not present in terrestrial organisms and therefore unlikely to have arisen by contamination. Moreover, the amino acids in this meteorite were optically inactive, consisting of D,L mixtures. This fact also indicated that the amino acids in the meteorite were not the product of contamination, since the amino acids found in earthly soil are optically active. Also found among the organic components of the Murchison meteorite are monocarboxylic and dicarboxylic acids, including malonic, succinic, and fumaric acids. Organic compounds have also been found in the Orgeuil meteorite, which fell in France about a century ago.

In the last few years radio-frequency spectroscopy has revealed that water, ammonia, and a large number of small organic molecules occur in interstellar dust clouds (Table 37-2). Among the latter are hydrogen cyanide, formaldehyde, and cyanoacetylene, which we have seen are important precursors of more complex organic compounds.

For many years there has been much speculation about the possible occurrence of life on other planets; indeed, the term *exobiology* has been coined to refer to extraterrestrial biology. It has become an important objective of the National Aeronautics and Space Administration to probe for signs of life on other planets and on the moon. Although the moon's surface appears to be completely anhydrous, traces of amino acids in lunar dust have been recorded. Among the planets of our solar system only Mars appears to afford the appropriate surface conditions to support a water-based life. Landings on Mars are scheduled for Viking spacecraft equipped with

Table 37-2 Some carbon compounds detected in interstellar space

Formaldehyde
Carbon monoxide
Hydrogen cyanide
Cyanoacetylene
Methanol
Formic acid
Formamide
Acetonitrile
Hydrogen isocyanide
Methylacetylene
Acetaldehyde
Thioformaldehyde
Methyleneimine

instrumentation to analyze the organic components of the surface and to determine whether living organisms exist or once existed on that planet.

That most of the common building-block biomolecules have been repeatedly found in simulated primitive-earth experiments under a great variety of conditions, in ancient rocks and shales, in meteorites, and in interstellar space clearly suggests that they may be ubiquitous and relatively stable products of chemical evolution, formed by energetically favored reaction pathways. Thus the basic biomolecules we know today may be the inevitable products of chemical evolution, not only on earth but wherever in the universe appropriate conditions exist.

We shall now survey the results of primitive-earth simulation experiments which have suggested possible ways in which the first primitive macromolecules, particularly proteins, enzymes and nucleic acids, were formed. Then we shall examine how they may have come together to yield the first "living" structures.

Prebiotic Condensing Agents

Once the primordial building-block molecules were formed by abiotic synthesis, the next step in their chemical evolution toward biomolecular systems must have been the formation of covalent bonds linking the building blocks to form more complex biomolecules, such as nucleotides, peptides, and lipids, and then polymers such as polysaccharides, polypeptides, and polynucleotides. But here we come to a paradox. The covalent linkages between the building blocks of such polymers are formed by condensation reactions, which entail removal of the elements of water from the monomeric units. However, peptide bonds, ester linkages, and glycosidic bonds are thermodynamically unstable in dilute aqueous systems; they tend to undergo hydrolysis with a large negative standard-free-energy change (page 397), so that at equilibrium only very small amounts of such linkages can exist. In order for primordial polypeptides, polysaccharides, and polynucleotides to have accumulated in any amount in the primordial sea or in localized aqueous systems, the rate of their formation must have exceeded the rate of their degradation.

There are only a few ways in which dehydrating condensation reactions between two building-block molecules, such as two amino acids, can take place with large yields. One is to carry out the reaction under anhydrous or nearly anhydrous conditions, particularly at temperatures above the boiling point of water. Another is to carry out the reaction with the aid of chemical condensing agents, compounds capable of extracting the elements of water from the molecules undergoing condensation. Both types of process may have been possible under primitive-earth conditions.

S. W. Fox, a leading investigator into the origins of life, has postulated that condensation reactions involved in the prebiotic synthesis of polymers were largely thermal in nature. As we shall see, he and his colleagues have shown that

Figure 37-5
Formation and action of condensing agents.

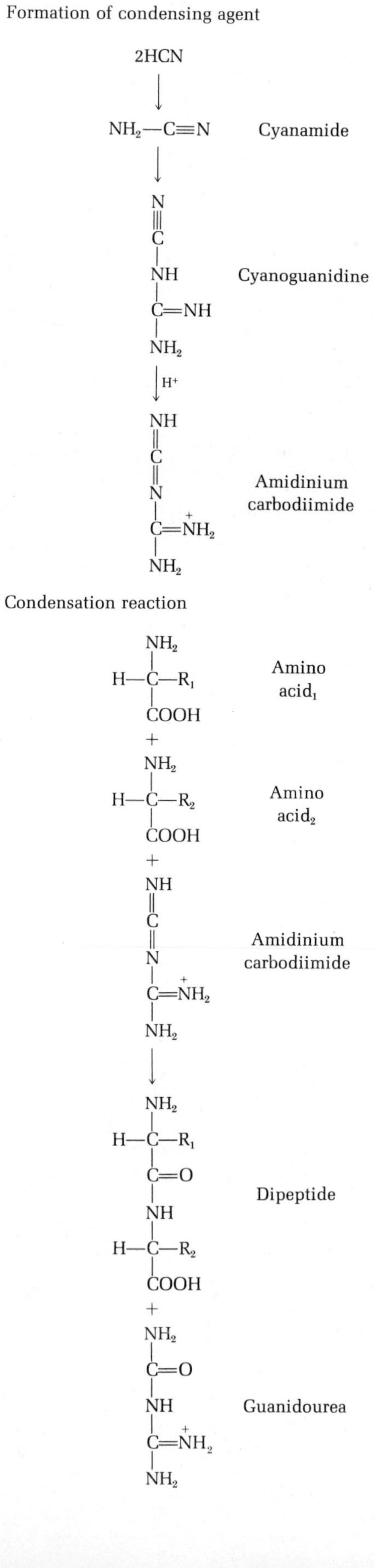

Figure 37-6
Some condensing agents formed in simulated primitive-earth experiments.

H—C≡N=O
Hydrogen cyanate

N≡C—C≡N
Cyanogen

HC≡C—C≡N
Cyanoacetylene

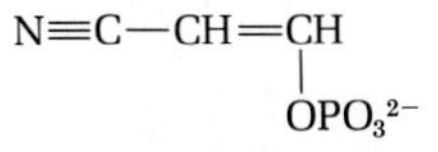

$$N≡C—CH=CH$$
$$|$$
$$OPO_3^{2-}$$
Cyanovinyl phosphate

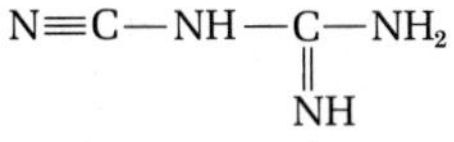

N≡C—NH—C—NH₂
‖
NH
Cyanoguanidine

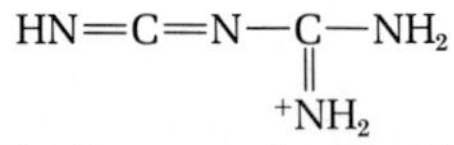

HN=C=N—C—NH₂
‖
⁺NH₂
Amidinium carbodiimide

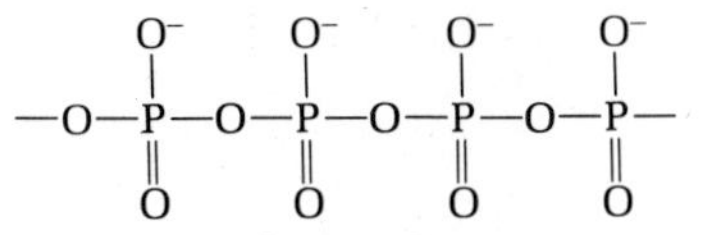

$$-O—\overset{O^-}{\underset{O}{\overset{|}{\underset{\|}{P}}}}—O—\overset{O^-}{\underset{O}{\overset{|}{\underset{\|}{P}}}}—O—\overset{O^-}{\underset{O}{\overset{|}{\underset{\|}{P}}}}—O—\overset{O^-}{\underset{O}{\overset{|}{\underset{\|}{P}}}}—$$
Polyphosphate

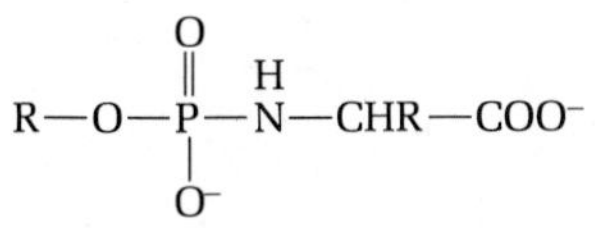

$$R—O—\overset{O}{\overset{\|}{\underset{O^-}{\underset{|}{P}}}}—\overset{H}{\underset{}{N}}—CHR—COO^-$$
An amino acid phosphoramidate

polypeptides may be formed simply by heating mixtures of amino acids at temperatures of about 130 to 180°C. Although temperatures this high may well have occurred near volcanoes, heating of organic compounds can be quite destructive, particularly for sugars. It therefore does not seem likely that thermal condensation was directly involved in the formation of such fragile biopolymers as polysaccharides or polynucleotides.

Because the living cells of today use a chemical condensing agent, the pyrophosphate group of ATP, to bring about the formation of such water-unstable linkages as glycosidic, ester, amide, and peptide bonds, it is attractive to think that chemical condensing agents participated in the prebiotic formation of the first nucleotides, polynucleotides, polypeptides, polysaccharides, and lipids. An example of a familiar chemical condensing agent is *dicyclohexyl carbodiimide* (page 119), often used to bring about the condensation of amino acids to yield peptides (page 118). M. Calvin and his colleagues have pointed out that hydrogen cyanide is a precursor of derivatives of carbodiimide. They have found that *cyanamide*, which is formed from hydrogen cyanide by application of radiant energy, dimerizes under primitive-earth conditions to yield *cyanoguanidine*, which is readily converted into a carbodiimide derivative. This condensing reagent promotes formation of simple peptides from amino acids, as shown in Figure 37-5. It also promotes formation of phosphate and acetate esters from alcohols, of pyrophosphate from orthophosphate, and of ADP from AMP and phosphate. *Cyanate* and *cyanogen* (Figure 37-6) are also effective condensing agents, as is *cyanovinyl phosphate*, formed by the action of phosphate on *cyanoacetylene*, one of the major products resulting from the action of electric discharges on mixtures of methane and nitrogen.

Another way of concentrating monomeric compounds and bringing about their condensation to form polymers is by adsorption on the surface of certain abundant minerals, particularly clay and the apatite minerals. One type of clay, montmorillonite, has been found by A. Katchalsky to promote a number of condensation reactions under very mild conditions, particularly the joining of amino acids to form long polypeptides with very high yields.

But most relevant to biological condensation reactions are polyphosphates and polyphosphate esters, which may have been forerunners of ATP, the most important biological condensing agent. Polyphosphates could easily have arisen from orthophosphate minerals in high-temperature zones near volcanoes or (under milder conditions) by the action of cyanoacetylene or cyanoguanidine; such reactions have been observed in primitive-earth simulation experiments.

Polyphosphates or their esters can bring about the dehydrating formation of peptides from amino acids on irradiation with ultraviolet light or on gentle heating. Polyphosphate esters also promote formation of AMP from adenine, ribose, and phosphate, of polynucleotides from mononucleotides, and of glucose polymers from glucose, as shown by G. Schramm, C. Ponnamperuma, and other investigators.

Polyphosphates have other desirable advantages from the biological point of view, for they react very slowly with water, i.e., are kinetically stable, but at the same time have a large negative value for $\Delta G°'$ of hydrolysis, i.e., are thermodynamically unstable. Very likely the polyphosphate group of ATP was selected as a biological condensing agent because of these chemical advantages. The adenosine moiety confers on ATP an easily recognized molecular "handle," by which it is bound to the active site of phosphate-transferring enzymes. An important evolutionary role of simple polyphosphates is also supported by the fact that many modern bacteria contain inorganic polyphosphate polymers (volutin or metachromatic granules, page 408), which function as a storage form of phosphate-bond energy.

L. E. Orgel and his colleagues have found that polyphosphates can be formed under very mild conditions by simply warming orthophosphate with urea and NH_4^+. Moreover, they have found that in the presence of Mg^{2+} such polyphosphates readily react with amino acids on evaporation, to give large yields of amino acid phosphoramidates (Figure 37-6); for example, with imidazole or histidine a phosphoramidate of a ring nitrogen is formed. Histidine residues are fundamental to the catalytic activity of many enzymes (page 224); indeed, the imidazole N-phosphate group is involved in some transphosphorylation reactions (page 458). Orgel and his colleagues have proposed that such phosphoramidates may have been early precursors of modern amino-group and carboxyl-group activating enzymes and that the intrinsic catalytic activity of the imidazole nucleus may have been exploited in early evolution of simple catalysts.

The Prebiotic Formation of Polypeptides

Polymers resembling polypeptides form when amino acids are heated, subjected to electric discharges, or treated with condensing agents such as polyphosphoric esters under simulated primitive-earth conditions. Fox and his colleagues have described polymers of amino acids which they have termed _proteinoids_, formed by heating amino acid mixtures at 130 to 180°C for a few hours or by warming them with polyphosphates at 50 to 60°C for longer periods. These amino acid polymers can be separated by chromatographic procedures into a relatively small number of species. Acidic, neutral, and basic proteinoids have been prepared in this manner. Proteinoids have particle weights which may approach 20,000 daltons; they may contain as many as 18 different amino acids in peptide linkage. Such proteinoids are not simply random polymers; their amino acid composition may differ significantly from the composition of the starting amino acid mixtures. With any given starting mixture the proteinoids produced are surprisingly constant in amino acid composition (Table 37-3). On hydrolysis with acids proteinoids yield free amino acids, as well as some non-amino acid components evidently formed as by-products in the thermal polymerization. Proteinoids are attacked by some proteolytic enzymes. Moreover, they show other properties

Table 37-3 Amino acid composition of a neutral soluble proteinoid prepared from an equimolar mixture of amino acids

	Mole percent
Lys	8.4
His	3.3
Arg	3.9
Asp	4.6
Glu	10.1
Thr	0.4
Ser	0.2
Pro	2.7
Gly	8.4
Ala	12.1
Val	9.6
Met	6.8
Ile	4.7
Leu	6.0
Cys	5.2
Tyr	5.1
Phe	4.5

Figure 37-7
The Akabori hypothesis for the origin of proteins.

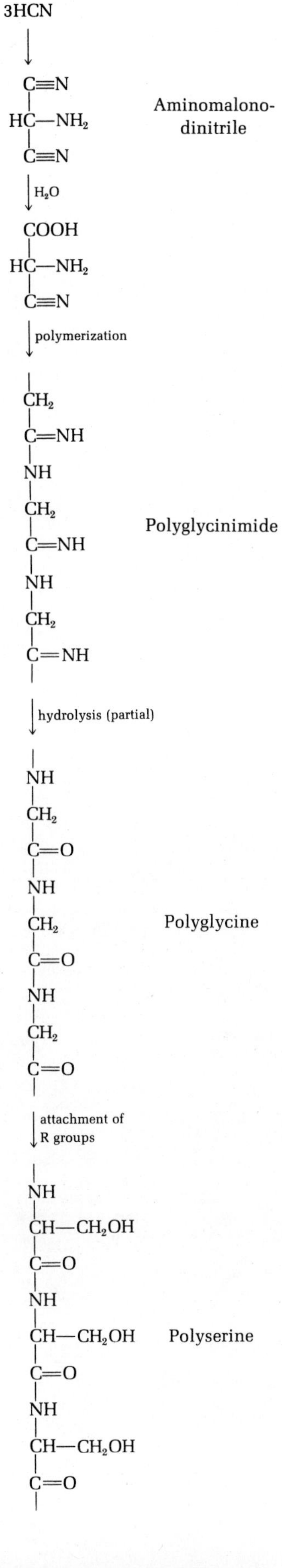

3HCN

|

C≡N
|
HC—NH₂   Aminomalono-
| dinitrile
C≡N

| H₂O

COOH
|
HC—NH₂
|
C≡N

| polymerization

|
CH₂
|
C=NH
|
NH
|
CH₂
| Polyglycinimide
C=NH
|
NH
|
CH₂
|
C=NH
|

| hydrolysis (partial)

|
NH
|
CH₂
|
C=O
|
NH
|
CH₂   Polyglycine
|
C=O
|
NH
|
CH₂
|
C=O
|

| attachment of
| R groups

|
NH
|
CH—CH₂OH
|
C=O
|
NH
|
CH—CH₂OH   Polyserine
|
C=O
|
NH
|
CH—CH₂OH
|
C=O
|

characteristic of proteins: they can be salted in or salted out, and they undergo isoelectric precipitation. The nonrandom nature of proteinoids is also indicated by the fact that their amino-terminal sequences do not reflect the composition of the entire polymer. Heme proteinoids have been prepared by mixing heme with the amino acid mixture before heating.

G. Krampitz and S. Fox have shown that incubation at pH 9 of a mixture of the aminoacyl adenylates of the common amino acids, which are enzyme-bound intermediates in the formation of aminoacyl-tRNAs (page 934), causes spontaneous formation of proteinoids of high molecular weight at mild temperatures. Since aminoacyl adenylates are readily formed from free adenylic acid and free amino acids in the presence of condensing agents, it appears possible that proteinoids may even be formed at low temperatures in dilute aqueous solutions by a reaction pathway resembling that taken by aminoacyl-tRNAs in present-day protein biosynthesis.

Other hypotheses have been proposed for the prebiotic origin of polypeptides. S. Akabori has suggested that the first polypeptides may have arisen by polymerization of monomeric units other than amino acids; for example, aminoacetonitrile, which is readily formed from ammonia, formaldehyde, and hydrogen cyanide (page 23). Polymerization of this monomer, followed by hydrolytic loss of ammonia, could have yielded polyglycine (Figure 37-7). The glycine residues of polyglycine could serve as the precursor of true proteins if the α hydrogen atoms of the glycine residues were replaced by appropriate R groups through secondary reactions.

Since peptide bonds are thermodynamically unstable in aqueous solutions, once a primitive proteinoid arose, it would be highly susceptible to hydrolytic breakdown in the warm primordial sea. Thus no single proteinoid molecule could be expected to survive for long. This fact raises a fundamental problem. It is difficult to see how any given proteinoid could have undergone residue-by-residue evolutionary improvement to an amino acid sequence better able to survive if each proteinoid molecule lasted but a short period and if there were no means of recording or replicating the amino acid sequence of the "better" proteinoids. This follows from mathematical considerations alone. In all the organisms known today there are only about 10^{12} different types of proteins, whereas over 10^{300} different types of proteins can theoretically be formed from 20 different amino acids. We shall return to this point.

The Prebiotic Origin of Enzymatic Activity

The water molecules of the early sea provided the most primitive and ubiquitous catalysts known, namely, H^+ and OH^- ions, which promote specific-acid or specific-base catalysis. Later, when the first organic acids and bases were formed abiotically, they yielded the first general-acid and general-base catalysts, for example, carboxyl groups and amino groups, respectively. Moreover, minerals also afforded the possibility of heterogeneous catalysis.

In the evolution of enzymes from simple precursors, it seems probable that catalytic capacity appeared first and that substrate specificity followed as a later evolutionary development. For example, general acids and general bases acting as catalysts have no substrate specificity; they enhance the rate of many different organic reactions involving proton exchanges. Moreover, it seems very likely that the first precursors of modern enzymes yielded only very small rate enhancements, perhaps only severalfold, compared with modern enzymes, which are capable of enhancing reaction rates anywhere from 10^8- to 10^{20}-fold.

The first primitive catalysts could have been short polypeptides containing an especially reactive α-ammonium group, an imidazole ring, or a carboxylate group functioning as a general-acid or general-base catalyst. The evolution of such a peptide into a more potent and a more specific catalyst may have involved the selection of those modifications of polypeptide structure which were better able to survive because they were more stable, had greater catalytic power, and thus were better able to modify their environment. High-molecular-weight proteinoids, like those described above, may have been the precursors of enzymes. Fox and his colleagues have detected several rate-enhancing activities in thermally formed proteinoids, including hydrolysis of esters, decarboxylation of α-keto acids (Figure 37-8), and deaminations. The esterase activity of certain proteinoids toward p-nitrophenyl acetate has been found to have an optimum pH, to exhibit saturation by substrate, i.e., Michaelis-Menten kinetics, to be destroyed by heat, and to depend on the presence of histidine residues. Zinc-containing proteinoids exhibit ATPase activity, and heme-containing proteinoids exhibit peroxidase activity.

Not much progress could have been made toward the evolution of catalysts with substrate specificity until long polypeptide chains arose which were capable of assuming specific three-dimensional globular conformations (page 222). Again, it would appear that the gradual residue-by-residue evolutionary improvement of polypeptide catalysts simply could not have taken place without some form of template capable of replicating amino acid sequences.

The Prebiotic Origin of Nucleotides and Nucleic Acids

As pointed out earlier (pages 24 and 1035), the organic building blocks of nucleotides, namely, pyrimidines, purines ribose, and 2-deoxyribose, have been shown to arise under simulated primitive-earth conditions. Moreover, nucleosides like adenosine and deoxyadenosine have also been detected in simulated primitive-earth experiments. When nucleosides and polyphosphates are heated or irradiated with ultraviolet light, mixtures of nucleotides are formed in which the phosphate group may be present on any one of the available carbon atoms of the ribose moiety. Not only 5'-adenylic acid

Figure 37-8
Decarboxylation of pyruvic acid catalyzed by a proteinoid.

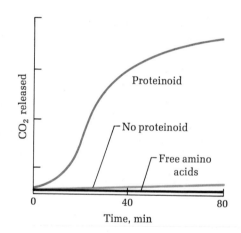

(AMP) but also ADP and ATP can be generated by heating or irradiating adenosine or mixtures of adenine and ribose in the presence of ethyl metaphosphate.

The next step in the chemical evolution of nucleic acids is the formation of internucleotide linkages between successive nucleotides. This also has been observed in simulated primitive-earth experiments in which nucleotides were heated at 50 to 65°C in the presence of polyphosphoric acid as condensing agent. Such abiotically formed oligonucleotides in most cases possess the 2′,5′ linkage; the 3′,5′ linkage does not form readily. Today's 3′,5′ linkage, which undoubtedly has significant advantages over the other possible types, may have arisen after the development of specific enzymes capable of enhancing the synthesis of 3′,5′ linkages in preference to others.

Abiotic Formation of Template Systems

We have seen that polypeptide and polynucleotide chains could well have been formed in the primordial broth. However, the question is: How can a specific sequence of monomers in such chains be recorded or replicated during the molecular evolution of polypeptides and nucleic acids? From the vantage point of today's knowledge of template function we know that nucleic acids function as templates but that proteins do not. Model experiments have shown that base pairing like that inherent in the DNA double helix can also be observed in much simpler systems. When solutions of simple derivatives of adenine and of uracil not containing ribose and phosphoric acid are mixed, adenine-uracil complexes form spontaneously, stabilized by hydrogen bonding. Such complexes are detectable by physicochemical methods; in fact, they can be crystallized from solution under some conditions. Similarly, cytosine and guanine derivatives will also form mixed complexes, but the pairs cytosine-adenine or uracil-guanine are far less stable. Similar physicochemical tests of all possible pairs of bases dissolved in anhydrous solvents show that adenine tends preferentially to complex with uracil and guanine with cytosine. Thus neither the sugar moiety nor the phosphodiester backbone of nucleic acids is required for specific pairing of bases, although the stability of such free A-U or G-C pairs is of course much lower than the base pairs in duplex nucleic acids. A given free mononucleotide will also associate in a preferential manner with specific nucleotide residues of a single-stranded polynucleotide in dilute solution. For example, when free adenosine 5′-phosphate is mixed with polyuridylic acid, the free AMP molecules associate with uridylic acid residues through hydrogen bonding between the complementary bases, to yield a helical structure having but one covalent backbone (Figure 37-9). Similarly, polycytidylic acid forms a complementary single-strand helix when mixed with free GMP. Thus a polynucleotide chain can act as a template to bind free complementary mononucleotides. It therefore becomes conceivable that prebiotic condensing agents might have formed internucleotide linkages between template-bound mononucleotides to yield a complementary polynucleotide.

Orgel and his colleagues have provided such a demonstration. They added a water-soluble carbodiimide condensing agent to a solution containing a single-stranded poly-U template molecule and free AMP. Not only was polyadenylic acid formed, but it was formed at a rate much higher than in the absence of the template (Figure 37-9). The complementary polynucleotide formed under these conditions has 2′,5′ internucleotide linkages. These and other experiments have shown that polynucleotides can serve as templates for nonenzymatic synthesis of complementary polynucleotides under abiotic primitive-earth conditions.

The Origin of Asymmetric Biomolecules

One of the distinguishing characteristics of living matter is that essentially all its organic compounds having one or more asymmetric carbon atoms occur in only one stereochemical configuration. Stereochemical asymmetry is not seen in inanimate matter except under laboratory conditions designed to produce compounds with asymmetric carbon atoms. Because amino acids were in all likelihood first formed from inorganic precursors in nonbiological reactions, they most probably occurred in the primordial broth as racemic mixtures. How, then, can we account for the fact that proteins today contain only L-amino acids?

As we have seen (page 128), an α helix will form from polypeptide chains containing either all L- or all D-amino acids. Since the L and D stereoisomers of any given amino acid are identical in structure and reactivity and differ only in the fact that they are nonsuperimposable mirror images of each other, biologically active proteins could in theory be made from either series. However, a stable α helix will not form from a polypeptide containing a random mixture of L and D stereoisomers. Hence there was probably a biological advantage in having the amino acids either all L or all D. It is now believed that L-amino acids were selected as the building blocks of proteins purely by chance, not because they had any intrinsic advantage over D-amino acids. This line of thought also suggests that all living organisms are derived from one cell line, since the proteins of all living forms possess L-amino acids.

Ultimately it is in the active site of enzymes capable of forming one specific stereoisomer from an optically inactive precursor that biological stereospecificity is preserved. Stereospecific catalysis of some nonbiological polymerization reactions has been observed in situations where the catalyst itself is an optical enantiomer.

The importance of optical specificity in the replication of nucleic acids has been shown in interesting experiments by Orgel and his colleagues. When polyuridylic acid in which all the ribose units were the D form was used as a template to promote the nonenzymatic condensation of hydrogen-bonded AMP molecules to form polyadenylic acid, the sub-

Figure 37-9
Arrangement of AMP molecules on a template of poly U. The formation of poly A is promoted by the poly-U template in the presence of a carbodiimide condensing agent.

Template function of poly U

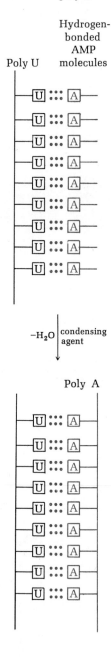

sequent condensation of the D stereoisomer of AMP was favored over that of the L stereoisomer. It therefore appears that once a polynucleotide template having optically homogenous monomer units is formed, it will preferentially direct the synthesis of another polynucleotide chain of the same optical type, even if the nucleotide pool from which the new chain is formed is a mixture of D and L forms.

Acceleration of poly-A formation by poly-U template

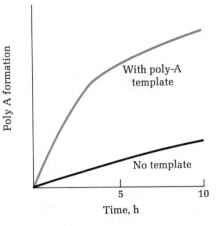

The Origin of "Life" vs. the Origin of Cells

We have seen how primitive polypeptides, polysaccharides, polynucleotides, catalysts, and replication templates could have arisen in the course of prebiotic chemical evolution. We now come to that critical moment in chemical evolution when the first semblance of "life" appeared, through the chance association of a number of abiotically formed macromolecular components to yield a unique system of enhanced survival value, capable of evolving toward a more complex structure, one better able to survive. However, the first structure possessing "life" was not necessarily a modern cell, complete with a membrane, a chromosome, ribosomes, enzymes, a metabolism, and the property of self-replication. The minimum requirement is that it could *potentially* evolve into a complete cell. It is at this point that a cluster of organic molecules, possibly a group of oligomers or polymers, associated with each other into a stabilized structure. Here we can readily imagine that such a more stable structure arose through hydrophobic interactions between the component molecules, so that their nonpolar surfaces associated together to form a micellelike structure, in a process made possible by the tendency of the surrounding water molecules to seek their position of maximum entropy (page 43). In this way a cluster or micelle with a hydrophilic outer surface and a partially hydrophobic interior could have formed. Such a structure may have included fatty acids or lipids, which spontaneously form membranes and micelles, or other molecules with hydrophobic zones, such as polypeptides rich in amino acids with hydrophobic R groups.

We cannot define "life" accurately enough to determine at what point it arose in the chain of events leading from preformed macromolecular components to a complete cell. However, it is generally agreed that a minimum requirement for life is one or more informational macromolecules capable of directing their own replication. One must then ask: Which had primacy in the origin of life, proteins or nucleic acids? One view suggests that the first protocells arose when primitive catalysts (presumably polypeptides) first became surrounded by a membrane or became incorporated in a gel-like matrix and that the resulting structure "learned" to maintain itself with a primitive metabolism. In this view the first cells functioned in the absence of nucleic acids and a genetic system, which they acquired later. Another view is that nucleic acids arose first and that they provided the information for the evolution of proteins. A third view is that both nucleic acids and proteins had to come together to form the first real precursor of a living cell. We shall now examine simulation experiments based on these hypotheses.

Coacervate Droplets

Oparin has suggested that the first precursors of cells, which he called *protobionts*, arose when a boundary or membrane formed around one or more macromolecules possessing catalytic activity, presumably proteins. He proposed that a cell phase could have arisen from the concentrated primordial broth by the process of *coacervation*, a physical phenomenon which often takes place in aqueous solutions of highly hydrated polymers. It is defined as the spontaneous separation of a continuous one-phase aqueous solution of a polymer into two aqueous phases, one having a relatively high polymer concentration and the other a relatively low concentration. The tendency to undergo coacervation is primarily a function of the molecular size of the polymer and of the extent to which its interstices can be penetrated by water. A coacervation process occurring in the primordial organic broth may have yielded microscopic droplets consisting of that phase in which the polymer concentration is high.

Oparin postulated further that some of these droplets may have entrapped a primitive catalyst, presumably a polypeptide, as well as substrate molecules, such as glucose or amino acids. Such coacervate droplets would thus possess a simple, one-reaction "metabolism." He postulated in addition that such a metabolizing droplet might interact further with the surrounding aqueous environment, to acquire other compounds that it could incorporate into its own structure, causing the droplet to grow. Upon reaching a size limited by physical considerations, the droplet would break up into smaller droplets, just as a coarse emulsion of oil in water disperses into a finer emulsion when shaken. Some of the progeny droplets so formed might retain catalyst molecules derived from the parent droplet and thus could "grow" another generation of droplets.

Oparin and his colleagues carried out a number of interesting model experiments with coacervate systems to illustrate the feasibility of his hypothesis. Under certain conditions concentrated aqueous solutions of polypeptides, polysaccharides, or RNA could be induced to form coacervate droplets having a volume of from 10^{-8} to 10^{-6} cm^3 and a polymer concentration of from 5 to 50 percent. Enzymes can be incorporated into such droplets (Figure 37-10). When coacervate droplets containing glycogen phosphorylase (page 433) were placed in a solution of glucose 1-phosphate, a starchlike polymer was formed in the droplets. If the droplets also contained β-amylase, which breaks down starch to maltose, they converted glucose 1-phosphate, obtained from the medium, into maltose, which then returned to the medium. Synthetic coacervate droplets could also be prepared containing the flavoprotein NADH dehydrogenase (page 488) and a reducible dye. Such droplets could accept reducing equivalents from external NADH and transfer them to the dye molecule, which was lib-

Figure 37-10
Primitive metabolism in synthetic coacervate droplets containing enzymes.

A droplet capable of synthesizing maltose from glucose 1-phosphate.

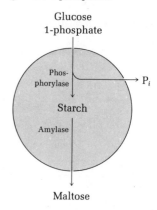

A droplet exhibiting electron transport.

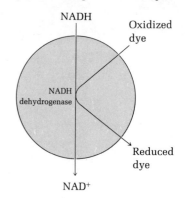

A coacervate droplet capable of transferring electrons from ascorbic acid via chlorophyll to a dye on illumination. Chl* represents chlorophyll in its excited state.

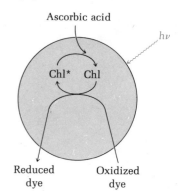

erated into the medium in reduced form. Photosynthetic droplets could be made by incorporating chlorophyll into a polymer; such droplets, when illuminated, promoted the reduction of dyes by ascorbic acid, a model of the Hill reaction (page 598).

Oparin's coacervate droplets are obviously models; they were made from biologically formed substances and were not formed under simulated primitive-earth conditions. Moreover, they do not provide for self-replication or evolution.

Proteinoid Microspheres

Self-forming, cell-like structures, called _microspheres_, have been described in great detail by Fox and his colleagues. These remarkable bodies arise spontaneously when hot concentrated solutions of thermally formed proteinoids are allowed to cool under appropriate conditions of pH and salt concentration. Microspheres are rather uniform spherical droplets about 2.0 μm in diameter (Figure 37-11). When the pH is properly adjusted, their outer boundary shows a double-layer structure resembling that of natural membranes; although microspheres contain no lipid, they do contain amino acids having nonpolar R groups, which may provide a lipidlike barrier. Microspheres appear to be rather stable at pH 3 to 7. When placed in hypertonic or hypotonic salt solutions microspheres will either shrink or swell, suggesting they have a semipermeable membrane surrounding an inner compartment containing entrapped salts and soluble proteinoids.

Microspheres show several types of cell-like behavior. They can be induced to undergo cleavage or division by suitable changes in pH or when exposed to $MgCl_2$. Moreover, if suspensions of microspheres are allowed to stand for 1 or 2 weeks, they undergo a budding process. Such buds may detach from the mother microsphere and form a second generation of microspheres (Figure 37-12). Proteinoid microspheres also give the appearance of conjugation. They can also be shown to accelerate chemical reactions when formed from proteinoids having catalytic activity.

Fox has pointed out that microspheres actually bear potential information since they are formed from proteinoids which are nonrandom in amino acid composition. The formation of proteinoids is in his view the first major step in the evolution of genetic information; the next step is the formation of the nonrandom structure of the microspheres.

In more recent work, Fox and his colleagues have shown that basic proteinoids, rich in lysine residues, selectively associate with the homopolynucleotides poly C and poly U but not with poly A or poly G. On the other hand, arginine-rich proteinoids associate selectively with poly A and poly G. In this manner, the information in proteinoids can be used to select polynucleotides. Moreover, it is striking that aminoacyl adenylates yield oligopeptides when incubated with proteinoid-polynucleotide complexes, which thus have some of the characteristics of ribosomes. Fox has suggested that proteinoids bearing this sort of primitive chemical informa-

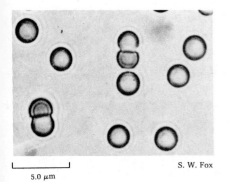

Figure 37-11
Proteinoid microspheres. Their average diameter is 1.9 μm.

5.0 μm

S. W. Fox

tion could have transferred it to a primitive nucleic acid; the specificity of interaction between certain proteinoids and polynucleotides suggests the beginnings of the genetic code.

Fox's comprehensive hypothesis for the origin of reproductive cells, which postulates the primacy of proteins as the first informational macromolecules, is outlined in Figure 37-13.

The Gene Hypothesis: Life without Proteins

An alternative hypothesis for the origin of life is based on the primacy of nucleic acids as the first informational macromolecules. This hypothesis was stated by the geneticist H. Muller in 1929. He proposed that life arose with the abiotic formation of one or more genes. Muller argued that the minimum properties of a living system, i.e., the capacity for metabolism and the capacity for self-replication are already *potentially* present in genes, which, when placed in the proper molecular environment within a cell, can code for the formation of daughter cells. When Muller first postulated this idea, the chemical nature of genes was unknown. Stated in modern terms, this hypothesis postulates that a nucleic acid molecule may possess the potential capacity for life by virtue of its ability to code for proteins, to undergo self-replication, and to undergo mutation. The acquisition of a boundary membrane and the development of catalysts are thus regarded as later evolutionary events. The gene hypothesis remained incompletely developed for some years, but with the newer knowledge of molecular genetics that has emerged, it has been more specifically elaborated by a number of investigators, especially N. W. Pirie, N. Horowitz, F. H. C. Crick, and L. E. Orgel.

Let us examine the three major lines of thought and evidence supporting the nucleic acid hypothesis. One arises from modern knowledge of the molecular structure and self-replication of viruses in host cells. Another arises from the wide range of biological functions nucleotides carry out in cells today. A third line of thought arises from the role of ribosomal RNA in protein synthesis.

The viruses, which are often referred to as structures "at the threshold of life," may conceivably represent the precursors of the first simple prokaryotic cells. The simplest, like tobacco mosaic virus, are little more than a complex of a single nucleic acid molecule, which bears the information, and a coat of recurring, identical protein monomers, which serves only as a protective sheath. The largest and most complex viruses, on the other hand, may contain a phospholipid membrane and one or several enzymatic activities; indeed, they do not differ much in size from the simplest prokaryotic cells. In the smallest viruses the nucleic acid contains only a few genes, but in the largest viruses, such as the vaccinia and pox viruses of animals, the nucleic acid has several hundred genes.

It has long been speculated that the largest known viruses may be very small cells. The largest animal viruses are similar in size and infectious behavior to a group of very small

Figure 37-12
Budding and replication of microspheres.

Buds on parent microspheres

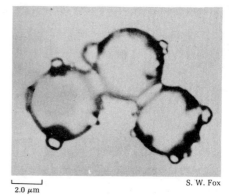

2.0 μm   S. W. Fox

Liberated buds

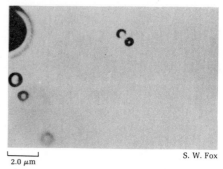

2.0 μm   S. W. Fox

Microspheres grown from a central, darkly stained bud

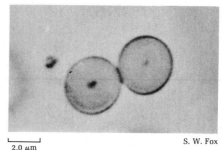

2.0 μm   S. W. Fox

Mature second-generation microsphere with bud

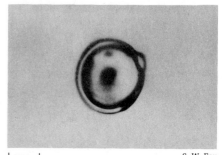

2.0 μm   S. W. Fox

Figure 37-13
Fox's hypothesis of the proteinoid origin of life.

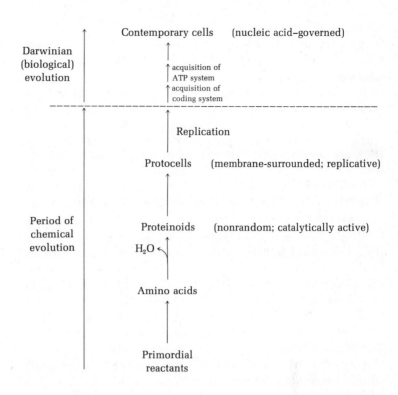

bacteria which can live only within specific host cells as parasites. These parasitic prokaryotes include the *Chlamydia,* among which are the infectious agents of the diseases *psittacosis* and *lymphogranuloma,* and the *Rickettsia,* somewhat larger organisms that cause tick fever and Rocky Mountain spotted fever. The causative agent of psittacosis (parrot fever) was for many years thought to be a virus. This agent, which cannot be cultured in free form independent of a host organism, has a particle weight of about 6 billion daltons, contains both DNA and RNA, and is capable of catalyzing some metabolic reactions. The psittacosis agent has now been reclassified as a bacterium on the basis of a more rigorous (and perhaps somewhat arbitrary) definition of viruses. According to this definition, a virus is an infectious agent with an intracellular phase in its life cycle which (1) reproduces from its nucleic acid alone, (2) contains only one type of nucleic acid, either RNA or DNA, (3) contains no ATP–ADP system for the transfer of metabolic energy, and (4) is unable to grow or undergo binary fission. However, the psittacosis organism is an extremely rudimentary and incompletely developed cell. Not only is it very small (0.2 to 0.5 μm) and simple in structure, but it entirely lacks an ATP–ADP system and phosphotransferases. It also lacks the enzymes required for glycolysis and the tricarboxylic acid cycle, although it does contain enzymes required in the biosynthesis of some of its components. Presumably this organism depends on the host cell for ATP energy and for its precursor biomolecules. Although the psittacosis organism is officially a bacterium, it could conceivably represent a transition form in the progressive evolution of a virus into true prokaryotic cells. On the other hand, as many, if not most, investigators believe, it may represent a mutational degeneration of a once complete bacterium, which lost genes irreversibly as it became dependent on the host cell.

Another line of reasoning supporting a central role of nucleic acids in the origin of life is suggested by the striking versatility of function of nucleotides in modern organisms. Nucleotides serve as the monomeric units in the structure of DNA and of the three major types of RNA. Specific nucleotides or nucleotide derivatives also serve as energy carriers (ATP, ADP), as hydrogen or electron carriers (NAD, NADP, FMN, FAD), as sugar carriers (UDP), as carriers of lipid components (CDP), and as acyl carriers (CoA). Thus nucleotides are important functional elements in all aspects of metabolism and energy transfer, as well as in the genetic apparatus. From nucleic acids may also be descended the characteristic phosphodiester linkages which join successive glycerol or ribitol residues to form the backbone of the teichoic acids (page 271) and which join glycerol and other alcohols to form the backbones of the phosphoglycerides (page 287). These relationships suggest very strongly that much of the fundamental metabolic and genetic machinery of the cell could have evolved or developed from nucleotides.

Perhaps we can now begin to see how a primitive form of life based on nucleotide chemistry may have arisen. The most primitive nucleic acids could have replicated themselves in the absence of enzymes, as we have seen, by the action of abiotically formed polyphosphates, the possible forerunners of ATP. The earliest forms of the pyridine-linked dehydrogenases may possibly have consisted of oligonucleotides of which one mononucleotide unit contained the reducible nicotinamide ring as a base, rather than one of the standard bases. It is thus quite plausible that a primitive nucleotide-based life could have existed in the absence of proteins. Relevant model reactions involving polynucleotides capable of base pairing with mononucleotides (and thus capable of replication of information) were described above. But it is quite clear that not much evolutionary progress could have been made without proteins, especially enzymes. How, then, did the first proteins arise if genes arose first?

This issue has recently been discussed in the light of modern developments in molecular genetics, particularly by Crick and Orgel. They suggest that the real answers to the origin of life may lie in the origin of ribosomes, of transfer RNA, and of the genetic code. They base their view on the circumstance that protein synthesis by ribosomes involves two types of nucleic acid that in the main do not serve a template function, namely, ribosomal RNA and transfer RNA. Transfer RNA does possess an anticodon triplet, but it makes up but a very small fraction of its total structure. Both tRNA and rRNA are single-strand polymers resembling globular proteins in that they possess complex folded three-dimensional structures, made possible by intrachain base pairing. These two forms of RNA are suggested to be the vital parts of a primordial system which "learned" how to make true informational proteins, not merely random polymers of amino acids. Thus the first primitive ribosomes were viewed as consisting entirely of RNA. (Parenthetically, we may note that the ribosomes of modern prokaryotic cells contain some 65 percent RNA, significantly more than those of eukaryotic cells, which evolved later.) Such a primitive proteinless ribosome may have been able to make the first ordered peptides with the participation of tRNA, a smaller molecule which was postulated to bind to the primordial ribosomal RNA, presumably by base pairing. Such a primitive tRNA may have had the ability to bind a given amino acid noncovalently, perhaps at or near the anticodon site, which might conceivably form a cagelike amino acid binding region. Indeed, much thought is now being given to the possible molecular or conformational relationships between specific base sequences of tRNAs and their corresponding amino acids. For example, why do the codons for all amino acids with hydrophobic R groups have U as the second letter? It is possible to imagine that tRNAs could bind the R group of certain amino acids noncovalently in a cavity of specific dimensions. It is also possible to imagine how a sequence of charged tRNA molecules might have become aligned on the surface of a ribosomal RNA molecule in such a way as to juxtapose the carboxyl group of one amino acid with the amino group of the next. If a prebiotic condensing agent, such as a polyphosphate, removed the elements of water from the properly aligned carboxyl and amino groups, a peptide bond could then arise. Thus the primitive peptide-forming machinery possibly consisted only of two different types of specifically folded RNA. At this stage in evolution, before the appearance of specific amino acid–activating enzymes, the association of the amino acid to the tRNA may have been noncovalent. Thus the system may have had very limited amino acid specificity. Nevertheless, the primitive forms of transfer RNA and ribosomal RNA may have performed one of the important functions of an enzyme molecule, namely, the provision of specific binding sites to position the loosely bound amino acid substrates in such a way as to allow them to interact in the presence of a nonenzymatic condensing agent, possibly a polyphosphate derivative.

The next step may have been the development of the first template for specifying the sequence of tRNAs and thus of the amino acids they carry. This template may have been provided by a loop of the primordial ribosomal RNA or by another RNA strand which could bind not only to the primitive tRNAs but also to the primitive ribosome, to constitute the first messenger RNA. This RNA may have coded for the sequence of one or more polypeptides which endowed the ribosomal apparatus with enhanced stability and activity; such polypeptides may have been the precursor of modern ribosomal proteins.

In this view, then, RNA, not DNA, represents the original protogene, the first informational macromolecule. Possibly DNA arose later than RNA to provide a more permanent engram, one capable of self-replication. Two facts from biochemistry support such a view. The discovery of RNA-directed DNA polymerases (reverse transcriptases, page 915) indicates that there need be no conceptual barrier in accepting RNA as the first informational nucleic acid. Moreover, we may recall that the direct precursors of the 2'-deoxyribonucleotide building blocks of DNA are the corresponding ribonucleotides (page 737).

Ultimately, the most compelling and overriding feature of the nucleic acid hypothesis, whatever its details, is that polynucleotides are capable of serving as templates in the absence of enzymes or proteins, in such a way that a complementary polynucleotide can be formed by an abiotic catalyst or condensing agent. Moreover, through the ability of polynucleotides to undergo mutation in the absence of proteins, as a result of such purely physical agencies as ultraviolet irradiation, a nucleic acid–based "life" can undergo evolutionary modification.

Attractive as these ideas might at first appear, they are subject to some major questions. The most crucial one is the origin of the genetic code. Apart from the fact that all hydrophobic amino acids are coded by triplets whose second base is U, there is little evidence that the coding triplets (page 962) bear any steric or chemical relationship to the amino acids for which they code. Model-building experiments have simply not revealed a satisfactory picture of the molecular correspondence between amino acids and their codons. For this reason it has been proposed that the genetic code is the result of a "frozen accident." It is this step in the molecular evolution of the genetic system for which there is yet no satisfactory model or theory. Indeed, Crick and Orgel have pointed out that it is not beyond reasonable possibility that genes and the genetic code may have been brought to earth by spaceship from some other body in the universe where intelligent life had already evolved. This is a throwback to the old hypothesis of "Panspermia," postulated at the turn of the twentieth century by the Swedish chemist S. Arrhenius, who proposed that life began on earth from seeds or sperm wafted from the outer reaches of the universe by celestial winds. Of course this idea is no answer to the problem, since one must then explain how life arose elsewhere.

The contrasting views of the proponents of proteins and the proponents of nucleic acids as set out above need not necessarily be mutually exclusive. Other scientists have argued with equal force that both primitive nucleic acids and proteins were required simultaneously in the origin of the single cell line that ultimately survived. Perhaps, as we learn more about the fundamental molecular interactions involved in the function of ribosomes during protein synthesis and of the structural relationships between the amino acids and their codons, we may yet see some molecular hint that will be the vital key to the problem.

Later Steps in Biochemical Evolution

Once a template system, a set of catalysts, and a surrounding membrane evolved, in whatever sequence, the process of cellular evolution becomes easier to comprehend. Presumably in the early stages of biological evolution, chemical evolution continued for some time, so that the first cells continued to select from the environment those organic molecules which enhanced their survival. Eventually the first anaerobic heterotrophs, which utilized organic compounds as sources of energy and carbon, depleted the primordial organic broth of nutrients. The first autotrophic cells, capable of using carbon dioxide, then arose, as a result of selection pressure, followed by the first photosynthetic cells, capable of utilizing light as energy source. Depletion of organic nitrogen compounds in the primitive primordial broth may also have led to the elaboration of the ability to fix atmospheric nitrogen. In the photosynthetic nitrogen-fixing blue-green algae we see the culmination of a metabolic adaption to a primitive sea which may have become completely devoid of organic carbon and nitrogen compounds. Long after oxygen first began to accumulate in the atmosphere, the first oxygen-consuming aerobic cells presumably arose, which possessed a selective advantage because they could extract more energy per molecule of exogenous nutrient, such as glucose, than anaerobic fermenting organisms. The evolution of aerobic respiration and oxidative phosphorylation completed the development of the basic repertoire of energy-yielding metabolic processes which survive in the living organisms of today. Although most evidence and current thought supports this sequence of events in the metabolic evolution of cells, we cannot exclude the possibility that some other sequence occurred. In fact, some investigators have proposed that autotrophic cells may have arisen first.

Only after these developments had taken place, which may have involved some 2,500 million years, or over one-half the earth's history, do we have our second great jump in cell evolution, the appearance of eukaryotic cells, which has rightfully been called one of the most significant events in biological evolution. Perhaps the third great step in biological evolution was the emergence of diploid-dominant eukaryotes some 700 million years ago. With these important steps in biological evolution a much wider range of cell differentiation became possible.

One hypothesis, which is given major support from recent investigations, holds that eukaryotic cells originated as the result of metabolic endosymbiosis. Presumably eukaryotes evolved from large anaerobic prokaryotic cells, whose fermentation capacity may no longer have been able to keep up with their energy demands. Such anaerobes may have engulfed small photosynthetic or heterotrophic prokaryotes, which became metabolic endosymbionts, the host cell furnishing glycolytic-fermentation capacity and the parasite either photosynthetic or respiratory capacity. Such endosymbionts may have been the forerunners of modern chloroplasts and mitochondria, respectively, whose ability to extract energy from sunlight or organic fuels may have endowed the host anaerobic cell with greatly enhanced survival value. Eukaryotes are much larger and more versatile cells, far better equipped to act on their environment, and thus modify it and capable of many modes of differentiation through their much larger genomes. Throughout the course of biological evolution the formation of nonlethal mutants, as the result of chemically or physically induced changes in the base sequence of the genome, made possible selection of those organisms best able to survive their contemporary conditions of life.

But the most striking of all the interlinked events which have led to the incredibly complex biological organisms we know today still remains at the divide where chemical evolution ends and biological evolution begins. Why and under what conditions does an ensemble of organic compounds become a self-organizing self-replicating system? The laws of chemistry and physics we know today do not forbid self-organization of matter; however, they provide no easy explanation for it. Our greatest task is to formulate in precise terms the molecular logic of self-organizing and self-replicating systems of organic compounds.

The questions raised by the origin of life need not remain philosophical speculations. They may well be approachable experimentally since, with modern knowledge of the properties of organic molecules and of biochemistry, it is possible to accelerate in the laboratory the chemical processes leading to, or involved in, the tendency of organic molecules to undergo self-organization. Living organisms may thus be an inevitable outcome of the evolution of self-organizing systems of organic molecules, given appropriate physical conditions. Indeed, they may be but stages in the further evolution of matter toward levels of organization that are still incomprehensible to us.

Summary

The earth was formed some 4,800 million years ago. During the first 1,500 million years organic compounds are believed to have been formed from atmospheric components such as H_2, NH_3, H_2O, and methane, under activation by ultraviolet light, lightning, shock waves, heat, and other forms of energy. Among these compounds were some 30 or 40 primordial biomolecules, including amino acids, purines, pyrimidines, pentoses, hexoses, fatty acids, and others. Such compounds or their precursors have also been detected in ancient fossils, in meteorites, and in interstellar space. Hydrogen cyanide, which is readily formed in simulated primitive-earth conditions, is a major precursor of nitrogenous organic molecules, via the formation of cyanamide, cyanoacetylene, and nitriles. Hydrogen cyanide is also a precursor of organic condensing agents, including cyanoguanidine, carbodiimides, and urea. The subsequent chemical evolution of organic molecules probably proceeded in a stepwise manner. The dehydrating condensation of amino acids to yield polypeptides can be brought about by heat, by organic condensing agents, by polyphosphates and phosphoramidates, or by clay and other mineral catalysts. Primitive polypeptides, called proteinoids, form readily when amino acid mixtures are heated above 100°C. Such proteinoids have a nonrandom amino acid composition, show a number of properties of modern proteins, and also have rudimentary catalytic activity. Presumably the first prebiotic "enzymes" had little specificity and low catalytic power; they may have utilized general-acid or general-base groups as catalytic centers. Nucleotides and polynucleotides also form under mild primitive-earth conditions, particularly in the presence of polyphosphates or phosphoramidates. Simple polynucleotides can serve as templates for the abiotic formation of complementary polynucleotides.

Primitive cell-like structures may have formed by the process of coacervation or by micelle formation driven by hydrophobic interactions, the tendency of the surrounding water molecules to seek the position of maximum entropy. A. I. Oparin has postulated the formation of coacervate droplets composed of polymers, into which primitive catalysts may have been incorporated. More recently S. W. Fox has described proteinoid microspheres, which exhibit many aspects of cell-like behavior. Such structures were visualized as developing in the absence of nucleic acids, which may have been acquired later. The other general hypothesis is that a primitive gene was required before proteins were added, an idea supported by the structure of modern viruses, the general importance of nucleotides in modern biochemistry, and the capacity for self-replication. In either case, some means for recording or coding and thus self-replication was required, as well as the capacity for further evolution. Development of the genetic code may have been the basic event in the origin of life. At the center of the problem is the process of self-organization of matter.

References

Books

CAIRNS-SMITH, A. G.: *The Life Puzzle*, University of Toronto Press, Toronto, 1971.

CALVIN, M.: *Chemical Evolution: Molecular Evolution towards the Origin of Living Systems on the Earth and Elsewhere*, Oxford University Press, London, 1969. Reconstruction of chemical evolution from molecular paleontology and from simulated primitive-earth experiments.

JUKES, T. H.: *Molecules and Evolution*, Columbia University Press, New York, 1966.

KENYON, D. H., and G. STEINMAN: *Biochemical Predestination*, McGraw-Hill, New York, 1959.

MARGULIS, L.: *Origin of Eukaryotic Cells*, Yale University Press, New Haven, Conn., 1970.

MILLER, S., and L. E. ORGEL: *The Origins of Life*, Prentice-Hall, Englewood Cliffs, N.J. 1973. Chemical problems in prebiotic organic synthesis are stressed.

MONOD, J.: *Chance and Necessity*, Knopf, New York, 1971, Philosophical discussion on molecular biology and the phenomenon of life.

OPARIN, A. I.: *Life: Its Origin, Nature, and Development*, Academic, New York, 1964.

ORGEL, L. E.: *The Origins of Life: Molecules and Natural Selection*, Wiley, New York, 1973.

ROHLFING, D. L., and A. I. OPARIN (eds.): *Molecular Evolution: Prebiological and Biological*, Plenum, New York, 1972. A volume honoring Sidney W. Fox.

SHKLOVSKII, I. S., and C. SAGAN: *Intelligent Life in the Universe*, Holden-Day, San Francisco, 1966. Origin of terrestrial and extraterrestrial life.

WOESE, C. R.: *The Genetic Code*, Harper & Row, New York, 1967. The origins of the genetic code and of living organisms are discussed.

WOOLDRIDGE, D. E.: *The Machinery of Life*, McGraw-Hill, New York, 1966. Unification of the naked-gene and protobiont hypotheses.

Articles

BARGHOORN, E. S.: "The Oldest Fossils," *Sci. Am.*, 224: 30–53 (1971).

BLACK, S.: "A Theory on the Origin of Life," *Adv. Enzymol.*, 38: 193–234 (1973). An interesting hypothesis and experiments on the origin of the genetic code.

CRICK, F. H. C.: "The Origin of the Genetic Code," *J. Mol. Biol.*, 38: 367–379 (1968). A provocative article emphasizing the origin of amino acid–transfer RNA specificity.

CRICK, F. H. C., and L. E. ORGEL: "Directed Panspermia," *Icarus*, 49: 341–348 (1973). The hypothesis that living organisms were deliberately transmitted to the earth by intelligent beings on another planet.

EGLINTON, G., and M. CALVIN: "Chemical Fossils," *Sci. Am.*, 216: 32–43 (1967). The development of organic geochemistry.

EIGEN, M.: "Self-Organization of Matter and the Evolution of Biological Macromolecules," *Naturwiss.*, 58: 465–523 (1971). An important theoretical paper (in English).

FERRIS, J. P.: "Cyanovinyl Phosphate: A Prebiological Phosphorylating Agent," *Science*, 161: 53–54 (1968).

FOX, S. W.: "Origin of the Cell: Experiments and Premises," *Naturwiss.*, 60: 359–369 (1973). Recent statement of Fox's views (in English).

LACEY, J. C., and K. M. PRUITT: "Origin of the Genetic Code," *Nature*, 223: 799–804 (1969). An attempt to provide a model for the interaction between nucleotides and polypeptides.

LOHRMANN, R., and L. E. ORGEL: "Prebiotic Activation Processes," *Nature*, 244: 418–420 (1973). Catalytic role of imidazole, urea, and Mg^{2+} in prebiotic condensation reactions.

OPARIN, A. I.: "The Origin of Life and the Origin of Enzymes," *Adv. Enzymol.*, 27: 347–380 (1965). The most recent statement of the protobiont hypothesis.

ORGEL, L. E.: "Evolution of the Genetic Apparatus," *J. Mol. Biol.*, 38: 381–393 (1968). A comparison of life without nucleic acids vs. life without proteins.

RAFF, R. A., and H. R. MAHLER: "The Nonsymbiotic Origin of Mitochondria," *Science*, 177: 575–582 (1972).

APPENDIXES

APPENDIX A

A Chronology of Biochemistry

The following represents a chronological listing of important events, experimental discoveries, or enunciation of ideas in the history of biochemistry. The entries since 1950 are somewhat arbitrarily chosen. They almost certainly fail to do justice to a number of important discoveries and to some individuals who have made significant contributions. This is due to the rapidly accelerating pace of modern biochemical research, to the much larger number of researchers, and to the often simultaneous discoveries made not only by two or more individuals but also by two or more groups of individuals.

Biochemistry had its earliest origins in speculations on the role of air in the utilization of food and on the nature of fermentation. Leonardo da Vinci (1452–1519) was among the first to compare animal nutrition to the burning of a candle, a line of reasoning that was further developed by van Helmont (1648). However, the real history of biochemistry did not begin until the late eighteenth century, when the science of chemistry began to take form.

1770–1774	Priestley discovered oxygen and showed it is consumed by animals and produced by plants.
1770–1786	Scheele isolated glycerol and citric, malic, lactic, and uric acids from natural sources.
1773	Rouelle isolated urea from urine.
1779–1796	Ingenhousz showed that light is required for oxygen production by green plants. He also proved that plants use carbon dioxide.
1780–1789	Lavoisier demonstrated that animals require oxygen, recognized that respiration is oxidation, first measured oxygen consumption by a human subject, and recognized that alcoholic fermentation is fundamentally a chemical process.
1783	Spallanzani deduced that protein digestion in the stomach is a chemical rather than mechanical process.
1804	Dalton enunciated the atomic theory.
1804	de Saussure carried out the first balance sheet for the stoichiometry of gas exchanges in photosynthesis.
1806	Vauquelin and Robiquet first isolated an amino acid, asparagine.
1810	Gay-Lussac deduced the equation for alcoholic fermentation.
1815	Biot discovered optical activity.
1828	Wohler synthesized the first organic compound from inorganic components: urea from lead cyanate and ammonia.
1830–1840	Liebig developed techniques of quantitative analysis and applied them to biological systems.
1833	Payen and Persoz purified diastase (amylase) of wheat, showed it to be heat-labile, and postulated the central importance of enzymes in biology.
1837	Berzelius postulated the catalytic nature of fermentation. He later identified lactic acid as a product of muscle activity.

1838	Schleiden and Schwann enunciated the cell theory.
1838	Mulder carried out the first systematic studies of proteins.
1842	Mayer enunciated the first law of thermodynamics and its applicability to living organisms.
1850–1855	Bernard isolated glycogen from the liver, showed it is converted into blood glucose, and discovered the process of gluconeogenesis.
1854–1864	Pasteur proved that fermentation is caused by microorganisms and demolished the spontaneous-generation hypothesis.
1857	Kölliker discovered mitochondria ("sarcosomes") in muscle.
1859	Darwin published *Origin of Species*.
1862	Sachs proved that starch is a product of photosynthesis.
1864	Hoppe-Seyler first crystallized a protein: hemoglobin.
1866	Mendel published his experiments leading to the principles of segregation and independent assortment of genes.
1869	Miescher discovered DNA.
1872	Pfluger proved oxygen is consumed by all the tissues of animals rather than by the blood or lungs alone.
1877	Kühne proposed the term enzyme and distinguished enzymes from bacteria.
1886	MacMunn discovered histohematins, later renamed cytochromes.
1890	Altmann described procedures for staining mitochondria, studied their distribution, and postulated them to have metabolic and genetic autonomy.
1893	Ostwald proved enzymes are catalysts.
1894	Emil Fischer demonstrated the specificity of enzymes and the lock-and-key relationship between enzyme and substrate.
1897	Bertrand coined the term coenzyme.
1897	Buchner discovered that alcoholic fermentation may occur in cell-free yeast extracts.
1897–1906	Eijkman proved that beriberi is a dietary-deficiency disease and that a water-soluble component of rice polishings can cure it.
1901–1904	Takamine and Aldrich, and also Abel, first isolated a hormone, epinephrine, and Stoltz achieved its synthesis.
1902	Emil Fischer and Hofmeister demonstrated that proteins are polypeptides.
1903	Neuberg first used the term biochemistry.
1905	Harden and Young showed the requirement of phosphate in alcoholic fermentation and concentrated the first coenzyme, cozymase, later shown to be NAD.
1905	Knoop deduced the β oxidation of fatty acids.
1907	Fletcher and Hopkins showed that lactic acid is formed quantitatively from glucose during anaerobic muscle contraction.
1909	Sörensen showed the effect of pH on enzyme action.
1911	Funk isolated crystals with vitamin B activity and coined the name vitamin.
1912	Neuberg proposed a chemical pathway for fermentation.
1912	Batelli and Stern discovered dehydrogenases.
1912	Warburg postulated a respiratory enzyme for the activation of

oxygen, discovered its inhibition by cyanide, and showed the requirement of iron in respiration.

1912–1922	Wieland showed the activation of hydrogen in dehydrogenation reactions.
1913	Michaelis and Menten developed a kinetic theory of enzyme action.
1913	Wilstatter and Stoll isolated and studied chlorophyll.
1914	Kendall isolated thyroxine.
1917	McCollum showed that xerophthalmia in rats is due to lack of vitamin A.
1922	Ruzicka recognized isoprene as the building block of many natural products.
1922	McCollum showed that lack of vitamin D causes rickets.
1922	Warburg and Negelein carried out the first measurements of the quantum efficiency of photosynthesis.
1923	Keilin rediscovered histohematins (cytochromes) and demonstrated changes in their oxidation state during respiratory activity.
1925	Briggs and Haldane made important refinements in the theory of enzyme kinetics.
1925–1930	Levene elucidated the structure of mononucleotides and showed they are building blocks of nucleic acids.
1925–1930	Svedberg invented the ultracentrifuge for determination of sedimentation rates of proteins.
1926	Sumner first crystallized an enzyme, urease, and proved it to be a protein.
1926	Jansen and Donath isolated vitamin B_1 (thiamine) from rice polishings.
1927	Muller, and also Stadler, demonstrated mutation of genes by x-rays.
1927	Windaus showed ergosterol is a precursor of vitamin D.
1928	Eggleton discovered phosphagen in muscle.
1928	Euler isolated carotene and showed it to have vitamin A activity.
1928–1932	Szent-Györgyi, and later Waugh and King, isolated ascorbic acid (vitamin C).
1928–1933	Warburg deduced the iron-porphyrin nature of the respiratory enzyme.
1929	Fiske and Subbarow isolated ATP and phosphocreatine from muscle extracts.
1930	Lundsgaard proved that muscles may contract in the absence of lactic acid formation.
1930–1933	Northrop isolated crystalline pepsin and trypsin and proved their protein nature.
1930–1935	Edsall and von Muralt isolated myosin from muscle.
1931	Engelhardt discovered that phosphorylation is coupled to respiration.
1932	Lohmann discovered the ATP-phosphocreatine reaction.
1932	Warburg and Christian discovered the "yellow enzyme," a flavoprotein.
1933	Keilin isolated cytochrome c and reconstituted electron transport in particulate heart preparations.

1933	Krebs and Henseleit discovered the urea cycle.
1933	Embden and also Meyerhof demonstrated crucial intermediates in the chemical pathway of glycolysis and fermentation.
1935	Williams and his colleagues deduced the structure of vitamin B_1.
1935	Rose discovered threonine, the last essential amino acid to have been recognized.
1935	Kuhn discovered that riboflavin (vitamin B_2) is a component of the "yellow enzyme."
1935	Schoenheimer and Rittenberg first used isotopes as tracers in the study of intermediary metabolism of carbohydrates and lipids.
1935	Stanley first crystallized a virus, tobacco mosaic virus.
1935	Szent-Gyorgyi showed the catalytic effect of dicarboxylic acids on respiration.
1935	Davson and Danielli postulated a model for the structure of cell membranes.
1935–1936	Warburg and Euler isolated and determined the structure and action of pyridine nucleotides.
1937	Krebs postulated the citric acid cycle.
1937	Lohmann and Schuster showed that thiamine is a component of the prosthetic group of pyruvate carboxylase.
1937–1938	Warburg showed how formation of ATP is coupled to the dehydrogenation of glyceraldehyde 3-phosphate.
1937–1941	Kalckar and Belitser independently carried out the first quantitative studies of oxidative phosphorylation.
1937	C. Cori and G. Cori began their incisive studies of glycogen phosphorylase.
1938	Hill found that cell-free suspensions of chloroplasts yield oxygen when illuminated in the presence of an electron acceptor.
1938	Braunstein and Kritzmann discovered transamination reactions.
1939	C. Cori and G. Cori demonstrated the reversible action of glycogen phosphorylase.
1939–1941	Lipmann postulated the central role of ATP in the energy-transfer cycle.
1939–1942	Engelhardt and Lyubimova discovered the ATPase activity of myosin.
1939–1946	Szent-Györgyi discovered actin and actomyosin.
1940	Beadle and Tatum deduced the one gene–one enzyme relationship.
1940–1943	Claude isolated and studied a mitochondrial fraction from liver.
1941–1944	Martin and Synge developed partition chromatography and applied it to amino acid analysis.
1942	Bloch and Rittenberg discovered that acetate is the precursor of cholesterol.
1943	A. Green and G. Cori crystallized muscle phosphorylase.
1943	Chance first applied sensitive spectrophotometric methods to enzyme-substrate interactions.
1943	Ochoa demonstrated that the P:O ratio of oxidative phosphorylation for the tricarboxylic acid cycle is 3.0.
1943–1947	Leloir and Munoz demonstrated fatty acid oxidation in cell-free

liver systems; Lehninger showed the requirement of ATP and the stoichiometry of fatty acid oxidation.

1944 Avery, MacLeod, and McCarty demonstrated that bacterial transformation is caused by DNA.

1945 Brand reported the first complete amino acid analysis of a protein, β-lactoglobulin, by chemical and microbiological methods.

1947–1950 Lipmann and Kaplan isolated and characterized coenzyme A.

1948 Leloir and his colleagues discovered the role of uridine nucleotides in carbohydrate metabolism.

1948 Hogeboom, Schneider, and Palade refined the differential-centrifugation method for cell fractionation.

1948 Calvin and Benson discovered that phosphoglyceric acid is an early intermediate in photosynthetic carbon dioxide fixation.

1948–1950 Kennedy and Lehninger discovered that the tricarboxylic acid cycle, fatty acid oxidation, and oxidative phosphorylation take place in mitochondria.

1949 Stein and Moore reported the complete amino acid analysis of β-lactoglobulin, determined by starch-column partition chromatography.

1949–1950 Sanger developed the 2,4-dinitrofluorobenzene method and Edman the phenylisothiocyanate procedure for identification of the N-terminal residues of peptides.

1950 Pauling and Corey proposed the α-helix structure for α-keratins.

1950–1953 Chargaff and colleagues discovered the base equivalences in DNA.

1951 Lehninger showed that electron transport from NADH to oxygen is the immediate energy source for oxidative phosphorylation.

1951 Lynen postulated the role of coenzyme A in fatty acid oxidation. Shortly after, the laboratories of Lynen, Green, and Ochoa isolated the enzymes of fatty acid oxidation.

1952–1953 Palade, Porter, and Sjostrand perfected thin sectioning and fixation methods for electron microscopy of intracellular structures.

1952–1954 Zamecnik and his colleagues discovered that ribonucleoprotein particles, later named ribosomes, are the site of protein synthesis.

1953 Sanger and Thompson completed the determination of the amino acid sequence of the A and B chains of insulin. Two years later Sanger and his colleagues reported the position of the disulfide cross-linkages.

1953 du Vigneaud carried out the first laboratory synthesis of the peptide hormones oxytocin and vasopressin.

1953 Horecker, Dickens, and also Racker elucidated the 6-phospho-gluconate pathway of glucose catabolism.

1953 Watson and Crick postulated the double-helical model of DNA structure.

1954 Arnon and colleagues discovered photosynthetic phosphorylation.

1954 Chance and Williams applied the oxygen electrode and difference spectrophotometry to study of the dynamics of electron transport in mitochondria.

1954–1958 Kennedy described the pathway of biosynthesis of triacylglycerols and phosphoglycerides and the role of cytidine nucleotides.

1955 Benzer carried out fine-structure genetic mapping and concluded that a gene has many mutable sites.

1955	Ochoa and Grunberg-Manago discovered polynucleotide phosphorylase.
1956	A. Kornberg discovered DNA polymerase.
1956	Umbarger reported that the end product isoleucine inhibits the first enzyme in its biosynthesis from threonine. Yates and Pardee reported the feedback inhibition of aspartate transcarbamoylase by cytidine triphosphate.
1956–1958	Anfinsen and White concluded that the three-dimensional conformation of proteins is specified by their amino acid sequence.
1957	Vogel, Magasanik, and others described genetic repression of enzyme synthesis.
1957	Sutherland discovered cyclic adenylic acid.
1957	Hoagland, Zamecnik, and Stephenson isolated transfer RNA and postulated its function.
1957	Skou discovered Na⁺K⁺-stimulated ATPase and postulated its role in transport of Na⁺ and K⁺ across the cell membrane.
1958–1959	S. B. Weiss, Hurwitz, and others discovered DNA-directed RNA polymerase.
1958	Crick enunciated the central dogma of molecular genetics.
1958	Meselson and Stahl provided experimental confirmation of the Watson-Crick model of semiconservative replication of DNA.
1958	Stein, Moore, and Spackman described the automatic amino acid analyzer, which greatly accelerated analysis of proteins.
1958	V. Ingram showed that normal and sickle-cell hemoglobin differ in a single amino acid residue in one of the chains.
1961–1968	Racker and his colleagues isolated F_1 ATPase from mitochondria and subsequently reconstituted oxidative phosphorylation in submitochondrial vesicles.
1950–1965	This period saw the elucidation of the major steps in the enzymatic pathways for the biosynthesis and degradation of amino acids, purines, pyrimidines, fatty acids, complex carbohydrates and lipids, and terpenoid compounds to which a great many investigators contributed, among whom were Ames, Baddiley, Bloch, Buchanan, Cantoni, Coon, B. D. Davis, D. E. Green, G. R. Greenberg, Gunsalus, Gurin, Handler, Hayaishi, Horecker, Kalckar, Kennedy, A. Kornberg, Lynen, Magasanik, Meister, Neuberger, Racker, Shemin, Sprinson, Strominger, and Weinhouse.
1960	Hirs, Moore, and Stein determined the amino acid sequence of ribonuclease; Anfinsen made important independent contributions.
1960	Kendrew reported the high-resolution x-ray analysis of the structure of sperm whale myoglobin.
1961	Jacob, Monod, and Changeux proposed a theory of the function and action of allosteric enzymes.
1961	Mitchell postulated the chemiosmotic hypothesis for the mechanism of oxidative and photosynthetic phosphorylation.
1961	Jacob and Monod proposed the operon hypothesis and postulated the function of messenger RNA.
1961	Nirenberg and Matthaei reported that polyuridylic acid codes for polyphenylalanine and thus opened the way to identification of the genetic code.
1961–1965	The laboratories of Nirenberg, Khorana, and Ochoa identified the genetic code words for the amino acids.

The Research Literature of Biochemistry

Some Annual Review Publications:

Advances in Carbohydrate Chemistry

Advances in Cell Biology

Advances in Comparative Physiology and
Biochemistry

Advances in Enzyme Regulation

Advances in Enzymology and Related Areas of
Molecular Biology

Advances in Immunology

Advances in Lipid Research

Advances in Protein Chemistry

Annual Review of Biochemistry

Annual Review of Genetics

Annual Review of Microbiology

Annual Review of Pharmacology

Annual Review of Physiology

Annual Review of Plant Physiology

Biochemical Society Symposia

Cold Spring Harbor Symposia in Quantitative
Biology

Essays in Biochemistry

Harvey Lectures

International Review of Cytology

Progress in Biophysics and Biophysical Chemistry

Progress in the Chemistry of Fats and Other Lipids

Progress in Nucleic Acid Research and Molecular
Biology

Vitamins and Hormones

Major Research Journals in Biochemistry and
Related Fields:

Accounts of Chemical Research

Analytical Biochemistry

Angewandte Chemie, International Edition in
English

Archives of Biochemistry and Biophysics

Bacteriological Reviews

Biochemical and Biophysical Research
Communications

Biochimie

Biochemical Journal

Biochemistry

Biochimica et Biophysica Acta

Biokhimiya (Russian; also translated into English)

Biological Reviews

Biopolymers

Bulletin de la Sociéte Chimie Biologique

Canadian Journal of Biochemistry

Canadian Journal of Microbiology

Cancer Research

Cell and Tissue Kinetics

Chemical Reviews

Chemistry and Physics of Lipids

Comparative Biochemistry and Physiology

Comptes Rendus de l'Acádemie des Sciences
(Paris), parts C and D

Comptes Rendus des Séances de la Société de
Biologie

European Journal of Biochemistry (formerly
Biochemische Zeitschrift)

Experimental Cell Research

Experientia

Federation Proceedings

Letters of the Federation of European Biochemical
Societies (FEBS Letters)

General Cytochemical Methods

Hoppe-Seyler's Zeitschrift für Physiologische
Chemie

Immunology

Indian Journal of Biochemistry

Journal of the American Chemical Society

Journal of the American Oil Chemists' Society

Journal of Bacteriology

Journal of Biochemistry (Tokyo)

Journal of Biological Chemistry

Journal of Cell Biology

Journal of Cell Science

Journal of Cellular Physiology

Journal of Chemical Education

Journal of Chromatography

Journal of Colloid and Interface Science

Journal of Electron Microscopy

Journal of Experimental Biology

Journal of General Microbiology

Journal of General Physiology

Journal of General Virology

Journal of Histochemistry and Cytochemistry

Journal of Immunology

Journal of Lipid Research

Journal of Membrane Biology

Journal of Molecular Biology

Journal of Neurochemistry

Journal of Pharmacology and Experimental Therapeutics

Journal of Theoretical Biology

Journal of Ultrastructure Research

Journal of Virology

Lipids

Macromolecules

Metabolism, Clinical and Experimental

Molecular and General Genetics

Molecular Pharmacology

Methods of Biochemical Analysis

Mutation Research

Nature

Naturwissenschaften

New England Journal of Medicine

Physiological Chemistry and Physics

Physiological Reviews

Plant Physiology

Proceedings of the National Academy of Sciences (U.S.)

Proceedings of the Royal Society (London), part B

Proceedings of the Society for Experimental Biology and Medicine

Protoplasma

Quarterly Review of Biology

Science

Steroids

Tetrahedron Letters

Virology

Zeitschrift für Naturforschung

Survey Works:

FLORKIN, M., and H. S. MASON (eds.): *Comparative Biochemistry*, Academic Press, Inc., New York, 7 vols., 1960ff.

FLORKIN, M., and E. H. STOTZ (eds.): *Comprehensive Biochemistry*, Elsevier Publishing Co., New York, 16 vols., 1967ff.

Collections of Biochemical Data, Constants, and Methods:

SOBER, H. A. (ed.): *Handbook of Biochemistry and Selected Data for Molecular Biology*, 2d ed., The Chemical Rubber Co., Cleveland, 1972.

DAWSON, R. M. C., D. C. ELLIOTT, W. H. ELLIOTT, and K. M. JONES (eds.): *Data for Biochemical Research*, Oxford University Press, London, 1959.

LONG C. (ed.): *Biochemist's Handbook*, D. Van Nostrand Company, Inc., Princeton, N.J., 1968.

DAYHOFF, M. O., and R. V. ECK: *Atlas of Protein Sequence and Structure*, National Biomedical Research Foundation, Silver Spring, Md., 5 vols. + 1 suppl., 1965ff.

Methods in Enzymology, Academic Press, Inc., New York. Multivolume series containing detailed descriptions of experimental methods.

APPENDIX C

Abbreviations Common in Biochemical Research Literature

A	Adenine or adenosine	FAD, FADH$_2$	Flavin adenine dinucleotide and its reduced form
ACP	Acyl carrier protein	FCCP	Carbonylcyanide p-trifluoro-methoxyphenylhydrazone
ACTH	Adrenocorticotrophic hormone		
ADH	Alcohol dehydrogenase	Fd	Ferredoxin
AMP, ADP, ATP	Adenosine 5'-mono-, di-, and triphosphate	FDNB (DNFB)	1-Fluoro-2,4-dinitrobenzene
		FDP	Fructose 1,6-diphosphate
cAMP	Cyclic AMP	FH$_2$, FH$_4$	Dihydro- and tetrahydrofolic acid
dAMP, dADP, dATP	2'-Deoxyadenosine 5'-mono-, di-, and triphosphate	fMet	N-Formylmethionine
Ala	Alanine	FMN, FMNH$_2$	Flavin mononucleotide and its reduced form
Arg	Arginine		
Asn	Asparagine	FP	Flavoprotein
Asp	Aspartic acid	Fru	D-Fructose
ATPase	Adenosine triphosphatase	G	Guanine or guanosine
C	Cytosine	Gal	D-Galactose
CD	Circular dichroism	GalNAc	N-Acetyl-D-galactosamine
Cer	Ceramide	GDH	Glutamate dehydrogenase
CMP, CDP, CTP	Cytidine mono-, di-, and triphosphate	GH	Growth hormone
		GLC	Gas-liquid chromatography
CM-cellulose	Carboxymethyl cellulose	Glc	D-Glucose
CoA, CoA-SH, acyl-CoA, acyl-S-CoA	Coenzyme A and its acyl derivatives	GlcNAc	N-Acetyl-D-glucosamine
		Gln	Glutamine
CoQ	Coenzyme Q; ubiquinone	Glu	Glutamic acid
Cys	Cysteine	Gly	Glycine
DEAE-cellulose	Diethylaminoethyl cellulose	cGMP	Cyclic GMP
DFP (sometimes DIFP)	Diisopropylphosphofluoridate	GMP, GDP, GTP	Guanosine mono-, di-, and triphosphate
DNA	Deoxyribonucleic acid	G3P	Glyceraldehyde 3-phosphate (3-phosphoglyceraldehyde)
DNase	Deoxyribonuclease		
DNP	2,4-Dinitrophenol	G6P	Glucose 6-phosphate
DOPA	Dihydroxyphenylalanine	GSH, GSSG	Glutathione and its oxidized form
DPN$^+$, DPNH	Same as NAD$^+$, NADH	Hb, HbO$_2$, HbCO	Hemoglobin, oxyhemoglobin, carbon monoxide hemoglobin
EC (followed by numbers)	Enzyme Commission, followed by numbers indicating formal classification of the enzyme		
		His	Histidine
		hnRNA	Heterogeneous nuclear RNA
EDTA	Ethylenediaminetetraacetic acid	Hyp	Hydroxyproline
EPR	Electron paramagnetic resonance	IgG	Immunoglobulin G
ESR	Electron spin resonance	Ile	Isoleucine
ETP	Electron transfer particle (from mitochondrial membrane)	IMP, IDP, ITP	Inosine mono-, di-, and triphosphate
FA	Fatty acid	LDH	Lactate dehydrogenase

Leu	Leucine	Rib	D-Ribose
LH	Luteinizing hormone	RNA	Ribonucleic acid
Lys	Lysine	mRNA	Messenger RNA
Mb, MbO$_2$	Myoglobin, oxymyoglobin	nRNA	Nuclear RNA
MDH	Malate dehydrogenase	rRNA	Ribosomal RNA
MtDNA	Mitochondrial DNA	tRNA	Transfer RNA
Met	Methionine	RNase	Ribonuclease
MetHb	Methemoglobin	RQ	Respiratory quotient
MSH	Melanocyte-stimulating hormone	Ser	Serine
NAD$^+$, NADH	Nicotinamide adenine dinucleotide (diphosphopyridine nucleotide) and its reduced form	T	Thymine or thymidine
		TH	Thyrotropic hormone
		Thr	Threonine
NADP$^+$, NADPH	Nicotinamide adenine dinucleotide phosphate (triphosphopyridine nucleotide) and its reduced form	TLC	Thin-layer chromatography
		TMP, TDP, TTP	Thymidine mono-, di-, and triphosphate
NAN, NANA	N-Acetylneuraminic acid	TMV	Tobacco mosaic virus
NEFA	Nonesterified fatty acid	TPCK	L-1-Tosyl-2-phenylethyl-chloromethylketone
NMN$^+$, NMNH	Nicotinamide mononucleotide and its reduced form	TPN$^+$, TPNH	Same as NADP$^+$, NADPH
NMR	Nuclear magnetic resonance	TPP	Thiamine pyrophosphate
OAA	Oxaloacetic acid	Tris	Tris(hydroxymethyl)amino-methane (a buffer component)
OD	Optical density		
ORD	Optical rotatory dispersion	Trp	Tryptophan
P$_i$	Inorganic phosphate	Tyr	Tyrosine
PAB or PABA	p-Aminobenzoic acid	U	Uracil or uridine
PEP	Phosphoenolpyruvate	UDP-gal	Uridine diphosphate galactose
3PG	3-Phosphoglycerate	UDP-glc	Uridine diphosphate glucose
PGA	Pteroylglutamic acid (folic acid)	UFA	Unesterified fatty acid
Phe	Phenylalanine	UMP, UDP, UTP	Uridine mono-, di-, and triphosphate
PMS	Phenazine methosulphate		
PP$_i$	Inorganic pyrophosphate	UV	Ultraviolet
ppGpp	Guanosine 5'-diphosphate-3'-diphosphate	Val	Valine
Pro	Proline		
PRPP	5-Phospho-α-D-ribose 1-pyrophosphate		

Unit Abbreviation and Prefixes and Physical Constants

A	Ampere		m	Molal concentration
Å	Angstrom		m	Meter
atm	Atmosphere		mg	Milligram
cal	Calorie		min	Minute
Ci	Curie		ml	Milliliter
cm	Centimeter		mm	Millimeter
dm	Decimeter		mm Hg	Millimeters of mercury pressure
$\mathscr{F}$	Faraday		mm H_2O	Millimeters of water pressure
g	Gram		mol	Mole
G	Gauss		mol wt	Molecular weight
h	Hour		N	Normal concentration
kat	Katal		nm	Nanometer
kcal	Kilocalorie		pH	$-\log H^+$ concentration
kJ	Kilojoule		pK'	$-\log K'$
K	Kelvin		ps	Picosecond
l	Liter		r	Revolution
ln	Logarithm to the base e		S	Svedberg unit
log	Logarithm to the base 10		Sf	Svedberg flotation unit
μm	Micrometer		STP	Standard temperature and pressure
μmol	Micromole		s	Second
M	Molar concentration		V	Volt

Prefixes Used in International System of Units

Multiple	Prefix	Abbreviation
10^{12}	tera	T
10^9	giga	G
10^6	mega	M
10^3	kilo	k
10^2	hecto	h
10^{-1}	deci	d
10^{-2}	centi	c
10^{-3}	milli	m
10^{-6}	micro	μ
10^{-9}	nano	n
10^{-12}	pico	p
10^{-15}	femto	f
10^{-18}	atto	a

Some Physical Constants

	Symbol or abbreviation	Value
Avogadro's number	N	6.022×10^{23} mol^{-1}
Curie	Ci	3.70×10^{10} disintegrations s^{-1}
Dalton	—	1.661×10^{-24} g
Faraday constant	$\mathscr{F}$	96,487 C mol^{-1} 23,062 cal V^{-1} mol^{-1}
Gas constant	R	8.314 J mol^{-1} K^{-1} 1.987 cal mol^{-1} K^{-1}
Planck's constant	h	6.626×10^{-34} J-s 1.584×10^{-34} cal-s
Velocity of light (vacuum)	c	2.997×10^{10} cm s^{-1}

Other Constants and Factors

$\pi = 3.1416$
$e = 2.718$
$\ln (\log_e) x = 2.303 \log_{10} x$
$1.0 \text{ J} = 0.239 \text{ cal} = 10^7 \text{ erg} = 1 \text{ W-s}$

International Atomic Weights

Element	Symbol	Atomic number	Atomic weight	Element	Symbol	Atomic number	Atomic weight
Aluminum	Al	13	26.98	Neodymium	Nd	60	144.27
Antimony	Sb	51	121.75	Neon	Ne	10	20.179
Argon	Ar	18	39.948	Nickel	Ni	28	58.71
Arsenic	As	33	74.92	Niobium	Nb	41	92.91
Barium	Ba	56	137.34	Nitrogen	N	7	14.0067
Beryllium	Be	4	9.012	Osmium	Os	76	190.2
Bismuth	Bi	83	208.98	Oxygen	O	8	16.000
Boron	B	5	10.81	Palladium	Pd	46	106.4
Bromine	Br	35	79.904	Phosphorus	P	15	30.974
Cadmium	Cd	48	112.40	Platinum	Pt	78	195.09
Calcium	Ca	20	40.08	Potassium	K	19	39.09
Carbon	C	6	12.01	Praseodymium	Pr	59	140.907
Cerium	Ce	58	140.12	Protactinium	Pa	91	231.036
Cesium	Cs	55	132.91	Radium	Ra	88	226.025
Chlorine	Cl	17	35.453	Radon	Rn	86	222
Chromium	Cr	24	52.00	Rhenium	Re	75	186.2
Cobalt	Co	27	58.93	Rhodium	Rh	45	102.91
Copper	Cu	29	63.55	Rubidium	Rb	37	85.467
Dysprosium	Dy	66	162.50	Ruthenium	Ru	44	101.07
Erbium	Er	68	167.26	Samarium	Sm	62	150.43
Europium	Eu	63	151.96	Scandium	Sc	21	44.96
Fluorine	F	9	18.998	Selenium	Se	34	78.96
Gadolinium	Gd	64	157.25	Silicon	Si	14	28.086
Gallium	Ga	31	69.72	Silver	Ag	47	107.868
Germanium	Ge	32	72.59	Sodium	Na	11	22.997
Gold	Au	79	196.97	Strontium	Sr	38	87.62
Hafnium	Hf	72	178.49	Sulfur	S	16	32.06
Helium	He	2	4.003	Tantalum	Ta	73	180.947
Holmium	Ho	67	164.93	Tellurium	Te	52	127.60
Hydrogen	H	1	1.0079	Terbium	Tb	65	158.93
Indium	In	49	114.82	Thallium	Tl	81	204.39
Iodine	I	53	126.905	Thorium	Th	90	232.038
Iridium	Ir	77	192.22	Thulium	Tm	69	169.934
Iron	Fe	26	55.84	Tin	Sn	50	118.69
Krypton	Kr	36	83.80	Titanium	Ti	22	47.90
Lanthanum	La	57	138.905	Tungsten	W	74	183.85
Lead	Pb	82	207.21	Uranium	U	92	238.03
Lithium	Li	3	6.941	Vanadium	V	23	50.941
Lutecium	Lu	71	174.97	Xenon	Xe	54	131.30
Magnesium	Mg	12	24.305	Ytterbium	Yb	70	173.04
Manganese	Mn	25	54.938	Yttrium	Y	39	88.906
Mercury	Hg	80	200.59	Zinc	Zn	30	65.38
Molybdenum	Mo	42	95.94	Zirconium	Zr	40	91.22

Logarithms

N	0	1	2	3	4	5	6	7	8	9
10	0000	0043	0086	0128	0170	0212	0253	0294	0334	0374
11	0414	0453	0492	0531	0569	0607	0645	0682	0719	0755
12	0792	0828	0864	0899	0934	0969	1004	1038	1072	1106
13	1139	1173	1206	1239	1271	1303	1335	1367	1399	1430
14	1461	1492	1523	1553	1584	1614	1644	1673	1703	1732
15	1761	1790	1818	1847	1875	1903	1931	1959	1987	2014
16	2041	2068	2095	2122	2148	2175	2201	2227	2253	2279
17	2304	2330	2355	2380	2405	2430	2455	2480	2504	2529
18	2553	2577	2601	2625	2648	2672	2695	2718	2742	2765
19	2788	2810	2833	2856	2878	2900	2923	2945	2967	2989
20	3010	3032	3054	3075	3096	3118	3139	3160	3181	3201
21	3222	3243	3263	3284	3304	3324	3345	3365	3385	3404
22	3424	3444	3464	3483	3502	3522	3541	3560	3579	3598
23	3617	3636	3655	3674	3692	3711	3729	3747	3766	3784
24	3802	3820	3838	3856	3874	3892	3909	3927	3945	3962
25	3979	3997	4014	4031	4048	4065	4082	4099	4116	4133
26	4150	4166	4183	4200	4216	4232	4249	4265	4281	4298
27	4314	4330	4346	4362	4378	4393	4409	4425	4440	4456
28	4472	4487	4502	4518	4533	4548	4564	4579	4594	4609
29	4624	4639	4654	4669	4683	4698	4713	4728	4742	4757
30	4771	4786	4800	4814	4829	4843	4857	4871	4886	4900
31	4914	4928	4942	4955	4969	4983	4997	5011	5024	5038
32	5051	5065	5079	5092	5105	5119	5132	5145	5159	5172
33	5185	5198	5211	5224	5237	5250	5263	5276	5289	5302
34	5315	5328	5340	5353	5366	5378	5391	5403	5416	5428
35	5441	5453	5465	5478	5490	5502	5514	5527	5539	5551
36	5563	5575	5587	5599	5611	5623	5635	5647	5658	5670
37	5682	5694	5705	5717	5729	5740	5752	5763	5775	5786
38	5798	5809	5821	5832	5843	5855	5866	5877	5888	5899
39	5911	5922	5933	5944	5955	5966	5977	5988	5999	6010
40	6021	6031	6042	6053	6064	6075	6085	6096	6107	6117
41	6128	6138	6149	6160	6170	6180	6191	6201	6212	6222
42	6232	6243	6253	6263	6274	6284	6294	6304	6314	6325
43	6335	6345	6355	6365	6375	6385	6395	6405	6415	6425
44	6435	6444	6454	6464	6474	6484	6493	6503	6513	6522
45	6532	6542	6551	6561	6571	6580	6590	6599	6609	6618
46	6628	6637	6646	6656	6665	6675	6684	6693	6702	6712
47	6721	6730	6739	6749	6758	6767	6776	6785	6794	6803
48	6812	6821	6830	6839	6848	6857	6866	6875	6884	6893
49	6902	6911	6920	6928	6937	6946	6955	6964	6972	6981
50	6990	6998	7007	7016	7024	7033	7042	7050	7059	7067
51	7076	7084	7093	7101	7110	7118	7126	7135	7143	7152
52	7160	7168	7177	7185	7193	7202	7210	7218	7226	7235
53	7243	7251	7259	7267	7275	7284	7292	7300	7308	7316
54	7324	7332	7340	7348	7356	7364	7372	7380	7388	7396
N	0	1	2	3	4	5	6	7	8	9

N	0	1	2	3	4	5	6	7	8	9
55	7404	7412	7419	7427	7435	7443	7451	7459	7466	7474
56	7482	7490	7497	7505	7513	7520	7528	7536	7543	7551
57	7559	7566	7574	7582	7589	7597	7604	7612	7619	7627
58	7634	7642	7649	7657	7664	7672	7679	7686	7694	7701
59	7709	7716	7723	7731	7738	7745	7752	7760	7767	7774
60	7782	7789	7796	7803	7810	7818	7825	7832	7839	7846
61	7853	7860	7868	7875	7882	7889	7896	7903	7910	7917
62	7924	7931	7938	7945	7952	7959	7966	7973	7980	7987
63	7993	8000	8007	8014	8021	8028	8035	8041	8048	8055
64	8062	8069	8075	8082	8089	8096	8102	8109	8116	8122
65	8129	8136	8142	8149	8156	8162	8169	8176	8182	8189
66	8195	8202	8209	8215	8222	8228	8235	8241	8248	8254
67	8261	8267	8274	8280	8287	8293	8299	8306	8312	8319
68	8325	8331	8338	8344	8351	8357	8363	8370	8376	8382
69	8388	8395	8401	8407	8414	8420	8426	8432	8439	8445
70	8451	8457	8463	8470	8476	8482	8488	8494	8500	8506
71	8513	8519	8525	8531	8537	8543	8549	8555	8561	8567
72	8573	8579	8585	8591	8597	8603	8609	8615	8621	8627
73	8633	8639	8645	8651	8657	8663	8669	8675	8681	8686
74	8692	8698	8704	8710	8716	8722	8727	8733	8739	8745
75	8751	8756	8762	8768	8774	8779	8785	8791	8797	8802
76	8808	8814	8820	8825	8831	8837	8842	8848	8854	8859
77	8865	8871	8876	8882	8887	8893	8899	8904	8910	8915
78	8921	8927	8932	8938	8943	8949	8954	8960	8965	8971
79	8976	8982	8987	8993	8998	9004	9009	9015	9020	9025
80	9031	9036	9042	9047	9053	9058	9063	9069	9074	9079
81	9085	9090	9096	9101	9106	9112	9117	9122	9128	9133
82	9138	9143	9149	9154	9159	9165	9170	9175	9180	9186
83	9191	9196	9201	9206	9212	9217	9222	9227	9232	9238
84	9243	9248	9253	9258	9263	9269	9274	9279	9284	9289
85	9294	9299	9304	9309	9315	9320	9325	9330	9335	9340
86	9345	9350	9355	9360	9365	9370	9375	9380	9385	9390
87	9395	9400	9405	9410	9415	9420	9425	9430	9435	9440
88	9445	9450	9455	9460	9465	9469	9474	9479	9484	9489
89	9494	9499	9504	9509	9513	9518	9523	9528	9533	9538
90	9542	9547	9552	9557	9562	9566	9571	9576	9581	9586
91	9590	9595	9600	9605	9609	9614	9619	9624	9628	9633
92	9638	9643	9647	9652	9657	9661	9666	9671	9675	9680
93	9685	9689	9694	9699	9703	9708	9713	9717	9722	9727
94	9731	9736	9741	9745	9750	9754	9759	9763	9768	9773
95	9777	9782	9786	9791	9795	9800	9805	9809	9814	9818
96	9823	9827	9832	9836	9841	9845	9850	9854	9859	9863
97	9868	9872	9877	9881	9886	9890	9894	9899	9903	9908
98	9912	9917	9921	9926	9930	9934	9939	9943	9948	9952
99	9956	9961	9965	9969	9974	9978	9983	9987	9991	9996
N	0	1	2	3	4	5	6	7	8	9

Solutions to Problems

Chapter 2

1 (a) 5.0; (b) 2.30; (c) 8.48; (d) 7.0;
  (e) 10.48; (f) 1.85
2 (a) 4.0×10^{-8} M; (b) 8.0×10^{-7} M;
  (c) 2.0×10^{-10} M; (d) 4.0×10^{-2} M;
  (e) 3.2×10^{-8} M; (f) 3.2×10^{-8} M
3 (a) 2.51×10^{-6}; (b) 5.60
4 (a) 0.0132; (b) 0.0417
6 (a) 0.0171; (b) 0.148; (c) 0.500
7 (a) 1.35×10^{-5} M; (b) 4.87
8 (a) 4.3×10^3; (b) 1.58; (c) phosphate
9 2.5
10 (a) 1.28 ml; (b) 10^{-5} M
11 (a) 2.89; (b) 6.31×10^{-11} M
12 $[A^{2-}] \approx K_2'$
13 (a) Add 22.43 ml 0.1 N NaOH to 27.57 ml
  0.1 N acetic acid; (b) 5.30
14 (a) 3.38; (b) 3.47
15 (a) 4.38; (b) 5.02
16 8.38
17 (a) 3.26; (b) 6.13×10^{-4} M
18 (a) 4.78×10^{-5} M; (b) 3.55×10^{-3} M;
  (c) 6.16×10^{-3} M; (d) 2.45×10^{-4} M

Chapter 4

1 Asn/Gln
2 Ornithine
3 Gly, 5.97; Ala, 6.02; Ser, 5.68; Thr, 6.53;
  Glu, 3.22
4 Gly: (a) +; (b) +; (c) +; (d) −
  Asp: (a) +; (b) +; (c) −; (d) −
  Lys: (a) +; (b) +; (c) +; (d) 0/−
  His: (a) +; (b) +; (c) +; (d) −
5 (a) Glu; (b) Lys, Arg; (c) Gly, Ala, Ser
6 Arg, Lys, Tyr, Ala, Glu, Asp
7 Ile, Ala, Ser, Glu
8 0.38 g
9 (a) 2.71; (b) 9.23
10 4.45
11 0.45°
12 (a) 0.152 M; (b) 0.182 M
13 0.64 mg Gly; 0.11 mg Phe
14 (a) tube 4, tube 3; (b) tube 1, 0.8 mg X,
  6.4 mg Y; tube 2, 9.6 mg X, 28.8 mg Y;
  tube 3, 38.4 mg X, 43.2 mg Y;
  tube 4, 51.2 mg X, 21.6 mg Y
15 (a) 1.44 mg; (b) 6.09×10^{-3} mg
16 (a) 3.50; (b) 0.0748 M
17 (a) 0.0843 M; (b) 0.0079 M; (c) 8

Chapter 5

1 (a) (1) Lys, Asp-Gly-Ala-Ala-Glu-Ser-Gly
      (2) Ala-Ala-His-Arg, Glu-Lys, Phe-Ile
      (3) Tyr-Cys-Lys, Ala-Arg, Arg, Gly
      (4) Phe-Ala-Glu-Ser-Ala-Gly
  (b) (1) 2,4-DNP-Lys, 2,4-DNP-Asp
      (2) 2,4-DNP-Ala, 2,4-DNP-Glu,
        2,4-DNP-Phe
      (3) 2,4-DNP-Tyr, 2,4-DNP-Ala, 2,4-DNP-Arg,
        2,4-DNP-Gly
      (4) 2,4-DNP-Phe
2 (a) Val-Ala-Lys-Glu-Glu-Phe, Val-Met-Tyr,

Cys-Glu-Trp, Met-Gly-Gly-Phe
  (b) Val-Ala-Lys-Glu-Glu-Phe, Val-homoserine
  lactone, Tyr, Cys-Glu-Trp, homoserine
  lactone, Gly-Gly-Phe
3 (a) Leu-Met-His-Tyr-Lys, Arg, Ser-Val-Cys-
  Ala-Lys, Asp-Gly-Ile-Phe-Ile
  (b) Leu-Met-His-Tyr-Lys, Arg, Ser,
  Val-Cys-Ala-Lys, Asp-Gly, Ile-Phe, Ile
4 (a) Chymotrypsin and thermolysin digestion;
  (b) chymotrypsin: Gly-Leu-Ser-Pro-Phe,
  His-Thr-Asp-Val-Ser-Ala-Ala-Trp,
  Gly-Glu-Val-Gly-Ala-His-Leu-Gly-Glu-Tyr,
  Gly-Ala-Glu-Ala-Thr-Glu; thermolysin: Gly,
  Leu-Ser-Pro-Phe-His-Thr-Asp, Val-Ser-Ala-
  Ala-Trp-Gly-Glu, Val-Gly-Ala-His,
  Leu-Gly-Glu-Tyr-Gly-Ala-Glu-Ala-Thr-Glu
5 Val-(Arg, Lys)-Pro-Gly
6 (a) C, C, C, A; (b) C, C, 0, A;
  (c) C, C, 0/A, A; (d) C, 0/C, A, A;
  (e) C, C, C, A
7 $a \approx e, b, c, d$
8 Changes in mobility toward anode:
  (a) decrease; (b) increase; (c) decrease;
  (d) increase; (e) decrease
9 Gly-Ile-Val-Glu-Glu-Cys-Cys-Ala-Ser-Val-Cys-
  Ser-Leu-Tyr-Glu-Leu-Glu-Asp-Tyr-Cys-Asp
10 Asp-Ala-Tyr-Glu-Lys-His-Ile-Glu-Val
11 Gln-Trp-Lys-Trp-Glu
12 Trp-Gly-Arg-Trp-Gly-Arg

Chapter 6

1 (a) 15.8 nm; (b) 37.9 nm
2 (a) 1 to 6; 8 to 15; 20 to 28; (b) 7; 16 to 19;
  (c) 13; 24
3 (a) 3,846 per 10^6 residues (b) 1.8
4 51%

Chapter 7

1 13,110
2 13,700
3 (a) 3,846 per 10^6 residues; (b) 1.8
4 13,000
5 494,000
6 (a) A; (b) C, A; (c) C, 0, A
7 (a) 5.85; (b) 8.25; (c) 4.90
8 Myosin, catalase, serum albumin, chymotrypsinogen, myoglobin, cytochrome c
9 52,000

Chapter 8

1 0.089 mM
2 1.5
3 (a) 2.31 s^{-1}; (b) 1.30 s
4 0.0031 s^{-1}
5 $[S] = 4K_M$
7 29,200
8 1.86×10^{-6} M
9 (a) Noncompetitive; both K_I values equal;
  (b) 2.44 mM; (c) 24.8 mM
10 (a) Competitive; (b) 5.4 mM; (c) 9.0 mM
11 (a) Uncompetitive; (b) 1.26 μM

12 (a) $\dfrac{1}{V_{max}}\left[K_M^B + \dfrac{K_S^A K_M^B}{[A]}\right]$; (b) $\dfrac{1}{V_{max}}\left[1 + \dfrac{K_M^A}{[A]}\right]$;

(c) $\dfrac{-K_M^A}{K_S^A K_M^B}$; (d) $\dfrac{1}{V_{max}}\left[1 - \dfrac{K_M^A}{K_S^A}\right]$

13 Double displacement
14 Ordered single displacement with NAD(H) leading
15 Random single displacement

Chapter 10

1 (a) 93.3°; (b) 52.7°; (c) 62.4°, 62.4°
2 (a) Pentaacetyl-α-D-galactose; (b) 2,3,4,6-tetra-O-methyl-α-D-glucose; 2,3,4,6-tetra-O-methyl-β-D-glucose
3 (a) D-Galactitol; (b) L-mannaric acid; (c) 1,3,4,6-tetra-O-methyl-N-dimethyl D-galactosamine
4 (a) 2,3,4,6-Tetra-O-methyl-D-galactose; 2,3,6-tri-O-methyl-D-glucose; (b) 2,3,4,6-tetra-O-methyl-D-glucose; 1,3,4,6-tetra-O-methyl-D-fructose
5 4-O-D-Glucosyl-6-O-D-galactosyl-D-glucose, or 6-O-D-glucosyl-4-O-D-glucosyl-D-galactose
6 Unbranched polymer of (6-O-D-glucosyl-4-O-D-glucosyl-) units
7 (a) 9.72%; (b) 10.3 glucose residues; (c) 49.7 mmol; (d) 12,300 glucose residues
8 820 glucose residues

Chapter 11

1 OPS, POS, OSP, SPO, SOP, PSO
2 (a) PPP, OOO, SSS, PPO, PPS, OOP, OOS, SSP, SPO, SSO, POP, PSP, OSO, SPS, SOS, OPO, POS, PSO, OPP, SPP, POO, SOO, PSS, OPS, OSS, SOP, OSP; (b) 18
3 Average of 16.7 carbons per fatty acid, assuming all are saturated
4 Phosphatidic acid; it is amphipathic and forms stable micelles in water; triacylglycerols do not
5 (a) A; (b) A; (c) O; (d) A; (e) C
6 (a) Glycerol, sodium stearate, sodium palmitate; (b) glycerol 3-phosphorylinositol, sodium stearate, sodium elaidate; (c) glycerol 3-phosphorylcholine, sodium palmitate, sodium oleate
7 (a) Stearic acid, oleic acid, glycerol, phosphoric acid, serine; (b) 1-palmitoyl-2-linoleyl phosphatidic acid, choline
8 (a) 1.13 g cm^{-3}; (b) sediment
9 Molar ratio lipids/protein = 41.7

Chapter 12

1 (a) 0.493; (b) 0.356; (c) 0.0594; (d) 1.037
2 (a) 2.78×10^{-5} M; (b) 9.78×10^{-6} M
3 AMP = 3.43×10^{-5} M; GMP = 1.06×10^{-5} M
4 (a) A; pU; pA; pA; pC; pU; (b) AU; AAC; U; (c) no effect; (d) an apurinic acid with the structure (5′)T-C--C-(3′), plus two molecules of the free bases adenine and guanine; (e) no effect
5 pGpCpCpApUpCpGpApC
6 ApUpCpApGpCpCpGpUpCpGpGp
7 (a) 1.77×10^{-5} M; (b) 1.30×10^{-5} M; (c) 8.00×10^{-6} M; (d) 1.41×10^{-5} M

Chapter 14

1 $t_{1/2} = 0.693/\lambda$
2 1.07×10^6
3 32.6 h
4 376 days
5 28.7% P; 71.3% S
6 1.46 mg ml^{-1}
7 1,163 ml

Chapter 15

1 (a) 10.1%; (b) 11.6%
2 $-275.68 - (-219.22) = -56.46$ kcal mol^{-1}
3 (a) 263; (b) 311
4 (a) -1.13 kcal mol^{-1}; (b) $+19.4$ kcal mol^{-1}; (c) $+11.1$ kcal mol^{-1}; (d) $+16.5$ kcal mol^{-1}
5 (a) -7.30 kcal mol^{-1}; (b) -8.66 kcal mol^{-1}; (c) -10.0 kcal mol^{-1}; (d) -11.4 kcal mol^{-1}
6 -12.3 kcal mol^{-1}
7 -1.3 kcal mol^{-1}
8 3.55×10^{-3} M
9 (a) [PEP] = [ADP] = 0.0178 mM; [Pyr] = [ATP] = 9.9822 mM (b) [PEP] = [ADP] = 0.0151 mM; [Pyr] = 5.98 mM; [ATP] = 11.98 mM
10 1.26 mM
11 (a) [ADP] = 0.0991 mM, [AMP] = 0.0009 mM; (b) $+3,344\%$

Chapter 16

1 (a) $-$; (b) $-$; (c) $+$; (d) $+$; (e) $+$; (f) $-$; (g) $-$; (h) $-$; (i) $+$
2 (a) and (b): D-fructose + 2ADP + $2P_i \longrightarrow$ 2 lactic acid + 2ATP + $2H_2O$
3 (a) Glycerol + NAD$^+$ + ADP + $P_i \longrightarrow$ lactic acid + NADH +H$^+$ + ATP + H_2O (b) L-Glycerol 3-phosphate + NAD$^+$ + 2ADP + $P_i \longrightarrow$ ethanol + CO_2 + NADH + H$^+$ + 2ATP + H_2O (c) D-Mannose + 2NAD$^+$ + $2P_i \longrightarrow$ 2 phosphoenolpyruvate + 2NADH + 2H$^+$ + $2H_2O$ (d) Sucrose + 5ADP + $5P_i \longrightarrow$ 4 ethanol + $4CO_2$ + 5ATP + $4H_2O$ (e) Glyceraldehyde + P_i + ADP $\longrightarrow$ ethanol + CO_2 + ATP + H_2O
4 Fructose 1,6-diphosphate + 2NAD$^+ \longrightarrow$ 2 phosphoenolpyruvate + 2NADH + 2H$^+$
5 190 mM glucose, 20 mM ethanol; increase phosphate to at least 380 mM
6 Fructose 1,6-diphosphate, dihydroxyacetone phosphate, D-glyceraldehyde 3-phosphate, D-glyceraldehyde, fructose 1-phosphate, acetaldehyde, 5-deoxyketopentose
7 (a) -25.84 kcal mol^{-1}; (b) -21.68 kcal mol^{-1}
8 -0.354 kcal mol^{-1}
9 (a) 0.79%; (b) 2.48%; (c) 7.63% (d) 22.1%; (e) 53.9%
10 -1.93 kcal mol^{-1}
11 (a) 0.30; (b) 1.50; (c) 5.10; (d) 7.58; (e) 11.3

Chapter 17

1 (a) 2; (b) 1; (c) 3; (d) 2
2 (a) Carbon atom 2 initially, then 1, 2, and 3; (b) carbon atom 2; (c) carbon atoms 2 and 3
3 (a) (1) HOOC—^{14}CH$_2$—CH(COOH)CH(OH)—COOH, (2) carbon atom 4, (3) carbon atom 2;

(b) (1) at carbon atoms marked (*):
HOOC—*CH$_2$—*CH(COOH)*CH(OH)COOH,
(2) carbon atoms 2, 3, and 4, (3) carbon atoms 1,
2, and 3; (c) (1) at carbon atoms marked (*):
HOOC—CH$_2$—CH(*COOH)CH(OH)*COOH,
(2) 1-^{14}C-α-ketoglutaric acid, (3) unlabeled,
(4) carbon atoms 1 and 4; (d) (1) at carbon
atoms marked (*): HOO*C—CH$_2$—
*CH(COOH)*CH(OH)COOH, (2) carbon atoms
1, 2, 3, and 4

4 (a) 0; (b) all; (c) 0

5 (a) 0; (b) 0; (c) one-half; (d) one-fourth;
(e) one-eighth

6 (a) Citric acid + 2NAD$^+$ + GDP + P $\longrightarrow$
succinic acid + 2CO$_2$ + 2NADH + 2H$^+$ + GTP
(b) 2 Pyruvic acid + 2H$_2$O + 3NAD$^+$ +
GDP + ATP $\longrightarrow$ succinic acid + 2CO$_2$ +
3NADH + 3H$^+$ + GTP + ADP
(c) Fumaric acid + AcCoA + 2H$_2$O +
3NAD$^+$ + GDP + P$_i$ $\longrightarrow$ succinic acid +
CoA + 2CO$_2$ + 3NADH + 3H$^+$ + GTP

7 (a) and (b) are the same as in Problem 6;
(c) 2 fumaric acid + 3H$_2$O + 5NAD$^+$ +
ADP + P$_i$ $\longrightarrow$ succinic acid +
4CO$_2$ + 5NADH + 5H$^+$ + ATP

8 Glucose + ATP + 2NADP$^+$ + H$_2$O $\longrightarrow$
ribose 5-phosphate + ADP + 2NADPH +
2H$^+$ + CO$_2$; carbon atom 2

9 5 Glucose 6-phosphate + 3ATP + 2H$_2$O $\longrightarrow$
6 ribose 5-phosphate + 3ADP + 2P$_i$

10 Glucose + 6NADP$^+$ + NAD$^+$ + 2H$_2$O + ADP +
P$_i$ $\longrightarrow$ pyruvate + 3CO$_2$ + 6NADPH +
NADH + 7H$^+$ + ATP

11 6 Ribose 5-phosphate + 16ADP + 10P$_i$ $\longrightarrow$
10 lactic acid + 16ATP + 10H$_2$O;
3-^{14}C-lactic acid, 1,3-^{14}C-lactic acid

12 −0.6 kcal mol^{-1}; $K' = 2.75\ M$

13 2.74 + 10^{-8} M

Chapter 18

1 (a) −0.35 V; (b) −0.32 V; (c) −0.29 V

2 a, d, e, b, c

3 a, d, b, c

4 c, d, f

5 (a) −2767 cal mol^{-1}; (b) −2814 cal mol^{-1};
(c) +7.10 kcal mol^{-1}; (d) −26.50 kcal mol^{-1}

6 (a) 6.2 μM; (b) 7.57 M

7 OAA = NADH = 0.0249 mM, NAD =
malate = 19.98 mM

8 OAA = 20 mM, malate = 30 mM

9 (a) +1.43 kcal mol^{-1}; (b) +13.33 kcal mol^{-1}

10 24-Hydroxycholecalciferol + NAD(P)H +
H$^+$ + O$_2$ $\longrightarrow$ 1,24-dihydroxycholecalciferol +
H$_2$O + NAD(P)$^+$

11 B

12 (a) Straight lines intersecting to the left of the
vertical axis; (b) they should be mutually
competitive

13 (a) NAD and NADH dehydrogenase reduced;
cytochromes b, c, a oxidized; (b) NAD,
NADH dehydrogenase and cytochrome b
reduced; cytochromes c, a oxidized; (c) all
reduced

Chapter 19

1 NADH + H$^+$ + $\frac{1}{2}$O$_2$ + 2ADP + 2P$_i$ $\longrightarrow$
NAD$^+$ + 3H$_2$O + 2ATP

2 NADH + H$^+$ + $\frac{1}{2}$O$_2$ + 3ADP + 3P$_i$ $\longrightarrow$
NAD$^+$ + 4H$_2$O + 3ATP

3 (a) 3.5; (b) 1.0; (c) 2.5; (d) 1.0

4 (a) Glyceraldehyde + 3O$_2$ + 17ADP + GDP +
18P$_i$ $\longrightarrow$ 3CO$_2$ + 17ATP + GTP + 21H$_2$O
(b) Pyruvic acid + 2$\frac{1}{2}$O$_2$ + 14ADP + GDP +
15P$_i$ $\longrightarrow$ 3CO$_2$ + 14ATP + GTP + 17H$_2$O
(c) Citric acid + 4$\frac{1}{2}$O$_2$ + 26ADP + GDP +
27P$_i$ $\longrightarrow$ 6CO$_2$ + 26ATP + GTP + 31H$_2$O
(d) Fumaric acid + 3O$_2$ + 18ADP +
18P$_i$ $\longrightarrow$ 4CO$_2$ + 18ATP + 20H$_2$O
(e) Oxaloacetic acid + 2$\frac{1}{2}$O$_2$ + 15ADP +
15P$_i$ $\longrightarrow$ 4CO$_2$ + 15ATP + 17H$_2$O

5 Fructose + 6O$_2$ + 2ADP + 4P$_i$ + 2GDP $\longrightarrow$
6CO$_2$ + 2ADP + 2GTP + 10H$_2$O

6 40.1%

7 40.4%

8 Aerobic conditions: phosphorylation
potential = 3.50; energy charge = 0.91;
anaerobic conditions: phosphorylation
potential = 0.54; energy charge = 0.83

Chapter 20

1 (a) Myristic acid + 6O$_2$ + 7CoA + 30ADP +
30P$_i$ $\longrightarrow$ 7 acetyl CoA + 36H$_2$O + 29ATP +
AMP + PP$_i$
(b) Arachidonic acid + 27O$_2$ + 157ADP +
157P$_i$ $\longrightarrow$ 20CO$_2$ + 172H$_2$O + 156ATP +
AMP + PP$_i$
(c) Palmitic acid + 7O$_2$ + 35ADP + 35P$_i$ $\longrightarrow$
4 acetoacetic acid + 37H$_2$O + 34ATP +
AMP + PP$_i$
(d) Monooleyldipalmitoylglycerol + 75O$_2$ +
430ADP + 430P$_i$ $\longrightarrow$ 53CO$_2$ + 477H$_2$O +
427ATP + 3AMP + 3PP$_i$
(e) Nonanoic acid + 12$\frac{1}{2}$O$_2$ + 72ADP +
72P$_i$ $\longrightarrow$ 9CO$_2$ + 80H$_2$O + 71ATP +
AMP + PP$_i$

2 (a) Carbon atom 1; (b) carbon atoms
1 and 4; (c) carbon atom 1

3 Carbon atoms 1, 3, and 5

Chapter 21

1 (a) Leucine; (b) isovaleryl-CoA dehydrogenase

2 (a) +8.76 kcal mol^{-1}; (b) −6.00 kcal mol^{-1}

3 (a) Phenylalanine + 10O$_2$ + 46ADP + 46P$_i$ $\longrightarrow$
9CO$_2$ + NH$_3$ + 45ATP + AMP + PP$_i$ + 45H$_2$O
(b) 2 Phenylalanine + 20O$_2$ + 90ADP +
88P$_i$ $\longrightarrow$ 17CO$_2$ + urea + 87ATP + 3AMP +
2PP$_i$ + 87H$_2$O

4 (a) 45 molecules; (b) 26.5 molecules

5 (1) amino nitrogen; (2) no labeling;
(3) no labeling; (5) α-carbon atom;
(6) β-carbon atom; (8) γ-carboxyl carbon
atom

6 2 Alanine + 4NAD$^+$ + 3ATP + 4H$_2$O $\longrightarrow$
acetoacetate + urea + CO$_2$ + 4NADH + 4H$^+$ +
2ADP + AMP + 4P$_i$

7 (a) γ-Carboxyl carbon atom; (b) γ-carboxyl
carbon atom; (c) carbon atom 2 or 3

8 22.4%

9 Glucose 6.33 ATP per carbon; butyric acid
7.00; alanine 6.00

10 (a) Lines intersecting to the left of the vertical
axis; (b) parallel lines

Chapter 22

1 −20.76 kcal mol^{-1}

2 35.2%

3 (a) 20.0%; (b) 37.5%

Chapter 23

1. (a) 2 Citric acid + $4NAD^+$ + $2FAD$ + $2ATP$ + $6H_2O \longrightarrow$ glucose + $6CO_2$ + $4NADH$ + $4H^+$ + $2FADH_2$ + $2ADP$ + $2P_i$

   (b) 2 Phenylalanine + $7O_2$ + $2NADPH$ + $2NADH$ + $4H^+$ + $2FAD$ + $2GTP$ + $2ATP$ + $6H_2O \longrightarrow$ glucose + 2 acetoacetate + $4CO_2$ + $2NH_3$ + $2NADP^+$ + $2NAD^+$ + $2FADH_2$ + $2GDP$ + $2ADP$ + $4P_i$

   (c) Palmitic acid + $11NAD^+$ + $11FAD$ + $5ATP$ + $4GTP$ + $27H_2O \longrightarrow$ 2 glucose + $4CO_2$ + $11NADH$ + $11H^+$ + $11FADH_2$ + $4ADP$ + AMP + $4GDP$ + PP_i + $8P_i$

   (d) 4 Leucine + $18NAD^+$ + $6FAD$ + $10ATP$ + $6GTP$ + $38H_2O \longrightarrow$ 3 glucose + $4NH_3$ + $6CO_2$ + $18NADH$ + $18H^+$ + $6FADH_2$ + $6ADP$ + $6GDP$ + $4AMP$ + $4PP_i$ + $12P_i$

2. Carbon atoms 1 and 6; carbon atoms, 2, 4, 6, 8, 10, 12, 14, and 16.

3. 2 Pyruvic acid + $2NADH$ + $2H^+$ + $4ATP$ + $2GTP$ + UTP + $6H_2O \longrightarrow$ glucose + $2NAD^+$ + $4ADP$ + $2GDP$ + UDP + $7P_i$, 8 high-energy phosphate bonds.

4. 2 Pyruvic acid + $NADPH$ + H^+ + NAD^+ + $4ATP$ + $2GTP$ + $UTP \longrightarrow$ ascorbic acid + $NADP^+$ + $NADH$ + H^+ + $4ADP$ + $2GDP$ + UDP + $7P_i$

5. (1) Carbon atoms 3 and 4; (2) carbon atoms 3, 4, and 5; (3) carbon atoms 1, 2, and 3

Chapter 24

1. 9 Acetyl-S-CoA + $16NADPH$ + $16H^+$ + $8ATP$ + $H_2O \longrightarrow$ stearic acid + $16NADP^+$ + $9CoASH$ + $8ADP$ + $8P_i$

2. 8 Pyruvate + $12NADPH$ + $4H^+$ + $7ATP$ + $8NAD^+$ + $H_2O \longrightarrow$ palmitoleic acid + $12NADP^+$ + $7ADP$ + $7P_i$ + $8CO_2$ + $8NADH$

3. Lauric acid + 3 acetyl-S-CoA + $6NADPH$ + $6H^+$ + $4ATP$ + $H_2O \longrightarrow$ stearic acid + $6NADP^+$ + $3CoASH$ + $4ADP$ + $4P_i$

4. $12\frac{1}{2}$ Glucose + $42NADPH$ + $42H^+$ + $48NAD^+$ + $FADH_2$ + $2ADP$ + $2P_i \longrightarrow$ tripalmitin + $24CO_2$ + $42NADP^+$ + $48NADH$ + $48H^+$ + FAD + $2ATP$ + $23H_2O$

5. 6 high-energy phosphate bonds

6. 12 high-energy phosphate bonds

7. 18 Acetic acid + $4O_2$ + $12NADPH$ + $12H^+$ + $36ATP$ + $9H_2O \longrightarrow$ cholesterol + $9CO_2$ + $12NADP^+$ + $18ADP$ + $18AMP$ + $18PP_i$ + $18P_i$; 36 high-energy phosphate bonds

8. 0.5 per carbon atom of glucose; 0.062 high-energy bonds are produced per carbon atom stored as palmitate

Chapter 25

1. Succinate + FH_4 + $2NAD^+$ + $NAD(P)H$ + H^+ + FAD + NH_3 + GTP + $H_2O \longrightarrow$ glycine + CO_2 + N^5,N^{10}-methylene FH_4 + $2NADH$ + $2H^+$ + $NAD(P)$ + $FADH_2$ + GDP + P_i

2. Isocitrate + NAD^+ + $NAD(P)H$ + H^+ + FAD + NH_3 + $2ATP$ + GDP + $H_2O \longrightarrow$ threonine + $2CO_2$ + $NADH$ + $NAD(P)^+$ + $FADH_2$ + $2ADP$ + P_i + GTP

3. Glycine + arginine $\longrightarrow$ NH_2—C—
   $\parallel$
   NH
   $NHCH_2COOH$ + $NH_2(CH_2)_3CHNH_2COOH$

NH_2—C—$NHCH_2COOH$ + methionine $\longrightarrow$
$\parallel$
NH
creatine + homocysteine

4. 5 high-energy phosphate bonds

5. Serine $\longrightarrow$ ethanolamine + CO_2; ethanolamine + 3 S-adenosylmethionine $\longrightarrow$ choline + 3 homocysteine + 3 methionine

6. Carbon 4 of the histidine moiety

7. $10CO_2$ + 2 nitrate + $31NADPH$ + $31H^+$ + $4NAD^+$ + $34ATP$ + $16H_2O \longrightarrow$ 2 glutamate + $31NADP^+$ + $4NADH$ + $4H^+$ + $34ADP$ + $34P_i$

8. α-Ketoglutarate + NH_3 + $3NAD(P)H$ + $3H^+ \longrightarrow$ proline + $3NAD(P)^+$ + $3H_2O$; α carbon atom

9. α-Ketoisovalerate + $[CH_2] \xrightarrow{\text{p-aminobenzoic acid}}$ α-ketopantoate; α-ketopantoate + $NAD(P)H$ + $H^+ \longrightarrow$ pantoate + $NADP^+$; aspartate $\longrightarrow$ β-alanine + CO_2; pantoate + β-alanine $\longrightarrow$ pantothenate

10. 2 high-energy phosphate bonds; (a) none; (b) carbon atom 2

Chapter 26

1. (a) Carbon atom 4 of purine ring; (b) carbon atom 4 of purine ring and carbon atom 2 of ribose moiety; (c) carbon atom 5 of purine ring; (d) nitrogen atom 1 of purine ring and amino nitrogen atom; (e) nitrogen atoms 3 and 9 of purine ring

2. (a) Carbon atom 6 of pyrimidine ring; (b) nitrogen atoms 1 and 3 and amino nitrogen; (c) nitrogen atoms 1 and 3 and amino nitrogen

3. 6 high-energy phosphate bonds

4. Oxaloacetate + ribose 5-phosphate + $3NH_3$ + NAD^+ + $NAD(P)H$ + H^+ + $5ATP \longrightarrow$ CTP + $NADH$ + H^+ + $NAD(P)^+$ + $4ADP$ + AMP + $4P_i$ + H_2O; 0.13 moles

5. Nicotinic acid + glucose + $NADP^+$ + NH_3 + $4ATP$ + $H_2O \longrightarrow$ $NADP^+$ + AMP + $2ADP$ + $2PP_i$ + $2NADPH$ + $2H^+$ + CO_2

6. Man: (a) nitrogen atom 3 of uric acid; (b) carbon atom 5 of uric acid; (c) nitrogen atoms of urea
   Turtle: (a) side chain —NH of allantoin; (b) ring carbonyl carbon of allantoin; (c) nitrogen atom of ammonia
   Amphibia: (a) nitrogen atoms of urea; (b) carboxyl carbon of glyoxylate; (c) nitrogen atoms of urea

Chapter 27

1. 139.0 cal

2. 188 s

3. (a) 4.9 l; (b) 8.67 mmol; (c) none; (d) 36.3 mmol

4. $1.9 + 10^{-6}$ mol g muscle^{-1} min^{-1}

5. 7.22×10^{-5} mol g muscle^{-1} min^{-1}

6. 1.41×10^{-3} mol g muscle^{-1} min^{-1}

7. $K' = 1.66 \times 10^2$

8. For phosphocreatine + ADP $\longrightarrow$ ATP + creatine: (a) -0.63 kcal mol^{-1}; (b) -0.64 kcal mol^{-1}

Chapter 28

1. 2.21 kcal (g ion)$^{-1}$

2. $2.24 \times 10^5/1$

3. (a) 51.6 μequivs h^{-1} mg^{-1}; (b) 57 K^+ ions

4. 9 Na^+ ions

5. 0.5%

6. pH gradient 5.35, 8.65

Chapter 30

1 (a) 1.0; (b) 1.33; (c) 0.90; (d) 0.693;
   (e) 0.857; (f) 0.703
2 Ethanol + 3O$_2$ + 18ADP + 18P$_i$ $\longrightarrow$
   2CO$_2$ + 21H$_2$O + 17ATP + AMP + PP$_i$
3 (a) 6.33; (b) 8.50; (c) 8.13; (d) 8.08;
   (e) 6.66; (f) 5.00
4 (a) Palmitic acid + 5O$_2$ + 23ADP + 23P$_i$ $\longrightarrow$
   4 β-hydroxybutyric acid + 22ATP + 22H$_2$O +
   AMP + PP$_i$
   (b) β-Hydroxybutyric acid + 4½O$_2$ + 27ADP +
   27P$_i$ $\longrightarrow$ 4CO$_2$ + 31H$_2$O + 26ATP +
   AMP + PP$_i$
5 Direct 130 ATP; via β-hydroxybutyrate
   126 ATP
6 −3 ATP

Chapter 31

1 2.21 + 10^6 daltons μm^{-1}
2 2,940
3 Little if any; base ratios show that M13 is
   single-stranded, although hairpin regions of
   base pairs might occur
4 590/1
5 (a) 30.6 nm, 6.03 × 10^4 daltons; (b) 106 nm,
   2.09 × 10^5 daltons; (c) 1840 nm,
   3.62 × 10^6 daltons
6 (1) (a) 1.748 g cm^{-3}; (b) 1.710 g cm^{-3};
   (c) 1.672 g cm^{-3}; (2) (a) 107°C; (b) 90.5°C;
   (c) 74.5°C; (3) (a) 50%; (b) 63%; (c) 77%
7 4.07 × 10^6
8 (a) 12.2; 2.32 × 10^9 mi; (b) 7.20 cm^3
9 (a) Double-stranded circle; (b) A = 1.3 × 10^6;
   B = 0.8 × 10^6; C = 0.7 × 10^6; D = 0.6 × 10^6;
   (c)

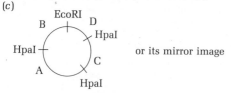

   or its mirror image

   (d) 160 nucleotide deletion, including the Hpa
   cleavage site between fragments A and B

Chapter 32

1 (a) 2,220 nucleotides per second; (b) 133 times
2 0.141 s
3 A 21%; G 29%; C 29%; U 21%
4 G 21.3%; C 21.3%; A 28.7%; T 28.7%
5 (a) 1.6 × 10^7 ATPs; (b) 4.44 × 10^5;
   (c) 8 × 10^6
6 1.5 × 10^8 ATPs; (b) 9.38 times
7 Nearest-neighbor base-frequency analysis; for
   parallel chains, TpC = ApG, TpG = ApC,
   CpT = GpA, etc.; for antiparallel chains,
   TpC = GpA, TpG = CpA, CpT = ApG, etc.
8 (a) Linear single strands of unit length and
   circular single strands of unit length;
   (b) single-stranded circles of unit length,
   single-stranded fragments greater than and
   less than unit length; (c) linear fragments
   greater than unit length; (d) unit-length
   circles and linear fragments less than unit
   length; (e) single-stranded circles; (f) the
   DNA forming the (+) strand is covalently

bound to the preexisting (+) strand and is
formed via continuous replication; the
nascent DNA forming the (−) strand is
synthesized discontinuously, i.e., via
Okazaki fragments

Chapter 33

1 (a) 5.2 s; (b) 36.9 times
2 414 phosphate bonds
3 (a) 3.2; (b) 10.4
4 (a) 1,000; (b) 27,300; (c) 5.45 × 10^7 ATPs;
   (d) 5.45 × 10^8 ATPs; (e) 7.5 × 10^8 ATPs;
   (f) 46.9 times; (g) tRNAs, amino acyl tRNA
   synthetases, protein factors, RNA
   polymerase, termination factors such as ρ,
   tRNA modification enzymes
5 (a) -E-Y-D-S- $\xrightarrow{\text{direction of transcription}}$; (b) mutant S,
   mutant D, mutant Y, mutant E

Chapter 34

1 (a) Gly-Gln-Ser-Leu-Leu-Ile; (b) Leu-
   Asp-Ala; (c) His-Asp-Ala-Cys-Cys-Tyr;
   (d) Met-Asp-Glu
2 a, b, c, d, consistent; e, f, g, inconsistent
3 HbI, transition; HbG$_{Honolulu}$, transversion;
   Hb$_{Norfolk}$, transition; HbM$_{Boston}$, transition;
   HbG$_{Phil}$, transition; HbO$_{Indonesia}$, transition.
4 (a) 5'TCGAGTAGCCGATGATCATCGTCGAC-
   GA3'; (b) 5'UCGAGUAGCCGAUGAUCA-
   UCGUCGACGA3'; (c) Ser-Ser-Ser-Arg;
   (d) Ser-Ser-Ala-Asp-Asp-His-Arg-Arg-Arg
5 (a) Yes; (b) UGG $\longrightarrow$ UUG $\longrightarrow$ CUG $\longrightarrow$
   CCG
6 (a) Lys; (b) Asn or Lys
7 GUA
8 Lys 150; Arg 30; Glu 30; Gly 6
9 (a) In the second codon; (b) first position;
   (c) AUGACCUUCAUAUGG;
   (d) Met-Ala-Phe-Ile-Trp

Chapter 35

1 (a) 0, 0, 0; (b) 13,800; 6,900; 0;
   (c) 1.5 × 10^5; 6,900; 34,500
2 3'-GCAGTCACTGC-5', 5'-CGTCAGTGACG-3'
3 CAP binding site, promoter
4 (a) histidinyl tRNA synthetase; (b) his
   repressor controls its own synthesis, and where
   it cannot bind to its own operator because the
   corepressor is absent, it is synthesized
   constitutively
5 (a) Deletion of all or most of lac operon;
   frame-shift mutation in structural genes;
   nonsense mutation in structural genes;
   altered codons producing nonfunctional
   enzymes; promoter mutation (inability to bind
   RNA polymerase); premature termination
   signal for transcription; mutation in CAP
   binding site preventing binding; mutation in
   i gene such that the repressor binds the
   operator even if the inducer is present
   (i^s mutation); (b) defect in adenyl cyclase;
   inability of CAP to bind cAMP; (c) (1)
   deletion of lac genes, (2) any mutation
   involving structural genes; premature
   termination signal for transcription, (3) i^s,
   (4) defective CAP binding site; (d) promoter
6 -REVILEDANDYDNADELIVER-

INDEX

Configurational and positional designations in chemical names (for example, 1,2-, α,β-, D-, L-, p-, cis-, erythro-) are disregarded in alphabetizing; the same or similar forms used as adjectives (α carbon atom, cis form) are not disregarded. Greek letters used as adjectives are alphabetized as if they were spelled out. Word division by space or hyphen is ignored; for example, alkyl ether acylglycerols *follows* alkylation, and cell wall *follows* cellubiose.